D0086299

ADVANCED REMOTE SENSING

ADVANCED REMOTE SENSING

Terrestrial Information Extraction and Applications

SECOND EDITION

Edited by

SHUNLIN LIANG
University of Maryland

JINDI WANG
Beijing Normal University

ELSEVIER

ACADEMIC PRESS
An imprint of Elsevier

Academic Press is an imprint of Elsevier
125 London Wall, London EC2Y 5AS, United Kingdom
525 B Street, Suite 1650, San Diego, CA 92101, United States
50 Hampshire Street, 5th Floor, Cambridge, MA 02139, United States
The Boulevard, Langford Lane, Kidlington, Oxford OX5 1GB, United Kingdom

Copyright © 2020 Elsevier inc. All rights reserved.

No part of this publication may be reproduced or transmitted in any form or by any means, electronic or mechanical, including photocopying, recording, or any information storage and retrieval system, without permission in writing from the publisher. Details on how to seek permission, further information about the Publisher's permissions policies and our arrangements with organizations such as the Copyright Clearance Center and the Copyright Licensing Agency, can be found at our website: www.elsevier.com/permissions.

This book and the individual contributions contained in it are protected under copyright by the Publisher (other than as may be noted herein).

Notices

Knowledge and best practice in this field are constantly changing. As new research and experience broaden our understanding, changes in research methods, professional practices, or medical treatment may become necessary.

Practitioners and researchers must always rely on their own experience and knowledge in evaluating and using any information, methods, compounds, or experiments described herein. In using such information or methods they should be mindful of their own safety and the safety of others, including parties for whom they have a professional responsibility.

To the fullest extent of the law, neither the Publisher nor the authors, contributors, or editors, assume any liability for any injury and/or damage to persons or property as a matter of products liability, negligence or otherwise, or from any use or operation of any methods, products, instructions, or ideas contained in the material herein.

Library of Congress Cataloging-in-Publication Data
A catalog record for this book is available from the Library of Congress

British Library Cataloguing-in-Publication Data
A catalogue record for this book is available from the British Library

ISBN: 978-0-12-815826-5

For information on all Academic Press publications visit our website at https://www.elsevier.com/books-and-journals

Publisher: Candice Janco
Acquisition Editor: Amy Shapiro
Editorial Project Manager: Lena Sparks
Production Project Manager: R.Vijay Bharath
Cover Designer: Matthew Limbert

Typeset by TNQ Technologies

Working together
to grow libraries in
developing countries

www.elsevier.com • www.bookaid.org

Contents

Contributors of the second edition

Yuqi Bai Center for Earth System Science, Tsinghua University, Beijing 100084, China, yuqibai@tsinghua.edu.cn

Jinshan Cao Collaborative Innovation Center for Geospatial Technology, Wuhan University, 129 Luoyu Road, Wuhan 430079, China, caojinshan0426@163.com

Erxue Chen Institute of Forest Resources Information Techniques, Chinese Academy of Forestry, Dongxiaofu No. 2, Xiangshan Road, Beijing 100091, China, chenerx@caf.ac.cn, chenerx@ifrit.ac.cn

Jun Chen State Key Laboratory of Remote Sensing Science, Jointly Sponsored by Beijing Normal University and Institute of Remote Sensing and Digital Earth of Chinese Academy of Sciences, 19 Xinjiekouwai Street, Beijing 100875, China; Beijing Engineering Research Center for Global Land Remote Sensing Products, Institute of Remote Sensing Science and Engineering, Faculty of Geographical Science, Beijing Normal University, 19 Xinjiekouwai Street, Beijing 100875, China, chenjun0903@qq.com

Jie Cheng State Key Laboratory of Remote Sensing Science, Jointly Sponsored by Beijing Normal University and Institute of Remote Sensing and Digital Earth of Chinese Academy of Sciences, 19 Xinjiekouwai Street, Beijing 100875, China; Beijing Engineering Research Center for Global Land Remote Sensing Products, Institute of Remote Sensing Science and Engineering, Faculty of Geographical Science, Beijing Normal University, 19 Xinjiekouwai Street, Beijing 100875, China, Jie_Cheng@bnu.edu.cn

Robert E. Dickinson Department of Geological Sciences, The University of Texas at Austin, Austin, TX 78712, USA, robted@jsg.utexas.edu

Cuicui Dou Nanjing Institute of Geography and Limnology, Chinese Academy of Sciences, 73 East Beijing Road, Nanjing 210008, China; School of Earth Sciences and Engineering, Hohai University, Nanjing, China, dccdou@163.com

Jinyang Du Numerical Terradynamic Simulation Group, W.A. Franke College of Forestry and Conservation, The University of Montana, Missoula, MT 59812, USA, jinyang.du@mso.umt.edu

Wenjie Fan Institute of RS and GIS, Peking University, 5 Yiheyuan Road, Beijing 100871, China, fanwj@pku.edu.cn

Hongliang Fang Institute of Geographic Sciences and Natural Resources Research, Chinese Academy of Sciences, 11A Datun Road, Beijing 100101, China, fanghl@lreis.ac.cn

Yi Fang School of Remote Sensing and Information Engineering, Wuhan University, 129 Luoyu Road, Wuhan 430079, China, fywhu@qq.com

Qiaoni Fu Nanjing Institute of Geography and Limnology, Chinese Academy of Sciences, 73 East Beijing Road, Nanjing 210008, China; School of Earth Sciences and Engineering, Hohai University, Nanjing, China

Shuai Gao Institute of Remote Sensing and Digital Earth, Chinese Academy of Sciences, Beijing 100101, China, gaoshuai@radi.ac.cn

Zhan Gao Faculty of Geographical Science, Beijing Normal University, 19 Xinjiekouwai Street, Beijing 100875, China, 1578090895@qq.com

Ruifang Guo Nanjing Institute of Geography and Limnology, Chinese Academy of Sciences, 73 East Beijing Road, Nanjing 210008, China, gr120206@126.com

Tao He   School of Remote Sensing and Information Engeering, Wuhan University, Luoyu Road No. 129, Wuhan 430079, China, taohers@whu.edu.cn

Wenli Huang   School of Resource and Environmental Sciences, Wuhan University, 129 Luoyu Road, Wuhan 430079, China, wenli.huang@whu.edu.cn

Shunping Ji   School of Remote Sensing and Information Engineering, Wuhan University, 129 Luoyu Road, Wuhan 430079, China, Jishunping2000@163.com

Kun Jia   State Key Laboratory of Remote Sensing Science, Jointly Sponsored by Beijing Normal University and Institute of Remote Sensing and Digital Earth of Chinese Academy of Sciences, 19 Xinjiekouwai Street, Beijing 100875, China;   Beijing Engineering Research Center for Global Land Remote Sensing Products, Institute of Remote Sensing Science and Engineering, Faculty of Geographical Science, Beijing Normal University, 19 Xinjiekouwai Street, Beijing 100875, China, jiakun@bnu.edu.cn

Bo Jiang   State Key Laboratory of Remote Sensing Science, Jointly Sponsored by Beijing Normal University and Institute of Remote Sensing and Digital Earth of Chinese Academy of Sciences, 19 Xinjiekouwai Street, Beijing 100875, China;   Beijing Engineering Research Center for Global Land Remote Sensing Products, Institute of Remote Sensing Science and Engineering, Faculty of Geographical Science, Beijing Normal University, 19 Xinjiekouwai Street, Beijing 100875, China, bojiang@bnu.edu.cn

Lingmei Jiang   State Key Laboratory of Remote Sensing Science, Jointly Sponsored by Beijing Normal University and Institute of Remote Sensing and Digital Earth of Chinese Academy of Sciences, 19 Xinjiekouwai Street, Beijing 100875, China;   Beijing Engineering Research Center for Global Land Remote Sensing Products, Institute of Remote Sensing Science and Engineering, Faculty of Geographical Science, Beijing Normal University, 19 Xinjiekouwai Street, Beijing 100875, China, jiang@bnu.edu.cn

Zengyuan Li   Institute of Forest Resources Information Techniques, Chinese Academy of Forestry, Dongxiaofu No. 2, Xiangshan Road, Beijing, 100091, China

Shunlin Liang   Department of Geographical Sciences, Univsersity of Maryland, College Park MD 20742, USA, sliang@umd.edu

Ming Lin   Tsinghua University, MengMinWei Scitech Building, S912, Beijing 100084, China, tj_linming@foxmail.com

Qiang Liu   State Key Laboratory of Remote Sensing Science, Jointly Sponsored by Beijing Normal University and Institute of Remote Sensing and Digital Earth of Chinese Academy of Sciences, 19 Xinjiekouwai Street, Beijing 100875, China;   College of Global Change and Earth System Science, Beijing Normal University, 19 Xinjiekouwai Street, Beijing 100875, China, toliuqiang@bnu.edu.cn

Suhong Liu   Faculty of Geographical Science, Beijing Normal University, 19 Xinjiekouwai Street, Beijing 100875, China, liush@bnu.edu.cn

Yaokai Liu   Academy of Opto-Electronics, Chinese Academy of Science, Beijing 100094, China, liuyk@aoe.ac.cn

Yuanbo Liu   Nanjing Institute of Geography and Limnology, Chinese Academy of Sciences, 73 East Beijing Road, Nanjing 210008, China, ybliu@niglas.ac.cn

Yufu Liu   Tsinghua University, MengMinWei Scitech Building, S917, Beijing 100084, China, liuyufu18@mails.tsinghua.edu.cn

Qian Ma   State Key Laboratory of Earth Surface Processes and Resource Ecology, College of Global Change and Earth System Science, Beijing Normal University, 19 Xinjiekouwai Street, Beijing 100875, China, maqian@bnu.edu.cn

Yuna Mao   State Key Laboratory of Earth Surface Processes and Resource Ecology, College of Global Change and Earth System Science, Beijing Normal University, 19 Xinjiekouwai Street, Beijing 100875, China, shanxian.08@163.com

Xiangcheng Meng   Faculty of Geographical Science, Beijing Normal University, 19 Xinjiekouwai Street, Beijing 100875, China, xiangchenmeng@yeah.net

Xihan Mu   State Key Laboratory of Remote Sensing Science, Jointly Sponsored by Beijing Normal University and Institute of Remote Sensing and Digital Earth of Chinese Academy of Sciences, 19 Xinjiekouwai Street, Beijing 100875, China;   Beijing Engineering Research Center for Global Land Remote

Sensing Products, Institute of Remote Sensing Science and Engineering, Faculty of Geographical Science, Beijing Normal University, 19 Xinjiekouwai Street, Beijing 100875, China, muxihan@bnu.edu.cn

Wenjian Ni Institute of remote sensing and digital earth, Chinese Academy of Sciences, A20 north, Datun road, Beijing 100101, China, niwj@radi.ac.cn

Zheng Niu Institute of Remote Sensing and Digital Earth, Chinese Academy of Sciences, 100101 Beijing, China, niuzheng@radi.ac.cn

Jinmei Pan State Key Laboratory of Remote Sensing Science, Jointly Sponsored by the Institute of Remote Sensing and Digital Earth of Chinese Academy of Sciences and Beijing Normal University, 100101 Beijing, China, panjm@aircas.ac.cn

Yong Pang Institute of Forest Resources Information Techniques, Chinese Academy of Forestry, Dongxiaofu No. 2, Xiangshan Road, Beijing, 100091, China, pangy@ifrit.ac.cn

Jingjing Peng Earth System Science Interdisciplinary Center, University of Maryland, College Park 20740 MD, USA, jingjingpeng89@gmail.com

Ying Qu School of Geographical Sciences, Northeast Normal University, Changchun 130024, China, quy100@nenu.edu.cn

Yonghua Qu State Key Laboratory of Remote Sensing Science, Jointly Sponsored by Beijing Normal University and Institute of Remote Sensing and Digital Earth of Chinese Academy of Sciences, 19 Xinjiekouwai Street, Beijing 100875, China; Beijing Engineering Research Center for Global Land Remote Sensing Products, Institute of Remote Sensing Science and Engineering, Faculty of Geographical Science, Beijing Normal University, 19 Xinjiekouwai Street, Beijing 100875, China, qyh@bnu.edu.cn

Jiancheng Shi State Key Laboratory of Remote Sensing Science, Jointly Sponsored by the Institute of Remote Sensing and Digital Earth of Chinese Academy of Sciences and Beijing Normal University, 100101 Beijing, China, shijc@radi.ac.cn

Jinling Song State Key Laboratory of Remote Sensing Science, Jointly Sponsored by Beijing Normal University and Institute of Remote Sensing and Digital Earth of Chinese Academy of Sciences, 19 Xinjiekouwai Street, Beijing 100875, China;

Beijing Engineering Research Center for Global Land Remote Sensing Products, Institute of Remote Sensing Science and Engineering, Faculty of Geographical Science, Beijing Normal University, 19 Xinjiekouwai Street, Beijing 100875, China, songjl@bnu.edu.cn

Wanjuan Song State Key Laboratory of Remote Sensing Science, Jointly Sponsored by Beijing Normal University and Institute of Remote Sensing and Digital Earth of Chinese Academy of Sciences, 19 Xinjiekouwai Street, Beijing 100875, China; Beijing Engineering Research Center for Global Land Remote Sensing Products, Institute of Remote Sensing Science and Engineering, Faculty of Geographical Science, Beijing Normal University, 19 Xinjiekouwai Street, Beijing 100875, China, songwanjuan@126.com

Guoqing Sun Department of Geographical Sciences, Univsersity of Maryland, College Park MD 20742, USA, guoqing.sun@gmail.com

Wanxiao Sun Department of Geography and Sustainable Planning, Grand Valley State University, 1 Campus Drive, Allendale, MI 49401-9403, USA, sunwa@gvsu.edu

Xin Tao Department of Geography, The State University of New York, Buffalo, NY 14261, USA, xintao@buffalo.edu

Xinpeng Tian CAS Key Laboratory of Coastal Environmental Processes and Ecological Remediation, Yantai Institute of Coastal Zone Research, Chinese Academy of Sciences, Yantai 264003, China, xptian@yic.ac.cn

Dongdong Wang Department of Geographical Sciences, Univsersity of Maryland, College Park MD 20742, USA, ddwang@umd.edu

Haoyu Wang Faculty of Geographical Science, Beijing Normal University, 19 Xinjiekouwai Street, Beijing 100875, China, why0925@mail.bnu.edu.cn

Jindi Wang State Key Laboratory of Remote Sensing Science, Jointly Sponsored by Beijing Normal University and Institute of Remote Sensing and Digital Earth of Chinese Academy of Sciences, 19 Xinjiekouwai Street, Beijing 100875, China; Beijing Engineering Research Center for Global Land Remote Sensing Products, Institute of Remote Sensing Science and Engineering, Faculty of Geographical

Science, Beijing Normal University, 19 Xinjiekou-wai Street, Beijing 100875, China, wangjd@bnu.edu.cn

Kaicun Wang   State Key Laboratory of Earth Surface Processes and Resource Ecology, College of Global Change and Earth System Science, Beijing Normal University, 19 Xinjiekouwai Street, Beijing 100875, China, kcwang@bnu.edu.cn

Wenhui Wang   I.M. Systems Group at NOAA/NES-DIS/STAR, 5200 Auth Road, Camp Springers, MD 20746, USA, wang.wenhui@gmail.com

Zhigang Wang   China Center for Resource Satellite Data and Applications, No. 5, Fengxian East Road, Beijing 100094, China, kevinwang2000@163.com

Jianguang Wen   Institute of Remote Sensing and Digital Earth of Chinese Academy of Sciences, Beijing 100101, China, wenjg@radi.ac.cn

Guiping Wu   Nanjing Institute of Geography and Limnology, Chinese Academy of Sciences, Nanjing 210008, China, gpwu@niglas.ac.cn

Zhiqiang Xiao   State Key Laboratory of Remote Sensing Science, Jointly Sponsored by Beijing Normal University and Institute of Remote Sensing and Digital Earth of Chinese Academy of Sciences, 19 Xinjiekouwai Street, Beijing 100875, China; Beijing Engineering Research Center for Global Land Remote Sensing Products, Institute of Remote Sensing Science and Engineering, Faculty of Geographical Science, Beijing Normal University, 19 Xinjiekouwai Street, Beijing 100875, China, zhqxiao@bnu.edu.cn

Chuan Xiong   Southwest Jiaotong University, Chengdu 611756, China, xiongchuan@swjtu.edu.cn

Chunyan Yan   China University of Geosciences, Beijing, 29 Xueyuan Road, Beijing 100083, China, 147583592@qq.com

Guangjian Yan   State Key Laboratory of Remote Sensing Science, Jointly Sponsored by Beijing Normal University and Institute of Remote Sensing and Digital Earth of Chinese Academy of Sciences, 19 Xinjiekouwai Street, Beijing 100875, China; Beijing Engineering Research Center for Global Land Remote Sensing Products, Institute of Remote Sensing Science and Engineering, Faculty of

Geographical Science, Beijing Normal University, 19 Xinjiekouwai Street, Beijing 100875, China, gjyan@bnu.edu.cn

Feng Yang   State Key Laboratory of Remote Sensing Science, Jointly Sponsored by Beijing Normal University and Institute of Remote Sensing and Digital Earth of Chinese Academy of Sciences, 19 Xinjiekouwai Street, Beijing 100875, China; Beijing Engineering Research Center for Global Land Remote Sensing Products, Institute of Remote Sensing Science and Engineering, Faculty of Geographical Science, Beijing Normal University, 19 Xinjiekouwai Street, Beijing 100875, China, yftaurus@mail.bnu.edu.cn

Wenping Yuan   School of Atmospheric Sciences, Sun Yat-sen University. No. 135, Xingang Xi Road, Guangzhou 510275, China, yuanwpcn@126.com

Xiuxiao Yuan   School of Remote Sensing and Information Engineering, Wuhan University, 129 Luoyu Road, Wuhan 430079, China, yuanxx@whu.edu.cn

Quan Zhang   Faculty of Geographical Science, Beijing Normal University, 19 Xinjiekouwai Street, Beijing 100875, China, zhangquanzq@126.com

Xiaotong Zhang   State Key Laboratory of Remote Sensing Science, Jointly Sponsored by Beijing Normal University and Institute of Remote Sensing and Digital Earth of Chinese Academy of Sciences, 19 Xinjiekouwai Street, Beijing 100875, China; Beijing Engineering Research Center for Global Land Remote Sensing Products, Institute of Remote Sensing Science and Engineering, Faculty of Geographical Science, Beijing Normal University, 19 Xinjiekouwai Street, Beijing 100875, China, xtngzhang@bnu.edu.cn

Zhiyu Zhang   Institute of remote sensing and digital earth, Chinese Academy of Sciences, A20 north, Datun road, Chaoyang District, Beijing 100101, China, zhangzy@irsa.ac.cn

Peisheng Zhao   George Mason University, 4400 University Drive, MSN 6E1, George Mason University, Fairfax, VA 22030, USA, pzhao@gmu.edu

Xiang Zhao   State Key Laboratory of Remote Sensing Science, Jointly Sponsored by Beijing Normal University and Institute of Remote Sensing and Digital

Earth of Chinese Academy of Sciences, 19 Xinjie-kouwai Street, Beijing 100875, China; Beijing Engineering Research Center for Global Land Remote Sensing Products, Institute of Remote Sensing Science and Engineering, Faculty of Geographical Science, Beijing Normal University, 19 Xinjiekouwai Street, Beijing 100875, China, zhaoxiang@bnu.edu.cn

Xiaosong Zhao   Nanjing Institute of Geography and Limnology, Chinese Academy of Sciences, 73 East Beijing Road, Nanjing 210008, China, xszhao@niglas.ac.cn

Yi Zheng   School of Atmospheric Sciences, Sun Yat-sen University. No. 135, Xingang Xi Road, Guangzhou 510275, China, zhengy263@mail2.sysu.edu.cn

Shugui Zhou   Faculty of Geographical Science, Beijing Normal University, 19 Xinjiekouwai Street, Beijing 100875, China, zhoushugui1990@msn.cn

Xiufang Zhu   Institute of Remote Sensing Science and Engineering, Faculty of Geographical Science, Beijing Normal University, 19 Xinjiekouwai Street, Beijing 100875, China, zhuxiufang@bnu.edu.cn

Foreword to the first edition

The Symposium on Quantitative Retrieval Algorithms in Remote Sensing was held in the summer of 2010 at Beijing Normal University. It was chaired by Professors Shunlin Liang and Xiaowen Li. During the Symposium, I stressed the roles of geography and remote-sensing science in the process of globalization. In the twenty-first century, world development has taken on three new characteristics: a constantly developing knowledge economy, constantly progressing globalization, and a widespread sustainable development theory. Since Earth science focuses on the relationships between human beings and the Earth's environment, it will significantly influence studies on the globalization process and sustainable development. For this reason, Earth science research in China should establish a much broader global outlook and extend its research perspective worldwide. Scientists should be much more concerned about global issues, multidisciplinary developments, and quantitative methods in the field of Earth science research.

Remote sensing is an important method of Earth observation. Satellite sensors can constantly observe the Earth's surface, and, with the development of remote-sensing science, it has become an important mechanism to determine spatial and temporal land-surface information quantitatively based on radiative transfer theory. High-level remote-sensing products are urgently needed to meet global changes and for many other applications. Generating these high-level products is challenging, however, and has become a hot research topic. Remote-sensing scientists, especially the young scientists among them, are fully aware of this and have accordingly paid more attention to quantitative methodology. Therefore, they are more eager to understand fundamental principles and practical algorithms.

Professors Xiaowen Li and Shunlin Liang are long-term explorers in the research field of quantitative remote sensing. They are not only world-known scientists but also tutors and friends trusted by many young scholars. Their most ardent wish is to satisfy the need of readers, especially young students, for knowledge of quantitative remote sensing.

To meet the needs of scientists and graduate students, Professors Shunlin Liang, Xiaowen Li, and Jindi Wang secured the collaboration of a group of scientists engaged in the frontiers of remote sensing in producing this book two years after the 2010 Symposium. The present volume introduces remote-sensing systems, remote-sensing models, the inversion algorithms of nearly 20 land-surface variables, and existing global products, all of which are state of the art. The book offers an extensive resource and reference that will help readers understand quantitative remote-sensing principles, communicate more effectively with other. Earth science researchers and promote the quantitative applications of remote sensing.

This book will be of significant value to both students and scientists worldwide, helping to promote better understanding of quantitative remote sensing and contributing to the further development of Earth science in the twenty-first century.

Guanhua Xu
Professor, Academician of Chinese Academy of Sciences and Former Minister of the Ministry of Science and Technology of the People's Republic of China

Preface to the first edition

As the technology of remote sensing has advanced over the last two decades, the scientific potential of the data that it produces has greatly improved. To better serve society's needs, the immense amounts of aggregated satellite data need to be transferred into high-level products in order to improve the predictive capabilities of global and regional models at different scales and to aid in decision making through various decision support systems. A general trend is that the data centers are distributing more high-level products rather than simply the raw satellite imagery.

An increasing number of researchers from a diverse set of academic and scientific disciplines are now routinely using remotely sensed data products, and the mathematical and physical sophistication of the techniques used to process and analyze these data have increased considerably. As a result, there is an urgent need for a reference book on the advanced methods and algorithms that are now available for extracting information from the huge volume of remotely sensed data, which are often buried in various journals and other sources. Such a book should be highly quantitative and rigorously technical; at the same time, it should be accessible to students at the upper undergraduate and first-year graduate student level.

To meet this critical demand, we have identified and organized a group of active research scientists to contribute chapters and sections drawn from their research expertise. Although this is an edited volume with multiple authors, it is well designed and integrated. The editors and authors have made great efforts to ensure the consistency and integrity of the text.

In addition to the introductory chapter, this book consists of five parts: (1) data processing methods and techniques; (2) estimation of land-surface radiation budget components; (3) estimation of biophysical and biochemical variables; (4) estimation of water cycle components; and (5) high-level product generation and application demonstrations. The titles and authors of the individual chapters are as follows:

Chapters	Titles	Authors
1	A Systematic View of Remote Sensing	S. Liang, J. Wang, B. Jiang
PART 1 Data Processing Methods and Techniques		
2	Geometric Processing and Positioning Techniques	X. Yuan, S. Ji, J. Cao, X. Yu
3	Compositing, Smoothing, and Gap-Filling Techniques	Z. Xiao
4	Data Fusion	J. Zhang, J. Yang
5	Atmospheric Correction of Optical Imagery	X. Zhao, X. Zhang, S. Liang

Continued

(*cont'd*)

(*cont'd*)

Chapter 1 presents introductory material and provides an overview of the book. From the system perspective, it briefly describes the essential components of the remote-sensing system, ranging from platforms and sensors, modeling approaches, and information extraction methods to applications.

Part 1 includes four chapters on data processing. Chapter 2 is the only chapter that presents the methods and techniques for handling geometric properties of remotely sensed data. These include the calibration of systematic errors, geometric correction, geometric registration, digital terrain model generation, and digital ortho-image generation.

Chapter 3 seeks to reconstruct spatial and temporal continuous high-quality imagery. As the temporal resolution of satellite observations greatly increases, images are more often contaminated by clouds and aerosols that partially or completely block the surface information. Two groups of techniques are presented. The first group deals with composite methods for aggregating the fine temporal resolution (say, daily) to the coarse resolution (say, weekly or monthly), and the second discusses smoothing and gapfilling methods to eliminate the impacts of clouds and aerosols at the same temporal resolution.

Chapter 4 introduces the basic principles and methods of data fusion for integrating multiple data sources on the pixel basis, which have different spatial resolutions, and are acquired from different spectra (optical, thermal, microwave). This chapter focuses mainly on low-level data products. (The methods for integrating high-level products are introduced in Chapter 22.)

Chapter 5 introduces methods for correcting the atmospheric effects of aerosols and water vapor on the optical imagery. Other atmospheric correction methods are discussed in Chapter 8 for thermal-IR data and in Part 4 for microwave data.

Part 2 focuses on estimation of surface radiation budget components. The surface radiation budget is characterized by all-wave net radiation (R_n) that is the sum of shortwave (S_n) and long-wave (L_n) net radiation

$$R_n = S_n + L_n = (s{\downarrow} - s{\uparrow}) + (L{\downarrow} - L{\uparrow})$$
$$= (1 - \alpha)S{\downarrow} + (L{\downarrow} - L{\uparrow})$$

where $S{\downarrow}$ is the downward shortwave radiation (discussed in Chapter 6), $S{\uparrow}$ is the upward shortwave radiation, α is the surface shortwave albedo (discussed in Chapter 7), $L{\downarrow}$ is the downward longwave radiation, and $L{\uparrow}$ is the upward longwave radiation. Longwave net radiation (L_n) can be also calculated by

$$L_n = \varepsilon L{\downarrow} - \varepsilon \sigma T_s^4$$

where σ is the Stefan-Boltzmann constant, ε is surface thermal broadband emissivity, and T_s is surface skin temperature. Estimation of ε and Ts is discussed in Chapter 8, and $L{\downarrow}$ and L_n are covered in Chapter 9.

Part 3 focuses on the estimation of biochemical and biophysical variables of plant canopy. Chapter 10 introduces the various methods for estimating plant biochemical variables, such as chlorophyll, water, protein, lignin and cellulose. The biophysical variables discussed in this book include leaf area index (LAI) in Chapter 11, the fraction of absorbed photosynthetically active radiation by green vegetation (FPAR) in Chapter 12, fractional vegetation cover in Chapter 13, vegetation height and vertical structure in Chapter 14, above-ground biomass in Chapter 15, and vegetation production in terms of gross primary production (GPP) and net primary production (NPP) in Chapter 16. Various inversion methods are introduced in this part, including optimization methods (Section 11.3.2), neural networks (Sections 11.3.3, 13.3.3 and 15.3.4), genetic algorithms (Section 11.3.4), Bayesian networks (Section 11.3.5), regression tree methods (Section 13.3.3), data assimilation methods (Section 11.4) and look-up table methods (Section 11.3.6). Part 3 also discusses multiple data sources besides optical imagery, such as Synthetic Aperture Radar (SAR) and Light Detection and Ranging (Lidar), and polarimetric InSAR data.

Part 4 is on estimation of water balance components. A general water balance equation is expressed by:

$$P = Q + E + \Delta S$$

where P is precipitation (discussed in Chapter 17), Q is runoff that is currently difficult to estimate from remote sensing, E is evapotranspiration (discussed in Chapter 18), and ΔS is the change in storage to which three chapters are related: soil moisture in Chapter 19, snow water equivalence in Chapter 20, and surface water storage in Chapter 21. In addition to optical and thermal data, microwave data are dealt with extensively in all chapters except in Chapter 18. The gravity data with the GRACE data are also briefly introduced in Chapter 21.

Part 5 deals with high-level product generation, integration, and application. Chapter 22 presents different methods for integrating high-level products of the same variable (e.g., LAI) that may be generated from different satellite data or different inversion algorithms. The data fusion methods for integrating low-level products are discussed in Chapter 4. Chapter 23 describes the typical procedures for producing high-level products from low-level satellite data and for developing a data management system that is used for effectively handling a large volume of satellite data. The last chapter demonstrates how remote-sensing data products can be used for land-cover and land-use change studies, particularly on mapping the extent of three major land-use types (urban, forest, and agriculture), detecting changes in these landuse types, and evaluating the environmental impacts of these land-use changes.

One important feature of this book is its focus on extracting land-surface information from satellite observations. All relevant chapters follow the same template: introduction to basic concepts and fundamental principles, review of practical algorithms with a comprehensive list of references, detailed descriptions of representative algorithms and case studies, surveys of current products, spatiotemporal variations of the variable, and identification of future research directions. The book includes almost 500 figures and tables, as well as 1700 references.

This book can serve as a text for upper-level undergraduate and graduate students in a variety of disciplines related to Earth observation. The entire book may be too lengthy for a one-semester or one quarter class, but most chapters in Parts 2—5 are relatively independent, and using a subset of them will be useful in such classes.

The text can also serve as a valuable reference book for anyone interested in the use and applications of remote-sensing data. Ideally, those using this book will have taken an introductory remote-sensing course, but we have written it at such a level that even those who have had little or no prior training in remote sensing can easily understand the overall development of this field.

Preface to the second edition

Since the first edition of this book was published in 2012, the field of remote sensing has experienced extensive growth and development. An updated text that examines and describes in detail this growth is now needed.

There are several remarkable trends. The first trend is the steadily increasing volume of remotely sensed data, driven by the growing number of satellites with higher spatial and temporal resolutions. For example, DigitalGlobe's satellite fleet currently generates 80TB per day of images. The constellations of smaller satellites, mostly operated by the commercial sector, provide high spatial and temporal resolutions imagery. Unmanned aerial vehicles (UAVs), platforms, and associated sensing technologies are now also collecting huge amounts of data for use in a variety of applications in a cost-effective manner.

The second trend is the widespread application of machine learning techniques that transform raw satellite observations into the values of various bio/geophysical variables. These methods, such as artificial neural network, support vector regression, random forest, and multivariate adaptive regression splines, are often based on extensive simulations of different radiative transfer models.

The third trend has been the gradual adaptation of cloud computing. It is essential to develop an infrastructure that connects global remotely sensed data collected and managed by various agencies and data centers located throughout the world. It will be a cost-effective approach for sharing, processing, archiving, and disseminating the massive size of remotely sensed data. The processing and analysis can be greatly enhanced by using a massive number of computing nodes through high-performance computing and high-throughput computing techniques.

Another trend is the generation of long-term consistent high-level satellite products that can be used directly by users for a variety of applications. The creation of long-term high-level land products leverages off the advantages of multi-source remote sensing data. It started from the NASA Earth Observing System (EOS) program in late 1980s. One of the product suites extensively discussed in this book is the Global Land Surface Satellite (GLASS) products, which are being distributed free of charge through the China National Data Sharing Infrastructure of Earth System Science (http://www.geodata.cn/thematicView/GLASS.html) and the University of Maryland (www.glass.umd.edu). The GLASS products have some unique features, one of which is long-term time series (from 1981 to present). Considerable efforts are also being made by the remote sensing community to develop the Climate Data Records (CDR) defined as the time series of measurements of sufficient length, consistency, and continuity to determine climate variability and change by the US National Research Council.

To incorporate state-of-the-art development of land remote sensing, this new book provides a major revision of the first edition by presenting new methods, new data products, and more applications. The chapter titles and author information are provided in the following table.

Chapters	Titles	Authors
1	A systematic view of remote sensing	S. Liang, J. Wang, B. Jiang
2	Geometric processing and positioning techniques	X. Yuan, J. Cao, S. Ji, Y. Fang
3	Compositing, smoothing, and gap-filling techniques	Z. Xiao
4	Atmospheric correction of optical imagery	X. Zhao, X. Tian, H. Wang, Q. Liu, S. Liang
5	Solar radiation	X. Zhang, S. Liang
6	Broadband albedo	Q. Liu, J. Wen, Y. Qu, T. He, J. Peng
7	Land surface temperature and thermal infrared emissivity	J. Cheng, S. Liang, X. Meng, Q. Zhang, S. Zhou
8	Surface longwave radiation budget	J. Cheng, W. Wang, S. Liang, F. Yang, S. Zhou
9	Canopy biochemical characteristics	Z. Niu, C. Yan, S. Gao
10	Leaf area index	H. Fang, Z. Xiao, Y. Qu, J. Song
11	Fraction of absorbed photosynthetically active radiation	X. Tao, Z. Xiao, W. Fan
12	Fractional vegetation cover	G. Yan, X. Mu, K. Jia, W. Song, Y. Liu, J. Chen, Z. Gao
13	Vegetation height and vertical structure	Y. Pang, W, Ni, Z. Li, W. Huang, E. Chen, G. Sun
14	Aboveground biomass	W. Ni, Y. Pang, Z. Zhang, W. Sun, S. Liang, E. Chen, G. Sun

(cont'd)

Chapters	Titles	Authors
15	Estimate of vegetation production of terrestrial ecosystem	W. Yuan, Y. Zheng
16	Precipitation	Y. Liu, R. Guo, Q. Fu, X. Zhao, C. Dou
17	Terrestrial evapotranspiration	K. Wang, R. Dickinson, Q. Ma, Y. Mao
18	Soil moisture contents	S. Liang, B. Jiang, T. He, X. Zhu
19	Snow water equivalent	L. Jiang, J. Du, J. Pan, C. Xiong, J. Shi
20	Water storage	G. Wu, Y. Liu
21	High-level land product integration methods	D. Wang
22	Data production and management system	Y. Bai, S. Liu, X. Zhao, Z. Wang, P. Zhao, Y. Liu, M. Lin
23	Urbanization: monitoring and impact assessment	X. Zhu, S. Liang
24	Remote sensing application in agriculture	X. Zhu, S. Liang
25	Forest cover changes: mapping and climatic impact assessment	B. Jiang, S. Liang

Most chapters have been considerably expanded and all have updated references. Chapter 1 provides a more comprehensive introduction to the remote sensing system and also serves as the "pointers" to various chapters of the book. Machine learning techniques are described in many chapters. We removed the chapter on data fusion but expanded the application section from one chapter to three (Chapters 23—25).

One of the first edition editors, Prof. Xiaowen Li, passed away in 2015. We lost a great colleague and friend. He made significant contributions to land remote sensing. For example, he was the primary developer of the well-known Li—Strahler geometric-optical vegetation reflectance model and pioneered in developing the simplified "kernels" modeling structure to characterize land surface directional reflectance that have been used for the MODIS surface albedo product generation and other applications. A full account of Prof. Li's lifetime achievements is available in a journal paper (Liu Q., et al., 2018. From Geometric-Optical Remote Sensing Modeling to Quantitative Remote Sensing Science—In Memory of Academician Xiaowen Li. *Remote Sensing* 10, 1764, 2018).

We would like to thank all the authors for their valuable contributions and are indebted to our many colleagues for their kind assistance in preparing this edition. Among them is Ms. Liulin Song who kept communicating with all contributors, managed all the documents, and applied for the permission of the copyright materials. Assistance from Dr. Hongmin Zhou is also greatly appreciated. Without their help, this project would probably never come to the end.

We also thank the editors and production personnel at Elsevier, particularly Ms. Lena Sparks, Editorial Project Manager, who has worked with us until the completion of this edition.

Lastly, we most appreciate the support of our families. To one and all, thank you!

This project was supported in part by the National Key Research and Development Program of China (Grant No. 2016YFA0600100), State Key Laboratory of Remote Sensing Science, and Beijing Engineering Research Center for Global Land Remote Sensing Products.

Shunlin Liang
Jindi Wang
October 2019

A systematic view of remote sensing

Advanced Remote Sensing, Second Edition
https://doi.org/10.1016/B978-0-12-815826-5.00001-5

1

© 2020 Elsevier Inc. All rights reserved.

Abstract

This chapter provides an overview of the remote sensing system, including the platform and sensor system, data transmission and ground receiving system, processing system of radiometric and geometric properties, analysis system for mapping category variables and generating high-level products of quantitative variables, product production and distribution system, product validation system, and remote sensing applications. It aims to present a complete picture of the state-of-the-art development of remote sensing techniques by linking different chapters in the rest of the book and filling in any possible gaps.

moves on to the data transmission and ground receiving system, the processing system for handling the geometric and radiometric properties of data, the analysis system for extracting information on both category and numerical variables of the Earth surface environment, the product generation and distribution system, the product validation system, and end-user applications. Applications largely define the data acquisition system, and end-users often need to validate the products to quantify their errors and uncertainties.

1.1 Introduction

We are living in a world where population is rapidly increasing, depleting natural resources, and experiencing the possible consequences of human-induced climate change. Our ability to meet these challenges partially depends on how well we understand the Earth system and use that information to guide our actions. Remote sensing is a tremendous source of information needed by policy-makers, resource managers, forecasters, and other users, and it has become increasingly vital for the effective and sustainable future management of the Earth. A remote sensing system consists of instrumentation, processing, and analysis designed to measure, monitor, and predict the physical, chemical, and biological aspects of the Earth system. Sophisticated new technologies have been developed to gather vast quantities of data, and the mathematical and physical sophistication of the techniques used to process and analyze the observed data has increased considerably.

The first chapter of the book aims to link diverse components to paint a full picture of a remote sensing system as illustrated in Fig. 1.1. It starts with a brief introduction to the platform and sensor system for acquiring data and then

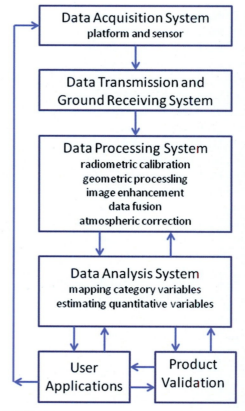

FIGURE 1.1   Key components of the remote sensing system.

1.2 Platform and sensor systems

The data acquisition system mainly consists of the sensor and the platform on which the sensor resides. The platform may be on the surface, in the air, or in space. A surface platform may be a ladder, tower, cherry picker, crane, building, or scaffolding that provides data used primarily for validation.

Aerial platforms include aircraft and balloons. Unmanned aerial systems (UAS), commonly known as a drone, have considerable potential to radically improve Earth observation by providing high spatial detail over relatively large areas in a cost-effective way and an entirely new capacity for enhanced temporal retrieval (Manfreda et al., 2018). In addition to the increasing availability of UAS and affordability, recent advances in sensor technologies and analytical capabilities have stimulated an explosion of interest from the remote sensing community. Increasing miniaturization allows multispectral, hyperspectral, and thermal imaging, as well as synthetic-aperture radar (SAR) and light detection and ranging (LiDAR) sensing to be conducted from UAS.

Spaceborne platforms are mainly satellites and space shuttles. As the landmark of spaceborne remote sensing, Landsat 1 was launched in 1972. Since then, there have been over 50 countries operating land remote sensing satellites. The Committee on Earth Observation Satellites (CEOS) database (http://database.eohandbook.com/database/missiontable.aspx) lists the current and future satellite missions and sensors. The following will mainly discuss the satellite remote sensing.

1.2.1 Geostationary satellites

A geostationary satellite is in an orbit that can only be achieved at an altitude very close to 35,786 km (22,236 miles) and which keeps the satellite fixed over one longitude at the equator. The satellite appears motionless at a fixed position in the sky to ground observers. There are several hundred communication satellites and several meteorological satellites in such an orbit. Fig. 1.2 illustrates a few typical meteorological satellites in the geostationary orbit relative to the polar-orbiting satellites.

US operational weather satellites include the Geostationary Operational Environmental Satellite (GOES) used for short-range warning and "now-casting" primarily to support the National Weather Service requirements. The procurement, design, and manufacturing of GOES are overseen by the National Aeronautics and Space Administration (NASA), while all operations of the satellites once in orbit are effected by the National Oceanic and Atmospheric Administration (NOAA). Before being launched, GOES satellites are designated by letters (-A, -B, -C). Once a GOES satellite is launched successfully, it is redesignated with a number (-1, -2, -3). Normally two GOES satellites are operational. Information on the GOES series is shown in Table 1.1. The third generation of GOES, the new GOES-R satellite series program, consisting of four satellites (from GOES-16), represents a significant improvement in spatial, temporal, and spectral observations over the capabilities of the previously operational GOES series. For example, the Advanced Baseline Imager (ABI) is the primary instrument on the GOES-R Series for imaging Earth's weather, oceans, and environment. The ABI provides three times more spectral information, four times the spatial resolution, and more than five times faster temporal coverage than the previous system.

European operational missions are currently operated by the European Organization for the Exploitation of Meteorological Satellites (EUMETSAT). EUMETSAT's geostationary satellite programs include the Meteosat First Generation system (up to Meteosat-7) from 1977 to 2017, four Meteosat Second Generation (MSG) satellites (MSG-1,2,3,4 or Meteosat-8,9,10,11) from 2004 to 2025, and six Meteosat

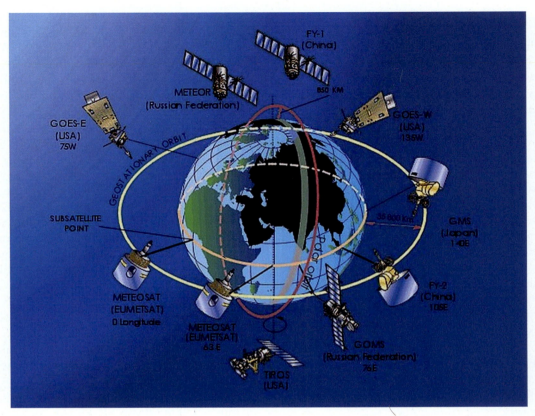

FIGURE 1.2   Illustration of the distribution of a few common geostationary satellites compared to the polar-orbiting satellites.

Third Generation (MTG) satellites from 2021 to 39. The MSG satellites carry an impressive pair of instruments: the Spinning Enhanced Visible and Infrared Imager (SEVIRI), which has the capacity to observe the Earth in 12 spectral channels and provide image data every half hour, and the Geostationary Earth Radiation Budget (GERB) instrument supporting climate studies.

The Japanese Geostationary Meteorological Satellite (GMS) series had five satellites from 1977. The Multifunctional Transport Satellites (MTSAT) are the successors to the GMS 1—5 satellite series. The MTSAT-2 from 2010 was also known as Himawari-7. Himawari-8 was operational from July 2015, and Himawari-9 started backup operation on March 2017. Both satellites

are located in orbit at around 140.7 degrees east and will observe the East Asia and Western Pacific regions for a period of 15 years. The Advanced Himawari Imager (AHI), similar to ABI, has six channel multispectral bands in the visible to near-infrared spectrum with 500m spatial resolution and provides full disk observations every 10 min and images of Japan every 2.5 min.

China has launched eight of the first-generation geostationary satellites named Fengyun (FY-2) from FY-2A to FY-2H since 1997. The second generation of geostationary meteorological satellites FY-4 was launched in December 2016, and multiple FY-4 satellites have been planned to provide service through 2037 when

TABLE 1.1   Information on GOES satellite series.

Satellites	Launch day	Status
1	October 16, 1975	Decommissioned
2	June 16, 1977	Decommissioned
3	June 16, 1978	Decommissioned
4	September 9, 1978	Decommissioned
5	May 22, 1981	Deactivated on July 18, 1990
6	April 28, 1983	Decommissioned
G	May 3, 1986	Failed to orbit
7	February 26, 1987	Used as a communications satellite; decommissioned 2012
8	April 13, 1994	Decommissioned 2004
9	May 23, 1995	Decommissioned 2007
10	April 25, 1997	Decommissioned 2009
11	May 3, 2000	Decommissioned 2011
12	July 23, 2001	Decommissioned 2013
13	May 24, 2006	On-orbit storage
14	June 27, 2009	On-orbit spare
15	March 4, 2010	Operational West backup
16 (GOES-R)	November 19, 2016	Currently operating as GOES East
17 (GOES-S)	March 1, 2017	Currently operating as GOES West
GOES-T	Planned to launch in 2020	
GOES-U	Planned to launch in 2024	

a successor program will be inaugurated. The Advanced Geosynchronous Radiation Imager (AGRI) aboard FY-4 is the corresponding version of ABI in the GOES-R series. It has 14 spectral bands, delivering full disk images every 15 min at a significantly improved resolution of 0.5–4 km.

1.2.2 Polar-orbiting satellites

Polar-orbiting satellites can provide an observational platform for the entire Earth surface, while their geostationary counterparts are limited to approximately 60 degrees of latitude of geostationary meteorological satellites at a fixed point over the Earth. Polar-orbiting satellites are able to circle the globe approximately once every 100 min. Most polar-orbiting Earth observation satellites, such as Terra, ENVISAT, and Landsat, have an altitude of about 800 km. They are in sun-synchronous orbits passing directly over a given spot on the ground at the same local time. A relatively low orbit allows detection and collection of data by instruments aboard a polar-orbiting satellite at a higher spatial resolution than from a geostationary satellite.

NASA has launched a series of polar-orbiting satellite missions with the ability to characterize the current state of the Earth system. The currently active satellites are illustrated in Fig. 1.3. All the missions fall into three types: exploratory, operational precursor and technology demonstration, and systematic.

Exploratory missions are designed to yield new scientific breakthroughs. Each exploratory satellite project is expected to be a one-time mission that can deliver conclusive scientific results addressing a focused set of scientific questions. In some cases, an exploratory mission may focus on a single pioneering measurement that opens a new window on the behavior of the Earth system. These missions are managed in the NASA Earth System Science program (ESSP). Examples include the Gravity Recovery and Climate Experiment (GRACE) and Cloud-SAT. GRACE data can be used for estimating soil moisture and surface/underground water (Section 20.4).

Operational precursor and technology demonstration missions enable major upgrades of existing operational observing systems. NASA is investing in innovative sensor technologies and developing more cost-effective

FIGURE 1.3   Illustration of the current NASA Earth observing satellites, downloaded from https://eospso.gsfc.nasa.gov/ in February 2019.

versions of its pioneer scientific instruments that can be used effectively by operational agencies. An example is the NMP EO-1 (New Millennium Program Earth Observing-1) mission launched on November 21, 2000, which includes three advanced land imaging sensors and five revolutionary crosscutting spacecraft technologies. The three sensors led to a new generation of lighter weight, higher performance, and lower cost Landsat-type Earth surface imaging instruments. The hyperspectral sensor Hyperion is the first of its kind to provide images of land surface in more than 220 spectral bands.

Systematic missions provide systematic measurements of key environmental variables that are essential to specify changes in forcings caused by factors outside the Earth system (e.g., changes in incident solar radiation) and to document the behavior of the major components of the Earth system. An example is the Earth Observing System (EOS) program. EOS is the centerpiece of NASA's recent Earth observation program. It was conceived in the 1980s and began to take shape in the early 1990s. It is composed of a series of satellites and sensors, a science component, and a data system supporting a coordinated series of polar-orbiting and low inclination satellites for long-term global observations of the land surface, biosphere, solid Earth, atmosphere, and oceans. Complete and still active EOS satellites are shown in Tables 1.2 and 1.3.

1.2.3 Overview of major satellite missions and programs

There exist 72 different government space agencies as of 2018, and 14 of those have launch capability. Six government space agencies have full launch capabilities, i.e., launch and recover multiple satellites, deploy cryogenic rocket

TABLE 1.2   Active EOS satellites as of April 2019.

Satellites	Launch day
Aqua	May 4, 2002
Aura	July 15, 2004
Cloud-aerosol LiDAR and infrared pathfinder Satellite observation (CALIPSO)	April 28, 2006
CloudSat	April 28, 2006
Cyclone global Navigation Satellite system (EVM-1) (CYGNSS)	December 15, 2016
Deep space climate observatory (DSCOVR)	February 11, 2015
ECOsystem spaceborne thermal radiometer experiment on space station (EVI-2) (ECOSTRESS)	June 29, 2018
Global ecosystem dynamics investigation LiDAR (EVI-2) (GEDI on ISS)	December 5, 2018
Global precipitation measurement core observatory (GPM Core)	February 27, 2014
Gravity recovery and climate experiment follow on (GRACE-FO)	May 22, 2018
Ice, cloud, and land Elevation Satellite-2 (ICESat-2)	September 15, 2018
Jason-3	January 17, 2016
Landsat 7	April 15, 1999
Landsat 8	February 11, 2013
Lightning imaging sensor on ISS (LIS on ISS)	February 19, 2017
Ocean surface topography Mission/Jason-2 (OSTM/Jason-2)	January 20, 2008
Orbiting carbon observatory 2 (OCO-2)	July 2, 2014
Quik Scatterometer (QuikSCAT)	June 19, 1999
Soil moisture active-passive (SMAP)	January 31, 2015
Solar radiation and climate experiment (SORCE)	January 25, 2003
Stratospheric aerosol and gas experiment III on ISS (SAGE III-ISS)	February 18, 2017

TABLE 1.2   Active EOS satellites as of April 2019.—cont'd

Satellites	Launch day
Terra	December 18, 1999
The global change observation mission-water (GCOM-W1)	May 18, 2012
Total solar irradiance spectral solar irradiance 1 (TSIS-1)	December 15, 2017

TABLE 1.3   Completed EOS satellites as of April 2019.

Satellites	Lunch day
Combined Release and Radiation Effects Satellite (CRRES)	July 25, 1990
Upper Atmosphere Research Satellite (UARS)	September 12, 1991
Atmospheric Laboratory of Applications and Science (ATLAS)	March 24, 1992
TOPEX/Poseidon	August 10, 1992
Spaceborne Imaging Radar-C (SIR-C)	April 19, 1994
Radar Satellite (RADARSAT)	November 4, 1995
Total Ozone Mapping Spectrometer-Earth Probe (TOMS-EP)	July 2, 1996
Advanced Earth Observing Satellite (ADEOS)	August 17, 1996
Orbview-2/SeaWiFS	August 1, 1997
Tropical Rainfall Measuring Mission (TRMM)	November 27, 1997
Tomographic Experiment using Radiative Recombinative ionospheric EUV and Radio Sources (TERRIERS)	May 18, 1999
Active Cavity Radiometer Irradiance Monitor Satellite (ACRIMSAT)	December 20, 1999
Challenging Mini-Satellite Payload (CHAMP)	July 15, 2000

(Continued)

TABLE 1.3   Completed EOS satellites as of April 2019.—cont'd

Satellites	Lunch day
Earth Observing-1 (EO-1)	November 21, 2000
Jason-1	December 7, 2001
Stratospheric Aerosol and Gas Experiment (SAGE III)	December 10, 2001
Gravity Recovery and Climate Experiment (GRACE)	March 17, 2002
SeaWinds (ADEOS II)	December 14, 2002
Ice, Cloud, and land Elevation Satellite (ICESat)	January 12, 2003
Polarization & Anisotropy of Reflectances for Atmospheric Sciences coupled with Observations from a LiDAR (PARASOL)	December 4, 2006
Aquarius	June 10, 2011
ISS-Rapid Scatterometer (ISS-RapidScat)	September 21, 2014
Cloud-Aerosol Transport System on ISS (CATS)	January 10, 2015

engines and operate space probes. They are the China National Space Administration (CNSA), the European Space Agency (ESA), the Indian Space Research Organization (ISRO), the Japan Aerospace Exploration Agency (JAXA), the NASA, and the Russian Federal Space Agency (RFSA or Roscosmos).

According to the Union of Concerned Scientists Database: As of November 30, 2018, there are 1957 Earth-orbiting satellites (US 849, China 284, Russia 152), 36% of these have a main purpose of either Earth Observation (EO) or Earth Science. Among active EO satellites (620) in 2017, a massive increase by 66% from the year 2016, their purposes can be grouped as following: 327 for optical imaging, 45 for radar imaging, 7 for infrared imaging, 7 for satellites, 64 for meteorology, and 60 for Earth Science.

The following will introduce the major satellite programs of the United States, Europe, and China.

1.2.3.1 USA

The United States has three major federal agencies involved in the EO satellites: NASA, NOAA, and US Geological Survey (USGS), but only NASA is responsible for launching all satellites for these agencies. The satellite missions managed by NASA have been briefly presented in Section 1.2.2, and USGS is currently managing the Landsat program. In the following, we will discuss the satellites operated by NOAA.

NOAA's operational environmental satellite system is composed of both geostationary and polar-orbiting satellites. GOES satellites, as discussed in Section 1.2.1, are mainly for national, regional, short-range warning, and "nowcasting." Complementing the GOES geostationary satellites are the polar-orbiting satellites known as Polar Operational Environmental Satellites (POES), Suomi National Polar-orbiting Partnership (S-NPP), and Joint Polar Satellite System (JPSS) for global, long-term forecasting and environmental monitoring. Both types of satellite are necessary for providing a complete global weather monitoring system.

The POES system includes the Advanced Very High Resolution Radiometer (AVHRR) and the Television Infrared Observation Satellite (TIROS) Operational Vertical Sounder (TOVS). The world's first meteorological satellite, TIROS, was launched on April 1, 1960 and demonstrated the advantage of mapping Earth's cloud cover from satellite altitudes.

On January 23, 1970, the first of the improved TIROS Operational Satellite (ITOS) was launched. Between December 11, 1970 and July 29, 1976, five ITOS satellites designated NOAA-1 through 5 were launched. From October 13, 1978 to July 23, 1981, satellites in the TIROS-N series were launched, where N represents the next generation of operational satellites. NOAA-6 and NOAA-7 were also launched during this time frame. On March 28, 1983, the first of the Advanced TIROS-N (or ATN) satellites, designated NOAA-8, was launched. NOAA continues to operate the ATN series of satellites today with improved instruments. Complementing the geostationary satellites are two NOAA polar-orbiting satellites, one crossing the equator at 7:30 a.m. local time and the other at 1:40 p.m. local time. The latest is NOAA-19, launched on February 6, 2009. NOAA-18 (PM secondary), NOAA-17 (AM backup), NOAA-16 (PM secondary), and NOAA-15 (AM secondary) all continue transmitting data as standby satellites. NOAA-19 is the "operational" PM primary satellite, and METOP-A, owned and operated by EUMETSAT, is the AM Primary satellite.

The first AVHRR sensor was a 4-channel radiometer, first carried on TIROS-N (launched October 1978). This was subsequently improved to a 5-channel instrument (AVHRR/2) that was initially carried on NOAA-7 (launched June 1981). The latest instrument version is AVHRR/3, with six channels, first carried on NOAA-15, launched in May 1998. Multiple global vegetation index datasets have been developed from NOAA-7 to now.

From 2011, NOAA has started the new JPSS program. JPSS is a collaborative program between the NOAA and NASA. This interagency effort is the latest generation of US polar-orbiting environmental satellites. The S-NPP satellite, launched in October 2011, is the predecessor to the JPSS series spacecraft and is considered the bridge between NOAA's legacy polar satellite fleet, NASA's EOS missions, and the JPSS constellation. S-NPP was constructed with a design life of 5 years but is still functioning normally. NOAA-20 (formerly JPSS-1), which launched into space on November 18, 2017, is the first spacecraft of NOAA's next generation of polar-orbiting satellites. Visible Infrared Imaging Radiometer Suite (VIIRS) is very similar to MODIS. NOAA-20 carries five similar instruments to the Suomi NPP. The following-on

satellites have been planned: JPSS-2 (2021), JPSS-3 (2026), and JPSS-4 (2031).

The Landsat missions have provided the long-term land surface observations at fine spatial resolutions (Fig. 1.4). On July 23, 1972, the Earth Resources Technology Satellite (ERTS-1) was launched and later renamed Landsat 1. The launches of Landsat 2, Landsat 3, Landsat 4, and Landsat 5 followed in 1975, 1978, 1982, and 1984, respectively. Landsat 5 provided the Thematic Mapper (TM) imagery for 28 years and 10 months; Landsat 6 failed to achieve orbit in 1993. Landsat 7 successfully launched in 1999 with the ETM + sensor, Landsat 8 in 2013, and both satellites continue to acquire data. The Landsat 9 satellite is expected to launch in December 2020.

1.2.3.2 Europe

The EUMETSAT Polar System mainly includes Metop that is a series of three polar-orbiting meteorological satellites: Metop-A (launched on October 19, 2006), Metop-B (launched on September 17, 2012), and Metop-C (launched on November 7, 2018). They are currently operated in unison.

The European Space Agency (ESA) has operated a series polar-orbiting satellite, such as CryoSat, Soil Moisture and Ocean Salinity (SMOS), and Envisat. Envisat was launched in 2002 with 10 instruments aboard but ended on April 08, 2012, following the unexpected loss of contact with the satellite. SMOS mission is a radio telescope in orbit, but pointing back to Earth not space. It was launched on November 2, 2009. CryoSat is Europe's first ice mission with an advanced radar altimeter specifically designed to monitor the most dynamic sections of Earth's cryosphere. CryoSat was launched on April 8, 2010.

ESA is developing the Sentinel satellite series. Each Sentinel mission is based on a constellation of two satellites to fulfill revisit and coverage requirements. These missions carry a range of technologies, such as radar and multispectral

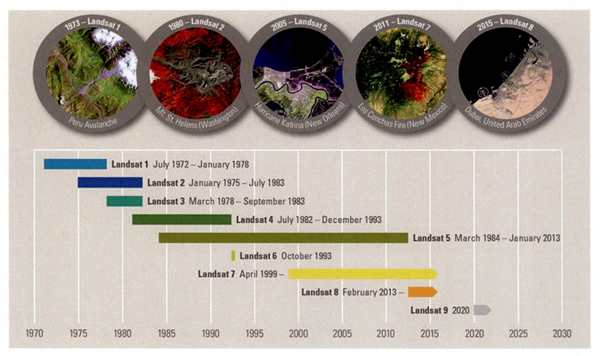

FIGURE 1.4   Landsat satellites timelines, downloaded from https://www.usgs.gov/land-resources/nli/landsat/ in February 2019.

imaging instruments. Sentinel-1 is a polar-orbiting, all-weather, day-and-night radar imaging mission for land and ocean service with Sentinel-1A launched on April 3, 2014 and Sentinel-1B on April 25, 2016. Sentinel-2 is a polar-orbiting, multispectral high-resolution imaging mission for land monitoring, with Sentinel-2A launched on June 23, 2015 and Sentinel-2B on March 7, 2017. Sentinel-3 is a multiinstrument mission to measure sea surface topography, sea- and land surface temperature, and ocean color and land color. Sentinel-3A was launched on February 16, 2016 and Sentinel-3B on April 25, 2018. Sentinel-4 is for atmospheric monitoring that will be embarked upon a Meteosat Third Generation-Sounder (MTG-S) satellite in geostationary orbit. Sentinel-5 Precursor (Sentinel-5P) is to provide timely data on a multitude of trace gases and aerosols. Sentinel-5P was launched on October 13, 2017. Sentinel-5 will monitor the atmosphere from polar orbit aboard a MetOp Second Generation satellite. Sentinel-6 carries a radar altimeter to measure global sea-surface height, primarily for operational oceanography and for climate studies.

Similar to the Landsat program, the SPOT program has also provided the long-term high-resolution satellite observations. The Landsat program has been mostly funded by the US government, but the SPOT program has been operating commercially. The SPOT satellites are summarized in Table 1.4. It is able to take stereo-pair images almost simultaneously to map surface topography.

1.2.3.3 China

China has developed several satellite series, such as meteorological satellite series Fengyun (FY), ocean satellite series Haiyang (HY), Earth resources satellite series Ziyuan (ZY),

TABLE 1.4 Overview of the SPOT satellites and data characteristics.

SPOT satellites	Launch date	Ending date	Spatial and spectral resolutions
1	February 22, 1986	December 31, 1990	One 10 m panchromatic band (0.51−0.73 µm); three 20 m multispectral bands: green (0.50−0.59), red (0.61−0.68 µm), near-infrared (0.79−0.89 µm)
2	January 22, 1990	July 2009	Same as SPOT 1
3	September 26, 1993	November 14, 1997	Same as SPOT-1
4	March 24, 1998	July 2013	One 10 m monospectral band (0.61−0.68 µm); three 20 m multispectral bands: green (0.50−0.59), red (0.61−0.68 µm), near-infrared (0.79−0.89 µm)
5	May 4, 2002	March 31, 2015	2.5/5 m panchromatic band; three 10 m multispectral bands: green (500−590 nm), red (610−680 nm), near-IR (780−890 nm) bands, and one 20m resolution on shortwave-infrared (1.58−1.75 µm)
6	September 9, 2012		One 1.5 m panchromatic band; four 6m multispectral bands: blue (450−525 nm), green (530−590 nm), red (625−695 nm), near-infrared (760−890 nm)
7	June 30, 2014		Same sensors as SPOT 6

environment and disaster monitoring small satellite constellation (HJ). Their launch times are shown in Fig. 1.5.

The resource ZY satellite series started with the China−Brazil Earth resource satellites (CBERS) jointly developed by China and Brazil with CBERS-1 launched in 1999 and CBERS-2 on October 21, 2003. ZY-1 02C was launched on December 22, 2011. ZY-3 is China's first high-resolution civilian optical transmission-type stereo mapping satellite that integrates the functions of surveying, mapping, and resources investigation. ZY-3 is equipped with two front and back view CCD cameras having the resolution better than 3.5 m, one CCD camera with the resolution better than 2.1 m, and one multispectral camera with the resolution better than 5.8 m. The swath is about 50 km.

The meteorological FY satellite series include both geostationary (FY2 and FY4) and polar-orbiting (FY-1 and FY-3) satellites. FY3 is the second generation of the Chinese meterological polar-orbiting satellites. FY-3A was launched on May 27, 2008 and carried 11 sensors. FY-3D is the latest one in the series launched on November 15, 2017. Additional satellites in this series have also been planned with FY-3E (2019), FY-3F (2019), and FY-3G (2022). FY-4A was launched on December 10, 2016, and additional five new FY-4's launches were also planned.

Gaofen (GF), meaning high resolution in Chinese, satellite series is part of the China High-Resolution Earth Observation System (CHEOS), an analog to Europe's Copernicus program of Sentinel Earth observation satellites. The first few satellites and some characteristics are shown in Table 1.5.

1.2.4 Small satellites and satellite constellations

All satellites can be classified into seven classes based on their masses (see Table 1.6). There is a trend in using small satellites for Earth observation for reducing the cost: heavier satellites require larger rockets with greater thrust, which also has greater cost to finance. Small satellites, also known as miniaturized satellites, are artificial satellites of low mass and size, usually under 500 kg. Most of these small satellites have been used with "mother" satellites that provide

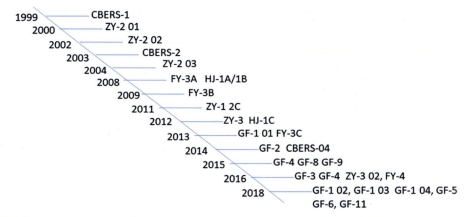

FIGURE 1.5   Major Chinese satellites relevant to land remote sensing and their launch times (Liang et al., 2018).

operating signals; however, more recent versions are operating independently. Femto satellites and other types of small satellites are beginning to revolutionize not only who can send satellite systems in space but also they have now given unprecedented access for data collection.

According to the Union of Concerned Scientists Database, all 620 active EO satellites in 2017 include 186 large satellites, 74 small satellites, 100 microsats, 215 Nanosats/CubeSats, and the remaining 45 satellites that do not have a launch mass specified. In particular, the number of Nanosats/CubeSats increased by 34.68% from 2016.

Satellite systems are generally transitioning from the single satellite model to the cooperative sensing approach. For missions requiring global or continuous coverage in real time or within a very short temporal period, there is a potential advantage in deploying a constellation of satellites. A satellite constellation is a group of satellites operating in a coordinated format. The well-known example is the global positioning system (GPS) constellation (Fig. 1.6).

Constellations of small satellites may offer a new approach to those science missions that would benefit from more frequent sampling by a larger number of lower cost sensors. Monitoring of time-varying phenomena such as cloud cover might be a good example of a mission where higher refresh rate at lower accuracy is a preferred approach.

The Planet Labs, a private company based in San Francisco, CA, USA, had launched 298 satellites, 150 of which were active, as of September 2018. The company is operating several Earth observation satellite constellations: *Flock*, *RapidEye*, and *Skysat*. The Flock constellation consists of the Dove Cubesats that weigh 4 kg (8.8 lb), $10 \times 10 \times 30$ cm (3.9 in $\times$ 3.9 in $\times$ 11.8 in) in length, width, and height. Each Dove satellite is tiny and has a lifespan of 1–3 years but can observe the Earth at 3–5 m spatial resolution.

The *RapidEye* constellation consists of five satellites producing 5-m (16 ft) resolution imagery that Planet acquired from the German company BlackBridge in 2015. The five satellites travel on the same orbital plane (at an altitude of 630 km) and together are capable of collecting over 4 million km of 5-m resolution, 5-band imagery every day in the blue (440–510 nm), green (520–590 nm), red (630–690 nm), red-edge (690–730 nm), and near infrared (760–880 nm).

The Skysat constellation, purchased from Google in 2017, is composed of CubeSat that can observe the Earth surface at a spatial resolution of 0.9 m in its 400–900 nm panchromatic band, making it the smallest satellite to be put

TABLE 1.5 Overview of the Gaofen first seven satellites and data characteristics.

GF-	Launch date	Notes
1	April 26, 2013	Two sensors: high-resolution cameras (HRC) and wide field imagers (WFI). HRC includes pan at 2 m and four multispectral bands (blue, green, red, and near-IR) at 8m with the swath of 68 km. WFI has similar four multispectral bands to HRC at 16m resolution with the swath of 830 km. Repeating cycle: ≤ 4 days at the equator
2	August 19, 2014	A single camera: one 1 m Pan and 4 m multispectral bands (blue, green, red, and near-IR). Swath: 45 km. Repeating cycle: ≤4 days at equator
3	September 8, 2016	A quad-polarization (vertical-vertical (VV); horizontalhorizontal (HH); vertical-horizontal (VH); horizontalvertical (HV)) C-Band SAR at 25 m spatial resolution, a 26-day repeat cycle
4	December 28, 2015	A geostationary satellite with a camera of 5 bands. The first four bands (blue, green, red, and near-IR) at 50 m resolution, the middle-IR (3.5–4.1 μm) at 400m resolution. Swath: 400 km.
5	September 5, 2018	Two hyperspectral/multispectral sensors for terrestrial earth observation and four atmospheric observation sensors: visible shortwave infrared hyperspectral camera, full-spectrum spectral imager, atmospheric aerosol multiangle polarization detector, atmospheric trace gases differential absorption spectrometer, main atmospheric greenhouse gases monitor, ultrahigh-resolution infrared atmospheric sounder
6	February 6, 2018	Similar to the GF-1 satellite, but using a different instrument suit, consisting of a 2/8 m resolution panchromatic/hyperspectral camera and a 16 m resolution wide angle camera
7	2019	Similar to ZY-3 with 3D topographical mapping

in orbit capable of such high-resolution imagery. The four multispectral bands have a spatial resolution of 2 m in blue (450–515 nm), green (515–595 nm), red (605–695 nm), and near-infrared (740–900 nm). As of September 2016, six SkySat satellites were launched. In October 2017, four additional Dove satellites were also launched being part of this constellation.

Besides the constellations with actual satellites simultaneously orbiting in space, the concept of virtual constellation has also been proposed. The CEOS defines virtual constellations as a *"set of space and ground segment capabilities that operate in a coordinated manner to meet a combined and common set of Earth Observation requirements."*

We are increasingly faced the challenging difficulty to address rapid changes in the global environment using data from single-satellite sensors or platforms due to the underlying limitations of data availability and tradeoffs that govern the design and implementation of existing satellite systems. Virtual constellations can principally be used to add value to Earth observation by combining sensors with similar spatial, spectral, temporal, and radiometric characteristics. Virtual constellations of planned and existing satellite sensors may help to overcome the limitation by combining existing observations to mitigate limitations of any one particular sensor. While multisensory applications are not new, the integration and harmonization of multisensor data is still challenging, requiring tremendous efforts of science and operational user communities.

The CEOS has formed seven virtual constellations so far to coordinate space-based, ground-based, and/or data delivery systems to meet a common set of requirements within a specific domain, including Atmospheric Composition (AC-VC), Land Surface Imaging (LSI-VC), Ocean Color Radiometry (OCR-VC), Ocean Surface Topography (OST-VC), Ocean Surface

TABLE 1.6 Classification of EO satellite system.

Satellites	Large satellite	Medium-sized satellite	Minisatellitemicrosat	Microsatellite or	Nanosatellite or nanosat	Picosatellite or picosat	Femtosatellite or femtosat
Mass	>1000 kg	500–1000 kg	100–500 kg	10–100 kg	1–10 kg	0.1–i kg	<0.1 kg

FIGURE 1.6 Global positioning system satellite constellation, downloaded from https://upload.wikimedia.org/wikipedia/commons/e/e2/GPS-constellation-3D-NOAA.jpg.

Vector Wind (OSVW-VC), Precipitation (P-VC), and Sea Surface Temperature (SST-VC). They leverage inter-Agency collaboration and partnerships to address observational gaps, sustain the routine collection of critical observations, and minimize duplication/overlaps, while maintaining the independence of individual CEOS Agency contributions.

1.2.5 Sensor types

The sensor technology has been well reviewed recently (Toth and Jozkow, 2016). There are two types of sensors: passive and active. Passive sensors detect natural radiation that is emitted by the object being viewed or reflected by the object from a source other than the instrument. Reflected sunlight is the most common external source of radiation sensed by passive sensors. Typical passive sensors include the following:

- Radiometer: An instrument that quantitatively measures the radiance of electromagnetic radiation in the visible, infrared, or microwave spectral region.
- Imaging radiometer: A radiometer that includes a scanning capability to provide a two-dimensional array of pixels from which an image may be produced. It is often called a scanner. Scanning can be performed

mechanically or electronically by using an array of detectors. Across-track scanners, scanning from one side of the sensor to the other across the platform flight direction using a rotating mirror, are called Whiskbroom Scanners, such as AVHRR. Alone-track scanners, scanning a swath with a linear array of charge-coupled devices (CCD) arranged perpendicular to the flight direction of the platform without using a mechanical rotation device, are called Pushbroom scanners, such as High Resolution Visible of SPOT and Advanced Land Imager of EO-1.

- Spectroradiometer: A radiometer that can measure the radiance in multiple spectral bands, such as the Moderate Resolution Imaging Spectroradiometer (MODIS) and the Multi-angle Imaging SpectroRadiometer (MISR).

Active sensors provide their own electromagnetic radiation to illuminate the scene they observe. They send a pulse of energy from the sensor to the scene and then receive the radiation that is reflected or backscattered from that scene. Typical active sensors include:

- Radar (Radio Detection and Ranging): A microwave radar that uses a transmitter operating at microwave frequencies to emit electromagnetic radiation and a directional antenna or receiver to measure the time of arrival of reflected or backscattered pulses of radiation from distant objects for determining the distance to the object.
- SAR: A side-looking radar imaging system that uses relative motion between an antenna and the Earth surface to synthesize a very long antenna by combining signals (echoes) received by the radar as it moves along its flight track for obtaining high spatial resolution imagery. There are multiple SAR systems in operation, and some examples can be seen in Section 13.4 for estimating vegetation canopy height and vertical

structure information, Section14.4.2 for estimating above-ground forest biomass, and Section 18.3.2.2 for mapping soil moisture.

- Interferometric synthetic-aperture radar (InSAR): A technique that compares two or more amplitude and phase images over the same geographic region received during different passes of the SAR platform at different times. InSAR can survey height information of the illuminated scene with cm-scale vertical resolution and 30-m pixel resolution, and covering areas 100×100 km (in standard beam modes). Examples include ERS-1 (1991), JERS-1 (1992), RADARSAT-1 and ERS-2 (1995), and ASAR (2002). While the majority of InSAR missions to date have utilized C-band sensors, recent missions such as ALOS PALSAR, TerraSAR-X, and COSMO SKYMED are expanding the available data in the L- and X-bands.
- Scatterometer: A high frequency microwave radar designed specifically to determine the normalized radar cross section of the surface. Over ocean surfaces, measurements of backscattered radiation in the microwave spectral region can be used to derive maps of surface wind speed and direction. It has also been used for mapping surface soil moisture and freeze/thaw states. Examples include the Advanced Microwave Instrument (AMI) of ERS-1 and ERS-2.
- LiDAR: An active optical sensor that uses a laser in the ultraviolet, visible, or near-infrared spectrum to transmit a light pulse and a receiver with sensitive detectors to measure the backscattered or reflected light. Distance to the object is determined by recording the time between the transmitted and backscattered pulses and using the speed of light to calculate the distance traveled. The details are given in Chapters 13 and 14.
- Laser Altimeter: A laser altimeter that uses a LiDAR to measure the height of the instrument platform above the surface. By independently knowing the height of the

platform with respect to the mean Earth's surface, the topography of the underlying surface can be determined. The Geoscience Laser Altimeter System (GLAS) of ICESat is a typical example of a space-based Laser Altimeter.

1.2.6 Data characteristics

The specifications of the platform and the sensor determine the resolutions of the remotely sensed data: spatial, spectral, temporal, and radiometric.

1.2.6.1 Spatial resolution

Spatial resolution is a measure of the smallest object that can be resolved by the sensor, or the ground area imaged for the instantaneous field of view (IFOV) of the sensor, or the linear dimension on the ground represented by each pixel. Fig. 1.7 shows the campus of the University of Maryland at College Park at four different spatial resolutions. Table 1.7 shows the spatial resolution of some common sensors.

1.2.6.2 Spectral resolution

The spectral resolution describes the number and width of spectral bands in a sensor system. Many sensor systems have a panchromatic band, which is one single wide band in the visible spectrum, and multispectral bands in the visible-near-IR or thermal-IR spectrum see (Table 1.7). Hyperspectral systems usually have hundreds of spectral narrow bands; for example, Hyperion on EO-1 satellite has 220 bands at 30-m spatial resolution.

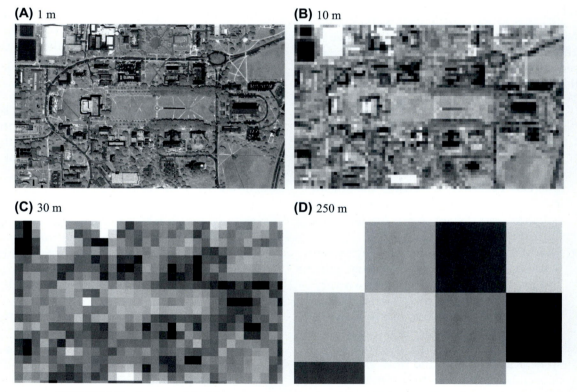

(A) 1 m   **(B)** 10 m   **(C)** 30 m   **(D)** 250 m

FIGURE 1.7   Campus of the University of Maryland at College Park at four Spatial resolutions.

TABLE 1.7   Characteristics of some commonly used satellite sensors.

	Satellite sensors	Spectral bands	Spatial resolution (m)	Radiometric resolution (bit)	Temporal resolution (day)	Temporal coverage
Coarse resolution (>1000 m)	POLDER	B1—B9	6000*7000	12	4	POLDER 1: October 1996 to June 1997 POLDER2: April to October 2003
Medium resolution (100 −1000 m)	MODIS	B1—B2	250	12	Daily	1999
		B3—B7	500			
		B8—B36	1000			
	AVHRR	B1—B5	1100 at nadir	10	Daily	
Fine resolution (5—100 m)	ALI/EO1					
	ASTER/Terra	B1	15	8		
		B2—B9	30			
		B11—B14	90	12		
	ETM+/Landsat 7	Pan	15	8	16	1999-
		B1—B5,B7	30			
		B6	60			
	HRV/SPOT5	Pan	2.5 or 5	8	26/2.4	2002-
		B1—B3	10			
		SW-IR	20			
Very high resolution (<5 m)	Ikonos	Panchromatic band	0.82 at nadir	11	3 days at 40 degrees latitude	1999-
		B1-b4	3.2 at nadir			
	Quickbird	Pan	0.61	11	1—3.5	2001-
		B1—B4	2.44			
	World view	Pan	0.5 at nadir	11	1.7—5.9	2007-
	Geoeye-1	Pan	1.41 at nadir	11	2.1—8.3 days at 40 degrees latitude	2008-
		B1—B4	1.65 at nadir			

1.2.6.3 Temporal resolution

Temporal resolution is a measure of the repeat cycle or frequency with which a sensor revisits the same part of the Earth's surface. The frequency characteristics are determined by the design of the satellite sensor and its orbit pattern.

The temporal resolutions of common sensors are also shown in Table 1.7.

1.2.6.4 Radiometric resolution

Radiometric resolution refers to the dynamic range, or the number of different output numbers

(A) 8 bits (256 levels)

(B) 4 bits (16 levels)

(C) 2 bits (4 levels)

(D) 1 bit (2 levels)

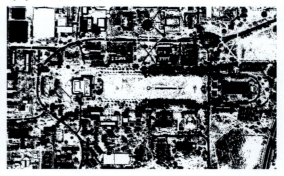

FIGURE 1.8   Campus of the University of Maryland at College Park at four radiometric resolutions.

in each band of data, and is determined by the number of bits into which the recorded radiation is divided. In 8-bit data, the digital numbers (DN) can range from 0 to 255 for each pixel (28 $^{1}/_{4}$ 256 total possible numbers). Obviously more bits results in higher radiometric accuracy of the sensor, as shown in Fig. 1.8. The radiometric resolutions of common sensors are shown in Table 1.7.

1.3 Data transmission and ground receiving system

There are three main options for transmitting data acquired by satellite sensors to the surface:

(1) the data can be directly transmitted to Earth if a Ground Receiving Station (GRS) is in the line of sight of the satellite;

(2) the data can be recorded on board the satellite for transmission to a GRS at a later time; and

(3) the data can also be relayed to the GRS through the Tracking and Data Relay Satellite (TDRS) System (TDRSS), which consists of a series of communications satellites in geosynchronous orbit. The data are transmitted from one satellite to another until they reach the appropriate GRS. NASA's TDRS started in the early 1970's and has evolved for three generations. The current TDRSS consists of 10 in-orbit satellites (four first generation, 3 s generation and two third generation satellites) distributed to provide near-constant communication links between the ground and orbiting satellites (e.g., Landsat).

There are two types of GRSs: fixed and mobile. Most GRSs are fixed, and Fig. 1.9 shows the locations of all currently active ground stations operated by the United States (only two in South Dakota and Aelaska) and International Cooperator ground station network for the direct downlink and distribution of Landsat 7 (L7) and/or Landsat 8 (L8) image data. As coverage of the globe by ground receiving stations is not complete, as seen from Fig. 1.9, the mobile station is an attractive solution to fill the holes and also an efficient means to perform acquisition in a remote location for a long period of time when a lot of images are needed for particular work (cartography of a region for example).

The ground receiving stations acquire, preprocess, archive, and process data. Their typical components and functions may include the data acquisition facility, the data processing facility, the value added facility, and user support services.

1.4 Data processing

A series of preprocessing tasks are needed to undertake before environmental information can be accurately extracted from remotely sensed data. Two types of preprocesses are conducted: radiometric processing and geometric processing. The radiometric processing may include sensor radiometric calibration, image enhancement (mostly filtering noises), atmospheric correction, and image fusion.

1.4.1 Radiometric calibration

Radiometric calibration is a process that converts recorded sensor voltages or digital numbers (DN) to an absolute scale of radiance or reflectance. Because outer space is such a harsh environment, the performance of all satellite sensors degrades over time. To achieve consistent and accurate measurements that can be used to detect climatic and environmental

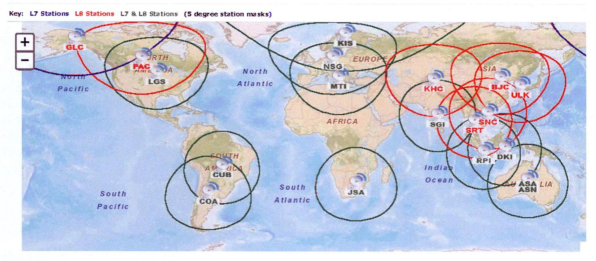

FIGURE 1.9 The locations of all active ground stations operated by our US and International Cooperator (IC) ground station network for the direct downlink and distribution of Landsat 8 and/or Landsat 9 data. The circles show the approximate area over which each station has the capability for direct reception of Landsat data. *Downloaded from https://landsat.usgs.gov/igs-network in February 2019.*

change, the digital numbers (DNs) need to be transformed into physical quantities.

Calibration measurements can be conducted in three stages: preflight, in-flight, and post-launch:

- Preflight calibration measures a sensor's radiometric properties before that sensor is sent into space. Preflight instrument calibration is performed at the instrument builder's facilities. The controllable and stable environment in the laboratory guarantees high calibration accuracy and precision
- In-flight calibration is usually performed on a routine basis with on-board calibration systems. More and more optical sensors have on-board calibration devices. For example, the AVHRR optical sensor does not have an on-board calibration capability, but the ETM+ has three on-board calibration devices: the Internal Calibrator, the Partial Aperture Solar Calibrator, and the Full Aperture Solar Calibrator. MODIS also has three dedicated calibration devices for the reflective bands: Solar Diffuser, Solar Diffuser Stability Monitor, and the Spectroradiometric Calibration Assembly. In addition, MODIS has two additional calibration techniques: looking at the Moon and at deep space. The MODIS sensor has such an onboard calibration system that promises an absolute error better than 2%.
- *Post-launch* calibration data have to be obtained from vicarious calibration techniques that typically make use of selected natural or artificial sites on the surface of the Earth. Prelaunch and onboard methods are better established, and postlaunch methods using invariant sites in vicarious calibration is becoming more popular with the changing design and demands of new instruments. Vicarious calibration using pseudo-invariant sites has become increasingly accepted as a fundamental postlaunch calibration method to monitor long-term performance of satellite

reflective solar sensors. There are several common desired characteristics of an invariant sited for example, temporal stability, spatial uniformity, little or nonvegetation, and relatively high surface reflectivity with approximately Lambertian reflectance. Commonly used sites include stable desert areas of the Sahara, Saudi Arabia, Sonoran, White Sand, and regions in Bolivia. By observing these sites with satellite sensor systems over extended periods of time, degradations (trends) in sensor responsivity can be monitored and quantified.

Some sensors have neither on-board calibration devices nor regular post-launch calibration, for example, AVHRR. One solution to calibrate the sensor is through *cross calibration*. MODIS as a well calibrated instrument (Xiong et al., 2018) has been used as a reference to calibrate other sensors using coincident observations of MODIS and the target sensors over the pseudo-invariant calibration sites. For example, Vermote and Kaufman (1995) proposed a cross-calibration method using a time series of MODIS and AVHRR data over a Saharan Desert site.

These two methods using data over the pseudo-invariant calibration sites provide absolute radiometric calibration of the sensors. Many sensors contain multiple detectors that have slightly different responsivities. As a result, the imagery produced by these sensors may contain a significant level of striping. One solution is to match the mean values of each detector over a period of times so that all detectors produce the relatively uniform values.

1.4.2 Geometric processing

No image acquired by sensors can perfectly represent the true spatial properties of the landscape. Many factors can also distort the geometric properties of remote sensing data, such as variations in the platform altitude, attitude and velocity, Earth rotation and curvature, surface

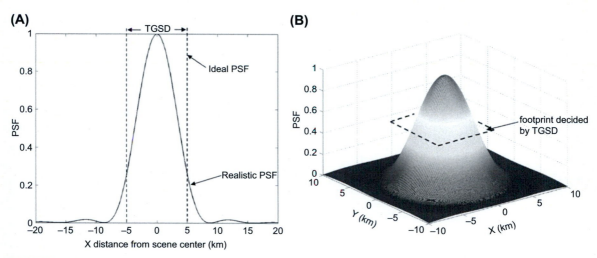

FIGURE 1.10 Point spread functions: (A) Realistic and ideal PSF model for simulating GEO radiances, (B) Sketch of EE in terms of the PSF, Dashed square shows the integral area for EE, which is determined by the TGSD (Zhang et al., 2006).

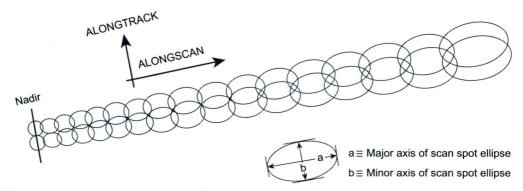

FIGURE 1.11 Sketch of pixel geometry for the AVHRR for adjacent scan lines to illustrate autocorrelation (Breaker, 1990).

relief displacement, and perspective projection. Some of these resulting distortions are systematic and can be corrected through analysis of sensor characteristics and platform ephemeris data, but others are random and have to be corrected by using ground control points (DCP).

In the sensor ground instantaneous field of view (IFOV), surface elements do not contribute to the pixel value equally, but rather, the central part contributes most to the pixel value. This kind of spatial effect is usually specified by the sensor point spread function (PSF) in the spatial domain, and the Fourier transform of the PSF is called the modulation transfer function (MTF), a precise measurement of details and contrast made in the frequency domain. The sensor PSF is often modeled as a Gaussian. Fig. 1.10 illustrates the PSF in two- and three-dimensions (Fig. 1.11), where TGSD is the threshold ground sample distance, which is the centroid-to-centroid distance between adjacent pixels.

The actual response function of the ground IFOV is often not square; for example, for MODIS, it is twice as wide cross-track as in-track because of time integration during scanning. For most whiskbroom scanners, such as

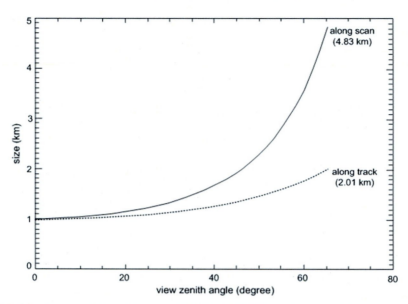

FIGURE 1.12   The actual size of the ground IFOV of AVHRR as a function of the view zenith angle.

AVHRR and MODIS, the actual size of the ground IFOV is a function of the scanning angle (see Fig. 1.12).

This topic is significant because level-1 radiance data or level-2 reflectance data should be corrected for geometric distortions before calculating geophysical parameters in order to obtain a truly absolute geophysical parameter. The details are discussed in Chapter 2.

1.4.3 Image quality enhancement

Imperfections or image artifacts are continuously caused by the instrument's electronics, dead or dying detectors, and downlink errors. Known artifacts include the scan-correlated shift, memory effect, modulation transfer function, and coherent noise. Dropped lines and in-operable detectors also exist as a result of decommutating errors and detector failure. Potential remnant artifacts include banding and striping. In the past, these effects were ignored or artificially removed using cosmetic algorithms during radiometric preprocessing. For example, dropped lines are usually filled with the values of the previous lines or the averages of the neighboring lines. Strips can be removed by using simple along-line convolution, high-pass filtering, and forward and reverse principal component transformations.

To assist human visual interpretation, various image enhancement techniques have been incorporated in many remote sensing digital image processing systems. These enhancement methods can be divided into spatial domain and frequency domain categories. In spatial domain techniques, we directly deal with the image pixels. Fig. 1.13 illustrates the effects of a linear enhancement technique. The pixel values are manipulated to achieve the desired enhancement. In frequency domain methods, the image is first transferred to the frequency domain. That is, the Fourier transform of the image is computed first, all the enhancement operations are performed on the Fourier transform of the image, and then the inverse Fourier transform is performed to obtain the resultant image.

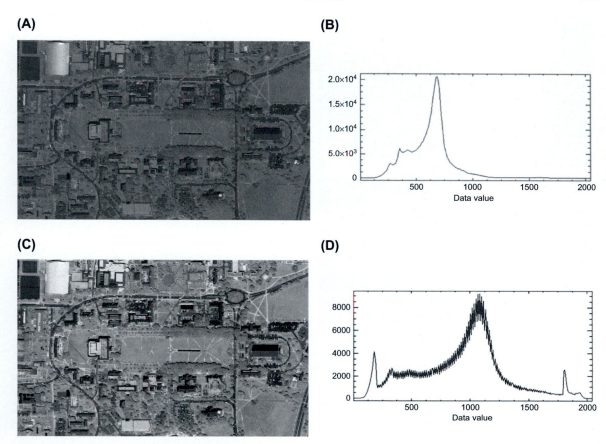

FIGURE 1.13 An example of linear enhancement: original image and its histogram (A) and (B); linearly enhanced image and its histogram (C) and (D). Note that the radiometric properties characterizing environmental conditions are artificially altered by image enhancement methods. Most image enhancement techniques for assisting visual interpretation should not be performed before quantitatively estimating biophysical variables. (A) Original image. (B) Histogram. (C) Enhanced image. (D) New histogram.

1.4.4 Atmospheric correction

Since the observed radiance recorded by a spaceborne or airborne sensor contains both atmospheric and surface information, atmospheric effects must be removed to estimate land surface biogeophysical variables, particularly from the reflective and thermal IR data, since microwave signals are not very sensitive to changes in atmospheric conditions.

Clouds in the atmosphere largely block Earth surface information and make most optical and thermal-IR imagery useless for terrestrial applications. Various cloud and shadow detection algorithms have been developed, and cloud mask is one of the high-level atmospheric products. However, this is still an active research area, and more effective and reliable algorithms are needed. For km coarse-resolution imagery (e.g., AVHRR and MODIS), there usually remain many cloudy or mixed cloudy pixels after applying the cloud mask. Various solutions have been used to address this issue. One solution relies on temporal compositing techniques, converting daily observations to weekly or

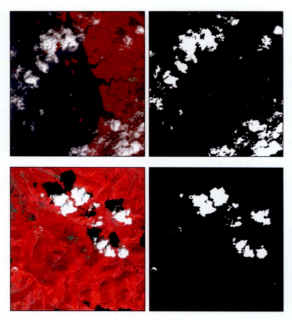

satellite imagery, different methods (Sun et al., 2017b), such as the threshold, radiative transfer and statistical methods have been developed to identify the clouds and the associated shadows. One example is shown in Fig. 1.14.

For optical imagery, both aerosol and water vapor scatter and absorb the radiation reflected from the surface. There are two approaches for atmospheric correction:

- The first assumes known atmospheric properties, usually the total amounts of aerosol and water vapor in the atmospheric column, which may be estimated from other sensors and/or other sources. Many atmospheric radiative transfer codes (e.g., MODTRAN, 6S) can be used to calculate the quantities required for atmospheric correction.
- The second relies only on the imagery itself without any external information.

FIGURE 1.14 Examples of cloud detection. The left column shows two false-color composite imagery of Landsat data and the detected clouds are shown in the right column (Sun et al., 2017b).

monthly data based on maximum vegetation index or other criteria; other solutions include replacement of these contaminated pixels using smoothing algorithms. This topic will be discussed in Chapter 3. For high-resolution

If the atmospheric information can be accurately estimated from other sources, the first approach is preferable, but quite often we are not that lucky. Fig. 1.15 shows the significant differences of MODIS surface reflectance before and after atmospheric correction (Liang et al., 2002). This topic will be discussed in Chapter 4.

(A)   (B)

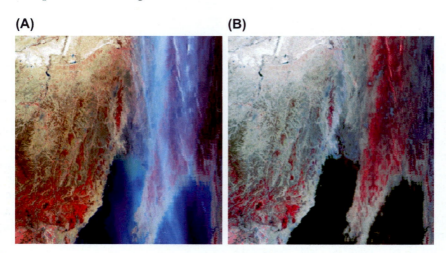

FIGURE 1.15   An example of atmospheric correction of MODIS imagery before (A) and after (B) (Gui et al., 2010). *From Liang, S., Fang, H., Chen, M., Shuey, C., Walthall, C., Daughtry, C., 2002. Atmospheric correction of landsat ETM+ land surface imagery: II. validation and applications. IEEE Trans. Geosci. Remote Sens. 40, 2736–2746. © 2002, IEEE.*

For thermal-IR imagery, if we can acquire atmospheric profile information (mainly temperature and water vapor) from sounding data, atmospheric correction is straightforward. The split-window approach based on two thermal-IR bands, when no such atmospheric profile information is available, is often used to estimate land surface temperature without atmospheric correction. The details are available in Chapter 7.

1.4.5 Image fusion and product integration

There are many cases where we need to integrate image data through image fusion techniques. Definitions of image fusion in the literature are very diverse. Image fusion can be viewed as a process that produces a single image from a set of input images. The fused image should have more complete information and is more useful for estimating land surface variables. It can improve both reliability by using redundant information and capability by using complementary information, as illustrated in Fig. 1.16.

Image fusion is not distinguished from image merging or image integration, which at the pixel level may be in many different forms, for example:

- Multitemporal images from the same or multiple sensors for change detection (e.g., merge TM images acquired at different times);

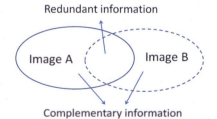

FIGURE 1.16   Illustration of image fusion.

- Multispatial images from the same or multiple sensors (e.g., merge ETM panchromatic and multispectral images);
- Multiple images of different spectral regions from the same or multiple sensors (e.g., merge SAR with optical imagery or visible bands with thermal bands);
- Remote sensing images with ancillary data (e.g., topographic map).

When the high-level satellite products are evaluated, it is surprising to see that most products are mainly generated from a single sensor. For example, the MODIS albedo product is mainly from MODIS data, which is also true for MISR, MERIS, etc. The same product from different satellite sensors may have different characteristics (e.g., spatial and temporal resolutions, accuracy). Instead of asking the user to pick the "best" product, we can generate a blended/integrated product from multiple-sensor products. Chapter 21 is devoted to addressing this topic in the example of leaf area index (LAI).

1.5 Mapping category variables

We are interested in two types of land surface variables: category and quantitative. The category variables represent the types of objects on the land surfaces and are usually mapped out through image classification. The purpose of image classification is to group together pixels that have similar properties into a finite set of classes. An example of a classified image is a land cover map. Fig. 1.17 is a global land cover map mapped from MODIS data. The key steps in the classification process are as follows:

(1) Definition of classification system (scheme): This depends on the objective and the characteristics of the remote sensing data. The purpose of such a scheme is to provide

(2) A framework for organizing and categorizing the information that can be

FIGURE 1.17 Global land cover classification map from MODIS.

MODIS IGBP Land Cover Legend

- Unclassified
- Water
- Evergreen Needleleaf Forest
- Evergreen Broadleaf Forest
- Deciduous Needleleaf Forest
- Deciduous Broadleaf Forest
- Mixed Forest
- Closed Shrublands
- Open Shrublands
- Woody Savannas
- Savannas
- Grasslands
- Permanent Wetlands
- Croplands
- Urban and Built-up
- Cropland/Natural Vegetation Mosaic
- Snow and Ice
- Barren or Sparsely Vegetated

extracted from the data. A number of classification schemes have been developed for mapping regional and global land cover and land use maps. The IGBP land cover classification system for global mapping using MODIS data is shown in Fig. 1.17.

(3) Selection of features: Classification is executed based on a series of features in the feature space. It divides the feature space into several classes based on a decision rule. Instead of using the original bands, they are often transformed into feature space to discriminate between the classes. Examples of features include various vegetation indexes, principal components and those from the Tasseled-Cap transformation, and other spatial, temporal, and angular features. The subset of features is selected to maximally distinguish different classes.

(4) Sampling of training data: Training is the process of defining the criteria by which these classes are recognized and is performed with either a supervised or an unsupervised method. Supervised training is closely controlled by the analyst, who selects pixels from each class based on high-resolution imagery, ground truth data, or maps, while unsupervised training is more computer-automated and enables the user to specify some parameters that the computer uses to uncover statistical patterns that are inherent in the data but do not necessarily correspond to classes in the classification scheme.

(5) Classification: A parametric or nonparametric decision rule, which is often called a classifier, is used to perform the actual sorting of pixels into distinct class values. There are various classifiers, such as the parallelepiped classifier, minimum distance classifier, maximum likelihood classifier, regression tree classifier, and support vector machine (SVM) classifier. They are compared with the training data so that an appropriate decision rule is selected for classification.

(6) Accuracy assessment: The classified results should be checked and verified for their accuracy and reliability. The training data are usually divided into two parts, one for training and the other for validation. In the evaluation of classification errors, a classification error matrix is typically formed, which is sometimes called a confusion matrix or contingency table.

The details of image classification techniques are not covered by this book, but the basic principles and progress can be found elsewhere (Dash and Ogutu, 2016; Lu and Weng, 2007). Some typical techniques for mapping land use types will be discussed in Chapters 23−25. At most spatial resolutions, the majority of pixels are mixed. If a pixel is required not just to be labeled as one of the cover types but to estimate the percentages of the cover types, it would be more challenging. How to estimate the fractional vegetation coverage within one pixel will be discussed in Chapter 12.

1.6 Estimating quantitative variables

To drive, calibrate, and validate the Earth process models and support various applications, high-level products of quantitative variables are much more desirable. How to generate these products is the main focus of this book. In the early stages of remote sensing technique development, visual interpretation was the approach commonly used for extracting land surface information. Statistical analysis later became a more common method for quantitatively estimating land surface information. As can be seen in the following chapters, various inversion techniques based on physically based surface radiation models have become the subject of mainstream research (Liang, 2007). It is necessary to provide an overview of these techniques. As many inversion algorithms are based on forward radiation modeling, let us first begin with that.

1.6.1 Forward radiation modeling

This is the process that links the pixel values of an image with surface characteristics through mathematical models (Liang, 2004). We will mainly present landscape generation, surface and atmosphere radiative transfer modeling, and sensor models.

1.6.1.1 Scene generation

Scene generation is a quantitative description of our understanding of the landscape. Strahler et al. (1986) identify two different scene models in remote sensing: H- and L-resolution models. H-resolution models are applicable where the elements of the scene are larger than the pixel size, and L-resolution models are applicable when the converse is true. H-resolution scenes can be generated using computer graphics techniques; for example, a vegetation canopy can be created with *Onyx* software (http://www.onyxtree.com/). L-resolution scenes can be generated using mathematical models or GIS (geographic information system) techniques.

1.6.1.2 Surface radiation modeling

Given landscape composition and its optical properties, we can predict the radiation field. Three types of models characterize the radiation field of the scene, and they are commonly used in optical remote sensing: geometric optical models, turbid-medium radiative transfer models, and computer simulation models.

In geometric optical models (Li and Strahler, 1985, 1986), canopy or soil is assumed to consist of geometric protrusions with prescribed shapes (e.g., cylinder, sphere, cones, ellipsoid, spheroid), dimensions, and optical properties that are distributed on a background surface in a defined manner (regularly or randomly distributed). The total pixel value is the weighted average of sunlit crown, sunlit ground, shadowed crown, and shadowed ground. Fig. 1.18 illustrates a simulated canopy field with an ellipsoid shape and the calculated sunlit and shadowed components.

Turbid-medium radiative transfer models treat surface elements (leaf or soil particle) as small absorbing and scattering particles with given optical properties, distributed randomly in the scene and oriented in given directions. In one-dimensional canopy models(Kuusk, 1995; Liang and Strahler, 1993b; Liang and Townshend, 1996; Verhoef, 1984), canopy elements are assumed to be randomly distributed, but three-dimensional RT models (Kuusk, 2018; Myneni et al., 1989) can take into account the structural information of the landscape, as shown in Fig. 1.19. The further development of geometric optical models has incorporated radiative transfer theory in calculating the individual sunlit/shadow components; the resulting models are often called hybrid models. In computer simulation models, the arrangement and orientation of scene elements are simulated on a computer, and the radiation properties are determined based on the radiosity equations (Borel et al., 1991; Huang et al., 2013; Qin and Gerstl, 2000) and/or Monte Carlo ray tracing (Disney et al., 2006; Gastellu-Etchegorry et al., 2015; Lewis, 1999; North, 1996; Qi et al., 2019) methods. Fig. 1.20 compares a photo of grass field and the simulated field using the Botanical Plant Modeling System (Lewis, 1999) based on the ray-tracing technique.

1.6.1.3 Atmospheric radiative transfer

The radiation at the Earth's surface is disturbed by the atmosphere before being captured by the sensor in the atmosphere (airborne sensors) or above the atmosphere (spaceborne sensors).

Atmospheric gases, aerosols, and clouds scatter and absorb the incoming solar radiation and the reflected and/or emitted radiation from the surface. As a result, the atmosphere greatly modulates the spectral dependence and spatial distribution of the surface radiation. The atmospheric radiative transfer theory is quite mature, and

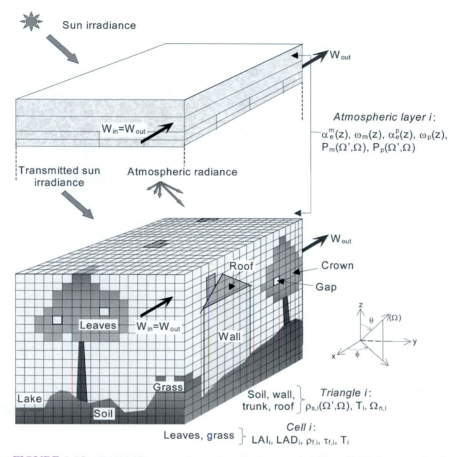

FIGURE 1.18 DART (discrete anisotropic radiative transfer) (Gastellu-Etchegorry, 2008).

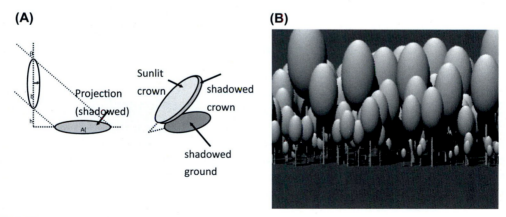

FIGURE 1.19 Principles of the geometric-optical model (A) for the canopy with an ellipsoid crown and the simulated canopy field (B).

(A)   **(B)**

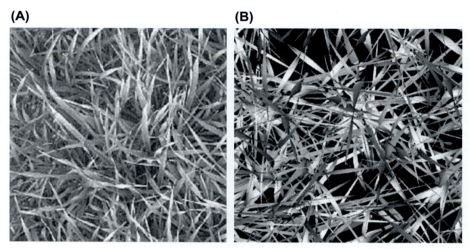

FIGURE 1.20   A photo of an actual canopy (A) and the simulated canopy field (B). *Courtesy from Dr. Mathias Disney at University College London.*

many computer software packages (e.g., MOD-TRAN, 6S) have been developed to enable us to calculate all necessary quantities, such as path radiance and transmittance.

1.6.1.4 Sensor modeling

As common detector materials do not respond across the entire optical spectrum, sensors have separate focal planes and noise mechanisms for each spectral region. The sensor model (Kerekes and Baum, 2005) can describe the effects of an imaging spectrometer on the spectral radiance mean and covariance statistics of a land surface. The input radiance statistics of every spectral channel are modified by electronic gain, radiometric noise sources, and relative calibration error to produce radiance signal statistics that represent the scene as imaged by the sensor.

The sensor model includes approximations for the spectral response functions and radiometric noise sources. The spectral response functions of each instrument can be measured and provided by the sensor manufacturers. Radiometric noise processes are modeled by adding variance to the diagonal entries of the spectral

covariance matrices. Radiometric noise sources come from the detector and electronics. Detector noise includes photon (shot) noise, thermal noise, and multiplexer/readout noise. Because detector parameters are often specified in terms of electrons, the noise terms are summed in a root sum squared sense in that domain before being converted to noise equivalent spectral radiance. The noise processes originating in the electronics include quantization noise, bit errors (in recording or transmitting the data), and noise arising within the electrical components.

Besides the sensor spectral response function and radiometric noise, the sensor model can also include spatial effects using PSF and MTF.

1.6.2 Inversion methods

Development of the inversion algorithms for estimating quantitative variables has a long history, as shown in Fig. 1.21. The early days focused on statistical methods and were then followed by physically based methods. Statistical methods are mainly based on a variety of vegetation indices through regression analysis. Physical algorithms rely on inverting surface

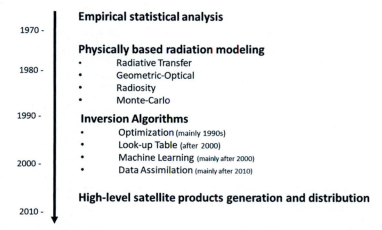

FIGURE 1.21 Major developmental milestones of quantitative remote sensing. *From Liang, S., Liu, Q., Yan, G., Shi, J., Kerekes, J.P., 2019. Foreword to the special issue on the recent progress in quantitative land remote sensing: modeling and estimation. IEEE J. Select. Topics Appl. Earth Obser. Remote Sens. 12, 391–395, ©2019, IEEE.*

radiation models. A new trend is to combine statistical and physical methods, and Section 1.6.2.4 on the direct estimation methods presents such an example.

1.6.2.1 Statistical analysis and machine learning techniques

Statistical models have proven to be very useful in various remote sensing applications. They are usually created using ground measurements. Because it is very expensive to collect extensive ground measurements under various conditions, the major weakness of models based on ground measurements is limited representation. An alternative solution is to simulate remotely sensed data using a physically based radiation model that may have been calibrated and validated by field measurements. The key modeling components have been discussed in Section 1.6.1.2.

Different statistical methods can be used to relate inputs and outputs of the model simulations. Besides the conventional multivariate regression analysis, different machine learning methods have been used, such as artificial neural networks (ANNs), SVM, self-organizing map (SOM), decision tree, random forest, case-based reasoning, neuro-fuzzy, genetic algorithm, and

the Multiple Adaptive Regression Spline (MARS). They are mainly used in two respects. The first is to simplify the complex physical model to execute the forward simulations in an inversion procedure. Complex models (such as atmospheric/surface radiative transfer models) are computationally expensive for repeated forward simulations. If the inversion process needs such a model, the machine learning technique can be used to replace it. The second is directly used for inversion. Based on the remote sensing data and the corresponding surface measurements as the input and output data pairs, a machine learning model can be established. Such input and output data pairs can also be generated by model simulations.

1.6.2.1.1 Artificial neural network

ANNs are complex computational models inspired by the human nervous system, which have the ability to learn from the training data to predict outcomes for new data. ANNs are probably the most widely used machine learning method for estimating land surface variables in the past decade, and use of ANNs has been steadily increasing.

There are many different types of ANNS for estimating land surface biogeophysical variables, such as multilayer perceptron (MLP), adaptive resonance theory, SOMs, radial basis function (RBF), and recurrent neural network, but MLP is mostly used based on the survey by Mas and Flores (2008). Although ANNs are considered to be "black boxes," we still have some flexibility, such as selecting input variables, determining appropriate characteristics for the training data, and optimizing the architecture of the network (number of layers and nodes) and the method to avoid overtraining.

Overfitting is often an issue. Training samples are usually erroneous due to observational uncertainties. Overfitting the training samples may degrade the prediction ability of an ANN model. An overtrained network is able to fit the training data precisely but fails for new datasets with subtle changes. There are some techniques to address this issue, such as the EBaLM-OTR (error back propagation and Levenberg—Marquardt algorithms for over training resilience) technique (Wijayasekara et al., 2011). This method uses k-fold cross-validation to determine the best architecture by dividing all data into three parts for training, validating, and testing each network architecture. Piotrowski and Napiorkowski (2013) evaluated multiple methods for catchment runoff modeling, such as the early stopping, the noise injection, the weight decay, and optimized approximation. These methods could be potentially used for estimating land surface parameters.

Determining network parameters (e.g., number of hidden layers and the number of neutrons in hidden layer) is often done by a trial-and-error method, which is empirical, is time-consuming, and results in the network configuration that may not be optimal. Some techniques have been proposed, such as the genetic algorithms (Castillo et al., 2000) and the particle swarm optimization (Da and Xiurun, 2005; Liu et al., 2015b), which could be potentially used for remote sensing inversion.

It is desirable to extract knowledge (e.g., explicit mathematical function) from trained neural networks for users to gain a better understanding of how the networks solve the nonlinear regression problems. Chan and Chan (2017) developed the piecewise linear artificial neural network (PWL-ANN) algorithm, which is to "open up" the black box of a trained neural network model so that rules in the form of linear equations are generated by approximating the sigmoid activation functions of the hidden neurons in an ANN. Setiono et al. (2002) described a method called rule extraction from function approximating neural networks (REF-ANN) for extracting rules from trained neural networks for nonlinear function approximation or regression.

ANNs that are composed of many layers are referred to deep learning. Deep learning has proven to be both a major breakthrough and an extremely powerful tool in applications where the target function is very complex and the datasets are large. It has become an increasingly important method in remote sensing (Zhang et al., 2016; Zhu et al., 2017a). The applications have been mostly for image classification and data fusion, but some studies on quantitative inversion of land surface variables have also been reported, such as soil moisture upscaling (Zhang et al., 2017a), land surface temperature (LST) (Yang et al., 2010), evapotranspiration (ET) (Chen et al., 2013), precipitation (Tao et al., 2016b), vegetation coverage (Jia et al., 2015), surface net radiation (Jiang et al., 2014), water quality optical parameters (Chen et al., 2014; Jamet et al., 2012), downward solar radiation (Tang et al., 2016), soil moisture (Xing et al., 2017), and so on.

1.6.2.1.2 Support vector machine

Support vector regression (SVR), which is also used in regression analysis, is a kind of kernel-based algorithm in machine learning. It minimizes the training error and the complexity of the model and uses the nonlinear kernel function that transforms the input data into a high-dimensional feature space. Compared with other machine learning methods, SVR is more suitable for high-dimensional nonlinear problems of small training samples.

Mountrakis et al. (2011) reviewed its wide applications in remote sensing, although more examples of applications are on image classification. Application examples of quantitative inversion include leaf nitrogen concentration in rice (Du et al., 2016; Sun et al., 2017a), LAI (Zhu et al., 2017b), evapotranspiration (Ke et al., 2016), nradiation (Jiang et al., 2016a), water depth and turbidity (Pan et al., 2015), grassland biomass (Marabel and Alvarez-Taboada, 2013), leaf nitrogen concentration (Omer et al., 2017; Wang et al., 2017; Yao et al., 2015), soil moisture (Ahmad et al., 2010), and other chemical parameters (Axelsson et al., 2013).

1.6.2.1.3 Regression tree

Regression trees and classification trees are often introduced together, but the main difference is to separate the type or estimate the value of the variable. The tree consists of a root node (containing all data), a set of internal nodes (splits), and a set of terminal nodes (leaves). A recursive segmentation algorithm is used to reduce the entropy within the class. The input data are hierarchical. The value of the internal node depends on the predicted average of each terminal node. Examples of quantitative inversion include biomass estimation (Blackard et al., 2008), forest structural parameters (Gomez et al., 2012; Mora et al., 2010), and forest ground coverage (Donmez et al., 2015).

1.6.2.1.4 Random forest

Random forest methods have been constructed with countless small regression trees to predict the variable. These small regression tree pruning are based on another random sample subset of the training dataset. Random forest methods can effectively overcome the problem of overfitting in regression, as well as "noise"

data and large datasets. Because random forest methods are less sensitive to the noise in the training dataset, it is better in the parameter estimation than the conventional regression tree method.

Random forest methods have been widely used for image classification (Belgiu and Dragut, 2016), but more and more is used in estimating quantitative parameters, such as grass nutrients and biomass (Ramoelo et al., 2015), nitrogen concentration in rice (Sun et al., 2017a), LAI (Beckschäfer et al., 2014; Li et al., 2017; Yuan et al., 2017; Zhu et al., 2017b), vegetation coverage (Halperin et al., 2016), biomass (Lopez-Serrano et al., 2016; Mutanga et al., 2012; Wang et al., 2016; Xia et al., 2018), surface temperature drop (Hutengs and Vohland, 2016), GPP (Tramontana et al., 2015), snow depth (Tinkham et al., 2014), precipitation (Kuhnlein et al., 2014), forest coverage and height (Ahmed et al., 2015), forest aboveground biomass (Karlson et al., 2015; Pflugmacher et al., 2014; Tanase et al., 2014), and other forest parameters (Garcia-Gutierrez et al., 2015).

RF is a robust nonlinear algorithm for predicting forest biophysical parameters. However, one drawback of RF regression algorithm when many input predictors are utilized is that it selects predictors that could be correlated to each other (Omer et al., 2016).

1.6.2.1.5 Multiple adaptive regression spline function

MARS is a nonlinear and nonparametric regression model. It is an extension of stepwise linear regression for adapting nonlinear regression. It is more flexible in establishing additive relationship or very strong relationship between the variables. It can be seen as an extended linear model that automatically simulates nonlinearity and the interaction among variables, so it has high computational efficiency.

Examples of applications include atmospheric correction to invert surface reflectance (Kuter et al., 2015), soil salinity (Nawar et al., 2014), land surface net radiation (Jiang et al., 2016a), aboveground biomass (Filippi et al., 2014), soil organic carbon content (Liess et al., 2016), chlorophyll concentration (Gholizadeh et al., 2015), and snow cover fraction (Kuter et al., 2018).

1.6.2.2 Optimization algorithms

Optimization inversion is the primary inversion method for estimating land surface variables in early days (Liang, 2004). It is based on a physically based model (such as canopy radiative transfer model) through the iterative method to continuously adjust the model parameters (x) so that the difference between the model calculated value $H(x)$ based on one or multiple radiative transfer (RT) models and satellite observations (y) is shrinking. The numerical difference is usually expressed by the following cost function $J(x)$. The iterative process is to minimize the cost (merit) function:

$$J(x) = (H(x) - y)^T R^{-1}(H(x) - y) + J_0 \quad (1.1)$$

where R is the covariance matrix of the observation errors, J_0 is the constrain term forcing that the estimates are as close to the background values (or ranges) as possible. These background values/ranges may come from the climatologies of existing satellite data, field measurements, or other priori knowledge.

There are some examples in using this approach. It has been widely used for estimating canopy and soil properties in the early years (Goel and Grier, 1987; Liang and Strahler 1993a, 1994) and land surface temperature (Liang, 2001). He et al. (2012) applied the optimization method to estimate aerosol optical depth and surface bidirectional reflectance distribution function (BRDF) parameters, which further lead to surface albedo calculation, from MODIS TOA observations. Zhang et al. (2018) further extended this method and estimated incident solar radiation as well.

Global land surface satellite products have not been generated by using the optimization approach. One of the major reasons is its computational cost because it iteratively runs the RT model in the inversion process. The recent efforts aim to develop various emulation techniques, i.e., using emulators that are computationally less expensive to replace the original RT model and provide an approximation of the model output trajectory. The applications of emulators have emerged in many disciplines (Castelletti et al., 2012; Conti et al., 2009; Lucia et al., 2004; Machac et al., 2016). There are two distinct emulation approaches: dynamic and statistical. A dynamic emulator is a simpler representation of the original dynamic model. For example, Xiao et al. (2015) replace the land surface dynamic model (e.g., dynamic vegetation model) that can be very complex by a simple differential equation characterized by LAI for estimating multiple land surface parameters simultaneously. Statistical emulators relate the input parameters to the model output statistically based on extensive model simulations by varying all key parameters. Verrelst et al. (2016) explored three machine learning regression algorithms (kernel ridge regression, neural networks, and Gaussian processes regression) to approximate the functions of three RT models (the leaf RT model PROSPECT-4, the canopy RT model PROSAIL, and the atmospheric RT model MODTRAN5) for global sensitivity analysis. It turns out the sensitivity study taking for over a month using the original RT model can be done in a few minutes. Gomez-Dans et al. (2016) developed the statistical Gaussian Process emulators: surrogate functions that accurately approximate canopy RT models (PROSAIL and SEMIDISCRETE) and atmospheric RT model (6S). The developed emulators can provide a fast and easy route to estimating the Jacobian of the original model, enabling the use of some optimization algorithms (e.g., efficient gradient descent methods). Rivera et al. (2015) developed a statistical emulator toolbox enabling multioutput machine learning regression algorithms available in MATLAB to approximate the RT model called SCOPE. Akbar and Moghaddam (2015) applied the global optimization method with a regulation term to estimate soil moisture from the combined active radar and passive radiometer microwave data at the same spatial resolution. The similar scheme is also applied for estimate soil moisture and roughness (Akbar et al., 2017).

1.6.2.3 Look-up table algorithms

Optimization algorithms are computationally expensive and very slow, performing the inversion process with a huge amount of remotely sensed data. The look-up table (LUT) approach has been used extensively to speed up the inversion process. It precomputes the model reflectance for a large range of combinations of parameter values. In this manner, the most computationally expensive aspect can be completed before the inversion is attempted, and the problem is reduced to searching a LUT for the modeled reflectance set that most resembles the measured set.

In an ordinary LUT approach, the dimensions of the table must be large enough to achieve high accuracy, which leads to much slower online searching. Moreover, many parameters must be fixed in the LUT method. To reduce the dimensions of the LUTs for rapid table searching, empirical functions are used to fit the LUT values so that a table searching procedure becomes a simple calculation of the local functions, or a simple linear regression is executed instead of table searching.

To reduce the dimensions of the LUTs for rapid table searching, Gastellu-Etchegorry et al. (2003) developed empirical functions to fit the LUT values so that a table searching procedure becomes a simple calculation of the local functions. Hedley et al. (2009) developed a so-called adaptive LUT method. It relies on a postprocessing step to organize the data into a binary space partitioning tree that facilitates an efficient inversion search algorithm.

Because of its simplicity and easy implementation, the LUT methods have continued to be widely used, and the recent examples include retrieving LAI (Banskota et al., 2015; Qu et al., 2014b; Tao et al., 2016a; Yang et al., 2017) and chlorophyll concentration (Darvishzadeh et al. 2008, 2012), crop biophysical parameters from LiDAR data (Ben Hmida et al., 2017), fraction of vegetation cover (Ding et al., 2016), grassland live fuel moisture content (Quan et al., 2016), soil moisture (Bai et al., 2017), snow depth (Che et al., 2016), surface broadband emissivity (Cheng et al., 2017), and estimating surface incident solar radiation (Zhang et al., 2014a).

1.6.2.4 Direct estimation methods

The direct estimation algorithms are very similar to the LUT algorithms, but the major difference is that the table searching is replaced by regression analysis (Liang, 2003; Liang et al., 1999). The first part is almost identical. The forward RT model is used to simulate TOA reflectance (radiance), but the inversion part relies on regression analysis that links the inverted parameters and the simulated reflectance/radiance instead of table searching. It is sometimes called hybrid inversion algorithm because it combines the RT model simulation (physical) with regression analysis (statistical). Fig. 1.22 illustrates the direct estimation method for estimating land surface albedo.

Liang et al. (2005) applied this method to estimate surface snow albedo from MODIS data using linear regression analysis at each angular bin in the solar illumination and sensor viewing geometry. The linear regression equations were training using the simulated data. This algorithm for estimating land surface albedo has been tested for a variety of remotely sensed data, such as MODIS (Wang et al., 2015c), Landsat data (He et al., 2018), AVIRIS (He et al., 2014b), and Chinese HJ (He et al., 2015a). In fact, the global land surface albedo products have been generated, such as the GLASS albedo product (Liang et al., 2013c; Qu et al. 2014a, 2016) from

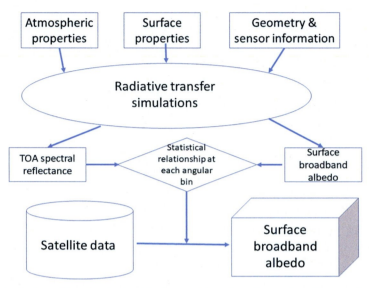

FIGURE 1.22   Illustration of the direct estimation method for estimating land surface albedo from satellite data.

MODIS data, and the VIIRS albedo product (Wang et al., 2013a; Zhou et al., 2016). Compare to the conventional albedo estimation algorithms, the direct estimation method does not require surface reflectance from atmospheric correction. Moreover, it also does not need an accumulation of observations for surface BRDF modeling during a certain period of time that cannot capture the rapid surface changes.

This approach has also been recently used to estimate surface shortwave net radiation (He et al., 2015b; Kim and Liang, 2010; Wang et al. 2014, 2015a,b) and TOA shortwave albedo and radiative flux (Wang and Liang 2016, 2017).

1.6.2.5 Data assimilation methods

The values of land surface variables, estimated using the methods from the previous sections from different sources, may not be physically consistent. Most techniques do not take advantage of observations acquired at different times and cannot handle observations with different spatial resolutions together. In particular, these techniques estimate only variables that significantly affect radiance received by the sensors. In many cases, the estimation of some variables not directly related to radiance is desirable.

Given the ill-posed nature of remote sensing inversion (the number of unknowns is far greater than the number of observations, and the multispectral data are highly correlated) and vast expansion of the amount of observation data, data fusion techniques, which simply register and combine datasets together from multiple sources, may be one solution. The data assimilation (DA) method allows use of all available information within a time window to estimate various unknowns of land surface models (Liang, 2007; Liang et al., 2013a; Liang and Qin, 2008). The information that can be incorporated includes observational data, existing pertinent a priori information, and, importantly, a dynamic model that describes the system of interest and encapsulates theoretical understanding.

A DA scheme commonly includes the following components: (1) a forward dynamic model that describes the time evolution of state variables such as surface temperature, soil moisture, and carbon stocks; (2) an observation model that relates the model estimates of state variables to satellite observations and vice versa; (3) an objective function that combines model estimates and observations along with any associated prior information and error structure; (4) an optimization scheme that adjusts forward model parameters or state variables to minimize the discrepancy between model estimates and satellite observations; and (5) error matrices that specify the uncertainty of the observations, model, and any background information (these are usually included in the objective function).

DA for inverting land surface parameters can integrate not only remote sensing data with different characteristics (multispectral, multiangular, and multitemporal) but also various measurement data and a prior knowledge (Lewis et al., 2012; Liang and Qin, 2008). The main difference with the optimization method described above is that it must rely on a dynamic equation that can be a mechanical physically based model, such as a crop growth model or a statistical model. The recent applications of DA may include soil moisture (Chen et al., 2015b; Fan et al., 2015; Han et al., 2015; Qin et al. 2013, 2015; Yang et al., 2016), hydrological parameters (Lei et al., 2014; Xie et al., 2014), heat fluxes (Wang et al., 2013b; Xu et al. 2014, 2015), carbon flux (Liu et al., 2015a), and crop yield (Cheng et al., 2016; Huang et al. 2015a,b, 2016).

A new remote sensing DA approach has been recently proposed to estimate an improved suite of land products simultaneously. The framework, as illustrated in Fig. 1.23, has several unique features compared to the conventional inversion methodology: making use of multiple satellite data, particularly time series observations; being able to incorporate *a prior* knowledge and various constraints; adapting the ensemble of multiple inversion algorithms for estimating the same land surface variable by taking advantage of the strengths of the individual algorithm. More importantly, this approach estimates land surface variables using both direct and indirect retrieval methods. The key parameters of the surface

radiative transfer model that are mostly sensitive to the observed radiance/reflectance can be directly estimated, and a group of other parameters (e.g., FAPAR, surface albedo) are calculated as an indirect inversion. A series of experiments have been conducted to demonstrate that this method is robust (Liu et al., 2014; Ma et al. 2017a,b, 2018; Shi et al. 2016, 2017; Xiao et al., 2015). However, the disadvantage of this method is computationally expensive and currently cannot be used for generating the global products. If the TOA observation data are directly assimilated (Shi et al., 2016), it can be worse. Computer emulation techniques may help (Gomez-Dans et al., 2016; Rivera et al., 2015).

1.6.2.6 Spatial and temporal scaling

Remote sensing data with different spatial resolutions have different temporal resolutions, generally speaking, high spatial resolution data with low temporal resolution. For example, Landsat has multispectral optical data at the spatial resolution of 30 m, but the satellite repeat cycle is 16 days and cloud contamination leads to much lower effective temporal resolution. In contrast, MODIS data have a higher temporal resolution but a lower spatial resolution. By combining data at different spatial resolutions, we will be able to generate new data products with high spatial and temporal resolutions.

Gao et al. (2006) proposed to generate a Landsat-like daily reflection at the spatial resolution of 30m from high spatial resolution low-frequency Landsat data and high-frequency low spatial resolution MODIS data using a new time-space adaptive reflectivity fusion model (STARFM) algorithm. Zhu et al. (2010) proposed the enhanced version of STARFM (ESTARFM). Similar ideas have also been applied to generate the LST products at the high spatial and temporal resolutions (Moosavi et al., 2015; Weng et al., 2014). These methods are also known in the literature as spatial and temporal data fusion methods (Chen et al., 2015a; Wu et al., 2015; Zhang et al., 2015).

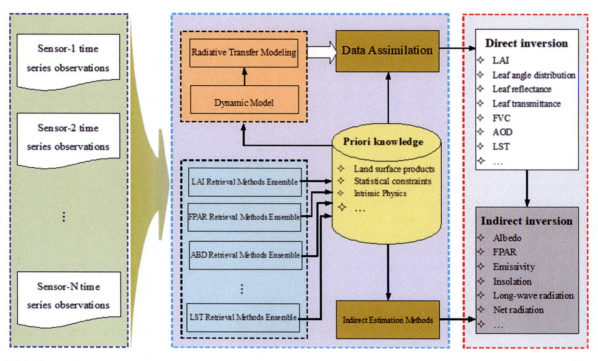

FIGURE 1.23 A data assimilation–based inversion framework for estimating a suite of land surface variables simultaneously.

If we need to generate the high-level products at multiple spatial resolutions, we need different methods. The first case is how to estimate from satellite data with multiple spatial resolutions that are acquired simultaneously. In theory, we can execute inversion separately. However, if we combine the observations at multiple spatial resolutions for joint inversion, relevant information at different scales will be passed to improve the inversion accuracy. For example, Jiang et al. (Jiang et al., 2016b) effectively improved the LAI inversion accuracy of ETM+ and MODIS data using the ensemble multiscale filter. De Vyver and Roulin (2009) used scale recursive estimation method to estimate precipitation. These multiscale estimation processes can be represented by a "tree" structure. All grids cover the same area, but each grid corresponds to a different spatial resolution. Each node is

associated with a node on a finer or coarser scale. The multiscale estimation method continues to loop up and down. Upward updates the information from fine to coarse scales and vice versa to transfer information from coarse to fine scales.

The second case is that the inverted satellite products of different scales may have systematic errors, spatial discontinuity, and uncertainty. Using fusion of multiscale data product, information on different scales is used to improve the accuracy and quality at each scale. He et al. (2014c) integrated land surface albedo products at three scales using multiresolution tree (MRT) method: MISR (1100m), MODIS (500m), and ETM+ (30m). They demonstrated that the multiscale data fusion method can fill the missing data and reduce the systematic error and uncertainty. This approach has also been applied to integrate the FAPAR products (Tao et al., 2017).

1.6.2.7 Regularization method

To overcome the "ill-posed" inversion problem, various regularization techniques have been recently proposed (Delahaies et al., 2017; Laurent et al., 2014) to increase the amount of information to enhance the stability and accuracy of the solution, such as making use of multisource data and a prior knowledge.

1.6.3 Use of multisource data

Modern remote sensing technology has evolved to acquire massive amounts of data at different spatial, temporal, and spectral resolutions using multiple sensors at various platforms for the monitoring of the Earth's environment. The geostationary satellite data can reveal the diurnal variation of the parameters due to high temporal resolution with relatively low spectral resolutions (few bands) in many systems, while the polar-orbiting satellite data usually have high spectral resolutions and global coverage but with low temporal resolution. Different sensors have their own advantages in different applications with different band settings, observing angles, and temporal and spatial resolutions. The integrative use of a variety of different satellite data is becoming a new trend for increasing the amount of known information to improve the accuracy of land parameter inversion.

Many research projects have shown that the accuracy of the products produced by integrating multisource remote sensing data is significantly higher than that of single sensor data. Generally, there are three types of integration according to data characteristics:

(1) Integrating data acquired using the similar principles at the similar spatial resolutions, for example, multispectral data from different sensors of polar-orbiting satellites and/or geostationary satellites: coarse resolution (MODIS/MERIS/VIIRS/MISR), fine resolution (ETM+/SPOT);

(2) Integrating data at different spatial resolutions;

(3) Integrating data acquired by sensors with different principles.

For example, Sun et al. (2011) demonstrated the potential of the combined use of LiDAR samples and SAR imagery for forest biomass mapping. A follow-up study further explored the use of integrated LiDAR, SAR, and multispectral imaging (Landsat) data for achieving the accuracy requirements of a global forest biomass mapping mission (Montesano et al., 2013). Other forest biomass mapping studies also demonstrated the value of integrating multispectral data and LiDAR data, which can effectively extend 3D information from discrete LiDAR measurements of forest structure across scales much larger than that of the LiDAR footprint (Pflugmacher et al., 2014; Zhang et al., 2014b) and so on.

1.6.4 Use of a prior knowledge

Any a prior information that affects the pixel values can help improve inversion accuracy and stability. Cui et al. (2014) demonstrated that the combination of a priori knowledge with a regularized inversion can robustly estimate surface albedo through the linear kernel-driven BRDF model for the cases where the number of observations is insufficient, or the angular distribution is poor.

Generation of a priori knowledge is the first step. A priori knowledge may come from measurement data, expert knowledge, or a variety of existing data products. The a priori knowledge for inversion can be generated by analyzing the surface measurements, the temporal and spatial distribution of the existing satellite data products, or the numerical model simulations. The a priori knowledge of the land surface products is mainly composed of statistical values, statistic constraints, and intrinsic physical correlations between the land surface variables. The statistical values include the annual mean and variance of the land surface products and their temporal and spatial distributions, for example, the empirical relationship between surface temperature, emissivity, broadband albedo, and LAI.

There are many mathematical equations that can be used to characterize the vegetation growth curve (Tsoularis and Wallace, 2002), such as the classical Fourier function to characterize the dynamic changes of NDVI (Hermance, 2007; Zhou et al., 2015) and surface temperature (Xu and Shen, 2013).

The logistic function has been used for LAI inversion (Qu et al., 2012), and the double logistic functions are also often used to characterize annual changes in vegetation indices (Beck et al., 2006; Fisher, 1994; Zhang et al., 2003).

Bayesian principles can be used for effective incorporation of a priori knowledge into the inversion process (Qu et al., 2014b; Shiklomanov et al., 2016; Varvia et al., 2018). Bayesian network method has been used to invert LAI and other parameters (Qu et al., 2014b; Quan et al., 2015; Zhang et al., 2012). Most of these studies using the Bayesian network approach assume that the inverted variables are independent, but Quan et al. (2015) take into account the correlation between variables. Laurent et al. (2014) estimated LAI and other parameters using Bayesian inversion algorithm by incorporating a priori knowledge and a coupled vegetation-atmosphere radiative transfer model.

Many satellite products have missing values and "noises" that need to be filled and smoothed before the statistical analysis. There have been studies that make use of a priori knowledge based on multiple satellite data products. For example, albedo (Fang et al., 2007; He et al., 2014a; Moody et al., 2005; Zhang et al., 2010), LAI (Fang et al. 2008b, 2013), FAPAR and fractional vegetation coverage (Verger et al., 2015 5350), land surface temperature (Bechtel, 2015), sea surface temperature (Banzon et al., 2014), ground radiation (Krahenmann et al., 2013; Posselt et al., 2014), aerosol optical properties

(Aznay et al., 2011), soil moisture (Owe et al., 2008), snow/ice covers (Husler et al., 2014), land cover (Broxton et al., 2014), and vegetation seasonal change (Verger et al., 2016).

1.6.5 Space—time constraints

Satellites observe the specific location of the Earth's surfaces periodically, and the current inversion algorithms estimate surface properties mainly using the data acquired at a specific time. However, many land variables are changing with time, such as LAI, surface temperature, and short-range surface shortwave radiation. Inversion relying on temporally discrete satellite observations cannot effectively capture the temporal variations. Therefore, one of the solutions in developing advanced inversion algorithms is to invert a temporal sequence of observations by using the coupled surface dynamic process model that describes the temporal variations of land surface variables with the radiative transfer model that links the surface state variables with radiance observed by the satellite sensors.

The time constraints apply to inverting the surface parameters from a time series of remote sensing data, assuming that some parameters are constant and some are temporally variable (Houborg et al., 2007 5397; Lauvernet et al., 2008 5398) or assuming that the parameters (such as LAI) follow a certain seasonal variation patterns or are temporally smooth (Quaife and Lewis, 2010).

The spatial constraints apply to a sequence of data at a small neighborhood. Assuming that some parameters are spatially uniform (e.g., same cover type), grouping these pixels can reduce the number of unknown variables (Atzberger, 2004; Laurent et al., 2013). For example, the 3×3 window was used to constrain the crop LAI inversion (Atzberger and Richter, 2012). In atmospheric correction, the entire image is often assumed to be a window with one set of atmospheric optical parameters, and sometimes the entire image is divided into multiple windows, each with the same atmospheric parameters to be estimated. Spatial smoothness is also used to constrain parameter inversion, as in the case of temporal smoothness constraints (Wang et al., 2008).

1.6.6 Algorithm ensemble

At present, the majority of satellite products have been generated by one "best" algorithm for each product. As different algorithms often have different characteristics, it is difficult to determine the "best" algorithm. As a result, there often exist systemic biases, and the accuracies of these products vary regionally under different conditions. Algorithm ensemble integrates the outputs of a group of algorithms and can greatly improve the inversion accuracy and stability.

Such an approach has been first applied to the development of the GLASS products (Liang et al., 2013c). GLASS land surface albedo product is generated in the integration of two estimation algorithms (Liu et al. 2013a,b). The GLASS ET product is based on integrating five algorithms (Yao et al., 2014).

The Bayesian Model Average (BMA) method has been proposed to integrate multiple downward and longwave radiation models (Wu et al., 2012).There are many other examples (Chen et al., 2015d; Kim et al., 2015; Shi and Liang, 2013a,b; Shi and Liang, 2014).

1.7 Production, archiving, and distribution of high-level products

Satellite observations can be converted into high-level bio/geophysical products using the inversion methods outlined above through a production system. Given satellite observations and the inversion algorithm, production of high-level products is not straightforward because of the huge amount of data. Creation of data information systems is essential. For example, NASA's principal Earth Science information system is the Earth Observing System Data and Information System (EOSDIS), which has been operational since August 1994. EOSDIS acquires, processes, archives, and distributes Earth Science data and information products created from satellite data that arrive at a rate of more than four trillion bytes (4 terabytes) per day. More and more information systems are supported by high-performance computing capabilities.

The various levels of data used by the EOSDIS are defined below. For some instruments, there will be no Level 1B product that is distinct from the Level 1A product. In these cases, the reference to Level 1B data can be assumed to refer to Level 1A data. Brief definitions follow:

- Level 0: Reconstructed, unprocessed instrument/payload data at full resolution; any and all communications artifacts (e.g., synchronization frames, communications headers, duplicate data removed). In most cases these data are provided by EDOS to a Distributed Active Archive Center (DAAC) as production datasets to produce higher level products.
- Level 1A: Reconstructed, unprocessed instrument data at full resolution, time-referenced, and annotated with ancillary information, including radiometric and geometric calibration coefficients, and georeferencing parameters (e.g., platform ephemeris, computed and appended but not applied to the Level 0 data).
- Level 1B: Level 1A data that have been processed to sensor units (not all instruments will have a Level 1B equivalent).
- Level 2: Derived geophysical variables at the same resolution and location as the Level 1 source data.
- Level 3: Variables mapped on uniform space—time grid scales, usually with some completeness and consistency.

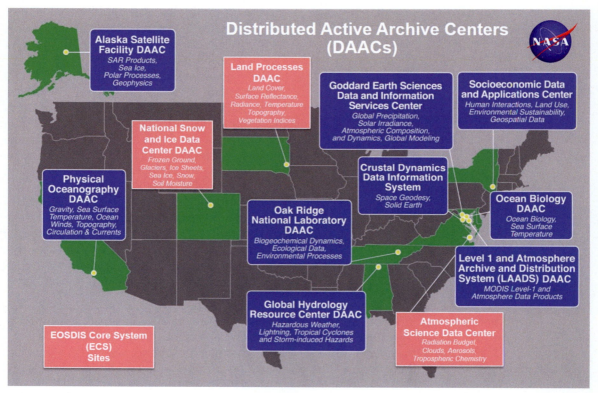

FIGURE 1.24   Distribution of the NASA DAAC, downloaded from NASA website NASA's Global Imagery Browse Services (GIBS), provides quick access to over 800 satellite imagery products, covering every part of the world. Most imagery is available within a few hours after satellite overpass, some products span almost 30 years, and the imagery can be rendered in the web client or GIS application.

- Level 4: Model output or results from analyses of lower level data (e.g., variables derived from multiple measurements).

Archiving and distribution of the huge amount of data and products are also challenging. As a user, we have several ways to search for EO data of interest.

NASA Earth Science information is archived at the DAACs located across the United States (Fig. 1.24). The DAACs specialize by topic area and make their data available to researchers around the world. Almost all data in EOSDIS are held online and accessed via ftp and https.

NASA's Land, Atmosphere Near real-time Capability for EOS (LANCE) provides near real-time imagery, and high-level products from several NASA instruments, including AIRS, AMSR2, LIS (ISS), MISR, MLS, MODIS, MOPITT, OMI, and VIIRS, are available in less than 3 h from satellite observations.

ESA distributes Earth observation data from ESA EO Missions, third party missions, ESA campaigns, the Copernicus space component, as well as sample and auxiliary data from a number of missions and instruments (https://earth.esa.int/web/guest/data-access), under different data policies and by various access mechanisms.

China's satellite data are archived and distributed through several agencies, such as the Chinese Meteorological Administration (http://www.cma.gov.cn/en2014/satellites/), mainly for meteorological satellite data, and China

Center for Resources Satellite Data and Application (http://www.cresda.com/EN/) mainly for other land satellite remote sensing data. The China National Data Sharing Infrastructure of Earth System Science (www.geodata.cn) distributes data products generated from major research projects in China, including surface measurements, high-level products from remotely sensed observations, and model simulations. It was created in 2004 and is being contributed to by more than 40 Chinese institutes including various local data centers.

CEOS International Directory Network (IDN) serves as a gateway to the world of Earth science data (https://idn.ceos.org/). The CEOS IDN is an international effort developed to assist researchers in locating information on available datasets.

In addition to the space agencies that usually produce the high-level products from a specific satellite data with a limited life span, some universities and research institutes led research teams are also generating the high-level satellite products from multiple satellite data. For example, the GLASS product suite initially included five products (Liang et al. 2013b,c), but it has been recently expanded into 12 products (see Table 1.8) from AVHRR, MODIS, and other satellite data. The GLASS products can be

TABLE 1.8 The list of the GLASS products and their characteristics as of 2019.

No	Product	Temporal range	Spatial range	Temporal resolution	Spatial resolution
1	Leaf area index	1981–2018	Global land surface	8 days	5 km before 2000, 1 km after 2000
2	Surface broadband albedo	1981–2018	Global	8 days	5 km before 2000, 1 km after 2000
3	Broadband emissivity	1981–2018	Global land surface	8 days	5 km before 2000, 1 km after 2000
4	Photosynthetically active radiation	2000–2018	Global land surface	Daily	5 km
5	Downward shortwave radiation	2000–2018	Global land surface	Daily	5 km
6	Surface longwave net radiation	1983, 1993, 2003, 2013	Global land surface	8 days	5 km before 2000, 1 km after 2000
7	Surface all-wave net radiation	2000–2018	Global land surface	8 days	5 km
S	Land surface temperature	1983, 1993, 2003, 2013	Global land surface	Instantaneous	5 km before 2000, 1 km after 2000
9	Fraction of absorbed photosynthetically active radiation	1981–2018	Global land surface	8 days	5 km before 2000, 1 km after 2000
10	Fractional vegetation cover	1981–2018	Global land surface	8 days	5 km before 2000, 1 km after 2000
11	Evapotranspiration	1981–2018	Global land surface	8 days	5 km before 2000, 1 km after 2000
12	Gross primary production	1981–2018	Global land surface	8 days	5 km before 2000, 1 km after 2000

freely downloaded from www.geodata.cn or www.glass.umd.edu.

More details on data, production, and distribution systems are presented in Chapter 22.

1.8 Product validation

One of the key components of information extraction is validation. Without a known accuracy, the high-level product cannot be used reliably and, therefore, has limited applicability. With various land products available, users need quantitative information on product uncertainties to select the most suitable product, or combination of products, for their specific needs. As remote sensing observations are generally merged with other sources of information or assimilated within process models, evaluation of product accuracy is required. Making quantified accuracy information available to the user can ultimately provide developers the necessary feedback for improving the products and can possibly provide methods for their fusion to construct a consistent long-term series of surface status.

Land product validation has to rely on ground measurements, which may be both time-consuming and very expensive. Because of its importance, such product validation must involve the efforts of the entire remote sensing community. Sharing the validation methodologies,

instruments, measured data, and results offer vital ways that will lead to success and progress. One of the critical issues is the mismatch between ground "point" measurements and kilometer-scale pixel values over the heterogeneous landscape. Upscaling "point" measurements using high-resolution remotely sensed data is the key to addressing this issue. Loew et al. (2017) provided a comprehensive review of state-of-the-art methods of satellite validation and documents their similarities and differences, and more details on validating different high-level land products are discussed in the relevant chapters.

1.9 Remote sensing applications

Remote sensing has generated comprehensive, near-real-time environmental data, information, and analyses. This important technology serves a wide range of users and empowers decision-makers to respond more effectively to the many environmental challenges facing us today. Remotely sensed information can be applied in different ways (Balsamo et al., 2018; Pasetto et al., 2018). We can group this information into the following categories, as illustrated in Fig. 1.25:

(1) Detection and monitoring. The long-term individual or combined high-level products

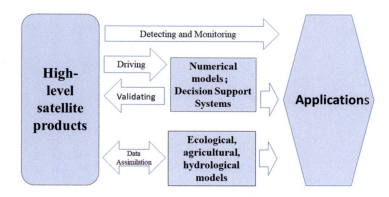

FIGURE 1.25　Different ways of applications of the high-level satellite products.

can be used for characterizing the surface dynamics resulted from natural variabilities and human activities, for example, LAI for the global "greening" trends (Zhu et al., 2016); snow cover and surface albedo for regional and global climate change (Chen et al., 2015c; He et al. 2013, 2014a).

(2) Driving numerical models and decision support systems. The satellite products have been used as inputs to various models, such as ecological models (Pasetto et al., 2018), hydrological models (McCabe et al., 2017), numerical weather prediction models (Fang et al., 2018), Earth system models (Balsamo et al., 2018; Simmons et al., 2016), or decision support systems (Mohammed et al., 2018; Murray et al., 2018; Rahman and Di, 2017).

(3) Validating model simulations. Numerical models may not actually use satellite products, but their simulation results are validated by satellite products, such as soil moisture (Gu et al., 2019), precipitation (Tapiador et al., 2018), temperature (Ouyang et al., 2018), and water storage (Zhang et al., 2017b).

(4) Assimilation into numerical models. The numerical models usually have many parameters that may be spatially variable that are difficult to predefine. DA approach can be used to adjust these parameters iteratively by matching the model simulated values with satellite products (Liang et al., 2013a; Liang and Qin, 2008). Thus, the models can generate the simulations with spatial-temporal continuity (satellite products are often irregular), such as soil moisture (Qin et al., 2009), or simulate the variables that cannot be directly observed or retrieved from remotely sensed data, such as carbon fluxes (Scholze et al., 2017) and crop yield (Fang et al., 2008a; Huang et al., 2019; Jin et al., 2018).

Fig. 1.26 depicts the linkage and flow of information from remote sensing observations and other in situ data to societal benefits. Data can be used for driving, calibrating, and validating models and decision support tools. The last four chapters of this book illustrate how different remote sensing data and products can be used for monitoring land cover and land use changes and assessing their environmental impacts.

GEO identified nine societal benefit areas in which there was recognition that clear societal benefits could be derived from a coordinated global observation system. As illustrated in Fig. 1.27, the nine societal benefit areas include, for example,

- Understanding environmental factors affecting human health and well-being.
- Improving management of energy resources.
- Understanding, assessing, predicting, mitigating, and adapting to climate variability and change.
- Improving water resource management through better understanding of the water cycle.
- Improving weather information, forecasting, and warning.

Many of these societal benefit areas are themselves complex clusters of issues, with multiple and varied stakeholders. In each area there are observational needs for numerous variables, with requirements for their accuracy, spatial and temporal resolution, and speed of delivery to the user. These societal benefit areas are now at widely varying levels of maturity with respect to establishing user needs, defining the observation requirements, and implementing coordinated systems.

1.10 Conclusion

Significant progress in engineering issues has enabled remarkable improvements in platform and sensor systems that are evidenced by many indicators, such as signal-to-noise ratios, resolutions, pointing accuracies, geometric and

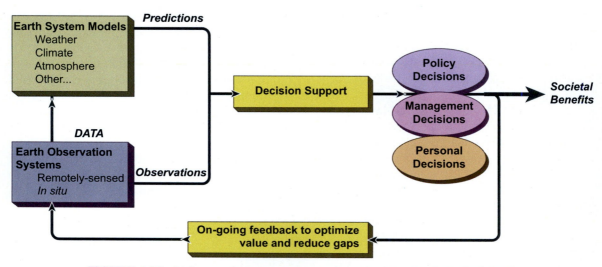

FIGURE 1.26   Linking earth observations with societal benefits (CENR/IWGEO, 2005).

FIGURE 1.27   GEOSS and applications (GEO, 2005).

spectroradiometric integrity, and calibration. The immense amounts of data available from satellite observations present great challenges. Considerable investments have been made in developing physical models to understand surface radiation regimes, and some of these models have been incorporated into useful algorithms for estimating land surface variables from satellite observations. However, development of realistic and computationally simplified surface radiation models mostly suitable for inversion of land surface variables from satellite data are still urgently required. Inversion of land surface parameters is generally a nonlinear ill-posed problem, and the use of more spatial and temporal constraints by incorporating a priori knowledge and integrating multiple-source data deserves further research.

Although remote sensing data products have been widely used, significant disconnects between remote sensing development and applications continue to exist. Some products developed by remote sensing scientists have not been widely used, and many variables required by land process models and decision support systems have not been generated. The product accuracy and application requirements may not always be consistent. Successful applications are not static but evolve as new sensor, data processing, and network technologies emerge. The improved linking of remote sensing science with applications would be most advantageous.

References

Ahmad, S., Kalra, A., Stephen, H., 2010. Estimating soil moisture using remote sensing data: a machine learning approach. Adv. Water Resour. 33, 69–80.

Ahmed, O.S., Franklin, S.E., Wulder, M.A., White, J.C., 2015. Characterizing stand-level forest canopy cover and height using Landsat time series, samples of airborne LiDAR, and the Random Forest algorithm. ISPRS J. Photogrammetry Remote Sens. 101, 89–101.

Akbar, R., Cosh, M.H., O'Neill, P.E., Entekhabi, D., Moghaddam, M., 2017. Combined radar-radiometer surface soil moisture and roughness estimation. IEEE Trans. Geosci. Remote Sens. 55, 4098–4110.

Akbar, R., Moghaddam, M., 2015. A combined active-passive soil moisture estimation algorithm with adaptive regularization in support of SMAP. IEEE Trans. Geosci. Remote Sens. 53, 3312–3324.

Atzberger, C., 2004. Object-based retrieval of biophysical canopy variables using artificial neural nets and radiative transfer models. Remote Sens. Environ. 93, 53–67.

Atzberger, C., Richter, K., 2012. Spatially constrained inversion of radiative transfer models for improved LAI mapping from future Sentinel-2 imagery. Remote Sens. Environ. 120, 208–218.

Axelsson, C., Skidmore, A.K., Schlerf, M., Fauzi, A., Verhoef, W., 2013. Hyperspectral analysis of mangrove foliar chemistry using PLSR and support vector regression. Int. J. Remote Sens. 34, 1724–1743.

Aznay, O., Zagolski, F., Santer, R., 2011. A new climatology for remote sensing over land based on the inherent optical properties. Int. J. Remote Sens. 32, 2851–2885.

Bai, X.J., He, B.B., Li, X., Zeng, J.Y., Wang, X., Wang, Z.L., Zeng, Y.J., Su, Z.B., 2017. First assessment of Sentinel-1A data for surface soil moisture estimations using a coupled water cloud model and advanced integral equation model over the Tibetan plateau. Remote Sens. 9, 20.

Balsamo, G., Agusti-Panareda, A., Albergel, C., Arduini, G., Beljaars, A., Bidlot, J., Bousserez, N., Boussetta, S., Brown, A., Buizza, R., Buontempo, C., Chevallier, F., Choulga, M., Cloke, H., Cronin, M., Dahoui, M., De Rosnay, P., Dirmeyer, P., Drusch, M., Dutra, E., Ek, M., Gentine, P., Hewitt, H., Keeley, S., Kerr, Y., Kumar, S., Lupu, C., Mahfouf, J.-F., McNorton, J., Mecklenburg, S., Mogensen, K., Muñoz-Sabater, J., Orth, R., Rabier, F., Reichle, R., Ruston, B., Pappenberger, F., Sandu, I., Seneviratne, S., Tietsche, S., Trigo, I., Uijlenhoet, R., Wedi, N., Woolway, R., Zeng, X., 2018. Satellite and in situ observations for advancing global earth surface modelling: a review. Remote Sens. 10, 2038.

Banskota, A., Serbin, S.P., Wynne, R.H., Thomas, V.A., Falkowski, M.J., Kayastha, N., Gastellu-Etchegorry, J.-P., Townsend, P.A., 2015. An LUT-based inversion of DART model to estimate forest LAI from hyperspectral data. IEEE J. Sel. Top. Appl. Earth Obs. Remote Sens. 8, 3147–3160.

Banzon, V.F., Reynolds, R.W., Stokes, D., Xue, Y., 2014. A 1/4 degrees-Spatial-Resolution daily sea surface temperature climatology based on a blended satellite and in situ analysis. J. Clim. 27, 8221–8228.

Bechtel, B., 2015. A new global climatology of annual land surface temperature. Remote Sens. 7, 2850–2870.

Beck, P.S.A., Atzberger, C., Hogda, K.A., Johansen, B., Skidmore, A.K., 2006. Improved monitoring of vegetation dynamics at very high latitudes: a new method using MODIS NDVI. Remote Sens. Environ. 100, 321–334.

Beckschäfer, P., Fehrmann, L., Harrison, R.D., Xu, J., Kleinn, C., 2014. Mapping Leaf Area Index in subtropical upland ecosystems using RapidEye imagery and the randomForest algorithm. iFor. Biogeosci. For. 7, 1.

Belgiu, M., Dragut, L., 2016. Random forest in remote sensing: a review of applications and future directions. ISPRS J. Photogrammetry Remote Sens. 114, 24–31.

Ben Hmida, S., Kallel, A., Gastellu-Etchegorry, J.P., Roujean, J.L., 2017. Crop biophysical properties estimation based on LiDAR full-waveform inversion using the DART RTM. IEEE J. Sel. Top. Appl. Earth Obs. Remote Sens. 10, 4853–4868.

Blackard, J.A., Finco, M.V., Helmer, E.H., Holden, G.R., Hoppus, M.L., Jacobs, D.M., Lister, A.J., Moisen, G.G., Nelson, M.D., Riemann, R., Ruefenacht, B., Salajanu, D., Weyermann, D.L., Winterberger, K.C., Brandeis, T.J., Czaplewski, R.L., McRoberts, R.E., Patterson, P.L., Tymcio, R.P., 2008. Mapping US forest biomass using nationwide forest inventory data and moderate resolution information. Remote Sens. Environ. 112, 1658–1677.

Borel, C.C., Gerstl, S.A.W., Powers, B.J., 1991. The radiosity method in optical remote sensing of structured 3-D surfaces. Remote Sens. Environ. 36, 13–44.

Breaker, L.C., 1990. Estimating and removing sensor-induced correlation from advanced very high resolution radiometer satellite data. J. Geophys. Res. 95, 9701–9711.

Broxton, P.D., Zeng, X.B., Sulla-Menashe, D., Troch, P.A., 2014. A global land cover climatology using MODIS data. J. Appl. Meteorol. Climatol. 53, 1593–1605.

Castelletti, A., Galelli, S., Ratto, M., Soncini-Sessa, R., Young, P.C., 2012. A general framework for Dynamic Emulation Modelling in environmental problems. Environ. Model. Softw 34, 5–18.

Castillo, P.A., Merelo, J., Prieto, A., Rivas, V., Romero, G., 2000. G-Prop: global optimization of multilayer perceptrons using GAs. Neurocomputing 35, 149–163.

CENR/IWGEO, 2005. Strategic plan for the U.S. integrated Earth observation system. In: N.S.a.T.C.C.o.E.a.n. Resources. National Science and Technology Council Committee on Environment and natural Resources, Washington DC, p. 149.

Chan, V., Chan, C., 2017. Towards developing the piece-wise linear neural network algorithm for rule extraction. Int. J. Cogn. Inf. Nat. Intell. 11, 17.

Che, T., Dai, L.Y., Zheng, X.M., Li, X.F., Zhao, K., 2016. Estimation of snow depth from passive microwave brightness temperature data in forest regions of northeast China. Remote Sens. Environ. 183, 334–349.

Chen, B., Huang, B., Xu, B., 2015a. Comparison of spatiotemporal fusion models: a review. Remote Sens. 7, 1798–1835.

Chen, J., Quan, W.T., Cui, T.W., Song, Q.J., Lin, C.S., 2014. Remote sensing of absorption and scattering coefficient using neural network model: development, validation, and application. Remote Sens. Environ. 149, 213–226.

Chen, W.J., Huang, C.L., Shen, H.F., Li, X., 2015b. Comparison of ensemble-based state and parameter estimation methods for soil moisture data assimilation. Adv. Water Resour. 86, 425–438.

Chen, X., Liang, S., Cao, Y., He, T., Wang, D., 2015c. Observed contrast changes in snow cover phenology in northern middle and high latitudes from 2001–2014. Sci. Rep. 5, 16820.

Chen, Y., Yuan, W.P., Xia, J.Z., Fisher, J.B., Dong, W.J., Zhang, X.T., Liang, S., Ye, A.Z., Cai, W.W., Feng, J.M.,

2015d. Using Bayesian model averaging to estimate terrestrial evapotranspiration in China. J. Hydrol. 528, 537–549.

Chen, Z.Q., Shi, R.H., Zhang, S.P., 2013. An artificial neural network approach to estimate evapotranspiration from remote sensing and AmeriFlux data. Front. Earth Sci. 7, 103–111.

Cheng, J., Liu, H., Liang, S., Nie, A., Liu, Q., Guo, Y., 2017. A framework for estimating the 30-m thermal-infrared broadband emissivity from Landsat surface-reflectance data. J. Geophys. Res. Atmospheres 122, 11405–11421.

Cheng, Z.Q., Meng, J.H., Wang, Y.M., 2016. Improving spring maize yield estimation at field scale by assimilating time-series HJ-1 CCD data into the WOFOST model using a new method with fast algorithms. Remote Sens. 8.

Conti, S., Gosling, J.P., Oakley, J.E., O'Hagan, A., 2009. Gaussian process emulation of dynamic computer codes. Biometrika 96, 663–676.

Cui, S.C., Yang, S.Z., Zhu, C.J., Wen, N., 2014. Remote sensing of surface reflective properties: role of regularization and a priori knowledge. Optik 125, 7106–7112.

Da, Y., Xiurun, G., 2005. An improved PSO-based ANN with simulated annealing technique. Neurocomputing 63, 527–533.

Darvishzadeh, R., Matkan, A.A., Ahangar, A.D., 2012. Inversion of a radiative transfer model for estimation of rice canopy chlorophyll content using a lookup-table Approach. IEEE J. Sel. Top. Appl. Earth Obs. Remote Sens. 5, 1222–1230.

Darvishzadeh, R., Skidmore, A., Schlerf, M., Atzberger, C., 2008. Inversion of a radiative transfer model for estimating vegetation LAI and chlorophyll in a heterogeneous grassland. Remote Sens. Environ. 112, 2592–2604.

Dash, J., Ogutu, B.O., 2016. Recent advances in space-borne optical remote sensing systems for monitoring global terrestrial ecosystems. Prog. Phys. Geogr. 40, 322–351.

Delahaies, S., Roulstone, I., Nichols, N., 2017. Constraining DALECv2 using multiple data streams and ecological constraints: analysis and application. Geosci. Model Dev. (GMD) 10, 2635–2650.

de Vyver, H.V., Roulin, E., 2009. Scale-recursive estimation for merging precipitation data from radar and microwave cross-track scanners. J. Geophys. Res. Atmospheres 114, 14.

Ding, Y.L., Zhang, H.Y., Li, Z.W., Xin, X.P., Zheng, X.M., Zhao, K., 2016. Comparison of fractional vegetation cover estimations using dimidiate pixel models and look-up table inversions of the PROSAIL model from Landsat 8 OLI data. J. Appl. Remote Sens. 10, 15.

Disney, M., Lewis, P., Saich, P., 2006. 3D modelling of forest canopy structure for remote sensing simulations in the optical and microwave domains. Remote Sens. Environ. 100, 114–132.

Donmez, C., Berberoglu, S., Erdogan, M.A., Tanriover, A.A., Cilek, A., 2015. Response of the regression tree model to high resolution remote sensing data for predicting percent tree cover in a Mediterranean ecosystem. Environ. Monit. Assess. 187, 12.

Du, L., Shi, S., Yang, J., Sun, J., Gong, W., 2016. Using different regression methods to estimate leaf nitrogen content in rice by fusing hyperspectral LiDAR data and laser-induced chlorophyll fluorescence data. Remote Sens. 8, 14.

Fan, L., Xiao, Q., Wen, J.G., Liu, Q., Jin, R., You, D.Q., Li, X.W., 2015. Mapping high-resolution soil moisture over heterogeneous cropland using multi-resource remote sensing and ground observations. Remote Sens. 7, 13273–13297.

Fang, H., Jiang, C., Li, W., Wei, S., Baret, F., Chen, J.M., Garcia-Haro, J., Liang, S., Liu, R., Myneni, R.B., Pinty, B., Xiao, Z., Zhu, Z., 2013. Characterization and intercomparison of global moderate resolution leaf area index (LAI) products: analysis of climatologies and theoretical uncertainties. J. Geophys. Res. Biogeosci. 118 https://doi.org/10.1002/jgrg.20051.

Fang, H., Kim, H., Liang, S., Schaaf, C., Strahler, A., Townshend, G.R.G., Dickinson, R., 2007. Developing a spatially continuous 1 km surface albedo data set over North America from Terra MODIS products. J. Geophys. Res. 112, D20206, 20210.21029/22006JD008377.

Fang, H., Liang, S., Hoogenboom, G., Teasdale, J., Cavigelli, M., 2008a. Crop yield estimation through assimilation of remotely sensed data into DSSAT-CERES. Int. J. Remote Sens. 29, 3011–3032.

Fang, H., Liang, S., Townshend, J., Dickinson, R., 2008b. Spatially and temporally continuous LAI data sets based on an new filtering method: examples from North America. Remote Sens. Environ. 112, 75–93.

Fang, L., Zhan, X.W., Hain, C.R., Liu, J.C., 2018. Impact of using near real-time green vegetation fraction in Noah land surface model of NOAA NCEP on numerical weather predictions. Adv. Meteorol. 12.

Filippi, A.M., Guneralp, I., Randall, J., 2014. Hyperspectral remote sensing of aboveground biomass on a river meander bend using multivariate adaptive regression splines and stochastic gradient boosting. Remote Sens. Lett. 5, 432–441.

Fisher, A., 1994. A model for the seasonal variations of vegetation indices in coarse resolution data and its inversion to extract crop parameters. Remote Sens. Environ. 48, 220–230.

Gao, F., Masek, J., Schwaller, M., Hall, F., 2006. On the blending of the Landsat and MODIS surface reflectance: predicting daily Landsat surface reflectance. IEEE Trans. Geosci. Remote Sens. 44, 2207–2218.

Garcia-Gutierrez, J., Martinez-Alvarez, F., Troncoso, A., Riquelme, J.C., 2015. A comparison of machine learning regression techniques for LiDAR-derived estimation of forest variables. Neurocomputing 167, 24–31.

Gastellu-Etchegorry, J.P., 2008. 3D modeling of satellite spectral images, radiation budget and energy budget of urban landscapes. Meteorol. Atmos. Phys. 102, 187–207.

Gastellu-Etchegorry, J.P., Gascon, F., Esteve, P., 2003. An interpolation procedure for generalizing a look-up table inversion method. Remote Sens. Environ. 87, 55–71.

Gastellu-Etchegorry, J.P., Yin, T.G., Lauret, N., Cajgfinger, T., Gregoire, T., Grau, E., Feret, J.B., Lopes, M., Guilleux, J., Dedieu, G., Malenovsky, Z., Cook, B.D., Morton, D., Rubio, J., Durrieu, S., Cazanave, G., Martin, E., Ristorcelli, T., 2015. Discrete anisotropic radiative transfer (DART 5) for modeling airborne and satellite spectroradiometer and LIDAR acquisitions of natural and urban landscapes. Remote Sens. 7, 1667–1701.

GEO, 2005. GEOSS 10-Year Implementation Plan Reference Document, p. 209.

Gholizadeh, H., Robeson, S.M., Rahman, A.F., 2015. Comparing the performance of multispectral vegetation indices and machine-learning algorithms for remote estimation of chlorophyll content: a case study in the Sundarbans mangrove forest. Int. J. Remote Sens. 36, 3114–3133.

Goel, N.S., Grier, T., 1987. Estimation of canopy parameters of row planted vegetation canopies using reflectance data for only four directions. Remote Sens. Environ. 21, 37–51.

Gomez-Dans, J.L., Lewis, P.E., Disney, M., 2016. Efficient emulation of radiative transfer codes using Gaussian processes and application to land surface parameter inferences. Remote Sens. 8, 32.

Gomez, C., Wulder, M.A., Montes, F., Delgado, J.A., 2012. Modeling forest structural parameters in the mediterranean pines of Central Spain using QuickBird-2 imagery and classification and regression tree analysis (CART). Remote Sens. 4, 135–159.

Gu, X.H., Li, J.F., Chen, Y.D., Kong, D.D., Liu, J.Y., 2019. Consistency and discrepancy of global surface soil moisture changes from multiple model-based data sets against satellite observations. J. Geophys. Res. Atmospheres 124, 1474–1495.

Gui, S., Liang, S., Wang, K., Li, L., Zhang, X., 2010. Assessment of three satellite-estimated land surface downwelling shortwave irradiance data sets. IEEE Geosci. Remote Sens. Lett. 7, 776–780.

Halperin, J., LeMay, V., Coops, N., Verchot, L., Marshall, P., Lochhead, K., 2016. Canopy cover estimation in miombo woodlands of Zambia: comparison of Landsat 8 OLI versus RapidEye imagery using parametric, nonparametric, and semiparametric methods. Remote Sens. Environ. 179, 170–182.

Han, X.J., Li, X., Rigon, R., Jin, R., Endrizzi, S., 2015. Soil moisture estimation by assimilating L-band microwave brightness temperature with geostatistics and observation localization. PLoS One 10, 20.

He, T., Liang, S., Song, D.-X., 2014a. Analysis of global land surface albedo climatology and spatial-temporal variation during 1981–2010 from multiple satellite products. J. Geophys. Res. Atmospheres 119, 10,281-210,298.

He, T., Liang, S., Wang, D., Cao, Y., Gao, F., Yu, Y., Feng, M., 2018. Evaluating land surface albedo estimation from Landsat MSS, TM, ETM +, and OLI data based on the unified direct estimation approach. Remote Sens. Environ. 204, 181–196.

He, T., Liang, S., Wang, D., Chen, X., Song, D.-X., Jiang, B., 2015a. Land surface albedo estimation from Chinese HJ satellite data based on the direct estimation approach. Remote Sens. 7, 5495–5510.

He, T., Liang, S., Wang, D., Shi, Q., Goulden, M.L., 2015b. Estimation of high-resolution land surface net shortwave radiation from AVIRIS data: algorithm development and preliminary results. Remote Sens. Environ. 167, 20–30.

He, T., Liang, S., wang, D., Shi, Q., Tao, X., 2014b. Estimation of high-resolution land surface shortwave albedo from AVIRIS data. IEEE J. Special Top. Appl. Earth Obs. Remote Sens. 7, 4919–4928.

He, T., Liang, S., Wang, D., Wu, H., Yu, Y., Wang, J., 2012. Estimation of surface albedo and reflectance from moderate resolution imaging spectroradiometer observations. Remote Sens. Environ. 119, 286–300.

He, T., Liang, S., Wang, D.D., Shuai, Y.M., Yu, Y.Y., 2014c. Fusion of satellite land surface albedo products across scales using a multiresolution tree method in the North Central United States. IEEE Trans. Geosci. Remote Sens. 52, 3428–3439.

He, T., Liang, S., Yu, Y., Liu, Q., Gao, F., 2013. Greenland surface albedo changes 1981-2012 from satellite observations. Environ. Res. Lett. 8, 044043. https://doi.org/044010. 041088/041748-049326/044048/044044/044043.

Hedley, J., Roelfsema, C., Phinn, S.R., 2009. Efficient radiative transfer model inversion for remote sensing applications. Remote Sens. Environ. 113, 2527–2532.

Hermance, J.F., 2007. Stabilizing high-order, non-classical harmonic analysis of NDVI data for average annual models by damping model roughness. Int. J. Remote Sens. 28, 2801–2819.

Houborg, R., Soegaard, H., Boegh, E., 2007. Combining vegetation index and model inversion methods for the extraction of key vegetation biophysical parameters using Terra and Aqua MODIS reflectance data. Remote Sens. Environ. 106, 39–58.

Huang, H.G., Qin, W.H., Liu, Q.H., 2013. RAPID: a Radiosity Applicable to Porous IndiviDual Objects for directional reflectance over complex vegetated scenes. Remote Sens. Environ. 132, 221–237.

Huang, J., Ma, H., Sedano, F., Lewis, P., Liang, S., Wu, Q., Su, W., Zhang, X., Zhu, D., 2019. Evaluation of regional estimates of winter wheat yield by assimilating three remotely sensed reflectance datasets into the coupled WOFOST–PROSAIL model. Eur. J. Agron. 102, 1–13.

Huang, J.X., Ma, H.Y., Su, W., Zhang, X.D., Huang, Y.B., Fan, J.L., Wu, W.B., 2015a. Jointly assimilating MODIS LAI and ET products into the SWAP model for winter wheat yield estimation. IEEE J. Sel. Top. Appl. Earth Obs. Remote Sens. 8, 4060–4071.

Huang, J.X., Sedano, F., Huang, Y.B., Ma, H.Y., Li, X.L., Liang, S., Tian, L.Y., Zhang, X.D., Fan, J.L., Wu, W.B., 2016. Assimilating a synthetic Kalman filter leaf area index series into the WOFOST model to improve regional winter wheat yield estimation. Agric. For. Meteorol. 216, 188–202.

Huang, J.X., Tian, L.Y., Liang, S., Ma, H.Y., Becker-Reshef, I., Huang, Y.B., Su, W., Zhang, X.D., Zhu, D.H., Wu, W.B., 2015b. Improving winter wheat yield estimation by assimilation of the leaf area index from Landsat TM and MODIS data into the WOFOST model. Agric. For. Meteorol. 204, 106–121.

Husler, F., Jonas, T., Riffler, M., Musial, J.P., Wunderle, S., 2014. A satellite-based snow cover climatology (1985-2011) for the European Alps derived from AVHRR data. Cryosphere 8, 73–90.

Hutengs, C., Vohland, M., 2016. Downscaling land surface temperatures at regional scales with random forest regression. Remote Sens. Environ. 178, 127–141.

Jamet, C., Loisel, H., Dessailly, D., 2012. Retrieval of the spectral diffuse attenuation coefficient K-d(lambda) in open and coastal ocean waters using a neural network inversion. J. Geophys. Res.-Oceans 117, 14.

Jia, K., Liang, S., Liu, S.H., Li, Y.W., Xiao, Z.Q., Yao, Y.J., Jiang, B., Zhao, X., Wang, X.X., Xu, S., Cui, J., 2015. Global land surface fractional vegetation cover estimation using general regression neural networks from MODIS surface reflectance. IEEE Trans. Geosci. Remote Sens. 53, 4787–4796.

Jiang, B., Liang, S., Ma, H., Zhang, X., Xiao, Z., Zhao, X., Jia, K., Yao, Y., Jia, A., 2016a. GLASS daytime all-wave net radiation product: algorithm development and preliminary validation. Remote Sens. 8, 222.

Jiang, B., Zhang, Y., Liang, S., Zhang, X., Xiao, Z., 2014. Surface daytime net radiation estimation using artificial neural networks. Remote Sens. 6, 11031–11050.

Jiang, J., Xiao, Z., Wang, J., Song, J., 2016b. Multiscale estimation of leaf area index from satellite observations based on an ensemble multiscale filter. Remote Sens. 8, 229.

Jin, X.L., Kumar, L., Li, Z.H., Feng, H.K., Xu, X.G., Yang, G.J., Wang, J.H., 2018. A review of data assimilation of remote sensing and crop models. Eur. J. Agron. 92, 141−152.

Karlson, M., Ostwald, M., Reese, H., Sanou, J., Tankoano, B., Mattsson, E., 2015. Mapping tree canopy cover and aboveground biomass in Sudano-Sahelian woodlands using Landsat 8 and random forest. Remote Sens. 7, 10017−10041.

Ke, Y.H., Im, J., Park, S., Gong, H.L., 2016. Downscaling of MODIS one kilometer evapotranspiration using Landsat-8 data and machine learning approaches. Remote Sens. 8, 26.

Kerekes, J.P., Baum, J.E., 2005. Full-spectrum spectral imaging system analytical model. IEEE Trans. Geosci. Remote Sens. 43, 571−580.

Kim, H., Liang, S., 2010. Development of a new hybrid method for estimating land surface shortwave net radiation from MODIS data. Remote Sens. Environ. 114, 2393−2402.

Kim, J., Mohanty, B.P., Shin, Y., 2015. Effective soil moisture estimate and its uncertainty using multimodel simulation based on Bayesian Model Averaging. J. Geophys. Res. Atmospheres 120, 8023−8042.

Krahenmann, S., Obregon, A., Muller, R., Trentmann, J., Ahrens, B., 2013. A satellite-based surface radiation climatology derived by combining climate data records and near-real-time data. Remote Sens. 5, 4693−4718.

Kuhnlein, M., Appelhans, T., Thies, B., Nauss, T., 2014. Improving the accuracy of rainfall rates from optical satellite sensors with machine learning − a random forests-based approach applied to MSG SEVIRI. Remote Sens. Environ. 141, 129−143.

Kuter, S., Akyurek, Z., Weber, G.W., 2018. Retrieval of fractional snow covered area from MODIS data by multivariate adaptive regression splines. Remote Sens. Environ. 205, 236−252.

Kuter, S., Weber, G.W., Akyurek, Z., Ozmen, A., 2015. Inversion of top of atmospheric reflectance values by conic multivariate adaptive regression splines. Inverse Prob. Sci. Eng. 23, 651−669.

Kuusk, A., 1995. A fast invertible canopy reflectance model. Remote Sens. Environ. 51, 342−350.

Kuusk, A., 2018. Canopy radiative transfer modeling. In: Chen, J. (Ed.), Terrestrial Ecosystems. In: Liang, S. (Ed.), Comprehensive Remote Sensing, vol. 3. Elsevier, pp. 9−22.

Laurent, V.C.E., Schaepman, M.E., Verhoef, W., Weyermann, J., Chávez, R.O., 2014. Bayesian object-based estimation of LAI and chlorophyll from a simulated Sentinel-2 top-of-atmosphere radiance image. Remote Sens. Environ. 140, 318−329.

Laurent, V.C.E., Verhoef, W., Damm, A., Schaepman, M.E., Clevers, J., 2013. A Bayesian object-based approach for estimating vegetation biophysical and biochemical variables from APEX at-sensor radiance data. Remote Sens. Environ. 139, 6−17.

Lauvernet, C., Baret, F., Hascoët, L., Buis, S., Le Dimet, F.-X., 2008. Multitemporal-patch ensemble inversion of coupled surface−atmosphere radiative transfer models for land surface characterization. Remote Sens. Environ. 112, 851−861.

Lei, F.N., Huang, C.L., Shen, H.F., Li, X., 2014. Improving the estimation of hydrological states in the SWAT model via the ensemble Kalman smoother: synthetic experiments for the Heihe River Basin in northwest China. Adv. Water Resour. 67, 32−45.

Lewis, P., 1999. Three-dimensional plant modelling for remote sensing simulation studies using the Botanical Plant Modelling System. Agronomie 19, 185−210.

Lewis, P., Gomez-Dans, J., Kaminski, T., Settle, J., Quaife, T., Gobron, N., Styles, J., Berger, M., 2012. An earth observation land data assimilation system (EO-LDAS). Remote Sens. Environ. 120, 219−235.

Li, X., Strahler, A., 1985. Geometric-optical modeling of a coniferous forest canopy. IEEE Trans. Geosci. Remote Sens. 23, 705−721.

Li, X., Strahler, A., 1986. Geometric-optical bi-directional reflectance modeling of a coniferous forest canopy. IEEE Trans. Geosci. Remote Sens. 24, 906−919.

Li, Z.W., Xin, X.P., Tang, H., Yang, F., Chen, B.R., Zhang, B.H., 2017. Estimating grassland LAI using the Random Forests approach and Landsat imagery in the meadow steppe of Hulunber, China. J. Integr. Agric. 16, 286−297.

Liang, S., 2001. An optimization algorithm for separating land surface temperature and emissivity from multispectral thermal infrared imagery. IEEE Trans. Geosci. Remote Sens. 39, 264−274.

Liang, S., 2003. A direct algorithm for estimating land surface broadband albedos from MODIS imagery. IEEE Trans. Geosci. Remote Sens. 41, 136−145.

Liang, S., 2004. Quantitative Remote Sensing of Land Surfaces. John Wiley & Sons, Inc, New York.

Liang, S., 2007. Recent developments in estimating land surface biogeophysical variables from optical remote sensing. Prog. Phys. Geogr. 31, 501−516.

Liang, S., Fang, H., Chen, M., Shuey, C., Walthall, C., Daughtry, C., 2002. Atmospheric correction of Landsat ETM+ land surface imagery: II. Validation and applications. IEEE Trans. Geosci. Remote Sens. 40, 2736−2746.

Liang, S., Li, X., Xie, X., 2013a. Land Surface Observation, Modeling and Data Assimilation. World Scientific.

Liang, S., Liu, Q., Yan, G., Shi, J., Kerekes, J.P., 2019. Foreword to the special issue on the recent progress in quantitative land remote sensing: modeling and estimation. IEEE J. Sel. Top. Appl. Earth Obs. Remote Sens. 12, 391−395.

Liang, S., Qin, J., 2008. Data assimilation methods for land surface variable estimation. In: Liang, S. (Ed.), Advances in Land Remote Sensing: System, Modeling, Inversion and Application. Springer, New York, pp. 313–339.

Liang, S., Shi, J., Yan, G., 2018. Recent progress in quantitative land remote sensing in China. Remote Sens. 10, 1490.

Liang, S., Strahler, A., Walthall, C., 1999. Retrieval of land surface albedo from satellite observations: a simulation study. J. Appl. Meteorol. 38, 712–725.

Liang, S., Strahler, A.H., 1993a. An analytic BRDF model of canopy radiative transfer and its inversion. IEEE Trans. Geosci. Remote Sens. 31, 1081–1092.

Liang, S., Strahler, A.H., 1993b. The calculation of the radiance distribution of the coupled atmosphere-canopy. IEEE Trans. Geosci. Remote Sens. 31, 491–502.

Liang, S., Strahler, A.H., 1994. Retrieval of surface BRDF from multiangle remotely sensed data. Remote Sens. Environ. 50, 18–30.

Liang, S., Stroeve, J., Box, J.E., 2005. Mapping daily snow/ice shortwave broadband albedo from Moderate Resolution Imaging Spectroradiometer (MODIS): the improved direct retrieval algorithm and validation with Greenland *in situ* measurement. J. Geophys. Res. Atmospheres 110. Art. No. D10109.

Liang, S., Townshend, J.R.G., 1996. A modified Hapke model for soil bidirectional reflectance. Remote Sens. Environ. 55, 1–10.

Liang, S., Zhang, X., Xiao, Z., Cheng, J., Liu, Q., Zhao, X., 2013b. Global LAnd Surface Satellite (GLASS) Products: Algorithms, Validation and Analysis. Springer.

Liang, S., Zhao, X., Yuan, W., Liu, S., Cheng, X., Xiao, Z., Zhang, X., Liu, Q., Cheng, J., Tang, H., Qu, Y.H., Bo, Y., Qu, Y., Ren, H., Yu, K., Townshend, J., 2013c. A long-term global LAnd surface satellite (GLASS) dataset for environmental studies. Int. J. Digital Earth 6, 5–33.

Liess, M., Schmidt, J., Glaser, B., 2016. Improving the spatial prediction of soil organic carbon stocks in a complex tropical mountain landscape by methodological specifications in machine learning approaches. PLoS One 11, 22.

Liu, M., He, H.L., Ren, X.L., Sun, X.M., Yu, G.R., Han, S.J., Wang, H.M., Zhou, G.Y., 2015a. The effects of constraining variables on parameter optimization in carbon and water flux modeling over different forest ecosystems. Ecol. Model. 303, 30–41.

Liu, N., Liu, Q., Wang, L., Liang, S., Wen, J., Qu, Y., Liu, S., 2013a. A statistics-based temporal filter algorithm to map spatiotemporally continuous shortwave albedo from MODIS data. Hydrol. Earth Syst. Sci. 17, 2121–2129. https://doi.org/2110.5194/hess-2117-2121-2013.

Liu, Q., Liang, S., Xiao, Z.Q., Fang, H.L., 2014. Retrieval of leaf area index using temporal, spectral, and angular information from multiple satellite data. Remote Sens. Environ. 145, 25–37.

Liu, Q., Wang, L., Qu, Y., Liu, N., Liu, S., Tang, H., Liang, S., 2013b. Priminary evaluation of the long-term GLASS albedo product. Int. J. Digital Earth 6, 69–95. https://doi.org/10.1080/17538947.17532013.17804601.

Liu, Y.M., Niu, B., Luo, Y.F., 2015b. Hybrid learning particle swarm optimizer with genetic disturbance. Neurocomputing 151, 1237–1247.

Loew, A., Bell, W., Brocca, L., Bulgin, C.E., Burdanowitz, J., Calbet, X., Donner, R.V., Ghent, D., Gruber, A., Kaminski, T., Kinzel, J., Klepp, C., Lambert, J.C., Schaepman-Strub, G., Schroder, M., Verhoelst, T., 2017. Validation practices for satellite-based Earth observation data across communities. Rev. Geophys. 55, 779–817.

Lopez-Serrano, P.M., Lopez-Sanchez, C.A., Alvarez-Gonzalez, J.G., Garcia-Gutierrez, J., 2016. A comparison of machine learning techniques applied to Landsat-5 TM spectral data for biomass estimation. Can. J. Remote Sens. 42, 690–705.

Lu, D., Weng, Q., 2007. A survey of image classification methods and techniques for improving classification performance. Int. J. Remote Sens. 28, 823–870.

Lucia, D.J., Beran, P.S., Silva, W.A., 2004. Reduced-order modeling: new approaches for computational physics. Prog. Aerosp. Sci. 40, 51–117.

Ma, H., Liang, S., Xiao, Z., Shi, H., 2017a. Simultaneous inversion of multiple land surface parameters from MODIS optical–thermal observations. ISPRS J. Photogrammetry Remote Sens. 128, 240–254.

Ma, H., Liang, S., Xiao, Z., Wang, D., 2018. Simultaneous estimation of multiple land surface parameters from VIIRS optical-thermal data. IEEE Geosci. Remote Sens. Lett. 15, 151–160.

Ma, H., Liu, Q., Liang, S., Xiao, Z., 2017b. Simultaneous estimation of leaf area index, fraction of absorbed photosynthetically active radiation and surface albedo from multiple-satellite data. IEEE Trans. Geosci. Remote Sens. 55, 4334–4354 doi:4310.1109/TGRS.2017.2691542.

Machac, D., Reichert, P., Albert, C., 2016. Emulation of dynamic simulators with application to hydrology. J. Comput. Phys. 313, 352–366.

Manfreda, S., McCabe, M.E., Miller, P.E., Lucas, R., Madrigal, V.P., Mallinis, G., Dor, E., Helman, D., Estes, L., Ciraolo, G., Mullerova, J., Tauro, F., de Lima, M.I., del Lima, J., Maltese, A., Frances, F., Caylor, K., Kohv, M., Perks, M., Ruiz-Perez, G., Su, Z., Vico, G., Toth, B., 2018. On the use of unmanned aerial systems for environmental monitoring. Remote Sens. 10, 28.

Marabel, M., Alvarez-Taboada, F., 2013. Spectroscopic determination of aboveground biomass in grasslands using spectral transformations, support vector machine and partial least squares regression. Sensors 13, 10027.

Mas, J.F., Flores, J.J., 2008. The application of artificial neural networks to the analysis of remotely sensed data. Int. J. Remote Sens. 29, 617–663.

McCabe, M.F., Rodell, M., Alsdorf, D.E., Miralles, D.G., Uijlenhoet, R., Wagner, W., Lucieer, A., Houborg, R., Verhoest, N.E.C., Franz, T.E., Shi, J.C., Gao, H.L., Wood, E.F., 2017. The future of Earth observation in hydrology. Hydrol. Earth Syst. Sci. 21, 3879–3914.

Mohammed, I.N., Bolten, J.D., Srinivasan, R., Lakshmi, V., 2018. Improved hydrological decision support system for the lower Mekong river basin using satellite-based earth observations. Remote Sens. 10, 17.

Montesano, P.M., Cook, B.D., Sun, G., Simard, M., Nelson, R.F., Ranson, K.J., Zhang, Z., Luthcke, S., 2013. Achieving accuracy requirements for forest biomass mapping: a spaceborne data fusion method for estimating forest biomass and LiDAR sampling error. Remote Sens. Environ. 130, 153–170.

Moody, E.G., King, M.D., Platnick, S., Schaaf, C.B., Gao, F., 2005. Spatially complete global spectral surface albedos: value-added datasets derived from terra MODIS land products. IEEE Trans. Geosci. Remote Sens. 43, 144–158.

Moosavi, V., Talebi, A., Mokhtari, M.H., Shamsi, S.R.F., Niazi, Y., 2015. A wavelet-artificial intelligence fusion approach (WAIFA) for blending Landsat and MODIS surface temperature. Remote Sens. Environ. 169, 243–254.

Mora, B., Wulder, M.A., White, J.C., 2010. Segment-constrained regression tree estimation of forest stand height from very high spatial resolution panchromatic imagery over a boreal environment. Remote Sens. Environ. 114, 2474–2484.

Mountrakis, G., Im, J., Ogole, C., 2011. Support vector machines in remote sensing: a review. ISPRS J. Photogrammetry Remote Sens. 66, 247–259.

Murray, N.J., Keith, D.A., Bland, L.M., Ferrari, R., Lyons, M.B., Lucas, R., Pettorelli, N., Nicholson, E., 2018. The role of satellite remote sensing in structured ecosystem risk assessments. Sci. Total Environ. 619, 249–257.

Mutanga, O., Adam, E., Cho, M.A., 2012. High density biomass estimation for wetland vegetation using WorldView-2 imagery and random forest regression algorithm. Int. J. Appl. Earth Obs. Geoinf. 18, 399–406.

Myneni, R.B., Ross, J., Asrar, G., 1989. A review on the theory of photon transport in leaf canopies. Agric. For. Meteorol. 45, 1–153.

Nawar, S., Buddenbaum, H., Hill, J., Kozak, J., 2014. Modeling and mapping of soil salinity with reflectance spectroscopy and Landsat data using two quantitative methods (PLSR and MARS). Remote Sens. 6, 10813–10834.

North, P.R.J., 1996. Three-dimensional forest light interaction model using a Monte Carlo method. IEEE Trans. Geosci. Remote Sens. 34, 946–956.

Omer, G., Mutanga, O., Abdel-Rahman, E.M., Adam, E., 2016. Empirical prediction of leaf area index (LAI) of endangered tree species in intact and fragmented indigenous forests ecosystems using WorldView-2 data and two robust machine learning algorithms. Remote Sens. 8, 26.

Omer, G., Mutanga, O., Abdel-Rahman, E.M., Peerbhay, K., Adam, E., 2017. Mapping leaf nitrogen and carbon concentrations of intact and fragmented indigenous forest ecosystems using empirical modeling techniques and WorldView-2 data. ISPRS J. Photogrammetry Remote Sens. 131, 26–39.

Ouyang, X.Y., Chen, D.M., Lei, Y.H., 2018. A generalized evaluation scheme for comparing temperature products from satellite observations, numerical weather model, and ground measurements over the Tibetan plateau. IEEE Trans. Geosci. Remote Sens. 56, 3876–3894.

Owe, M., de Jeu, R., Holmes, T., 2008. Multisensor historical climatology of satellite-derived global land surface moisture. J. Geophys. Res. 113, F01002.

Pan, Z.G., Glennie, C., Legleiter, C., Overstreet, B., 2015. Estimation of water depths and turbidity from hyperspectral imagery using support vector regression. IEEE Geosci. Remote Sens. Lett. 12, 2165–2169.

Pasetto, D., Arenas-Castro, S., Bustamante, J., Casagrandi, R., Chrysoulakis, N., Cord, A.F., Dittrich, A., Domingo-Marimon, C., Serafy, G., Karnieli, A., Kordelas, G.A., Manakos, I., Mari, L., Monteiro, A., Palazzi, E., Poursanidis, D., Rinaldo, A., Terzago, S., Ziemba, A., Ziv, G., 2018. Integration of satellite remote sensing data in ecosystem modelling at local scales: practices and trends. Methods Ecol. Evol. 9, 1810–1821.

Pflugmacher, D., Cohen, W.B., Kennedy, R.E., Yang, Z., 2014. Using Landsat-derived disturbance and recovery history and lidar to map forest biomass dynamics. Remote Sens. Environ. 151, 124–137.

Piotrowski, A.P., Napiorkowski, J.J., 2013. A comparison of methods to avoid overfitting in neural networks training in the case of catchment runoff modelling. J. Hydrol. 476, 97–111.

Posselt, R., Mueller, R., Trentmann, J., Stockli, R., Liniger, M.A., 2014. A surface radiation climatology across two Meteosat satellite generations. Remote Sens. Environ. 142, 103–110.

Qi, J., Xie, D., Yin, T., Yan, G., Gastellu-Etchegorry, J.-P., Li, L., Zhang, W., Mu, X., Norford, L.K., 2019. LESS: LargE-Scale remote sensing data and image simulation framework over heterogeneous 3D scenes. Remote Sens. Environ. 221, 695–706.

Qin, J., Liang, S., Yang, K., Kaihotsu, I., Liu, R.G., Koike, T., 2009. Simultaneous estimation of both soil moisture and model parameters using particle filtering method through the assimilation of microwave signal. J. Geophys. Res. Atmospheres 114, 13.

Qin, J., Yang, K., Lu, N., Chen, Y.Y., Zhao, L., Han, M.L., 2013. Spatial upscaling of in-situ soil moisture measurements based on MODIS-derived apparent thermal inertia. Remote Sens. Environ. 138, 1–9.

Qin, J., Zhao, L., Chen, Y.Y., Yang, K., Yang, Y.P., Chen, Z.Q., Lu, H., 2015. Inter-comparison of spatial upscaling methods for evaluation of satellite-based soil moisture. J. Hydrol. 523, 170–178.

Qin, W.H., Gerstl, S.A.W., 2000. 3-D scene modeling of semi-desert vegetation cover and its radiation regime. Remote Sens. Environ. 74, 145–162.

Qu, Y., Liang, S., Liu, Q., Li, X., Feng, Y., Liu, S., 2016. Estimating Arctic sea-ice shortwave albedo from MODIS data. Remote Sens. Environ. 186, 32–46.

Qu, Y., Liu, Q., Liang, S., Wang, L., Liu, N., Liu, S., 2014a. Improved direct-estimation algorithm for mapping daily land-surface broadband albedo from MODIS data. IEEE Trans. Geosci. Remote Sens. 52, 907–919.

Qu, Y., Zhang, Y., Xue, H., 2014b. Retrieval of 30-m-resolution leaf area index from China HJ-1 CCD Data and MODIS Products through a dynamic bayesian network. IEEE J. Sel. Top. Appl. Earth Obs. Remote Sens. 7, 222–228.

Qu, Y.H., Zhang, Y.Z., Wang, J.D., 2012. A dynamic Bayesian network data fusion algorithm for estimating leaf area index using time-series data from in situ measurement to remote sensing observations. Int. J. Remote Sens. 33, 1106–1125.

Quaife, T., Lewis, P., 2010. Temporal constraints on linear BRDF model parameters. IEEE Trans. Geosci. Remote Sens. 48, 2445–2450.

Quan, X.W., He, B.B., Li, X., 2015. A bayesian network-based method to alleviate the ill-posed inverse problem: a case study on leaf area index and canopy water content retrieval. IEEE Trans. Geosci. Remote Sens. 53, 6507–6517.

Quan, X.W., He, B.B., Li, X., Liao, Z.M., 2016. Retrieval of grassland live fuel moisture content by parameterizing radiative transfer model with interval estimated LAI. IEEE J. Sel. Top. Appl. Earth Obs. Remote Sens. 9, 910–920.

Rahman, M.S., Di, L.P., 2017. The state of the art of space-borne remote sensing in flood management. Nat. Hazards 85, 1223–1248.

Ramoelo, A., Cho, M.A., Mathieu, R., Madonsela, S., van de Kerchove, R., Kaszta, Z., Wolff, E., 2015. Monitoring grass nutrients and biomass as indicators of rangeland quality and quantity using random forest modelling and World View-2 data. Int. J. Appl. Earth Obs. Geoinf. 43, 43–54.

Rivera, J.P., Verrelst, J., Gomez-Dans, J., Munoz-Mari, J., Moreno, J., Camps-Valls, G., 2015. An emulator toolbox to approximate radiative transfer models with statistical learning. Remote Sens. 7, 9347–9370.

Scholze, M., Buchwitz, M., Dorigo, W., Guanter, L., Quegan, S., 2017. Reviews and syntheses: systematic Earth observations for use in terrestrial carbon cycle data assimilation systems. Biogeosci. Discuss. 1–49.

Setiono, R., Wee Kheng, L., Zurada, J.M., 2002. Extraction of rules from artificial neural networks for nonlinear regression. IEEE Trans. Neural Netw. 13, 564–577.

Shi, H., Xiao, Z., Liang, S., Ma, H., 2017. A method for consistent estimation of multiple land surface parameters from MODIS top-of-atmosphere time series data. IEEE Trans. Geosci. Remote Sens. 55, 5158–5173.

Shi, H., Xiao, Z., Liang, S., Zhang, X., 2016. Consistent estimation of multiple parameters from MODIS top of atmosphere reflectance data using a coupled soil-canopy-atmosphere radiative transfer model. Remote Sens. Environ. 184, 40–57.

Shi, Q., Liang, S., 2013a. Characterizing the surface radiation budget over the Tibetan Plateau with ground-measured, reanalysis, and remote sensing data sets: 1. Methodology. J. Geophys. Res. Atmospheres 118, 9642–9657.

Shi, Q., Liang, S., 2013b. Characterizing the surface radiation budget over the Tibetan Plateau with ground-measured, reanalysis, and remote sensing data sets: 2. Spatiotemporal analysis. J. Geophys. Res.: Atmospheres 118, 8921–8934.

Shi, Q., Liang, S., 2014. Surface sensible and latent heat fluxes over the Tibetan Plateau from ground measurements, reanalysis, and satellite data. Atmos. Chem. Phys. 14, 5659–5677.

Shiklomanov, A.N., Dietze, M.C., Viskari, T., Townsend, P.A., Serbin, S.P., 2016. Quantifying the influences of spectral resolution on uncertainty in leaf trait estimates through a Bayesian approach to RTM inversion. Remote Sens. Environ. 183, 226–238.

Simmons, A., Fellous, J.L., Ramaswamy, V., Trenberth, K., Study Team Comm Space, R., 2016. Observation and integrated Earth-system science: a roadmap for 2016–2025. Adv. Space Res. 57, 2037–2103.

Strahler, A.H., Woodcock, C.E., Smith, J.A., 1986. On the nature of models in remote sensing. Remote Sens. Environ. 20, 121–139.

Sun, G., Ranson, K.J., Guo, Z., Zhang, Z., Montesano, P., Kimes, D., 2011. Forest biomass mapping from lidar and radar synergies. Remote Sens. Environ. 115, 2906–2916.

Sun, J., Yang, J., Shi, S., Chen, B.W., Du, L., Gong, W., Song, S.L., 2017a. Estimating rice leaf nitrogen concentration: influence of regression algorithms based on passive and active leaf reflectance. Remote Sens. 9, 15.

Sun, L., Mi, X.T., Wei, J., Wang, J., Tian, X.P., Yu, H.Y., Gan, P., 2017b. A cloud detection algorithm-generating method for remote sensing data at visible to short-wave infrared wavelengths. ISPRS J. Photogrammetry Remote Sens. 124, 70—88.

Tanase, M.A., Panciera, R., Lowell, K., Tian, S., Hacker, J.M., Walker, J.P., 2014. Airborne multi-temporal L-band polarimetric SAR data for biomass estimation in semi-arid forests. Remote Sens. Environ. 145, 93—104.

Tang, W.J., Qin, J., Yang, K., Liu, S.M., Lu, N., Niu, X.L., 2016. Retrieving high-resolution surface solar radiation with cloud parameters derived by combining MODIS and MTSAT data. Atmos. Chem. Phys. 16, 2543—2557.

Tao, L.L., Li, J., Jiang, J.B., Chen, X., 2016a. Leaf area index inversion of winter wheat using modified water-cloud model. IEEE Geosci. Remote Sens. Lett. 13, 816—820.

Tao, X., Liang, S., He, T., Wang, D., 2017. Integration of satellite fraction of absorbed photosynthetically active radiation products: method and validation. IEEE Trans. Geosci. Remote Sens. (revised).

Tao, Y.M., Gao, X.G., Hsu, K.L., Sorooshian, S., Ihler, A., 2016b. A deep neural network modeling framework to reduce bias in satellite precipitation products. J. Hydrometeorol. 17, 931—945.

Tapiador, F.J., Navarro, A., Jimenez, A., Moreno, R., Garcia-Ortega, E., 2018. Discrepancies with satellite observations in the spatial structure of global precipitation as derived from global climate models. Q. J. R. Meteorol. Soc. 144, 419—435.

Tinkham, W.T., Smith, A.M.S., Marshall, H.P., Link, T.E., Falkowski, M.J., Winstral, A.H., 2014. Quantifying spatial distribution of snow depth errors from LiDAR using Random Forest. Remote Sens. Environ. 141, 105—115.

Toth, C., Jozkow, G., 2016. Remote sensing platforms and sensors: a survey. ISPRS J. Photogrammetry Remote Sens. 115, 22—36.

Tramontana, G., Ichii, K., Camps-Valls, G., Tomelleri, E., Papale, D., 2015. Uncertainty analysis of gross primary production upscaling using Random Forests, remote sensing and eddy covariance data. Remote Sens. Environ. 168, 360—373.

Tsoularis, A., Wallace, J., 2002. Analysis of logistic growth models. Math. Biosci. 179, 21—55.

Varvia, P., Rautiainen, M., Seppänen, A., 2018. Bayesian estimation of seasonal course of canopy leaf area index from hyperspectral satellite data. J. Quant. Spectrosc. Radiat. Transf. 208, 19—28.

Verger, A., Baret, F., Weiss, M., Filella, I., Peñuelas, J., 2015. GEOCLIM: a global climatology of LAI, FAPAR, and FCOVER from VEGETATION observations for 1999—2010. Remote Sens. Environ. 166, 126—137.

Verger, A., Filella, I., Baret, F., Peñuelas, J., 2016. Vegetation baseline phenology from kilometric global LAI satellite products. Remote Sens. Environ. 178, 1—14.

Verhoef, W., 1984. Light scattering by leaf layers with application to canopy reflectance modeling: the SAIL model. Remote Sens. Environ. 16, 125—141.

Vermote, E., Kaufman, Y.J., 1995. Absolute calibration of AVHRR visible and near-infrared channels using ocean and cloud views. Int. J. Remote Sens. 16, 2317—2340.

Verrelst, J., Sabater, N., Rivera, J.P., Munoz-Mari, J., Vicent, J., Camps-Valls, G., Moreno, J., 2016. Emulation of leaf, canopy and atmosphere radiative transfer models for fast global sensitivity analysis. Remote Sens. 8, 27.

Wang, D., Liang, S., 2016. Estimating high-resolution top of atmosphere albedo from Moderate Resolution Imaging Spectroradiometer data. Remote Sens. Environ. 178, 93—103.

Wang, D., Liang, S., 2017. Estimating top-of-atmosphere daily reflected shortwave radiation flux over land from MODIS data. IEEE Trans. Geosci. Remote Sens. 55, 4022—4031.

Wang, D., Liang, S., He, T., Cao, Y., Jiang, B., 2015a. Surface shortwave net radiation estimation from FengYun-3 MERSI data. Remote Sens. 7, 6224—6239.

Wang, D., Liang, S., He, T., Shi, Q., 2015b. Estimation of daily surface shortwave net radiation from the combined MODIS data. IEEE Trans. Geosci. Remote Sens. 53, 5519—5529.

Wang, D., Liang, S., He, T., Yu, Y., 2013a. Direct estimation of land surface albedo from VIIRS data: algorithm improvement and preliminary validation. J. Geophys. Res. 118, 12577—12586.

Wang, D., Liang, S., He, T., Yu, Y.Y., Schaaf, C., Wang, Z.S., 2015c. Estimating daily mean land surface albedo from MODIS data. J. Geophys. Res. Atmospheres 120, 4825—4841.

Wang, D., Liang, S., Tao, H., 2014. Mapping high-resolution surface shortwave net radiation from Landsat data. IEEE Geosci. Remote Sens. Lett. 11, 459—463.

Wang, K., Tang, R.L., Li, Z.L., 2013b. Comparison of integrating LAS/MODIS data into a land surface model for improved estimation of surface variables through data assimilation. Int. J. Remote Sens. 34, 3193—3207.

Wang, L.A., Zhou, X.D., Zhu, X.K., Dong, Z.D., Guo, W.S., 2016. Estimation of biomass in wheat using random forest regression algorithm and remote sensing data. Crop J. 4, 212—219.

Wang, L.A., Zhou, X.D., Zhu, X.K., Guo, W.S., 2017. Estimation of leaf nitrogen concentration in wheat using the MK-SVR algorithm and satellite remote sensing data. Comput. Electron. Agric. 140, 327—337.

Wang, Y., Yang, C., Li, X., 2008. Regularizing kernel-based BRDF model inversion method for ill-posed land surface parameter retrieval using smoothness constraint. J. Geophys. Res. Atmospheres 113.

Weng, Q.H., Fu, P., Gao, F., 2014. Generating daily land surface temperature at Landsat resolution by fusing Landsat and MODIS data. Remote Sens. Environ. 145, 55–67.

Wijayasekara, D., Manic, M., Sabharwall, P., Utgikar, V., 2011. Optimal artificial neural network architecture selection for performance prediction of compact heat exchanger with the EBaLM-OTR technique. Nucl. Eng. Des. 241, 2549–2557.

Wu, H., Zhang, X., Liang, S., Yang, H., Zhou, H., 2012. Estimation of clear-sky land surface longwave radiation from MODIS data products by merging multiple models. J. Geophys. Res. 117, D22107 doi:22110.21029/22012JD017567.

Wu, P.H., Shen, H.F., Zhang, L.P., Gottsche, F.M., 2015. Integrated fusion of multi-scale polar-orbiting and geostationary satellite observations for the mapping of high spatial and temporal resolution land surface temperature. Remote Sens. Environ. 156, 169–181.

Xia, J.Z., Ma, M.N., Liang, T.G., Wu, C.Y., Yang, Y.H., Zhang, L., Zhang, Y.J., Yuan, W.P., 2018. Estimates of grassland biomass and turnover time on the Tibetan Plateau. Environ. Res. Lett. 13, 12.

Xiao, Z., Liang, S., Wang, J., Xie, D., Song, J., Fensholt, R., 2015. A framework for the simultaneous estimation of leaf area index, fraction of absorbed photosynthetically active radiation and albedo from MODIS time series data. IEEE Trans. Geosci. Remote Sens. 53, 3178–3197.

Xie, X., Meng, S., Liang, S., Yao, Y., 2014. Improving streamflow predictions at ungauged locations with real-time updating: application of an EnKF-based state-parameter estimation strategy. Hydrol. Earth Syst. Sci. 18, 3923–3936.

Xing, C.J., Chen, N.C., Zhang, X., Gong, J.Y., 2017. A machine learning based reconstruction method for satellite remote sensing of soil moisture images with in situ observations. Remote Sens. 9, 24.

Xiong, X.X., Angal, A., Barnes, W.L., Chen, H., Chiang, V., Geng, X., Li, Y.H., Twedt, K., Wang, Z.P., Wilson, T., Wu, A.S., 2018. Updates of moderate resolution imaging spectroradiometer on-orbit calibration uncertainty assessments. J. Appl. Remote Sens. 12, 18.

Xu, T., Bateni, S., Liang, S., 2015. Estimating turbulent heat fluxes with a weak-constraint data assimilation scheme: a case study (HiWATER-MUSOEXE). IEEE Geosci. Remote Sens. Lett. 12, 68–72.

Xu, T., Bateni, S.M., Liang, S., Entekhabi, D., Mao, K., 2014. Estimation of surface turbulent heat fluxes via variational assimilation of sequences of land surface temperatures from Geostationary Operational Environmental Satellites. J. Geophys. Res. Atmospheres 119, 10,780-710,798, doi:710.1002/2014JD021814.

Xu, Y.M., Shen, Y., 2013. Reconstruction of the land surface temperature time series using harmonic analysis. Comput. Geosci. 61, 126–132.

Yang, B., Knyazikhin, Y., Mottus, M., Rautiainen, M., Stenberg, P., Yan, L., Chen, C., Yan, K., Choi, S., Park, T., Myneni, R.B., 2017. Estimation of leaf area index and its sunlit portion from DSCOVR EPIC data: theoretical basis. Remote Sens. Environ. 198, 69–84.

Yang, G.J., Pu, R.L., Huang, W.J., Wang, J.H., Zhao, C.J., 2010. A novel method to estimate subpixel temperature by fusing solar-reflective and thermal-infrared remote-sensing data with an artificial neural network. IEEE Trans. Geosci. Remote Sens. 48, 2170–2178.

Yang, K., Zhu, L., Chen, Y.Y., Zhao, L., Qin, J., Lu, H., Tang, W.J., Han, M.L., Ding, B.H., Fang, N., 2016. Land surface model calibration through microwave data assimilation for improving soil moisture simulations. J. Hydrol. 533, 266–276.

Yao, X., Huang, Y., Shang, G.Y., Zhou, C., Cheng, T., Tian, Y.C., Cao, W.X., Zhu, Y., 2015. Evaluation of six algorithms to monitor wheat leaf nitrogen concentration. Remote Sens. 7, 14939–14966.

Yao, Y., Liang, S., Li, X., Hong, Y., Fisher, J.B., Zhang, N., Chen, J., Cheng, J., Zhao, S., Zhang, X., Jiang, B., Sun, L., Jia, K., Wang, K., Chen, Y., Mu, Q., Feng, F., 2014. Bayesian multimodel estimation of global terrestrial latent heat flux from eddy covariance, meteorological, and satellite observations. J. Geophys. Res. Atmospheres 119, 2013JD020864.

Yuan, H., Yang, G., Li, C., Wang, Y., Liu, J., Yu, H., Feng, H., Xu, B., Zhao, X., Yang, X., 2017. Retrieving soybean leaf area index from unmanned aerial vehicle hyperspectral remote sensing: analysis of RF, ANN, and SVM regression models. Remote Sens. 9, 309.

Zhang, P., Li, J., Olson, E., Schmit, T.J., Li, J.L., Menzel, W.P., 2006. Impact of point spread function on infrared radiances from geostationary satellites. IEEE Trans. Geosci. Remote Sens. 44, 2176–2183.

Zhang, D.Y., Zhang, W., Huang, W., Hong, Z.M., Meng, L.K., 2017a. Upscaling of surface soil moisture using a deep learning model with VIIRS RDR. ISPRS Int. J. Geo-Inf. 6, 20.

Zhang, H.K.K., Huang, B., Zhang, M., Cao, K., Yu, L., 2015. A generalization of spatial and temporal fusion methods for remotely sensed surface parameters. Int. J. Remote Sens. 36, 4411–4445.

Zhang, L.J., Dobslaw, H., Stacke, T., Guntner, A., Dill, R., Thomas, M., 2017b. Validation of terrestrial water storage variations as simulated by different global numerical models with GRACE satellite observations. Hydrol. Earth Syst. Sci. 21, 821–837.

Zhang, L.P., Zhang, L.F., Du, B., 2016. Deep learning for remote sensing data a technical tutorial on the state of the art. IEEE Geosci. Remote Sens. Mag. 4, 22–40.

Zhang, X., Liang, S., Wang, K., Li, L., Gui, S., 2010. Analysis of global land surface shortwave broadband Albedo from multiple data sources. IEEE J. Special Top. Appl. Earth Obs. Remote Sens. 3, 296–305.

Zhang, X., Liang, S., Zhou, G., Wu, H., Zhao, X., 2014a. Generating Global LAnd Surface Satellite incident shortwave radiation and photosynthetically active radiation products from multiple satellite data. Remote Sens. Environ. 152, 318–332.

Zhang, X.Y., Friedl, M.A., Schaaf, C.B., Strahler, A.H., Hodges, J.C.F., Gao, F., Reed, B.C., Huete, A., 2003. Monitoring vegetation phenology using MODIS. Remote Sens. Environ. 84, 471–475.

Zhang, Y., He, T., Liang, S., Wang, D., Yu, Y., 2018. Estimation of all-sky instantaneous surface incident shortwave radiation from Moderate Resolution Imaging Spectroradiometer data using optimization method. Remote Sens. Environ. 209, 468–479.

Zhang, Y., Liang, S., Sun, G., 2014b. Mapping forest biomass with GLAS and MODIS data over Northeast China. IEEE J. Special Top. Appl. Earth Obs. Remote Sens. 7, 140–152.

Zhang, Y., Qu, Y., Wang, J., Liang, S., 2012. Estimating leaf area index from MODIS and surface meteorological data using a dynamic Bayesian network. Remote Sens. Environ. 127, 30–43.

Zhou, J., Jia, L., Menenti, M., 2015. Reconstruction of global MODIS NDVI time series: performance of harmonic ANalysis of time series (HANTS). Remote Sens. Environ. 163, 217–228.

Zhou, Y., Wang, D., Liang, S., Yu, Y., He, T., 2016. Assessment of the Suomi NPP VIIRS land surface albedo data using station measurements and high-resolution albedo maps. Remote Sens. 8, 137.

Zhu, X.L., Chen, J., Gao, F., Chen, X.H., Masek, J.G., 2010. An enhanced spatial and temporal adaptive reflectance fusion model for complex heterogeneous regions. Remote Sens. Environ. 114, 2610–2623.

Zhu, X.X., Tuia, D., Mou, L., Xia, G.S., Zhang, L., Xu, F., Fraundorfer, F., 2017a. Deep learning in remote sensing: a comprehensive review and list of resources. IEEE Geosci. Remote Sens. Mag. 5, 8–36.

Zhu, Y.H., Liu, K., Liu, L., Myint, S.W., Wang, S.G., Liu, H.X., He, Z., 2017b. Exploring the potential of WorldView-2 red-edge band-based vegetation indices for estimation of mangrove leaf area index with machine learning algorithms. Remote Sens. 9, 20.

Zhu, Z., Piao, S., Myneni, R.B., Huang, M., Zeng, Z., Canadell, J.G., Ciais, P., Sitch, S., Friedlingstein, P., Arneth, A., Cao, C., Cheng, L., Kato, E., Koven, C., Li, Y., Lian, X., Liu, Y., Liu, R., Mao, J., Pan, Y., Peng, S., Penuelas, J., Poulter, B., Pugh, T.A.M., Stocker, B.D., Viovy, N., Wang, X., Wang, Y., Xiao, Z., Yang, H., Zaehle, S., Zeng, N., 2016. Greening of the earth and its drivers. Nat. Clim. Chang. 6, 791–795.

Geometric processing and positioning techniques

© 2020 Elsevier Inc. All rights reserved.

Abstract

The geometric processing and object positioning of remote sensing imagery are techniques used to map a remote sensing image into a specified object space coordinate system according to the geometric processing model of a satellite sensor with a few ground control points to acquire spatial information about the object in the area covered by the image. An investigation of the theories and methods of geometric processing as well as the object positioning of remote sensing imagery is important for accurate geometric processing of satellite remote sensing imagery and the faster generation of image products in different levels. This chapter mainly discusses the calibration of systematic errors in remote sensing imagery, geometric correction, geometric registration, digital terrain model generation, and digital orthoimage generation. The basic principles and methods of the relevant techniques will be introduced systematically.

The geometric processing and object positioning of remote sensing imagery are techniques used to map a remote sensing image into a specified object space coordinate system according to the geometric processing model of a satellite sensor with a few ground control points (GCPs) to acquire spatial information about the object in the area covered by the image. An investigation of the theories and methods of geometric processing as well as the object positioning of remote sensing imagery is important for accurate geometric processing of satellite remote sensing imagery and the faster generation of image products in different levels.

This chapter mainly discusses the calibration of systematic errors in remote sensing imagery, geometric correction, geometric registration, digital terrain model (DTM) generation, and digital orthoimage generation. The basic principles and methods of the relevant techniques will be introduced systematically.

2.1 Overview

The geometric processing and object positioning of remote sensing imagery are important components of the extraction of object spatial information in an area covered by a satellite remote sensing imagery, the core techniques of which are the establishment of an appropriate geometric processing model of the linear array push-broom sensors and the correct solution of the parameters in the model. The two-dimensional satellite remote sensing imagery is acquired by a one-dimensional linear array scanner in the push-broom mode, and each scan-line image is in the central projection. The imaging time, position, and attitude of each scan line differs from those of the other scan lines, thus producing an imaging mechanism that differs from that of the traditional frame-sensor imagery. Therefore, traditional theories of the geometric processing and object positioning of remote sensing imagery can no longer be utilized. For this reason, scientists have developed many different rigorous geometric models of satellite remote sensing imagery according to different assumptions about the variation rules of the sensor position and attitude as well as the settings of specific parameters.

In all the types of rigorous geometric processing models, the basic principle is still based on collinearity equations. These equations have a rigorous theory and high geometric processing accuracy but suffer from a complicated form, geometric processing difficulty, and the requirement for the users to have a strong background in photogrammetry.

To use satellite imagery conveniently and retain the confidentiality of some specific parameters, the empirical geometric processing model, which uses general mathematical functions to describe the geometric relationship between the image space coordinates and the corresponding object space coordinates, is employed. Models of this type do not require interior and exterior orientation elements, and they have a simple form. Furthermore, the model parameters can be solved easily.

The most common empirical geometric processing models include the general polynomial model, the direct linear transformation (DLT) model, the affine transformation model and the rational function model (RFM). The RFM is an expansion of the first three models. It is a more general and perfect empirical geometric processing model. Because the IKONOS high-resolution satellite imagery used in the United States employs the RFM to replace the rigorous geometric processing model based on the collinearity condition equation for the geometric processing and object positioning of satellite imagery, the RFM has gained considerable attention, and it has been studied by many scholars. The RFM can achieve a very high object positioning accuracy.

Tao and Hu (2001) developed a method for solving the rational polynomial coefficients (RPCs) based on the least squares adjustment. They used one SPOT image and one aerial image for the experiment, and they showed that the RFM with denominators has a higher accuracy than the RFM without denominators. After an experiment using a pair of SPOT images and a pair of NAPP images, Yang (2000) found that the RFM in the third order (or even in the second order with different denominators) could replace the rigorous geometric processing model for SPOT imagery and that the RFM in the first order is adequate for aerial images. Di et al. (2003) investigated a method for improving the RFM accuracy in both the object space and the image space using GCPs. Fraser et al. (2006) studied the bundle block adjustment based on the RFM. Grodecki and Dial (2001) confirmed that the RFM could replace the rigorous geometric processing model for image rectification, orthoimage generation, and three-dimensional object reconstruction, etc., in the geometric processing of satellite imagery (Zhang et al., 2012).

Currently, many suppliers of high-resolution remote sensing imagery provide RPCs to users as a standard positioning assistance dataset along with the images. The existing commercial remote sensing image—processing software (such as ERDAS, PCI, and ENVI) already includes an RPC-based geometric processing module for satellite imagery. The RFM has become a generalized geometric processing and object positioning model for high-resolution satellite imagery, and researchers are further investigating the accuracy of RPC solutions, error propagation properties, adjustment optimization schemes, and other applications.

In addition, because satellite imagery is influenced by various factors during the imaging process, image data inevitably contain a series of errors that produce geometric distortions. To extract accurate and reliable terrestrial information from satellite remote sensing imagery, the preliminary correction of the systematic imaging errors is a necessary and important step before geometric processing; digital elevation model (DEM) generation or DOM generation based on the above geometric processing models can be implemented. This step can eliminate the influence of image distortions on the accuracy of the terrestrial information extracted from satellite imagery. However, the research results on this aspect of imaging remain limited.

2.2 In-orbit geometric calibration of satellite remote sensing imagery

2.2.1 Systematic error sources of satellite remote sensing imagery

Remote sensing imagery acquired by different sensors has a series of geometric distortions corresponding to self-geometric characteristics, which include the attitude variation of the satellite during flight; the image displacement caused by the curvature of the Earth; the image displacement caused by atmospheric refraction; the image slippage caused by the Earth's rotation; the sensor capability, such as the structural errors of the CCD array; and the projection mode that the user ultimately selects.

Toutin (2004a) suggested that the sources of systematic errors that produce geometric distortions of satellite remote sensing imagery can be classified into two types: errors from the observation devices (the remote sensing platform, the sensor and other measuring devices of the imaging system such as gyroscope and star trackers) and errors that arise from the observed objects (such as atmosphere, Earth). The details of these error sources are listed in Table 2.1.

Most error sources listed in Table 2.1 can be predicted, and the image geometric distortions caused by those sources are systematic. Some errors, especially the image distortions related to the measuring instrument, can be corrected by

TABLE 2.1   Classification of error sources of high-resolution remote sensing imagery.

Class	Subclass	Error source
Errors from image acquisition system	Platform	Platform velocity variation; platform attitude variation
	Sensor	The sensor scan rate variation; scanning side view angle variation
	Measure instrument	Clock error and time is not synchronized
Errors from observed object	Atmosphere	Refraction
	Earth	Earth curvature; Earth rotation; terrestrial factors; etc.
	Map projection	Earth to ellipsoid and ellipsoid to map projection transformation

the satellite ground station or the parameters provided by the image suppliers. The other errors, such as image distortions caused by atmospheric conditions, are often difficult to correct because it is difficult to obtain information about the time, position, and weather conditions related to the atmospheric window at the time of image acquisition. The following error sources influence remote sensing imagery in specific ways:

(1) the altitude of the remote sensing platform, camera focal length, and terrain relief mainly influence the pixel interval of the imagery;
(2) the variation of the sensor attitude (pitch, roll and yaw) and side-look angle influence the image shape;
(3) the velocity of the remote sensing platform influences the image line interval and causes scan-line fracture or overlap;
(4) the side-look angle, the curvature of the Earth, and terrain relief cause variation in the image resolution and generate image parallax in the direction of the scan line.

In addition, projection errors exist that are caused by using a reference ellipsoid to replace the geoid and projection of the reference ellipsoid onto the tangent plane.

In the geometric processing and object positioning of satellite remote sensing imagery, image geometric distortions can be expressed by one or several functions of the image coordinates. The changing parts of the systematic errors can be considered as a time sequence in the adjustment of the stochastic model or simply as a random error, and they can be eliminated with GCPs.

2.2.1.1 The Earth curvature correction

The surface of the Earth is curved, but the datum in the geometric processing of a remote sensing imagery is a tangential plane to the Earth's surface. Therefore, the imaging position of a ground point $P(X, Y, Z)$ is actually point

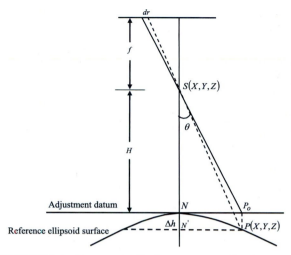

FIGURE 2.1 Image point displacement caused by the curvature of the Earth.

P_0, and it generates offset dr in the remote sensing image (Fig. 2.1).

Without considering the influence of the image inclination, the correction of the image point displacement caused by the curvature of the Earth in the radial direction is

$$\delta = \frac{H}{2Rf^2}r^3 \tag{2.1}$$

where r is the radius vector when the photo nadir point is taken as the pole, and $r = \sqrt{x'^2 + y'^2}$. Here, x' and y' are the image point coordinates; f is the camera focal length; H is the altitude of the camera station; and R is the curvature radius of the Earth.

Therefore, the following corrections are applied to the image coordinate offsets in the x and y directions:

$$\begin{cases} \delta_x = \dfrac{x'}{r}\delta = \dfrac{Hr^2}{2Rf^2}x' \\[2mm] \delta_y = \dfrac{y'}{r}\delta = \dfrac{Hr^2}{2Rf^2}y' \end{cases} \tag{2.2}$$

The influence of the curvature of the Earth can be neglected when the nadir view angle is small.

From Fig. 2.1, we can conclude that the correction value of the image coordinate offset for satellite imagery acquired by a linear array sensor in the push-broom mode becomes larger along with the distance between the image points and the projection center. This enlargement may generate additional upper and lower errors. In addition, to consider the influence of the Earth's curvature strictly on the positioning accuracy of satellite imagery, it is preferable that the triangulation be performed in the geocentric Cartesian coordinate system or the tangent plane coordinate system. With the emergence and popularity of high-resolution remote sensing imagery, the camera focal length has become longer, the altitude of the remote sensing platform has become higher, and the influence of the Earth's curvature on the image point coordinates has become smaller. Taking CBERS-02B as an example, the altitude of the satellite is 780 km, and the camera focal length is 3.396 m. According to Eq. (2.2), the maximum image coordinate offset caused by the curvature of the Earth on the edge of the satellite imagery is only 1/4 pixel, which is equivalent to the measurement error of the image point coordinates and can therefore be ignored.

The image point displacement caused by terrain relief is similar to that caused by the curvature of the Earth. Calculating the differentials of collinearity equation

$$\begin{cases} x - x_0 = -f\dfrac{a_1(X - X_S) + b_1(Y - Y_S) + c_1(Z - Z_S)}{a_3(X - X_S) + b_3(Y - Y_S) + c_3(Z - Z_S)} \\ y - y_0 = -f\dfrac{a_2(X - X_S) + b_2(Y - Y_S) + c_2(Z - Z_S)}{a_3(X - X_S) + b_3(Y - Y_S) + c_3(Z - Z_S)} \end{cases}$$
$$(2.3)$$

We can obtain the following formula:

$$\begin{cases} dx = -f\dfrac{\overline{Z}c_1 - \overline{X}c_3}{\overline{Z}^2}dZ \\ dy = -f\dfrac{\overline{Z}c_2 - \overline{Y}c_3}{\overline{Z}^2}dZ \end{cases}$$
$$(2.4)$$

where dx and dy are the image point displacements caused by the terrain relief;

$$\overline{X} = a_1(X - X_S) + b_1(Y - Y_S) + c_1(Z - Z_S);$$

$$\overline{Y} = a_2(X - X_S) + b_2(Y - Y_S) + c_2(Z - Z_S);$$

$$\overline{Z} = a_3(X - X_S) + b_3(Y - Y_S) + c_3(Z - Z_S).$$

In projective imagery, the direction of the image point displacement caused by terrain relief is away from principal point, in contrast with its effect on Radar imagery.

2.2.1.2 Atmospheric refraction correction

The atmosphere is a nonuniform medium, and the refractive index of electromagnetic wave transmission in the atmosphere changes with the height above the Earth. Therefore, the propagation path of an electromagnetic wave is not a straight line. This nonlinear propagation destroys the collinear relationship between the ground point, the perspective center, and the corresponding image point at the imaging time, causing image point offsets. The following is a general model for the correction of the influences of atmospheric refraction on projective imagery:

$$\Delta r = \frac{n_H(n - n_H)}{n(n + n_H)} \cdot \left(r + \frac{r^3}{f^2}\right) \qquad (2.5)$$

where n and n_H are the air refractive indices on the ground and at the photo station, respectively.

Assuming that $K = \frac{n_H(n - n_H)}{n(n + n_H)}$, the corrections of the image point displacement caused by atmospheric refraction in the x and y directions are, respectively,

$$\begin{cases} \Delta x = K\left(x + \dfrac{xr^2}{f^2}\right) \\ \Delta y = K\left(y + \dfrac{yr^2}{f^2}\right) \end{cases}$$
$$(2.6)$$

The coefficient K can be solved according to the following formula:

$$K = \left(\frac{2410H}{H^2 - 6H + 250} - \frac{2410H}{h^2 - 6h + 250} \cdot \frac{h}{H} \right) \cdot 10^{-6}$$

$$(2.7)$$

where H is the altitude of the remote sensing platform (unit: km) and h is the average elevation of the area covered by the image (unit: km).

Assuming that the average elevation of the area covered by the image is 0, the coefficients K are only related to the altitude of the remote sensing platforms, i.e.,

$$K = \frac{2410H}{H^2 - 6H + 250} \cdot 10^{-6} = \frac{2410}{H + \dfrac{250}{H} - 6} \cdot 10^{-6}$$

$$(2.8)$$

Therefore, the correction for the image point displacement caused by atmospheric refraction is

$$D_r = \left(\frac{0.113(P_1 - P_2)H}{H - h} - \left(\frac{2410H}{H^2 - H \cdot 6 + 250} - \frac{2410h}{h^2 - h \cdot 6 + 250} \right) \frac{r^3}{r + f^2} \right) \cdot 10^{-6}$$

$$(2.9)$$

where P_1 and P_2 are the air pressures on the ground and at the altitude of the sensor, respectively, $P_1 = e^{6.94 - 0.125H}$ and $P_2 = e^{6.94 - 0.125h}$.

Eq. (2.9) is only applicable to the correction of image point displacement in traditional aerial photography in which the flight altitude is below 12 km. This formula is not applicable to satellite sensor platforms, whose orbital altitudes are at several hundred kilometers. In particular, when the ground elevation reaches more than 4000 m, the correction for the image point displacement cannot be calculated according to Eq. (2.9) and should be calculated based on the principle of atmospheric imaging stratification. In Fig. 2.2, the altitude of the satellite sensor platform is usually above several hundred kilometers, and the image point displacement caused by the atmospheric refraction is only a few micrometers, which is far less than the value calculated by Eq. (2.9). Therefore, this error source can be neglected when processing satellite remote sensing imagery.

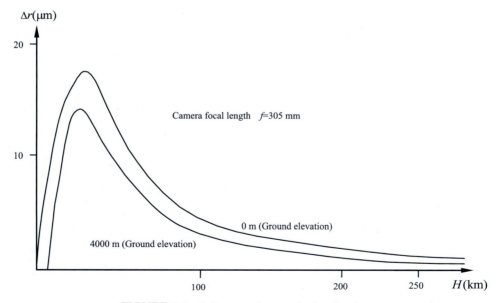

FIGURE 2.2   Influences of atmospheric refraction.

2.2.1.3 The Earth rotation correction

The Earth's rotation can cause parallel rupture for satellite imagery acquired by linear array sensors in push-broom mode. In particular, when the satellite moves from north to south while the Earth rotates from west to east, the Earth's rotation leads the projection of each scan-line image on the ground; thus, these images have a westward migration that produces an image point displacement because each scan line of the satellite imagery has a different imaging time. For satellite imagery acquired by a linear array sensor in push-broom mode, a single scan-line image is not influenced by the Earth's rotation. However, because the imaging time of each scan-line image is different, the image point displacement caused by the Earth's rotation influences a series of scan-line images, as shown in Fig. 2.3.

The parallel image dislocation caused by the Earth's rotation, Δy_e, is given by the following formula:

$$\Delta y_e = t_e \cdot v_\phi \qquad (2.10)$$

where t_e is the total acquisition time of the image, which can be obtained from the satellite ephemeris file; v_ϕ is the linear velocity of the Earth's rotation at this latitude, $v_\phi = 40000 \cos B/24$.

Here B is the latitude of the center of the area covered by the image.

As an example, the Wuhan area is covered by a CBERS-02B image. The maximum momentum error Δy in the image space, calculated according to Eq. (2.10), is approximately 900 pixels if the perimeter of the Earth at the equator is assumed to be 40,000 km, the camera focal length is 3.396 m, the altitude of the satellite is 780 km. Based on these assumptions, this error source should be considered when the satellite remote sensing imagery is processed.

2.2.1.4 The CCD manufacture error correction

Current optical imaging satellites usually have one or more linear array CCD sensors. In general, the sensors are calibrated in a laboratory before the satellite launches to determine the relative and absolute position of the CCD, the principal point coordinates, the focal length, and other parameters. However, excessive acceleration changes CCD position during the satellite launch process, and the focal length of the linear array CCD sensor changes, along with the other instrument parameters, when the satellite moves. The images acquired by this satellite sensor have geometric distortions that influence their geometric positioning. Therefore, the sensor should be calibrated using GCPs or a geometric calibration field after the satellite launches.

Because of the technical limitation of CCD sensor production, a single linear array CCD sensor is not adequate to cover the entire imaging area. Therefore, many existing CCD linear arrays usually utilize multiple TDI CCD mosaic techniques to obtain a large image (Baltsavias et al., 2006), as shown in Fig. 2.4.

TDI CCD is a new type of CCD that has a plane structure and a linear array output. TDI CCD has a multiseries time delay integration function so that a single scanned object on the ground can be exposed repeatedly in the integration direction, and the exposure value can be output after integration processing. This process

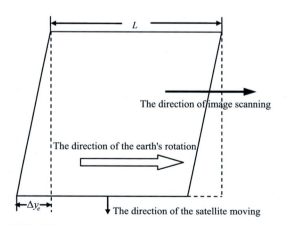

FIGURE 2.3 Image parallel dislocations caused by the Earth's rotation.

(A)

(B)

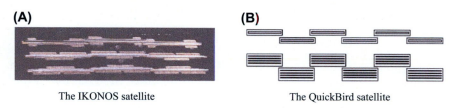

The IKONOS satellite   The QuickBird satellite

FIGURE 2.4   The structure of CCD linear arrays.

can guarantee that the camera segment retains better imaging performance in the near-infrared spectrum and under other low-light situations.

CCD devices always have a package structure and a specific geometric size, and the actual pixel number of the CCD sensor is greater than the effective pixel number. Therefore, if two CCD components are directly joined together, then gaps must exist between the two components that form an imaging blind spot and destroy the image data. At present, the two common methods of joining CCDs are optical joints and mechanical joints.

Fig. 2.5 shows a schematic diagram of a CCD cross-joint. Although the scan-line image acquired by each CCD is not in a straight line, a single continuous image can still be obtained by joining these scan-line images together; however, the initial and end positions of the entire image exhibit an image-staggering phenomenon. In practice, we can appropriately expand the area covered by the image and remove the excess staggered image-by-image processing technology to fulfill the application requirements.

In comparison to optical joints, staggered mechanical joints do not introduce a prism; consequently, color shading does not exist, and the energy is not dissipated. In addition, staggered mechanical joints are not affected by the length of the prism, and they can be used for all types of optical systems. Because the CCD is joined together in a staggered mode and placed within the focal plane, it is difficult to arrange the inspection components.

Both optical and mechanical joints produce many geometric and systematic problems:

(1) sensors may have different principal lengths;
(2) sensors may be rotated relative to the line of the image plane;
(3) sensors may be rotated relative to the image plane; and
(4) displacement may exist in the image plane.

Different locations of the CCD linear arrays produce different imaging angles (shown in Fig. 2.6). At a certain reference ground elevation, H_0, a single image can be fitted well to the plane. However, if the elevation of the area covered by the image changes substantially, then the difference in the CCD line array locations produces geometric displacement errors. For example, a

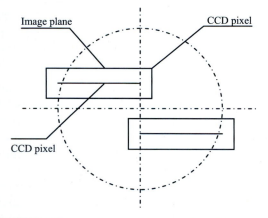

FIGURE 2.5   A schematic diagram of a CCD staggered joint.

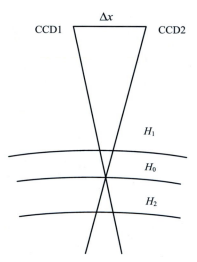

FIGURE 2.6   Image point displacement caused by CCD linear arrays at different locations at different altitudes.

displacement of 1 pixel will be caused by a terrain relief of 450 m relative to the reference plane for IRS-1C/1D imagery. For QuickBird imagery, only a terrain relief of more than 2.8 km will cause a 1 pixel displacement. Nowadays, errors caused by CCD joints are mainly corrected with the in-orbit geometric calibration method using calibration field. The details of this method will be introduced in Section 2.2.2.

2.2.2 In-orbit geometric calibration model

To establish the in-orbit geometric calibration model of satellite remote sensing imagery, we should first establish the rigorous geometric processing model. Owing to the difference of the physical structure of satellite sensors and the operating principle of position and attitude determination systems, the rigorous geometric processing models of different satellite remote sensing images are slightly different. In this section, we only take the ZiYuan-3 satellite in China

for example and introduce its in-orbit geometric calibration model.

The rigorous geometric processing model actually describes a series of space coordinate transformations of a certain point, including the image space coordinate system, the sensor coordinate system, the satellite body coordinate system, the attitude determination reference coordinate system, the space-fixed inertial reference coordinate system, and the Earth-fixed ground reference coordinate system. In the actual process, the J2000 celestial coordinate system is often employed as the space-fixed inertial reference coordinate system, and the WGS 84 geocentric coordinate system is often employed as the Earth-fixed ground reference coordinate system. Therefore, the rigorous geometric processing model can be expressed as follows:

$$
\begin{bmatrix} X \\ Y \\ Z \end{bmatrix}_{WGS84} = \begin{bmatrix} X \\ Y \\ Z \end{bmatrix}_{GPS}
$$

$$
+ m\mathbf{R}_{J2000}^{WGS84}\mathbf{R}_{Star}^{J2000}\left(\mathbf{R}_{Star}^{Body}\right)^{\mathrm{T}}\mathbf{R}_{Sensor}^{Body}\begin{bmatrix} \tan(\psi_Y) \\ \tan(\psi_X) \\ -1 \end{bmatrix}f
$$

(2.11)

where $(X, Y, Z)_{WGS84}^{\mathrm{T}}$ and $(X, Y, Z)_{WGS84}^{\mathrm{T}}$ are the coordinates of ground point P and the satellite position S in the WGS 84 geocentric coordinate system; m is the scale factor; $\mathbf{R}_{J2000}^{WGS84}$ is the J2000-to-WGS 84 rotation matrix; $\mathbf{R}_{Star}^{J2000}$ is the rotation matrix from the attitude determination reference coordinate system to the J2000 celestial coordinate system; $\mathbf{R}_{Star}^{Body}$ and $\mathbf{R}_{Sensor}^{Body}$ represent the rotation matrices from the attitude determination reference coordinate system and the sensor coordinate system to the satellite body coordinate system; (ψ_Y, ψ_X) are the CCD-detector look angles in the sensor coordinate system; and f is the focal length.

To establish the geometric calibration model based on CCD-detector look angles, the following equation firstly needs to be defined:

$$\begin{bmatrix} x \\ y \\ z \end{bmatrix} = \left(\mathbf{R}_{Star}^{Body}\right)^T \mathbf{R}_{Sensor}^{Body} \begin{bmatrix} \tan(\psi_Y) \\ \tan(\psi_X) \\ -1 \end{bmatrix} f \quad (2.12)$$

where $(x, y, z)^T$ represents the coordinates of each CCD detector in the attitude determination reference coordinate system.

With reference to Eq. (2.12), the following equation can be defined further:

$$\begin{cases} \tan(\psi_Y') = -\dfrac{x}{z} \\ \tan(\psi_X') = -\dfrac{y}{z} \end{cases} \quad (2.13)$$

where (ψ_Y', ψ_X') represent the look angles of each CCD detector in the attitude determination reference coordinate system.

Dividing the left and right parts of Eq. (2.12) by $-z$, Eq. (2.14) can be obtained:

$$\begin{bmatrix} -x/z \\ -y/z \\ -1 \end{bmatrix} = \begin{bmatrix} \tan(\psi_y') \\ \tan(\psi_x') \\ -1 \end{bmatrix}$$

$$= -\frac{f}{z}\left(\mathbf{R}_{Star}^{Body}\right)^T \mathbf{R}_{Sensor}^{Body} \begin{bmatrix} \tan(\psi_y) \\ \tan(\psi_x) \\ -1 \end{bmatrix} \quad (2.14)$$

From Eq. (2.14), it can be seen that the look angles (ψ_Y', ψ_X') can describe the comprehensive effects of the exterior calibration parameters $\mathbf{R}_{Star}^{Body}$ and $\mathbf{R}_{Sensor}^{Body}$ and the interior calibration parameters (ψ_Y, ψ_X) and f on the direct georeferencing of satellite images. On the other hand, when the satellite orbits the Earth, the parameters $\mathbf{R}_{Star}^{Body}$, $\mathbf{R}_{Sensor}^{Body}$, (ψ_Y, ψ_X), and f change very little and can be considered stable over a short period of time (for example, 3 months). Therefore, the look angles (ψ_Y', ψ_X') can be also

considered stable over a short period of time, which can provide possibilities for calibrating the CCD-detector look angles in the attitude determination reference coordinate system.

Substituting Eq. (2.14) into Eq. (2.11), Eq. (2.15) can be obtained:

$$\begin{bmatrix} X \\ Y \\ Z \end{bmatrix}_{WGS84} = \begin{bmatrix} X \\ Y \\ Z \end{bmatrix}_{GPS}$$

$$+ \lambda \mathbf{R}_{J2000}^{WGS84} \mathbf{R}_{Star}^{J2000} \begin{bmatrix} \tan(\psi_Y') \\ \tan(\psi_X') \\ -1 \end{bmatrix} \quad (2.15)$$

where λ is a scale factor and $\lambda = -mz$.

Transforming Eq. (2.5), we can obtain

$$\begin{bmatrix} \tan(\psi_Y') \\ \tan(\psi_X') \\ -1 \end{bmatrix} = \frac{1}{\lambda}\left(\mathbf{R}_{J2000}^{WGS84} \mathbf{R}_{Star}^{J2000}\right)^T \left\{ \begin{bmatrix} X \\ Y \\ Z \end{bmatrix}_{WGS84} - \begin{bmatrix} X \\ Y \\ Z \end{bmatrix}_{GPS} \right\}$$

$$(2.16)$$

Eq. (2.16) is the in-orbit geometric calibration model of high-resolution satellite imagery based on CCD-detector look angles. For high-resolution satellite sensors, a linear CCD array often has many detectors. If we calculate the look angles of each detector, no fewer than one GCP in each image column should be available. In the actual process, a polynomial model is often employed as the CCD-detector look angle model as follows (Wang et al., 2014; Zhang et al., 2014; Cao et al., 2015):

$$\begin{cases} \psi_Y' = a_0 + a_1 N + a_2 N^2 + a_3 N^3 + \cdots \\ \psi_X' = b_0 + b_1 N + b_2 N^2 + b_3 N^3 + \cdots \end{cases} \quad (2.17)$$

where $(a_0, a_1, a_2, a_3, \cdots b_0, b_1, b_2, b_3, \cdots)$ are polynomial coefficients; N is the CCD-detector number.

When sufficient and evenly distributed GCPs in the image-covered area are available, we can calculate the calibration parameters with the least squares adjustment. The specific algorithm flow is as follows:

(1) According to the third formulation in Eq. (2.16), the scale factor λ is firstly calculated with GCPs;
(2) According to Eq. (2.17), the CCD-detector look angles (ψ'_Y, ψ'_X) corresponding to the GCPs in the attitude determination reference coordinate system are calculated;
(3) According to Eq. (2.17), the following error equation is established:

$$\mathbf{V} = \mathbf{A}\mathbf{x} - \mathbf{l} \qquad (2.18)$$

where

$$\mathbf{V} = \begin{bmatrix} v_{\psi'_Y} \\ v_{\psi'_X} \end{bmatrix};$$

$$\mathbf{A} = \begin{bmatrix} 1 & N & N^2 & N^3 & \cdots & 0 & 0 & 0 & 0 & 0 \\ 0 & 0 & 0 & 0 & 0 & 1 & N & N^2 & N^3 & \cdots \end{bmatrix};$$

$$\mathbf{x} = \begin{bmatrix} a_0 & a_1 & a_2 & a_3 & \cdots & b_0 & b_1 & b_2 & b_3 & \cdots \end{bmatrix}^{\mathrm{T}};$$

$$\mathbf{l} = \begin{bmatrix} \psi'_Y \\ \psi'_X \end{bmatrix}$$

(4) According to the least squares adjustment, the unknown parameter $\mathbf{x}$ in Eq. (2.18) is calculated as follows:

$$\mathbf{x} = (\mathbf{A}^{\mathrm{T}}\mathbf{A})^{-1}\mathbf{A}^{\mathrm{T}}\mathbf{l} \qquad (2.19)$$

(5) According to Eq. (2.17), the look angles of each CCD detector in the attitude determination reference coordinate system are calculated.

It can be seen from the above equations that the error equation established in this section is linear when we calculate the calibration parameters. There is no need to calculate the calibration parameters iteratively or take the laboratory calibration value as the initial value of the calibration parameters, so the solution process is very convenient.

2.2.3 In-orbit geometric calibration of the ZiYuan-3 satellite

In this section, the in-orbit geometric calibration of the ZiYuan-3 satellite is first performed with the highly precise GCPs in the Songshan calibration field. The obtained calibration parameters are then used to perform the georeferencing of the other ZiYuan-3 images to evaluate the effectiveness and feasibility of the introduced calibration method in Section 2.2.2.

The ZiYuan-3 satellite is mainly used for 1: 50,000 scale stereo mapping as well as updating some elements in 1:25,000 scale topographic maps. The ZiYuan-3 satellite is equipped with three independent panchromatic cameras for forward (FWD), nadir (NAD), and backward (BWD) views. The NAD camera has a 2.1 m ground sample distance (GSD) and 51 km ground swath. Both the FWD and BWD cameras have a 3.5 m GSD and 52 km ground swath width. The designed look angles of the FWD and BWD cameras are, respectively, +22 degrees and −22 degrees from the NAD camera, forming a large base-to-height ratio of 0.8. The general parameters of each dataset are depicted in Table 2.2, and the distribution of the GCPs is shown in Fig. 2.7.

2.2.3.1 Calibration accuracy analysis of the CCD-detector look angles

The ZiYuan-3 FWD, NAD, and BWD cameras are all linear array CCD cameras. Therefore,

TABLE 2.2   General parameters of each dataset.

Area	Songshan	Luoyang	Nanyang	Taiyuan	Bellegarde
Acquisition date	2012.2.3	2012.1.24	2012.2.3	2012.3.13	2012.2.29
Terrain relief/m	86−1130	88−370	61−138	743−1545	54−245
Number of ground control points	48	24	28	17	9

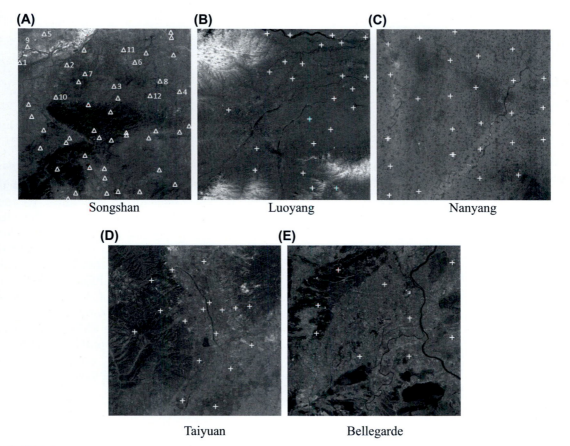

FIGURE 2.7   Distribution of the ground control points in (A) Songshan, (B) Luoyang, (C) Nanyang, (D) Taiyuan, and (E) Bellegarde.

when performing in-orbit geometric calibration, the CCD detectors corresponding to the GCPs should be distributed evenly within the linear array. Based on this principle, three different GCP layouts in the Songshan area were selected to perform geometric calibration; the remaining GCPs were considered as check points. The selected GCPs are denoted as △ and numbered from 1 to 9 in Fig. 2.7A. In each GCP layout, the maximum error and the root mean square error (RMSE) of the CCD-detector look angles corresponding to the check points are listed in Table 2.3.

TABLE 2.3   The calibration accuracy of the FWD, NAD, and BWD cameras.

Camera	Number of ground control points	Number of check points	Maximum error/″		RMSE/″	
			ψ'_y	ψ'_x	ψ'_y	ψ'_x
FWD	5	43	−1.822	−1.913	0.899	0.911
	7	41	−1.712	1.837	0.836	0.976
	9	39	−1.724	−2.196	0.852	0.770
NAD	5	43	−2.145	−1.283	1.020	0.583
	7	41	−2.390	−1.379	0.970	0.597
	9	39	−2.363	−1.506	1.010	0.591
BWD	5	43	2.524	−2.260	1.122	1.050
	7	41	−2.076	−2.625	0.947	0.880
	9	39	−2.262	−2.699	1.000	0.925

When we perform the in-orbit geometric calibration of the ZiYuan-3 cameras with the introduced calibration method in Section 2.2.2, the unknown parameters are actually the coefficients of the CCD-detector look angle model. From the results in Table 2.3, we can see that the calibration accuracy of the CCD-detector look angles obtained with five GCPs reaches ±1.0″. The calibration accuracy no longer improves significantly when the number of GCPs increases. It means that the introduced calibration method can not only eliminate the correlation between the calibration parameters but also reduce the number of required GCPs in the in-orbit geometric calibration of the ZiYuan-3 satellite.

2.2.3.2 Direct georeferencing accuracy analysis

To further evaluate the effectiveness of the obtained CCD-detector look angles, we performed the georeferencing of the ZiYuan-3 NAD image in the Songshan area. The look angles of each CCD detector in the attitude determination reference coordinate system were first calculated with different GCPs layouts. The georeferencing of the NAD image was then performed according to Eq. (2.15). The maximum error and the RMSE of the residual errors of the check points are listed in Table 2.4.

From the results in Table 2.4, we can draw the following conclusions.

TABLE 2.4   The georeferencing results of the ZiYuan-3 NAD image in the Songshan area.

Calibration	Number of ground control points	Number of check points	Maximum/m			RMSE/m		
			X	Y	Planimetry	X	Y	Planimetry
Before	0	48	772.255	671.104	1022.145	765.757	621.666	986.333
After	5	43	−2.897	−4.976	5.522	1.279	1.957	2.338
	7	41	−3.112	−4.249	4.946	1.361	1.532	2.049
	9	39	−3.396	−4.193	4.869	1.332	1.661	2.129

Note: "Before calibration" means that the georeferencing was performed according to Eq. (2.11).

First, the georeferencing accuracy of the ZiYuan-3 NAD image obtained before calibration reaches up to about ±1 km in planimetry. This demonstrates that the values of some orientation parameters have changed during the satellite launch and differ a lot from the values calibrated in laboratory before the satellite launch.

Second, when five GCPs in the image-covered area are available, the look angles of each CCD detector can be calculated with the introduced calibration method. The systematic errors in the orientation parameters, such as $\mathbf{R}_{Star}^{Body}$ and $\mathbf{R}_{Sensor}^{Body}$, can be effectively eliminated. The georeferencing accuracy of the ZiYuan-3 images can be thereby improved. For the NAD image in the Songshan area, the georeferencing accuracy is improved to ±2.338 m, which is corresponding to about ±1 pixel. The georeferencing accuracy of the NAD image is no longer improved significantly as the number of GCPs increases. The georeferencing accuracy is improved by only ±0.209 m in planimetry as the number of GCPs increases from five to nine.

2.2.3.3 Extrapolated georeferencing accuracy analysis

A topic of greater concern for the geometric calibration is the geometric performance of the calibrated CCD-detector look angles in the extrapolated georeferencing of the ZiYuan-3 images covering other areas. Therefore, we performed the georeferencing of the ZiYuan-3 NAD images covering the Luoyang, Nanyang, Taiyuan, and Bellegarde areas with the CCD-detector look angles calibrated with five GCPs in the Songshan area. The maximum error and the RMSE of the residual errors of the check points are listed in Table 2.5.

From the results in Table 2.5, we can see that the georeferencing accuracy of the ZiYuan-3 NAD images after calibration is improved significantly. The georeferencing accuracy of the image covering the Luoyang area is improved from ±1005.534 m to ±8.133 m in planimetry, and the accuracy in the other areas also reaches better than ±10 m. It demonstrates that the camera installation errors and the CCD-detector look angle errors are the major errors that affect the georeferencing accuracy of the ZiYuan-3 satellite images. It is feasible to geometrically calibrate the look angles of each CCD detector in the attitude determination reference coordinate system with the introduced method in Section 2.2.2. The introduced calibration method can effectively eliminate the influence of the camera installation errors and the CCD-detector look

TABLE 2.5 The georeferencing results of the ZiYuan-3 NAD images.

Area	Calibration	Number of check points	Maximum/m			RMSE/m		
			X	Y	Planimetry	X	Y	Planimetry
Luoyang	Before	24	934.527	430.075	1028.739	927.104	389.327	1005.534
	After	24	9.597	8.207	11.118	6.159	5.311	**8.133**
Nanyang	Before	28	783.342	646.714	1014.351	773.924	608.874	984.727
	After	28	−5.555	6.439	6.555	2.077	2.707	**3.412**
Taiyuan	Before	17	782.802	669.247	1029.889	771.619	631.003	996.775
	After	17	−8.264	−9.973	12.419	6.102	7.440	**9.623**
Bellegarde	Before	9	475.397	−1351.071	1432.269	423.522	1330.623	1396.399
	After	9	5.600	8.104	9.796	3.328	5.981	**6.845**

angle errors on the georeferencing of the ZiYuan-3 satellite images. The georeferencing accuracy of the ZiYuan-3 images can be improved significantly.

From the above, the introduced calibration method is simple and feasible. Five evenly distributed GCPs are sufficient to calculate the look angles of each CCD detector, and the values of the camera installation angles and the CCD-detector look angles calibrated in laboratory are unnecessary. After the geometric calibration, the georeferencing accuracy of the ZiYuan-3 satellite images can be improved significantly. Of course, we can see from Eq. (2.16) that the position and attitude determination accuracy and the GCP measurement accuracy are the major factors that affect the in-orbit geometric calibration. For the ZiYuan-3 satellite, the position determination accuracy can reach a level of centimeter, and the attitude determination accuracy can reach a level of second of arc. Therefore, the introduced geometric calibration method can achieve a satisfied calibration accuracy.

2.3 Geometric rectification of a single remote sensing image

The geometric rectification of satellite remote sensing imagery is the process of projecting one original image into a specific reference coordinate system, eliminating the geometric distortions of the original image and producing a new image that satisfies the requirements of the map projection or the graphical expression. This process consists of the following two main steps: the first step is the transformation of the image coordinates, which requires the transformation of the image coordinates into the map projection coordinates or the object space coordinates; the second step is the resampling of the gray value after the transformation of the image coordinates. The process of the geometric

rectification of remote sensing imagery proceeds as follows:

(1) determine an appropriate geometric processing model between the image space coordinates and the object space coordinates according to the imaging mode of the remote sensing imagery;
(2) confirm the geometric rectification formulas according to the above geometric processing models;
(3) implement adjustments to solve the model parameters according to the coordinates of the GCPs and the corresponding image points and evaluate the accuracy;
(4) implement the geometric transformation and resampling of the original image.

2.3.1 Image geometric rectification models

The emergence of CCD linear array sensors enabled the implementation of high-precision object positioning and the surveying and mapping of large-scale topography with satellite remote sensing imagery. The establishment of image geometric rectification models is the foundation of the geometric processing of satellite imagery, and it represents the mathematical relationship between the three-dimensional object space coordinates of the ground points and the two-dimensional image coordinates of the corresponding image points. Because the characteristics of various sensors are different, there are many differences among the different geometric processing models in terms of strictness, complexity, and accuracy. At present, the common geometric processing models utilized for remote sensing imagery can be classified into two general types: rigorous geometric processing models and empirical geometric processing models.

2.3.1.1 Rigorous geometric processing models

For satellite imagery acquired by linear array sensors in push-broom mode, the entire image is generated by scanning successive single lines along the sensor's flight direction. Each scan-line image and the scanned objects on the ground satisfy a strict center projection relationship (that is, the scanned objects, the perspective center of the lens, and the corresponding image points satisfy a strict center projection relationship).

Additionally, each scan-line image has its own exterior orientation elements. Because of the sensor's imaging mechanism and the CCD linear array's center projection imaging characteristics in the row direction, a rigorous geometric processing model is established according to the strict mathematical relationship by which the scanned objects, the perspective center of lens, and the corresponding image points occupy the same straight line at a specific imaging time, as shown in Eq. (2.20) (Poli, 2004).

Because each scan-line image has its own exterior orientation elements, we call Eq. (2.20) the extended collinear condition equation to distinguish it from the traditional collinear condition equation. This rigorous geometric processing model provides a more accurate representation of the geometric imaging characteristics of a single linear array sensor, and the theory is strict. The geometric processing accuracy of the high-resolution remote sensing imagery according to this model is also improved as follows:

$$\begin{bmatrix} X \\ Y \\ Z \end{bmatrix} = \lambda \mathbf{M}_t \begin{bmatrix} x \\ 0 \\ -f \end{bmatrix} + \begin{bmatrix} X_{St} \\ Y_{St} \\ Z_{St} \end{bmatrix} \quad (2.20)$$

where (X, Y, Z) are the coordinates of the ground points in object space; (X_{St}, Y_{St}, Z_{St}) are the coordinates of the perspective center in object space at a certain time t; λ is a ratio factor; $\mathbf{M}_t$ is a rotation matrix constructed by the rotation angles

$(\phi_t, \omega_t, \kappa_t)$ between the object space coordinate system and the corresponding scan-line image coordinate system at time t; and the element values of the matrix are as follows:

$$\begin{cases} a_1 = \cos\phi_t \cos\kappa_t - \sin\phi_t \sin\omega_t \sin\kappa_t \\ a_2 = -\cos\phi_t \sin\kappa_t - \sin\phi_t \sin\omega_t \cos\kappa_t \\ a_3 = -\sin\phi_t \cos\omega_t \\ b_1 = \cos\omega_t \sin\kappa_t \\ b_2 = \cos\omega_t \cos\kappa_t \\ b_3 = -\sin\omega_t \\ c_1 = \sin\phi_t \cos\kappa_t + \cos\phi_t \sin\omega_t \sin\kappa_t \\ c_2 = -\sin\phi_t \sin\kappa_t + \cos\phi_t \sin\omega_t \cos\kappa_t \\ c_3 = \cos\phi_t \cos\omega_t \end{cases}$$

$$(2.21)$$

Because remote sensing platforms (such as satellites and spacecraft) move steadily in the aerospace, the imaging interval between the adjacent scan lines is very short (generally on the order of a millisecond). Therefore, the exterior orientation elements of each scan-line image are usually expressed as a polynomial function of the imaging time t in actual applications:

$$\begin{cases} X_{S_t} = m_0 + m_1 t + m_2 t^2 + \cdots + m_n t^n \\ Y_{S_t} = n_0 + n_1 t + n_2 t^2 + \cdots + n_n t^n \\ Z_{S_t} = s_0 + s_1 t + s_2 t^2 + \cdots + s_n t^n \\ \phi_t = d_0 + d_1 t + d_2 t^2 + \cdots + d_n t^n \\ \omega_t = e_0 + e_1 t + e_2 t^2 + \cdots + e_n t^n \\ \kappa_t = f_0 + f_1 t + f_2 t^2 + \cdots + f_n t^n \end{cases} \quad (2.22)$$

where $(Xs_t, Ys_t, Zs_t, \phi_t, \omega_t, \kappa_t)$ are the exterior orientation elements of the scan-line image at time t; $(m_i, n_i, s_i, d_i, e_i, f_i)(i = 0, 1, \cdots n)$ are the polynomial coefficients of the exterior orientation elements, which can be considered to be the n-order rate of every exterior orientation element; and t is the corresponding imaging moment of each scan-line image, which can be determined by the imaging time of the central scan-line image and the image scan interval. The specific form is given by the following formula:

$$t = t_c + lsp \cdot (y - y_c) \qquad (2.23)$$

where t_c is the imaging time of the central scan-line image, lsp is the image scan interval, y is the line coordinate of any line in the image, and y_c is the line coordinate of the central scan-line image.

The polynomial order can be selected flexibly, according to the specific application. Normally, for middle-to-high-orbit satellite remote sensing images (e.g., SPOT), satellites move more steadily in their orbits, and the influences of various perturbation forces are reduced; consequently, the linear or quadratic polynomial model is adequate to fit their orbit equations.

For low-orbit satellite remote sensing images (e.g., QuickBird, IKONOS), the satellite orbits are affected, for example, by the atmosphere, radiation, and gravitational forces; therefore, a quadratic or higher-order polynomial model should be used to fit their orbit equations.

However, in the case of ill conditions of the coefficient matrix of normal equations when solving the polynomial coefficients, a higher polynomial order will produce an oscillation phenomenon between the values of the exterior orientation elements that were calculated using the polynomial coefficients and the true values. Additionally, this higher polynomial order will not provide a better fit to the orbit equations. Therefore, the polynomial order should be selected according to the actual application when implementing the geometrical processing of satellite remote sensing imagery acquired by linear array sensors.

When implementing the spatial resectioning of satellite remote sensing imagery, Eq. (2.20) can be transformed into the following formula:

$$\begin{cases} x = -f\dfrac{a_1(X - X_{St}) + b_1(Y - Y_{St}) + c_1(Z - Z_{St})}{a_3(X - X_{St}) + b_3(Y - Y_{St}) + c_3(Z - Z_{St})} \\[3mm] 0 = -f\dfrac{a_2(X - X_{St}) + b_2(Y - Y_{St}) + c_2(Z - Z_{St})}{a_3(X - X_{St}) + b_3(Y - Y_{St}) + c_3(Z - Z_{St})} \end{cases}$$
$$(2.24)$$

By substituting Eqs. (2.21) and (2.22) into Eq. (2.24) and expanding it to the first order by Taylor series, the linearized error equations of the image space resectioning can be obtained and have the following matrix form:

$$\mathbf{V_x} = \mathbf{A_1 t_1} + \mathbf{A_2 t_2} - \mathbf{L_x}, \qquad \mathbf{E} \qquad (2.25)$$

where $\mathbf{V_x} = [v_x \quad v_y]^T$ is the correction vector for the observed image point coordinates;

$\mathbf{t_1} = [\Delta m_0 \quad \Delta n_0 \quad \Delta s_0 \quad \Delta m_1 \quad \Delta n_1 \quad \Delta s_1 \cdots]^T$ is the correction vector of the polynomial coefficients of the line exterior orientation elements;

$\mathbf{t_2} = [\Delta d_0 \quad \Delta e_0 \quad \Delta f_0 \quad \Delta d_1 \quad \Delta e_1 \quad \Delta f_1 \cdots]^T$ is the correction vector of the polynomial coefficients of the angle exterior orientation elements; and

$\mathbf{A_1}$ and $\mathbf{A_2}$ are the coefficient matrices of the unknowns $\mathbf{t_1}$ and $\mathbf{t_2}$, respectively (i.e., they are the first-order partial derivatives of the unknowns in each observation equation), and they have the following forms:

$$\mathbf{A_1} = \begin{bmatrix} \dfrac{\partial x}{\partial m_0} & \dfrac{\partial x}{\partial n_0} & \dfrac{\partial x}{\partial s_0} & \dfrac{\partial x}{\partial m_1} & \dfrac{\partial x}{\partial n_1} & \dfrac{\partial x}{\partial s_1} & \cdots \\[3mm] \dfrac{\partial y}{\partial m_0} & \dfrac{\partial y}{\partial n_0} & \dfrac{\partial y}{\partial s_0} & \dfrac{\partial y}{\partial m_1} & \dfrac{\partial y}{\partial n_1} & \dfrac{\partial y}{\partial s_1} & \cdots \end{bmatrix}$$

$$= \begin{bmatrix} \dfrac{\partial x}{\partial X_{Si}} & \dfrac{\partial x}{\partial Y_{Si}} & \dfrac{\partial x}{\partial Z_{Si}} & t\dfrac{\partial x}{\partial X_{Si}} & t\dfrac{\partial x}{\partial Y_{Si}} & t\dfrac{\partial x}{\partial Z_{Si}} & \cdots \\[3mm] \dfrac{\partial y}{\partial X_{Si}} & \dfrac{\partial y}{\partial Y_{Si}} & \dfrac{\partial y}{\partial Z_{Si}} & t\dfrac{\partial y}{\partial X_{Si}} & t\dfrac{\partial y}{\partial Y_{Si}} & t\dfrac{\partial y}{\partial Z_{Si}} & \cdots \end{bmatrix}$$

and,

$$\mathbf{A_2} = \begin{bmatrix} \dfrac{\partial x}{\partial d_0} & \dfrac{\partial x}{\partial e_0} & \dfrac{\partial x}{\partial f_0} & \dfrac{\partial x}{\partial d_1} & \dfrac{\partial x}{\partial e_1} & \dfrac{\partial x}{\partial f_1} & \cdots \\[3mm] \dfrac{\partial y}{\partial d_0} & \dfrac{\partial y}{\partial e_0} & \dfrac{\partial y}{\partial f_0} & \dfrac{\partial y}{\partial d_1} & \dfrac{\partial y}{\partial e_1} & \dfrac{\partial y}{\partial f_1} & \cdots \end{bmatrix}$$

$$= \begin{bmatrix} \dfrac{\partial x}{\partial \phi_i} & \dfrac{\partial x}{\partial \omega_i} & \dfrac{\partial x}{\partial \kappa_i} & t\dfrac{\partial x}{\partial \phi_i} & t\dfrac{\partial x}{\partial \omega_i} & t\dfrac{\partial x}{\partial \kappa_i} & \cdots \\[3mm] \dfrac{\partial y}{\partial \phi_i} & \dfrac{\partial y}{\partial \omega_i} & \dfrac{\partial y}{\partial \kappa_i} & t\dfrac{\partial y}{\partial \phi_i} & t\dfrac{\partial y}{\partial \omega_i} & t\dfrac{\partial y}{\partial \kappa_i} & \cdots \end{bmatrix},$$

in which $\mathbf{E}$ is the unit matrix, and $\mathbf{L_x}$ is the residual vector of the observations of the image point coordinates.

According to the principle of least squares adjustment, the matrix form of the normal equations can be obtained as follows:

$$\begin{bmatrix} \mathbf{A}_1^T\mathbf{A}_1 & \mathbf{A}_1^T\mathbf{A}_2 \\ \mathbf{A}_2^T\mathbf{A}_1 & \mathbf{A}_2^T\mathbf{A}_2 \end{bmatrix} \begin{bmatrix} \mathbf{t}_1 \\ \mathbf{t}_2 \end{bmatrix} = \begin{bmatrix} \mathbf{A}_1^T\mathbf{L}_X \\ \mathbf{A}_2^T\mathbf{L}_X \end{bmatrix} \qquad (2.26)$$

Once the coordinates of one GCP are measured, a pair of error equations with the form of Eq. (2.25) can be obtained, and every unknown parameter can be solved according to the principle of least squares adjustment. When quadratic polynomials are used to fit the variations of the exterior orientation elements, a total of 18 unknowns must be solved. Theoretically, at least 9 GCPs are necessary; otherwise, the normal equations will be rank-defected, and the unknowns cannot be solved. However, even with enough GCPs available, the solutions of the unknowns are still unstable, and they can deviate far from the true values as a result of the strong correlations between the unknowns because of the narrow field of view of the sensors.

2.3.1.2 Empirical geometric processing models

Empirical geometric processing models avoid the geometric imaging process of sensors and use general mathematical functions to represent the mathematical relationship between the three-dimensional object space coordinates of the ground points and the two-dimensional image space coordinates of the corresponding image points. This type of model is simple in form and suitable for various remote sensors. Additionally, the geometric parameters of the sensors in the imaging process are not necessary for these geometric models. For a long time, these models have been used as practical models in the geometrical processing of satellite remote sensing imagery. Common empirical geometric processing models mainly include the general polynomial model, the DLT model, the affine transformation model, and the RFM.

2.3.1.2.1 The general polynomial model

The general polynomial model is a simple empirical geometric processing model that ignores the specific imaging mechanism of the satellite remote sensing imagery and mathematically simulates the image distortions directly. The model considers the image distortions to be the combined effects caused by zoom, translation, rotation, affinity, partial twisting, bending, and other more advanced basic distortions, and it utilizes a proper polynomial to describe the mathematical relationship between the object space coordinates (X, Y, Z) of the ground points and the image space coordinates (x, y) of the corresponding image points as follows:

$$\begin{cases} x = a_0 + a_1 X + a_2 Y + a_3 X^2 + a_4 XY + a_5 Y^2 + a_6 X^3 + a_7 X^2 Y + a_8 XY^2 + a_9 Y^3 + \cdots \\ y = b_0 + b_1 X + b_2 Y + b_3 X^2 + b_4 XY + b_5 Y^2 + b_6 X^3 + b_7 X^2 Y + b_8 XY^2 + b_9 Y^3 + \cdots \end{cases}$$
$$(2.27)$$

The general polynomial model has a simple form and low computation requirements, but it ignores the influence of terrain relief on the accuracy of image geometric processing. Therefore, this model is suitable only for implementing geometric processing on those satellite images with smaller geometric distortions. For areas with large terrain reliefs, especially when the sensor's lateral angle is large, the positioning accuracy of Eq. (2.27) decreases significantly. Hence, the ground elevation values can be introduced, and an improved general polynomial model can be obtained as follows:

$$\begin{cases} x = a_0 + a_1 X + a_2 Y + a_3 Z + a_4 X^2 + a_5 Y^2 + a_6 Z^2 + a_7 XY + a_8 XZ + a_9 YZ + \cdots \\ y = b_0 + b_1 X + b_2 Y + b_3 Z + b_4 X^2 + b_5 Y^2 + b_6 Z^2 + b_7 XY + b_8 XZ + b_9 YZ + \cdots \end{cases}$$
$$(2.28)$$

The accuracy of the image geometric processing based on the improved general polynomial model is improved by considering the variation of the terrain relief, but it is still affected by the number, distribution, and accuracy of the GCPs and the actual terrain. When using polynomial models to implement image positioning, the fitting accuracy at the GCPs is very high, whereas the interpolation values of the other ground points may deviate from their true values (that is, an oscillation phenomenon may occur at some ground points).

2.3.1.2.2 The direct linear transformation model

The DLT model directly establishes a mathematical relationship between the image space coordinates (x, y) and the object space coordinates (X, Y, Z) of the corresponding ground points with the following specific form:

$$\begin{cases} x = \dfrac{L_1X + L_2Y + L_3Z + L_4}{L_9X + L_{10}Y + L_{11}Z + 1} \\[2mm] y = \dfrac{L_5X + L_6Y + L_7Z + L_8}{L_9X + L_{10}Y + L_{11}Z + 1} \end{cases} \quad (2.29)$$

The model contains 11 transformation parameters that can be transformed strictly and mutually with the interior and exterior orientation elements in the traditional collinear condition equations. However, the use of this model to implement the geometric processing of high-resolution remote sensing imagery is an approximation because the model does not consider the variation of the exterior orientation elements with the imaging time of each scan line of the remote sensing imagery nor does it equate the images acquired in the dynamic push-broom mode to the images acquired in the static frame model. Therefore, the geometric processing accuracy of satellite imagery using this model is theoretically lower than that achieved with the rigorous geometric processing model based on the collinear condition equations.

To describe the imaging characteristics of CCD linear array sensors in aerospace photogrammetry, Okamoto (1999) and Wang (1999) proposed an extended direct linear transformation (EDLT) model and a self-calibration direct linear transformation (SDLT) model based on the DLT model, respectively, which are shown in Eqs. (2.30) and (2.31):

$$\begin{cases} x = \dfrac{L_1X + L_2Y + L_3Z + L_4}{L_9X + L_{10}Y + L_{11}Z + 1} + L_{12}x^2 \\[2mm] y = \dfrac{L_5X + L_6Y + L_7Z + L_8}{L_9X + L_{10}Y + L_{11}Z + 1} + L_{13}xy \end{cases} \quad (2.30)$$

$$\begin{cases} x = \dfrac{L_1X + L_2Y + L_3Z + L_4}{L_9X + L_{10}Y + L_{11}Z + 1} \\[2mm] y = \dfrac{L_5X + L_6Y + L_7Z + L_8}{L_9X + L_{10}Y + L_{11}Z + 1} + L_{12}xy \end{cases} \quad (2.31)$$

The EDLT model and the SDLT model improve the DLT model under different assumptions, and they add items to correct the image point coordinates. In comparison with DLT, these two models do not significantly increase the amount of calculation, but their positioning accuracy is significantly improved.

2.3.1.2.3 The affine transformation model

Based on the theory of parallel projection, the affine transformation model uses the affine transformation parameters $A_1 \sim A_8$ to establish the mathematical relationship between the image space coordinates (x, y) of the image points and the object space coordinates (X, Y, Z) of the corresponding ground points as follows:

$$\begin{cases} x = A_1X + A_2Y + A_3Z + A_4 \\ y = A_5X + A_6Y + A_7Z + A_8 \end{cases} \quad (2.32)$$

Existing research has shown that improved accuracy can be achieved with this model to implement the approximate geometric processing of satellite remote sensing imagery when the field of view of the linear array sensors is narrow or the satellite images have already received preliminary geometric calibration and resampling by the image providers.

2.3.1.2.4 The rational function model

The RFM is an extended expression of the general polynomial model, the DLT model, and the affine transformation model, and it is a more generalized form of various empirical geometric processing models for remote sensing imagery. The RFM has a simple form, and it can be applied to all types of remote sensors, including new aviation and aerospace sensors. In addition, the RFM does not require a number of the geometric parameters that influence the satellite imaging process, such as the satellite orbit ephemeris, the attitude angles of the sensors, and the physical characteristic parameters of the sensors and imaging mode. Furthermore, the RPCs have no definite physical meanings; thus, the core information about the sensors remains confidential. Existing studies have shown that the RFM can replace the rigorous geometric processing model for image rectification, ortho-image generation, three-dimensional reconstruction, and other applications in the geometric processing of satellite remote sensing imagery acquired by a single linear array sensor in push-broom mode (Tao and Hu, 2002).

The RFM represents image point coordinates (r, c) as a ratio of polynomials of the ground point coordinates (X, Y, Z) (OGC, 1999):

$$\begin{cases} r_n = \dfrac{p_1(X_n, Y_n, Z_n)}{p_2(X_n, Y_n, Z_n)} \\[2mm] c_n = \dfrac{p_3(X_n, Y_n, Z_n)}{p_4(X_n, Y_n, Z_n)} \end{cases} \quad (2.33)$$

where (X_n, Y_n, Z_n) and (r_n, c_n) are the normalized image coordinates and the object space coordinates, respectively. Their values occupy the range from -1.0 to $+1.0$.

The RFM adopts the normalized coordinates to improve the solution stability of each coefficient in the model and to reduce the rounding errors introduced by an excessive quantity differential of data in the calculation process. The normalization formula is

$$\begin{cases} X_n = \dfrac{X - X_o}{X_s} \\[2mm] Y_n = \dfrac{Y - Y_o}{Y_s} \\[2mm] Z_n = \dfrac{Z - Z_o}{Z_s} \end{cases} \quad (2.34)$$

$$\begin{cases} r_n = \dfrac{r - r_o}{r_s} \\[2mm] c_n = \dfrac{c - c_o}{c_s} \end{cases} \quad (2.35)$$

where $(X_0, Y_0, Z_0, r_0, c_0)$ are the translation parameters of normalization and $(X_s, Y_s, Z_s, r_s, c_s)$ are ratio parameters of normalization. The specific formulae are as follows:

$$\begin{cases} X_o = \dfrac{\sum X}{m} \\[2mm] Y_o = \dfrac{\sum Y}{m} \\[2mm] Z_o = \dfrac{\sum Z}{m} \\[2mm] r_o = \dfrac{\sum r}{m} \\[2mm] c_o = \dfrac{\sum c}{m} \end{cases} \quad (2.36)$$

$$\begin{cases} X_s = \max(|X_{max} - X_o|, |X_{min} - X_o|) \\ Y_s = \max(|Y_{max} - Y_o|, |Y_{min} - Y_o|) \\ Z_s = \max(|Z_{max} - Z_o|, |Z_{min} - Z_o|) \\ r_s = \max(|r_{max} - r_o|, |r_{min} - r_o|) \\ c_s = \max(|c_{max} - c_o|, |c_{min} - c_o|) \end{cases} \quad (2.37)$$

where m is the number of control points.

In Eq. (2.33), neither the maximum order of each coordinate component X_n, Y_n, Z_n in each polynomial p_i ($i = 1, 2, 3, 4$) nor the total order of all three coordinate components exceeds 3. Each polynomial form is (for convenience, the subscript n is omitted in the following text) as follows:

$$\begin{aligned} p &= \sum_{i=0}^{m1} \sum_{j=0}^{m2} \sum_{k=0}^{m3} a_{ijk} X^i Y^j Z^k \\ &= a_0 + a_1 Z + a_2 Y + a_3 X + a_4 ZY \\ &\quad + a_{14} Y^2 X + a_{15} Z X^2 + a_{16} Y X^2 + \\ &\quad a_{17} Z^3 + a_{18} Y^3 + a_{19} X^3 \end{aligned} \quad (2.38)$$

where $a_i (i = 0, 1, \cdots, 19)$ are the RPCs.

In the RFM, the distortions caused by the optical projection can generally be described by the ratios of the first-order terms. The distortions caused by the Earth's curvature, by atmospheric refraction, and by lens distortion can be represented by the second-order terms, and unknown distortions of the high-order components, such as camera vibration, can be represented by the third-order items (Toutin, 2004b).

When solving RPCs, Eq. (2.33) can be transformed into the following formula:

$$\begin{cases} F_r = p_1(X,Y,Z) - r{\cdot}p_2(X,Y,Z) = 0 \\ F_c = p_3(X,Y,Z) - c{\cdot}p_4(X,Y,Z) = 0 \end{cases} \quad (2.39)$$

Therefore, the matrix form of the error equation is

$$\mathbf{V} = \mathbf{BX} - \mathbf{L}, \quad \mathbf{P} \quad\quad (2.40)$$

where

$$\mathbf{B} = \begin{bmatrix} \dfrac{\partial F_r}{\partial a_i} & \dfrac{\partial F_r}{\partial b_j} & \dfrac{\partial F_r}{\partial c_i} & \dfrac{\partial F_r}{\partial d_j} \\[2mm] \dfrac{\partial F_c}{\partial a_i} & \dfrac{\partial F_c}{\partial b_j} & \dfrac{\partial F_c}{\partial c_i} & \dfrac{\partial F_c}{\partial d_j} \end{bmatrix},$$

$$(i = 0, 1, \cdots, 19; j = 1, 2, \cdots 19);$$

$$\mathbf{X} = [a_0, \cdots, a_{19}, b_1, \cdots, b_{19}, c_0, \cdots, c_{19}, d_1, \cdots d_{19}]^{\mathrm{T}};$$

$$\mathbf{L} = [r, c]^{\mathrm{T}}; \text{ and}$$

$\mathbf{P}$ is a weight matrix that is generally assumed to be a unit matrix.

Thus, the following normal equation can be obtained:

$$\mathbf{B}^{\mathrm{T}}\mathbf{PBX} = \mathbf{B}^{\mathrm{T}}\mathbf{PL} \quad\quad (2.41)$$

In Eq. (2.41), there are 78 unknown parameters. If an image covers more than 39 GCPs, it is possible to solve the RPCs according to the principle of the least squares adjustment as follows:

$$\mathbf{X} = \left(\mathbf{B}^{\mathrm{T}}\mathbf{PB}\right)^{-1}\mathbf{B}^{\mathrm{T}}\mathbf{PL} \quad\quad (2.42)$$

The error equation obtained from the linear RFM is also linear, and the solutions of the RPCs can be obtained without iterations or the initial values of the RPCs according to the least squares adjustment. However, an uneven distribution of GCPs or the overparameterization of the model still causes the normal equation singular or rank defective and leads to unstable solutions of the RPCs. In this case, ridge estimation can be employed to solve the RPCs, and their solutions based on ridge estimation are as follows:

$$\mathbf{X} = \left(\mathbf{B}^{\mathrm{T}}\mathbf{PB} + k\mathbf{E}\right)^{-1}\mathbf{B}^{\mathrm{T}}\mathbf{PL} \quad\quad (2.43)$$

where k is a ridge parameter that is generally a positive decimal.

2.3.2 Layout of ground control points

When implementing the geometric rectification of satellite imagery, the parameters of the geometric processing models should be solved first according to the coordinates of the GCPs and the image space coordinates of their corresponding image points. However, the number, distribution, and accuracy of the GCPs directly influence the accuracy of the image geometric rectification. For the rigorous geometric processing model, when a linear polynomial is used to fit the variations of the exterior orientation elements, there are 12 unknown parameters in total. In theory, at least six GCPs are necessary to solve these 12 unknowns; otherwise, the normal equations will be rank defected or singular, and the unknown parameters cannot be solved.

For the RFM, at least 39 GCPs are necessary to solve the 78 RPCs, but in most cases, it is difficult or impossible to obtain this many GCPs. Therefore, when utilizing the RFM to implement the geometric rectification of satellite remote sensing imagery, the terrain-independent scenario is usually adopted for the first solution of the RPCs (Tao and Hu, 2001); then, a few GCPs are used to eliminate the systematic errors in the RPCs (Di et al., 2003; Fraser and Hanley, 2005) so that the RFM can replace the rigorous geometric processing model for the geometric processing of satellite imagery.

2.3.2.1 Principles of selecting ground control points

GCPs are generally divided into two classes: artificial marked points and distinct target points. When laying out artificial marked points, to conveniently interpret and measure the coordinates of the GCPs in the images and ensure their accuracy, the sizes of the GCPs are generally determined by the image scales, and the artificial marked points contrast with the surrounding objects. Distinct target points refer to those natural points that cannot be easily destroyed and can be accurately identified in images, such as house corners, road corner points, the intersections of nearly orthogonal line feature objects, and fixed-point feature objects.

Because a single satellite remote sensing image usually covers a large area and the resolution of the image is limited, distinct target points are often used as GCPs to implement image geometric processing; thus, the number of GCPs should be determined by the rectification model.

2.3.2.2 Distribution requirements of ground control points

In addition to their number and accuracy, the distribution of GCPs is an important factor that affects the accuracy of image geometric rectification. Using the same number of GCPs but with different distributions, the accuracy of the image geometric rectification may vary considerably. Therefore, when selecting GCPs in satellite remote sensing imagery, we should ensure that the GCPs are as evenly distributed over the entire image as possible; additionally, the terrain characteristics should be considered so that a sufficient altitude difference is maintained between the GCPs. When the number of GCPs is limited, the layout of the GCPs should be at the image corners or along the periphery of the image to the greatest extent possible.

2.3.3 Image resampling

When implementing image geometric rectification, the coordinates of the image to be rectified should be projected onto the original image using the geometric rectification models, allowing the coordinates of the sampled points to be obtained. Normally, the coordinates of the sampled points are not integers, and the contributions of the gray values of neighboring points with integer coordinates should be accumulated to constitute a new gray value at any such point. This process is called image gray resampling. There are three primary methods of image gray resampling: the nearest-neighbor interpolation, the bilinear interpolation, and the bicubic convolution interpolation.

2.3.3.1 Nearest-neighbor interpolation

The nearest-neighbor interpolation essentially takes the gray I_N of the known pixel N nearest to the sampled points $P(x, y)$ as the resampling gray. The resampling function is

$$W(x_c, y_c) = 1, \quad (x_c = x_N, y_c = y_N) \qquad (2.44)$$

The resampling gray is

$$I_p = W(x_c, y_c) \cdot I_N = I_N \qquad (2.45)$$

where $x_N = \text{int}(x_p + 0.5)$; $y_N = \text{int}(y_p + 0.5)$.

The nearest-neighbor interpolation is the simplest method, and it has better radiation fidelity. However, this method may cause image point displacement within a pixel, and the geometric accuracy of this method is poorer than those of the other two methods.

2.3.3.2 Bilinear interpolation

The bilinear interpolation resampling function is approximately equivalent to the sinc function and can use a linear function of the triangle shown in Fig. 2.8 to express the resampling formula as follows:

$$W(x_c) = 1 - |x_c|, \quad (0 \le |x_c| \le 1) \qquad (2.46)$$

When implementing the bilinear interpolation, the gray values of four known pixels around the sampled point p are required for the calculation.

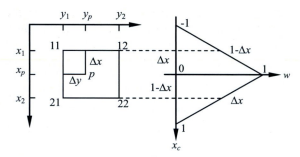

FIGURE 2.8 Gray resampling using bilinear interpolation.

The gray value of the sampled point p is

$$I_p = \mathbf{W}_x \cdot \mathbf{I} \cdot \mathbf{W}_y^T = \begin{bmatrix} W_{x_1} & W_{x_2} \end{bmatrix} \begin{bmatrix} I_{11} & I_{12} \\ I_{21} & I_{22} \end{bmatrix} \begin{bmatrix} W_{y_1} \\ W_{y_2} \end{bmatrix}$$

(2.47)

where $W_{x_1} = 1 - \Delta x$; $W_{x_2} = \Delta x$; $W_{y_1} = 1 - \Delta y$; $W_{y_2} = \Delta y$.

The bilinear interpolation calculation is relatively simple, and its gray sampling accuracy is relatively high; consequently, it is often used in practice. However, the sampled image is slightly blurred.

2.3.3.3 Bicubic convolution

Bicubic convolution uses a cubic resampling function to approximate the sinc function (shown in Fig. 2.9):

$$\begin{cases} W(x_c) = 1 - 2x_c^2 + |x_c|^3, & 0 \le |x_c| \le 1 \\ W(x_c) = 4 - 8|x_c| + 5x_c^2 - |x_c|^3, & 1 \le |x_c| \le 2 \\ W(x_c) = 0, & |x_c| > 2 \end{cases}$$

(2.48)

where x_c is the coordinate of neighboring pixel x, with the sampled point p as its origin, and the pixel interval is 1.

When Eq. (2.48) is used in the y direction on the image, the symbol x should be substituted by the symbol y.

Assuming that p is the sampled point, the coordinate differences Δx and Δy from point p to the nearest-neighbor point (22 in the upper left direction) are both small values, i.e.,

$$\begin{cases} \Delta x = x_p - \text{int}(x_p) = x_p - x_{22} \\ \Delta y = y_p - \text{int}(y_p) = y_p - y_{22} \end{cases}$$

(2.49)

When using the cubic convolution to resample the gray value of point p, the known gray values I_{ij} of 4×4 neighboring points (i, j) $(i = 1,2,3,4; j = 1,2,3,4)$ are required.

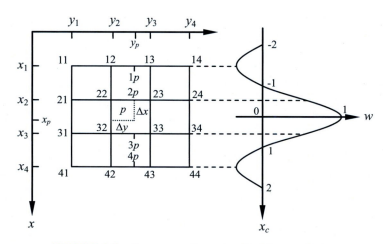

FIGURE 2.9 Gray resampling using bicubic convolution.

The gray value of the sampled point p is

$$I_p = \mathbf{W}_x \cdot \mathbf{I} \cdot \mathbf{W}_y^T \qquad (2.50)$$

where

$$\mathbf{W}_x = [\, W_{x_1} \quad W_{x_2} \quad W_{x_3} \quad W_{x_4} \,];$$

$$\mathbf{W}_y = [\, W_{y_1} \quad W_{y_2} \quad W_{y_3} \quad W_{y_4} \,];$$

$$\mathbf{I} = \begin{bmatrix} I_{11} & I_{12} & I_{13} & I_{14} \\ I_{21} & I_{22} & I_{23} & I_{24} \\ I_{31} & I_{32} & I_{33} & I_{34} \\ I_{41} & I_{42} & I_{43} & I_{44} \end{bmatrix};$$

$$\begin{aligned} W_{x_1} &= -\Delta x + 2\Delta x^2 - \Delta x^3, & W_{y_1} &= -\Delta y + 2\Delta y^2 - \Delta y^3 \\ W_{x_2} &= 1 - 2\Delta x^2 + \Delta x^3, & W_{y_2} &= 1 - 2\Delta y^2 + \Delta y^3 \\ W_{x_3} &= \Delta x + \Delta x^2 - \Delta x^3, & W_{y_3} &= \Delta y + \Delta y^2 - \Delta y^3 \\ W_{x_4} &= -\Delta x^2 + \Delta x^3, & W_{y_4} &= -\Delta y^2 + \Delta y^3 \end{aligned}$$

The bicubic convolution resampling has a high accuracy, but its calculation is complex and time consuming.

2.3.4 Accuracy evaluation

After the geometric rectification of satellite remote sensing imagery, the accuracy of the geometric rectification should be evaluated quantitatively so that image users can choose different images according to their accuracy requirements when extracting geographic information from satellite imagery. Geometric rectification accuracy is employed to measure the differences between the position of a point on the rectified image and its real position, and it is determined by the following formula:

$$\begin{cases} m_X = \sqrt{\dfrac{\sum\limits_{i=1}^{N} (X_T - X_R)_i^2}{N}} \\[2em] m_Y = \sqrt{\dfrac{\sum\limits_{i=1}^{N} (Y_T - Y_R)_i^2}{N}} \\[2em] m_P = \sqrt{m_X^2 + m_Y^2} \end{cases} \qquad (2.51)$$

where m_X and m_Y are the accuracies of the rectified image in the X direction and the Y direction, respectively; m_P is the planimetric accuracy of the rectified image; (X_T, Y_T) and (X_R, Y_R) are the true coordinates of the check points and their rectified coordinates on the rectified image, respectively; and N is the number of check points used to evaluated the accuracy.

2.4 Geometric registration of satellite remote sensing imagery

With the development of sensor technology, remote sensing imagery has become increasingly redundant. Satellite remote sensing imagery of a particular area can be acquired by different sensors at different times, and it can contain different spectral ranges. To fully utilize the advantages of various remote sensing images, we must implement image fusion for thematic map production, automatic computer classification, disaster monitoring, and other applications. However, before implementing image fusion, we should ensure the geometric consistency of these images; consequently, we must first implement image registration.

Image geometric registration refers to the projection of two or more images that cover the same area and are acquired by different sensors (with different views and at different times) into a single coordinate system based on the transformation parameters and the production of the best match at the pixel level.

2.4.1 Automatic extraction of image registration points

The extraction of conjugate features is the first step in solving the transformation parameters between different remote sensing images. Conjugate features include point features, linear features, and regional features. There is no fixed mathematical representation of feature extraction based on linear features and regional features; therefore, this section introduces only feature extraction based on point features.

2.4.1.1 Image matching based on image gray

2.4.1.1.1 Correlation coefficient matching

Gray correlation was originally used for image matching, which is also called digital correlation. Digital correlation utilizes the correlation coefficient of two image blocks to evaluate their similarity; then, a pair of conjugate points can be determined by setting the threshold of the similarity measure.

Digital correlation is often implemented in two-dimensional space. First, a reference point should be selected in the reference image, and a pixel array with $m \times n$ pixels, whose center is the reference point, should also be selected as a target window. Then, the conjugate point can be searched in the image to be matched. Here, the image region that may contain the conjugate point corresponding to the reference point to be matched is estimated in advance, and a searching window containing $k \times l$ pixels ($k > m$, $l > n$) is established. Digital correlation sequentially selects a matching window with $m \times n$ pixels. With its window center at point (c, r) in the searching window, it calculates the similarity measure ρ between the target window and the matching window. If the similarity measure is the maximum, and it is larger than the given similarity threshold, the center pixel of the matching window is considered to be the conjugate point corresponding to the reference point.

One-dimensional matching is often used in image matching; in this case, n equals 1 (that is, the procedure of searching conjugate points occurs along a line).

The correlation coefficient is the most common similarity measure, as Eq. (2.52) shows. Because it is a linear invariant of gray, it cannot be influenced by the overall radiation variation.

$$\rho(c, r) = \frac{\sum\limits_{i=1}^{m} \sum\limits_{j=1}^{n} \left(g_{i,j} - \bar{g}\right)\left(g'_{i+r,j+c} - \bar{g}'_{r,c}\right)}{\sqrt{\sum\limits_{i=1}^{m} \sum\limits_{j=1}^{n} \left(g_{i,j} - \bar{g}\right)^2 \cdot \sum\limits_{i=1}^{m} \sum\limits_{j=1}^{n} \left(g'_{i+r,j+c} - \bar{g}'_{r,c}\right)^2}}$$

$$(2.52)$$

where $\bar{g}'_{r,c} = \frac{1}{m \cdot n} \sum\limits_{i=1}^{m} \sum\limits_{j=1}^{n} g'_{i+r,j+c}$ is the average gray value of the matching window and $\bar{g} =$

$\frac{1}{m \cdot n} \sum\limits_{i=1}^{m} \sum\limits_{j=1}^{n} g_{i,j}$ is the average gray value of the target window.

2.4.1.1.2 Least squares matching

In the 1980s, Professor Ackermann (1983), in Germany, proposed a new image-matching method, least squares matching (LSM). In this method, the square sums of the absolute differences of the gray values between the target window and the corresponding matching window should be minimal. The least squares principle is applied to image matching through a series of derivations.

Both radiant distortions and geometric distortions exist between two images acquired at different times. The matching window is relatively small; therefore, only geometric distortions described by the first-order polynomial are usually considered, as Eq. (2.53) shows:

$$\begin{cases} x_2 = a_0 + a_1 x + a_2 y \\ y_2 = b_0 + b_1 x + b_2 y \end{cases} \quad (2.53)$$

Assuming that the radiant distortion between two images is linear, Eq. (2.54) can be established according to the gray values of any pair of points in two matching windows. The equation contains two radiant distortion parameters, h_0 and h_1, and six geometric distortion parameters, a_0, a_1, a_2, b_0, b_1, and b_2.

$$g_1(x, y) = h_0$$
$$+ h_1 g_2(a_0 + a_1 x + a_2 y, \ b_0 + b_1 x + b_2 y) \quad (2.54)$$

By taking the radiant and geometric distortion parameters as unknown parameters and linearizing Eq. (2.54), we obtain

$$v = c_1 \Delta h_0 + c_2 \Delta h_1 + c_3 \Delta a_0 + c_4 \Delta a_1 + c_5 \Delta a_2$$
$$+ c_6 \Delta b_0 + c_7 \Delta b_1 + c_8 \Delta b_2 + g_2 - g_1$$

$$(2.55)$$

where

$$c_1 = 1; \quad c_2 = g_2; \quad c_3 = \frac{\partial g_2}{\partial a_0} = \dot{g}_x;$$

$$c_4 = \frac{\partial g_2}{\partial a_1} = x\dot{g}_x; \quad c_5 = \frac{\partial g_2}{\partial a_2} = y\dot{g}_x; \quad c_6 = \frac{\partial g_2}{\partial b_0} = \dot{g}_y;$$

$$c_7 = \frac{\partial g_2}{\partial b_1} = x\dot{g}_y; \quad c_8 = \frac{\partial g_2}{\partial b_2} = y\dot{g}_y.$$

Then, the matrix form of the normal equation can be obtained as follows:

$$(\mathbf{C}^T\mathbf{C})\mathbf{X} = (\mathbf{C}^T\mathbf{L}) \tag{2.56}$$

where

$$\mathbf{C} = [c_1 \quad c_2 \quad c_3 \quad c_4 \quad c_5 \quad c_6 \quad c_7 \quad c_8]^T;$$
$$\mathbf{X} = [\Delta h_0 \quad \Delta h_1 \quad \Delta a_0 \quad \Delta a_1 \quad \Delta a_2 \quad \Delta b_0 \quad \Delta b_1 \quad \Delta b_2]^T;$$
$$\mathbf{L} = [g_2 - g_1].$$

If there are more than eight conjugate points in the two matching windows, then each pair of points can be used to establish one error equation, such as Eq. (2.55). By combining these error equations and applying the least squares principle, solutions for the unknown parameters can be obtained as follows:

$$\mathbf{X} = (\mathbf{C}^T\mathbf{C})^{-1}(\mathbf{C}^T\mathbf{L}) \tag{2.57}$$

The unknown parameters should be solved iteratively, and their initial values can be defined as $h_0 = 0, h_1 = 1, a_0 = 0, a_1 = 1, a_2 = b_0 = b_1 = 0, b_2 = 1$. The correction values for the deformation parameters are calculated according to the pixel array of the right image after the geometric and radiant corrections. After obtaining the correction values X for the parameters, the deformation parameters are calculated according to the following algorithm. Assuming that $h_0^{i-1}, h_1^{i-1}, a_0^{i-1}, a_1^{i-1}, a_2^{i-1}, b_0^{i-1}, b_1^{i-1}, b_2^{i-1}$ are the previous deformation parameters and $\Delta h_0^i, \Delta h_1^i, \Delta a_0^i, \Delta a_1^i, \Delta a_2^i, \Delta b_0^i, \Delta b_1^i, \Delta b_2^i$ are the correction values after the ith iteration of this

time, the following equation can be obtained relative to the geometric deformation parameters:

$$\begin{bmatrix} 1 \\ x_2 \\ y_2 \end{bmatrix} = \begin{bmatrix} 1 & 0 & 0 \\ a_0^i & a_1^i & a_2^i \\ b_0^i & b_1^i & b_2^i \end{bmatrix} \begin{bmatrix} 1 \\ x \\ y \end{bmatrix}$$

$$= \begin{bmatrix} 1 & 0 & 0 \\ \Delta a_0^i & 1 + \Delta a_1^i & \Delta a_2^i \\ \Delta b_0^i & \Delta b_1^i & 1 + \Delta b_2^i \end{bmatrix}$$

$$\times \begin{bmatrix} 1 & 0 & 0 \\ a_0^{i-1} & a_1^{i-1} & a_2^{i-1} \\ b_0^{i-1} & b_1^{i-1} & b_2^{i-1} \end{bmatrix} \begin{bmatrix} 1 \\ x \\ y \end{bmatrix} \tag{2.58}$$

Therefore,

$$\begin{cases} a_0^i = a_0^{i-1} + \Delta a_0^i + a_0^{i-1}\Delta a_1^i + b_0^{i-1}\Delta a_2^i \\ a_1^i = a_1^{i-1} + a_1^{i-1}\Delta a_1^i + b_1^{i-1}\Delta a_2^i \\ a_2^i = a_2^{i-1} + a_2^{i-1}\Delta a_1^i + b_2^{i-1}\Delta a_2^i \\ b_0^i = b_0^{i-1} + \Delta b_0^i + a_0^{i-1}\Delta b_1^i + b_0^{i-1}\Delta b_2^i \\ b_1^i = b_1^{i-1} + a_1^{i-1}\Delta b_1^i + b_1^{i-1}\Delta b_2^i \\ b_2^i = b_2^{i-1} + a_2^{i-1}\Delta b_1^i + b_2^{i-1}\Delta b_2^i \end{cases} \tag{2.59}$$

For the radiant deformation parameters, we can obtain

$$\begin{bmatrix} 1 \\ g_1 \end{bmatrix} = \begin{bmatrix} 1 & 0 \\ \Delta h_0^i & 1 + \Delta h_1^i \end{bmatrix} \begin{bmatrix} 1 & 0 \\ h_0^{i-1} & h_1^{i-1} \end{bmatrix} \begin{bmatrix} 1 \\ g_2 \end{bmatrix} \tag{2.60}$$

Therefore,

$$\begin{cases} h_0^i = h_0^{i-1} + \Delta h_0^i + h_0^{i-1}\Delta h_1^i \\ h_1^i = h_1^{i-1} + h_1^{i-1}\Delta h_1^i \end{cases} \tag{2.61}$$

All of the deformation parameter values are obtained after each iteration, the coefficient values of the unknown parameters in Eq. (2.55) are recalculated, and the correction values of the parameters are solved again according to Eq. (2.57). These iterations are repeated until the correlation coefficient between g_1 and g_2 is smaller than the previous one or the correction values of the geometric deformation parameters (especially the displacement corrections a_0 and b_0) are smaller than the given threshold, producing all of the deformation parameter values.

In image matching based on image gray, the relative positions of the images to be matched should be known to generate epipolar images so that a better image-matching result can be obtained. Fig. 2.10 shows the matching results of a stereo pair of images acquired by an Ultra-Cam UCD digital camera after using correlation coefficient matching and LSM. The cross lines are the matched conjugate points between the two images.

2.4.1.2 Image matching based on features

Image matching based on features is usually more complicated than image matching based on image gray, and it relies on the extraction and description of the features. Image matching based on features usually includes four steps:

(1) the features are detected. The features of interest are often sectional extreme values caused by gray variations, and they often contain distinguished structure information.

(2) the features are described. Feature parameters or feature vectors are established to reflect the main differences between the various types of feature-based image-matching algorithms. The selection of feature parameters determines the features that should be used in image matching and the features that should be neglected.

(3) the features are matched, and the candidate points are obtained. According to the similarity between feature vectors, feature-based image matching takes a given distance function, such as the Euclidean distance, block distance, or Mahalanobis distance, as the similarity measure of the features and implements image matching.

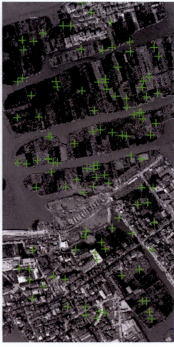

FIGURE 2.10   The matching results of a stereo pair of images.

(4) mismatched conjugate points are eliminated. Regardless of the description of the feature parameters and the similarity measures used in image matching, the existence of mismatched conjugate points is inevitable. After image matching, the mismatched conjugate points in the set of candidate points can be eliminated according to the geometric constraint information between the images. The epipolar constraint is the most commonly used geometric constraint. In the following, we use the SIFT feature operator as an example to introduce feature-based image matching (Lowe, 2004). The main steps are described below.

2.4.1.2.1 Detecting the extreme values of the scale space

The concept of the image scale space should be introduced to establish the SIFT operator. Eq. (2.62) gives the definition of the image scale space $L(x,y,\sigma)$, which is the convolution of the scale-changing Gaussian function G and the image I.

$$L(x,y,\sigma) = G(x,y,\sigma) \otimes I(x,y) \quad (2.62)$$

The definition of Gaussian function is

$$G(x,y,\sigma) = \frac{1}{2\pi\sigma^2}e^{\frac{-(x^2+y^2)}{2\sigma^2}} \quad (2.63)$$

where (x, y) are the image coordinates and σ is the factor of the scale space, which determines the smoothing extent of the images. The large scale corresponds to the general features of the images and the small scale corresponds to the detailed features of the images.

After establishing the scale space, we can detect the points with extreme values. First, we should establish the DOG (difference of Gaussian) pyramid images:

$$D(x,y,\sigma) = (G(x,y,k\sigma) - G(x,y,\sigma)) \otimes I(x,y)$$
$$= L(x,y,k\sigma) - L(x,y,\sigma)$$
$$(2.64)$$

If the value of a point is the maximum or minimum value in the neighboring 26 pixels (which contains 8 pixels in the same layer as the point, 9 pixels in the layer above, and 9 pixels in the layer below), then the point should be taken as a key point, as shown in Fig. 2.11.

2.4.1.2.2 Direction distribution of key points

Using the gradient direction distribution of the neighboring pixels of the key points to specify the direction parameters for each key point, the operator may have the characteristic of rotation invariance.

$$\begin{cases} m(x,y) = \sqrt{(L(x+1,y)-L(x-1,y))^2 + (L(x,y+1)-L(x,y-1))^2} \\ \theta(x,y) = \tan^{-1}\left(\frac{L(x,y+1)-L(x,y-1)}{L(x+1,y)-L(x-1,y)}\right) \end{cases}$$
$$(2.65)$$

Eq. (2.65) is used to calculate the gradient modulus and the gradient direction at the point (x, y). In the formula, the scale used by L is also the scale of each key point itself. During an actual calculation, the sampling is usually performed in the window that contains the key point and uses the histogram to count the gradient directions of the neighboring pixels of

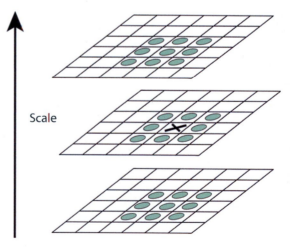

FIGURE 2.11 The symbol X represents the key point (Lowe, 2004).

the key point. The range of the gradient histogram is 0°−360 degrees, and every 10 degrees represents one direction so that there are 36 directions in all. The peak of the histogram represents the main gradient direction in the neighborhood of the feature point, and this direction is used as the direction of the key point.

2.4.1.2.3 The description of feature points

During the construction of the feature-description parameters, the local area surrounding the feature points should be rotated clockwise by angle θ to ensure that the local area has the characteristic of rotation invariance. In the region after rotation, the 16×16 rectangle windows with the key points at their center are evenly divided into 16 4×4 subwindows (in Fig. 2.12, the size is 2×2). Then, the accumulated values of the gradient in 8 directions can be calculated for each subwindow, and the gradient direction histograms can be drawn. Subsequently, 128 accumulated values of the gradient and the vector that they define can be obtained and used as the description parameters of the feature point.

Following the above process, the vectors of the SIFT operator can remove the influences caused by geometric distortions, such as scale variations and rotations. The illumination influence can be further decreased by normalizing these vectors.

2.4.1.2.4 Feature matching

After the vectors of the SIFT operator in the two images are generated, the Euclidean distance can be taken as the similarity measure of the key points of the two images. Then, the first and second nearest key points to each key point in the reference image (according to the Euclidean distance of the feature description vectors) can be found in the image to be matched. In the two key points, if the ratio of the nearest distance to the second nearest distance is smaller than the given threshold, then the matching is considered to be successful.

2.4.1.2.5 The elimination of mismatched conjugate points

Geometric constraints between the images are usually employed to eliminate the mismatched conjugate points. For satellite remote sensing imagery, the approximate epipolar model, the homograph matrix model, or the affine transformation model can be used as the geometric constraint model, and the RANSAC algorithm (Fischler and Bolles, 1981) or the authority-selected iteration algorithm can be used to eliminate the mismatched points.

In comparison to correlation coefficient matching, feature-based image matching can obtain a better image-matching result when the relative position of the images to be matched is

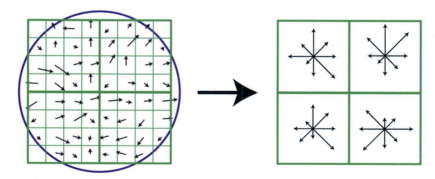

FIGURE 2.12   Feature vectors constituted by gradient direction histograms (Lowe, 2004).

FIGURE 2.13   The matching results by feature-based image matching.

unknown. In Fig. 2.13, the left image and the right image are the image-matching results of an aerial image and a SPOT-5 image by the SIFT operator, respectively, and the initial orientation parameters of the two images are unknown. Therefore, a better image-matching result can still be obtained by the SIFT operator even when the differences in the image resolution and the relative rotation angle between the two images to be matched are large. However, it is difficult to find conjugate points using correlation coefficient matching for these images.

2.4.2 Mathematical models of image registration

The mathematical model of image registration describes the geometric mapping relationship between two images. It is difficult to establish the physical meaning of the corresponding relationship between two images at the subpixel level (which is mainly affected by the parallaxes caused by terrain relief and the different elevations of objects). Therefore, the geometric fitting model is often used instead of the physical model to implement geometric registration between two images. Geometric registration models can be classified into linear transformation models and nonlinear transformation models, and the two types of models will be introduced below.

2.4.2.1 Linear transformation models

The affine transformation model, which is shown in Eq. (2.53), is the most common linear transformation model. This model ignores the influences caused by terrain relief and image projection modes, and it turns the image registration between two images into a complete transformation between two image planes. In the affine transformation model, six parameters containing the rotation, translation, and scale parameters between two images are solved. More than three conjugate points are required to solve the affine transformation coefficients in Eq. (2.53).

2.4.2.2 Nonlinear transformation models

The general polynomial model (of more than one order) and the small-panel correction model are the most common nonlinear transformation models. Compared with the affine transformation model, polynomial models with high orders can also simulate some high-order deformations other than rotation, translation, and scaling. However, polynomial models also have shortcomings: more conjugate points are required, and the artificial selection of these conjugate points increases the work required to set up the model. Furthermore, the solutions of the parameters are unstable.

In fact, polynomial models cannot always simulate the unknown deformations accurately because there are strong correlations between the polynomial coefficients. As the order of the polynomial increases, its solutions become increasingly unstable. Therefore, only the second- and third-order polynomial models are utilized in actual applications. The second-order polynomial model is given as follows:

$$\begin{cases} x_2 = a_0 + a_1 x + a_2 y + a_3 x^2 + a_4 y^2 + a_5 xy \\ y_2 = b_0 + b_1 x + b_2 y + b_3 x^2 + b_4 y^2 + b_5 xy \end{cases}$$

$$(2.66)$$

The geometric distortions caused by terrain reliefs differ in mountainous areas and along the boundaries between plain areas and mountainous areas; consequently, using the general polynomial model to implement image registration over the entire image cannot usually meet the accuracy requirements. In these cases, the method of simulating local deformations should be used (that is, the whole image should be divided into blocks, and image registration should be implemented block by block).

In the small-panel correction model, the whole image is divided into a number of small panels, and every panel is registered individually. The premise of small-panel correction is that the boundary points of these panels should be registered precisely. The most common panel shape is triangular. When implementing image registration, we can first divide the entire image into triangles according to the numerous conjugate points acquired by image matching; then, each triangle panel can be registered according to the affine transformation model. The triangles can be generated by the Delaunay triangle irregular–network algorithm.

2.5 Construction of a digital terrain model

2.5.1 The concept of the DEM and structure of the model

The DTM was originally used to design highways. The DTM is defined as an array of orderly values that are used to describe the spatial distribution of various properties of the Earth's surface (for example, mainly geomorphic information, object information, natural resource and environment information, and economical information). When the DTM is used only to describe and express spatial information, such as terrain relief and elevations, it is also called a DEM.

The DEM has many expression forms, the most common two of which are the regular grid DEM and the triangulated irregular network (TIN) DEM.

2.5.1.1 The regular grid DEM

The regular grid divides the ground into a series of regular grid cells, and each cell is expressed by the elevation of an object point (shown in Fig. 2.14), as follows:

$$DEM = \{Z_{ij}\}, \quad (i = 1, 2, \cdots, m; \ j = 1, 2, \cdots, n)$$

$$(2.67)$$

With limited storage, the regular grid DEM can easily be compressed. Additionally, it is suitable for computer management, making it the most common form. The DEM data in many countries are supplied in the form of regular grid, such as the USGS DEM in the United States and the DEM data below the scale of 1:50,000 in China. Table 2.6 lists some reference information about the commonly used DEM database.

There are also obvious shortcomings of the regular grid DEM. For example, it always causes

57	82	84	83	81	88
55	54	53	82	87	83
59	57	55	56	89	84
35	59	58	52	57	63
33	35	30	34	64	64
87	89	34	65	66	68

FIGURE 2.14   The regular grid digital elevation model.

TABLE 2.6 Some common used digital elevation model (DEM) databases.

Name	Format of DEM	Resolution	Coverage	Date source	Time
SRTM1/SRTM3	Longitude and latitude grid	1″/3″	Total land on Earth	SRTM	2002
1:50,000 DEM in China	Kilometer grid	25 m	China	Relief maps	2002
ASTER_GDEM	Longitude and latitude grid	1″/3″	Total land on Earth	ASTER	2009
GMTED2010	Longitude and latitude grid	7.5″/15″/30″	Total land on Earth	Multisources	2011
WorldDEM	Longitude and latitude grid	12 m	Total land on Earth	TanDEM-X	2014

large data redundancy when it is used to describe simple and flat ground with a fixed resolution, and it cannot express topography accurately when it is used to describe complicated and undulating ground. Therefore, various types of feature data, such as ridgelines, valley lines, and break lines, must be added to improve the topographical expression accuracy when using a regular grid DEM to describe the ground.

2.5.1.2 The triangulated irregular network

The TIN utilizes the original sample points to constitute many nonoverlapping triangles that cover the entire region according to a set of rules. The ground surface is described approximately with these triangles (shown in Fig. 2.15).

Because of the irregularity of the TIN, the organization, storage, and application of its data are more complicated than that of the regular grid DEM. In addition to storing the elevation of the points, the planimetric position and the

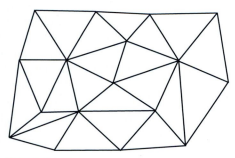

FIGURE 2.15 The triangulated irregular network digital elevation model.

topological relationship between adjacent triangles are recorded.

The construction of the TIN is also an important step, and the criterion for triangulation division is often used to construct the nonoverlapping triangles based on the discrete sampling points. Delaunay is the most common triangulation algorithm. However, the radiation scanning algorithm, the simulated annealing algorithm, and algorithms based on mathematical morphology are also used.

2.5.2 Preprocessing of DEM data

Terrain data should be previously available to establish a DEM. Terrain data are mainly composed of elevation data. In some situations, various types of topographical feature lines, such as ridgelines, valley lines, and break lines, are also included. After acquiring the source data, the DEM data should be preprocessed, which includes data format transformation, coordinate system transformation, gross error detection, filtering, and data division. Because this book mainly discusses satellite remote sensing imagery, only original data collection, gross error detection, and filtering are introduced in the following text.

2.5.2.1 DEM data collection

Currently, the data sources used to generate a DEM mainly include digital topographic maps, photogrammetric and remote sensing data, ground measurement data, LiDAR data, and similar sources. Aerial photogrammetry is always the primary method and the most valuable data source for topographical map generation and updating. Land satellite remote sensing imagery, which can rapidly obtain DEM data over a large region, is also an effective data source. However, because of the low spatial resolution of the satellite sensors, the images acquired by Landsat and SPOT can only be used to generate a small-scale DEM, whereas the images acquired by QuickBird, IKONOS, and SAR can be used to generate a DEM with high accuracy and high resolution.

Table 2.7 gives the relationship between the DEM resolution and the scale of the original data. In addition to aerial images and midscale and large-scale topographical maps, high-resolution remote sensing imagery is an important data source used to generate fine topography—scale DEMs.

2.5.2.2 Blunder detection of original data

Regardless of the data collection method used, the original DEM data inevitably contain some errors. When preprocessing the data, errors should be eliminated to prevent them from spreading and enlarging during the DEM generation procedure. Based on the characteristics of the errors, the errors in the original DEM data can be classified into three categories: random error, systematic error, and gross error. For the DEM data acquired from remote sensing imagery, gross errors are the main error source; therefore, they should be eliminated during data preprocessing. The most common methods of eliminating gross errors are stereoscopic manual visual inspection and gross error detection based on the fitting curved surface or trend surface.

2.5.2.2.1 The stereoscopic manual visual inspection method

For the discrete points acquired from stereo remote sensing images, a TIN is usually employed to generate a three-dimensional terrain surface model. Under the three-dimensional visual environment, gross errors can be detected by human—computer interaction. However, this method requires that the software platform be capable of constructing a three-dimensional net, scanning, and allowing a quick interactive response; furthermore, the workers must be experienced.

2.5.2.2.2 Gross error detection based on the fitting curved surface

This method assumes that variations in the Earth's surface are consistent with some natural tendency, and that they can be described by a continuous and smooth curved surface. The curved surface expresses the macroscopic variation tendency of the topography. If the elevation observation of one sampling point is far away from the surface, the point is assumed to be affected by gross error. Therefore, using the fitting curved surface to detect gross errors includes two main steps: the selection of the fitting curved surface and the determination of the threshold.

Any continuous surface can be approximated by a higher-order polynomial, but the solution

TABLE 2.7   Relationship between digital elevation model (DEM) resolution and scale of original data (Hutchinson and Gallant, 2000).

Scale	DEM resolution	Original data
Fine topo-scale	5—50 m	Aerial images, topographic map of 1:5000 to 1:50,000
Coarse topo-scale	50—200 m	Aerial images, topographic map of 1:50,000 to 1:200,000
Mesoscale	0.2—5 km	Topographic map of 1:100,000 to 1:250,000
Macroscale	5—500 km	Topographic map of 1:250,000 to 1:1,000,000, national control points of elevation and plane

of the higher-order polynomial parameters is usually unstable. Therefore, a second- or third-order polynomial is usually employed for real applications. For the determination of the threshold, a statistical method is often used (that is, the RMSE of the original DEM points is obtained, and 2 or 3 times the RMSE can be used as the threshold). In fact, the variation in the Earth's surface cannot be fully described by the statistical method; thus, selecting the appropriate threshold has a great impact on the result of gross error detection.

Generally speaking, the method based on the fitting of a curved surface can detect most of the data points that contain gross error, but this method has greater uncertainty as well. Therefore, the method of stereoscopic manual visual inspection should be used for further analysis.

2.5.2.3 Filtering the source data

The DEM source data obtained from a stereo pair of images is usually in the form of a three-dimensional intensive and discrete point cloud, and the density, distribution, and accuracy of the point cloud may affect the accuracy and rationality of the description of the terrain by a DEM. Therefore, it is necessary to filter the source data with a filtering algorithm to improve the accuracy and rationality of the terrain expression.

Nearest-neighbor sampling, mean filtering, median filtering based on a moving window, and low-pass filtering based on the frequency domain are the common data-filtering methods. Many experiments have shown that implementing an appropriate filter can improve the accuracy of the terrain description by a DEM. However, filtering does not work in all cases. When accidental errors are the majority error source, rather than systematic or gross errors, implementing a filter can effectively improve the data quality. However, the effect is not obvious, and it does not always improve the data quality.

2.5.3 Interpolating of DEM data

DEM data interpolation is an important step in DEM generation and application. When generating a DEM, the elevation of each grid point from the known planimetric position should be interpolated from the discrete points. In DEM applications, the elevations of the nongrid points are usually interpolated from the elevations of the grid points. There are many interpolation methods for DEM data, and each method has its own advantages and disadvantages. Therefore, different interpolation methods should be selected according to different situations. This section introduces only the interpolation method based on the moving curved surface.

The following are the main steps of the moving curved surface interpolation method:

(1) For each grid point $P(X_p, Y_p)$ in the DEM, the data points $P_i(X_i, Y_i)$ in the corresponding grid blocks can be retrieved from the total data points, and the origin of the coordinate system can be moved onto the grid points:

$$\begin{cases} \overline{X}_i = X_i - X_p \\ \overline{Y}_i = Y_i - Y_p \end{cases} \tag{2.68}$$

(2) Taking the unknown point P as the center of a circle and R as the radius of the circle to select adjacent data points (shown in Fig. 2.16), the data points that fall inside the circle are selected. The number of selected points depends on the utilized local fitting function, and the number of selected points must be more than 6 when using a quadratic function to implement the interpolation. When the distance between the data point $P_i(X_i, Y_i)$ and the unknown point $P(X_p, Y_p)$ satisfies

$$d_i = \sqrt{\overline{X}_i^2 + \overline{Y}_i^2} < R \tag{2.69}$$

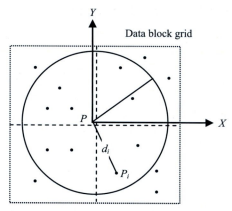

FIGURE 2.16   Selected data points in the circle with a radius of R and P as the center.

the point P_i is selected. If the number of selected points is not adequate, then R should be increased until the number of points meets the requirements.

(3) The observation equation is established. If a quadric surface is chosen as the fitting surface, the surface equation is as follows:

$$Z = AX^2 + BXY + CY^2 + DX + EY + F \quad (2.70)$$

The observation equation is established from the data point $P_i(X_i, Y_i)$ as follows:

$$v_i = \overline{X}_i^2 A + \overline{X}_i \overline{Y}_i B + \overline{Y}_i^2 C + \overline{X}_i D + \overline{Y}_i E + F - Z_i \quad (2.71)$$

(4) The weight w_i of each data point is set. The determination of w_i is related to the distance d_i between the data point P_i and the unknown point P. For smaller values of d_i, d_i has greater influence and greater weight. In contrast, greater values of d_i have smaller weight. The following formulas are commonly used to determine the weight:

$$w_i = \frac{1}{d_i^2} \quad or \quad w_i = \left(\frac{R - d_i}{d_i} \right) \quad (2.72)$$

(5) The normal equation is established and solved. According to the principle of the least squares adjustment, the following solution of the normal equation is established according to Eq. (2.72):

$$\mathbf{X} = (\mathbf{M}^T \mathbf{W} \mathbf{M})^{-1} \mathbf{M}^T \mathbf{W} \mathbf{Z} \quad (2.73)$$

where $\mathbf{M}$ is the design matrix constructed by the coefficients in Eq. (2.71), $\mathbf{Z}$ is a constant vector, and $\mathbf{W}$ is the weight matrix of data points.

The coefficient F of Eq. (2.70) is the interpolated elevation Zp of the unknown point because $\overline{X}_p = 0, \overline{Y}_p = 0$.

2.6 Orthoimage production

Remote sensing imagery can accurately represent the surface scenery, and it is more intuitive than a topographical map. However, aerial and satellite images are in center or multicenter projections, and the images have geometric distortions caused by image incline and terrain relief. Consequently, these images are not a simple deflation of the surface scenery, and they are not completely similar with the surface scenery. Therefore, in actual applications, remote sensing imagery should usually be rectified into an orthoimage in an orthographic projection. This procedure is called orthoimage production. Because the orthoimage production procedure divides the entire image into small blocks and then rectifies the image block by block, orthoimage production is also known as digital differential rectification.

2.6.1 Digital differential rectification of frame perspective imagery

2.6.1.1 Principles of digital differential rectification

The basic task of digital differential rectification is to achieve the mapping transformation between two two-dimensional images. In the rectification procedure, the geometric relationship between the original image and the orthoimage should first be determined. The coordinates of one pixel in the image before and after the rectification are assumed to be (x, y) and (X, Y), respectively, and they are assumed to satisfy the following mapping relationship:

$$\begin{cases} x = f_x(X, Y) \\ y = f_y(X, Y) \end{cases} \tag{2.74}$$

or

$$\begin{cases} X = \phi_x(x, y) \\ Y = \phi_y(x, y) \end{cases} \tag{2.75}$$

Eq. (2.74) is used to calculate the image point coordinates (x, y) of the original image according to the image point coordinates (X, Y) of the orthoimage. This calculation is called the inverse or indirect method of rectification. Eq. (2.75) works in the opposite way. This formula is used to calculate the point coordinates in the orthoimage according to the image point coordinates of the original image, and it is called the forward or direct method of rectification. The following section will introduce the operation

procedure for rectifying the perspective image into an orthoimage using the inverse and forward methods.

2.6.1.2 Digital differential rectification based on the inverse method

The processing flow of digital differential rectification based on the inverse method is as follows (shown in Fig. 2.17):

(1) The object space coordinates are calculated.

Assuming that the coordinates of any point P in the orthoimage are(X', Y'), the coordinates of the left-bottom point in the orthoimage are(X_0, Y_0), and the scale denominator of the orthoimage is M; then, the following are the object–space coordinates of point P:

$$\begin{cases} X = X_0 + MX' \\ Y = Y_0 + MY' \end{cases} \tag{2.76}$$

(2) The image space coordinates are calculated.

According to the rectification Eq. (2.74) of the inverse method, the corresponding image space coordinates of point P in the original image can

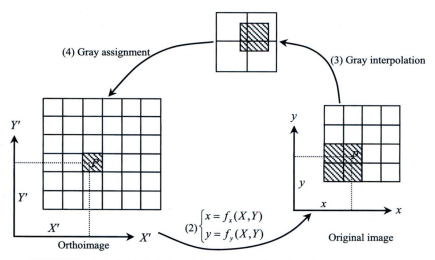

FIGURE 2.17 Digital differential rectification based on the inverse method.

be calculated. For the frame perspective imagery, the rectification formula of the inverse method is given by collinearity equations as follows:

$$\begin{cases} x - x_0 = -f\dfrac{a_1(X - X_S) + b_1(Y - Y_S) + c_1(Z - Z_S)}{a_3(X - X_S) + b_3(Y - Y_S) + c_3(Z - Z_S)} \\[2mm] y - y_0 = -f\dfrac{a_2(X - X_S) + b_2(Y - Y_S) + c_2(Z - Z_S)}{a_3(X - X_S) + b_3(Y - Y_S) + c_3(Z - Z_S)} \end{cases}$$
$$(2.77)$$

where Z is the elevation of point P and can usually be interpolated by DEM.

(3) The grays are resampled.

After obtaining the image space coordinates of point P, the coordinates should be transformed into image coordinates according to the interior orientation parameters. Here, the image coordinates may not be located in the pixel center, in which case, it is necessary to implement gray resampling. Generally, the gray value $g(x, y)$ of point $p(x, y)$ can be interpolated according to the method of bilinear interpolation.

(4) The gray values are reassigned.

The gray value of image point p is assigned to the corresponding pixel $P(X', Y')$ in the orthoimage as follows:

$$G(x, y) = g(x, y) \qquad (2.78)$$

By implementing the above operations for every pixel in the orthoimage in turn, the orthorectified digital image can be obtained.

2.6.1.3 Digital differential rectification based on the forward method

According to the forward Eq. (2.75) of rectification, digital differential rectification based on the forward method calculates the image space coordinates of each point in the rectified image with the image space coordinates of the corresponding point in the original image (shown in Fig. 2.18).

For the perspective imagery, the forward formula is the following converted form of the collinearity equations:

$$\begin{cases} X = Z\dfrac{a_1 x + a_2 y - a_3 f}{c_1 x + c_2 y - c_3 f} \\[2mm] Y = Z\dfrac{b_1 x + b_2 y - b_3 f}{c_1 x + c_2 y - c_3 f} \end{cases} \qquad (2.79)$$

When calculating the image point coordinates in the orthoimage according to Eq. (2.79), the elevation Z of the point should be known, but it is a function of the unknown coordinates (X, Y). Therefore, an initial elevation Z_0 should be set according to the DEM when implementing orthoimage rectification. After obtaining (X, Y),

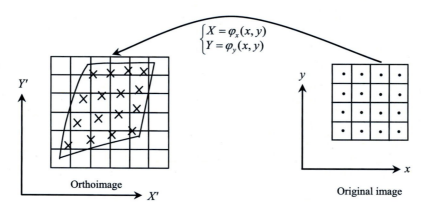

$$\begin{cases} X = \varphi_x(x, y) \\ Y = \varphi_y(x, y) \end{cases}$$

FIGURE 2.18   Digital differential rectification based on the forward method.

a new elevation Z should be interpolated again according to the DEM. Iterations are performed in this way, and the accurate coordinates of the point in the orthoimage can be obtained. In addition to the iterative solution, the image points obtained by this method are in an irregular distribution. No point may exist in some pixels, whereas several repeated points may exist in other pixels. Therefore, orthoimage rectification usually adopts the inverse method because of the above shortcomings of the forward method.

2.6.2 Digital differential rectification of linear array remote sensing imagery

The rigorous geometric processing model and the RFM are often used as the rectification models to implement the digital differential rectification of the satellite remote sensing imagery acquired by a linear array in push-broom mode. When the accuracy requirement of the orthoimage product is not high, the empirical geometric processing models, such as the general polynomial model and the affine transformation model, can also be used as the rectification models. At present, digital differential rectification usually utilizes the inverse method; therefore, this section introduces only the principle of the digital differential rectification of satellite remote sensing imagery based on the inverse method.

2.6.2.1 Digital differential rectification based on the rigorous geometric processing model

Points in each scan line of the satellite remote sensing imagery acquired by a linear array in push-broom mode have a strict perspective relationship with the corresponding objects on the ground. However, the satellite orbit parameters and attitude parameters of each scan-line image are different, and the orbit parameters and attitude parameters of the scan line that contains the image point corresponding to the objects on the ground should be calculated first when calculating the image point coordinates based on the rigorous geometric processing model. Therefore, the basic steps of the digital differential rectification based on the forward method are as follows:

(1) the scan time t corresponding to the scan line y in the image is calculated from Eq. (2.23);
(2) the exterior orientation elements of the scan-line image corresponding to the scan time t are calculated according to Eq. (2.22);
(3) each element value of the rotation matrix is calculated according to Eq. (2.21);
(4) the image space coordinate y' of the image point corresponding to the given ground point is calculated according to the second equation of Eq. (2.24);
(5) steps (1)–(4) are repeated by taking $y + y'$ as the new value of y. The iteration is thought to be convergent when y' is smaller than the given threshold (e.g., 0.001 pixels). Then, the image space coordinate x of the image point can be obtained by the exterior orientation elements according to the first equation of Eq. (2.24). If y' is larger than the given threshold, the iteration should be repeated until y' is smaller than the given threshold; next, we can obtain the image space coordinates (x, y) of the image point corresponding to the ground point.

The procedure described above requires many iterations to converge, and the initial coordinate y of the image point in the image space coordinate system must be known. In the digital differential rectification procedure, an accurate initial coordinate of the image point is never available; therefore, the coordinate y of the center scan line is usually taken as the initial value of the image point.

For each point in the orthoimage, the image space coordinates of the corresponding point in the original image can be calculated according to the steps described above. After resampling for each image point, the digital image after orthorectification can be obtained.

2.6.2.2 Digital differential rectification based on the RFM

The procedure of digital differential rectification for linear array satellite remote sensing imagery based on the RFM is similar to that of the traditional frame perspective imagery. The procedures differ only in that Eq. (2.33) should be substituted by Eq. (2.77).

For the digital differential rectification of high-resolution satellite remote sensing imagery covering a large ground area, implementing the orthorectification based on the rigorous geometric processing model is time consuming. However, using the RFM to substitute the strict geometric processing model, the image space coordinates of the image point corresponding to the ground point can be directly obtained according to Eq. (2.33), and many iterative calculations can be avoided. Therefore, the digital differential rectification can be completed within an acceptable length of time.

2.6.3 The orthoimage mosaic

The orthoimage is obtained by digital differential rectification of a single original image. In practical applications, it is necessary to join each single orthoimage together to generate a large orthoimage according to the map or designated areas; this operation is called the orthoimage mosaic. The orthoimage mosaic is the procedure of joining mutual adjacent orthoimages together into a unified orthoimage. To achieve a good visual effect, the new generated orthoimage should be seamless in geometry and color. Therefore, the procedure of orthoimage mosaic requires not only the geometrical

joining of the images but also the procedure of image dodging and tone balance, in which mosaic line searching is a key step.

2.6.3.1 Image dodging and tone balance

In the remote sensing image acquisition procedure, the interference of the internal and external environmental factors causes the tonality, intensity, and contrast of the different interior areas of a single remote sensing image to differ to a certain extent. Some areas are lighter, whereas other areas are darker (generally, the central areas are lighter, and the surrounding areas are darker), and a single ground object may have different intensity values in different areas of the image. Furthermore, the tones of different remote sensing images covering the same ground area may not be balanced (that is, a single ground object may have different intensity values in adjacent images). Therefore, to obtain a seamless orthoimage, image dodging and tone balance should be implemented both in a single image and between different images.

2.6.3.1.1 Image dodging in a single remote sensing image

The procedure of image dodging in a single remote sensing image can be divided into four steps: image preprocessing, background image generation, MASK operation, and image postprocessing.

(1) For the image-dodging processing of colorful remote sensing imagery, the color space transformation of the image should be performed first. This process entails the transformation of the color image in the RGB space to the image in the IHS space. After extracting the I component of the image in the IHS space, the image-dodging processing of I can be implemented separately. This procedure is similar to that of a black and white image.

(2) The original image is resampled to make it smaller; then, the frequency spectrum can be

obtained by a Fourier transformation. A Gaussian low-pass filter is applied in the frequency domain, and a small-sized background image can be obtained using the inverse Fourier transformation. When the resulting image is enlarged to its original size, the background image is obtained.

(3) According to the principle of MASK image dodging, an image with an uneven distribution can be considered to be the result of the superposition of one image with the illumination evenly distributed and a background image. This representation can be described using the following mathematical model:

$$I'(x,y) = I(x,y) + B(x,y) \qquad (2.80)$$

where $I'(x,y)$ represents the actual image with the illumination unevenly distributed, $I(x,y)$ represents the image with the illumination evenly distributed in ideal conditions, and $B(x,y)$ represents the background image.

By subtracting the original image from the background image, an image with an improved intensity distribution can be obtained. This processing method is called the MASK operation, which has the following mathematical expression:

$$I_{out} = I_{in} - I_{blur} + offset \qquad (2.81)$$

where I_{out} is the image with an improved intensity distribution, I_{in} is the original image, I_{blur} is the background image, and $offset$ is the offset.

Introducing an offset to Eq. (2.81) distributes the pixel values of the processed images within the grayscale range. In addition, the value of the offset determines the average intensity of the resulting image. To maintain the average intensity of the original image, the value of the offset can be set equal to the average intensity of the original image.

(4) The dynamic range of gray values of the remote sensing images processed using the MASK operation is reduced, and the overall contrast of the resulting images is decreased. Then, the resulting images can be processed using the subsection linear stretching method to enhance their overall contrast.

The above processing is carried out on the I component of the IHS space for color remote sensing images. The component I′, which has an evenly distributed illumination, can be obtained after the above process. Then, by replacing the component I in the IHS space with the component I′ and implementing the inverse Fourier transformation from the IHS color space to the RGB color space, a color remote sensing image with improved light distribution can be obtained.

When a single remote sensing image is dodged with the Mask method, the contrast of the resulting image may be decreased. To solve this problem, researchers improved the Mask method. The contrast of the resulting image can be retained to a certain extent. However, when an obvious gradient mutation region exists in the original image, the phenomenon of the halo and the distorted gray will appear in the corresponding region in the resulting image. In response to this deficiency, Yuan et al. (2014) expanded and swallowed the high-contrast region in the original image. The distorted gray at the junction of special regions can be effectively eliminated, and the texture clarity of the dark region in the original image can be significantly enhanced.

2.6.3.1.2 Tone balance among different images

The tone-balance processing among different remote sensing images can be completed using the tone-balance method based on the Wallis filter.

The Wallis filter is a relatively special filter that has the capability of self-adaptation. This filter can enhance the contrast of the original signal while simultaneously reducing the noise. The Wallis filter is a type of local image transformation, and it can approximately equalize the variance and the gray average at different locations of the image (that is, it not only increases the contrast in image areas where the contrast is small but also decreases the contrast in image areas where the contrast is large, enhancing the weak gray information in the image). When the Wallis filter is applied to remote sensing images, the average and the variance of the image gray can be mapped to the given gray value and variance.

The general mathematical expression of the Wallis transformation is

$$f(x,y) = (g(x,y) - m_g)\frac{cv_f}{cv_g + (1-c)v_f} + bm_f$$
$$+ (1-b)m_g$$

$$(2.82)$$

where $g(x,y)$ is the gray value of the original image; $f(x,y)$ is the gray value of the resulting image after the Wallis transformation; mg and vg are the average and the standard deviation of the gray of the original local image, respectively; mf and vf are the target value of the average and the standard deviation of the resulting image, respectively; $c \in [0,1]$ is an expanding constant of the image variance; and $b \in [0, 1]$ is the intensity coefficient of the image. The average of the image is mapped to mf when $b \to 1$, and the average of the image is mapped to mg when $b \to 0$.

Eq. (2.82) can also be expressed as

$$f(x,y) = g(x,y)r_1 + r_0 \qquad (2.83)$$

where

$$r_1 = \frac{cv_f}{cv_g + (1-c)v_f}$$

and $r_0 = bm_f + (1 - b - r_1)m_g$. Parameters r_1 and r_0 are a multiplicative coefficient and an additive coefficient, respectively.

In the typical Wallis filter, $c = 1$ and $b = 1$; therefore, Eq. (2.82) can be expressed as

$$f(x,y) = (g(x,y) - m_g)\frac{v_f}{v_g} + m_f \qquad (2.84)$$

Here,

$$\begin{cases} r_1 = \dfrac{v_f}{v_g} \\ r_0 = m_f - r_1 m_g \end{cases} \qquad (2.85)$$

The processing flow of the tone-balance operation among different remote sensing images based on the Wallis filter is as follows:

(1) The data are prepared. Before implementing tone-balance processing among different remote sensing images, each image must be processed individually by image dodging. The inconsistent tonality and contrast of each image will influence the subsequent tone-balance processing results among the different images.

(2) The standard parameters are determined. An image with relatively ideal tonality and contrast is selected from the images to be processed, and the average and standard deviation of the image are statistically calculated. These values can be considered to be the target average and standard deviation. Additionally, these values can be treated as the target average and standard deviation when statistically calculating the average and the standard deviation of each image to be processed.

(3) The coefficients of Wallis filter are calculated. Setting $c = 1$ and $b = 1$, the average and the standard deviation of each image can be approximately consistent. Then, the multiplicative coefficient and the additive coefficient can be calculated according to Eq. (2.85).

(4) Wallis filter processing is implemented. Each image is processed by the Wallis filter according to Eq. (2.83).

(5) The overlapping image areas are processed further. After the above processing, each image has self-consistent tonality and contrast, but the overlapping areas between the adjacent images may still have some color differences, which may influence the color effect of the whole image after performing the image mosaic. To eliminate these color differences, further tone-balance processing of the overlapped image areas should be implemented. Because the overlapping image areas have the same distribution of ground objects and the shape of the image histograms should be roughly the same, the tone-balance processing of the overlapping image areas can be implemented based on the method of histogram matching to further improve the color effect of the entire image after the image mosaic.

Wallis filter is essentially a linear transformation of pixel values of each color channel in an image. Although the linear transform can be used to correct the color consistency of multiple remote sensing images, the cascade transmission mode of the linear transform has the inherent defect of cumulative error, which determines that the conventional Wallis filter cannot be effectively applied to the color consistency correction of large scenes. On the basis of Wallis filter, researchers presented a tone-balance method based on the least squares block adjustment. This method first calculates the tone information (gray average and gray variance) of a single orthophoto image and the common overlapping area of its adjacent images. Then, on the basis of the requirement that the overlapping regions of the tone-balanced orthophoto images should have the same or similar tone, the tone adjustment parameters of each image are totally calculated according to the least squares adjustment. In addition, the weighted Wallis filter can also avoid the cumulative error caused by spatial transmission and achieve the color consistency correction of multiple remote sensing images in large areas (Fan, 2017).

2.6.3.2 Image mosaic

The orthoimage mosaic refers to the process of joining many orthoimages that share overlapping areas into a large orthoimage and clipping the orthoimage according to the figure profile to generate the orthoimage maps. Determining the optimal mosaic line in the effective areas of the overlapping image is a key technology of the image mosaic. A reasonable mosaic line should avoid features above the ground, such as roofs and tree crowns. It should also avoid areas with large color contrasts to improve the geometric quality of the mosaic images and the efficiency of orthoimage production.

2.6.3.2.1 Mosaic line searching

The DEM used in the orthoimage rectification procedure does not consider objects above the ground (such as houses and trees), and areas such as roofs and the tree crowns cannot be rectified into the correct positions; thus, projective differences are generated. Because of the central projection characteristics, the size and the direction of the projections differ among images. Obviously, a "seamless" mosaic is impossible to realize if the mosaic line passes through areas that contain large projective differences. Mosaic line searching can be implemented in the differential image of the images to be jointed together to avoid areas with large contrasts. The differential image can be obtained according to the following formula:

$$g(i,j) = g_1(i_1,j_1) - g_2(i_2,j_2) \qquad (2.86)$$

where $g(i, j)$ is the pixel gray in the differential image and $g_1(i_1, j_1)$ and $g_2(i_2, j_2)$ are the gray of corresponding pixels in the first and second images to be jointed, respectively. Areas with large contrasts between the images will become highlighted areas in the differential image. Therefore, the principle of mosaic line searching requires that pixels with large gray values be avoided

The searching scope of the starting point

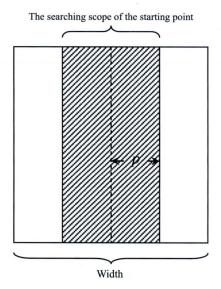

Width

FIGURE 2.19   A selection schematic diagram of the starting point of the mosaic line.

and that pixels with small gray values in the differential image be selected.

There are many algorithms for mosaic line searching. Here, we introduce a relatively simple random greedy method with the following steps:

(1) The starting point of the mosaic line is selected.

The projective differences are relatively small in the middle part of the orthoimage; therefore, the selected mosaic line should be as close to the middle as possible. As shown in Fig. 2.19, taking the search from the top to bottom as an example, we can randomly select a pixel point as the starting point in the middle of the first line in the differential image. The searching scope is $[(0.5 - p) \cdot \text{width}, (0.5 + p) \cdot \text{width}]$,

here p is the percentage of the interval scope and the width is the width of the differential images. The scope of the hatched lines in Fig. 2.19 is the selection scope of the starting point of the mosaic line.

(2) The next path point is selected.

After the starting point is determined, the next path point can be selected using a greedy search method (that is, the pixel with the minimum gray value in the candidate pixels can be selected as the next path point). As shown in Fig. 2.20, the candidate pixels are the n pixels in the searching direction ahead. Taking the direction from top to bottom as the searching direction and assuming that the jth pixel is the path point in the ith line, the optional pixels in the $(i+1)^{\text{th}}$ line should be those n pixels between j-$n/2$ and $j + n/2$, and the value of n is usually 3, 5, or 7.

(3) The sum of the gray values is calculated.

After the last line of the image is searched, the gray values of each pixel on the path are accumulated, and the sum of the gray values of each pixel on the path can be obtained.

(4) The mosaic line is searched many times, and the optimal path is selected.

Because the starting point of the mosaic line is random, the path obtained is also random and unreliable. Therefore, it is necessary to select different starting points many times and to repeat the above process. For each instance, when the current optimal path is obtained, it is necessary to compare it with the historical optimal path. If the accumulated sum of the gray values of the current path is smaller than

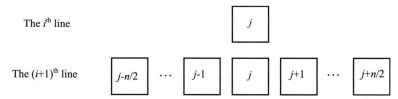

FIGURE 2.20   A selection schematic diagram of the next path point.

that of the historical optimal path, then the current path should be substituted for the historical optimal path; otherwise, the next instance of searching should be continued until the search times equal the given times. The searching times are usually between 1000 and 10,000, and they can be set according to the actual situation.

2.6.3.2.2 Image filling

The effective filling area of each image can be obtained using the polygon intersection method after the mosaic line is generated. There are many image filling methods, and only the boundary-marked filling method is introduced here.

The boundary-marked filling method can be divided into two steps:

(1) the scan-line transformation of each boundary of the filling area is implemented (that is, the pixels on the boundary of the area are marked); and

(2) the area is filled row by row. A mark parameter is set for each line, and the initial value of the parameter is set to lie outside the area. The value of the parameter should be set to lie inside the area when the boundaries of the area are met, and the value should be set to lie outside the area when the boundaries are met again. When the value of the parameter is set to lie inside the area, the gray information of the corresponding image areas should be assigned to the image to be filled.

2.7 Summary

This chapter introduces geometric processing and positioning techniques for satellite remote sensing imagery, including systematic error calibration, geometric rectification, geometric registration, DEM generation, and DOM generation. The core technique is the establishment of a suitable geometric processing model for satellite imagery and the precise solution of all unknown parameters of the model.

In actual applications, users can select a suitable geometric processing model to implement the geometric processing of satellite imagery according to different requirements. When the accuracy requirement of the image products is not high, the geometric processing model whose parameters can be solved easily and whose computation demand is small can be used, such as the general polynomial model, the DLT model, or the affine transformation model. In contrast, when the accuracy requirement of the image product is relatively high, the expanded collinearity equation model or the RFM should be used.

Compared with the digital differential rectification of satellite imagery based on the expanded collinearity equation model, the digital differential rectification based on the RFM can achieve a consistent geometric accuracy, and the iterative calculations can be avoided so that its computation demand is far lower than that of the expanded collinearity equation model. Therefore, if the RPCs are available, the RFM should be employed as often as possible to implement the digital differential rectification of satellite imagery.

To extract geographic spatial information from satellite imagery, the generation of image products at different levels is an essential step, and the geometric accuracy of these products influences a series of subsequent analyses and decisions. Therefore, generating various image products at different levels with high accuracy from satellite imagery is of ongoing research interest. Some progress has been made, but further research is still necessary.

For example, for image geometric rectification, a certain number of GCPs should be available to solve the parameters of the geometric rectification model. However, in some areas with complicated terrain or poor environments, it is expensive to obtain GCPs, and in some reconnaissance areas, it is impossible to obtain them. In this case, implementing satellite image geometric rectification with high accuracy but without GCPs is still difficult or impossible.

Additionally, the method of image matching using dense points is often employed to automatically generate a DEM according to several satellite remote sensing images covering the same ground area. However, due to the problems caused by image obstruction, poor texture, tree coverage, and repeated texture, it is difficult to match the dense point cloud with high accuracy and high reliability according to this method; thus, it is impossible to generate a DEM automatically for these areas. Moreover, searching for the optimal mosaic line automatically and avoiding artificial buildings (e.g., houses) during the orthoimage mosaic process remains an important research topic.

2.8 Questions

1. Which factors affect the in-orbit geometric calibration accuracy of satellite remote sensing imagery?
2. What are the advantages and disadvantages of the rigorous geometric processing models and empirical geometric processing models used in the geometric rectification of satellite remote sensing imagery?
3. Which factors affect the geometric rectification accuracy of satellite remote sensing imagery?
4. Which factors affect the geometric registration accuracy of satellite remote sensing imagery?
5. What are the advantages and disadvantages of the regular grid DEM and the TIN DEM?
6. What are the advantages and disadvantages of the digital differential rectification based on the inverse method and the digital differential rectification based on the forward method?
7. What are the differences between the digital differential rectification of frame perspective imagery and the digital differential rectification of linear array remote sensing imagery?

8. Which important factors should be considered in the orthoimage mosaic of satellite remote sensing imagery?

References

Ackermann, F., 1983. High precision digital image correlation. In: Proceedings of the 39th Photogrammetric Week (Stuttgart, Germany).

Baltsavias, E., Zhang, L., Eisenbeiss, H., 2006. DSM generation and interior orientation determination of IKONOS images using a testfield in Switzerland. Photogramm. Fernerkund. GeoInf. 1, 41–54.

Cao, J., Yuan, X., Gong, J., 2015. In-orbit geometric calibration and validation of ZY-3 three-line cameras based on CCD-detector look angles. Photogramm. Rec. 30 (150), 211–226.

Di, K.C., Ma, R.J., Li, R.X., 2003. Rational functions and potential for rigorous sensor model recovery. Photogramm. Eng. Remote Sens. 69 (1), 33–41.

Fan, C., Chen, X., Zhong, L., Zhou, M., Shi, Y., Duan, Y., 2017. Improved Wallis dodging algorithm for large-scale super-resolution reconstruction remote sensing images. Sensors 17 (3), 623.

Fischler, M.A., Bolles, R.C., 1981. Random sample consensus: a paradigm for model fitting with applications to image analysis and automated cartography. CACM 24 (6), 381–395.

Fraser, C.S., Dial, G., Grodecki, J., 2006. Sensor orientation via RPCs. ISPRS J. Photogrammetry Remote Sens. 60 (3), 182–194.

Fraser, C.S., Hanley, H.B., 2005. Bias-compensated RPCs for sensor orientation of high-resolution satellite imagery. Photogramm. Eng. Remote Sens. 71 (8), 909–915.

Grodecki, J., Dial, G., 2001. IKONOS geometric accuracy. In: Proceedings of ISPRS Working Groups on High Resolution Mapping from Space 2001 (Hanover, Germany).

Hutchinson, M.F., Gallant, J.C., 2000. Digital Elevation Models and Representation of Terrain Shape. John Wiley and Sons, New York.

Lowe, D.G., 2004. Distinctive image features from scale-invariant key points. Int. J. Comput. Vis. 60 (2), 91–110.

OGC (OpenGIS Consortium), 1999. The OpenGIS Abstract Specification~Topic7: The Earth Imagery Case. http://portal.opengeospatial.org/files/?artifact_id=892.

Okamoto, A., 1999. Geometric characteristics of alternative triangulation models for satellite imagery. In: Proceedings of 1999 ASPRS Annual Conference. Oregon, USA.

Poli, D., 2004. Orientation of satellite and airborne imagery from multi-line pushbroom sensors with A rigorous sensor model. In: International Archives of Photogrammetry and Remote Sensing. Istanbul, Turkey.

Tao, C.V., Hu, Y., 2001. A comprehensive study of the rational function model for photogrammetric processing. Photogramm. Eng. Remote Sens. 67 (12), 1347−1357.

Tao, C.V., Hu, Y., 2002. 3D reconstruction methods based on the rational function model. Photogramm. Eng. Remote Sens. 68 (7), 705−714.

Toutin, T., 2004a. Geometric processing of remote sensing images: models, algorithms and methods. Int. J. Remote Sens. 25 (10), 1893−1924.

Toutin, T., 2004b. Spatio-triangulation with multi-sensor VIR/SAR images. IEEE Trans. Geosci. Remote Sens. 42 (10), 2096−2103.

Wang, M., Yang, B., Hu, F., Zang, X., 2014. On-orbit geometric calibration model and its applications for high-resolution optical satellite imagery. Remote Sens. 6, 4391−4408.

Wang, Y.N., 1999. Automated triangulation of linear scanner imagery. In: Proceedings of ISPRS Work Groups on "Sensors and Mapping from 1999" (Hanover, Germany).

Yang, X.H., 2000. Accuracy of rational function approximation in photogrammetry. In: Proceedings of 2000 ASPRS Annual Conference (Washington DC, USA).

Yuan, X., Han, Y., Fang, Y., 2014. Improved Mask dodging algorithm for aerial imagery. J. Remote Sens. 18 (3), 630−641.

Zhang, G., Jiang, Y., Li, D., Huang, W., Pan, H., Tang, X., Zhu, X., 2014. In-orbit geometric calibration and validation of ZY-3 linear array sensor. Photogramm. Rec. 29 (145), 68−88.

Zhang, Y., Lu, Y., Wang, L., Huang, X., 2012. A new approach on optimization of the rational function model of high-resolution satellite imagery. IEEE Trans. Geosci. Remote Sens. 50 (7), 2758−2764.

Further reading

Jacobsen, K., 1997. Calibration of IRS-1C Pan Camera. German Society for Photogrammetry and Remote Sensing, pp. 163−170.

Compositing, smoothing, and gap-filling techniques

Abstract

Because of the influences of cloud cover, seasonal snow, and many other factors, time series of land surface parameters extracted from remote sensing data often suffer from discontinuities and missing data, which have seriously restricted the application of extracted land surface parameters to research in global change and other fields. This chapter introduces several different techniques for compositing, smoothing, and gap filling of remotely sensed time series data, which

generate land surface parameter products that are temporally and spatially continuous.

A multitemporal compositing method is designed to select the optimal remote sensing data as the pixel value in a composite time window, according to certain criteria. This is a common processing method used to obtain cloud-free and spatially continuous images. Section 3.1 of this chapter gives a detailed description of several commonly used compositing algorithms. Here, time series data smoothing and gap-filling methods are applied to reconstruct high-

© 2020 Elsevier Inc. All rights reserved.

quality time series data with temporal continuity and spatial integrity, thereby eliminating the influences of cloud. Section 3.2 introduces several typical smoothing and gap-filling algorithms for time series data.

3.1 Multitemporal compositing techniques

Because of the influences of clouds, aerosols, and other factors, single-phase remote sensing images are often subject to missing data, seriously limiting the application of extracted land surface parameters to research in global change and many other fields. Through multitemporal remote sensing data compositing, the influences of these factors can be reduced or eliminated (Holben, 1986; Cihlar et al., 1994).

Multitemporal compositing uses set criteria to select the pixel values with the highest quality among multiscene matching data within a specific time range. For different application targets, a variety of data compositing criteria have been proposed, each with the purpose of eliminating the influences of cloud cover, aerosols, and other factors as much as possible (Cihlar et al., 1994; Qi and Kerr, 1997). Several typical data compositing algorithms are briefly introduced in this section.

3.1.1 Maximum vegetation index composite

Vegetation index is usually obtained by linear or nonlinear combination operations on remotely sensed red and near-infrared (NIR) reflectance data, which are simple and effective parameters for characterizing vegetation cover and growth status. The maximum vegetation index composite chooses the maximum value of the vegetation index within a given time range as the selection criterion for the remote sensing data.

The maximum value composite (MVC), first proposed by Holben (1986), was applied to the maximum normalized difference vegetation index (NDVI) to synthesize multitemporal AVHRR data. Based on the multiscene matching satellite observations within a given time range, the corresponding NDVI images are calculated,

and then a comparison is performed pixel by pixel to choose the remote sensing data with maximum NDVI. This method is simple and easily implemented, which is why it has been widely applied to monitor changes in global vegetation cover (Illera et al., 1996; Kasischke et al., 1993; Peters et al., 2002; Potter and Brooks, 2000). However, MVC tends to select data that are far away from the subsatellite point in practical applications (Cihlar et al., 1994), yielding poor performance when eliminating cloud cover for certain vegetation types (Sousa et al., 2002).

3.1.2 Minimum band reflectance composite

To avoid some of the problems associated with MVC, selection based on the minimum band reflectance was proposed by Qi and Kerr (1997). Generally, these methods involve selecting the minimum blue band reflectance (mBlue) (Vermote and Vermeulen, 1999) and minimum red band reflectance (mRed) (Cabral et al., 2003; Chuvieco et al., 2005).

In the red band, the reflectance of cloud is significantly higher than that of vegetation. If the minimum reflectance of the red channel is selected, the probability of selecting cloud-contaminated pixel values should be reduced. However, this method still retains a large number of cloud-contaminated pixel values because the reflectance of the red channel in cloud-shadow areas is also very low (Qi and Kerr, 1997).

To eliminate the influences of cloud shadows, selection of the third lowest near-infrared reflectance (NIRm3) and the third darkest value (Darkm3) has been suggested by some scholars. Selection of the third lowest values is based on the assumption that there is a low likelihood of a cloud shadow falling on a given pixel more than twice, over a period of 1 month (Cabral et al., 2003).

The composite results of daily VEGETATION data obtained by four compositing techniques are shown in Fig. 3.1. All algorithms performed similarly with respect to cloud elimination. However, mRed retained cloud shadows, and

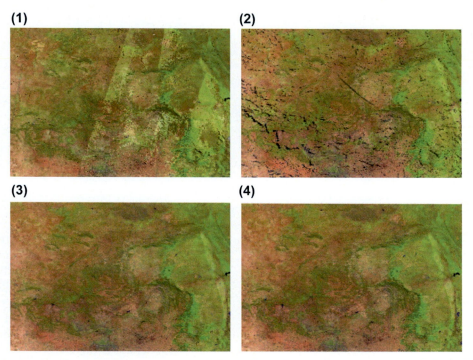

FIGURE 3.1 Comparison of different compositing techniques: (1) maximum value composite, (2) mRed, (3) NIRm3, and (4) Darkm3 (Cabral et al., 2003).

MVC yielded highly heterogeneous images because of the different acquisition geometries of neighboring pixels (Cabral et al., 2003). Comparatively speaking, the NIRm3 and Darkm3 were able to generate high-quality images which appeared to be visually smoother, while effectively eliminating the influences of clouds and cloud shadows.

3.1.3 Maximum surface temperature composite

For burned land mapping, the maximum surface temperature (MaxTs) is an accurate criterion to composite remote sensing data. This is because the temperature in recently burned areas is higher than that of the surrounding unaffected vegetation and is also higher than those of clouds and cloud shadows (Pereira et al., 1999).

In composites of MODIS and AVHRR data, the MaxTs criterion has achieved good results in distinguishing burned areas from the surrounding vegetation. Fig. 3.2 shows the MODIS multitemporal composites generated by MVC and MaxTs over an area affected by a large forest fire. By comparing the results of the two methods, it can be seen that the MaxTs composite can distinguish burned areas from the surrounding vegetation more clearly than the MVC method.

3.1.4 Mixing criteria compositing

In multitemporal compositing, the use of mixing criteria to select remote sensing data was proposed by Qi et al. (1993). The above criteria for choosing maximum vegetation index, maximum surface temperature, and maximum view zenith angle can be combined to form a

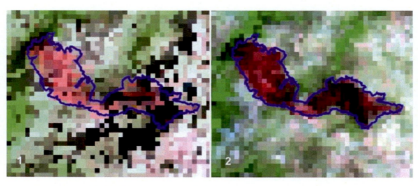

FIGURE 3.2 Composite images of MODIS multitemporal data over a burned area generated with different criteria: (1) maximum value composite and (2) MaxTs (Chuvieco et al., 2005).

variety of data compositing techniques (Cabral et al., 2003; Carreiras and Pereira, 2005; Carreiras et al., 2003; Cihlar et al., 1994).

Fig. 3.3 shows the results of applying several criteria for multitemporal compositing to MODIS multitemporal images collected in September 2003. Visually, all the results appear to be significantly better than the daily observation data, but clouds and cloud shadows were still present in the composite results for MaxNDVI, MinR1, and MinR2. This means that MaxNDVI, MinR1, and MinR2 were unable to properly eliminate the cloud-contaminated data. The MaxTs method yielded the best composite results, with nearly all clouds and cloud shadows eliminated from the daily remote sensing data. The compositing techniques that combine criteria for choosing maximum temperature and minimum view zenith angle or minimum NIR reflectance have been demonstrated to be capable of eliminating all clouds and cloud shadows.

3.1.5 MODIS vegetation index compositing technique

In the MVC approach, data far away from the subsatellite point are more likely to be selected than those closer by. Therefore, the selection of remote sensing data with a minimum view zenith angle has been proposed (Chuvieco

et al., 2005). However, before data compositing, strict data screening is required to eliminate cloud-contaminated data. To avoid the selection of data far away from the subsatellite point, a bidirectional reflectance distribution function (BRDF) model correction is adopted to synthesize MODIS vegetation indices (Van Leeuwen et al., 1999; Schaaf et al., 2002).

Fig. 3.4 shows the flowchart for the MODIS vegetation index compositing algorithm. Based on cloud identification and other information, high-quality directional reflectance data are selected within a 16-day time window. When the number of high-quality directional reflectance data within the time window is greater than 5, the Walthall BRDF model is used for fitting the reflectance of each band. The Walthall BRDF model can be expressed by Eq. (3.1) (Walthall et al., 1985) as follows:

$$\rho_\lambda(\theta_v, \phi_s, \phi_v) = a_\lambda \theta_v^2 + b_\lambda \theta_v \cos(\phi_v - \phi_s) + c_\lambda$$
(3.1)

where a_λ, b_λ, and c_λ are model parameters determined by least squares fitting. The nadir reflectance for each pixel within a 16-day interval can then be derived.

When the number of cloud-free data during a 16-day period is less than 5, the constrained view angle, maximum value composite (CV-MVC) approach is adopted to select the highest NDVI

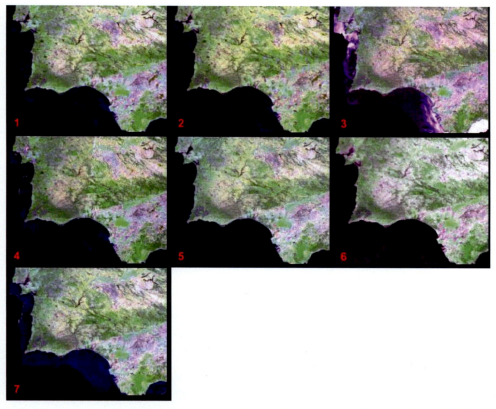

FIGURE 3.3 Composite results for MODIS multitemporal images in September 2003. (1) MinR1, (2) MinR2, (3) MaxTs, (4) MaxNDVI, (5) MinAZMaxTs, (6) MinR2MaxTs, and (7) MaxTsMinR2 (Chuvieco et al., 2005).

based on two cloud-free pixels with their view angles closest to nadir. First, all cloud-free reflectance data are sorted within the time window by view zenith angle, and then two cloud-free reflectance data with view angles closest to the view zenith angle are selected to calculate NDVI. The reflectance data with maximum NDVI values are selected as the composite results within the time window. If only one cloud-free observation exists within the time window, this observation will be automatically selected to represent the optimal value within the composite time. If 16-day observation data are all subject to cloud contamination, then the vegetation indices are calculated for all days, and the pixel value with the maximum NDVI among all the observations is selected.

The MODIS vegetation index compositing algorithm was applied to 1 year of daily AVHRR data to generate 23 consecutive global composites of the NDVI. Fig. 3.5 shows the results of the NDVI composite from August 13 to 28, 1989. It can be seen from Fig. 3.6 that, over the entire time profile, NDVI values obtained by the MODIS vegetation index compositing technique are slightly smaller than those obtained by MVC.

3.2 Time series data smoothing and gap filling

If the composite time period is too long, the time compositing technology is unable to reflect

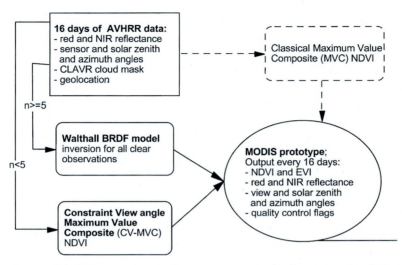

FIGURE 3.4　Flowchart for the MODIS vegetation index compositing algorithm applied to AVHRR test datasets (Van Leeuwen et al., 1999).

FIGURE 3.5　Global normalized difference vegetation index image using the MODIS vegetation index compositing algorithm (Van Leeuwen et al., 1999).

real changes in land surface parameters; if it is too short, the influences of clouds cannot be effectively eliminated, especially in cloudy areas. The smoothing and gap-filling method for time series data aims to reconstruct high-quality time series data and eliminate the influences of clouds. To date, a variety of smoothing and gap-filling methods for time series of remote sensing data have been developed (Viovy et al.,

1992; Hermance, 2007; Julien et al., 2006; Roerink et al., 2000; Moody et al., 2005). Generally speaking, these methods can be divided into two categories: the first is smoothing and gap filling in the time domain, mainly by asymmetric Gaussian function fitting (Jonsson and Eklundh, 2002), weighted least squares linear regression (Sellers et al., 1994), or a Savitzky–Golay (SG) filter (Chen et al., 2004; Cao et al., 2018); the other is

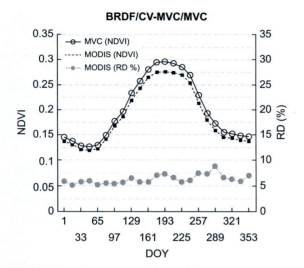

FIGURE 3.6 Comparison between global means of normalized difference vegetation index obtained by MVC and the MODIS vegetation index compositing algorithm (Van Leeuwen et al., 1999).

smoothing and gap filling in the frequency domain, e.g., the fitting method with a Fourier function (Roerink et al., 2000).

This section will focus on several common smoothing and gap-filling methods for time series data.

3.2.1 Curve fitting method

Based on least squares fitting, multiple fitting methods for the outer envelope of NDVI time series data were proposed by Jönsson and Eklundh (2004). These methods mainly include adaptive SG filtering, an asymmetric Gaussian function and double logistic function fitting.

Given time series data $(t_i, y_i), i = 1, 2, ..., N$, the parameter c in the fitting model $f(t; \mathbf{c})$ can be obtained through minimizing the following cost function:

$$\chi^2 = \sum_{i=1}^{N} [\omega_i (f(t_i; \mathbf{c}) - y_i)]^2 \qquad (3.2)$$

where ω_i is the weight of the ith data point.

Note that the atmospheric effects on surface reflectances generally cause increases in reflectance in the red band rather than in the NIR band, resulting in decreases in the retrieved LAI (Fernandes et al., 2003; Chen et al., 2006). To account for negatively biased noise, the fitting is implemented in two steps (Jönsson and Eklundh, 2002). The first step minimizes the cost function when all data points have the same weights. It is generally believed that data points with values greater than those simulated by the fitting model in the first step are less affected by noise and are thus of higher quality. Therefore, in the second step, the minimization is repeated with the weight of low-quality data points decreased by a factor of 10. This yields the smooth envelope curve for the NDVI time series, as shown in Fig. 3.7. In Fig. 3.7, thin solid lines indicate the NDVI time series; dotted lines indicate the fitting curves after the first iteration; and thick solid lines indicate the NDVI envelope curve after the two-step iteration.

3.2.1.1 Adaptive SG filtering

To suppress noise and to smooth data, a polynomial equation $f(t) = c_1 + c_2 t + c_3 t^2$ is used for fitting $2n + 1$ data within a sliding

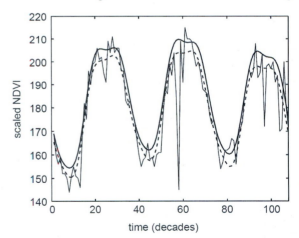

FIGURE 3.7 The upper envelope of normalized difference vegetation index time series data obtained by two-step fitting (Jönsson and Eklundh, 2004).

window at each position i of the time series data $(t_i, y_i), i = 1, 2, \ldots, N$. Data are equally spaced in SG filtering (Savitzky, and Golay, 1964). Then, after fitting the polynomial equation, y_i is replaced by the simulated value g_i at position i. Considering the negative bias in remote sensing data noise, the above two steps are undertaken to obtain the smooth envelope curve of time series data.

The sliding window size n not only determines the smoothness of the results of adaptive SG filtering but also greatly influences the ability to fit rapidly changing time series data by an adaptive SG filter. Fig. 3.8 shows the filtering results obtained with sliding windows of five and three points, respectively. It can be seen that with sliding window of 5, some anomalous peak values are not fitted properly; meanwhile, with the sliding window of 3, the filtering results closely follow the anomalous peak values.

3.2.1.2 Asymmetric Gaussian function and double logistic function fitting

In both of the function fitting methods, the fitting model usually has the following form:

$$f(t; \mathbf{c}) = c_1 + c_2 g(t; \mathbf{x}) \tag{3.3}$$

where $\mathbf{x}$ is the parameter of the basis function $g(t; \mathbf{x})$.

In asymmetric Gaussian function fitting, the following basis function is selected:

$$g(t; x_1, x_2, \cdots, x_5) = \begin{cases} \exp\left[-\left(\dfrac{t - x_1}{x_2}\right)^{x_3}\right] & 当 t > x_1 \\ \exp\left[-\left(\dfrac{x_1 - t}{x_4}\right)^{x_5}\right] & 当 t < x_1 \end{cases} \tag{3.4}$$

where x_1 is the position at which the basis function reaches its maximum or minimum; x_2 and x_3 are the width and flatness of the right basis function, respectively; and similarly x_4 and x_5 are the width and flatness of the left basis function, respectively. To ensure the smooth results of model simulation corresponding to time series data, parameters x_2, x_3, x_4, and x_5 should lie within specified ranges.

In double logistic function fitting, the basis function is as follows:

$$g(t; x_1, x_2, x_3, x_4) = \frac{1}{1 + \exp\left(\dfrac{x_1 - t}{x_2}\right)}$$

$$- \frac{1}{1 + \exp\left(\dfrac{x_3 - t}{x_4}\right)} \tag{3.5}$$

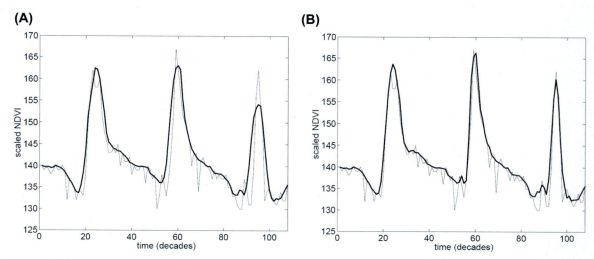

(A) scaled NDVI vs time (decades)

(B) scaled NDVI vs time (decades)

FIGURE 3.8 Results of adaptive SG filtering with different sliding window lengths (Jönsson and Eklundh, 2004). (A) the sliding window of 5 and (B) the sliding window of 3.

where x_1 is the position of the left inflection point and x_2 is the slope of the left inflection point of the basis function; similarly, x_3 is the position of the right inflection point and x_4 is the slope of the right inflection point. As in the asymmetric Gaussian function fitting, specified value ranges are also required for parameters to ensure smooth results in the model simulation.

MODIS LAI time series data were used for comparing the above fitting methods by Gao et al. (2008) (Fig. 3.9). The asymmetric Gaussian function fitting was applied to smoothing and gap filling of the MODIS LAI product to generate high-quality LAI products with temporal and spatial continuity.

3.2.2 Ecosystem-dependent temporal interpolation technique

Within a small region, pixels of the same ecological category show similar growth and phenological behaviors, which is the basis for such categorization in describing how vegetation parameters of an ecological category change over time. Each ecological category enters consecutive phenological stages (green-up, full maturity, senescence, and dormancy) at approximately the same time. Therefore, the parameters

at pixel level in the region will represent very similar ecological behaviors. However, under different growth conditions, discrepancies can be found between pixels in terms of the magnitude of ecological behavior variations. Based on these inherent properties, Moody et al. (2005) proposed an ecosystem-dependent temporal interpolation technique.

This method mainly uses pixel-level and regional ecosystem—dependent phenological behavioral curves to obtain high-quality time series data of pixel parameters. Using this algorithm, Moody et al. (2005) interpolated the missing data in the standard MOD43B3 albedo product. Fig. 3.10 shows the MOD43B3 white-sky albedo product over four seasons. Fig. 3.11 shows the albedo product after interpolation. The albedo product obtained using ecosystem-dependent temporal interpolation is not only able to maintain the same quality as that of the original data but also yields the statistical properties of the filled data and an estimate of the quality of the processed data.

Ding et al. (2017) improved the ecosystem-dependent temporal interpolation technique to address the uncertainties in interpolating missing data in satellite-derived LAI time series for heterogeneous landscape, particularly

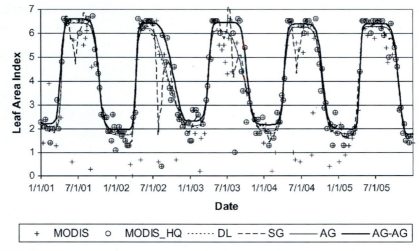

FIGURE 3.9   Comparison of results obtained by several fitting methods (Gao et al., 2008).

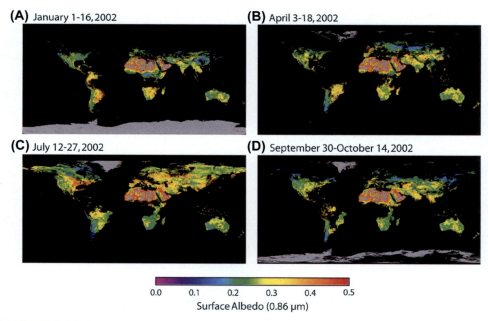

FIGURE 3.10   Global MOD43B3 white-sky albedo for four 16-day periods: (A) January 1—16, (B) April 3—18, (C) July 12—27, and (D) September 30—October 14, 2002 (Moody et al., 2005).

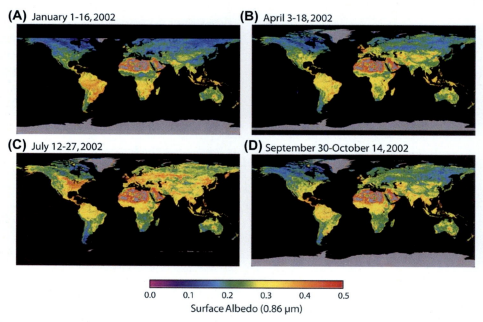

FIGURE 3.11   Spatially complete white-sky albedo after the temporal interpolation technique was applied to four 16-day periods: (A) January 1—16, (B) April 3—18, (C) July 12—27, and (D) September 30—October 14, 2002 (Moody et al., 2005).

heterogeneous grasslands. The improved algorithm can more accurately predict missing LAI value for different seasons and proportions of missing data (Ding et al., 2017).

3.2.3 Temporal spatial filter algorithm

The MODIS standard LAI product is not continuous in time and space. To fill gaps and improve product quality, Fang et al. (2008) proposed a temporal spatial filter (TSF) algorithm to integrate both the spatial and temporal characteristics of different plant functional types and generate continuous LAI data. The purpose of the TSF algorithm is to process pixels with low-quality and missing data in the original products (quality control identification (QC) > 32). The algorithm flowchart is shown in Fig. 3.12 and comprises three main steps:

(1) Calculate the background value. For vegetation types with the same function, the pixel with missing data is first supplemented by a multiyear mean. If the pixel has no value for all years, the pixel does not have a multiyear mean and instead the pixel's background value is calculated by vegetation continuous field—ecological curve fitting (VCF-ECF). The new VCF-ECF emphasizes the dependence of time series data for each target pixel on regional VCF-dependent phenological behaviors, thereby maintaining the temporal and spatial integrity at pixel level.

(2) Calculate the observed value. The original MODIS LAI data are treated as observed values. When $32 \leq QC < 128$ (LAI is obtained by an alternate algorithm), the original MODIS LAI value can be considered as an observed value. The same quality identification is adopted to calculate the LAI variance of vegetation types with the same function. When $QC > 128$, there will be no retrieved value of LAI, and local time filtering is needed to acquire the observed

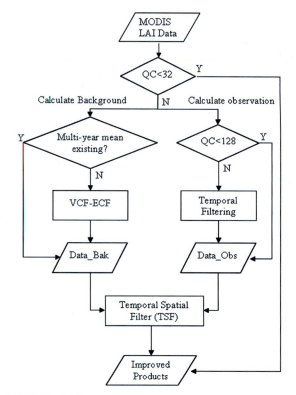

FIGURE 3.12 Temporal spatial filter flowchart. *Sourced from Fang, H., Liang, S., Townshend, J.R., Dickinson, R.E., 2008. Spatially and temporally continuous LAI data sets based on and integrated filtering method: examples from North America. Remote Sens. Environ. 112, 75–93.*

value of LAI in an adjacent time. If no desired QC value can be found in an adjacent time, then filtering, such as SG filtering (see Section 3.2.1.1), is used to calculate the observed value from the background values.

(3) Time-space filtering. The target value x_a is calculated using the background value x_b and the observed value x_o is obtained by the above two steps as follows:

$$x_a(r_i) = \frac{E_b^{-2}x_b(r_i) + E_o^{-2}w(r_i, r_j)\{x_b(r_i) + [x_o(r_j) - x_b(r_j)]\}}{E_b^{-2} + E_o^{-2}}$$

(3.6)

where E_b^2 and E_o^2 are the error variances of the background and observed values. Assuming that errors in the background and observed values are homogeneous and not spatially related, E_b^2 and E_o^2 are considered as locally independent. $w(r_i, r_j)$ is the weighting function, which is dependent on the time distance $d_{i,j}$ between point r_i and point r_j:

$$w(r_i, r_j) = \max\left(0, \frac{R^2 - d_{i,j}^2}{R^2 + d_{i,j}^2}\right) \qquad (3.7)$$

where R is a prescribed influence radius. When processing MODIS LAI data, Fang et al. (2008) set the influence radius as 16.

Fang et al. (2008) employed the TSF algorithm to process the MODIS LAI product and generated a continuous LAI product for North America. Fig. 3.13A shows the MODIS LAI product for North America on the 225th day of 2000. Because of cloud cover, especially at high latitudes, the MODIS LAI product has a lower quality. Processed LAI data are shown in Fig. 3.13B. It can be seen that the LAI product has improved spatial integrity after time-space filtering. These results confirm that the temporal and spatial continuity of the LAI product after TSF filtering is significantly improved.

The TSF algorithm can also be applied when processing other land surface parameters. Fang et al. (2007) used the TSF method when filtering the MODIS albedo product to generate a 1-km resolution albedo product with spatial continuity in North America.

3.2.4 Smoothing and gap-filling algorithm based on the wavelet transform

The wavelet transform (WT) can be used to analyze signals in time−frequency space and reduce noise, while retaining the important components in the original signals. In the past 20 years, WT has become a very effective tool in signal processing. Currently, WT is widely used in numerous remote sensing applications,

such as image fusion (see Chapter 4, Ranchina et al., 2003), the extraction of hyperspectral data characteristics (Pu and Gong, 2004), and phenological detection (Sakamoto et al., 2005).

Lu et al. (2007) proposed an algorithm for applying the WT to eliminate noise in time series observations and to fill missing data. First, based on quality identification and blue band data, linear interpolation is performed on the time series data; then, the time series data are decomposed into different time scales, and a set of adjacent time scales with the highest correlation is used to construct the new time series data.

Figs. 3.14 and 3.15 show time series data for the MODIS LAI and albedo products before and after WT, respectively. Compared with the original data, the reconstructed LAI and albedo time series data are relatively smooth, have no major fluctuations, and are more consistent with the variations of these parameters in nature.

3.2.5 Time series surface reflectance reconstruction

Currently, several datasets based on AVHRR data were produced using different data processing methods (James and Kalluri, 1994; Los et al., 2000; Pedelty et al., 2007; Tucker et al., 2005; Pinzon and Tucker, 2014). These datasets, with various spatial and temporal resolutions, have different temporal extent. Among them, the representative dataset is now being routinely generated by the Land Long-Term Data Record (LTDR) project, the goal of which is to produce a global land surface climate data record (CDR) from the AVHRR, the Moderate Resolution Imaging Spectroradiometer (MODIS), and the Visible Infrared Imaging Radiometer Suite (VIIRS) instruments using a MODIS-like operational production approach (Pedelty et al., 2007). The LTDR project has created daily surface reflectance and NDVI products from AVHRR observations.

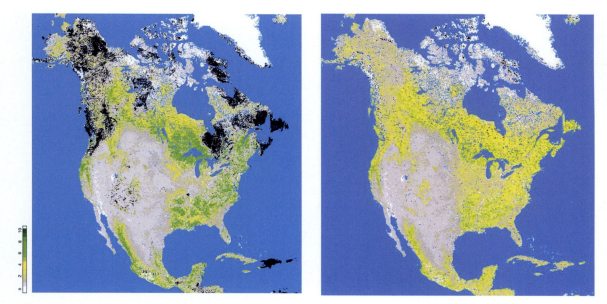

FIGURE 3.13   Comparison of MODIS LAI products before and after temporal spatial filter (TSF) filtering: (A) MODIS LAI (Day 225, 2000) and (B) LAI map after TSF filtering.

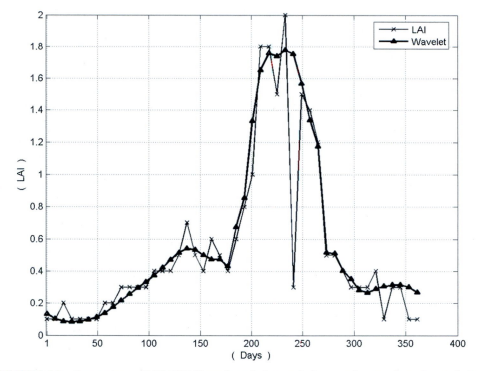

FIGURE 3.14   Comparison of MODIS LAI products before and after wavelet transform (Lu et al., 2007).

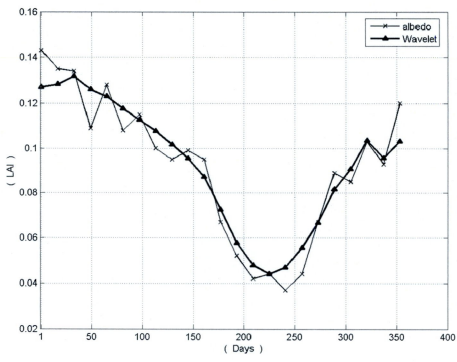

FIGURE 3.15   Comparison of MODIS albedo products before and after wavelet transform (Lu et al., 2007).

Although the quality of the LTDR AVHRR products was greatly improved over earlier versions (Nagol et al., 2014; Sobrino and Julien, 2016; Tian et al., 2015), some cloudy or partially cloudy pixels remained in the surface reflectance and NDVI products, and on some occasions, several days of surface reflectance and NDVI values were missing from the time series. Residual cloud and aerosol contamination in the surface reflectance and NDVI products significantly limits their applications of land surface monitoring and results in temporal and spatial inconsistencies in subsequent downstream products. Hence, it has been widely recognized that consistent and gap-filled land surface reflectance and NDVI time series are fundamentally important.

As mentioned above, a variety of methods were developed to reconstruct temporal NDVI trajectories (Chen et al., 2004; Hermance, 2007;

Julien et al., 2006; Lu et al., 2007; Viovy et al., 1992; Xiao et al., 2015). However, there are only a few ways to reconstruct surface reflectance time series. Tang et al. (2013) developed a time series cloud detection (TSCD) algorithm to remove cloud contamination in the MODIS surface reflectance product (MOD09A1) and fill in missing pixels. The TSCD algorithm performs very well, particularly when the land surface is stable or changing only slowly (Tang et al., 2013). However, the TSCD algorithm is dependent on surface reflectance in the blue band and other auxiliary information, which makes it difficult to reprocess the AVHRR surface reflectance data.

In view of the shortcomings of currently available NDVI and surface reflectance reconstruction algorithms, Xiao et al. (2015) developed a temporally continuous vegetation indices–based land surface reflectance reconstruction (VIRR) method to reconstruct time series of surface reflectance. The daily LTDR AVHRR surface reflectance data were first aggregated into 8-day intervals to maintain a temporal resolution consistent with the MODIS and VIIRS surface reflectance data and the aggregated 8-day surface reflectance data were used to calculate NDVI. Compared with surface reflectance, NDVI exhibits obvious seasonal changes that make it relatively easy to reconstruct temporally continuous NDVI. Considering the negative bias of NDVI when surface reflectance values are contaminated by clouds, NDVI upper envelopes were constructed. The cloud-contaminated surface reflectance values were identified using the NDVI time series and its upper envelope. Then the surface reflectance time series was further reconstructed from cloud-free surface reflectance values by incorporating the NDVI upper envelopes as constraints. Fig. 3.16 shows the general process of the VIRR method to reconstruct temporally continuous NDVI and surface reflectance from LTDR AVHRR data. Detailed descriptions of the important parts of the VIRR algorithm are given below.

3.2.5.1 Surface reflectance screening

An initial screening process was applied to the daily LTDR AVHRR surface reflectance data to determine whether the surface reflectance values were valid. The characteristic feature of vegetated target spectra is minimum reflectance in the red band due to strong absorption of photosynthetically active radiation by chlorophyll in green leaves and maximum reflectance in the NIR band because of strong reflection by leaf cells. However, the NIR reflectance in cloud-contaminated or snow-covered pixels is lower than reflectance in the red band (Luo et al., 2008). Generally, cloud contamination and snow cover are considered as noise for retrieval of land surface parameters such as LAI and FAPAR. Therefore, for each data point, the surface reflectance values were assumed to be invalid if the reflectance in the red band was

higher than the NIR reflectance. At the same time, the two-band enhanced vegetation index (EVI2) should be lower than the NDVI under normal conditions (Jiang et al., 2008; Zhang, 2015). If the EVI2 values were higher than the NDVI values, the surface reflectance values were also assumed to be invalid. In this study, only valid surface reflectance values were used for compositing.

3.2.5.2 Surface reflectance composition

The daily surface reflectance data were formed into 8-day composite images from 3 days before through 4 days after each composite time. If the number of data points with valid surface reflectance values over an 8-day compositing period was two or more, the CV-MVC method was used. The view zenith angles were taken for all data points with valid surface reflectance values over the 8-day compositing period, after which these view zenith angles were sorted in ascending order. Then, the NDVI values were computed for the data points with the two smallest view zenith angles, after which the surface reflectance with the higher NDVI value was selected (Van Leeuwen et al., 1999). If only one valid surface reflectance value was available over the composite period, this surface reflectance value was automatically selected to represent the period. If all surface reflectance values during the 8-day period were invalid, a multiyear mean was calculated to fill the gap.

3.2.5.3 NDVI reconstruction

Composite surface reflectance data are still susceptible to spurious data points. Therefore, the NDVI values derived from surface reflectance are also temporally discontinuous. Considering the negative bias of NDVI when surface reflectance values are contaminated by clouds, a penalized least squares regression based on a three-dimensional discrete cosine transform (DCT-PLS) developed by Garcia (2010) was used to reconstruct the NDVI upper envelopes.

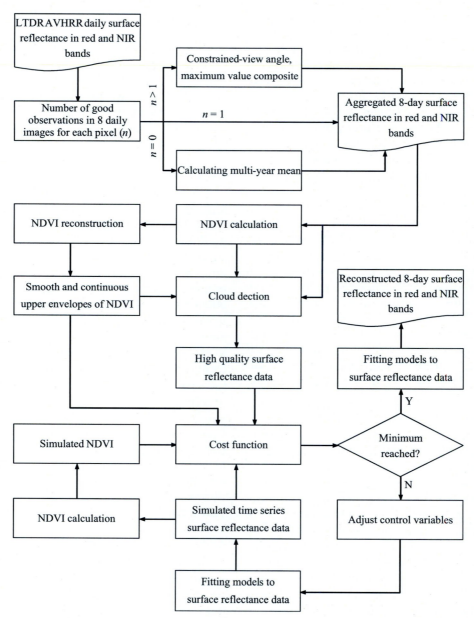

FIGURE 3.16   Schematic description of the method used to reconstruct normalized difference vegetation index and surface reflectance time series from LTDR AVHRR.

Let X be a vector that contains a series of equally spaced NDVI values $x_i, i = 1, 2, \cdots, n$. Let $\mathbf{W}$ be the diagonal matrix that contains the weights, w_i, corresponding to the NDVI value x_i. The DCT-PLS consists in minimizing the following functional to find the best smoothing estimate, $\widehat{X}$, of X (Garcia, 2010).

$$F\left(\widehat{X}\right) = \left\| \mathbf{W}^{1/2}\left(\widehat{X} - \mathbf{X}\right) \right\|^2 + s\|\mathbf{D}\widehat{X}\|^2 \quad (3.8)$$

where D is the Laplace operator, and s is a real positive scalar that controls the degree of smoothing. Minimizing on $\widehat{X}$ can be achieved using an iterative procedure. With type-2 DCT and inverse discrete cosine transform (IDCT), the $\widehat{X}$ can be expressed as

$$\widehat{X}_{\{k+1\}} = \mathbf{IDCT}\left(\Gamma \, \mathbf{DCT}\left(\mathbf{W}\left(\mathbf{X} - \widehat{X}_{\{k\}} \right) + \widehat{X}_{\{k\}} \right) \right) \quad (3.9)$$

where $\widehat{X}_{\{k\}}$ refers to $\widehat{X}$ calculated at the kth iteration step, and Γ is a diagonal matrix with the components

$$\Gamma_{i,i} = \left(1 + s(2 - 2\cos((i-1)\pi/n))^2 \right)^{-1} \quad (3.10)$$

Note that atmospheric effects on surface reflectance generally cause an increase in reflectance in the red band rather than in the NIR band, resulting in decreases in the NDVI. To account for negatively biased noise, we used the weight function

$$w_i = \begin{cases} 1.0 & u_i > 0.0 \\ \left(1.0 - \left(\dfrac{u_i}{4.685} \right)^2 \right)^2 & -1.0 < \dfrac{u_i}{4.685} < 0.0 \\ 0.0 & \dfrac{u_i}{4.685} < -1.0 \end{cases}$$

$$(3.11)$$

where u_i is the studentized residual, calculated by

$$u_i = r_i \left(1.4826 \ \mathrm{MAD}(\mathbf{r}) \sqrt{1 - \frac{\sqrt{1 + \sqrt{1 + 16\,s}}}{\sqrt{2}\sqrt{1 + 16\,s}}} \right)^{-1} \quad (3.12)$$

where $r_i = x_i - \widehat{x}_i$ is the residual of the ith observation, and MAD denotes the median absolute deviation (Rousseeuw and Croux, 1993). This iteratively reweighted procedure leads to a smoothed curve adapted to the upper envelope of the NDVI in the time series.

3.2.5.4 Cloud detection of surface reflectance

Atmospheric effects on surface reflectance generally cause negatively biased noise within the vegetation index values. The smoothing method described above was first used to calculate the upper envelopes of NDVI, denoted by NDVI_Env. If NDVI values at each time satisfy the following conditions, the surface reflectance data were deemed to be contaminated by clouds or other factors. Otherwise, they were considered to be cloud free and of high quality.

$$|\mathrm{NDVI} - \mathrm{NDVI\_Env}| > \alpha \times \mathrm{NDVI\_Env} \quad (3.13)$$

where α is a threshold and was set to 0.4 in this study.

The method was applied to check the time series surface reflectance, and any data points assessed as contaminated were filtered out.

3.2.5.5 Surface reflectance reconstruction

The surface reflectance time series was reconstructed according to cloud-free surface reflectance values by incorporating the reconstructed NDVI upper envelopes as constraints.

Let $(t_i, \mathbf{I}_i), i = 1, 2, \cdots, m$ be a series of data points, where t_i is time and $\mathbf{I}_i$ is the surface reflectance in the various bands. In this study, only surface reflectance in the red and NIR bands were reconstructed. Therefore, $\mathbf{I}_i = \left(\rho_i^{\text{red}}, \rho_i^{\text{NIR}}\right)^T$, where ρ_i^{red} and ρ_i^{NIR} are the surface reflectance for the ith position in the red and NIR bands, respectively. For each data point $\mathbf{I}_i$, a quadratic polynomial function, $f(t) = at^2 + bt + c$, was fitted to all $2n + 1$ surface reflectance values in each band in a moving time window using the least squares method. An optimization method searched for the trajectories of the quadratic polynomial functions that best fitted the surface reflectance in the given time window, i.e., minimizing

$$
\begin{aligned}
J(X_i) = &\sum_{j=-n}^{n} \left(f_{\text{red}}(t_{i+j}) - \rho_{i+j}^{\text{red}}\right)^2 \\
&+ \sum_{j=-n}^{n} \left(f_{\text{NIR}}(t_{i+j}) - \rho_{i+j}^{\text{NIR}}\right)^2 \\
&+ \sum_{j=-n}^{n} \left(\text{NDVI\_Sim}_{i+j} - \text{NDVI\_Env}_{i+j}\right)^2
\end{aligned}
$$

$$(3.14)$$

where $f_{\text{red}}(t_i)$ and $f_{\text{NIR}}(t_i)$ are the quadratic polynomial functions for the red and NIR bands, respectively; $X_i = (a_{\text{red}}, b_{\text{red}}, c_{\text{red}}, a_{\text{NIR}}, b_{\text{NIR}}, c_{\text{NIR}})^T$ is a control vector consisting of the coefficients of the quadratic polynomial functions; NDVI_Sim_i is a simulated NDVI calculated using the values of the quadratic polynomial functions; and n is the size of the moving time window. In this study, n was set to 2. Then the reconstructed surface reflectance values at the ith position were set to the values of these quadratic polynomial functions according to the optimal values of the control variables. Detailed information about the reconstruction method has been given by Xiao et al. (2015).

Xiao et al. (2017) used the VIRR method to generate GLASS AVHRR NDVI and surface reflectance products from 1982 to 2015. To illustrate the performance of the VIRR algorithm, time series of the GLASS AVHRR NDVI and surface reflectance products were compared with

those of the LTDR AVHRR NDVI and surface reflectance products over a sample of sites with different biome types, and the GLASS AVHRR NDVI product was also compared with the reconstructed NDVI from LTDR AVHRR data using the SG filter (denoted by SG AVHRR) over these sites. In addition, the GLASS AVHRR surface reflectance product was compared with the LTDR AVHRR surface reflectance product in space.

Fig. 3.17 shows time series of the GLASS AVHRR NDVI and surface reflectance, the LTDR AVHRR NDVI and surface reflectance over a sample of sites with different biome types for 1982−2015. And time series of the SG AVHRR NDVI are also shown in Fig. 3.17 for comparison to evaluate the performance of the refined VIRR algorithm. There were gaps in the LTDR AVHRR surface reflectance in late 1994 and 2000. Fig. 3.17 demonstrates that the refined VIRR algorithm provides reasonable NDVI and surface reflectance values for the missing pixels over these sites, and the SG AVHRR NDVI also shows complete temporal profiles. Over these sites, most of the LTDR AVHRR surface reflectance values in the red band are greater than 0.2, and the corresponding NDVI values are around zero, which may result from serious contamination due to residual clouds. The refined VIRR algorithm successfully removed surface reflectance values contaminated by residual clouds and reconstructed temporally continuous time series of NDVI and land surface reflectance (Fig. 3.17). Excellent agreement was achieved between the time series of GLASS AVHRR NDVI and the upper envelope of the time series NDVI calculated from the LTDR AVHRR surface reflectance product, and the time series of GLASS AVHRR surface reflectance in the red and NIR bands were in good agreement with the lower envelopes of the LTDR AVHRR surface reflectance over these sites.

The GLASS AVHRR and SG AVHRR NDVI profiles demonstrated excellent agreement for these years except over the Yucheng site.

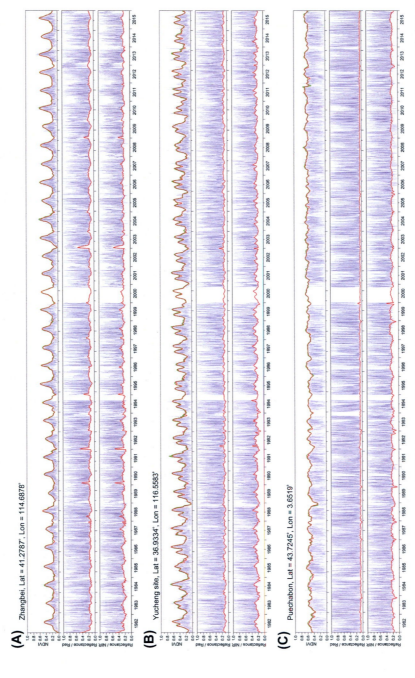

FIGURE 3.17 Time series of normalized difference vegetation index (NDVI) and surface reflectance in the red and near-infrared bands from GLASS AVHRR and LTDR AVHRR for the (A) Zhangbei, (B) Yucheng, (C) Puechabon, (D) Larose, (E) Wankama, and (F) Turco sites from 1982 to 2015. For comparison, the reconstructed NDVI profiles using the SG filter over these sites are also shown.

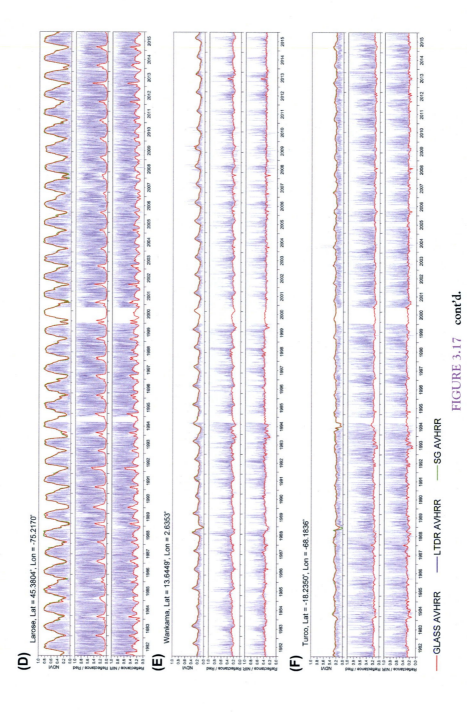

FIGURE 3.17 cont'd.

The biome type for Yucheng site is cropland with double annual vegetation seasons. The SG AVHRR NDVI profile over this site was not able to follow sudden decrease of underlying data values between two growing seasons for some years.

Examples of the reconstructed surface reflectance are shown in Fig. 3.18. Fig. 3.18A shows an RGB image of the GLASS AVHRR surface reflectance on January 9, 2010, which was composited from the LDTR AVHRR daily surface reflectance from January 6 to 13, 2010, and then reconstructed using the refined VIRR method. For comparison, an RGB image of the LDTR AVHRR surface reflectance on January 9, 2010, is shown in Fig. 3.18B. The images were produced with an RGB color scheme that uses the red band (as red color), the NIR band (as green color), and the red band (as blue color), and the same color enhancement was used. In Fig. 3.18B, the LDTR AVHRR surface reflectance is contaminated with residual clouds, and large clouds can in fact be seen over the image. The RGB image in Fig. 3.18A shows that the refined VIRR method effectively removed the residual clouds in the LDTR AVHRR surface reflectance. Fig. 3.18C shows an RGB image of the GLASS AVHRR surface reflectance on July 12, 2010, and an RGB image of the LDTR AVHRR surface reflectance on July 12, 2010, is shown in Fig. 3.18D. The images were produced with the same RGB color scheme and color enhancement as in Fig. 3.18A. In Fig. 3.18D, most areas of the image were covered by residual clouds. The refined VIRR algorithm effectively identified these residual clouds based on the temporally continuous NDVI. The RGB image of the GLASS AVHRR surface reflectance from the refined VIRR method (Fig. 3.18C) shows that almost all contamination due to clouds was removed.

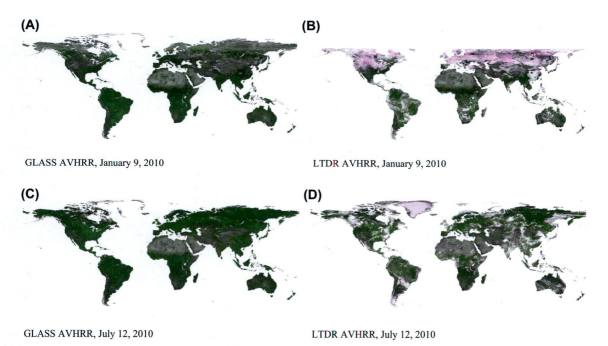

(A)

GLASS AVHRR, January 9, 2010

(B)

LTDR AVHRR, January 9, 2010

(C)

GLASS AVHRR, July 12, 2010

(D)

LTDR AVHRR, July 12, 2010

FIGURE 3.18 RGB images of the GLASS AVHRR and LTDR AVHRR surface reflectance on January 9 and July 12, 2010. The RGB images were produced with an RGB color scheme that used the red band (as red color), the near-infrared band (as green color), and the red band (as blue color), and the same color enhancement was used.

The VIRR method is a fully automated procedure and provides fast smoothing of data. Therefore, the method is particularly suitable for operational applications with a long time series of global data. Another advantage of the refined VIRR method is that it simultaneously reconstructs time series of surface reflectance in the red and NIR bands, which avoids inconsistent estimation of surface reflectance in the different bands.

3.3 Summary

Time series of remotely sensed land surface parameters act as important model inputs when studying global environmental change. However, noise in the data impairs the characterization and prediction performance of models. Through multiphase data synthesis, smoothing, and gap filling, time series of land surface parameters can be reconstructed to improve the quality of the existing land surface parameter products.

On the other hand, these methods have their respective shortcomings. For instance, empirical analysis is required to determine the smoothing window size in SG filtering. Current synthesis products have lower atmospheric interference than the originals, but atmospheric influences are not totally eliminated during the synthesis process. Thus, these compositing techniques may not be applicable for improving land surface parameter products in areas with data missing over long time periods (such as tropical rain forests).

References

Cabral, A., De Vasconcelos, M.J.P., Pereira, J.M.C., Bartholome, E., Mayaux, P., 2003. Multi-temporal compositing approaches for SPOT-4 VEGETATION. Int. J. Remote Sens. 24, 3343–3350.

Cao, R.Y., Chen, Y., Shen, M.G., Chen, J., Zhou, J., Wang, C., Yang, W., 2018. A simple method to improve the quality of NDVI time-series data by integrating spatiotemporal information with the Savitzky-Golay filter. Remote Sens. Environ. 217, 244–257.

Carreiras, J.M.B., Pereira, J.M.C., 2005. SPOT-4 VEGETATION multitemporal compositing for land cover change studies over tropical regions. Int. J. Remote Sens. 26, 1323–1346.

Carreiras, J.M.B., Pereira, J.M.C., Shimabukuro, Y.E., Stroppiana, D., 2003. Evaluation of compositing algorithms over the Brazilian Amazon using SPOT-4 VEGETATION data. Int. J. Remote Sens. 24, 3427–3440.

Chen, J., Jonsson, P., Tamura, M., Gu, Z., Matsushita, B., Eklundh, L., 2004. "A simple method for reconstructing a high-quality NDVI time-series data set based on the Savitzky–Golay filter. Remote Sens. Environ. 91, 332–344.

Chen, J.M., Deng, F., Chen, M.Z., 2006. Locally adjusted cubic-spline capping for reconstructing seasonal trajectories of a satellite-derived surface parameter. IEEE Trans. Geosci. Remote Sens. 44 (8), 2230–2238.

Chuvieco, E., Ventura, G., Martin, M.O., Gomez, I., 2005. Assessment of multitemporal compositing techniques of MODIS and AVHRR images for burned land mapping. Remote Sens. Environ. 94, 450–462.

Cihlar, J., Manak, D., D'Iorio, M., 1994. Evaluation of compositing algorithms for AVHRR over land. IEEE Trans. Geosci. Remote Sens. 32, 427–437.

Ding, C., Liu, X.N., Huang, F., 2017. Temporal interpolation of satellite-derived leaf area index time series by introducing spatial-temporal constraints for heterogeneous grasslands. Remote Sens. 9, 12.

Fang, H., Liang, S., Townshend, J.R., Dickinson, R.E., 2008. Spatially and temporally continuous LAI data sets based on and integrated filtering method: examples from North America. Remote Sens. Environ. 112, 75–93.

Fang, H., Liang, S., Kim, H., Townshend, J.R., Schaaf, C.L., Strahler, A.H., Dickinson, R.E., 2007. Developing a spatially continuous 1 km surface albedo data set over North America from Terra MODIS products. J. Geophys. Res. Atmos. 112, D20206.

Fernandes, R., Butson, C., Leblanc, S.G., Latifovic, R., 2003. Landsat-5 TM and Landsat-7 ETM+based accuracy assessment of leaf area index products for Canada derived from SPOT-4 VEGETATION data. Can. J. Remote Sens. 29 (2), 241–258.

Gao, F., Morisette, J.T., Wolfe, R.E., Ederer, G., Pedelty, J., Masuoka, E., Myneni, R., Tan, B., Nightingale, J., 2008. An algorithm to produce temporally and spatially continuous MODIS-LAI time series. IEEE Geosci. Remote Sens. Lett. 5 (1), 60–64.

Garcia, D., 2010. Robust smoothing of gridded data in one and higher dimensions with missing values. Comput. Stat. Data Anal. 54, 1167–1178.

Hermance, J.F., 2007. Stabilizing high-order, non-classical harmonic analysis of NDVI data for average annual models by damping model roughness. Int. J. Remote Sens. 28, 2801–2819.

Holben, B.N., 1986. Characteristics of maximum-value composite images for temporal AVHRR data. Int. J. Remote Sens. 7, 1435–1445.

Illera, P., Fernández, A., Delgado, J.A., 1996. Temporal evolution of the NDVI as an indicator of forest fire danger. Int. J. Remote Sens. 17 (6), 1093–1105.

James, M.E., Kalluri, S.N.V., 1994. The Pathfinder AVHRR land data set: an improved coarse resolution data set for terrestrial monitoring. Int. J. Remote Sens. 15 (17), 3347–3363.

Jiang, Z.Y., Huete, A.R., Didan, K., Miura, T., 2008. Development of a two-band enhanced vegetation index without a blue band. Remote Sens. Environ. 112, 3833–3845.

Jönsson, P., Eklundh, L., 2002. Seasonality extraction by function fitting to timeseries of satellite sensor data. IEEE Trans. Geosci. Remote Sens. 40 (8), 1824–1832.

Jönsson, P., Eklundh, L., 2004. TIMESAT—a program for analyzing time-series of satellite sensor data. Comput. Geosci. 30, 833–845.

Julien, Y., Sobrino, J.A., Verhoef, W., 2006. Changes in land surface temperatures and NDVI values over Europe between 1982 and 1999. Remote Sens. Environ. 103, 43–55.

Kasischke, E.S., French, N.H.F., Harrell, P., Christensen, N.L., Ustin, S.L., Barry, D., 1993. Monitoring of wildfires in Boreal Forests using large area AVHRR NDVI composite image data. Remote Sens. Environ. 45, 61–71.

Los, S.O., Collatz, G.J., Sellers, P.J., Malmstrom, C.M., Pollack, N.H., DeFries, R.S., et al., 2000. A global 9-yr biophysical land surface dataset from NOAA AVHRR data. J. Hydrometeorol. 1 (2), 183–199.

Lu, X., Liu, R., Liu, J., Liang, S., 2007. Removal of noise by wavelet method to generate high quality temporal data of terrestrial MODIS products. Photogramm. Eng. Remote Sens. 73, 1129–1139.

Luo, Y., Trishchenko, A.P., Khlopenkov, K.V., 2008. Developing clear-sky, cloud and cloud shadow mask for producing clear-sky composites at 250-meter spatial resolution for the seven MODIS land bands over Canada and North America. Remote Sens. Environ. 112, 4167–4185.

Moody, E.G., King, M.D., Platnick, S., Schaaf, C.B., Gao, F., 2005. Spatially complete global spectral surface albedos: value-add datasets derived from Terra MODIS land products. IEEE Trans. Geosci. Remote Sens. 43, 144–157.

Nagol, J.R., Vermote, E.F., Prince, S.D., 2014. Quantification of impact of orbital drift on inter-annual trends in AVHRR NDVI data. Remote Sens. 6, 6680–6687.

Pedelty, J., Vermote, E., Devadiga, F.S., Roy, D., Schaaf, C., Privette, J., et al., 2007. Generating a long-term land data record from the AVHRR and MODIS instruments.

In: IEEE International Geoscience and Remote Sensing Symposium, pp. 1021–1025.

Pereira, J.M.C., Sa, A.C.L., Sousa, A.M.O., Martín, M.P., Chuvieco, E., 1999. Regional-scale burnt area mapping in Southern Europe using NOAA-AVHRR 1 km data. In: Chuvieco, E. (Ed.), Remote Sensing of Large Wildfires in the European Mediterranean Basin. Berlin Springer-Verlag, pp. 139–155.

Peters, A.J., Walter-Shea, E.A., Ji, L., Viña, A., Hayes, M., Svodoba, M.D., 2002. Drought monitoring with NDVI-based standardized vegetation index. Photogramm. Eng. Remote Sens. 62 (1), 71–75.

Pinzon, J., Tucker, C., 2014. A non-stationary 1981-2012 AVHRR NDVI3g time series. Remote Sens. 6, 6929–6960.

Potter, C.S., Brooks, V., 2000. Global analysis of empirical relations between annual climate and seasonality of NDVI. Int. J. Remote Sens. 19 (15), 2921–2948.

Pu, R.L., Gong, P., 2004. Wavelet transform applied to EO-1 hyperspectral data for forest LAI and crown closure mapping. Remote Sens. Environ. 91, 212–224.

Qi, J., Huete, A.R., Hood, J., Kerr, Y., 1993. Compositing multitemporal remote sensing data sets. Pecora 12, 206–213. Sioux Falls7 American Society of Photogrammetry and Remote Sensing.

Qi, J., Kerr, Y., 1997. On current compositing algorithms. Remote Sens. Rev. 15, 235–256.

Ranchina, T., Aiazzib, B., Alparonec, L., Barontib, S., Wald, L., 2003. Image fusion-The ARSIS concept and some successful implementation schemes. ISPRS J. Photogrammetry Remote Sens. 58, 4–18.

Roerink, G.J., Menenti, M., Verhoef, W., 2000. Reconstructing cloudfree NDVI composites using Fourier analysis of time series. Int. J. Remote Sens. 21, 1911–1917.

Rousseeuw, P.J., Croux, C., 1993. Alternatives to the median absolute deviation. J. Am. Stat. Assoc. 88, 1273–1283.

Sakamoto, T., Yokozawa, M., Toritani, H., Shibayama, M., Ishitsuka, N., Ohno, H., 2005. A crop phenology detection method using time-series MODIS data. Remote Sens. Environ. 96, 366–374.

Savitzky, A., Golay, M.J.E., 1964. Smoothing and differentiation of data by simplified least squares procedures. Anal. Chem. 36, 1627–1639.

Schaaf, C.B., Gao, F., Strahler, A.H., Lucht, W., Li, X., Tsang, T., et al., 2002. First operational BRDF, albedo and nadir reflectance products from MODIS. Remote Sens. Environ. 83, 135–148.

Sellers, P., Tucker, C., Collatz, G., Los, S., Justice, C., Dazlich, D., Randall, D., 1994. A global 1 by 1 NDVI data set for climate studies. Part 2: the generation of global fields of terrestrial biophysical parameters from the NDVI. Int. J. Remote Sens. 15, 3519–3546.

Sobrino, J.A., Julien, Y., 2016. Exploring the validity of the long-term data record V4 database for land surface monitoring. IEEE J. Sel. Top. Appl. Earth Obs. Remote Sens. 9, 3607–3614.

Sousa, A.M.O., Pereira, J.M.C., Silva, J.M.N., 2002. Evaluating the performance of multitemporal image compositing algorithms for burned area analysis. Int. J. Remote Sens. 24, 1219—1236.

Tang, H., Yu, K., Hagolle, O., Jiang, K., Geng, X., Zhao, Y., 2013. Cloud detection method based on a time series of MODIS surface reflectance images. Int. J. Digit. Earth 6, 157—171.

Tian, F., Fensholt, R., Verbesselt, J., Grogan, K., Horion, S., Wang, Y., 2015. Evaluating temporal consistency of long-term global NDVI datasets for trend analysis. Remote Sens. Environ. 163, 326—340.

Tucker, C.J., Pinzon, J.E., Brown, M.E., Slayback, D.A., Pak, E.W., Mahoney, R., et al., 2005. An extended AVHRR 8-km NDVI dataset compatible with MODIS and SPOT vegetation NDVI data. Int. J. Remote Sens. 26, 4485—4498.

Van Leeuwen, W.J.D., Huete, A.R., Laing, T.W., 1999. MODIS vegetation index compositing approach: a prototype with AVHRR data. Remote Sens. Environ. 69, 264—280.

Vermote, E.F., Vermeulen, A., 1999. Atmospheric Correction Algorithm: Spectral Reflectances (MOD09). MODIS Algorithm Theoretical Basis Document.

Viovy, N., Arino, O., Belward, A.S., 1992. The best index slope extraction (bise)—a method for reducing noise in NDVI time-series. Int. J. Remote Sens. 13, 1585—1590.

Walthall, C.L., Norman, J.M., Welles, J.M., Campbell, G., Blad, B.L., 1985. Simple equation to approximate the bi-directional reflectance from vegetative canopies and bare soil surfaces. Appl. Opt. 24, 383—387.

Xiao, Z., Liang, S., Wang, T., Liu, Q., 2015. Reconstruction of satellite-retrieved land-surface reflectance based on temporally-continuous vegetation indices. Remote Sens. 7, 9844—9864.

Xiao, Z., Liang, S., Tian, X., Jia, K., Yao, Y., Jiang, B., 2017. Reconstruction of long-term temporally continuous NDVI and surface reflectance from AVHRR data. IEEE J. Sel. Top. Appl. Earth Obs. Remote Sens. 10 (12), 5551—5568.

Zhang, X., 2015. Reconstruction of a complete global time series of daily vegetation index trajectory from long-term AVHRR data. Remote Sens. Environ. 156, 457—472.

C H A P T E R

4

Atmospheric correction of optical imagery

© 2020 Elsevier Inc. All rights reserved.

Abstract

Quantitative remote sensing is an inevitable trend in the application of remote sensing in the 21st century. Its core role is to establish a quantitative relationship between the electromagnetic spectrum and the information obtained by the sensor and then use electromagnetic wave information to quantitatively detect all kinds of surface characteristic parameters. Quantitative remote sensing is based on the precise calibration of sensors and the atmospheric correction of remote sensing data. This chapter introduces the atmospheric effects on the process of remote sensing imaging and atmospheric correction to specifically discuss the utilization of atmospheric correction in processing remote sensing data. Section 1 presents the background of atmospheric effects. Section 2 discusses various algorithms for the estimation of aerosol properties, which are based on spectral, temporal, angular, spatial and polarization characteristics. Section 3 introduces the algorithms used for estimating atmospheric moisture content. Then, the influences of other atmospheric components and commonly used atmospheric correction models are illustrated. Finally, an application of atmospheric correction is presented.

4.1 Atmospheric effects

4.1.1 Atmospheric characterization in a quantitative remote sensing model

An optical remote sensing system can be divided into five subsystems (Fig. 4.1): the scene radiative transfer model, the atmospheric radiative transfer model, the navigation system, the sensor system, and the mapping and binning system. The scene radiative transfer model describes the relationship between surface radiative signals and surface characteristic parameters.

The atmospheric radiative transfer model characterizes the atmospheric impacts on surface radiative signals that are received by the remote sensors. The navigation system mainly involves the surface imaging systems of carried satellites, aircraft, and other sensor platforms. The sensor system mainly includes the sensor spectral response, spatial response, band division, noise treatment, and digital processing, and the mapping and binning system mainly involves projection transformation and imagery resampling.

The atmospheric radiative transfer model is a critical link that connects surface characteristic parameters with the signals received by remote sensors. This model is also considered to be an important factor that influences the quantitative retrieval of surface characteristic parameters using remote sensing imagery.

4.1.2 Atmospheric composition

The atmosphere is a medium consisting of a variety of gases and aerosols between satellite sensors and the Earth's surface. When electromagnetic waves are transmitted from the atmosphere to the Earth's surface and from the Earth's surface to sensors, the atmosphere is the channel through which they are transmitted. The lower atmosphere has a great influence on the transmission of electromagnetic radiation, mainly by changing the distribution of the radiation spectrum and energy intensity. The upper atmosphere (atmospheric height over 100 km) has little influence on electromagnetic radiation transmission in common remote sensing windows and can be neglected.

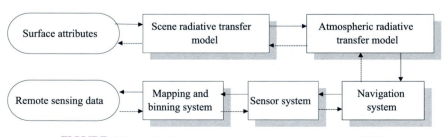

FIGURE 4.1   Optical remote sensing system models (Liang, 2004).

The atmosphere is a mixture of various gases and suspended liquid and solid impurities. An atmosphere free of water vapor and impurities is called dry-clean air. The atmosphere at Earth's surface is mainly composed of oxygen, nitrogen, and several inert gases, which together account for 99.9% of the total air. Excluding carbon dioxide and water vapor ozone, other compositions of these gases are stable in the troposphere. The water vapor and carbon dioxide concentrations vary with region, season, and weather conditions. In most cases, water vapor and carbon dioxide account for 0%–4% and 0.033% of the atmosphere, respectively, as shown in Table 4.1.

The large numbers of solid and liquid particles suspended in the atmosphere are called aerosols, which typically include haze, smoke, and fog. Aerosols are mainly produced by volcanic eruptions; ground dust; sandstorms; forest fires; various acidic particles; and industrial, transportation, construction, agricultural, and other production and living activities.

TABLE 4.1 Major atmospheric constituents.

Stable compositions		Variable compositions	
Composition	Volume ratio (%)	Composition	Volume ratio (%)
N_2	78.084	H_2O	0.04
O_2	20.948	O_3	12×10^{-4}
Ar	0.934	SO_2	0.001×10^{-4}
CO_2	0.033	NO_2	0.001×10^{-4}
Ne	18.18×10^{-4}	NH_3	0.001×10^{-4}
He	5.24×10^{-4}	NO	0.0005×10^{-4}
Kr	1.14×10^{-4}	H_2S	0.00005×10^{-4}
Xe	0.089×10^{-4}	Nitric acid vapor	Trace
H_2	0.5×10^{-4}		
CH_4	1.5×10^{-4}		
N_2O	0.27×10^{-4}		
CO	0.19×10^{-4}		

These aerosols can be categorized by particle size. Those with a particle size between 5.0×10^{-3} and 0.2 µm are called Aitken nuclei, those with a particle size from 0.2 to 1 µm are called large particles, and those with a particle size greater than 1 µm are called giant particles. In the troposphere, the aerosol concentration is subject to exponential decay with increasing height; in the stratosphere, the aerosol concentration is relatively stable.

4.1.3 Interaction between electromagnetic waves and the atmosphere

Atmospheric gas molecules, aerosols, cloud droplets, ice crystals, and other particles contain many charged electrons and protons. These charges are unevenly distributed in particles. When an electromagnetic wave touches a particle, the charges are forced to vibrate under the excitation of the electromagnetic wave, and then secondary electromagnetic waves are emitted in all directions. Secondary electromagnetic waves generated in this way are referred to as scattering. The energy stimulated by electromagnetic waves in particles does not transform into secondary electromagnetic waves scattering outward. Some electromagnetic energy is transformed into heat energy, and this process is called absorption. If polarization effects are disregarded and the particles can be considered to be isotropic, then the one-dimensional radiative transfer equation can be expressed as follows:

$$\mu \frac{dL(\tau, \mu, \varphi)}{d\tau} = L(\tau, \mu, \varphi) - \frac{\omega}{4\pi} \int_0^{2\pi}$$

$$\times \int_{-1}^{1} L(\tau, \mu_i, \varphi_i) P(\mu, \varphi, \mu_i, \varphi_i) d\mu_i d\varphi_i \quad (4.1)$$

where L is the radiance; (μ, φ) are the observation angle coordinates, where $\mu = \cos(\theta)$ and θ is the zenith angle; φ is the relative azimuth between the solar direction and the observation direction; τ is the aerosol optical thickness (AOT), which is defined by $\tau = \int_0^z \sigma_e(z)dz$, where $\sigma_e(z)$ is the extinction coefficient; ω is the single scattering albedo, which describes the probability that scattering will occur in a photonic collision with particles of that medium; and $P(\mu, \phi, \mu_i, \phi_i)$ is the scattering phase function describing the probability of photon scattering from other directions.

Molecular absorption refers to the conversion of radiant energy into oscillation energy that is molecularly excited. Scattering refers to the redistribution of incident energy in other directions, and the total effect shifts incident energy away from the scatter.

The atmospheric molecules that absorb electromagnetic waves mainly include O_2, O_3, H_2O, CO_2, CH_4, and N_2O. Of these molecules, ozone, carbon dioxide, and water vapor absorb solar radiation energy the most effectively. Ozone has a strong absorption band in the ultraviolet region (0.22−0.32 μm) and in the 0.6, 4.7, 9.6, and 14 μm spectral regions. Carbon dioxide is mainly distributed in the lower atmosphere, with absorption bands in the 1.4, 1.6, 2.0, 2.7, 4.3, 4.8, 5.2, and 15.0 μm spectral regions, of which the bands at 2.7, 4.3, and 15 μm are strong absorption bands. Water vapor has absorption bands at 0.94, 1.1, 1.38, 1.87, 2.7, 3.2, and 6.3 μm, which cover almost the entire infrared band.

Scattering redistributes energy in the incident direction toward other directions, and the total effect results in the transfer of energy away from the incident direction. Atmospheric scattering has a great influence on the transmission of electromagnetic waves. It reduces the intensity of direct sunlight and weakens the radiation of electromagnetic waves reaching the ground or that radiating outward from the ground. In addition, it also changes the direction of solar radiation and produces diffuse sky scattering light (also known as sky light or sky radiation), which enhances radiation at the ground and the "brightness" of the atmosphere itself.

The atmospheric scattering of electromagnetic waves is mainly divided into two categories: selective scattering, which is further divided into Rayleigh scattering and Mie scattering, and nonselective scattering. When the diameter of an atmospheric particle causing scattering is far smaller than the wavelength of the incident electromagnetic waves, Rayleigh scattering occurs. The scattering of visible light by oxygen, nitrogen, and other gas molecules in the atmosphere belongs to this category. The Rayleigh scattering intensity is inversely proportional to the fourth power of the wavelength, and forward scattering has equal intensity to backscattering. Rayleigh scattering is one of the main causes of radiometric distortion and blurring of remote sensing shortwave imagery. When the diameter of an atmospheric particle causing scattering is equal to the wavelength of the incident waves, Mie scattering occurs. Scattering by small particles and other aerosols suspended in the atmosphere belongs to this category. Forward Mie scattering is usually far greater than Mie backscattering. Under general atmospheric conditions, Rayleigh scattering plays a dominant role; however, when Mie scattering is superimposed on Rayleigh scattering, the sky becomes gloomy. When the diameter of the atmospheric particle causing scattering is much larger than the wavelength of the incident waves, nonselective scattering occurs, and its scattering intensity is not related to the wavelength. Scattering by clouds, fog, water droplets, and dust in the atmosphere falls into this category.

The physical matter in the atmosphere attenuates electromagnetic signals due to absorption, refraction, and other phenomena. Atmospheric effects are among the key issues in quantitative remote sensing, and they can interfere with the quantitative application of remote sensing. The removal of this interference is necessary to quantitatively use remotely sensed data.

4.1.4 Major aspects of atmospheric correction

Atmospheric correction mainly consists of two parts: the estimation of atmospheric parameters and the retrieval of surface reflectance. If the surface is Lambertian and all of the atmospheric parameters are known, then remote sensing imagery can be calculated to directly retrieve surface reflectance. Based on the radiative transfer theory and assuming that the target is a uniform Lambertian surface, the radiance received by a sensor at the top of the atmosphere (TOA) can be expressed as follows:

$$L = L_0 + \frac{\rho}{1 - s\rho} \cdot \frac{TF_d}{\pi} \qquad (4.2)$$

where L_0 is the atmospheric radiation path in the case of no surface reflection, T is the transmittance from the surface to the sensor, s is the atmospheric spherical reflectance, ρ is the surface target reflectance, and F_d is the downward radiation flux reaching the surface. According to the equation, the radiance received by the sensors is given by L, L_0, and s, and TF_d/π can be calculated by the radiative transfer model and used to calculate the surface reflectance.

As shown in Fig. 4.2, MODTRAN (see Section 4.4.1) is used to simulate the apparent radiance within the wavelength range of 0.4–2.5 μm. For the simulation, vegetation is used for the surface reflectance, the atmosphere is set to have midlatitude summer conditions, the aerosol

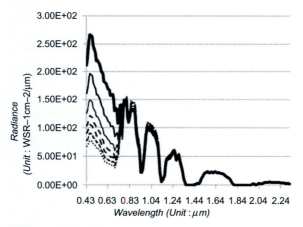

FIGURE 4.2 Changes in vegetation spectral radiance at different visibilities (the visibility is set equal to 2, 5, 10, 16, 23, 35, and 60 km for the respective curves from top to bottom).

type is set to a rural type, the water vapor content remains constant, and the visibility varies from 2 to 60 km. As indicated in the figure, visibility has the greatest impact on the visible spectrum. The radiance of the visible band gradually decreases with increasing visibility.

Solar electromagnetic radiation travels through the atmosphere from space and interacts with the atmosphere at the surface. Then, it is reflected from the surface and passed through the atmosphere until the signal is received by satellite sensors. This process is very complicated. The photoelectric system characteristics, solar altitude, and terrain and atmospheric conditions of the remote sensing sensors themselves cause distortion of the spectral radiance. In addition to particle scattering, the main absorbers include water vapor, ozone, oxygen, and aerosols. Molecular scattering and absorption by ozone, oxygen, and other gases are relatively easy to correct because the concentrations of these elements are relatively stable in time and space; however, estimating the aerosol and water vapor parameters is rather difficult. Therefore, the removal of the impacts of aerosols and water vapor is the main component of atmospheric correction. The atmospheric correction algorithms used in typical satellite sensor images are shown in Table 4.2.

It is very difficult to obtain surface information accurately by removing atmospheric effects from remote sensing images through atmospheric correction. There are two main aspects that affect the accuracy of atmospheric correction.

4.1.4.1 Internal sensor error

Ideally, the electromagnetic radiation of each band recorded by the remote sensing sensor system can accurately record the radiation information leaving the target object. However, there are many kinds of noise (errors) that enter the data acquisition system. These noises received by sensors are independent of signals, which will seriously affect the signal-to-noise ratio of remote sensing images. The distortion caused

TABLE 4.2   Atmospheric correction algorithms and product information of typical satellite sensors.

Satellite/Sensor	Algorithms	Time range	Spatial resolution	Band	References
Terra/MODIS; Aqua/MODIS	–	2000– 2002–	250/500/1000 m	Bands 1–7	Vermote et al. (2002)
Terra/MISR	–	2000–	1100 m		
Suomi NPP/ VIIRS	–	2012–	500/1000 m 0.05 degrees	Bands 1–11	Justice et al. (2013)
NOAA/AVHRR	–	1981–	0.05 degrees	Bands 1–3	Tanre et al. (1992)
Landsat4, 5/TM; Landsat5/ETM+; Landsat8/OLI	EDAPSLEDAPS; LaSRC	Landsat 4 TM: 1982–93; Landsat 5 TM: 1984–2012; Landsat 7 ETM+: 1999 –; Landsat 8 OLI: 2013 –	30 m	TM/ETM+: Bands 1–5, 7; OLI: Bands 1–7	Schmidt et al. (2013), Vermote et al. (2016)
Sentinel-2 MSI; SENTINEL-3 Synergy	Sen2Cor –	2015.06–; 2016.10–	10/20/60 m 300 m	Bands 1–12; S1–S4, S6	Malenovsky et al. (2012)
FY3A/MERSI & VIRR– FY3A/VIRR		FY3A/MERSI & VIRR: 2009.06 –2014.09 FY3C/VIRR: 2014–	250/1000 and 1100 m 1100 m		Fan et al. (2016)

by sensor sensitivity is mainly caused by the characteristics of its optical system or photoelectric conversion system. With the increase in the service life of the sensor, the sensitivity of the photoelectric system to the signal will be attenuated to varying degrees.

4.1.4.2 External errors caused by environmental factors

Even if the remote sensing system works normally, the acquired data may still have radiation errors. The atmosphere and terrain are two of the most important environmental attenuation sources causing radiation errors. When estimating the influence of atmospheric scattering and absorption on satellite imagery using the atmospheric radiation transfer model, the atmospheric effects can be eliminated by determining the atmospheric parameters, and the atmospheric correction of the imagery can be realized. However, it is very difficult to obtain the eigenvalues of the atmospheric condition in images simultaneously. Specifically, for historical satellite data, atmospheric information at that time was often unknown.

The slope and aspect of terrain can also cause radiation errors, especially when the study area is completely shadowed, which greatly affects the radiance value of its pixels. The purpose of slope and aspect corrections is to remove the change in illumination caused by terrain. After correction, two objects with the same reflection characteristics have the same radiance value in the image, although the aspects are different. In remote sensing images with a wide range of mountain areas, corrections to the slope and aspect effects are indispensable if surface reflectance is to be accurately obtained.

There have been many studies on atmospheric correction in remote sensing images, which have led to a large number of atmospheric radiation transfer programs (models). According to the different characteristics of the sensor, such as the difference in spectral settings and the imaging mechanism, a variety of atmospheric correction methods have emerged. However, for any image, due to many factors and uncertainties, it is almost impossible to correct every pixel perfectly. In the follow-up section of this chapter, we will mainly introduce the estimation of aerosol properties and water vapor content and briefly introduce the removal methods of other atmospheric components.

4.2 Correcting the aerosol impact

Atmospheric aerosols refer to multidispersed bodies consisting of small particles suspended in the atmosphere, whose scale ranges between 0.001 and 10 μm. Atmospheric aerosols have a considerable impact on the global climate. Through their absorption and scattering of solar and infrared radiation, they change the radiation budget of the Earth—atmosphere system. By absorbing and scattering radiation, aerosols can directly interfere with the signal reception of optical sensors. The estimation of aerosol attributes is mainly used to estimate the AOT. The algorithms used to determine the spatial variation in the AOT are fully described by the spectral, temporal, angular, spatial, and polarization signatures, along with the other information in this section.

4.2.1 Spectral information—based correction method

The development of multispectral and hyperspectral sensors has led to atmospheric correction methods based on spectral characteristics. The dark target method is the earliest classic algorithm based on spectral characteristics. This method assumes that some bands are affected by aerosols more significantly than other bands. The bands that are minimally affected by aerosols are used to determine the surface reflectance, which is then used to estimate the AOT of the bands sensitive to aerosols. This method has a long history and is one of the most commonly used atmospheric correction methods presently available.

For the Moderate-Resolution Imaging Spectroradiometer (MODIS) and Medium-Resolution Imaging Spectrometer (MERIS) data, the dark target method is also used to estimate the AOT for atmospheric correction (Santer et al., 1999; Vermote et al., 2002). The dark object method is mainly employed for processing imagery where dense vegetation is most often seen as a dark object, and its application to remote sensing imagery without dark targets is difficult.

4.2.1.1 Midinfrared dark target method

The basic principle of this method is that the dark target pixels of the images to be corrected (such as dense vegetation images) and the linear relationship between the 2.1 μm midinfrared (MIR) band and the red and blue band reflectance are used to calculate the AOT (Kaufman et al. 1997a,b). In the case of Lambertian surface reflection, the atmospheric property is uniform, and multiple atmospheric scattering and the diffuse reflection of neighboring pixels can be ignored. Therefore, the apparent reflectance of a dark target in the MIR band is minimally affected by the atmosphere, and it can be approximated as surface reflectance. Based on the linear relationship between this reflectance and the blue and red band reflectance, the surface object reflectance in the blue and red bands can be calculated. Then, the AOT in the red and blue bands can be estimated. According to the aerosol model, the optical thicknesses of the other bands can also be determined. The dark target method is applicable to both multi-spectral remote sensing imagery and hyperspectral imagery (Liang and Fang, 2004; Zhao et al., 2008).

Taking Landsat TM imagery as an example, the procedures to apply MIR dark target method processing (Liang et al., 1997) are shown below:

* **Identification of the dark object:** by setting a low reflectance threshold (e.g., 0.05), a dark target pixel is found in the MIR (band 7) imagery with a wavelength of approximately 2.1 μm, and its pixel reflectance is denoted as $\rho_{2.1}$. Additionally, to extract dense vegetation as a dark object, a vegetation index is often used to assist in determining the vegetation density;
* **Calculation of the surface reflectance in the visible band:** the following statistical linear relationship is used to calculate the surface reflectance in the blue (band 1) and red (band 3) bands:

$$\rho_{\text{red}} = 0.5\rho_{2.1}$$
$$\rho_{\text{blue}} = 0.25\rho_{2.1} \tag{4.3}$$

* **Calculation of the AOT:** according to the surface reflectance in the red and blue bands and the remote sensing apparent reflectance, the radiative transfer model can be used to calculate the AOT in the red and blue bands. Next, through a function of the optical thickness, the AOTs of the other bands can be determined:

$$\tau_i = a\lambda_i^{-b} \tag{4.4}$$

where τ_i and λ_i are the AOT and the central wavelength of Channel i ($1 \leq i \leq 5$), respectively. Parameters a and b can be obtained by estimating the optical thicknesses in the blue and

red bands, respectively. Parameter b is often called the Angstrom exponent.

* **Calculation of the surface reflectance:** For nondense vegetation pixels in the imagery, the interpolation technique using a moving window is adopted to estimate the spatial distribution of the AOT; then, the surface reflectance of all pixels can be calculated.

Fig. 4.3 shows an application case of this method through a comparison between an original color composite image (bands 1, 2, and 3) and the image after atmospheric correction. This TM image was acquired on February 26, 1990, at the central longitude of 63°45′19″ and at the central latitude of 17°20′39″. Bolivia, South America, was included in this image, with a resolution of 800 × 800 pixels. The figure clearly shows that many fuzzy zones were removed after atmospheric correction, and the surface characteristics concealed by the fuzzy atmosphere were recovered well. Although there were some clouds in the image, the overall contrast of the image was improved.

There are many methods for AOT inversion based on MODIS data, among which the dark object method is the most widely used. After decades of development, MODIS aerosol products have undergone several versions of updates. Currently, NASA's official website offers both C5 and C6 versions of aerosol products. The spatial resolution of the C5 version is 10 km, and that of the C6 version is 3 km (Remer et al., 2013) and 10 km. Compared with C5, the C6 version has made great improvements in aerosol type and product quality control. The algorithm of the MODIS dark target includes the following steps (Levy et al., 2007b).

4.2.1.1.1 Selection of dark pixels

For MODIS terrestrial aerosol inversion, determining dark pixels and determining the relationship between the red and blue bands of

(A) **(B)**

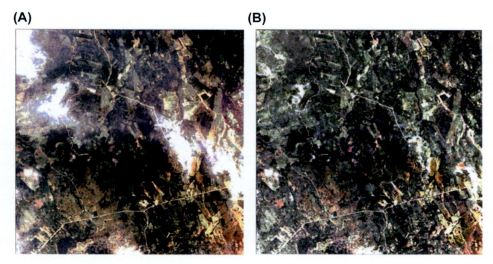

FIGURE 4.3   Application case using the dark target method for atmospheric correction. (A) The image before correction and (B) the image after correction (Liang et al. 1997).

dark pixels and shortwave infrared band surface reflectance is an important technical step. First, a 10*10 km spatial range is selected from the image. The number of pixels in the range varies with the spatial resolutions of different bands. After removing all cloud, ice, and snow and continental water pixels in this area, the pixels with apparent reflectance greater than 0.01 and less than 0.25 at 2.1 μm are selected, and then the pixels with reflectance higher than 50% and lower than 20% in the red band are removed. The remaining pixels can be considered dark pixels in the 10*10 km space range.

4.2.1.1.2 Determining the surface reflectance of dark pixels

Previous studies have shown that the relationship between the reflectance of dense vegetation at 2.1 μm and that at 0.47 and 0.66 μm is related not only to the scattering angle but also to the vegetation density (Gatebe et al., 2001; Remer et al., 2001). Therefore, by adding the scattering angle and shortwave infrared normalized vegetation index as auxiliary data, the apparent

reflectance at the 2.1 μm band can be used to estimate the visible reflectance at red and blue bands (Levy et al., 2007b).

4.2.1.1.3 Determination of the aerosol model

In aerosol inversion, it is necessary to determine the aerosol model, but it is very difficult to determine the model accurately. The terrestrial aerosol model varies greatly with seasons and regions, so it is often necessary to assume the aerosol model in the inverted area based on empirical assumptions. The early method to determine aerosol type was to utilize the difference in aerosols between the red and near-infrared (NIR) bands and to use a simple ratio method to determine aerosol type (Kaufman et al., 1997a). However, due to the complexity of aerosol composition and the uncertainty of surface reflectance, this method is difficult to effectively express the real optical properties of aerosols. At present, the dataset of terrestrial aerosol models in different seasons and regions of the world is usually established by clustering based on ground-based AERONET observations

(Levy et al., 2007a). The dataset divides the aerosol models into three types: strong absorption, weak absorption, and moderate absorption, each of which is composed of coarse and fine aerosol particles (Levy et al., 2013).

4.2.1.1.4 Calculating aerosol optical thickness over dark targets

At present, NASA has released global atmospheric aerosol products (MOD04/MYD04) using the MODIS dark target algorithm for terrestrial aerosol inversion. The data have been updated to version C6.1 after algorithm updates. At the same time, the accuracy of the algorithm has been greatly improved, especially in the bright urban area (Tian et al., 2018a). The aerosol information can be obtained in most other areas, except some deserts (such as the Saharan desert) and snow-covered regions of Greenland (Levy et al., 2013).

4.2.1.2 Near-infrared dark target method

In the traditional dark target method for atmospheric correction, data from the MIR band (MIR, 2.1 μm) are required. However, a large number of surface observation satellite sensors provide data only for the four bands from the visible to the NIR spectrum (400–1000 μm) at present. Because of the lack of a MIR band, the conventional dark target method cannot be applied. An atmospheric correction algorithm involving dark targets in the visible band and the 0.85 μm NIR band was proposed (Richter et al., 2006) for correcting TM and ETM + images.

Compared with the MIR dark target method, the NIR dark target method is characterized by the use of the linear relationship between the dense vegetation reflectance in the 0.85 μm spectral region and that in the red band reflectance (Richter et al., 2006):

$$\rho_{red} = \alpha \rho_{nir} = \frac{\rho_{nir}}{10} \tag{4.6}$$

After the dark object is identified, the surface reflectance in the red band of the dark target area

is obtained according to the above equation; the atmospheric visibility of the red band can then be determined based on the relationship between the surface reflectance and the apparent radiance.

Through comparison with the atmospheric correction results using the MIR dark target algorithm, this algorithm is found to have a standard deviation of less than 0.005, indicating that this algorithm is reliable and can be applied to the atmospheric correction of other sensor data.

4.2.1.3 Deep blue method

The surface reflectance of most objects in the deep blue (DB) band (412 nm) ranges from 0 to 0.1; therefore, the contributions of these surface reflectances to the apparent reflectance are significantly lower than those in other bands. This increases the sensitivity of the satellite sensor signals to aerosols to a certain extent. Therefore, it is an effective method to retrieve the AOT by using the apparent reflectance of the DB band observed by satellite. Accordingly, the problem with surface and atmospheric decoupling in aerosol inversions can be solved by building prior knowledge from reflectance data (Hsu et al., 2004, 2006). Because this algorithm uses a MODIS DB band, it is also called DB. This algorithm is used to retrieve the AOT from remote sensing images in nondense vegetation areas, and it has been improved to process aerosol products in MODIS C5 and C6 versions (Hsu et al., 2013). In addition, considering the influence of BRDF characteristics on land surfaces, the AOT inversion accuracy in bright areas has been improved on the basis of the DB algorithm (Tian et al., 2018b). The basic process of the DB algorithm in MODIS aerosol products can be summarized by the following four steps (Hsu et al. 2013).

4.2.1.3.1 Pixel selection

In the first step, the dark blue algorithm is similar to the dark target method. Before inversion, the cloud- and snow-covered pixels need to be removed. In general, the threshold method is used to detect cloud pixels in the 0.412, 0.650, 1.38, 11, and 12 µm bands. The normalized difference snow index, 0.555 and 0.86 µm reflectance data, and 11 µm brightness temperature data were used for ice and snow pixels, while the MODIS surface cover products were used for water screening.

4.2.1.3.2 Determination of surface reflectance

For a pixel in an image, the dark blue algorithm divides the surface into three types according to the MODIS surface cover products and vegetation cover: (1) natural vegetation coverage, (2) urban areas, and (3) arid/semiarid areas. For different regions, three different methods are used to determine the surface reflectance. Similar to the dark target method, the correlations among the 0.65, 0.47, and 2.1 µm models are established in vegetation-covered areas, while in arid/semiarid areas, the long time series MODIS data are used to construct a prior surface reflectance database with a resolution of 0.1×0.1. Cloud detection and aerosol inversion are realized in urban areas using a hybrid method. At the same time, AERONET and remote sensing data are used to determine the BRDF angle and shape information under different NDVI values.

4.2.1.3.3 Determining aerosol models

The aerosol model selection method with the dark blue algorithm is similar to that with MODIS version C6, where only a few modifications have been made to dust areas. The results show that brightness temperature differences of 8.6 and 11 microns can be used to detect mineral dust particles with strong absorption in the blue band (Hansell et al., 2007). Therefore, the dark target algorithm in the C6 version optimizes the determination of the strong absorption aerosol region (Hsu et al., 2013).

4.2.1.3.4 Obtaining aerosol optical thickness

Similar to the dark target algorithm, NASA has launched a worldwide production release of aerosol products based on the dark blue algorithm. Correspondingly, the latest product version is C6.1. It is worth mentioning here that for highlighted urban surfaces, the precision difference between the C6.1 and C6 versions of dark blue products is small, and the new version has not improved significantly. The existing results show that aerosol information can be obtained in both low-reflective vegetation areas and high-reflective desert areas, but it is not applicable in snow-covered areas. At the same time, its spatial distribution trend is consistent with that of dark target products.

4.2.2 Temporal information—based correction method

With the improved time resolution of the satellite data, the atmospheric correction method based on time series has been proposed. This method determines the aerosol information of a pixel by assuming that the surface reflectance is constant in a certain time interval, and that only some of the values of the pixel are significantly affected by aerosols. The values that are minimally affected by aerosols are used to determine the surface reflectance, which is used to estimate the AOTs of the values that are greatly affected by aerosols. This method requires the data to have a higher time resolution; therefore, it is applicable to most data with a moderate resolution.

The linear regression method, as a traditional atmospheric correction method based on time series imagery, is first introduced in this section. Then, the atmospheric correction algorithm based on temporal characteristics, which has been improved by Liang, for MODIS reflectance products can be implemented (Liang et al., 2006).

4.2.2.1 Linear regression method

Under the assumption that the surface reflectances of some pixels in an image are stable over a time series, differences in the remote sensing apparent reflectance mainly reflect changes in the atmospheric conditions. Based on the apparent reflectance of these "invariant" pixels, a linear relationship between the images at different times can be established to remove the differences caused by atmospheric interference. This method is a relative atmospheric correction. If the imagery is acquired in synchronization with a corresponding surface reflectance measurement, then this process becomes more accurate.

It is assumed that N "invariant" pixels can all be identified from M images acquired at different times. If a clear image is selected, e.g., reference image J, then other images can be corrected based on the linear regression of the N pixel values in image J (see Fig. 4.4).

L_j is the radiance of reference image J, and L_i is the radiance of image I. The corresponding equation for processing is shown below:

$$L_j^k = a_i^k + b_i^k L_i^k \qquad (4.7)$$

Two coefficients, a and b, are generated in the linear regression analysis on the radiance of N pixels in each band K for each image. These two coefficients are used to correct all other pixels in band k of image I. Thus, invariable pixels with different brightnesses are required in each band. If only one type of pixel exists in the image with little change in brightness, then considerable errors occur in the linear transformation.

4.2.2.2 Improved multitemporal imaging method

The dark target method has been used for the atmospheric correction of MODIS reflectance products. This algorithm is only applied to imagery with dense vegetation. In the case of no dense vegetation, the atmospheric correction of MODIS products is not accurate. Liang et al. proposed an improved method for the estimation of AOT based on the higher time resolution of MODIS data and the assumption that the surface reflectance of the surface object is fixed over a specified time interval (Liang et al., 2006). The process of this method is shown in Fig. 4.5. This algorithm mainly detects the cleanest observation of each pixel over time series images in a certain time interval (least affected by the atmosphere). Then, using a lookup table, the AOT of the hazy image is calculated. This algorithm produces better correction results for images with no dense vegetation, especially in the case of "bright target" imagery.

This algorithm was used to estimate the AOT of a MODIS scene acquired on November 11, 2002, which was greatly affected by aerosols. The retrieved AOT in the blue band is shown in Fig. 4.6.

4.2.3 Angular information—based correction method

There are many airborne and aerospace sensors that observe the Earth's surface from various angles simultaneously, such as the Multiangle Imaging Spectroradiometer (MISR). The MISR can acquire global observational data every 9 days, with a spatial resolution of 275—1100 m. The MISR sensor consists of nine cameras, and

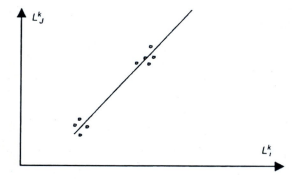

FIGURE 4.4 Illustration of the linear relationship between the "invariant" pixels in two images acquired at two different times.

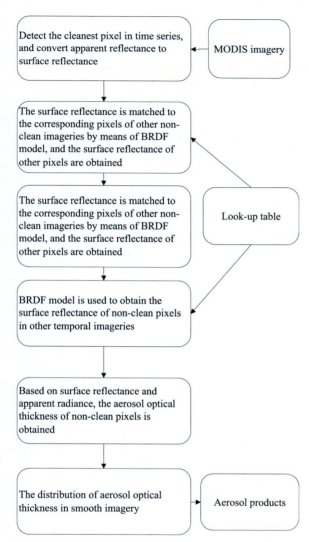

FIGURE 4.5 Flowchart of the multitemporal correction algorithm (Liang et al. 2006).

are more sensitive to the impact of aerosols, which facilitates the use of multiangle data for atmospheric correction. Consequently, this method takes advantage of the fact that aerosol impacts are different for images acquired from various angles.

In the case of the vertical observations of the land surface, brighter surface targets and a shorter atmospheric path make haze detection significantly more difficult. However, in the case of surface targets observed at large angles, the atmospheric path becomes longer, thus increasing the intensity of the path radiance of haze. As a result, haze that is difficult to detect at nadir is more easily detected in large-angle observations (Martonchik et al., 2004).

The MISR sensor on the Terra satellite has an imaging function that performs at nine observation angles. When the zenith angle is large, the haze observation is more obvious than the zenith condition, which makes it convenient to calculate the AOT. Further use of angular information in atmospheric correction is addressed elsewhere (Liang, 2004).

Atmospheric correction can also be carried out by matching the spatial characteristics between clear and hazy regions, such as the histogram matching method and the cluster matching method. In the histogram matching method, it is assumed that a clear area and a hazy area due to aerosol scattering exist simultaneously in a single image, and the surface reflectance of the clear area can be easily determined by removing Rayleigh scattering, minimal aerosol scattering, and gas absorption. If the histogram of the hazy region is matched with the histogram of the clear region, we can estimate the AOT and remove the impact of aerosol scattering (Richter, 1996). However, if the landscapes of the clear and hazy regions are not the same, then their histograms of surface reflectance should not be the same. For example, the histogram of the urban region should be quite different from that of a rural region.

each camera covers four bands: the blue band (446 nm), the green band (558 nm), the red band (672 nm), and the NIR band (867 nm). The nine cameras have surface observation angles of −70.5, −60.0, −45.6, −26.1, 26.1, 0, 45.6, 60.0, and 70.5 degrees, of which the positive values indicate forward observations and the negative values indicate backward observations (Diner et al., 1999). Generally, larger observation angles

(A)   **(B)**

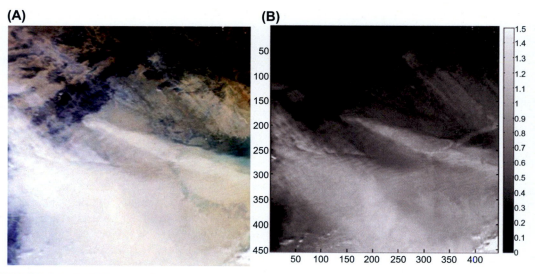

FIGURE 4.6   MODIS image (left) and the estimated aerosol content in the blue band (right) (Liang et al. 2006).

However, the histograms of each surface type should be similar. As an improvement, the cluster matching method first identified the clusters (surface types) using images at longer wavelengths that are less sensitive to aerosols and then matched the histograms of these clusters across the study area (Liang et al. 2001, 2002). This method can address different landscapes and even small hazy patches. The pixel cluster matching method was used in the atmospheric correction of TM imagery, taking into account the adjacency effect correction, and was also applied to other satellite images, such as MODIS and the sea wide field-of-view sensor (SeaWiFS) (Liang et al., 2002).

4.2.4 Polarization information—based correction method

In the visible band, atmospheric scattering has strong polarization, while surface reflection exhibits low polarization. The polarized radiation observed by satellites is sensitive to the size of the aerosol particles and their refractive index but insensitive to surface changes. The combined use of polarization and radiation information can better determine aerosol optical properties. At present, the running remote sensing polarization sensor is POLDER (POLarization and Directionality of Earth Reflectance), developed by the French Space Agency. POLDER provides radiation, polarization, and multiangle information for aerosol retrieval; consequently, a series of polarization-based atmospheric correction algorithms have been developed (Deuzé et al., 2001). POLDER data were used to determine the AOT over the Sea of Japan (Kawata et al., 2000). Moreover, the AOT determined by individual uses of reflectance or polarization reflectance was compared with that determined by the combined method. The data validation indicates that the combined method demonstrates better performance in the determination of the AOT (Kawata et al., 2000). POLDER data were used by Sano et al. to evaluate aerosol properties over land and the global distribution of the Angstrom exponent (Sano, 2004). These authors found that the year-round

AOTs are relatively large in Central Africa and Southeast Asia.

4.2.5 Multisensor cooperative inversion algorithm

To make full use of the advantages of different sensors in extracting aerosol inversion parameters, many researchers have proposed and used the method of the multisensor cooperative inversion of aerosols. For example, using the GOME-ATSR2 and CIAMACHY-AATSR sensors, a multisensor collaborative retrieval method for AOT and aerosol type (Holzer-Popp et al., 2002) was implemented; using multiangle MISR data to extract surface information, the results were used to retrieve aerosol information from multiband MODIS data (Vermote et al., 2007) and MODIS data from the Terra and Aqua satellites in three bands. To solve the radiation transfer equation, the cooperative inversion method for aerosols from two MODIS satellites was realized (Tang et al., 2005). The multisensor cooperative inversion method provides more known parameters for aerosol inversion by using existing remote sensing data sources and effectively solves the problem of solving the "ill-conditioned equation" in aerosol inversion to a certain extent. However, this method is based on multisensor data at the same time or at short intervals, which requires strict conditions for data acquisition and contains too many approximations and assumptions. The error in real-time data processing is transmitted to the aerosol inversion, which makes it difficult to effectively use in the operational inversion of terrestrial aerosols.

4.2.6 Joint inversion of atmospheric surface parameters

Accurate atmospheric correction requires accurate atmospheric parameters, and the remote sensing retrieval of atmospheric parameters (especially aerosol parameters) requires surface reflectance spectral information, which is contradicting. Several of the aerosol estimation methods introduced above simplify surface reflection, such as the Lambertian surface assumption. In fact, neglecting the bidirectional reflectance of a surface may lead to large aerosol parameter inversion errors. Therefore, researchers propose a joint surface—atmosphere optimization inversion method that simultaneously solves the atmospheric aerosol parameters and surface bidirectional reflectance parameters. For example, a joint inversion study was conducted for MSG/SEVIRI data from geostationary meteorological satellites (Govaerts et al., 2010; Wagner et al., 2010).

Such methods (He et al., 2012; Zhang et al., 2018) first need to construct a surface—atmosphere coupled radiation transfer model. The RPV model (Rahman et al., 1993) and 6S model (Vermote et al., 2006) can be used to characterize the bidirectional reflectance of a surface. Then, the model is inverted by the optimization method, and the surface parameters (bidirectional reflectance model parameters) and atmospheric parameters (AOT) are obtained simultaneously. The optimization method is used to obtain a set (or groups) of optimal estimates of the parameters to be estimated within a reasonable range to minimize the cost function. The cost function designed here is as follows:

$$
\begin{aligned}
J(x) &= (y_m(x) - y_o)S_y^{-1}(y_m(x) - y_o)^T \\
&+ (x - x_b)S_b^{-1}(x - x_b)^T \\
&= J_y + J_x
\end{aligned}
\tag{4.8}
$$

here x represents the joint vector of the surface and atmospheric parameters. y_0 represents the observed apparent reflectance at the top of the atmospheric. $y_m(x)$ predicts the apparent reflectance at the top of the atmospheric after x is fed into the surface—atmosphere coupled radiation transfer model introduced earlier, x_b is the a priori knowledge mean of the model parameters, S_y is the covariance matrix of the apparent reflectance error, and S_b is the a priori knowledge of the model parameters in the covariance matrix. In short, the cost function is divided into two parts: J_y is the fitting effect of the model prediction on apparent reflectance, and J_x is the degree of conformity between the model parameters and a priori knowledge.

4.3 Correcting the impact of water vapor

In addition to the impact of aerosols, atmospheric water vapor is a major factor affecting remote sensing imagery. Therefore, in the quantitative application of NIR and MIR remote sensing imagery, the impact of water vapor absorption in the atmosphere must be removed. As shown in Fig. 4.7, MODTRAN is used to simulate the apparent radiation within the wavelength range of 0.4−2.2 μm, in which vegetation reflectance is used as the background value. The atmosphere type is set to midlatitude summer conditions, the aerosol type is set to rural, the visibility is set to 23 km, and the column water vapor content is increased from 0.2 to 9.0 g/cm^2. The curves indicate the simulation results as the water vapor content changes. The water

vapor shows significant absorption features in the 0.94 and 1.14 μm spectral regions. Moreover, as the water vapor content increases, the absorption features of the water vapor become more pronounced.

The water vapor content in the atmospheric column is generally obtained through a water vapor absorption channel. At present, many algorithms can be used to estimate the atmospheric column water vapor content, including the narrowband and wideband ratio methods (Frouin et al., 1990), the continuum interpolated band ratio method (Bruegge et al., 1992; Kaufman and Gao, 1992) (CIBR), the 3-channel ratio method (Gao and Kaufman, 2002), the curve-fitting method (Gao and Goetz, 1990), the atmospheric precorrected differential absorption method (Schlapfer et al., 1998) (APDA), the smoothness test method (Qu et al., 2003), and the linear interpolated regression ratio method that evolved from the CIBR (Schlapfer et al., 1996).

If the band ratio method is directly used to estimate the water vapor content (Bruegge et al., 1992; Frouin et al., 1990; Schlapfer et al., 1996), the impact of the atmospheric path radiation is not removed, and the decrease in the surface reflectance leads to a greater ratio due to the impact of atmospheric path radiation. Therefore,

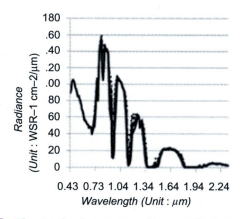 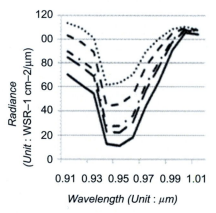

FIGURE 4.7 The simulated apparent radiance as the water vapor content changes (climate type, midlatitude summer climate; aerosol type, rural; visibility, 23 km; water vapor content, increasing from 0.2 to 9 g/cm^2 from top to bottom).

these algorithms are more suitable for regions with high background reflectance. In the low-reflectance regions, the water vapor estimation accuracy is reduced. The water vapor content is usually underestimated for darker surfaces (Gao and Goetz, 1990). Because of the removal of the impact of atmospheric path radiation, the estimation accuracy of the water vapor content in a low-background region is somewhat improved if the APDA method is used (Schlapfer et al., 1998).

More methods for estimating the column water vapor content, such as the differential absorption method and the split-window algorithm, are discussed in Section 4, Chapter 6, of *Quantitative Remote Sensing* of Land Surfaces (Liang, 2004).

4.4 Correcting the impacts of other constituents

The correction methods used to remove the impacts of atmospheric aerosols and water vapor have been introduced in the previous sections. In addition to aerosols and water vapor, the Earth's atmosphere includes a mixture of nitrogen, oxygen, argon, and carbon dioxide (see Table 4.1). These mixed gases are also called dry-clean gases. Of the components of atmospheric gas, aerosols and water vapor are the most important and most active elements, dramatically changing with time and space. Because of the prevailing convection and turbulence in the atmosphere, these gases at different heights in different layers are fully exchanged and mixed. Therefore, the percentages of various compositions of these dry-clean gases remain stable and constant from the surface to a height of 85 km.

For the atmospheric correction of remote sensing imagery in the visible band, the extraction and correction of parameters are required for aerosol, water vapor, and other variable compositions based on imagery. However, because the contents of some invariable gases in the atmosphere change insignificantly with time

and space, they can be set through constant observed or calculated values for different regions, heights, and seasons to remove their impacts.

4.5 Commonly used models and software

For most atmospheric correction algorithms, it is necessary to solve the radiative transfer equation and establish lookup tables for rapid atmospheric correction, resulting in a series of atmospheric correction models based on the theory of atmospheric radiative transfer. These models include MODTRAN (Berk et al., 2008), 6S (Vermote et al., 2006), and other atmospheric radiation approximate calculation models. Most of these models and software are intended for professionals. Their operation is complex; consequently, there have been many interface-based atmospheric correction software packages, such as the atmospheric correction modules FLAASH, ACTOR, and ACORN. Based on MODTRAN, these models are constructed from a large number of lookup tables for the convenient and rapid atmospheric correction of a variety of sensor data. This section first describes two commonly used radiative transfer model software packages in which the forward calculation results can be used for atmospheric corrections. Several interface-based software packages for fast atmospheric correction are briefly covered in the following section.

4.5.1 MODTRAN model

MODTRAN (Moderate-Resolution Atmospheric Transmittance and Radiance Code) is the software used to calculate atmospheric transmittance and the radiative transfer algorithm with a moderate spectral resolution (Berk et al., 2008). MODTRAN can be used to calculate atmospheric transmittance, atmospheric background radiation (upward and downward atmospheric radiation), the radiance of single solar or lunar scattering, direct solar irradiance, etc. MODTRAN was based on LOWTRAN, with a spectral resolution of $1 \, cm^{-1}$. Several important databases have been updated in MODTRAN 5.0, which greatly improves the calculation accuracy (Berk et al., 2005). The accuracies of Rayleigh scattering and the complex refractive index are also improved, and the related options for calculating the azimuth of the solar scattering contribution are added in DISORT. In addition, seven kinds of land surface BRDF models are introduced so that the simulations are not limited to Lambertian surfaces. More accurate calculations of transmittance and radiation can significantly enhance the analysis of hyperspectral imagery.

The input parameters for the operation of MODTRAN can be divided into five types, all of which are determined by compiling CARD1 into CARD5:

- Controlling and operating parameters
  These parameters must determine what type of radiative transfer model is selected and whether multiple scattering calculations are required. These settings are mainly performed in CARD1, and CARD5 provides the options for the repeat calculations.
- Atmospheric parameters
  The atmospheric type is determined through the options in CARD1, while other specific parameters, including aerosols, are mainly selected through the options in CARD2.
- Surface parameters
  Initial options for the surface parameter settings are provided in CARD1, and the specific surface parameter settings are applied in CARD4 according to the parameters set in CARD1.
- Observation geometry
  The geometric type of atmospheric path is determined in CARD1; CARD3 is the input option for the observation geometry parameters, in which various methods are

combined to achieve the parameter input. Users can select the input method according to their own circumstances.

- Sensor parameters
  CARD1A has the option to allow or disallow the input of the sensor channel response function; the file name of the channel response function is input in CARD1A3, and the wavelength range for the simulation calculation is input in CARD4.
  The main output result of MODTRAN is the simulated apparent radiance. Based on the surface reflectance and atmospheric parameter input, the apparent radiances for specific atmospheric parameters can be obtained. According to the radiative transfer equation, the parameters required for atmospheric correction can be solved, enabling the construction of lookup tables for atmospheric correction.

4.5.2 6S Model

The 6S Model (Second Simulation of the Satellite Signal in the Solar Spectrum) is compiled with the FORTRAN programming language and is applied to the simulation calculation of the atmospheric radiative transfer model in the solar reflection band (0.25–4 µm). The latest approximation and successive approximation algorithms are used in this model to calculate scattering and absorption, in which the parameter input is improved to provide more accurate simulation. Based on the assumption of a cloudless sky, many issues are considered in this model, including absorption by water vapor, carbon dioxide, ozone, and oxygen; molecular and aerosol scattering; and nonuniform surface bidirectional reflectance. Among these considerations, gas absorption is calculated with a spectral interval of $10 \, cm^{-1}$ and a spectral integral step of 2.5 nm; therefore, this method is mostly used to process multiangular remote sensing data in the visible and NIR bands (Vermote et al., 2006).

4.5.3 FLAASH

FLAASH (Fast Line-of-sight Atmospheric Analysis of Spectral Hypercubes) is an atmospheric correction module developed by the Spectral Sciences Institute of Optical Imaging with the support of the US Gas Dynamics Lab (Adler-Golden et al., 1998), and it has been integrated into ENVI. The module can be used for the rapid atmospheric correction analysis of Landsat, SPOT, AVHRR, ASTER, MODIS, MERIS, AATSR, IRS, and other multispectral and hyperspectral data, aerial imagery, and hyperspectral images with user-defined formats. This module can also be applied to effectively remove the impacts of the atmosphere, illumination, and other factors on the surface reflectance, producing more accurate values of the reflectance, radiance, surface temperature, and other real physical parameters of the models. The FLAASH module is directly combined with the MODTRAN4 atmospheric radiative transfer code, enabling the direct selection of standard MODTRAN atmospheric models and aerosol types in relation to imagery for use and, hence, for the calculation of surface reflectance. The FLAASH module can be applied to the correction of the adjacency effect and the visibility calculation for the entire image. In addition, FLAASH can be used to generate the classified imagery of cirrus and other thin clouds for spectral smoothing (Flaash, 2009).

4.5.4 ACTOR

ATCOR is an ERDAS IMAGINE module for atmospheric correction and haze removal, which is applied to the correction of atmospheric impacts on the spectral reflectance of surfaces and the removal of the impacts of thin clouds and haze. This module can be used to correct both flat images in the imaged area and images with greater height changes, which requires a digital elevation model (DEM) of the imaged area. Atmospheric conditions change with weather and the solar elevation angle, which inevitably alters the spectral reflectance of the surface object. Such changes can prevent a direct comparison of images acquired at different times or by different sensors. The ATCOR module can be applied to atmospheric correction for the removal of such interferences. ATCOR includes two submodules, ATCOR2 (2D) and ATCOR3 (3D) (Richter, 2010a,b).

4.5.5 ACORN

ACORN is an imagery-based atmospheric correction software developed by ImSpec LLC in Colorado, USA, that can be used to calculate radiation values and surface reflectance within the wavelength range of 350—2500 nm (AnalyticalImaging and GeophysicsLLC, 2002). ACORN provides a series of atmospheric correction strategies, including the empirical method and a method based on the radiative transfer theory, which can be used for the atmospheric correction of hyperspectral and multispectral reflectance in a variety of atmospheric correction models.

4.6 Application of GF-1 WFV atmospheric correction

The Gaofen-1 satellite (GF-1) was successfully launched into orbit at the Jiuquan Satellite Launch Center in April 2013. The satellite uses a sun-synchronous orbit. The average height of the track is approximately 644 km. It belongs to the optical imaging remote sensing satellite and has a design life of 5—8 years. The GF-1 satellite is equipped with four medium-resolution cameras, and the subsatellite point can achieve a spatial resolution of 16 m. There are four bands: blue, green, red, and NIR. The width of the field-of-view of each camera can reach 200 km, and the observation data of four cameras (WFV1, WFV2, WFV3, and WFV4) can obtain observations with an 800 km width. The satellite features a high resolution, wide coverage, and 4-day temporal coverage in China.

To better quantify the GFV satellite WFV data and eliminate the influence of the atmosphere on satellite signals, it is important to complete high-precision atmospheric correction. Because of the problems of insufficient auxiliary data and band settings, the WFV data atmospheric correction mainly has the following difficulties.

- A limited band leads to insufficient access to aerosol information
  The WFV sensor has fewer bands and does not have a 2.1 μm shortwave infrared band, similar to the MODIS sensor. Therefore, the aerosol dark pixel inversion algorithm cannot be simply used, which makes it difficult to perform high-precision atmospheric correction.
- Insufficient angle auxiliary information
  The GF-1 satellite is equipped with four sensors, WFV1, WFV2, WFV3, and WFV4, and the image width can reach 800 km. Its metadata .xml document only provides four types of angle information (sun angle and azimuth and sensor zenith angle and azimuth) of the image center point. For WFV wide-format cameras, the difference in observed geometric angles between the image edge pixels and center point pixels can cause large errors in quantitative applications. Therefore, the observation geometry of each cell also needs to be calculated.
- Data calibration problem
  Because of the hardware problems of the WFV sensor itself, its radiation measurement capability cannot be stable for 1 year. As the

instrument ages, the sensor has a large calibration error (Yang et al., 2015). The wide coverage of the WFV is made up of four cameras, which also poses a greater challenge for scaling four cameras to the same measurement.

Focusing on the above problems, a general model of atmospheric correction for the WFV camera is constructed. The basic process is described below.

4.6.1 Radiation calibration

The GF WFV camera radiometric calibration work is undertaken by the China Centre for Resources Satellite Data and Application. The radiometric calibration results are issued once a year. The calibration results of the GF WFV data are listed in Table 4.3. Excluding 2013, the calibration formula is as follows:

$$L_e(\lambda_e) = \text{Gain} \cdot \text{DN} + \text{Offset} \quad (4.9)$$

where $L_e(\lambda_e)$ is the radiance, DN is the digital number of the satellite observations, Gain is the calibration slope, and Offset is the absolute calibration coefficient offset.

The data calibration formula for 2013 is different. It uses the following formula to complete the calibration:

$$L_e(\lambda_e) = \frac{\text{DN} - \text{Offset}}{\text{Gain}} \quad (4.10)$$

4.6.2 Geometric correction and angle-assisted data calculation

Geometric correction is a key technology for remote sensing image processing and application. The usual sensor models are based on collinear conditional equations and establish a strict imaging geometry model. However, the limiting attitude control of the satellite during imaging results in a very complex imaging geometry. When building a model, the exact

TABLE 4.3 Summary of the GF WFV camera radiation calibration coefficients.

Year	Sensor	B1		B2		B3		B4	
		Gain	Offset	Gain	Offset	Gain	Offset	Gain	Offset
2016	WFV 1	0.1843		0.1477		0.122		0.1365	
	WFV 2	0.1929		0.154		0.1349		0.1359	
	WFV 3	0.1753		0.1565		0.148		0.1322	
	WFV 4	0.1973		0.1714		0.15		0.1572	
2015	WFV 1	0.1816		0.156		0.1412		0.1368	
	WFV 2	0.1684		0.1527		0.1373		0.1263	
	WFV 3	0.177		0.1589		0.1385		0.1344	
	WFV 4	0.1886		0.1645		0.1467		0.1378	
2014	WFV 1	0.2004		0.1733		0.1745		0.1713	
	WFV 2	0.1648		0.1383		0.1514		0.16	
	WFV 3	0.1243		0.1122		0.1257		0.1497	
	WFV 4	0.1563		0.1391		0.1462		0.1435	
2013	WFV 1	5.851	0.0039	6.014	0.0125	5.82	0.0071	5.35	0.0369
	WFV 2	7.153	0.0047	6.823	0.0193	6.239	0.0334	6.235	0.0235
	WFV 3	8.368	0.003	9.451	0.0429	7.010	0.0226	6.992	0.0217
	WFV 4	7.474	0.0274	8.996	0.0011	7.711	0.0117	7.462	0.005

orbit, attitude, sensor imaging parameters, and imaging method of the satellite must be known.

The GF WFV data released by the China Resources Satellite Application Center are divided into two categories: preprocessed radiation-corrected imagery (L1A grade) and preprocessed geometric-corrected imagery (L2A grade). The L1A level is an image data product that has undergone processing, such as data analysis and uniformization radiation correction. It provides rational polynomial coefficients (RPCs). The L2A-level data product is an image product generated by geometric correction and map projection based on L1A-level data and has image projection information. Currently, there are few L2A-level data released. For a large number of L1A-level data, geometric correction of GF WFV data can be implemented based on the RPC model based on the RPC parameter file (*.rpb) provided. When the geometric correction is implemented, the pixel-by-pixel angle calculation is simultaneously completed based on the geometric relationship.

4.6.3 Atmospheric parameter acquisition

The four bands of the WFV sensor are less affected by atmospheric water vapor, especially in the visible light band. It is very difficult to obtain atmospheric water vapor content information directly from WFV data. Therefore, the method adopted here is to perform extraction from the MODIS atmospheric water vapor product on the same date as the image to be corrected.

Compared with atmospheric molecules, the spatial and temporal distributions of aerosols over land are quite different. How to obtain aerosol information for the whole scene is very important for atmospheric correction. Considering that the WFV lacks a shortwave infrared band of approximately 2.1 μm, the surface directional reflection of high-resolution images is more pronounced. Therefore, an improved dark blue algorithm is used to obtain atmospheric aerosol information. The basic idea is to use the existing mature MODIS BRDF parameter product dataset (MOD43B1) to accurately obtain the 500 m resolution surface reflectance information under the GF-1 WFV observation geometry based on the nuclear drive model to support GF-1. This method is used to support the AOT inversion of the GF-1 WFV data. The inversion results are shown in Fig. 4.8.

4.6.4 Atmospheric correction

Atmospheric correction of the radiative transfer method involves simulating the relationship between the atmospheric parameters of the satellite synchronization and the true reflectivity of the surface by simulating the radiation transmission process between the atmospheric—surface remote sensor. After determining the atmospheric parameter information, we can separate the respective contributions of gas at the ground from the radiation information obtained by the sensor and then obtain the surface reflectivity. The basic idea of atmospheric correction based on the radiation transmission model is to simulate the propagation of electromagnetic waves in the geogas system under various atmospheric parameter conditions (atmospheric AOT), atmospheric modes (related to atmospheric gas parameters), surface elevations, and satellite observation geometries (satellite zenith angle, solar zenith angle, and relative azimuth) through the MODTRAN model. Using the results of the simulation, a lookup table of atmospheric radiation transmission parameters is constructed in conjunction with the spectral response function of the satellite sensor.

Fig. 4.9 shows the comparison of true color composite maps before and after WFV data atmospheric correction. The figure shows that atmospheric correction better removes the atmospheric effects, especially atmospheric aerosol scattering, and improves the clarity of the image.

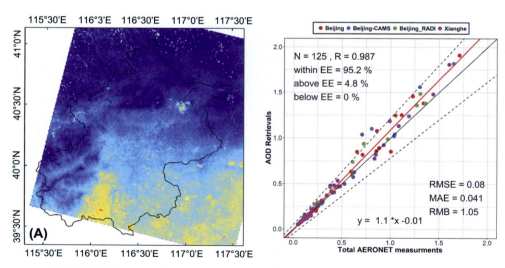

FIGURE 4.8   Aerosol optical thickness inversion and verification results in Beijing.

FIGURE 4.9 GF-1 WFV data true color composite map (left: before atmospheric correction; right: after atmospheric correction).

4.7 Conclusions

Most of the practical algorithms for the estimation of land surface variables are based on remotely sensed surface reflectance. Atmospheric correction is a process in which the TOA radiance received by sensors is converted to surface reflectance. Temporal and spatial dynamic changes in the two main atmospheric parameters (aerosols and water vapor) make atmospheric correction difficult. Aerosol distribution mainly affects the shortwave signals, while water vapor affects the NIR signals. This chapter describes several atmospheric correction methods for the estimation of aerosol content from spectral, temporal, angular, spatial, and polarization signatures. These methods are applicable to remote sensing data in various imaging models. The advantages, disadvantages, and applicabilities of each algorithm are briefly evaluated, and the methods for correcting the impact of water vapor content are described. Finally, the models and software commonly used in atmospheric correction are discussed.

Although a variety of atmospheric correction algorithms have been introduced, no algorithm is universally applicable. For example, in the dark target method, the presence of dark targets is required in the image, such as dense vegetation, and surface reflectance of the dark target in the MIR band is required to define a linear relationship with the surface reflectance of the dark target in the visible band. Thus, the dark target method is not applicable to images with no dark targets. Therefore, appropriate atmospheric correction algorithms should be selected based on the aims, requirements, and conditions of the specific research.

References

Adler-Golden, S., Berk, A., Bernstein, L., Richtsmeier, S., Acharya, P., Matthew, M., Anderson, G., Allred, C., Jeong, L., Chetwynd, J., 1998. FLAASH, a MODTRAN4 atmospheric correction package for hyperspectral data retrievals and simulations. In: Proc. 7th Ann. JPL Airborne Earth Science Workshop. JPL Publication Pasadena, CA, pp. 9–14.

AnalyticalImaging, GeophysicsLLC, 2001. ACORN User's Guide Stand Alone Version: Analytical Imaging and Geophysics LLC, p. 64.

Berk, A., Anderson, G., Acharya, P., Modtran, E.S., 2008. 5.2.0.0 User's Manual Air Force Res. Lab., Space Veh. Directorate, Air Force Materiel Command, Bedford, MA, USA, pp. 01731–03010.

Berk, A., et al., 2005. MODTRAN 5: A reformulated atmospheric band model with auxiliary species and practical multiple scattering options: Update. In: Shen, S.S., Lewis, P.E. (Eds.), Algorithms and Technologies for Multispectral, Hyperspectral, and Ultraspectral Imagery XI. SPIE, Bellingham, Wash, pp. 662–667. https://doi.org/10.1117/12.606026.

Bruegge, C.J., Conel, J.E., Green, R.O., Margolis, J.S., Holm, R.G., Toon, G., 1992. Water-vapor column abundance retrievals during fife. J. Geophys. Res. 97, 18759-11878.

Deuzé, J.L., Bréon, F.M., Devaux, C., Goloub, P., Herman, M., Lafrance, B., Maignan, F., Marchand, A., Nadal, F., Perry, G., Tanré, D., 2001. Remote sensing of aerosols over land surfaces from POLDER-ADEOS-1 polarized measurements. J. Geophys. Res. 106, 4913–4926.

Diner, D., Martonchik, J., Borel, C., Gerstl, S., Gordon, H., Myneni, R., 1999. Level 2 Surface Retrieval Algorithm Theoretical Basis Document. NASA/JPL, JPL D-11401, Rev. D.

Fan, C., Guang, J., Xue, Y., Di, A., She, L., Che, Y., 2016. An atmospheric correction algorithm for FY3/MERSI data over land in China. In: Geoscience and Remote Sensing Symposium (IGARSS), 2016 IEEE International. IEEE, pp. 4067–4070.

Flaash, U.G., 2009. Atmospheric Correction Module: QUAC and Flaash User Guide v. 4.7. ITT Visual Information Solutions Inc., Boulder, CO, USA.

Frouin, R., Deschamps, P.Y., Lecomte, P., 1990. Determination from space of atmospheric total water vapor amounts by differential absorption near 940nm: theory and airborne verification. J. Appl. Meteorol. 29, 448–459.

Gao, B.C., Goetz, A., 1990. Determination of total column water vapor in the atmosphere at high spatial resolution from AVIRIS data using spectral curve fitting and band ratioing techniques. Proc. SPIE 1298, 138–149.

Gao, B.C., Kaufman, Y.J., 2002. The MODIS Near-IR Water Vapor Algorithm, Algorithm Technical Background Document. ATBD-MOD05, NASA.

Gatebe, C.K., King, M.D., Tsay, S.C., Ji, Q., Arnold, G.T., Li, J.Y., 2001. Sensitivity of off-nadir zenith angles to correlation between visible and near-infrared reflectance for use in remote sensing of aerosol over land. IEEE Trans. Geosci. Remote Sens. 39, 805–819.

Govaerts, Y.M., Wagner, S., Lattanzio, A., Watts, P., 2010. Joint retrieval of surface reflectance and aerosol optical depth from MSG/SEVIRI observations with an optimal estimation approach: 1. Theory. J. Geophys. Res. Atmos. 115.

Hansell, R.A., Ou, S.C., Liou, K.N., Roskovensky, J.K., Tsay, S.C., Hsu, C., Ji, Q., 2007. Simultaneous detection/separation of mineral dust and cirrus clouds using MODIS thermal infrared window data. Geophys. Res. Lett. 34.

He, T., Liang, S.L., Wang, D.D., Wu, H.Y., Yu, Y.Y., Wang, J.D., 2012. Estimation of surface albedo and directional reflectance from Moderate Resolution Imaging Spectroradiometer (MODIS) observations. Remote Sens. Environ 119, 286–300.

Holzer-Popp, T., Schroedter, M., Gesell, G., 2002. Retrieving aerosol optical depth and type in the boundary layer over land and ocean from simultaneous GOME spectrometer and ATSR-2 radiometer measurements, 1. Method description. J. Geophys. Res. Atmos. 107.

Hsu, N.C., Jeong, M.J., Bettenhausen, C., Sayer, A.M., Hansell, R., Seftor, C.S., Huang, J., Tsay, S.C., 2013. Enhanced deep blue aerosol retrieval algorithm: the second generation. J. Geophys. Res. Atmos. 118, 9296–9315.

Hsu, N.C., Tsay, S.C., King, M.D., Herman, J.R., 2004. Aerosol properties over bright-reflecting source regions. IEEE Trans. Geosci. Remote Sens. 42, 557–569.

Hsu, N.C., Tsay, S.C., King, M.D., Herman, J.R., 2006. Deep blue retrievals of Asian aerosol properties during ACE-Asia. IEEE Trans. Geosci. Remote Sens. 44, 3180–3195.

Justice, C.O., Roman, M.O., Csiszar, I., Vermote, E.F., Wolfe, R.E., Hook, S.J., Friedl, M., Wang, Z.S., Schaaf, C.B., Miura, T., Tschudi, M., Riggs, G., Hall, D.K., Lyapustin, A.I., Devadiga, S., Davidson, C., Masuoka, E.J., 2013. Land and cryosphere products from Suomi NPP VIIRS: overview and status. J. Geophys. Res. Atmos. 118, 9753–9765.

Kaufman, Y.J., Gao, B.C., 1992. Remote sensing of water vapor in the near IR from EOS/MODIS. IEEE Trans. Geosci. Remote Sens. 30, 871–884.

Kaufman, Y.J., Tanre, D., Remer, L.A., Vermote, E.F., Chu, A., Holben, B.N., 1997a. Operational remote sensing of tropospheric aerosol over land from EOS moderate resolution imaging spectroradiometer. J. Geophys. Res. Atmos. 102, 17051–17067.

Kaufman, Y.J., Wald, A.E., Remer, L.A., Gao, B.-C., Rong-Rong, L., Flynn, L., 1997b. The MODIS 2.1-μm channel-correlation with visible reflectance for use in remote sensing of aerosol. IEEE Trans. Geosci. Remote Sens. 35, 12.

Kawata, Y., Izumiya, T., Yamazaki, A., 2000. The estimation of aerosol optical parameters from ADEOS/POLDER data. Appl. Math. Comput. 116, 197–215.

Levy, R.C., Mattoo, S., Munchak, L.A., Remer, L.A., Sayer, A.M., Patadia, F., Hsu, N.C., 2013. The Collection 6 MODIS aerosol products over land and ocean. Atmos. Meas. Tech. 6, 2989–3034.

Levy, R.C., Remer, L.A., Dubovik, O., 2007a. Global aerosol optical properties and application to Moderate Resolution Imaging Spectroradiometer aerosol retrieval over land. J. Geophys. Res. Atmos. 112.

Levy, R.C., Remer, L.A., Mattoo, S., Vermote, E.F., Kaufman, Y.J., 2007b. Second-generation operational algorithm: retrieval of aerosol properties over land from inversion of Moderate Resolution Imaging Spectroradiometer spectral reflectance. J. Geophys. Res. Atmos. 112.

Liang, S.L., 2004. Quantitative Remote Sensing of Land Surfaces. Wiley-Interscience.

Liang, S.L., Fang, Fallash-Adl, H., Kalluri, S., JaJa, J., Kaufman, Y.J., Townshend, J.R.G., 1997. An operational atmospheric correction algorithm for landsat Thematic Mapper imagery over the land. J. Geophys. Remote Sens. 40, 2736–2746.

Liang, S.L., Fang, H.L., 2004. An improved atmospheric correction algorithm for hyperspectral remotely sensed imagery. IEEE Geosci. Remote Sens. Lett. 1, 112–117.

Liang, S.L., Fang, H.L., Chen, M.Z., 2001. Atmospheric correction of landsat ETM+ land surface imagery - Part I: methods. IEEE Trans. Geosci. Remote Sens. 39, 2490–2498.

Liang, S.L., Fang, H.L., Morisette, J.T., Chen, M.Z., Shuey, C.J., Walthall, C.L., Daughtry, C.S.T., 2002. Atmospheric correction of landsat ETM plus land surface imagery - Part II: validation and applications. IEEE Trans. Geosci. Remote Sens. 40, 2736–2746.

Liang, S.L., Zhong, B., Fang, H.L., 2006. Improved estimation of aerosol optical depth from MODIS imagery over land surfaces. Remote Sens. Environ. 104, 416–425.

Malenovsky, Z., Rott, H., Cihlar, J., Schaepman, M.E., Garcia-Santos, G., Fernandes, R., Berger, M., 2012. Sentinels for science: potential of Sentinel-1, -2, and -3 missions for scientific observations of ocean, cryosphere, and land. Remote Sens. Environ. 120, 91–101.

Martonchik, J.V., Diner, D.J., Kahn, R., Gaitley, B., Holben, B.N., 2004. Comparison of MISR and AERONET aerosol optical depths over desert sites. Geophys. Res. Lett. 31, L16102.

Qu, Z., Kindel, B.C., Goetz, A.F.H., 2003. The high accuracy atmospheric correction for hyperspectral data (HATCH) model. IEEE Trans. Geosci. Remote Sens. 41, 1223–1231.

Rahman, H., Verstraete, M.M., Pinty, B., 1993. Coupled surface-atmosphere reflectance (csar) model .1. Model description and inversion on synthetic data. J. Geophys. Res. Atmos. 98, 20779–20789.

Remer, L.A., Mattoo, S., Levy, R.C., Munchak, L.A., 2013. MODIS 3 km aerosol product: algorithm and global perspective. Atmos. Meas. Tech. 6, 1829–1844.

Remer, L.A., Wald, A.E., Kaufman, Y.J., 2001. Angular and seasonal variation of spectral surface reflectance ratios: implications for the remote sensing of aerosol over land. IEEE Trans. Geosci. Remote Sens. 39, 275–283.

Richter, R., 1996. Atmospheric correction of satellite data with haze removal including a haze/clear transition region. Comput. Geosci. 22, 675–681.

Richter, R., 2010a. ATCOR-2/3 user guide version 7.1. In: DLR - German Aerospace Center, Remote Sensing Data Center.

Richter, R., 2010b. ATCOR-4 user guide version 5.1. In: DLR - German Aerospace Center, Remote Sensing Data Center.

Richter, R., Schlapfer, D., Muller, A., 2006. An automatic atmospheric correction algorithm for visible/NIR imagery. Int. J. Remote Sens. 27, 2077–2085.

Sano, I., 2004. Optical thickness and Angstrom exponent of aerosols over the land and ocean from space-borne polarimetric data. Adv. Space Res. 34, 833–837.

Santer, R., Carrere, V., Dubuisson, P., Roger, J.C., 1999. Atmospheric correction over land for MERIS. Int. J. Remote Sens. 20, 1819–1840.

Schlapfer, D., Borel, C.C., Keller, J., Itten, K.I., 1998. Atmospheric precorrected differential absorption technique to retrieve columnar water vapor. Remote Sens. Environ. 65, 353–366.

Schlapfer, D., Keller, J., Itten, K.I., 1996. Imaging spectrometry of tropospheric ozone and water vapor. In: Parlow, E. (Ed.), The 15th EARSeL Symposium Basel, pp. 439–446. Rotterdam.

Schmidt, G., Jenkerson, C., Masek, J., Vermote, E., Gao, F., 2013. Landsat ecosystem disturbance adaptive processing system (LEDAPS) algorithm description. In: US Geological Survey.

Tang, J., Xue, Y., Yu, T., Guan, Y., 2005. Aerosol optical thickness determination by exploiting the synergy of TERRA and AQUA MODIS. Remote Sens. Environ. 94, 327–334.

Tanre, D., Holben, B.N., Kaufman, Y.J., 1992. Atmospheric correction algorithm for Noaa-Avhrr products - theory and application. IEEE Trans. Geosci. Remote Sens. 30, 231–248.

Tian, X., Liu, Q., Li, X., Wei, J., 2018a. Validation and comparison of MODIS C6. 1 and C6 aerosol products over Beijing, China. Remote Sens. 10, 2021.

Tian, X., Liu, S., Sun, L., Liu, Q., 2018b. Retrieval of aerosol optical depth in the arid or semiarid region of Northern Xinjiang, China. Remote Sens. 10, 197.

Vermote, E., Justice, C., Claverie, M., Franch, B., 2016. Preliminary analysis of the performance of the Landsat 8/OLI land surface reflectance product. Remote Sens. Environ. 185, 46–56.

Vermote, E., et al., 2006. Second Simulation of a Satellite Signal in the Solar Spectrum—Vector (6SV); 6S User Guide Version 3. NASA Goddard Space Flight Center, Greenbelt, MD, USA.

Vermote, E.F., Roger, J.C., Sinyuk, A., Saleous, N., Dubovik, O., 2007. Fusion of MODIS-MISR aerosol inversion for estimation of aerosol absorption. Remote Sens. Environ. 107, 81–89.

Vermote, E.F., Saleous, N.Z.E., Justice, C.O., 2002. Atmospheric correction of MODIS data in the visible to middle infrared: first results. Remote Sens. Environ. 83, 15.

Wagner, S.C., Govaerts, Y.M., Lattanzio, A., 2010. Joint retrieval of surface reflectance and aerosol optical depth from MSG/SEVIRI observations with an optimal estimation approach: 2. Implementation and evaluation. J. Geophys. Res. Atmos. 115.

Yang, A.X., Zhong, B., Lv, W.B., Wu, S.L., Liu, Q.H., 2015. Cross-calibration of GF-1/WFV over a desert site using landsat-8/OLI imagery and ZY-3/TLC data. Remote Sens. 7, 10763—10787.

Zhao, X., Liang, S.L., Liu, S.H., Wang, J.D., Qin, J., Li, Q., Li, X.W., 2008. Improvement of dark object method in atmospheric correction of hyperspectral remotely sensed data. Sci. China Ser. D Earth Sci. 51, 349—356.

Zhang, Y., He, T., Liang, S.L., Wang, D.D., Yu, Y.Y., 2018. Estimation of all-sky instantaneous surface incident short-wave radiation from Moderate Resolution Imaging Spectroradiometer data using optimization method. Remote Sens. Environ 209, 468—479.

Solar radiation

Abstract

This chapter introduces the principles and algorithms for estimating downward surface shortwave radiation and photosynthetically active radiation (PAR) from remotely sensed data. The advantages and disadvantages of the existing algorithms are also briefly discussed. Section 5.1 briefly introduces basic concepts, such as the solar radiation spectrum, solar constant, and shortwave radiation and PAR. Section 5.2 presents the current observation networks of global solar radiation, including global energy balance archive, baseline surface radiation network, surface radiation budget network, and FLUXNET. Section 5.3 introduces the related solar radiation estimation methods. Section 5.4 introduces the current existing solar radiation products from remote sensing observations. This chapter ends with a short summary.

© 2020 Elsevier Inc. All rights reserved.

5.1 Basic concepts

5.1.1 Solar radiation spectrum

Energy is transmitted in three different ways: conduction, convection, and radiation. In conduction and convection, molecules function as the necessary media because energy cannot be transmitted in a vacuum. However, most of the large distance between the sun and the Earth is vacuum, with the exception of their atmospheres; therefore, solar energy cannot be transmitted to Earth by conduction and convection. Instead, radiation continuously transmits energy from the sun to the Earth, providing a major energy source for life on Earth. Solar radiation has many functions. For instance, the radiation absorption by land surface varies with latitude, so the north–south exchange of cold and warm air masses results in atmospheric circulation. Furthermore, the surface temperature increases with the increase in solar radiation absorbed by the surface, which facilitates the vertical movement of the atmosphere, resulting in convective weather. In addition, the formation of clouds, fog, rain, and snow requires solar radiation; thus, solar radiation is a worthwhile topic for deep investigation.

The solar radiation spectrum shows the solar radiation distribution by wavelength. Despite the broad wavelength range of solar radiation, very little energy is emitted as very long- or very short-wavelength radiation; the majority of the energy is concentrated within the wavelength range of 250–2500 nm, accounting for 99% of the total solar radiation. The visible, infrared, and ultraviolet bands contain 50%, 44%, and 6% of the energy, respectively. The peak radiation wavelength is approximately 480 nm. As the solar light penetrates the atmosphere, the solar radiation is attenuated by the atmosphere through absorption, scattering, and reflection.

5.1.2 Solar constant

The change in top-of-atmosphere (TOA) irradiance is mainly dependent on the distance between the sun and the Earth. Given the solar irradiance at the average Earth–sun distance,

$\overline{E_0}$, solar irradiance can be calculated using Formula 5.1 (Duffle and Beckman, 1980; Liang, 2004):

$$E_0 = \overline{E_0}(1 + 0.033 * \cos(2\pi d_n / 365)) \quad (5.1)$$

where d_n is the day number in reference to a year, ranging from 1 to 365. A more precise formula is

$$E_o = \overline{E_0}[1.00.00128 \sin \chi + 0.000719 \cos 2\chi + 0.000077 \sin 2\chi] \quad (5.2)$$

(Liang, 2004; Spencer, 1971), where $\chi = 2\pi(d_n - 1)/365$. In most cases, the difference between the results of these two formulas is negligible (Liang, 2004).

The solar constant is generally the integration of irradiance over the full wavelength range, calculated by Formula 5.3:

$$I_0 = \int_0^\infty \overline{E_0}(\lambda)d\lambda \quad (5.3)$$

The average monitored solar constant was approximately 1369 W/m^2 ±0.25% during the 1980s (Hartmann et al., 1999). The Moderate Resolution Transmission (MODTRAN) radiative-transfer software (Anderson et al., 1999) provides TOA radiance datasets derived from different data sources corresponding to various solar constants, such as 1362.12, 1359.75, 1368.00 and 1376.23 W/m^2 (Liang, 2004). The latest most accurate value of total solar irradiance during the 2008 solar minimum period is 1360.8 ± 0.5 W/m^2 according to measurements from the total irradiance monitor on NASA's Solar Radiation and Climate Experiment and a series of new radiometric laboratory tests (Kopp and Lean, 2011).

5.1.3 Shortwave radiation and photosynthetically active radiation

Surface shortwave radiation is referred to total solar irradiance with wavelengths in the range of 300–3000 nm and is usually defined as

$$I_g = \int_{0.3\mu m}^{3.0\mu m} I(\lambda)d\lambda \quad (5.4)$$

where $I(\lambda)$ is the spectral irradiance.

Photosynthetically active radiation (PAR) is the solar radiation within the wavelength range of 400–700 nm.

$$\text{PAR} = \int_{0.4\ \mu m}^{0.7\ \mu m} I(\lambda)d\lambda \qquad (5.5)$$

PAR constitutes the basic source of energy for biomass by controlling the photosynthetic rate of organisms on land, thus directly affecting plant growth. In addition, PAR is a key climatic resource because it influences the substance and energy exchange between the land and atmosphere (Li et al., 1997). A wide range of land surface ecosystem models, including many biological models and land surface–atmosphere models, simulate biological dynamics and the interactions between the global carbon cycle and water cycle. In nearly all of these models, photosynthesis adjusts the water and carbon exchange between the vegetation canopy and atmosphere (Liang et al., 2006), and incident PAR is one of the most important input parameters for these models.

PAR is the visible part of DSSR and is correlated with the overall DSSR. The specific correlation can be expressed by

$$\text{PAR} = I_g \cdot \xi \qquad (5.6)$$

where I_g is the shortwave radiation reaching at surface, and ξ is the coefficient for the proportion of PAR in shortwave radiation I_g. Although ξ is generally stable (around 0.5), it is not a constant.

5.1.4 Attenuation of solar radiation

Because of the Earth's atmosphere, solar radiation is attenuated by phenomena such as atmospheric absorption and scattering as well as cloud reflectance before reaching the land surface (Fig. 5.1). Generally, when solar radiation penetrates the atmosphere, water vapor and carbon dioxide absorb the infrared energy, while the ozone layer absorbs the UV energy. The short-wavelength radiation in the visible waveband is scattered by aerosol and air molecules;

thus, only some of the solar radiation penetrates the atmosphere and reaches the land surface. The total amount of radiation reaching the horizontal land surface, I_g, is composed of direct solar radiation, scattered solar radiation, and radiation resulting from surface reflectance, as expressed by the formula below:

$$I_g = I_b * \cos(\theta) + I_{as} + I_r \qquad (5.7)$$

$$I_d = \int I_b * \sin(\theta) \qquad (5.8)$$

where I_b is the surface direct solar radiation in the orthogonal direction; I_d is the direct solar radiation reaching the horizontal surface; I_{as} is the scattered radiation; I_r is the radiation reflected by the surface; and θ is the incident angle, i.e., the solar zenith angle or the complement of the solar elevation angle.

5.1.5 Earth radiation budget

The Earth radiation budget (ERB) is a concept for understanding the energy the Earth receives from the sun and the Earth emits back to outer space. The ERB brings into imbalance as the Earth receives more energy from the sun. As a result, the Earth emits more infrared energy. If the Earth emits more infrared energy than it absorbs, the Earth cools. The Earth emits less energy as it cools. Then, the radiation budget bring into balance again. In a word, the energy received from the sun heats the Earth, meanwhile, the emitted energy lowers the surface temperature. The ERB reflects the relationships between incoming solar radiations and outgoing radiation, which is illustrated in Fig. 5.1.

The surface radiation budget involves the following variables: downward shortwave radiation, upward shortwave radiation, downward longwave radiation, upward longwave radiation, and net radiation, as shown in Formula 5.9. The net surface shortwave and longwave radiations, S_n and L_n, respectively, are summed to obtain the net surface radiation:

$$R_n = S_n + L_n = (S\downarrow - S\uparrow) + (L\downarrow - L\uparrow)$$
$$= (1-a)S\downarrow + (L\downarrow - L\uparrow) \qquad (5.9)$$

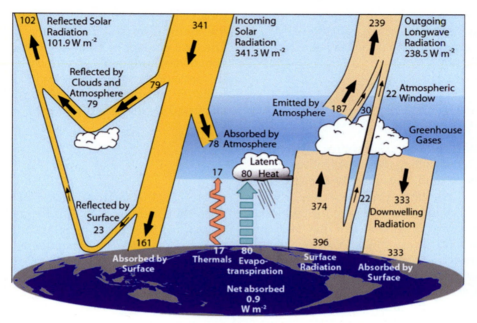

FIGURE 5.1 The global annual mean Earth's energy budget for 2000–05 (W/m²). The broad arrows indicate the schematic flow of energy in proportion to their importance (Trenberth and Fasullo, 2012).

where $S\downarrow$ is the downward shortwave radiation, $S\uparrow$ is the upward shortwave radiation, α is the surface shortwave albedo, $L\downarrow$ is the downward longwave radiation, and $L\uparrow$ is the upward longwave radiation.

5.2 Observation network of land surface radiation

There are several radiation networks for radiation measurements such as the global energy balance archive (GEBA) (Gilgen and Ohmura, 1999; Liang et al., 2010), the baseline surface radiation network (BSRN) (Liang et al., 2010; Ohmura et al., 1998), the surface radiation budget network (SURFRAD) (Augustine et al. 2000, 2005; Liang et al., 2010), and FLUXNET (Baldocchi et al., 2001; Liang et al., 2010). The data obtained by these observation networks can be used to verify the solar radiation estimation models or the accuracy of atmospheric radiative–transfer simulation models.

Instruments for radiation observation can be divided by purpose and requirements into pyranometers, pyrgeometers, net pyrradiometers, and PAR meters. Pyranometers are the most common instrument for measuring total surface radiation. Pyrgeometers measure upward and downward longwave radiation. General pyrradiometers measure shortwave radiation and longwave radiation, while net pyrradiometers measure upward and downward radiations. PAR meters measure PAR.

The GEBA database collects surface solar radiation data globally and is maintained by ETH Zurich. Observation data of the surface radiation budget data have been collected at more than 2000 stations over more than 250,000 months since 1950. Gilgen et al. (1998) estimated and assessed the precision of the monthly and yearly mean downward solar radiation data from GEBA. The root mean square errors (RMSE) of the monthly and yearly data were 5% and 2%, respectively.

The BSRN is an observation network that provides surface solar radiation data through continuous global-scale observations; many of its stations having been in operation since 1992. Of the 59 total observation stations, at least 40 stations are now in operation. The observation precisions of the BSRN are 5, 2, and 5 W/m^2 for solar shortwave radiation, direct solar radiation, and scattered solar radiation, respectively (Ohmura et al., 1998). Fig. 5.2 shows the geographical distribution of the existing BSRN stations and future stations.

The SURFRAD observation network was established to better understand the global surface radiation budget and investigate the mechanism of climate change by NOAA in 1993. The purpose of the SURFRAD network is to provide precise, continuous, long-term observation data regarding the surface solar radiation for the entire United States. The solar and infrared are the primary measurements at the SURFRAD network, ancillary observations include direct and diffuse solar, PAR, spectral solar, and meteorological parameters. To date, SURFRAD has seven stations and the data can be downloaded from the Internet (http:// www.srb.noaa.gov). The relative error of the measurements is between ±2 and ± 5%. Table 5.1 shows the detailed information of the SURFRAD stations.

FLUXNET is a global observation network of microclimate towers that measures carbon dioxide, water vapor, and energy exchange between the atmosphere and land surface using the eddy covariance method. FLUXNET contains a series of regional networks, including AmeriFlux, CarboEuropeIP, AsiaFlux, KoFlux, OzFlux, Flux-Canada, and ChinaFlux. From 1996 to July 2009, 500 observation towers measured the fluxes and related parameters on a long-term basis. Because different instruments were used in different regions and countries, no unified criterion exists for assessing the precision of surface solar radiation observations for the FLUXNET sites. Fig. 5.3 indicates the distribution of existing FLUXNET towers.

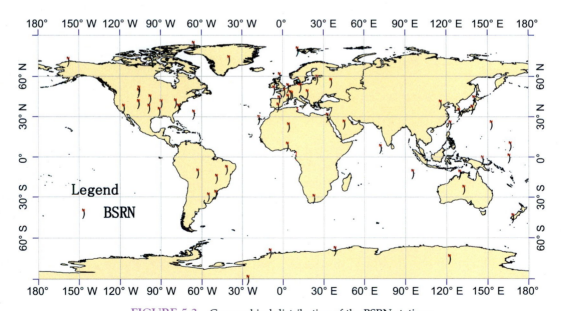

FIGURE 5.2   Geographical distribution of the BSRN stations.

TABLE 5.1   Detailed information on the SURFRAD stations.

Name	Longitude/latitude	Elevation (m)	Vegetation cover type
Bondville, IL	40.05°N, 88.37°W	213	Cropland
Boulder, CO	40.13°N, 105.24°W	1689	Grassland
Fort Peck, MT	48.31°N, 105.10°W	634	Grassland
Sioux Falls, SD	43.73°N, 96.62°W	473	Grassland
Penn State, PA	40.72°N, 77.93°W	376	Cropland
Goodwin Creek, MS	34.25°N, 89.87°W	98	Pasture
Desert Rock, NV	36.63°N, 116.02°W	1007	Desert

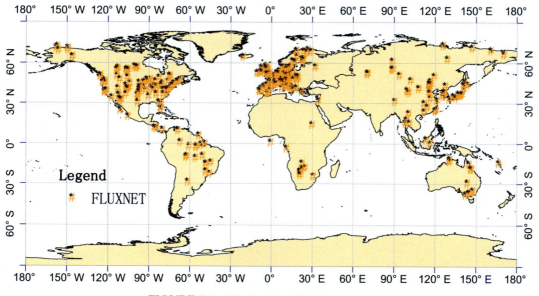

FIGURE 5.3   Distribution of FLUXNET stations.

In addition to the large-scale global radiation observation network introduced in this section, many regional radiation observation networks exist, including the atmospheric radiation measurement, whose data can be accessed at http://www.arm.gov/. The purpose of the GEWEX Asia Monsoon Experiment (GAME/AAN) is to improve the accuracy of Asian monsoon simulations by studying the role of Asian monsoons in the global energy and water budgets (Yang et al., 2008). The Greenland Climate Network (GC-net) has more than 20 stations observing surface solar radiation and relevant meteorological parameters in Greenland with an observation precision of 5%−15% (Liang et al., 2010). The Aerosol Robotic Network (Aeronet) is the land surface aerosol observation network built by NASA PHOTONS, with some of the stations observing solar radiation. In the Amazon, the surface solar radiation is overestimated by approximately 6−13 W/m²Wm^{-2}.

5.3 Surface radiation estimation based on satellite remote sensing and GCM

One of the most practical and reliable method for estimating the solar radiation is based on remote sensing because remote sensing observations are superior to station-based observations for acquiring the features of both regional and global coverages (Liang, 2004). Many algorithms have been developed to calculate the amount of solar radiation at surface; however, few global land surface radiation products exist due to the limited applicability of the current algorithms. This section discusses the advantages, disadvantages, and applicability of the existing algorithms for estimating surface radiation.

Since the 1960s, meteorological satellites have greatly expanded our knowledge of surface energy and radiation budgets. Many radiometers and sensors have been developed to measure or observe surface solar radiation, such as the ERB sensor carried on Nimbus-7 (Jacobowitz and Tighe, 1984), the Clouds and the Earth's Radiant Energy System (CERES) sensor carried on three satellites (Barkstrom and Smith, 1986), and the Geostationary Earth Radiation Budget sensor carried on Meteosat-8 and Meteosat-9 (Harries et al., 2005). Multiband sensors are also used to generate surface radiation products, such as the Spinning Enhanced Visible and Infrared Imager (SEVIRI) sensor carried on the METEOSAT Second Generation satellite and the GOES-R ABI (Laszlo et al., 2008) and Moderate Resolution Imaging Spectroradiometer (MODIS) sensors (Liang et al., 2006).

There are currently many approaches for estimating the amount of solar radiation at surface, including the radiative-transfer model; parameterization methods; empirical models; lookup table method, and also machine learning methods. The MODTRAN radiative-transfer model is representative of the first method; however, its computational complexity makes it unsuitable for estimating high spatial and

temporal resolution global surface radiation flux directly. Parameterization method can be divided into spectral and broadband models. In spectral models, the amount of solar radiation at surface in each waveband is calculated, and the total surface solar radiation is then obtained by integration. The broadband models directly calculate the total solar radiation using various models. Compared with broadband models, spectral models have certain disadvantages, such as computational complexity. Many spectral models have been introduced and have been tested (Gueymard, 1995; Iqbal, 1983; Van Laake and Sanchez-Azofeifa, 2004). An even greater number of surface solar radiation research efforts have used the relatively simple broadband model (Bird and Hulstrom, 1981a,b; Gueymard, 1993a,b; Huang et al., 2018; Iqbal, 1983; Maxwell, 1998; Pinker and Laszlo, 1992; Qin et al., 2015; Ryu et al., 2008; Tang et al., 2016; Yang et al., 2000). The estimation precision of this algorithm depends on the model selection, the quality of the input parameters, instrument calibration, the spatial and temporal resolution of the data, and the time difference between the remote sensing data acquisition and surface data measurement (Wang and Pinker, 2009). Most of the broadband models estimate only shortwave radiation, and models that directly estimate PAR are less common. PAR is estimated in many models by conversion from the shortwave radiation model to the PAR directly, although the conversion relationship is subject to the influence of multiple factors. Therefore, the direct use of constants for conversion inevitably causes potential errors. Traditional empirical approaches can also be used to estimate surface solar radiation, such as the empirical regression method and the relative sunshine duration—based estimation of surface solar radiation. The disadvantages of these approaches are their limited universality and the limitations of their use of ground observation data. Liang et al. (2006) presented a lookup table—based surface solar radiation estimation method, which overcomes such disadvantages as the large number and complexity of input parameters in previous estimation models. In addition, this method requires only the parameters observed by remote sensing, instead of the retrieved parameters of cloud and aerosol properties.

Remote sensing—based surface solar radiation retrieval method is strongly influenced by the calibration precision of the sensors. Schmetz (1989) and Pinker et al. (1995) found that the precision error of estimation of instantaneous incoming shortwave radiation based on geostationary satellite was between 10% and 15%, the error associated with hourly estimations of the surface solar radiation under clear-sky conditions is 5%—10%, and the error associated with all-sky estimations could reach as high as 15% —30%. Therefore, the estimation precision of surface solar radiation must be improved (Liang et al., 2010). The following section will thoroughly discuss the methods and principles available for estimating surface solar radiation.

5.3.1 Empirical model

5.3.1.1 Simple empirical model

5.3.1.1.1 Lacis and Hansen model

According to the model proposed by Lacis and Hansen (1974), the global irradiance is directly calculated instead of estimating the direct, diffuse, and reflected components separately. The equations are as follows:

$$I_g = I_0 \cos(\theta)\big($$

$$\times \big(0.647 - r'_s - a_0\big) / \big(1 - 0.0685 r_g\big) + 0.353 - a_w\big) \tag{5.10}$$

$$r'_s = 0.28/(1 + 6.43\cos(\theta)) \tag{5.11}$$

$$a_0 = 1 - T_O \tag{5.12}$$

where I_0 is the extraterrestrial solar irradiance, r_g is the ground albedo, and θ is the solar zenith

angle; the transmittance of ozone absorptance, T_O, is calculated using Eq. (5.42). a_w is obtained using Eq. (5.45) with the following correction:

$$X_w = X_w(p/1013.25)^{0.75}(273/T)^{0.5} \tag{5.13}$$

5.3.1.1.2 Gueymard model

This model is a purely mathematical model and does not estimate total radiation by first calculating the atmospheric transmission of solar radiation and then calculating the direct solar radiation and scattered solar radiation, as in other models (Gueymard, 1993b). The downward surface shortwave radiation (DSSR) is directly calculated using the formula below:

$$I_g = I_0 F_{bg}(w) F_g(p, \beta) \sum_{j=0}^{4} g_j \sin^j h \tag{5.14}$$

where $F_{bg}(w)$ is the function of water vapor and $F_g(p, \beta)$, whose value is 1.0 at sea level, is a function of station pressure p and the Angstrom turbidity coefficient β.

$$F_g(p, \beta) = 1 + (0.0752 - 0.107\beta)(1 - p/p_0) \tag{5.15}$$

$$p / p_0 = \exp(0.00177 - 0.11963z - 0.00136z^2) \tag{5.16}$$

where z is surface elevation in kilometers, p_0 is the pressure at sea level, and g_j is a function of the atmospheric turbidity coefficient, calculated by

$$T = \ln(1 + 10\beta) \tag{5.17}$$

$$g_j = \sum_{k=0}^{3} d_{kj} T^k \quad \text{where } j = 1 - 4 \text{ and } g_0$$

$$= 0.006 \tag{5.18}$$

Table 5.2 lists the specific values of d_{kj} (Gueymard, 1993b).

TABLE 5.2 Values of d_{kj} used in Formula 5.18.

j	1	2	3	4
d_{0j}	0.38702	1.35369	−1.59816	0.66864
d_{1j}	−3.8625	1.53300	−1.90377	0.80172
d_{2j}	0.09234	−1.07736	1.63113	−0.75795
d_{3j}	0	0.23728	−0.38770	0.18895

5.3.1.2 Relative sunshine duration model

The method of estimating surface solar radiation based on relative sunshine duration can be considered an empirical model. Many studies focus on the sunshine duration−based estimation of surface solar radiation (Falayi et al., 2008; Hanna and Siam, 1981; Kumar et al., 2001; Safari and Gasore, 2009; Telahun, 1987). The most widely used method is that of Angstrom (1924), who presented a linear relationship between the ratio of average daily global radiation to the corresponding value on a completely clear day and the average daily sunshine duration to the possible sunshine duration. The expression for calculating daily solar radiation is as follows:

$$H = H_0 \left(a + b \frac{n}{N} \right) \qquad (5.19)$$

where H and H_0 are, respectively, the monthly mean daily global radiation and the daily extraterrestrial radiation on a horizontal surface, n and N are, respectively, the monthly average sunshine duration and the monthly average maximum possible daily sunshine duration, and a and b are empirically derived regression constants. The daily extraterrestrial radiation can be estimated using Eq. (5.20).

$$H_0 = \frac{24}{\pi} I_0 \left(\cos(lat)\cos \delta \sin w_s + \frac{\pi}{180} w_s \sin \lambda \sin \delta \right) \qquad (5.20)$$

where lat is the latitude of the site, δ is the solar declination, and w_s is the mean sunrise hour angle for the given month, which can be estimated by Eq. (5.21).

$$w_s = \cos^{-1}(-\tan(lat)\tan \delta) \qquad (5.21)$$

The maximum possible daily sunshine duration, N, can be obtained by Eq. (5.22).

$$N = \frac{2}{15} w_s \qquad (5.22)$$

The relationship between monthly mean daily global radiation and the daily extraterrestrial radiation on a horizontal surface can also be expressed in quadratic, third degree, and logarithmic functions (Akinoğ;lu and Ecevit, 1990; Almorox and Hontoria, 2004; Ampratwum, 1999; Ertekin and Yaldiz, 2000). Other methods can also be used to estimate the surface downward shortwave radiation, such as the algorithm using the temperature, humidity, and precipitation measured at the station as proposed by Thornton and Running (1999).

5.3.2 Parameterization method

Another commonly applied method of estimating surface solar radiation is the parameterization method, i.e., establishing a physical model to simulate the direct interaction between solar radiation and the atmosphere. When penetrating the atmosphere, solar radiation is partly absorbed by water vapor, gas, or ozone or it is partly absorbed and scattered by aerosol. Through a careful calculation of the amount of radiation absorbed, reflected, and scattered, the amount of solar radiation at surface can be obtained.

At present, most of the surface solar radiation estimation models operate under clear-sky rather than cloudy-sky conditions. The physical simulation of clouds is still a key factor to solar radiation estimation using remotely sensed data. This section presents the parameterization model and the algorithm used to estimate surface solar radiation under clear-sky and cloudy-sky conditions, respectively.

5.3.2.1 Inputs for parameterization models

There are more than a dozen models for estimating solar radiation at surface under clear-sky conditions. These models are more complex and require many parameters, such as surface and atmospheric parameters, as inputs to estimate solar radiation at surface. Atmospheric parameters include water vapor content, ozone content, aerosol optical thickness, and the Angstrom turbidity coefficient.

Regarding the acquisition method for surface parameters, the MODIS sensor carried by Terra and Aqua provides many land surface and atmospheric products with extensively verified and assessed precision (Gao and Kaufman, 2003; Kahn et al., 2007; Liu et al., 2009). In this section, several broadband models are selected to estimate surface solar radiation with greatly reduced complexity compared with radiative-transfer software.

Table 5.3 compares the land surface parameters required by various broadband models under clear-sky conditions. These basic parameters can be obtained from MODIS land surface or atmospheric products. The detailed information of MODIS products used in these models is presented in Table 5.4.

5.3.2.2 Clear-sky model

This section introduces the broadband and spectral models used to estimate the amount of shortwave radiation at surface under clear-sky conditions.

5.3.2.2.1 Broadband model

In most of the broadband transmission-based parameterization models, the atmospheric absorption of direct solar radiation is calculated by transmittance T_t according to the following formula:

$$I_d = I_0 \cos \theta * T_t \qquad (5.23)$$

TABLE 5.3 Parameters required by broadband models under clear-sky conditions.

Model	θ	w	u_o	τ_a	β	α	p	r_g	T
Bird (Bird and Hulstrom, 1981b)	✓	✓	✓	✓			✓	✓	
Davies and Hay (1979)	✓	✓	✓				✓	✓	
Hoyt (1978)	✓	✓	✓	✓				✓	
Lacis and Hansen (1974)	✓	✓	✓				✓	✓	✓
Choudhury (1982)	✓	✓	✓			✓		✓	
CPCR2 Gueymard (1989)	✓	✓	✓		✓	✓	✓		

Parameters: θ, solar zenith angle; w, water vapor content; u_o, ozone content; τ_a, aerosol optical thickness; β, Angstrom turbidity coefficient; α, Angstrom coefficient; p, pressure at the station; r_g, surface albedo; T, surface temperature.

where I_0 is the extraterrestrial solar radiation with the sun—Earth distance correction by Formula 5.1, and T_t is the sum of transmittances generated by the atmospheric attenuation of solar radiation.

$$T_t = T_R T_A T_O T_W T_G T_N \qquad (5.24)$$

where T_R, T_A, T_O, T_W, T_G, and T_N represent Rayleigh scattering, aerosol absorptance and scattering, absorption by ozone, absorption by water vapor, absorption by mixed gas, and transmittance of NO_2, respectively. Some of the broadband models do not consider T_N because the effect of NO_2 on radiation is very small.

5.3.2.2.1.1 Modified Bird model   Bird's model is based on the comparison of SOLTRAN3 and SOLTRAN4 (Bird and Hulstrom, 1981b) and

TABLE 5.4 MODIS products required by parameterization models.

MODIS product	Product	Version	Spatial resolution	Parameter
Geographical location product	MOD03	5	1 km	Solar zenith angle
	MYD03			
Land surface product	MCD43B3	5	1 km	Surface albedo
	(Terra and Aqua)			
Aerosol product	MOD04	5	10 km	Aerosol optical thickness Atmospheric turbidity coefficient
	MYD04			
Water vapor product	MOD05	5	1 km	Water vapor
	MYD05			
Atmospheric product	MOD07	5	5 km	Ozone
	MYD07			
Cloud product	MOD06 MYD06	5	1 and 5 km	Cloud mask (1 km); surface temperature (5 km); surface pressure (5 km)

has been evaluated in several studies (Gueymard, 2003a,b; Ryu et al., 2008). With Bird's model, direct normal irradiance can be obtained as

$$I_d = I_0 \cos(\theta)0.9662 T_A T_R T_G T_O T_W \quad (5.25)$$

Diffuse radiation can be expressed as

$$I_{as} = I_0 \cos(\theta)0.79 T_O T_W T_G T_{AA}$$
$$\frac{0.5(1 - T_R) + B_a(1 - T_{AS})}{1 - M + M^{1.02}} \quad (5.26)$$

where T_{AA} is the transmittance of aerosol absorption, T_{AS} is the transmittance of aerosol scattering, and B_a is the forward scattering ratio due to the aerosols and is treated as a constant, namely 0.84 (Annear and Wells, 2007). M is the optical air mass.

$$M = \left(\cos(\theta) + 0.15(93.884 - \theta)^{-1.25} \right)^{-1} \quad (5.27)$$

The irradiance from the multiple reflections between the ground and sky is calculated as

$$I_r = r_g r_s (I_b + I_d)/(1 - r_g r_s) \quad (5.28)$$

where r_s and r_g are the atmospheric and ground albedo, respectively, which can be obtained from the MODIS land product (MCD43B3). The atmospheric albedo, r_s, can be calculated as

$$r_s = 0.0685 + (1 - B_a)\left(1 - \frac{T_A}{T_{AA}}\right) \quad (5.29)$$

The aerosol transmittance of absorptance, T_{AA}, was calculated by Bird and Hulstrom (1981b) using the formula

$$T_{AA} = 1 - K_l(1 - M + M^{1.06})(1 - T_A) \quad (5.30)$$

where K_l is an empirical absorptance coefficient. Bird and Hulstrom (1981b) recommended that the coefficient be set to 0.1.

The transmittance of aerosol absorption and scattering, T_A, was calculated by the formula

$$T_A = \exp\left(-\tau_a^{0.873}\left(1 + \tau_a - \tau_a^{0.7088}\right)M^{0.9108}\right)$$
(5.31)

In the original publication, the average aerosol optical depth, τ_a, was defined from the spectral depth at 0.38 and 0.50 μm (Bird and Hulstrom, 1981b; Gueymard, 2003a). Here, the MODIS aerosol optical depth is directly used instead. The transmittance of aerosol scattering can be obtained from

$$T_{AS} = T_A/T_{AA}$$
(5.32)

The transmittance of Rayleigh scattering, T_R, was calculated by Bird and Hulstrom (1981b) using

$$T_R = \exp\left(-0.0903M_p^{0.84}\left(1 + M_p - M_p^{1.01}\right)\right)$$
(5.33)

where M_p is the pressure-corrected air mass, which can be obtained by

$$M_p = M_p/1013.25$$
(5.34)

The transmittance of water vapor, T_W, was calculated by

$$T_w = 1$$
$$-\ 2.4959X_w\left(\left(1 + 79.034X_w\right)^{0.6828} + 6.385X_w\right)^{-1}$$
(5.35)

where X_w is the precipitable water content in a slanted path, which was calculated as

$$X_w = wM$$
(5.36)

The transmittance of ozone absorptance, T_O, was calculated by Bird and Hulstrom (1981b) as

$$T_O = 1 - 0.1611X_o(1 + 139.48X_o)^{-0.3035}$$
$$-\ 0.002715X_o\left(1 + 0.044X_o + 0.0003X_o^2\right)^{-1}$$
(5.37)

where X_o is the ozone amount in a slanted path and can be calculated by Bird and Hulstrom (1981b) as

$$X_o = u_oM$$
(5.38)

where u_o is the ozone amount.

The transmittance of mixed gases (carbon dioxide and oxygen), T_G, can be calculated by Bird and Hulstrom (1981b) as

$$T_G = \exp\left(-0.0127M_p^{0.26}\right)$$
(5.39)

The validation results of the estimated DSSR of the Modified Bird model using MODIS products from Terra under clear-sky conditions at the Bondville station from 2003 to 2005 is shown in Fig. 5.4.

5.3.2.2.2 Spectral model

Under clear-sky conditions, the spectral model to estimate spectral radiation is much more complex than the broadband model. In the spectral model, the transmittance of ozone, water vapor, aerosol, and mixed gas is calculated for various wavelength ranges, and shortwave radiation is obtained by integration within a certain wavelength range. The advantage of spectral models over broadband models is that radiation can be calculated at any wavelength or in any wavelength range. However, due to this computational complexity, the spectral model requires longer calculation times to estimate surface solar radiation than does the broadband model. Two relatively simple spectral models, reported by Iqbal (1983) and Gueymard (1995), will be introduced.

5.3.2.2.2.1 Iqbal spectral model Similar to general broadband models, spectral models are also generated by multiplying the TOA spectral irradiance by the atmospheric transmission of solar radiation. The TOA spectral irradiance is dependent on the sun–Earth distance. Spectral

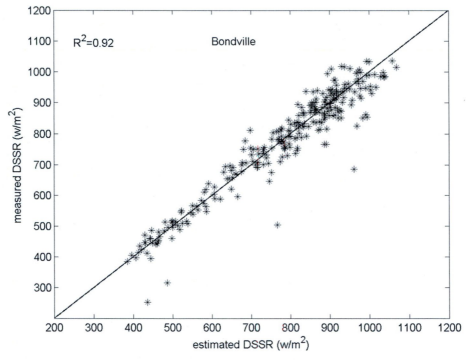

FIGURE 5.4 The validation results of the estimated DSSR of the Modified Bird model using MODIS products from Terra under clear-sky conditions at the Bondville station from 2003 to 05.

models divide the solar radiation at surface into direct solar radiation and scattered solar radiation. In the Iqbal (1983), the atmospheric attenuation of solar radiation is divided into the following parts:

$$T_{t\lambda} = T_{O\lambda}T_{R\lambda}T_{G\lambda}T_{W\lambda}T_{A\lambda} \qquad (5.40)$$

where $T_{O\lambda}$, $T_{R\lambda}$, $T_{G\lambda}$, $T_{W\lambda}$, and $T_{A\lambda}$ represent the transmission coefficients of ozone absorption, Raleigh scattering, absorption by mixed gas, absorption by water vapor, and aerosol scattering at different wavelengths, respectively. These transmission coefficients can be calculated using the following formulas:

$$T_{O\lambda} = \exp[-k_{o\lambda}u_0M_0] \qquad (5.41)$$

$$T_{R\lambda} = \exp[-0.008735\lambda^{-4.08}M_p] \qquad (5.42)$$

$$T_{G\lambda} = \exp\left[-1.4k_{g\lambda}m_0 \, / \, (1 + 118.93k_{g\lambda}M_0)^{0.45}\right] \qquad (5.43)$$

$$T_{W\lambda} = \exp\left[-0.2385k_{w\lambda}wM_0 \, / \, (1 + 20.07k_{w\lambda}wM_0)^{0.45}\right] \qquad (5.44)$$

$$T_{A\lambda} = \exp[-\beta\lambda^{-1.3}M_p] \qquad (5.45)$$

where λ is the wavelength of direct solar radiation (μm); u_0 and w(cm) represent the ozone and water vapor contents, respectively; β is the Angstrom turbidity coefficient; and $k_{g\lambda}$, $k_{w\lambda}$, and $k_{o\lambda}$ are the coefficients of solar radiation for mixed gas, water vapor, and ozone, respectively, in terms of wavelength (for specific values, please refer to Iqbal (1983)). M is the relative air mass, and M_p is the air mass corrected by pressure, both of which are calculated according to the formulas below:

$$M = 1/\cos\theta \quad \theta <= 60° \tag{5.46}$$

$$M = 1/\left(\cos\theta + 0.15(93.385 - \theta)^{-1.253}\right) \quad \theta$$
$$> 60° \tag{5.47}$$

$$M_p = M_p/1013.25 \tag{5.48}$$

where θ is the solar zenith angle. Transmittances at different wavelengths are calculated using the formulas above, and the amount of solar radiation at surface can be obtained by wavelength integration.

To determine the scattered solar radiation using the Iqbal spectral model, the following formula is used:

$$F = 0.9302 \cos\theta^{0.2556} \tag{5.49}$$

$$I_{as\lambda} = E_{0\lambda}\cos\theta T_{O\lambda}T_{w\lambda}[0.5T_{A\lambda}(1 - T_{R\lambda}) + [Fw_0 T_{R\lambda}(1 - T_{A\lambda})]] \tag{5.50}$$

where w_0 is single-scattering albedo, which is derived from the specific aerosol type given by MODIS aerosol products (Kaufman and Tanre' 1998; Van Laake and Sanchez-Azofeifa, 2004).

The left side of Eq. 5.50 describes scattered solar radiation caused by Rayleigh scattering of direct solar radiation, while the right side describes scattered solar radiation caused by aerosol scattering of direct solar radiation. Fig. 5.5 shows the verification of the spectral model at the Desert Rock and Boulder stations, which has reduced applicability due to the computational complexity of this model.

5.3.2.3 Cloudy-sky model

Compared with clear-sky surface solar radiation estimation models, the radiation estimation under cloudy-sky conditions is much more complex, with large estimation uncertainties. Generally, there are far fewer clear-sky days than cloudy-sky days, especially in Beijing, China, and in the Amazon area in South America, where pollution is a serious problem.

In the cloudy-sky model, the algorithm used to estimate the total solar radiation is slightly

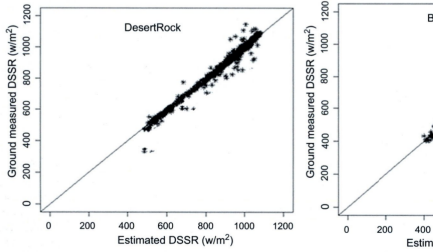
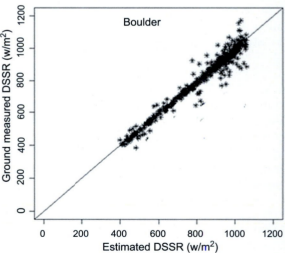

FIGURE 5.5   Verification of the Iqbal spectral model using MODIS products (left) from Terra and (right) from Aqua at SURFRAD stations under clear-sky conditions from 2003 to 05.

similar to that used in the clear-sky model. The cloudy-sky model can be divided into the simple broadband and complex narrowband models, which are introduced in detail as follows.

5.3.2.3.1 Broadband model

The broadband model used for cloudy-sky conditions, proposed by Choudhury (1982), is simple with readily accessible parameters. The surface solar radiation on a cloudy day can be known if the cloud coverage, solar zenith angle, cloud optical thickness, and land surface albedo are given. The specific calculation is as follows:

$$I_{gcld} = I_0(1 - F_{cld} + F_{cld}T_{cld})/(1 - F_{cld}r_g x_{cld})$$
(5.51)

$$T_{cld} = (0.97(2 + 3\cos\theta))/(4 + 0.6\tau_{cld})$$
(5.52)

$$x_{cld} = 0.6\tau_{cld}/(4 + 0.6\tau_{cld})$$
(5.53)

where I_0 is the extraterrestrial solar irradiance, I_{gcld} is the total amount of shortwave radiation under the cloudy-sky condition, F_{cld} is the fraction of cloud cover, τ_{cld} is the cloud optical thickness, and r_g is the surface albedo, all of which can be derived from MODIS products. Fig. 5.6 shows the verification of the combined Bird clear-sky model and broadband cloudy-sky model at the SURFRAD station based on MODIS products from Terra of 2003–05. Blue represents a clear sky, while red represents a cloudy sky.

5.3.2.3.2 Dual-band model

Under cloudy-sky conditions, part of the solar radiation will be reflected back into space by the top of the cloud, while the other part penetrates the cloud and reaches the land surface. Using the clear-sky model to calculate the amount of radiation reaching the top of the cloud, the cloud reflectance and transmittance of solar radiation can be determined. Thus, the amount of shortwave radiation at surface is known. When using this method, Stephens (Stephens, 1978a,b) divided the shortwave radiation into two wavelength ranges: 300–750 nm and 750–4000 nm.

Assuming that the sunlight falls on the top of a cloud at the solar zenith angle θ and that the lower boundary of the cloud is a nonreflecting boundary, the reflectance Re and transmittance T_r of the cloud with an optical thickness of τ_{cld} can be calculated using the following formula (Coakley and Chylek, 1975):

(1) For nonabsorptive media, $w_0 = 1$.

$$\text{Re}(u) = \frac{\beta(\mu)\tau_{cld}/\mu}{1 + \beta(\mu)\tau_{cld}/\mu}$$
(5.54)

$$T_r(\mu) = 1 - \text{Re}(\mu)$$
(5.55)

(2) For absorptive media, $w_0 < 1$.

$$\text{Re}(\mu) = (u^2 - 1)\exp(\tau_{eff}) - \exp(-\tau_{eff})/R$$
(5.56)

$$T_r(\mu) = 4u/R$$
(5.57)

$$u^2 = [1 - w_0 + 2\beta w_0]/(1 - w_0)$$
(5.58)

$$\tau_{eff} = \{(1 - w_0)[1 - w_0 + 2\beta(\mu)w_0]\}^{1/2}\tau_{cld}/\mu$$
(5.59)

$$R = (u + 1)^2 \exp(\tau_{eff}) - (u - 1)^2 \exp(-\tau_{eff})$$
(5.60)

where $\mu = \cos\theta$; θ is the solar zenith angle; τ_{cld} is the cloud optical thickness; w_0 is the single-scattering albedo; and $\beta(\mu)$ is the backward scattering of incident solar radiation, which can be calculated by linear interpolation using the lookup table of cloud optical thicknesses and solar zenith angles (Stephens et al., 1984; Van Laake and Sanchez-Azofeifa, 2004).

This estimation algorithm is based on two assumptions: first, the cloud has a uniform distribution and is isotropic; and second, both ozone absorption and water vapor absorption occur at the upper layer of the cloud. Fig. 5.7 shows the results from verification of the dual-band

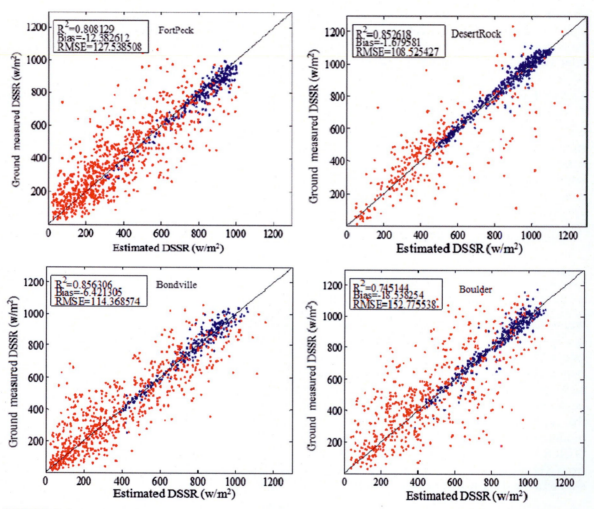

FIGURE 5.6   Validation results of the clear-sky and broadband cloudy-sky models at the SURFRAD station based on MODIS products from Terra of 2003–05.

model using MODIS products from Terra under cloudy-sky conditions at two SURFRAD stations from 2003 to 2005.

The parameterization method relies heavily on the precision of input parameters. The precision of MODIS products used as input parameters would affect the retrieval precision of DSSR and PAR. Therefore, the direct use of this method might not be suitable for generating global radiation products.

5.3.3 Lookup table method

The lookup table–based estimation of the amount of solar radiation at surface is introduced in this section. Section 5.3.2 shows the estimation of the amount of solar radiation using the parameterization method, which requires many input parameters, such as aerosol optical thickness and ozone content. In many cases, however, these input parameters might be unavailable or

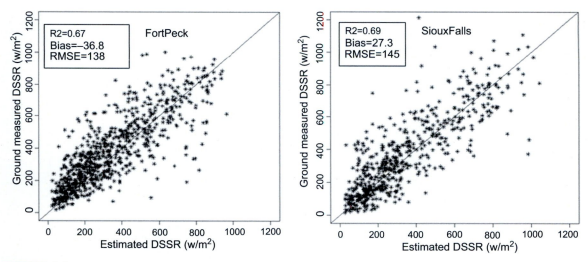

FIGURE 5.7 Verification of the dual-band cloudy-sky model using MODIS products from Terra at two SURFRAD sites from 2003 to 05.

have insufficient precision for the model, resulting in poor universality of the general radiation parameterization model. The algorithm for the lookup table—based surface solar radiation estimation does not require complex input parameters, and the input parameters, which consist of remote sensing observed data and geometric parameters.

The procedures for the lookup table—based estimation of DSSR and PAR are illustrated in Fig. 5.8.

TOA radiance detected by the sensor provides information on the path radiance and the radiance caused by backward scattering by particles and atmospheric molecules. Path radiance is determined by the optical properties of the atmosphere; thus, it is independent of the surface conditions. The other component is the radiation that reaches the land surface through the atmosphere and is reflected back to the atmosphere and sensed by the sensor.

Assuming that the land surface is Lambertian with a uniform surface, TOA radiance can be expressed by the following formula (Liang et al., 2006; Vermote et al., 1997):

$$I_{\text{TOA}}(\mu, \mu_v, \phi) = I_0(\mu, \mu_v, \phi)$$
$$+ \frac{r_g(\mu, \mu_v, \phi)}{1 - r_g(\mu, \mu_v, \phi) r_s} \mu \overline{E_0} \gamma(\mu) \gamma(\mu_v) \quad (5.61)$$

where $I_{\text{TOA}}(\mu, \mu_v, \phi)$ is the spectral radiance, which is received by the sensor for a given illuminating/viewing geometry: solar zenith angle is θ ($\mu = \cos \theta$), viewing zenith angle is θ_v ($\mu_v = \cos \theta_v$), and relative azimuth angle is ϕ. In Formula 5.61, the first item on the right is $I_0(\mu, \mu_v, \phi)$, which is the path radiance, and the second item, which is the radiation reflected by the land surface, is dependent on the atmospheric conditions. The relevant parameters include $\gamma(\mu)$, transmittance from the sun to the land surface; $\gamma(\mu_v)$, transmittance from the land surface to the sensor; r_s, atmospheric spherical albedo; $r_g(\mu, \mu_v, \phi)$, surface reflectance; and $\overline{E_0}$, solar spectral irradiance.

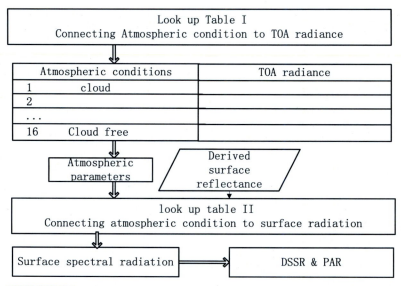

FIGURE 5.8   Flowchart used to estimate DSSR and PAR using a lookup table.

For a uniform Lambertian surface, the downward surface spectral flux can be expressed as follows (Liang et al., 2006; Vermote et al., 1997):

$$F(\mu) = F_0(\mu) + \frac{r_g(\mu, \mu_v, \phi)r_s}{1 - r_g(\mu, \mu_v, \phi)r_s} \mu \overline{E_0}\gamma(\mu) \quad (5.62)$$

where the solar zenith angle is θ ($\mu = \cos\theta$), $F_0(\mu)$ is the downward spectral flux excluding radiation reflected by the surface, $r_g(\mu, \mu_v, \phi)$ is the surface reflectance, r_s is the atmospheric spherical albedo, $\overline{E_0}$ is the solar spectral irradiance, and $\gamma(\mu)$ is the total transmittance.

The optical properties of the atmosphere modulate the variation in TOA radiance, DSSR and PAR. One specific atmospheric condition will result in a DSSR or PAR value and TOA radiance. Thus, N atmospheric conditions correspond to N pairs of DSSR or PAR and TOA radiance. Given a sufficient change in the value of N, DSSR and PAR can be derived from TOA radiance. DSSR and PAR can clearly be directly determined from TOA radiance without requiring any other atmospheric parameters. By dividing the atmospheric conditions into cloudy-sky and clear-sky scenarios and setting the cloud absorption coefficient and visibility for different conditions, a parameter set including various atmospheric conditions corresponding to visibilities that range from complete cloud coverage to clear skies is formed. Using MODTRAN, DSSR and PAR under different observation conditions and atmospheric conditions are simulated. In this way, the lookup table is established to estimate DSSR and PAR. The parameter configuration for the cloud absorption coefficient and visibility is given in the next section.

Liang et al. (2006) presented the lookup table method to estimate PAR at different geometric information and atmospheric observation conditions based on MODTRAN4. It can simulate and output upward TOA radiance and land surface downward spectral radiation for any sensor and then calculate PAR through integration. The 400–700 nm spectral range can be used to estimate PAR. This lookup table method can be extended to the shortwave band for DSSR estimating.

In MODTRAN4, the model parameters of specified atmospheric optical properties can be divided into four types: the geometric information; the atmospheric composition; the optical properties of aerosol; and the optical properties of clouds. TOA radiance and surface solar radiation vary with observation geometry conditions. The change in TOA radiance and surface solar radiation was simulated under as many observation conditions as possible, as shown in Table 5.7.

Elevation is also another factor to estimate DSSR and PAR. The simplest method involves using atmospheric radiative–transfer software for simulation and then establishing a statistical regression–based relation between elevation and DSSR and PAR. The elevation can then be corrected. However, this method does not consider the ozone, water vapor, and aerosol changes with elevation. However, for the lookup table–based method, the elevation can be treated as a dimension in the lookup table. The elevation-change parameters selected for the simulation are shown in Table 5.5.

In simulation, the default values of water vapor and ozone content in MODTRAN4 are directly used. The parameter configuration also involves aerosol and cloud properties. Table 5.6 provides the aerosol information used in the simulation. Table 5.7 presents the cloud parameters used in the simulation. The influences of water vapor on the DSSR estimation cannot be ignored, thus the estimated DSSR value should be corrected using the empirical function for the lookup method. Water vapor correction is quite simple. The solar radiation transmittance under the action of actual water vapor content is calculated and compared with the transmittance corresponding to the fixed water vapor content in the lookup table method, yielding a correction factor. Multiplying the radiation estimated by the lookup table method with the correction factor, the corrected radiation value is obtained.

By setting different surface reflectances according to Formula 5.61, the relation of TOA radiance $I_{TOA}(\mu, \mu_v, \phi)$ to three variables, $I_0(\mu, \mu_v, \phi)$, r_s, and $\mu\overline{E}_0\gamma(\mu)\gamma(\mu_v)$, is established. The first lookup table can then be built according to different atmospheric conditions and includes nine basic variables: cloud absorption coefficient, atmospheric visibility, solar zenith angle, viewing zenith angle, relative azimuth angle, elevation, $I_0(\mu, \mu_v, \phi)$, r_s, and $\mu\overline{E}_0\gamma(\mu)\gamma(\mu_v)$, as shown in Table 5.8.

A second lookup table can be established using the same method according to Formula 5.62. The association between the surface radiation and surface reflectance is built through four variables: $F_0(\mu)$, r_s, $\mu\overline{E}_0\gamma(\mu)$, and $F_d(\mu)$, where $F_d(\mu)$ is the scattered radiation. The second lookup table is shown in Table 5.9.

To establish the relationship between the first and second lookup tables, the relation of TOA radiance to surface radiation in the visible band for each sensor is generated, i.e., according to the first lookup table, the association between the TOA radiance and atmospheric condition index is determined. Meanwhile, the association between the atmospheric condition index and surface radiation is established based on the second lookup table. The specific computational methods are as follows:

TABLE 5.5 Parameter configurations of observation geometry information and elevation used in MODTRAN4 simulations.

Solar zenith angle	0°,10°, 20°, 30°, 40°, 50°, 55°, 60°, 70°, 80°, 85°, 90°
Viewing zenith angle	0°, 10°, 20°, 30°, 45°, 65°, 85°
Relative azimuth angle	0°, 30°, 60°, 90°, 120°, 150°, 180°
Elevation (km)	0.000, 1.500, 3.000, 4.500, 5.900

TABLE 5.6   Aerosol type and visibility in MOD-TRAN simulations (km).

	Rural aerosol	Tropospheric aerosol	Urban aerosol
Visibility (km)	5	5	5
	10	10	10
	20	20	20
	30	30	30
	100	100	100
	300	300	300

(1) To derive the surface reflectance using specific algorithm or directly using relevant products.

(2) The TOA radiance under all atmospheric conditions, ranging from the clearest sky to the cloudiest sky, is calculated using the derived reflectance. The sensed TOA radiance by different sensors and simulated TOA radiance values in lookup table are compared under all atmospheric conditions to determine the atmospheric condition index.

TABLE 5.7   Cloud absorption coefficient (km^{-1}) at 550 nm and relevant configurations in MODTRAN simulations.

	Altostratus clouds	Stratus clouds	Stratus/stratocumulus clouds	Nimbostratus clouds
Absorption coefficient (km^{-1})	1	1	1	1
	5	5	3	5
	7	20	10	10
	15	50	15	30
	20	56.9	38.7	45
	50			
	90			
	128			
Thickness (km)	0.6	0.67	1.34	0.5
Base height (km)	2.4	0.33	0.66	0.16

TABLE 5.8   Lookup table 1: association between the top-of-atmosphere radiance and atmospheric condition index.

Cloud absorption coefficient Visibility (atmospheric condition)	Solar zenith angle	Viewing zenith angle	Relative azimuth angle	Elevation	$I_0(\mu, \mu_v, \phi)$	r_s	$\mu\overline{E_0}\gamma(\mu)\gamma(\mu_v)$
...	...	...	...	...	...	...	...
...	...	...	...	...	...	...	...

TABLE 5.9 Lookup table 2: association between the surface radiation and atmospheric condition index.

Cloud absorption coefficient Visibility (atmospheric condition)		Solar zenith angle	Elevation	$F_0(\mu)$	r_s	$\mu\overline{E_0}\gamma(\mu)$	$F_d(\mu)$
…	…	…		…	…	…	…
…	…	…		…	…	…	…

(3) By referring to lookup table 2 and using the derived atmospheric condition index, the surface solar radiation is calculated by interpolation.

The Global LAnd Surface Satellite (GLASS) DSSR and PAR products were developed based on the improved lookup table method from 2008 to 2010 at a 5-km spatial resolution and a 3-h temporal resolution, the first global radiation products at such high resolutions, from multiple polar-orbiting and geostationary satellite data, including the MODIS, the Meteosat Second Generation SEVIRI, the Multi-functional Transport Satellite (MTSAT)-1R, and the Geostationary Operational Environmental Satellite (GOES) Imager. The GLASS products utilize both the polar-orbiting and geostationary satellite to generate the global DSSR and PAR products, as illustrated in Fig. 5.9, which is an instantaneous fusion product of DSSR in a gridded equal-angle projection with a 5-km spatial resolution and a 3-h temporal resolution. Ground-based measurement data from 34 sites are used to validate the improved algorithm and the GLASS products. The validation results of the instantaneous DSSR and PAR products at all validation sites are notably good with coefficient of determination values of 0.83 and 0.84, respectively, and RMSE values of 115.0 and 49.0 W/m², respectively.

5.3.4 Machine learning methods

Besides the methods mentioned above, machine learning methods are also an alternative way to estimate DSSR (Lam et al., 2008; Tang et al., 2016; Wandera et al., 2017; Wei et al., 2019; Yang et al., 2018; Zhou et al., 2017). The widely used machine learning methods include gradient boosting regression tree (GBRT), random forest (RF), multivariate adaptive

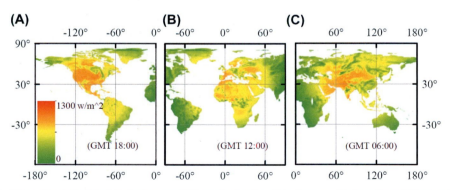

FIGURE 5.9 The estimated global instantaneous ISR product using data from multiple polar-orbiting and geostationary satellites on June 27, 2008 (Zhang et al., 2014).

regression spline (MARS), and artificial neural network (ANN), and so on.

The GBRT method, also known as gradient boosting decision tree, is based on iterative optimization (Wang et al., 2016). GBRT can be used for regression and classification problems. The GBRT method can automatically find nonlinear interactions through decision trees (Johnson and Zhang, 2014) and has a better predictive capacity than a single decision tree. Assuming $\{y_i, x_i\}_1^N$ is training data, where x is a set of predictor variables, y is the target variable, and N is the number of training samples. The GBRT method constructs M different individual decision trees $h(x;a_1)$, $h(x;a_2)$, ..., $h(x;a_M)$, and then $h(x;a_m)$ can be used as the basis function to express the approximation function $f(x)$ as follows (Ding et al., 2016):

$$
\begin{cases}
f(x) = \sum_{m=1}^{M} f_m(x) = \sum_{m=1}^{M} \beta_m h(x;a_m) \\
\\
h(x;a_m) = \sum_{j=1}^{J} \gamma_{jm} I(x \in R_{jm}), \text{ where } I = 1
\end{cases}
$$

$$
\text{if } x \in R_{jm}; \ I = 0, \quad \text{otherwise}
$$

(5.63)

where β_m and α_m represent the weight and classifier parameter of each decision tree. A specified loss function $L(y, f(x))$ is used to estimate the prediction performance of these two parameters. Each tree partitions the input space into J regions $R_{1m}, R_{2m}, ..., R_{jm}$ and each R_{jm} predicts the constant γ_{jm}. Fig. 5.10 shows the GBRT process flow (Zhang and Haghani, 2015). In this study, the GBRT method is implemented on the Python platform using the Ensemble module in the scikit-learn toolbox (Pedregosa et al., 2012).

The performance of the GBRT method is mainly influenced by learning rate, n-estimators, max-depth, and subsample. The learning rate parameter, which shrinks the contribution of each tree, may need to be small to avoid overfitting. However, a small learning rate parameter could lead to a high computational cost of applications. The n-estimators parameter is the number of boosting stages to perform. A larger n-estimators parameter usually provides better performance for the training dataset. However, model complexity increases with increasing n-estimators, leading to poor prediction performance. The max-depth parameter is the maximum depth for individual regression estimator, which limits the number of nodes in the tree. The subsample parameter is the fraction of samples to be used for fitting individual

Initialize $f_0(x) = \arg\min_\rho \sum_{i=1}^{N} L(y_i, \rho)$

For m = 1 to M do
  For i = 1 to n do
    Compute the negative gradient

$$
\tilde{y}_{im} = -\left[\frac{\partial L(y_i, f(x_i))}{\partial f(x_i)}\right]_{f(x)=f_{m-1}(x-1)}
$$

End;
Fit a regression tree $h(x;a_m)$ to predict the targets $\tilde{y}_{im}$ from covariates x_i for all training data

Compute a gradient descent step size as $\rho_m = \arg\min_\rho \sum_{i=1}^{n} L(y_i, f_{m-1}(x_i) + \rho h(x_i;a_m))$

Update the model as $f_m(x) = f_{m-1}(x) + \rho_m h(x_i;a_m)$
End;
Output the final model $f_M(x)$

FIGURE 5.10   The main process flow of the GBRT method.

base learner. Choosing a subsample parameter smaller than 1.0 can reduce variance but increases bias. This parameter can help prevent overfitting (Fig. 5.11).

Wu et al. tried to estimate DSSR at a spatial resolution of 5 km and a temporal resolution of 1 day using Advanced Very High Resolution Radiometer (AVHRR) data based on four machine learning methods which includes GBRT, RF, MARS, and ANN. For model construction, the ground measurements of daily DSSR data at CDC/CMA stations from 2001 to 03 were used as the target variables. The dataset was divided into clear-sky and cloudy-sky datasets according to the cloud mask from the AVHRR cloudy- and clear-sky radiation properties dataset. Eight variables extracted from the AVHRR data were used as predictor variables under clear-sky conditions, including solar zenith angle, viewing zenith angle, relative azimuth angle, TOA shortwave broadband albedo, spectral reflectance of channels 1 and 2, and

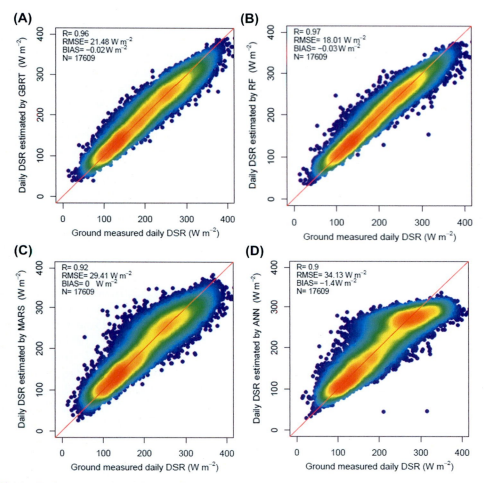

FIGURE 5.11 Evaluation results of the training dataset's daily DSSR estimates under clear-sky conditions based on the (A) GBRT, (B) RF, (C) MARS, and (D) ANN method against ground measurements. N is the number of total data points (Wei et al., 2019).

brightness temperature of channels 4 and 5. In addition to these eight variables, cloud optical depth was added as a predictor variable under cloudy-sky conditions. To construct the model, the dataset is randomly divided into two groups: 80% to train the model and the remaining 20% to validate the trained model under both clear-sky and cloudy-sky conditions. The optimal parameters providing the highest average R^2 were selected by k-fold cross validation in the training data. After determining the optimal parameters, the optimal models were used to estimate DSSR.

The DSSR estimates based on four machine learning methods were evaluated using ground measurements at 96 sites over China. The measurements were collected from the Climate Data Center of the Chinese Meteorological Administration (CDC/CMA) from 2001 to 2003. The evaluation results showed that the GBRT method performed best at both daily and monthly time scales under both clear- and cloudy-sky conditions. The validation results at the daily time scale showed an overall RMSE of $30.34 \, \text{W/m}^2$ and an R value of 0.90 under clear-sky conditions, whereas these values were $42.03 \, \text{W/m}^2$ and 0.86, respectively, under cloudy-sky conditions (Fig. 5.12).

Although the GBRT method can automatically find nonlinear interaction via decision tree learning and has been successful for many research applications, the GBRT method also has some disadvantages. First, the GBRT method does not have an explicit regularization. Second, the step size of learning rate parameter, which is regarded as implicit regularization, may need to be small to avoid overfitting. However, the small learning rate parameter implies a high computational cost of applications. Third, the regression tree learner is treated as a black box. The other machine learning methods also have overfitting issues (Wei et al., 2019). The RF method has the overfitting issue due to very complex trees, especially in high-dimensional noisy datasets (Salles et al., 2018). Excessive number of basis functions may cause an overfitting issue in MARS (Koc and Bozdogan, 2015). The ANN method also has the risk of overfitting, especially when the size of the network is large (Hagiwara and Fukunaizu, 2008).

5.4 Current existing products and long-term variations

5.4.1 Existing products and evaluation

Currently, four representative global satellite DSSR products are widely used, including the Global Energy and Water Cycle Experiment—Surface Radiation Budget (GEWEX-SRB) (Pinker and Laszlo, 1992) at 3-h temporal resolution and 1° spatial resolution, the International Satellite Cloud Climatology Project—Flux Data (ISCCP-FD) at 3-h intervals and 2.5° spatial resolution (Zhang et al., 2004), the University of Maryland (UMD)/Shortwave Radiation Budget (SRB) product with 0.5° spatial resolution (Ma and Pinker, 2012), and the CERES-EBAF with 1° spatial resolution (Kato et al. 2012, 2013). The Detailed Information on these four datasets is presented in the following sections and summarized in Table 5.10.

The primary GEWEX-SRB algorithm is an updated version of the University of Maryland algorithm (Pinker and Laszlo, 1992). The major GEWEX-SRB (V3.0) inputs include cloud parameters derived from the International Satellite Cloud Climatology Project DX data products (Rossow and Schiffer, 1999), temperature and moisture profiles, and ozone-column data. GEWEX-SRB V3.0 estimated DSSR on the basis of the cloud fraction, atmospheric composition, background aerosol, and spectral albedo, with the TOA measured cloudy- and clear-sky radiances acting as a constraint through radiative-transfer computations.

The ISCCP-FD data are an update of the ISCCP-FC data, which were generated mainly using International Satellite Cloud Climatology Project—C1 input data (Zhang et al., 1995). The

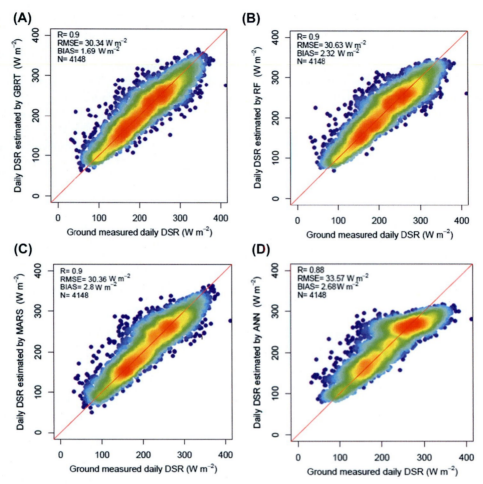

FIGURE 5.12 Evaluation results of the validation dataset's daily DSSR estimates under clear-sky conditions based on the (A) GBRT, (B) RF, (C) MARS, and (D) ANN method against ground measurements. *N* is the number of total data points (Wei et al., 2019).

TABLE 5.10 Summary of satellite-derived DSSR products.

Dataset	Spatial resolution	Temporal resolution	Temporal range
GEWEX-SRB V3.0	1°	3 Hourly, daily, monthly	1983.7–2007.12
ISCCP-FD	~280 km	3 Hourly	1983.7–2009.12
CERES-EBAF	1°	Monthly	2000.3–2013.3
UMD-SRB V3.3.3	0.5°	Monthly	1983.7–2007.6

ISCCP-FD calculated DSSR using a more advanced NASA Goddard Institute for Space Studies (GISS) radiative-transfer model and improved ISCCP cloud climatology obtained from the ISCCP-D1 data (Rossow and Schiffer 1999) and ancillary datasets with 3-h interval, 2.5° spatial resolution, and 5 pressure levels.

The UMD-SRB V3.3.3 dataset by Ma and Pinker (2012) is a new version of the UMD-SRB dataset. The previous version, V3.3, was generated at 2.5° spatial resolution by converting TOA radiance into broadband albedo to infer atmospheric transmissivity and to estimate DSSR, whereas UMD-SRB V3.3.3 calculates DSSR by matching radiative-transfer computations and satellite observations based on a lookup table, which contains the values of atmospheric transmissivity and reflectivity in five broadband intervals as a function of solar zenith angle, water vapor and ozone amount, aerosol single-scattering albedo, asymmetry factor, optical depth, and cloud optical depth using radiative-transfer computations in five broadband intervals (Pinker and Laszlo, 1992).

The CERES-EBAF incorporates more accurate cloud information than that in reanalysis. The CERES-EBAF DSSR values were obtained using cloud and aerosol properties derived from instruments on the A-train Constellation based on a radiative-transfer model of CERES with k-distribution and correlated-k for radiation (FLCKKR) with a two stream approximation using the independent column approximation considering observational constraints of TOA irradiance from CERES (Kato et al., 2013). Besides the cloud and aerosol properties data from the A-train satellite, the other inputs included temperature and humidity profiles, ozone amounts, ocean spectral surface albedo, TOA albedo, and emissivities.

Four satellite estimates of DSSR, including the GEWEX-SRB V3.0, the ISCCP-FD, the UMD-SRB V3.3.3 product, and the CERE-EBAF, were evaluated using comprehensive ground measurements at 1151 sites around the world from the GEBA and the China Meteorological Administration (CMA) by Zhang et al. (2015). It was found that satellite estimates of DSSR agreed better with surface measurements at monthly than at daily time scale. The mean bias of monthly mean estimates averaged over all four satellite-derived products was 8.4 W/m^2 at GEBA sites, with a range from 5 to 10.9 $W/m^2 wm^{-2}$ depending on the specific satellite-derived product, and 12.4 $W/m^2 wm^{-2}$ at CMA sites, with a range from 8.1 to 18.3 $W/m^2 wm^{-2}$. The monthly DSSR of these four datasets correlate very well with direct surface measurements. The CERES-FSW DSSR data showed the best performance among these four datasets, with overall correlation coefficients (R) of 0.97 and 0.95 for GEBA and CMA respectively, positive biases of 5.0 and 8.1 $W/m^2 wm^{-2}$, and RMSE values of 18.8 and 20.5 W/m^2.

5.4.2 Temporal and spatial patterns of solar radiation

Surface solar radiation shows a certain trend at the global scale in terms of spatial and temporal variations. The total yearly surface solar radiation absorbed by land surfaces is always greater in low-latitude areas than in high-latitude areas, while the seasonal variation in high-latitude areas is more significant than that in low-latitude areas. Theoretically, on the vernal and autumnal equinoxes, the equator is exposed to direct sunlight and thus receives the most radiation on the planet, while the radiation level progressively decreases toward the Polar Regions. Considering clouds and other factors, this trend might differ under real conditions.

In summer, the daytime in the northern hemisphere is long and increases in regions at higher latitudes. The Arctic Circle even has a polar day. High-latitude regions receive lower solar radiation due to the effect of the solar elevation angle, but the total amount of solar radiation that they receive is large as a result of a longer duration of sunlight. In winter, the daytime length in the southern hemisphere increases, and higher latitudes in southern hemisphere gives rise to longer daytimes. Polar days occur south of the Antarctic Circle. During this period, the radiation received by the Antarctic Circle and high-latitude areas reaches its maximum. Fig. 5.13 shows the pattern of the temporal and spatial distributions of monthly accumulated solar radiation in March, June, September, and December of 2007 based on the surface solar radiation data provided by GEWEX.

A fundamental determinant of climate and life on the Earth is the solar radiation incident at the Earth's surface. Any change in this precious energy source affects our lives profoundly. Global dimming refers to a decrease in the amounts of sunlight and energy reaching the Earth compared with those amounts in a previous time period. Conversely, global brightening refers to an increase in the amount of sunlight and energy reaching the Earth compared with a previous time period. Currently, the two phenomena cannot be explained by solar changes; rather, the atmospheric composition change constitutes the major reason for global dimming and global brightening.

The analysis of various types of long-term radiation observation data shows that DSSR exhibits a particular variation pattern (Ohmura, 2009; Pinker et al., 2005; Wild et al., 2005). By 1990, it was found that the surface downward shortwave radiation recorded by most of the

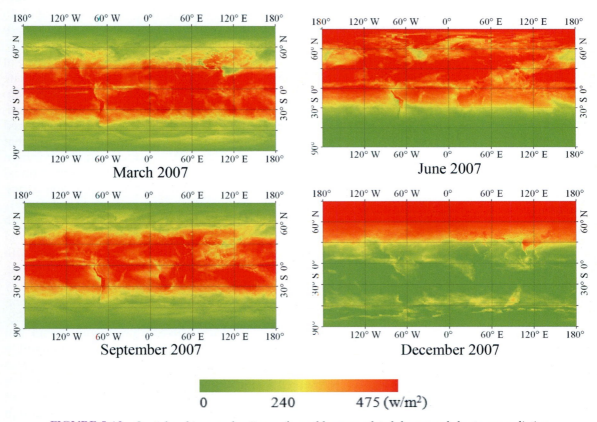

FIGURE 5.13 Spatial and temporal patterns of monthly accumulated downward shortwave radiation.

stations had a declining trend, which is indicative of global dimming. The analysis of the latest observation data indicated that, starting in late 1980s, the downward shortwave radiation showed an increasing trend, which is indicative of global brightening (Wild et al., 2005). Fig. 5.14 shows the temporal variation of atmospheric transmittance observed by stations in different areas under clear-sky conditions (Wild et al., 2005). Fig. 5.15 shows GEBA global radiation flux anomalies, estimated shortwave cloud cover radiative effect (CCRE) anomalies, and residual anomalies after removing estimated shortwave CCRE from GEBA global radiation (Norris and Wild, 2007).

Studies of global brightening and dimming, based on either GEBA observational data or global meteorological observational data, cannot avoid yielding errors due to the limited number of observation stations. Comparatively speaking, the use of remote sensing data in the study of long-term radiation variation at the global scale will produce results that are more

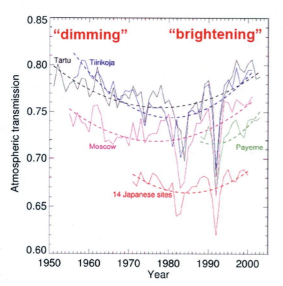

FIGURE 5.14 Variation of atmospheric transmittance at stations in various regions (Wild et al., 2005).

accurate due to its large spatial coverage. Nevertheless, global land surface radiation products still suffer from issues related to satellite substitution, sensor calibration, satellite observation geometry information, and long-term comprehensive aerosol observation. Evan et al. (2007) noted that the current ISCCP data are not suitable for the analysis of long-term variations, especially the variation trend. Moreover, the current GCM is unable to simulate the interannual variation trend.

Through the analysis of long-term satellite data, Pinker et al. (2005) found that the surface solar radiation between 1983 and 2001 increased at a rate of 0.16 W/m$^2 Wm^{-2}$ annually based on GEWEX and ISCCP satellite retrieved data. The increasing trend could be divided into two stages: decreasing between 1983 and 90 and increasing between 1990 and 2001. Fig. 5.16 shows the trend in the variation of global surface radiation fluxes from 1983 to 2001.

Global dimming and global brightening are of considerable significance in the study of climate change and water circulation. The fourth evaluation report of the IPCC indicated that the continental and global surface temperature showed a slightly decreasing trend from 1950 to 70 that reversed after 1980 especially in the northern hemisphere, which is similar to the interannual variation trend of surface solar radiation (Liang et al., 2010).

Zhang et al. (2015) also found that both GEWEX-SRB V3.0 and ISCCP-FD showed similar trends at the global scale but with different magnitudes. A significant dimming was found between 1984 and 91, followed by brightening from 1992 to 2000, and then by a significant dimming over 2001−07. The CERES-EBAF product showed a brightening trend, but not significantly since 2000. The variability from satellite estimates at pixel level was also analyzed (Fig. 5.17). The results are comparable with previous studies based on observed DSSR at the surface for specific regions, although

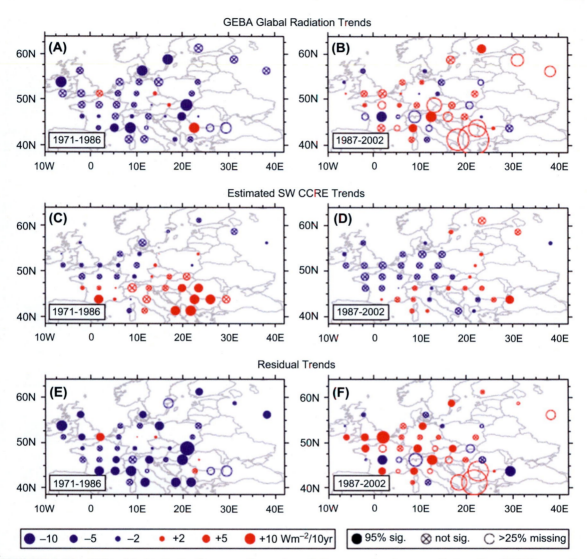

FIGURE 5.15 GEBA global radiation flux anomalies, estimated shortwave cloud cover radiative effect (CCRE) anomalies, and residual anomalies after removing estimated shortwave CCRE from GEBA global radiation (Norris and Wild, 2007).

some inconsistencies still exist and the magnitudes of the variations should be further quantified. As shown in Fig. 5.18, Zhang et al. (2015) also examined the contributions of clouds versus aerosols to the variations of DSSR and found that clouds and aerosols primarily determine the long-term variations of DSSR from satellite observations.

5.5 Summary

This chapter introduced some of the basic concepts of solar radiation, typical surface solar radiation observation networks at the global scale, and typical approaches to estimate surface solar radiation based on remote sensing observations. These approaches include the empirical

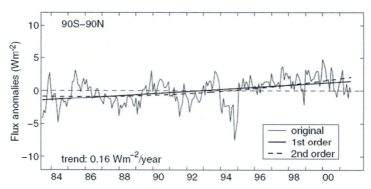

FIGURE 5.16 Trend in the variation of global surface radiation fluxes from 1983 to 2001 (Pinker et al., 2005).

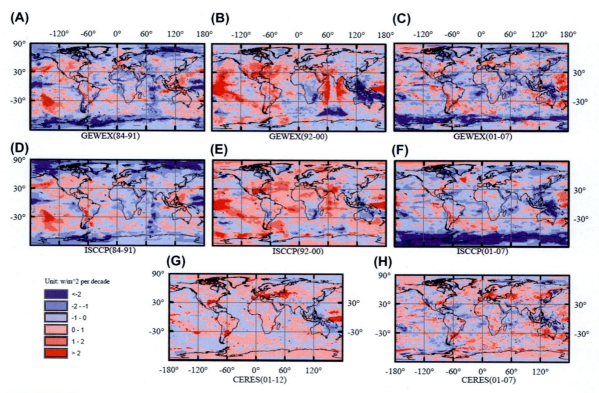

FIGURE 5.17 Segmental trend analysis of R_s from satellite observations by GEWEX-SRB, ISCCP-FD, and CERES-EBAF for each pixel: (A), (B), and (C) are the trends from GEWEX-SRB during three time periods, (D), (E), and (F) are the trends from ISCCP-FD during three time periods, (G) and (H) are the trends from CERES-EBAF during two time periods. Units are W/m² per decade.

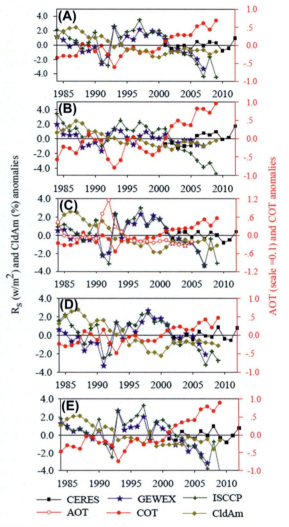

FIGURE 5.18 Long-term variability of cloud amount (CldAm), aerosol optical depth (AOT) and cloud optical depth (COT), and DSSR for CERES-EBAF, GEWEX-SRB V3.0, and ISCCP-FD: (A) globe, (B) land, (C) ocean, (D) northern hemisphere, (E) southern hemisphere.

For the general parameterization models, the results obtained using a specific algorithm are dependent on not only the model itself but also on the quality of the parameters because multiple surface or atmospheric parameters are taken as the input parameters, and their quality is, in turn, determined by retrieval precision and resolution. Therefore, the application of the parameterization method is limited. Compared with the parameterization method, the lookup table retrieval method requires fewer input parameters and is relatively simple. The lookup table method is one of the best choices to generate long-term radiation products at a global or large regional scale. The topographic should be further considered in the lookup table method. Overfitting is one common issue in the machine learning methods.

Numerous studies have substantiated the findings of significant decadal DSSR variability observed both at sites distributed around the world and in specific regions. However, current satellite DSSR products show different temporal trends. Much efforts must be conducted for solar radiation estimation in the future.

It would be desirable to obtain more accurate input data, including cloud-property parameters, aerosol and atmospheric profiles, and advanced radiative computation models. This would improve the accuracy of radiation estimates. Moreover, more surface measurements, including clouds and aerosols, should be collected, which would enable more quantitative analyses in the future.

methods, the parameterization methods, the lookup table methods, and also the machine learning methods. In addition, the retrieval results from various models were preliminarily validated, and the advantages and disadvantages of the existing methods were reported.

Nomenclature

M_A	Aerosol optical mass
β	Angstrom turbidity coefficient
α	Angstrom wavelength coefficient
τ_a	Broadband aerosol optical depth
τ_{cld}	Cloud optical depth
I_{gcld}	Cloudy-sky total (global) solar irradiance on a horizontal surface

ξ	Coefficient for the proportion of photosynthetically active radiation in shortwave radiation
I'_r	Corrected value of the direct surface radiation and scattered radiation by multiple reflections of the adjacent pixels (W/m^2)
μ	Cosine of solar zenith angle
μ_v	Cosine of view zenith angle
H_0	Daily extraterrestrial radiation on a horizontal surface (W/m^2)
I_b	Direct solar irradiance in the orthogonal direction (W/m^2)
I_d	Direct solar irradiance on a horizontal surface (W/m^2)
I_0	Extraterrestrial solar irradiance over the full wavelength (W/m^2)
B_a	Forward scattering ratio due to the aerosols
F_{cld}	Fraction of cloud coverage
w_s	Mean sunrise hour angle for the given month
u_i	Mixed gas mass (CO_2, CO, N_2O, CH_4, or O_2)
N	Monthly average maximum possible daily sunshine duration
n	Monthly average sunshine duration
H	Monthly mean daily global radiation (W/m^2)
M	Optical air mass
u_o	Ozone amount (cm)
M_o	Ozone optical mass
M_p	Pressure-corrected air mass
I_r	Radiation reflected by the surface (W/m^2)
ϕ	Relative azimuth angle
w_0	Single-scattering albedo
lat	Site latitude
r_s	Sky or atmospheric albedo
ϕ_s	Solar azimuth angle
δ	Solar declination
I_{as}	Solar irradiance on a horizontal surface due to atmospheric scattering (W/m^2)
θ	Solar zenith angle
E_0	Spectral solar irradiance ($W/m^2/\mu m$)
$\overline{E_0}$	Spectral solar irradiance at the average sun–Earth distance ($W/m^2/\mu m$)
r_g	Surface albedo
z	Surface elevation (km)
p	Surface pressure (mb or hPa)
r_g	(μ, μ_v, ϕ) Surface reflectance
T	Surface temperature (K)
I_g	Total (global) solar irradiance on a horizontal surface (W/m^2)
I_{topo}	Total (global) solar irradiance on a horizontal surface after topographic correction (W/m^2)
T_t	Total transmittance of atmosphere
$\gamma(\mu_v)$	Transmittance from the surface to the sensor
$\gamma(\mu)$	Transmittance in the solar illumination direction
T_G	Transmittance of absorptance of mixed gases
T_{AA}	Transmittance of aerosol absorptance

T_A	Transmittance of aerosol absorptance and scattering
T_{AS}	Transmittance of aerosol scattering
T_N	Transmittance of NO_2
T_O	Transmittance of ozone absorptance
T_O	Transmittance of ozone absorptance and scattering
T_R	Transmittance of Rayleigh scattering
T_W	Transmittance of water vapor absorptance
θ_v	View zenith angle
w	Water vapor amount (cm)
M_w	Water vapor optical mass

Acknowledgements

This chapter was partly supported by National Basic Research Program of China (2015CB953701).

References

Akinoğlu, B.G., Ecevit, A., 1990. Construction of a quadratic model using modified Ångstrom coefficients to estimate global solar radiation. Sol. Energy 45, 85–92.

Almorox, J., Hontoria, C., 2004. Global solar radiation estimation using sunshine duration in Spain. Energy Convers. Manag. 45, 1529–1535.

Ampratwum, D., 1999. Estimation of solar radiation from the number of sunshine hours. Appl. Energy 63, 161–167.

Anderson, G.P., Berk, A., Acharya, P.K., Matthew, M.W., Bernstein, L.S., James, H., Chetwynd, J., Dothe, H., Adler-Golden, S.M., Ratkowski, A.J., Felde, G.W., Gardner, J.A., Hoke, M.L., Richtsmeier, S.C., Pukall, B., Mello, J.B., Jeong, L.S., 1999. MODTRAN4: radiative transfer modeling for remote sensing. Proc. SPIE 3866.

Angstrom, A., 1924. Solar and terrestrial radiation. Report to the international commission for solar research on actinometric investigations of solar and atmospheric radiation. Q. J. R. Meteorol. Soc. 50, 121–126.

Annear, R.L., Wells, S.A., 2007. A comparison of five models for estimating clear-sky solar radiation. Water Resour. Res. 43, W10415.

Augustine, J.A., Deluisi, J., Long, C.N., 2000. SURFRAD-a national surface radiation budget network for atmospheric research. Bull. Am. Meteorol. Soc. 81, 2341–2357.

Augustine, J.A., Hodges, G.B., Cornwall, C.R., Michalsky, J.J., Medina, C.I., 2005. An update on SURFRAD—The GCOS surface radiation budget network for the continental United States. J. Atmos. Ocean. Technol. 22, 1460–1472.

Baldocchi, D., Falge, E., Gu, L., Olson, R., Hollinger, D., Running, S., Anthoni, P., Bernhofer, C., Davis, K., Evans, R., 2001. FLUXNET: a new tool to study the temporal and spatial variability of ecosystem-scale carbon dioxide, water Vapor, and energy flux densities. Bull. Am. Meteorol. Soc. 82, 2415–2434.

Barkstrom, B.R., Smith, G.L., 1986. The earth radiation budget experiment: science and implementation. Rev. Geophys. 24, 379–390.

Bird, R.E., Hulstrom, R.L., 1981a. Review, evaluation, and improvement of direct irradiance models. ASME Trans. J.Sol. Energy Eng. 103, 182–192.

Bird, R.E., Hulstrom, R.L., 1981b. A simplified clear sky model for direct and diffuse insolation on horizontal surfaces. Technical Report No. SERI/TR-642-761. In: Golden, Colorado: Solar Energy Research Institute.

Choudhury, B., 1982. A parameterized model for global insolation under partially cloudy skies. Sol. Energy 29, 479–486.

Coakley, J.A., Chylek, P., 1975. The two-stream approximation in radiative transfer Including the angle of the incident radiation. J. Atmos. Sci. 32, 409–418.

Davies, J.A., Hay, J.E., 1979. Calculation of the solar radiation incident on a horizontal surface. In: John, L. (Ed.), Proceedings of the First Canadian Solar Radiation Data Workshop. Downsview (ON): Canadian Atmospheric Environment Service (Now Environment Canada); Proceddings, First Canadian Solar Radiation Data Workshop.

Ding, C., Wang, D.G., Ma, X.L., Li, H.Y., 2016. Predicting short-term subway ridership and prioritizing its influential factors using gradient boosting decision trees. Sustainability 8, 1100.

Duffle, J.A., Beckman, W.A., 1980. Solar Engineering of Thermal Processes. Wiely, New York.

Ertekin, C., Yaldiz, O., 2000. Comparison of some existing models for estimating global solar radiation for Antalya (Turkey). Energy Convers. Manag. 41, 311–330.

Evan, A.T., Heidinger, A.K., Vimont, D.J., 2007. Arguments against a physical long-term trend in global ISCCP cloud amounts. Geophys. Res. Lett. 34, L04701.

Falayi, E.O., Adepitan, J.O., Rabiu, A.B., 2008. Empirical models for the correlation of global solar radiation with meteorological data for Iseyin, Nigeria. Int. J. Phys. Sci. 3, 210–216.

Gao, B.C., Kaufman, Y.J., 2003. Water vapor retrievals using moderate resolution Imaging spectroradiometer (MODIS) near-infrared channels. J. Geophys. Res. Atmos. 108, 4389.

Gilgen, H., Ohmura, A., 1999. The global energy balance archive. Bull. Am. Meteorol. Soc. 80, 831–850.

Gilgen, H., Wild, M., Ohmura, A., 1998. Means and trends of shortwave irradiance at the surface estimated from global energy balance archive data. J. Clim. 11, 2042–2061.

Gueymard, C., 1989. A two-band model for the calculation of clear sky solar irradiance, illuminance, and photosynthetically active radiation at the earth's surface. Sol. Energy 43, 253–265.

Gueymard, C., 1993a. Critical analysis and performance assessment of clear sky solar irradiance models using theoretical and measured data. Sol. Energy 51, 121–138.

Gueymard, C., 1993b. Mathemratically integrable parameterization of clear-sky beam and global irradiances and its use in daily irradiation applications. Sol. Energy 50, 385–397.

Gueymard, C., 1995. SMARTS2, a simple model of atmospheric radiative transfer of sunshine: algorithms and performance assessment. In: Florida Solar Enegy Center, Cocoa, FL.

Gueymard, C.A., 2003a. Direct solar transmittance and irradiance predictions with broadband models. Part I: detailed theoretical performance assessment. Sol. Energy 74, 355–379.

Gueymard, C.A., 2003b. Direct solar transmittance and irradiance predictions with broadband models: Part II: validation with high-quality measurements. Sol. Energy 74, 381–395.

Hagiwara, K., Fukunaizu, K., 2008. Relation between weight size and degree of over-fitting in neural network regression. Neural Netw. 21, 48–58.

Hanna, L.W., Siam, N., 1981. The empirical relation between sunshine and radiation and its use in estimating evaporation in North East England. Int. J. Climatol. 1, 11–19.

Harries, J.E., Russell, J.E., Hanafin, J.A., Brindley, H., Futyan, J., Rufus, J., Kellock, S., Matthews, G., Wrigley, R., Last, A., Mueller, J., Mossavati, R., Ashmall, J., Sawyer, E., Parker, D., Caldwell, M., Allan, P.M., Smith, A., Bates, M.J., Coan, B., Stewart, B.C., Lepine, D.R., Cornwall, L.A., Corney, D.R., Ricketts, M.J., Drummond, D., Smart, D., Cutler, R., Dewitte, S., Clerbaux, N., Gonzalez, L., Ipe, A., Bertrand, C., Joukoff, A., Crommelynck, D., Nelms, N., Llewellyn-Jones, D.T., Butcher, G., Smith, G.L., Szewczyk, Z.P., Mlynczak, P.E., Slingo, A., Allan, P., Ringer, M.A., 2005. The geostationary earth radiation budget projectAmerican meteorological society. Bull. Am. Meteorol. Soc. 86, 945–960.

Hartmann, D.L., Bretherton, C.S., Charlock, T.P., Chou, M.D., Genio, A.D., Dickinson, R.E., Fu, R., Houze, R.A., King, M.D., Lau, K.M., Leovy, C.B., Sorooshian, S., Washburne, J., Wielicki, B., Willson, R.C., 1999. Radiation, clouds, water vapor, precipitation, and atmospheric circulation. In: EOS Science Plan. NASA GSFC, pp. 39–114.

Hoyt, D.V., 1978. A model for the calculation of solar global insolation. Sol. Energy 21, 27–35.

Huang, G., Liang, S., Lu, N., Ma, M., Wang, D., 2018. Towards a broadband parameterization scheme for estimating surface solar irradiance: development and preliminary results on MODIS products. J. Geophys. Res. Atmos. 123 https://doi.org/10.1029/2018JD028905.

Iqbal, M., 1983. An Introduction to Solar Radiation. Academic Press, Toronto.

Jacobowitz, H., Tighe, R., 1984. The earth radiation budget derived from the Nimbus 7 ERB experiment. J. Geophys. Res. 89, 4997–5010.

Johnson, R., Zhang, T., 2014. Learning nonlinear functions using regularized Greedy forest. IEEE Trans. Pattern Anal. Mach. Intell. 36, 942–954.

Kahn, R.A., Garay, M.J., Nelson, D.L., Yau, K.K., Bull, M.A., Gaitley, B.J., Martonchik, J.V., Levy, R.C., 2007. Satellite-derived aerosol optical depth over dark water from MISR and MODIS: comparisons with AERONET and implications for climatological studies. J. Geophys. Res. Atmos. 112, D18205.

Kato, S., Loeb, N., Rutan, D., Rose, F., Sun-Mack, S., Miller, W., Chen, Y., 2012. Uncertainty estimate of surface irradiances computed with MODIS-, CALIPSO-, and CloudSat-derived cloud and aerosol properties. Surv. Geophys. 33, 395–412.

Kato, S., Loeb, N.G., Rose, F.G., Doelling, D.R., Rutan, D.A., Caldwell, T.E., Yu, L., Weller, R.A., 2013. Surface irradiances consistent with CERES-derived top-of-atmosphere shortwave and longwave irradiances. J. Clim. 26, 2719–2740.

Kaufman, Y.J., Tanre, D., 1998. Algorithm for Remote Sensing of Tropospheric Aerosols from MODIS. Algorithm Theoretical Basis Document. NASA Goddard Space Flight Center.

Koc, E.K., Bozdogan, H., 2015. Model selection in multivariate adaptive regression splines (MARS) using information complexity as the fitness function. Mach. Learn. 101, 35–58.

Kopp, G., Lean, J., 2011. A new, lower value of total solar irradiance: evidence and climate significance. Geophys. Res. Lett. 38, L01706.

Kumar, N.M., Kumar, P.V.H., Rao, P.R., 2001. An empirical model for estimating hourly solar radiation over the Indian seas during summer monsoon season. Indian J. Marione Sci. 30, 123–131.

Lacis, A.A., Hansen, J.E., 1974. A parameterization for the absorption of solar radiation in the earth's atmosphere. J. Atmos. Sci. 31, 118–133.

Lam, J.C., Wan, K.K.W., Yang, L., 2008. Solar radiation modelling using ANNs for different climates in China. Energy Convers. Manag. 49, 1080–1090.

Laszlo, I., Ciren, P., Liu, H., Kondragunta, S., Tarpley, J.D., Goldberg, M.D., 2008. Remote sensing of aerosol and radiation from geostationary satellites. Adv. Space Res. 41, 1882–1893.

Li, Z., Moreau, L., Cihlar, J., 1997. Estimation of photosynthetically active radiation absorbed at the surface. J. Geophys. Res. 102, 29717–29727.

Liang, S., 2004. Quantitative Remote Sensing of Land Surfaces. Wiley, Hoboken, New Jersey.

Liang, S., Zheng, T., Liu, R.G., Fang, H., Tsay, S.C., Running, S., 2006. Estimation of incident photosynthetically active radiation from Moderate Resolution Imaging Spectrometer data. J. Geophys. Res. Atmos. 111, D15208.

Liang, S., Wang, K., Zhang, X., Wild, M., 2010. Review on estimation of land surface radiation and energy budgets from ground measurement, remote sensing and model simulations. IEEE J. Sel. Top. Appl. Earth Obs. Remote Sens. 3, 225–240.

Liu, J.C., Schaaf, C., Strahler, A., Jiao, Z.T., Shuai, Y.M., Zhang, Q.L., Roman, M., Augustine, J.A., Dutton, E.G., 2009. Validation of Moderate resolution imaging spectroradiometer (MODIS) albedo retrieval algorithm: dependence of albedo on solar zenith angle. J. Geophys. Res. Atmos. 114, D01106.

Ma, Y., Pinker, R.T., 2012. Modeling shortwave radiative fluxes from satellites. J. Geophys. Res. Atmos. 117, D23202.

Maxwell, E.L., 1998. METSTAT–the solar radiation model used in the production of the National Solar Radiation Data Base (NSRDB). Sol. Energy 62, 263–279.

Norris, J.R., Wild, M., 2007. Trends in aerosol radiative effects over Europe inferred from observed cloud cover, solar "dimming"; and solar "brightening". J. Geophys. Res. 112, D08214.

Ohmura, A., 2009. Observed decadal variations in surface solar radiation and their causes. J. Geophys. Res. 114, D00D05.

Ohmura, A., Dutton, E.G., Forgan, B., Fröhlich, C., Gilgen, H., Hegner, H., Heimo, A., König-Langlo, G., McArthur, B., Müller, G., Philipona, R., Pinker, R., Whitlock, C.H., Dehne, K., Wild, M., 1998. Baseline surface radiation network (BSRN/WCRP):new precision radiometry for climate research. Bull. Am. Meteorol. Soc. 79, 2115–2136.

Pedregosa, F., Gramfort, A., Michel, V., Thirion, B., Grisel, O., Blondel, M., Prettenhofer, P., Weiss, R., Dubourg, V., Vanderplas, J., 2012. Scikit-learn: machine learning in Python. J. Mach. Learn. Res. 12, 2825–2830.

Pinker, R.T., Laszlo, I., 1992. Modeling surface solar irradiance for satellite applications on a global scale. J. Appl. Meteorol. 31, 194–211.

Pinker, R.T., Frouin, R., Li, Z., 1995. A review of satellite methods to derive surface shortwave irradiance. Remote Sens. Environ. 51, 108–124.

Pinker, R.T., Zhang, B., Dutton, E.G., 2005. Do satellites detect trends in surface solar radiation? Science 308, 850–854.

Qin, J., Tang, W.J., Yang, K., Lu, N., Niu, X.L., Liang, S.L., 2015. An efficient physically based parameterization to derive surface solar irradiance based on satellite atmospheric products. J. Geophys. Res. Atmos. 120, 4975–4988. https://doi.org/10.1002/2015JD023097.

Rossow, W.B., Schiffer, R.A., 1999. Advances in understanding clouds from ISCCP. Bull. Am. Meteorol. Soc. 80, 2261–2288.

Ryu, Y., Kang, S., Moon, S.K., Kim, J., 2008. Evaluation of land surface radiation balance derived from moderate resolution imaging spectroradiometer (MODIS) over complex terrain and heterogeneous landscape on clear sky days. Agric. For. Meteorol. 148, 1538–1552.

Safari, B., Gasore, J., 2009. Estimation of global solar radiation in Rwanda using empirical models. Asian J. Sci. Res. 2, 68–75.

Salles, T., Goncalves, M., Rodrigues, V., Rocha, L., 2018. Improving random forests by neighborhood projection for effective text classification. Inf. Syst. 77, 1–21.

Schmetz, J., 1989. Towards a surface radiation climatology: retrieval of downward irradiances from satellites. Atmos. Res. 23, 287–321.

Spencer, J.W., 1971. Fourier series representation of the position of the sun. Search 2, 172–172.

Stephens, G.L., 1978a. Radiation profiles in extended water clouds. I. theory. J. Atmos. Sci. 35, 2111–2122.

Stephens, G.L., 1978b. Radiation profiles in extended water clouds. II: parameterization schemes. J. Atmos. Sci. 35, 2123–2132.

Stephens, G.L., Ackerman, S., Smith, E.A., 1984. A shortwave parameterization revised to improve cloud absorption. J. Atmos. Sci. 41, 687–690.

Tang, W.J., Qin, J., Yang, K., Liu, S.M., Lu, N., Niu, X.L., 2016. Retrieving high-resolution surface solar radiation with cloud parameters derived by combining MODIS and MTSAT data. Atmos. Chem. Phys. 16, 2543–2557.

Telahun, Y., 1987. Estimation of global solar radiation from sunshine hours, geographical and meteorological parameters. Sol. Wind Technol. 4, 127–130.

Thornton, P.E., Running, S.W., 1999. An improved algorithm for estimating incident daily solar radiation from measurements of temperature, humidity, and precipitation. Agric. For. Meteorol. 93, 211–228.

Trenberth, K.E., Fasullo, J.T., 2012. Tracking earth's energy: from El Niño to global warming. Surv. Geophys. https://doi.org/10.1007/s10712-10011-19150-10712.

Van Laake, P.E., Sanchez-Azofeifa, G.A., 2004. Simplified atmospheric radiative transfer modelling for estimating incident PAR using MODIS atmosphere products. Remote Sens. Environ. 91, 98–113.

Vermote, E., Tanré, D., Deuzé, J.L., Herman, M., Morcrette, J.J., 1997. Second simulation of the satellite signal in the solar spectrum, 6S: an overview. IEEE Trans. Geosci. Remote Sens. 35, 675–686.

Wandera, L., Mallick, K., Kiely, G., Roupsard, O., Peichl, M., Magliulo, V., 2017. Upscaling instantaneous to daily evapotranspiration using modelled daily shortwave radiation for remote sensing applications: an artificial neural network approach. Hydrol. Earth Syst. Sci. 21, 197–215.

Wang, H., Pinker, R.T., 2009. Shortwave radiative fluxes from MODIS: model development and implementation. J. Geophys. Res. Atmos. 114, D20201.

Wang, Y.Z., Feng, D.W., LI, D.S., Chen, X.Y., Zhac, Y.X., Niu, X., 2016. A Mobile Recommendation System Based on Logistic Regression and Gradient Boosting Decision Trees. 2016 International Joint Conference on Neural Networks. Ieee, New York, pp. 1896–1902.

Wei, Y., Zhang, X., Hou, N., Zhang, W., Jia, K., Yao, Y., 2019. Estimation of surface downward shortwave radiation over China from AVHRR data based on four machine learning methods. Sol. Energy 177, 32–46.

Wild, M., Gilgen, H., Roesch, A., Ohmura, A., Long, C.N., Dutton, E.G., Forgan, B., Kallis, A., Russak, V., Tsvetkov, A., 2005. From dimming to brightening: decadal changes in solar radiation at earth's surface. Science 308, 847–850.

Yang, K., Huang, G.W., Tamai, N., 2000. A hybrid model for estimating global solar radiation. Sol. Energy 70, 13–22.

Yang, K., Pinker, R.T., Ma, Y., Koike, T., Wonsick, M.M., Cox, S.J., Zhang, Y., Stackhouse, P., 2008. Evaluation of satellite estimates of downward shortwave radiation over the Tibetan Plateau. J. Geophys. Res. 113, D17204.

Yang, L., Zhang, X.T., Liang, S.L., Yao, Y.J., Jia, K., Jia, A.L., 2018. Estimating surface downward shortwave radiation over China based on the gradient boosting decision tree method. Remote Sens. 10, 185.

Zhang, Y.R., Haghani, A., 2015. A gradient boosting method to improve travel time prediction. Transp. Res. C Emerg. Technol. 58, 308–324.

Zhang, Y.C., Rossow, W.B., Lacis, A.A., 1995. Calculation of surface and top of atmosphere radiative fluxes from physical quantities based on ISCCP data sets: 1. Method and sensitivity to input data uncertainties. J. Geophys. Res. Atmos. 100, 1149–1165.

Zhang, Y., Rossow, W.B., Lacis, A.A., Oinas, V., Mishchenko, M.I., 2004. Calculation of radiative fluxes from the surface to top of atmosphere based on ISCCP and other global data sets: refinements of the radiative transfer model and the input data. J. Geophys. Res. Atmos. 109, D19105.

Zhang, X., Liang, S., Zhou, G., Wu, H., Zhao, X., 2014. Generating Global LAnd Surface Satellite incident shortwave radiation and photosynthetically active radiation products from multiple satellite data. Remote Sens. Environ. 152, 318–332.

Zhang, X., Liang, S., Wild, M., Jiang, B., 2015. Analysis of surface incident shortwave radiation from four satellite products. Remote Sens. Environ. 165, 186–202.

Zhou, Q.T., Flores, A., Glenn, N.F., Walters, R., Hang, B., 2017. A machine learning approach to estimation of downward solar radiation from satellite-derived data products: an application over a semi-arid ecosystem in the US. PLoS One 12, 19.

Advanced Remote Sensing, Second Edition
https://doi.org/10.1016/B978-0-12-815826-5.00006-4

© 2020 Elsevier Inc. All rights reserved.

Abstract

Albedo is a key parameter that is widely used in land surface energy balance studies, mid-to-long-term weather prediction, and global climate change investigation. Remote sensing is an effective way to map the land surface albedo on both the regional and global scales. This chapter first presents the consideration of light incidence and reflection angles to the calculation of surface albedo. Then, the retrieval method of spectral albedo based on bidirectional reflectance models inversion and narrowband-to-broadband albedo conversion is discussed. The next section describes the direct-estimation approaches that retrieve surface broadband albedo directly from single-angle surface reflectance or top-of-atmosphere reflectance, which can be easily adopted in operational global albedo production. After that, the available global land surface albedo remote sensing products and application examples of phenology analysis based on global surface albedo products is briefly reviewed. In the end, a short discussion is given on the cutting-edge problems in global surface albedo estimation.

Albedo is a key parameter that is widely used in land surface energy balance studies, mid-to-long-term weather prediction, and global climate change investigation (Dickinson, 1995). It is defined as the ratio of the surface-reflected irradiance to the incident irradiance. Because albedo quantifies the capacity of a surface to reflect solar radiation, it is one of the main driving factors of the energy balance and interaction between land surface and atmosphere. An increase in albedo leads to a decrease of net radiation, decreasing the latent heat and sensible heat and weakening the low-level convergence. As a result, cloud formation and precipitation decreases. Soil moisture may also decrease, further increasing surface albedo and forming a positive feedback mechanism. The reduction of cloud cover causes an increase in solar radiation and further increases the net radiation, which introduces a negative feedback. Albedo plays a key role in the maintenance of a stable state between the positive and negative feedback mechanisms. Therefore, it is a fundamental component in determining the surface radiation budget and atmospheric motion during global and regional climatic change. The spatial and temporal variations of the surface albedo are affected by natural processes, such as soil moisture, vegetation dynamics, and snow coverage, in addition to land utilization and other human activities (Gao et al., 2005).

Remote sensing is an effective way to map the land surface albedo on both the regional and global scales. Early climate applications require an albedo-estimation accuracy of ± 0.05 (Henderson-Sellers and Wilson, 1983). According to Global Climate Observing System, the expected accuracy of surface albedo products should be greater than 0.01 (or 5% relative accuracy) to meet application needs. Additionally, the spatial resolution should be better than 1 km, and the temporal resolution should be 1 day (GCOS, 2006). This set of parameters is the goal of quantitative remote sensing inversion of surface albedo.

This chapter first presents the consideration of light incidence and reflection angles to the calculation of surface albedo (Section 6.1). Then, the retrieval method of spectral albedo based on bidirectional reflectance models inversion, and narrowband-to-broadband albedo conversion is discussed in Section 6.2., and Section 6.3 describes the direct-estimation approaches that retrieve surface broadband albedo directly from single-angle surface reflectance or top-of-atmosphere (TOA) reflectance, which can be easily adopted in operational global albedo production. The following two sections introduce the available global land surface albedo remote sensing products (Section 6.4) and application examples of phenology analysis based on global surface albedo products (Section 6.5). Finally, a short discussion is given on the cutting-edge problems in global surface albedo estimation.

6.1 Land surface bidirectional reflectance modeling

6.1.1 Definition of land surface bidirectional reflectances and broadband albedo

Land surface albedo quantifies the solar radiation reflectivity of the land surface, which is characterized by its variation in the spectral and directional domains. To simplify this concept,

the directional characteristics of monochromatic light are discussed first.

6.1.1.1 Bidirectional reflectance distribution function

Light incident on an ideal smooth surface is reflected specularly, while that on an ideal rough surface is reflected diffusely. However, natural terrain surfaces are neither ideally smooth nor ideally rough. The notation of bidirectional reflectance is based on the dependence of the solar radiation reflectivity of the land surface on both the incident solar direction and the sensor viewing direction. The following definition of the bidirectional reflectance distribution function (BRDF) was given by Nicodemus et al., 1977:

$$\text{BRDF} = f_r(\theta_i, \phi_i; \theta_r, \phi_r; \lambda) = \frac{dL_r(\theta_i, \phi_i; \theta_r, \phi_r; \lambda)}{dE_i(\theta_i, \phi_i; \theta_r, \phi_r; \lambda)}$$

(6.1)

where θ_i is the solar zenith angle, ϕ_i is the solar azimuth angle, θ_r is the viewing zenith angle, and ϕ_r is the viewing azimuth angle. The differential quantity $dE_i(\theta_i, \phi_i; \theta_r, \phi_r; \lambda)$ denotes the increment of the spectral radiance in an incoming beam per unit solid angle and per unit area of the surface, while $dL_r(\theta_i, \phi_i; \theta_r, \phi_r; \lambda)$ characterizes the corresponding reflected spectral radiance from the surface to the sensor viewing direction (Fig. 6.1). The BRDF is expressed in units of $1/\text{sr}$. The wavelength symbols are usually omitted, provided that this omission will not cause confusion, and the expression is simplified to

$$\text{BRDF} = f_r(\theta_i, \phi_i; \theta_r, \phi_r) = \frac{dL_r(\theta_i, \phi_i; \theta_r, \phi_r)}{dE_i(\theta_i, \phi_i; \theta_r, \phi_r)}$$

(6.2)

The above definition is based on the infinitesimal area of the surface and the infinitesimal solid angle. This rigorous mathematical basis endows the BRDF with an explicit physical connotation such that it is intrinsic to the surface properties of the specific location and independent of the measurement conditions. However,

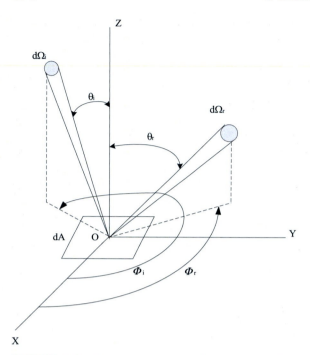

FIGURE 6.1 The configuration of bidirectional reflectance distribution function parameters.

because of the difficulties encountered during practical measurement of the BRDF, people tend to use other physical quantities whose definitions are based on the BRDF.

6.1.1.2 The definition of BRDF, reflectance, reflectance factor, and albedo

The ratio of the radiant existence to the irradiance defines the reflectance. Rigorously, reflectance is defined for monochromatic wavelengths, which is known as the spectral reflectance. In practice, the radiant flux is always measured in a wavelength interval, or so-called spectral band. So long as the wavelength interval is small, the band reflectance can also be defined. The reflectance, ρ, is a dimensionless physical quantity with a value in the inclusive interval from 0 to 1.

In comparison, the reflectance factor is the ratio of the radiant flux reflected by a surface to

that reflected by an ideal (lossless) and diffuse (Lambertian) standard surface of the same reflected-beam geometry, in the same wavelength range and irradiated under the same conditions. The reflectance factor (R) is a dimensionless physical quantity defined per unit wavelength. Unlike the reflectance, the reflectance factor can take values greater than 1.

The reflectance and the reflectance factor are both ratios, but they differ in two ways. The denominator of the reflectance is the incoming irradiance, while that of the reflectance factor is the reflected radiant flux of an ideal diffuse surface. Additionally, the numerator of the reflectance is the integral reflected radiant flux for a hemispherical space, while that of the reflectance factor is the reflected radiant flux in a specific direction.

The albedo is defined as the ratio between the reflected energy and the incident energy over a unit area, from which it can be inferred that there is some similarity between the albedo and the reflectance. In fact, the albedo of a single wavelength is identical to the reflectance. However, the albedo is usually defined over a certain wavelength range, such as the visible light albedo or the shortwave albedo.

The reflectance, reflectance factor, and albedo are all measurable physical quantities. Because reflectance and albedo must be measured hemispherically, the detector (for example, an integrating sphere or a pyranometer) should be able to conduct cosine integration. In contrast, the measurement of the reflectance factor is specific to the direction of the emergent ray; accordingly, a directional detector (for example, a spectrometer) is required. To determine the emergent light of an ideal diffuse surface, a reference panel is usually adopted to approximate the ideal diffuse standard surface.

As mentioned above, the BRDF is a description of a surface's intrinsic properties and is independent of the measurement conditions. However, the reflectance, reflectance factor, and albedo are related to the measurement conditions, such as the directional distribution of the incident radiance. In the case of the broadband albedo, the spectral variation of the incident radiance should also be considered. Some other physical quantities, such as directional—hemispherical reflectance (DHR), BHR, bidirectional reflectance factor (BRF) and hemispherical—directional reflectance factor (HDRF), are further defined according to the varied directional distribution of the incoming radiance and the observation method, all of which can be obtained by integrating the BRDF.

Typically, the measured surface is assumed to be a flat and homogeneous plane; therefore, the size of the area has no effect on the measurement result. Because each pixel of a remote sensing -image usually covers a certain area of a heterogeneous surface, the BRDF cannot be effectively defined for remote sensing pixel; however, the definitions of the reflectance, reflectance factor, and albedo remain valid.

Some of the literature does not adhere to this strict definition of these physical quantities but follows less rigorous naming conventions. The most common example is referring to the measured BRF data as the BRDF. In fact, these two quantities not only have different physical meanings but their values also differ by a factor of π, even for Lambertian surfaces. Some of the ambiguous usage of terms, such as interchange of the reflectance factor and reflectance or referring to the narrowband albedo as the reflectance, can be tolerated within certain contexts, e.g., when the observed target is homogeneous or Lambertian. However, these concepts must be standardized and carefully utilized in serious research investigations. The detailed explanations of the definitions of these reflectance quantities are presented in the literature (Schaepman-Strub et al., 2006; Martonchik et al., 2000; Liang and Strahler, 2000).

6.1.1.3 Definitions of relative physical quantities

The reflectivity of natural terrain is related to both the incident and observation geometry of the ray. The commonly employed observation geometries are specific direction observation and hemispheric integration. The definition of reflectance (albedo) is based on the radiant flux (energy); therefore, this term can only be used to describe hemispherically integrated observations, while specific direction observations correspond to the reflectance factor.

The three commonly used incident geometries are natural irradiance, ideal parallel radiation, and ideal diffuse radiation. Accordingly, there are six possible combinations of the incident and observation geometries: bidirectional, diffuse hemispherical-directional, hemispherical-directional, directional-hemispherical, directional-diffuse hemispherical, and bihemispherical.

6.1.1.3.1 Bidirectional reflectance factor

The BRF is the measurable physical quantity that is closest to the BRDF. It is defined as the ratio of the reflected radiant flux from the target surface area to the equivalent flux from an ideal diffuse surface in the same location under an identical view geometry and parallel illumination.

$$\text{BRF} = R(\theta_i, \phi_i; \theta_r, \phi_r) = \frac{d\Phi_r(\theta_i, \phi_i; \theta_r, \phi_r)}{d\Phi_r^{id}(\theta_i, \phi_i)} \quad (6.3)$$

where $d\Phi_r(\theta_i, \phi_i; \theta_r, \phi_r)$ is the differential radiant flux in the viewing direction, which has the following relationship with the differential radiance:

$$d\Phi_r(\theta_i, \phi_i; \theta_r, \phi_r) = \cos\theta_r \sin\theta_r dL_r(\theta_i, \phi_i; \theta_r, \phi_r) d\theta_r d\phi_r dA \quad (6.4)$$

The quantity $d\Phi_r^{id}(\theta_i, \phi_i)$ in Eq. (6.3) is the differential radiant flux reflected to the view direction from an ideal diffuse surface. Because the

Lambertian surface has no angular dependence, the view zenith and azimuth angles are omitted. For an ideal standard surface, all of the energy of the incoming beam is reflected in a diffusive way; therefore,

$$d\Phi_r^{id}(\theta_i, \phi_i) = d\Phi_i(\theta_i, \phi_i) = dE_i(\theta_i, \phi_i)dA \quad (6.5)$$

Thus, the relationship between the BRF and the BRDF can be expressed as follows:

$$\text{BRF} = R(\theta_i, \phi_i; \theta_r, \phi_r) = \frac{f_r(\theta_i, \phi_i; \theta_r, \phi_r)}{f_r^{id}(\theta_i, \phi_i)}$$

$$= \pi f_r(\theta_i, \phi_i; \theta_r, \phi_r) \quad (6.6)$$

i.e., the BRF of an ideal surface can be expressed as the BRDF multiplied by π. The instantaneous field of view (IFOV) of the instrument may integrate over a certain viewing solid angle rather than an infinitesimal solid angle. Because the BRDF of a natural scene is continuous, the angular influence of IFOV is often neglected, but it should not be neglected for large IFOV sensor measurements.

6.1.1.3.2 Diffuse hemispherical–directional reflectance factor

The concept of the diffuse hemispherical–directional reflectance factor (HDRF_diff) is similar to the definition of the BRF, but the incoming radiance is ideally isotropic over the entire hemisphere. The HDRF_diff can be expressed as the integral of the BRDF over the incoming hemisphere,

$$\text{HDRF\_diff} = R(2\pi; \theta_r, \phi_r) = \frac{d\Phi_r(2\pi; \theta_r, \phi_r)}{d\Phi_r^{id}(2\pi)}$$

$$= \int_0^{2\pi} \int_0^{\pi/2} f_r(\theta_i, \phi_i; \theta_r, \phi_r) \sin\theta_i \cos\theta_i d\theta_i d\phi_i$$

$$(6.7)$$

6.1.1.3.3 Hemispherical–directional reflectance factor

The definition of the HDRF is also similar to that of the BRF, but it permits realistic angular distributions of the illumination (Strub et al., 2003).

In practice, it is often assumed that the natural irradiance is composed of a direct component and a diffuse component.

$$\text{HDRF} = R(\theta_i, \phi_i, 2\pi; \theta_r, \phi_r) = \frac{d\Phi_r(\theta_i, \phi_i, 2\pi; \theta_r, \phi_r)}{d\Phi_r^{id}(\theta_i, \phi_i, 2\pi)}$$

$$= \frac{\int_0^{2\pi} \int_0^{\pi/2} f_r(\theta_i, \phi_i; \theta_r, \phi_r)\sin\theta_i \cos\theta_i L_i(\theta_i, \phi_i; \theta_r, \phi_r)d\theta_i d\phi_i}{\int_0^{2\pi} \int_0^{\pi/2} \sin\theta_i \cos\theta_i L_i(\theta_i, \phi_i; \theta_r, \phi_r)d\theta_i d\phi_i}$$

$$(6.8)$$

If we divide the irradiance into a direct part (with angles θ_0, ϕ_0) and a diffuse part, we may continue in the following ay:

$$\text{HDRF} = R(\theta_i, \phi_i, 2\pi; \theta_r, \phi_r)$$

$$= (1-s)\pi f_r(\theta_0, \phi_0; \theta_r, \phi_r) + s \int_0^{2\pi} \int_0^{\pi/2} f_r(\theta_i, \phi_i; \theta_r, \phi_r)\sin\theta_i \cos\theta_i d\theta_i d\phi_i$$

$$= (1-s)\text{BRF}(\theta_0, \phi_0; \theta_r, \phi_r) + s\text{HDRF\_diff}(\theta_r, \phi_r)$$

$$(6.9)$$

where s corresponds to the fractional amount of the diffuse radiant flux:

$$s = \frac{L_i^{diff}}{\left(1/\pi\right)E_{dir}(\theta_0, \phi_0) + L_i^{diff}} \quad (6.10)$$

6.1.1.3.4 Directional–hemispherical reflectance

The DHR is the ratio of the radiant flux (for light reflected by a unit surface area into the view hemisphere) to the illumination radiant flux when the surface is illuminated with a parallel beam of light from a single direction.

$$\text{DHR} = \rho(\theta_i, \phi_i; 2\pi) = \frac{d\Phi_r(\theta_i, \phi_i; 2\pi)}{d\Phi_i(\theta_i, \phi_i)}$$

$$= \int_0^{2\pi} \int_0^{\pi/2} f_r(\theta_i, \phi_i; \theta_r, \phi_r)\sin\theta_r$$

$$\cos\theta_r d\theta_r d\phi_r$$

$$(6.11)$$

Eq. (6.11) corresponds to the definition of black-sky albedo (BSA) for a monochromatic incoming beam (Lucht et al., 2000).

6.1.1.3.5 Diffuse hemispherical–hemispherical reflectance

The diffuse hemispherical–hemispherical reflectance (BHR_diff) is the ratio of the radiant flux reflected from a unit surface area into the whole hemisphere to the isotropic incident radiant flux. It is equal to the integral of the BRDF over the incoming and outgoing hemispheres or the integration of the DHR over the incoming hemisphere.

$$\text{BHR\_diff} = \rho(2\pi; 2\pi) = \frac{d\Phi_r(2\pi; 2\pi)}{d\Phi_i(2\pi)}$$

$$= \int_0^{2\pi} \int_0^{\pi/2} \rho(\theta_i, \phi_i; 2\pi)\sin\theta_i$$

$$\cos\theta_i d\theta_i d\phi_i \tag{6.12}$$

Eq. (6.12) can also be interpreted as the so-called white-sky albedo (WSA) in the moderate resolution imaging spectroradiometer (MODIS) product suite for a monochromatic incoming beam.

6.1.1.3.6 Hemispherical–hemispherical reflectance

The concept of the BHR is similar to that of the BHR_diff, but it permits realistic angular distributions of illumination. Similar to the HDRF, the incident radiance flux is usually assumed to be composed of a direct component and a diffuse component in practice, although the definition of the BHR allows any directional distribution of the incident radiance flux.

$$\text{BHR} = \rho(\theta_i, \phi_i, 2\pi; 2\pi) = \frac{d\Phi_r(\theta_i, \phi_i, 2\pi; 2\pi)}{d\Phi_i(\theta_i, \phi_i, 2\pi)}$$

$$= \frac{\int_0^{2\pi} \int_0^{\pi/2} \rho(\theta_i, \phi_i; 2\pi)L_i(\theta_i, \phi_i)\sin\theta_i \cos\theta_i d\theta_i d\phi_i}{\int_0^{2\pi} \int_0^{\pi/2} L_i(\theta_i, \phi_i)\sin\theta_i \cos\theta_i d\theta_i d\phi_i}$$

$$\tag{6.13}$$

When the irradiance is divided into a direct part (with angles θ_0, ϕ_0) and a diffuse part and s corresponds to the fractional amount of the diffuse radiant flux,

$$\text{BHR} = (1 - s)\text{DRF}(\theta_0, \phi_0; 2\pi)$$

$$+ s\text{BHR\_diff}(2\pi; 2\pi) \tag{6.14}$$

Eq. (6.13) is also called the blue-sky albedo or the actual albedo for ambient monochromatic illumination.

6.1.1.3.7 Broadband albedo

The DHR (BSA), the BHR_diff (WSA), and the BHR (blue-sky albedo) are also called spectral albedos, which are represented by $a(\theta, \lambda)$. They are expressed as a function of wavelength and the geometry of the incident beam, which can be ideal direct radiation, ideal diffuse radiation, or natural irradiance.

The albedo on a certain spectral band can be determined by integrating the spectral albedo over the corresponding wavelength range, weighted by the spectral response function.

$$A(\theta, \lambda) = \frac{\int_{\lambda 1}^{\lambda 2} \alpha(\theta, \lambda)F_d(\theta, \lambda)d\lambda}{\int_{\lambda 1}^{\lambda 2} F_d(\theta, \lambda)d\lambda} \tag{6.15}$$

Here, $A(\theta, \lambda)$ denotes the band albedo; F_d is the incident downward solar radiant flux at the bottom of the atmosphere, which can be expressed as a function of the wavelength and the incidence angle; and λ_1 and λ_2 are the start and end wavelengths of the integration, respectively.

The following band albedos are commonly applied to remote sensing. The narrowband albedo corresponds to a specific sensor in a certain spectral channel, e.g., the albedo at band 1 of TM or the albedo at band 2 of MODIS. The shortwave albedo is the most important albedo in the study of the surface energy budget because solar radiation is mainly distributed over the shortwave band. In the literatures, the most frequently used wavelength ranges for shortwave albedo are 0.25—0.5 μm, which covers a wider wavelength region of the solar radiation distribution, and 0.3—3 μm, which corresponds to the wavelength range of the pyrheliometer. The difference between the two wavelength ranges is minimal because the incoming solar radiation at the wavelength beyond 0.3—3 μm approaches zero. The visible and near-infrared albedos, corresponding to the wavelength ranges 0.3—0.7 μm and 0.7—3.0 μm, respectively, are useful to distinguish the apparent differences between the reflective characteristics of plants and soil. The shortwave, visible, and near-infrared albedos can be conceptualized as a broadband albedo, but the term "broadband albedo" most commonly refers to the shortwave albedo.

6.1.2 Observations data of surface bidirectional reflection

6.1.2.1 Laboratory and field observations

The phenomenon of bidirectional reflection caught scientists' attention first in measurements of light reflected from man-made materials and other small samples and later from natural terrain. Some data from the observation of natural land surface are introduced below.

The main instruments used in the field measurement of bidirectional reflection are field goniometric devices, spectrometers, and reference boards. As mentioned above, according to its rigorous definition, the BRDF cannot be measured; therefore, the measured quantity that is commonly called as BRDF is actually either the BRF or the HDRF, reflecting the anisotropic characteristics of the surface. Ideally, the irradiance of the target object and of an ideal diffuse reference board are measured without diffuse light from the sky, using a spectrometer that has a small field of view (FOV), and the ratio of the irradiances is the BRF. However, because natural incident light is composed of direct light and diffuse light, this measurement actually

corresponds to the HDRF. Under clear-sky conditions, the fractional of direct radiant flux generally exceeds 80% of the total incident radiant flux, so the discrepancy between the BRF and the HDRF can be ignored with minimal error. However, the literature also indicates that the difference between HDRF and BRF must be considered when the incident radiance has a large diffuse component; extra procedures for measuring and data processing are recommended to convert HDRF to BRF.

Different field goniometric devices (herein called "field goniometers") have different levels of automaticity. Two representative fully automatic field goniometer systems are the Swiss Field—Goniometer System (FIGOS) (Sandmeier and Itten, 1999) and the Sandmeier Field Goniometer (SFG) constructed by NASA (Sandmeier, 2000). Both systems allow the spectroradiometer to be positioned at any view zenith and azimuth angle using a motor. The completion of a multiangle observation cycle (including 11 zenith angles and 18 azimuth angles) requires approximately 10—20 min. Fig. 6.2 shows the automatic field goniometer designed by Chinese scientists, which is less complex and cheaper than FIGOS. The device uses a "crane boom" structure to control the view zenith angle, and the view azimuth angle varies with the rotation of the base on the arc rail. The height of the device can be dynamically adjusted to follow the growth of a plant. Such designs have reduced the weight of the system and its disturbance of the natural illumination. Because expensive full-automatic field goniometers are quite bulky, they are also replaced by simple apparatus in many field campaigns. The measurement of the bidirectional quantities of barren soil can be performed indoors with prepared samples and smaller observation device under illumination by artificial light (Wu et al., 2010) (Fig. 6.3). These goniometer devices still have many problems, such as the destructive effect to the observed target and its surroundings during the installation of the goniometer devices. If

FIGURE 6.2   The automatic field goniometer designed by Chinese scientists (Liu et al., 2002).

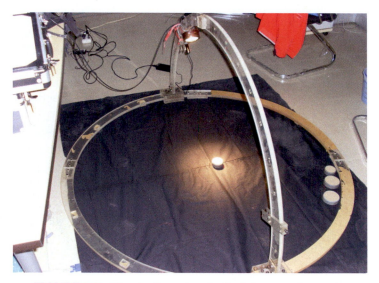

FIGURE 6.3   The simple goniometer for indoor measurements.

we enlarge the visual area of the spectrometer while maintaining the small field angle, the observation height of the spectrometer must be increased accordingly, which is technologically impractical.

A complete set of multiangle observations requires a long measurement time; the change in solar angle and illumination conditions during this period impedes the standardization of the multiangle dataset. To shorten the measurement time, the measurement can be reduced to several typical sample planes. The bidirectional reflective characteristics are more evident on the principal plane, which is defined as the plane that is determined by the incident solar beam and the surface normal. Generally, movement of the observing instrument from the backward direction to the forward direction causes the viewing zenith angle to change gradually from 90° backward to a nadir of 0° and then forward to 90°, while the viewing azimuth angle changes accordingly from 0° (corresponding to the backward direction) to 180°. During field observation, the bidirectional reflection distribution on the principal plane is given measurement priority if the distribution of the entire hemisphere is difficult to collect.

6.1.2.1.1 Bidirectional reflective characteristics of vegetation canopy

Fig. 6.4 shows the profile of the measured bidirectional reflectance of a corn canopy in the principal plane, measured at the Yingke Oasis Experimental Station in the city of Zhangye on June 22, 2008. The measurement result is actually the HDRF, which was acquired by an ASD Field-Spec Pro FR 2500 spectrometer and the field goniometers designed by Beijing Normal University. The negative view zenith angles denote backward observation, while the positive ones correspond to forward observation. Four planes were observed: the principal plane, the perpendicular principal plane, the plane along the row, and the plane across the row. The maximal view zenith angle was 60°, which was approached in increments of 10°. The measured

target was corn fields, which have an LAI of approximately 2.5. One of the main features of vegetation reflection is that the near-infrared reflection is significantly greater than the red band. The second feature is that the bidirectional reflection distribution in the red band has a mound-like shape, which is chiefly caused by surface scattering, while the result in the near-infrared band presents a shallow bowl shape that is mainly caused by volumetric scattering. The third feature is the existence of a hotspot, which refers to a phenomenon of increased reflectance when the viewing direction coincides with the solar illumination direction. In Fig. 6.4, the hotspot direction is at 22° in the backscattering direction. As shown in the profile, the bidirectional reflectance increases when the view zenith angle is between backward 20 and 30°. However, a much higher sampling resolution is needed to reveal the details of the hotspot in the reflectance profile. Because measurements in the hotspot direction are hampered by the sensor shadow, the hotspot phenomenon is difficult to observe in practice.

6.1.2.1.2 Bidirectional reflective characteristics of bare soil

Fig. 6.5 presents the principal plane distribution profiles of HDRF measured at the second

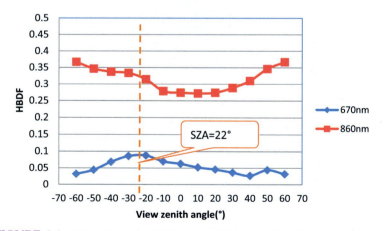

FIGURE 6.4   Bidirectional reflection of vegetation canopy in the principal plane.

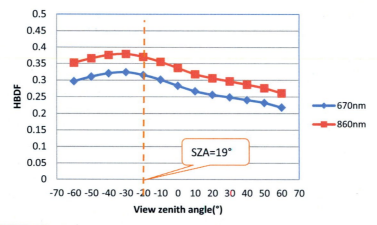

FIGURE 6.5 Bidirectional reflection characteristics of bare soil in the principal plane.

experimental site, in Zhangye city in the Huazhaizi desert on June 22, 2008, with ASD FieldSpec Pro FR 2500 spectrometer and simple field goniometers. The principal and perpendicular principal planes were measured in increments of 10° to a maximal view zenith angle of 60°. The measured target was the desert surface with less than 2% coverage of xerophytic vegetation. The difference of the soil reflectance in the visible and near-infrared bands is less significant than the difference for vegetation, and the directional variations are similar. The hotspot direction occurs at approximately negative 19°; however, according to

Fig. 6.5, the maximum bidirectional reflectance occurs at negative 30°, which illustrates that many other factors also influence the position of the peak value.

6.1.2.1.3 Bidirectional reflective characteristics of ice/snow

Fig. 6.6 shows the HDRF profiles in principal plane measured at the Binggou site in Qilian County on March 23, 2008, with an ASD Field-Spec Pro FR 2500 spectrometer and rotary field goniometers. During the multiangle observation of snow, the view azimuth angle was increased in 10° increments, and the view zenith angle

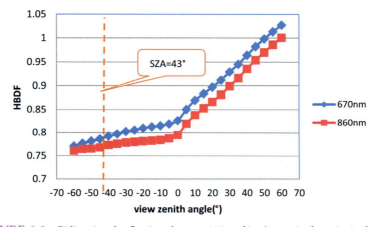

FIGURE 6.6 Bidirectional reflection characteristics of ice/snow in the principal plane.

was increased in 5° increments (to a maximum of 60°). The results indicate that the reflectances of the snowfield in both the visible and NIR bands are high, and both bands exhibit almost identical directional characteristics. The hotspot phenomenon is not apparent in the figure, and the forward reflectance of the snow is greater than its backward reflectance, which is in contrast to the reflectances of vegetation and bare soil, for which backscattering dominates. Fig. 6.6 shows that the maximum value of the measured HDRF of the snowfield exceeds 1, which is possible according to the definition of the reflectance factor. Although we habitually call the measured data as BRDF, the actual physical variable corresponding to the measurement should be considered during data analysis and interpretation.

6.1.2.2 Remote sensing observation data

Generally speaking, the means of acquiring multiangle remote sensing data can be grouped into the following three categories:

The first category is similar to the ground-measuring method. The sensor keeps focused on the same ground object during its movement along the track. It obtains multiangular along-track observations by adjusting its own inclination. CHRIS aboard PROBA is a typical spaceborne sensor that utilizes this observation method. It is able to obtain observations at five different angles by adjusting the inclination of the platform. The satellite can also carry several cameras to obtain images from various angles instead of altering the satellite attitude; a typical example of this configuration is MISR aboard Terra, which acquires observations at nine different angles using nine cameras with a maximum viewing zenith angle of 70°.

The second category employs a panel camera to form wide-angle images and then extracts repeated observations of the same surface target from either multiple adjacent images of the same track or multiple images from adjacent tracks. All of these images can supply observations of different viewing geometries. POLDER aboard PARASOL, which has a ground resolution of approximately 6 km, is a typical spaceborne sensor with such an imaging design. The images from one track of POLDER have up to 14 different angles, of which the maximum zenith is 70°.

The third technique uses a wide-angle scanning sensor, and the repeated observations of the same surface target are captured from multiple adjacent tracks. Many mid-to-low-resolution remote sensors, such as NOAA-AVHRR, Terra-MODIS, SPOT VEGETATION, and FY3-VIRR/MERSI, have scanning angles exceed 50° and are able to cover most of the global surface in a single day. Therefore, the data obtained on different dates for the same pixel can be used to compose a multiangle observation dataset. This method requires several days to obtain sufficient directional samples, during which the change of the surface state is not negligible. Therefore, these sensors are not strictly multiangle sensors. On the other hand, because of their large viewing angle, the surface bidirectional reflective characteristics should be considered when using the data from these sensors quantitatively.

The multiangle observations acquired with the first method have good time synchronism, and the viewing angles are all located within the orbital plane. The time synchronism of the third method is weaker, and its viewing angles are located near the plane perpendicular to the orbital plane of the satellite. Fig. 6.7 shows an example of the distribution of the solar/view angles of a multiangle dataset acquired by MODIS and MISR in the view hemisphere. The view angles of the data acquired with both methods are limited and not evenly distributed.

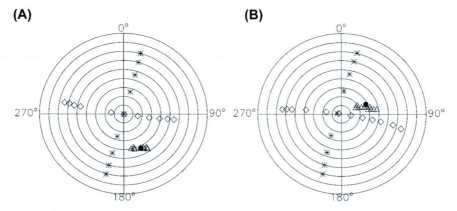

FIGURE 6.7 The hemispherical distribution of solar/view angles of MODIS and MISR data (A) January, Sahara; (B) February, South Africa; the polar radius and angle represent the zenith and azimuth, respectively; the diamond and asterisk patterns represent the observation directions of MODIS and MISR, respectively; and the triangle and round dot patterns represent the solar directions of MODIS and MISR, respectively (Jin et al., 2002a,b).

Concerning the second method described above, multiangle observations can be obtained in both the along-track (real-time) and cross-track (multiday, within the overlap of images from adjacent tracks) directions with the advantages of quantity and even distribution. However, because of its wide-angle imaging with panel array cameras, the spatial resolution of the data is usually low. The BRDF dataset extracted from POLDER data by the POSTEL Thematic Center of the European Space Agency is a typical bidirectional reflectance observation dataset that covers various land types with the best available quality, which is actually the HDRF after atmospheric correction. Every dataset is a collection of cloud-free directional samples of a certain pixel within a month. The following figure (Fig. 6.8) illustrates a POLDER-BRDF dataset acquired in July 2006 at (−27.08, 123.75) of an Open Shrublands land type (IGBP_class 7). The dataset is composed of the data of several tracks, and the data of each track are distributed along an arc extending from the upper left to the lower right of the viewing hemisphere. The purple points in the figure, indicating solar direction, illustrate that the observations near the solar direction have larger reflectance (the hotspot phenomenon) (Fig. 6.8).

In the application of multiangle models and inversions, people always want the bidirectional reflectance dataset contains a large number of directional samples evenly distributed over the hemisphere and in shorter acquisition time. However, there is a limitation on the directional samples available from a single sensor or single satellite. Therefore, the utilization of multiple sensors onboard multiple satellites has the advantage of producing more evenly distributed observations over a shorter time, which is beneficial to the inversion with multiangle remote sensing models. Some studies have illustrated that the inversion of BRDF and albedo with combined MODIS and MISR data is helpful in the extraction of surface bidirectional reflectance information (Lucht, 1998; Jin, 2002c). However, the band discrepancies of different sensors and the error in their geometric and radiometric calibration are obstacles to the combination of data from multiple sensors.

Compared with satellite remote sensing, airborne remote sensing has the advantage of flexibility and high resolution; consequently, airborne multiangle remote sensors have always been an important data source for surface bidirectional reflectance modeling and inversion

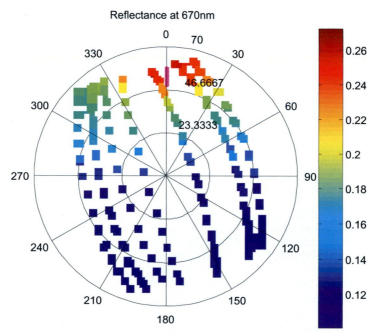

FIGURE 6.8 The angular distribution and reflectances at 670 nm of a typical POLDER-BRDF dataset.

studies. The airborne sensors used to capture bidirectional reflectance data include ASAS (Irons et al., 1991), AirMISR (Diner et al., 1998), AMTIS (Wang, 2000), etc. All these three sensors utilize the along-track alternation of the camera inclination method (the first method mentioned above) to obtain multiangle data. Some other airborne sensors, like AirPOLDER (Leroy et al., 2001), DuncanTech camera (Chopping et al., 2003), and WiDAS (Li Xin et al., 2009), capture multiangle images by panel array cameras with wide-angle lenses.

6.1.3 Surface bidirectional reflectance model

6.1.3.1 Physical model

The physical models of bidirectional reflectance deduce the relationship between the surface bidirectional reflectance and surface parameters from the physical processes that occur between the incident light and the actual surface, and the model parameters have explicit physical significances (Jacquemoud et al., 2000). Taking the vegetation canopy reflectance as an example, the physical models can be divided into four categories: radiative transfer (RT) models, geometric optical (GO) models, geometric—radiative transfer (GO-RT) mixed models, and computer simulation models.

6.1.3.1.1 Radiative transfer models

The theoretical basis of RT models is the theory of RT in a chaotic medium, which is based on the transfer process of radiation in the horizontal-homogeneous layers of canopy. The RT equation is solved to predict the interaction between the radiation and the canopy and explain the radiation transfer mechanism in the vegetation canopy. Next, the directional and spectral characteristics of the absorption, transmission, and reflection are obtained in relation to the incident radiation on the canopy and the underlying surface, which is the basic principle of all bidirectional reflectance models.

The RT equation is the core of the theory; it describes the transfer characteristics of electromagnetic waves in a horizontal-homogeneous and perpendicular-inhomogeneous medium. If polarization effects are ignored, the RT equation can be expressed as follows:

$$\frac{\partial I(\tau, s)}{\partial \tau} = - I(\tau, s)$$
$$+ (1/4\pi) \int P(s, s') I(\tau, s') d\omega' + \varepsilon(r, s)/\sigma\rho$$

$$(6.16)$$

where $I(\tau, s)$ is the radiance, $P(s, s')$ is the phase function, τ is the optical depth, and ε represents emission from the interior of the vegetation canopy. The horizontal-homogenous vegetation canopy can be divided into several thin layers. Within each of these layers is a randomly distributed tiny scatterer. Therefore, the RT characteristics can be described by introducing the optical depth, the scattering phase function, etc. The solution of the RT equation must adopt approximate or numerical solutions. For example, KM theory (Kubelka and Munk, 1931) presumes that the radiation at each canopy level is specified in terms of the downward and upward fluxes and incident and reflective direct parallel fluxes, thus defining a group of differential equations that simplify the RT equations.

$$dE_-/d(-\tau) = -(\alpha + \gamma)E_- + \gamma E_+ + s_1 F_- + s_2 F_+$$
$$dE_+/d(\tau) = -(\alpha + \gamma)E_+ + \gamma E_- + s_1 F_+ + s_2 F_-$$
$$dF_-/d(-\tau) = -(K + s_1 + s_2)F_-$$
$$dF_+/d(\tau) = -(K + s_1 + s_2)F_+$$

$$(6.17)$$

where E_- and E_+ represent the downward and upward scattering fluxes, which are determined by the absorption coefficient α and reflective coefficient γ, respectively, and F_- and F_+ represent the incident and reflective parallel fluxes,

which are described by the absorption coefficient K and the forward and backward scattering coefficients s_1 and s_2, respectively. The essence of KM theory is to approximate the complex differential—integration process with four linear partial differential equations.

The main canopy bidirectional reflectance models based on RT theory include analytical models based on KM theory, such as the Suits model (Suits, 1972), the SAIL model (Verhoef, 1984), and the Kuusk model (Kuusk, 1985), which considers the hotspot effect; the independent discrete model proposed by Idso and Wit (1970), which was further developed into the Goudriaan model (Goudriaan, 1977), the Cupid model (Norman et al., 1985), and several others; and the three-dimensional RT model considering the heterogeneous structure of the canopy (Disney et al., 2006), etc.

6.1.3.1.2 Geometric optical models

The RT model is based on the scattering of small particles. GO models are built on "scene synthesis technology." This method assumes that the reflectance of a pixel is an area-weighted sum of the reflectance signatures of four components: the sunlit crown, the sunlit background, the shadowed crown, and the shadowed background. Jackson and Palmer (1972) proposed a four-component model of row crops; Li and Strahler (1986, 1988, 1992) calculated the variation of the four-component area with the solar/view angles based on the parameters of sparse forest structure, including tree density, canopy size, and height, and proposed the bidirectional reflectance model of a natural forest. Jupp et al. (1986) proposed a GO model suitable for multilayer canopies. Strahler and Jupp (1990) explained the bidirectional characteristics of discontinuous vegetation at two different scales: the canopy scale and the leaf scale. In later developments, Li and Strahler (1992) included the ellipsoidal crown shapes and the effects of mutual shadowing to develop a GO model suitable for forests with high degrees of canopy closure.

The most representative GO model is the Li–Strahler pure GO model (Li and Strahler 1986), which can be described as follows:

$$L_s = K_g L_g + K_c L_c + K_t L_t + K_z L_z \qquad (6.18)$$

where L_s is the total radiance of the pixel, L_g, L_c, L_t, L_z are the radiance of the four components (the sunlit crown, the sunlit background, the shadowed crown, and the shadowed background, respectively), and K_g, K_c, K_t, K_z are the area proportions of the corresponding components, which can be described as functions of the solar/view angles and the canopy structure parameters. The focus of this model is the calculation of the proportions of components and their individual radiances.

The advantage of the GO model is its ability for the description of surface scattering characteristics and the hotspot effect of discrete vegetation, such as sparse forests and orchard gardens.

6.1.3.1.3 Geometric—radiative transfer mixed model

The GO and RT models have advantages on different scales. Based on the pure GO model and discontinuous vegetation gap probability, Li et al. (1995) calculated the contribution of multiple scattering to each component's radiance with the RT method and simulated the reflectance of the sunlit and shadowed components of two layers. They combined the GO and RT models with gap probability theory and developed the GO-RT mixed model, which has obtained good results in the calculation of forest albedo and bidirectional reflectance under different solar heights.

Because the GO-RT mixed model combines the advantages of both the RT and GO models, it has been widely used in the scientific community. Typical representations include the four-scale model (Chen and Leblanc, 1997), the GeoSAIL model (Huemmrich, 2001), and FLIM (Rosema et al., 1992).

6.1.3.1.4 Real scene computer simulation model

Computer simulation models of real scenes use the high-speed calculation and graphic and image processing abilities of modern computers to simulate real vegetation canopies and other scene structures, building models by tracing photon scattering on the canopy surface. Typical modeling methods include the Monte Carlo ray tracing method (Ross and Marshak, 1988; 1989), the radiosity method (Borel et al., 1991), and discrete anisotropic radiative transfer model (Gastelluetchegorry et al., 1996).

The computer simulation model can solve complex RT equations through modern computer technology and simulate the actual structure of vegetation under various viewing conditions, which provides an effective means for the study and validation of other models.

6.1.3.2 Empirical model

The empirical models, also called statistical models, depict the shape of bidirectional reflectance with several mathematical functions, which are usually without explicit physical meanings. Typical empirical models include the Minnaert model (Minnaert, 1941), the Shibayama model (Shibayama, 1985), the Walthall model (Walthall et al., 1985), and the modified Walthall model (Liang et al., 1994; Danaher et al., 2002).

Because empirical models are statistical descriptions or relative analyses of observations, they have many advantages, including their simplicity and calculation speed. However, the development of an empirical model requires a large amount of measured data, and the models vary by vegetation and soil type, which limits the applicability of the model. In addition, because of the lack of a logical relationship between the parameters at different spectral bands, the number of model parameters increases with the number of spectral bands, which causes difficulties during inversion (Goel, 1988).

6.1.3.2.1 Minnaert model

The Minnaert (1941) is one of the earliest empirical models that have been used in planetary astronomy to describe the lunar surface's bidirectional reflectance. The model is simple and reciprocal but only able to roughly approximate the reflectance of the Earth's surface.

$$\rho(\theta_i, \theta_r, \varphi) = \rho_L \frac{(k+1)}{2}(\cos \theta_i \cos \theta_r)^{(k-1)} \quad (6.19)$$

where ρ is the surface reflectance, ρ_L is the surface albedo, i represents the incident angle, and r represents the emergent angle. The Minnaert constant, represented by k, is a parameter that describes the non-Lambertian characteristics of the surface and has a value from 0 to 1. When $k = 1$, we obtain the Lambertian reflectance formula. For dark surfaces, k is approximately 0.5; for brighter surfaces, k increases. It is close to 1 when the surface is very bright.

6.1.3.2.2 Shibayama model

Shibayama and Wiegand (1985) proposed a linear empirical model satisfying the reciprocal principle.

$$\rho(\theta_i, \theta_r, \varphi) = a + b \sin \theta_r + c \sin \theta_r \sin \frac{\varphi}{2}$$
$$+ d \frac{\sin \theta_r}{\cos \theta_i} \quad (6.20)$$

where a, b, c, and d are fitting coefficients.

6.1.3.2.3 Walthall model and modified Walthall model

Walthall et al. (1985) proposed a functional relationship between the bidirectional reflectance and the view zenith angle, solar zenith angle, and relative azimuth angle based on the simulation data of the canopy reflectance distributions of 18 soybeans under different conditions defined by three solar zenith angles, two spectral bands (visible and near-infrared), and three LAI levels.

$$\rho(\theta_i, \theta_r, \varphi) = a\theta_r^2 + b\theta_r \cos \cos(\varphi_r - \varphi_i) + c \quad (6.21)$$

where a, b, and c are model coefficients to be determined. This model is mainly used for soil,

but it also produces good simulation results for soybean bidirectional reflectance, which is described in a paper by Walthall et al. (1985).

Nilson and Kuusk (1989) have modified this model to make it reciprocal:

$$\rho(\theta_i, \theta_r, \varphi) = a\theta_i^2\theta_r^2 + b(\theta_i^2 + \theta_r^2) + c\theta_i\theta_r \cos \varphi + d \quad (6.22)$$

where a, b, c, and d are fitting coefficients.

6.1.3.3 Semiempirical models

Semiempirical models represent a compromise between empirical models and physical models. They reduce the complexity of physical models by applying approximations and simplifications, while retaining the physical meaning.

6.1.3.3.1 Kernel-driven model

The kernel-driven model is currently the most commonly used semiempirical model. It is a combination of approximations of the RT model of a turbid medium and the GO model of a discrete canopy. The model contains an isotropic kernel, a volume scattering kernel, and a GO kernel (Roujean et al., 1992). The kernel functions have physical meanings; thus, they can provide insight into the mechanism of the surface bidirectional phenomenon. Compared to the physical model, the kernel-driven model is simple and easy to operate in routine production. For instance, AMBRALS (Algorithm for MODIS Bidirectional Reflectance Anisotropies of the Land Surface) for MODIS products has adopted the kernel-driven model.

$$R(\theta_i, \theta_r, \varphi; \lambda) = f_{iso}(\lambda)k_{iso} + f_{geo}(\lambda)k_{geo}(\theta_i, \theta_r, \varphi)$$
$$+ f_{vol}(\lambda)k_{vol}(\theta_i, \theta_r, \varphi) \quad (6.23)$$

The above equation is the common expression of the kernel-driven model in which k_{iso} is the isotropic kernel function, which is usually treated as a constant equal to 1; k_{geo} and k_{vol} are the GO kernel and the volume scattering kernel, respectively, which are functions of the incident/reflective angle and are independent of wavelength; f_{iso}, f_{geo}, and f_{vol} are the corresponding coefficients, which are functions of the wavelength but independent of angle.

The kernel-driven model can be composed of different kernel functions. The currently available kernels are described in the following subsections.

6.1.3.3.1.1 RossThick kernel This kernel was proposed by Roujean et al. (1992) as an approximation of the Ross RT theory (Ross, 1981) to describe the bidirectional reflectance of a dense vegetation canopy. The model presumes that the background and the scattering elements all produce Lambertian reflections, and that the orientations of the scatterers are randomly distributed. The simplified kernel function is a single-scattering solution, and multiple scattering is not considered. The function is normalized to 0 when the solar/view zenith angles are all 0. The RossThick kernel function is expressed as follows:

$$k_{thick}(\theta_i, \theta_r, \varphi) = \frac{\left(\frac{\pi}{2} - \xi\right)\cos\xi + \sin\xi}{\cos\theta_i + \cos\theta_r} - \frac{\pi}{4}$$

$$(6.24)$$

where ξ is the phase angle, $\cos\xi = \cos\theta_i \cos\theta_r + \sin\theta_i \sin\theta_r \cos\varphi$.

6.1.3.3.1.2 RossThin kernel This kernel, proposed by Wanner et al. (1995), is useful for the description of volume scattering in canopy with low LAI. The kernel function is described as follows:

$$k_{thin} = \frac{\left(\frac{\pi}{2} - \xi\right)\cos\xi + \sin\xi}{\cos\theta_i \cos\theta_r} - \frac{\pi}{2} \qquad (6.25)$$

6.1.3.3.1.3 RossHotspot kernel (modified Ross-Thick kernel) Maignan et al. (2004) considered

the hotspot effect of the canopy and proposed a new volume scattering kernel:

$$k_{thickM}(\theta_i, \theta_r, \varphi) = \frac{4}{3\pi} \frac{1}{\cos\theta_i + \cos\theta_r} \Bigg[$$

$$\times \left(\frac{\pi}{2} - \xi\right)\cos\xi + \sin\xi \Bigg] \left(1 + \left(1 + \frac{\xi}{\xi_0}\right)^{-1}\right) - \frac{1}{3}$$

$$(6.26)$$

where ξ_0 is a characteristic angle that reflects the ratio of the size of the scatters within to the vertical height of the vegetation canopy. To reduce the number of model parameters, ξ_0 is set to 1.5°.

6.1.3.3.1.4 LiSparse and LiSparseR kernels The LiSparse kernel (Wanner et al., 1995), which is a simplified form of the GO model, is applied to sparse canopies distributed on Lambertian surfaces. The kernel function is expressed as follows:

$$k_{sparse}(\theta_i, \theta_r, \varphi) = O(\theta_i, \theta_r, \varphi) - \sec\theta_i - \sec\theta_r$$

$$+ \frac{1}{2}(1 + \cos\xi)\sec\theta_r$$

$$(6.27)$$

Within the model,

$$O(\theta_i, \theta_r, \phi) = \frac{1}{\pi}(t - \sin t \cos t)(\sec\theta_i + \sec\theta_r)$$

$$\cos t = \frac{h}{b} \frac{\sqrt{D^2 + (\tan\theta_i \tan\theta_r \sin\varphi)}}{\sec\theta_i + \sec\theta_r} D$$

$$= \sqrt{\tan^2\theta_i + \tan^2\theta_r - 2\tan\theta_i \tan\theta_r \cos\phi}$$

$$\cos\xi = \cos\theta_i \cos\theta_r + \sin\theta_i \sin\theta_r \cos\phi$$

where h/b is the ratio of the crown ellipsoid's vertical to horizontal radii, commonly valued as 2.0.

Usually, the range of the solar zenith of satellite multiangle observations is too small to robustly derive bidirectional models, which produces large errors when the observation is extrapolated to other solar zenith angles. If the model satisfies the reciprocal principle (that is, the surface bidirectional reflectance is invariant when the solar angle and view angle are exchanged), the error is mitigated. Thus, the LiSparseR kernel is proposed as a modified LiSparse kernel, according to the reciprocal principle (Lucht, 1998).

$$k_{sparseR}(\theta_i, \theta_r, \varphi) = O(\theta_i, \theta_r, \varphi) - \sec\theta_i - \sec\theta_r$$
$$+ \frac{1}{2}(1 + \cos\xi)\sec\theta_i \sec\theta_r$$

$$(6.28)$$

6.1.3.3.1.5 LiDense kernel
Wanner et al. (1995) proposed a simplified description of the GO model for dense vegetation canopies. The kernel function has the following form:

$$k_{dense}(\theta_i, \theta_r, \varphi) = \frac{(1 + \cos\xi)\sec\theta_r}{\sec\theta_r + \sec\theta_v - O(\theta_i, \theta_r, \varphi)} - 2$$

$$(6.29)$$

6.1.3.3.1.6 LiTransit kernel
To mitigate the extrapolation error of the LiSparse kernel for large solar zenith angles, Li et al. (2000), (Gao et al., 2001) pointed out that the reciprocal principle does not always hold for bidirectional reflectance at the pixel scale and proposed a nonreciprocal solution scheme. The kernel is based on the idea that the observed canopy gap probability decreases when the solar/view zenith angles increase and the LiDense kernel should be used to replace the LiSparse kernel to describe canopy bidirectional reflectance at large solar/view zeniths (that is, the LiSparse kernel should be used at smaller zenith angles and transition to the LiDense kernel at larger zenith angles). The kernel functions are described as follows:

$$k_{Transit} = \begin{cases} k_{Sparse}, & B \leq 2 \\ k_{Dense} = \frac{2}{B}k_{Sparse}, & B > 2 \end{cases}$$

$$(6.30)$$

where
$$B(\theta_i, \theta_v, \phi) = \sec(\theta_i) + \sec(\theta_r) - O(\theta_i, \theta_r, \varphi).$$

6.1.3.3.1.7 Roujean geometric kernel
Roujean modeled the scattering characteristics of low vegetation covered surfaces as vertical opaque protrusions uniformly distributed on a plane. The kernel function has the following form:

$$k_{roujean}(\theta_i, \theta_r, \varphi) = \frac{1}{2\pi}[(\pi - \varphi)\cos\varphi + \sin\varphi]\tan\theta_i\tan\theta_r$$
$$- \frac{1}{\pi}\left(\tan\theta_i + \tan\theta_r\right.$$
$$\left. + \sqrt{\tan\theta_i^2 + \tan\theta_r^2 - 2\tan\theta_i\tan\theta_r\cos\varphi}\right)$$

$$(6.31)$$

6.1.3.3.2 RPV model

To describe the characteristics of the atmosphere—surface coupled bidirectional reflectance, Rahman et al. (1993) introduced a nonlinear empirical three-parameter model for the vegetation canopy, which is abbreviated as the RPV (Rahman-Pinty-Verstraete) model. The model considers the hotspot effect and satisfies reciprocal principle. It is basically described as follows:

$$\rho(\theta_i, \theta_r, \varphi) = \rho_0 M(\theta_i, \theta_r, a)F(\xi, \Theta)H(\rho_0, \theta_i, \theta_r, \varphi)$$

$$(6.32)$$

$$M(\theta_i, \theta_r, a) = (\cos\theta_i \cos\theta_r)^{a-1}(\cos\theta_i + \cos\theta_r)^{a-1}$$

$$F(\xi, \Theta) = \frac{1 - \Theta^2}{\left(1 + 2\Theta\cos g + \Theta^2\right)^{3/2}}$$

$$H(\rho_0, \theta_i, \theta_r, \varphi) = 1 + \frac{1 - \rho_0}{1 + G}$$

$$G = \sqrt{\tan^2\theta_i + \tan^2\theta_r - 2\tan\theta_i \tan\theta_r \cos\varphi}$$

where ξ is the phase angle, and Θ is an empirical parameter that describes the scattering phase function. $\Theta > 0$ means that the forward scattering is dominant, and $\Theta < 0$ means backward scattering dominates.

Martonchik et al. (1998) replaced the Henyey—Greenstein function in the model with an exponential function and proposed the modified MRPV model.

$$F(\xi, b) = \exp(-b \cos \xi) \qquad (6.33)$$

where b is an empirical parameter describing the phase function.

6.2 The albedo-estimation method based on bidirectional reflectance model inversion

6.2.1 Inversion of the bidirectional reflectance model and derivation of narrowband albedo

The most-recognized albedo retrieving algorithm currently in operation is the AMBRALS, which was proposed for MODIS BRDF/albedo product. The albedo calculation procedures of AMBRALS consists of three steps: (1) atmospheric correction is applied to the observed upward radiation at the top of the atmosphere to calculate the HDRF at the surface; (2) the kernel-driven model parameters are inverted by fitting clear-sky observations, and the BSA and WSA are obtained after the integration of the model-predicted BRDF; and (3) based on the solar radiation spectra and the response function of each band of the sensor, the broadband BSA and WSA are obtained after conversion from narrowband to broadband.

The fitting of the surface bidirectional reflectance model and the calculation of the narrowband albedo are introduced here first. In the calculation, the narrowband albedo is considered as approximately equal to the center wavelength spectral albedo.

6.2.1.1 Bidirectional reflectance model and data fitting

AMBRALS is based on the kernel-driven linear bidirectional reflectance model (Eq. 6.34), which uses a combination of the RossThick (VOL kernel) and LiSparseR (GO kernel) kernels

in the operational production of the global BRDF/albedo. The kernel-driven model is used for mass data processing because of its brevity, high speed, and strong data-fitting ability.

Through the least squares method, the optimal f_k are inverted by fitting the observations $\rho(\wedge)$ (that is, the reflectance observations $\rho(\wedge)$ under the angle conditions of θ_i, θ_r, and φ are known), and the analytical solutions of the equations can be obtained by minimizing the following error function:

$$e^2(\wedge) = \frac{1}{d} \sum_n \frac{(\rho(\theta_{i,n}, \theta_{r,n}, \varphi_n, \wedge) - R(\theta_{i,n}, \theta_{r,n}, \varphi_n))^2}{\omega_n}$$

$$(6.34)$$

where $\wedge$ represents the model parameters, which are $\{f_k | k = 1, \ldots, 3\}$; d is the degree of freedom, which corresponds to the number of observations minus the number of kernel coefficients f_k; and ω_n is the weight factor of the corresponding observation. After the construction of the kernel coefficients, the bidirectional reflectance under any solar incident angle and viewing conditions can be extrapolated with the kernel-driven model.

6.2.1.2 Albedo from integration of bidirectional reflectance

There is an explicit mathematical relationship between the spectral albedo and the BRDF (that is, the black-sky spectral albedo $\alpha_{b\lambda}(\theta_i)$ is the integration of the BRDF within the 2π observational hemisphere, while the white-sky spectral albedo $\alpha_{w\lambda}$ is the integration of the black-sky spectral albedo $\alpha_{b\lambda}(\theta_i)$ within the 2π incident hemisphere). For quicker calculation, the directional–hemispherical integral $h_k(\theta_i, \lambda)$ and the bihemispherical integral $H_k(\lambda)$ are defined by Eqs. (6.37) and (6.38), respectively. Based on the linear characteristics of the kernel-driven model, $\alpha_{b\lambda}(\theta_i)$ and $\alpha_{w\lambda}$ can be expressed as the weighted average of $h_k(\theta_i, \lambda)$ and $H_k(\lambda)$, where the weights are the inverted kernel coefficients (Eqs. 6.35 and 6.36). The actual spectral albedo

$\alpha_{a\lambda}(\theta_i)$ can then be obtained by the weighted average of $\alpha_{b\lambda}(\theta_i)$ and $\alpha_{w\lambda}$, where the weights are the proportions of skylight in the total radiation (s) and solar direct light in total radiation ($1 - s$) (Eq. 6.39). In practice, to void multiple integration and release system resources, the integration kernels are precomputed. For $h_k(\theta_i, \lambda)$, according to the variation of the viewing geometry, the lookup tables from the incident solar zenith angle to the kernel integrals are prebuilt or obtain the approximating functions of incident solar zenith angle and directional–hemispherical integral kernels (Eq. 6.40). $H_k(\lambda)$ are the constant values computed by numerical integration.

$$\alpha_{b\lambda}(\theta_i) = \sum_k f_k(\lambda) h_k(\theta_i, \lambda) \quad (6.35)$$

$$\alpha_{w\lambda} = \sum_k f_k(\lambda) H_k(\lambda) \quad (6.36)$$

$$h_k(\theta_i, \lambda) = \frac{1}{\pi} \int_0^{2\pi} \int_0^{\frac{\pi}{2}} [K_k(\theta_i, \theta_v, \psi, \lambda)$$
$$\times] \sin \theta_v \cos \theta_v d\theta_v d\psi \quad (6.37)$$

$$H_k(\lambda) = 2 \int_0^{\frac{\pi}{2}} h_k(\theta_i, \lambda) \sin \theta_i \cos \theta_i d\theta_i \quad (6.38)$$

$$\alpha_{a\lambda}(\theta_i) = s\alpha_{w\lambda} + (1 - s)\alpha_{b\lambda}(\theta_i) \quad (6.39)$$

Table 6.1 lists some integration values of the common isotropic (k_iso), surface scattering (k_geo), and volume scattering (k_vol) kernels. By convention, the integration of the BSA under

TABLE 6.1 Coefficients in the calculation of black-sky and white-sky albedos for common kernel functions.

Kernel name	g_k	g_{0k}	g_{1k}	g_{2k}
Isotropic	1	1	0	0
LiSparseR	−1.377622	−1.284909	−0.166314	0.041840
RossThick	0.189184	−0.007574	−0.070987	0.307588
RossHotspot	0.0952955	0.010939	−0.024966	0.132210

different solar zenith angles uses the cubic polynomial approximation.

$$h_k(\theta) = g_{0k} + g_{1k}\theta^2 + g_{2k}\theta^3 \quad (6.40)$$

In summary, the narrowband WSA is given by

$$\alpha_{ws}(\lambda) = f_{iso}(\lambda) g_{iso} + f_{vol}(\lambda) g_{vol} + f_{geo}(\lambda) g_{geo} \quad (6.41)$$

and the BSA is given by

$$\alpha_{bs}(\theta, \lambda) = f_{iso}(\lambda) \left(g_{0iso} + g_{1iso}\theta^2 + g_{2iso}\theta^3 \right)$$
$$+ f_{vol}(\lambda) \left(g_{0vol} + g_{1vol}\theta^2 + g_{2vol}\theta^3 \right)$$
$$+ f_{geo}(\lambda) \left(g_{0geo} + g_{1geo}\theta^2 + g_{2geo}\theta^3 \right) \quad (6.42)$$

6.2.2 Narrowband-to-broadband albedo conversion

The surface broadband albedo is the ratio of the surface upward radiation flux (F_u) to the downward radiation flux (F_d) within a certain wavelength range.

$$A(\theta, \varLambda) = \frac{F_u(\varLambda)}{F_d(\theta, \varLambda)} = \frac{\int_\varLambda F_d(\theta, \lambda) \alpha(\theta, \lambda) d\lambda}{\int_\varLambda F_d(\theta, \lambda) d\lambda} \quad (6.43)$$

where $\varLambda$ is the wavelength range from λ_1 to λ_2. If $\varLambda$ is set in the range of $0.25 \sim 5.0 \, \mu m$, $\alpha(\theta_i, \varLambda)$ is shortwave albedo, while the ranges $0.4 \sim 0.7 \, \mu m$ and $0.7 \sim 5.0 \, \mu m$ correspond to the visible and near-infrared albedo, respectively.

The surface broadband albedo depends on the atmospheric conditions in addition to the surface properties. The downward radiant flux distribution at the bottom of the atmosphere is the weighting function for the conversion of the spectral albedos to broadband albedos, and different atmospheric conditions have different downward flux distributions. Thus, the surface broadband albedos derived from remotely sensed data under one specific solar zenith angle and atmospheric condition may not be applicable to other conditions because the spectral distribution of the downward radiation flux can vary. When studying surface albedo, the inherent albedo (such as the frequently used BSA and WSA) should be separated from the apparent albedo (the observed albedo, which is influenced by the atmospheric conditions). The inherent albedo is entirely independent of the atmospheric conditions, and the apparent albedos are equivalent to those measured by albedometers or pyranometers in the field (Liang et al., 1999). If the inherent albedos are provided, users can transform them into the apparent albedos under any desired atmospheric conditions (Lucht et al., 2000). When the atmospheric downward fluxes are known, their integration with the inherent spectral albedo produces much more accurate broadband albedo products. However, users may not want to implement such a procedure in many practical applications. Given the narrowband albedos, they are likely to want to predict the average broadband albedos under general atmospheric conditions.

The common approximation algorithm to convert narrowband albedos to broadband albedo can be expressed as follows:

$$A = c_0 + \sum_{i=1}^{n} c_i \alpha_i \qquad (6.44)$$

where A is the surface broadband albedo, α_i is the narrowband albedo of band i, and c_i is the conversion coefficient, which can be calculated through either field measurements of certain surface types or model simulations. In practice, it is impractical to develop a universal formula based only on ground measurements because collecting extensive datasets for different atmospheric and surface conditions is expensive. Model simulation is a better approach to the development of universal conversion formulas, and ground measurements are valuable for validation.

Liang (2001) has provided simple conversion coefficients for calculating the average land surface broadband albedos from a variety of narrowband sensors under common atmospheric and surface conditions. These results are based on extensive RT simulations using the Santa Barbara DISORT Atmospheric Radiative Transfer code (Ricchiazzi et al., 1998). Compared with previous studies, a larger set of surface reflectance data were employed in the model simulations. A total of 256 surface reflectance spectra were employed, including soil (43), vegetation canopy (115), water (13), wetland and beach sand (4), snow and frost (27), city (26), road (15), rock (4), and other cover types (9), covering a wide range of typical surface types. Eleven atmospheric visibility levels (2, 5, 10, 15, 20, 25, 30, 50, 70, 100, and 150 km) were used for different aerosol loadings, and five atmospheric profiles of the MODTRAN defaults were used (tropical, midlatitude winter, subarctic summer, subarctic winter, and US62). These profiles also represent different concentrations of water vapor and other gas profiles. A range of nine solar zenith angles was simulated from 0 to 80° in 10° increments. After the downward radiant flux (including the direct and skylight scattering) was determined, the narrowband spectral albedo was calculated by integrating the downward radiant flux and the surface reflectance spectrum. The broadband albedo was obtained from the ratio of the surface broadband upward radiant flux to the downward radiant flux. Furthermore, the conversion coefficients of the nine sensors, including ALI, ASTER, AVHRR, GOES, Landsat7 ETM+, MISR, MODIS, POLDER, and SPOT VEGETATION, were obtained through regression analysis. Recently, Peng et al. (2017) analyzed more than 7000 spectral measurements for typical land cover types, mostly vegetation, and suggested NDVI-related NTB conversion coefficients for snow-free land surface.

6.2.2.1 Vegetation and soil

The following conversion coefficients from the narrowband albedo of typical satellite sensor to the broadband albedo, which can be used to estimate broadband albedo under ordinary atmospheric conditions, were published by Liang Shunlin et al., in 2001.

$$A^{\text{ASTER}} = 0.484\alpha_1 + 0.335\alpha_3 - 0.324\alpha_5$$
$$+ 0.551\alpha_6 + 0.305\alpha_8 - 0.367\alpha_9 - 0.0015$$

$$A^{\text{AVHRR}} = -0.3376\alpha_1^2 - 0.2707\alpha_2^2 + 0.7074\alpha_1\alpha_2$$
$$+ 0.2915\alpha_1 + 0.5256\alpha_2 + 0.0035$$

$$A^{\text{GOES}} = 0.0759 + 0.7712\alpha$$

$$A^{\text{ETM+}} = 0.356\alpha_1 + 0.130\alpha_3 + 0.373\alpha_4 + 0.085\alpha_5$$
$$+ 0.072\alpha_7 - 0.0018$$

$$A^{\text{MISR}} = 0.126\alpha_2 + 0.343\alpha_3 + 0.451\alpha_4 + 0.0037$$

$$A^{\text{MODIS}} = 0.160\alpha_1 + 0.291\alpha_2 + 0.243\alpha_3$$
$$+ 0.116\alpha_4 + 0.112\alpha_5 + 0.081\alpha_7 - 0.0015$$

$$A^{\text{POLDER}} = 0.112\alpha_1 + 0.388\alpha_2 - 0.266\alpha_3$$
$$+ 0.668\alpha_4 + 0.0019$$

$$A^{\text{VEGETATION}} = 0.3512\alpha_1 + 0.1629\alpha_2 + 0.3415\alpha_3$$
$$+ 0.1651\alpha_4$$

6.2.2.2 Snow cover

Because the visible-band reflectance of a snow-covered surface is quite high, it is better to use separate coefficients for narrowband-to-broadband albedo conversion to ensure accurate conversion for snow/ice albedos. Stroeve proposed the following conversion coefficients for MODIS in 2004 (Stroeve and Box, 2004):

$$A^{\text{MODIS}} = -0.0093 + 0.1574\alpha_1 + 0.2789\alpha_2$$
$$+ 0.3829\alpha_3 + 0.1131\alpha_5 + 0.0694\alpha_7$$

6.3 The direct estimation of surface albedo

6.3.1 Overview of the direct-estimation method

Although the albedo-estimation method based on the inversion of the bidirectional reflectance model has clear and explicit physical significance, the algorithm processing flow is quite complicated, and it requires a large amount of calculations. In addition, the accuracy of the final albedo products depends on the performance of all of these processes. The uncertainties associated with each procedure may be canceled out, but they are also likely to accumulate. For instance, the AMBRALS algorithm of MODIS is sensitive to the uncertainty of the atmospheric correction. It was also assumed that the surface bidirectional reflectance characteristics would remain constant for 16 days in the model inversion; however, this assumption does not always hold (e.g., snowfall, snowmelt, harvest), which often leads to a failure in the parameter inversion of the bidirectional reflectance model, and it further affects the estimation accuracy of the surface broadband albedo.

The goal of the direct estimation of the land surface broadband albedos is to discard the complicated multistep inversion process and relate the TOA narrowband bidirectional reflectance to the land surface broadband albedos with a linear regression function that replaces the three physics-based steps of atmospheric correction, narrowband BRDF modeling, and narrowband-to-broadband albedo conversion. The algorithm becomes simpler and more efficient, and it is suitable for the production of global albedo products. Because the atmospheric correction is not necessary, the chain of data processing is shortened, and the inversion accuracy does not rely on the accuracy of the aerosol estimation. The surface broadband albedo can be estimated from a single observation, which makes it possible to provide daily surface albedo products using MODIS data, thus enabling the characterization of rapid changes of the land surface albedo.

The idea of the direct estimation of surface broadband albedo can be traced to earlier studies that linearly related the TOA reflectance to the surface broadband albedos (Chen and Ohring, 1984; Pinker, 1985; Koepke and Kriebel, 1987). Those earlier studies mainly focus on the connection between the planetary shortwave albedo and the surface shortwave albedo. Liang et al. (1999) suggested linking the narrowband TOA reflectance of MODIS with the surface broadband albedos and brought significant improvements. As these algorithms assumed Lambertian surface in the atmospheric RT simulation, without considering the impact of surface reflectance anisotropy on the algorithm, Liang et al. (2005) made some improvements in his developed algorithm. The improvements considered the impact of surface reflectance anisotropy on the inversion results by simulating the bidirectional reflectances of snow/ice with the DISORT model.

After a training database has been built by RT simulation, the relationship between the TOA reflectances and the land—surface broadband albedos can be developed using an angular bin regression method. In earlier studies, Liang et al. (1999) obtained this relationship using the neural network (NN) method. In a subsequent study, Liang et al.,(2003) used the projection pursuit regression method. However, these two studies did not consider the anisotropy of the land surface. Consequently, Liang et al. (2005) further improved this method by dividing the solar/view geometry space into angular bins and calculating the regression coefficients of each angular bin using a linear regression method. This approach has been adopted for the estimation of the daily land—surface albedo of the Greenland ice sheet from MODIS data. Validation of the algorithm against station measurements on Greenland ice showed good results.

In the GLASS (Global Land Surface Satellite) system developed by the global surface characteristics parameters production and application research group at Beijing Normal University, two direct-estimation methods have been adopted: the broadband albedo direct inversion algorithm based on the MODIS surface bidirectional reflectance (AB1—Angular Bin 1) and an algorithm based on the MODIS TOA bidirectional reflectance (AB2—Angular Bin 2). Similar algorithm has also been applied to data acquired by MISR (He et al., 2017), which is also onboard EOS-terra; and data acquired by Visible Infrared Imaging Radiometer Suite (VIIRS) (Wang et al., 2013), which is the next-generation sensor to replace MODIS; and data from lots of relative high—resolution sensors such as the Chinese HJ-CCD (He et al., 2015; Wen et al., 2015), the Landsat sensors, MSS, TM, ETM +, and OLI (He et al., 2018). In the following, we take the AB1 and AB2 algorithms as examples to introduce the direct-estimation method.

6.3.2 Albedo-estimation method based on surface bidirectional reflectance data

6.3.2.1 General concept

A multivariant linear regression relationship is assumed between the surface broadband albedo and the surface bidirectional reflectances in MODIS visible and near-infrared spectral bands, which can be expressed by the following formula:

$$A = c_0(\theta_i, \theta_r, \varphi) + \sum_{i=1}^{n} c_i(\theta_i, \theta_r, \varphi)\rho_i(\theta_i, \theta_r, \varphi)$$

(6.45)

where A represents the surface albedo (more specifically, the shortwave [0.3—3 μm] WSA or BSA at a certain solar zenith angle); c_i are the regression coefficients; and ρ_i are the surface directional reflectances at band i, which are all functions of solar/view zenith angles.

The first step of the surface albedo inversion is the calculation of the regression coefficients, c_i. Because of the bidirectional reflectance of the natural land surface, the regression coefficients c_i change with the solar/view angles. For the convenience of numerical calculation, the solar/view geometry space was divided into a three-dimensional grid comprising the solar zenith, the view zenith, and the relative azimuth, in which each grid element is called an angular bin. Reducing the grid size would improve the precision but increase the use of computer resources; therefore, we have to seek a balance between these considerations. The solar zenith angle of the centers of the angular bins varies from 0 to $80°$ in $4°$ increments (i.e., $0°$, $4°$,..., $80°$). The view zenith angle of the centers of the angular bins varies from 0 to $64°$ in $4°$ increments (i.e., $0°$, $4°$, ..., $64°$). The relative azimuth angle of the centers of the angular bins varies from 0 to $180°$ in $20°$ increments (i.e., $0°$, 20, ..., $180°$).

The AB1 algorithm is used to build the linear regression relationship between the surface bidirectional reflectances on the MODIS bands and the broadband albedo for each grid element, which has the advantages of simple computation, low input—data requirements, and full consideration of the surface bidirectional and spectral characteristics.

6.3.2.2 Building the training dataset

The regression coefficients of the direct-estimation approach are obtained using training data; therefore, a training dataset containing the multiband bidirectional reflectance and the broadband albedo is needed. The quality and representativeness of this training dataset strongly influences the performance of the AB1 algorithm. As a production algorithm for the global albedo product, the AB1 algorithm must consider the various surface types in different regions. The work load required to simulate all surface types with the RT model would be too great; therefore, similar to the method of Cui et al. (2009), a training dataset that presents

various global surfaces was built based on the POLDER-BRDF dataset. Although the POLDER-BRDF observations cover many different observation angles, it remains impossible to ensure enough observations within each angular bin. Consequently, the POLDER-BRDF database must be interpolated and extrapolated to build the complete training dataset. It can then be integrated to obtain the broadband albedo.

6.3.2.2.1 The fitting and interpolation method of the POLDER-BRDF database

The PARASOL (Polarization & Anisotropy of Reflectances for Atmospheric Sciences coupled with Observations from a Lidar), launched by the Centre National d'Etudes Spatiales on October 18, 2004, hosts the third-generation POLDER (Polarization and Directionality of the Earth's reflectances), which can collect global polarization and directional data for the reflected solar radiation from the atmosphere and the Earth surface. The spatial resolution of the POLDER product is 6 km × 7 km, providing abundant angular, spectral, and polarization information. These data are the latest currently available multiangle satellite remote sensing data and provide much information about the Earth surface, atmospheric aerosols, and clouds. The imaging characteristics of the POLDER sensor enable it to obtain observation data for as many as 14 angles, up to $60°$, in every track. Each sample of the dataset is composited using the monthly observed data. POLDER was identified as a satellite sensor that can provide comprehensive surface bidirectional reflectance characteristics. The POLDER Level 3 BRDF database contains the surface reflectance observations of all tracks during a month, after cloud removal and atmospheric correction. The data processing algorithm fully considers the impacts of cloud detection, molecule absorption correction, and aerosols in the stratosphere and troposphere.

The linear kernel-driven model is used in the POLDER Level 3 algorithm to predict the surface bidirectional reflectance characteristics, which uses the LiSparseR and modified RossThick (identified here as RossHotspot Section 6.1.3.3.1.3) kernel functions. For each POLDER-BRDF multiangle observation dataset, the kernel-driven model coefficients were fitted by the least squares method and used in the model to calculate the bidirectional reflectance at the central point of each angular bin.

The kernel-driven model is also used in the calculation of the surface albedo corresponding to each dataset; here, the broadband albedo in the 0.3–5 μm band range is represented by the WSA and BSA. Because the BSA is a function of the solar zenith angle, we have calculated all BSAs between 0 and 80°, in 5° intervals. As a feature of the kernel-driven model, the narrowband albedo is the weighted sum of the integration of the kernel functions, with the kernel coefficients as weights (detailed in Section 6.2.1). The conversion algorithm from the narrowband albedo to the broadband albedo is introduced in Section 6.2.2 and is applied in different forms to ice/snow and snow-free surfaces.

6.3.2.2.2 Land cover classification

The bidirectional reflectance characteristics vary by land cover type. Although the AB1 surface albedo inversion algorithm is based on regression, which can adapt to the variation of the bidirectional reflectance by dividing the solar/view geometry space into a three-dimensional grid, the linear regression model has approximation error; therefore, it is necessary to introduce the land cover type information and subdivide the training samples to reduce the uncertainty of the linear regression model.

If global land cover data (such as the MODIS land use/land cover products) are used to support the albedo inversion, there will be more input of the algorithm, thus decreases its applicability. Furthermore, new problems must be considered: first, error exists in the global land cover products, especially error caused by the mixed pixels in the 1-km resolution data; second,

the surface albedo is a rapidly changing physical parameter, but the land cover types are subjective judgment results according to the surface long-term cover state and are not all consistent over time. For instance, the albedo of cropland changes significantly after snowfall but its land cover type is still cropland.

Therefore, we used a relatively simple classification scheme which is directly based on remote sensing observations. Basically, we divide the POLDER-BRDF datasets into three classes: vegetation, bare soil, and snow/ice. The following classification criteria are used: (1) for each observation, if the NDVI value is greater than 0.2, the pixel is categorized as vegetation; (2) if the reflectance of the blue or red channel is greater than 0.3, the pixel is categorized as ice/snow; and (3) the remaining pixels are categorized as bare soil.

In the production of the training dataset, the average reflectance and average NDVI of each spectral band are calculated for use as the basis of this classification. To maintain the consistency of the calculated albedo of each category during the transition, we have designated intermediate classes. For instance, the pixels whose corresponding NDVI is between 0.18 and 0.24 would be categorized as belonging to "intermediate class A," and the pixels whose reflectance in the blue band is between 0.24 and 0.4 would be categorized as belonging to "intermediate class B." The other pixels are temporarily categorized into pure vegetation, pure bare soil, and pure ice/snow.

The POLDER-BRDF datasets are categorized into five classes, resulting in 4737 pure vegetation datasets, 2401 pure soil datasets, 627 pure ice/snow datasets, 1136 intermediate class A datasets, and 123 intermediate class B datasets. During the regression analysis, the classification was further aggregated into three classes: vegetation (pure vegetation + intermediate class A), soil (pure soil + intermediate class A + intermediate class B), and snow/ice (pure snow/ice + intermediate class B). In the angular bin regression procedure,

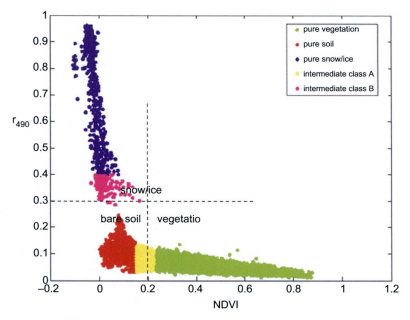

FIGURE 6.9   Scatter diagram of the average reflectance in the blue band and the average NDVI of the POLDER-BRDF dataset. (Blue, red, green, yellow, and purple pixels are pure snow/ice, pure soil, pure vegetation, intermediate class A, and intermediate class B, respectively).

the datasets of the different classes were analyzed separately. In this way, proper overlapping guarantees that the regression result has desirable consistency in the transition regions of the different land cover types. Fig. 6.9 shows the scatter diagram of the five categories of training data in the NDVI-R490 feature space, in which r_{490} represents the average reflectance in the blue band.

6.3.2.2.3 Band conversions from POLDER to MODIS

Because the training datasets are obtained by POLDER but the inversion algorithm is applied to MODIS data, the relationship between the surface reflectance observations of the two sensors at each band must be built for band conversions. The principle of band conversion relies on the premise that certain rules exist in the spectra of the surface features, and the surface reflectances in different bands correlate with each other. This premise has been widely adopted in the Earth observation studies. A continuous spectrum of the typical surface is collected first, and the corresponding reflectance is calculated through the spectral response functions in each POLDER and MODIS band. Then, the linear conversion relationship is built according to the statistical properties of the data in these bands.

A set of 493 field-measured spectral data of typical land covers, including crops, natural vegetation, soil, sand, water, snow and ice, were used to derive the band conversion coefficients, which are listed in Table 6.2.

The root mean square error (RMSE) values for band conversion of the first four bands of MODIS are smaller than those of the latter three bands; therefore, a greater uncertainty was added to the band conversion of the latter three bands, which is considered when the albedo regression equations are built.

TABLE 6.2 The band conversion coefficients from POLDER to MODIS.

Band name	MODIS-b1	MODIS-b2	MODIS-b3	MODIS-b4	MODIS-b5	MODIS-b6	MODIS-b7
POLDER-b1	0.024593	0.032882	0.912575	0.207528	−0.356931	−1.039259	−1.153855
POLDER-b2	0.306280	−0.036396	0.143223	0.613727	−0.015207	−0.441590	−0.504465
POLDER-b3	0.691690	−0.011160	−0.060149	0.121219	0.418777	1.540050	1.822514
POLDER-b4	−0.044711	0.299618	0.005727	0.119292	−0.117775	−0.186372	−0.114003
POLDER-b5	−0.005402	0.645341	0.021614	−0.026334	−0.580513	−0.792087	−0.776048
POLDER-b6	0.030160	0.081619	−0.026446	−0.043885	1.488419	1.267713	0.918089
Offset	0.004258	0.001039	−0.007456	−0.000374	0.022993	0.070931	0.070192
RMSE	0.003089	0.003553	0.004188	0.004286	0.020661	0.041961	0.047242

6.3.2.3 Regression method

The above Eq. (6.45) describes the relationship between the surface multiband BRF and the broadband albedo, in which the regression coefficients are undetermined. For each solar/view angular bin, a group of regression coefficients, i.e., $c_i | i = 0, \ldots, n$, must be estimated according to the training data.

To solve the equations through the linear least squares method, the equations can be written in matrix form:

$$Y = AX \qquad (6.46)$$

where X is the matrix comprising the multiband reflectances in the training data, which has dimensions of $(n + 1) \times m$. The number of MODIS correlated bands is given by $n = 7$, m is the number of training data in the grid, and Y is the matrix comprising the albedo in the training data and has a degree of $18 \times m$. Because the WSA and BSA from 0 to 80° in 5° increments must be calculated, there are 18 different albedo values. A is the matrix of the regression coefficients, which has dimensions of $(n + 1) \times 18$.

The linear least squares solution A^* is expressed as follows:

$$A^* = (X^T X)^{-1} X^T Y \qquad (6.47)$$

The linear least squares method normally has a simple form and provides good fitting results. However, if intercorrelations exist in the training dataset, a robust full retrieval cannot be made. Numerical tests indicate that if A^* is calculated from seven MODIS bands by the simple least square solution, the result is sensitive to the noise in the MODIS spectral data.

The six POLDER bands are all positioned in the visible and near-infrared range, with wavelengths less than 1020 nm; however, the shortwave albedo corresponds to the range from 300 to 5000 nm, so POLDER cannot supply surface spectral information in the midinfrared spectral range. The first seven MODIS spectral bands are distributed in the visible and near-infrared range; thus, they are suitable for use in albedo calculation. However, the uncertainties of the fifth, sixth, and seventh MODIS bands in the band conversion process are large and the inversion is sensitive to noise when all seven MODIS bands are included. On the other hand, because the last three MODIS bands contain information about the surface albedo in midinfrared spectral range, they should be retained in the inversion. Thus, we adopted another technique: a stable solution can be acquired by adding the simulated data noise into the regression algorithm. Specifically, X is the training data produced by the POLDER data after conversion to the MODIS wavebands. We presume that the training data are accurate but that the observations contain noise; therefore, statistical random noise is added to $\mathbf{X}$, which becomes $\widetilde{\mathbf{X}}$. The anti-noise least square solution under this design is expressed as follows, which is the least-error solution to be applied to observations with noise:

$$\mathbf{A}^* = \left(\widetilde{\mathbf{X}}^T\widetilde{\mathbf{X}}\right)^{-1}\widetilde{\mathbf{X}}^T\mathbf{Y} \tag{6.48}$$

The procedure of adding noise to the data is not actually implemented when estimating $\mathbf{A}^*$ because the specific noise in limited data cannot represent the statistics of noise. Because $\widetilde{\mathbf{X}}^T\widetilde{\mathbf{X}}$ and $\widetilde{\mathbf{X}}^T\mathbf{Y}$ can be directly calculated from X^TX and X^TY, it is possible to bypass generating $\widetilde{\mathbf{X}}$. It is presumed that the noise of the MODIS bands all average to 0, while their covariance matrix is Δ. Then,

$$\widetilde{\mathbf{X}}^T\widetilde{\mathbf{X}} = \mathbf{X}^T\mathbf{X} + m\Delta, \widetilde{\mathbf{X}}^T\mathbf{Y} = \mathbf{X}^T\mathbf{Y} \tag{6.49}$$

Therefore, the anti-noise least square solution can be calculated by

$$\mathbf{A}^* = \left(\mathbf{X}^T\mathbf{X} + m\Delta\right)^{-1}\mathbf{X}^T\mathbf{Y} \tag{6.50}$$

In the current algorithm, we have made simple estimation to the data noise in MODIS bands, and set Δ as a diagonal matrix in which the diagonal elements are the variances of the noise of each band. Two main factors are embodied in Δ: the residual uncertainty of the atmospheric correction and the uncertainty introduced by band conversion from POLDER to MODIS. At present, the noises of the first and second bands of MODIS are believed to be small enough that their standard variance can be set to 0.01; the third and fourth bands are prone to be influenced by aerosol error, so the standard variance of their noises is valued at 0.02. Uncertainty has been introduced to the band conversion of the simulation data at the fifth to seventh bands; thus, the corresponding standard variance is set to 0.04. It is sometimes difficult to strictly estimate the constraints, such as data noise, in an inversion method. However, even if the constraints are not accurate, usually their negative effect on the inversion results will not be significant, but the benefit to the stability of the inversion will be apparent.

To review the regression results, the statistics of inversion errors of the WSA and BSA at the 45° solar zenith angle were calculated based on the training data (which is derived from the POLDER data after interpolation and band conversion to MODIS). The average RMSE of the 50,061 angular bins are listed in Table 6.3, which shows that the fitting effect of the training data with the regression formula is best for surfaces covered with vegetation, i.e., the residual is within 0.01. As the average albedo increases for bare soil and snow surfaces, the fitting residual also increases, while the relative error is within 6%. The regression RMSE of WSA has the same variation trend as the BSA.

6.3.2.4 The results of the AB1 algorithm

To demonstrate the effect of algorithm AB1 on the inversion of the surface albedo, we have

TABLE 6.3 The residual statistics of the AB1 algorithm applied to the training data.

Category of training data	Average RMSE of WSA	Average RMSE of BSA at 45° solar zenith angle	Relative error of WSA	Relative error of BSA at 45° solar zenith angle
Vegetation	0.0078	0.0063	4.91%	4.25%
Bare soil	0.0118	0.0103	5.12%	4.67%
Ice/snow	0.0248	0.0199	3.85%	3.19%

chosen the clear-sky MODIS surface reflectance product (MOD09GA) of Fort-Peck station in North America from 2000 to 2006 and calculated the WSA and BSA at local noon with the AB1 algorithm. From these results, the blue-sky albedo (actual albedo) was calculated through a simple weighted sum with the skylight fraction factor $s = 0.3$. The comparison of this result and the surface-measured albedo at the flux stations is shown in Fig. 6.10. Accordingly, the inversion result of MODIS standard albedo inversion algorithm (MCD43B3) is also illustrated. This station is located at (47.3079°N, 105.101°W) with a grassland terrain. The figure shows that the AB1 algorithm result and the MCD43B3 product both reflect the temporal variation of the surface-measured albedo. However, as the effects of cloud on the images have not been completely removed, the data noise cannot be ignored, especially in the case of the large winter fluctuations

of the surface albedo. This fluctuation partly occurs because the separation of cloud from snow is quite difficult and partly reflects the actual change of the surface state. Compared with the MCD43B3 products, the temporal resolution of the albedo extracted with the AB1 algorithm is much higher, and the fluctuation is more significant. On the other hand, we can see that the surface-measured data also have apparent fluctuation, which may be explained by the impact of weather because the albedo is correlated with light rays and measurement conditions, according to its definition.

6.3.3 The TOA reflectance—based method

The albedo inversion algorithm AB2 can use the TOA directional reflectance routinely acquired by the MODIS onboard Terra/Aqua

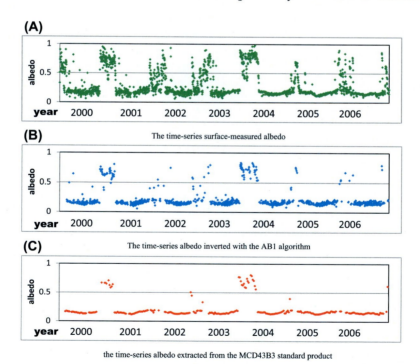

FIGURE 6.10   The time series albedo at Fort-Peck station in North America from 2000 to 06. (A) Surface measurements; (B) AB1 algorithm inversion result; and (C) extracted values from MCD43B3 albedo products.

platform to directly invert the surface broadband albedo, which is independent of the atmospheric correction; thus, atmospheric correction difficulties and errors are evaded. The main difference between the AB1 and AB2 algorithms is that the simulation of the atmospheric RT is contained in the forward modeling of the AB2 algorithm. Both algorithms have the similar processing procedures, except for the atmospheric correction part. Because these steps are identical, they are not reintroduced in this section; instead, this section focuses on the atmospheric RT simulation and linear regression method.

6.3.3.1 Atmospheric radiative transfer simulation

To build the linear regression relationship between the TOA narrowband bidirectional reflectance and the surface broadband albedo, a training dataset representing various surface bidirectional reflectance characteristics and atmospheric conditions must be built in algorithm AB2. More specifically, based on the training dataset (the surface bidirectional reflectance) of algorithm AB1, the atmospheric RT model 6S (Second Simulation of a Satellite Signal in the Solar Spectrum) was applied to the simulation of the TOA apparent reflectance to generate a training dataset that covers the bidirectional reflectance characteristics under various atmospheric and surface conditions. As the calculation requirements of the atmospheric RT is quite large, it is usually approximated with an analytical formula based on the physical process, of which the parameters are obtained from the simulation result of the 6S model. Qin et al. 2001 proposed an approximate formula based on the non-Lambertian surface to calculate the TOA apparent reflectance.

$$\rho^*(i,v) = \rho_0(i,r)$$
$$+ \frac{\mathbf{T}(i) \cdot \mathbf{R}(i,r) \cdot \mathbf{T}(r) - t_{dd}(i) \cdot t_{dd}(r) \cdot |\mathbf{R}(i,r)| \cdot \bar{\rho}}{1 - r_{hh}\bar{\rho}}$$
(6.51)

where $\rho^*(i,r)$ is the TOA apparent reflectance, $\rho_0(i,r)$ is the path reflectance of atmosphere (that is, the part of the solar downward radiation that is scattered by the atmosphere and entered the satellite sensors), $\bar{\rho}$ is the hemispherical albedo of the atmosphere, $\mathbf{T}$ is the matrix of the atmospheric transmittance, r_{hh} is the surface diffuse bihemispherical reflectance (i.e., the WSA), and $\mathbf{R}$ is the surface reflectance matrix. The two types of parameters in the formula are mutually independent. The first type reflects the intrinsic properties of the atmospheric, and the second reflects the characteristics of the surface directional reflectance.

The matrix of the atmospheric transmittance, $\mathbf{T}$, can be described by the following formula:

$$\mathbf{T}(i) = [t_{dd}(i)t_{dh}(i)] \qquad (6.52)$$

$$\mathbf{T}(r) = \begin{bmatrix} t_{dd}(r) \\ t_{hd}(r) \end{bmatrix} \qquad (6.53)$$

where t_{dd} is the directional transmittance, t_{dh} is the directional–hemispherical transmittance, and t_{hd} is the hemispherical–directional transmittance. Here, subscripts d and h are abbreviations for "directional" and "hemispherical," respectively.

For the surface, the formula of the reflectance matrix is as follows:

$$\mathbf{R}(i,v) = \begin{bmatrix} r_{dd}(i,v)r_{dh}(i) \\ r_{hd}(v)r_{hh} \end{bmatrix} \qquad (6.54)$$

where r_{dd} is the surface bidirectional reflectance factor; r_{dh} and r_{hd} are the directional–hemisphere and hemisphere–directional albedo, respectively; and r_{hh} is the diffuse hemispherical–hemispherical albedo.

The parameters adopted by the simulation of the RT in the atmosphere are shown in Table 6.4, in which the input parameters for the 6S code are set as follows: six types of atmosphere (tropical, midlatitude summer, midlatitude winter, subarctic summer, subarctic winter, and US62 standard) and six types of aerosol (continental, maritime, urban, desert, biomass, and haze).

TABLE 6.4   The input parameters for the atmospheric simulation with 6S code.

Atmospheric parameters	Parameter settings
Atmosphere type	Tropical, midlatitude summer, midlatitude winter, subarctic summer, subarctic winter, and US62 standard
Aerosol type	Continental, maritime, urban, desert, biomass, and haze
Optical depth of aerosol	0.01, 0.05, 0.1, 0.2
Elevation of target (km)	0, 0.5, 1.0, 1.5, 2.0, 2.5, 3
Solar zenith angle (°)	0, 4, 8, …, 76, 80
View zenith angle (°)	0, 4, 8, …, 60, 64
Relative azimuth angle (°)	0, 20, 40, …, 160, 180

Haze is a user-defined aerosol type, for which the respective fractions of four aerosol particles—dust, water soluble, soot, and oceanic—are 15%, 75%, 10%, and 0%. The amount of water vapor is set to the default value, and the AOD (aerosol optical depth) at 550 nm can assume values of 0.01, 0.05, 0.1, and 0.2, which describe conditions from clear to relatively turbid aerosol loading. The target altitude varies from 0 to 3.5 km in 0.5 km increments. The simulation of the solar/view zenith angle is described in Section 6.3.2.

After the TOA apparent reflectance and the corresponding surface albedo dataset are built, the empirical correlation between the TOA apparent reflectance and the surface broadband albedo is built through the regression analysis described by Formula (6.49). The wavelength ranges of the four surface directional reflectance bands in the POLDER-BRDF dataset (b1, 490 nm; b2, 565 nm; b3, 670 nm; b4, 765 nm; b5, 865 nm; and b6, 1020 nm) are more consistent with the first four MODIS bands (b1, 648 nm; b2, 859 nm; b3, 466 nm; and b4, 554 nm), which have less RMSE in the band conversion, than

the latter three MODIS bands (b5, 1244 nm; b6, 1631 nm; and b7, 2119 nm). Furthermore, the latter three MODIS bands are apt to be influenced by the absorption of water vapor in the atmosphere. Therefore, we chose the first four MODIS bands as the input of the TOA directional reflectance in the regression analysis.

The linear regression coefficients are obtained by least squares regression, in which R^2 and the RMSE are used to evaluate the stability of the regression model. The statistical results illustrate that the RMSE values of the linear regression between the TOA reflectances under various atmosphere and aerosol types and surface broadband albedos are almost below 0.01 and the R^2 values are almost above 0.9. When the angular bins with 32° solar zenith angle, 56° view zenith angle, and 100° relative azimuth angle are chosen for analysis and testing, the R^2 of the regression result slightly decreases from 0.978 to 0.959 after the simulation of the atmospheric RT, while the RMSE slightly increased from 0.006 to 0.008. This insignificant difference indicates good linear regression correlation between the TOA directional reflectance and the surface broadband albedo; thus, the use of the noncorrected TOA reflectance can achieve similar albedo estimation results to those obtained by the surface reflectance. Algorithm AB2 is independent of the atmospheric correction, which leads to the simplification of the data processing and the avoidance of possible errors introduced by estimation of the atmospheric parameters or inadequate atmospheric correction.

6.3.3.2 The result of the AB2 algorithm

Fig. 6.11 shows the validation results of this algorithm with the measurements at surface radiation observation stations, in which the estimated albedo (blue points) has good consistency with the measured albedo (red points) at the Fork Peck station from 2003 to 06. Compared with the surface measurements, the albedo calculated by algorithm AB2 has a more sporadic distribution. The inversion error in the ice area is also

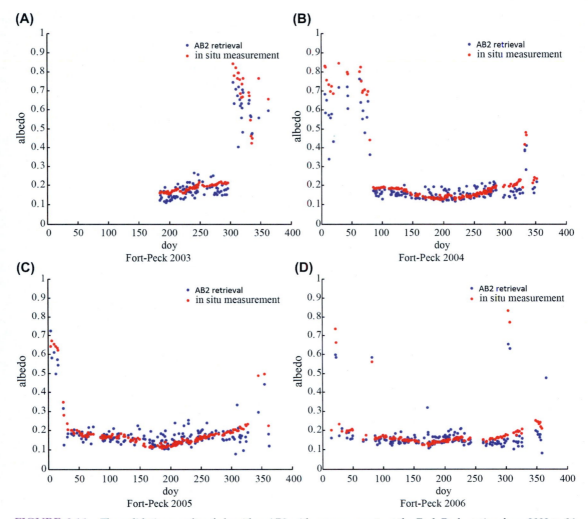

FIGURE 6.11 The validation results of algorithm AB2 with measurements at the Fork Peck station from 2003 to 06.

larger than that of the vegetation and soil areas. This algorithm can produce daily surface albedo, which responds better to snowfall and snowmelt; thus, it can reflect the temporal variation of the albedo caused by snowfall.

6.4 Global land surface albedo products and validation

6.4.1 Global surface albedo products from satellites

Many satellite albedo products have been operationally produced and published at a range of spatial (from 250 m to 20 km) and temporal (from daily to monthly) scales (Schaaf et al., 2008). The surface bidirectional reflectance/albedo inversion algorithm based on the kernel-driven model is currently the most widely used method and has been well applied to the surface albedo products of MODIS. The POLDER and PARASOL series of sensors, which have better multiangle observation capabilities but lower spatial resolutions (6 km), also have released distinguishing global albedo products. Some other remote sensing sensors, such as the geostationary meteorological satellites Meteosat and MSG, and the polar orbital satellite sensors, e.g., AVHRR, VEGETATION, MERIS, and VIIRS, also have albedo products that cover different spatial scopes. The data from FY-3 series of Chinese satellites have also been applied to regional and global surface albedo inversion. Albedo can also be derived from low spatial—resolution and wideband sensors, such as CERES and ERBE, which were designed for cloud or radiation flux observations.

The quantitative products from single source data are limited in terms of spatial/temporal coverage and data quality. So, several recent albedo products utilize multisource remote sensing data. For example, the GLASS and GLOBALBEDO products use multisource data in different terms to achieve long-term temporal coverage, while the MuSyQ try to increase the temporal resolution of albedo product by combining multisource data in the same term for model inversion. Although the use of multisource data helps to improve the spatial/temporal resolution of albedo projects, the problem of band discrepancy between the wavebands of different sensors has not been satisfactorily solved. In addition, the influences of topographic effects and scaling effects on surface albedo (Davidson et al., 2004; Li et al., 2000) have been ignored in the current operational albedo inversion algorithm.

6.4.1.1 MODIS albedo

MODIS is the main sensor board on the NASA EOS (Earth Observation System) series of satellites. Terra/MODIS began to collect data in 2000, and Aqua/MODIS started in 2003.

MODIS albedo products, including the MOD43 series product derived from the Terra data, the MYD43 series product from the Aqua data, and the MCD43 series from both Terra and Aqua, are inverted from multiangle and multiband observations collected from multiple days using the semiempirical kernel-driven bidirectional reflectance model (Gao et al., 2005; Schaaf et al., 2002; Lucht et al., 2000). These products comprise one of the most abundant and detailed databases of global albedo. The following table lists the current categories of MODIS albedo products (Table 6.5).

TABLE 6.5 The albedo product categories of MODIS released by NASA.

Product name	Product category	Classes of grids	Spatial resolution	Temporal resolution
MOD/MYD/MCD43A3	Albedo	Tile	500 m	16 Day
MOD/MYD/MCD43B3	Albedo	Tile	1000 m	16 Day
MOD/MYD/MCD43C3	Albedo	CMG	0.05°	16 Day
MOD/MYD/MCD43A1	BRDF-Albedo Model Parameters	Tile	500 m	16 Day
MOD/MYD/MCD43B1	BRDF-Albedo Model Parameters	Tile	0.05°	16 Day
MOD/MYD/MCD43C1	BF-Albedo Model Parameters	CMG	5600 m	16 Day
MOD/MYD/MCD43A2	BRDF-Albedo Quality	Tile	500 m	16 Day
MOD/MYD/MCD43B2	BRDF-Albedo Quality	Tile	1000 m	16 Day
MOD/MYD/MCD43C2	BRDF-Albedo Snow-free Quality	Tile	0.05°	16 Day
MOD/MYD/MCD43A4	Nadir BRDF—Adjusted Reflectance	Tile	500 m	16 Day
MOD/MYD/MCD43B4	Nadir BRDF—Adjusted Reflectance	Tile	1000 m	16 Day
MOD/MYD/MCD43C4	Nadir BRDF—Adjusted Reflectance	CMG	0.05°	16 Day

Tile and CMG represent two grids based on different projections of the MODIS products. Tile presents a grid of sinusoidal projection which divide the Earth's surface into 36 * 18 publishing units, and CMG (Climate Modeling Grid) presents globally equal longitude—latitude grids (0.05°, 1°, etc.).

As an example, the MCD43A3 Albedo Product (MODIS/Terra and Aqua Albedo Daily L3 Global 500 m SIN Grid V006) provides both the WSA and BSA (at local solar noon) for MODIS bands one to seven and for three broadbands (0.3—0.7 μm, 0.7—5.0 μm, and 0.3—5.0 μm). In its latest released version, i.e., Collection 6, the temporal granularity has been updated to daily, while in previous version it is 8 days.

6.4.1.2 POLDER albedo

POLDER (Polarization and Directionality of the Earth's Reflectances) is one of the main sensors of the A-train plan, which was a joint effort between France and the United States. The sensor is capable of polarized and along-track multiangle observation in the visible and near-infrared bands.

The POLDER albedo product (Leroy et al., 1997; Maignan et al., 2004; Bacour and Breon, 2005) was provided in three periods: 1996.11—1997.06, 2003.04—2003.10, and 2005.07—2010.08. The corresponding POLDER sensors are onboard ADEOS-1, ADEOS-2, and PARASOL, respectively. Although all the three versions of POLDER albedo products adopt the sinusoidal project, 1/12° spatial resolution, and 10-days temporal granularity, they have difference in their publishing format.

6.4.1.3 VIIRS albedo

VIIRS, as the successor of MODIS, is the next-generation optical sensor in the middle resolution. The first VIIRS sensor was launched onboard NPP satellite in October 2011.

There are actually two sets of VIIRS albedo products derived from different algorithms. The one released by NASA is inherited from MODIS BRDF/albedo algorithm (Liu et al., 2017), resulting in a series of complex products, i.e., BRDF model parameters, narrowband albedos, broadband albedos, NBARs, with spatial resolution of 500 m/1 km/0.05°, and temporal granularity of 1 day. The other one is release by NOAA. It is generated with the direction estimation method (Wang et al., 2013; Zhou et al., 2016), resulting in only the broadband blue-sky albedo product which keeps the same swath projection and resolution as the raw VIIRS data.

6.4.1.4 Meteosat albedo

The Meteosat series Earth geostationary meteorological satellite belongs to the European Organisation for the Exploitation of Meteorological Satellites. Two generations of satellites, the MFG (Meteosat First Generation) and MSG (Meteosat Second Generation), have been successfully launched to date.

The long-term Meteosat surface albedo (MSA) product (Pinty et al., 2000; Govaerts et al., 2004, 2006), which is generated using MFG data, covers two regions: the area centered at 0° longitude from 1982 to 2006 and the Indian Ocean Area centered at 63°E longitude from 1998 to the present. These two products are generated from the satellite data in 10 days, with a spatial resolution of 3 km, and include the WSA and BSA in visible and near-infrared broadband (0.4—1.1 μm).

The MSG albedo product is produced with the data from the SEVIRI sensor. It is generated in 5-day composite period and in a single-day rolling frequency. The data product is in a spatial resolution of 3 km, including three broadband albedos of the visible, near-infrared, and shortwave infrared.

6.4.1.5 CLARA-SAL

The time series from the Clouds, Albedo, and Radiation-Surface Albedo (CLARA-SAL) provide the global albedo derived from AVHRR sensors (Karlsson et al., 2017). As the newest

CLARA-SAL covers all Earth's surface, both land and ocean included, and a nearly continuous temporal range of 34 years (1982–2015), it serves as the unique data source for long-term albedo change studies. It has a relatively low spatial resolution of 0.25°, and two modes of temporal resolution, i.e., 5-day average and monthly average. Only BSA is provided in the CLARA-SAL product.

6.4.1.6 CERES albedo

The CERES sensors, which are designed for radiation flux observation by the CERES (Clouds and the Earth's Radiant Energy System) project, operate on board TRMM, EOS-terra, EOS-aqua, and S-NPP satellites. Based on CERES observations, the developed Energy Balanced and Filled (EBAF) dataset includes upwarding and downwarding longwave and shortwave radiation flux in spatial resolution of 1°, and temporal resolution of monthly average, from 2000 to date (Wielicki et al., 1998). The CERES albedo is derived as the ratio of EBAF upwarding and downwarding shortwave flux.

6.4.1.7 GLOBALBEDO

The GLOBALBEDO is a project supported by ESA, aiming at generating land surface albedo at global scale with data from European satellites such as AATSR, SPOT4-VEGETATION, SPOT5-VEGETATION2, and MERIS. It adopted a combined model inversion strategy to generate the broadband albedo at the visible, near-infrared, and shortwave bands, with spatial resolution of 1 km/0.05°/0.5°, temporal resolution of 8-days/1-month, and temporal coverage from 1998 to 2011 (Lewis et al. 2001; Muller et al., 2012). One merit of the GLOBAL-BEDO product is that it is gap filled.

6.4.1.8 GLASS albedo

The GLASS albedo is produced by the research group in Beijing Normal University under the support of the National High Technology Research and Development Program of China.

It is generated with the direct-estimation method, and then be postprocessed with the statistics-based temporal filtering algorithm to fill gaps and reduce noises (Qu et al., 2014; Liu et al. 2013a,b). It used two data sources, MODIS and AVHRR. The albedo from MODIS data covers range of 2000–15, with spatial resolution of 1 km/0.05° and temporal granularity of 8 days. The albedo from AVHRR data covers range of 1981–2015, with spatial resolution of 0.05° and temporal granularity of 8 days. Compared to CLARA-SAL, the GLASS albedo also has the longest temporal coverage, but with higher spatial–temporal resolution and gapless feature; thus, it is an ideal data source for long-term analysis of land surface processes.

6.4.1.9 MuSyQ albedo

MuSyQ (Multi-source data Synergized Quantitative remote sensing production system) is another project supported by the National High Technology Research and Development Program of China. The MuSyQ albedo is generated by combined model inversion with data from MODIS and FY3-MESSI, which is the major optical sensor on the FY3 series of Chinese meteorological satellites (Wen et al., 2017). As the result of using multisource data, MuSyQ albedo enhanced its temporal resolution, with 10-day composition window and 5-day granularity. It is composed of BSA and WSA in visible, near-infrared, and shortwave band, and in sinusoidal projection and 1 km spatial resolution.

6.4.2 Issues in validating the remote sensing albedo products

6.4.2.1 The scale matching method in the validation of land surface albedo products

As an important indicator and driving factor for the land surface energy budget, the albedo product is widely concerned, and its quality and authenticity need to be verified by validation. The aim of validation is to compare the remote sensing—derived data products with relatively more accurate reference values, which usually comes from ground measurement, airborne observation, or high-resolution satellite images, and to assess the accuracy, most importantly, to let data users know the potential and limitations of these products. The reference value to validate albedo product may come from the worldwide network of tower-based flux measuring instruments. For example, there are over 500 sites in the FLUXNET, and a great part of them provide incoming and outgoing shortwave radiation measurements in half hour interval; and BSRN has more than 40 sites measuring incoming and outgoing shortwave radiation in every minute. The GC-Net was set up in Greenland in 1999 to measure meteorological and radiation flux in 18 automatic weather stations distributed around Greenland, and the provided incoming and outgoing shortwave flux can be used for validation of albedo in snow/ice surface.

Lots of research works use the incoming and outgoing flux measurements of pyranometers for albedo validation. However, the footprint of tower-based pyranometer corresponds to spatial scale of several meters to several scores of meters, which is much smaller than the area represented by the low-resolution remote sensing pixel. The mismatch in spatial scale may bring uncertainty to the reference values, especially in case of heterogeneous scene, and result in misleading validation conclusions. So, a rigorous validation strategy must deal with the scale matching problem.

One approach for scale matching is to select data from homogeneous sites for validation purpose. In this case, surface measurements with small FOVs are able to represent the footprint of coarse-resolution pixels. Susaki et al. (2007) demonstrated that differences in the albedo spatial resolution have negligible effects when the measurement site can be assumed to be homogeneous. However, completely homogeneous sites at the scale of coarse-resolution pixels are rare, spatial heterogeneity is almost always encountered in the validation activities of albedo product.

Another approach is to choose those measurements in which the FOV is representative of the satellite pixel footprint (Barnsley et al., 2000; Román et al., 2009; Cescatti et al., 2012). These measurements are usually gathered from the tops of towers of sufficient height, so that the radius of FOV of pyranometers can reach to more than 100 m. The representativeness of tower measurements is assessed by analyzing the spatial variation in the vicinity of the tower (Lucht et al., 2000; Román et al., 2009). However, the number of sites that satisfy the criteria of representativeness is still limited, and the distribution of these sites are not uniform, and cannot cover all the major land cover types.

In this consideration, the third approach is recommended, which utilizes high-resolution remote sensing imagery to upscale the point-scale measurements to coarse resolutions, and is applicable for more heterogeneous areas (Liang et al., 2002; Wang et al., 2004). Even in the homogeneous or representative sites as the first and second approaches selects, the upscaling approach can also be adopted to estimate more accurate reference values for validation by adjusting the ground-based observation data according to the high-resolution imagery. The earliest prototype of this method used high-resolution images to derive the fractional ground cover of different scene elements and assigned typical albedos to each scene element. The coarse-resolution albedo is subsequently predicted using a linear mixture model. The limitations of this approach include the effects of land cover classification errors and the representativeness of the albedo values. Liang et al., (2002) improved the inversion method for high-resolution albedo and directly used inverted albedo from the high-resolution images as the upscaling bridge. This multiscale validation procedure was summarized by Zhang et al. (2010a,b) as1 performing "two matches": first, the in situ measurements are matched with the high-resolution remote sensing products; then, the high-resolution remote sensing products are matched with the coarse-resolution product. When both matches are tested and approved, the coarse-resolution product is validated. The advantage of this method is the introduction of the high-resolution albedo information. Consequently, the validation results are affected by the uncertainties in the inverted high-resolution albedo.

The framework for the multiscale validation of the coarse-resolution albedo products is shown in Fig. 6.12. Briefly, three major steps are involved in estimating the coarse-resolution reference value for validation: (1) collecting the in situ measurements; (2) calibrating the high-resolution albedo maps with the in situ measurements; and (3) aggregating the high-resolution albedo maps to coarse-resolution and assessing the uncertainty level of each coarse-resolution pixel using the

coarea high-resolution pixels. So, the coarse-resolution reference value can be compared to the coarse-resolution albedo product and access the accuracy of the later. In this framework, several procedures are designed to control the uncertainties in the validation based on the analysis of the uncertainty sources and their propagation. For the in situ measurements, a sampling scheme is employed to catch the heterogeneity on both the 30 m and the 1 km scales. Each coarse-resolution pixel is scored according to its total uncertainty resulting from geometric misalignment and nonlinearity in the albedo upscaling. Then, the candidate pixels with acceptable uncertainty are used in the validation of the albedo products.

The solid-color boxes identify the steps that might introduce uncertainties into the validation, and the gradient-filled boxes identify the solutions to the uncertainties that result from the previous step.

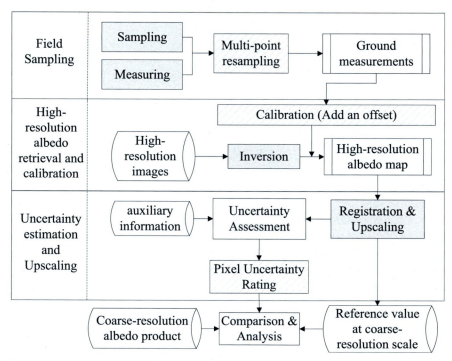

FIGURE 6.12   The proposed multivalidation procedures for the coarse-resolution albedo products (Peng et al., 2015).

6.4.2.2 Uncertainties in the validation and their assessment

The key issue in validating satellite albedo products is estimating their reference values at the pixel scale. Using the multiscale validation framework, the reference value is the aggregated coarea high-resolution albedo. However, a reference value might not always be a satisfactory approximation of the true surface albedo because of the uncertainties in the validation procedure.

Uncertainties originate from both the data source and the validation process. In terms of data, both field measurements and high-resolution albedo maps are subject to errors. The errors within in situ measurements come from imperfect instruments, inappropriate operations, and insufficient samples. Conversely, proper instruments, strict operation protocols, and reasonable sampling schemes reduce field measurement uncertainties. The errors in the high-resolution albedo products originate from instrumental limitations, radiometric calibrations, geometric mismatches, atmospheric corrections, and retrieval algorithms. The data errors are composed of bias and random errors. Field measurement bias can be reduced by proper instrument adjustment before use and careful operation, whereas random errors in field measurements can be reduced with repeated measurements. Random errors in high-resolution albedo products are reduced during the upscaling (i.e., averaging) process, whereas bias is preserved. Therefore, bias in high-resolution albedo products should be corrected for as much as possible. For this purpose, field measurements are referenced to calibrate the bias in high-resolution albedo products.

Scale effects and geometric mismatches between two images introduce uncertainty in the upscaling process. As Wu and Li (2009) summarized, scale effects "refer to the contrast of information or the different characteristics at different scales," and their primary causes are surface heterogeneity and the linearity/nonlinearity of the retrieving algorithms. In this context, the uncertainty caused by the scale effect is the combined result of spatial heterogeneity and the retrieval model. As previous studies have revealed, the linearity/nonlinearity of the retrieval model is important for scale effects. Albedo is usually retrieved using a linear model. However, even a linear model can result in a scale effect when nonuniform illumination (shading or sheltering) or multiple scattering (multiple collisions between light rays and the in situ surface) is involved because these effects change the amount of reflected radiation. Nonuniform illumination incurs a scale effect only if intrapixel heterogeneity exists, whereas multiple scattering incurs a scale effect even with a uniform surface (Li et al., 2000). Topology is the primary cause for nonuniform illumination and multiple scattering between 30 m and 1 km.

In contrast, the uncertainty from geometric mismatching is independent of retrieval algorithms. Two types of geometric mismatching exist: registration error, which results from incorrect operations and is beyond the scope of this paper, and intrinsic uncertainty, which cannot be avoided. Because the basic unit of a remote sensing image is the pixel, and resampling occurs several times in albedo product generation, it is difficult to improve the subpixel registration accuracy of a coarse-resolution pixel, thus, the resampling brings intrinsic uncertainty to the geometric matching. Some of the other uncertainties become decreased after upscaling, whereas the uncertainty caused by geometric mismatching directly adds to the final total uncertainty. So, to diminish the geometric uncertainty is very crucial to a successful validation.

The aforementioned uncertainties affect the accuracy of the reference values. The validation strategy reduces the uncertainties as much as possible and assesses the residual uncertainty. Because the uncertainty sources are independent, summing the squared uncertainties of different sources and mechanisms is a practical method for estimating the final uncertainty in the reference values. The following equation is proposed to conceptually represent the sources and propagation of uncertainties as well as to estimate the total uncertainty in a reference value:

$$\varepsilon_{tot}^2 = \varepsilon_{grd}^2 + \varepsilon_{high}^2 + \varepsilon_{gm}^2 + \varepsilon_{scale}^2 \qquad (6.55)$$

where ε_{tot}^2 is the uncertainty in the reference albedo value for each coarse-resolution pixel calculated from the upscaled high-resolution albedo. The four items on the right side of the equation estimate the uncertainty derived from the in situ measurements, the high-resolution albedo image, the geometric matching, and upscaling, respectively. The detailed formula for estimating these four items can be found in the reference (Peng et al., 2015).

6.4.2.3 The issue of albedo scaling in the mountainous areas

The scale effect in land surface albedo is most prominent in mountainous areas. If the surface is flat, the albedo of low-resolution pixel is the area-weighted average of albedos of all the subpixel factions. As the weighted average is a linear operator, the scaling from high-resolution albedo map to low-resolution albedo map is also a liner transformation in flat surface. In contrast, when the surface is rugged, as in the mountainous area, the incident radiance is not uniform anymore. So, the weight function for albedo averaging is no longer proportional to area, and the scaling of albedo is no longer the simple linear transformation. Wen discussed the scale effect of surface albedo caused by terrain (Wen et al., 2009) and proposed a topology-based scaling factor to solve the issues in upscaling albedo in mountainous areas. The scaling factor is calculated based on the slope, aspect, and mutual shadowing of the microsurfaces inside the coarse-resolution pixel.

Fig. 6.13 illustrates the model-simulated albedo of remote sensing pixel with different solar angle and different resolution of the pixel, on a scene in which the average slope of terrain is 30.7°. The albedo of all the microsurfaces is 0.30. So, if there is no scale effect, the pixel-scale albedo should also be 0.30. However, due to the effect of terrain, the pixel-scale albedo actually distributed in the range of 0.19–0.24, as the pixel size increase from 600 to 2400 m.

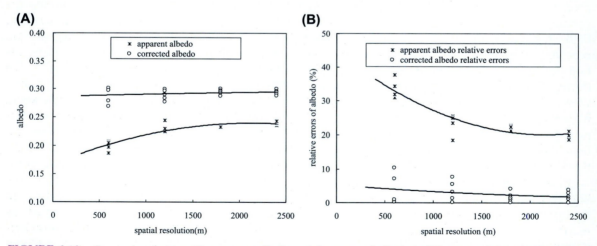

FIGURE 6.13 Comparison between the apparent albedo and the corrected albedo in different spatial resolution and with different solar zenith angle. (A) Apparent albedo and corrected albedo and (B) relative errors of apparent albedo and corrected albedo compared to 0.30.

After applying the scaling factor in the upscaling process, the corrected pixel-scale albedos approach the microsurfaces albedo, i.e., 0.3, and the residual difference is smaller when the resolution of pixel decreases (Fig. 6.13A). Fig. 6.13B presents the relative RMSE of the corrected coarse-pixel albedo. As the result of the effect of microsurface topography, the apparent albedos differ from the microsurface albedo, and the error is larger with higher resolution of the pixel. This error from scale effect can be as large as 30% when the pixel size is 600 m, and decrease to 20% when the pixel size approaches 2400 m. In comparison, the albedos corrected with the scaling factor show much smaller RMSE, which mostly distributed below 0.5, indicating that the scaling factor effectively described the scale effect in rugged terrain. The RMSE after scale correction becomes larger (close to 10%) when the solar zenith angle is large.

6.5 Temporal and spatial analysis of the global land surface albedo

6.5.1 The method to calculate regional average and monthly average albedo

Albedo significantly influences the surface net radiation and plays an important role in global and regional meteorological modeling, for which surface albedo data products can provide helpful information. The land surface albedo is determined by many factors, such as vegetation cover, snow/ice, and the characteristics of clouds and aerosols in the atmosphere (Jin et al., 2002a,b; Zhou et al., 2003; Greenfell and Perovich, 2004), which cause seasonal changes and regional differences. Therefore, many meteorology studies have investigated the land surface albedo and its changing trend on the global or regional scale.

Most of the remote sensing albedo products only provide BSA and WSA, which need to be converted to blue-sky albedo (also called actual albedo) for the purpose of climatological analysis. The conversion formula can be referred to 6.14, or it can be commonly written as follows:

$$\alpha = (1-s)\alpha_{BSA} + s\alpha_{WSA} \qquad (6.56)$$

where α denotes for blue-sky albedo in shortwave band, α_{BSA} and α_{WSA} denote for BSA and WSA in shortwave band, and s is the ratio of diffuse downwarding radiant flux to the total downwarding flux. Because the difference between BSA and WSA is very small for most of the snow-free land surface, it is omitted in some of the application, which means we can simply use the BSA for analysis, or use a constant s for calculating the blue-sky albedo. However, in the following part of this chapter, the spatial and temporal distribution of s is derived from monthly average flux data from the NCEP (National Centers for Environmental Prediction) dataset for more accurate calculation of blue-sky albedo.

Climatological analysis is usually performed on large spatial and temporal scale, which means it uses data of low spatial and temporal resolution. The way to calculate average albedo in low temporal resolution, e.g., monthly or annual, from the commonly available daily or 8-day product, is to use the downwarding radiant flux as weight. To calculate the spatial average albedo, it is also necessary to use the downwarding radiant flux in each pixels as weight. Besides, the variation of area in each pixel should also be considered if the projection of the albedo product is not equal-area projection. For example, when we calculate global average albedo from the 0.05° product in CMG projection, if the area distortion was not considered, there would be a positive bias in the global average albedo because the pixels in polar area were given too much weight. In Zhang's research on global albedo, he used the solar radiation and pixel area together as the weight to calculate the regional and global average albedo from MODIS product (Zhang et al., 2010a,b). The average albedo $\bar{\alpha}$ is defined as follows:

$$\bar{\alpha} = \frac{\sum F_d * \omega_{area} * \alpha_i}{\sum F_d * \omega_{area}} \qquad (6.57)$$

where α_i is the albedo of each pixel, F_d is the weight of incoming radiance at surface, and ω_{area} represents the area weight of the pixel. The summation can be performed in the spatial domain, or in the spatial and temporal domain, thus the monthly or annual average albedo of the region can be derived. The incoming radiance at surface can be estimated from statistics of the global climate dataset. For example, the MERRA (Modern-Era Retrospective Analysis for Research and Applications) dataset issued by NASA has a relatively high spatial resolution of $1/2° \times 2/3°$, its monthly downwarding radiant flux data is adopted by He et al. 2014 as radiance weight.

Many validation and application research works indicate that the MODIS albedo product (MCD43A3) has good accuracy as well as relatively high spatial/temporal resolution. It is often adopted as reference to verify other albedo product. Zhang et al. 2010a,b had analyzed the global and regional average albedos derived from MODIS product and compared them to that derived from model-simulated dataset such as ISCCP and GEWEX. He et al. 2014 also compared GLASS and GLOBALBEDO products with MODIS albedo product and performed climatological analysis. In the following part, we briefly illustrate the spatial and temporal distribution of albedo based on their results.

6.5.2 Temporal variation of global albedo

Table 6.6 lists the seasonal averaged and yearly averaged BSA and WSA from MODIS data from 2000 to 08 of different spatial domains, including the global, southern hemisphere, and northern hemisphere. The multiyear averages of the global MODIS BSA and WSA are 0.235 and 0.245, while the corresponding standard deviation is 0.001. The estimated values without considering their area weight are 0.30 and 0.31. This later result is slightly larger than the true values because it overrates the contribution of ice/snow cover at high latitudes to the average albedo.

The surface albedo reaches its maximum value in winter and its minimum value in summer because snow in winter increases the albedo. The seasonal changes of albedo between 2000 and 08 are also shown in Table 6.6, in which spring refers to the period from March to May, and the remaining periods can be inferred in the same manner. Fig. 6.14 demonstrates the monthly changes of the BSA and WSA in global, southern

TABLE 6.6 The seasonal changes of the global, southern hemisphere, and northern hemisphere average shortwave albedo from 2000 to 08.

Area	Type of albedo	Yearly average of albedo	Seasonal albedo			
			Spring	Summer	Autumn	Winter
North hemisphere	BSA	0.225 ± 0.002	0.251 ± 0.003	0.202 ± 0.003	0.209 ± 0.003	0.243 ± 0.003
	WSA	0.235 ± 0.001	0.262 ± 0.003	0.216 ± 0.003	0.217 ± 0.003	0.248 ± 0.003
South hemisphere	BSA	0.255 ± 0.004	0.174 ± 0.004	0.148 ± 0.002	0.268 ± 0.008	0.361 ± 0.004
	WSA	0.264 ± 0.004	0.184 ± 0.004	0.157 ± 0.002	0.277 ± 0.008	0.37 ± 0.004
Global	BSA	0.235 ± 0.001	0.231 ± 0.002	0.19 ± 0.002	0.233 ± 0.003	0.302 ± 0.002
	WSA	0.235 ± 0.001	0.242 ± 0.002	0.203 ± 0.002	0.242 ± 0.003	0.309 ± 0.002

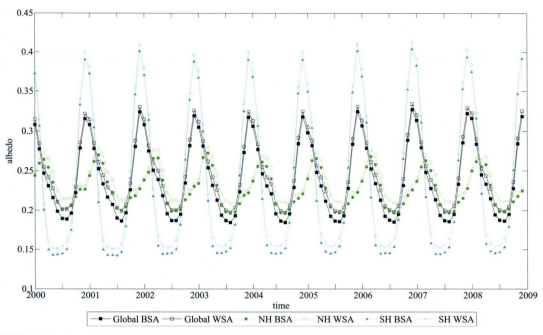

FIGURE 6.14   The changes of the monthly averages of the global, southern hemisphere, and northern hemisphere short-wave albedo during 2000—08.

hemisphere, and northern hemisphere, which clearly reveal that the pattern of albedo seasonal changes was consistent during those 9 years.

The difference between the BSA and WSA is notably larger in the summer than in the winter, which may be caused by the fact that the surface bidirectional reflectance characteristics are more apparent in summer, when plants are thick, than in winter, when the land surface is covered by bare soil and ice/snow. The yearly average of the WSA is also higher than that of the BSA by approximately 0.01.

During the winter in the northern hemisphere, the sun zenith angle is too large and the duration of sunshine is too short or there is no daytime at all. Therefore, there is no MODIS albedo product in the high-latitude areas of the northern hemisphere from December to the following February. Similarly, in high-latitude areas of the southern hemisphere, there is no MODIS albedo product from May to July.

6.5.3 The surface albedo of different latitudinal zones

Fig. 6.15 illustrates the seasonal changes of the BSA and WSA of different latitudinal zones. The following features can be inferred

(1) There are only slight changes in the annual and monthly averages of the albedo between 30°N and 50°S.

(2) The area with the largest seasonal variations of albedo is between 30°N and 90°N, which occurs because of the phenological changes of the vegetation. The regional variations are also quite significant between 50°S and 70°S because of the obvious seasonal changes of ice/snow cover.

(3) The area with the minimum albedo variation is between 10°S and 10°N.

(4) The seasonal changes of the surface albedo of different latitudinal zones in the northern

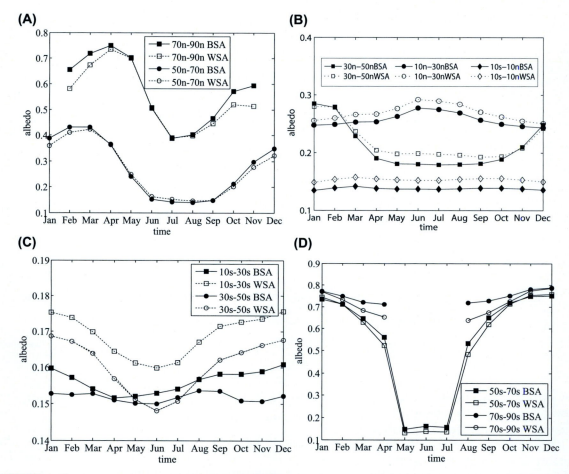

FIGURE 6.15 Seasonal changes of the MODIS BSA and WSA of different latitudinal zones (A) Latitude 50°N~90°N; (B) Latitude 10°S~50°N; (C) Latitude 50°S~10°S; (D) Latitude 90°S~50°S.

hemisphere are larger than those in southern hemisphere.

(5) Invalid albedo value in the MODIS product occurs in the area between 70°N and 90°N and between 70°S and 90°S because the incoming solar radiance is weak, and the reflected radiance is to weak a signal to be quantified by the satellite sensors.

6.5.4 The comparison of different albedo products

He et al. (2014) compared nine albedo datasets in the global scale, among them, five datasets

have been introduced in Section 6.4.1, and the main parameters of the other four datasets are listed in Table 6.7.

Fig. 6.16 presents the multiyear monthly average albedo in north hemisphere, south hemisphere, and whole globe, from different albedo datasets. As can be seen from the global values in Fig. 6.16A, most of the satellite albedo products agree relatively well and can likely satisfy the accuracy requirements for global climate applications, with differences less than 0.02. ISCCP climatological albedos were found to be considerably underestimated, particularly from June to September. This confirms the finding of

TABLE 6.7 The major parameters of four global-scale albedo datasets.

Dataset name	Full name	Spatial resolution	Time span	Reference paper
ISCCP	International Satellite Cloud Climatology Project	2.5°	1983—2009	Zhang et al. (2004)
GEWEX	Global Energy and Water Exchanges Project	1°	1983—2007	Pinker and Laszlo (1992)
ERBE	Earth Radiation Budget Experiment	0.25°	1985—89	Li and Garand, 1994
MERIS	Medium-Resolution Imaging Spectrometer	0.25°	2002—06	Popp et al. (2011)

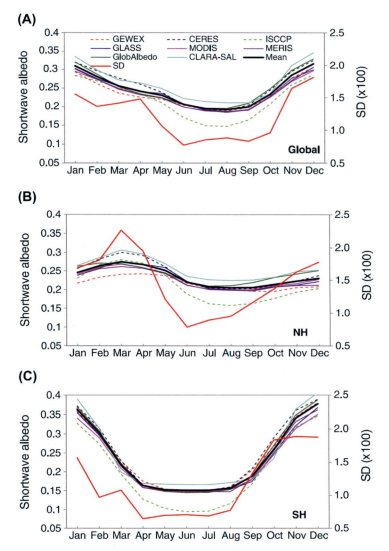

FIGURE 6.16 Monthly climatological surface shortwave albedo derived from satellite-based albedo datasets for (A) the globe, (B) the northern hemisphere, and (c) the southern hemisphere.

Stackhouse et al. (2012), the same paper also pointed out that GEWEX had significant underestimation in albedo over snow/ice surfaces. The CLARA-SAL albedo product tends to have the highest values in most seasons for both hemispheres. There are three reasons for this overestimation. First, the CLARA-SAL product includes sea ice albedo estimates, which may affect our regional aggregates. Its coarse resolution (0.05°) is likely to result in mixed pixels from sea ice and land cover. It matched the data from CERES very well, which is also a coarse-resolution dataset with pixels mixed from sea ice and land covers. Second, the CLARA-SAL product provides only the BSA, which is typically with higher value than the actual "blue-sky" albedo, if the solar zenith angle is large. Third, the weaker cloud detection ability of the AVHRR sensors makes it more likely for cloud pixels to be misidentified as snow/ice pixels at high-latitude area, leading to overestimation of surface albedo.

Differences between the albedo datasets were found to be larger for the northern hemisphere than for the southern hemisphere. This is likely because seasonal snow cover is more extensive in the Northern than in the southern hemisphere. Disagreement on albedo for the seasonal snow area could be attributed to a high degree of sensitivity of the land surface albedo to the different spatial resolutions (e.g., partial/subpixel snowmelt) and temporal composition strategies used in the different albedo datasets. In addition, changes in plant phenology, poleward expansion of tree line, and other climate-related variations are more prominent and complex and in the northern hemisphere they increase the uncertainties in the climatological comparison of surface albedo datasets. Differences in absolute sensor calibration and narrowband-to-broadband conversion may also lead to biases in climatological surface albedo values derived from these datasets.

The zonal mean albedo was also calculated to help further identify differences between the various albedo datasets. Fig. 6.17 shows the zonal mean albedo of 70°N–80°S in January and 80°N–70°S in July using 10° intervals. Most albedo datasets agree well with each other in middle-to-low latitudes, except for ISCCP albedos, which were considerably underestimated as discussed before. We also found that GEWEX surface albedos were underestimated in winter at latitudes higher than 50°. For the latitudinal zones from 40°S to 60°S, the small land surface area and complex terrain may lead to the difference in the different albedo datasets.

The albedo datasets were found to be substantially more inconsistent in winter at high latitudes than that in summer or low latitudes. As errors in albedo estimations tend to increase with solar zenith, differences in the handling of observations with large solar zenith angle in surface anisotropy modeling may lead to the differences in albedo estimation for these datasets. The cutoff angle and angle normalization in satellite products may also contribute to the large difference in albedo for high-latitude regions.

On the context that MODIS albedo is the most accurate dataset among all the nine datasets, GLASS albedo has the best agreement with MODIS albedo. A slightly larger difference among them was found in snow cover transition region 40°N–60°N in January, which is probably a result of their differences in snow albedo retrieval algorithms. The MODIS albedo algorithm tends to generate a snow-free albedo if the majority of observations in the temporal composite window (16 days) is free of snow, whereas the GLASS albedo product uses a temporal filter to smooth the daily albedo and then calculate the mean albedo in its composite window (17 days), which is more likely to have included the albedo from ephemeral snow and result in a higher albedo than the MODIS product.

MERIS albedo climatology also showed consistent values with MODIS albedo in terms of zonal averages, except for the high northern latitudes (>50°N) in January. This is because

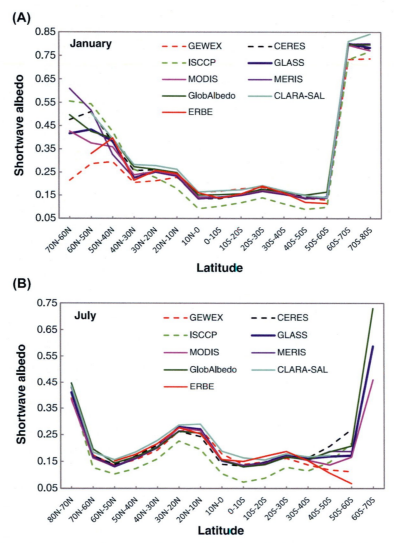

FIGURE 6.17 Climatological surface shortwave albedo at different latitudes from nine global dataset. (A) January and (B) July.

the MERIS algorithm relies on the MODIS aerosol product for atmospheric correction and uses MODIS surface anisotropy information in its retrieval procedure. The overestimation of MERIS albedo in high-latitude winter has also been reported in Fischer et al. (2009), which was, however, attributed to the inaccurate atmospheric correction.

GLOBALBEDO albedo also matched well with MODIS data partly because it used the MODIS surface anisotropy product in the gap-filling postprocess to generate a spatially complete dataset. This gap filling, however, may cause some uncertainties for events such as ephemeral snow and short-term drought. The overestimation of GLOBALBEDO at high-latitude winters has been attributed to the difficulty in snow detection (Muller, 2013).

The overestimation of CLARA-SAL albedo may be attributed to the constant-AOD setting in CLARA-SAL algorithm. In addition, CLARA-SAL provides the temporal-averaged albedo on a monthly basis using the instantaneous albedo values while most of the other satellite albedo products are normalized to the local solar noon. Therefore, it is possible that CLARA-SAL may have an overestimation of surface albedo compared with other satellite products. The ERBE dataset only provides albedo in the middle-to-low latitude, and it matched quite well with other datasets in this regions.

6.5.5 Surface albedo of different land types

Albedo climatologies of different land types have been widely used in climate modeling studies. Based on the resolution, as well as consistency, of global albedo datasets, three datasets, i.e., GLASS, GLOBALBEDO, and MODIS, are selected for building albedo climatologies of typical land cover types during the period 2001−10. For this purpose, a multiyear stable land cover map is generated from the annual MODIS land cover data (MCD12C1).

Fig. 6.18 presents the monthly average albedo of all three datasets on different land types and different latitudinal zones. For most of the cover types between 40°N and 40°S, results show that climatological surface albedo values do not change much across seasons. Data in the zones 20°N−20°S are not plotted because they have small seasonal variations, and the differences among those three datasets are very small. The magnitude of seasonal albedo variation is much larger for middle-to-high latitudes, mainly because snowfall and snowmelt can significantly affect surface reflectivity. Plant phenology and soil moisture also contribute to the differences in this variation at different latitudes. The effects

of snow vary over the different cover types, for example, the pixels with sparse vegetation cover (e.g., shrub and grass) tend to have much higher surface albedos than densely vegetated pixels (e.g., forest) in high-latitude winters (Wu et al., 2012).

6.5.6 Change trend of annual average albedo

GLASS and CLARA-SAL datasets provide the longest continuous time series of albedo data over the globe. But the resolution of CLARA-SAL is relatively low, in the meanwhile, the GLASS product is in good quality and free of gap, so here the GLASS albedo product is adopted to study to long-term change trend of global albedo. According to the above analysis, the land area mainly distributed in the north hemisphere, and there is also more variation in albedo in the north hemisphere than that of south hemisphere. So, the following analysis adopts the GLASS albedo in north hemisphere as example to illustrate the annual albedo change trend.

We have removed the seasonal variation trend from the temporal series of monthly averaged albedos to compare the nonperiodic variations of the global monthly average albedo from 1981 to 2010. The albedo without the seasonal trend is determined by the following formula:

$$\Delta\alpha(yy, mm) = \alpha(yy, mm) - \overline{\alpha}(mm) \qquad (6.58)$$

where $\alpha(yy, mm)$ represents the monthly average albedo in "yy" (year) and "mm" (month), and $\overline{\alpha}(mm)$ represents the average value of several years in a particular "mm" (month).

Fig. 6.19 presents the anomalies of monthly land surface albedo in the north hemisphere in January and July, and the snow cover area in the same spatial and temporal domain. It is found that the July surface albedo decreased at a rate of 0.0013/decade from 1981 to 2010,

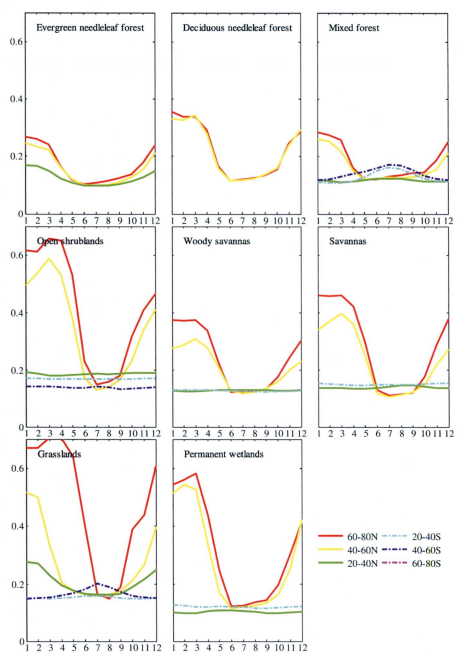

FIGURE 6.18   Monthly averages of albedo climatologies at different latitudinal zones, derived from MODIS, GLASS, and GlobAlbedo products for eight land cover types. *X axis* represents the month; *y axis* represents the surface albedo.

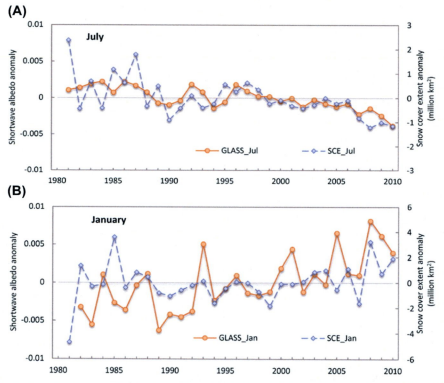

FIGURE 6.19 Surface albedo anomalies for the northern hemisphere in July and January, from the GLASS albedo product and snow cover extent anomalies for 1981—2010. (A) July and (B) January.

according to the GLASS albedo product. Hand in hand with the surface albedo decrease, the snow cover extent (SCE) decreased at a rate of 5.46×10^5 km^2/decade. The correlation coefficient between GLASS surface albedo anomalies and SCE anomalies (taking the year of 2000 data as a reference) was 0.61 for the whole time period; it increased to 0.77 for the period of 2000—10, as data after 2000 were believed to have improved sensitivity to surface changes particularly during transition periods over snow surfaces. Besides the SCE decrease, changes in snow morphology have been reported to be able to contribute approximately one-third of the changes in snow albedo based on model simulations, which, however, is currently difficult to be accurately quantified by remote sensing data on a global basis.

We also found that for the northern hemisphere, surface albedo in winter changed in a direction opposite to that observed in summer. GLASS albedo product showed an increase in surface albedo at a rate of 0.0029/decade. At the same time, SCE increased by 4.4×10^5 km^2/decade. The correlation between albedo anomalies and SCE is 0.44, which increased to 0.56 when only the data for the period of 2000—10 were used. One possible reason for this relationship is that global warming increased atmospheric water vapor content in winter and consequently more and/or heavier snowfall on land surface in winter in the northern latitudes.

Compared with the relatively larger correlation in July, the decreased sensitivity of satellite albedo to SCE changes in January can result from the reduced spatial coverage at high latitudes in winter, and the differences in satellite albedo data processing add uncertainty in this analysis. Because of the limitations of satellite albedo products in high northern latitudes in winter, further efforts are urged to investigate the reason of the albedo's decreasing trend and poor correlation of satellite albedo and SCE.

Temporal trends in surface albedo vary from place to place. Fig. 6.20 shows the trend of surface albedo per pixel during 2000–10. Climate change has resulted in early snowmelt and/or vegetation onset over boreal areas, which may have caused the dramatic decrease in surface albedo, particularly over forest areas in early spring. In addition, a significant albedo decrease (>0.05/decade) in early spring in northern Russia was found in this study, which was due to the snow masking effects of vegetation (Loranty et al., 2014); this can be attributed to the northward expansion of the boreal forest in the northern tundra area in Russia, which has resulted from the recent climate warming.

A decreasing summer land surface albedo has been found in the Greenland ice sheet since the 1980s, which is consistent with former research (He et al., 2013). Although the extent of the albedo-decreasing area is not quite significant, as shown in Fig. 6.20C, its associated snow/ice albedo feedback with the adjacent sea ice area could bring amplified effects to global climate change.

Observed changes in surface albedo may also be the result of soil moisture changes, especially in semiarid areas. It was found by Dorigo et al. (2012) that observed albedo decrease and increase in southwest Africa and central to west Australia, as can be seen in Fig. 6.20B,C, could be largely explained by soil moisture increase and decrease, respectively (particularly from March to May). Major soil moisture variations are mainly precipitation driven, especially in the arid/semiarid areas. If a drier condition becomes more frequent and/or extensive in the future, which leads to soil moisture decrease,

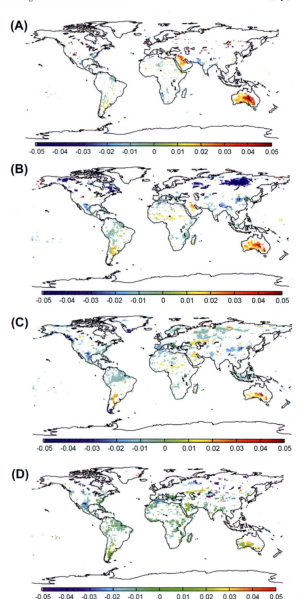

FIGURE 6.20 Distribution of surface albedo change trend in 2000–10. (A) Winter: December–February; (B) Spring: Mar–May, (C) Summer: June–August, and (D) Autumn: September–November. Trend is statistically significant at the 95% confidence level. Trend is not calculated over the Antarctica.

vegetation cover may decrease accompanied by increase in surface albedo. This may result in reduced evapotranspiration and further contribute to the decrease in precipitation, which will in turn exacerbate the drought condition.

6.6 Problems and prospects in the study of broadband albedo

Although the studies in albedo inversion algorithms and surface albedo products have achieved great success, some problems still exist. Not only the accuracy, temporal resolution, and data integrity of albedo product should to be further enhanced but also the research on extreme surfaces, e.g., snow/ice, ocean, and mountain, is in urgent need, and the validation work on albedo product should be pushed forward to longer and larger extend.

The frequently used albedo inversion algorithm based on the kernel-driven bidirectional reflectance model mainly applies to vegetation. There lacks appropriate albedo-estimation model in many other land cover types. For example, in high-latitude areas, the large albedo of snow cover in winter significantly influences the surface energy balance. Some studies have indicated that the MODIS albedo product underestimates the albedo in the polar regions (Wang and Zender, 2010), and this issue is still open to discuss. In contrast to the numerous studies and products on land surface albedo, the studies of albedo over the vast area of ocean are surprisingly scarce. Currently, the only published global albedo dataset that covers ocean is the CLARA-SAL, which is in low resolution and only provides BSA. One of the reasons for the lack of products on ocean is that the ocean surface albedo, affected by solar angle and wind speed, exhibits intense variation in the temporal domain, so the ocean surface albedo requires totally different production procedure as well as form of product from the land surface albedo (Feng et al., 2016). Another important but

neglected land type is the mountainous area, which covers a quarter of the total land surface, and even two-third of the territory of China. Currently, the effect of rugged terrain on albedo (Davidson and Wang, 2004; Wen et al., 2009) is ignored in all operational algorithm of albedo product. Even the theoretic and experimental research works seldom consider the terrain effect on albedo. The difficult point on rugged terrain albedo research is the scale effect and its derivative issue of how to adequately define albedo in albedo measurement and application in rugged terrain (Wen et al., 2018). Thus, although there exists physical RT model on rugged terrain, it is still difficult to build concise and invertible BRDF/albedo model. To increase the overall accuracy of global albedo estimation, it is imperative to put more strength on the research about BRDF/albedo model on the surface of these extreme areas.

To increase the temporal resolution and accuracy of albedo product, it is necessary to increase the available information in the inversion, and to use observation data from multiple sensor is major resort to increase information. One way to deal with multisensor data first uses the direct inversion algorithm, which can calculate albedo based on each single-angle observations, and then merges the albedo-estimation results from different sensors according to their uncertainty estimation. Another approach combines all observations from multiple sensors in the inversion of BRDF model, and then derives albedo from BRDF model through angular integration and narrow-to-broadband conversion. To successfully perform inversion of the BRDF model, it is prerequisite to collect multiangular data in sufficient number as well as quality, which may require a long data-accumulation time if only single source of data is available, e.g., the 16-day composition time for MODIS albedo product. This results in the relatively low actual temporal resolution in current albedo product and leads to the poor ability to reflect variation of albedo during the snow season or fast vegetation grow season. With the rapid development of remote sensing techniques, an increasing number of sensors are running on orbit, which can provide sufficient number of evenly distributed directional observations in a short composition time (Wen et al., 2017). However, to use multisensor data, we still face unsolved scientific and technical problems such as the inconsistent band design and spatial resolution, as well as inadequate calibration or other quality drawbacks in the multiple sensors. The albedo products inverted from remote sensing data are influenced by weather factors, such as clouds and aerosols, which cause gaps in products and inconvenience for users. So, developing the gap-filling algorithm, which involves spatial–temporal autocorrelation and a priori knowledge, is also an indispensable part in albedo researches.

Like other quantitative remote sensing products, surface albedo products require strict and massive validation. Currently, the albedo observation data on snow/ice, ocean, and mountain surfaces are far from sufficient to support the validation of albedo product in these areas. The commonly used albedo validation data are measurements from flux towers or automatic weather stations. The spatial resolutions of these measurements differ dramatically from the pixels of typical low-resolution remote sensing data (for example, 1 km); consequently, there are large uncertainties when the surface-measured data are directly used to validate pixel-scale albedo products. Although several scholars have reported albedo measurement methods for inhomogeneous areas and scaling methods based on high-resolution remote sensing images, in general, scaling problems is not yet solved in the validation of albedo products. The issue of temporal normalization of albedo also needs to be studied. The field-measured albedo is continuous, which exhibits fluctuation with the variation of solar angles and atmospheric state. However, the albedo needed in climate model or other researches are daily or even monthly average values, and

the remote sensing albedo products are instantaneous, or normalized into the local solar noon BSA and WSA. The discrepancy in the temporal setting introduces uncertainty in the surface albedo validation as well as energy balance estimation.

In summary, we try to put forward 10 questions that need be attended to in the study of surface albedo.

1. How to build model for snow/ice surface as well as its mix pixel and more accurately estimate surface in polar areas?
2. How to design and generate albedo product that can depict the temporal variation, especially diurnal variation of ocean surface albedo?
3. How the terrains in subpixel scale influence BRDF/albedo in the pixel scale, and how to parameterize this effect?
4. How to deal with the different spectral setting of multiple sensors in the combined inversion?
5. How to evaluate the information gain from multisensor data, especially from those relatively low-quality data?
6. How to merge remote sensing datasets of different resolution, ground observations, and a priori knowledge to reduce noise and fill gaps in the global albedo product?
7. How to evaluate and quantify the uncertainty in albedo product from multiple sources in the albedo merging?
8. How to carry out measurement of surface albedo, as well as BRDF, in the extreme environment of polar, ocean, and mountainous areas?
9. How to perform spatial scale matching or scale conversion between ground-based albedo measurement and satellite product?
10. How to evaluate the variation of albedo in the diurnal cycle and perform temporal normalization to albedo products accordingly?

References

Bacour, C., Bréon, F.M., 2005. Variability of land surface BRDFs. Remote Sens. Environ. 98, 80—95.

Barnsley, M., Hobson, P., Hyman, A., Lucht, W., Muller, J., Strahler, A., 2000. Characterizing the spatial variability of broadband albedo in a semidesert environment for MODIS validation. Remote Sens. Environ. 74 (1), 58—68.

Borel, C.C., Gerstl, S., Powers, B.J., 1991. The radiosity method in optical remote sensing of structured 3-D surfaces. Remote Sens. Environ. 36 (1), 13—44.

Cescatti, A., Marcolla, B., Santhana Vannan, S.K., Pan, J.Y., Román, M.O., Yang, X., Ciais, P., Cook, R.B., Law, B.E., Matteucci, G., Migliavacca, M., Moors, E., Richardson, A.D., Seufert, G., Schaaf, C.B., 2012. Intercomparison of MODIS albedo retrievals and in situ measurements across the global FLUXNET network. Remote Sens. Environ. 121, 323—334.

Chen, J.M., Leblanc, S., 1997. A 4-scale bidirectional reflection model based on canopy architecture. IEEE Trans. Geosci. Remote Sens. 35, 1316—1337.

Chen, T., Ohring, G., 1984. On the relationship bet ween clear-sky planetary and surface albedos. J. Atmos. Sci. 41, 156—158.

Chopping, M.J., Rango, A., Havstad, K.M., Schiebe, F.R., Ritchie, J.C., Schmugge, T.J., French, A., McKee, L., Davis, R.M., 2003. Canopy attributes of Chihuahuan Desert grassland and transition communities derived from multi-angular 0.65μm airborne imagery. Remote Sens. Environ. 85 (3), 339—354.

Cui, Y., Mitomi, Y., Takamura, T., 2009. An empirical anisotropy correction model for estimating land surface Albedo for radiation budget studies. Remote Sens. Environ. 113, 24—39.

Danaher, T.J., 2002. An Empirical BRDF Correction for Landsat TM and ETMt Imagery. Brisbane, Australia: Proceedings of the 11th. Australasian Remote Sensing and Photogrammetry Conference, September.

Davidson, A., Wang, S., 2004. The effects of sampling resolution on the surface albedos of dominant land cover types in the north American boreal region. Remote Sens. Environ. 93, 211—224.

Dickinson, R.E., 1995. Land processes in climate models. Remote Sens. Environ. 55 (1), 27—38.

Diner, D.J., Barge, L.M., Bruegge, C.J., Chrien, T.G., Conel, J.E., Eastwood, M.L., Garcia, J.D., Hernandez, M.A., Kurzweil, C.G., Ledeboer, W.C., Pignatano, N.D., Sarture, C.M., Smith, B.G., 1998. The Airborne Multi-angle SpectroRadiometer (AirMISR): instrument description and first results. IEEE Trans. Geosci. Remote Sens. 36, 1339—1349.

Disney, M., Lewis, P., Saich, P., 2006. 3D modelling of forest canopy structure for remote sensing simulations in the optical and microwave domains. Remote Sens. Environ. 100 (1), 114—132.

Dorigo, W., Jeu, R.D., Chung, D., Parinussa, R., Liu, Y., Wagner, W., et al., 2012. Evaluating global trends (1988—2010) in harmonized multi-satellite surface soil moisture. Geophys. Res. Lett. 39 (18), 18405.

Feng, Y.B., Liu, Q., Qu, Y., Liang, S.L., 2016. Estimation of the ocean water albedo from remote sensing and meteorological reanalysis data. IEEE Trans. Geosci. Remote Sens. 2 (54), 850—868. https://doi.org/10.1109/TGRS.2015.2468054.

Fischer, J., Preusker, R., Muller, J., Zuhlke, M., 2009. ALBEDOMAP-validation Report. Available at: http://www.brockmann-consult.de/albedomap/pdf/MERIS-AlbedoMap-Validation-1.0.pdf.

Gao, F., Li, X., Strahler, A.H., Schaaf, C., 2001. Evaluation of the LiTransit kernel for BRDF modeling. Remote Sens. Rev. 19, 205—224.

Gao, F., Schaaf, C., Strahler, A., Roesch, A., Lucht, W., Dickinson, R., 2005. The MODIS BRDF/albedo climate modeling grid products and the variability of albedo for major global vegetation types. J. Geophys. Res. 110, D01104. https://doi.org/10.1029/2004JD00519.

Gastelluetchegorry, J.P., Demarez, V., Pinel, V., Zagolski, F., 1996. Modeling radiative transfer in heterogeneous 3-d vegetation canopies. Remote Sens. Environ. 58 (58), 131—156.

GCOS, 2006. Satellite- Based Products for Climate. Supplemental Details to the Satellite- Based Component of the "Implementation Plan for the Global Observing System for Climate in Support of the UNFCCC", GCOS-107 (WMO/TD No. 1338). 90p.

Goel, N.S., 1988. Models of vegetation canopy reflectance and their use in estimation of biophysical parameters from reflectance data. Remote Sens. Rev. 4 (1), 1—212.

Goudriaan, J., 1977. Crop Micrometeorology: A Simulation Study. Simulation Monographs. Pudoc, Wageningen.

Govaerts, Y., Lattanzio, A., Pinty, B., Schmertz, J., 2004. Consistent surface albedo retrieved from two adjacent geostationary satellite. Geophys. Res. Lett. 31, L15201. https://doi.org/10.1029/2004GL020418 03.

Govaerts, Y.P., Taberner, M., Lattanzio, A., 2006. Spectral conversion of surface albedo derived from Meteosat first generation observations. IEEE Geosci. Remote Sens. Lett. 3, 23—27. https://doi.org/10.1109/LGRS.2005.854202.

Greenfell, T.C., Perovich, D.K., 2004. Seasonal and spatial evolution of albedo in a snow-ice-land-ocean environment. J. Geophys. Res. 109, C01001.

He, T., Liang, S., Yu, Y., Wang, D., Gao, Liu, Q., 2013. Greenland surface albedo changes in July 1981−2012 from satellite observations. Environ. Res. Lett. 8 (4), 044043. https://doi.org/10.1088/1748-9326/8/4/044043.

He, T., Liang, S., Song, D., 2014. Analysis of global land surface albedo climatology and spatial-temporal variation during 1981−2010 from multiple satellite products. J. Geophys. Res. Atmos. 119 (10) https://doi.org/10.1002/2014JD021667, 281−10,298.

He, T., Liang, S., Wang, D., Chen, X., Song, D.-X., Jiang, B., 2015. Land surface albedo estimation from Chinese HJ satellite data based on the direct estimation approach. Remote Sens. 7, 5495−5510.

He, T., Liang, S., Wang, D., 2017. Direct estimation of land surface albedo from simultaneous MISR data. IEEE Trans. Geosci. Remote Sens. 55, 2605−2617.

He, T., Liang, S., Wang, D., Cao, Y., Gao, F., Yu, Y., Feng, M., 2018. Evaluating land surface albedo estimation from Landsat MSS, TM, ETM +, and OLI data based on the unified direct estimation approach. Remote Sens. Environ. 204, 181−196.

Henderson-Sellers, A., Wilson, M.F., 1983. Surface albedo data for climatic modeling. Rev. Geophys. 21 (8), 1743−1778. https://doi.org/10.1029/RG021i008p01743.

Huemmrich, K.F., 2001. The GeoSail model: a simple addition to the SAIL model to describe discontinuous canopy reflectance. Remote Sens. Environ. 75, 423−431.

Idso, S.B., Wit, C.T., 1970. Light relations in plant canopies. Appl. Opt. 9 (1), 177−184.

Irons, J.R., Ranson, K.J., Irish, R.R., Huezel, F.G., 1991. An off-nadir-pointing imaging spectroradiometer for terrestrial ecosystem studies. IEEE Trans. Geosci. Remote Sens. 29, 66−74.

Jackson, J.E., Palmer, J.W., 1972. Interception of light model hedgerow orchards in relation to latitude, time of year and hedgerow configuration and orientation. J. Appl. Ecol. 9 (2), 341−358.

Jacquemoud, S., Bacour, C., Poilvé, H., Frangi, J.P., 2000. Comparison of four radiative transfer models to simulate plant canopies reflectance: direct and inverse mode. Remote Sensing of Environment 74 (3), 471−481.

Jin, Y., Gao, F., Schaaf, C.B., Li, X., Strahler, A.H., Bruegge, C.J., Martonchik, J.V., 2002a. Improving MODIS surface BRDF/Albedo retrievals with MISR observations. IEEE Trans. Geosci. Remote Sens. 40 (7), 1,593−1,604.

Jin, Y., Gao, F., Schaaf, C.B., Li, X., Strahler, A.H., Bruegge, C.J., Martonchik, J.V., 2002. Improving MODIS surface BRDF/Albedo retrievals with MISR observations. IEEE Trans. Geosci. Remote Sens. 40, 604.

Jin, Y., Schaaf, C.B., Gao, F., Li, X., Strahler, A.H., Zeng, X., Dickinson, R.E., 2002b. How does snow impact the albedo of vegetated land surface as analyzed with MODIS data? Geophys. Res. Lett. 29, 1374.

Jupp, D.L.B., Walker, J., Penridge, L.K., 1986. Interpretation of vegetation structure in Landsat MSS imagery: a case study in disturbed semi-arid eucalypt woodlands. Part 2. Model-based analysis. J. Environ. Manag. 23 (1), 35−57.

Karlsson, K.G., Anttila, K., Trentmann, J., Stengel, M., Fokke Meirink, J., Devasthale, A., et al., 2017. Clara-a2: the second edition of the CM SAF cloud and radiation data record from 34 years of global AVHRR data. Atmos. Chem. Phys. 17 (9), 5809−5828.

Koepke, P., Kriebel, K.T., 1987. Improvements in the short-wave cloud-free radiation budget accuracy, part I: numerical study including surface anisotropy. J. Clim. Appl. Meteorol. 26, 374−395.

Kubelka, P.J., Munk, F., 1931. Ein Beitrag zur Optik der Farbanstriche. Z. Tech. Physik 12, 593−601.

Kuusk, A., 1985. The hotspot effect of uniform vegetative cover. Sov. J. Remote Sens. 3, 645−658.

Leroy, M., Deuze, J.L., Breon, F.M., Hautecoeur, O., Herman, M., Buriez, J.C., Tanre, D., Bouffies, S., Chazette, P., Roujean, J.L., 1997. Retrieval of atmospheric properties and surface bidirectional reflectances over land from POLDER/ADEOS. J. Geophys. Res. 102, 17023−17037.

Leroy, M., Hautecoeur, O., Ponchaut, F., Alonso, L., Moreno, J., 2001. The digital airborne spectrometer experiment (DAISEX). In: Wooding, M., Harris, R.A. (Eds.), Proceedings of the Workshop Held July. European Space Agency, ESA SP-499, p. 13.

Lewis, P., Brockmann, C., Danne, O., et al., 2001. GlobAlbedo Algorithm Theoretical Basis Document: Version 3.0. 323pp. Available from: http://www.globalbedo.org.

Li, X., Strahlar, A., Woodcock, C.E., 1995. A hybrid geometric optical-radiative transfer approach for modeling albedo and directional reflectance of discontinuous canopies. IEEE Trans. Geosci. Remote Sens. 33, 466−480.

Li, Z., Garand, L., 1994. Estimation of surface albedo from space: a parameterization for global application. J. Geophys. Res. Atmos. 99 (D4), 8335−8350.

Li, X., Strahler, A.H., 1986. Geometric-Optical bidirectional reflectance modeling of a conifer forest canopy. IEEE Trans. Geosci. Remote Sens. 24 (6), 906−919.

Li, X., Strahler, A.H., 1988. Modeling the gap probability of a discontinuous vegetation canopy. IEEE Trans. Geosci. Remote Sens. 26 (2), 161−170.

Li, X., Strahler, A.H., 1992. Geometric-optical bidirectional reflectance modeling of the discrete crown vegetation canopy: effect of crown shape and mutual shadowing. IEEE Trans. Geosci. Remote Sens. 30 (2), 276−292.

Li, X., Wang, J., Strahler, A.H., 2000. Scale effects and scaling up by geometric optical model. Sci. China Ser. E 43, 17–22.

Li, X., Li, X.W., Li, Z.Y., et al., 2009. Watershed Allied Telemetry Experimental Research. J. Geophys. Res. Atmos. 114, D22103. https://doi.org/10.1029/2008JD011590.

Liang, S., 2001. Narrowband to broadband conversions of land surface albedo. Remote Sens. Environ. 76, 213–238.

Land surface bidirectional reflectance distribution function (BRDF): recent advances and future prospects. In: Liang, S., Strahler, A. (Eds.), Remote Sens. Rev. 18, 83–551.

Liang, S., Strahler, A.H., 1994. Retrieval of surface BRDF from multiangle remotely sensed data. Remote Sensing of Environment 50, 18–30.

Liang, S., Strahler, A., Walthall, C., 1999. Retrieval of land surface albedo from satellite observations: a simulation study. J. Appl. Meteorol. 38, 712–725.

Liang, S., Fang, H., Chen, M., Walthall, C., Daughtry, C., Morisette, J., Schaaf, C., Strahler, A., 2002. Validating MODIS land surface reflectance and albedo products: methods and preliminary results. Remote Sens. Environ. 83 (1–2), 149–162.

Liang, S., Shuey, C., Russ, A., Fang, H., Chen, M., Walthall, C., Daughtry, C., 2003. Narrowband to broadband conversions of land surface albedo: II. Validation. Remote Sens. Environ. 84 (1), 25–41.

Liang, S., Stroeve, J., Box, J., May 26, 2005. Mapping daily snow shortwave broadband albedo from MODIS: the improved direct estimation algorithm and validation. J. Geophys. Res. 110 (D10). Art. No. D10109.

Liu, Q.H., Li, X.W., Chen, L.F., 2002. Field campaign for quantitative remote sensing in Beijing. Proc. IGARSS02 3133–3135.

Liu, Q., Liu, Q., Wang, L.Z., et al., 2013a. A statistics-based temporal filter algorithm to map spatiotemporally continuous shortwave albedo from MODIS data. Hydrol. Earth Syst. Sci. 17 (6), 2121–2129.

Liu, Q., Wang, L.Z., Qu, Y., et al., 2013b. Preliminary evaluation of the long-term GLASS albedo product. Int. J. Digit. Earth 6 (Suppl. 1), 5–33.

Liu, Y., Wang, Z., Sun, Q., et al., 2017. Evaluation of the VIIRS BRDF, Albedo and NBAR products suite and an assessment of continuity with the long term MODIS record. Remote Sens. Environ. 201, 256–274.

Loranty, M.M., Berner, L.T., Goetz, S.J., Jin, Y.F., Randerson, J.T., 2014. Vegetation controls on northern high latitude snow-albedo feedback: observations and CMIP5 model simulations. Glob. Chang. Biol. 20 (2), 594–606. https://doi.org/10.1111/Gcb.12391.

Lucht, W., 1998. Expected retrieval accuracies of bidirectional reflectance and albedo from EOS-MODIS and MISR angular sampling. J. Geophys. Res. 103 (8), 763-8,778.

Lucht, W., Schaaf, C.B., Strahler, A.H., 2000. An algorithm for the retrieval of albedo from space using semiempirical BRDF models. IEEE Trans. Geosci. Remote Sens. 38, 977–998.

Maignan, F., Bréon, F.M., Lacaze, R., 2004. Bidirectional reflectance of Earth targets: evaluation of analytical models using a large set of spaceborne measurements with emphasis on the Hot Spot. Remote Sens. Environ. 90 (2), 210–220.

Martonchik, J.V., Diner, D.J., Pinty, B., Verstraete, M.M., Myneni, R.B., Knyazikhin, Y., Gordon, H.R., 1998. Determination of land and ocean reflective, radiative, and biophysical properties using multiangle imaging. IEEE Trans. Geosci. Remote Sens. 36, 1266–1281.

Martonchik, J.V., Bruegge, C.J., Strahler, A., 2000. A review of reflectance nomenclature used in remote sensing. Remote Sens. Rev. 19, 9–20.

Minnaert, M., 1941. The reciprocity principle in lunar photometry. Astrophys. J. 93, 403–410.

Muller, J.-P., 2013. GlobAlbedo Final Validation Report. University College London. Available at: http://www.globalbedo.org/docs/GlobAlbedo_FVR_V1_2_web.pdf.

Muller, J.-P., et al., 2012. The ESA GlobAlbedo project for mapping the Earth's land surface albedo for 15 years from European sensors. In: IEEE Geoscience and Remote Sensing Symposium (IGARSS) 2012. IEEE, Munich, Germany.

Nicodemus, F.E., et al., 1977. Geometric Considerations and Nomenclature for Reflectance. Monograph 161. National Bureau of Standards (US).

Nilson, T., Kuusk, A., 1989. A reflectance model for the homogeneous plant canopy and its inversion. Remote Sens. Environ. 27, 157–167.

Norman, J.M., Welles, J.M., Walter-Shea, E.A., 1985. Constrasts among bidirectional reflectances of leaves, canopies, and soils. IEEE Trans. Geosci. Remote Sens. 23, 659–668.

Peng, J.J., Qiang, L., Wen, J.G., Liu, Q.H., Yong, T., Wang, L.Z., et al., 2015. Multi-scale validation strategy for satellite albedo products and its uncertainty analysis. Sci. China Earth Sci. 58 (4), 573–588.

Peng, S., Wen, J.G., Xiao, Q., et al., 2017. Multi-staged NDVI dependent snow-free land-surface shortwave albedo narrowband-to-broadband (NTB) coefficients and their sensitivity analysis. Remote Sens. 9, 93. https://doi.org/10.3390/rs9010093.

Pinker, R.T., 1985. Determination of surface albedo from satellite. Adv. Space Res. 5, 333–343.

Pinker, R.T., Laszlo, I., 1992. Modeling surface solar irradiance for satellite applications on a global scale. J. Appl. Meteorol. 31 (2), 194–211.

Pinty, B., Roveda, F., Verstraete, M.M., Gobron, N., Govaerts, Y., Martonchik, J., Diner, D., Kahn, R., 2000. Surface albedo retrieval from METEOS AT — Part 1: theory. J. Geophys. Res. 105, 18099–18112.

Popp, C., Wang, P., Brunner, D., Stammes, P., Zhou, Y., Grzegorski, M., 2011. MERIS albedo climatology for FRESCO+O-2 A-band cloud retrieval. Atmos. Meas. Tech. 4 (3), 463–483. https://doi.org/10.5194/amt-4-463-2011.

Qin, W.H., Herman, J.R., Ahmad, Z., 2001. A fast, accurate algorithm to account for non-Lambertian surface effects on TOA radiance. J. Geophys. Res. Atmos. 106 (D19), 22671–22684.

Qu, Y., Liu, Q., Liang, S., et al., 2014. Direct-estimation algorithm for mapping daily land-surface broadband Albedo from MODIS data. IEEE Trans. Geosci. Remote Sens. 52 (2), 907–919.

Rahman, H., Pinty, B., Verstraete, M.M., 1993. Coupled surface-atmosphere reflectance (CSAR) model, 2, Semiempirical surface model usable with NOAA advanced very high resolution radiometer data. J. Geophys. Res. Atmos. 98, 20-20.

Ricchiazzi, P., Yang, S., Gautier, C., Sowle, D., 1998. SBDART :A research and teaching software tool for plane parallel radiative transfer in the earth's atmosphere. Bull. Atm. Meteorol. Soc. 79, 2101–2114.

Román, M.O., Schaaf, C.B., Woodcock, C.E., Strahler, A.H., Yang, X., Braswell, R.H., Curtis, P.S., Davis, K.J., Dragoni, D., Goulden, M.L., 2009. The MODIS (Collection V005) BRDF/albedo product: assessment of spatial representativeness over forested landscapes. Remote Sens. Environ. 113 (11), 2476–2498.

Rosema, A., Verhoef, W., Noorbergen, H., Borgesius, J.J., 1992. A new forest light interaction model in support of forest monitoring. Remote Sens. Environ. 42, 23–41.

Ross, J., 1981. The Radiation Regime and the Architecture of Plant Stands. Dr. W. Junk Publ., The Netherlands.

Ross, J.K., Marshak, A.L., 1988. Calculation of canopy bidirectional reflectance using the Monte Carlo method. Remote Sens. Environ. 24 (2), 213–225.

Ross, J., Marshak, A., 1989. The influence of leaf orientation and the specular component of leaf reflectance on the canopy bidirectional reflectance. Remote Sens. Environ. 27 (3), 251–260.

Roujean, J.L., Leroy, M., Deschamps, P.Y., 1992. A bidirectional reflectance model of the earth's surface for the correction of remote sensing data. J. Geophys. Res. 97 (D18), 20455–20468.

Sandmeier, S.R., 2000. Acquisition of bidirectional reflectance factor data with field goniometers. Remote Sens. Environ. 73 (3), 257–269.

Sandmeier, S.R., Itten, K.I., 1999. A field goniometer system (FIGOS) for acquisition of hyperspectral BRDF data. IEEE Trans. Geosci. Remote Sens. 37 (2), 978–986.

Schaaf, C., Gao, F., Strahler, A.H., Lucht, W., Li, X., Tsang, T., Strugnell, N.C., Zhang, X., Jin, Y., Muller, J.P., Lewis, P., Barnsley, M., Hobson, P., Disney, M., Roberts, G., Dunderdale, M., Doll, C., d'Entremont, R., Hu, B., Liang, S., Privette, J.L., Roy, D., 2002. First operational BRDF, albedo and nadir reflectance products from MODIS. Remote Sens. Environ. 83, 135–148.

Schaaf, C., Martonchik, J., Pinty, B., et al., 2008. Retrieval of Surface Albedo from Satellite Sensors, Advances in Land Remote Sensing: System, Modeling, Inversion and Application. Springer.

Schaepman-Strub, G., Schaepman, M.E., Painter, T.H., Dangel, S., Martonchik, J.V., 2006. Reflectance quantities in optical remote sensing–definitions and case studies. Remote Sens. Environ. 103 (1), 27–42. https://doi.org/10.1016/j.rse.2006.03.002.

Shibayama, M., Wiegand, C.L., 1985. View azimuth and zenith, and solar angle effects on wheat canopy reflectance. Remote Sens. Environ. 18, 91–103.

Stackhouse, P.W., Cox, S.J., Mikovitz, J.C., Rossow, W.B., Zhang, Y.C., Hinkelman, L.M., Pinker, R.T., Raschke, E., Kinne, S., 2012. Chapter 4: surface radiation budget. In: Rep., World Climate Research Programme. NASA.

Strahler, A.H., Jupp, D., 1990. Modeling bidirectional reflectance of forests and woodlands using Boolean models and geometric optics. Remote Sens. Environ. 34 (3), 153–166.

Stroeve, J., Box, J.E., et al., 2004. Accuracy assessment of the MODIS 16-day albedo product for snow: comparisons with Greenland in situ measurements. Remote Sens. Environ. 94 (1), 46–60.

Strub, G., et al., 2003. Evaluation of spectrodirectional Alfalfa canopy data acquired during DAISEX'99. IEEE Transactions on Geoscience and Remote Sensing 41, 1034–1042.

Suits, G.H., 1972. The calculation of the directional reflectance of vegetative canopy. Remote Sens. Environ. 2, 117–175.

Susaki, J., Yasuoka, Y., Kajiwara, K., Honda, Y., Hara, K., 2007. Validation of MODIS albedo products of paddy fields in Japan. IEEE Trans. Geosci. Remote Sens. 45 (1), 206–217.

Verhoef, W., 1984. Light scattering by leaf layers with application to canopy reflectance modeling: the SAIL model. Remote Sens. Environ. 16, 125—141.

Walthall, C.L., et al., 1985. Simple equation to approximate the bidirectional reflectance from vegetative canopies and bare soil surfaces. Appl. Opt. 24 (3), 383—387.

Wang, J.F., 2000. An airborne multi-angle TIR/VNIR imaging system. Remote Sens. Rev. 19 (1—4), 161—170.

Wang, X., Zender, C.S., 2010. MODIS snow albedo bias at high solar zenith angles relative to theory and to in situ observations in Greenland. Remote Sens. Environ. 114, 563—575.

Wang, K., Liu, J., Zhou, X., Sparrow, M., Ma, M., Sun, Z., Jiang, W., 2004. Validation of the MODIS global land surface albedo product using ground measurements in a semidesert region on the Tibetan Plateau. J. Geophys. Res. 109 (D5), D05107.

Wang, D., Liang, S., He, T., et al., 2013. Direct estimation of land surface albedo from VIIRS data: algorithm improvement and preliminary validation. J. Geophys. Res. Atmos. 118 (22), 12,577—12,586.

Wanner, W., Li, X., Strahler, A.H., 1995. On the derivation of kernels for kernel-driven models of bidirectional reflectance. J. Geophys. Res. 100, 077—089.

Wen, J.G., Liu, Q., Liu, Q.H., Xiao, Q., Li, X.W., 2009. Scale effect and scale correction of land-surface albedo in rugged terrain. Int. J. Remote Sens. 30 (20), 5397—5420.

Wen, J.G., Liu, Q., Tang, Y., Dou, B.C., You, D.Q., Xiao, Q., Liu, Q.H., Li, X.W., 2015. Modeling land surface reflectance coupled BRDF for HJ-1/CCD data of rugged terrain in Heihe river basin. China. IEEE J. Sel. Top. Appl. Earth Obs. Remote Sens. 8 (4), 1506—1518.

Wen, J., Dou, B., You, D., et al., 2017. Forward a small-timescale BRDF/Albedo by multisensor combined BRDF inversion model. IEEE Trans. Geosci. Remote Sens. 55 (2), 683—697.

Wen, J., Liu, Q., Xiao, Q., Liu, Q., You, D., Hao, D., et al., 2018. Characterizing land surface anisotropic reflectance over rugged terrain: a review of concepts and recent developments. Remote Sensing 10 (3).

Wielicki, B., Barkstrom, B.R., Baum, B., et al., 1998. Clouds and the earth's radiant energy system (CERES): algorithm overview. Bull. Am. Meteorol. Soc. 36 (4), 1127—1141.

Wu, H., Li, Z.L., 2009. Scale issues in remote sensing: a review on analysis, processing and modeling. Sensors 9 (3), 1768—1793.

Wu, Y.Z., Lu, H.Y., Liu, Q., 2010. Inversion of the asymmetry factor for desert areas of China. Sci. China Earth Sci. 53 (4), 561—567. https://doi.org/10.1007/s11430-010-0026-y.

Wu, H., Liang, S., Tong, L., He, T., Yu, Y., 2012. Bidirectional reflectance for multiple snow-covered land types from MISR products. IEEE Geosci. Remote Sens. Lett. 9 (5), 994—998. https://doi.org/10.1109/LGRS.2012.2187041.

Zhang, Y.C., Rossow, W.B., Lacis, A.A., Oinas, V., Mishchenko, M.I., 2004. Calculation of radiative fluxes from the surface to top of atmosphere based on ISCCP and other global data sets: refinements of the radiative transfer model and the input data. J. Geophys. Res. 109, D19105. https://doi.org/10.1029/2003JD004457.

Zhang, X.T., Liang, S.L., Wang, K.C., Li, L., Gui, S., 2010a. Analysis of global land surface shortwave broadband Albedo from multiple data sources. J. Sel. Top. Earth Obs. Remote Sens. 3 (3), 296—305.

Zhang, R., Tian, J., Li, Z., et al., 2010b. Principles and methods for the validation of quantitative remote sensing products. Sci. China Earth Sci. 53, 741—751.

Zhou, L., Dickinson, R.E., Tian, Y., Zeng, X., Dai, Y., Yang, Z.L., Schaaf, C.B., Gao, F., Jin, Y., Strahler, A., Myneni, R.B., Yu, H., Wu, W., Shaikh, M., 2003. Comparison of seasonal and spatial variations of albedos from moderate-resolution imaging spectroradiometer (MODIS) and common land model. J. Geophys. Res. 108, 4488.

Zhou, Y., Wang, D., Liang, S., Yu, Y., He, T., 2016. Assessment of the Suomi NPP VIIRS Land Surface Albedo Data Using Station Measurements and High-Resolution Albedo Maps. Remote Sensing 8, 137.

Land surface temperature and thermal infrared emissivity

© 2020 Elsevier Inc. All rights reserved.

Abstract

Land surface temperature (LST) and land surface emissivity (LSE) determine the longwave radiation in land surface radiation and energy budgets and are the key input parameters in climatic, hydrological, ecological, and biogeochemical models. Section 7.1 provides the traditional definitions of temperature and emissivity as well as several definitions for the temperature and emissivity of heterogeneous and nonisothermal mixed pixels using remote sensing pixel scales. Section 7.2 illustrates the algorithm used to retrieve the average LST, including the retrieval algorithms designed for thermal infrared data and passive microwave data. Section 7.3 first presents two methods for LSE measurements and then introduces typical LSE retrieval algorithms. Finally, a formula for converting a narrowband emissivity into a broadband emissivity is described. Section 7.4 lists the most frequently used LST and LSE products. Section 7.5 briefly describes the fusion of thermal infrared LST and microwave LST.

7.1 The definitions of land surface temperature and land surface emissivity

For homogenous and isothermal objects, the definitions of temperature and emissivity can be formulated in accordance with those in classical physics. However, at the remote sensing pixel scale, homogenous pixels are difficult to find, except for large-area bodies of water, deserts, and rich grasslands; conversely, mixed pixels are more pervasive (Li et al., 1999). Given the heterogeneous and nonisothermal features of mixed pixels, applying the definitions of temperature and emissivity from classical physics to mixed pixels is impracticable. Thus, defining the temperature and emissivity of mixed pixels using thermal infrared remote sensing is a scientific problem that requires a solution. Below, we review the existing definitions of temperature and emissivity.

7.1.1 The definition of land surface temperature

7.1.1.1 Thermodynamic or kinetic temperature (Norman and Becker, 1995)

The actual temperature of the surface of an object can be measured with a high-precision thermometer by contacting the object. Thermodynamic temperature is a macroscopic quantity that is constant throughout any group of subsystems that are in thermodynamic equilibrium. The conditions for thermodynamic equilibrium in a system composed of several subsystems can be obtained by maximizing the energy of the entropy pairs in each subsystem. For the entire system, the equilibrium of each subsystem is achieved when the energy differentiation in each entropy pair remains unchanged:

$$\frac{\partial S}{\partial E} = \frac{1}{T} \quad (7.1)$$

where S is entropy, E is energy in J, and T is the kinetic temperature in degrees K. In the above equation, the partial differentiation of E with respect to S mainly suggests that other parameters remain unchanged in this state. Although total entropy maximization is not used to illustrate the right side of this equation, this definition is consistent with the definition of the absolute temperature of an ideal gas:

$$PV = NkT \quad (7.2)$$

where P is the air pressure, usually in kPa or atm; V is the volume, usually in L; N is the number of molecules in mol; and $k = 1.38 \times 10^{-23}$ J/K, which is the Boltzmann constant. T is the thermodynamic temperature or absolute temperature in degrees K. The thermodynamic temperature can be explained as a kinetic temperature from a statistical perspective and is defined based on the average kinetic energy of particles at the microscopic scale. The average kinetic energy of a single-atom gas particle system without rotation and vibration is expressed as follows:

$$\frac{1}{2}m<v^2> = \frac{3}{2}kT \quad (7.3)$$

where m is the particle weight, $<v^2>$ is the mean value of the square of the particle velocity, and k is the Boltzmann constant. The temperature defined above is known as the "mobile kinetic temperature"; however, "kinetic temperature" is more commonly used. A more general definition for kinetic temperature must consider rotational kinetic energy. For a more detailed discussion, please refer to Present's monograph (Present, 1958).

7.1.1.2 Brightness temperature

Brightness temperature has been widely used in ground-based thermal infrared measurements and thermal infrared remote sensing applications. When the radiance of an object equals that of a certain blackbody, the physical temperature of this blackbody is defined as the brightness temperature of the object. Brightness temperature has the dimensions of temperature but lacks the physical meaning of temperature.

7.1.1.3 Radiometric temperature (Becker and Li, 1995)

The radiometric temperature is defined by L_λ, i.e., the radiance emitted by a land surface. Let ε_λ be the land surface emissivity (LSE); R_λ is the radiation value measured by a radiometer and is then approximated as follows:

$$R_\lambda = L_\lambda + (1 - \varepsilon_\lambda)R_{u,\lambda} \quad (7.4)$$

where $L_\lambda = \varepsilon_\lambda B_\lambda(T)$ in W/m^2/sr/μm, T is the land surface radiometric temperature in degrees K, $R_{u,\lambda}$ is the atmospheric downward radiation in W/m^2/sr/μm, and B_λ is the Planck function.

$$B_\lambda(T) = \frac{2hc^2}{\pi\lambda^5\left(e^{\frac{hc}{k\lambda T}} - 1\right)} = \frac{C_1}{\lambda^5\left(e^{\frac{C_2}{\lambda T}} - 1\right)} \quad (7.5)$$

where $h = 6.626 \times 10^{-34}$ J s, $c = 2.99,793 \times 10^8$ m/s, $k = 1.3806 \times 10^{-23}$ J/K, and $C_2 = hc/k = 14388$ μm·K.

The radiometric temperature is the temperature of a blackbody with the radiance of $L_\lambda/\varepsilon_\lambda$, that is

$$T = B^{-1}[L_\lambda / \varepsilon_\lambda] = \frac{C_2}{k\ln\left(\frac{C_1}{\lambda^5(L_\lambda/\varepsilon_\lambda)} + 1\right)} \quad (7.6)$$

where B^{-1} is the inverse function of the Planck function, and ln is the natural logarithm.

7.1.1.4 Equivalent or average temperature

For heterogeneous and nonisothermal pixels, even if each subpixel is a blackbody, the overall radiation behavior of the pixel cannot be described by an isothermal blackbody in the thermal infrared domain (8–14 μm). In this range, the concept of equivalent temperature, or average temperature, should be applied.

Most of the algorithms for estimating temperature and emissivity by remote sensing do not consider the heterogeneous and nonisothermal features of land surfaces. Assuming that a land surface is isothermal or approximately isothermal in the wavelength range of the sensor, the equivalent temperature or average temperature of the pixel can be retrieved from the sensor's radiometric measurements (Gillespie et al., 1998; Wan and Dozier, 1996).

7.1.2 Definition of land surface temperature

7.1.2.1 Spectral emissivity

Spectral emissivity is the ratio of the radiation energy of the object to the radiation energy of the isothermal blackbody, as expressed by the following:

$$\varepsilon(\lambda, T) = \frac{M(\lambda, T)}{M_b(\lambda, T)} \quad (7.7)$$

where $M(\lambda, T)$ is the object's exitance in W/m^2, and $M_b(\lambda, T)$ is the isothermal blackbody's exitance. $\varepsilon(\lambda, T)$, which depicts the object's capacity for emitting thermal radiation, is determined by the object's composition and physical status. For most land surface categories, emissivity is directional. At the remote sensing pixel scale, the heterogeneous and nonisothermal features of pixels render the above definition inapplicable. As a result, LSE must be redefined on the remote sensing pixel scale.

7.1.2.2 e-Emissivity (Norman and Becker, 1995)

e-Emissivity is defined as the ratio of the total radiation of a natural object's surface to the blackbody radiation with an identical temperature distribution. When the pixel has N components,

$$\varepsilon_{e,i}(\theta, \phi) = \frac{\sum\limits_{k=1}^{N} a_k \varepsilon_{r,i,k}(\theta, \phi) T_{R,i,k}^n(\theta, \phi)}{\sum\limits_{k=1}^{N} a_k T_{R,i,k}^n(\theta, \phi)} \quad (7.8)$$

where a_k is the normalized-area proportion of component K, $\varepsilon_{r,i,k}(\theta, \phi)$ is the emissivity of each component in the direction of (θ, ϕ), and $T_{R,i,k}^n(\theta, \phi)$ is the approximation of the exponential function of blackbody radiation. Thus, e-emissivity is a function of an object's component temperatures.

7.1.2.3 r-Emissivity (Norman and Becker, 1995)

r-Emissivity can be used to express the area weighting of a component's emissivity, which has no relation with the component temperature. If a pixel has N components, then

$$\varepsilon_{r,i}(\theta, \phi) = \sum\limits_{k=1}^{N} a_k \varepsilon_{r,i,k}(\theta, \phi) \quad (7.9)$$

The above two emissivity types do not incorporate the contributions that result from multiple scattering by the components. Wan and Dozier (1996) defined two equivalent emissivity types similar to the method of Norman and Becker:

$$\overline{\varepsilon_i} = \frac{\int_{\lambda_1}^{\lambda_2} f(\lambda)[a_i \varepsilon_i(\lambda) B(\lambda, T_1) + a_2 \varepsilon_2(\lambda) B(\lambda, T_2)] d\lambda}{\int_{\lambda_1}^{\lambda_2} f(\lambda)[B(\lambda, T_1) + B(\lambda, T_2)] d\lambda} \quad (7.10)$$

and

$$\overline{\varepsilon_i} = \frac{\int_{\lambda_1}^{\lambda_2} f(\lambda)[a_1 \varepsilon_1(\lambda) + a_2 \varepsilon_2(\lambda)] d\lambda}{\int_{\lambda_1}^{\lambda_2} f(\lambda) d\lambda} \quad (7.11)$$

where $\overline{\varepsilon_i}$ is the equivalent emissivity for channel i; λ_1 and λ_2 are the upper and lower boundaries of channel i, respectively; $f(\lambda)$ is the response function for channel i; a_i is the proportional area of the component; ε_i is the component emissivity; B is the Planck function; and T_i is the temperature of the component.

7.1.2.4 Equivalent emissivity for a nonisothermal surface (Li et al., 1999)

For pixels with a 3D structure, we have

$$\varepsilon_0 = \varepsilon_{BRDF} + \Delta\varepsilon(T|T_0) \quad (7.12)$$

where $\varepsilon_{\mathrm{BRDF}} = \bar{\varepsilon} + \Delta\varepsilon_{multi}$, i.e., the emissivity at the same temperature; T_0 is the reference temperature; $\bar{\varepsilon}$ is the emissivity of the pixel material; and $\Delta\varepsilon_{multi}$ is the contribution to $\varepsilon_{\mathrm{BRDF}}$ resulting from multiple scattering among the components of the pixel. $\Delta\varepsilon(T|T_0)$ is the apparent emissivity increment resulting from the 3D structure of the pixel and the nonisothermal feature of the pixel.

7.1.2.5 Component effective emissivity

Chen et al. (2004) proposed a definition for a component's effective emissivity. For nonisothermal pixels with two components, the component effective emissivity is defined as follows:

$$\varepsilon_{e1} = a_1(\theta)\varepsilon_1(\theta) + \Delta\varepsilon_{1s}(\theta) \tag{7.13}$$

and

$$\varepsilon_{e2} = a_2(\theta)\varepsilon_2(\theta) + \Delta\varepsilon_{2s}(\theta) \tag{7.14}$$

where $a_1(\theta)$ and $a_2(\theta)$ are the area proportions of the two components; θ is the observation angle; $\varepsilon_1(\theta)$ and $\varepsilon_2(\theta)$ are the component emissivities; and $\Delta\varepsilon_{1s}(\theta)$ and $\Delta\varepsilon_{2s}(\theta)$ are regarded as emissivity increments resulting from the multiple scattering of the two components' thermal radiations, which is related strictly to the geometric structure of the object, the optical features of the components, and the observation angle and is unrelated to the component temperature. The thermal radiation of the pixel is expressed as follows:

$$L(\theta) = \varepsilon_{e1}(\theta)B(T_1) + \varepsilon_{e2}(\theta)B(T_2) \tag{7.15}$$

where T_1 and T_2 are the temperatures of the two components. Although several definitions have been proposed for the emissivities of heterogeneous and nonisothermal pixels, methods for acquiring these emissivities from a remote sensor's radiometric measurements have yet to be investigated. Cheng et al. (2008a) analyzed trends in the variation of the radiometric temperature of nonisothermal planar pixels with two components. Assuming that the radiometric temperature remains constant in the narrowband emission, the r-emissivity of the planar mixed pixels can be retrieved from the simulated thermal infrared hyperspectral data; a method

for acquiring the r-emissivity of the entire thermal infrared region is discussed below.

7.2 The estimation of average land surface temperature

Although several definitions of land surface temperature (LST) are listed in Section 7.1.1, the average temperature of a pixel is the only quantity that can be acquired by remote sensing. For conciseness, this temperature is hereinafter referred to as the LST. At the global and regional scales, LST is mainly detected by infrared and passive microwave sensors. LST detection by infrared thermal sensors provides a greater spatial resolution, ranging from 90 m for ASTER to 1 km for MODIS to several dozen kilometers for meteorological satellites. As a result of the limited penetration capacity of infrared radiation, infrared LST can be acquired only during clear sky conditions. In contrast, microwave radiation is only slightly affected by atmospheric influences, which allows LST to be acquired in all weather conditions. Compared with infrared sensors, passive microwave sensors have a lower spatial resolution and lower precision. However, neither of these two types of sensors can achieve the 1K retrieval precision required in practical applications (Wang et al., 2008). For example, the MODIS global LST product is the most reliable LST product, with a precision of 1K verified only for homogenous surfaces, such as water bodies and sandy land (Wan et al. 2002, 2004). For global land surfaces, homogeneity on a scale of 1 km is extremely rare. Thus, the retrieval precision for microwave sensors is far lower than 1K.

Table 7.1 shows the channels, spectral ranges, and algorithms used for LST retrieval by typical infrared sensors. Depending on the channel(s) used by the sensors, there are three LST retrieval algorithms that may be applied: the single-channel algorithm, the split-window algorithm, and the multichannel algorithm. Table 7.2 lists the feature parameters of passive microwave sensors relevant for LST retrieval.

TABLE 7.1 Typical infrared sensors for land surface temperature (LST) retrieval.

Sensor	Channel(s)	Spectral range (μm)	Algorithm
ETM+/ Landsat 7	6	10.4−12.5	Single-channel algorithm[a]
AVHRR/ NOAA	3, 4, 5	3.55−3.93 10.30−11.30 11.50−12.50	Split-window algorithm[b] TISI algorithm[c]
MODIS/ EOS	20, 22, 23, 29, 31, 32, 33	3.66−3.84 3.929−3.989 4.02−4.08 8.4−8.7 10.78−11.28 11.77−12.77 13.185−13.485	Split-window algorithm[b] Day/night algorithm[c]
ASTER/ EOS	10, 11, 12, 13, 14	8.125−8.475 8.475−8.825 8.925−9.275 10.25−10.95 10.95−11.65	TES algorithm[d]
AATSR/ ENVISAT	6, 7	Central wavelength: 10.85 and 12.0 Channel width: 0.9 and 1.0	Split-window algorithm[b]
ABI/ GOES-R	14, 15	Central wavelength: 11.2 and 12.3	Split-window algorithm[b]
SEVIRI/ MSG	9, 10	Central wavelength: 10.8 and 12.0	Split-window algorithm[b]
IRMSS/ CBRES	9	10.4−12.5	Single-channel algorithm[a]
MERSI/FY-5 3		Central wavelength: 11.25	Single-channel algorithm[a]
IRS/HJ-1	4	10.5−12.5	Multichannel algorithm[c]
S-VISSR/ FY-2	IR1, IR2	10.3−11.3 11.5−12.5	Split-window algorithm[a]

TABLE 7.1 Typical infrared sensors for land surface temperature (LST) retrieval.—cont'd

Sensor	Channel(s)	Spectral range (μm)	Algorithm
VIRR/FY-3	3, 4, 5	3.55−3.93 10.3−11.3 11.5−12.5	Split-window algorithm[b] TISI algorithm[c]

[a] please refer to Section 7.2.1.
[b] refer to Section 7.2.2.
[c] refer to Section 7.2.3.
[d] refer to Section 7.3.4.

TABLE 7.2 Passive microwave sensors for land surface temperature (LST) retrieval.

Sensor	Frequency (GHz)	37 GHz spatial resolution (km)	Width (km)	Launch date
SMMR (Nimbus-7)	6.6, 10.7, 18, 21, 37	18 × 27	780	1978
SSM/I (DMSP)	19.35, 22.235, 37, 85.5	28 × 37	1400	1987
AMSR-E (EOS Aqua)	6.925, 10.65, 18.7, 23.8, 36.5, 89	8 × 14	1445	2002

7.2.1 Single-channel algorithms

The spectral radiance received by a thermal infrared sensor in the direction of (θ_r, φ_r) can be expressed as follows:

$$L_\lambda(\theta_r, \varphi_r) = \left(\begin{array}{l} \varepsilon_\lambda(\theta_r, \varphi_r)B_\lambda(T_s) + \\ \displaystyle\int_{2\pi} \rho_{b,\lambda}(\theta_i, \varphi_i, \theta_r, \varphi_r)L_{\downarrow,\lambda}(\theta_i, \varphi_i)\cos\theta_i d\Omega_i + \\ \rho_{b,\lambda}(\theta_i, \varphi_i, \theta_r, \varphi_r)E_{sun,\lambda}(\theta_s) \end{array} \right)$$
$$\tau_{\uparrow,\lambda}(\theta_r, \varphi_r) + L_{\uparrow,\lambda}(\theta_r, \varphi_r)$$

$$(7.16)$$

where θ_r and φ_r are the zenith angle and azimuth angle of the sensor, respectively, and θ_i and φ_i are the zenith angle and azimuth angle of the downward atmospheric radiance, respectively. $\varepsilon_\lambda(\theta_r, \phi_r)$ is the LSE, and $B_\lambda(T_s)$ is the blackbody radiation at the temperature T_s. $\rho_{b,\lambda}(\theta_i, \phi_i, \theta_r, \phi_r)$ is the BRDF of the land surface, θ_s is the solar zenith angle, $\tau_{\uparrow,\lambda}(\theta_r, \varphi_r)$ is the transmittance of the entire atmosphere from the land surface to the sensor, and $L_{\uparrow,\lambda}(\theta_r, \varphi_r)$ is the atmospheric upward radiance. Scattered solar radiation is not considered in this case. The spectral resolution capacity of the sensor is limited, and each channel has a certain width. Through the convolution of Eq. (7.16) with a channel response function, an expression for the spectral radiance in each channel is obtained. Assuming that the land surface is Lambertian and ignores the solar irradiance, the radiative transfer equation for the thermal infrared is given as follows:

$$L_i = \left[\varepsilon_i B_i(T_s) + (1 - \varepsilon_i)L_{d,i}\right]\tau_i + L_{u,i} \quad (7.17)$$

where $L_{u,i}$ and $L_{d,i}$ are the atmospheric upward radiance and atmospheric downward radiance, respectively (or, hereinafter, the atmospheric upward radiation and atmospheric downward radiation).

LSE and the atmospheric parameters ($L_{u,i}$, $L_{d,i}$, and τ_i) must be known before LST can be solved from L_i. For sensors with only a thermal infrared channel such as TM/ETM+, retrieving LST using only radiometric measurements is difficult. In ideal conditions, the LSE and the atmospheric temperature and humidity profile are also known. Therefore, the LST can be solved according to Eq. (7.17) based on atmospheric parameters that are obtained from a radiative transfer equation. This method is thus known as the radiative transfer equation method. In practice, however, obtaining the atmospheric temperature and humidity profiles to determine LSE is difficult. The radiative transfer equation method, the single-channel algorithm, and generalized single-window algorithm are among various parameterization schemes that can be used to determine the atmospheric parameters and LSE.

7.2.1.1 The radiative transfer equation method

The three atmospheric parameters in Eq. (7.17), namely, upward radiance, downward radiance, and transmittance, can be simulated using an atmospheric transfer equation, provided that the atmospheric temperature and humidity profiles are known. Assuming that the LSE is known, the LST can be obtained by an inversion of the Planck function. However, the real-time atmospheric temperature and humidity profiles that are required for atmospheric simulations are generally unavailable; thus, reanalysis products are used as surrogate radiosondes to perform atmospheric corrections of the thermal infrared band. For example, Barsi et al. (2003) and Tardy et al. (2016) developed the atmospheric correction tools for the Landsat satellite series using the National Centers for Environmental Prediction (NCEP) and the European Centre for Medium-Range Weather Forecasts (ECMWF) Interim Reanalysis (ERA-Interim), respectively. Meng and Cheng (2018) compared the accuracy of eight global reanalysis products, namely, NCEP/FNL, NCEP/DOE Reanalysis2, MERRA-3, MERRA-6, MERRA2-3, MERRA2-6, JRA-55, and ERA-Interim, which provides a guide for selecting the proper reanalysis products for thermal infrared atmospheric correction. Their results indicated that MERRA-6 and ERA-Interim are more accurate than the other reanalysis products for different water vapor contents and surface elevations. The overall biases and root mean square errors (RMSEs) between simulated LSTs and actual LSTs are all less than 0.2 and 1.1K.

Moreover, Meng et al. (2017) proposed a new algorithm to estimate LSE, which is used for improving the accuracy of FY-3C/MERSI LST derived by the radiative transfer equation method. The new algorithm includes three parts: a lookup table (LUT)—based method to determine the emissivity of a vegetated surface, an empirical method to derive the emissivity of bare soil from the Global LAnd Surface Satellite (GLASS) broadband emissivity (BBE) product, and the angular dependent atmospheric correction. The details are as follows:

(1) The three input parameters of the LUT are leaf emissivity, soil emissivity, and LAI. Leaf emissivity was calculated from the measurements by the methods of Pandya et al. (2013) and Wang et al. (2012) and the ASTER and MODIS spectral library for five composited vegetation land cover types based on MCD12Q1. The emissivity of the soil background underneath the vegetation canopy was estimated from the mean GLASS BBE in the nonvegetation cover season. LAI was obtained from the GLASS LAI product.

(2) The regression coefficients between GLASS BBE and FY-3C/MERSI emissivity for bare soils are determined by emissivity spectra from the ASTER spectral library, MODIS spectral library, and the measured soil emissivity from Wang et al. (2012).

(3) The SeeBor V5.0 profiles were used to establish the relationships between three atmospheric parameters (atmospheric transmittance, atmospheric upward radiance, and atmospheric downward radiance) observed at different view zenith angles and three atmospheric parameters observed at the nadir.

The validation result shows that the absolute bias of LST is less than 1K, and the standard deviation and RMSE of LST are both less than 1.95K, which indicates that the new LSE algorithm improves the accuracy of LSE and, therefore, the LST for sensors with broad spectral ranges.

7.2.1.2 The single-channel algorithm

Qin and Karnieli (2001) proposed a single-window algorithm for retrieving LST from TM data. Based on the mean value theorem (McMillin, 1975), the average atmospheric temperature T_a is introduced to approximately express the atmospheric upward radiance and the atmospheric downward radiance. Assuming that the average temperature of the atmospheric upward radiance and the average temperature of the atmospheric downward radiance are equal, a linear approximation of the Planck function can be calculated at room temperature. An expression for the LST is then obtained:

$$T_s = [a(1 - C - D) + (b(1 - C - D) + C + D)T_6 - DT_a]/C \qquad (7.18)$$

where T_6 is the brightness temperature of the sixth TM channel, $a = -67.355351$, $b = 0.458606$, $C = \varepsilon_6 \tau_6$, and $D = (1 - \tau_6)[1 + \tau_6(1 - \varepsilon_6)]$. ε_6 and τ_6 are the LSE and atmospheric transmittance of the sixth channel, respectively.

This algorithm requires only three parameters, the LSE, atmospheric transmittance, and average temperature, and the last two can be estimated from the atmospheric temperature and humidity profiles or the observations at meteorological stations. The shortcoming of this algorithm is that the determination of the empirical equations for estimating the atmospheric transmittance and the average temperature use only the standard atmospheric profile data. Because standard atmospheric profile data represent the statistical results of large samples, the actual atmospheric conditions cannot be reflected, which limits the applicability of this algorithm.

7.2.1.3 Generalized single-window algorithms

Jiménez-Muñoz and Sobrino (2003) proposed a generalized single-window algorithm that can be applied for LST retrieval from any type of thermal infrared data or from TM6 data. In this algorithm, LST is expressed as follows:

$$T_s = \gamma \left\{ \frac{1}{\varepsilon} \left[\varphi_1 L_\lambda^{at-sensor} + \varphi_2 \right] + \varphi_3 \right\} + \delta \qquad (7.19)$$

$$\gamma = \left\{ \frac{c_2 B_\lambda(T_o)}{T_o^2} \left[\frac{\lambda^4}{c_1} B_\lambda(T_o) + \lambda^{-1} \right] \right\}^{-1} \qquad (7.20)$$

and

$$\delta = -\gamma B_\lambda(T_o) + 1 \qquad (7.21)$$

where $L_\lambda^{at-sensor}$ is the at-sensor radiance, ε is the land surface emissivity, λ is the equivalent wavelength, $C_1 = 1.19104 \times 10^8 \, W/m^2/sr/\mu m^4$, and $C_2 = 14388 \, \mu m \cdot K$. T_o is the reference temperature, which is usually the brightness temperature corresponding to the at-sensor radiance. φ_1, φ_2, and φ_3 are simple functions of atmospheric water content w, and their expressions can be obtained by fitting.

Compared with the single-channel algorithm, the input parameters of the generalized single-window algorithm are only the LSE and atmospheric water content, thereby making this algorithm more attractive. Jiménez-Muñoz et al. (2009) modified this algorithm by (1) reintroducing four atmospheric profile databases while regressing atmospheric function coefficients and (2) recalculating the atmospheric function coefficients for Landsat 4, 5, and 7. The modified algorithm was evaluated using simulated data. The results indicated that for atmospheric water contents in the range of 0.5–2 g/cm², satisfactory results were produced with this algorithm, with an error of 1–2K. However, for an atmospheric water content of 3 g/cm², the error became unacceptable. Cristóbal et al. (2009) made further improvements to the generalized single-channel algorithm proposed by Jiménez-Muñoz and Sobrino. When computing the atmospheric functions φ_1, φ_2, and φ_3, a near-surface temperature was incorporated as an input parameter. Landsat TM/ETM+ data were used to verify the improved algorithm. The results revealed that the simultaneous use of atmospheric water content and the near-surface temperature to retrieve LST increased the retrieval precision. RMSE reached 0.9K, whereas the RMSE was 1.5K for the algorithm in which only the atmospheric water content was involved.

Precision is the primary concern for the users of all three algorithms. Sobrino and Romaguera (2004) conducted a comparative analysis of these three algorithms and found that when real-time atmospheric profile data were used, the RMSE of the radiative transfer equation method was 0.6K, whereas the RMSE values for both the single-window algorithm and generalized single-channel algorithm were 0.9K. When no real-time atmospheric profile data were involved, the radiative transfer equation method was no longer applicable; the RMSEs of the single-window algorithm and the generalized single-channel algorithm were 2 and 0.9K, respectively. Duan et al. (2008) evaluated two single-channel algorithms using simulated HJ-1B data and found that the generalized single-channel algorithm is superior to the single-window algorithm. The FY-3C/MERSI LST validation results of Meng et al. (2017) indicated that the accuracy of the radiative transfer equation method is slightly higher than the generalized single-channel algorithm, with their RMSE values being 1.77 and 1.89K, respectively.

7.2.2 Split-window algorithms for thermal infrared sensors

McMillin (1975) proposed a split-window algorithm for sea surface temperature estimations from remote sensing data. By taking full advantage of the different atmospheric absorption features of the two channels in the atmospheric window (10.5–12.5 μm), the atmospheric influence is removed through a combination of brightness temperatures (mainly linear) in the two channels. This algorithm includes the following basic assumptions: (1) seawater can be approximated as the blackbody with an emissivity equal to 1; (2) the atmospheric window absorption is weak, and the water vapor absorption coefficient is approximately constant; and (3) the Plank function can be approximated with a first-order Taylor series expansion around the central wavelength. The typical split-window algorithm can be expressed as follows:

$$T_s = a_0 + a_1 T_i + a_2 T_j \qquad (7.22)$$

where a_i ($i = 1, 2$) is a coefficient, and T_i and T_j are the brightness temperatures of the two channels. Because of its simple form, this split-window algorithm is frequently applied as an empirical equation. In fact, the derivation of the split-window algorithm is highly complex. For the detailed derivation process, please refer to Prata et al. (1995). Based on the a priori knowledge of the sea surface temperature and the atmospheric temperature and humidity profiles, large quantities of samples are randomly generated; the brightness temperature of the corresponding channel is simulated using an atmospheric radiative transfer equation (such as LOWTRAN, MODTRAN), and the coefficients in Eq. (7.22) are then obtained through statistical regression. The regression coefficients derived by different researchers vary slightly (Wan and Dozier, 1996).

Because of the homogenous nature of seawater, its emissivity is relatively stable and approaches unity. The split-window algorithm has been successfully applied to sea surface temperature retrievals with a retrieval precision of 0.3K (Niclos et al., 2007). Many researchers have also applied the split-window algorithm in land surface temperature retrievals. Compared with sea surfaces, land surfaces are much more complicated. As a result of the 3D structure of land surfaces, numerous factors with considerable uncertainty affect the temporal and spatial variation of LSE; therefore, a land surface cannot be approximated as a blackbody. Various equations for split-window algorithms have been derived by numerous researchers for use in LST retrievals (Becker and Li, 1990b; Caselles et al., 1997; Price, 1984; Sobrino et al. 1994; Ulivieri et al., 1992; Vidal, 1991; Wan and Dozier, 1996; Yu et al., 2009). However, only a few typical algorithms are briefly introduced in this section for the sake of brevity.

Price (1984) was the first to apply the sea surface temperature split-window algorithm to land surface temperature retrievals. In this algorithm, the land surface temperature is expressed as follows:

$$T_s = [T_4 + 3.33(T_4 - T_5)]\left(\frac{3.5 + \varepsilon_4}{4.5}\right)$$
$$+ 0.75 T_5(\varepsilon_4 - \varepsilon_5) \tag{7.23}$$

where T_4 and T_5 are the brightness temperatures of the fourth and fifth channels of the AVHRR, respectively, and ε_4 and ε_5 are the corresponding LSEs.

Becker and Li (1995) introduced the atmospheric water content (the total water content of the atmospheric column) into their previously proposed local split-window algorithm (Becker and Li, 1990b), making this algorithm applicable to the majority of atmospheric conditions.

$$T_s = A_0 + P\frac{T_4 + T_5}{2} + M\frac{T_4 - T_5}{2} \tag{7.24}$$

$$A_0 = -7.49 - 0.407w \tag{7.25}$$

$$P = 1.03 + (0.211 - 0.031 \cos\theta \cdot w)(1 - \varepsilon_4)$$
$$- (0.37 - 0.074w)(\varepsilon_4 - \varepsilon_5) \tag{7.26}$$

$$M = 4.25 + 0.56w + (3.41 + 1.59w)(1 - \varepsilon_4)$$
$$- (23.58 - 3.89w)(\varepsilon_4 - \varepsilon_5) \tag{7.27}$$

Aiming at retrieving land surface temperatures from MODIS data, Wan and Dozier proposed a generalized split-window algorithm (Wan and Dozier, 1996). Compared with other split-window algorithms for land surface temperature retrievals, this generalized split-window algorithm is an operational algorithm that has been applied for generating MODIS global LST products and has been extensively verified.

$$T_s = C + \left(A_1 + A_2\frac{1-\varepsilon}{\varepsilon} + A_3\frac{\Delta\varepsilon}{\varepsilon^2}\right)\frac{T_{31} + T_{32}}{2}$$
$$+ \left(B_1 + B_2\frac{1-\varepsilon}{\varepsilon} + B_3\frac{\Delta\varepsilon}{\varepsilon^2}\right)\frac{T_{31} - T_{32}}{2}$$

$$(7.28)$$

where A_i, B_i, and C are coefficients; $\varepsilon = (\varepsilon_{31} + \varepsilon_{32})/2$ and $\Delta\varepsilon = \varepsilon_{31} - \varepsilon_{32}$; and ε_{31} and ε_{32} are the emissivities of MODIS channels 31 and 32, respectively.

The split-window algorithms cited above are not an inclusive list of all reported algorithms. For a specific thermal infrared sensor, determining the algorithm that is the most widely applicable with a higher precision can be achieved only through numerous ground verifications and comparisons. Yu et al., in the process of developing an LST retrieval algorithm for ABI/GOES-R, analyzed the precision of nine split-window algorithms using simulated data. Their results indicated that the split-window algorithm developed by Ulivieri et al. (1992) has a simple form, is insensitive to variations in emissivity and water vapor, and proved to be the optimal algorithm for ABI/GOES-R. In addition, matched ground-measured temperatures from SURFRAD and GOES-8 data were used to evaluate the optimal algorithm, and the results showed that this algorithm achieved the precision of 2.3K required by the GOES-R program (Yu et al., 2009).

7.2.3 Multichannel algorithms

For sensors with multiple thermal infrared channels, such as MODIS, ASTER, and AVHRR, LST and LSE can be estimated by multichannel algorithms. The algorithms introduced in this section can retrieve LSE and LST simultaneously. The LSE estimation methods introduced below (Section 7.3) can also be used to estimate the LST because only one unknown (LST) remains after the LSE is determined.

7.2.3.1 The temperature-independent spectral index method

Becker and Li defined temperature-independent spectral indices (TISI), which are used to retrieve LST and LSE from the day and night observations on the NOAA AVHRR channels 3, 4, and 5 (Becker and Li, 1990a). For night observations, the TISI are defined as follows:

$$\text{TISI}_n = M\frac{L_3(T_{g3n})}{L_4(T_{g4n})^{\alpha_4} L_5(T_{g5n})} \qquad (7.29)$$

where $L_3(T_{g3n})$, $L_4(T_{g4n})$, and $L_5(T_{g5n})$ are the radiances in the night observations in AVHRR channels 3, 4, and 5, respectively, and T_{g3n}, T_{g4n}, and T_{g5n} are the corresponding land surface brightness temperatures. M is a known constant, and α_4 is used to eliminate the influence of LST on the TISI. For their specific expressions, please refer to their respective references. Note that the following assumption is made:

$$\text{TISI}_n \cong \text{TISIE}_n = \frac{\varepsilon_3}{\varepsilon_4^{\alpha_4}\varepsilon_5} \qquad (7.30)$$

where ε_3, ε_4, and ε_5 are the emissivities of AVHRR channels 3, 4, and 5, respectively.

For day observations, the TISIs are defined as follows:

$$\text{TISI}_d = M\frac{L_3(T_{g3d})}{L_4(T_{g4d})^{\alpha_4} L_5(T_{g5d})} \qquad (7.31)$$

where $L_3(T_{g3n})$, $L_4(T_{g4n})$, and $L_5(T_{g5n})$ are the radiances in the day observations in AVHRR channels 3, 4, and 5, respectively, and T_{g3n}, T_{g4n}, and T_{g5n} are the corresponding land surface brightness temperatures. In day observations, the solar radiation and the land surface thermal radiation in channel 3 are of the same order of magnitude and must be considered. Multiple scattering is neglected. The radiance of channel 3 is expressed as follows:

$$L_3(T_{g3d}) = D_3(T_{g3d}) + \rho_3(\theta_s, \theta)R_{g3}^s(\theta_s)\cos(\theta_s)$$

$$(7.32)$$

where $D_3(T_{g3d})$ is the observed radiance of channel 3 without the contribution of solar radiation, $\rho_3(\theta_s, \theta)$ is the land surface bidirectional reflectance, and $R^s_{g3}(\theta_s)$ is the solar radiation reaching the land surface. The TISI in day observations are expressed as follows:

$$\text{TISI}_d = \text{TISIE}_d + M \frac{\rho_3(\theta_s, \theta) R^s_{g3}(\theta_s) \cos(\theta_s)}{L_4(T_{g4d})^{\alpha_4} L_5(T_{g5d})}$$

(7.33)

Assuming that the field TISIs in day observations are constant, i.e., $\text{TISIE}_e = \text{TISIE}_d$, then the land surface bidirectional reflectance is obtained. The directional emissivity of the channel is then obtained from calculations based on Kirchhoff's law:

$$\rho_3(\theta_s, \theta) = \frac{(\text{TISI}_d - \text{TISI}_n) L_4(T_{g4d})^{\alpha_4} L_5(T_{g5d})}{M R^s_{g3}(\theta_s) \cos(\theta_s)}$$

(7.34)

and

$$\varepsilon_3(\theta) = 1 - \frac{\pi \rho_3(\theta_s, \theta)}{f_3(\theta_s, \theta)}$$

(7.35)

where $f_3(\theta_s, \theta)$ is an angular form factor. The directional emissivities of channels 4 and 5 can be calculated from of the emissivity of channel 3. The emissivities of channels 4 and 5 are then used in the local split-window algorithm (Becker and Li, 1990b) to calculate the LST.

Several researchers including Li (Li et al., 2000; Nerry et al., 1998; Petitcolin and Vermote, 2002), improved this algorithm, making it more theoretically sound. However, this algorithm is excessively complex and involves many assumptions, which has constrained its promotion and widespread application.

7.2.3.2 The MODIS day/night algorithm

Wan and Li (1997) proposed a physical algorithm to simultaneously retrieve LST and LSE using MODIS day and night observations, making a day/night algorithm operational for the first time. The key assumptions involved in this algorithm are as follows:

(1) the LSEs are identical in day and night observations, and the land surface is Lambertian;
(2) the bidirectional reflectance factors in the midinfrared thermal channels are identical; and
(3) the MODIS atmospheric sounding channels and the corresponding retrieval algorithms provide the atmospheric temperature and humidity profiles; the profile shapes are accurate and can be described with two parameters (viz., the temperature and water content of the lower layer of the atmosphere).

Based on day/night observations of the 7 infrared MODIS channels (channels 20, 22, 23, 29, 31, 32 and 33), 14 equations are constructed for the solution of 14 land surface and atmospheric parameters (namely, the daytime and nighttime LSTs, the LSEs of the 7 channels, the daytime and nighttime temperatures and water vapor contents of the lower layer of the atmosphere, and the bidirectional reflectance factor).

Theoretically, if the equation set is adequately posed, a unique solution exists. In actual retrieval, however, there are many uncertain factors affecting the precision of the algorithm. With 14 equations to be solved, the computation process is especially complex; moreover, a solution is possible only when an atmospheric model is used to determine several parameters. There are, however, substantial variations in the climate of a given area between the day and night observations; e.g., there may be clouds at night although the sky is clear in the daytime. Additionally, because of variations in satellite orbits, geometric corrections are required to match the two scenes in day and night observations. However, errors inevitably arise as a result of changes in pixel values caused by their resampling during geometric corrections.

7.2.3.3 *The integrated retrieval algorithm*

Ma et al. (2000) proposed an integrated retrieval algorithm for the MODIS airborne simulator (MAS), also known as the two-step retrieval algorithm, which was then applied to MODIS data (Ma et al., 2002). There are 50 channels in the range of 0.47–14.17 μm for MAS, of which 19 channels are identical to those used by MODIS; 11 channels with wavelengths greater than 3 μm are identical to those of MODIS. The integrated retrieval algorithm allows for the simultaneous retrieval of atmospheric temperature and LSE and LST humidity profiles.

Assuming that the LSE remains constant within the wavelength ranges of 3–5 μm and 8–14.5 μm, the atmospheric temperature and humidity profile can be discretized into 40 layers. The number of unknown parameters to be retrieved plus the LST total 83. However, MAS has only 20 channels (MODIS has 16), resulting in an underdetermined equation set. By adopting eigenvectors to represent the temperature and humidity profiles (Smith and Woolf, 1976), these profiles can be represented by three and five vectors, respectively. Through this method, the number of parameters to be retrieved is reduced from 83 to 11, and the previously underdetermined set is converted into an overdetermined set.

A two-step solution method is then used in the parameter retrieval. In the first step, the initial values of the parameters to be retrieved are obtained by a regression method; in the second step, the initial values are adjusted using the Tikhonov regularization method (Hansen, 1998) to obtain a regularized solution. The Newton method is then applied in a further correction of the regularized solution to obtain the final result.

The results in the literature (Ma et al. 2000, 2002) indicate that the integrated retrieval algorithm successfully retrieves atmospheric temperature and humidity profiles, the LSE and LST. As a result of the limited information content provided by the observation data and the complexity of the atmosphere and the land surface, this algorithm cannot be applied to generate LST products on a global scale; however, it has made considerable progress in terms of structuring retrieval algorithms.

7.2.3.4 *Algorithms for hyperspectral data from meteorological satellites*

To date, spaceborne hyperspectral satellite data for acquiring LST and thermal infrared emissivity remain unavailable. Spaceborne hyperspectral thermal infrared sensors that detect temperature and humidity profiles and atmospheric composition provide operational products. For nadir-viewing sensors, the received thermal radiation signals come from land surface and atmospheric thermal radiation. Interference information should be isolated during retrievals of land surface or atmospheric information, both of which are unknown. For this reason, the simultaneous retrieval of land surface parameters and atmospheric parameters is often chosen; thus, LST and LSE are usually regarded as by-products of hyperspectral meteorological satellites.

In terms of algorithms, statistical regression (Smith and Woolf, 1976) is usually employed to obtain the initial values of LST, LSE, and atmospheric conditions (temperature and humidity profiles and atmospheric composition). An iteration method is then employed for optimization, thereby obtaining an optimal estimate for each parameter. Susskind et al. presented algorithms for calculating the land surface parameters and atmospheric parameters for AIRS/AMSU/HSB data (Susskind et al., 2003), and Li and Li (2008) presented a physical algorithm for the simultaneous retrieval of LST, LSE, and temperature and humidity profiles from AIRS data.

7.2.4 Microwave methods

With full cloud coverage or partial cloud coverage, thermal infrared remote sensing cannot acquire LST information. In contrast, passive microwave sensors can penetrate clouds and are subject to little atmospheric interference. Based on the Planck blackbody radiation principle and the Rayleigh—Jeans approximation, the microwave radiance of ordinary surface features has a simple linear relationship with the real temperature. In this sense, utilizing passive microwave data to retrieve LST is essential.

The PMW radiations observed by the satellite sensors can be described by the following radiative transfer equation:

$$B_p(T_{bp}) = \tau_p e_p B_p(T_s) + (1 - e_p)\tau_p B_p(T_{a\ down})$$
$$+ B_p(T_{a\ up})$$

$$(7.36)$$

where subscript p represents the frequency; T_{bp} and T_s are the brightness temperature and LST in K, respectively; $B_p(T_{bp})$ and $B_p(T_s)$ are the radiation at the sensor and the radiation at the land surface in W/m^2/sr/Hz, respectively; τ_p is the atmospheric transmissivity; e_p is the land surface emissivity; T_a is the average atmospheric temperature; and $B_p(T_{a\cdot down})$ and $Bp(T_{a\ up})$ are the downward and upward atmospheric radiations in W/m^2/Sr/Hz, respectively. According to the Rayleigh—Jeans approximation, Eq. (7.36) can be rewritten as follows:

$$T_{bp} = \tau_p e_p T_s + (1 - e_p)\tau_p T_{a\ down} + T_{a\ up} \quad (7.37)$$

Eq. (7.37) clearly shows the four parameters that are necessary for deriving LST, namely, τ_p, ep, $T_{a\ down}$, and $T_{a\ up}$. For low-frequency channels, the atmospheric effects are negligible, and the LST can be calculated by dividing the brightness temperature by emissivity. However, the effects of the atmosphere should be taken into account in higher frequency channels to obtain more accurate LST estimations.

However, in microwave frequencies, factors affecting LSE are complex and difficult to determine, resulting in dramatic temporal and spatial variations in the LSE. Because of the low spatial resolution of microwave remote sensors, the acquisition of ground-measured data is very difficult. Furthermore, little research has been performed on the surface self-emission mechanism of microwaves. Accordingly, few LST retrieval algorithms have been developed for passive microwave data. Most of the algorithms apply the linear or nonlinear combinations of single or multi-microwave channels. In the following paragraphs, LST retrieval algorithms for the sensors listed in Table 7.2 are briefly reviewed.

Owe and Van de Griend (2001) found that in semiarid districts, the 37 GHz vertically polarized SMMR brightness temperature has a strong linear relationship with the LST and can therefore be applied to LST retrieval. Guha and Lakshmi (2004) used 6.6, 10.7, and 18 GHz SMMR data to retrieve soil moisture and LST in the middle and southern part of the United States. A comparison was made with NCEP reanalysis products, and the results indicated that the SMMR data allowed for qualitative predictions of the seasonal cycle of hydrological changes of land surfaces.

Mcfarland et al. (1990), using 19 and 37 GHz mean polarization difference data and the brightness temperature difference for the 85v and 37v channels, classified land surfaces into cropland/rangeland, wet soil, and dry soil. Using statistical regression, the LST was effectively approximated for the three soil types. Weng and Grody (1998) developed a physical algorithm for LST retrieval from the SSM/I 19.35 and 22.23 GHz brightness temperatures. This algorithm adopted adjacent channels to eliminate the variation in LSE, so it eliminated the need for information on land surface classifications. Test results with simulation data showed that the RMSE of the LST retrieval was 3.8K, and the RMSE for the SSM/I data was 4.4K. Fily et al. (2003) found that horizontally polarized and vertically polarized emissivities of snow surfaces and iceless land surfaces (at 19 and 37 GHz, respectively, for SSM/I) displayed strong linear relationships. The authors derived an expression for LST retrieval from

SSM/I data for the subpolar regions of Canada. A comparison of the simultaneously measured air temperature, the LST and the temperature retrieved from infrared sensor data indicated that the RMSE ranged from 2 to 3.5°C, and the deviation was approximately 1−3°.

Based on the land surface and atmospheric radiative transfer equations, Njoku and Li (1999) proposed an algorithm for retrieving land surface parameters (soil moisture, vegetation water content, and land surface temperature) from AMSR 6−18 GHz data. Except for bare soil, the algorithm achieved a precision of 2° for LST retrieval when tested with simulation data. When the algorithm was applied to SMMR data in the Sahel, Africa, the standard deviation between the retrieval results and the output of the NCEP model was 2.7°. Gao et al. (2008) retrieved the LST from AMSR-E data for the Amazon rainforest. The retrieved LST had a strong correlation with the LST measured at the meteorological stations. Taking AMSR-E as an example, the following methods introduce the process of establishing the LST retrieval algorithm for microwave sensors based on empirical statistical models:

Holmes et al. (2009) selected 17 sites that represent different vegetation and climate types from the FLUXNET flux tower network around the world. These sites have high-quality observations of longwave radiation, sensible heat flux, and near-surface air temperature. The sensible heat flux observations of the site in 2005 were used to determine the typical longwave emissivity of different vegetation type zones, and then the LST of each site was calculated by Stefan—Boltzmann's law. With the premise that the 37 GHz vertically polarized brightness temperature of AMSR-E is more suitable for retrieving LST than the other microwave channels, a simple linear regression between the 37 GHz vertically polarized brightness temperature ($T_{B,\ 27V}$) and the LST (T_s) was established to retrieve the LST for the 17 vegetation zones. The equation is as follows:

$$T_s = 1.11 T_{B,27V} - 15.2, \quad T_{B,27V} > 259.8 \quad (7.38)$$

The parameters such as water content, single scattering albedo, vegetation canopy roughness, and wilting point were then input into the radiation transmission model to evaluate the accuracy of this algorithm. The accuracy of derived T_s can reach 2K in the forest area and 3.5K in the sparse vegetation area, respectively.

Few studies have been conducted on LST retrieval from passive microwave data, and the precision of their LST retrievals from passive data was low compared with using thermal infrared data. To fully utilize the advantages of passive microwave data in LST retrievals, the microwave radiation mechanism requires further extensive study.

7.3 LSE estimation methods

For most land surface types in the natural world, emissivity is directional. Indirect measurement methods can measure emissivities at a zenith angle of 10°. Direct measurement methods can theoretically measure emissivities at zenith angles of 0−90° because of their greater flexibility. In LSE remote sensing retrievals, land surfaces are generally assumed as Lambertian, and directionality is neglected.

7.3.1 Emissivity measurement methods

There are two categories of emissivity measurement methods: indirect and direct measurement methods. Indirect measurement methods (Salisbury et al., 1994), based on Kirchhoff's law (for opaque objects, $\varepsilon_\lambda = 1 - \rho_\lambda$, where ρ_λ is the directional hemispheric reflectance), can obtain the emissivity in infrared channels by measuring ρ_λ. The measurement of directional hemispheric reflectance requires an active radiation source, and bidirectional reflectance in each direction is measured in a 2π space. The directional hemispheric reflectance is then obtained from the bidirectional reflectance using an integrating sphere. In laboratory conditions, the use of an integrating sphere based on the reciprocity principle can easily realize an approximate measurement of ρ_λ. Because of the limitations in working conditions and dimensions, integrating spheres are currently only applicable to the measurements of small-scale specimens rather than the emissivities of targets in natural conditions. Indirect measurement methods can achieve high-precision emissivities and are therefore frequently applied in measuring the precision of emissivities obtained through other methods. Earlier work by Salisbury et al. (1994) indicated that the absolute precision of an emissivity obtained by indirect measurement is less than 0.01. Infrared spectrum information for surface features in the ASTER spectral library (Baldridge et al., 2009) and the MODIS UCSB spectral library (Snyder and Wan, 1998) were obtained through indirect measurements.

The principle of direct measurement methods is based on the definition of emissivity: $\varepsilon_\lambda = L_\lambda / L_{b\lambda}(T_s)$. Here, only the object surface temperature is required for calculating emissivity. However, with this method, the following two problems must be resolved:

(1) The precise acquisition of object surface temperature. The conventional contact temperature measurement breaks the thermal equilibrium mechanism of the original surface through the contact of a temperature-sensing element with the object surface, resulting in considerable error.

(2) Because the target object has a temperature similar to the surrounding environment, part of the environmental radiation reflected by the object, together with the radiation from the target itself, is sensed by the sensor. Therefore, this part of the radiation must be subtracted from the radiance measured by the sensor.

The "emissivity box" method has been extensively applied since its introduction. In this method, a bottomless box is used to cover the target; the inside wall of the box has a high reflectance, and two types of materials with different properties are used to manufacture the cover plate. One material has a high reflectance, whereas the other has a high emissivity. By the alternate use of the two types of cover plates, the environmental radiation is varied, thereby achieving the solution of the thermal infrared radiation equation. This method, entailing no direct temperature measurement, provides a strong control of environmental radiation with widespread applicability. However, the measurement precision of this method is limited by the properties of the material of the inner wall of the box. Other methods applied in field measurements include the shading method and the reference-board method (Korb et al., 1996; Wan et al., 1994). The shading method assumes that the background radiation is isotropic, and the land surface is Lambertian, resulting in certain limitations to its application. In the reference-board method, the emissivity of the reference board should be known and stable.

Direct measurement methods, indirect measurement methods, and combinations of passive and active measurement methods each have advantages and disadvantages. For indoor and field measurements of LSE, the appropriate measuring method can only be chosen depending on specific targets and experimental conditions. At present, indirect measurement methods are mainly used in laboratory measurements of LSE; in field measurements, the radiation barrel method is generally used. With the development of temperature/emissivity separation algorithms (Cheng et al., 2010; Wang et al., 2008), hyperspectral Fourier transform spectrometers and gilded reference baffle boards have been used to acquire surface features and environmental radiation information, thus enabling the LSE spectrum and the LST to be extracted.

7.3.2 Classification-based methods

In split-window algorithms for land surfaces, the LSE is assumed to be known. In practice, the LSE varies tremendously in time and space, especially for soil emissivity, which makes obtaining accurate LSE values difficult on the pixel scale. Snyder and Wan (1998), using laboratory-measured LSE spectra in a kernel-driven BRDF model, conducted a fitting of the emissivities of 14 types of surface features in accordance with the IGBP classification system. By combining MODIS land cover products, the corresponding relationships between LSE and land cover were established. Using an LUT, the emissivity of each pixel can be obtained for use in LST retrieval using a split-window algorithm (Wan and Dozier, 1996). The emissivity of water bodies, ice/snow, and vegetation is stable with little variation, and they are therefore suitable for this classification method. The emissivities of other land cover types, such as bare rock and soil, vary enormously, and the assignment of constant values will inevitably result in greater error. Because of the uncertainty of the classifications themselves, especially for mixed pixels and transitions from one type to another, the LSE obtained through this classification method suffers from discontinuity.

7.3.3 NDVI-based methods

Based on Botswana's radiometer-measured data, which correspond to the AVHRR channel, Van de Griend and Owe (1993) found that the LSE has a strong logarithmic relationship with NDVI.

$$\varepsilon = 1.0094 + 0.047 \ln(\text{NDVI}) \qquad (7.39)$$

Through further study of the relationship between LSE and NDVI, Olioso (1995) noted that this dependence is closely related to soil emissivity, the effective emissivity of leaves, canopy structure, the optical features of leaves, solar position, and the proportion of soil penetrated by sunlight but is insensitive to observational geometric conditions.

Valor and Caselles (1996) used a vegetation cover method to calculate LSE and applied it to more complex mixed pixels to obtain satisfactory results. The specific equation is shown below:

$$\varepsilon = \varepsilon_v P_v + \varepsilon_g(1 - P_v) + \mathrm{d}\varepsilon \qquad (7.40)$$

where ε_v is the emissivity of vegetation, ε_g is the emissivity of a land surface with nonvegetation cover, usually soil, P_v is vegetation coverage, and $\mathrm{d}\varepsilon$ is the contribution of multiple scatterings caused by the internal geometric structure of the pixel.

Sobrino et al. (2001) proposed an NDVI threshold method for AVHRR data; here, NDVI was used to distinguish between vegetated areas and nonvegetated areas, and the emissivities of the two types of areas were thus obtained. Mathematical expressions for the AVHRR, MODIS, SEVIRS, AASTR, and TM data are also given (Sobrino et al., 2008) as presented below:

$$\varepsilon_\lambda = \begin{cases} a_\lambda + a_\lambda \rho_{red} & \text{NDVI} < \text{NDVI}_s \\ \varepsilon_{v\lambda} P_v + \varepsilon_{s\lambda}(1 - P_v) + C_\lambda & \text{NDVI}_s \leq \text{NDVI} \\ & \leq \text{NDVI}_v \\ \varepsilon_{v\lambda} & \text{NDVI} > \text{NDVI}_v \end{cases}$$

$$(7.41)$$

where ρ_{red} is the soil reflectance in the red channel, $\varepsilon_{s\lambda}$ is the emissivity of soil, $\varepsilon_{v\lambda}$ is the emissivity of vegetation, NDVI_s is the NDVI of soil, NDVI_v is the NDVI of vegetation, $P_v = \left(\frac{\text{NDVI}-\text{NDVI}_s}{\text{NDVI}_v-\text{NDVI}_s}\right)^2$ is the fractional vegetation cover, and C_λ is a correction factor. When $\text{NDVI} < \text{NDVI}_s$, the pixel represents soil; when $\text{NDVI}_s \leq \text{NDVI} \leq \text{NDVI}_v$, the pixel is a mixture of vegetation and soil, i.e., partial vegetation cover; when $\text{NDVI} > \text{NDVI}_v$, the pixel represents vegetation. In Eq. (7.38), a discontinuity of emissivity occurs when $\text{NDVI} = \text{NDVI}_s$ or $\text{NDVI} = \text{NDVI}_v$. ε_λ and ρ_{red} have a poor relationship in some soil specimens. The equation can then be simplified as follows:

$$\varepsilon_{\lambda} = \begin{cases} \varepsilon_{s\lambda} & \text{NDVI} < \text{NDVI}_s \\ \varepsilon_{v\lambda}P_v + \varepsilon_{s\lambda}(1 - P_v) + C_{\lambda} & \text{NDVI}_s \leq \text{NDVI} \\ & \leq \text{NDVI}_v \\ \varepsilon_{v\lambda} & \text{NDVI} > \text{NDVI}_v \end{cases}$$

(7.42)

Although the NDVI threshold method contains several improvements compared with other classification methods, it still fails to indicate considerable changes in the LSE, especially for areas without vegetation cover.

7.3.4 Multichannel methods

The methods for determining LSE based on classification and NDVI mainly involve visible and near-infrared spectral information for land surfaces. In contrast, multichannel algorithms to determine LSE mainly use infrared spectral information. The coupling of surface emissivity and temperature makes their separation from the sensor's radiometric measurements an essentially ill-posed problem, in which N equations are involved to solve $N + 1$ unknowns; thus, the equation set is not complete (Liang, 2004). Specific strategies must be employed (for instance, establishing redundant equations and reducing the number of parameters to be retrieved) to make the equation set complete. In attempts to resolve this problem, previous researchers have proposed many algorithms (Becker and Li, 1990a; Gillespie et al., 1998; Liang, 2001; Wan and Li, 1997). The prerequisites for these algorithms include the completion of atmospheric corrections, i.e., the completion of the removal of the influences of atmospheric transmittance and the radiation path and a known atmospheric downward radiance. Thus,

$$L_{g,i} = \varepsilon_i B_i(T_s) + (1 - \varepsilon_i)L_{d,i} \quad (7.43)$$

where $L_{g,i}$ is the ground-leaving radiance, which can be regarded as the received radiance when the sensor is placed on a land surface.

7.3.4.1 The normalized emissivity method

The normalized emissivity method (NEM) is also known as the blackbody spectrum curve-fitting method. In this algorithm, the channel with the highest emissivity is not fixed, and the most suitable channel can be selected as the channel with the highest emissivity. Therefore, this algorithm has the advantage of flexibility and is applicable to the more complex surface feature spectrum (Gillespie 1985). In this algorithm, the maximum emissivity is fixed, regardless of which channel has the maximum emissivity. Superior results can be achieved by first separating vegetation from rocks based on an image classification and then assigning the pixels different respective maximum emissivities. However, such a classification scheme often results in discontinuities between the two types of land cover.

Hook et al. (1992) conducted experiments on the spectra of 81 types of substances, and 58% of the temperatures retrieved by the NEM differed from the true values by less than 1K. The value of the temperature retrieved by the reference channel method was only 21%. Clearly, the precision of the NEM was much higher. However, the maximum emissivity of a single value is not applicable for all types of surface features; thus, greater error arises for a gray body, whereas a high-precision retrieval is ensured for geological targets. The NEM results mainly depend on the accuracy of the assumed maximum emissivity.

7.3.4.2 The α residual method

The α residual method, which was first proposed by Kealy and Gabel (1990), takes advantage of the Wien approximation in the Planck law and omits the "item −1" in the denominator. For a blackbody at 300K, using the Wien approximation at the wavelength of 1 μm incurs an error of approximately 1%.

An expression for temperature composed of emissivity and radiation is obtained by applying a logarithmic process to radiation emitted by the object. An expression for α residual is obtained by subtracting the average temperature for N channels from the temperature for one channel.

$$\alpha_i = \lambda_i \ln \varepsilon_i - \mu_\alpha \qquad (7.44)$$

where $\mu_\alpha = \frac{1}{N} \sum_{k=1}^{N} \lambda_k \ln \varepsilon_k$. α_i can be calculated from ground-measured surface radiation; according to the known thermal infrared emissivity spectra of various surface features, the relationship between μ_α and the spectral contrast (characterized by the variance of α) can be fitted, and μ_α can thus be determined. Next, from the definition of the α residual, the emissivity of each channel, ε_i, can be obtained. Finally, the temperature of each channel is computed according to the Planck function, and the mean temperature value is used as the optimal estimate of the temperature.

With the application of the Wien approximation, a systematic error between the emissivity spectrum and the true value inevitably occurs, and this error changes with temperature. Moreover, mathematically speaking, the α residual method is more complex than most of the other methods.

7.3.4.3 The MMD method

Matsunaga (1994) described an empirical relationship between the channel average emissivity and the spectral emissivity contrast (the difference between the minimum emissivity and the maximum emissivity, MMD) and used it to extract emissivity information from surface features. In the first step, an initial guess value for the emissivity is obtained from the measured radiation by a certain method and then adjusted according to the fitted empirical relationship. The adjusted emissivity, combined with the measured radiation, is then used to calculate the temperature for each channel. The mean value

is taken as the target temperature. Iterations are performed until the difference between the target temperatures that are calculated in two adjacent loops is smaller than the noise equivalent temperature difference of the device. The emissivities of surface features are obtained based on the ultimate target temperature and radiometric measurements. The precision of this algorithm mainly depends on the accuracy of the empirical relationship and the noise level of the device.

7.3.4.4 The TES algorithm for ASTER

The TES algorithm for ASTER thermal infrared data combines the advantages of the NEM method, the spectral contrast method, and the MMD method and adds certain external constraints. Through optimization, a stepwise refinement is achieved (Gillespie et al., 1998). The TES algorithm consists of four modules:

(1) The NEM module: By assuming a maximum emissivity, the initial values of the temperature and the emissivity are obtained, and the influence of the atmospheric downward radiation is removed. Here, the major improvement is that the determination of the maximum emissivity is an iterative process in which the maximum emissivity is adjustable. Therefore, errors resulting from the use of a single maximum emissivity to obtain an initial emissivity value and the separation of the atmospheric downward radiation can be reduced.
(2) The ratio module: In this module, the emissivity ratio is the emissivity obtained in the NEM module divided by its mean value. This calculation reduces the sensitivity of the emissivity ratio to the initial temperature.
(3) The MMD module: In the MMD module, the core of the TES algorithm, the ultimate emissivity, and temperature are determined according to an empirical relationship of the minimum emissivity with the difference between the maximum and minimum spectral emissivity ratios.

(4) The quality evaluation module: The reliability of the temperature and emissivity obtained is reported here.

The ASTER TES algorithm—based temperature retrieval has an error of approximately ± 1.5K; the error for the channel emissivity is approximately ± 0.015. A potentially uncertain factor for this algorithm is the empirical relationship between the emissivity and spectral contrast. Specifically, for a gray body with MMD < 0.03, the algorithm precision is so low that it nearly becomes invalid (Jiménez-Muñoz et al. 2006). Payan and Royer (2004) assessed the application of the TES algorithm by using hyperspectral thermal infrared data for land surfaces. The spectra of approximately 490 types of surface features, including soil, were selected from the ASTER spectral library for simulation. The RMSE of the retrieved emissivity was approximately 0.02, which was mainly attributed to the uncertainty of the NEM module.

7.3.4.5 Optimization methods

Liang (2001) formulated an optimization algorithm for the simultaneous estimation of LSE and LST. The key part of this algorithm is the establishment of an empirical constraint (or an empirical relationship). For MODIS, the empirical relationship can be expressed as follows:

$$\begin{aligned}
\widehat{\varepsilon}_{\min} =\ & 0.067 + 0.319\varepsilon_{20} + 0.232\varepsilon_{22} + 0.271\varepsilon_{23} \\
& + 0.381\varepsilon_{29} + 0.280\varepsilon_{31} + 0.261\varepsilon_{32} - 0.583\varepsilon_{rangge} \\
& - 0.822\varepsilon_{med}
\end{aligned}$$

$$(7.45)$$

where ε_{ij} is the emissivity of MODIS channel ij, ε_{rangge} is the range of absolute emissivities for the six channels, and ε_{med} is the midvalue. For ASTER, the empirical relationship can be expressed as follows:

$$\begin{aligned}
\widehat{\varepsilon}_{\min} =\ & 0.101 + 0.3098\varepsilon_{10} + 0.2352\varepsilon_{11} \\
& + 0.3477\varepsilon_{12} + 0.2458\varepsilon_{13} + 0.2862\varepsilon_{14} \\
& - 0.5406\varepsilon_{rangge} - 0.4411\varepsilon_{med}
\end{aligned} \qquad (7.46)$$

These symbols have the same meanings as those of the MODIS expression. A cost function with constraints is then established and iteratively solved using a nonlinear optimization method.

A series of numerical tests was performed to assess the new algorithm. The algorithm validity was proved with the following major features: (1) extremely different empirical equations; (2) the feasibility of the regularization method; (3) the definition of additional a priori knowledge, which is naturally integrated into the retrieval algorithm; and (4) the application of an optimized inversion algorithm with a solid foundation in computational mathematics.

The analysis of the principles and features of various thermal infrared remote sensing—based multichannel LST and LSE extraction algorithms indicates the following: (1) atmospheric corrections must be supported by other sensors or independent atmospheric data obtained from actual measurements; and (2) to convert an underdetermined problem into an effectively posed problem, some algorithms make several assumptions and approximations regarding emissivity, whereas others utilize empirical relationships between emissivity and its spectral variation. These assumptions and relationships can only be properly applied to certain land cover types.

7.3.5 Retrieval algorithms for hyperspectral data

7.3.5.1 The iterative spectrally smooth temperature and emissivity separation algorithm

The iterative spectrally smooth temperature and emissivity separation algorithm proposed by Borel is used for temperature and emissivity separations from hyperspectral thermal infrared data (Borel, 1998). The LST is estimated based on the assumption that the thermal infrared emissivity spectrum of a natural land surface is smoother than the atmospheric downward radiance spectrum. If the estimated LST deviates

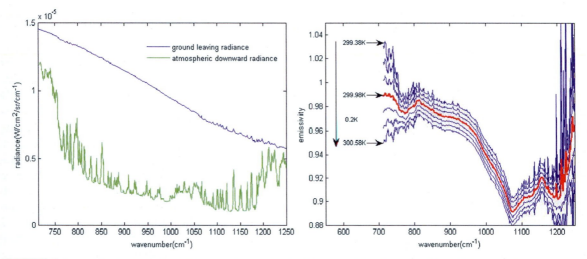

FIGURE 7.1 An illustration of the ISS temperature and emissivity separation algorithm. The left diagram shows simulated soil and atmospheric downward radiance curves. The right diagram shows emissivity curves corresponding to different soil temperatures. The actual land surface temperature (LST) is 300K.

from the true value, the emissivity spectrum calculated according to Eq. (7.44) is saw-toothed, where the atmospheric emission line is located. This phenomenon is termed the "atmospheric downward radiance residue," the degree of which (smoothness) is described by defining an index in a form similar to Eq. (7.45). The temperature corresponding to the emissivity with the minimum smoothness index is considered as the optimal LST. Eq. (7.44) is used to calculate the LSE spectrum. Fig. 7.1 shows the atmospheric downward radiance residues with different LSE estimates.

$$\varepsilon_i = \frac{\left(L_{g,i} - L_{d,i}\right)}{\left(B_i(T_s) - L_{d,i}\right)} \qquad (7.47)$$

and

$$S = \sum_{i=2}^{N-1} \left\{ \varepsilon_i - \frac{\varepsilon_{i-1} + \varepsilon_i + \varepsilon_{i+1}}{3} \right\}^2 \qquad (7.48)$$

where ε_i is the emissivity, $L_{g,i}$ is the ground-leaving radiance, $L_{d,i}$ is the atmospheric downward radiance, $B_i(T_s)$ is the Planck function, T_s is the LST, and i is the channel.

The ISS algorithm does not involve an empirical relationship; thus, it avoids errors in temperature and emissivity separations resulting from the uncertainties of empirical relationships. Here, atmospheric spectral characteristics are utilized, eliminating the requirement for the prior separation of the atmospheric downward radiation contribution. Therefore, the dissemination of the atmospheric downward radiation separation error in the extraction of emissivity information is precluded.

Other temperature and emissivity separation algorithms (Xie, 1993; Knuteson et al., 2004) similar to the ISS algorithm have also been proposed; these algorithms are almost identical, with only slight modifications. Ingram and Henry (2001) thoroughly analyzed the sensitivity of the ISS algorithm to model error and device noise. Their results indicated that this algorithm can achieve a high precision when using hyperspectral data. Studies conducted by Cheng et al. (2008b) suggested that the ISS algorithm may be invalid in certain cases. The invalidity is caused by singular emissivities generated during the emissivity calculation process and cannot be overcome by the ISS algorithm itself. Cheng et al. also proposed the use of a multilayer perceptron neural network—based algorithm for the simultaneous retrieval of the LST and the LSE spectrum, which may serve as a beneficial complement to the ISS algorithm. Cheng et al. (2010) proposed a new stepwise refining temperature and emissivity separation algorithm, which solved the problem of algorithm invalidity caused by singular emissivities and displayed high LSE and LST retrieval precisions.

7.3.5.2 Correlation-based algorithms

A prerequisite for the temperature and emissivity separation algorithms reported in the literature is the assumption that the pixels are internally isothermal. For the ASTER TES algorithm (Gillespie et al., 1998) and the MODIS day/night LST algorithm (Wan and Li, 1997), only a mean equivalent temperature is available for the entire thermal infrared spectral region. In fact, nonisothermal pixels and mixed pixels are widespread natural phenomena. In times of direct sunlight, temperature differences between components can reach over 20K. If an equivalent temperature is defined for the entire thermal infrared spectral region, then the r-emissivity of mixed pixels in each channel is impossible to obtain from radiation measurements and the equivalent temperature. Thus, this problem requires a compromise between accurately calculating the r-emissivity and temperature. From the perspective of practical applications, we expect that r-emissivity, which is closely associated with the physical and chemical properties of land surfaces, can be obtained by thermal infrared sensors with a high spectral resolution and broad spectral coverage.

Cheng et al. (2008a) proposed a correlation-based temperature and emissivity separation algorithm (CBTES) in which the correlation between the equivalent atmospheric downward radiance and the LSE is treated as the basis for LST optimization. In this algorithm, the LSE and the equivalent atmospheric downward radiance are considered as n-dimensional vectors X and Y, respectively. By taking the maximum brightness temperature corresponding to the ground-leaving radiance as the center and the noise equivalent temperature difference of the sensor as the increment, a series of LSTs is generated. The correlation between the emissivity X_i corresponding to each LST and the equivalent atmospheric downward radiance Y is calculated according to Eq. (7.46). The absolute value is adopted, and the LST corresponding to the minimum correlation is the optimal LST estimate, which, together with radiometric measurements, is then used to calculate the LSE. Fig. 7.2 illustrates the relationship between the atmospheric downward radiance and the LSE.

$$corr(i) = \frac{X_i \cdot Y}{\|X_i\| \|Y\|}, \quad X_i \in R^n, Y \in R^n \quad (7.49)$$

and

$$\text{optimal } T = T_i|_{\min(abs(corr(i)))} \quad (7.50)$$

where $\bullet$ is the vector inner product, $\|\|$ is the vector modulus, abs is the adoption of the absolute value, and min is the adoption of the minimum value.

Two typical types of nonisothermal pixels are considered. One is mixed pixels (such as a mixed system of soil and vegetation) with considerable temperature variations during direct sunlight; the other is bare soil, or Gobi, where sunlit and shaded areas with considerable temperature differences are formed in response to a blockage inside the pixel. Based on the assumptions that each nonisothermal pixel consists of two Lambertian components and that the atmospheric conditions inside the pixels are the same, a simulation is performed to analyze the variation of radiometric temperature in the spectral range of 714—1250 cm^{-1} (8—14 μm) when the emissivity is defined as the r-emissivity. Additionally, assuming that the radiometric temperature is approximately fixed within a narrow spectral range, the equivalent temperatures of nonisothermal pixels within a narrow spectral

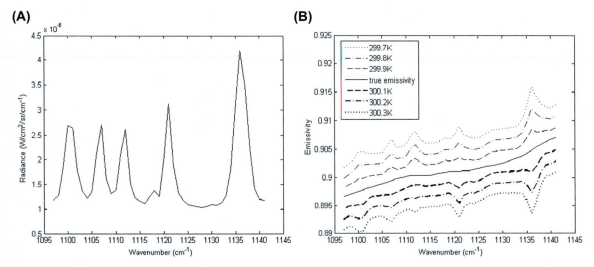

FIGURE 7.2   The relationship of the atmospheric downward radiation to the land surface emissivity (LSE): (A) the curve of the equivalent atmospheric downward radiance; (B) emissivity curves corresponding to different soil temperatures. *From Cheng, J., Liu, Q., Li, X., Qing, X., Liu, Q., Du, Y., 2008. Correlation-based temperature and emissivity separation algorithm. Sci. China Ser. D Earth Sci. 51 363–372, copyright © 2008 with permission from Science Press.*

TABLE 7.3   The parameters for nonisothermal pixels in the simulation.

Temperature difference/K	Component temperature/K		Component emissivity	Component area
10	T1 = 289.2	T2 = 299.2	$\varepsilon_1 = \varepsilon_2 = \varepsilon_{\text{soil}}$	a1 = a2 = 0.5
15	T1 = 289.2	T2 = 304.2	$\varepsilon_1 = \varepsilon_2 = \varepsilon_{\text{soil}}$	a1 = a2 = 0.5
20	T1 = 289.2	T2 = 309.2	$\varepsilon_1 = \varepsilon_2 = \varepsilon_{\text{soil}}$	a1 = a2 = 0.5

range are retrieved using the CBTES algorithm. Vegetation emissivity is derived from the mean value of vegetation emissivity from the ASTER spectral library (http://speclib.jpl.nasa.gov), and soil emissivity is derived from the mean value of soil emissivity from the MODIS UCSB spectral library (http://g.icess.ucsb.edu/modis/EMIS/html/em.html). Table 7.3 lists the parameters used for nonisothermal soil mixed pixels in the simulation.

Fig. 7.3 presents comparisons of the retrieved radiometric temperature and the r-emissivity and their true values. The retrieved values show similar trends of variation with respect to the true values, indicating the validity of the

CBTES algorithm in extracting r-emissivity from nonisothermal planar pixels.

7.3.5.3 Downward radiance residue index algorithms

Wang et al. (2008) proposed a downward radiation residue index method. Similar to the above two algorithms, the downward radiance residue index (DRRI) was established to describe the spectral residue of the atmospheric downward radiation in the estimated emissivity spectrum, as expressed by Eq. (7.48):

$$\text{DRRI} = \varepsilon_2 - \left(\frac{v_3 - v_2}{v_3 - v_1}\varepsilon_1 + \frac{v_2 - v_1}{v_3 - v_1}\varepsilon_3\right) \quad (7.51)$$

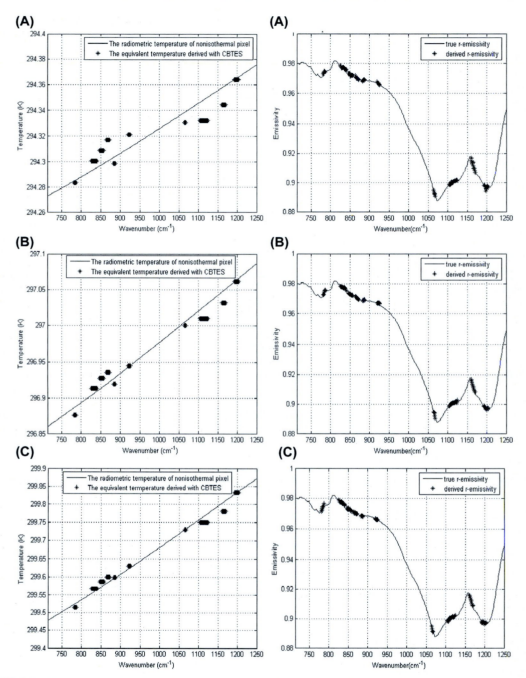

FIGURE 7.3   Variation trends of the radiometric temperature of nonisothermal pixels at different soil temperatures and the equivalent temperature and r-emissivity retrieved by the correlation-based temperature and emissivity separation algorithm (CBTES) algorithm: (A) a temperature difference of 10K; (B) a temperature difference of 15K; and (C) a temperature difference of 20 (K)

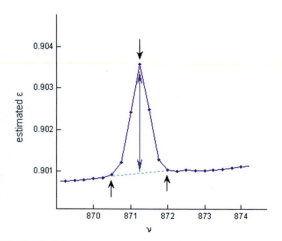

FIGURE 7.4 The spectral characteristics of the atmospheric downward radiation residue and the downward radiance residue index (DRRI) component.

where v_1, v_2, and v_3 are three adjacent channels, of which v_2 is located where atmospheric emissivity occurs; and ε_1, ε_2, and ε_3 are the corresponding emissivity estimates. Fig. 7.4 is a schematic diagram illustrating the principle of the algorithm. With an inaccurate LST estimate, the atmospheric emission line residue is included in the computed emissivity, as shown by the curve in the diagram. Calculating the DRRI facilitates the evaluation of the precision of the LST estimate. The atmospheric downward radiation residue can be found in several places on the entire estimated emissivity curve. N multiples of v_1, v_2, and v_3 are selected for accumulation to obtain the final DRRI. The emissivity corresponding to the temperature with the minimum DRRI is the desired LSE. This case is also applicable for the temperature.

The downward radiation residue index method has many advantages compared with the spectral smoothing method (Borel, 1998). For example, fewer channels are selected, reducing computation time, which facilitates the processing of large quantities of satellite-borne remote sensing data. Moreover, the downward radiation residue index method successfully prevents the occurrence of singular values and can therefore be adapted to various climatic conditions (Wang et al., 2008).

7.3.5.4 Multiscale wavelet–based temperature and emissivity separation algorithm

The commonly adopted assumption in the temperature and emissivity separation algorithm for hyperspectral data is that the thermal infrared emissivity of a natural surface is smoother than the atmospheric downward radiance spectrum. If an estimated LST deviates from its true value, the calculated emissivity is saw-toothed at the atmospheric emission line. This phenomenon is termed as an "atmospheric downward radiance residue," which is primarily determined by the degree of temperature deviation. Thus, a temperature and emissivity separation can be regarded as a process of seeking a suitable surface temperature that can remove the influence of the atmosphere downward radiance on the derived emissivity. Zhou and Cheng (2018) proposed a multiscale wavelet–based temperature and emissivity separation algorithm (MSWTES). The algorithm considers that when the surface emissivity is calculated by an incorrect surface temperature, the atmospheric downwelling signal will remain in the obtained surface emissivity, and the multiscale wavelet can transform the emissivity spectrum from the spatial domain to the frequency domain and reconstruct its low-frequency part and high-frequency part. The wavelet energy of the low-frequency part is mainly contributed by the surface emission. The wavelet energy of the high-frequency part is mainly contributed by the downward radiation of the atmosphere. The ratio of high-frequency energy to low-frequency energy of surface emissivity spectrum can be used to measure the residual level of the atmospheric radiance in the emissivity spectrum. Multiscale wavelet decomposition can be expressed by the following formula:

$$f(t) = \sum_{k=-\infty}^{\infty} c(k)\varphi_k(t) + \sum_{j=0}^{\infty} \sum_{k=-\infty}^{\infty} d(j,k)\psi_{j,k}(t)$$

(7.52)

where $\varphi_k(t)$ is the scale function and $\psi_{j,k}(t)$ is the wavelet function, and the signal $f(t)$ is the ground-leaving radiance spectrum or the emissivity spectrum. The first term of the above equation is the low-scale approximation of the signal, while the second term is the detailed or the high-frequency component of the signal. Using the $1095-1145$ cm^{-1} spectral region as an example, we decompose the ground-leaving radiance into different levels by multiscale wavelets and reconstruct the high-frequency component of ground-leaving radiance. The comparison between the high-frequency reconstruction part of the surface emission radiation and the downward radiation of the atmosphere is shown in the figure below.

Fig. 7.5 shows that the surface emitted radiance is a low-frequency signal and the atmosphere downward radiance is a high-frequency signal. Assuming that the frequency range of the emissivity spectrum calculated by Eq. (7.44) is consistent with the frequency range of the surface emission radiation, we can obtain the optimal decomposition scale by comparing the correlation coefficient between the atmospheric downradiation and the surface-expected radiation reconstructed after decomposition. The ratio of high-frequency energy to low-frequency energy can be used to measure the residual level

of atmospheric downradiation signals in the estimated surface emissivity.

MSWTES algorithm is implemented in three steps.

Step (1): Determine the scale of decomposition. The scale of decomposition is mainly determined by the frequency range of the downward radiation of the atmosphere inside the surface emission radiation. Wavelet decomposition is performed on the surface emission radiation, and the high-frequency portion is reconstructed. The correlation coefficients of the high-frequency part of the surface emission (D1 + D2 + …) and the downward radiation of the atmosphere are compared. When the correlation coefficient reaches the maximum, the corresponding decomposition scale is the optimal scale (Fig. 7.6).

Step (2): Determine the optimal surface temperature. Taking the $850-900$ cm^{-1} spectral interval as an example, the average bright temperature of the interval is first calculated and used as the initial value of the surface temperature. At the same time, with the brightness temperature as the center, ±20K is the search range, and the optimal temperature is obtained

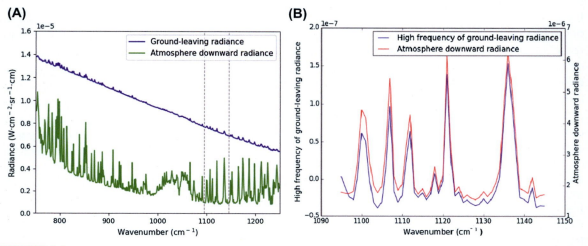

FIGURE 7.5 Contrast of spectral characteristics. (A) The ground-leaving radiance versus atmosphere downward radiance in $750-1250$ cm^{-1}; (B) reconstructed high-frequency component of ground-leaving radiance and atmosphere downward radiance in $1095-1145$ cm^{-1}. *From Zhou S, Cheng J. 2018. A multi-scale wavelet-based temperature and emissivity separation algorithm for hyperspectral thermal infrared data. Int. J. Remote Sens. 39 (22), 8092–8112, copyright © 2018 with permission from Taylor & Francis Group.*

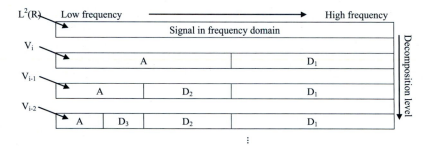

FIGURE 7.6 Multiscale wavelet decomposition of the ground-leaving radiance spectrum. *From Zhou S, Cheng J. 2018. A multi-scale wavelet-based temperature and emissivity separation algorithm for hyperspectral thermal infrared data. Int. J. Remote Sens. 39 (22), 8092–8112, copyright © 2018 with permission from Taylor & Francis Group.*

by iteration. The ratio of the high-frequency signal energy to the low-frequency signal energy is used as the cost function. When the energy of the atmospheric downlink signal contained in the solved emissivity spectrum is the lowest, the cost function takes the minimum value. The formula for the cost function is as follows:

$$\text{Loss} = \frac{\sum\limits_{k}\sum\limits_{j=1}^{L} d_{j,k}^2}{\sum\limits_{k} a_k^2} \tag{7.53}$$

where *Loss* is the cost function; $d_{j,k}$ is the high-frequency wavelet coefficient at level i and translation k; L is the decomposition level; and a_k represents the low-frequency scale coefficient.

Step (3) Repeat step 1 and step 2 for 776–854 cm^{-1}, 1095–1145 cm^{-1}, 1196–1248 cm^{-1}, and 750–1250 cm^{-1}. The derived surface temperatures are averaged to obtain the final estimated surface temperature (Fig. 7.7).

French scientists Laurent Poutier and Francoise Nerry conducted field experiments in 2004 to measure radiation data from soil, rock, and man-made materials and measured the direction of the samples using hemispherical reflectometry in the laboratory (Kanani et al., 2007). The emissivity of each sample can be calculated according to Kirchhoff's law. Fig. 7.8

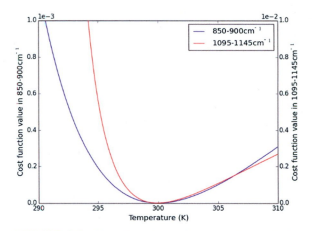

FIGURE 7.7 Variation of cost function value with the estimated temperature in both the 850–900 cm^{-1} and 1095–1145 cm^{-1} spectral regions. *From Zhou S, Cheng J. 2018. A multi-scale wavelet-based temperature and emissivity separation algorithm for hyperspectral thermal infrared data. Int. J. Remote Sens. 39 (22), 8092–8112, copyright © 2018 with permission from Taylor & Francis Group.*

shows the emissivity of 10 samples, and Fig. 7.9 shows the RMSE of the emissivity inversion, which is compared with the inversion results of the ISS algorithm. The results of MSWTES are superior to the ISS algorithm.

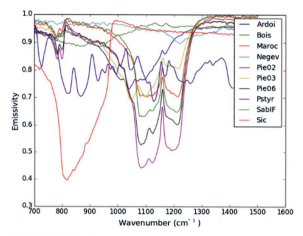

FIGURE 7.8   The laboratory measured directional emissivity spectra of 10 samples. *From Zhou S, Cheng J. 2018. A multi-scale wavelet-based temperature and emissivity separation algorithm for hyperspectral thermal infrared data. Int. J. Remote Sens. 39 (22), 8092—8112, copyright © 2018 with permission from Taylor & Francis Group.*

7.3.6 The calculation of the surface longwave broadband emissivity

The surface BBE is the key parameter for estimating the net surface longwave radiation. Generally, a 10% error in the BBE results in a longwave net radiation bias of 15—20 W/m² (Ogawa and Schmugge, 2004). BBE varies greatly with space, especially in nonvegetated areas. Therefore, inputting BBE as a constant into a model (such as GCM) entails potential uncertainty. Although remote sensing can provide a more realistic BBE, existing thermal infrared sensors can only provide several discrete narrow-band emissivities within the spectral range of 3—14 μm, which cannot meet the requirement for calculating net surface longwave radiation. In this section, the emissivity spectra of water, snow, and minerals at 1—200 μm were simulated using radiative transfer tools, which were used to investigate the suitable BBE spectral domain for estimating surface longwave net radiation. In addition, ASTER and MODIS data are taken as examples to verify the methods for converting narrowband emissivity into BBE.

The surface longwave net radiation (L_n) is the difference between the surface upwelling long-wave radiation and surface downward longwave radiation, which can be expressed as follows:

$$L_n = \int_{\lambda 1}^{\lambda 2} \varepsilon_\lambda [B(T_s) + \rho_\lambda L_{a\lambda}]d\lambda - \int_{\lambda 1}^{\lambda 2} L_{a\lambda} d\lambda \quad (7.54)$$

where ε_λ is the surface spectral emissivity, $B(T_s)$ is the Planck function at temperature T_s, ρ_λ is the directional hemispherical spectral reflectance, and $L_{a\lambda}$ is the surface downward longwave radiation. The lower and upper limit of wavelengths are $\lambda 1 = 0$ and $\lambda 2 = \infty$, respectively. Assuming that the surface is at thermodynamic equilibrium and follows Kirchhoff's law, then (7.54) can be expressed as follows:

$$L_n = \int_{\lambda 1}^{\lambda 2} \varepsilon_\lambda [B(T_s) - L_{a\lambda}]d\lambda \quad (7.55)$$

Assuming ε_λ is independent of $B(T_s)$ and $L_{a\lambda}$, and neglecting the bandpass effect, (7.55) can be formulated as (7.56). Although this assumption will affect the accuracy of surface longwave net radiation calculations, the research of Cheng et al. (2013) indicated that the bias and root mean square (RMS) of replacing (7.55) with (7.56) were 0.55 and 2.31 W/m², respectively.

$$L_n = \varepsilon_{BB}(\sigma T_s^4 - L_a); \quad \text{with } L_a$$

$$= \int_{\lambda 1}^{\lambda 2} L_{a\lambda} d\lambda; \quad \text{and } \varepsilon_{BB} = \frac{\int_{\lambda_1}^{\lambda_2} \varepsilon_\lambda B_\lambda(T_s)d\lambda}{\int_{\lambda_1}^{\lambda_2} B_\lambda(T_s)d\lambda}$$

$$(7.56)$$

where σ is the Stefan—Boltzmann's constant (5.67×10^{-8} W/m²/K⁴). L_a is the wavelength integration of $L_{a\lambda}$, and ε_{BB} is the BBE. Theoretically, the spectral range of integration in (7.56) should be $0-\infty$ μm. However, we cannot determine ε_λ and $L_{a\lambda}$ over such a broad spectral range using modern sensors or simulation tools. Currently, the spectral range of 4—100 μm is

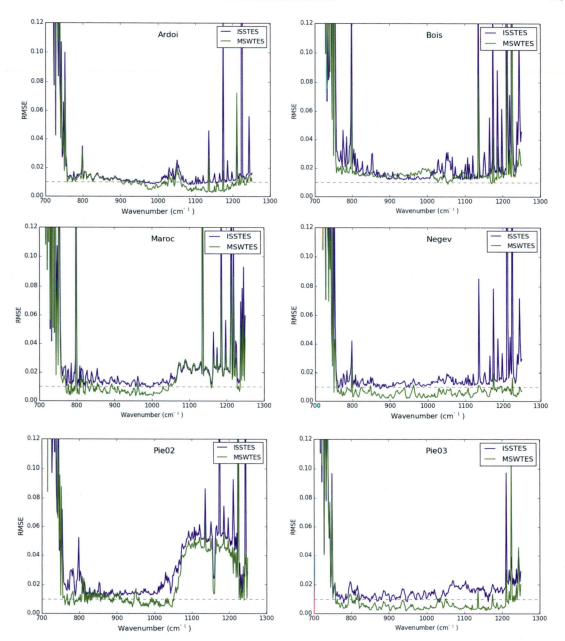

FIGURE 7.9 Root mean square error (RMSE) of an emissivity retrieval for 10 samples using the multiscale wavelet–based temperature and emissivity separation algorithm (MSWTES) and ISSTES algorithms. The *dashed line* corresponds to an RMSE of 0.01. *From Zhou S, Cheng J. 2018. A multi-scale wavelet-based temperature and emissivity separation algorithm for hyperspectral thermal infrared data. Int. J. Remote Sens. 39 (22), 8092–8112, copyright © 2018 with permission from Taylor & Francis Group.*

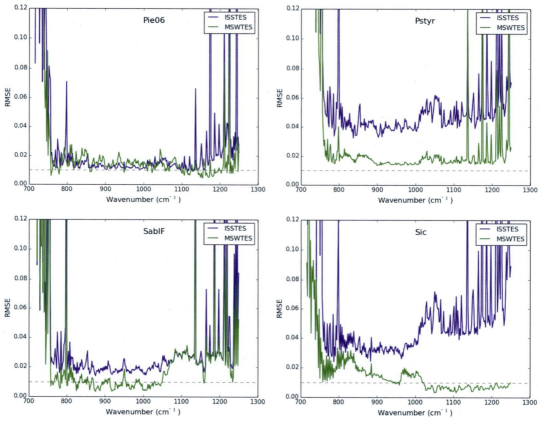

FIGURE 7.9   cont'd.

adopted when estimating the surface longwave radiation budget. The spectral range of 4−100 µm accounts for 99.5% of the total radiation of a 300K blackbody, whereas the spectral range of 1−200 µm accounts for 99.92%. The accuracy of all-wavelength surface longwave net radiation using the spectral domain 1−200 µm would then be higher than using the spectral domain of 4−100 µm. Calculating the all-wavelength L_n in real applications is impractical, so we investigated the accuracy of L_n in the 4−100 µm spectral domain and other spectral domains (e.g., 3−100, 2.5−100, 2.5−200, and 1−200 µm) to replace the all-wavelength L_n.

(1) The emissivity spectra of minerals, water, and snow/ice was simulated with modern radiative transfer tools.

(2) The latest Thermodynamic Initial Guess Retrieval (TIGR) database was used to simulate atmospheric downward radiance in the 0.2−1000 µm spectral range. LST were specified as Ta, Ta + 8, and Ta + 15, respectively, where Ta is the near-surface air temperature of the TIGR atmosphere profiles.

(3) All-wavelength L_n and the L_n for other spectral domains were calculated using (7.55).

TABLE 7.4   Bias and root mean square (RMS) of L_n estimated obtained using different spectral domains.

Spectral domain (μm)	Bias(W/m²)			RMS(W/m²)		
	T_a	$T_a + 8$	$T_a + 15$	T_a	$T_a + 8$	$T_a + 15$
4−100	4.212	3.979	3.725	5.387	5.140	4.871
3−100	4.460	4.388	4.322	5.679	5.618	5.562
2.5−100	4.459	4.393	4.336	5.676	5.624	5.578
2.5−200	0.928	0.920	0.912	0.993	0.984	0.977
1−200	0.929	0.921	0.914	0.993	0.986	0.979

TABLE 7.5   Bias and root mean square (RMS) when estimating L_n at 2.5−200 μm using the broadband emissivity (BBE) for different spectral domains.

Spectral domain (μm)	Bias (W/m²)			RMS (W/m²)		
	T_a	$T_a + 8$	$T_a + 15$	T_a	$T_a + 8$	$T_a + 15$
3−14	0.07	0.387	0.687	0.863	0.990	1.292
8−12	−0.046	−0.254	−0.451	1.383	1.585	2.070
8−13.5	0000	0.001	0.002	0.978	1.120	1.453
8−14	0.010	0.054	0.096	0.911	1.045	1.373

The biases and RMSs of the L_n between different spectral ranges and the full spectrum calculated by the above method are shown in Table 7.4. The biases of the 4−100, 3−100, and 2.5−100 μm spectral domains ranged from 3.725 to 4.46 W/m², and the RMS lies between 4.971 and 5.679 W/m². The bias and RMS of the 2.5−200 μm spectral domain were slightly smaller than the 1−200 μm spectral domain. The average bias and RMS of the 2.5−200 μm spectral domain were 0.92 and 0.985 W/m², respectively, which were smaller than the 4−100 μm spectral domain by 3.052 and 4.148 W/m², respectively. Therefore, L_n calculated from the 2.5−200 μm spectral domain is superior to the other spectral domain calculations.

Existing thermal infrared sensors can only provide several discrete narrowband emissivities within a spectral range of 3−14 μm. Many studies have been carried out with the aim of estimating BBE in different spectral domains (e.g., 3−14, 8−12, 8−13.5, and 8−14 μm), and the accuracy of BBE for different spectral domains must be systematically quantified when calculating L_n. We first calculated BBE for the 2.5−200, 3−14, 8−12, 8−13.5, and 8−14 μm spectral domains using (7.56).

The accuracy for each spectral domain is presented in Table 7.5. Although the 3−14 μm spectral domain had the lowest RMS, the RMS difference between the 3−14 μm spectral domain and other domains was not significant. In addition, the 3−14 μm spectral domain had the largest bias. The 8−12 μm spectral domain had the largest RMS and a negative bias whose absolute value was only lower than the 3−14 μm spectral domain. Spectral domains 3−14 and 8−12 μm were inappropriate for calculating L_n for the 2.5−200 μm spectral domain. The 8−13.5 μm spectral domain had a minimum bias and an acceptable RMS. The average bias and RMS for this domain were 0.001 and 1.184 W/m², respectively. The bias of the 8−14 μm spectral domain was larger the 8−13.5 μm spectral domain, and the RMS of the 8−14 μm spectral domain was slightly lower than in the 8−13.5 μm spectral domain. Therefore, the BBE for spectral domain 8−13.5 μm was best for calculating the L_n of the 2.5−200 μm spectral domain. Using these simulation data, we investigated the accuracy of the 2.5−200 μm L_n calculated by the 8−13.5 μm BBE when replacing all-wavelength L_n. In this case, the average bias and RMS were 1.473 and 2.746 W/m², respectively.

Based on the spectral emissivity ε_λ of a land surface, the BBE is defined as follows (Ogawa et al. 2002):

$$\varepsilon_{\lambda_1 - \lambda_2} = \frac{\int_{\lambda_1}^{\lambda_2} \varepsilon(\lambda) B(\lambda, T) d\lambda}{\int_{\lambda_1}^{\lambda_2} B(\lambda, T) d\lambda} \quad (7.57)$$

where $B(\lambda, T)$ is the Planck function, and T is the surface temperature. Prior research indicates that when T increases from 270 to 300K, the

BBE varies by only 0.005. Therefore, T can be taken as a constant, such as 300K λ_1 and λ_2 are the upper and lower wavelength limits, respectively, for BBE. Theoretically, calculating a radiation budget requires the broadband emissivities at wavelengths of 3 μm to infinity, whereas remote sensing can provide only narrowband emissivities within the 3−14 μm spectral range. Ogawa and Schmugge (2004) and Cheng et al. (2013) suggested that the highest precision for longwave net radiation can be achieved by using BBE within the wavelength range of 8–13.5 μm with a clear sky. Moreover, for more extensive applications, methods for calculating the BBE within the wavelength ranges of 3–14 μm, 8–12 μm, 8–14 μm, and 8–13.5 μm have also been introduced.

If a spectral response function $f(\lambda)$ for the MODIS or ASTER thermal infrared channels is introduced into the above equation, then the narrowband emissivity can be expressed as follows (Ogawa et al. 2002):

$$\varepsilon_{ch} = \frac{\int_{\lambda_1}^{\lambda_2} \varepsilon(\lambda)B(\lambda, T)f(\lambda)d\lambda}{\int_{\lambda_1}^{\lambda_2} B(\lambda, T)f(\lambda)d\lambda} \quad (7.58)$$

where λ_1 and λ_2 are the wavelength response ranges for the MODIS and ASTER thermal infrared channels, respectively. Given the broadband emissivities and narrowband emissivities, the conversion between them can be generally expressed in linear form as follows:

$$\varepsilon_{\lambda_1-\lambda_2} = \sum_{ch=1}^{N} a_{ch}\varepsilon_{ch} + c \quad (7.59)$$

where a_{ch} is the conversion coefficient for each narrowband, c is a constant, and N is the number of narrowbands. MODIS provides six narrowband emissivities within the wavelength range of 3−14 μm, whereas ASTER has five narrowband emissivities in the 8−14 μm spectral range. Information related to these narrowbands is listed in Table 7.6.

In total, 424 emissivity spectra, including 240 emissivity spectra from the ASTER spectral library (Baldridge et al., 2009), 109 emissivity spectra from the MODIS UCSB spectral library (Snyder and Wan, 1998; Snyder et al., 1998), and 75 soil emissivity spectra that were measured outdoors were selected for calculating the BBE and narrow emissivity in Eqs. (7.57) and (7.58). The surface types included soil, vegetation, rock, water body, ice, and snow. Because the data selected from the ASTER spectral library are directional hemispheric reflectance values, they were converted into emissivities according to Kirchhoff's law. That is, at thermal equilibrium, the relationship between the emissivity and reflectance can be expressed as $\varepsilon_\lambda = 1 - \rho_\lambda$. Based on the above spectral data, a regression analysis was conducted to obtain all the conversion coefficients for MODIS and ASTER, which are shown in Eqs. (7.60) and (7.61). The R^2 and RMS for ASTER were 0.983 and 0.005, respectively. The R^2 and RMS for MODIS were 0.932 and 0.010, respectively.

$$\varepsilon_{bb\_ast} = 0.197 + 0.025\varepsilon_{10} + 0.057\varepsilon_{11} + 0.237\varepsilon_{12}$$
$$+ 0.333\varepsilon_{13} + 0.146\varepsilon_{14}$$

$$(7.60)$$

$$\varepsilon_{bb\_mod} = 0.095 + 0.329\varepsilon_{29} + 0.572\varepsilon_{31} \quad (7.61)$$

where ε_{bb_ast} is the ASTER BBE, $\varepsilon_{10} - \varepsilon_{14}$ are the five ASTER narrowband emissivities, ε_{bb_mod} is the MODIS BBE, and ε_{29} and ε_{31} are the MODIS narrowband emissivities for channels 29 and 31. The conversion formulas were tested via leaving one-out cross-validation. The 423 samples were used to establish the conversion function, and the remaining one sample was used for validation. This process was repeated 424 times to cover the entire set of samples. Each time, we obtained one conversion formula, as well as the corresponding R^2 and one bias between the predicted BBE and the true BBE.

The mean error and mean RMS and R^2 calculated by the ASTER conversion function were 0, 0.005, and 0.983 ± 0.0001, respectively. The corresponding values for MODIS were 0, 0.010, and 0.932 ± 0.0005, respectively. The coefficients of the conversion function were 0.197 ± 0.0, 0.025 ± 0.0, 0.057 ± 0.0, 0.237 ± 0.0, 0.333 ± 0.0, and 0.146 ± 0.0 for ASTER and 0.095 ± 0.002, 0.329 ± 0.0, and 0.572 ± 0.002 for MODIS.

7.3.7 The retrieval of the surface longwave broadband emissivity

According to Kirchhoff's law, a complementary relationship between the surface emissivity and reflectance exists, which allows the emissivity to be calculated from the reflectance. However, deriving the temperature and/or emissivity for a thermal infrared sensor with one or two channels is difficult. Although the emissivity is fitted as a linear function of red reflectance using the spectra in the ASTER spectral library, the physical connection between the surface emissive and reflective variables is unclear (Cheng et al., 2017). This section uses the simulation results of the Hapke radiative transfer model to investigate the feasibility of calculating the surface longwave BBE from reflective variables.

The formulas of the Hapke reflective and emissive models are provided as below with the assumption that the particles emit and scatter isotropically:

$$r_{hd}(e) = (1 - \gamma)/(1 + 2\gamma\mu)$$
$$\varepsilon_d(e) = \gamma \frac{1 + 2\mu}{1 + 2\gamma\mu} \tag{7.62}$$

where $r_{hd}(e)$ and $\varepsilon_d(e)$ are hemisphere-directional reflectance and directional emissivity, respectively; $\mu = \cos(e)$ and e is the viewing zenith angle; and $\gamma = \sqrt{1 - w}$ and w is the particle single scattering albedo.

The optical constants of the various minerals and the effective radius of the particles are input into the Mie Code to produce Mie parameters, and the Mie parameters are then input into the Hapke radiative transfer model to produce the simulated nadir-viewing surface reflectance and emissivity. The simulated surface reflectance spectra were convolved with the filter functions of seven optical Landsat TM5 channels to generate the corresponding channel reflectance and emissivity. The linear relationship between the emissivity and red reflectance was very weak, but a significant multiple linear relationship was found:

$$\varepsilon = 0.988 - 0.496R_1 + 1.186R_2 - 0.737R_3 \tag{7.63}$$

where ε is the emissivity of channel 6, and R_i is the Landsat reflectance for channel i. The simulated surface reflectance spectra were also convolved with the filter functions of seven optical MODIS channels to produce the corresponding spectral albedos, assuming that surface is Lambertian. According to the research of Cheng et al. (2013), the surface BBE can be calculated from the simulated emissivity spectra's assigned surface temperature of 300K. After obtaining the surface spectral albedos and BBE, a statistically significant multiple linear function was derived:

$$\varepsilon_{bb} = 0.989 - 0.292\alpha_1 - 0.377\alpha_2 + 0.688\alpha_4$$
$$- 0.027\alpha_7 \tag{7.64}$$

where ε_{bb} is the BBE, and α_i is the MODIS spectral albedo for channel i. The determination coefficient is 0.753. The bias and RMSE are 1×10^{-6} and 0.003, respectively. Clearly, a physical linkage exists between the surface emissive and reflective variables.

On the above basis, GLASS BBE products were generated based on MODIS data from 2000 to 2010. To date, this is the world's first BBE product with the highest spatial resolution (1 km) and time resolution (8 days). Fig. 7.10 shows a schematic diagram of the global BBE on January 1, 2003.

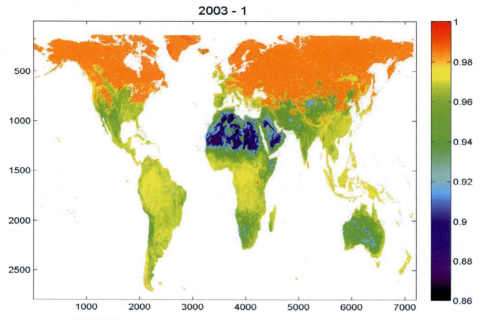

FIGURE 7.10   The global broadband emissivity on January 1, 2003.

According to the pixel NDVI, the surface is divided into five categories to respectively calculate each BBE; they are water, ice/snow, bare soil $(0 < \text{NDVI} \leq 0.156)$, vegetated surfaces $(\text{NDVI} > 0.156)$, and a transition zone $(0.1 < \text{NDVI} < 0.2)$. The water and ice/snow BBE was set to 0.985 based on the ASTER and MODIS spectral library, and the snow BBE was simulated using the radiative transfer mode (Cheng et al., 2010). For the bare soil and transition zone, a linear function between the ASTER BBE and seven MODIS spectral black-sky albedos was established to invert their respective BBEs (Cheng and Liang 2014a). When NDVI < 0.1 or 0.2 < NDVI, the algorithm of the bare soil or vegetated surfaces is used to calculate the respective BBE. In the overlap between the bare soil and transition zone, BBEs are the average of the values calculated by the formulas of bare soil and transition zone, respectively. Similarly, BBEs of the overlap between the transition zone and vegetated surfaces $(0.15 < \text{NDVI} < 0.2)$ are calculated by averaging the results of their formulas.

TABLE 7.6   Spectral ranges for the Moderate Resolution Imaging Spectrometer (MODIS) and Advanced Spaceborne Thermal Emission and Reflection Radiometer (ASTER) thermal infrared channels.

MODIS		ASTER	
Channel	Spectral range	Channel	Spectral range
29	8.40–8.70	10	8.13–8.48
31	10.8–11.3	11	8.48–8.83
32	11.8–12.3	12	8.93–9.28
–	–	13	10.25–10.95
–	–	14	10.95–11.65

The satellite data used to retrieve BBE from bare soil and transition zone are (1) the MODIS albedo product (MCD43B3) and its quality control (QC) data (MCD43B2), (2) the MODIS NDVI product (MOD13A2), and (3) the ASTER emissivity product (AST05). The requisite auxiliary data were the soil taxonomy, which has 12

TABLE 7.7   Bias and root mean square error (RMSE) for each of the eight soil orders in bare soils.

Soil order	Alfisols	Aridisols	Entisols	Gelisols	Inceptisols	Mollisols	Oxisols	Vertisols
Sample number	34	12,547	33,025	34,009	513	2821	11	157
Bias	0.004	−0.002	0.002	−0.003	−0.008	−0.006	−0.004	−0.001
RMSE	0.010	0.011	0.011	0.012	0.019	0.009	0.016	0.008

soil orders with a spatial resolution of approximately 0.0333°. The ideal situation is to derive one formula using all the extracted BBE—albedo pairs. The reasons are (1) the spatial resolution of the soil taxonomy is much coarser than the satellite data and (2) the accuracy of the soil taxonomy is relatively low. We derived one formula using all the extracted BBE—albedo pairs for bare soil. However, this method did not work well when all soil orders were combined. Thus, we derived three formulas, i.e., one formula for andisols, one formula for ultisols, and one formula for the remaining eight soil orders. The formulas are shown as follows, and the biases (RMSE) for each of the eight soil orders in bare soils are shown in Table 7.7.

$$\varepsilon_{BB\_s1} = 0.963 + 0.643a_1 - 1.011a_3 - 0.137a_7\cdots$$
$$(7.65)$$

$$\varepsilon_{BB\_s2} = 0.976 + 0.138a_1 + 0.040a_2 + 0.264a_3$$
$$- 0.383a_4 + 0.031a_6 - 0.124a_7\cdots$$
$$(7.66)$$

$$\varepsilon_{BB\_s3} = 0.953 - 0.827a_1 + 0.447a_2 + 0.570a_3$$
$$- 0.041a_4 + 0.130a_5 + 0.006a_6 - 0.153a_7\cdots$$
$$(7.67)$$

where ε_{BB_s1}, ε_{BB_s2}, and ε_{BB_s3} are the BBE for andisols, ultisols, and the remaining eight soil orders, respectively.

In sparsely vegetated areas, determining whether the pixels relate to bare soil or vegetated surface is difficult. Thus, the variation in the BBE for these pixels would be larger than the actual variation. We propose specifying a transition zone to mitigate the BBE difference between the bare soil and vegetated pixels by using the NDVI and giving a BBE estimation method for the transition zone. In this study, pixels with an NDVI ranging from 0.1 to 0.2 were labeled as transition zone pixels. If the NDVI is between 0.1 and 0.156, its BBE is the average of the BBEs calculated using the formula for bare soil and the transition zone. If the NDVI lies between 0.156 and 0.2, its BBE is the average of the BBEs calculated using the formulas for the transition zone and vegetation. Similar to the method used for bare soil, we derived three formulas, i.e., one formula for andisols, one formula for vertisols, and one formula for the remaining eight soil orders. The formulas are expressed as follows and the biases (RMSE) for each of the eight soil orders in the transition zones are shown in Table 7.8.

$$\varepsilon_{BB\_t1} = 1.006 - 0.339a_2 + 0.142a_7\cdots \quad (7.68)$$

TABLE 7.8   Bias and root mean square error (RMSE) for each of the eight soil orders in transition zones.

Soil order	Alfisols	Aridisols	Entisols	Gelisols	Inceptisols	Mollisols	Oxisols	Vertisols
Sample number	75	22,927	19,977	26,013	944	5418	12	1730
Bias	0.008	−0.002	0.007	−0.001	0.005	−0.008	0.008	0.004
RMSE	0.012	0.010	0.015	0.010	0.016	0.014	0.010	0.007

$$\varepsilon_{BB\_t2} = 0.964 + 0.195a_1 + 0.256a_2 - 0.745a_3$$
$$+ 0.099a_6 - 0.300a_7 \cdots$$

$$(7.69)$$

$$\varepsilon_{BB\_t3} = 0.954 - 0.782a_1 + 0.345a_2 - 0.7760a_3$$
$$- 0.111a_4 + 0.056a_5 + 0.080a_6 - 0.131a_7 \cdots$$

$$(7.70)$$

where ε_{BB_s1}, ε_{BB_s2}, and ε_{BB_s3} are the BBEs for andisols, vertisols, and the remaining eight soil orders, respectively.

The BBE of a vegetated surface can be derived from the LUT constructed by the 4SAIL radiative transfer model provided with the BBEs of leaves, soil background, and LAI. Table 7.9 gives the BBE values (8–13.5 μm) for six composited land cover types. We averaged the emissivity spectra from different sources for composited IGBP forest, grassland, and cropland classes, and we then calculated the corresponding BBE individually. Directionality was ignored in the calculation of BBE due to a lack of directional emissivity measurements. We calculated the mean BBE of each 8-day period for different soil types using the GLASS BBE from 2001 to 2010; the resulting value was designated the BBE of the soil background. The LAI was extracted from the GLASS LAI product. To make the BBE LUT more flexible, the variation ranges for three principal model inputs were set as follows: the leaf BBE ranges from 0.935 to 0.995 and has an interval of 0.01; the soil BBE varies from 0.71 to 0.99 and has an interval of 0.01; and the LAI ranges from 0 to 6.0 and has an interval of 0.5. According to Verhoef et al. (2007), the difference between the 4SAIL-modeled directional emissivity values using four LIDFs (planophile, plagiophile, spherical, and erectophile) is within 0.005 when the LAI ranges of 0.5–4. Therefore, a spherical distribution function was adopted. The spectral range was set to 8–13.5 μm in the simulation. We first used 4SAIL to model the directional BBE for

TABLE 7.9   Leaf broadband emissivity (BBE) values for six composited vegetation land cover types.

Composited class	IGBP class	BBE	Leaf emissivity
Forest	1, 2, 3, 4, 5	0.9771	Mean of three live canopy BBEs from the ASTER spectral library and 24 leaf BBEs from the MODIS spectral library
Grassland	10	0.9785	Mean of green grass BBE from the ASTER spectral library and Ref. Pandya et al. (2013)
Cropland	12, 14	0.9627	Mean of maize BBE, sorghum BBE, pearl millet BBE from Ref. Pandya et al. (2013); wheat BBE of Li et al. (2013), and *Acer rubrum* BBE from Luz and Crowley (2007)
Savanna	8, 9	0.9778	50% Forest + 50% grassland
Shrub land	6, 7	0.9771	Forest
Other types	16, 254	0.9785	Mean value of the above five types

1—evergreen needleleaf forest; 2—evergreen broadleaf forest; 3—deciduous needleleaf forest; 4—deciduous broadleaf forest; 5—mixed forest; 6—closed shrublands; 7—open shrublands; 8—woody savannas; 9—savannas; 10—grasslands; 12—croplands; 14—cropland/natural vegetation; 16—barren or sparsely vegetated; 254—unclassified.

view zenith angles ranging from 0 to 85° at 5° intervals. Then, the directional BBE was integrated to produce hemispherical BBE. In total, there are 2710 situations represented in the constructed hemispherical BBE LUT.

The research of Cheng et al. (2014b) indicated that the average difference between the estimated

BBE and the measured BBE was 0.016 when using field measurements in China for BBE validation. The MODIS BBE derived from the Version 5 emissivity product was larger than the estimated BBE, and the biases were −0.008 and −0.010 for the bare soil and transition zone, respectively, while the RMSEs were 0.026 and 0.023, respectively. The MODIS BBE derived from the Version 4.1 emissivity product was less than the estimated BBE, and the biases were 0.001 and 0.003 for the bare soil and transition zone, respectively, while the RMSEs were 0.015 and 0.013, respectively. The estimated BBE was also in accordance with the ASTER BBE, i.e., the biases were −0.001 and 0.001 for the bare soil and transition zone, while the RMSEs were 0.012 and 0.011, respectively. Cheng et al. (2016) validated the BBE over the

vegetated surface, and the accuracy of the new method exceeds 0.005 over fully vegetated surfaces. Compared with the BBE calculated using ASTER and MODIS11C2 V5 emissivity products, the new BBE correctly describes the seasonal variation of vegetation abundance, while the other two products failed to characterize this variation.

Figs. 7.11 and 7.12 show the comparison results between NAALSED BBE and GLASS BBE for the summer season and winter season, respectively. Visually, the GLASS BBE was more complete than the NAALSED BBE, especially for the winter season. There was almost no missing data in the GLASS BBE, while there were many gaps in the NAALSED BBE. The spatial patterns of NAALSED BBE and GLASS BBE were very similar. In general, GLASS BBE

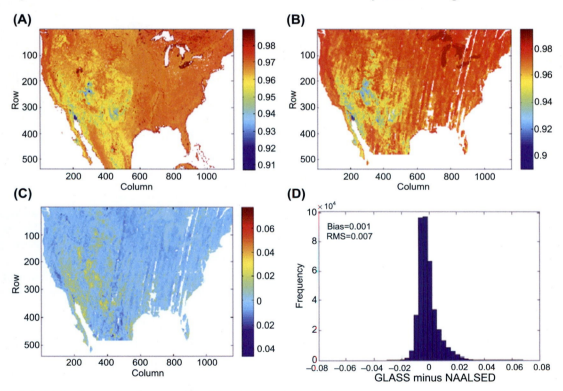

FIGURE 7.11 Comparison between NAALSED BBE and GLASS BBE for the summer season. (A) GLASS BBE, (B) NAALSED BBE, (C) the difference between GLASS BBE and NAALSED BBE, and (D) the histogram showing the difference. *From Cheng J., Liang S.L., Yao Y.J., Ren B.Y., Shi L.P., Liu H., 2014. A comparative study of three land surface broadband emissivity datasets from satellite data, Remote Sens. 6, 111–134.*

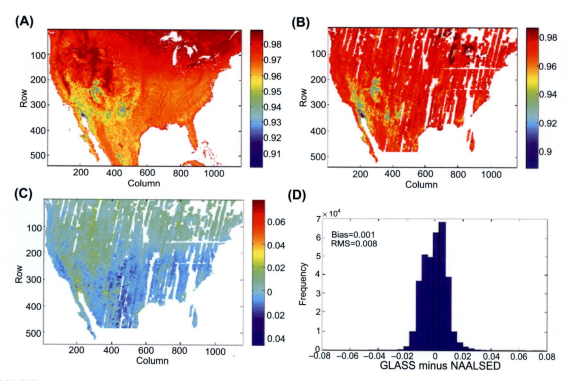

FIGURE 7.12   Comparison between NAALSED BBE and GLASS BBE for the winter season. (A) GLASS BBE, (B) NAALSED BBE, (C) the difference between GLASS BBE and NAALSED BBE, and (D) the histogram showing the difference. *From Cheng J., Liang S.L., Yao Y.J., Ren B.Y., Shi L.P., Liu H., 2014. A comparative study of three land surface broadband emissivity datasets from satellite data, Remote Sens. 6, 111—134.*

and NAALSED BBE were in accordance. The bias and RMSE were −0.001 and 0.007 for the summer season, respectively, and −0.001 and 0.008 for the winter season, respectively. GLASS BBE products therefore have a higher overall precision and can provide data support for related research.

7.4 LSE and LST products

Sections 7.2 and 7.3 describe LST and LSE retrieval algorithms. Despite the numerous available algorithms, few algorithms can retrieve LST and LSE on a large scale and thereby generate LST and LSE products. As a result of complex and variable land surface conditions, the accuracy of assumptions and

approximations related to these algorithms is challenged. Typically, the assumptions and approximations established on a local scale are false at larger scales. Additionally, algorithm complexity and operational efficiency are the major factors limiting the generation of products on a large scale. Tables 7.10 and 7.11 list typical LST and LSE products and their characteristics.

7.5 Fusion of land surface temperature products

Currently, the remote sensing data used in the retrieval of LST products are generally located in the thermal infrared spectral range. The thermal infrared LST retrieval algorithm is only effective

TABLE 7.10 Typical land surface temperature (LST) products.

Product name	Sensor	Retrieval algorithm	Spatial resolution	Temporal resolution	Spatial coverage	Temporal coverage
MODIS LST*	MODIS/EOS	Split-window algorithm Day/night algorithm	1 km/6 km	1 d	Global	2000—present
AVHRR LST	AVHRR/ NOAA14	Split-window algorithm	8 km	1 d	Africa	1995—2000
AVHRR LST	AVHRR/NOAA	Split-window algorithm	1.1 km	1 d	Global	1998—2007
AATSR LST	AATSR/ ENVISAT	Split-window algorithm	1 km	3 d	Global	2004—present
ASTER LST	ASTER/TERRA	TES	90 m	16 d	Global	2000—present
MVIRI LST	MVIRI/ METEOSAT	Neural network	5 km	30 min	Europe/Africa	1999—2005
SEVIRI LST	SEVIRI/MSG	Split-window algorithm	3 km	15 min	Europe/Africa/South America	2006—present

Only the basic MODIS LST products are listed. Other types of LST products can be derived from the basic products. For more details, please refer to https://lpdaac.usgs.gov/lpdaac/products/modis_products_table.

TABLE 7.11 Typical land surface emissivity (LSE) products.

Product's name	Sensor	Retrieval algorithm	Spatial resolution	Temporal resolution	Spatial coverage	Temporal coverage
MODIS LSE	MODIS/EOS	Day/night algorithm	6 km	1 d	Global	2000—present
ASTER LSE	ASTER/TERRA	TES	90 m	16 d	Global	2000—present
GLASS LSE[a]	MODIS/EOS AVHRR/NOAA	Regression method	1 km/5 km	8 d	Global	1985—present

[a] GLASS LSE is a broadband emissivity product in the 8—13.5 and 3—14 μm windows that is retrieved from the AVHRR and MODIS visible and near-infrared products, respectively.

during clear sky conditions. These products cannot reach the full surface coverage because of cloud coverage. More than 60% of MODIS LST products are affected by cloud cover (Chen et al., 2011). Dynamic monitoring of resource environments at regional or global scales requires complete and continuous LST information, while existing LST products cannot meet the requirements of these applications. In contrast, microwave radiation is much less affected by the atmosphere and can penetrate clouds and reach the land surface, which leads to a higher spatial integrity of the microwave data. Many algorithms for microwave LST retrieval have also been developed (see Section 7.2.4).

To fill the gaps in thermal infrared LST products, researchers have begun to blend microwave and thermal infrared data to obtain LST with all-weather coverage in recent years.

(Wang et al., 2014) first tried to blend the MODIS and AMSR-E LST data, in which the MODIS cloud coverage ratio was treated as the weight to calculate the LST in the fused pixel. The fusion formula is as follows:

$$LST_{cloud} = \frac{LST_{AMSR-E} - LST_{MODIS} \cdot (1 - C)}{C}$$

(7.71)

where LST_{cloud} is the estimated AMSR-E LST with cloud coverage. LST_{AMSR-E} and LST_{MODIS} are the AMSR-E and MODIS LSTs obtained under clear sky conditions, respectively, and C is the ratio of the cloud amount in a single AMSR-E pixel. However, the spatial resolution and accuracy of the microwave LST are much lower than the thermal infrared LST due to the limitation of microwave sensors. For example, the spatial resolution of the AMSR-E microwave sensor data is 25 km, and the accuracy of the retrieved microwave LST is approximately 5–6K from the different algorithms. Therefore, although the LST calculated by the weight ratio of the cloud amount reaches the aim of providing all-weather coverage, the accuracy of LST in the gap filling area is not substantially improved, and a grid effect also exists in these areas.

To solve these problems, Kou et al. (2016) proposed a method using the Bayesian maximum entropy to fuse MODIS and AMSR-E LST data and verified the method on the Tibetan Plateau. The accuracy of blended LST data is between 2.31 and 4.53K. This fusion method uses the geostatistical interpolation method to eliminate the grid effect and reflect the details of the spatial variation of LST in the filling area to a certain degree; however, Kou et al. (2016) did not solve the problems of low spatial resolution and the low accuracy of AMSR-E LST. Furthermore, Duan et al. (2017) proposed a space–time interpolation model fusion method for MODIS and AMSR-E LST data based on Eq. (7.71). The

AMSR-E LST was downscaled, and its accuracy was improved before fusion. This method was verified by taking the extent of China in 2009 as an example. The LST error in clear skies is approximately 2K, while the LST error in cloud coverage is between 3.5 and 4.4K.

Currently, the research on the fusion of thermal infrared and microwave LST data is still in the preliminary stage. Because of the large-scale difference between MODIS and AMSR-E data, the scaling of AMSR-E LST data becomes a key issue during the fusion process; however, the existing fusion algorithms do not adequately address the scaling problem. In addition, the development of a perfect fusion method is also a task that requires in-depth study. Therefore, although the fusion of thermal infrared and microwave data is currently a very promising method for generating spatially complete LST products, further research is needed.

7.6 Summary

This chapter reviews remote sensing estimation algorithms for LST and LSE. Neither thermal infrared sensors nor passive microwave sensors currently achieve the temperature retrieval precision of 1K required for practical applications. Additional research is required to improve LST remote sensing retrieval precision and to provide seamless LST products with the desired temporal resolution on the global scale. Synergies between thermal infrared and microwave models and retrieval algorithms are a likely future research focus. Temperature and emissivity separation algorithms for hyperspectral data are expected to generate high-precision LSE spectra, and the relevant studies on this subject are in their initial stages. However, the existing algorithms fail to meet the requirements for generating LST and LSE products and must be verified by satellite-borne data.

Acronyms

AATSR	Advanced Along-Track Scanning Radiometer
AMSR-E	Advanced Microwave Scanning Radiometer-Enhanced
ASTER	Advanced Spaceborne Thermal Emission and Reflection Radiometer
AVHRR	Advanced Very High-Resolution Radiometer
BBE	Broadband emissivity
CBRES	China–Brazil Earth Resources Satellite
DMSP	Defense Meteorological Satellite Program
EOS	Earth Observing System
ETM	Enhanced Thematic Mapper
FY-2	Feng Yun 2
FY-3	Feng Yun 3
GLASS	Global LAnd Surface Satellite
HJ-1	Huan Jing 1
IRMSS	Infrared Multispectral Scanner
IRS	Infrared Scanner
MERSI	MEdium Resolution Spectral Imager
METEOSAT	Meteorology Satellite
METEOSAT	Meteorology Satellite
MODIS	Moderate Resolution Imaging Spectrometer
MODTRAN	MODerate resolution TRANSmittance
MVIRI	Meteosat visible and infrared imager
MVIRI	Meteosat visible and infrared Imager
NOAA	National Oceanic and Atmospheric Administration
S-VISSR	Stretched Visible and Infrared Spin Scan Radiometer
SAIL	Scattering by Arbitrarily Inclined Leaves
SMMR	Scanning Multichannel Microwave Radiometer
SSM/I	Special Sensor Microwave/Imager
SURFRAD	SURface RADiation Network
TM	Thematic Mapper
VIRR	Visible and infrared radiometer

References

Baldridge, A.M., Hook, S.J., Grove, C.I., Rivera, G., 2009. The ASTER spectral library version 2.0. Remote Sens. Environ. 113, 711–715.

Barsi, J.A., Barker, J.L., Schott, J.R., 2003. An atmospheric correction parameter calculator for a single thermal band earth-sensing instrument. In: Geoscience and Remote Sensing Symposium, 2003. IGARSS'03. Proceedings. 2003 IEEE International, vol. 5, pp. 3014–3016.

Becker, F., Li, Z.-L., 1990a. Temperature-independent spectral indices in thermal infrared bands. Remote Sens. Environ. 32, 17–33.

Becker, F., Li, Z.-L., 1990b. Toward a local split window method over land surface. Int. J. Remote Sens. 11, 369–393.

Becker, F., Li, Z.-L., 1995. Surface temperature and emissivity at various scales: definition, measurement and related problems. Remote Sens. Rev. 12, 225–253.

Borel, C.C., 1998. Surface emissivity and temperature retrieval for a hyperspectral sensor. Proc. IEEE Conf. Geosci. Remote Sens. 504–509.

Caselles, V., Coll, C., Valor, E., 1997. Land surface temperature determination in the whole Hapex Sahell area from AVHRR data. Int. J. Remote Sens. 18, 1009–1027.

Chen, L.F., Liu, Q.H., Chen, S., Tang, Y., Zhong, B., 2004. Definition of component effective emissivity for heterogeneous and non-isothermal surfaces and its approximate calculation. Int. J. Remote Sens. 25, 231–244.

Chen, S.S., Chen, X.Z., Chen, W.Q., Su, Y.X., Li, D., 2011. A simple retrieval method of land surface temperature from AMSR-E passive microwave data—a case study over Southern China during the strong snow disaster of 2008. Int. J. Appl. Earth Obs. Geoinf. 13 (1), 140–151.

Cheng, J., Liang, S., Wang, J., Li, X., 2010. A stepwise refining algorithm of temperature and emissivity separation for hyperspectral thermal infrared data. IEEE Trans. Geosci. Remote Sens. 48, 1588–1597.

Cheng, J., Liu, Q., Li, X., Qing, X., Liu, Q., Du, Y., 2008a. Correlation-based temperature and emissivity separation algorithm. Sci. China Ser. D Earth Sci. 51, 363–372.

Cheng, J., Xiao, Q., Li, X., Liu, Q.-H., Du, Y.-M., 2008b. Multi-Layer perceptron neural network based algorithm for simultaneous retrieving temperature and emissivity from hyperspectral FTIR data. Spectrosc. Spectr. Anal. 28, 780–783.

Cheng, J., Liang, S., 2014. Estimating the broadband longwave emissivity of global bare soil from the MODIS shortwave albedo product. J. Geophys. Res. Atmos. 119, 614–634.

Cheng, J., Liang, S.L., Yao, Y.J., Ren, B.Y., Shi, L.P., Liu, H., 2014. A comparative study of three land surface broadband emissivity datasets from satellite data. Remote Sens. 6, 111–134.

Cheng, J., Liang, S., Yao, Y., Zhang, X., 2013. Estimating the optimal broadband emissivity spectral range for calculating surface longwave net radiation. IEEE Geosci. Remote Sens. Lett. 10 (2), 401–405.

Cheng, J., Liang, S., Verhoef, W., Shi, L., Liu, Q., 2016. Estimating the hemispherical broadband longwave emissivity of global vegetated surfaces using a radiative transfer model. IEEE Trans. Geosci. Remote Sens. 54 (2), 905–917.

Cheng, J., Liang, S., Nie, A., Liu, Q., 2017. Is there a physical linkage between surface emissive and reflective variables over non-vegetated surfaces? J. Indian Soc. Remote Sens. 1–6.

Cristobal, J., Jimenez-Munoz, J.C., Sobrino, J.A., Ninyerola, M., Pons, X., 2009. Improvements in land surface temperature retrieval from the landsat series thermal band using water vapor and air temperature. J. Geophys. Res. 114 https://doi.org/10.1029/2008JD010616.

Duan, S.B., Yan, G.J., Qian, Y.G., Li, Z.L., Jiang, X.G., Li, X.W., 2008. Single channel algorithms for retrieving land surface temperature from HJ-1B simulated data. Prog. Nat. Sci. 18 (9), 1001–1008.

Duan, S.B., Li, Z.L., Leng, P., 2017. A framework for the retrieval of all-weather land surface temperature at a high spatial resolution from polar-orbiting thermal infrared and passive microwave data. Remote Sens. Environ. 195, 107–117.

Fily, M., Royer, A., Goita, K., Prigent, C., 2003. A simple retrieval method for land surface temperature and fraction of water surface determination from satellite microwave brightness temperatures in sub-arctic areas. Remote Sens. Environ. 85, 328–338.

Gao, H., Fu, R., Dickinson, R.E., Juarez, R.I.N., 2008. A practical method for retrieving land surface temperature from AMSR-E over the amazon forest. IEEE Trans. Geosci. Remote Sens. 46, 193–199.

Gillespie, A.R., 1985. Lithologic mapping of silicate rocks using TIMS. In: The TIMS Data Users' Workshop, 86-38. JPL Publication, pp. 29–44.

Gillespie, A.R., Rokugawa, S., Matsunaga, T., Cothern, J.S., Hook, S.J., Kahle, A.B., 1998. A temperature and emissivity separation algorithm for advanced spaceborne thermal emission and reflection radiometer (ASTER) images. IEEE Trans. Geosci. Remote Sens. 36, 1113–1126.

Van de Griend, A.A., Owe, M., 1993. On the relationship between thermal emissivity and the normalized difference vegetation index for natural surfaces. Int. J. Remote Sens. 14, 1119–1131.

Guha, A., Lakshmi, V., 2004. Use of the scanning multichannel microwave radiometer (SMMR) to retrieve soil moisture and surface temperature over the central United States. IEEE Trans. Geosci. Remote Sens. 42, 1482–1494.

Hansen, P.C., 1998. Rand-deficient and Discrete Ill-Posed Problems. Society for Industrial and applied Mathematics, Philadephia, USA.

Holmes, T., De Jeu, R., Owe, M., et al., 2009. Land surface temperature from Ka band (37 GHz) passive microwave observations. J. Geophys. Res. Atmos. 114 (D4).

Hook, S.J., Gabell, A.R., Green, A.A., Kealy, P.S., 1992. A comparison of techniques for extracting emissivity information from thermal infrared data for geological studies. Remote Sens. Environ. 42, 123–135.

Ingram, P.M., Muse, A.H., 2001. Sensitivity of iterative spectrally smooth temperature/Emissivity to algorithmic assumption and measurement noise. IEEE Trans. Geosci. Remote Sens. 39, 2158–2167.

Jiménez-Muñoz, J.C., Sobrino, J.A., Gillespie, A., Sabol, D., Gustafson, W.T., 2006. Improved land surface emissivities over agricultural areas using ASTER NDVI. Remote Sens. Environ. 103 (4), 474–487.

Jiménez-Muñoz, J.C., Cristobal, J., Sobrino, J.A., Soria, G., Ninyerola, M., Pons, X., 2009. Revision of the single-channel algorithm for land surface temperature retrieval from Landsat Thermal-Infrared data. IEEE Trans. Geosci. Remote Sens. 47, 339–349.

Jiménez-Muñoz, J.C., Sobrino, J.A., 2003. A generalized single-channel method for retrieving land surface temperature from remote sensing data. J. Geophys. Res. 108 (D22).

Kanani, K., Poutier, l., nerry, F., stoll, M.-p., 2007. Directional effects consideration to improve out-doors emissivity retrieval in teh 3-13 um domain. Opt. Express 15 (19), 12464–12482.

Knuteson, R.O., Best, F.A., DeSlover, D.H., Osborne, B.J., Revercomb, H.E., Sr, W.L.S., 2004. Infrared land surface temperature remote sensing using high spectral resolution aircraft observations. Adv. Space Res. 33, 1114–1119.

Korb, A.R., Dybwad, P., Wadsworth, W., Salisbury, J.W., 1996. Portable fourier transform infrared spectrometer for field measurements of radiance and emissivity. Appl. Opt. 35, 1679–1692.

Kou, X., Jiang, L., Bo, Y., Yan, S., Chai, L., 2016. Estimation of land surface temperature through blending MODIS and AMSR-E data with the Bayesian maximum entropy method. Remote Sens. 8 (2), 105.

Li, H., Liu, Q., Du, Y., Jiang, J., Wang, H., 2013. Evaluation of the ncep and modis atmospheric products for single channel land surface temperature retrieval with ground measurements: a case study of hj-1b irs data. IEEE J. Sel. Top. Appl. Earth Obs. Remote Sens. 6 (3), 1399–1408.

Li, J., Li, J.L., 2008. Derivation of a global hyperspectral resolution surface emissivity spectra from advanced infrared sounder radiance measurements. Geophys. Res. Lett. 35, L15807. https://doi.org/10.1029/2008GL034559.

Li, X., Strahler, A.H., Friedl, M.A., 1999. A conceptual model for effective directional emissivity from nonisothermal surfaces. IEEE Trans. Geosci. Remote Sens. 37, 2508–2517.

Li, Z.-L., Petitcolin, F., Zhang, R., 2000. A physically based algorithm for land surface emissivity retrieval from combined mid-infrared and thermal infrared data. Sci. China Ser. D Earth Sci. 43, 23–33.

Liang, S., 2001. An optimization algorithm for separating land surface temperature and emissivity from multispectral thermal infrared imagery. IEEE Trans. Geosci. Remote Sens. 39, 264–274.

Liang, S., 2004. Quantitative Remote Sensing of Land Surface. John Wiley and Sons, Inc, Jew Jersey.

Luz, B.R.D., Crowley, J.K., 2007. Spectral reflectance and emissivity features of broad leaf plants: prospects for remote sensing in the thermal infrared (8.0–14.0μm) ☆. Remote Sens. Environ. 109 (4), 393–405.

Ma, X.L., Wan, Z., Moeller, C.C., Menzel, W.P., Gumley, L.E., 2002. Simultaneous retrieval of atmospheric profiles, land-surface temperature, and surface emissivity from Moderate-Resolution Imaging Spectrometer thermal infrared data: extension of a two-step physical algorithm. Appl. Opt. 41, 909–924.

Ma, X.L., Wan, Z., Moller, C.C., Menzel, W.P., Gumley, L.E., Zhang, Y.L., 2000. Retrieval of geophysical parameters from Moderate Resolution Imaging Sepctrometer thermal infrared data: evaluation of a two-step physical algorithm. Appl. Opt. 39, 3537–3550.

Matsunaga, T., 1994. A temperature-emissivity separation method using an empirical relationship between the mean, the maximum, the minimum of the thermal infrared emissivity sepctrum. J. Remote Sens. Soc. Jpn. 14, 230–241.

Mcfarland, M.J., Miller, R.L., Neale, C.M.U., 1990. Land surface temperature derived from the SSM/I passive microwave brightness temperatures. IEEE Trans. Geosci. Remote Sens. 28, 839–845.

McMillin, L.M., 1975. Estimation of sea surface temperature from two infrared window measurements with different absorption. J. Geophys. Res. 20, 11587–11601.

Meng, X., Cheng, J., Liang, S., 2017. Estimating land surface temperature from feng yun-3C/MERSI data using a new land surface emissivity scheme. Remote Sens. 9 (12), 1247.

Meng, X., Cheng, J., 2018. Evaluating eight global reanalysis products for atmospheric correction of thermal infrared sensor—application to landsat 8 TIRS10 data. Remote Sens. 10 (3), 474.

Nerry, F., Petitcolin, F., Stoll, M.P., 1998. Bidirectional reflectivity in AVHRR channel 3: application to a region in northern Africa. Remote Sens. Environ. 66, 298–316.

Niclos, R., Caselles, V., Coll, C., Valor, E., 2007. Determination of sea surface temperature at large observation angles using an angular and emissivity dependent split-window equation. Remote Sens. Environ. 111, 107–121.

Njoku, E.G., Li, L., 1999. Retrieval of land surface parameters using passive microwave measurments at 6-18 GHz. IEEE Trans. Geosci. Remote Sens. 37, 79–93.

Norman, J.M., Becker, F., 1995. Terminology in thermal infrared remote sensing of natural surfaces. Remote Sens. Rev. 12, 159–173.

Ogawa, K., Schmugge, T., Jacob, F., French, A., 2002. Estimation of broadband land surface emissivity from multispectral thermal infrared remote sensing. Agronomie 22 (6), 695–696.

Ogawa, K., Schmugge, T., 2004. Mapping surface broadband emissivity of the sahara desert using ASTER and MODIS data. Earth Interact. 8, 1–14.

Olioso, A., 1995. Simulating the relationship between thermal emissivity and the normalized difference vegetation index. Int. J. Remote Sens. 16, 3211–3216.

Owe, M., Van de Griend, A.A., 2001. On the relationship between thermodynamic surface temperature and high-frequency (37 GHz) vertically polarized brightness under semi-arid conditions. Int. J. Remote Sens. 22, 3521–3532.

Pandya, M.R., Shah, D.B., Trivedi, H.J., Lunagaria, M.M., Pandey, V., Panigrahy, S., et al., 2013. Field measurements of plant emissivity spectra: an experimental study on remote sensing of vegetation in the thermal infrared region. J. Indian Soc. Remote Sens. 41 (4), 787–796.

Payan, v., Royer, A., 2004. Analysis of temperature and emissivity separation (TES) algorithm applicability and sensitivity. Int. J. Remote Sens. 25, 15–37.

Petitcolin, F., Vermote, E., 2002. Land surface reflectance, emissivity and temperature from MODIS middle and thermal infrared data. Remote Sens. Environ. 83, 112–134.

Prata, A.J., Caselles, V., Coll, C., Sobrino, J.A., Ottle, C., 1995. Thermal remote sensing of land surface temperature from satellites: current status and future prospects. Remote Sens. Environ. 12, 175–224.

Present, R.D., 1958. Kinetic Theory of Gases. McGraw-Hill Book Co, NY.

Price, J.C., 1984. Land surface temperature measurements from the split window channels of the NOAA 7 advanced very high resolution radiometer. J. Geophys. Res. 89, 7231–7237.

Qin, Z., Karnieli, A., 2001. A mono-window algorithm for retrieving land surface temperature from Landsat TM data and its application to the Israel-Egypt border region. Int. J. Remote Sens. 22, 3719–3746.

Salisbury, J.W., Wald, A., A'aria, D.M., 1994. Thermal-infrared remote sensing and Kirchhoff's law 1. laboratory measurements. J. Geophys. Res. 99, 11897–11911.

Smith, W.L., Woolf, H.M., 1976. The use of eigenvectors of statistical covariance matrices for interpreting satellite sounding radiometer observations. J. Atmos. Sci. 33, 1127–1140.

Snyder, W.C., Wan, Z., 1998. BRDF modles to predict spectral reflectance and emissivity in the thermal infrared. IEEE Trans. Geosci. Remote Sens. 36, 214–225.

Snyder, W.C., Wan, Z., Zhang, Z., feng, Y.-Z., 1998. Classification-based emissivity for land surface temperature measurement from space. Int. J. Remote Sens. 19, 2753–2774.

Sobrino, J.A., Li, Z.L., Stoll, M.P., Becker, F., 1994. Improvements in the split-window technique for land surface temperature determination. IEEE Trans. Geosci. Remote Sens. 32 (2), 243–253.

Sobrino, J.A., Jiménez-Muñoz, J.C., Sòria, G., Romaguera, M., Guanter, L., Moreno, J., Plaza, A., Martinez, P., 2008. Land surface emissivity retrieval from different VNIR and TIR sensors. IEEE Trans. Geosci. Remote Sens. 46, 316–327.

Sobrino, J.A., Raissouni, N., Li, Z.-L., 2001. A comparative study of land surface emissivity retrieval using NOAA data. Remote Sens. Environ. 75, 256–266.

Sobrino, J.A., Romaguera, M., 2004. Land surface temperature retrieval from MSG1-SEVIRI data. Remote Sens. Environ. 92, 247–254.

Susskind, J., Barnet, C.D., Blaisdell, J.M., 2003. Retrieval of atmospheric and surface parameters from AIRS/AMSU/HSB data in the presence of clouds. IEEE Trans. Geosci. Remote Sens. 41, 390–409.

Tardy, B., Rivalland, V., Huc, M., Hagolle, O., Marcq, S., Boulet, G., 2016. A software tool for atmospheric correction and surface temperature estimation of Landsat infrared thermal data. Remote Sens. 8 (9), 696.

Ulivieri, C., Castronouvo, M.M., Francioni, R., Cardillo, A., 1992. A SW algorithm for estimating land surface temperature from satellites. Adv. Space Res. 14, 59–65.

Valor, E., Caselles, V., 1996. Mapping land surface emissivity from NDVI: application to European, African, and south American areas. Remote Sens. Environ. 57, 167–184.

Verhoef, W., Jia, L., Xiao, Q., Su, Z., 2007. Unified optical-thermal four-stream radiative transfer theory for homogeneous vegetation canopies. IEEE Trans. Geosci. Remote Sens. 45 (6), 1808–1822.

Vidal, A., 1991. Atmospheric and emissivity correction of land surface temperature measured from satellite using ground measurements or satellite data. Int. J. Remote Sens. 12, 2449–2460.

Wan, Z., Dozier, J., 1996. A generalized split-window algorithm for retrieving land-surface temperature form space. IEEE Trans. Geosci. Remote Sens. 34, 892–905.

Wan, Z., Li, Z.-L., 1997. A Physics-based algorithm for retrieving land-surface emissivity and temperature from EOS/MODIS data. IEEE Trans. Geosci. Remote Sens. 35, 980–996.

Wan, Z., Ng, D., Dozier, J., 1994. Spectral emissivity measurements of land-surface materials and related radiative transfer simulations. Adv. Space Res. 14, 91–94.

Wan, Z., Zhang, Y., Zhang, Q.C., Li, Z.-L., 2004. Quality assessment and validation of the MODIS global land surface temperature. Int. J. Remote Sens. 25, 261–274.

Wan, Z., Zhang, Y.L., Zhang, Q.C., Li, Z.-L., 2002. Validation of the land surface temperature products retrieved from terra moderate resolution imaging sepctrometer data. Remote Sens. Environ. 83, 163–180.

Wang, H.S., Li, H., Cao, B., Du, Y., Xiao, Q., Liu, Q., 2012. HIWATER: Dataset of Thermal Infrared Spectrum

Observed by BOMEM MR304 in the Middle Reaches of the Heihe River Basin. Institute of Remote Sensing Applications, Chinese Academy of Sciences, Beijing, China.

Wang, W., Liang, S., Meyer, T., 2008. Validating MODIS land surface temperature products using long-term nighttime ground measurements. Remote Sens. Environ. 112, 623–635.

Wang, X., Ouyang, X.Y., Tang, B.-H., Li, Z.-L., Zhang, R., 2008. A new method for temperature/emissivity separation from hyperspectral thermal infrared data. In: IGARSS'09, pp. 286–289.

Weng, F., Grody, N.C., 1998. Physical retrieval of land surface temperature using the special sensor microwave imager. J. Geophys. Res. 103, 8839–8848.

Xie, R., 1993. Retrieving surface temperature and emissivity from high spectral resolution radiance observations. In: Atmospheric Sciences. University of Wisconsin.

Yu, Y., Tarpley, D., Privette, J.L., Goldberg, M.D., Raja, M.K.R.V., Vinnikov, K.Y., Xu, H., 2009. Developing algorithm for operational GOES-R land surface temperature product. IEEE Trans. Geosci. Remote Sens. 47, 936–951.

Zhou, S., Cheng, J., 2018. A multi-scale wavelet-based temperature and emissivity separation algorithm for hyperspectral thermal infrared data. Int. J. Remote Sens. 39 (22), 8092–8112.

Further reading

Chen, L.F., Zhuang, J.L., Xu, X.R., 2000. Concept and verification of effective specific radiation rate of thermal radiation of non-isothermal mixed pixels. Chin. Sci. Bull. 45, 22–28.

Cheng, J., Liang, S., Liu, Q., Li, X., 2011. Temperature and emissivity separation from ground-based MIR hyperspectral data. IEEE Trans. Geosci. Remote Sens. 49, 1473–1484.

Cheng, J., Liu, Q.H., Li, X.W., 2007. Retrieval of atmospheric trace gases by spaceborne hyperspectral thermal infrared sensors. Remote Sens. Inf. 90–97.

Colwell, R., Simonett, D., Ulaby, F. (Eds.), 1983. Manual of Remote Sensing Interpretation and Applications, second ed., vol. 2. Am. Soc. of Photogramm., Falls Church, Va.

Li, X.W., Wang, J.D., 1999. Definition of surface emissivity for non-isothermal pixels. Chin. Sci. Bull. 44, 1612–1617.

Li, Z.-L., Becker, F., 1993. Feasibility of land surface temperature and emissivity determination from AVHRR data. Remote Sens. Environ. 43, 67–85.

Tang, S.H., Li, X.W., Wang, J.D., Zhu, Q.J., Zhang, L.H., 2006. An improved TES algorithm based on revised ALPHA difference spectrum. Chin. Sci. Seri. D Geosci. 36 (7), 663–671.

Tian, G.L., 2006. Thermal Infrared Remote Sensing. Electronic Industry Press, Beijing.

Wang, X.H., Qiu, S., Jiang, X.G., Ouyang, X.Y., Li, Z.L., 2010. Research on retrieving surface temperature and specific emissivity from hyperspectral thermal infrared data. Arid Reg. Geogr. 33 (3), 419–426.

Wang, T., Shi, J., Yan, G., Zhao, T., Ji, D., Xiong, C., July 2014. Recovering land surface temperature under cloudy skies for potentially deriving surface emitted longwave radiation by fusing MODIS and AMSR-E measurements. In: Geoscience and Remote Sensing Symposium (IGARSS), 2014 IEEE International. IEEE, pp. 1805–1808.

Xu, X.R., 2005. Remote Sensing Physics. Peking University Press, Beijing.

Zhang, R.H., 2009. Quantitative Thermal Infrared Remote Sensing Model and Ground Experiment Basis. Science Press, Beijing.

CHAPTER

8

Surface longwave radiation budget

Advanced Remote Sensing, Second Edition
https://doi.org/10.1016/B978-0-12-815826-5.00008-8

297

© 2020 Elsevier Inc. All rights reserved.

Abstract

This chapter focuses on the estimation of the surface longwave radiation budget (4–100 µm), which includes surface downward longwave radiation, surface upwelling longwave radiation, and surface net longwave radiation. Section 8.1 is dedicated to the background and methods for estimating the surface downward longwave radiation, Section 8.2 presents methods for estimating the surface upwelling longwave radiation, Section 8.3 presents methods for estimating the surface net longwave radiation, and Section 8.4 introduces the long-term ground measurement networks used to validate the surface longwave radiation budget retrieval and the existing surface longwave radiation budget products. A summary is given in Section 8.5.

8.1 Surface downward longwave radiation

8.1.1 Background

Surface downward longwave radiation is the result of atmospheric absorption, emission, and scattering. It is a direct measure of the radiative heating of the surface by the atmosphere. The clear-sky surface downward longwave radiation depends on the vertical profiles of the atmospheric temperature, moisture, and presence of other gases (Ellingson, 1995; Ellingson et al., 1993; Lee and Ellingson, 2002):

$$F_d = 2\pi \int_{\lambda_1}^{\lambda_2} \int_0^1 I_\lambda(z = 0, -\mu)\mu d\mu d\lambda \qquad (8.1)$$

where λ_1 and λ_2 describe the spectral range of the surface downward longwave radiation; λ is the wavelength; z is the altitude; $\mu = \cos(\theta)$, where θ is the local zenith angle; and $I_\lambda(z = 0, -\mu)$ is the downward spectral radiance at the surface. Under clear-sky conditions, I_λ can be expressed as follows:

$$I_{\lambda,clear}(z = 0, -\mu) = -\int_0^{Z_t} B(T_z)\frac{\partial T_\lambda(0, z; -\mu)}{\partial z}dz$$

$$(8.2)$$

where $B(T_z)$ is the Planck function evaluated at altitude z, Z_t is the altitude of the satellite, and T_λ is the transmittance from the surface to the altitude z.

The surface downward longwave radiation is dominated by the radiation from a shallow layer close to the surface of the Earth. The atmosphere above 500 m from the surface only accounts for 16%–20% of the total surface downward longwave radiation. The contribution of the lowest 10 m of the atmosphere accounts for 32%–36% of the total surface downward longwave radiation (Schmetz, 1989). Previous studies have indicated that the atmospheric temperature and moistures profiles are the most important parameters for estimating the clear-sky surface downward longwave radiation. It is sufficient to use the climatological CO_2 and O_3 mass mixing ratios because variations in the mixing ratios of these two gases have small impacts on the surface downward longwave radiation. A 50% change in the mixing ratio of the two species only modifies the surface downward longwave radiation by $1 \, W/m^2$ (Smith and Wolfe, 1983). The cloud base height, cloud base temperature, cloud cover, and cloud emissivity are important parameters under cloudy-sky conditions. Cloud contributions arise mainly from the atmospheric window channels (Schmetz, 1989).

During the past several decades, significant effort has been devoted to estimating the surface downward longwave radiation. Comprehensive reviews of these studies are available from the literature (Niemela et al., 2001; Flerchinger et al., 2009; Diak et al., 2004; Ellingson, 1995; Schmetz, 1989). The methods described in these studies can be divided into three categories: profile-based (physical) methods, hybrid methods, and meteorological parameter–based methods. The profile-based methods calculate the surface downward longwave radiation using a radiative transfer model and satellite-derived or radiosonde atmospheric profiles. In the hybrid methods, the surface downward longwave radiation and top-of-atmosphere (TOA) spectral radiance for a particular sensor are usually first simulated using a radiative transfer model and a large number of atmospheric profiles. Empirical relationships between the surface downward longwave radiation and the TOA radiance or the brightness temperature are then established using statistical analysis. The physics that govern surface downward longwave radiation are embedded in the radiative transfer simulation process. Meteorological parameter–based models estimate the surface downward longwave radiation using screen-level ($\sim 2 \, m$ above the surface) air temperature and moisture measurements. Representative methods from each category will be given in the following subsections.

8.1.2 Profile-based methods

The profile-based methods are straightforward, i.e., they calculate the surface downward longwave radiation using either a radiative transfer model or highly parameterized

equations and atmospheric temperature/moisture profiles. The merit of profile-based methods is their basis in physics. One of their major disadvantages is that errors in the input parameters (atmosphere profiles and cloud parameters) affect the accuracy of the derived surface downward longwave radiation. Moreover, atmospheric profiles are expensive to obtain, and they are not always available.

Darnell et al. (1983) used a radiative transfer model with input data from the NOAA Television and InfraRed Observation Satellite (TIROS) Operational Vertical Sounder (TOVS) products (temperature profile, precipitable water, cloud cover, and effective cloud-top height) to calculate the surface downward longwave radiation for both clear and cloudy skies. In their follow-up study, Darnell et al. (1986) conducted an extensive comparison of their detailed retrieval method with monthly averaged ground measurements at four US sites for a 1-year period. Their validation results showed that the standard error is $10 \, W/m^2$ and R^2 is 0.98.

Frouin et al. (1988) developed a radiative transfer technique for estimating the surface downward longwave radiation over the ocean using TOVS temperature and moisture profiles and cloud parameters (fractional cloud coverage, cloud emissivity, and cloud-top and base height) obtained from Geostationary Operational Environmental Satellite system (GOES) Visible and Infrared Spin Scan Radiometer (VISSR) data. Four methods with different sophistication levels were investigated in the study. In Method A, which was the most refined method, the cloud reflectance in the visible band was determined first. Then, the cloud liquid water path, downward and upward cloud emissivities, cloud fraction, and cloud-top altitude were estimated. Finally, the cloud base altitude was derived from the cloud-top altitude and the cloud liquid water path. In Method B, the downward emissivity and the geometrical thickness of the clouds were obtained from the cloud liquid water path by an empirical formula. In Method C, a constant cloud thickness (0.5 km) was assumed. In Method D, the effect of clouds was only parameterized as a function of the fractional cloud cover. The first two methods are not applicable at night. The validation results showed that the most sophisticated method produced the best result, with a correlation coefficient of 0.73 and standard error of 20.6 W/m² for half-hourly comparisons and 0.83 and 15.7 W/m² for daily comparisons. However, the standard error of the simplest method was only 4 W/m² larger than that of the most sophisticated method. Therefore, the authors recommended the simple methods (C and D) for global applications.

The Clouds and the Earth's Radiant Energy System (CERES) science team adopted three plans for estimating the surface downward longwave radiation. Plan A and Plan C are hybrid-based methods that will be described in Section 8.1.3. In Plan B, the total-sky surface downward longwave radiation was estimated using parameterized equations (Gupta et al., 1997b) that were originally developed for computing the surface downward longwave radiation globally using TOVS meteorological data (Gupta et al., 1997a; Gupta, 1989; Gupta and Wilber, 1992). The CERES science team adapted Gupta (1989) approach to the estimation of surface longwave fluxes using other meteorological data (Gupta et al., 1997b). The input parameters to the algorithm were the surface temperature and emissivity, the temperature and humidity profiles, the fractional cloud amounts, and the cloud-top height from the Meteorology Ozone and Aerosol archival product, the visible infrared scanner, and the Moderate Resolution Imaging Spectroradiometer (MODIS). The cloud base heights and water vapor burden below the cloud base were derived from the above variables. The surface downward longwave radiation was calculated using the following equation:

$$F_d = F_{d,c} + \sum C_{21} A_{c1} \quad (8.3)$$

where $F_{d,c}$ is the clear-sky surface downward longwave radiation, and C_{21} and A_{c1} are the cloud forcing factor and fractional cloud amount for each cloud layer, respectively. The clear-sky

surface downward longwave radiation is computed using the following equation:

$$F_{d,c} = \left(A_0 + A_1 V + A_2 V^2 + A_3 V^3\right) T_e^{3.7} \quad (8.4)$$

where $V = \ln(W)$ (W is the atmospheric water vapor burden) and T_e is the atmospheric effective emitting temperature, which is calculated as follows:

$$T_e = k_s T_s + k_1 T_1 + k_2 T_2 \quad (8.5)$$

where T_1 and T_2 are the mean temperatures of the first (surface-800 hPa) and second (800−680 hPa) atmospheric layers next to the surface. k_s, k_1, and k_2 were determined from the sensitivity analysis. The cloud forcing factor for a particular layer was represented as follows:

$$C_{21} = T_{cb}^4 / \left(B_0 + B_1 W_c + B_2 W_c^2 + B_3 W_c^3\right) \quad (8.6)$$

where T_{cb} is the cloud base temperature and W_c is the water vapor burden below the cloud base for the layer under consideration. The two variables were computed using the following procedures:

1. Cloud base pressure (P_{cb}) was derived by combining the cloud-top pressure with climatological estimates of the cloud thickness.
2. T_{cb} was obtained by matching P_{cb} against the temperature profile.
3. W_c was computed from the humidity profile.

8.1.3 Hybrid methods

Hybrid methods are based on extensive radiative transfer simulations and statistical analysis. Compared to the profile-based methods, this category of methods is less sensitive to errors in the atmospheric profiles. Smith and Wolfe (1983) used a linear regression analysis on 1200 in situ soundings to obtain relations between the VISSR Atmosphere Sounder (VAS) TOA radiance and the surface longwave radiation budget at a pressure level of 1000 hPa for clear-sky and cloudy-sky conditions. They found that the window channels were the most important predictors in both cases. Their statistical clear-sky models explain 98.1% of the variances in the simulated database, with a standard error of 10.3 W/m². Morcrette and Deschamps (1986) estimated the surface downward longwave radiation using regression equations from the second High-Resolution Infrared Sounder (HIRS/2) TOA radiance under clear-sky conditions. The results were compared with hourly ground measurements over three sites in Western Europe, and the standard errors ranged from 1630 W/m². Lee and Ellingson (2002) developed nonlinear surface downward longwave radiation models using HIRS/2 data for both clear-sky and cloudy-sky conditions.

Wang et al. (2009c; 2010; 2009) proposed a general framework of hybrid methods for estimating both surface downward and upwelling longwave

radiation using MODIS and GOES data under clear-sky conditions. Section 8.1.3.1 introduces this general framework. The MODIS and GOES surface downward longwave radiation models developed on the basis of this general framework will be presented in Sections 8.1.3.2 and 8.1.3.3. The hybrid methods adopted by the CERES science team will be presented in Section 8.1.3.4.

8.1.3.1 The general framework of the hybrid methods

The general framework of hybrid methods (Wang and Liang, 2009c, 2010; Wang et al., 2009) (see Fig. 8.1) consists of two steps. The first step is to generate simulated databases using extensive radiative transfer simulation. Surface downward (or upwelling) longwave radiation and TOA radiance for a particular satellite instrument are simulated using a radiative

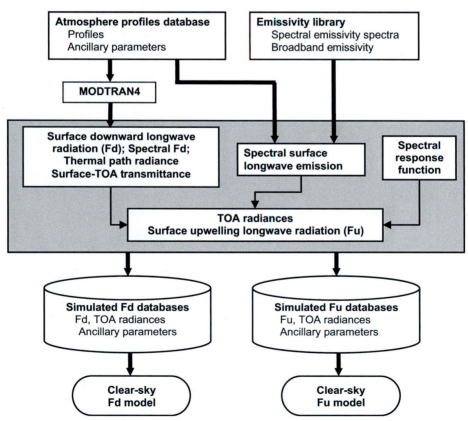

FIGURE 8.1   Flowchart of the general framework of hybrid methods.

transfer model and a large number of clear-sky atmospheric profiles. The physics that govern the surface longwave radiation budget are embedded in the radiative transfer simulation processes. The second step is to conduct a statistical analysis of the simulated databases to derive models. Previous studies (Diak et al., 1996; Liang, 2003; Liang et al., 2005) have applied similar ideas to determine the land surface broadband albedo and surface insolation.

Simulated databases were generated using radiative transfer code, the atmospheric profile database, and the emissivity library by the following steps:

1. The spectral surface downward longwave radiation, thermal path radiance, and transmittance were simulated for each profile using the Moderate Resolution Transmittance Code Version 4 (MODTRAN4) (Berk et al., 1999). The path radiance and transmittance were calculated at five sensor view zenith angles (VZAs) (0, 15, 30, 45, and 60°). The surface temperature was assigned based on the surface air temperature from the temperature profile.
2. The surface downward longwave radiation was simulated for each profile using MODTRAN4.
3. The spectral and integrated surface longwave emissions were calculated for each profile using the Planck function. Eleven surface temperatures were assigned to each profile to simulate the surface conditions for different types of land cover under similar atmospheric conditions. The differences between the surface temperatures and the surface air temperatures were $-10, -8, -6, -4, -2, 0, 2, 4, 6, 8,$ and $10°C$. The University of California Santa Barbara or the John Hopkins University emissivity library (ASTER, 1999; Wan, 1999) was used to account for the surface emissivity effect (see Chapters 5 and 6 for further information).
4. The surface upwelling longwave radiation was synthesized based on Eq. (8.25) in Section 8.2. The broadband emissivity was derived for each emissivity spectrum.

5. The TOA radiance was synthesized based on Eq. (8.7):

$$L = \int_{\lambda_1}^{\lambda_2} \left((L_{\downarrow\lambda} (1 - \varepsilon_\lambda) + \varepsilon_\lambda B(T_s)) \tau_\lambda + L_{p\lambda} \right) SRF d\lambda$$

$$(8.7)$$

where λ is the wavelength, $L_{\downarrow\lambda}$ is the spectral surface downward longwave radiation, ε_λ is the emissivity, T_s is the surface temperature, $B(T_s)$ is the Planck function, τ_λ is the surface-TOA transmittance, $L_{p\lambda}$ is the thermal path radiance, SRF is the spectral response function of a particular sensor channel, and λ_1 and λ_2 specify the spectral range of the channel.

Table 8.1 shows the structure of the simulated databases. The separate simulated databases for the surface downward and upwelling longwave radiation were produced using the atmospheric profile databases and emissivity libraries described above. This framework assumes a Lambertian surface. In addition to the TOA radiance and the surface downward/upwelling longwave radiation, the broadband emissivity, surface elevation, surface temperature, surface air temperature, and column water vapor (CWV) corresponding to each profile are also

TABLE 8.1   The structure of the simulated databases used for developing the surface downward or upwelling longwave radiation models.

Fields	0°	15°	30°	45°	60°
TOA radiance (Channels 1 … n)					
Surface downward or upwelling longwave radiation	Independent of sensor view zenith angle				
Surface elevation (H)					
Surface temperature (T_s)					
Surface air temperature (T_{air})					
Column water vapor					
Broadband emissivity (ε)					

stored in the simulated databases. Only the TOA radiance is dependent on the sensor VZAs among all fields in the database.

A radiative transfer simulation requires representative atmospheric profiles and emissivity spectra of different surfaces. Theoretically, the employment of a greater number of representative atmospheric profiles and emissivity spectra leads to the derivation of a better model. However, one must limit the size of the simulated databases because radiative transfer simulation is time consuming, and the data handling capacity of statistical software packages is restricted. The clear-sky surface downward longwave radiation is dominated by the near-surface atmospheric temperature and moisture conditions, but it is not sensitive to the surface emissivity (ε). Accordingly, the simulated surface downward longwave radiation databases can be generated using a larger atmospheric profile database and a smaller emissivity library. The surface upwelling longwave radiation is dominated by the surface temperature (T_s) and ε, but it is not sensitive to the atmospheric conditions. Therefore, a smaller atmospheric profile database and a larger emissivity library can be used for the simulation of surface upwelling longwave radiation databases.

After the radiative transfer simulation, the surface downward or upwelling longwave radiation models can be derived using regression techniques. Stepwise regression can be used to identify the ideal channels and nonlinear terms used for estimating the surface downward or upwelling longwave radiation. Separate regression coefficients for the five fixed sensor VZAs are estimated for each model. The values at an arbitrary VZA are calculated using linear interpolation.

8.1.3.2 Clear-sky surface downward longwave radiation model for MODIS

8.1.3.2.1 Surface downward longwave radiation model in North America

Eq. (8.8) shows Wang and Liang's (2009c) linear model for estimating the surface downward longwave radiation from MODIS observations based on the general framework:

$$F_d = a_0 + \sum a_i L_i + bH \tag{8.8}$$

where a_0, a_i, and b are the regression coefficients; L_i are the MODIS TOA radiances ($i = 27, 28, 29, 31, 32, 33, 34$); and H is the surface elevation. A total of 10 sets of coefficients corresponding to 5 VZAs (0, 15, 30, 45, and 60°) and 2 observation times (day and night) were estimated. The linear model can account for more than 92% of the variations in the simulated databases, with biases of zero and standard errors of less than 16.50 W/m^2.

Stepwise regression was used to identify the optimal channel combinations on the basis of the standard errors (standard deviation of the difference between the predicted and simulated surface downward longwave radiation values) and the R^2 values of each combination. The results indicated that the MODIS channels 27−29 and 31−34 were the most valuable for predicting the surface downward longwave radiation. This result is consistent with the physics that govern the surface downward longwave radiation: 27 and 29 are water vapor channels; 33 and 34 are near-surface air temperature profile channels; and 29, 31, and 32 are used for retrieving T_s. The surface pressure is also an important factor in the estimation of the surface downward longwave radiation because of the effect of pressure broadening of the spectral lines (Flerchinger et al., 2009; Lee and Ellingson, 2002). In this model, the surface elevation (H) was used as a proxy for the surface pressure to account for the surface pressure effect.

The residual analysis indicated that the surface downward longwave radiation tends to be underestimated under high temperature/moisture conditions and overestimated under low temperature conditions. A nonlinear surface downward longwave radiation model was developed to account for this nonlinear effect.

TABLE 8.2   Nonlinear MODIS clear-sky surface downward longwave radiation model regression coefficients.

	Day					Night				
	0°	15°	30°	45°	60°	0°	15°	30°	45°	60°
$a0$	150.204	153.149	162.142	180.911	214.228	84.143	87.069	95.437	112.646	142.438
$a1$	4.453	4.344	3.909	3.119	2.129	5.365	5.274	4.899	4.184	3.049
$a2$	−1.740	−1.800	−1.989	−2.411	−3.279	−1.782	−1.833	−1.993	−2.374	−3.199
$a3$	−21.030	−20.367	−18.460	−14.022	−3.723	−15.508	−14.870	−13.068	−8.880	0.425
$a4$	32.217	31.676	30.225	−26.553	16.927	27.077	26.520	25.066	21.511	13.061
$b1$	−150.869	−154.969	−167.043	−192.689	−239.237	−106.529	−110.082	−119.872	−140.713	−177.342
$b2$	33.176	34.007	35.638	40.589	53.681	62.673	63.050	63.200	64.904	69.793
$b3$	−26.812	−25.894	−22.376	−16.065	−6.780	−40.546	−39.727	36.611	−30.986	−21.948
$c1$	−1.911	−1.907	−1.902	−1.914	−1.987	−1.984	−1.977	−1.966	−1.962	−2.001

The same set of predictor variables used in the linear model was also used in the nonlinear model:

$$\text{LWDN} = L_{Tair}\left(a_0 + a_1 L_{27} + a_2 L_{29} + a_3 L_{33} \right.$$

$$\left. + a_4 L_{34} + b_1 \frac{L_{32}}{L_{31}} + b_2 \frac{L_{33}}{L_{32}} + b_3 \frac{L_{28}}{L_{31}} + c_1 H \right)$$

$$(8.9)$$

where L_{Tair} represents the surface air temperature (it is equal to L_{31} during the night and L_{32} during the day); a_i, b_i, and c_1 are regression coefficients; and H is the surface elevation. The regression coefficients of the nonlinear models are given in Table 8.2. The three ratio terms represent the effect of water vapor on the surface downward longwave radiation. The signs of the b_i values are consistent with the physics because $\frac{L_{32}}{L_{31}}$ and $\frac{L_{28}}{L_{31}}$ are negatively correlated to the CWV and $\frac{L_{33}}{L_{32}}$ is positively correlated to the CWV. The negative c_1 values are also consistent with the surface pressure effect. The nonlinear model can account for more than 93% of the variations in the simulated databases, which have biases of zero and standard errors of less than 14.90 (night) and 15.20 (day) W/m^2.

Wang and Liang (2009c) applied the nonlinear model to both MODIS Terra and Aqua TOA radiance and validated the model using ground data from six Surface Radiation Budget Network (SURFRAD) sites. The average root mean square errors (RMSEs) of the nonlinear models were 17.60 (Terra, see Fig. 8.2) and 16.17 W/m^2 (Aqua). Gui et al. (2010) evaluated the nonlinear model using 1 year (2003) of ground measurements from 15 sites located in North America, the Tibetan Plateau, Southeast Asia, and Japan with different land cover types and surface elevations. Their results show that the average bias among the 15 sites is −9.2 W/m^2, and the average standard deviation is 21.0 W/m^2. Larger errors may exist among the Tibet sites (which have elevations > 4000 m).

8.1.3.2.2 Surface downward longwave radiation model in globe

In the previous section, Wang and Liang (2009c) introduced a hybrid algorithm for estimating surface longwave radiation on the North America area. If the method is extended to the global scale, a lot of work is needed, such as building an atmospheric profile library. The temperature and humidity profiles have large temporal and spatial variations, and it is very difficult to ensure the representativeness of the constructed atmospheric profiles on a global scale. Therefore, this section developed a new method to estimate surface downward

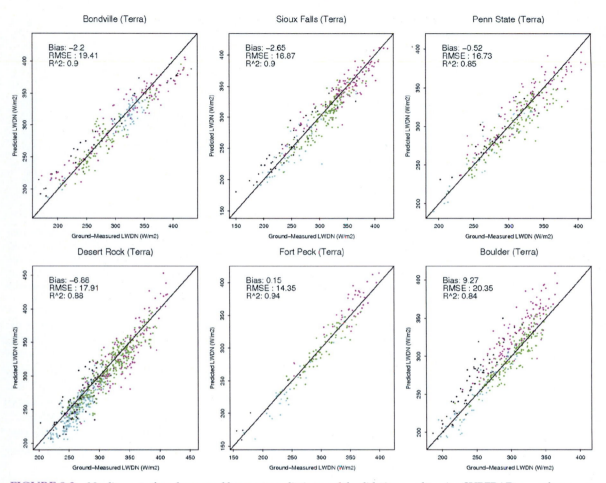

FIGURE 8.2 Nonlinear surface downward longwave radiation model validation results using SURFRAD ground measurements and MODIS Terra data.

longwave radiation using the main atmospheric parameters and avoid the construction of the global atmospheric profile library.

Cheng et al. (2017) developed an efficient hybrid method for estimating 1-km instantaneous clear-sky surface downward longwave radiation (LWDN) from MODIS thermal infrared observations and the MODIS near-infrared CWV data product:

$$LWDN = a_0 + a_1 LWUP + a_2 \log(1 + w)$$
$$+ a_3 \log(1 + w)^2 + a_4 Rad29 \qquad (8.10)$$

where LWUP is the surface upwelling longwave radiation, w is the total CWV, and Rad29 is the MODIS TOA radiance for channel 29.

The primary considerations motivating the use of this formulation are as follows:

1. LWUP is the result of Earth–atmosphere interactions and equals the sum of the surface's self-emission and the LWDN reflected from the surface. Compared to land–surface temperature (LST), a commonly used indicator of Earth–atmosphere interactions (Gupta et al., 2010; Zhou et al., 2007), the hybrid method for estimating LWUP is easy to implement and has a high accuracy (Nie et al., 2016; Wang et al., 2009), while the LST estimate is prone to contamination by many factors such as clouds (Wan, 2008) and occasionally has large

uncertainties (Wang et al., 2008). LWUP was adopted as a proxy of air temperature at screening level to predict LWDN.

2. Water vapor is the major greenhouse gas in the atmosphere and shows the most prominent variation over short time scales. The CWV (also called total precipitable water vapor) and its vertical distribution are critical to LWDN. Previously, CWV was used as one of the predictors to calculate LWDN (Gupta et al., 2010; Zhou et al., 2007) and has been used for calculating the atmospheric downward and upwelling radiance at the window region in the single channel algorithm for LST retrieval (Jimenez Munoz and Sobrino, 2003; Qin et al., 2001). Therefore, we used the same form of CWV as that used by Zhou et al. (2007) to predict LWDN.

3. The weighting function of MODIS water vapor sounding channel 29 peaks at the surface. The TOA radiance of MODIS channel 29 can represent the water vapor at the surface to some extent. Thus, we used the TOA radiance from MODIS channel 29 to characterize the water vapor in the lower atmosphere in this study.

One of the inputs to the proposed hybrid method is the LWUP, whose accuracy will certainly affect the accuracy of the LWDN estimates. Among the sites used, 32 sites have LWUP measurements. The hybrid method for estimating LWUP was validated using these data. The validation results are shown in Fig. 8.3. When the LWUP was smaller than 500 W/m^2, no obvious underestimation or overestimation appeared in the scatterplot, while a few samples with positive bias were noted when the LWUP values were larger than 500 W/m^2. These samples belonged to FLUXNET sites US-SRM and US-Wkg. The exact reasons for this inconsistency are not clear. The bias and RMSE values were 0.569 and 24.291 W/m^2, which are consistent with the results of validation during the algorithm development stage (Cheng and Liang, 2016).

The study validated the proposed method based on 62 globally distributed sites from six

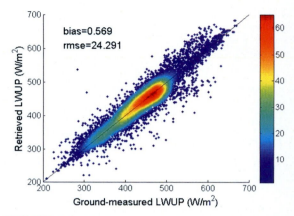

FIGURE 8.3 Validation results of the hybrid method for LWUP estimation. *From Cheng, J., et al. 2017. An efficient hybrid method for estimating clear-sky surface downward longwave radiation from MODIS data. J. Geophys. Res. Atmos. 122 (5), 2616–2630. Copyright © 2017 with permission from American Geophysical Union.*

observational networks (see Fig. 8.4). The validated results showed that the correlation coefficient was 0.951, and the bias and RMSE were 0.081 and 20.929 W/m^2. The LWDN values were overestimated at the low end of the LWDN range, where the ground-measured LWDN values were approximately 150 W/m^2.

At the same time, the method proposed in this study is overestimated in high-altitude areas

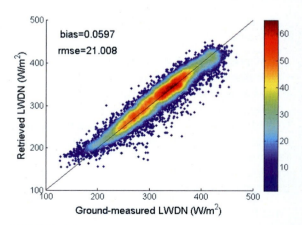

FIGURE 8.4 Validation results of the proposed hybrid method. *From Cheng, J., et al. 2017. An efficient hybrid method for estimating clear-sky surface downward longwave radiation from MODIS data. J. Geophys. Res. Atmos. 122 (5), 2616–2630. Copyright © 2017 with permission from American Geophysical Union.*

TABLE 8.3 Results of comparing Eqs. 8.10 and 8.11 for Sites at High Elevation and With the af and am Climatic Types (Cheng et al., 2017).

Site type	High elevation	af	am
No. of sites	5	3	2
Bias (W/m^2)	9.407(0.924)	13.16 (4.824)	26.743 (20.299)
RMSE (W/m^2)	23.919(19.895)	26.339 (22.858)	30.353 (26.3834)

The values in the parentheses show the results of applying Eq. (8.11). *From Cheng, J., et al., 2017. An efficient hybrid method for estimating clear-sky surface downward longwave radiation from MODIS data. J. Geophys. Res. Atmos. 122 (5), 2616–2630. Copyright © 2017 with permission from American Geophysical Union.*

(CWV < 0.5 g/cm^2); in tropical climates (high CWV), there is an underestimation. To solve these problems, Cheng et al. (2017) proposed a complementary approach:

$$LWDN = 283.157cwv^{0.245} \quad (8.11)$$

The proposed method was verified using observation points Qinghai Flux (37.01°, 101.33°) and D105AWS (33.06°, 91.94°). The results show that the complementary method overcomes the overestimation of high altitude, with bias and RMSE from 9.407 and 23.919 W/m^2 to −0.924 and 19.895 W/m^2, respectively (see Table 8.3).

8.1.3.3 Clear-sky surface downward longwave radiation models for GOES Sounders and GOES-R ABI

GOES Sounders provide similar longwave channels to those provided by MODIS. Following the general framework, GOES Sounder channels 4−6 (lower tropospheric and near-surface air temperature), 7−8 (T_s), and 10−12 (atmospheric water vapor content) were considered for the development of the surface downward longwave radiation model. Eq. (8.12) shows the GOES Sounder nonlinear surface downward longwave radiation model:

$$F_d = a_0 + L_7\left(a_1 + a_2 L_{12} + a_3 L_{10} + a_4 L_5 + a_5 L_4 \right.$$
$$\left. + a_6 \frac{L_7}{L_8} + a_7 \frac{L_5}{L_7} + a_8 \frac{L_{11}}{L_8} + a_9 H \right)$$
$$(8.12)$$

where F_d is the surface downward longwave radiation, L_i is the TOA radiance of channel i (W/m^2/μm/sr), a_j ($j = 0, 9$) is the regression coefficient, and H is the surface elevation. The surface pressure effect on the surface downward longwave radiation was considered using Wang and Liang's method (2009c) (using H as a proxy for the surface pressure). The nonlinear model explains more than 93% of the variations in the simulated databases, with biases of zero and standard errors (e) of less than 16 W/m^2 ($\sim 6\%$ under typical conditions). H explains $\sim 1.5\%$ of the variations. The three ratio terms represent the CWV and explain $\sim 1\%$ of the variations. Table 8.4 shows the GOES-12 Sounder nonlinear surface downward longwave radiation model fitting results. The model was evaluated using ground data from one full year from four SURF-RAD sites, which had RMSEs of less than 22.50 W/m^2 at all four sites.

Wang and Liang (2010) also investigated the feasibility of applying the hybrid method framework to the GOES-R Advanced Baseline Imager data. Their results indicated that the surface downward longwave radiation may not be

TABLE 8.4 GOES-12 Sounder nonlinear surface downward longwave radiation model fitting results.

VZA	0°	15°	30°	45°	60°
a_0	−44.8722	−51.6357	−72.1654	−120.3777	−228.1702
a_1	96.3724	97.1171	99.3169	106.5137	126.8047
a_2	2.6209	2.5206	2.1459	1.5720	0.9204
a_3	1.4669	1.4733	1.5315	1.6141	1.8386
a_4	−27.1931	−26.9319	−26.3887	−25.1835	−22.9124
a_5	35.4864	35.1178	34.3086	32.0538	25.9863
a_6	−92.2204	−93.7095	−97.6262	−108.2041	−131.6922
a_7	43.5084	46.8090	56.2904	79.3296	131.6080
a_8	−27.0000	−26.7329	−25.9316	−24.5630	−22.4952
a_9	−1.9935	−1.9968	−2.0179	−2.0717	−2.1967
SE(e)	14.78	14.82	14.91	15.13	15.63
R^2	0.936	0.936	0.935	0.933	0.930

estimated accurately using the ABI data following the general framework. ABI has only one temperature sounding channel at 13.3 μm (with weighting function peaks at the surface) and no channel at a longer wavelength. The atmospheric radiation above 500 m contributes ~20% of the total flux at the surface (Schmetz, 1989). Wang and Liang's MODIS study (2009c) and the GOES Sounder results (2010) indicate that a lower tropospheric temperature sounding channel at ~13.64 μm (with weighting function peaks at ~2 km above the surface) is necessary to provide atmospheric temperature information from above 500 m. Wang and Liang modified the GOES-12 Sounder to use channels that are only available in the GOES-R ABI. The modified model explains ~1% less of the variation in the simulated database, primarily due to the lack of the 13.64 μm channel. The ground validation results show that this modified model has large RMSEs (>40 W/m^2) and biases (>20 W/m^2) under high Ta conditions.

8.1.3.4 Surface downward longwave radiation hybrid models for CERES

Section 8.1.2 introduced the CERES science team's Plan B surface downward longwave radiation algorithm, which is a profile-based method. The CERES Plan A and Plan C longwave algorithms belong to the hybrid category. CERES measures the broadband TOA radiances in three channels: total (0.2—100 μm), shortwave (0.2—5 μm), and longwave (8—12 μm). The Plan A algorithm (Inamdar and Ramanathan, 1997a,b) is designed for broadband sensors, which is different from other methods introduced in this chapter (for narrowband sensors). This algorithm uses the TOA broadband longwave fluxes (which are derived from the broadband radiance) and other correlated meteorological variables, including the total CWV (w), the surface skin temperature, and the near-surface air temperature, to estimate the surface downward longwave radiation under clear-sky conditions. In this algorithm, the TOA and surface longwave radiations are simulated

using a radiative transfer model, which uses soundings from ships as input. The window (8—14 μm) and nonwindow models of the downward radiation are fitted separately, and the regression coefficients vary with the geographical region (Kratz et al., 2010):

$$F_d = F_{d,win} + F_{d,nw} \qquad (8.13)$$

The surface downward longwave radiation at the window region ($F_{d,win}$) is estimated using the total surface upwelling longwave radiation (F_u, see Section 8.2), the window surface upwelling longwave radiation ($F_{u,win}$), the window upward longwave TOA radiation ($F_{toa,win}$), w, the near-surface air temperature (θ_0), and the air temperature at pressure P_e (θ_a, $P_e = (P_s/2000.0 + 0.45)P_s$), where P_s is the surface pressure in hPa (see Eqs. 8.14). $c_1 - c_6$ are regression coefficients. The emission from the atmospheric water vapor is expressed as a linear term in w because the water vapor absorption does not saturate in the window region. The term $\ln\left(\frac{F_{toa,win}}{F_{u,win}}\right)$ represents the optical depth.

$$
\begin{aligned}
F_{d,win} = {} & c_1\left(F_{u,win} - F_{toa,win}\right) \\
& + \left[c_2 w + c_3 \ln\left(\frac{F_{toa,win}}{F_{u,win}}\right) + c_4\theta_0 + c_5\theta_a\right]F_{toa,win} \\
& + c_6 F_u
\end{aligned}
$$
$$(8.14)$$

The nonwindow surface downward longwave radiation ($F_{d,nw}$) is estimated using F_u, the nonwindow surface upward longwave radiation ($F_{u,nw}$), the nonwindow upward TOA longwave radiation ($F_{toa,nw}$), θ_0, and θ_a (see Eq. 8.15). In the nonwindow region, the water vapor emission to the surface is expressed as $\ln(w)$ because the water vapor absorption in this region tends to have a logarithmic limit.

$$
\begin{aligned}
F_{d,nw} = {} & c_7\left(F_{u,nw} - F_{toa,nw}\right) \\
& + \left[c_8 \ln(w) + c_9\theta_0 + c_{10}\theta_a\right]F_{toa,nw} + c_{11}F_u
\end{aligned}
$$
$$(8.15)$$

The CERES science team adopted the algorithm developed by Zhou et al. (2007) as its Plan C longwave algorithm to maintain two independent algorithms after the replacement of the CERES window channel in future CERES instruments. The Plan C algorithm estimates the clear-sky and cloudy-sky components separately and determines the all-sky surface downward longwave radiation using Eqs. 8.16—8.18:

$$F_{d,all} = F_{d,clr}f_{clr} + F_{d,cld}(1 - f_{clr}) \quad (8.16)$$

$$F_{d,clr} = a_0 + a_1 F_u + a_2 \ln(1 + \text{PWV})$$
$$+ a_3[\ln(1 + \text{PWV})]^2 \quad (8.17)$$

$$F_{d,cld} = b_0 + b_1 F_u + b_2 \ln(1 + \text{PWV})$$
$$+ b_3[\ln(1 + \text{PWV})]^2 + b_4(1 + \text{LWP})$$
$$+ b_5(1 + \text{IWP}) \quad (8.18)$$

where $F_{d,all}$, $F_{d,clr}$, and $F_{d,cld}$ are the all-sky, clear-sky, and cloudy-sky surface downward radiation values, respectively; F_u is the surface upwelling longwave radiation; PWV is the column precipitable water vapor; LWP is the cloud liquid water path; IWP is the ice water path; and a_0–a_3 and b_0–b_5 are regression coefficients. The clear-sky and cloudy-sky models were formulated based on detailed studies of the radiative transfer models and observational data. Zhou et al. (2007) also assumed that the surface downward longwave radiation consists of window and nonwindow (H_2O and CO_2 bands) components. In the nonwindow region, the surface downward longwave radiation is approximately equivalent to a blackbody of atmospheric emitting temperature. PWV and LWP dominate the window region, which involves the water vapor continuum absorption and the effective cloud base emission. LWP is used as a proxy for the cloud base height.

8.1.4 Meteorological parameter—based methods

Satellite and radiosonde observations of the vertical temperature and moisture profiles are not always available. In addition, the surface downward longwave radiation is decoupled from the satellite observations (TOA radiances) under overcast conditions. Many parameterizations have been developed to use surface meteorological measurements, including the screen-level air temperature, vapor pressure, and fractional cloud cover, to estimate the surface downward longwave radiation (Choi et al., 2008; Dilley and O'Brien, 1998; Flerchinger et al., 2009; Idso and Jackson, 1969; Iziomon et al., 2003; Kjaersgaard et al., 2007b; Niemela et al., 2001; Prata, 1996).

8.1.4.1 Bayesian model averaging

In this section, the Bayesian model averaging (BMA) method is introduced to integrate selected bulk formulas for predicting SDLR under all-sky condition for the first time. The BMA method uses the posterior probability as the weight to conduct a weighted average of the possible individual models (Raftery et al., 2005; Hoeting et al., 1999). The posterior probability reflects the models' predictive performance and can be derived in the training period (Raftery et al., 2005).

We employ SDLR_p and SDLR_n denote the predicted and observed SDLR_s at a given time, respectively $f_1, f_1, \ldots, f_n$ is an ensemble of n bulk formulas used to predict SDLR_p. According to the total probability formula, the predictive probability density function (PDF) of SDLR_p based on the multibulk formulas ensemble can be calculated as

$$p(\text{SDLR}_p | f_1, f_1, \ldots, f_n) = \sum_{i=1}^{n} p(\text{SDLR}_p | f_i)$$
$$p(f_i | \text{SDLR}_p)$$

$$(8.19)$$

where $p(f_i|\text{SDLR}_n)$ is the forecast PDF of bulk formula f_i alone, and $p(f_i|\text{SDLR}_n)$ is the posterior probability of bulk formula f_i, which can reflect how well the bulk formula f_i fits the observations. The PDF of BMA is a weighted average of the conditional PDFs of each individual bulk formula, weighted by their posterior model probabilities. The posterior probabilities of all single bulk formulas add up to one, i.e., $\sum_{i=1}^{n} p(f_i|\text{SDLR}_o) = 1$, and they can be viewed as weights w_i. Equation (8.19) can be rearranged as follows:

$$\text{p}(\text{SDLR}_p|f_1, f_1, ..., f_n) = \sum_{i=1}^{n} w_i p(\text{SDLR}_p|f_i)$$

(8.20)

Assuming the conditional PDF of $SDLR_p$ is normally distributed with mean value of E and variance σ^2. Denote the $g(\cdot)$ is used to represent the associated Gaussian PDF of SDLR_p.

$$E(\text{SDLR}_p|f_i) = g(\text{SDLR}_p|\{E_i, \sigma_i^2\})$$

(8.21)

$$\text{p}(\text{SDLR}_p|f_1, f_1, ..., f_n) = \sum_{i=1}^{n} w_i p(\text{SDLR}_p|\{E_i, \sigma_i^2\})$$

(8.22)

The optimal estimation of SDLR by the BMA method is to derive the conditional expected value of SDLR_p, which can be written as follows:

$$E(\text{SDLR}_p|f_1, f_1, ..., f_n) = \sum_{i=1}^{n} w_i E_i)$$

(8.23)

As a result, the key problem is obtaining the posterior probabilities w_i of each model, which makes the estimated SDLR_p closest to the measurement SDLR_n. The w_i can be estimated by maximum likelihood from the training data (Fisher, 1922). The EM algorithm is iterative and alternates between the Expectation (or E) step and the Maximization (or M) step to find the maximum likelihood value (Mclachlan and Krishnan, 2007). Firstly, we assume that all the bulk formulas have the same weight,

then the Expectation step can be written as follows:

$$z_{i,t} = \frac{w_i g(\text{SDLR}_{o,t}|\{E_{i,t}, \sigma_i^2\})}{\sum_{i=1}^{n} w_i g(\text{SDLR}_{o,t}|\{E_{i,t}, \sigma_i^2\})}$$

(8.24)

where the $z_{i,t}$ denotes the probability that the bulk formula i to be the best forecaster at training data t. The M step will estimate the w_i using the $z_{i,t}$ calculated in E step:

$$w_i = \frac{1}{N} \sum_{t} z_{i,t}$$

(8.25)

$$\sigma_i^2 = \frac{1}{N} \sum_{t} \sum_{i=1}^{n} z_{i,t} (E_{i,t} - \text{SDLR}_{o,t})^2$$

(8.26)

where N is the number of observations in the training dataset, and n is the number of bulk formulas. The E and M steps are iterated to convergence, and then the w_i for each bulk formula is obtained.

8.1.4.2 Clear-sky parameterizations

The radiation from a shallow layer close to the surface of the Earth contributes the majority of the surface downward longwave radiation. Meteorological parameter—based clear-sky downward longwave radiation models typically share a general form based on the Stefan—Boltzmann equation:

$$F_{d,c} = \varepsilon_{a,clear} \sigma T_a^4$$

(8.27)

where $F_{d,c}$ is the clear-sky surface downward longwave radiation, σ is the Stefan—Boltzmann constant, T_a is the screen-level air temperature, and $\varepsilon_{a,clear}$ is the atmospheric emissivity. $\varepsilon_{a,clear}$ is usually estimated using T_a, the screen-level vapor pressure (e_0, which is derived from the screen-level dew point temperature), or the precipitable water (w). Table 8.5 summarizes the $\varepsilon_{a,clear}$ models proposed by different investigators.

Compared to the profile-based and hybrid methods introduced in Sections 8.1.2 and 8.1.3, meteorological parameter—based models are usually site-specific, and they are affected by

TABLE 8.5   Parameterizations for estimating the clear-sky atmospheric emissivity in Eq. (8.27). e_0 is the water vapor pressure, T_a is the screen-level air temperature (K), and w is the precipitable water.

Source	Clear-sky emissivity model
Angstrom (1918)	$\varepsilon_{a,clear} = a - b \times 10^{-ce_0}$
Brunt (1932)	$\varepsilon_{a,clear} = 0.52 + 0.065\sqrt{e_0}$
Idso and Jackson (1969)	$\varepsilon_{a,clear} = 1 - 0.261 \exp\left\{ -7.77 \times 10^{-4}(T_a - 273)^2 \right\}$
Brutsaert (1975)	$\varepsilon_{a,clear} = 1.24(e_0/T_a)^{1/7}$
Satterlund (1979)	$\varepsilon_{a,clear} = 1.08\left\{ 1 - \exp\left(-e_0^{T_a/2016}\right) \right\}$
Idso (1981)	$\varepsilon_{a,clear} = 0.7 + 5.95 \times 10^{-5}e_0 \exp(1500/T_a)$
Prata (1996)	$\varepsilon_{a,clear} = 1 - (1+w)\exp(-\sqrt{1.2 + 3.0w})$
Dilley and O'Brien (1998)	$\varepsilon_{a,clear} = 1 - \exp\left[-1.66\left(2.232 - 1.875(T_a/273.16) + 0.7365(w/25)^{0.5}\right) \right]$
Niemela et al. (2001)	$\varepsilon_{a,clear} = 0.72 + 0.009(e_0 - 0.2)$ for $e_0 \geq 0.2$ $\varepsilon_{a,clear} = 0.72 - 0.076(e_0 - 0.2)$ for $e_0 < 0.2$
Iziomon et al. (2003)	$\varepsilon_{a,clear} = 1 - X\exp(-Ye_0/T_a)$ $X = 0.35$, $Y = 10.0$ for lowland sites $X = 0.43$, $Y = 11.5$ for highland sites

geographical locations and local atmospheric conditions (Choi et al., 2008). For example, Brutsaert's (1975) model performs well under warm and wet conditions (Prata, 1996), and Satterlund's (1979) formulation provides better results under conditions with temperatures of less than 0°C. The atmospheric emissivity estimated by Idso's model (1981) is too high under very dry conditions (Niemela et al., 2001; Prata, 1996). Moreover, the coefficients for the meteorological parameter–based models may vary significantly with site. Different coefficients have been reported for the widely used clear-sky atmospheric emissivity model developed by Brunt (1932), with variability as large as 32% (Iziomon et al., 2003; Monteith and Szeicz, 1961; Swinbank, 1963). The accuracy of meteorological parameter–based models is typically 5–30 W/m² , depending on the suitability of the model to the local atmospheric conditions and the data used for model calibration.

Some clear-sky meteorological parameter–based models are not based on the atmospheric emissivity. Swinbank (1963) argued that only the very lowest layers of water vapor affect the surface downward longwave radiation due to the strong absorption lines. He introduced a simple model that only uses the screen-level temperature:

$$F_{d,c} = 5.31 \times 10^{-14}T_a^6 \qquad (8.28)$$

Dilley and O'Brien (1998) reported that the screen-level temperature is a better indicator of the mass of water vapor than the surface vapor pressure, and they incorporated both the screen-level temperature and the precipitable water (kg/m²) terms in their model:

$$F_{d,c} = 59.38 + 113.7(T_a/273.16)^6 + 96.96(w/25)^{0.5} \qquad (8.29)$$

An elevation correction should be included in simple meteorological parameter–based models if the study site is not located at sea level (Flerchinger et al., 2009). Marks and Dozier (1979) suggested an elevation correction to Brutsaert's (1975) clear-sky model:

$$\varepsilon_{a,c,ele} = 1.723\left(e_0'/T_a'\right)^{1/7} \times (P_0/P_{sl}) \qquad (8.30)$$

TABLE 8.6 Typical parameterization scheme for surface downward longwave radiation in clear skies.

Source	Clear-sky model
Brunt (1932)	$\text{LWDR} = \left(a_1 + b_1 e_a^{1/2}\right)\sigma T_a^4$
Swinbank (1963)	$\text{LWDR} = \left(a_2 T_a^2\right)\sigma T_a^4$
Idso and Jackson (1969)	$\text{LWDR} = \left(1 - a_3 \exp\left[b_3(273 - T_a)^2\right]\right)\sigma T_a^4$
Brutsaert (1975)	$\text{LWDR} = \left(a_4 \left(\frac{e_a}{T_a}\right)^{b_4}\right)\sigma T_a^4$
Idso (1981)	$\text{LWDR} = \left(a_5 + b_5 e_a \exp\left[\frac{1500}{T_a}\right]\right)\sigma T_a^4$
Prata (1996)	$\text{LWDR} = \left(1 - \left[\left(1 + 46.5\left(\frac{e_a}{T_a}\right)\right)\exp\left(-\left(a_6 + b_6 46.5\left(\frac{e_a}{T_a}\right)\right)^{\frac{1}{2}}\right)\right]\right)\sigma T_a^4$
Carmona et al. (2014)	$\text{LWDR} = \left(a_7 + b_7 T_a + d_7 Rh\right)\sigma T_a^4$

From Guo, Y., Cheng, J., Liang, S., 2018. Comprehensive assessment of parameterization methods for estimating clear-sky surface downward longwave radiation. Theor. Appl. Climatol. (6), 1–14. Copyright © 2017 with permission from Springer-Verlag GmbH Austria, part of Springer Nature.

where T_a' and e_0' are the screen-level temperature and vapor pressure, respectively, adjusted to their sea-level equivalents; P_0 is the surface pressure (in hPa) at the site; and P_{sl} is the sea-level pressure. Deacon (1970) proposed an elevation correction to Swinbank's (1963) clear-sky model:

$$F_{d,c,ele} = F_{d,c} - 3.5 \times 10^{-5} z\sigma T_a^4 \qquad (8.31)$$

where z is the elevation (in meters).

Therefore, to make full use of the advantages of various parametric models and obtain the stability of the developed parametric models at large regional scales or global scales, Guo et al. (2018) selected seven typical parametric schemes, and these parametric schemes are shown in Table 8.6. These seven parameterization scheme coefficients were corrected using the site data 2/3 described in Guo et al. (2018), and the corrected parametric models were validated using the remaining 1/3 of the observation data. The verification results showed that Swinbank (1963) and Idso and Jackson (1969) are significantly lower in accuracy than the other five. Therefore, this section uses BMA method to integrate five parameterization schemes with higher accuracy to obtain surface downward

longwave radiation integration results. The specific verification results are shown in Section 8.1.4.4.

Besides, the clear sky was identified using a cloud fraction that was calculated using the following equation (Crawford and Duchon, 1999):

$$c = 1 - \frac{S_W\downarrow}{S_W\uparrow_0} \qquad (8.32)$$

where $S_W\downarrow$ is the observed downward shortwave radiation, $S_W\uparrow_0$ is the theoretical value of downward shortwave radiation; when $c < 0.05$, we define the situation as clear sky. Note that nighttime data were excluded because $S_W\downarrow$ was not available. The input parameter e_a (in hPa) was calculated by

$$e_a = e_s \left(\frac{RH}{100}\right) = \left(6.108 \exp\left[\frac{17.27 T_a}{T_a + 273.3}\right]\right) \left(\frac{RH}{100}\right)$$

$$(8.33)$$

where e_s (in hPa) is the saturation vapor pressure. T_a and relative humidity (RH) are in degrees centigrade and percentage, respectively. The derived clear-sky data for each site were randomly divided into two parts.

8.1.4.3 All-sky parameterizations

The presence of clouds increases the atmospheric longwave radiation at the surface because the radiation from water vapor and other gases in the lower atmosphere is supplemented by the cloud emission in the wavelength that the gas emission lacks. Most all-sky models modify the clear-sky atmospheric emissivity using the cloud cover. Table 8.7 summarizes the all-sky atmospheric emissivity models developed by different investigators. The accuracies of all-sky models are similar to those of the clear-sky models. In addition, both the all-sky and clear-sky model coefficients require local calibration in the same manner.

TABLE 8.7   All-sky atmospheric emissivity parameterizations. c is the fractional cloud cover.

Source	All-sky model
Brunt (1932)	$\varepsilon_{a,cloudy} = (1 + 0.22c)\varepsilon_{a,clear}$
Jacobs (1978)	$\varepsilon_{a,cloudy} = (1 + 0.26c)\varepsilon_{a,clear}$
Maykut and Church (1973)	$\varepsilon_{a,cloudy} = (1 + 0.22c^{2.75})\varepsilon_{a,clear}$
Sugita and Brutsaert (1993)	$\varepsilon_{a,cloudy} = (1 + uc^v)\varepsilon_{a,clear}$ u and v are coefficients
Unsworth and Monteith (1975)	$\varepsilon_{a,cloudy} = (1 - 0.84c)\varepsilon_{a,clear} + 0.84c$
Crawford and Duchon (1999)	$\varepsilon_{a,cloudy} = (1 - c) + c\varepsilon_{a,clear}$
Lhomme et al. (2007)	$\varepsilon_{a,cloudy} = 1.18 \times (1.37 - 0.34c)\left(\frac{e_0}{T_a}\right)^{1/7}$
Iziomon et al. (2003)	$\varepsilon_{a,cloudy} = 1 - X \exp(-Ye_0/T_a)(1 + ZN^2)$ N is the cloud cover (in Okta) $X = 0.35$, $Y = 10.0$, $Z = 0.0035$ for lowland sites $X = 0.43$, $Y = 11.5$, $Z = 0.0050$ for highland sites

Kimball et al. (1982) developed an all-sky model based on Idso's (1981) clear-sky equation, assuming that clouds contribute additional radiation and that the atmosphere is opaque at wavelengths outside the 8–14 μm window region. This model supports a maximum of four layers of clouds:

$$F_{d,cloudy} = F_{d,clear} + \sum_i \tau_{8i} c_i \varepsilon_i f_{8i} \sigma T_{ci}^4 \qquad (8.34)$$

$$\tau_{8i} = 1 - \varepsilon_{8zi}(1.4 - 0.4\varepsilon_{8zi})$$

$$\varepsilon_{8zi} = 0.24 + 2.98 \times 10^{-6} e_0^2 \exp(3000/T_a)$$

$$f_{8i} = -0.6732 + 0.6240 \times 10^{-2} T_{ci} - 0.9140 \times 10^{-5} T_{ci}^2$$

where τ_{8i} is the atmospheric transmittance in the 8–14 μm window, ε_{8zi} is the 18–14 μm atmospheric emittance in the zenith direction, and f_{8i} is the fraction of the blackbody radiation emitted in the 8–14 μm window at cloud temperature. c_i, ε_i, and T_{ci} are the fractional cloud cover, cloud emittance, and cloud base temperature at level i, respectively.

Diak et al. (2000) proposed a method to estimate the total-sky surface downward longwave radiation:

$$F_d = F_{d,c} + (1 - \varepsilon_{a,c})c\sigma T_c^4 \qquad (8.35)$$

where $F_{d,c}$ is the clear-sky surface downward longwave radiation, $\varepsilon_{a,c}$ is the clear-sky emissivity, c is the effective cloud fraction derived from the real-time GOES cloud product, and T_c is the cloud temperature. $F_{d,c}$ and ε_a are estimated using Prata's (1996) clear-sky equation.

The fractional cloud cover is required to estimate the surface downward longwave radiation under all-sky conditions. The fractional cloud cover can be obtained from either visual observations or satellite data. Several satellite-derived cloud cover products are available (such as MODIS and GOES cloud products). During the day, the fractional cloud cover can also be estimated using the incoming surface shortwave radiation and the theoretical incoming clear-sky surface shortwave radiation (Crawford and Duchon, 1999; Lhomme et al., 2007, see Eq. 8.32).

Wang and Liang (2009b) evaluated two widely accepted meteorological parameter—based methods, namely the Brunt (1932) equation and the Brutsaert (1975) equation, and they estimated the global total-sky atmospheric downward longwave radiation using meteorological observations at 36 globally distributed sites. The Crawford and Duchon (1999) parameterization was chosen to calculate the cloud effect on the downward longwave radiation. Their results suggest that the two methods could be applied to most of the Earth's land surfaces. The Brunt (1932) equation is more accurate for elevations less than 1000 m, while the Brutsaert (1975) equation is more accurate for elevations greater than 1000 m. Overall, the instantaneous surface downward longwave radiation under all-sky conditions is estimated to have an average bias of 2 W/m^2 (0.6%), an average standard deviation of 20 W/m^2 (6%), and an average

correlation coefficient of 0.86; the daily downward longwave radiation under all-sky conditions is estimated to have a standard deviation of 12 W/m^2 (3.7%) and an average correlation coefficient of 0.93. Wang and Liang (2009b) applied the two methods to the globally available meteorological observations from 3200 stations and estimated the decadal variation in the downward longwave radiation. Their results showed that the daily downward longwave radiation increased at an average rate of 2.2 W/m^2 per decade from 1973 to 2008 (see Fig. 8.5). This rising trend is primarily the result of increases in the air temperature, atmospheric water vapor, and CO_2 concentration.

Similarly, to make full use of the advantages of various parametric models and obtain the stability of the developed parametric models at large regional scales or global scales, Cheng et al. (2019) selected seven typical cloudy parameterization schemes. Their specific forms are shown in Table 8.8. Similar to the clear-sky estimation method, firstly, the effective data of the cloudy sky are extracted according to the cloud

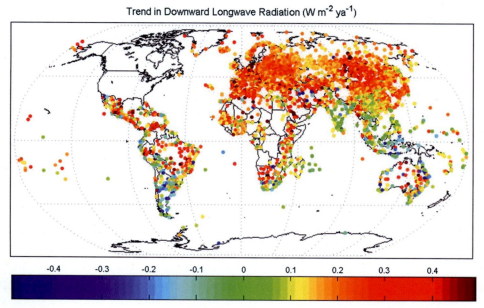

FIGURE 8.5   Linear trend of daily downward longwave radiation at 3200 global stations (1973—2008). Each point in the figure represents one station, and the colors of the points indicate the trend values (Wang and Liang, 2009b).

TABLE 8.8 Seven cloudy-sky bulk formulas for surface downward longwave radiation (SDLR) estimation.

Bulk formulas	Equation
Jacobs (1978)	$SDLR = SDLR_{clr}(1 + a_1 c)$
Lhomme et al. (2007)	$SDLR = SDLR_{clr}(a_2 + b_2 c)$
Maykut and Church (1973)	$SDLR = SDLR_{clr}(1 + a_3 c^{b_3})$
Konzelmann et al. (1994)	$SDLR = SDLR_{clr}(1 - a_4 c^{b_4}) + d_4 c^{e_4} \sigma T^4$
Crawford and Duchon (1999)	$SDLR = SDLR_{clr}(1 - c) + c\sigma T^4$
Carmona et al. (2014)	$SDLR = [(a_6 + b_6 T + d_6 RH)(1 - c) + c]\sigma T^4$
Carmona et al. (2014)	$SDLR = (a_7 + b_7 T + d_7 RH + e_7 c)\sigma T^4$

Carmona et al. (2014) denote the first and second multiple linear regression models of Carmona et al. (2014), respectively. The selected seven bulk formulas are abbreviated as Jacobs, Lhomme, MC, Konzelmann, Crawford, Carmona et al. (2014), respectively. a, b, d, and e are the coefficients, $SDLR_{clr}$ is the clear-sky SDLR, T is the screen-level temperature (unit: K), σ is the Stephen–Boltzmann constant ($5.67 \cdot 10^{-8}$ W/m^2/K^4), RH is the relative humidity (%), and c is the cloud fraction.
From Cheng, J., Yang, F., Guo, Y., 2019. A comparative study of bulk parameterization schemes for estimating cloudy-sky surface downward longwave radiation. Remote Sens. 11 (5), 528. Copyright © 2019 with permission from MDPI.

amount, and it is judged to be cloudy when $c > 0.05$. The models were corrected using the site observation data 2/3 listed in the text, and then the models were validated using the remaining 1/3 of the site observations. The verification results showed that the accuracy of these seven models is very high, so at the same time, we integrated the seven parameterization schemes to obtain surface downward longwave radiation using the BMA method. The specific verification results were shown in Section 8.1.4.4.

8.1.4.4 Verification based on ground measurement data

8.1.4.4.1 Verification in clear skies

Because the coefficients of each parametric model are derived from the data of a particular

region, and have certain limitations, the coefficients should be calibrated according to the state of the study area before use. The global coefficient correction results are shown in Table 8.9. In general, the coefficient difference is relatively large. Except for the Swinbank (1963) and Prata (1996) two parameterization schemes, the coefficient variation is less than 15%, and other coefficient changes are above 15%. When using the parameterization scheme to estimate the surface downward longwave radiation, it is very necessary to perform coefficient correction according to the study area.

The data of all sites are combined to obtain the overall verification accuracy. Fig. 8.6 is a comparison verification scatterplot of the surface downward longwave radiation observation value with seven parameterization schemes and BMA estimation values, and Table 8.10 is the statistical result corresponding to Fig. 8.7. It can be seen that the bias obtained from the original coefficient is $-16.96 \sim 15.99$ W/m^2, the RMSE is $22.23 \sim 36.65$ W/m^2, and the bias after the correction coefficient is $-4.53 \sim 0.01$ W/m^2. The RMSE is $20.35 \sim 34.38$ W/m^2, and the verification accuracy is greatly improved. With the exception of Idso and Jackson (1969), other parameterization schemes are underestimated. Because Swinbank (1963) and Idso and Jackson (1969) two parameterization schemes only consider temperature factors and do not consider humidity, their RMSE is significantly higher than other parameterization schemes, and the accuracy is worse than other parametric schemes. This is consistent with many existing studies (Duarte et al., 2006; Kjaersgaard et al., 2007a,b; Kruk et al., 2010). Carmona et al. (2014) has the highest accuracy, with a bias of -0.11 W/m^2 and an RMSE of 20.35 W/m^2, followed by Idso (1981); Prata (1996); Brunt (1932), and Brutsaert (1975). The BMA integration results are similar to Carmona et al. (2014), but BMA is more stable than other parametric schemes.

TABLE 8.9 Comparison of original and adjusted coefficient values for seven clear-sky parameterization methods.

Parameterization schemes	Coefficients	Adjusted coefficients	Original coefficients	Relative difference (%)
Brunt (1932)	$a1$	0.63	0.52	21.88
	$b1$	0.04	0.07	−34.46
Swinbank (1963)	$a2$	9.01×10^{-6}	9.37×10^{-6}	−3.83
Idso and Jackson (1969)	$a3$	0.26	0.26	−1.88
	$b3$	-2.90×10^{-4}	-7.77×10^{-4}	−62.67
Brutsaert (1975)	$a4$	1.05	1.24	−15.68
	$b4$	0.09	1/7	−38.53
Idso (1981)	$a5$	0.68	0.7	−2.34
	$b5$	4.69×10^{-5}	5.95×10^{-5}	−21.23
Prata (1996)	$a6$	1.35	1.2	12.26
	$b6$	2.78	3	−7.55
Carmona et al. (2014)	$a7$	−0.44	−0.34	28.62
	$b7$	3.7×10^{-3}	3.36×10^{-3}	10.12
	$d7$	2.7×10^{-3}	1.94×10^{-3}	39.18

From Guo, Y., Cheng, J., Liang, S., 2018. Comprehensive assessment of parameterization methods for estimating clear-sky surface downward longwave radiation. Theor. Appl. Climatol. (6), 1–14. Copyright © 2017 with permission from Springer-Verlag GmbH Austria, part of Springer Nature.

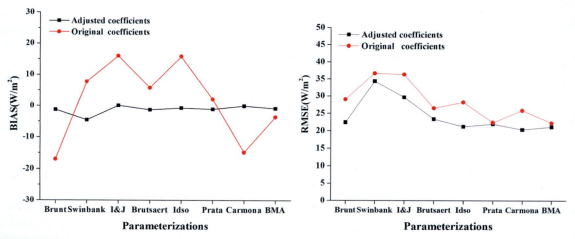

FIGURE 8.6 Accuracy of parameterization schemes and the BMA method under clear sky. *From Guo, Y., Cheng, J., Liang, S., 2018. Comprehensive assessment of parameterization methods for estimating clear-sky surface downward longwave radiation. Theor. Appl. Climatol. (6), 1–14. Copyright © 2017 with permission from Springer-Verlag GmbH Austria, part of Springer Nature.condition.*

TABLE 8.10 Statistical results of seven parameterization models and BMA using original and adjusted coefficients.

Parameterization schemes	Adjusted coefficients			Original coefficients		
	Bias	RMSE	R^2	Bias	RMSE	R^2
Brunt (1932)	−1.27	22.41	0.91	−16.96	29.05	0.91
Swinbank (1963)	−4.53	34.38	0.83	7.72	36.65	0.83
Idso and Jackson (1969)	0.01	29.62	0.81	15.99	36.31	0.83
Brutsaert (1975)	−1.36	23.35	0.9	5.77	26.5	0.91
Idso (1981)	−0.79	21.21	0.92	15.82	28.19	0.92
Prata (1996)	−1.13	21.96	0.91	2.05	22.41	0.92
Carmona et al. (2014)	−0.11	20.35	0.92	−14.81	25.85	0.91
BMA	−0.89	21.13	0.92	−3.60	22.23	0.92

From Guo, Y., Cheng, J., Liang, S., 2018. Comprehensive assessment of parameterization methods for estimating clear-sky surface downward longwave radiation. Theor. Appl. Climatol. (6), 1–14. Copyright © 2017 with permission from Springer-Verlag GmbH Austria, part of Springer Nature.

8.1.4.4.2 Verification in cloudy skies

Similar to the clear-air coefficient correction method, the data are divided into two parts, and the data of 3/2 are used for coefficient correction, and the global coefficients of six parameterization schemes on cloudy days are obtained, as shown in Table 8.11, where bulk formula from Crawford and Duchon (1999) has no variable coefficients, thus it is not listed in Table 8.11 where Crawford and Duchon (1999) and Konzelmann et al. (1994) have a special case with a model coefficient of 1, with no variable coefficients. It can be seen from the results that the change of the parameters of the cloudy parameterization scheme is more than 30% except for Lhomme et al. (2007), and more than half of the coefficient changes are greater than 50%.

The accuracy of seven kinds of parameterization schemes and BMA is shown in Fig. 8.7. Table 8.12 shows the accuracy statistics. The bias before correction is 29.76 W/m², the RMSE is up to 47.15 W/m², whereas the bias after coefficient correction is all within 10 W/m², the RMSE are within 30 W/m², suggesting that the accuracy has been greatly improved. Except for Carmona et al. (2014) and Crawford and Duchon (1999), others have different degrees of underestimation, as opposed to clear-sky conditions. Carmona et al. (2014) has the highest accuracy, with bias at 0 W/m² and RMSE at 20.13 W/m². The BMA integration results are similar to Carmona et al. (2014), but BMA is more stable than other parametric schemes.

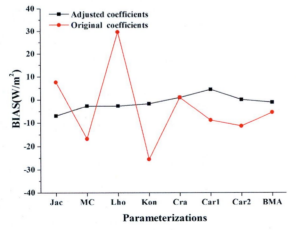

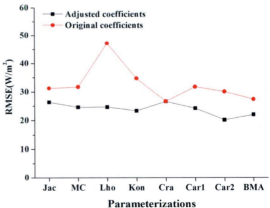

FIGURE 8.7 Accuracy of parameterization schemes and the BMA method under cloudy-sky condition (Jac, Lho, Kon, Cra, Car1, and Car2 represent Jacobs, Lhomme, Konzelmann, Crawford and Duchon, Carmona et al. (2014), respectively). *From Cheng, J., Yang, F., Guo, Y., 2019. A comparative study of bulk parameterization schemes for estimating cloudy-sky surface downward longwave radiation. Remote Sens., 11 (5), 528. Copyright © 2019 with permission from MDPI.*

TABLE 8.11 Variation of coefficient values for six cloudy-sky bulk formulas before and after calibration.

Bulk formula	Coefficients	Original coefficients	Adjusted coefficients	Relative difference (%)
Jacobs (1978)	$a1$	0.26	0.17	−34.62
Maykut and Church (1973)	$a2$	0.22	0.12	−45.45
	$b2$	2.75	0.30	−89.09
Lhomme et al. (2007)	$a3$	1.03	1.06	2.91
	$b3$	0.34	0.07	−79.41
Konzelmann et al. (1994)	$a4$	1.00	0.29	−71.00
	$b4$	4.00	0.23	−94.25
	$d4$	0.95	0.34	−64.29
	$e4$	4.00	0.27	−93.25
Carmona et al. (2014)	$a6$	−0.88	0.33	−137.5
	$b6$	5.20×10^{-3}	8.05×10^{-4}	−84.512
	$d6$	2.02×10^{-3}	3.90×10^{-3}	93.07
Carmona et al. (2014)	$a7$	−0.34	0.55	−261.77
	$b7$	3.36×10^{-3}	5.05×10^{-4}	−84.97
	$d7$	1.94×10^{-3}	2.57×10^{-3}	32.47
	$e7$	0.21	5.60×10^{-2}	−73.71

From Cheng, J., Yang, F., Guo, Y., 2019. A comparative study of bulk parameterization schemes for estimating cloudy-sky surface downward longwave radiation. Remote Sens. 11 (5), 528. Copyright © 2019 with permission from MDPI.

As mentioned above, in this section, we explored the parameterization method and BMA method for estimating surface downward longwave radiation. This section is divided into three parts: introduction of the BMA principle and parameterization method, introduction of the widely used parameterization method under clear-sky and all-sky conditions, and the accuracy analysis and verification of seven kinds of models. Through the verification and analysis of selected parameterization schemes using the site observation data described in Guo et al. (2018) and Cheng et al. (2019), we can draw the following conclusions:

(1) When using the parameterization method, the model coefficients must be calibrated first;

(2) Different parameterization methods have certain regional applicability, and the accuracy of surface downward longwave radiation estimate using BMA to integrate various parameterization methods is at least not worse than that using other parameterization methods alone;

(3) Under clear-sky conditions, Carmona et al. (2014) has the highest precision, followed by Idso (1981), Prata (1996), Brunt (1932), and Brutsaert (1975). The BMA integration method and Carmona et al. (2014) have similar precision. Carmona et al. (2014) has the highest precision under all-sky conditions, followed by Konzelmann et al., (1994), Lhomme et al. (2007), and Maykut and Church (1973).

TABLE 8.12 Variation of coefficient values for six cloudy-sky bulk formulas before and after calibration.

Bulk formula	Coefficients	Original coefficients	Adjusted coefficients	Relative difference (%)
Jacobs (1978)	$a1$	0.26	0.17	−0.36
Maykut and Church (1973)	$a2$	0.22	0.12	−0.44
	$b2$	2.75	0.30	−0.89
Lhomme et al. (2007)	$a3$	1.03	1.06	0.03
	$b3$	0.34	0.07	−0.80
Konzelmann et al. (1994)	$a4$	1	0.29	−0.71
	$b4$	4	0.23	−0.94
	$d4$	0.952	0.34	−0.64
	$e4$	4	0.27	−0.93
Carmona et al. (2014)	$a6$	−0.88	0.33	−1.38
	$b6$	$5.2 \cdot 10^{-3}$	$8.05 \cdot 10^{-4}$	−0.85
	$d6$	$2.02 \cdot 10^{-3}$	$3.90 \cdot 10^{-3}$	0.93
Carmona et al. (2014)	$a7$	−0.34	0.55	−2.61
	$b7$	$3.36 \cdot 10^{-3}$	$5.05 \cdot 10^{-4}$	−0.85
	$d7$	$1.94 \cdot 10^{-3}$	$2.57 \cdot 10^{-3}$	0.32
	$e7$	0.213	$5.6 \cdot 10^{-2}$	−0.74

From Cheng, J., Yang, F., Guo, Y., 2019. A comparative study of bulk parameterization schemes for estimating cloudy-sky surface downward longwave radiation. Remote Sens. 11 (5), 528. Copyright © 2019 with permission from MDPI.

8.2 Surface upwelling longwave radiation

Surface upwelling longwave radiation, which is dominated by the surface temperature (see Chapter 7), is an indicator of how warm the Earth's surface is. This section focuses on methods used to estimate the surface upwelling longwave radiation. Three methods will be presented. Section 8.2.1 describes the widely used temperature-emissivity method, and Section 8.2.2 presents the linear and ANN model methods developed using the general framework of hybrid methods (see Section 8.1.3).

8.2.1 Temperature-emissivity method

Theoretically, the surface upwelling longwave radiation consists of two components: the surface longwave emission and the reflected surface downward longwave radiation (Liang, 2004):

$$F_u = \varepsilon \int_{\lambda_1}^{\lambda_2} \pi B(T_s) d\lambda + (1 - \varepsilon) F_d \qquad (8.36)$$

where F_u is the surface upwelling longwave radiation, ε is the surface broadband emissivity, T_s is the surface temperature, $B(T_s)$ is Planck's function, λ_1 and λ_2 specify the spectral range of surface upwelling longwave radiation, and F_d is the surface downward longwave radiation. Three parameters are required to accurately estimate the surface upwelling longwave radiation: T_s, ε, and the surface downward longwave radiation. T_s and ε can be estimated using the methods presented in Chapter 8, and the surface downward longwave radiation can be estimated using methods given in Section 8.1.

The surface upwelling longwave radiation is dominated by the surface longwave emission. During the past several decades, it has generally been believed that the surface upwelling longwave radiation can be accurately estimated from satellite-derived surface temperature and emissivity products. However, while the sea surface temperature can be estimated with high accuracy (~ 0.5K) and the sea surface emissivity is mostly uniform, estimating the LST and emissivity remains challenging. The accuracy of the available satellite-derived LST products is approximately 1–3K, and the accuracy of the land surface emissivity remains poorly known.

8.2.2 Hybrid methods

The clear-sky TOA radiance contains information regarding the surface temperature, emissivity, and surface downward longwave radiation. The hybrid method derives the surface upwelling longwave radiation directly from the satellite TOA radiance or BT without separately estimating the three variables on the right-hand side of Eq. (8.25). The surface upwelling longwave radiation and surface downward longwave radiation have the same order of magnitude. The two terms on the right side of Eq. (8.25) have opposite signs, which partly mitigates the errors in the surface emissivity (Diak et al., 2000). The advantage of this method is that the problem of separating the LST and emissivity is bypassed. As a result, a more accurate estimation of the surface upwelling longwave radiation may be achieved.

The hybrid method has mainly been used to estimate the surface downward longwave radiation (Inamdar and Ramanathan, 1997a; Lee, 1993; Lee and Ellingson, 2002; Morcrette and Deschamps, 1986; Smith and Wolfe, 1983). Smith and Wolfe (1983) used a hybrid method to estimate the surface upwelling and downward longwave radiation at 1000 hPa pressure level using the NOAA geostationary satellites VISSR

VAS TOA radiance. Meerkoetter and Grassl (1984) used the hybrid method to estimate the surface upwelling and net longwave radiation using split-window radiance data from the Advanced Very High-Resolution Radiometer (AVHRR). However, these studies focused on estimating the surface longwave radiation over sea surfaces, and constant emissivity was usually assumed.

Based on the framework of the hybrid model presented in Section 8.1.3, Cheng et al. (2016) established a model for estimating upwelling longwave radiation from the surface under clear skies using MODIS data. They explicitly considered the emissivity effect in the radiation transfer simulation process. The first and second subsections below introduce a linear model based on MODIS data and a dynamic learning neural network (DLNN) model. Section 8.2.2.3 presents a surface upwelling longwave radiation estimation model based on Visible Infrared Imaging Radiometer Suite (VIIRS)/S-NPP data. Section 8.2.2.4 describes the surface upwelling longwave radiation model based on GOES Sounder and GOES-R ABI data.

8.2.2.1 MODIS linear surface upwelling longwave radiation model

The estimation of surface longwave radiation using the hybrid model involves two steps: (1) establish a training database. For a specific sensor, using a large number of representative atmospheric profile data and typical ground emissivity spectra, the surface upwelling longwave radiation and TOA radiance can be simulated by the radiation transfer model. (2) Analyze the database and establish a model between multichannel TOA radiance and surface longwave radiation. The analysis results show that the surface emission contributes the most to the surface longwave radiation. The channel 29, 31, and 32 of MODIS are sensitive to the change of surface temperature and are in the atmospheric window, which are most suitable to estimate the surface longwave radiation and also consistent with the physical process of upwelling longwave radiation. Under the framework of the above hybrid model, Cheng et al. proposed a linear model for estimating global surface upwelling longwave radiation with consideration of physical laws and higher computing efficiency. The linear model formula is as follows:

$$\text{LWUP} = a_0 + a_1 L_{29} + a_2 L_{31} + a_3 L_{32} \quad (8.37)$$

where a_0, a_1, a_2, and a_3 are regression coefficients and L_{29}, L_{31}, and L_{32} are the TOA radiances for MODIS channels 29, 31, and 32, respectively. The above formulas are fitted by training data, and the coefficients can be obtained by least square method.

The global land surface was divided into three subregions according to their latitudes, i.e., a low-latitude region ($0° - 30°$N, $0° - 30°$S), a midlatitude region ($30°-60°$N, $30° - 60°$S), and a high-latitude region ($60° - 90°$N, $60° - 90°$S). Generate surface upwelling longwave radiation and channel radiance for different regions and different zenith angles. The simulation data are calculated by the atmospheric radiation transfer model MODTRAN, where the atmospheric profile and the surface emissivity are two important input parameters. The AIRS atmospheric profile products are used here. AIRS atmospheric temperature profile has 28 atmospheric temperature layers, the corresponding air pressure is from 1100 to 0.1 mb, while the water vapor has 14 layers, and the corresponding air pressure is from 1100 to 50 mb. Cheng collected the 2 years global AIRS L2 products from 2007 to 2008, including 41,724, 35,487, 2842 for low-latitude, midlatitude and high-latitude, respectively. The LST was generated with the bottom layer temperature and the difference to cover the possible variation of actual LST. For example, the difference for the midlatitude region lies between -15 and 20K. The LST was assigned as the bottom layer temperature plus $[-15, 20]$K, with a step of 5K. Land surface emissivity was obtained from the Advanced

Spaceborne Thermal Emission and Reflection Radiometer (ASTER) spectral library and MODIS UCSB spectral library. Representative soil, vegetation, snow/ice, and water emissivity spectra were selected to construct the emissivity database. In addition, mixed pixels were also considered by area weighted average pure component emissivity spectra linearly. In total, 84 emissivity spectra representing various natural surfaces were used. The emissivity spectrum in the spectral region beyond the span of the spectral library was extrapolated from the original emissivity spectra. The above parameters are input into MODTRAN to calculate the surface upwelling longwave radiation and atmospheric parameters (atmospheric upwelling radiation, downward radiation, atmospheric transmittance). The MODIS TOA radiance can be obtained by using the following formulas:

$$L_{\text{i}} = \frac{\int_{\lambda_1}^{\lambda_2} ((\varepsilon_\lambda B(T_s) + (1 - \varepsilon_\lambda) L_{\downarrow \lambda}) \tau_\lambda + L_{\uparrow \lambda}) f_i(\lambda) d\lambda}{\int_{\lambda_1}^{\lambda_2} f_i(\lambda) d\lambda}$$

$$(8.38)$$

where L_i is the MODIS TOA radiance for channels 29, 31, and 32, ε_λ is the emissivity spectra, $L_{\downarrow \lambda}$ is the spectral LWDN, $L_{\uparrow \lambda}$ is the thermal path radiance, $f_i(\lambda)$ is the response function for channels 29, 31, and 32, and λ_1, λ_2 are the spectral range for a specific channel.

After generating the simulation data, the model coefficients are fitted for different regions and different observation angles respectively. The fitting results of the model are shown in Table 8.13. Statistics show that the linear model can explain 97.9% variance of simulation data, and the RMSE is less than 12.31 W/m².

8.2.2.2 Dynamic learning neural network model

Neural network technology has been proved to have advantages in modeling of nonlinear problems. Neural network does not need prior

TABLE 8.13 Summary of fitting results of linear model.

	Θ	a_0	a_1	a_2	a_3	R^2	Bias	RMSE
Low-latitude region	0°	118.807	−1.236	155.740	−126.281	0.991	0.00	7.87
	15°	121.078	−1.182	158.025	−129.038	0.991	0.00	7.99
	30°	128.588	−0.884	165.195	−137.866	0.990	0.00	8.46
	45°	144.119	0.348	178.241	−154.825	0.988	0.00	9.51
	60°	176.288	6.153	198.059	−185.369	0.979	0.00	12.31
Midlatitude region	0°	98.654	−1.460	138.154	−104.873	0.994	0.01	6.32
	15°	100.396	−1.505	140.500	−107.528	0.994	0.01	6.42
	30°	106.164	−1.566	147.916	−116.038	0.993	0.00	6.76
	45°	118.150	−1.252	161.760	−132.508	0.991	0.00	7.47
	60°	143.546	1.590	185.170	−163.217	0.987	0.00	9.33
High-latitude region	0°	74.506	−6.201	114.816	−73.069	0.996	0.00	4.78
	15°	48.974	4.817	18.136	20.384	0.999	0.00	1.76
	30°	48.918	4.695	19.121	19.476	0.999	0.00	1.81
	45°	48.897	4.442	21.289	17.455	0.999	0.00	1.93
	60°	49.262	3.829	26.592	12.446	0.999	0.00	2.20

From Cheng, J., Liang, S., 2016. Global Estimates for high-spatial-resolution clear-sky land surface upwelling longwave radiation from MODIS data. IEEE Trans. Geosci. Remote Sens. 1—15, © 2016, IEEE.

knowledge about the statistical distribution of data and the relationship between input and output variables. It can directly establish the relationship between input and output variables through training set. Cheng et al. (2016) used a DLNN to establish a model for estimating surface upwelling longwave radiation. DLNN is a multilayer perceptron that improves the structure by applying Kalman filtering technology to the learning process. The DLNN updates the weights globally, avoiding backpropagation and causing the learning process to be too long and improving the learning speed. In addition, the DLNN has features such as global minimization, convergence guarantee, and built-in weighting function. The input parameters of the DLNN model are the same as the linear model, which is the TOA radiance of the MODIS channel 29, 31 and 32, and the output is the upwelling longwave radiation of the surface. The DLNN model is able to explain the variance of the simulated database over 99.2%. The sensor's viewing angle ranges from 0 to 60°, and the RMSE varies from 1.45 to 7.76 W/m². From the training results, the RMSE of the DLNN is smaller than the linear model.

The linear and DLNN models were validated using SURFRAD, ASRCOP, and GAME-AAN upwelling longwave radiation observations. The location distribution of the 19 stations is shown in Fig. 8.8, and the verification results are shown in Table 8.14:

As can be seen from the table, the linear model is more accurate than the DLNN model. For the SURFRAD, ASRCOP, and GAME-AAN observation networks, the biases of linear model were −4.99, 1.06, and 2.49 W/m², respectively, and the RMSEs were 13.47, 17.61, and

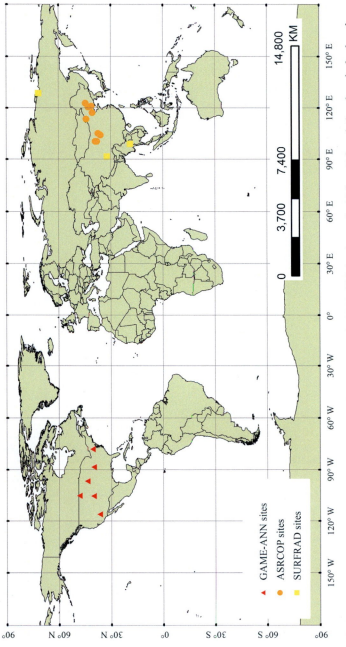

FIGURE 8.8 Distribution of 19 validation sites. *From Cheng, J., Liang, S., 2016. Global estimates for high-spatial-resolution clear-sky land surface upwelling longwave radiation from MODIS data. IEEE Trans. Geosci. Remote Sens. 1–15, © 2016, IEEE.*

TABLE 8.14   Validation results for Linear model and DLNN model (W/m^2).

Network	Site name	No. of Observation	Linear models		DLNN models	
			Bias	RMSE	Bias	RMSE
SURFRAD	Bondville	545	0.89	16.51	−0.37	18.88
	Boulder	683	−0.67	15.00	−1.09	18.23
	Desert Rock	1192	−16.2	17.5	−14.12	19.44
	Fort Peck	616	−1.66	11.52	−4.10	14.34
	Penn State	384	−0.74	7.70	−3.08	10.08
	Sioux Falls	583	−8.58	12.60	−10.84	16.05
ASRCOP	Arou	148	13.66	26.29	11.8	27.05
	Dongsu	122	−3.68	12.71	−5.78	13.89
	Jingzhou	76	3.92	11.11	−2.71	10.56
	Miyun	49	8.36	13.35	6.55	13.96
	Naiman	41	−2.07	10.76	−4.43	11.94
	Shapotou	51	−11.06	33.83	−12.38	34.45
	Tongyu grass	71	−4.82	14.20	−6.99	15.26
	Tongyu crop	77	−3.09	12.41	−5.26	13.18
	Yinke	73	11.86	23.15	10.78	22.68
	Yuzhong	113	−2.53	18.27	−3.49	20.18
GAME-AAN	Amdo	220	−3.51	15.85	−4.91	17.86
	Kogma	32	8.75	35.62	8.46	35.53
	Tiksi	28	2.22	34.52	0.58	31.95

From Cheng, J., Liang, S., 2016. Global Estimates for high-spatial-resolution clear-sky land surface upwelling longwave radiation from MODIS data. IEEE Trans. Geosci. Remote Sens. 1–15, © 2016, IEEE.

28.67 W/m^2. According to the simulation data fitting results, DLNN is more suitable to fit the nonlinear relationship between LWUP and TOA, but the training data is noise-free, and there are problems such as instrument noise and calibration error in the actual observation of satellite. These problems may be amplified by the DLNN, which ultimately causes the prediction result to be worse than the linear model. In addition, the linear model is much more efficient than the DLNN model, so the linear model is more suitable for estimating global surface upwelling longwave radiation.

8.2.2.3 Surface upwelling longwave radiation models for VIIRS

In the past few years, NASA has operated many satellite systems, including the well-known EOS system (MODIS, ASTER, etc.). Now NASA is building a new satellite system, the Joint Polar Satellite System (JPSS), in which S-NPP is one of the most important satellites in the satellite system, and VIIRS (Visible infrared Imaging Radiometer) is mounted on the S-NPP satellite. VIIRS is an extension and improvement of the high-resolution radiometer AVHRR and

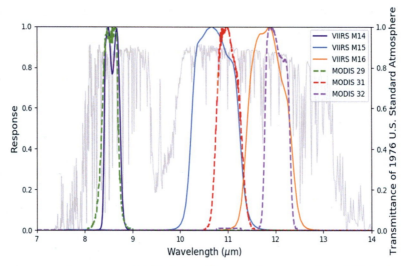

FIGURE 8.9 Relative spectral responses of Visible Infrared Imaging Radiometer Suite (VIIRS) channels M14, M15, and M16, and Moderate Resolution Imaging Spectroradiometer (MODIS) channels 29, 31, and 32. The gray line represents the transmittance of the 1976 U.S. Standard Atmosphere. *From Zhou, S., Cheng, J., 2018. Estimation of high spatial-resolution clear-sky land surface-upwelling longwave radiation from VIIRS/S-NPP data. Remote Sens. 10 (2), 253. Copyright © 2018 with permission from MDPI.*

the resolution imaging spectrometer MODIS in the Earth observation series. It has become a key bridge to ensure the long-term continuation of climate data records on a global scale.

The spatial resolution of the midresolution band of the VIIRS is 750 m, and the thermal infrared bands M14, M15, and M16 are both in the atmospheric window and similar to the spectral ranges of MODIS 29, 31, and 32 (see Fig. 8.9).

The establishment of the VIIRS upwelling radiation model is similar to MODIS. The linear model can be written as follows:

$$LWUP = a_0 + a_1 M14 + a_2 M15 + a_3 M16 \quad (8.39)$$

where a_0, a_1, a_2, and a_3 are coefficients; and M14, M15, and M16 are TOA radiance for moderate resolution bands 14, 15, and 16 of the VIIRS. The linear model is fitted for each subregion and VZA. The above coefficients can be obtained by fitting the training data, and the construction of the training data is similar to the previous one. The training results of the model are shown in Table 8.15 below.

It can be seen from the table that the linear model can account for more than 97.7%, 98.5%, and 99.1% of the variation of the LWUP in the simulation database for the low-latitude, middle-latitude, and high-latitude regions, respectively. The bias is zero, and the RMSE ranges from 5.27 to 13.02 W/m². The RMSE in the low-latitude region is larger than that in the middle-latitude and high-latitude regions. In addition, the RMSE for the larger VZA is greater than that with the smaller VZA.

The model was verified by the SURFRAD observation. The results show that the average deviation and RMSE of the derived LWUP for seven SURFRAD stations are −4.59 and 16.15 W/m², respectively. The verification results are shown in Fig. 8.10.

As can be seen from Figs. 8.6, the verification results are basically concentrated on the 1:1 line, and the night is more concentrated than during the day. This may be because the temperature field at night is more uniform than during the day. Compared to other sites, the Desert Rock,

TABLE 8.15   Fitting results of linear models.

Angle	a_0	a_1	a_2	a_3	R^2	Bias	RMSE
Low-latitude region							
0°	124.404	2.687	119.530	−93.350	0.989	0.00	8.82
15°	126.927	2.833	121.603	−95.997	0.988	0.00	8.97
30°	135.126	3.434	128.092	−104.459	0.988	0.00	9.13
45°	151.431	5.290	139.829	−120.664	0.985	0.00	10.46
60°	182.429	12.293	157.379	−149.538	0.977	0.00	13.02
Midlatitude region							
0°	99.959	1.747	104.644	−73.428	0.993	0.00	6.94
15°	101.853	1.769	106.772	−75.933	0.992	0.00	7.04
30°	108.090	1.922	113.550	−84.018	0.992	0.00	7.39
45°	120.822	2.647	126.401	−99.870	0.990	0.00	8.11
60°	146.517	6.157	148.690	−129.866	0.985	0.00	9.79
High-latitude region							
0°	77.525	0.915	87.049	−50.963	0.995	0.00	5.27
15°	79.219	1.588	88.103	−52.734	0.995	0.00	5.16
30°	82.928	1.339	94.582	−59.759	0.995	0.00	5.41
45°	90.741	1.020	107.407	−73.892	0.994	0.00	5.91
60°	107.699	1.298	132.253	−102.344	0.991	0.00	6.95

From Zhou S, Cheng J. Estimation of high spatial-resolution clear-sky land surface-upwelling longwave radiation from VIIRS/S-NPP data. Remote Sens. 2018, 10 (2), 253. Copyright © 2018 with permission from MDPI.

NV, site has a significant underestimation. Wang et al. (2009) thought that cirrus pollution is the main reason for the underestimation of longwave radiation at this site.

To verify this inference, we downloaded 3 years of ground observations from the Atmospheric Radiation Measurement SGP C1 site. The downloaded data include upwelling longwave radiation, cloud bottom height, and other ancillary data. Cloud bottom height information is obtained using Micropulse Lidar (MPL). The MPL is a ground-based lidar that determines the height of the cloud floor by emitting a laser pulse into the sky and then receiving the backscattered signal from the receiver. First, the VIIRS cloud mask is used to filter out the pixels in the nonclear sky. Based on this, the VIIRS clear-sky pixels are further screened according to the cloud bottom height information of the ground MPL. A VIIRS pixel with no apparent backscatter is considered a true clear-sky pixel, while a cell with one or more cloud height information is considered a nonclear-sky pixel. The LWUP is calculated by the linear hybrid model developed above, and the calculation result is shown in Fig. 8.11. It can be seen from the figure that the deviation of the cloudy sky and the RMSE are −0.65 and 16.04 W/m², respectively, while the deviation of the sunny day and the RMSE are −2.94 and 15.62 W/m². Therefore, the cloud mask of VIIRS cannot identify all the clouds, which leads to the underestimation of LWUP inversed by VIIRS.

The obvious underestimation of LWUP at Desert Rock, NV, may also be related to Broadband Emissivity (BBE) of the surface. Taking the developed linear model for the midlatitude region at nadir view as an example, we investigated the effects of surface BBE on the accuracy of LWUP estimation. The bias and RMSE of the fitting liner model at nadir are zero and 6.94 W/m². We calculated the average bias and RMSE for each emissivity spectrum as well as the corresponding BBE. The relationship between BBE versus bias is shown in Fig. 8.12; 78% samples have a negative bias when their BBE is less than 0.966% and 88%; BBE has a positive bias when BBE is equal to or larger than 0.966. It is very likely to produce negative bias at Desert Rock, NV, from the point of the developed linear models, because its BBE is less than 0.966. We have also investigated the relationship between the fitting residual and LST with the same data, and no significant overestimated or underestimated trend is found. Thus, cloud and surface BBE are two primary factors that affect the accuracy of the LWUP estimate.

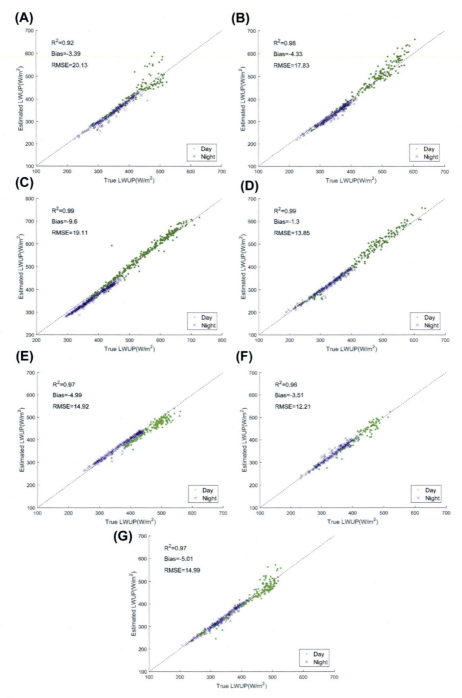

FIGURE 8.10 Validation results of the linear model at SURFRAD sites. (A) Bondville, IL; (B) Boulder, CO; (C) Desert Rock, NV; (D) Fort Peck, MT; (E) Goodwin Creek, MS; (F) Penn State, PA; (G) Sioux Falls, SD. *From Zhou, S., Cheng, J., 2018. Estimation of high spatial-resolution clear-sky land surface-upwelling longwave radiation from VIIRS/S-NPP data. Remote Sens. 10 (2), 253. Copyright © 2018 with permission from MDPI.*

(A)

(B)

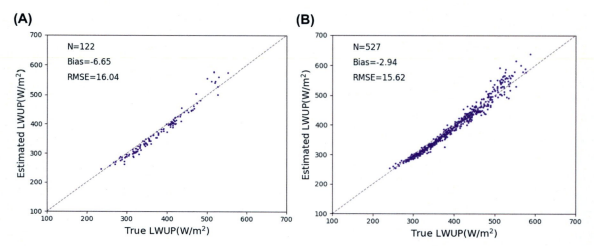

FIGURE 8.11 Validation results of the linear model at the Atmospheric Radiation Measurement (ARM) program Southern Great Plains (SGP) C1 site. (A) Pixels that were identified as clear sky by the VIIRS cloud mask, whereas cloudy by ground-based lidar; (B) pixels that were identified as clear sky by both the VIIRS Cloud Mask and the lidar. *From Zhou, S., Cheng, J., 2018. Estimation of high spatial-resolution clear-sky land surface-upwelling longwave radiation from VIIRS/S-NPP data. Remote Sens. 10 (2), 253. Copyright © 2018 with permission from MDPI].*

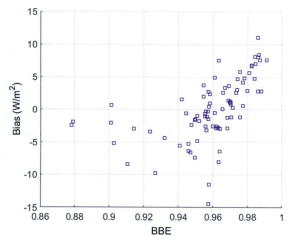

FIGURE 8.12 The relationships between broadband emissivity (BBE) versus bias of retrieved LWUP. *From Zhou, S., Cheng, J., 2018. Estimation of high spatial-resolution clear-sky land surface-upwelling longwave radiation from VIIRS/S-NPP data. Remote Sens. 10 (2), 253. Copyright © 2018 with permission from MDPI.*

8.2.2.4 Surface upwelling longwave radiation models for GOES Sounders and GOES-R ABI

The radiative transfer simulations and statistical analyses for GOES Sounders and GOES-R ABI are similar to those for MODIS. Eq. (8)–(40)

TABLE 8.16 GOES-12 Sounder linear surface upwelling longwave radiation model fitting results.

VZA	0°	15°	30°	45°	60°
b_0	124.9927	125.9401	128.9878	135.2046	148.1727
b_1	−130.4156	−132.0319	−137.2860	−148.0604	−170.4925
b_2	153.7796	155.1242	159.4967	168.4509	187.0587
b_3	4.6375	4.8304	5.4884	6.9761	10.5636
SE(e)	2.99	3.04	3.21	3.64	4.95
R^2	0.999	0.999	0.999	0.998	0.997

shows the linear model developed to estimate the surface upwelling longwave radiation using the GOES Sounder channels 10, 8, and 7, which are comparable to MODIS channels 29, 31, and 32, respectively (Wang and Liang, 2010):

$$F_u = b_0 + b_1 L_7 + b_2 L_8 + b_3 L_{10} \quad (8.40)$$

where F_u is the surface upwelling longwave radiation and b_j ($j = 0,3$) are regression coefficients. Table 8.16 summarizes the GOES-12 Sounder surface upwelling longwave radiation model fitting results. The model accounts for more than 99% of the variations in the simulated databases, with biases of zero and standard errors of less than 5.00 W/m^2.

Eq. (8.41) shows the preliminary ABI (full system) surface upwelling longwave radiation model:

$$F_{u,GOES-R} = d_0 + d_1 L_{11} + d_2 L_{13} + d_3 L_{14} + d_4 L_{15}$$

(8.41)

The ABI model fitting results are similar to those of the GOES-12 Sounder and MODIS. No concerns exist regarding the surface upwelling longwave radiation model because all channels used in the MODIS and GOES Sounder surface

upwelling longwave radiation models are available in the ABI. Moreover, the ABI has an additional window channel at 10.35 μm that can also be used to estimate the surface upwelling longwave radiation.

All MODIS and GOES-12 Sounder models presented in this section were evaluated using ground measurements. Wang et al. (2010, 2009) validated the models using ground data from the SURFRAD sites. For MODIS, the RMSEs of the ANN model method are smaller than those of the other two methods at all sites. The average RMSEs of the ANN model method are 15.89 W/m^2 (Terra, see Fig. 8.13) and 14.57 W/m^2 (Aqua);

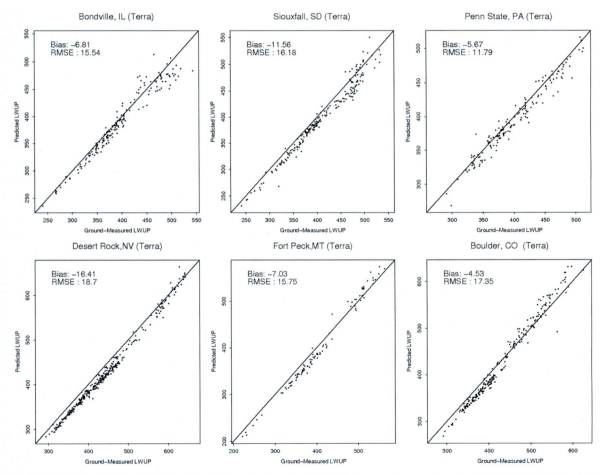

FIGURE 8.13 ANN model method validation results using SURFRAD ground measurements and MODIS Terra top-of-atmosphere radiance data.

the average biases are $-8.67\,\mathrm{W/m^2}$ (Terra) and $-7.21\,\mathrm{W/m^2}$ (Aqua). The biases and RMSEs for Aqua are $\sim 1.3\,\mathrm{W/m^2}$ smaller than those of Terra. The biases and RMSEs of the ANN model method are $5\,\mathrm{W/m^2}$ smaller than those of the temperature-emissivity method and $\sim 2.5\,\mathrm{W/m^2}$ smaller than those of the linear model method. For the GOES-12 Sounder linear model, the RMSEs are less than $21\,\mathrm{W/m^2}$ at four SURF-RAD sites. Gui et al. (2010) validated the MODIS linear surface model using ground measurements from a 1-year period at 15 North American, Tibetan Plateau, Southeast Asian, and Japanese sites, with an average bias of $-9.2\,\mathrm{W/m^2}$ and a standard deviation of $21.0\,\mathrm{W/m^2}$.

8.3 Surface net longwave radiation

Surface net longwave radiation (SNLR) is the difference between the surface downward and upwelling longwave radiation. SNLR is dominated by longwave radiation at night and at most times of the year in the polar regions. SNLR is an important part of radiant energy balance, and its accurate estimation is of great significance to weather forecast, climate simulation, and land surface process simulation results. This section will introduce the estimation of the SNLR under clear-sky conditions and the SNLR under cloudy conditions.

8.3.1 Estimation of surface net longwave radiation in clear sky

At present, there are two methods for estimating surface net longwave net radiation: one is the component mode. More specifically, the methods described in Sections 8.1.3 and 8.2.2 can be used to predict surface downward longwave radiation (SDLR) and surface upwelling longwave radiation (SULR); the difference between the SDLR and the SULR is SNLR using the component mode. The second method is the integrated mode. According to Formula 8.40, we can directly establish the relationship between SNLR and the CWV using the data in Cheng et al. (2017):

$$\mathrm{LWNT} = 73.133 - 0.698 * \mathrm{LWUP} + 130.726$$
$$* \ln(1 + \mathrm{PWV}) - 15.335 * [\ln(1 + \mathrm{PWV})]^2$$
$$(8.42)$$

At the same time, we verified accuracies of two SNLR estimation methods (component mode and integrated mode). The verification results are shown in Fig. 8.14. The bias and RMSE

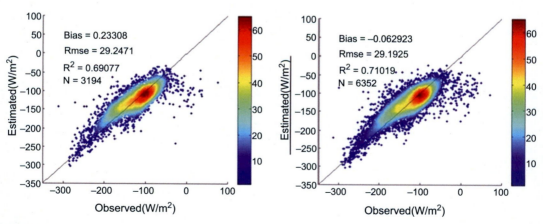

FIGURE 8.14   The verification results of surface net longwave radiation: the component mode result (the left figure) and the integrated mode result (the right figure).

of the component mode and the integrated mode are 0.23308 and 29.2461, −0.082923, and 29.9152 W/m², respectively.

All of the above methods are based on MODIS data. However, MODIS data are only provided after 2000, which makes it impossible to estimate surface net longwave radiation before 2000 with this set of schemes. In response to this problem, we proposed the following solutions:

(1) The MERRA-2 reanalysis data can provide the near-surface temperature and humidity data to calculate SDLR using parameterization schemes before 2000. From the analysis results of the parameterization method by Guo et al. (2018), it can be seen that the BMA integrated parameterization method was relatively stable under different climate and surface cover conditions. Therefore, MERRA-2 can be used as the input data, and based on the analysis results of the parameterization method by Guo et al. (2018), the remote sensing estimation of SDLR before 2000 can be realized by integrating the parameterization method.

(2) Based on AVHRR data, the surface temperature before 2000 can be inverted. The GLASS Broad Band Emissivity (BBE) data can provide the emissivity data before 2000. Thus, considering GLASS BBE data and AVHRR surface temperature as input data and based on Boltzmann's law, SULR can be estimated using the temperature-emissivity method before 2000.

In summary, the MERRA-2 reanalysis data are used as the input data, and the SDLR is calculated by the integrated parameterization method; the temperature data of the AVHRR inversion and the GLASS BBE data are used as input data, and the SULR is calculated by the temperature-emissivity method. The difference between the two is to obtain a remote sensing estimate of SNLR before 2000, which makes up for the data vacancy 2000 years ago.

8.3.2 Estimation of surface net longwave radiation in cloudy sky

The SNLR is an important part of surface radiation budget. Significant progress has been made in the estimation of clear-sky SNLR. However, estimating the cloudy-sky SNLR remains a major challenge. Therefore, this section attempts to develop a linear method and a multivariate adaptive regression spline (MARS) model to estimate SNLR from the cloudy sky (Guo et al., 2018).

8.3.2.1 Methods

The surface net radiation is the sum of surface net shortwave radiation and SNLR, which can be expressed as follows:

$$R_n = R_{nS} + R_{nL} \tag{8.43}$$

According to Eq. (8.43), SNLR can be expressed as follows:

$$R_{nL} = R_n - R_{ns} \tag{8.44}$$

Previous studies have shown that surface net radiation can be estimated directly from surface shortwave radiation. Kaminsky and Dubayah (1997) explored the relationship between surface net radiation and shortwave fluxes in central Canada, and they suggested that a single linear equation using surface net shortwave radiation can be used to estimate surface net radiation. Alados et al. (2003) did similar work in a semiarid area. Therefore, the right-hand side of Eq. (8.44) can be expressed only by R_{ns}. This means that we can estimate SNLR by R_{ns}. As surface shortwave net radiation can be precisely estimated under all weather conditions, we explore the relationship between SNLR and surface net shortwave radiation to estimate SNLR under a cloudy sky. This allows measurements of cloud information to be bypassed, including cloud-top temperature, cloud-top height, and cloud optical depth, which is unstable. A linear model and a MARS model were used to predict SNLR. In addition, NDVI was also incorporated into the developed models to qualify the effect of vegetation on the estimation of SNLR.

8.3.2.1.1 Linear model

The linear model is expressed as follows:

$$R_{nL} = a_1 R_{nS} + b_1 \qquad (8.45)$$

$$R_{nS} = (1 - \alpha) R_{Sd} \qquad (8.46)$$

When the NDVI was considered, the linear model becomes the following:

$$R_{nL} = a_2 R_{nS} + b_2 \text{NDVI} + c_2 \qquad (8.47)$$

where a_1, b_1, a_2, b_2, and c_2 are regression coefficients; α is the shortwave broadband albedo derived from GLASS ABD; R_{Sd} is the surface shortwave downward radiation derived from GLASS DSR; and NDVI is derived from MOD13A2.

To fit the above equations, spatial and temporal matching between satellite data and site measurements was implemented through three steps: (1) spatial matching. The location of each site was determined by matching the site's coordinates and the geolocation fields of the images. (2) Cloudy-sky identification. The cloudy-sky condition of the sites was determined by cloud fraction c. The calculation for c was referenced from Carmona et al. (2014). When c was greater than 0.05, the sky was considered cloudy. (3) Temporal matching. The timing of the satellite data and measurements was matched using the nearest-neighbor method to obtain time-matched data. After the above processes, more than 90,000 samples were obtained. The samples were randomly divided into two parts. Two-thirds of the samples were used to fit the coefficients in Eqs. 8.45 and 8.47 by linear regression fitting, and one-third of the samples were used to validate the developed models.

8.3.2.1.2 MARS model

The MARS is a form of regression analysis introduced by Friedman (1991). MARS builds models in the following form:

$$f(x) = \sum_{i=1}^{k} c_i B_i(x) \qquad (8.48)$$

where $B_i(x)$ is basic function, which is a constant or has the form $\max(0, x - const)$ or $\max(0, const - x)$, and c_i is a constant coefficient.

MARS is a nonparametric regression technique in which nonlinear responses between variables are described by a series of linear segments of different slopes, each of which is fitted using a basic function. Breaks between segments are defined by a knot in a model that initially overfits the data and then is simplified using a backward/forward stepwise cross-validation procedure to identify the terms to be retained in the final model. This produces continuous models with continuous derivatives and has more power and flexibility to model relationships that are nearly additive or involve interactions between at most a few variables.

MARS was applied for SDLR estimation in this study. To explore the nonlinearity among SDLR, net shortwave radiation, and NDVI, we used MARS to establish the nonlinear relationship using the same samples as in Section (1). This was implemented on the MATLAB platform with a tool called *ARESLab*. First, we use the function *aresev* to select the number of basic functions using cross-validation. Figure 8.15 shows the change in mean square error (MSE) for models of each fold as the number of basic functions increases. The 10 pink dotted lines show the MSE for models of each fold. The pink solid line is the mean MSE for each model size. The circle and vertical dashed lines are at the minimum of the solid lines, which is the optimum number of basic functions. As shown in Fig. 8.15, the number of basic functions was set to 11. Then, we use the two part samples from Section (1) to develop and validate the MARS model.

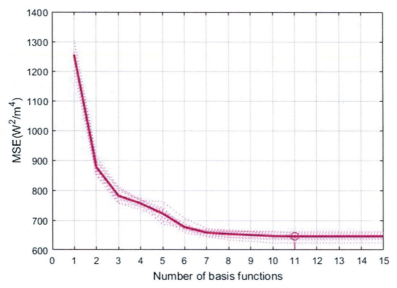

FIGURE 8.15 Determination of the optimum number of basic functions. *From Guo, Y., Cheng, J., 2018. Feasibility of estimating cloudy-sky surface longwave net radiation using satellite-derived surface shortwave net radiation. Romote Sens. 10 (4), 596. Copyright © 2018 with permission from MDPI.*

8.3.2.2 Results

8.3.2.2.1 Validation of the linear model

Using the method described in Section 8.3.2.1, we obtained 92,220 samples. The coefficients in Eqs. (8.45) were fitted by linear regression using two-thirds of the samples:

$$R_{nL} = -0.12R_{nS} - 11.74 \qquad (8.49)$$

The bias and RMSE were used as the primary indicators of accuracy. In addition to bias and RMSE, the determination coefficient (R^2) was also used as an indicator to test the performance of the developed model. The training results are shown in Fig. 8.16A. The R^2 was 0.34, and the bias and RMSE were 0.006 and 30.22 W/m^2, respectively. N is the number of samples for training purposes. Then, an evaluation of the model's accuracy was implemented using the remaining one-third of samples that were extracted. The validation results are shown in Fig. 8.16B. N is the number of samples for validation purposes. The R^2 was 0.34, and the bias and RMSE were 0.02 and 30.19 W/m^2,

respectively. The SDLR values were overestimated at the low end of the SDLR range, where the ground-measured SDLR values were approximately -150 W/m^2, while the SDLR values were underestimated at the high end of the SDLR range (ground-measured SDLR values were approximately 0). In addition, there is a ceiling for the linear model. This maybe because the surface shortwave net radiation is not enough to estimate SDLR.

In addition to clouds, many factors can affect the surface radiation budget, including near-surface air temperature and humidity, soil moisture, and land cover type. However, most of these parameters cannot be obtained directly by optical-thermal satellite under cloudy conditions. The NDVI from MOD13 was used to provide additional land surface information. Then, we obtained a new linear model (LM-NDVI) with two variables as follows:

$$R_{nL} = -0.12R_{nS} + 28.11\text{NDVI} - 23.76 \quad (8.50)$$

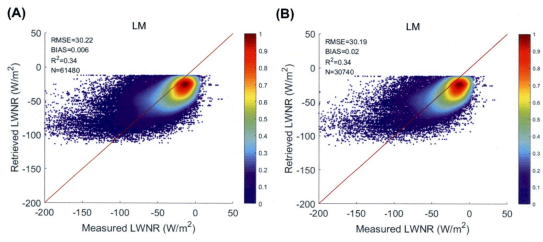

FIGURE 8.16 Accuracy of linear model: (A) training results, (B) validation results. *From Guo, Y., Cheng, J., 2018. Feasibility of estimating cloudy-sky surface longwave net radiation using satellite-derived surface shortwave net radiation. Romote Sens. 10 (4), 596. Copyright © 2018 with permission from MDPI.*

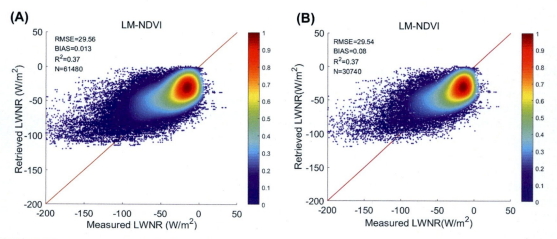

FIGURE 8.17 Accuracy of LM-NDVI model: (A) training results, (B) validation results (2). *From Guo, Y., Cheng, J., 2018. Feasibility of estimating cloudy-sky surface longwave net radiation using satellite-derived surface shortwave net radiation. Romote Sens. 10 (4), 596. Copyright © 2018 with permission from MDPI.*

The training results are shown in Fig. 8.17A. The R^2 was 0.37, and the bias and RMSE were 0.013 and 29.56 W/m², respectively. Then, one-third of the samples were extracted and used to validate the LM-NDVI model. The evaluation results of LM-NDVI are shown in Fig. 8.17B. The R^2 was 0.37, and the bias and RMSE were 0.08 and 29.54 W/m², respectively. Compared with the linear model, the LM-NDVI model did not significantly improve, but the distribution of scatters seems more reasonable. There is no obvious upper limit for LM-NDVI.

8.3.2.2.2 Validation of MARS model

The MARS model was trained with the same samples used in Section 4.1. Fig. 8.18A shows the training results of the MARS model. The validation process was performed to evaluate the MARS model, whose results are shown in Fig. 8.18B. The performance of the MARS model is slightly better than the linear model. The determination coefficient is 0.36, and the bias and RMSE values are 0.02 and 29.86 W/m², respectively, for the validation results. Similar to the linear model, the obvious maximum and minimum values of the retrieved SDLR are -20 and -120, respectively.

For the same reason as in Section 4.1, NDVI was considered and inputted to the MARS model for new SDLR predictions (MARS-NDVI model). Fig. 8.19 plots the scatter density of the ground measurements against the MARS-NDVI model-predicted SDLR using training

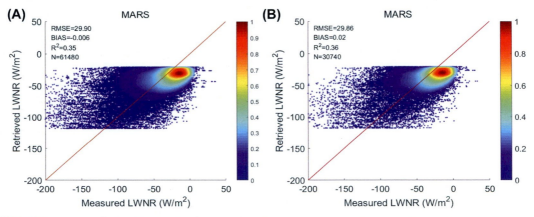

FIGURE 8.18 Accuracy of MARS model: (A) training results, (B) validation results. *From Guo, Y., Cheng, J., 2018. Feasibility of estimating cloudy-sky surface longwave net radiation using satellite-derived surface shortwave net radiation. Romote Sens. 10 (4), 596. Copyright © 2018 with permission from MDPI.*

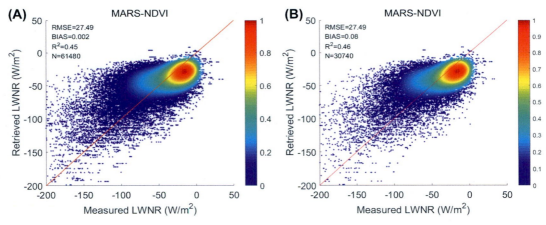

FIGURE 8.19 Accuracy of MARS-NDVI model: (A) training results, (B) validation results. *From Guo, Y., Cheng, J., 2018. Feasibility of estimating cloudy-sky surface longwave net radiation using satellite-derived surface shortwave net radiation. Romote Sens. 10 (4), 596. Copyright © 2018 with permission from MDPI.*

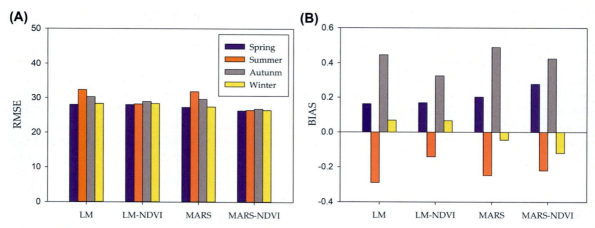

FIGURE 8.20 Validation results of four developed models for spring, summer, autumn, and winter: (A) RMSE, (B) bias. *From Guo, Y., Cheng, J., 2018. Feasibility of estimating cloudy-sky surface longwave net radiation using satellite-derived surface short-wave net radiation. Remote Sens. 10 (4), 596. Copyright © 2018 with permission from MDPI.*

TABLE 8.17   Summary statistics of four developed models for spring, summer, autumn, and winter.

Seasons	No. of samples	Linear model		LM-NDVI		MARS		MARS-NDVI	
		RMSE	Bias	RMSE	Bias	RMSE	Bias	RMSE	Bias
Spring	9275	28.05	0.16	28.04	0.17	27.25	0.20	26.30	0.28
Summer	9287	32.37	−0.29	28.24	−0.14	31.84	−0.25	26.48	−0.22
Autumn	7194	30.31	0.45	28.94	0.33	29.62	0.49	26.86	0.43
Winter	4111	28.35	0.07	28.36	0.07	27.44	−0.05	26.45	−0.12

From Guo, Y., Cheng, J., 2018. Feasibility of estimating cloudy-sky surface longwave net radiation using satellite-derived surface shortwave net radiation. Remote Sens. 10 (4), 596. Copyright © 2018 with permission from MDPI.

samples and the validation samples, respectively. The validation results are almost the same as the training results. This indicates that the training samples are sufficient. The validation results showed that the SDLR predictions are well related to the ground measurements with a determination coefficient of 0.46 and small bias and RMSE values of 0.08 and 27.49 W/m², respectively. Moreover, the MARS-NDVI model performs much better than the MARS model, indicating that the NDVI is an unneglectable factor and needs to be incorporated into the SDLR estimations. Meanwhile, the MARS-NDVI model has a better performance

than the LM-NDVI model, which indicates that the nonlinear model is more robust than the linear model.

Vegetation has obvious seasonal variations and may have certain effects on different SDLR estimation models. Therefore, to further investigate the adaptabilities of different SDLR estimation models among different seasons, the samples were divided into four parts: spring, summer, autumn, and winter to test four developed models. The results are shown in Fig. 8.20 and summarized in Table 8.17. The results showed that the SDLR predictions have obvious seasonal differences without considering the

NDVI. Both the linear model and MARS perform the worst in summer with the largest RMSE of greater than 30 W/m^2, while these models have better performances in winter. This is because less surface shortwave net radiation was transformed into longwave radiation due to the transpiration of the vegetation. The correlation between surface shortwave net radiation and SDLR weakened with the increase in vegetation coverage. Therefore, the seasonal differences are obviously decreased when considering NDVI.

The validation accuracy of the developed MARS-NDVI model is comparable to the accuracy of the general model of Zhou et al. (2007), which is reported to have a bias and RMSE of -2.31 and 29.25 W/m^2, respectively, at nine AmeriFlux sites. The developed MARS-NDVI model is a promising method for estimating SDLR using current satellite products (e.g., GLASS SDR and ABD, MODIS NDVI).

8.3.3 Global surface net longwave radiation product generation

According to the above estimation algorithm, the production program is programmed, and the data of MOD02, MOD03, MOD05, and MOD11 in 2003 are used to invert the global clear-sky instantaneous longwave net radiation with a spatial resolution of 1 KM and a temporal resolution of instantaneous. The program flow is as follows:

(1) Get the MOD021KM data and obtain the corresponding MOD03, MOD05, and MOD11 data;
(2) Get the radiance values of the three bands of band29, band31, and band32, MOD03 latitude and longitude information and angle information, MOD05 water vapor data, and QC of MOD11;
(3) From pixel-by-pixel cycle, when the pixel is clear (QC judgment of MOD11), different

coefficients are selected according to different latitudes to obtain the surface upwelling longwave radiation; the surface downward longwave radiation is obtained from surface upwelling longwave radiation and water vapor data, and the difference between the two is surface net longwave radiation;
(4) The calculation results are saved in HDF standard format.

It can be seen from the above inversion results (Fig. 8.21) that due to the influence of the cloud, the lack of data will seriously affect the user's use. Therefore, it is necessary to study the estimation of the cloudy-sky surface net longwave radiation and fill in the missing data.

8.4 Ground validation networks and existing satellite-derived surface longwave radiation budget products

Quantifying the surface longwave radiation budget with high accuracy is a fundamental prerequisite for reliable weather prediction, climate simulation, and land surface modeling (Wild et al., 2001). The meteorological, hydrological, and agricultural research communities require an accuracy of $5-10 \text{ W/m}^2$ for the surface longwave radiation budget retrieved from satellite data at spatial resolutions from 25 to 100 km and temporal resolutions from 3 h to 1 day (CEOS and WMO, 2000; GCOS, 2006; GEWEX, 2002). However, there is no global surface longwave radiation budget dataset that meets these requirements. Section 8.4.1 presents the existing surface longwave radiation budget products derived from satellite data, and Section 8.4.2 gives some examples that described temporal and spatial variation analysis of surface longwave radiation.

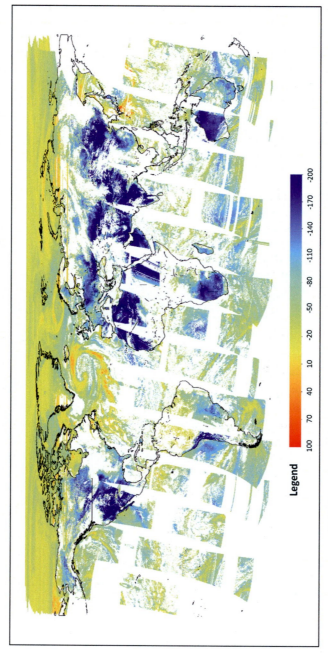

FIGURE 8.21   Global surface net longwave radiation results on the 150th day of 2003.

8.4.1 Existing surface longwave radiation budget products

Four major long-term surface longwave radiation budget datasets are currently available. The first dataset is derived from CERES, which is onboard the NASA Earth Observing System (EOS) Terra and Aqua satellites and the Tropical Rainfall Measuring Mission (TRMM) satellite (Inamdar and Ramanathan, 1997; Wielicki 1998; Gupta et al., 1997). The second long-term surface longwave radiation budget dataset is provided by the NASA WCRP/GEWEX project using data from the GOES (ASDC, 2006). The third dataset is an 18-year surface longwave radiation budget dataset derived from International Satellite Cloud Climatology Project (ISCCP) data (Zhang and Rossow, 2002; Zhang et al., 1995, 2004). The fourth dataset is a 22-year surface downward longwave radiation dataset for the Arctic (Francis and Secora, 2004). Table 8.18 summarizes the spatial resolution, temporal coverage, satellite instrument, instrument footprint, and stated accuracy associated with each dataset. Gui et al. (2010) evaluated the CERES, WCRP/GEWEX, and ISCCP surface longwave radiation. The results show that the CERES datasets perform the best overall; however, all of the datasets are biased. Readers may consult this paper for more details.

Surface longwave radiation budget data are also available from Global Climate Model (GCM) simulations. However, most GCMs have simplified land—surface emissivity, and considerable differences exist between model outputs and satellite-derived products. Figs. 8.22 and 8.23 compare the average monthly surface downward and upwelling longwave radiation budgets from ISCCP, GEWEX, and GCMs in IPCC AR4.

8.4.2 Spatiotemporal variation analysis of surface downward longwave radiation

Many scholars have analyzed the temporal and spatial distribution characteristics of surface downward longwave radiation on the global or regional scale based on current longwave radiation products (including satellite products and reanalysis data) and have also obtained many valuable results (Kato et al., 2011; Wang and Dickinson 2013, Wang et al., 2018). For example, Wang and Dickinson (2013) analyzed spatiotemporal variations of the global surface downward longwave radiation at long-term scale based on two satellite products (CERES SYN and GEWEX SRB) and six reanalysis data (ERA-Interim, ERA-40, CSFR, MERRA, NCEP, and JRA-25). It has been found that although there are some differences in the estimated radiation values of

TABLE 8.18 Summary of existing surface longwave radiation budget datasets.

	CERES	WCRP/GEWEX	ISCCP	Arctic dataset
Available products and spatial resolution	Surface downward and upwelling longwave radiation 1°	Surface downward and upwelling longwave radiation 1°	Surface downward and upwelling longwave radiation 280 km	Surface downward longwave radiation 100 km
Temporal coverage	1998—present	1983—2005	1983—2001	1979—2002
Satellite	TRMM, EOS Terra/Aqua	GOES	TOVS	TOVS
Instrument footprint (km)	20	10 × 40	40	40
Stated accuracy of monthly averages (W/m²)	21	33.6	20—25	30

8. Surface longwave radiation budget

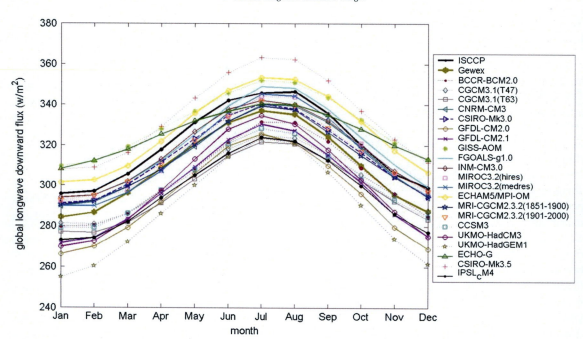

FIGURE 8.22   Monthly averages of longwave downward radiation from two satellite products (ISCCP and GEWEX) and different Global Climate Models in the IPCC AR4 (Liang et al., 2010).

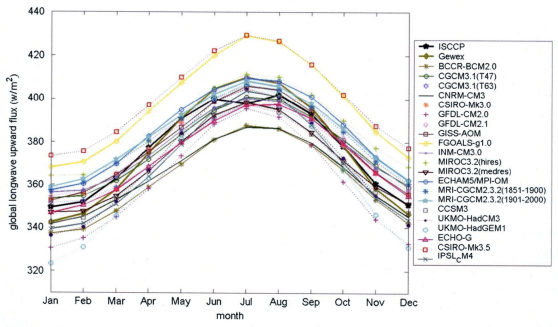

FIGURE 8.23   Monthly averages of longwave upwelling radiation from two satellite products (ISCCP and GEWEX) and different Global Climate Models in the IPCC AR4 (Liang et al., 2010).

different data sources, the various surface downward longwave radiation data on the global, ocean, and land have basically increased since 1979 (as shown in Fig. 8.24). This may be due to the increasing greenhouse gases, which has caused global warming (Prata, 2008; Wang and Liang, 2009a). It has also been found that the variation of surface downward longwave radiation is smaller than that from the ocean, which is consistent with the observations of the sites over the past 10 years (Trenberth, 2011). The study also evaluated spatially reanalysis data and satellite products. It is believed that the CERES SYN surface downward longwave radiation products have the highest accuracy and minimum deviation, while ERA-Interim has the best estimate of surface downward longwave radiation in desert areas. Therefore, the spatial distribution map of surface downward longwave radiation from 2003 to 2010 was produced based on CERES SYN and ERA-Interim products (see Fig. 8.25). From the ocean to the land, the surface downward longwave radiation shows a downward trend.

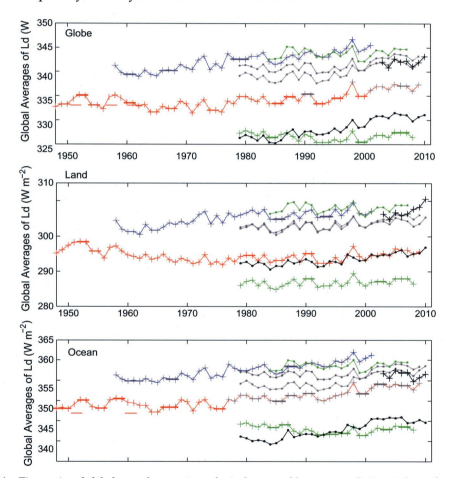

FIGURE 8.24 Time series of global annual mean atmospheric downward longwave radiation at the surface (Ld) averaged over (A) global (upper), (B) land (middle), and (C) ocean (bottom): NCEP (red cross), JRA-25 (green cross), ERA-40 (blue cross), CERES (black cross), CSFR (*red dot*), ERA-Interim (*blue dot*), GEWEX SRB (*green dot*), and MERRA (*black dot*). *From Wang, K., Dickinson, R. E., 2013. Global atmospheric downward longwave radiation at the surface from ground-based observations, satellite retrievals, and reanalyses. Rev. Geophys. 51 (2), 150–185. Copyright © 2013 with permission from American Geophysical Union.*

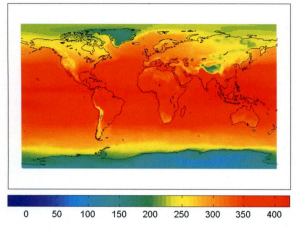

FIGURE 8.25 Multiyear mean surface downward long-wave radiation at the surface (Ld) averaged CERES SYN products from 2003 to 2010, with Ld of ERA-Interim over the deserts (unit: W/m²). *From Wang, K., Dickinson, R. E., 2013. Global atmospheric downward longwave radiation at the surface from ground-based observations, satellite retrievals, and reanalyses. Rev. Geophys. 51 (2), 150–185. Copyright © 2013 with permission from American Geophysical Union.*

For example, Wang et al. (2017) analyzed spatial distribution characteristics of surface downward longwave radiation (SDLR) over high-altitude areas (Qinghai-Tibet Plateau) based on SDLR products produced by the new algorithm and the existing SDLR products (CERES-SSF, ERA-Interim, MERRA and NCEP-CFSR). The comparison results showed that the five radiation products show a similar spatial distribution pattern in the study area, that is, the radiation values in the Tibetan Plateau and the northwest high altitude areas are lower, and the areas outside the area have higher radiation values (see Fig. 8.26). This can be explained by the fact that atmospheric temperature (including cloud base temperature) and humidity decrease with altitude; therefore, due to its close relationship with near-surface atmospheric temperature and humidity, surface downward longwave radiation is correspondingly reduced. However, the differences between these products are still evident. In particular, the details of the changes from many SDLRs clearly indicate

that the radiation texture of the newly proposed method is clearer, while other products are very smooth in spatial texture.

In general, we have found that a variety of reanalysis data and satellite products have a certain degree of uncertainty in the temporal and spatial variation pattern of SDLR. Therefore, when analyzing the temporal and spatial variation pattern of SDLR, it is better to combine remotely sensed products of the current development algorithm and reanalysis data. Even we can use Bayesian averaging method to integrate multiple product results so that the analysis results are more consistent and more convincing.

8.5 Summary

This chapter discussed methods for estimating the surface longwave radiation budget, and it consisted of three major sections. The first section presented methods for estimating the surface downward longwave radiation, the second section discussed methods for estimating the surface upwelling longwave radiation, and the third section presented methods for estimating the surface net longwave radiation.

In recent decades, many studies have been devoted to the estimation of the surface downward longwave radiation. Three categories of surface downward longwave radiation budget methods have been summarized: (1) profile-based methods, (2) hybrid methods, and (3) meteorological parameter—based methods. Profile-based methods and hybrid methods can be potentially applied at a global scale using satellite observations. Meteorological parameter—based models are usually site-specific and require local calibration.

Three methods were presented for the estimation of the surface upwelling longwave radiation: (1) the temperature-emissivity method, (2) the linear model method, and (3) the ANN model method. The temperature-emissivity method has been widely used to estimate the

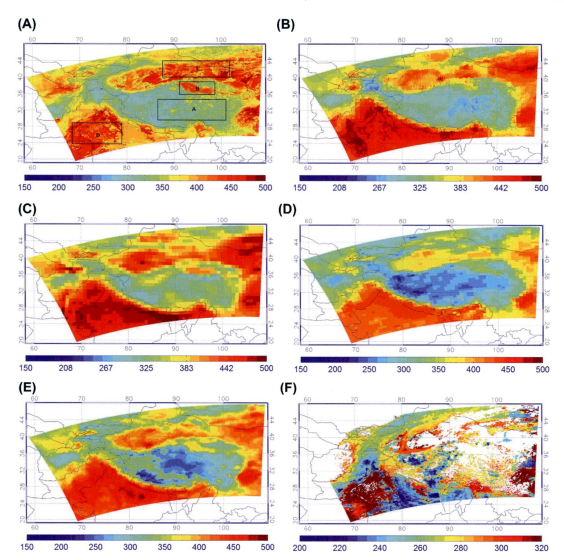

FIGURE 8.26 SDLR maps (W/m²) of the High Mountain Asia area on July 29, 2010 (A): proposed, (B): CERES-SSF, (C): ERA-Interim, (D): MERRA, (E): NCEP-CFSR, and (F): cloud base temperature (K). *From Wang et al. 2018. Remote. Sens. Environ. Copyright © 2018 with permission from Elsevier Inc.*

surface upwelling longwave radiation budget; however, unity or constant surface emissivity is usually assumed. The linear and ANN model methods derive the surface upwelling longwave radiation directly from the satellite TOA radiance data. The advantage of these two methods is that the problem of separating the LST and emissivity is bypassed; therefore, a more accurate estimate of the surface upwelling longwave radiation may be achieved.

The surface net longwave radiation is simply the difference between the surface downward and upwelling longwave radiation values. It can be derived by first estimating the surface downward and upwelling longwave radiation individually or it can be estimated directly using the hybrid methods. The incoming surface short-wave radiation and meteorological parameters are also used to estimate the net surface longwave radiation and the net surface all-wave radiation.

Acknowledgments

This work was funded by the National Basic Research Program of China via grant2015CB953701 and the National Natural Science Foundation of China via grant 41771365.

References

Alados, I., et al., 2003. Relationship between net radiation and solar radiation for semi-arid shrub-land. Agric. For. Meteorol. 116 (3–4), 0–227.

Angstrom, 1918. A study of the radiation of the atmosphere. Smithson. Misc. Collect. 65, 1–159.

ASDC, 2006. Radiation Budget: Atmosphere Science Data Center.

ASTER, 1999. ASTER Spectra Library. Jet Propulsion Laboratory.

Berk, A., et al., 1999. MODTRAN4 User's Manual. Hanscom AFB, MA (Air Force Research Laboratory, Space Vehicles Directorate, and Air Force Materiel Command).

Brunt, D., 1932. Notes on radiation in the atmosphere. Q. J. R. Meteorol. Soc. 58, 389–420.

Brutsaert, W., 1975. On a derivable formula for long-wave radiation from clear skies. Water Resour. Res. 11 (5), 742–744.

Carmona, F., et al., 2014. Estimation of daytime downward longwave radiation under clear and cloudy skies conditions over a sub-humid region. Theor. Appl. Climatol. 115 (1–2), 281–295.

CEOS, WMO, 2000. CEOS/WMO Online Database: Satellite Systems and Requirements: The Committee on Earth Observation Satellites and the World Meteorological Organization.

Cheng, et al., 2016. IEEE Trans. Geosci. Remote Sens.

Cheng, J., Liang, S., 2016. Global estimates for high-spatial-resolution clear-sky land surface upwelling longwave radiation from MODIS data. IEEE Trans. Geosci. Remote Sens. 1–15.

Cheng, J., et al., 2017. An efficient hybrid method for estimating clear-sky surface downward longwave radiation from MODIS data. J. Geophys. Res. Atmos. 122 (5), 2616–2630.

Cheng, J., Yang, F., Guo, Y., 2019. A comparative study of bulk parameterization schemes for estimating cloudy-sky surface downward longwave radiation. Remote Sens. 11 (5), 528.

Choi, M., et al., 2008. Assessment of clear and cloudy sky parameterizations for daily downwelling longwave radiation over different land surfaces in Florida, USA. Geophys. Res. Lett. 35 (20), L20402.

Crawford, T.M., Duchon, C.E., 1999. An improved parameterization for estimating effective atmospheric emissivity for use in calculating daytime downwelling longwave radiation. J. Appl. Meteorol. 38 (4), 474–480.

Darnell, W.L., et al., 1983. Downward longwave radiation at the surface from satellite measurements. J. Appl. Meteorol. 22 (11), 1956–1960.

Darnell, W.L., et al., 1986. Downward longwave surface radiation from sun-synchronous satellite data: validation of methodology. J. Appl. Meteorol. 25 (7), 1012–1021.

Deacon, E.L., 1970. The derivation of Swinbank's long-wave radiation formula. Q. J. R. Meteorol. Soc. 96 (408), 313–319.

Diak, G.R., et al., 2000. Satellite-based estimates of longwave radiation for agricultural applications. Agric. For. Meteorol. 103 (4), 349–355.

Diak, G.R., Mecikalski, J.R., Anderson, M.C., Norman, J.M., Kustas, W.P., Torn, R.D., Dewolf, R.L., 2004. Estimating land surface energy budgets from space: review and current efforts at the university of wisconsin—madison and USDA ARS. Bull.Amer.Meteor.Soc 85 (1), 65–78.

Diak, G.R., et al., 1996. A note on first estimates of surface insolation from GOES-8 visible satellite data. Agric. For. Meteorol. 82 (1–4), 219–226.

Dilley, A.C., O'Brien, D.M., 1998. Estimating downward clear sky long-wave irradiance at the surface from screen temperature and precipitable water. Q. J. R. Meteorol. Soc. 124 (549), 1391–1401.

Duarte, H.F., Dias, N.L., Maggiotto, S.R., 2006. Assessing daytime downward longwave radiation estimates for clear and cloudy skies in Southern Brazil. Agric. For. Meteorol. 139 (3-4), 171–181.

Ellingson, R.G., Wiscombe, W.J., DeLuisi, J., Kunde, V., Melfi, H., Murcray, D., Smith, W., 1993. The SPECTral Radiation Experiment (SPECTRE): Clear-sky observations and their use in ICRCCM and ITRA. Proc. of the Int. Radiation Symp. 451–453.

Ellingson, R.G., 1995. Surface longwave fluxes from satellite observations: a critical review. Remote Sens. Environ. 51 (1), 89–97.

Fisher, R.A., 1922. On the mathematical foundations of theoretical statistics. Phil. Trans. Roy. Soc. A 222.

Flerchinger, G.N., et al., 2009. Comparison of algorithms for incoming atmospheric long-wave radiation. Water Resour. Res. 45 (3), W03423.

Francis, J., Secora, J., 2004. A 22-year dataset of surface long-wave fluxes in the Arctic. In: Fourteenth ARM Science Team Meeting Proceedings. Albuquerque, New Mexico.

Friedman, J.H., 1991. Multivariate adaptive regression splines. Ann. Stat. 19 (1), 1—67.

Frouin, R., et al., 1988. Downward longwave irradiance at the ocean surface from satellite data: methodology and in situ validation. J. Geophys. Res. 93 (C1), 597—619.

GCOS, 2006. Systematic Observation Requirements for Satellite-Based Products for Climate-Supplemental Details to the Satellite-Based Component of the Implementation Plan for the Global Observing System for Climate in Support of the UNFCCC, p. 90 (The Global Climate Observation System).

GEWEX, 2002. Global Energy and Water Cycle Experiment: Phase I: GEWEX.

Gui, S., et al., 2010. Evaluation of satellite-estimated surface longwave radiation using ground-based observations. J. Geophys. Res. 115, D18214.

Guo, Y., Cheng, J., 2018. Feasibility of estimating cloudy-sky surface longwave net radiation using satellite-derived surface shortwave net radiation. Remote Sens 10 (4), 596.

Guo, Y., Cheng, J., Liang, S., 2018. Comprehensive assessment of parameterization methods for estimating clear-sky surface downward longwave radiation. Theor. Appl. Climatol. (6), 1—14.

Gupta, R.K., et al., 1997a. Estimation of surface temperature over agriculture region. Adv. Space Res. 19 (3), 503—506.

Gupta, S.K., 1989. A parameterization for longwave surface radiation from sun-synchronous satellite data. J. Clim. 2 (4), 305—320.

Gupta, S.K., et al., 1997b. Clouds and the Earth's Radiant Energy System (CERES) Algorithm Theoretical Basis Document: An Algorithm for Longwave Surface Radiation Budget for Total Skies (Subsystem 4.6.3).

Gupta, S.K., Wilber, A.C., 1992. Longwave surface radiation over the globe from satellite data: an error analysis. Int. J. Remote Sens. 14 (1), 95—114.

Gupta, S.K., et al., 2010. Improvement of surface longwave flux algorithms used in CERES processing. J. Appl. Meteorol. Climatol. 49 (7), 1579—1589.

Hoeting, J.A., et al., 1999. Bayesian model averaging: a tutorial. Stat. Sci. 14 (4), 382—401.

Idso, S.B., 1981. A set of equations for full spectrum and 8- to 14-um and 10.5- to 12.5-um thermal radiation from cloudless skies. Water Resour. Res. 17 (2), 295—304.

Idso, S.B., Jackson, R.D., 1969. Thermal radiation from the atmosphere. J. Geophys. Res. 74 (23), 5397—5403.

Inamdar, A.K., Ramanathan, V., 1997a. Clouds and the Earth's Radiant Energy System (CERES) Algorithm Theoretical Basis Document: Estimation of Longwave Surface Radiation Budget from CERES (Subsystem 4.6.2).

Inamdar, A.K., Ramanathan, V., 1997b. On monitoring the atmospheric greenhouse effect from space. Tellus Ser. B Chem. Phys. Meteorol. 49 (2), 216—230.

Iziomon, M.G., et al., 2003. Downward atmospheric longwave irradiance under clear and cloudy skies: measurement and parameterization. J. Atmos. Sol. Terr. Phys. 65, 1107—1116.

Jacobs, J.D., 1978. Radiation climate of Broughton Island. In: Barry, R.G., J.D, J. (Eds.), Energy Budget Studies in Relation to Fast-Ice Breakup Processes in Davis Strait. University of Colorado, Boulder, CO, pp. 105—120 (Institute of Arctic and Alp.Research. Occas).

Jiménez-MuNoz, Juan, C., 2003. A generalized single-channel method for retrieving land surface temperature from remote sensing data. J. Geophys. Res. 108 (D22), 4688.

Kaminsky, K.Z., Dubayah, R., 1997. Estimation of surface net radiation in the boreal forest and northern prairie from shortwave flux measurements. J. Geophys. Res. Atmos. 102 (D24), 29707—29716.

Kato, S., et al., 2011. Improvements of top-of-atmosphere and surface irradiance computations with CALIPSO-, CloudSat-, and MODIS-derived cloud and aerosol properties. J. Geophys. Res. Atmos. 116 (D19), D19209.

Kimball, B.A., et al., 1982. A model of thermal radiation from partly cloudy and overcast skies. Water Resour. Res. 18 (4), 931—936.

Kjaersgaard, J., et al., 2007a. Long-term comparisons of net radiation calculation schemes. Boundary-Layer Meteorol. 123 (3), 417—431.

Kjaersgaard, J.H., et al., 2007b. Comparison of models for calculating daytime long-wave irradiance using long term data set. Agric. For. Meteorol. 143 (1—2), 49—63.

Kratz, D.P., et al., 2010. Validation of the CERES edition 2B surface-only flux algorithms. J. App. Meteorol. Climatol. 49 (1), 164—180.

Konzelmann, T., van de Wal, R.S.W., Greuell, W., Bintanja, R., Henneken, E.A.C., Abe-Ouchi, A., 1994. Parameterization of global and longwave incoming radiation for the greenland ice sheet. Glob. Planet. Chang. 9, 143—164.

Kruk, N.S., Vendrame, Í.F., Da Rocha, H.R., Chou, S.C., Cabral, O., 2010. Downward longwave radiation estimates for clear and all-sky conditions in the Sertãozinho region of São Paulo, Brazil. Theor. Appl. Climatol. 99 (1-2), 115—123.

Lee, H.-T., 1993. Development of a Statistical Technique for Estimating the Downward Longwave Radiation at the Surface from Satellite Observations. Dept. of Meteorology, College Park, Maryland, p. 150 (University of Maryland).

Lee, H.-T., Ellingson, R.G., 2002. Development of a nonlinear statistical method for estimating the downward longwave radiation at the surface from satellite observations. J. Atmos. Ocean. Technol. 19 (10), 1500—1515.

Lhomme, J.P., et al., 2007. Estimating downward long-wave radiation on the Andean Altiplano. Agric. For. Meteorol. 145 (3–4), 139–148.

Liang, S., 2003. A direct algorithm for estimating land surface broadband albedos from MODIS imagery. IEEE Trans. Geosci. Remote Sens. 41 (1), 136–145.

Liang, S., 2004. Quantitative Remote Sensing of Land Surfaces. John Wiley & Sons, Jew Jersey.

Liang, S., et al., 2005. Mapping daily snow shortwave broadband albedo from MODIS: the improved direct estimation algorithm and validation. J. Geophys. Res. 110, D10109.

Liang, S., et al., 2010. Review on estimation of land surface radiation and energy budgets from ground measurement, remote sensing and model simulations. IEEE J. Sel. Top. Appl. Earth Obs. Remote Sens. 3 (3), 225–240.

Marks, D., Dozier, J., 1979. A clear-sky longwave radiation model for remote alpine areas. Theor. Appl. Climatol. 27 (2), 159–187.

Maykut, G.A., Church, P.E., 1973. Radiation climate of barrow, Alaska. J. Appl. Meteorol. 12 (620–628).

Mclachlan, G.J., Krishnan, T., 2007. The EM Algorithm and Extensions, second ed.

Meerkoetter, H., Grassl, H., 1984. Longwave net flux at the ground from radiance at the top, IRS '84 current problems in atmospheric radiation. In: Proceedings of the International Radiation Symposium. Perugia, Italy.

Monteith, J.L., Szeicz, G., 1961. The radiation balance of bare soil and vegetation. Q. J. R. Meteorol. Soc. 87 (372), 159–170.

Morcrette, J.J., Deschamps, P.Y., 1986. Downward longwave radiation at the surface in clear sky atmospheres: comparison of measured, satellite-derived and calculated fluxes. In: Proc. ISLSCP Conf, pp. 257–261. Rome, ESA SO-248M Darmstadt, Germany.

Nie, A., Qiang, L., Jie, C., 2016. Estimating clear-sky land surface longwave upwelling radiation from MODIS data using a hybrid method. Int. J. Remote Sens. 37 (8), 1747–1761.

Niemela, S., et al., 2001. Comparison of surface radiative flux parameterizations: part I: longwave radiation. Atmos. Res. 58 (1), 1–18.

Prata, A.J., 1996. A new long-wave formula for estimating downward clear-sky radiation at the surface. Q. J. R. Meteorol. Soc. 122 (533), 1127–1151.

Prata, F., 2008. The climatological record of clear-sky longwave radiation at the Earth's surface: evidence for water vapour feedback? Int. J. Remote Sens. 29 (17–18), 5247–5263.

Qin, Z., Karnieli, A., Berliner, P., 2001. A mono-window algorithm for retrieving land surface temperature from Landsat TM data and its application to the Israel-Egypt border region. Int. J. Remote Sens. 22 (18), 3719–3746.

Raftery, A.E., et al., 2005. Using bayesian model averaging to calibrate forecast ensembles. Mon. Weather Rev. 133 (5), 1155–1174.

Satterlund, D.R., 1979. An improved equation for estimating long-wave radiation from the atmosphere. Water Resour. Res. 15 (6), 1649–1650.

Schmetz, J., 1989. Towards a surface radiation climatology: retrieval of downward irradiances from satellites. Atmos. Res. 23 (3–4), 287–321.

Smith, W.L., Wolfe, H.M., 1983. Geostationary Satellite Sounder (VAS) Observations of Longwave Radiation Flux, the Satellite Systems to Measure Radiation Budget Parameters and Climate Change Signal. International Radiation Commission, Igls, Austria.

Sugita, M., Brutsaert, W., 1993. Cloud effect in the estimation of instantaneous downward longwave radiation. Water Resour. Res. 29 (3), 599–605.

Swinbank, W.C., 1963. Long-wave radiation from clear skies. Q. J. R. Meteorol. Soc. 89 (381), 339–348.

Trenberth, K.E., 2011. Changes in precipitation with climate change. Clim. Res. 47 (47), 123–138.

Unsworth, M.H., Monteith, J.L., 1975. Long-wave radiation at the ground I. Angular distribution of incoming radiation. Q. J. R. Meteorol. Soc. 101 (427), 13–24.

Wan, Z., 1999. MODIS Land-Surface Temperature Algorithm Theoretical Basis Document: Version 3.3. University of California, Santa Barbara, CA.

Wan, Z., 2008. New refinements and validation of the collection-6 MODIS land-surface temperature/emissivity product. Remote Sens. Environ. 112 (1), 59–74.

Wang, K., Dickinson, R.E., 2013. Global atmospheric downward longwave radiation at the surface from ground-based observations, satellite retrievals, and reanalyses. Rev. Geophys. 51 (2), 150–185.

Wang, K., Liang, S., 2009a. Estimation of daytime net radiation from shortwave radiation measurements and meteorological observations. J. Appl. Meteorol. Climatol. 48 (3), 634–643.

Wang, K., Liang, S., 2009b. Global atmospheric downward longwave radiation over land surface under all-sky conditions from 1973 to 2008. J. Geophys. Res. 114, D19101.

Wang, T., Shi, J., 2017. Cloudy-sky Longwave Downward Radiation Estimation by Combining MODIS and AIRS/AMSU Measurements//. Agu Fall Meeting.

Wang, W., Liang, S., 2009c. Estimation of high-spatial resolution clear-sky longwave downward and net radiation over land surfaces from MODIS data. Remote Sens. Environ. 113 (4), 745–754.

Wang, W., Liang, S., 2010. A method for estimating clear-sky instantaneous land-surface longwave radiation with GOES sounder and GOES-R ABI data. IEEE Geosci. Remote Sens. Lett. 7 (3), 708–712.

Wang, W., Liang, S., Meyers, T., 2008. Validating MODIS land surface temperature products using long-term nighttime ground measurements. Remote Sens. Environ. 112 (3), 623–635.

Wang, W., et al., 2009. Estimating high spatial resolution clear-sky land surface upwelling longwave radiation from MODIS Data. IEEE Trans. Geosci. Remote Sens. 47 (5), 1559–1570.

Wang, T., Shi, J., Yu, Y., Husi, L., Gao, B., Zhou, W., Chen, L., 2018. Cloudy-sky land surface longwave downward radiation (LWDR) estimation by integrating MODIS and AIRS/AMSU measurements. Remote Sens. Environ. 205, 100–111.

Wild, M., et al., 2001. Evaluation of downward longwave radiation in general circulation models. J. Clim. 14, 3227–3238.

Wielicki, B.A., 1998. Clouds and the Earth's radiant energy system (CERES) : an earth observing system experiment. Bull. Am. Meteorol. Soc. 36 (4), 1127–1141.

Zhang, Y.-C., Rossow, W.B., 2002. New ISCCP global radiative flux data products. GEWEX News 12 (4), 7.

Zhang, Y.-C., et al., 1995. Calculation of surface and top of atmosphere radiative fluxes from physical quantities based on ISCCP data sets: 1. Method and sensitivity to input data uncertainties. J. Geophys. Res. 100 (D1), 1149–1165.

Zhang, Y., et al., 2004. Calculation of radiative fluxes from the surface to top of atmosphere based on ISCCP and other global data sets: refinements of the radiative transfer model and the input data. J. Geophys. Res. 109, D19105.

Zhou, S., Cheng, J., 2018. Estimation of high spatial-resolution clear-sky land surface-upwelling longwave radiation from VIIRS/S-NPP data. Remote Sens. 10 (2), 253.

Zhou, Y., et al., 2007. An improved algorithm for retrieving surface downwelling longwave radiation from satellite measurements. J. Geophys. Res. 112, D15102.

Further reading

Allen, R.G., et al., 1998. Crop Evapotranspiration: Guidelines for Computing Crop Water Requirements. Irr. & Drain, Rome, Italy (UN-FAO).

Augustine, J.A., et al., 2000. SURFRAD—a national surface radiation budget network for atmospheric research. Bull. Am. Meteorol. Soc. 81 (10), 2341–2357.

Augustine, J.A., et al., 2005. An update on SURFRAD—the GCOS surface radiation budget network for the continental United States. J. Atmos. Ocean. Technol. 22 (10), 1460–1472.

Baldridge, A.M., Hook, S.J., Grove, C.I., et al., 2009. The ASTER spectral library version 2.0. Remote Sens. Environ. 113 (4), 711–715.

Fang, H., et al., 2007. Developing a spatially continuous 1 km surface albedo data set over North America from Terra MODIS products. J. Geophys. Res. 112, D20206.

FAO, 1990. Annex V. FAO Penman-Monteith Formula. Foot and Agriculture Organization of the United Nations, Rome, Italy.

Gupta, S.K., et al., 2004. Validation of parameterized algorithms used to derive TRMM—CERES surface radiative fluxes. J. Atmos. Ocean. Technol. 21 (5), 742–752.

Hansen, S., 2000. Markvandsbalance, Appendix A: Estimation of Net Radiation, Technical Note. The Royal Veterinary and Agricultural University, Copenhagen, 28pp.

Insightful, 2005. S-PLUS 7 for UNIX User Guide.

Irmak, S., et al., 2003. Predicting daily net radiation using minimum climatological data. J. Irrig. Drain. Eng. 129 (4), 256–269.

Kjaersgaard, J., et al., 2009. Comparison of the performance of net radiation calculation models. Theor. Appl. Climatol. 98 (1), 57–66.

Snyder, W.C., Wan, Z., Zhang, Y., et al., 1998. Classification-based emissivity for land surface temperature measurement from space. Int. J. Remote Sens. 19 (14), 2753–2774.

CHAPTER

9

Canopy biochemical characteristics

© 2020 Elsevier Inc. All rights reserved.

Abstract

Plants contain biochemicals such as chlorophyll, water, protein, lignin, and cellulose. The accurate estimation of biochemical concentrations in vegetation canopies is important to understanding ecosystem functions on different scales. This chapter introduces methods and models for the extraction of plant biochemical parameters using remote sensing. Section 9.1 reviews the principles and methods of extracting plant biochemical concentrations using remote sensing. Section 9.2 describes the empirical and semiempirical extraction methods that provide the many different types of experimental data used to extract biochemical concentrations on the leaf and canopy scales. Section 9.3 focuses on the application of various radiative transfer models and relevant retrieval algorithms to the extraction of plant biochemical parameters. Section 9.4 focuses on the study of extracting vertical distribution of vegetation biochemical components based on hyperspectral lidar. Section 9.5 summarizes this chapter.

Plants contain biochemicals such as chlorophyll, water, protein, lignin, and cellulose. The accurate estimation of biochemical concentrations in vegetation canopies is important to understanding ecosystem functions on different scales.

This chapter introduces methods and models for the extraction of plant biochemical parameters using remote sensing. Section 9.1 reviews the principles and methods of extracting plant biochemical concentrations using remote sensing.

Section 9.2 describes the empirical and semiempirical extraction methods that provide the many different types of experimental data used to extract biochemical concentrations on the leaf and canopy scales. Section 9.3 focuses on the application of various radiative transfer models and relevant retrieval algorithms to the extraction of plant biochemical parameters. Section 9.4 focuses on the study of extracting vertical distribution of vegetation biochemical components based on hyperspectral lidar. Section 9.5 summarizes this chapter.

9.1 Overview of principles and methods

This section briefly introduces plant biochemicals and their spectral characteristics and investigates the feasibility of using hyperspectral remote sensing to extract plant biochemical parameters. Two categories of methods commonly used in the extraction of plant biochemical parameters are introduced: empirical and semiempirical extraction and the radiative transfer model. A detailed account of the respective characteristics of these two categories is given to compare their advantages and disadvantages and to discuss relevant developmental trends and possible directions for improvement.

9.1.1 Remote sensing of plant biochemical parameters

Plant biochemicals are key parameters in many ecological processes, such as photosynthesis, respiration, transpiration, and decomposition and also convey stress information. The extraction of plant biochemical concentrations is important for the fields of agricultural economics, ecosystem balance, and environmental safety. However, the traditional methods for the extraction of plant biochemical concentrations are often destructive or time-delayed, which restricts their application.

The National Aeronautics and Space Administration launched the Accelerated Canopy Chemistry Program (ACCP) in 1991 to provide a theoretical and empirical basis for the use of remote sensing data to extract nitrogen and lignin concentrations in ecosystems. In this program, a large amount of biochemical and spectral data were obtained, and research on the extraction of biochemical concentrations using remote sensing was significantly promoted. Soon afterward, many European organizations and research institutions initiated large-scale tests on similar topics. Since the Ninth Five-Year Plan Period (1996–2000), China has also approved related research projects in the National Key Basic Research Program (973) and the National High Technology Research and Development Program of China (863) with a focus on precision agriculture and global change research, which has greatly promoted the development of related disciplines in China.

Since the 1990s, remote sensor hardware technology has undergone constant development and improvement, as have multispectral, hyperspectral, and even ultraspectral sensors, laying the foundation for the remote sensing retrieval of biochemical parameters on different scales. Airborne and spaceborne sensors obtain remote sensing images on different scales, with spectral resolutions ranging from several nanometers to dozens of nanometers and spatial resolutions

from a few meters to hundreds of meters. As early as the 1980s, researchers had realized that the spectral characteristics of vegetation could be used to extract biochemical parameters; however, extensive quantitative research on the extraction of plant biochemical concentrations did not begin until the emergence of hyperspectral imaging technology. Empirical and semiempirical approaches were the first to be proposed, but physical-model inversion methods have attracted increasing attention as ways to overcome the limitations of these earlier approaches.

Different plant biochemicals selectively absorb electromagnetic radiation in different wavelengths because of their different elemental compositions and chemical structures. Infrared radiation is absorbed by molecular vibration, while visible absorption corresponds to electronic transitions. For example, water, the predominant biochemical in plant leaves, has two distinct absorption bands near 1400 and 1940 nm, which are in the shortwave infrared region. The locations of these bands are determined by molecular vibrations, which are in turn dictated by the molecular structure. For chlorophyll *a* and *b*, because of electronic transitions inside the molecules, distinct absorption bands can be observed in the blue and red regions of the spectrum. The widths of the blue and red absorption bands are approximately 90 nm and less than 50 nm, respectively. However, unlike water, these molecules exhibit almost no absorption in the near-infrared and shortwave-infrared regions. The absorption characteristics of carbon-containing substances such as cellulose, hemicellulose, lignin, protein, and carbohydrates in leaves are more complex; the width of each band is usually less than 100 nm, and the absorption is significantly weaker.

When using the traditional wideband sensor to study biochemicals, the sensor's nanoscale spectral resolution is usually greater than the width of the biochemical absorption peaks. In this case, it is only possible to obtain the average reflectance in this range. Moreover, the continuous spectral of the research object cannot be obtained because of the wideband sensor's discrete bands, which greatly limits its applications in biochemical research.

It is clear that improved spectral resolution and the acquisition of continuous spectral curve could better reflect the detailed absorption behavior of biochemicals, which would be of great significance to biochemical research. Imaging spectroscopy, first established in the 1980s, focuses on the study of the acquisition of image data of very narrow and continuous spectra in the ultraviolet, visible, near-infrared, and infrared regions of the electromagnetic spectrum. This field of study effectively promotes the remote sensing retrieval of plant biochemical information.

9.1.1.1 Leaf structure and its biological, physical, and chemical properties

The nutrient production of plants through photosynthesis affects the leaf radiation signal and thus the corresponding canopy radiation signal. A healthy leaf needs three elements to produce nutrients: CO_2, water, and light energy. CO_2 in the air and water supplied by the root system

are the basic raw materials of photosynthesis, which is initiated in the presence of light energy. The leaf is the major organ of photosynthesis. Fig. 9.1 shows an illustrated profile and a microscopic image of a typical green leaf. Because of the differences among plant species and environmental conditions, leaf cellular structures vary greatly. Atmospheric CO_2 enters the internal structure of the leaf mainly through stomata on the lower epidermis. The stomata are surrounded by guard cells, which can either swell or shrink. When they swell, the stomata open to admit CO_2.

The stratum corneum is the outermost layer of the epidermis of the leaf. It can diffuse light but reflects very little of it. In some cases, a layer of fluff "hair" will grow from the upper and lower epidermis when the leaves are under direct sunlight. These "hairs" are beneficial because they can reduce the sunlight intensity on the plant. However, visible light and near-infrared light still penetrate the stratum corneum and upper epidermis to reach the mesophyll cells in the palisade tissue and sponge tissue.

In common green leaves, photosynthesis occurs in two types of mesophyll cells: those in palisade tissue and those in sponge tissue. Palisade and sponge tissue cells contain

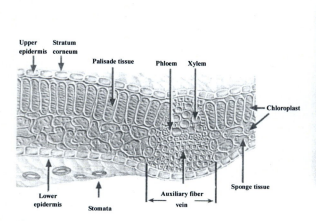
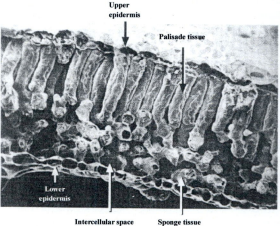

FIGURE 9.1   Illustrated profile (left) and real microscopic profile (right) of a typical green leaf.

chloroplasts, and chloroplasts contain chlorophyll. Palisade tissue cells, which face toward the upper surface, generally contain more chloroplasts than do sponge tissue cells. Therefore, the upper surface of the leaf is greener than the lower surface. When a leaf is exposed to sunlight, the biochemical molecules in the mesophyll cells will absorb some wavelengths, reflecting others. This absorption results in a high-energy state, or excited state. Based on the different molecular absorption or reflection characteristics of the biochemicals, their concentrations in leaves can be deduced from the leaves' reflection spectra.

The sponge tissue in leaves controls the amount of energy reflected by the near-infrared reflection. Sponge tissue lies below the palisade tissue and consists of cells and intercellular spaces. Oxygen and carbon dioxide exchange here during photosynthesis and respiration. The large amount of near-infrared energy that is reflected by leaves is due to multiple scattering between cell walls and air gaps.

9.1.1.2 Spectral characteristics of biochemicals

From the above introduction, we can see that the influence of leaf structure and biochemical concentration on incident radiation is expressed in the reflection spectrum. In fact, each biochemical component has a characteristic absorption spectral curve, as shown in Fig. 9.2. Curran (1989) listed 42 absorption characteristics (Table 9.1) of leaf biochemicals in the visible and near-infrared regions. The spectrum (such as a

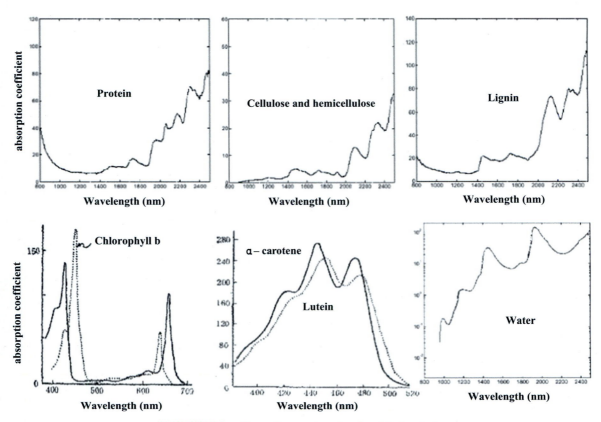

FIGURE 9.2　Absorption spectra of various biochemicals.

TABLE 9.1   Visible and near-infrared absorption characteristics of leaf biochemicals.

Wavelength (μm)	Electron transition or bond vibration	Chemical(s)	Remote sensing considerations
0.43	Electronic transition	Chlorophyll *a*	Atmospheric scattering
0.46	Electronic transition	Chlorophyll *b*	
0.64	Electronic transition	Chlorophyll *b*	
0.66	Electronic transition	Chlorophyll *a*	
0.91	C—H stretch, third overtone	Protein	
0.93	C—H stretch, third overtone	Oil	
0.97	O—H bend, first overtone	Water, starch	
0.99	O—H stretch, second overtone	Starch	
1.02	N—H stretch	Protein	
1.04	C—H stretch, C—H deformation	Oil	
1.12	C—H stretch, second overtone	Lignin	
1.20	O—H bend, first overtone	Water, cellulose, starch, lignin	
1.40	O—H bend, first overtone	Water	
1.42	C—H stretch, C—H deformation	Lignin	
1.45	O—H stretch, first overtone	Starch, sugar	Atmospheric absorption
	C—H stretch C—H deformation	Lignin, water	
1.49	O—H stretch, first overtone	Cellulose, sugar	
1.51	N—H stretch, first overtone	Protein, nitrogen	
1.53	O—H stretch, first overtone	Starch	
1.54	O—H stretch, first overtone	Starch, cellulose	
1.58	O—H stretch, first overtone	Starch, sugar	
1.69	C—H stretch, first overtone	Lignin, starch, protein, nitrogen	
1.78	C—H stretch, first overtone/ O—H stretch/H—O—H deformation	Cellulose, sugar, starch	
1.82	O—H stretch/C—O stretch, second overtone	Cellulose	
1.90	O—H stretch, C—O stretch	Starch	
1.94	O—H stretch, O—H deformation	Water, lignin, protein, nitrogen, starch, cellulose	Atmospheric absorption

TABLE 9.1   Visible and near-infrared absorption characteristics of leaf biochemicals.—cont'd

Wavelength (μm)	Electron transition or bond vibration	Chemical(s)	Remote sensing considerations
1.96	O—H stretch/O—H bend	Sugar, starch	
1.98	N—H asymmetry	Protein	
2.00	O—H deformation C—O deformation	Starch	
2.06	N=H bend, second overtone/ N=H bend/N—H stretch	Protein, nitrogen	
2.08	O—H stretch/O—H deformation	Sugar, starch	
2.10	O=H bend/C—O stretch/ C—O—C stretch, third overtone	Starch, cellulose	
2.13	N—H stretch	Protein	
2.18	N—H bend, second overtone/ C—H stretch/C—O stretch C=O stretch/C—N stretch	Protein, nitrogen	
2.24	C—H stretch	Protein	Rapid decrease in signal-to-noise ratio of sensor
2.25	O—H stretch, O—H deformation	Starch	
2.27	C—H stretch/O—H stretch CH$_2$ bend/CH$_2$ stretch	Cellulose, starch, sugar	
2.28	C—H stretch, CH$_2$ deformation	Starch, cellulose	
2.30	N—H stretch, C=O stretch, C—H bend, second overtone	Protein, nitrogen	
2.31	C—H bend, second overtone	Oil	
2.32	C—H stretch/CH$_2$ deformation	Starch	
2.34	C—H stretch/O—H deformation/ C—H deformation/O—H stretch	Cellulose	
2.35	CH$_2$ bend, second overtone C—H deformation, second overtone	Cellulose, protein, nitrogen	

leaf reflectance spectrum, with which we are familiar) provides information on the biochemical concentrations in the leaves. For instance, as chlorophyll concentration increases, the vegetation's red edge (the maximal value of the first-order derivative of the reflectance spectrum from 680 to 750 nm) will move toward the red end, also known as red shift. Thus, researchers can combine vegetation reflectance data and biochemical information to extract biochemical concentrations.

In Fig. 9.2, the vegetation components' characteristic absorption spectral curves are all continuous. The extraction of plant biochemical concentrations requires the information encoded in subtle changes in vegetation spectra, but the spectral resolutions of actual remote sensors vary. Fig. 9.3 shows the vegetation band

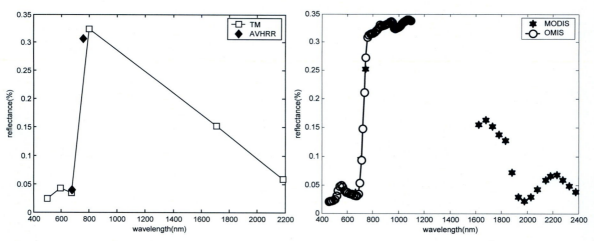

FIGURE 9.3   Vegetation spectral curves of different sensors.

reflectances of four different sensors: AVHRR, TM, MODIS (in the visible and near-infrared channels), and OMIS (operational modulator image system, developed by the Shanghai Institute of Technical Physics, Chinese Academy of Sciences, in the range 400–1100 nm). The figure shows that sensors with higher spectral resolutions provide more information. AVHRR only provides values for two wide channels; thus, an AVHRR spectrum cannot be used to extract biochemical concentrations. The TM sensor has five channels in the optical spectrum, and its spectral curve can provide certain vegetation information (such as green peaks and red valleys). However, its spectral resolution is low. The spectral resolution of MODIS is higher than TM in the range 400–2500 nm, and it has 9 visible and near-infrared channels and 16 middle-infrared channels, which provide more continuous spectral information. OMIS has 64 channels below 1100 nm, and the obtained vegetation spectral curves can be considered continuous.

The development of hyperspectral remote sensing allows quantitative biochemical information to be extracted from vegetation. Since the 1990s, hyperspectral remote sensing technology has developed rapidly. Hyperspectral remote sensing is capable of subdividing a specific

spectral region and capturing accurate and continuous spectral information of surface features, enabling the identification and analysis of detailed spectral information.

9.1.2 Introduction to theories and methods

Methods for extracting plant biochemical parameters can be classified into two categories: empirical and semiempirical methods and physical-model inversion methods.

The "empirical and semiempirical method" category is diagrammed in Fig. 9.4A. The correlation between biochemical concentration (y) and spectral factor (reflectance or its variants, spectral index, etc.) (x) is determined to establish a statistical model. If x is given, the biochemical concentration (y) can be obtained using this model. The "physical-model inversion method" category is diagrammed in Fig. 9.4B. Using this physical model, the vegetation spectrum can be simulated if the input parameters are given. This process is called a feedforward process or forward modeling. If the vegetation spectrum is known, we can retrieve the model parameters through the

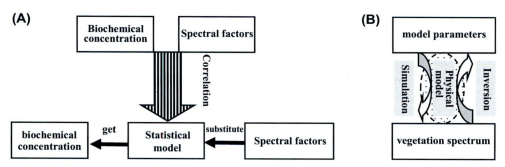

FIGURE 9.4   Two categories of methods for extracting biochemical parameters: (A) empirical and semiempirical methods and (B) physical model inversion methods.

corresponding feed backward process. Normally, the biochemical parameters are the input for the leaf physical model. Through retrieval, we can obtain the biochemical concentrations.

Both methods have advantages and disadvantages. Empirical and semiempirical methods are simple and easy to apply but lack robustness because of the limitations of the methods themselves, the difficulty to apply in different place and time, etc. A model retrieval method has definite physical meaning and can be used at any place and time within the assumed range of the model. However, this type of method is time consuming because of accuracy limitations and the speed of the model's algorithm. More accurate models and faster algorithms are needed to make this type of method advantageous. Therefore, research on the two methods must be conducted in parallel at present and in the near future.

9.1.2.1 Empirical and semiempirical methods

One important contributor to empirical modeling methods is laboratory-based near-infrared spectroscopy (NIRS, Marten et al., 1989), which was used to extract plant biochemicals immediately after its emergence (Curran et al., 1992; Peñuelas et al., 1995; Jacquemoud et al., 1995). Generally, stepwise multiple regression is used to determine the spectral region in which the reflectance and/or its variants (usually derivative spectroscopic data) are most closely related to the biochemical concentrations from the training samples. Regression equations are then established to determine the biochemical concentrations of unknown samples.

NIRS is quite satisfactory under well-controlled laboratory conditions. However, the use of remote sensing data introduces many interference factors, including variations in solar radiation intensity and angle and the influence of the observation environment, canopy structure, surface, and atmosphere. In such cases, robustness and portability might be lost. Grossman et al. (1996) and Dawson et al. (1999) carefully used leaf reflectance to test stepwise multiple regression in which leaf carbon, nitrogen, lignin, cellulose, dry matter, and water concentrations were quantitatively estimated. Their research revealed that the high correlation coefficients obtained from stepwise multiple regression of biochemical data and reflectance spectra were questionable for the following reasons: the data used were selected randomly from a database; the spectral regions selected from this database were different from those used by other researchers; and the dependence on the selected data and the given absorption characteristics failed to explain the biochemical information in the database. These researchers noted that special attention must be paid when using stepwise multiple linear regression for the reflectance spectra of fresh leaves because the selected spectral regions seemed to be independent of the biochemical absorption characteristics.

Despite its disadvantages, NIRS is both simple and easy to apply. Many attempts have been made to improve this method for application to remote sensing data on different scales. Kokaly and Clark (1999) and Kokaly (2001) proposed an improved method that used spectra to estimate plant nitrogen, lignin, and cellulose concentrations. They used the continuum-removed reflectance spectrum, which was normalized by the convolution depth. Next, in combination with stepwise multiple linear regression, the empirical method was established using a laboratory spectrum of dry leaves to determine a group of wavelengths that were highly relevant to leaf biochemical characteristics. Good results were obtained when these wavelengths were applied separately in the biochemical data at the seven observation points. In addition, the regression equations derived from the data samples were applied to the remaining samples to estimate biochemical concentrations. The concentrations of lignin and cellulose were better estimated for new forest leaf data than for nonforest leaf data. For nitrogen concentration, the estimation effect of all samples was quite satisfactory when using the convolution depth with the normalized area. The continuity of this method is reflected in that it can be used in independent datasets and different plant species. Curran et al. (2001) verified this method, finding that it exhibited good performance under certain conditions.

The index method is another type of semiempirical method. It has been long known that the reflectance spectrum of vegetation in the visible region is mainly determined by chlorophyll concentration (Thomas and Gaussman, 1977). An important aspect of this semiempirical method is the development of multiple indices that are highly relevant to biochemical concentration. The classic method is to combine several narrowband reflectances into an index associated with a specific vegetation characteristic, such as normalized difference vegetation index (NDVI). These indices are used to estimate chlorophyll concentration (Yoder and Daley, 1990; Chappelle et al., 1992; Gitelson and Merzlyak, 1996; Lichtenthaler et al., 1996) and water concentration (Hunt et al., 1987, Hunt and Rock, 1989; Inoue et al., 1993; Penuelas et al., 1993). In fact, the Minolta SPAD-502 chlorophyll meter, which is commonly used to extract immediate chlorophyll concentrations, is based on this principle. The red edge parameter is a widely applied index that is closely related to vegetation functions and chlorophyll concentration. It has already been used to estimate chlorophyll concentration (Horler et al., 1983; Curran et al., 1991). We also classify the red edge parameter as a spectral index, so as to distinguish it from the empirical method, which simply uses reflectance or its variants and biochemical component regression.

However, similarly to empirical methods, the index method usually lacks robustness. It is subject to significant background interference, especially when used on the canopy scale. In the mathematical and physical sense, Verstraete and Pinty (1996a) elaborated how to design an optimal spectral index and proposed a general design principle. They noted that a spectral index could be designed to be highly sensitive to a specific interference factor and weakly dependent on other interference factors but not exclusively sensitive to a specific interference factor. Therefore, in the extraction of biochemical information, an important principle of spectral index development is minimizing the index's sensitivity to interference factors, such as lower surface and atmospheric influence, while maximizing its sensitivity to biochemicals (Huete, 1988; Baret et al., 1994; Verstraete et al., 1996b). Many spectral indices have been proposed and applied to remote sensing data on different scales. These indices are under continuous improvement and development. Presently, this topic is still a very active research area (Chappelle et al., 1992; Gitelson et al., 1996; Bisun, 1998; Blackburn, 1998; Thenkabail et al., 2000; Haboudane et al., 2002).

9.1.2.2 Radiative transfer models

The premise of the retrieval of biochemical parameters by manipulating a model is that the model uses biochemical information as input parameters. A physical model for leaves considers the physical mechanism for light-leaf interaction and leaf structure, describing in detail the light transmission process in leaves. Several physical models for leaves are introduced below.

9.1.2.2.1 N-stream models

These models are based on K-M theory, assuming leaves to be thick plates of scattering and absorption materials (Fig. 9.5). The N-stream equation is a simplification of the radiative transfer equation. A simple analytical solution for leaf reflectance and transmittance can be obtained by solving these equations.

A two-stream model (Allen and Richardson, 1968) and a four-stream model (Fukshanky et al., 1991; Richter and Fukshansky, 1996) have been successfully used to describe the feed-forward process of leaf radiative transfer and to calculate leaf optical parameters, such as scattering and absorption. Yamada and Fujimura (1991) later proposed a more complex model, which divided leaves into four parallel layers: the upper epidermis, palisade tissue, sponge tissue, and lower epidermis. K-M theory was applied in each layer using different parameters;

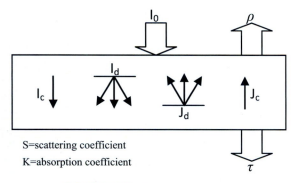

S=scattering coefficient
K=absorption coefficient

FIGURE 9.5   Two-stream model.

leaf reflectance and transmittance, i.e., scattering and absorption coefficient functions, were derived through the combination of solutions and reasonable boundary conditions. A more in-depth research program was performed to associate the absorption coefficient in the visible spectral region with chlorophyll concentration. Through retrieval, these models could be used in nondestructive measurements of photosynthetic pigments.

9.1.2.2.2 Random model

Tucker and Garratt (1977) first proposed a random model. The radiative transfer was simulated using a Markov chain (Fig. 9.6). They divided the leaf into two independent tissues: the palisade tissue and the sponge tissue. Four radiative states (scattering, reflection, absorption, and transmittance) were defined, as was the transition probability from one radiative state to another. The probability was determined from the optical properties of the leaf material. An initial vector expressing the incident radiation was allowed to reach a steady state through the iteration of state transitions. Thus, the leaf reflectance and transmittance were obtained.

This method was improved by several researchers. The SLOP model is one such improvement (Maier et al., 1999). Ma et al. (1990) described the leaf as a flat plate consisting of water, on the surface of which irregular spherical particles were randomly distributed. Ganapol et al. (1998), who put forward LEAFMOD, drew an analogy between the leaf and evenly mixed biochemical substances full of scattered and absorbed light. This model was suitable for the simulation of leaf optical properties.

9.1.2.2.3 Ray tracing model

Among the existing models, only the ray tracing model can describe the complex internal structure of leaves observed under a microscope (Fig. 9.7). The model must describe single cells and their arrangement in tissues in detail; the optical constants of the leaf materials (cell wall,

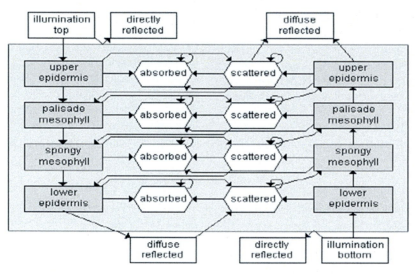

FIGURE 9.6   Random model diagram.

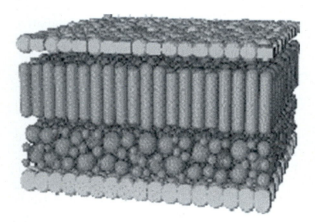

FIGURE 9.7   Ray tracing model.

cytoplasm, pigment, stoma, etc.) are still undefined. Reflection, refraction, and absorption laws can be used to determine the transmission processes of single photons falling on leaves. Once the directions of a sufficient number of photons are simulated, the radiation transmission in the leaves can be estimated from a statistical standpoint. The earliest studies were on the cell level, especially epidermal cells; the shape of the epidermal cells of some plants allows these cells to act as a lens, concentrating the light in the upper, chloroplast-rich area of the palisade tissue.

Some research has focused on the transmission path of light in the entire leaf. Allen et al. (1973) simulated the spectrum of albino maple leaves through 100 circular arcs (intercellular space and cell wall) in the two media. This model was used to test the mirror reflection and scattering characteristics of the cell wall. Kumar and Silva (1973) found that this model underestimated the reflectance of the near-infrared reflection platform and overestimated the transmittance in some cases. They also found that the actual spectrum could be better represented by including two more media, the cytoplasm and the chloroplast, thus increasing internal scattering. All of these methods ignored the characterization of the absorption phenomena outside the near-infrared region. In these models, the leaf was often described as a 2D object, although the 3D structure of the leaf is essential for physiological functions (for example, the diffusion of CO_2, H_2O, and O_2) and light scattering (Vogelmann et al., 1993). The 3D ray tracing model RAYTRAN was developed accounting for the above considerations (Govaerts et al., 1996). This model requires that the 3D description of the internal leaf structure complies with the anatomical morphology and physiological characteristics of the leaves.

9.1.2.2.4 Plate model

The original plate model treated the leaf as an absorption plate with a Lambertian surface and required the leaf refractive index and absorption coefficient. This model has been successfully applied to simulate the tight structure of corn leaves (no air or intercellular space).

However, many leaf structures differ significantly from corn leaves. The application of this model was later extended to nontight structures by regarding a leaf as N plates separated by $N-1$ air gaps. N could also be a real number (i.e., N was not necessarily an integer). The PROSPECT (Jacquemoud and Baret, 1990) model was developed on this basis. The model assumes that the luminous flux inside the leaf is isotropic, and ρ_{90} and τ_{90} are the reflectance and transmittance of each layer, respectively. For isotropic incident light, $R_{N,90}$ and $T_{N,90}$ are given by

$$\frac{R_{N,90}}{b_{90}^N - b_{90}^{-N}} = \frac{T_{N,90}}{a_{90} - a_{90}^{-1}} = \frac{1}{a_{90}b_{90}^N - a_{90}^{-1}b_{90}^{-N}}$$

$$(9.1)$$

in which

$$a_{90} = \left(1 + \rho_{90}^2 - \tau_{90}^2 + \delta_{90}\right)/(2\rho_{90})$$

$$b_{90} = \left(1 - \rho_{90}^2 + \tau_{90}^2 + \delta_{90}\right)/(2\tau_{90})$$

$$\delta_{90} = \sqrt{\left(\tau_{90}^2 - \rho_{90}^2 - 1\right)^2 - 4\rho_{90}^2}$$

In Eq. (9.1), the leaf optical properties are closely related to N-structural parameters. Theoretically speaking, N is a parameter related to the leaf cell arrangement; ρ_{90} and τ_{90} are related to the refractive index n (≈ 1.4, associated with wavelength) and the transmission coefficient τ of each layer. τ is closely related to the absorption coefficient, which in turn is determined by the various biochemical concentrations.

9.1.2.2.5 Conifer leaf model LIBERTY

All of the models above were developed specifically for broadleaf structures, which have extended plane and obviously different layers. But the conifer leaf has distinctive structures: it lacks an evident palisade tissue and its profile contains nearly all of spherical cells. The characterization and simulation of the elements of a conifer leaf are still hindered by many unsolved problems. The shapes and sizes of conifer leaves make spectral measurement difficult too. Even in the laboratory, it is difficult to determine the reflectance and transmittance of a single leaf. Therefore, the slat model is not suitable for the conifer leaf, which restricts the development of conifer leaf models. The LIBERTY model (Dawson et al., 1998) was developed specifically for the conifer leaf. It simulates the spectral characteristics of conifer foliage or a single leaf, with an excellent ability to accurately simulate the spectra of dry and wet foliage.

The LIBERTY model assumes leaf cells to be spherical particles. The leaf surface consists of several layers of equally spaced spherical cells of uniform size. The spherical cells are separated by air gaps. A quasi-infinitely thick medium is formed because of cell agglomeration. The LIBERTY model needs six input parameters: average cell diameter d, intercellular space characterizing the upward radiation components in cells X_u, base absorption a_b, albino absorption a_a, conifer leaf thickness t, and biochemical concentration C. The biochemicals considered include chlorophyll, water, protein, lignin, and cellulose.

9.2 Empirical and semiempirical methods

The amount of biochemical concentration affects the leaf spectrum and hence the reflectance of the entire canopy. These two parameters are closely linked, and efforts have been made to establish relationships between biochemical concentration and the spectrum or one or more of its variants (derivative, logarithm, spectral index) to estimate concentrations. This simple approach has aroused the interest of many researchers.

In Section 9.2.1, a stepwise multiple regression method is introduced to extract cellulose, lignin, carbon, and nitrogen concentrations on the leaf scale. Considering that many studies focus on chlorophyll, which is the most important plant biochemical, methods of its extraction by empirical and semiempirical methods on the leaf and canopy scales are introduced in Section 9.2.2. The applications of the above theories are explored using experimental data and self-testing data.

9.2.1 Extraction of biochemical concentration on the leaf scale

This section focuses on the extraction of cellulose, lignin, carbon, and nitrogen concentrations. The leaf spectra and biochemical data used in this section come from ACCP experimental data (Aber and Martin, 1999). All tree leaf types (basswood leaf, white oak leaf, sugar maple, red maple, red oak, red wood, red pine, white pine, tamarack pines, American beech) are included, and a continuous separation method is adopted for analyzing cellulose and lignin concentrations (Newman94). The combustion method is used to obtain carbon and nitrogen concentrations. A PerkinElmer 2400 spectrophotometer is used for the leaf spectral measurements. The ACCP experimental data used in this study are listed in Table 9.2. The leaf reflectance spectra are measured in the range 400–2498 nm with an

interval of 2 nm. The spectrum is resampled to 10 nm for analysis.

9.2.1.1 Cellulose concentration

Stepwise multiple regression uses a linear combination of independent variables x_i to fit the observed dependent variable set Y:

$$Y = a_0 + \sum_{i=1}^{N} a_i x_i \quad (9.2)$$

where Y is a specific biochemical concentration; a_0 is a constant; x_i is the reflectance or one of its variants (first-order derivative) in band i; and a_i (i = 1, N) is the regression coefficient of the ith band.

Stepwise multiple linear regression is conducted over the cellulose concentration in the BHI region with reflectance as the independent variable and again with the first-order derivative of reflectance as the independent variable. Table 9.3 shows the selected wavelengths and coefficients with reflectance and its first-order derivative as the independent variable.

When reflectance is the independent variable, 2250, 1890, 1490, 2280, 2340, and 2090 nm are just or close to the absorption wavelengths of the cellulose molecules. When the first-order derivative of reflectance is the independent variable, 2320, 1790, and 2250 nm are the absorption wavelengths of the cellulose molecules. The other identified wavelengths may be the unknown absorption wavelengths of cellulose molecules or are caused by randomness.

The two regression equations obtained using the data of the BHI region are substituted into

TABLE 9.2 Simple list of experimental data used.

Sampling area(abbreviated form)	Leaf type	Sample size	Biochemical types
Blackhawk Island(BHI)	Dry leaves	182	Cellulose, lignin, total carbon, total nitrogen
Howland (HOW)	Dry leaves	187	Cellulose, lignin, total carbon, total nitrogen
Harvard Forest (HF)	Dry leaves	187	Cellulose, lignin, total carbon, total nitrogen
Jasper Ridge (JR)	Fresh leaves	40	Cellulose, lignin, total carbon, total nitrogen

TABLE 9.3   Leaf cellulose concentration regression relation and selected bands.

	Reflectance		First-order derivative of reflectance
a_0	28.862	a_0	44.631
Band i nm	Coefficient a_i	Band i nm	Coefficient a_i
2130	658.476	2320	−20,895.564
2410	2,508.026	2300	−24,385.654
2250	−376.588	1680	−15,278.368
2440	−2,088.553	600	3,686.841
1890	436.844	2420	−10,197.646
1490	−314.556	850	−9,854.551
2280	1,042.409	1790	−28,876.059
2010	−1,139.557	2080	41,889.11
660	256.934	700	−998.719
2340	−1,441.54	2460	−41,827.108
2090	272.333	2250	6,590.467
680	−171.765	2480	25,451.451
2230	−380.955		
2020	767.107		

the HOW and HF region samples. The leaf cellulose concentrations of these two regions are then estimated and compared with the real values (Fig. 9.8). A satisfactory regression result is obtained for the correction samples (BHI region), regardless of the independent variable, reflectance, or first-order derivative of reflectance. The error is approximately 3%. The cellulose concentrations of the validation samples (HOW and HF) can also be properly estimated, with errors of 14.48%, 14.65% and 10.07%, 9.07%. These results show that it is possible to derive a general estimation model for dry leaf samples.

9.2.1.2 Lignin concentration

Similarly, two stepwise multiple linear regressions are conducted over lignin concentration in

the BHI region, using either reflectance or the first-order derivative of reflectance as the independent variable. Table 9.4 shows the selected wavelengths and coefficients with reflectance and the first-order derivative of reflectance as the independent variable.

In the regression using reflectance as the independent variable, only 1430 and 1700 nm are the absorption wavelengths of the lignin molecules; in the regression using the first-order derivative of reflectance as the independent variable, 1450, 1630, 1940, 1370, and 1190 nm are the absorption wavelengths of the lignin molecules.

The two regression equations obtained using the data from the BHI region are substituted into the HOW and HF region samples. The leaf lignin concentrations of the two regions are then estimated and compared with the real values (Fig. 9.9). A satisfactory regression result is obtained for the correction samples (BHI region) regardless of whether the independent variable is reflectance or its first-order derivative, with an error of approximately 5%. For the HOW validation sample, the regression equations obtained using reflectance and the first-order derivative of reflectance produced errors of 31.77% and 13.44%, respectively. For the HF region, however, the two equations both achieve a satisfactory estimation result, with errors of 12.06% and 10.04%, respectively.

9.2.1.3 Carbon concentration

Two stepwise multiple linear regressions are conducted over leaf carbon concentration in the BHI region, one using reflectance and the other using the first-order derivative of reflectance as the independent variable. As shown in Table 9.5, carbon is mainly distributed in such substances as cellulose, lignin, and starch. For the regression equation, no matter whether the independent variable is reflectance or the first-order derivative of reflectance, most of the selected bands are within the absorption bands of the aforesaid substances. However, some bands are still randomly selected in.

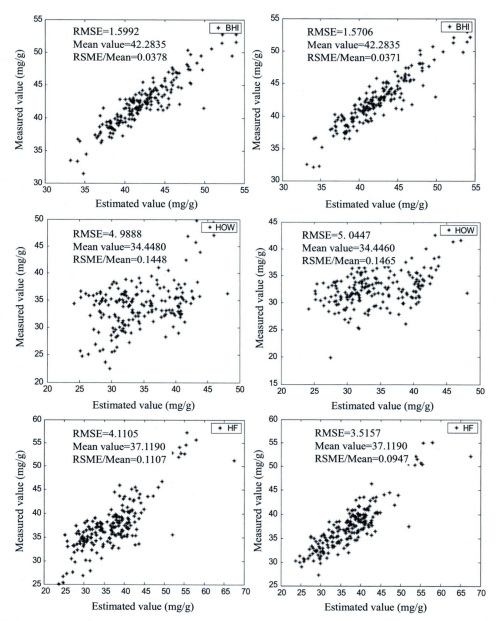

FIGURE 9.8   Test of regression equations for leaf cellulose concentration in three regions (left: reflectance is the independent variable; right: the first-order derivative of reflectance is the independent variable).

The two regression equations obtained using the data from the BHI region are substituted into the HOW and HF region samples. The leaf carbon concentrations of the two regions are then estimated and compared with the real values. From Fig. 9.10, it can be seen that a satisfactory regression result is obtained for the correction samples (BHI region), regardless of

TABLE 9.4 Leaf lignin concentration regression relation and selected bands.

	Reflectance		First-order derivative of reflectance
a_0	28.706	a_0	35.265
Band i nm	Coefficient a_i	Band i nm	Coefficient a_i
2300	−1354.39	2370	−33,132.118
2320	1950.590	2420	38,257.306
1430	−257.738	1580	13,674.827
2380	−1809.06	1450	10,686.943
2260	−1260.58	2210	26,456.241
650	−57.189	1630	21,175.473
2310	1133.209	1940	4,686.527
2230	2165.731	790	8,043.826
2210	−2200.30	1810	−27,402.878
1370	93.656	1370	3,657.470
2430	314.154	2160	11,054.804
2170	1466.705	1190	8,562.878
2150	−734.812	2390	16,785.976
1700	204.359		
2480	350.061		

the independent variable, reflectance, or first-order derivative of reflectance. The overall errors are only 0.71% and 0.51%, respectively. For the validation samples (HOW and HF regions), satisfactory estimation results are also obtained, with overall errors of only 2.61%, 1.62% and 2.20%, 1.36%.

9.2.1.4 Nitrogen concentration

Two stepwise multiple linear regressions are conducted over leaf nitrogen concentration in the BHI region, using either reflectance or the first-order derivative of reflectance as the independent variable. As shown in Table 9.6, four absorption bands of protein molecules are selected when reflectance is the independent variable; five absorption bands of protein molecules are selected when the first-order derivative of reflectance is the independent variable. Most other bands are not.

The two regression equations obtained using the data from the BHI region are substituted into the HOW and HF region samples. Then the leaf nitrogen concentrations of the two regions are estimated and compared with the real values. From Fig. 9.11, it can be seen that a satisfactory regression result is obtained for the correction samples (BHI region), regardless of the independent variable, reflectance, or first-order derivative of reflectance; the overall errors are only 4.56% and 3.31%, respectively. For the validation samples (HOW and HF regions), satisfactory estimation results are also obtained, with overall errors of only 15.79%, 14.70% and 11.47%, 13.18%.

The above regressions of the cellulose, lignin, total carbon, and total nitrogen concentrations of dry leaves and the leaf spectra and tests show that the concentrations of these biochemicals in dry leaves can be properly estimated using stepwise multiple linear regression using either reflectance or the first-order derivative of reflectance as the independent variable; moreover, the established relationship can be applied to other samples. The use of the first-order derivative of reflectance as the independent variable removes some random factors, so the regression and test results with this independent variable are better than that with reflectance. However, some of the selected bands corresponding to the components do not fall within the absorption bands of related substances. The sources of some of these bands cannot be fully explained; some might be the absorption bands of unknown substances, while some others might be selected randomly, depending on the nature of the method used. When these prediction relationships are applied to fresh leaves, the estimated results are not satisfactory. This result may be related to sample size, but the main reason lies in the

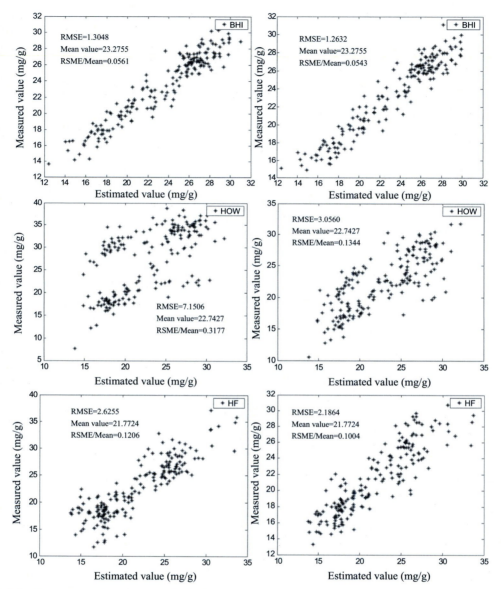

FIGURE 9.9 Test of regression equations for leaf lignin concentration in three regions (left: reflectance is the independent variable; right: the first-order derivative of reflectance is the independent variable).

water present in fresh leaves. Because of the presence of water, many substances have weaker effects, contributing weaker components. Therefore, further study is necessary on how to effectively eliminate the effect of water while increasing the effect of these components.

9.2.2 Extraction of chlorophyll concentration

Chlorophyll is the most important biochemical in the determination of plant growth conditions. Here, we introduce the methods for the

TABLE 9.5 Leaf carbon concentration regression relation and selected bands.

Reflectance		First-order derivative of reflectance	
a_0	44.402	a_0	49.164
Band i nm	Coefficient a_i	Band i nm	Coefficient a_i
1720	−138.915	1580	3984.504
1790	129.109	1940	3760.091
1480	98.774	550	1485.997
2060	322.160	1870	2735.807
2150	−380.731	2430	8329.378
2380	−45.154	430	−1883.896
1550	−493.709	1240	−3926.871
1610	331.447	1730	3340.442
2220	205.722	1700	1421.357
2240	−118.428	1200	3617.996
2120	76.902	2200	5307.829
		1650	−1589.747
		2070	−1917.720
		520	−516.603
		1590	3799.942
		1260	−4698.241
		2310	3108.189
		2400	3591.095
		1520	3043.164
		2370	−2808.535

extraction of the chlorophyll concentration from reflectance spectra data on the leaf and canopy scales separately. The data used in this section come from agronomic sampling of different varieties of summer maize under different nitrogen treatments and from canopy and leaf spectral measurements on the farm at the Beijing Academy of Agriculture and Forestry Sciences from July to September 2003. The physiological morphology parameters and biochemicals were also measured at the same time. An ASD2500 hyperspectrometer was used for the canopy spectral measurements, and a Licor1800-12s integrating sphere externally connected to the ASD spectrometer was used for the leaf spectral measurements. The measurements covered different growth stages of summer maize.

9.2.2.1 Spectral index

The spectral index is commonly used to study the relationship between biological indicators (such as pigment concentration and chlorophyll fluorescence) based on physiological conditions obtained from the field or laboratories and plant reflectance. The spectral index is the combination of the reflectances of some specific bands and is related to the leaf pigmentation or photosynthesis and the stress state of the vegetation.

Many spectral indices associated with vegetation functions are on the leaf scale rather than the canopy scale. Their relationships with pigment concentration or chlorophyll fluorescence can be easily observed. For instance, Gamon et al. (1997) proposed the reflectance combination of PRI (531 and 570 nm: PRI $= (R_{531} − R_{570})/(R_{531} + R_{570})$) as a physiological reflectance index, which was associated with the epoxidation of lutein ring pigment and the photosynthetic efficiency of a canopy under nitrogen stress. This index could be used as the intermediate index between the radiation utilization efficiency of leaves and that of a completely illuminated canopy instead of a covered canopy. However, if the canopy structure was not a problem, the relative photosynthetic efficiency also could be obtained by remote sensing.

Horler et al. (1983) and some recent leaf and canopy studies explored the relationship between the changes in red edge position and slope and chlorophyll concentration. During the process from positive photosynthesis of healthy leaves to aging because of chlorophyll loss, the red edge position and slope will change.

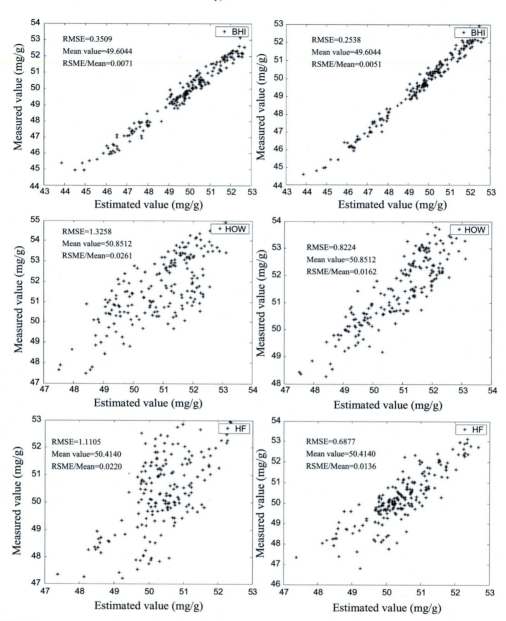

FIGURE 9.10   Test of regression equations for leaf carbon concentration in three regions (left: reflectance is the independent variable; right: the first-order derivative of reflectance is the independent variable).

The movement of the red edge position toward the blue end is called blue shift. According to some studies, blue shift is caused by a decreased concentration of chlorophyll *b* (Rock et al., 1988).

To better retrieve leaf/canopy chlorophyll concentrations, the establishment and application of efficient spectral indices on a theoretical basis is important. This method is an important part of retrieval using empirical or semiempirical methods. According to spectral regions and relevant parameters, spectral indices with potential values can be divided into four categories (Zarco-Tejada et al., 1999 a,b), listed in Table 9.7.

TABLE 9.6 Leaf nitrogen concentration regression relation and selected bands.

Reflectance		First-order derivative of reflectance	
a_0	3.087	a_0	2.869
Band i nm	Coefficient a_i	Band i nm	Coefficient a_i
1450	47.898	2140	−1209.906
2040	−159.654	2060	940.667
2030	102.770	2250	380.397
2480	56.822	2280	1328.752
2180	−22.362	410	−638.220
1510	−29.422	1960	−1679.859
660	−21.375	2470	1073.881
1640	41.277	1620	1143.008
680	17.600	1910	−648.372
1980	−10.092	1350	886.454
1400	−22.404	2420	3274.417
		2410	−1626.489
		790	708.796
		680	135.718
		1030	1277.322
		1870	−1247.685
		1500	−1441.210
		2020	−682.345
		850	−450.175
		1840	1397.505
		1280	−1125.733

9.2.2.2 Chlorophyll concentration on the leaf scale

Using the measured spectral data of maize and chlorophyll concentration data, the correlation between each spectral index and chlorophyll concentration and the differences between the two representation methods of chlorophyll concentration (quality/dry weight and quality/area) are investigated.

The results show that chlorophyll concentration has an obvious correlation with most of the spectral indices within a concentration range. However, the correlation of the representation method of the chlorophyll concentration that uses quality/area as its dimension with each spectral index is more significant than that of the representation that uses quality/dry weight as its dimension. This finding reveals that quality/area should be adopted as the dimension when remote sensing data are used to extract chlorophyll concentration, which can be explained from a theoretical standpoint. According to radiative transfer theory, light extinction in a leaf includes scattering and absorption, depending on the optical path the light takes in the leaf, and the optical path is directly related to the scattering and absorption substance concentrations.

Theoretically speaking, most of the spectral indices within a certain range can be used to estimate the leaf chlorophyll concentration. As an individual index is not sensitive to vegetation types, it is possible to establish a general chlorophyll concentration estimation model suitable for all types of leaves. If the sample types are fixed, the result of using the established regression relation established to extract the chlorophyll concentration of vegetation leaves will be improved.

9.2.2.3 Chlorophyll concentration of crops on the canopy scale

The canopy spectra detected by remote sensing are the mixed spectra of vegetation and soil. The most direct approach initially used was to establish a correlation between the canopy spectral reflectance or one of its variants and the chlorophyll concentrations of some samples; this relationship was then applied to predict the chlorophyll concentrations of other

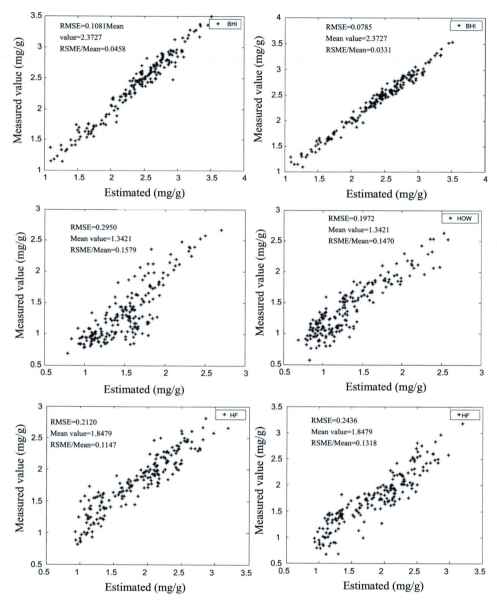

FIGURE 9.11   Test of regression equations for leaf nitrogen concentration in three regions (left: reflectance is the independent variable; right: the first-order derivative of reflectance is the independent variable).

samples. It was found that this method was influenced by many factors, making finding a universal relationship difficult. As a consequence, the prediction equations varied with time and place. The established relationship was unable to provide theoretical explanations because the establishment of the relationship between mixed spectrum and vegetation

parameters was influenced by so many different factors.

As some of the scattering and absorption mechanisms in vegetation are taken into account, the spectral index has a certain physical meaning. As we can see, it is possible to identify a universal relationship at the leaf scale. At the canopy scale, however, the effect is not

TABLE 9.7   Results of unbiased data-parameter retrieval using the PROSPECT model.

Categories	Name	Equation
Visible ratio	NPCI	$(R_{680} - R_{430})/(R_{680} + R_{430})$
	NPQI	$(R_{415} - R_{435})/(R_{415} + R_{435})$
	PRI1	$(R_{531} - R_{570})/(R_{531} + R_{570})$
	PRI2	$(R_{550} - R_{531})/(R_{550} + R_{531})$
	PRI3	$(R_{570} - R_{539})/(R_{570} + R_{539})$
	SRPI	R_{430}/R_{680}
	Carter1	R_{695}/R_{420}
	Green index	R_{554}/R_{677}
	Lic1	R_{440}/R_{690}
	Area covered by reflectances in 450–680 nm	$AR = \int_{450}^{680} R$
Visible/near-infrared ratio	NDVI	$(R_{774} - R_{677})/(R_{774} + R_{677})$
	Lic2	$(R_{800} - R_{680})/(R_{800} + R_{680})$
	Lic3	(R_{440}/R_{740})
	SIPI	$(R_{800} - R_{450})/(R_{800} + R_{650})$
	MCARI	$[(R_{700} - R_{670}) - 0.2(R_{700} - R_{550})] * (R_{700}/R_{670})$
	TCARI	$3[(R_{700} - R_{670}) - 0.2(R_{700} - R_{550}) * (R_{700}/R_{670})]$
	SR	R_{774}/R_{677}
	PSSRa	R_{800}/R_{680}
	PSSRb	R_{800}/R_{635}
Red-edge reflectance	Vog1	$(R_{734} - R_{747})/(R_{715} - R_{720})$
	Vog2	$(R_{734} - R_{747})/(R_{715} - R_{726})$
	Vog3	R_{740}/R_{720}
	GM	R_{750}/R_{700}
	Carter2	R_{695}/R_{760}
	CI	$R_{675} * R_{690}/(R_{683}^2)$
Red-edge derivative	D	D_{715}/D_{705}
	DPR1	$D_{\lambda p}/D_{\lambda p+12}$
	DPR2	$D_{\lambda p}/D_{\lambda p+22}$
	DP21	$D_{\lambda p}/D_{703}$
	DP22	$D_{\lambda p}/D_{720}$
	Area covered by derivative spectra in red-edge region	$AD = \int_{680}^{760} D$

satisfactory when the relationship between a certain spectral index and the chlorophyll concentration based on some number of samples is used to extract the chlorophyll concentrations of other samples because of background noise. Therefore, we must find a method that can eliminate this background noise.

Recent studies reveal that the combination of SAVI (soil-adjusted vegetation index, Huete, 1988) with a specific spectral index greatly reduces the noise. For example, OSAVI (optimized soil-adjusted vegetation index, Rondeaux et al., 1996) is a type of SAVI that is a combination of reflectances at 800 and 670 nm, defined as $OSAVI = (1 + 0.16)(R_{800} - R_{670})/(R_{800} + R_{670} + 0.16)$. By its definition, the determination of OSAVI requires no soil information or other types of information, in contrast to other indices, which rely on actual soil spectral characteristics. Moreover, it is also the best at removing the influence of the soil background for most crops.

Leaf area index (LAI) can somewhat reflect the ratio of vegetation to soil in a remote sensing pixel. Therefore, when a spectral index is used to retrieve chlorophyll concentration, the presence of the soil background effect can be converted into the influence of LAI on the established relationship. It is possible, at least in theory, to establish a universal relationship with chlorophyll concentration (independent of LAI) by combining certain spectral indices with SAVI.

The spectral index TCARI is an improvement on MCARI (see definition in Section 9.2.2.1). Haboudane et al. (2002) found that although the influences of LAI on the relationships between two indices, TCARI and OSAVI, and the chlorophyll concentration were significant, TCARI and OSAVI can be combined to isolate the influences of LAI and the soil background on the estimation of the chlorophyll concentration. Fig. 9.12 shows the distribution of TCARI and OSAVI at different chlorophyll concentrations and LAI.

The most important information present in Fig. 9.12 is the distribution of chlorophyll concentration in OSAVI-TCARI space. For all coverages (LAIs), the chlorophyll concentration is distributed along the curves with the same origin, with high values near the x-axis (OSAVI) and low values near the y-axis (TCARI). In addition, the points representing the same chlorophyll concentration but different LAI are

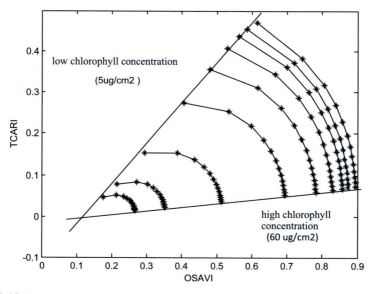

FIGURE 9.12 Distribution of TCARI and OSAVI at different chlorophyll concentrations and LAI.

distributed along a straight line with bare soil as the origin. The isolines of chlorophyll concentration intersect near the origin of the scatter diagram and travel from the center as coverage (LAI) increases. To show this phenomenon more clearly, only two isolines, one corresponding to a low chlorophyll value (5 μg/cm^2) and one corresponding to a high chlorophyll value (60 μg/cm^2), are listed in Fig. 9.12. This figure shows that the chlorophyll concentration is associated with only the slope of the isoline, not LAI. The slope of the isoline decreases as chlorophyll concentration increases. In other words, the slope of the TCARI/OSAVI curve can be used to remove the influence of LAI and extract the chlorophyll concentration.

In fact, taking the ratio of TCARI to OSAVI significantly reduces its sensitivity to LAI while retaining its high sensitivity to changes in chlorophyll concentration. This ratio has a generally corresponding relationship with chlorophyll concentration within a wide range of LAI (0.3—8). This relationship is especially close when the chlorophyll concentration is within the range 15—60 μg/cm^2. On the basis of the above analysis, Haboudane put forward a

prediction relation between the chlorophyll concentration and the TCARI/OSAVI ratio. As shown in Fig. 9.13, the y-axis represents chlorophyll concentration and the x-axis represents TCARI/OSAVI. This relation is believed to be universal for the extraction of the canopy chlorophyll concentration of crops.

An important task yet to be performed is the improvement of Haboudane's relation. The distribution of chlorophyll concentration in TCARI/OSAVI space and the slopes of the isolines is associated with chlorophyll concentration. The point at which these isolines intersect is close but not equal to the coordinate origin. The relation established by Haboudane implies that the intersection point is the origin or that its influence is ignored. An intersection point can be found in the figure, and it is assumed to be O′ (0.11, 0.01). Assuming that chlorophyll concentration is related to the slope of the straight line with O′ as the origin, the relation between chlorophyll concentration and (TCARI-(0.01))/(OSAVI-0.11) can be plotted, as in Fig. 9.14. Compared with Fig. 9.13, the samples in Fig. 9.14 are more concentrated, i.e., less sensitive to the influence of LAI and soil. Therefore, it is the influence of the intersection point

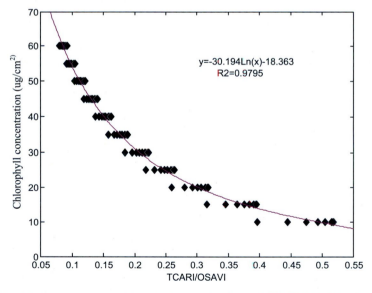

FIGURE 9.13   Relationship between canopy chlorophyll concentration and TCARI/OSAVI ratio without considering the points of intersection.

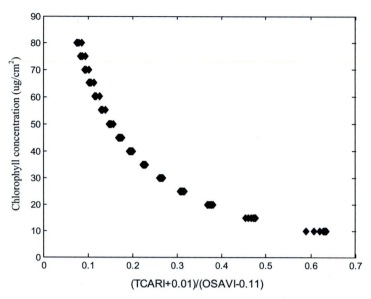

FIGURE 9.14   Relationship between canopy chlorophyll concentration and TCARI/OSAVI ratio with the intersection point considered.

on the relation between chlorophyll concentration and the combination of TCARI and OSAVI should be considered. In practical modeling, the coordinates of the intersection point O′ are assumed to be unknown for the sake of accuracy and are obtained through sample fitting.

In addition, the chlorophyll concentration and TCARI/OSAVI ratio distribution indicate that it is more appropriate to express the relation between chlorophyll concentration and the TCARI/OSAVI ratio as a reciprocal function (Fig. 9.14). Therefore, the relation between the chlorophyll concentration and TCARI and OSAVI should be fitted using Eq. (9.3):

$$y = \frac{a\prime}{\dfrac{TCARI - b\prime}{OSAVI - c\prime} - d\prime} + e \quad (9.3)$$

where y is the chlorophyll concentration and b′ and c′ are the x and y coordinates of the intersection point of the isolines, respectively. For convenience in least squares fitting, the equation can be simplified as follows:

$$y = (a * OSAVI - b)/(TCARI - c * OSAVI - d) + e \quad (9.4)$$

Eq. (9.4) is used to fit the regression relation between chlorophyll concentration and TCARI and OSAVI. Fig. 9.15 shows the relation between the chlorophyll concentrations of the samples and the fitted chlorophyll concentrations; the regression relation and the values corresponding to a, b, c, d, and e in Eq. (9.4) are presented.

The relation for the extraction of the canopy chlorophyll concentration established by Haboudane is used to extract the chlorophyll concentration of maize at each sample point using canopy spectral data measured at the Beijing Academy of Agriculture and Forestry Sciences; the results are compared with the measured values. The results show that the improved estimation relation is better than that proposed by Haboudane for the extraction of the chlorophyll concentration of maize in terms of the correlation between the estimated chlorophyll concentration and the measured chlorophyll concentration and the ratio of the RMSE between the estimated value and measured value to the real mean value (RMSE/mean, representing the overall error). The data used in modeling are all samples

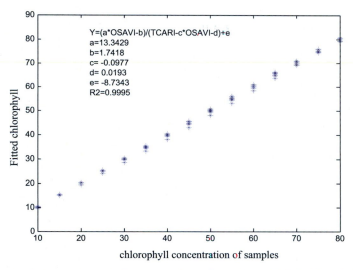

FIGURE 9.15   Relationship between chlorophyll concentrations of samples and fitted chlorophyll concentrations and the estimation of canopy chlorophyll concentration.

obtained through a theoretical model. Therefore, the improved prediction relation can be applied to the estimation of the canopy chlorophyll concentration of maize in other areas. Considering the fact that many crops are categorized as gramineous monocotyledons with similar mesophyll structures, this relation may be suitable for the extraction of the canopy chlorophyll concentration of gramineous crops in general. However, further validation with more measured data remains to be done.

9.3 Extraction using physical models

A leaf reflectance model is needed for the extraction of biochemical parameters on both the canopy and the leaf scales, and the two retrieval methods share some similar features. In both methods, a cost function is constructed, through the minimization of which, using some chosen optimization methods, the estimated parameter value can be obtained. This approach thus requires the rational selection of a cost function, the optimization of said cost function, and the selection of an accurate and time-efficient retrieval algorithm. All of these considerations directly affect whether the retrieval method can be applied in a feasible manner.

Hyperspectral data provide the quasi-continuous spectrum of the vegetation, enabling biochemical parameter retrieval, but the abundant information provided by hyperspectral data contains redundancies. The use of all bands in the extraction will greatly increase the calculation time. Therefore, one must consider how to minimize the number of bands used while maximizing the amount of necessary information to eliminate redundant information.

9.3.1 Overview of the retrieval methods

A vegetation remote sensing physical model is a nonlinear expression of input parameters; it is not necessarily an analytic expression. How, then, do we obtain the surface parameters required (model input parameters) from remote sensing data through retrieval? This type of problem is actually a nonlinear optimization problem. A successful retrieval requires the combination of three factors: a good cost function, a good optimized algorithm, and a good retrieval strategy.

9.3.1.1 The cost function in retrieval

The cost function in Eq. (9.5a) is the form most often used in the traditional retrieval algorithm because it is simple and intuitive. In the cost function, Y_M is the model-simulated value and Y_{obs} is the observed value, which represent the canopy reflectance and the leaf spectrum, respectively; S represents other parameters of the model; and X represents the biochemical parameters for retrieval. The form itself is able to clarify the purpose of retrieval, that is, to make the simulated value close to the real value as possible.

$$COST(X) = \sum [Y_M - Y_{obs}(S;X)]^2 \qquad (9.5a)$$

In consideration of the fact that the contribution of each observed value to the cost function may be different, researchers are sometimes more willing to use a cost function such as that shown in Eq. (9.5b). This function differs from Eq. (9.5a) in that all observed values are all the weighted values, although the total weight is 1. The weight value is specified artificially, or obtained through statistics from some observation data.

$$COST(X) = \sum W_k [Y_{M,k} - Y_{obs,k}(S_k;X)]^2$$
$$(9.5b)$$

The selection of the actual cost function should be based on the specific conditions. This chapter discusses the selection of several cost functions with a focus on analysis of Bayesian retrieval.

Bayesian retrieval gives constraints in the form of a prior probability distribution from the perspective of probability theory and provides retrieval results in the form of a post probability distribution. With regard to other methods, Bayesian theory has the following features:

① prior probability has a definite physical meaning, which provides the approach to introduce other sources of information into the equation solution;
② when there is noise or the equation is not yet determined, an exact solution for the equation is impossible. Bayesian theory explicitly notes that the retrieval result is a probability distribution, and the function of retrieval is to increase the amount of information carried by

the post probability distribution compared with the prior probability distribution information. Therefore, Bayesian retrieval not only is the ideal method for solving the problem when the equation is not determined but also can improve retrieval accuracy and stability for noisy data given some prior knowledge of the parameters.

Bayesian inference is the theoretical basis of Bayesian retrieval. The Bayesian method was introduced and used in geology, atmospheric studies, and other fields very early, and the principle and method of the application of Bayesian inference in geology have been discussed in many books (Menke, 1984; Tarantola, 1987). However, the application of Bayesian retrieval in remote sensing began only a little over 10 years ago (Li et al., 1998; Moulin et al., 2003), mainly because of doubt regarding the existence of prior knowledge (i.e., the prior probability distribution). Prior knowledge refers to all of the information about the unknown parameters we can gather, excluding the equation. In fact, any solution for the retrieval at all implies the use of certain prior knowledge, and a solution in which no prior knowledge is used is no more than a surface phenomenon. For decades, people have accumulated massive amounts of observation data and built up various databases, all of which can serve as prior knowledge. When the research object of remote sensing is the Earth rather than any other planet, the prior estimation of any parameter at a specific time and a specific location on the ground is not far from the real value.

Simply, Bayesian inference can be expressed as follows:

$$P(X|Y) = \frac{P(X)P(Y|X)}{\int P(x)P(Y|x)dx} \qquad (9.6)$$

in which $P(X|Y)$ represents the conditional probability of the occurrence of X when the known event is Y; $P(Y|X)$ represents the conditional probability of the occurrence of Y when the known event is X; and $P(X)$ is the so-called prior probability, that is to say, the estimation of the occurrence probability of X when Y is unknown and the common denominator is the normalization factor.

To apply Bayesian inference to retrieval, the method we use is to treat X as the unknown parameter value and Y as the observed data. The corresponding parameter value when the value of P(X|Y) is at its maximum is called the maximal post probability solution of retrieval, which is represented by X^* here. If any noise and the prior probability distribution of the unknown parameters are Gaussian, the maximal post probability solution is the extreme value point of the cost function below:

$$\text{COST } (X) = (Y_M - Y_{obs})^T \Sigma^{-1} (Y_M - Y_{obs})$$
$$+ \left(X - \overline{X_{prior}}\right)^T \Delta^{-1} \left(X - \overline{X_{prior}}\right)$$

$$(9.7)$$

in which Y_M represents the simulated value of the model of the canopy or leaf spectrum Y; Y_{obs} represents the observed value; Σ represents the covariance of the noise distribution; $\overline{X_{prior}}$ represents the mean value of the prior distribution of the unknown parameters (biochemicals); and Δ is the covariance matrix of the prior distribution of the unknown parameters.

9.3.1.2 Retrieval algorithm

The ultimate goal of physical-model retrieval is to minimize the cost function, and the parameter value when the cost function is minimized is the retrieval solution. Biochemical information retrieval is obtained through the minimization of the cost function using a canopy or leaf model and the measured spectral data. Because both canopy and leaf models are a generic function of biochemical parameters, an accurate analytical solution for the model parameters cannot be obtained simply as the solution to a linear equation. Instead, we usually use an iterative optimization method to obtain an acceptable estimated solution by minimizing the cost function. The process of optimization can be expressed as in the flow chart in Fig. 9.16:

① An initial estimated value X_0 of model parameter X is given and substituted into the model to calculate the simulated value;

② The cost function COST (X_0) is calculated using the observed data and the result of ①. If the cost function is less than a given threshold, the retrieval is over; otherwise, the process enters step ③;

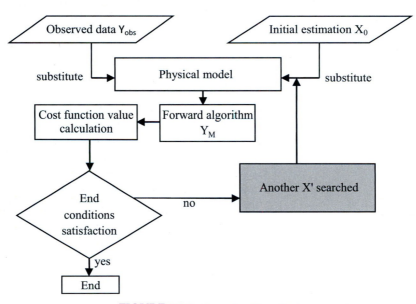

FIGURE 9.16 Inversion flow chart.

③ Another X_0 is found through some search method to replace X_0, and steps ① and ② are repeated;

④ Step ③ is repeated until either the cost function becomes less than a given threshold value or the iteration time reaches a predefined value.

The virtual box in the flow chart is actually an important optimization step, and the algorithm used determines the search speed and accuracy and whether a minimal value can be found. SIMPLEX (Lagarias et al., 1998) is a traditional iteration algorithm popular in the optimization field. Because it is simple and does not require the derivation information of the optimization function, it has been widely used. Most optimization algorithms will adopt the SIMPLEX algorithm, although the performances of some other algorithms are also discussed.

9.3.1.3 Retrieval strategy

The parameters to be retrieved may be large in number. Some may be sensitive, while others may not be. It is also possible that the retrieval model is a combination of several models. In this case, the following parameters will constitute the criteria for retrieval strategy selection: whether the retrieval of all parameters is conducted simultaneously; whether some insensitive parameters are fixed; and whether the retrieval of the final model is conducted directly or the retrieval of several component models is conducted separately.

We can conduct the extraction of biochemical component information at both the leaf and canopy scales. The biochemical parameters are usually the direct input parameters of the leaf model, while the canopy biochemical parameters are coupled into the canopy model through the leaf model. Thus, it is obvious that for the leaf scale, we can directly invert the biochemical parameters; for the canopy, we can directly invert the biochemical information through the coupling model, and if the component spectrum is available or can be obtained through extraction, what remains is simply the retrieval from the leaf model. The extraction for the two scales is therefore both connected and independent. The leaf model retrieval is actually the basis of all extraction processes. The extraction of biochemical information at the leaf and canopy scales and present problems are analyzed below.

9.3.2 Leaf-scale biochemical parameter retrieval

The retrieval from a theoretical model of leaf biochemicals can be conducted through a cost-function retrieval algorithm using a certain optical physical leaf model. In the extraction process, the sensitivity of each model parameter to the retrieval and the degree of influence of noise on the accuracy of the retrieval results should be evaluated to understand the importance of prior knowledge to the retrieval.

9.3.2.1 Unbiased data retrieval

To use real measured remote sensing data to invert biochemical component information, simulated unbiased data can be used for a retrieval experiment and a discussion of the retrieval degree of the model and the sensitivity of the model parameters. The method is as follows: first, select a model and specify the input parameters of this model; then use the model to simulate the leaf spectrum; finally, use the simulated spectrum as the real value (also called the observed value) to invert the "forgotten" input parameters. Researchers (Maire et al., 2004) have used similar methods to study the retrieval strategy for chlorophyll concentration. The Goel and Strebel retrieval strategy (Goel and Strebel, 1983) can also be used as a reference: change the form of the cost function; change the number of parameters and observed values; use the combined parameters or combined spectrum; and use different optimization methods and prior knowledge.

The extraction abilities of two models, PROSPECT and LIBERTY, are analyzed below.

9.3.2.1.1 PROSPECT model retrieval

As seen in the above introduction, PROSPECT requires two types of parameters: a structure parameter N and biochemical concentrations. If chlorophyll, water, protein, and cellulose + lignin, the concentrations of which are c_{ab}, c_w, c_p and c_{cl} respectively, are included as the biochemicals of interest, there will be five parameters. The prior knowledge of $[N, c_{ab}, c_p, c_w, c_{cl}]$ is as follows (Jacquemoud, 1996):

Lower bound = $[1, 16.5 \, \mu g/cm^2, 0.0046 \, cm,$
$0.00,048 \, g/cm^2, 0.00,034 \, g/cm^2]$
Upper bound = $[5, 85.5 \, \mu g/cm^2, 0.0405 \, cm,$
$0.00,172 \, g/cm^2, 0.0085 \, g/cm^2]$
Mean value = $[2, 48.6 \, \mu g/cm^2, 0.0115 \, cm,$
$0.00,096 \, g/cm^2, 0.00,168 \, g/cm^2]$
Standard deviation = $[1, 15.2 \, \mu g/cm^2,$
$0.0067 \, cm, 0.00,029 \, g/cm^2, 0.00,129 \, g/cm^2]$

Within the range of the lower and upper bounds, a 100-parameter set is generated at random following a Gaussian distribution; the spectrum data of 100 leaves in 421 bands from 400 to 2500 nm in 5-nm intervals are simulated using the PROSPECT model. Each leaf spectrum includes reflectance and transmittance. These simulated leaf spectra are taken as observed values for biochemical component parameter extraction.

The retrieval of the five parameters, including the mesophyll structure parameter N, is conducted simultaneously in the experiment. As the data are unbiased, prior knowledge is not added to the cost function; that is, we select the form given by Eq. (9.5a). The SIMPLEX iterative

optimization algorithm is selected, as it is a commonly used retrieval algorithm.

The retrieval results are shown in Table 9.8. The RMSE (root mean square error)/mean value represents the overall average error of the retrieval. It can be seen that, for unbiased leaf data, the real parameters are properly represented using the PROSPECT model retrieval method, providing a basis for biochemical concentration extraction from a measured spectrum.

9.3.2.1.2 LIBERTY model retrieval

If the four biochemicals (chlorophyll, water, protein, and lignin + cellulose) are taken into consideration, including the five nonbiochemical parameters (average cell diameter D, cell gap x_u, thickness of the needle leaf t, bond absorption a_b, albino absorption a_a), the LIBERTY model has nine input parameters.

The spectrum data of 100 leaves in 421 bands in a range of 400—2500 nm in 5-nm intervals are simulated using the LIBERTY model. Each leaf spectrum includes the reflectance of infinitely thick leaves and the reflectance and transmittance of a single leaf. The extraction results of the nine parameters with these simulated leaf spectra as the observed values are shown in Table 9.9. The table shows that errors in the model retrieval results can be tolerated. For the four biochemical parameters, the chlorophyll error is 13%; the water error is 15%; the protein error is approximately 20%; and the lignin and cellulose error is 16%. However, the retrieval result is not ideal for unbiased data.

To accurately invert biochemical concentrations, the retrieval parameters must be reduced to create certain constraints on the extraction

TABLE 9.8 Results of unbiased-data parameter retrieval using the PROSPECT model.

Retrieval parameters	N	c_{ab} ($\mu g/cm^2$)	c_w (cm)	c_p (g/cm^2)	c_{cl} (g/cm^2)
Mean value	1.98	48.6	0.0118	0.00094	0.0017
RMSE/mean value	6e-011	8e-010	1e-009	1e-009	8e-009

TABLE 9.9 Results of all-parameter extraction using the LIBERTY model.

Retrieval parameter	D	x_u	t	a_b	a_a	c_{ab} ($\mu g/cm^2$)	c_w (cm)	c_p (g/cm^2)	c_{cl} (g/cm^2)
Mean value	45.8	0.03	1.60	0.0006	5.15	51.2	0.0116	0.00059	0.0017
RMSE	7.87	0.001	0.007	0.0001	0.76	6.75	0.00169	0.00012	0.00027
RMSE/mean value	0.17	0.04	0.004	0.14	0.15	0.13	0.15	0.20	0.16

TABLE 9.10 Results of biochemical concentration extraction using the LIBERTY model.

Retrieval parameters	c_{ab} ($\mu g/cm^2$)	c_w (cm)	c_p (g/cm^2)	c_{cl} (g/cm^2)
Mean value	51.54486	0.01155689	0.000609918	0.001599231
RMSE	0.3556004	0.0000287294	0.000046136	6.84E-09
RMSE/mean value	0.0069	0.0025	0.0756	0

process. If the nonbiochemical parameters are known and fixed, the extraction of only the biochemical parameters is enough to obtain an accurate retrieval, with results as shown in Table 9.10. Therefore, because the LIBERTY model parameters are so large in number, the accurate retrieval of biochemical parameters relies heavily on the accurate prior knowledge of other parameters.

9.3.2.2 Extraction from noisy data

Real-observed data are all affected by some kind of noise. Even with data correction, it is not possible to eliminate all of the noise, so the influence of the noise on the retrieval must also be considered. Among the various types of noise, random noise refers to irregular noise that occurs at random, such as spot noise. Here, the random noise is assumed to be additive noise with a Gaussian distribution of zero mean value. System noise, generally with the multiplicative nature, refers to an entire drift of the data because of system error. For instance, system noise due to sensor calibration error causes an overall shift, higher or lower, of the observed data compared with the real values.

One hundred parameter combinations are selected at random within the bound range of

each parameter using the PROSPECT model to simulate the leaf spectrum. Next, a Gaussian random noise of zero mean and standard deviation of 0.01, 0.05, or 0.1 and a system noise with an error of 5%, 10%, or 15% are added into these data. A cost function with the form of Eq. (9.5a), namely that with no prior knowledge is used to invert the noisy data of the full wave band. Table 9.11 lists the biochemical concentration retrieval results in the presence of different levels of noise. The first line represents the random noise level, and the first column represents the system noise level. The four biochemical values represent chlorophyll, water, protein, and cellulose and lignin.

It can be seen that the performance of each biochemical parameter with different noise levels is uneven. In broad terms, chlorophyll and water can be classified together, as can protein and cellulose and lignin.

When the system noise is low, water and chlorophyll can both obtain ideal retrieval results. Even if the random noise level is higher (i.e., the standard deviation is 0.1), the retrieval error is only approximately 8%. In contrast, protein and cellulose and lignin are very sensitive to random noise, and the retrieval already fails for moderate noise (i.e., a standard deviation of 0.05).

TABLE 9.11   The retrieval results of biochemicals with different combined noise levels.

System noise	Biochemicals	Random noise 0.01		0.05		0.1	
		RMSE	RMSE/mean value	RMSE	RMSE/mean value	RMSE	RMSE/mean value
5%	Cab	2.4986	0.0521	3.4149	0.0712	3.8575	0.0804
	Cw	0.0005	0.0431	0.0007	0.0616	0.0008	0.0720
	Cp	0.0009	0.8792	0.0009	0.9108	0.0009	0.9203
	Ccl	0.0011	0.6519	0.0012	0.7126	0.0014	0.8213
10%	Cab	5.7194	0.1193	7.0091	0.1462	7.7808	0.1623
	Cw	0.0011	0.0985	0.0014	0.1261	0.0016	0.1462
	Cp	0.0018	1.8918	0.0019	1.9986	0.0021	2.1650
	Ccl	0.0033	1.9586	0.0037	2.1532	0.0046	2.6958
15%	Cab	9.9874	0.2083511	10.5121	0.2401	12.7429	0.2658
	Cw	0.0023	0.2111035	0.0024	0.2251	0.0025	0.2341
	Cp	0.0028	2.8779056	0.0028	2.8727	0.0029	2.9872
	Ccl	0.0059	3.481261	0.0062	3.6498	0.0066	3.8529

When the system noise is larger, the retrieval results for the chlorophyll and water as well as the protein and cellulose and lignin are seriously affected. Even if there is no random noise and the system noise level is low (5% deviation), the retrieval of protein and cellulose and lignin fully fails, while the retrieval errors for chlorophyll and water are low, only 5% and 4%. When the system noise increases to 10%, the retrieval errors for chlorophyll and water increase to approximately 14% and 12%, respectively, and when the system noise increases to 15%, the extraction of chlorophyll and water is almost impossible.

As shown here, chlorophyll, water, protein, and cellulose and lignin have clearly different resistances to noise. Random noise has essentially no influence on chlorophyll and water,

while the influence of the system noise is stronger. However, both types of noise can seriously affect the retrieval accuracy for protein and cellulose and lignin.

These results can be explained as follows. On one hand, it is assumed that the random noise has a Gaussian distribution (mean 0) for chlorophyll and water because the retrieval is conducted over the full wave bands. As a consequence, the influences of the random noise on chlorophyll and water retrieval are statistically offset between each band. The same "offset effect" does not affect protein or cellulose and lignin because among the four biochemicals, the water content is the most sensitive, followed by chlorophyll; cellulose and lignin and chlorophyll have similar sensitivity, and the sensitivity of protein is the smallest. So why are the retrieval results of cellulose and lignin and chlorophyll, which have nearly the same sensitivity, quite different? Consider the sensitivity of each component in each band. For bands below 800 nm, only chlorophyll has a significant sensitivity to the spectrum, while the sensitivities of the other three biochemicals, such as water, are negligible. For bands above 800 nm, chlorophyll loses its sensitivity to the spectrum, while the other biochemicals gain this sensitivity. Among the three nonchlorophyll parameters, the sensitivity of water is far greater than that of the other two parameters in each band. As the action of the water covers up the influence of the protein and cellulose and lignin, it is understandable that the retrieval results for protein and cellulose and lignin have a very weak resistance to noise. This result also means that it is, in fact, not possible to extract these two biochemicals via remote sensing data retrieval. Jacquemoud et al. (1996) proposed that the remaining components after the removal of pigment and water, including protein, cellulose and lignin, be classified into one component: dry matter.

Table 9.11 also shows that the system noise has a very obvious influence on the retrieval results of all four biochemicals. For higher levels of system noise, the retrieval results for chlorophyll and water content are also not ideal. Therefore, we assume that for remote sensing data with system noise, we may improve the retrieval accuracy by changing the cost function if the noise attributes are known. For example, system noise is equivalent to an overall "translation" of the real data, which does not change the data "shape." With regard to the spectrum, the real spectrum and the noise spectrum in a two-dimensional plane are two parallel curves with an inevitable relevance of 100%. Therefore, for parameter retrieval from such noise-affected data, it may be more appropriate for us to use the correlation coefficients of the observed spectrum and model simulation spectrum as the cost functions for biochemical retrieval.

9.3.2.3 Extraction from observed data

We used the observed data of the LOPEX93 Database (Hosgood et al., 1995) to test the retrieval of leaf-scale biochemicals. The LOPEX database contains 70 leaf samples, which represent 50 types of woody plants and herbs. As the data reflect the diversity of internal leaf structure and pigment concentration as well as water content and other biochemicals, the spectral characteristics of the leaves also vary within a wide range. We adopted 62 samples of fresh leaf data to conduct this research. The leaf spectrum data were sampled from 400 to 2500 nm hemisphere reflectance and transmittance with a sampling interval of 1 nm. The spectrum of each sample was the average spectrum of five leaves of the same type, and the subdivided spectra evenly corresponded to 5 nm sampling intervals.

At the leaf scale, the PROSPECT model was used to extract three biochemical parameters: chlorophyll, water, and dry matter. The retrieval process was realized by minimizing the cost function in Eq. (9.5a), and the SIMPLEX iterative optimization algorithm was again used to minimize this function.

The retrieval of biochemical concentrations and structure parameters was conducted simultaneously using the spectrum data sampled over entire spectrum, that is, from 400 to 2500 nm. The results are shown in Figs. 9.17—9.19, in which the retrieval results for the three biochemical parameters are presented: chlorophyll, water, and dry matter. It can be seen that the retrieval results of the chlorophyll and water contents are quite good. For the chlorophyll concentration, the correlation coefficient between the model retrieval value and the measured value is 0.88 and the RMSE is 7.14 $\mu g/cm^2$, but the average real value is 48.79 $\mu g/cm^2$, an average retrieval accuracy of 85%. For the water content, the correlation coefficient between the model retrieval value and the measured value is 0.97 and the RMSE is 0.0021 cm, but the average real value is

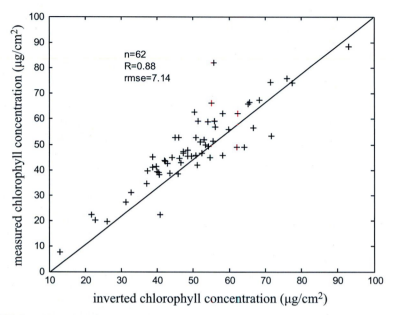

FIGURE 9.17   The relationship between the inverted and measured chlorophyll content.

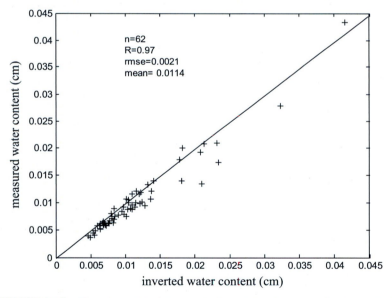

FIGURE 9.18   The relationship between the inverted and measured water content.

0.0114 cm, an average retrieval accuracy of 82%. For dry matter content, however, the retrieval result is worse: the correlation coefficient is only 0.58; the RMSE is 0.0032 g/cm^2; and the average value is 0.0055, which is not sufficiently accurate. In other words, the retrieval was not successful.

Why, then, were the retrieval results for chlorophyll and water quite good, while those

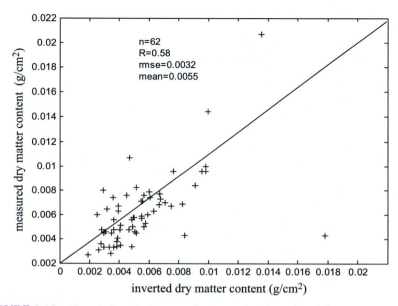

FIGURE 9.19   The relationship between the inverted and measured dry matter content.

for the dry matter were not? Was it because the parameters involved in the retrieval are less sensitive, leading to the failure of the retrieval? We performed an experiment to test this hypothesis: the other parameters were fixed to the real values, and only the dry matter content was inverted.

However, the retrieval results were still not ideal. The absorption spectra of the components might explain the failure of the retrieval. Fig. 9.20 shows the absorption spectra of the three components. The figure shows that the absorption spectrum of the dry matter is very weak. In the visible-light range, its action is completely obscured by that of the chlorophyll, while its absorption spectrum in the near-infrared range is far less than that of the water. Therefore, there is no range of the leaf spectrum in which its contribution is dominant. In addition, the model's sensitivity to water is far greater than that to dry matter, so the retrieval of the dry matter information of fresh leaves is quite difficult.

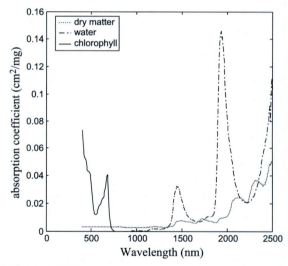

FIGURE 9.20   The absorption spectra of chlorophyll, water, and dry matter.

9.3.3 Canopy-scale biochemical parameter retrieval

The radiative transfer model SAILHT is used for the canopy reflectance model in this section

(Verhoef, 1984; Kuusk, 1991). Supposing that the canopy is uniformly distributed, this model describes four ascending and descending fluxes in a uniform canopy. The model parameters include the following: leaf reflectance $\rho(\lambda)$ and transmittance $\tau(\lambda)$, LAI, mean leaf angle (MLA) θ_l, spot size s, soil reflectance $\rho_s(\lambda)$, and horizontal visibility Vis (used for calculating the scattering component of solar radiation). The direction spectrum is simulated by changing the measuring conditions: solar zenith angle θ_s and azimuth angle φ_s, view zenith angle θ_v and azimuth angle φ_v. If PROSPECT is embedded into the SAILH model to provide its input parameters to the leaf spectrum, the parameters that the entire PROSPECT + SAILH model requires can be categorized as follows:

- Canopy physiological and biochemical parameters: chlorophyll concentration c_{ab}, water content c_w, mesophyll structure parameter N, LAI, MLA θ_l, and spot size parameter s;
- The soil reflectance $\rho_s(\lambda)$ (supposed to be Lambertian);
- External parameters: view zenith angle θ_v and azimuth angle φ_v, solar zenith angle θ_s and azimuth angle φ_s and horizontal visibility Vis.

As you can see, the leaf spectrum is an input parameter of the SAILH model. Therefore, after the theoretical coupling of the PROSPECT model and the SAILH model, a feedforward process can be used to simulate the dichroic reflectance spectrum of the canopy R, and the reverse process can be used to invert the SAILH model parameters. The biochemical parameters are extracted through retrieval from the coupling model.

9.3.3.1 Retrieval from simulated data: multiple-phase retrieval of biochemical parameters at the canopy scale

We used our canopy model SAILH and substituted the reflectance and transmittance of

TABLE 9.12 SAILH model simulation parameter settings.

LAI	θ_i	P_s	S	Vis(km)	θ_s	φ_s	θ_v	φ_v
2	57	Measured	0.25	50	30°	0°	60°	0°

the LOPEX leaf samples. Other parameters (see Table 9.12) were set to simulate the canopy spectrum, and the biochemical concentration retrieval was conducted with this spectrum as "the measured data."

Firstly, the PROSPECT model was coupled to the SAILH model. The retrieval of biochemical parameters was performed on the canopy spectrum directly, and other parameters were fixed as the real values. Fig. 9.21 shows the inverted chlorophyll concentration and water content within the acceptable iteration times and search field values. Although the retrieval of the chlorophyll concentration obtained satisfactory results for some samples, the inverted chlorophyll concentrations in many samples are still around the initial value. Therefore, the retrieval was not successful. The accuracy of the water content retrieval is acceptable but far less accurate than the retrieval results at the leaf scale.

The results show that it is more difficult to couple the PROSPECT and SAILH model and directly invert leaf data from the canopy data than to simply invert the PROSPECT model, and the degree of sensitivity of the parameters being retrieved is different, which leads to unsatisfactory retrieval results for some parameters.

9.3.3.2 Retrieval of biochemical parameters from real-observed data

We conducted Bayesian retrieval of biochemical concentrations using the canopy maize spectrum and physiological and biochemical data observed in the Beijing Academy of Agricultural and Forestry Sciences and the PROSPECT + SAILH models. We chose the maize

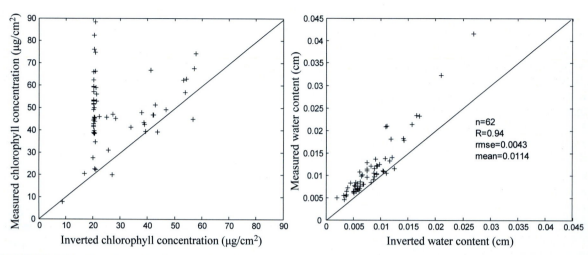

FIGURE 9.21   The relationship between the chlorophyll and water content directly inverted from the canopy spectrum and the real value.

spectrum and physiological data of August 20 as our experimental data. The solar zenith angle was calculated according to the observation time. The relative azimuth was zero because of vertical observation. For the soil spectrum, we chose the measured spectrum from that day. Meteorological data were the primary source of horizontal visibility. The other six canopy physiological and biochemical parameters were as follows: chlorophyll concentration c_{ab}, water content c_w, mesophyll structure parameter N, LAI, MLA θ_l, and spot size parameter s. These parameters were treated as unknown variables, and their retrieval was conducted simultaneously. The retrieval was conducted with two different parameter settings: ① the prior mean and variance were set to the measured values and the noise level was estimated to have a variance of 0.0005; ② the noise level was estimated to have a variance of 0.0005, and others were same to ①.

For the former noise-level setting, the retrieval results were affected by actual data noise and other sensitive parameters at the canopy scale;

most of the retrieval results for the water content and chlorophyll concentration were at the border; and the retrieval of LAI, although not at the border, still exhibited a large gap compared with the actual value.

For the latter noise-level setting, the accuracy of the retrieval results was greatly improved because of the constraints of prior knowledge. Although the retrieval results were still not ideal, the results manifested an essential difference from those obtained at the former noise-level setting, for which the measured data noise effect was completely not inverted. These results reaffirmed the importance of prior knowledge.

9.3.4 The influence of spectral resolution and band selection

Hyperspectral data have redundancy, while multispectral data have insufficiency to accurately invert extract biochemical parameters (chlorophyll, water). Knowledge of the required spectral resolution must be available and the number of bands necessary for biochemical parameter retrieval forms the basis for relevant remote sensing work. For this reason, we conducted a study of band selection for biochemical component retrieval for many types of remote sensing data, including hyperspectral data, and analyzed the significance of the selected bands.

9.3.4.1 The influence of spectral resolution on the retrieval of biochemicals

NOAA-AVHRR has two bands in the visible and near-infrared spectrum, with center wavelengths of 680 and 760 nm; TM has six bands, with center wavelengths of 500, 595, 677.5, 800, 1707.5, and 2187.5 nm. MODIS has 25 bands in the visible and near-infrared spectrum, with center wavelengths of 547, 664, 707, 745, 786, 834, 875, 910, 945, 1623, 1680, 1730, 1780, 1830, 1880, 1930, 1980, 2030, 2080, 2142, 2180, 2230, 2280, 2330, and 2380 nm. OMIS has 64 bands for which the wavelengths are shorter than 1100 nm, and the spectral resolution is 10 nm. The spectral resolution of the four types of sensors increases gradually. This fact allows us to analyze the influence of spectral resolution on biochemical retrieval. We still use the leaf spectrum from 400 to 2500 nm with 5 nm intervals simulated using the PROSPECT simulation model with some added random noise. The retrieval results for chlorophyll concentration and water content using the data from various sensors are compared by resampling the sensor response functions of AVHRR, TM, MODIS, and OMIS as equivalent leaf spectra from various sensors.

Table 9.13 shows the retrieval results for chlorophyll and water contents obtained from the equivalent spectra of the various sensors. The usability evaluation of the various sensors with respect to biochemical parameter retrieval is given in the table.

The table shows that the NOAA-AVHRR sensor only has two bands: one in the visible and the other in the near-infrared. The spectral resolution is only 100 nm. As a result, the retrieval of biochemical information is almost

TABLE 9.13 The influence of spectral resolution on biochemical parameter retrieval.

Sensor	Retrieval results (RMSE/mean value of c_{ab}, c_w)	√ Evaluation
AVHRR	0.34987, 0.28096	×
TM	0.07553, 0.14131	⅄
MODIS	0.04223, 0.03315	√
OMIS	0.01912, 0.09338	√

impossible. In fact, these results can be explained by the mathematical requirements of the retrieval process. The two bands are equivalent to only two equations, and there are three unknown parameters (leaf structure parameter, chlorophyll, and water). Thus, it is an underdetermined equation, and the retrieval is not successful. For the TM sensor, the spectral resolution is improved with respect to AVHRR, and there are four more bands available. As a result, the retrieval results are clearly superior to those obtained from AVHRR, but the retrieval results are still not ideal considering that the noise we used is small. Thus, TM should be used to extract biochemical information only in the absence of any other choice; MODIS has 25 bands with spectral resolution from 10 to 50 nm, and the retrieval results for chlorophyll and water content using the MODIS spectral data are therefore more ideal; the retrieval results using OMIS data are good, as expected.

This study reveals that the retrieval of biochemical parameters depends highly on spectral resolution, and multiple-spectrum or hyperspectral remote sensing is a necessary condition.

9.3.4.2 Band selection specifically for the retrieval of biochemicals

According to the above analysis, we have learned that the retrieval results for biochemical parameters using MODIS and OMIS data are

similar. We also know that the correlation between the bands of hyperspectral data is very
strong, and redundant information is present.
If we wished to use remote data retrieval for a
practical application, the calculation overhead
would be an important consideration, and too
many bands would increase the redundant information and the computation time. Therefore, it is
necessary to perform band selection. Here, we
still superpose random noise of standard deviation 0.01 and system noise of 5% deviation on
the 100-leaf simulated spectrum. In the following
steps, the band selections for the full-band data
and the equivalent data of MODIS are conducted
specifically for the retrieval of the chlorophyll
concentration and the water content separately
(as shown in Tables 9.14—9.17):

① First, the retrieval of biochemicals in each
band is conducted separately, and the bands
with the lowest RMSE are selected.
② Next, we fix the bands selected in ①. The
retrieval of biochemicals from the remaining

bands and the fixed bands is conducted, and
the bands with the lowest RMSE are selected
again.
③ Once 10 bands have been selected, the band
selection is stopped.

Fig. 9.22 shows the change in RMSE/mean
value with the increase of selected bands when
bands are selected using spectrum data of 421
bands. The figure shows that for both chlorophyll
and water content, the inverted RMSE/mean
value decreases as the number of selected bands
increases, and the decrease of the inverted
RMSE/mean value slows or stops once the number of selected bands has reached a certain point.
For the band selection of the full wave bands, the
decrease of the RMSE/mean value of the chlorophyll concentration or water content retrieval
band decreases more slowly once the third band
has been selected. The selection results of the
biochemical retrieval bands of the simulated
MODIS data are similar. This finding reveals
that a limited number of bands are sufficient for

TABLE 9.14   The band selection results for the retrieval of chlorophyll concentration using full-band data.

Selected order	1	2	3	4	5
Wavelength (nm)	630	1080	585	705	565
RMSE/mean value	0.1512	0.0408	0.0337	0.0311	0.0320
Selected order	6	7	8	9	10
Wavelength (nm)	1215	1090	700	1085	570
RMSE/mean value	0.0319	0.0328	0.0317	0.0300	0.0319

TABLE 9.15   The band selection results for the retrieval of chlorophyll concentration using MODIS equivalent data.

Selected order	1	2	3	4	5
MODIS wavelength (nm)	547	834	707	745	786
RMSE/mean value	0.4258	0.0534	0.0501	0.0479	0.0479
Selected order	6	7	8	9	10
MODIS wavelength (nm)	664	875	910	945	1623
RMSE/mean value	0.0478	0.0478	0.0477	0.0477	0.0477

TABLE 9.16 The band selection results for the retrieval of water content using full-band data.

Selected order	1	2	3	4	5
Wavelength(nm)	1900	935	1405	1400	1905
RMSE/mean value	0.2431	0.0488	0.0412	0.0306	0.0243
Selected order	6	7	8	9	10
Wavelength(nm)	770	1405	1410	1400	1415
RMSE/mean value	0.0188	0.0189	0.0182	0.0248	0.0186

TABLE 9.17 The band selection results for retrieval of water content using MODIS equivalent data.

Selected order	1	2	3	4	5
MODIS wavelength (nm)	1880	786	1830	1623	1780
RMSE/mean value	0.8087	0.1652	0.0696	0.0261	0.0261
Selected order	6	7	8	9	10
MODIS wavelength (nm)	834	1680	875	1730	910
RMSE/mean value	0.0261	0.0174	0.0174	0.0174	0.0174

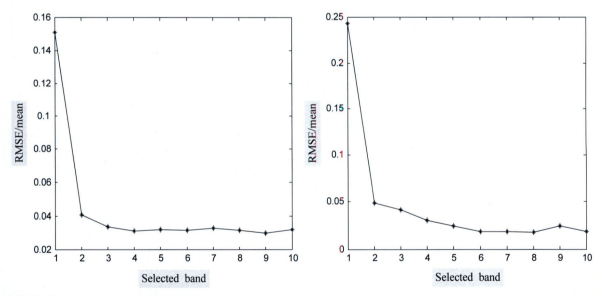

FIGURE 9.22 The change in RMSE/mean value with the increase in number of selected bands (left: chlorophyll; right: water).

the retrieval of biochemical information when using multispectral or hyperspectral data. If many bands are used, there will be redundant information, which will produce undesirable effects. Therefore, it is necessary to select bands (Fig. 9.23).

Let us analyze the meaning of the band selection for chlorophyll and water retrieval. For the

409	426	442	458
474	491	507	523
540	556	572	589
605	621	637	653
670	686	703	719
735	751	768	784
800	816	833	849
865	882	898	914

FIGURE 9.23   Hyperspectral lidar prototype and wavelength (left: prototype; right: central wavelength nm).

chlorophyll concentration, 630 nm, red light, was the first band selected when using the full wave band data to select bands; 1080 nm, in the near-infrared, was the second band selected; and the third band was green light. When using MODIS equivalent data for band selection, the first selected band was green light, followed by near-infrared and then red light. This result is consistent with the biological properties of chlorophyll. As we know, chlorophyll has an absorption characteristic for red light but a stronger reflection characteristic for green light. In other words, chlorophyll is very sensitive to both red and green light. For water content, the first selected band

was 1900 nm, followed by 935 nm and then 1405 nm when using the full wave band data for band selection; of the bands selected using MODIS equivalent data, the first corresponded to 1880 nm, the second corresponded to 786, and the third corresponded to 1830 nm. Both the first and the third selected bands were consistent with water absorption characteristics (Fig. 9.24).

We found that the second selected band for both chlorophyll and water was a near-infrared band. Their sensitivities show that chlorophyll and water both have the weakest sensitivity in this area, which seems to contradict the explanation above. In fact, this contradiction is resolved when we consider the sensitivity analysis of dry matter content. In the near-infrared region, chlorophyll and water are very insensitive, but dry matter is relatively sensitive and is not covered by the action of the water and chlorophyll. Additionally, the retrieval we conducted can be regarded as multiple-parameter retrieval, because the retrieval is conducted over all parameters, although the only ones we consider are chlorophyll and water. Thus, only when the retrieval of each parameter is "accurate" is the error at

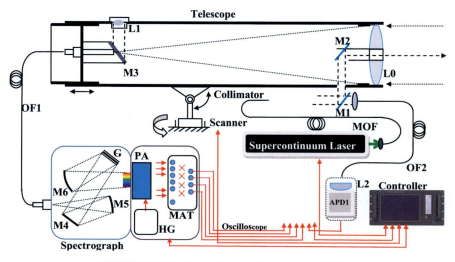

FIGURE 9.24   Optical path of hyperspectral lidar.

a minimum. It is therefore understandable why the second selected band is always in the near-infrared region when the band selection for the retrieval of chlorophyll and water content is conducted.

Starting from the fourth band, the selected bands were secondary sensitive bands or near-infrared bands for both chlorophyll and water. This result is in line with the theoretical analysis. In addition, as the number of selected bands increased, the error decreased very slowly. The significance of these later selected bands was therefore not obvious.

9.4 Extraction of vertical distribution of biochemical components in vegetation using hyperspectral lidar

The full-waveform lidar echoes with hyperspectral attributes can be obtained by hyperspectral lidar, which provides a new remote sensing method for detecting the three-dimensional distribution of vegetation biochemical characteristics. This section briefly introduces the latest progress of vertical detection of vegetation characteristics using hyperspectral lidar, discusses the feasibility of extracting vertical distribution of biochemical components using hyperspectral lidar, and introduces the research of terrestrial hyperspectral lidar instrument and its data processing flow. Finally, we carry out detection research based on instrument for extracting vertical distribution information of chlorophyll and carotene and give the inversion results, and the development trend of this method is discussed.

9.4.1 Study on vertical extraction of vegetation characteristics using hyperspectral lidar

In areas rich in biological species, natural vegetation communities usually present three-dimensional distribution of tree–shrub–grass. There are great differences in light, water, and nutrients at different heights. Therefore, detecting three-dimensional distribution characteristics is not only the need of accurate ecosystem monitoring and management but also of great significance for early warning of pests and diseases, and prediction of ecosystem evolution (Wang et al., 2004, Asner et al., 2014), but the existing remote sensing methods have encountered great difficulties in solving these problems.

For the complex vertical distribution of vegetation, the signal received by passive optical sensors usually contains contributions from different heights. In fact, it is a two-dimensional accumulation under the three-dimensional distribution. In the actual imaging process, the three-dimensional object become two-dimensional, which brings greater uncertainty to the inversion of vegetation parameters. Although multiangle observation can obtain the optical signal of vegetation from different angles, and then obtain the vertical distribution information of chlorophyll and other biochemical components (Zhao et al., 2006), it is difficult to obtain the most important angles such as "hot spots" and "cold spots" on aeronautical and satellite scales, and it is also difficult to directly apply it to inversion of vertical stratified distribution on satellite scale. Although lidar has been widely used in the extraction of vegetation structure parameters, it mainly uses its ranging features and is limited by its few spectral features. Therefore, the fusion of the two sensors has attracted wide attention, but the study showed that there are many shortcomings and errors in the fusion of the two sensors, no matter in the preacquisition or postprocessing of data. To further combine the advantages of the two sensors, a direct method is to use an instrument and equipment to directly collect the point cloud or waveform data of lidar with spectral attribute information for structure and biochemical parameters extraction.

The hyperspectral lidar combines the hyperspectral observation ability of passive optics and the vertical detection characteristics of lidar. It has the ability to detect the fine structure and spectrum of the canopy (Wing et al., 2012; Hopkinson et al., 2016). Therefore, it has great advantages. It does not depend on sunlight and is not affected by the observed geometry. It can form accurate point observation ability and reduce the effect of mixed pixels. At present, many countries have developed and applied hyperspectral lidar. For example, Morsdorf et al. (2009) have used four-band instruments of Edinburgh University to diagnose the vertical distribution of vegetation physiological parameters. The trees of different ages simulated by ecological process models have been diagnosed, and the effects of different ages on NDVI and PRI profiles of vegetation have been studied. Suomalainen et al. (2011) found that the vertical distribution of the spectrum can also be used for automatic detection and classification of targets. Vertical geometry and spectral information can be obtained simultaneously through one observation. Hakala et al. (2012) studied the deforested *Pinus yunnanensis* indoors. The results showed that different band combinations could be used not only to distinguish canopy branches and leaves but also to detect the differences of leaf water content, chlorophyll, and other biochemical components at different canopy heights. Rall et al. (2004) firstly proposed the concept of active vegetation index based on two-band instrument and realized the horizontal observation of deciduous vegetation leaves by low power laser diode, it showed that characteristic changes of chlorophyll content in vegetation can be detected in the band near 680 nm. Wei et al. (2012) used a four-band instrument to study the changes in the concentration of fine biochemical components in leaves. Li et al. (2014) used the hyperspectral lidar prototype developed by the Institute of Remote Sensing Earth to the extraction of chlorophyll, carotene, and nitrogen from the leaves of Safflower Sheep's Foot and *Lagerstroemia indica*. Compared with laboratory analysis, related coefficients were 0.85, 0.71, and 0.51.

The above research shows that the hyperspectral lidar instrument, combined with the characteristics of hyperspectral and lidar sensors, can detect the physiological processes directly related to vegetation physiological structure and photosynthesis and has a strong scientific value and application potential in the detection of complex surface vegetation characteristics. However, for the inversion of vertical distribution of biochemical components, there is no specific method based on hyperspectral lidar. Gao et al. (2018) have carried out the research of vertical detection of biochemical components in vegetation based on hyperspectral lidar. The difference of laser-reflected signals in different bands caused by biochemical components has been studied, and the inversion method of vertical distribution of biochemical components based on this instrument has been proposed. The calibration and processing methods adapted to the characteristics of the instrument are also proposed.

9.4.2 Experiment and data processing of hyperspectral lidar instruments

9.4.2.1 Hyperspectral lidar instruments

Sun et al. (2014) developed a 32-band hyperspectral lidar prototype in the State Key Laboratory of Remote Sensing Science. The prototype covers a range of wavelengths from visible to near infrared (Fig. 9.23). The equipment is mainly composed of two-dimensional scanning platform, supercontinuum pulse laser source, transmitting and receiving system fixed on the scanning platform, multichannel full waveform measuring device, and control center (Niu et al., 2015). The scanning system adopts two-dimensional turntable scheme PTU-D48 E (FLIR Systems Inc). The minimum angle steps of horizontal and vertical design are 0.026° and 0.0°, respectively. The horizontal scanning displacement corresponding to the distance of 10 m is about 4—5 mm. The scanning system and the main control system are connected by serial port and follow the communication protocol of the turntable.

The pulse width of the supercontinuum spectrum pulse laser source (NKT SuperK Compact) is 1—2 ns, the peak power is about 20 KW, and the pulse frequency is 20—40 KHz. The "white light" laser emitted by the supercontinuum spectrum laser passes through the collimator and becomes a convergent beam (divergence angle is less than 5 mrad). The beam is refracted twice through two mirrors (M1, M2) and enters the optical axis of the telescope and emits. The mirror M1 is a microtransparent mirror. Most of the light in the laser beam is reflected into the mirror barrel, and a few of the light is projected through the optical fiber into the emission detection sensor APD1. The acquisition of the emission beam is realized, and the collected signal is used to trigger a measurement. The scattered light of the detected target is collected by achromatic refractive telescope (focal length 400 mm, aperture 80 mm), and collected by optical fibers at the focal point. Then, 32 independent bands are generated by grating spectrometer. After splitting, the emitted light is projected onto the corresponding linear detector (PA), and the echo detection of four bands is realized at the same time. The detector uses a linear PMT photomultiplier array (Hamamatsu Photon Technology Inc.) with a sensitive wavelength ranging from 400 to 1000 nm and a rise time of 0.6 ns. The output signal of photodetector is sampled and stored by high-speed oscilloscope (DPO5204B, Tektronix Inc.) to produce high-resolution waveform echo. The scanning process is controlled by the host computer to generate X- and Y-axis motion (Fig. 9.24).

The prototype system can measure flight time and echo spectrum by postprocessing the recorded waveforms of each channel. The two-axis turntable drives the optical part to scan, and a series of echo signals of different wavelengths can be obtained.

9.4.2.2 Experiments and data processing

Torch flower (*Kniphofia uvaria*) growing in loose and fertile sandy loam soil is used in the experiments. It is mostly clustered in lawn or beside rocks. In the experiment, hyperspectral lidar scans at a distance of 7 m from its top of it.

In the experiment, the horizontal and vertical steps of hyperspectral lidar scanning are 80 and 160, so a total of 12,800 points are scanned. In the scanning process, firstly, the center of the scanning target is aligned, the turntable moves to the upper left corner of the scanning range, then rotates to the right. The rotation angle is a horizontal angle. For each step, 32 measurements are made. After repeating to the designated position, the angle of an inclined step is rotated downwards, then turning left to the designated position and stopping.

The hyperspectral lidar scans the ground objects and records 32 bands of echo waveform information at each point. The typical echo waveform is shown in Fig. 9.25, the left one is a waveform, and the right one is two waveforms.

For the 32-band echo waveform files recorded, data cleaning is carried out first, including renaming the files, deleting duplicate data, etc. Then noise removal is carried out according to the frequency of signal occurrence, and then band segmentation is carried out to obtain the information of each band, which is saved separately. For each waveform, the threshold range of transmitting and receiving signals is determined by the energy distribution of transmitting and receiving waveforms. For example, all transmitting waveforms are counted, and the position where the maximum intensity of transmitting waveforms appears is counted. Thus, the approximate Gauss distribution is obtained. The threshold range of transmitting waveform position is determined by using three times Sigma as the threshold. According to the threshold range of transmitting waveform and receiving waveform, the signal of threshold distance range is extracted. The maximum value

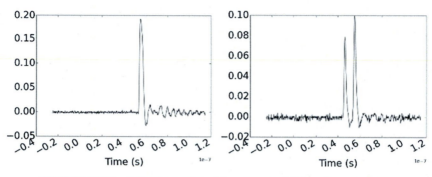

FIGURE 9.25   703 nm wavelength hyperspectral lidar echo waveform.

of Gauss decomposition is used as the strength of the waveform and its position is recorded. The intensity of transmitting waveform is corrected by dividing the intensity of receiving waveform by the intensity of transmitting waveform. Finally, the intensity of correcting at a certain position of 32 waveforms is obtained in turn and stored in the way of point cloud. Finally, various parameters are extracted by the algorithm. The typical processing flow is shown in Fig. 9.26.

9.4.2.3 Hyperspectral lidar point cloud data

According to the scanning results of lidar instrument and the basic data processing, the hyperspectral lidar point cloud data are obtained. Fig. 9.27 is the elevation distribution of point cloud data. The top part is close to the instrument and the bottom part is far from the instrument. It can be seen from the figure that the hyperspectral lidar can accurately distinguish the different height of the leaves in the range of 40 cm.

According to the data processing method introduced above, the echo waveforms of 32 bands of lidar are fitted, and the peak value by Gauss-fitting is taken as the echo intensity value at this point. The echo intensity value is divided by the emission intensity value to obtain the spectral calibration echo intensity information of 32 bands. Fig. 9.28 is 800, 670, 572, and

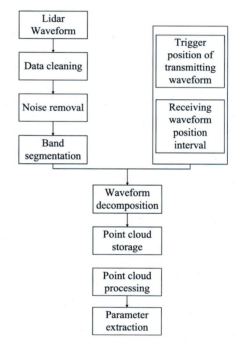

FIGURE 9.26   Data processing flow of hyperspectral lidar.

523 nm point clouds of lidar echo intensity, respectively. From the figure, we can see that red and green leaves have strong reflection to 800 nm laser, red leaves have strong reflection to 670 nm laser, while green leaves have weak reflection. There are obvious differences between 572 and 523 nm laser point clouds at green leaves.

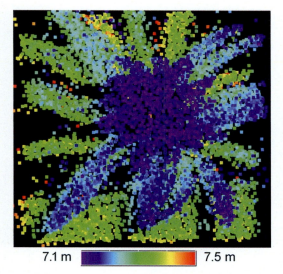

7.1 m ▨▨▨▨▨▨▨▨ 7.5 m

FIGURE 9.27   Elevation distribution of hyperspectral lidar point cloud.

9.4.3 Inversion method and results of vertical distribution of vegetation biochemical components

9.4.3.1 Relationship between biochemical components and hyperspectral lidar

After scanning, destructive sampling of the leaves was carried out and the extraction method was used in the laboratory. Fig. 9.29 is the hyperspectral lidar echo intensity curve of the leaves at different positions of the torch flower. The contents of chlorophyll a, b, carotenoids, nitrogen, and water in leaves were obtained by laboratory tests of sampling samples. The relationship between biochemical components and lidar spectra was established. In the study, NDVI and PRI were mainly used to establish the relationship with biochemical components of

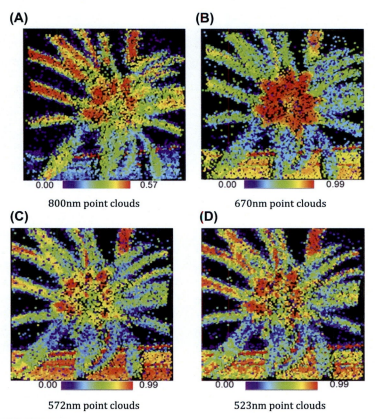

(A)

0.00 ▨▨▨▨ 0.57

800nm point clouds

(B)

0.00 ▨▨▨▨ 0.99

670nm point clouds

(C)

0.00 ▨▨▨▨ 0.99

572nm point clouds

(D)

0.00 ▨▨▨▨ 0.99

523nm point clouds

FIGURE 9.28   Point clouds of hyperspectral lidar with different wavelengths.

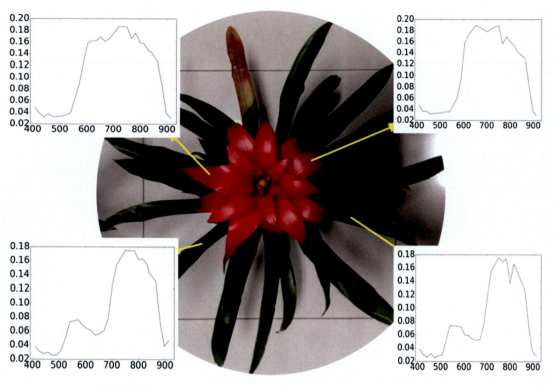

FIGURE 9.29   Hyperspectral lidar reflection spectrum of torch flower.

vegetation. The total number of modeling samples was 18, including 9 red leaves and 9 green leaves. The correlation was as shown in Table 9.18. It is shown as follows:

$$\text{NDVI} = (\rho_{800} - \rho_{670})/(\rho_{800} + \rho_{670}) \quad (9.8)$$

$$\text{PRI} = (\rho_{572} - \rho_{523})/(\rho_{572} + \rho_{523}) \quad (9.9)$$

It can be seen from Fig. 9.29 that the hyperspectral lidar echo intensity curves of red leaves and green leaves are obviously different. The red light band near 670 nm of red leaves has strong reflection, while the green leaves have obvious valleys in this band. The study shows that the existence of the valleys is due to the strong absorption of chlorophyll in this band. Thus, hyperspectral laser thunder can be seen. The intensity curve can reflect the change of chlorophyll content in vegetation (Fig. 9.30).

9.4.3.2 Vertical distribution of vegetation index of hyperspectral lidar

The spectral reflection intensity of point clouds is processed to obtain the point cloud data of lidar vegetation index. Fig. 9.30 is the NDVI point cloud and PRI point cloud. Based on the point cloud data of vegetation index, the vertical distribution of vegetation index is calculated. The right one is the corresponding curve of vegetation index along the height distribution. The NDVI value is obviously lower at the top of red leaves (7.1 m), but higher at the middle of 7.2 m. PRI vegetation index did not change significantly in the vertical direction (Fig. 9.31).

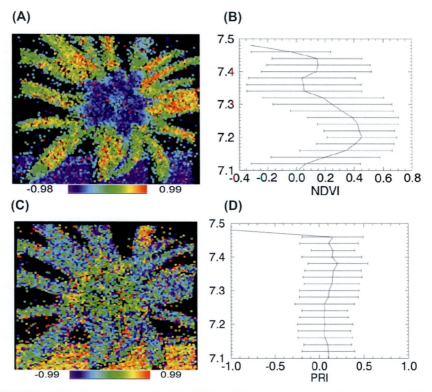

FIGURE 9.30   Point clouds (800, 670, PRI) at different wavelengths of hyperspectral lidar.

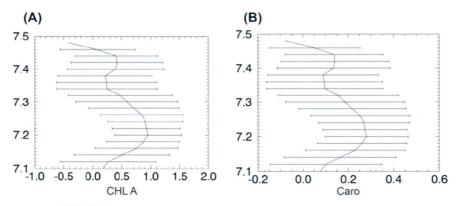

FIGURE 9.31   Vertical distribution of chlorophyll *a* and carotenoids.

9.4.3.3 Vertical distribution of biochemical components

Based on the relationship between vegetation index and biochemical components of hyperspectral lidar (Table 9.18), point clouds of

vegetation index can be calculated according to formulas to obtain point clouds of various biochemical parameters. Taking flower above as an example, the changes of chlorophyll *a* and carotenoid (mg/g) biochemical components

TABLE 9.18   Relationship between vegetation index and biochemical components of hyperspectral lidar.

	Formula	R Square
Chlorophyll *a*	$Y = 1.78 \times NDVI + 0.15$	0.86
	$Y = -4.42 \times PRI - 0.01$	0.67
	$Y = 0.39 \times NDVI + 1.78 \times PRI + 0.23$	0.84
Chlorophyll *b*	$Y = 0.62 \times NDVI + 0.06$	0.88
	$Y = -1.54 \times PRI + 0.01$	0.68
	$Y = 0.83 \times NDVI + 0.61 \times PRI + 0.23$	0.85
Carotenoid	$Y = 0.47 \times NDVI + 0.07$	0.90
	$Y = -1.16 \times PRI + 0.03$	0.71
	$Y = 0.60 \times NDVI + 0.39 \times PRI + 0.09$	0.88

along height were calculated. At the top (7.1 m), the content of chlorophyll *a* was lower than 0.5 and carotene was lower than 0.2, while at the middle (7.2 m), the content of chlorophyll *a* and carotenoid was higher, which was in vertical direction with the red and green leaves of vegetation. The distribution of chlorophyll and carotene in red leaves at the top was low, while that in green leaves at the middle was high (Fig. 9.31).

In this study, the ground hyperspectral lidar instrument developed by our laboratory is used to carry out indoor scanning with torch flower as an example. The basic flow of data processing is proposed for the 32-band lidar point cloud data. Then, according to the relationship between biochemical components and spectra, the vertical distribution of biochemical components (chlorophyll *a*, carotene) in three-dimensional space is obtained, which is in good agreement with laboratory results. The research shows that the new hyperspectral lidar can provide support for accurate three-dimensional vegetation modeling and quantitative remote sensing inversion.

The hyperspectral lidar instrument can measure the spectrum at different locations and can measure the biochemical components at different locations in detail. It avoids the influence of structural parameters on the inversion of biochemical components. Compared with multiangle passive optical observation, the hyperspectral lidar instrument can provide more effective observation means. Compared with passive hyperspectral remote sensing, the instrument can obtain hyperspectral information at different locations, thus avoiding the interference of spectral information at different altitudes. Compared with lidar sensors, it can provide abundant spectral information, so it can establish links with physical properties. Previous studies have shown that with the improvement of hardware level and the development of more experiments, hyperspectral lidar can obtain more and higher quality data, which has great application prospects in the detection of vegetation physiological and biochemical characteristics.

9.5 Summary

A comprehensive introduction focusing on plant biochemical extraction using remote sensing is presented in this chapter, involving the following aspects: empirical and semiempirical statistical regression extraction methods and physical-model retrieval methods; both leaf and canopy scales; theory and practice; and application to model formation. Through the analysis of the above results, we can come to the following conclusions:

(1) The empirical and semiempirical statistical regression method for the extraction of plant biochemicals is a simple method and thus may be applied in remote sensing.
Although a good general estimation model for vulnerable components (cellulose, lignin, etc.) can be obtained at the dry leaf scale, for fresh leaves, the influence of water covers their action. Therefore, the problem of how to develop an extraction method for biochemicals in fresh leaf data or even the canopy must still be solved in this field.
Using a spectroscopic index to extract vegetation chlorophyll concentration has a certain theoretical meaning and also has potential applications. At the leaf scale, most spectral indices can be used to obtain some general regression equations for particular leaf types, which explain why a good result can be obtained from the extraction of the chlorophyll concentration using these indices. However, because of the sensitivity of these indices to leaf types, the estimation model obtained from one set of data may be ineffective when applied to other samples. However, some individual spectral indices also show a weak insensitivity to leaf types.
After we use the leaf chlorophyll concentration formed using the GM index to test the estimated model through experimental data, we find that the estimated value is consistent with the actual value and the estimated model is an applicable model.
At the canopy scale, the influence of background factors can be effectively removed by combining the spectroscopic index TCARI and the soil-adjustable index OSAVI. The estimation model is improved and proven to be able to extract crop canopy chlorophyll concentrations through verification with experimental data.

(2) As the plant biochemical component information extraction method through physical-model retrieval has a clear physical meaning, it will remain a primary concern of study in this field into the near future.
The radiative transfer method is constrained by the invertibility of the model. Therefore, only those models that can be inverted are our objects. The leaf model PROSPECT is a model with complete invertibility, but the accurate retrieval of biochemical parameters using the LIBERTY model requires accurate prior knowledge of other parameters.
For a specific application, different cost functions for retrieval must be selected depending on the nature of the data. Because of the system noise in data, using the correlation coefficient as the cost function can clearly reduce the sensitivity of the retrieval to system noise; Bayesian retrieval makes full use of prior knowledge, and the retrieval results are subject to the joint control of random noise in the data and prior knowledge. For data with significant noise effects, Bayesian retrieval can be used to properly constrain the retrieval results. The actual retrieval problems are often quite complex. These complex problems can be simplified using multiple-phase retrieval. To invert a canopy model to obtain a leaf spectrum and then invert the leaf spectrum to obtain the required parameters is the practice of such a theory; the real application of the extraction of biochemical parameters through retrieval from a physical model is still subject to the constraints of the retrieval algorithm's accuracy and the amount of calculation.
Hyperspectral data contain redundancies, so band selection is necessary. Band selection aiming at the retrieval of biochemical parameters reveals that, past a certain point, a larger number of bands does not imply a more accurate retrieval of biochemical component information because each

additional band improves the accuracy by a negligible amount. Therefore, only a small, optimal number of bands is needed for the actual retrieval.

Using physical-model retrieval to extract biochemical parameters is limited by the accuracy of the model and the parameters considered. Therefore, the development of an accurate and easy retrieval model is the fundamental hurdle to the improvement of this method.

(3) To improve the accuracy and practicality of plant biochemical component retrieval using hyperspectral remote sensing data is an important goal in the future.

It should be noted that the actual retrieval of plant biochemical information using hyperspectral images is not presented in this chapter. Compared with ground data, hyperspectral images might have more uncertainty, including not only image-calibration and atmospheric-correction error but also problems of mixed pixels caused by enlargement of the observational scale; the latter concern introduces a mixed-pixel decomposition problem. Calibration and atmospheric-correction error is essentially system error, which has little impact on the chlorophyll concentration estimation models established using a spectral index. Most system error will be eliminated during this process, which is an advantage of such a model because these models are combinations of addition, subtraction, multiplication, and division.

As system error has significant impact on parameter retrieval from physically based models, a prerequisite for model retrieval is to obtain perfect calibrated images. In addition, time is also a major consideration for model retrieval, which is one of the motivating reasons to perform optimized algorithm selection. Certainly, the application of models and methods is our ultimate goal. Only in this way can the advantages of remote sensing truly be displayed; many researchers have attempted to estimate plant biochemical component information using spaceborne and airborne hyperspectral remote sensing data (Huang et al., 2004; Yilmaz et al., 2008). Future research should work toward the improvement of the estimated retrieval accuracy and the satisfaction of practical requirements.

(4) Hyperspectral lidar is of great significance to the development of vegetation quantitative remote sensing.

The hyperspectral lidar ranges from visible light to near infrared can not only measure the height and density of vegetation in single-band lidar but also have multispectral narrowband information. It can detect the biochemical characteristics of vegetation and obtain the vertical distribution of vegetation nutrients and moisture. Using the hyperspectral lidar, the vertical distribution of biochemical components (chlorophyll *a*, carotene) in three-dimensional space could be obtained, which is in good agreement with laboratory results. The new hyperspectral lidar can provide support for accurate three-dimensional vegetation modeling and quantitative remote sensing inversion.

References

Aber, J.D., Martin, M., 1999. Leaf Chemistry, 1992-1993 (ACCP). Data set. Available on-line. http://www.daac. ornl.gov. From Oak Ridge National Laboratory Distributed Active Archive Center, Oak Ridge, TN, U.S.A).

Allen, W.A., Richardson, A.J., 1968. Interaction of light with a plant canopy. J. Opt. Soc. Am. 58, 1023—1028.

Allen, W.A., Gausman, H.W., Richardson, A.J., 1973. Willstätter-Stoll theory of leaf reflectance evaluation by ray tracing. Appl. Opt. 12, 2448—2453.

Asner, G.P., Anderson, C.B., Martin, R.E., Knapp, D.E., Tupayachi, R., Sinca, F., Malhi, F., 2014. Landscape-scale changes in forest structure and functional traits along an Andes-to-Amazon elevation gradient. Biogeosciences 11 (3), 843—856.

Baret, F., Vanderbilt, V.C., Steven, M.D., Jacquemoud, S., 1994. Use of spectral analogy to evaluate canopy reflectance sensitivity to leaf optical properties. Remote Sens. Environ. 48, 253–260.

Bisun, D., 1998. Remote sensing of chlorophyll a, chlorophyll b, chlorophyll a+b, and total carotenoid content in Eucalyptus leaves. Remote Sens. Environ. 66, 111–121.

Blackburn, G.A., 1998. Quantifying chlorophylls and carotenoids at leaf and canopy scales: an evaluation of some hyperspectral approaches. Remote Sens. Environ. 66, 273–285.

Chappelle, E.W., Kim, M.S., McMurtrey, J.E., 1992. Ratio analysis of reflectance spectra (RARS): an algorithm for the remote estimation of the concentrations of chlorophyll A, chlorophyll B and the carotenoids in soybean leaves. Remote Sens. Environ. 39, 239–247.

Curran, P.J., 1989. Remote sensing of foliar chemistry. Remote Sens. Environ. 30, 271–278.

Curran, P.J., Dungan, J.L., Macler, B.A., Plummer, S.E., 1991. The effect of a red leaf pigment on the relationship between red-edge and chlorophyll concentrations. Remote Sens. Environ. 35, 69–75.

Curran, P.J., Dungan, J.L., Macler, B.A., Plummer, S.E., Peterson, D.L., 1992. Reflectance spectroscopy of fresh whole leaves for the estimation of chemical composition. Remote Sens. Environ. 39, 153–166.

Curran, P.J., Dungan, J.L., Peterson, D.L., 2001. Estimating the foliar biochemical concentration of leaves with reflectance spectrometry: testing the Kokaly and Clark methodologies. Remote Sens. Environ. 76, 349–359.

Dawson, T.P., Curran, P.J., Plummer, S.E., 1998. LIBERTY: modeling the effects of leaf biochemistry on reflectance spectra. Remote Sens. Environ. 65, 50–60.

Dawson, T.P., Curran, P.J., North, P.R.J., Plummer, S.E., 1999. The propagation of foliar biochemical absorption features in forest canopy reflectance: a theoretical analysis. Remote Sens. Environ. 67, 147–159.

Fukshanky, L., Fukshansky-Kazarinova, N., Remisowsky, A.M.V., 1991. Estimation of optical parameters in a living tissue by solving the inverse problem of the multiflux radiative transfer. Appl. Opt. 30, 3145–3153.

Gamon, J.A., Serrano, L., Surfus, J.S., 1997. The photochemical reflectance index: an optical indicator of photosynthetic radiation-use efficiency across species, functional types, and nutrient levels. Oecologia 112, 492–501.

Ganapol, B., Johnson, L., Hammer, P., Hlavka, C.A., Peterson, D.L., 1998. LEAFMOD: a new within-leaf radiative transfer model. Remote Sens. Environ. 6, 182–193.

Gao, S., Niu, Z., Sun, G., Qin, Y.C., Li, W., Tian, H.F., 2018. Vertical distribution inversion of biochemical parameters using hyperspectral LiDAR. Journal of Remote Sensing 22 (5), 737–744.

Gitelson, A.A., Merzlyak, M.N., 1996. Signature analysis of leaf reflectance spectra: algorithm development for remote sensing of chlorophyll. J. Plant Physiol. 148, 494–500.

Gitelson, A.A., Merzlyak, M.N., Grits, Y., 1996. Novel algorithms for remote sensing of chlorophyll content in higher plants. In: Proceedings of the 1996 International Geoscience and Remote Sensing Symposium. Nebraska, Lincoln, pp. 2355–2357.

Goel, N.S., Strebel, D.E., 1983. Retrieval of vegetation canopy reflectance models for estimating agronomic variables. Remote Sens. Environ. 13, 487–507.

Govaerts, Y.M., Jacquemoud, S., Verstraete, M.M., Ustin, S.L., 1996. Three-dimensional radiative transfer modeling in a dicotyledon leaf. Appl. Opt. 35, 6585–6598.

Grossman, Y.L., Ustin, S.L., Jacquemoud, S., Sanderson, E.W., Schmuck, G., Verdebout, J., 1996. Critique of stepwise multiple linear regression for the extraction of leaf biochemistry information from leaf reflectance data. Remote Sens. Environ. 56, 182–193.

Haboudane, D., Miller, J.R., Tremblay, N., Zarco-Tejadad, P.J., Dextraze, L., 2002. Integrated narrow-band vegetation indices for prediction of crop chlorophyll content for application to accuracy agriculture. Remote Sens. Environ. 81, 416–426.

Hakala, T., Suomalainen, J., Kaasalainen, S., Chen, Y., 2012. Full waveform hyperspectral LiDAR for terrestrial laser scanning. Opt. Express 20 (7), 7119–7127.

Hopkinson, C., Chasmer, L., Gynan, C., Mahoney, C., Sitar, M., 2016. Multisensor and Multispectral LiDAR Characterization and Classification of a Forest Environment. Canadian Journal of Remote Sensing 42 (5), 501–520.

Horler, D.N.H., Dockray, M., Barber, J., 1983. The red edge of plant leaf reflectance. Int. J. Remote Sens. 4, 278–288.

Hosgood, B., Jacquemoud, S., Andreoli, G., Verdebout, J., Pedrini, A., Schmuck, G., 1995. Leaf Optical Properties Experiment 93 (LOPEX93) Report EUR-16096-EN[R]. European Commission, Joint Research Centre. Institute for Remote Ssnsing Applications, Ispra, Italy.

Huang, Z., Turner, B.J., Dury, S.J., Wallis, I.R., Foley, W.J., 2004. Estimating foliage nitrogen concentration from HYMAP data using continuum removal analysis. Remote Sens. Environ. 93, 18–29.

Huete, A.R., 1988. A soil-adjusted vegetation index (SAVI). Remote Sens. Environ. 25, 295–309.

Hunt, E.R., Rock, B.N., 1989. Detection of changes in leaf water content using near and middle-infrared reflectances. Remote Sens. Environ. 30, 43–54.

Hunt, E.R., Rock, B.N., Nobel, P.S., 1987. Measurement of leaf relative water content by infrared reflectance. Remote Sens. Environ. 22, 429–435.

Inoue, Y., Morinaga, S., Shibayama, M., 1993. Non-destructive estimation of water status of intact crop leaves based on spectral reflectance measurements. Jpn. J. Crop Sci. 62, 462–469.

Jacquemoud, S., Baret, F., 1990. PROSPECT: a model of leaf optical properties. Remote Sens. Environ. 34, 75–91.

Jacquemoud, S., Verdebout, J., Schmuck, G., Andreoli, G., Hosgood, B., 1995. Investigation of leaf biochemistry by statistics. Remote Sens. Environ. 54, 180–188.

Jacquemoud, S., Ustin, S.L., Verdebout, J., Schmuck, G., Andreoli, G., Hosgood, B., 1996. Estimating leaf biochemistry using the PROSPECT leaf optical properties model. Remote Sens. Environ. 56, 194–202.

Kokaly, R.F., 2001. Investigating a physical basis for spectroscopic estimates of leaf nitrogen concentration. Remote Sens. Environ. 75, 153–161.

Kokaly, R.F., Clark, R.N., 1999. Spectroscopic determination of leaf biochemistry using band-depth analysis of absorption features and stepwise multiple linear regression. Remote Sens. Environ. 67, 267–287.

Kumar, R., Silva, L., 1973. Light ray tracing through a leaf cross section. Appl. Opt. 12, 2950–2954.

Kuusk, A., 1991. The hot-spot effect in plant canopy reflectance. In: Ross, R.B., Ross, J. (Eds.), Photon–Vegetation Interaction: Applications in Optical Remote Sensing and Plant Ecology. Springer-Verlag, Heidelberg, pp. 139–159.

Lagarias, J.C., Reeds, J.A., Wright, M.H., Wright, P.E., 1998. Convergence properties of the Nelder-Mead simplex method in low dimensions. SIAM J. Optim. 9, 112–147.

Li, X., Wang, J., Hu, B., Strahler, A.H., 1998. On utilization of a priori knowledge in inversion of remote sensing models. Sci. China, Ser.D 46, 580–585.

Li, W., Sun, G., Niu, Z., Gao, S., Qiao, H., 2014. Estimation of biochemical content of leaves using a novel hyperspectral full-waveform LiDAR system. Remote Sens. Lett. 5, 693–702.

Lichtenthaler, H.K., Gitelson, A., Lang, M., 1996. Nondestructive determination of chlorophyll content of leaves of a green and an aurea mutant of tobacco by reflectance measurements. J. Plant Physiol. 148, 483–493.

Ma, Q., Ishimaru, A., Phu, P., Yasuo, K., 1990. Transmission, reflection, and depolarization of an optical wave for a single leaf. IEEE Trans. Geosci. Remote Sens. 28, 865–872.

Maier, S.W., Lüdeker, W., Günther, K.P., 1999. SLOP: a revised version of the stochastic model for leaf optical properties. Remote Sens. Environ. 68, 273–280.

Maire, G., Francois, C., Dufrene, E., 2004. Towards universal broad leaf chlorophyll indices using PROSPECT simulated database and hyperspectral reflectance measurements. Remote Sens. Environ. 89, 1–28.

Marten, G., Shenk, J., Barton II, F.E., 1989. Near Infrared Reflectance spectroscopy(NIRS): Analysis of Forage Quality, U. S. Dept. of Agric. Handbook, vol. 643. USDA, Washington, DC.

Menke, W., 1984. Geophysical Data Analysis: Discrete Inverse Theory. Academic Press, New York.

Morsdorf, F., Nichol, C., Malthus, T., Woodhouse, I.H., 2009. Assessing forest structural and physiological information content of multi-spectral LiDAR waveforms by radiative transfer modelling. Remote Sensing of Environment 113 (10), 2152–2163.

Moulin, S., Guerif, M., Baret, F., 2003. Retrieval of biophysical variables on N-treatments of a wheat crop using hyperspectral measurements. In: Proc. of International Geoscience and Remote Sensing Symposium, pp. 21–25 (Toulouse, France).

Niu, Z., Xu, Z., Sun, G., Huang, W., Wang, L., Feng, M., Li, W., He, W., Gao, S., 2015. Design of a new multispectral waveform LiDAR instrument to monitor vegetation. IEEE Geosci. Remote Sens. Lett. 12 (7), 1506–1510.

Peñuelas, J., Filella, I., Biel, C., Serrano, L., Save, A., 1993. The reflectance at the 950–970 nm region as an indicator of plant water status. Int. J. Remote Sens. 14, 1887–1905.

Peñuelas, J., Baret, F., Fillella, I., 1995. Semi-empirical indices to assess carotenoids/chlorophyll a ratio from leaf spectral reflectance. Photosynthetica 31, 221–230.

Rall, J.A.R., Knox, R.G., 2004. Spectral ratio biospheric lidar. IEEE Int. Geosci. Remote Sens. Symp. 3, 1951–1954.

Richter, T., Fukshansky, L., 1996. Optics of a bifacial leaf: 1. A novel combined procedure for deriving the optical parameters. Photochem. Photobiol. 63, 507–516.

Rock, B.N., Hoshizaki, T., Miller, J.R., 1988. Comparison of in situ and airborne spectral measurements of the blue shift associated with forest decline. Remote Sens. Environ. 24, 109–127.

Rondeaux, G., Steven, M., Baret, F., 1996. Optimization of soil-adjusted vegetation indices. Remote Sens. Environ. 55, 95–107.

Sun, G., Niu, Z., Gao, S., Huang, W., Wang, L., Li, W., Feng, M., 2014. 32-channel hyperspectral waveform LiDAR instrument to monitor vegetation: design and initial performance trials. Proc. SPIE-Int. Soc. Opt. Eng. 9263, 926331-926331-7.

Suomalainen, J., Hakala, T., Kaartinen, H., Räikkönen, E., Kaasalainen, S., 2011. Demonstration of a virtual active hyperspectral LiDAR in automated point cloud classification. ISPRS J. Photogram. Remote Sens. 66 (5), 637–641.

Tarantola, A., 1987. Inverse Problem Theory: Method for Data Fitting and Model Parameter Estimation. Elservier Science Publishers, Amsterdam.

Thenkabail, P.S., Smith, R.B., De Pauw, E., 2000. Hyperspectral vegetation indices and their relationships with agricultural crop characteristics. Remote Sens. Environ. 71, 158–182.

Thomas, J.R., Gaussman, H.W., 1977. Leaf reflectance vs. leaf chlorophyll and carotenoid concentrations for eight crops. Agron. J. 69, 799–802.

Tucker, C.J., Garratt, M.M., 1977. Leaf optical system modeled as a stochastic process. Appl. Opt. 16, 635–642.

Verhoef, W., 1984. Light scattering by leaf layers with application to canopy reflectance modeling: the SAIL model. Remote Sens. Environ. 16, 125–141.

Verstraete, M.M., Pinty, B., 1996. Designing optimal spectral indices for remote sensing applications. IEEE Trans. Geosci. Remote Sens. 34, 1254–1265.

Verstraete, M.M., Pinty, B., Myneni, R.B., 1996. Potential and limitations of information extraction on the terrestrial biosphere from satellite remote sensing. Remote Sens. Environ. 58, 201–214.

Vogelmann, J.E., Rock, B.N., Moss, D.M., 1993. Red edge spectral measurements from sugar maple leaves. Int. J. Remote Sens. 14, 1563–1575.

Wang, J., Wang, Z., Huang, W., Ma, Z., Liu, L., Zhao, C., 2004. Vertical distribution and spectral response of nitrogen in winter wheat canopy. J. Remote Sens. 8 (4), 309–316.

Wei, G., Shalei, S., Bo, Z., Shuo, S., Faquan, L., Xuewu, C., 2012. Multi-wavelength canopy LiDAR for remote sensing of vegetation: Design and system performance. ISPRS Journal of Photogrammetry and Remote Sensing 69, 1–9.

Wing, B.M., Ritchie, M.W., Boston, K., Cohen, W.B., Gitelman, A., Olsen, M.J., 2012. Prediction of understory vegetation cover with airborne lidar in an interior ponderosa pine forest. Remote Sensing of Environment 124, 730–741.

Yamada, N., Fujimura, S., 1991. Nondestructive measurement of chlorophyll pigment content in plant leaves from three-color reflectance and transmittance. Appl. Opt. 30, 3964–3973.

Yilmaz, M.T., Hunt Jr., E.R., Jackson, T.J., 2008. Remote sensing of vegetation water content from equivalent water thickness using satellite imagery. Remote Sens. Environ. 112, 2514–2522.

Yoder, B.J., Daley, L.S., 1990. Development of a visible spectroscopic method for detemining chlorophyll a and b in vivo in leaf samples. Spectroscopy 5, 44–50.

Zarco-Tejada, P.J., Miller, J.R., Mohammed, G.H., Noland, T.L., Sampson, P.H., 1999a. Canopy optical indices from infinite reflectance and canopy reflectance models for forest condition monitoring: applications to hyperspectral CASI data. In: Proceedings of the 1999 IEEE International Geoscience and Remote Sensing Symposium, pp. 1878–1881 (Hamburg, Germany).

Zarco-Tejada, P.J., Miller, J.R., Mohammed, G.H., Noland, T.L., Sampson, P.H., 1999b. Optical indices as bioindicators of forest condition from hyperspectral CASI data. In: Proceedings 19th Symposium of the European Association of Remote Sensing Laboratories (EARSel). Valladolid, Spain.

Zhao, C., Huang, W., Wang, J., Liu, L., Song, X., Ma, Z., Li, C., 2006. Inversion of vertical distribution of chlorophyll content in winter wheat using multi-angle spectral information. J. Agric. Eng. 22 (6), 104–109.

Further reading

Campbell, J.B., 1987. Introduction to Remote Sensing. Published by the. Guilford Press, New York, London.

Jago, R.A., Cutler, M.E.J., Curran, P.J., 1999. Estimating canopy chlorophyll concentration from field and airborne spectra. Remote Sens. Environ. 68, 217–224.

CHAPTER

10

Leaf area index

OUTLINE

© 2020 Elsevier Inc. All rights reserved.

Abstract

Leaf area index (LAI) indicates the amount of leaf area in an ecosystem. LAI is a critical parameter for understanding terrestrial ecological, hydrological, and biogeochemical processes. This chapter first introduces field methods for measuring LAI, including direct and indirect methods, which can be divided into contact methods and noncontact optical methods. LAI can be estimated from remote sensing data using either statistical or physical methods. The statistical methods use the empirical relationship between the LAI and surface reflectance or vegetation indices. The physical methods determine the LAI based on radiative light transfer processes within the canopy. Various retrieval methods, including neural network methods, genetic algorithms, Bayesian networks, and lookup table methods, are discussed. A real-time LAI estimation method that uses the data assimilation approach is introduced. The chapters describes how LAI is retrieved from the light detection and ranging technology based on a forest model. Some major global moderate-resolution LAI products, such as Moderate Resolution Imaging Spectroradiometer, GEOV1, GLASS, GLOBMAP, ECOCLIMAP, and CCRS, are illustrated. The global LAI climatology is presented at the end of the chapter.

Leaf area index (LAI) is a critical parameter in many land surface vegetation and climate models that simulate the carbon and water cycles. The vegetation canopy structure, including the LAI, directly influences the radiative transfer process of sunlight in vegetation and, therefore, determines the radiometric characteristics of the top of the canopy (TOC), such as reflectance. Remote sensing methods estimate the canopy LAI based on its relationship with TOC radiometric information, such as the statistical relationship between the TOC reflectance and the LAI, or from the outputs of physical canopy radiative transfer models.

This chapter first introduces the LAI definitions and its measurement methods in section 10.1. The statistical and physical LAI retrieval methods are introduced in Sections 10.2 and 10.3, respectively. Several commonly used retrieval algorithms, such as the neural network (NN) algorithms, genetic algorithms (GAs), Bayesian networks, and the lookup table (LUT)

methods, are described. Section 10.4 introduces the more recent data assimilation retrieval method. Section 10.5 describes the LAI retrieval from the light detection and ranging (lidar) technology. The last section presents the major global moderate-resolution LAI products and discusses LAI the climatology characteristics.

10.1 Definitions

LAI is defined as the total one-sided green leaf area per unit of ground surface (Chen and Cihlar, 1996). LAI indicates the area of ground occupied by plants and is an important structural property of vegetation. Because leaf surfaces are the primary sites of energy and mass exchange, important processes such as canopy interception, evapotranspiration, and gross photosynthesis are directly proportional to the LAI.

Along with the green LAI, there are several subtly different definitions of LAI (Fang et al., 2019). The total LAI (or the all-sided LAI) is based on the total area outside the leaves, accounting for leaf shape, per unit ground area. The effective LAI (Le) is calculated from the canopy gap fraction, assuming that the foliage spatial distribution is random (Chen and Cihlar, 1996):

$$Le = 2 \int_{0}^{\pi/2} \ln[1/p(\theta)]\cos\theta \sin\theta d\theta \qquad (10.1)$$

where $p(\theta)$ is the proportion of gaps in the foliage in the view direction with a zenith angle of θ.

For needle-leaf forests, the formula for calculating the green LAI (or the needle area index, L) is expressed as (Chen, 1996)

$$L = (1 - \alpha) \times L_e \times \gamma_E/\Omega_E \qquad (10.2)$$

where γ_E is the needle-to-shoot area ratio, which quantifies the effect of foliage clumping within shoots; Ω_E is the element clumping index, which quantifies the effect of foliage clumping at scales larger than shoots; and α is the woody-to-total area ratio used to remove the contribution of the supporting woody material from the total area including foliage, branches, and tree trunks, all of which affect ground-based optical measurements.

Two methods are used in ground-based LAI measurements: the direct method measures the LAI in a direct way, whereas the indirect method derives the LAI from other more easily measurable parameters.

10.1.1 Direct leaf area index measurement

LAI can be assessed directly by harvesting leaves and measuring their area in the field or in the laboratory. This assessment can be performed using either destructive sampling or litter traps. Comparatively, the destructive sampling method is more appropriate for short-stature ecosystems, such as grasslands, agricultural crops, and tundra, whereas litter traps are more appropriate for forests.

Leaf area meters (e.g., Li-3000, Licor, Nebraska, USA, see Fig. 10.1) are common instruments for this type of measurement. Alternatively, leaf area can be calculated in the laboratory through the specific leaf area (SLA), which is the fresh leaf area in square centimeters per gram of dry foliage mass. The SLA is first determined, and then the projected leaf area is measured with an image analyzer. The SLA and total dry

FIGURE 10.1 Li-3000 Area Meter. ©*LI-COR Biosciences, used by permission.*

mass of each foliage age class are multiplied and totaled to calculate the LAI for the canopy.

Direct sampling methods have some drawbacks. For example, these methods are usually labor intensive for areas that are necessarily large to adequately characterize the spatial heterogeneity. Furthermore, one type of direct sampling method, the litter-fall method, is suitable for green summer vegetation but not for evergreens. Nevertheless, direct sampling methods are often considered the most accurate way to measure LAI and are often implemented as calibration tools for indirect measurement techniques.

10.1.2 Indirect leaf area index estimation

Indirect methods of determining the LAI are generally faster and more convenient than direct methods. Indirect methods are conducted with a standard strategy. Multiple readings are taken for each plot, and sensors are placed at sites systematically selected on transects. LAI can also be obtained with a nonstandard strategy, in which readings are taken at a single point per plot, and the distance and orientation from a subject tree are standardized to reduce variability. Indirect methods of estimating the LAI can be divided into two categories: (1) indirect contact methods and (2) indirect optical methods.

10.1.2.1 Indirect contact method

In forest stands, direct estimates of the vegetative LAI are often derived from the allometric relationship between the leaf area per tree and the diameter at breast height (DBH). An allometric equation relates the stem diameter to the leaf area:

$$\log y = a + b \log(x) \tag{10.3}$$

where y is the leaf area in square meters, and x is the stem diameter in centimeters at DBH. The coefficients a and b vary with plant species, tree size, nutrient availability, and fertilization. When accurate LAI estimates are needed, site-specific allometric equations should be developed.

The point quadrats method is also used to measure leaf area indirectly. In this method, the vegetation canopy is pierced with a long thin needle at a known elevation (i.e., the angle between the needle and the horizontal plane when vertically projected) and azimuth angle (i.e., the bearing of the needle from north when horizontally projected). The number of hits or contacts of the needle with "green" canopy elements are then counted. The vegetation LAI is determined from

$$Ni = L \times Ki \qquad (10.4)$$

where Ni is the number of contacts of the needle with the vegetation at elevation i, and Ki is the extinction coefficient at elevation i. The point quadrats method is suitable for short vegetation with large leaves but not for forests.

10.1.2.2 Indirect optical method

The indirect optical method calculates the LAI using measurements of canopy transmittance. Canopy transmittance is converted to the LAI using the Beer—Lambert law:

$$L = -\ln(Qi / Q0)/K \qquad (10.5)$$

where K is the light extinction coefficient, Qi/Q0 is the canopy transmittance, Qi is the below-canopy photosynthetic active radiation (PAR), and Q0 is the average total incoming PAR. In practice, it is usually assumed that leaf inclination angles are spherically distributed and that the foliage is distributed randomly in space.

LAI-2000 (and the successor LAI-2200) is one of the most commonly used instruments to measure LAI (Fig. 10.2). The ratio of each ring's above- and below-canopy radiation value is referred to as the canopy gap fraction for each detector. The conversion of gap fraction data to LAI values assumes that only light from the sky is detected by the sensor beneath the canopy. Thus, the measurements should be made on a cloudy day or at sunset or sunrise.

Hemispherical canopy photography acquires photographs through a hemispherical (fisheye) lens oriented toward the zenith beneath the

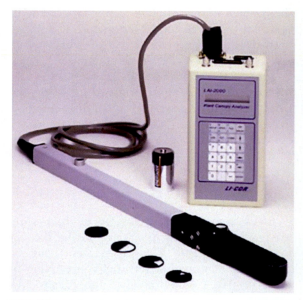

FIGURE 10.2   The LAI-2000 Canopy Analyzer. ©*LI-COR Biosciences, used by permission.*

canopy or placed above the canopy looking downward. Hemispherical photographs show a complete view of all sky directions, with the zenith in the center of the image and the horizons at the edges (Fig. 10.3). Traditionally, analog hemispherical photography was used to determine LAI. Digital cameras are now available with a large number of pixels and high radiometric image quality.

The Sunfleck Ceptometer (METER Group, Inc. USA, see Fig. 10.4) is a line quantum sensor that uses 80 individual sensors on a probe and control unit. This instrument strictly measures the sun-fleck fraction or the quantity of PAR radiation using a probe under a canopy in an open field. Because of the large variability between measurements, a sufficient number of measurements must be made to obtain a reliable and representative result. Moreover, this technique is not suitable in coniferous forests as a result of penumbral effects on the sun-fleck fraction.

The Tracing Radiation and Architecture of Canopies (TRAC) instrument (Third Wave Engineering, Ontario, Canada, see Fig. 10. 5) accounts

FIGURE 10.3   A hemispherical image (Fang et al., 2014).

FIGURE 10.5   The TRAC. *Image courtesy from Leblanc, S.G., Chen, J.M., Kwong, M., 2002. Tracing Radiation and Architecture of Canopies TRAC Manual Version 2.1.3. Natural Resources Canada.*

FIGURE 10.4   The AccuPAR. *Image courtesy of METER.*

for the canopy gap fraction and the canopy gap size distribution. This instrument is carried by hand by a person walking at a steady pace. Using the solar beam as a probe, it uses three photosensitive sensors to transmit direct light at high frequencies. The clumping index obtained from the TRAC can be used to convert the effective LAI to the LAI.

The indirect methods must take into account the influence of shoot structure, dead branches, and stems. One of the basic assumptions of the indirect methods is that the foliage is black and randomly distributed. In reality, no real canopy conforms exactly to this assumption. Thus, all the optical instruments that indirectly estimate the LAI actually estimate the effective LAI, Le, when the foliage in a canopy is not randomly distributed (i.e., clumped). In general, ground LAI measurements usually focus on a specific study site or small patches of vegetation.

10.2 Statistical methods

Spectroradiometer measurements have demonstrated that surface parameters, such as the LAI, are related to radiometric measurements. Thus, LAI can be estimated from radiometric information through an empirical relationship or a physical model inversion. The following section discusses general methods to estimate the LAI from remotely sensed data. In comparison to the ground measurement method, the remote sensing method provides a practical method to estimate the LAI of a large area with high temporal coverage (Myneni et al., 1997).

Vegetation indices are designed to maximize the sensitivity to vegetation characteristics while minimizing confounding factors such as soil background reflectance and directional and atmospheric effects. The most commonly used vegetation indices use red and near-infrared (NIR) canopy reflectances or radiances. These reflectances or radiances are combined in the form of ratios such as the ratio vegetation index (RVI) or the normalized difference (NDVI).

$$RVI = \rho_{RED}/\rho_{NIR}$$
$$NDVI = (\rho_{NIR} - \rho_{RED})/(\rho_{NIR} + \rho_{RED}) \quad (10.6)$$

where ρ_{RED} and ρ_{NIR} represent spectral reflectances in the red and NIR regions, respectively.

These indices enhance the contrast between soil and vegetation but minimize the effects of illumination conditions. However, they are sensitive to the optical properties of the soil background. For a given amount of vegetation, darker soil substrates result in higher vegetation index (VI) values. The soil adjusted vegetation index (SAVI) is used to suppress the soil effect (Huete, 1988):

$$SAVI = (\rho_{NIR} - \rho_{RED})/(\rho_{NIR} + \rho_{RED} + C)(1 + C) \quad (10.7)$$

The constant C is introduced to minimize soil-brightness influences. C can vary from zero to infinity as a function of the canopy density. If $C = 0$, then SAVI is equivalent to NDVI.

The enhanced vegetation index (EVI) was developed to optimize the vegetation signal with improved sensitivity in high-biomass regions (Huete et al., 2002):

$$EVI = G \frac{\rho_{NIR} - \rho_{RED}}{\rho_{NIR} + C_1 \times \rho_{Red} - C_2 \times \rho_{Blue} + X} \quad (10.8)$$

where ρ_{Blue} is the surface reflectance in the blue band; C_1 and C_2 are the coefficients of the aerosol

resistance term; X is the canopy background adjustment factor; and G is the gain factor.

A common procedure for estimating LAI is to establish an empirical relationship between VI and the LAI by statistically fitting observed LAI values to the corresponding VI values. Among the proposed LAI-VI relations are the following forms (Qi et al., 1994):

$$L = Ax^3 + Bx^2 + Cx + D$$
$$L = A + Bx^c \quad (10.9)$$
$$L = -1/2A \ln(1 - x)$$

where x is either the VI or the reflectances derived from remotely sensed data, and the coefficients A, B, C, and D are empirical parameters that vary by vegetation type. Given a set of coefficients, the equations can be applied to remotely sensed images to map the spatial LAI distributions.

Generally, the vegetation indices approach a saturation level asymptotically for LAI values ranging from 2 to 6, depending on the type of VI used, the type of vegetation, and the experimental conditions. The advantage of this approach is its simplicity and ease of computation, but its major limitation is that no LAI-VI equation is universally applicable to diverse vegetation types because the empirical coefficients depend primarily on the vegetation type. To operationally use the VI approach, an LAI-VI equation must be established for each vegetation type, such as coniferous, deciduous, mixed forests, and nonforest vegetation (Myneni et al., 1997). However, the development of these equations requires substantial LAI measurements and corresponding remote sensing data.

10.3 Canopy model inversion methods

Canopy reflectance models relate fundamental surface parameters, such as LAI and leaf optical properties, to scene reflectance values for a given sun-surface-sensor geometry. These models can be grouped according to four categories: parametric, geometrical, turbid medium, and computer simulation models. Parametric models are based on simple mathematical functions of reflectance distribution. The other models involve physical radiative transfer processes in the canopy. These models have been used to estimate canopy morphological and optical properties.

10.3.1 Radiative transfer modeling

Radiative transfer theory has been recognized as the principal modeling method that accounts for radiation in the atmosphere, and it is also widely used for land surface modeling (e.g., vegetation canopy, snow, and soil) (Liang, 2004). Generally, continuous vegetation is approximated by a horizontally homogenous and vertically layered model. The interaction between light and continuous vegetation consists of absorption and scattering; therefore, radiative transfer theory studies concerning atmospheric physics and particle transfer can be translated into studies of the bidirectional reflectance distribution function (BRDF) related to continuous vegetation. Radiative transfer models describing the interaction between the canopy and incident radiation were developed from the transfer mechanism of radiation through a turbid medium. In these models, the canopy is assumed to be horizontally uniform, with plane parallel distinct layers, above a horizontal ground surface. Each layer consists of vegetation elements that are small and randomly distributed. These elements and ground are characterized by scattering and absorbing optical properties. The canopy architecture is described at the leaf level through the LAI and leaf angle distribution. Properties such as leaf size and the spatial location of the leaf elements are ignored (Goel and Thompson, 2000). The advantage of the radiative transfer model is that it considers multiple scattering and is important for identifying

homogeneous vegetation, especially in infrared and microwave bands. However, even in the case of homogeneous vegetation, only numerical solutions can be obtained from differential equations in 3D space; establishing a clear analytic expression between vegetation structure and BRDF is difficult.

A classic textbook on canopy radiative transfer modeling was written by Ross (1981). There have also been several reviews on this topic (e.g., Goel and Grier, 1988; Myneni et al., 1989, 1990; Qin and Liang, 2000; Strahler, 1997). An in-depth review of mathematical modeling development was compiled by Myneni and Ross (1991). The one-dimensional radiative transfer equation of a flat homogeneous canopy for radiance I at the direction (U) is given by

$$-\mu\frac{\partial I(\tau,\Omega)}{\partial x} + h(\tau,\Omega)G(\Omega)I(\tau,\Omega)$$
$$= \frac{1}{\pi}\int_0^{2\pi}\int_{-1}^1 \Gamma(\Omega',\Omega)I(\tau,\Omega'd\Omega') \tag{10.10}$$

where $h(\tau,\Omega)$ is an empirical correction function to account for the variation of the extinction coefficient. The optical depth s can be defined by the leaf area density $\mu_1(z)$ as $\tau(z) = \int_0^z u_1(z)dz$; the geometry function $G(\Omega)$ is usually defined to represent the mean projection of a unit foliage area in the direction Ω characterized by the zenith angle m and the azimuth angle $\varnothing$. And in the equation, $\Gamma(\Omega,\Omega')$ is the area scattering phase function, which is consisting of both diffuse and specular components (Liang, 2004).

In recent decades, a great deal of work has been conducted throughout the world, including the establishment of theories and methods related to radiative transfer models, the acquisition of field observation data and remote sensing data related to a wide range of vegetation canopies, and the verification and application of models. Since 1999, RAMI (Radiation Transfer Model Intercomparison) has compared more than 10 BRDF models, and the fourth time of RAMI was called in 2010. The models involved in the comparison include SAILH (Verhoef, 1984; Verhoef, 1985), GORT (Li et al., 1995), discrete anisotropic radiative transfer (DART) (Gastellu-Etchegorry et al., 1996), and others (Pinty et al., 2004). Each type of model has a distinctive application scope. The SAIL model is a representative radiative transfer model. In recent years, Gastellu-Etchegorry et al. extended the SAIL model by considering the anisotropy of the vegetation canopy and soil background. Myneni et al. (1990) then developed a 3D radiative transfer model for the canopy, and Nilson and Kuusk (1989) developed a two-layer canopy reflectance model for forests. These models have been applied in land surface parameter retrieval from satellite data. The most typical application is generating the MODIS (Moderate Resolution Imaging Spectroradiometer) LAI product from a 3D radiative transfer model.

10.3.1.1 A brief introduction to the models

The SAIL model (scattering by arbitrarily inclined leaves) is one of the classic models (Verhoef, 1984), extended from the Suits model (Suits, 1972) of Verhoef and Bunnik (Verhoef, 1981). This model, which uses the radiative transfer equation to describe the spectral and directional reflectance of the vegetation canopy, assumes that the canopy is a mixture of horizontal, homogenous, and infinitely extending isotropic leaves with randomly distributed positions. The SAIL model is used to simultaneously solve for the intensity of the upward and downward scattered light within the canopy. For the given structural and environmental parameters of the canopy, the canopy reflectance can be calculated at any solar height, observation direction, and the proportion of the scattered light in the sky. The model simulation can satisfactorily reflect the anisotropic reflectance of vegetation in various optical bands.

Over the past 20 years, the SAIL model has been constantly improved (Verhoef, 1984; Verhoef and Bach, 2003). The original SAIL model failed to represent the hotspot effect. Kuusk (1991b) introduced a hotspot effect parameter into the SAIL model to obtain SAILH, a model that effectively simulates the hotspot effect caused by the single light-scattering effect of leaf shoots. Jupp and Strahler (1991) applied the Boolean theory in 1991 to calculate mutual overshadowing among leaves of continuous vegetation and discussed the pattern of the hotspot effect in continuous vegetation. These two attempts considerably improved the SAIL model and its simulation precision near hotspots. Huemmrich (2001) developed a Geo-SAIL model by integrating a geometric-optical model and the SAIL model in 2000 (Huemmrich, 2001). The GeoSAIL model addresses the applicability problem for noncontinuous vegetation and can be applied to simulate the reflectance of noncontinuous vegetation canopies. Bacour et al. (2002) developed the ProSAIL model, which temporally coupled a leaf radiative transfer model (PROSPECT) and a canopy radiative transfer model (SAIL) (Bacour, 2002). By combining SAIL model parameters and biophysical composition parameters (e.g., chlorophyll and leaf water content of leaves), more precise spectral feature parameters of leaves can be obtained.

10.3.1.2 SAILH model—based simulation

The SAILH model needs seven input parameters:

(1) LAI;
(2) Average leaf angle (ALA);
(3) SL;
(4) Leaf reflectance;
(5) Leaf transmittance;
(6) Soil reflectance;
(7) Horizontal visibility.

LAI represents the canopy structural features; the leaf hemisphere reflectance and

TABLE 10.1 Parameter configurations in the SAILH model.

Parameter	Standard value	Scope range
Solar zenith angle	30 degrees	0–70 degrees, step length by 10 degrees
Relative azimuth angle	0 and 180 degrees	0–90 degrees, step length by 15 degrees
Leaf area index	4	3–6, step length by 0.5
Hotspot parameter	0.2	0.0, 0.01, 0.05, 0.1, 0.15, 0.2, 0.3, 0.4
Leaf angle distribution	Inclined	Flat-plate, vertical, inclined, extreme, spherical

transmittance represent leaf spectral features; the soil reflectance represents background spectral features; and horizontal visibility represents atmospheric conditions. Parameter configurations are shown in Table 10.1. Using these parameters, we obtained the BRDFs for the visible and NIR bands in the principal plane for various LAI values and zenith angles. The LAI of winter wheat and variations in the principal plane for the visible and NIR bands with the viewing zenith angle were obtained, as shown in Fig. 10.6. Additionally, the SAILH model satisfactorily reflects the vegetation hotspot effect.

The composition reflectance data needed by the model were the experimental data collected from the spectral database of the Beijing Normal University. For ground measurements of the

canopy spectra for winter wheat, the SE 590 portable field spectrometer (Spectron Engineering Inc., USA) was employed. The wavelength scope of the spectrometer was 400–1100 nm, and the spectral resolution was approximately 3 nm. Each group of measurements provided spectral data in 252 wavelengths. The spectrometer's field of view was 15 degrees, and the average reflectance of the reference board was 50%. For the reflectance and transmittance measurements of the leaves and stems of winter wheat, an SE590 spectrometer and an external integrating sphere (1800-12S, External Integrating Sphere, LI-COR., USA) were used, with wavelengths ranging from 400 to 1100 nm. All input parameters in the SAILH model were accurately measured in the dataset: LAI 2.2912, ALA 48.23 degrees, leaf size 5.8–21.4 cm, and

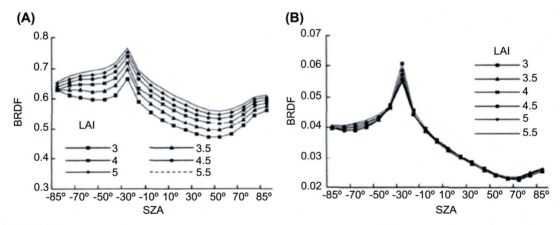

FIGURE 10.6 Variation of canopy reflectance for thenear-infrared and red bands in the principal plane, with different LAI values and viewing zenith angles, simulated using the SAILH model for winter wheat.

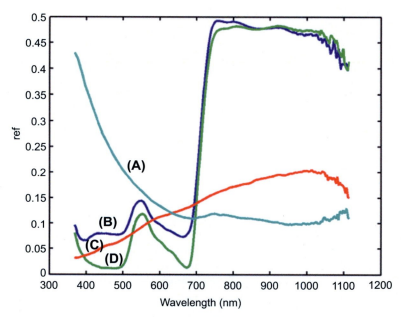

FIGURE 10.7   The input spectral parameters in the SAILH model. The ratio of scattered light in each waveband to scattered light from the sky (cyan) (A), leaf reflectance (blue) (B), soil reflectance (red) (C), and leaf transmittance (green) (D).

average canopy height 73.5 cm. The input spectral parameters for SAILH are shown in Fig. 10.7.

In the SAILH model, the canopy reflectance is also simulated (Fig. 10.8). Simulated and measured canopy reflectance values were similar, especially in the visible band, where the difference was below 0.02. In the NIR band, the simulated reflectance value was slightly higher than the measured reflectance, with the maximum difference below 0.1. Using the ground-measured

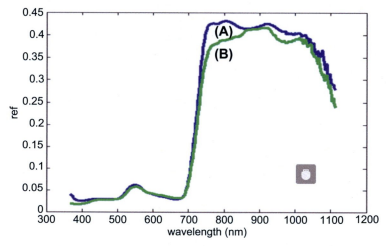

FIGURE 10.8   A comparison between (A) the canopy reflectance simulated by the SAILH model and (B) the ground-measured reflectance.

input parameters, the relationship between the ground-measured canopy spectra (spectra B in Fig. 10.8) and the leaf (spectra A in Fig. 10.8) spectra (spectra A in Fig. 10.8) for winter wheat can be represented by the SAILH model. Thus, the simulation capacity of the SAILH model for the spectral features of the vegetation canopy was verified.

10.3.1.3 The 3D radiative transfer model

In the 3D radiative transfer model, vegetation in the 3D space is divided into a finite number of unit-scale cubes, and the incident radiation is discretized, depending on the zenith angle and azimuth angles at the time of incidence. The most striking feature of this model is that the radiation of the preceding cube is treated as the second radiation source of the subsequent cube depending on its outgoing direction. At the same time, multiple scattering is taken into account; thus, the radiation fields of all cubes are properly associated to obtain the radiation distribution of the entire space. For horizontally homogenous media or noncontinuous canopies, such as row crops and orchards with isolated tree crowns, the canopy closure is usually uncertain. Kimes (1991) proposed a canopy reflectance model to deal with the heterogeneities resulting from the uncertain canopy closure of plants. The canopy is divided into matrices of rectangle units based on the assumption that radiative transfer is limited to finite directions. Gastellu-Etchegorry et al. (1996) extended this method by overcoming its disadvantages and establishing DART model (Liang, 2004). Myneni et al. (1990) conducted research on establishing 3D radiative transfer equations and their solutions. Theoretically, 3D canopy radiative transfer equations can be used to deal with nonhomogeneous canopies of any form (Myneni et al., 1990).

A 3D radiative transfer model was applied to retrieve the MODIS LAI and to generate the temporally continuous MODIS LAI products with continuous time series.

10.3.2 Optimization techniques

To calculate canopy biophysical parameters, such as LAI, the canopy reflectance model must be solved. Model inversion has been mainly applied over the directional distribution of reflectance. The inversion of a canopy radiative transfer model is usually ill posed, and the equation solution does not depend continuously on the data (Kimes et al., 2000). Because the canopy radiative transfer model is not an analytic function, an optimization step and a search for the solution must be performed numerically and iteratively.

The general reflectance model inversion problem can be stated as follows: given a set of empirical reflectance measurements, the set of canopy biophysical variables must be determined so that the computed reflectances best fit the empirical reflectances. The fit of the empirical data is determined by a merit function, ε^2:

$$\varepsilon^2 = \sum_{i=1}^{n} \sum_{j=1}^{B} W_i \left(r_{ij} - \widehat{r}_{ij} \right)^2 \qquad (10.11)$$

where r_{ij} is the observed directional reflectance for a given viewing and solar angle geometry; $\widehat{r}_{ij}$ is the simulation model estimate; n is the number of reflectance samples; B is the number of spectral bands; and W_i represents the weight. A penalty function can be used to limit the independent parameter space to physically possible values. The ability to correctly determine the target parameter space through model inversion, therefore, depends on the dataset $\widehat{r}_{ij}$, the degree to which the model represents physical reality and the chosen optimization algorithm's ability to minimize ε^2 over the parameter space.

10.3.2.1 Minimization in one or multiple dimensions

The minimization problem is simple if there is only one variable (1D). Multidimensional optimization problems can be broken down into a series of 1D minimizations. For 1D minimization, a search space in which a minimum exists can be easily defined. If we can find three points a, b, and c, with a<b<c and f(a)>f(b)<f(c), at least one minimum point must exist in the interval (a, c). The points a, b, and c are said to bracket the minimum.

An elegant and robust method of locating a minimum in such a bracket is the Golden Section Search, which involves evaluating a function at some point x in the larger of the two intervals (a,b) and (b,c). If f(x)<f(b), then x replaces the midpoint b, and b becomes an end point. If f(x)>f(b), then b remains the midpoint, and x replaces one of the end points. In both cases, the width of the bracketing interval is decreased, and the minima position will be better defined (Fig. 10.9). The procedure is then repeated until a desired width tolerance is achieved. If the new test point x is chosen to be a proportion of $(3 - \sqrt{5})/2$ (hence, the name Golden Section) along the larger subinterval, measured from the midpoint b, then the width of the full interval (a,c) will decrease at an optimal rate (Press et al., 1992a).

The Golden Section Search requires no information about the derivative of a function. If such information is available, it can be used to predict the best choice for the new point x in the above algorithm, leading to faster convergence (Press et al., 1992a).

10.3.2.2 Nonderivative and derivative methods

We distinguish between methods that use no derivatives (e.g., function values, such as the reflectance method) and methods that use only first derivatives (e.g., slope or force methods) or the second derivatives (e.g., curvature or force-constant methods). Optimization methods that do not require any derivative of a function are generally easy to implement but require the most steps to reach the minimum. These methods may work well in special cases when the function is random in character or the

variables are essentially uncorrelated. Examples of these methods are the simplex minimization, GAs, NNs, and simulated annealing methods. More detailed descriptions of these methods are provided in other sections. The methods that require derivatives are usually preferred for optimization purpose.

Methods that use only the first derivatives are often used in the preliminary minimization step for very large systems. The first derivative indicates the downhill direction and suggests the size of the steps that should be taken when stepping down the hill (e.g., large steps on a steep slope or small steps on flatter areas that are hoped to be near the minimum). Methods such as the steepest-descent approach and a variety of conjugate gradient minimization algorithms belong to this category. The search starts at an arbitrary point and then slides down the gradient until a sufficiently close solution is obtained. The iterative procedure of the steepest-descent algorithm in 1D takes the form

$$x_{k+1} = x_k - \alpha_k \nabla f(x_k),$$

where α_k is the step size, and ∇ is the gradient or the partial derivative of the function in the search direction.

The minima of a function can either be global (i.e., over the entire search space) or local (i.e., over some small neighborhood). We are usually the most interested in finding the global optimum, but this can be difficult. The choice of a particular optimization algorithm depends on the mathematical properties of the function to be minimized. Some of these methods, including the downhill simplex method (Privette et al., 1994), the conjugate direction set method (Kuusk, 1991a), and the quasi-Newton method (Pinty et al., 1990), have been used to determine LAI. These methods are generally available in standard software libraries (Press et al., 1992b) and are commonly used for complex, nonlinear formulations such as those used in canopy reflectance modeling.

10.3.3 Neural networks

NN is a computationally efficient machine learning method with a high interpolation capacity. NNs can be used to efficiently and accurately approximate complex nonlinear functions. Research has indicated that levels of retrieval precision similar to those from physical models can be achieved by NNs trained with simulated data from radiative transfer models. Using NN retrieval, the CYCLOPES program, sponsored by the European Union, has produced global LAI products using NN retrieval.

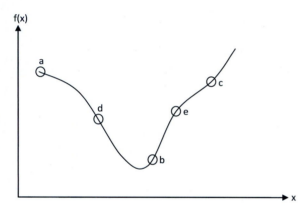

FIGURE 10.9 The Golden Section Search. Initial bracket (a, b, c) becomes (d, b, c) and (d, b, e).

10.3.3.1 CYCLOPES leaf area index algorithm

In CYCLOPES LAI products, the simulated data obtained from the radiative transfer model PROSPECT + SAIL are used to train NNs. The inputs of NNs consist of the median of the solar zenith angle (θ_s) and the surface reflectance in the red band (B2), the NIR band (B3), and the shortwave infrared band (SWIR) in solar zenith angle observations; the output is the LAI (L). The network structure is shown in Fig. 10.10 0.

After simulated data are normalized for training, the NN is trained by the L-M optimization algorithm. Next, the normalized vegetation reflectance, which is in the red, NIR, and SWIRs, is corrected for radiance, cloud screening, atmospheric conditions, and BRDF. The result is combined with the median of the solar zenith angle before being input into the NN to produce 10-day composite LAI products.

Fig. 10.11 shows CYCLOPES LAI images of France from August 2002 (Fig. 10.11A) and August 2003 (Fig. 10.11B). In contrast to the LAI from 2002, the 2003 LAI value was clearly lower, which could be attributed to the tremendous drought in Western Europe in 2003.

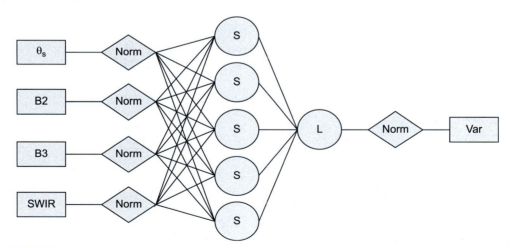

FIGURE 10.10   The neural network structure of the retrieval algorithm for CYCLOPES LAI products.

(A) **(B)**

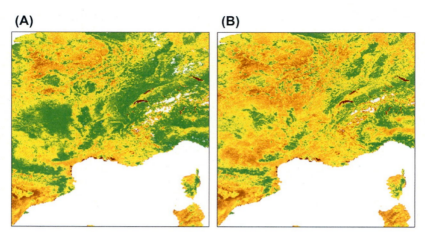

FIGURE 10.11   The CYCLOPES LAI of France in (A) August 2002 and (B) August 2003 (http://postel.mediasfrance.org/).

10.3.3.2 GLASS LAI algorithm

Xiao et al. (2014) developed a method to retrieve LAI profiles from MODIS reflectance time series data using general regression neural networks (GRNNs).

The GRNNs, which consist of an input layer, a hidden layer, a summation layer, and an output layer, were proposed by Specht (1991). The network structure of GRNNs is shown in Fig. 10.12. The expression for the output layer of a GRNN that uses a Gauss kernel function can be written as

$$\mathbf{Y}'(\mathbf{X}) = \frac{\sum\limits_{i=1}^{n} \mathbf{Y}^i \exp\left(-\dfrac{D_i^2}{2\sigma^2}\right)}{\sum\limits_{i=1}^{n} \exp\left(-\dfrac{D_i^2}{2\sigma^2}\right)} \qquad (10.12)$$

where $D_i^2 = (\mathbf{X} - \mathbf{X}^i)^T(\mathbf{X} - \mathbf{X}^i)$; $\mathbf{X}^i$ and $\mathbf{Y}^i$ ($i = 1$, 2, ..., n) are the input and output of the ith sample, respectively; n is the number of samples; $\mathbf{X}$

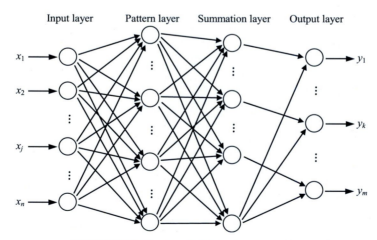

FIGURE 10.12   The GRNN network structure.

is the input vector; $\mathbf{Y}'(\mathbf{X})$ is the output when the predicted input is $\mathbf{X}$; and σ is the parameter that controls the smoothing process of the fitted results. GRNN has the general approximation characteristics of smooth functions. As long as a sufficient amount of data is available, high-precision approximations can be achieved. Even with a small number of multidimensional samples, this algorithm is still effective. GRNNs, which can be rapidly trained without iterations, easily and accurately describe the features of dynamic systems.

The GRNNs were trained using the fused time series LAI values from MODIS and CYCLOPES LAI products and the preprocessed time series MODIS reflectance over the BELMANIP sites. The trained GRNNs were then used to retrieve LAI from the preprocessed MODIS reflectance data. The inputs of the GRNNs include annual reflectance time series data in the red (R) and NIR bands, that is, the input vector $\mathbf{X} = (R_1, R_2, \cdots, R_{46}, NIR_1, NIR_2, \cdots, NIR_{46})^T$ includes 92 vectors. The output is the LAI time series of the corresponding year, $\mathbf{Y} = (LAI_1, LAI_2, \cdots, LAI_{46})^T$, which includes 46 vectors.

Because of its excellent performance in retrieving LAI from MODIS reflectance time series data, the method was extended to estimate LAI profiles from AVHRR reflectance time series data (Xiao et al., 2016). The fused LAI values from the CYCLOPES and MODIS LAI products were aggregated to 0.05 degrees spatial resolution using the spatial-average method. Then, the aggregated LAI time series values and the corresponding reprocessed AVHRR reflectance values of the BELMANIP sites were used to train GRNNs (Xiao et al., 2016). The reprocessed AVHRR reflectance values from an entire year were inputted to the GRNNs to estimate the 1-year LAI profiles.

The retrieval method described above was used to generate a long time series of Global LAnd Surface Satellite (GLASS) LAI product from Advanced Very High—Resolution Radiometer (AVHRR) and MODIS reflectance data. Fig. 10.13 shows the GLASS LAI profiles for different vegetation types (Xiao et al., 2014). In general, the GLASS LAI profiles obtained by the GRNNs (denoted by GRNN LAI) are consistent with the seasonal variation trends of MODIS LAI (denoted by MOD LAI) and CYCLOPES LAI (denoted by CYC LAI). The GLASS LAI profiles are comparatively smooth, with continuous time series variations. The MODIS LAI profiles, however, fluctuate radically, especially during growing seasons. Direct comparison with ground measurements shows that the accuracy of the GLASS LAI product is superior to the accuracy of the MODIS, CYCLOPES, and GEOV1 LAI products (Xiao et al., 2016, 2017).

10.3.4 Genetic algorithms

10.3.4.1 Introduction

GAs attempt to minimize functions using an approach analogous to evolution and natural selection (Davis, 1991; Goldberg, 1989). A solution in the search space is encoded as a chromosome composed of N genes (parameters). A population of chromosomes (possible solutions) is maintained for each iteration. New solutions are generated by combining existing solutions. Optimal solutions evolve through the iterative production of new generations of chromosomes in which good solutions are combined and bad ones discarded.

A chromosome is a string of bits formed by concatenating the bit strings that represent each of the n parameters. The number of bits used to encode each parameter depends on the desired tolerance. Table 10.2 shows the evolution of a population of two chromosomes, each with two genes. Two parent chromosomes in generation k are cut at a random crossover location, and the opposing sections are combined to form two children in generation K+1. In addition, a small number of mutations are introduced, randomly changing chromosome bits. A simple GA model

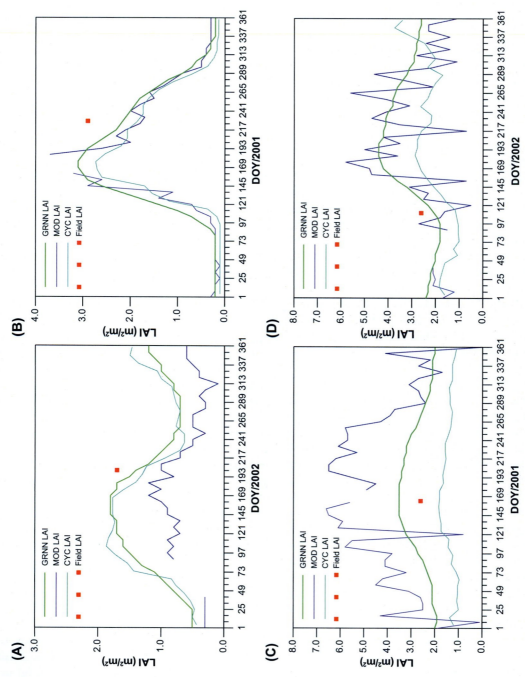

FIGURE 10.13　The profiles of GLASS, MODIS, and CYCLOPES LAI products for different vegetation types. (A) Crops, (B) grassland, (C) broad-leaved forest, and (D) needle-leaved forest.

TABLE 10.2 Chromosome evolution through cross-over, rejoining, and mutation from generation k to generation k+1.

	Generation K	
Genes and chromosomes	Chromosome 1	1001010 1100011
	Chromosome 2	0011100 0011010
Crossover	Chromosome 1	10010101100011
	Chromosome 2	00111000011010
Rejoining	Chromosome 1	10010000011010
	Chromosome 2	00111101100011
Mutation	Chromosome 1	10010000101010
	Chromosome 2	00111101100000
Generation K+1		
Chromosomes and genes	Chromosome 1	1001000 0101010
	Chromosome 2	0011110 1100000

includes reproduction, crossover, and mutation. These genetic operations alter the composition of "offspring" during reproduction. A complete GA requires other parameters such as population size and the probabilities of applying genetic operators.

A typical GA in the optimization process comprises a number of ad hoc steps:

(1) the parameter search space is determined;
(2) an arbitrary encoding algorithm is developed to establish a one-to-one relationship between a set of solutions (represented by chromosomes) and the discrete points of the search space;
(3) a trial set of parameters known as the initial population is randomly generated;
(4) high-performance parameters are selected according to the objective function, known as natural selection;
(5) the solutions from the above steps are taken and used to form a new population through mating and mutation; the solutions are selected based on their fitness and form new solutions (offspring); and

(6) steps (4) and (5) are repeated until convergence is reached.

10.3.4.2 The application of the GA in LAI retrieval

GAs have been applied to a variety of remote sensing optimization problems in recent years. Fang et al. (2003) applied the GA technique and a Markov chain reflectance model (MCRM, Kuusk, 1995b, 2001) to retrieve the surface LAI from both field-measured reflectances and atmospherically corrected Landsat ETM+ data. A genetic software called GENESIS (Grefenstette, 1990) was used. GENESIS is easy to use, and it provides default parameter settings that are robust for a variety of applications. In the GA, each chromosome was represented by real numbers. Each gene (free parameter) takes a range of floating point values with a user-defined output format.

Inputs of the forward MCRM are summarized in Table 10.3. The solar zenith angle θi represents the values at which the ETM+ data are acquired. The leaf water content and leaf dry matter content, including protein, cellulose, and lignin, are from (Jacquemoud et al., 1996). For the ETM+ data, only the nadir viewing angle was considered. In this case, the sensitivity of the inversion to the hotspot parameter S_L ($=0.15$) was low. Two leaf angle distribution parameters are set to zero ($e = 0$; $\theta m = 0$), assuming a spherical leaf orientation. Thus, there is no dependence on the leaf angle θm (Kuusk, 1995a). Six free parameters were identified: LAI, Sz, Cab, N, rs1, and rs2. Their effective ranges are displayed in Table 10.3.

For accurate inversions, the number of free parameters that significantly impact the canopy reflectance should be minimal (Kimes et al., 2000). In practice, the number of genes was reduced by fixing the parameters that change less rapidly. The fixed values represent the general conditions of the study area in the inversion experiments. Sz was the most stable variable in the test site. With the NIR value only, a stable

TABLE 10.3 The parameters needed to run the radiative transfer model, Markov chain reflectance model (MCRM). The free parameters are marked with an asterisk.

Parameters	Symbol	Values
External parameters		
Solar zenith angle (°)	θ_i	27.8, 46.6, 31.4, and 30.2
Angstrom turbidity factor	τ	0.1
Canopy structure parameters		
Leaf area index*	LAI	0–10.0
Leaf linear dimension/canopy height ratio	S_L	0.15
Markov parameter describing clumping*	S_z	0.4–1
Eccentricity of the leaf angle distribution	e	0.0
Mean leaf angle of the elliptical LAD	θ_m	0.0
Leaf spectral and directional properties		
Chlorophyll AB concentration ($\mu g/cm^2$)*	C_{ab}	20–90
Leaf equivalent water thickness (cm)	C_w	0.01
Leaf protein content (g/cm^2)	C_p	0.001
Leaf cellulose and lignin content (g/cm^2)	C_c	0.002
Leaf structure parameter*	N	1–3
Soil spectral and directional properties (Price 1990)		
Weight of the first Price function*	r_{s1}	0–1.0
Weight of the second Price function*	r_{s2}	−1.0–1.0
Weight of the third and fourth Price function	r_{s3}, r_{s4}	0.0

Sz value (0.8) can be achieved with the lowest CV (0.1286). Other parameters, such as Cab and N, change more rapidly over the area (Fang et al., 2003). In this case, Sz = 0.8 was chosen to represent the general status of the study area, and it was fixed in later inversions. After Sz was fixed, a similar GA optimization procedure was conducted to fix the subsequent Cab (50) and N (1.8) values; thus, the number of genes was reduced from 6 to 5, 4, and 3. When the number of free parameters is decreased, the retrieved Cab and N values may differ from their values in the previous inversion.

A typical GA output is shown in Table 10.4 for one point. Column ε^2 provides the number of local minimum values, which was 10 in this study, whereas the italicized row represents the global minimum ε^2. Each LAI–GA point in the figures is denoted as the value at the local minimum of the merit function.

The GA optimization method provides an alternative to inverting the RT models in remote sensing. The advantages of GA are twofold. First, it scans all the initial conditions and provides several possible solutions for the detailed examination of the global optimum solution, thereby avoiding the inaccuracies introduced by traditional minimization algorithms. Second, it runs only the forward RT model in a constrained parameter space and is straightforward

TABLE 10.4 An example of a genetic algorithm output for six genes.

LAI-GA	S_z	C_{ab}	N	r_{s1}	r_{s2}	ε^2	Generations	Number trials
2.53	0.95	65	3.41	0.06	−.02	8.16e−03	26	791
2.53	0.43	67	3.19	0.02	−0.08	7.81e−03	30	900
2.53	0.96	68	3.41	0.06	−0.02	8.22e−03	27	826
2.61	0.44	68	2.58	0.02	−0.08	7.69e−03	24	750
2.58	0.95	67	3.48	0.03	−0.02	8.09e−03	31	929
2.59	0.94	65	3.5	0.02	−0.08	7.90e−03	22	686
2.58	0.95	69	3.5	0.02	−0.08	7.94e−03	21	658
2.61	0.44	68	2.58	0.02	−0.08	7.70e−03	32	963
2.59	0.96	48	3.5	0.02	−0.08	8.28e−03	23	722
2.58	0.95	69	3.48	0.02	−0.08	7.97e−03	17	542

in the optimization process. In this study, most of the computational time was used in the GA optimization process, although reducing the number of free parameters helped. In the GA, the initial condition space must be scanned, and a large number of iterations are needed for convergence toward appropriate solutions. To solve this problem, more efficient GA optimization algorithms and GA-RT coupling methods are needed. To operationally map LAI values from satellite data, the computational time must be radically reduced before this method can be applied regionally.

10.3.5 Bayesian networks

10.3.5.1 A brief introduction to Bayesian networks

Bayesian retrieval algorithms, based on Bayes' theorem, have long been the most commonly used retrieval algorithms in geophysics (Klaus Mosegaard, 2002). In obtaining remote sensing data-based land surface parameters, the land surface parameters and observation data are regarded as random variables. The retrieval process is illustrated in

Fig. 10.14. The distribution of land surface parameters in the parameter data space (A) is inferred from that in parameter data space (B). Formula (10.13) is an expression of Bayes' theorem.

$$p(A|B = b_i) = \frac{p(A)p(B = b_i|A)}{p(B = b_i)}$$

$$= \frac{p(A)p(B = b_i|A)}{\sum_j p(A = a_j)p(B = b_i|A = a_j)} \quad (10.13)$$

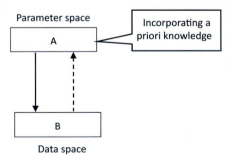

FIGURE 10.14 An illustration of Bayes' theorem and the Bayesian retrieval algorithm.

Here, A and B are the parameter space and the data space, respectively; a_j and b_i are the random events occurring in spaces A and B and corresponding to the retrieval parameter values and observation data, respectively; p(A) in the numerator is the a priori distribution of parameters; p(B|A) describes the probability density distribution of the model and observation data error; and the denominator, which is not related to parameter variations, is a normalization function.

Bayesian networks, which are mathematically based on Bayesian quantification schemes, integrate graphics and probability knowledge. A directed acyclic graph is used to depict the mutual dependence among the variables. Each node in the network represents a random variable, whereas the directed arcs connecting the nodes denote dependence among variables.

By introducing more constraining information on the land surface parameters, such as land cover type and growth period, to serve as auxiliary parameters for retrieval, Fig. 10.15 extends the Bayesian network shown in Fig. 10.14.

The joint probability density of random variables A, B, and C can be calculated by Formula (10.14):

$$p(C, A, B) = p(C, A) * p(B|C, A)$$
$$= p(C) * p(A|C) * p(B|C, A) \qquad (10.14)$$

Given the conditional independence assumption in Bayesian networks, which states that given A, B, and C are conditionally independent, we have

$$p(B|C, A) = p(B|A) \qquad (10.15)$$

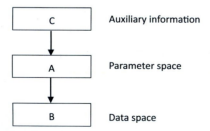

FIGURE 10.15   A conceptual diagram for a Bayesian network.

Formula (10.15) can be rewritten as

$$p(C, A, B) = p(C, A) * p(B|C, A)$$
$$= p(C) * p(A|C) * p(B|A) \qquad (10.16)$$

In Bayesian network—based parameter retrievals, the distribution of posteriori probability densities for parameter A is obtained after acquiring observation data and the matching parameter information. According to Bayes' theorem, the following formula is obtained:

$$p(A|B = b_i, C = c_k) = \frac{p(C = c_k)p(A|C = c_k)p(B = b_i|A)}{\sum_j p(C = c_k)p(A = a_j|C = c_k)p(B = b_i|A = a_j)}$$

$$(10.17)$$

The difference between Formulas (10.13) and (10.17) is that the priori knowledge $p(A)$ in Formula (10.13) becomes $p(C = c_k)p(A|C = c_k)$ in Formula (10.17), where $p(C = c_k)$ denotes the probability distribution of factors that affect the parameters. These factors include time, position, elevation, land use and other auxiliary information, and the parameter information, which are observed simultaneously with ground measurements. $p(A|C = c_k)$ is the distribution of the probability densities of the parameters to be retrieved after acquiring the aforementioned information. The information in the two items can be derived from the long-term accumulation of Earth observation data and the quantified relationship between them, $p(A|C)$, which can be acquired by machine learning or statistical method.

An example of the LAI retrieval in remote sensing data—based land surface parameter retrievals is demonstrated in Qu, Wang et al. (2008), and the corresponding network structure is presented in Fig. 10.16. The computational formula for Bayesian network—based parameter retrieval is given below. The variable T represents the growth period, such as the growth time for crops; A is the ALA; and V1 and V2 are the relative azimuth angle (VAzimuth) and the viewing zenith angle (VZenith), respectively. Other variables are represented by the capital initial letters of the corresponding nodes. Lowercase letters are used to represent the specific

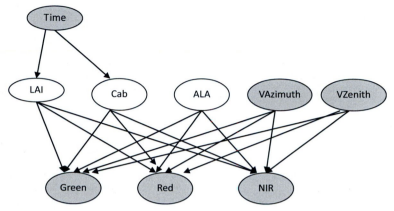

FIGURE 10.16   A Bayesian network model for retrieving LAI and Cab. (The shaded nodes represent the observable parameters, and the white nodes represent the free parameters).

values of the corresponding variables. Assume that the reflectance values in the three wave bands, which are obtained under the condition that $V1 = v1$ and $V2 = v2$, are g, r, and n, respectively. Then, in the LAI retrieval, the posteriori probability for $LAI = l$ is

where *Const* denotes the constant factor and $p(L = l|T)$ is the conditional probability distribution of LAI for the given growth period. The last factor indicates the data fitting capacity under the given conditions. Therefore, the Bayesian retrieval algorithm can integrate the information

$$p(L = l|T = t, V1 = v1, V2 = v2, G = g, R = r, N = n)$$

$$= \frac{p(L = l|T = t) \sum_{\{C,A\}} p(G = g, R = r, N = n|L = l, C, A, V1 = v1, V2 = v2)}{\sum_{\{L\}}\left[p(L|T = t) \sum_{\{C,A\}} p(G = g, R = r, N = n|L, C, A, V1 = v1, V2 = v2)\right]} \quad (10.18)$$

Because the denominator in Formula (10.18) is not associated with the variation of parameter L, the above formula can be written as

from three aspects: the a priori distribution of parameters, the physical model in remote

$$p(L = l|T = t, V1 = v1, V2 = v2, G = g, R = r, N = n)$$
$$= Const * p(L = l|T = t) \sum_{\{C,A\}} p(G = g, R = r, N = n|L = l, C, A, V1 = v1, V2 = v2) \quad (10.19)$$

sensing techniques, and the parameter information provided by observation data.

10.3.5.2 *The application of Bayesian network in LAI retrieval*

The Bayesian network—based retrieval scheme was tested using measured canopy reflectance and the LAI data for winter wheat at multiple time points. In this dataset, multiangle ground observation data collected on April 2, 12, 17, and 21 in 2001 year, were selected to estimate LAI and Cab values. The measurement time was 9:00—11:00 a.m. (local time). Five zenith angles in principal plane observations at 55, 25, 0, −25, and −55 degrees were chosen; the negative sign indicates forward observations. The measurement period covered the wheat phenology of standing, jointing, and flagging stages. The results of this retrieval, which used measured land surface data, are shown in Fig. 10.17.

The measured LAI and C_{ab} values were the mean measurements at multiple points on each observation date in the plot. Because leaf biochemical components and observation spectra were obtained from different data source, there was a certain time lag between the measurements of the two types of data. Therefore, the true parameter values were based on the assumption that the environmental variable of the measured target did not vary excessively within the time interval between the measurements of the two types of data. This factor could be one of the major reasons for the considerable error in the C_{ab} retrieval.

The results of this retrieval using land surface data indicate that the estimated LAI values were similar to the measured values, with a root mean square error (RMSE) of 0.22. However, the error in the estimated values of C_{ab} was relatively larger, with an RMSE of 4.0. The largest error occurred on April 17, and the second-largest

error occurred on April 21, with absolute errors of 11 and 9 µg/cm^2, respectively. The C_{ab} retrieval results show the same pattern as the simulated data results. Thus, for larger C_{ab}, generally above 50 µg/cm^2, the parameter retrieval results are unreliable.

Bayesian network—based land surface parameter retrievals, based on Bayes' theorem and integrated with graph theory, can graphically and intuitively describe the relationships among variables. In remote sensing retrievals, the uncertainties in remote sensing and ground-measured data enable Bayesian networks to satisfactorily describe the uncertainties of geoscience data. As the theory of remote sensing—based retrievals develops, Bayesian networks will be used in an increasing number of applications in remote sensing—based retrievals. When considering the dynamic variations of land surface vegetation parameters, the Bayesian network—based retrieval algorithms can be applied to estimate the time series features of land surface parameters. Thus, algorithms for dynamic Bayesian network—based time series parameter estimates can be developed.

10.3.6 Lookup table methods

LUT is a data array with precalculated items that represent matching input and output values. The aim of a LUT is to replace complex computation processes with a simpler array-indexing operation. LUTs establish a correspondence between canopy reflectances and various parameter values, such as viewing geometry, canopy structure parameters, leaf reflectance and directional properties, and soil background spectral and directional properties. LUTs are usually precomputed for a large range of BRDF parameter value combinations, and the most computationally intensive model simulation is completed before the inversion. Thus, the

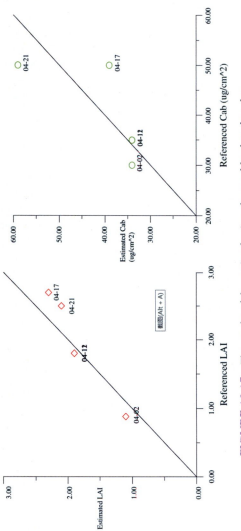

FIGURE 10.17   The results of the retrieval using observed land surface data.

inversion problem is greatly expedited by the use of a LUT for the modeled reflectance set. The inverse problem is formulated as follows: given element d∈D, find all p∈P for which f(p) = d. Here, d is the measured reflectance, and p represents the retrieved parameters. The canopy RT model f() is used to precompute elements of the space D by running the model for all canopy realizations P. Thus, with more elements in space P, real canopies can be more accurately represented. A simple two-step procedure for LUT generation and optimization is described below.

The ideal way to create large LUTs is to discretize more input values and run the RT model for all cases. We use the (Fang and Liang, 2005) study to demonstrate the procedure for constructing a LUT based on the MCRM. The MCRM was run with variable solar zenith angles (SZA = 20, 30, 35, 40, 45, 50, 55, 60, 65, 70, 75 degrees), view zenith angles (0−90 degrees in increments of 10 degrees), relative azimuth angles (0−180 degrees in increments of 15 degrees), LAI (0.1−10 in increments of 0.1), and different soil reflectance index (SRI) values (0.01, 0.05, 0.1, 0.15, 0.2, 0.25, 0.3, 0.4, 0.5, 0.6, 0.8, 1.0). The PROSPECT model (Jacquemoud and Baret, 1990) was used to simulate leaf optical properties. Leaf biophysical parameters were specified by the leaf chlorophyll A + B concentration (Cab = 10−90 in increments of 10 $\mu g/cm^2$), the Markov parameter (0.4−0.9 in increments of 0.1), and the effective number of elementary layers (1.0−3.0 in increments of 0.5). The leaf orientation was assumed to be spherical. An example of the output MODIS red nadir and NIR reflectances in the principle plane for different LAI and SRI values is shown in Fig. 10.18. When the solar zenith angle is low (30 degrees), the red reflectance is low and insensitive to LAI changes. When the LAI increases, the red reflectance decreases slowly, whereas the NIR reflectance increases. The red reflectance

decreases and the NIR reflectance increases in relation to increases in the LAI when SZA = 50 degrees. The hotspot effect is obvious for both solar zenith angles.

For the LUT inversion, only search operations are needed to identify the parameter combinations that yield the best fit between measured and LUT spectra. The LUT is sorted according to a cost function, which is a simple RMSE between the measured and modeled spectra:

$$RMSE = \sqrt{\frac{1}{n}\sum_{i=1}^{n}\left(R_{measured,i} - R_{LUT,i}\right)} \quad (10.20)$$

where $R_{measured,i}$ is the measured reflectance at wavelength i; $R_{LUT,i}$ is the simulated reflectance at wavelength i stored in the LUT; and n is the number of wavelengths.

The LUT solution is not always the optimal solution because it may not be unique, making it an ill-posed problem. Therefore, the distribution of the set of variables that provides the smallest RMSE is considered as the solution and is represented by its median values. For some variables, the distribution of the best solution can be so large that any value in the range of variation of the variable can be taken. Under such conditions, the median value represents the center of the range instead of the high-frequency values. To prevent widely spread solutions, constraints can be imposed by accounting for prior information and reducing the LUT. The median from the best 10, 20, 40, and 100 combinations can be treated as the solution.

LUT is very accurate, but it may remain computationally expensive in some specific cases. In recent years, machine learning tools were also used to replace the complex model. Gomez-Dans et al. (2016) used the Gaussian process regression (GPR) algorithm to fit the PRO-SAIL model and the PROSAIL+6S model, finding it accurately approximated radiative transfer models (RTMs) with a small set of the training dataset. They called this approach emulation and the machine learning model emulator. Verrelst et al. (2016) tested kernel ridge regression, NN, and GPR as emulators for PROSPECT-4, PROSAIL, and MODTRAN5 and showed very promising results in replacing RTMs. Besides the monospectral emulation, the emulators are also suitable for approximating canopy reflectance and sun-induced fluorescence spectra, by introducing the dimensionality reduction method (Gomez-Dans et al., 2016; Verrelst et al., 2017).

NNs and LUT methods are commonly used in LAI retrieval. Both techniques build on the relationship between biophysical parameters and the simulated canopy reflectance. The forward radiative transfer simulation is complicated, and the simulation's accuracy depends on the physical models. GAs have the advantage of identifying the global optima; however, their use is limited because of computation costs. Bayesian network models have the potential to be used for LAI retrieval from remote sensing data.

10.4 Data assimilation methods

Many algorithms have been developed for remote sensing data—based LAI estimation methods, including the estimation method, which is based on the statistical relationship between the LAI and the spectral VI; the method that uses physical model retrievals; and other nonparametric methods. Each method has advantages and disadvantages. Model retrieval algorithms are based on physics, with extensive applications in a wide range of domains. Therefore, in remote sensing data—based LAI estimations from remote sensing data, retrievals based on radiative transfer models are gaining popularity. All of these methods use data measured at single time points to retrieve biological parameters. This approach limits the availability of information,

which further leads to time series discontinuities and low precision levels.

The inversion problem is essentially an ill-posed inversion because of its multiple solutions and uncertainty. Therefore, acquiring as much information as possible is necessary to improve the precision of land surface parameter estimates. One feasible method, the data assimilation technique, which is based on reasonable descriptions of the dynamic variations of biological parameters and ion time series from multiple sites, can improve the quality of variable biological profile retrievals.

Generally, two types of assimilation algorithms are used: variational methods, such as the 4D-variational method, and sequential assimilation methods, such as the Kalman filter (KF).

10.4.1 Variational assimilation methods

In 4DVAR data assimilation, an attempt is made to find an evolution of model states that most closely matches observations given during the assimilation interval. Lauvernet et al. (2008) developed a multitemporal-patch ensemble inversion scheme to account for the spatial and temporal constraints in the inversion process. Koetz et al. (2005) proposed a method to estimate LAI based on multitemporal remote sensing observations. The inversion was initially solved exclusively based on radiometric information for each observation, and a simple semi-mechanistic canopy structure dynamic model was used to fit over the radiometrically estimated LAI. Then, the fitted LAI was integrated in the inversion as a priori information. Dente et al. (2008) presented a method to assimilate LAI retrieved from Environmental Satellite (ENVISAT), Advanced Synthetic Aperture Radar, and Medium-spectral Resolution Imaging Spectrometer (MERIS) data into the crop environment resource synthesis wheat crop growth model with the objective of improving the accuracy of the wheat yield predictions at catchment scale. A variational assimilation algorithm has been applied to minimize the difference between simulated and remotely sensed LAI and to determine the optimal set of input parameters. Liu et al. (2008) proposed a data assimilation approach that combines the satellite observations of MODIS albedo with a dynamic leaf model. The approach optimizes the dynamic model parameters such that the difference between the estimated surface reflectances based on the modeled leaf area and those of satellite observations is minimized, and the results show that the seasonal cycle of the directly retrieved leaf areas is smooth and consistent with both observations and current understanding of processes controlling leaf area dynamics. Xiao et al. (2009) developed a temporally integrated inversion method to estimate LAI from time series MODIS reflectance data. This method is briefly introduced in this section.

The variational assimilation algorithm can easily integrate dynamic model information and remote sensing time series data to retrieve land surface parameters. N observation data points are given in the assimilation window, $\{\mathbf{y}_k | k = 0, 1, 2, ..., N\}$, and the variational assimilation algorithm is used to solve for the optimal estimated value of cost function $J(\mathbf{x})$.

$$J(\mathbf{x}) = \frac{1}{2} \sum_{i=1}^{n} (\mathbf{y}_i - H_i[\alpha, LAI_i(\theta)])^T \mathbf{R}_i^{-1}$$

$$(\mathbf{y}_i - H_i[\alpha, LAI_i(\theta)]) + \frac{1}{2}(\mathbf{x} - \mathbf{x}_b)^T \mathbf{B}^{-1}(\mathbf{x} - \mathbf{x}_b)$$

$$(10.21)$$

where $\mathbf{x}$ is the coupling model parameter subject to optimal estimation, includes the radiative transfer model parameter α and the dynamic model parameter θ; $\mathbf{x}_b$ is the a priori information of these parameters; $\mathbf{y}_i$ is the observation data of the sensor; $H_i(\cdot)$ is the radiative transfer mode model; $\mathbf{B}$ and $\mathbf{R}$ are the error covariances of the a priori information and the observation data,

respectively; and $LAI_i(\theta)$ describes the pattern of temporal LAI variations. In this study, the following empirical model was used:

$$LAI_t(\theta) = vb + \frac{k}{1 + \exp(-c(t - p))}$$
$$- \frac{k + vb - ve}{1 + \exp(-d(t - q))}$$

$$(10.22)$$

where k, c, d, p, q, vb, and ve are model parameters.

Using the above-described variational method to integrate MODIS time series reflectance data, LAI values are retrieved. Fig 10.19 shows the LAI temporal profiles retrieved by the temporally integrated inversion method (denoted by LAI-TS) at the center pixel of the Bondville site in 2001. The Bondville site is an agricultural site in the Midwestern part of the United States. The field at the site was continuous no-till with alternating years of soybean and maize crops. In 2001, the crop was maize with a maximum LAI of 4.38 and a corresponding height of 2.4 m. To better compare and analyze the retrieved results, MODIS LAI values and LAI time series retrieved by the routine method (denoted by LAI-SP) are also shown in Fig. 10.19. Clearly, the MODIS LAI values at this pixel have markedly underestimated field measurements in the crop growing season. Fluctuations, particularly during the crop growing season, result from the difficulty of acquiring cloud free images in this time period due to the high humidity in the atmosphere. For days 169 and 177, MODIS LAI values are missing. The LAI time series retrieved by the routine method also show similar change characteristics, and there is an abrupt drop at day 201. By comparison, the temporally integrated inversion method can remove noise shown as abrupt rises or drops. Moreover, the accuracy of the retrieved LAI has been significantly improved over the MODIS LAI product as compared with the field-measured LAI data.

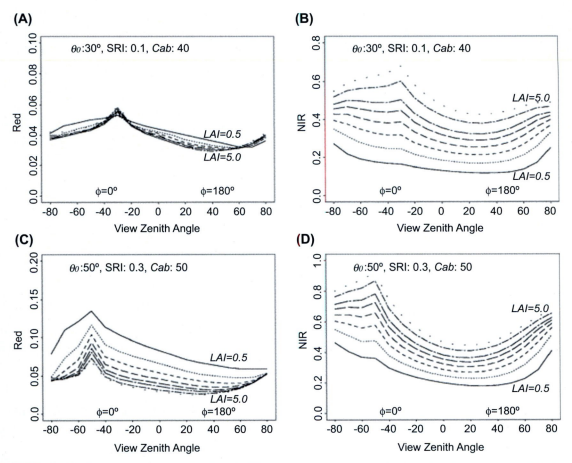

FIGURE 10.18 Simulations of MODIS red and NIR reflectances for various viewing directions are shown. The figure shows reflectances in the principal plane with varying viewing zenith angles and LAI values. (A) and (B) are the red and NIR reflectances, respectively, for SZA (30 degrees), SRI (0.1), and Cab (40). (C) and (D) are the red and NIR reflectances, respectively, for SZA (50 degrees), SRI (0.4), and Cab (60). LAI = {0.5, 1.0, 1.5, 2.0, 2.5, 3.0, 4.0, and 5.0}.

10.4.2 The sequential data assimilation algorithm

The sequential data assimilation algorithms only consider observations made in the past until the time of analysis. Based on the ECOCLIMAP land cover classification, Samain et al. (2008) demonstrated the advantage in using a KF to improve the spatial coherence and time consistency of the surface BRDF. Xiao et al. (2011) proposed a method in which MODIS time series reflectance data (MOD09A1) are used to achieve real-time retrieval. As new observations arrive, the ensemble Kalman filter (EnKF) is used to update LAI recursively by combining predictions from the dynamic model and MODIS reflectance data. In the absence of new observations, the biophysical variables can be propagated using the dynamic model. This section introduces the EnKF-based method for iterative LAI retrieval.

The EnKF method is a novel sequential data assimilation algorithm that is mainly applied to

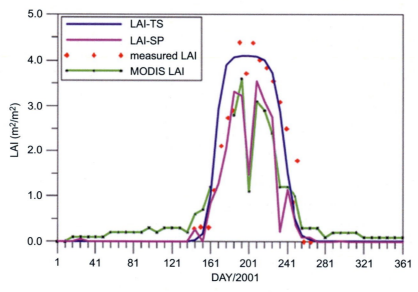

FIGURE 10.19   Retrieved LAI temporal profiles at the center pixel of the Bondville site in 2001.

nonlinear models. $\mathbf{x}$ is defined as an n-dimensional state vector, and the matrix $\mathbf{A} \in \Re^{n \times N}$ is composed of the set of N state vectors, $\mathbf{A} = (\mathbf{x}_1, \mathbf{x}_2, \dots, \mathbf{x}_N)$. Let $\overline{\mathbf{A}} \in \Re^{n \times N}$ be the mean matrix of the set. The ensemble perturbation matrix $\mathbf{A}' \in \Re^{n \times N}$ is defined as $\mathbf{A}' = \mathbf{A} - \overline{\mathbf{A}}$, and the standard analysis equation can be expressed as

$$\mathbf{A}^a = \mathbf{A} \\ + \mathbf{A}'\widehat{\mathbf{A}}'^{\mathrm{T}}\widehat{\mathbf{H}}^{\mathrm{T}}\left(\widehat{\mathbf{H}}\widehat{A}'\widehat{\mathbf{A}}'^{\mathrm{T}}\widehat{\mathbf{H}}^{\mathrm{T}} + \mathbf{R}\right)^{-1}\left(\mathbf{D} - \widehat{\mathbf{H}}\widehat{A}\right)$$
$$(10.23)$$

where $\mathbf{H} \in \Re^{m \times n}$ is the observation operator that describes the relationship between a state vector and an observation, and $\mathbf{R} \in \Re^{m \times m}$ is the observation error covariance matrix.

In Formula (10.24), $\mathbf{H}$ is assumed to be a linear observation operator. When MODIS reflectance data are assimilated using a process model, the canopy radiative transfer model $\mathbf{h}(\cdot)$ is a nonlinear function of the state vector, and the analysis equation in Formula (10.24) is no longer applicable. By augmenting the state vector, $\widehat{\mathbf{x}}^{\mathrm{T}} = [\mathbf{x}^{\mathrm{T}}, \mathbf{h}^{\mathrm{T}}(\mathbf{x})]$, $\widehat{\mathbf{A}} \in \Re^{\hat{n} \times N}$ is defined as the set matrix of the augmented state vector,

and $\widehat{\mathbf{A}}' \in \Re^{\hat{n} \times N}$ is defined as the ensemble perturbation matrix of the state vector. The analysis equation

$$\mathbf{A}^a = \mathbf{A} \\ + \mathbf{A}'\mathbf{A}'^{\mathrm{T}}\mathbf{H}^{\mathrm{T}}\left(\mathbf{H}A'\mathbf{A}'^{\mathrm{T}}\mathbf{H}^{\mathrm{T}} + \mathbf{R}\right)^{-1}(\mathbf{D} - \mathbf{H}A)$$
$$(10.24)$$

can then be used to solve the nonlinear data assimilation problem for the observation operator.

The EnKF-based iterative update algorithm is illustrated in Fig. 10.20. After the preprocessing of MODIS LAI products with a self-adaptive SG filter, the annual and interannual LAI variation patterns can be statistically analyzed. Once the LAI variation trend information has been acquired by the SARIMA model, the process model describing the dynamic LAI variation patterns can be constructed. Based on the prediction information of SARIMA model, the process model is built as follows:

$$\mathrm{LAI}_t = F_t \times \mathrm{LAI}_{t-1} \qquad (10.25)$$

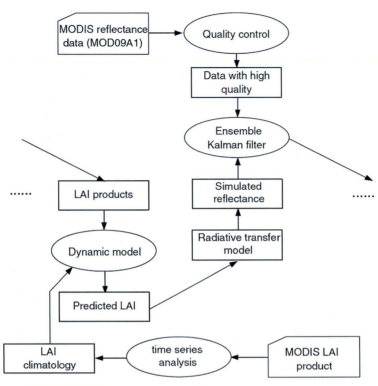

FIGURE 10.20   A flowchart for the real-time retrieval algorithm.

where

$$F_t = 1 + \frac{1}{\text{LAI}_t^{\text{clim}} + \varepsilon} \times \frac{d\text{LAI}_t^{\text{clim}}}{dt} \qquad (10.26)$$

where $\text{LAI}_t^{\text{clim}}$ is the temporal profile from LAI climatology, and $\varepsilon = 10^{-3}$ prevents a null denominator. The model incorporates the canopy LAI climatology. Predicated LAI from the dynamic model was input into the radiative transfer model to simulate reflectance of different bands, which was then compared with MODIS reflectance to correct the LAI predicted using the EnKF techniques.

Based on MODIS data from the BELMANIP sites, the real-time LAI retrieval algorithm was tested. Fig. 10.21 shows the results retrieved by the real-time inversion method in 2002 at the Sud-Ouest site. The site features croplands according to the IGBP classification. The land

cover is composed mainly of crops (corn, soybean, and sunflower) and grasslands. The LAI climatology for the year 2002 is forecasted by the SARIMA model from the upper envelope of 2000–2001 MODIS LAI data, which describes the general change tendency of LAI in 2002 (see Fig. 10.21A). Based on the climatology, a dynamic model is constructed to evolve LAI in time and then used to provide the short-range forecast of LAI. Forecasts from the dynamic model are used with the EnKF techniques to estimate real-time LAI from the time series MODIS reflectance data.

For a better evaluation of the quality of the retrieved LAI values, Fig. 10.21B shows the time series of LAI values using the real-time LAI retrieval algorithm, MODIS LAI values, and the BELMANIP mean LAI for July 2, 2002. The real-time retrieved and MODIS LAI profiles

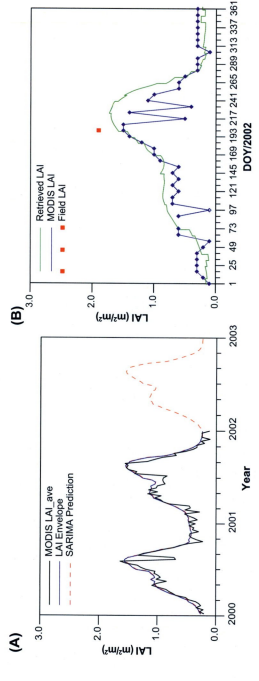

FIGURE 10.21   (A) LAI predicted by the SARIMA model based on the outer envelop of the MODIS LAI; (B) retrieved LAI temporal profiles by the real-time inversion method at the Sud-Ouest site in 2002.

are in very good agreement at the beginning and the end of the growing season. During the peak of the growing season, the MODIS LAI profile shows dramatic fluctuations, while the retrieved LAI profile from our new method is relatively smooth. Comparatively speaking, the real-time retrieved LAI estimate is closer to BELMANIP mean LAI, than MODIS LAI values.

10.5 LAI retrieval from lidar data

The return waveform of lidar is a successive signal that is a product of interaction between lidar system and vegetation. The physical model that links lidar system and vegetation can help us have a better understanding of their interaction mechanisms and reveal the relationship between the signal and vegetation variables (Ni-Meister et al., 2001; Sun and Ranson, 2000). Most studies concerning retrieving LAI from lidar data and physical model are based on gap theories.

10.5.1 Retrieving LAI from FAVD

Sun and Ranson (2000) built a model to retrieve LAI based on RT theories and 3D model. The model was built based on RT theories and 3D model. In this model, the forest canopy is described as a turbid scattering medium that is parameterized by its foliage area volume density (FAVD), μ, the Ross-Nilson G-factor, and the foliage reflectance, r_{leaf}. The background (soil) within the forest scene is parameterized by its reflectance, r_{soil}.

Because lidar is working under hotspot conditions, the bidirectional gap fraction p of the canopy in the direction Ω_i, at the canopy depth z, can be expressed as

$$p(z, \Omega_i) = \exp\left(-\int_0^z u_L(z')G(z', \Omega_i)dz'\right) \quad (10.27)$$

where $u_L(z')$ is the density of a one-sized leaf area at depth z', $G(z', \Omega_i)$ is the G-factor, the

mean projection of unit foliage area at a depth z' in the direction Ω_i.

The emitting laser pulse was divided into n narrow pulse (Iemit(i), i = 1 ... n), with the duration time set to the time resolution of the system, and the canopy was divided into m layers from top to bottom, with the thickness Δz the same as the vertical range resolution of the system. Supposing the lidar return pulse begins at a time when the first subemitting pulse reaches the first canopy layer, then, when the ith subpulse reaches the jth subcanopy layers, the time delay is

$$t(i, j) = \frac{2(i + j) \times \Delta z}{c} \quad (10.28)$$

In addition, the returned energy is

$$I(i, j) = I_{emit}(i) \times R_j \times E_{j-1} \quad (10.29)$$

where $R_j = \Gamma \times u_j \times \omega$ is the backscatter factor of the jth canopy, ω is a parameter determined by the lidar system, Γ is the scattering phase function of the canopy ($\Gamma = r_{leaf} \times G/\pi$), and E_{j-1} is the extinction above this canopy layer

$$\left(E_{j-1} = \exp\left(-\sum_{k=1}^{j-1} (u \times G)_k \times \Delta z\right)\right).$$

When the ith subpulse reaches the m+1 soil layer, the returned energy is calculated from

$$I(i, m + 1) = I_{emit}(i) \times r_{soil} \times \omega \times E_m \quad (10.30)$$

To set the lidar system parameter, ω, the soil return sub-Gaussian wave extracted return waveform was used. So, in this model, the calibration of the lidar system is not necessary. The return signal is recorded in m + n digital bins according to the time delay.

Based on this model, Ma et al. (2015) inversed the FAVD and LAI value of Dayekou forest site. The results were shown in Figs. 10.22–10.23.

10.5.2 Retrieving leaf area index from gap fraction

Ni-Meister et al. (2001) developed a method to derive gap fraction from lidar full waveforms through GORT model. The basic assumption is that gap fraction is the reverse of vertical canopy profile as laser energy can only penetrate into the lower canopy layer or ground through gaps. Using this relationship, vertical gap fraction can be calculated from lidar waveform through

$$P(z) = 1 - \frac{R_v(z)}{R_v(0)} \frac{1}{1 + \frac{\rho_v}{\rho_g} \times \frac{R_g}{R_v(0)}} \quad (10.31)$$

where $P(z)$ denotes the probability that there is a gap above height z in a canopy, and $R_v(z)$, $R_v(0)$, and R_g represent the integrated laser energy returns from canopy to height z, from canopy to ground and ground return, respectively. The parameters of ρ_v and ρ_g represent the canopy and ground reflectance.

Effective LAI can be calculated using the following:

$$LAI_{cum}(z) = -\int_{z0}^{z} \frac{1}{G} \times \frac{d \log P(z)}{dz} dz \quad (10.32)$$

where $z0$ is the height location of the canopy bottom. The projection coefficient G is typically set to 0.5 assuming a random foliage distribution within the canopy crown.

Based on this model, Tang retrieved vertical LAI profiles over tropical rain forests using waveform lidar at La Selva, Costa Rica (Tang et al., 2012), derived and validated LAI at multiple spatial scales through lidar remote sensing (Tang et al., 2014a), and then retrieved large-scale LAI and vertical foliage profile from spaceborne waveform lidar (Tang et al., 2014b). Marselis calculated the LAI value through gap fraction integration at different height intervals. According to the differences of LAI value of some certain height intervals, the total LAI value, canopy height, and vegetation coverage, Marselis et al. (2018) studied the distribution of different kinds of vegetation and made the map of the study area, shown in Figs. 10.24—10.25.

10.6 Global and regional leaf area index products

10.6.1 Major global moderate-resolution leaf area index products

With the launch of remote sensing satellites, the data acquired by moderate-resolution

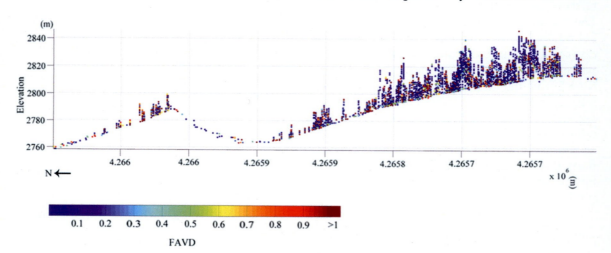

FIGURE 10.22   FAVD distribution in Dayekou forest site area.

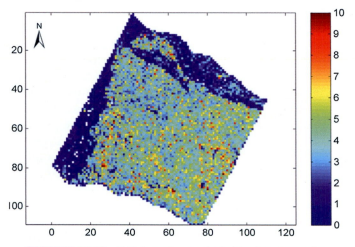

FIGURE 10.23   LAI inversion results in Dayekou forest site.

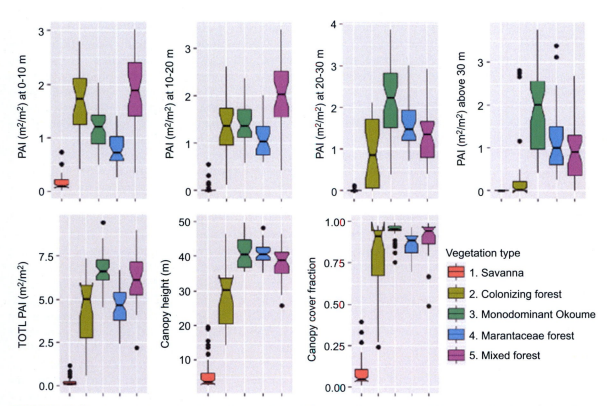

FIGURE 10.24   Seven indexes describing characteristics of the vegetation structure of five vegetation types (colored), ordered by successional stage.

spectral sensors have become the major data source for generating global LAI products. These data have been used to generate multiple types of global and regional LAI products (Fang et al., 2019). The MODIS, GLASS, and GLOB-MAP LAI products were derived from the TERRA-AQUA satellites (from 2000 to the present). Three types of global LAI products have been generated from SPOT/VEGETATION sensor data: the GLOBCARBON LAI product, the CYCLOPES LAI product, and the GEOV1 LAI product. The CCRS LAI product, a regional product, has also been generated from the SPOT/VEGETATION sensor data. The ECO-CLIMAP LAI product is derived from National Oceanic and Atmospheric Administration (NOAA)/AVHRR NDVI information. Several time-limited LAI products, such as the Polarization and Directionality of the Earth's Reflectance (POLDER) LAI product and the MERIS LAI product, and coverage-limited LAI products, such as the MISR LAI product and the MSG/SEVIRI LAI product, have also been developed.

The MODIS LAI product, produced by the TERRA-AQUA satellites from 2000 to the present, is the LAI product with the most extensive applications. Through constant improvements, the fifth version, based on the true LAI, has been developed. This product, available at http://wist.echo.nasa.gov, has a spatial resolution of 1 km, a temporal resolution of 8 days, and a sinusoidal projection. The retrieval algorithms for the MODIS LAI include a main algorithm and an alternate algorithm. The main algorithm uses a geobotanical chart as a priori knowledge and divides global vegetation into eight categories (Yang et al., 2006). This algorithm employs different input parameters for different land cover types and uses a 3D radiative transfer model for forward predictions. In this algorithm, different observation geometries and reflectances in various wave bands for different types of soil are precomputed to form a table. The retrieval is then

conducted using a LUT (Knyazikhin et al. 1998a,b). During the retrieval, only the observed reflectances for different wave bands are compared with the simulated reflectances. After a threshold has been set, the mean values and the variances of the LAI can be calculated. The former are treated as the retrieval results, and the latter are treated as uncertainty values. If the main algorithm fails, the alternate algorithm is used. This algorithm is based on the empirical relationships between different LAI types, and NDVI values are used to calculate the LAI. A lower level of precision is achieved compared with the main algorithm.

The GEOV1 LAI product has been available since 1998 from http://www.geoland2.eu/. The product is provided in a plate carrée projection at 1/112 spatial resolution and a 10-day frequency. The GEOV1 LAI product was derived from SPOT/VEGETATION sensor data using backpropagation NNs that were trained by fused and scaled "best estimates" of LAI from the MODIS and CYCLOPES products and the SPOT/VEGETATION nadir surface reflectance values over the BELMANIP (BEnchmark Land Multisite ANalysis and Intercomparison of Products) network of sites.

The GLASS LAI product (Xiao et al., 2014, 2016), a global LAI product with long time series, is generated and released by the Center for Global Change Data Processing and Analysis of Beijing Normal University (http://www.bnu-datacenter.com/). It can also be downloaded from the University of Maryland (http://www.glass.umd.edu.). The GLASS LAI product has a temporal resolution of 8 days. It includes LAI retrievals derived from MODIS surface reflectance data (denoted by GLASS MODIS) and LAI retrievals generated from the long-term data record AVHRR reflectance data (denoted by GLASS AVHRR). The GLASS MODIS LAI product is provided in a sinusoidal projection at a spatial resolution of 1 km and spans from 2000 to present, while the GLASS AVHRR LAI product is provided in a geographic latitude/longitude projection at a spatial resolution of 0.05 degrees (approximately 5 km at the equator) and spans from 1981 to present. Details of the GLASS LAI algorithm are specified in Section 10.3.3.2. Figs. 10.25 and 10.26 displays spatial distribution maps of global mean LAI for the GLASS LAI product in January and July during 2001−10.

The GLOBMAP LAI product provides a 0.08 degrees spatial resolution on geographic grid and spans the period from 1981 to 2016 (http://www.globalmapping.org/globalLAI/). For 2000−16, the GLOBMAP LAI product was generated from MODIS surface reflectance data based on the GLOBCARBON LAI retrieval algorithm (Deng et al., 2006). It was provided in a temporal resolution of 8 days. For 1981−99, The GLOBMAP LAI product was generated from AVHRR observations based on the pixel-level relationships between the AVHRR data

(GIMMS NDVI) and the LAI values retrieved from the MODIS surface reflectance data (Liu et al., 2012). It was provided in a temporal resolution of half month. The contaminated pixels in the GLOBMAP LAI product were filled by locally adjusted cubic spline capping approach (Chen et al., 2006).

The ECOCLIMAP dataset provides the mean values of biological and physical variables for land surface simulations, including those for the LAI. The dataset, available at http://www. cnrm.meteo.fr/gmme/PROJETS/ ECOCLIMAP/page_ecoclimap.htm, uses a ground sampling distance of 1/120 degrees and a step length of 1 month. This dataset also uses the Plate Carrée projection. Based on a global classification scheme, the ECOCLIMAP data, which is used with several land cover maps and a world climate distribution, divide the globe into 15 major land cover classes. For each type of land cover, LAI variations are

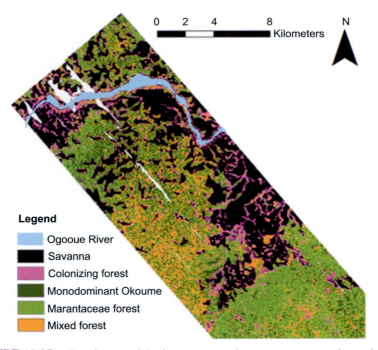

FIGURE 10.25 Classification of the five successional vegetation types in the study area.

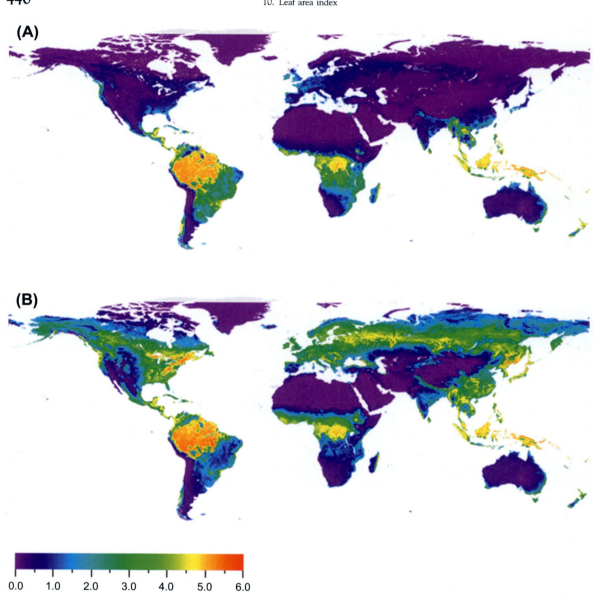

FIGURE 10.26 Global mean LAI maps for the GLASS LAI product in (A) January and (B) July over 2001–2010.

determined by field surveys. This product takes into account the vegetation aggregation effect at the plant and canopy scales. It is representative of green leaves and includes the lower layer of the forest. For each pixel in the ECOCLIMAP grid, an NDVI product is synthesized on a monthly basis into the corresponding pixel category using global NOAA/AVHRR products with a one-year cycle. The time trajectory of the LAI is regulated by the LAI maximum and minimum. The advantage of this algorithm is the low rate of spatial LAI variations for each vegetation type. The ECOCLIMAP LAI is a mean value.

The CCRS LAI product, a regional product derived from SPOT/VEGETATION sensor reflectance data for Canada, is generated by the Canada Center for Remote Sensing (Fernandes et al., 2003). This product is normalized to general geometry conditions, with a ground sampling distance of 1 km and a step length of 10 d (from 1998 to the present). It uses the Lambert conical projection, and its algorithm is based on the empirical relationship between the measured LAI and the VI corresponding to seven land cover types in Canada. This algorithm is applied in Canada by deriving regional land cover maps based on SPOT/VEGETATION data.

10.6.2 Leaf area index climatology

Satellite remote sensing provides a unique way to obtain LAI values over large areas. An example of the global LAI is given in Fig. 10.27. This LAI map was produced from the MODIS onboard the Terra and Aqua satellites. The LUT method is used by the MODIS science team to estimate the LAI from a three-dimensional radiative transfer model. The MODIS LAI product is a 1-km global data product that is updated once every 8-day period throughout the calendar year. The Earth Resources Observation Systems Data Center Distributed Active Archive Center makes it available free of charge. Major landscape units are clearly distinguishable in Fig. 10.27. The forest LAI values are generally higher than those for grasses and agricultural crops. The LAI is the lowest for urban, snow, and desert areas.

In addition to MODIS, other satellite sensors can provide global LAI maps. The CYCLOPES project generates global LAI products and the associated uncertainty information from the SPOT/VEGETATION sensor at 1/112 degrees (approximately 1 km at the equator) resolution. The monthly 1-km GLOBCARBON LAI product is derived from the combined use of the

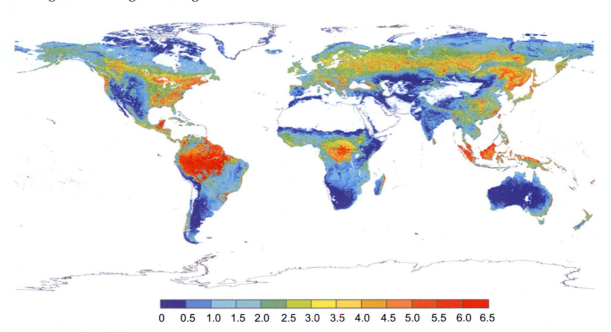

FIGURE 10.27 Global LAI produced by MODIS/Terra + Aqua Collection six in July 2017.

SPOT/VEGETATION and ENVISAT/ATSR instruments. NOAA AVHRR data, which are used all over the world, are unique for long-term surface change analyses and global research. The Terra Multiangle Imaging Spectroradiometer, Envisat MERIS, and ADEOS POLDER are notable satellite instruments that provide global LAI products with various spatial and temporal resolutions.

Fig. 10.28 shows an example of the global monthly mean LAI for the major biome types.

The monthly data from the 8-day MODIS/Terra + Aqua version 6 LAI product were combined for the period from 2003 to 2017. The LAI shows a clear intraannual vegetation cycle for areas other than evergreen broadleaf forests. Deciduous forests showed the greatest yearly LAI undulation. The PFT standard deviation is higher in the summer and lower in the winter. The LAI increase from the springtime to the summertime typically ranges from 1.0 to 2.0 for shrubs, grasses, and crops.

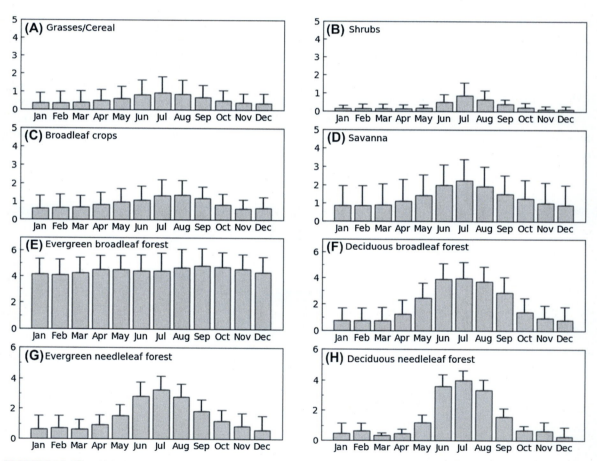

FIGURE 10.28   Global monthly mean LAI for different biome types derived from MODIS/Terra + Aqua LAI product from 2003 to 2017. The standard deviation is also included.

10.7 Summary

LAI is a critical parameter for describing terrestrial ecosystem. Global LAI products have been generated through remote sensing techniques for a variety of applications. To better meet user requirements, further studies are necessary to improve data continuity and the spatial and temporal stability of existing LAI products. Assessing the accuracy of remote sensing LAI products is critical to ensuring that they are used effectively in various disciplines (Morisette, 2006). A series of validation studies have been conducted at local, regional, and global scales. Data assimilation methods have been used to estimate the real-time LAI effectively (Xiao et al., 2011), which could be applied to generate long-term LAI products systematically.

References

Bacour, C., Jacquemoud, S., Tourbier, Y., Dechambre, M., Frangi, J.-P., 2002. Design and analysis of numerical experiments to compare four canopy reflectance models. Remote Sens. Environ. 79, 72–83.

Chen, J.M., 1996. Optically-based methods for measuring seasonal variation of leaf area index in boreal conifer stands. Agric. For. Meteorol. 80, 135–163.

Chen, J.M., Deng, F., Chen, M., 2006. Locally adjusted cubic-spline capping for reconstructing seasonal trajectories of a satellite-derived surface parameter. IEEE Trans. Geosci. Remote Sens. 44, 2230–2238.

Chen, J.M., Cihlar, J., 1996. Retrieving leaf area index for boreal conifer forests using Landsat TM images. Remote Sens. Environ. 55, 153–162.

Davis, L., 1991. Handbook of Genetic Algorithms. Van Nostrand Reinhold, New York.

Deng, F., Chen, J.M., Plummer, S., Chen, M.Z., Pisek, J., 2006. Algorithm for global leafarea index retrieval using satellite imagery. IEEE Trans. Geosci. Remote Sens. 44 (8), 2219–2229.

Dente, L., Satalino, G., Mattia, F., Rinaldi, M., 2008. Assimilation of leaf area index derived from ASAR and MERIS data into CERES-wheat model to map wheat yield. Remote Sens. Environ. 112, 1395–1407.

Fang, H., Baret, F., Plummer, S., Schaepman-Strub, G., 2019. An overview of global leaf area index (LAI): methods, products, validation, and applications. Rev. Geophys. https://doi.org/10.1029/2018RG000608.

Fang, H., Liang, S., 2005. A hybrid inversion method for mapping leaf area index from MODIS data: experiments and application to broadleaf and needleleaf canopies. Remote Sens. Environ. 94, 405–424.

Fang, H., Liang, S., Kuusk, A., 2003. Retrieving leaf area index using a genetic algorithm with a canopy radiative transfer model. Remote Sens. Environ. 85, 257–270.

Fang, H., Liang, S., 2014. Leaf Area Index Models. Reference Module in Earth Systems and Environmental Sciences. Elsevier, Oxford. ISBN: 978-0-12-409548-9.

Fernandes, R.A., Butson, C., Leblanc, S.G., Latifovic, R., 2003. Landsat-5 and Landsat-7 ETM+ based accuracy assessment of leaf area index products for Canada derived from SPOT-4 VEGETATION data. Can. J. Remote Sens. 29, 241–258.

Gastellu-Etchegorry, J.P., Demarez, V., Pinel, V., Zagolski, F., 1996. Modeling radiative transfer in heterogeneous 3-D vegetation canopies. Remote Sens. Environ. 58, p131–156.

Goel, N.S., Grier, T., 1988. Estimation of canopy parameters for inhomogeneous vegetation canopies from reflectance data: III. Trim: a model for radiative transfer in heterogeneous three-dimensional canopies. Remote Sens. Environ. 25, 255–293.

Goel, N.S., Thompson, R.L., 2000. A snapshot of canopy reflectance models and a universal model for the radia? tion regime. Remote Sens. Rev. 18, 197–225.

Goldberg, D.E., 1989. Genetic Algorithms in Search, Optimization and Machine Learning. Addison-Wesley, Reading, MA.

Gomez-Dans, J.L., Lewis, P.E., Disney, M., February 2016. Efficient emulation of radiative transfer codes using Gaussian processes and Application' to land surface parameter inferences. Remote Sens. 8 (2), 119.

Grefenstette, J., 1990. A User's Guide to GENESIS from. http: www.ail.nrl.navy.mil/galist/src.

Huemmrich, K.F., 2001. The GeoSail model: a simple addition to the SAIL model to describe discontinuous canopy reflectance. Remote Sens. Environ 75, 423-421.

Huete, A., Didan, K., Miura, T., Rodriguez, E.P., Gao, X., Ferreira, L.G., 2002. Overview of the radiometric and biophysical performance of the MODIS vegetation indices. Remote Sens. Environ. 83, 195–213.

Huete, A.R., 1988. A soil-adjusted vegetation index (SAVI). Remote Sens. Environ. 25, 295–309.

Jacquemoud, S., Baret, F., 1990. PROSPECT: a model of leaf optical properties spectra. Remote Sens. Environ. 34, 75–91.

Jacquemoud, S., Ustin, S.L., Verdebout, J., Schmuck, G., Andreoli, G., Hosgood, B., 1996. Estimating leaf biochemistry using the PROSPECT leaf optical properties model. Remote Sens. Environ. 56, 194–202.

Jupp, D.L.B., Strahler, A.H., 1991. A hotspot model for leaf canopies. Remote Sens. Environ. 38, 193–210.

Kimes, D.S., 1991. Radiative transfer in homogeneous and heterogeneous vegetation canopies. In: Myneni, R.B., Ross, J. (Eds.), Photon-Vegetation Interactions: Applications in Optical Remote Sensing and Plant Ecology. Springer-Verlag, Berlin, pp. 339–388.

Kimes, D.S., Knyazikhin, Y., Privette, J.L., Abuelgasim, A.A., Gao, F., 2000. Inversion methods for physically-based models. Remote Sens. Rev. 18, 381–440.

Klaus Mosegaard, A.T., 2002. Probabilistic Approach to Inverse Problems. Academic Press, Paris, France.

Knyazikhin, Y., Martonchik, J.V., Diner, D.J., Myneni, R.B., Verstrate, M.M., Pinty, B., Gobron, N., 1998a. Estimation of vegetation canopy leaf area index and fraction of absorbed photosynthetically active radiation from atmosphere-corrected MISR data. J. Geophys. Res. 103, 32,239-232,256.

Knyazikhin, Y., Martonchik, J.V., Myneni, R.B., Diner, D.J., Running, S.W., 1998b. Synergistic algorithm for estimating vegetation canopy leaf area index and fraction of absorbed photosynthetically active radiation from MODIS and MISR data. J. Geophys. Res. (103) 32,257-232,276.

Koetz, B., Baret, F., Poilve, H., Hill, J., 2005. Use of coupled canopy structure dynamicand radiative transfer models to estimate biophysical canopy characteristics. Remote Sens. Environ. 95 (1), 115–124.

Kuusk, A., 1985. The hot spot effect of a uniform vegetative cover. Soy. J. Remote Sens. 3, 645–658.

Kuusk, A., 1991a. Determination of vegetation canopy parameters from optical measurements. Remote Sens. Environ. 37, 207–218.

Kuusk, A., 1991b. The hot spot effect in plant canopy reflectance. In: Myneni, R.B., Ross, J. (Eds.), Photon-Vegetation Interactions. Applications in Optical Remote Sensing and Plant Ecology. Springer-Verlag, Berlin, pp. 139–159.

Kuusk, A., 1995a. A fast invertible canopy reflectance model. Remote Sens. Environ. 51, 342–350.

Kuusk, A., 1995b. A Markov chain model of canopy reflectance. Agric. For. Meteorol. 76, 221–236.

Kuusk, A., 2001. A two-layer canopy reflectance model. J. Quant. Spectrosc. Radiat. Transf. 71, 1–9.

Lauvernet, C., Baret, F., Haucoet, L., Buis, S., Dimet, F.X.L., 2008. Multitemporalpatchensemble inversion of coupled surface-atmosphere radiative transfer modelsfor land surface characterization. Remote Sens Environ 112 (3), 851–861.

Li, X., Woodcock, C., Davis, R., 1995. A hybrid geometric optical-radiative transfer approach for modeling albedo and directional reflectance of discontinuous canopies. IEEE Trans. Geosci. Remote Sens. 33, 466–480.

Liang, S., 2004. Quantitative Remote Sensing of Land Surfaces. John Wiley and Sons, Inc, New York.

Liu, Y., Liu, R.G., Chen, J.M., 2012. Retrospective retrieval of long-term consistent globalleaf area index (1981–2011) from combined Avhrr and Modis data. J. Geophys. Res.Biogeosci. 117, G04003.

Liu, Q., Gu, L., Dickinson, R.E., Tian, Y., Zhou, L., Post, W.M., 2008. Assimilation ofsatellite reflectance data into a dynamical leaf model to infer seasonally varying leafareas for climate and carbon models. J Geophys. Res. 113, D19113.

Ma, H., Song, J.L., Wang, J.D., 2015. Forest canopy LAI and vertical FAVD profile inversion from airborne full-waveform LiDAR data based on a radiative transfer model. Remote Sens. 7, 1897–1914.

Marselis, S.M., Tang, H., Armston, J.D., Calders, K., Labriere, N., Dubayah, R., 2018. Distinguishing vegetation types with airborne waveform lidar data in a tropical forest-savanna mosaic: a case study in Lope National Park, Gabon. Remote Sens. Environ. 216, 626–634.

Morisette, J.T., Baret, F., Privette, J.L., Myneni, R.B., Nickeson, J.E., Garrigues, S., Shabanov, N.V., Weiss, M., Fernandes, R.A., Leblanc, S.G., Kalacska, M., Sanchez-Azofeifa, G.A., Chubey, M., Rivard, B., Stenberg, P., Rautiainen, M., Voipio, P., Manninen, T., Pilant, A.N., Lewis, T.E., Liames, J.S., Colombo, R., Meroni, M., Busetto, L., Cohen, W.B., Turner, D.P., Warner, E.D., Petersen, G.W., Seufert, G., Cook, R., 2006. Validation of global moderate-resolution LAI products: a framework proposed within the CEOS land product validation subgroup. IEEE Trans. Geosci. Remote Sens. 44, 1804–1817.

Myneni, R.B., Asrar, G., Gerstl, S.A.W., 1990. Radiative transfer in three dimensional leaf canopies. Transp. Theory Stat. Phys. 19, 205–250.

Myneni, R.B., Ross, J., 1991. Photon-Vegetation Interactions: Applications in Optical Remote Sensing and Plant Ecology. Springer-Verlag., New York.

Myneni, R.B., Ramakrishna, R., Nemani, R., Running, S.W., 1997. Estimation of global leaf area index and absorbed par using radiative transfer models. IEEE Trans. Geosci. Remote Sens. 35, 1380–1393.

Myneni, R.B., Ross, J., Asrar, G., 1989. A review on the theory of photon transport in leaf canopies. Agric. Forest Meteorol. 45, 1–153.

Ni-Meister, W., Jupp, D.L.B., Dubayah, R., 2001. Modeling lidar waveforms in heterogeneous and discrete canopies. IEEE Trans. Geosci. Remote Sens. 39, 1943–1958.

Nilson, T., Kuusk, A., 1989. A reflectance model for the homogeneous plant canopy and its inversion. Remote Sens. Environ. 27, 157–167.

Pinty, B., et al., 2004. Radiation transfer model Intercomparison(RAMI) exercise: results from the second phase. J. Geophsical Res. 109, 1–19.

Pinty, B., Verstraete, M.M., Dickinson, R.E., 1990. A physical model for the bidirectional reflectance of vegetation canopies – Part 2: inversion and validation. J. Geophys. Res. B 95, 11767–11775.

Press, W.H., Teukolsky, S.A., Vetterling, W.T., Flannery, B.P., 1992a. Numerical Recipes in C: The Art of Scientific Computing. Cambridge University Press, New York.

Press, W.H., Teukolsky, S.A., Vetterling, W.T., Flannery, B.P., 1992b. Numerical Recipes in Fortran 77: The Art of Scientific Computing. Cambridge University Press, New York.

Privette, J.L., Myneni, R.B., Tucker, C.J., Emery, W.J., 1994. Invertibility of a 1-D discrete ordinates canopy reflectance model. Remote Sens. Environ. 48, 89–105.

Qin, W., Liang, S., 2000. Plane Parallel Canopy Radiation Transfer Modeling and Applications : Recent Advances and Future Directions [J]. Remote Sens. Rev. 18, 281–305.

Qi, J., Chehbouni, A., Huete, A.R., Kerr, Y., Sorooshian, S., 1994. A modified soil adjusted vegetation index (MSAVI). Remote Sens. Environ. 48, 119–126.

Qu, Y., Wang, J., Wan, H., Li, X., Zhou, G., 2008. A Bayesian network algorithm for retrieving the characterization of land surface vegetation. Remote Sens. Environ. 112, 613–622.

Ross, J., 1981. The Radiation Regime and Architecture of Plant Stands. Dr. W. Junk Publishers.

Samain, O., Roujeana, J.L., Geiger, B., 2008. Use of a Kalman filter for the retrieval of surface BRDF coefficients with a time-evolving model based on the ECOCLIMAP land cover classification. Remote Sens. Environ. 112, 1337–1346.

Specht, D.F., 1991. A general regression neural network. IEEE Trans. Neural Netw. 2 (6), 568–576.

Suits, G.H., 1972. The calculation of the directional reflectance of vegetation canopy. Remote Sens. Environ. 2, 117–175.

Sun, G.Q., Ranson, K.J., 2000. Modeling lidar returns from forest canopies. IEEE Trans. Geosci. Remote Sens. 38, 2617–2626.

Tang, H., Dubayah, R., Swatantran, A., Hofton, M., Sheldon, S., Clark, D.B., Blair, B., 2012. Retrieval of vertical LAI profiles over tropical rain forests using waveform lidar at La Selva, Costa Rica. Remote Sens. Environ. 124, 242–250.

Tang, H., Brolly, M., Zhao, F., Strahler, A.H., Schaaf, C.L., Ganguly, S., Zhang, G., Dubayah, R., 2014a. Deriving and validating Leaf Area Index (LAI) at multiple spatial scales through lidar remote sensing: A case study in Sierra National Forest, CA. Remote Sens. Environ. 143, 131–141.

Tang, H., Dubayah, R., Brolly, M., Ganguly, S., Zhang, G., 2014b. Large-scale retrieval of leaf area index and vertical foliage profile from the spaceborne waveform lidar (GLAS/ICESat). Remote Sens. Environ. 154, 8–18.

Verhoef, W., 1984. Light scattering by leaf layers with application to canopy reflectance modeling: the SAIL Model. Remote Sens. Environ. 16, 125–141.

Verhoef, W., 1985. Earth observation modeling based on layer scattering matrices. Remote Sens. Environ. 17 (2), 165–178.

Verhoef, W., Bach, H., 2003. Remote sensing data assimilation using coupled radiative transfer models. Phys. Chem. Earth 28, 3–13.

Verhoef W, N.J.J.B., 1981. Influence of crop geometry on multispectral reflectance determined by the use of canopy reflectance models. In: International Coll. On Spectral Signatures of Objects in Remote Sensing, Avignon. France.

Verrelst, J., Sabater, N., Rivera, J.P., Munoz-Marĩí, J., Vicent, J., Camps-Valls, G., Moreno, J., Aug. 2016. Emulation of leaf, canopy and atmosphere radiative transfer models for fast global sensitivity analysis. Remote Sens. 8 (8), 673.

Verrelst, J., Rivera Caicedo, J.P., Munoz-Marĩí, J., Camps-Valls, G., Moreno, J., Sep. 2017. SCOPE-based emulators for fast generation of synthetic canopy reflectance and sun-induced fluorescence spectra. Remote Sens. 9 (9), 927.

Xiao, Z., Liang, S., Wang, J., Song, J., Wu, X., 2009. A temporally integrated inversion method for estimating leaf area index from MODIS data. IEEE Trans. Geosci. Remote Sens. 47 (8), 2536–2545.

Xiao, Z., Liang, S., Wang, J., Jiang, B., Li, X., 2011. Real-time retrieval of leaf area index from MODIS time series data. Remote Sens. Environ. 115 (1), 97–106.

Xiao, Z., Liang, S., Wang, J., Chen, P., Yin, X., Zhang, L., Song, J., 2014. Use of general regression neural networks for generating the GLASS leaf area index product from time series MODIS surface reflectance. IEEE Trans. Geosci. Remote Sens. 52 (1), 209–223.

Xiao, Z., Liang, S., Wang, J., Xiang, Y., Zhao, X., 2016. Long time series global land surface satellite (GLASS) leaf area index product derived from MODIS and AVHRR data. IEEE Trans. Geosci. Remote Sens. 54, 5301–5318.

Xiao, Z., Liang, S., Jiang, B., 2017. Evaluation of four long time-series global leaf area index products. Agric. For. Meteorol. 246, 218–230.

Yang, W., Tan, B., Huang, D., Rautiainen, M., Shabanov, N.V., Wang, Y., Privette, J.L., Huemmrich, K., Fensholt, F., Sandholt, R.I., Weiss, M., Ahl, D.E., Gower, S.T., Nemani, R.R., Knyazikhin, Y., Myneni, R.B., 2006. MODIS leaf area index products: from validation to algorithm improvement. IEEE Trans. Geosci. Remote Sens. 44 (7), 1885–1898.

Fraction of absorbed photosynthetically active radiation

Abstract

The fraction of absorbed photosynthetically active radiation (FAPAR) characterizes the energy absorption capacity of vegetation canopy. It is a basic physiological variable describing the vegetation structure and related material and energy exchange processes and is an important parameter for estimating the net primary production of terrestrial ecosystem using a remote sensing–based method. This chapter first discusses some related concepts and then introduces the principles and current status of FAPAR estimation. At present, the remote sensing–based methods for estimating FAPAR can be divided into two categories: empirical methods and methods using radiative

transfer models or other physical models. Section 11.3 introduces FAPAR products as well as their intercomparison results. Case studies of the study area near the AmeriFlux sites are given in Section 11.4.

The fraction of absorbed photosynthetically active radiation (FAPAR) characterizes the energy absorption capacity of vegetation canopy. It is a basic physiological variable describing the vegetation structure and related material and energy exchange processes and is an important parameter for estimating the net primary production of terrestrial ecosystem using a

© 2020 Elsevier Inc. All rights reserved.

remote sensing—based method. This chapter first discusses some related concepts and then introduces the principles and current status of FAPAR estimation. At present, the remote sensing—based methods for estimating FAPAR can be divided into two categories: empirical methods and methods using radiative transfer (RT) models or other physical models. Section 11.3 introduces FAPAR products as well as their intercomparison results. Case studies of the study area near the AmeriFlux sites are given in Section 11.4.

11.1 Introduction

Vegetation plays a key role in the global energy balance, carbon cycle, and water budget of the Earth by controlling the exchanges between the lower atmosphere and the continental biosphere. Vegetation photosynthesis is responsible for the conversion of about 50 PgC/yr[1] of atmospheric CO_2 into biomass, which represents about 10% of the atmospheric carbon content (Carrer et al., 2013). Land-use changes, mainly attributed to deforestation, have led to an emission level of 1.7 PgC/yr[1] in the tropics, offsetting by a small amount of uptake of about 0.1 PgC in temperate and boreal areas—thereby producing a net source of around 1.6 PgC/yr[1] (Houghton, 1995). One of the most

important factors to monitor vegetation growth is the distribution of the FAPAR within vegetation as it constrains the photosynthesis rate through the energy absorbed by the vegetation. The FAPAR is the fraction of incoming solar radiation absorbed by plants in the 400—700 nm spectral range (Liang et al., 2012; Sellers et al., 1997). The definition of FAPAR can apply to vegetation only. It does not include incident solar radiation reflected by vegetation and solar radiation absorbed by the background (including soil, lichens, and litter under the forest), but must include the proportion reflected from the background and absorbed by vegetation. Given the boundaries of the vegetation, atmosphere, and soil, the solar radiation related to the canopy includes the following terms (Gobron et al., 2006a): (incoming solar flux)$I_{TOC}^{\downarrow}$, (flux to the ground)$I_{Ground}^{\downarrow}$, (flux from the ground)$I_{Ground}^{\uparrow}$, and (outgoing solar flux)$I_{TOC}^{\uparrow}$. These terms are shown in Fig. 11.1. Note that irradiance and flux are not distinguished here for convenience. FAPAR can be calculated by

$$FAPAR = \left(I_{TOC}^{\downarrow} - I_{Ground}^{\downarrow} + I_{Ground}^{\uparrow} - I_{TOC}^{\uparrow}\right)/I_{TOC}^{\downarrow}$$

(11.1)

FAPAR is one of the 50 essential climate variables recognized by the UN Global Climate Observing System (GCOS, 2011) and is a critical input parameter in biogeophysical and biogeochemical processes described in many climate

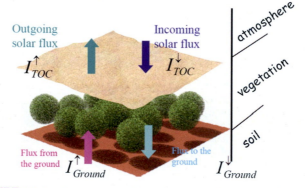

FIGURE 11.1   Composition of the solar energy that accesses the canopy.

and ecological models, e.g., Community Land Model, Community Earth System Model, and crop growth models (Bonan et al., 2002; Kaminski et al., 2012; Maselli et al., 2008; Tian et al., 2004). The Moderate Resolution Imaging Spectroradiometer (MODIS) FAPAR product (MOD15) is a critical input for MODIS evapotranspiration (MOD16), in addition to gross (GPP) and net primary production (NPP) products (MOD17) (Liang et al., 2012). A 10% increase in FAPAR would result in equal amounts of GPP, NPP, and carbon sink increases.

Despite the existence of the aforementioned numbers, the spatial distributions of carbon sources and sinks remain a core question and a subject of debate for the broad scientific community. In this regard, an improved representation of vegetation status in the ecological modeling is desirable. The reliable estimates of GPP, NPP, and carbon flux depend on high FAPAR input accuracy. An accuracy of 10% or 0.05 in FAPAR is considered acceptable in agronomical and other applications (GCOS, 2011).

The direct validation of satellite FAPAR products with ground measurements has generated some encouraging results, particularly when compared with previous versions of FAPAR products. The MODIS Collection 4 FAPAR product has been validated with ground measurements to demonstrate an accuracy of 0.2 (Baret et al., 2007; Fensholt et al., 2004; Huemmrich et al., 2005; Olofsson and Eklundh, 2007; Steinberg et al., 2006; Turner et al., 2005; Weiss et al., 2007; Yang et al., 2006), and the MODIS Collection 5 FAPAR product presents an improved accuracy to around 0.1 (Baret et al., 2013; Camacho et al., 2013; Martinez et al., 2013; McCallum et al., 2010; Pickett-Heaps et al., 2014; Xiao et al., 2015a). This improvement could be the result of a new stochastic RT model, which adequately captures the 3D effects of foliage clumping and species mixtures of natural ecosystems (Kanniah et al., 2009). The Multiangle Imaging SpectroRadiometer (MISR) FAPAR product exhibits performance similar to that of the MODIS C5 FAPAR product. However, the MODIS and MISR FAPAR products might show overestimation at certain sites. For example, Martinez et al. (2013) reported that MODIS tends to provide high values in cultivated areas and Mediterranean forests, such as the Puéchabon. The MODIS FAPAR product may also have positive bias for very low FAPAR values. A similar overestimation problem has been detected in MISR FAPAR data, with a positive bias as large as 0.16 in broadleaf forests (Hu et al., 2007). In addition to the overestimation problem, underestimations have been detected in the MODIS Collection 4 FAPAR product for certain sites in Switzerland (Olofsson and Eklundh, 2007). Overall, the current FAPAR products are close to the accuracy requirement but further improvements are still needed (Tao et al., 2015).

11.2 FAPAR estimation method

FAPAR can be derived from field measurements at the point scale, but the monitoring network of in situ measurements is insufficient for global coverage. Satellite sensors efficiently acquire land surface information at regional and global scales, providing new opportunities for monitoring biophysical parameters (Liang et al., 2012). The estimation of FAPAR from optical remote sensing is based on physical models or empirical relationships (Liang, 2007).

Empirical relationships between FAPAR and observations or derivatives from observations are established without knowledge of the underlying physical mechanism in the RT process. Therefore, simplicity is the primary advantage (Gobron et al., 1999). However, no unique relationship between FAPAR and the vegetation index is universally applicable to all conditions because canopy reflectance is also dependent on other factors such as geometrical measurement and spatial resolution (Asrar et al., 1992; Friedl, 1997). Moreover, the relationship between FAPAR and the vegetation index such as the normalized difference vegetation index (NDVI) is quite sensitive to the reflectance of background material (Asrar et al., 1992). Physical models analyze the interactions between solar radiation and vegetation canopies and reveal cause−effect relationships (Pinty et al., 2011; Widlowski et al., 2007). They are generally applicable to most conditions including over different land covers and during different time periods, although they require complex parameterizations.

Physical models for the retrieval of biophysical characteristics from reflected radiation of canopy can be divided into several classes (Liang, 2004): RT, geometric-optical, hybrid, and Monte Carlo, in addition to other computer simulations. The pure geometric-optical model considers only single scattering within the canopy, whereas an RT model also includes multiple scattering. Monte Carlo models and computer simulations are based on RT principles but are executed following random events rather than explicit formulae and therefore are computationally intensive. They may be used as surrogate truths to evaluate other RT and geometric-optical models (Widlowski, 2010; Widlowski et al., 2007).

In addition to the retrieval model performance, the determinants of FAPAR accuracy can be traced to the accuracy of such input parameter as leaf area index (LAI), soil background reflectance, and fractional canopy cover. LAI is one of the most important parameters in the determination of FAPAR, and its accuracy directly influences that of FAPAR. A 10% change in tree LAI could account for a 55% change in FAPAR (Asner et al., 1998). The collection of soil background reflectance is important for guaranteeing that the simulated reflectance can cover the entire set of observed surface reflectance data (Fang et al., 2012; Knyazikhin et al., 1998b; Shabanov et al., 2005). Otherwise, saturation of the relationship between FAPAR and surface reflectance may occur; very high FAPAR values are not reliable (Weiss et al., 2007). The correct estimation of

FAPAR also relies on that of fractional canopy cover, the underestimation of which might cause unrealistically high FAPAR values (Kanniah et al., 2009).

11.2.1 Empirical methods

As mentioned above, FAPAR is closely related to the vegetation canopy structure such as LAI and is also affected by the solar and viewing angles. To date, many studies have proposed various empirical algorithms for this parameter estimation. Because of the good correlation between FAPAR and LAI, in some biogeochemical process models, FAPAR is calculated as a function of LAI and extinction coefficient (Ruimy et al., 1994). Wiegand et al. (1992) obtained the empirical relationship between FAPAR and LAI as follows:

$$FAPAR = 1 - e^{-LAI}, \ R^2 = 0.952, \\ RMSE = 0.054 \quad (11.2)$$

Casanova et al. (1998) also mentioned that because the fraction of the photosynthetically active radiation (PAR) penetrating through the canopy to the incoming PAR has an exponentially decreasing relationship with LAI, the FAPAR can be expressed as follows:

$$FAPAR = 1 - e^{-K \times LAI}, \\ \text{where } K \text{ is the extinction coefficient} \quad (11.3)$$

However, because such methods need to obtain the leaf area index first, and also need to determine the extinction coefficient of the canopy, it is not commonly used in empirical relationship based methods.

The FAPAR can also be obtained by establishing its empirical relationship with the vegetation index. It can be retrieved by establishing a regression equation between field-measure FAPAR and vegetation indices such as NDVI, which can be calculated from the original image, reflectance image, or image after atmospheric correction. Although this method is convenient and flexible, the absorption of light by plants varies with the seasonal phase change of light reflection by plants, which is susceptible to various factors such as vegetation type, growth stage, and site environment. Thus, the applications of this type of model are limited.

Studies have shown that under certain conditions, there is a linear relationship between FAPAR and NDVI (Asrar et al., 1984; Goward and Huemmrich, 1992; Sellers, 1985). Myneni and Williams (1994) studied the relationship between FAPAR and NDVI under different canopy, soil, and atmospheric conditions using the RT method. They found that the relationship between FAPAR and NDVI is sensitive to the background, atmosphere, and canopy bidirectional reflectance characteristics. If the study is limited to the vicinity of the subsatellite point, the effects of atmospheric and bidirectional reflectance characteristics can be neglected. In the case of soil with moderate reflectance, the background effect can be neglected. Therefore, Myneni and Williams believe that the linear relationship between FAPAR and NDVI is tenable when the solar zenith angle is less than 60°, the observation angle of the nearby subsatellite point is less than 30°, the soil background reflectance is moderate (NDVI is about 0.12), and the atmospheric optical thickness is less than 0.65 at 550 nm.

Roujean and Breon (1995) used the SAIL model to simulate the radiation transfer inside the canopy and surface reflectance. The relationships between FAPAR and NDVI under different solar zenith angles, viewing zenith angles, and relative azimuth angles were studied. They found the relationship will improve when the solar zenith angle or viewing zenith angle increases. The reason is that the ray path increases and the background influence decreases, but it also brings the saturation problem of NDVI to higher LAI.

The backup algorithm, which Myneni et al. (2002) used in MODIS FAPAR product, is also based on the empirical NDVI algorithm. In the Vegetation Photosynthesis Model, the FAPAR is a function of the MODIS Enhanced Vegetation Index. The difference between EVI and NDVI is that the EVI adds blue reflectance based on the NDVI (Huete et al., 1997). In the CASA model, the FAPAR algorithm uses the linear stretching mode of the normalized vegetation index NDVI, that is, the simple ratio SR = $(1 + \text{NDVI})/(1-\text{NDVI})$, and the calculation of FAPAR is as follows (Potter et al., 1993):

$$FAPAR = \min\left(\frac{SR - SR_{min}}{SR_{max} - SR_{min}}, 0.95\right) \quad (11.4)$$

FAPAR in the Glo-PEN model is a linear function of SR (Prince and Goward, 1995):

$$FAPAR = \frac{SR - SR_{min}}{SR_{max} - SR_{min}}$$
$$(FAPAR_{max} - FAPAR_{min}) \quad (11.5)$$

Compared with NDVI and SR, the difference vegetation index (DVI) (see Eq. 11.6) can minimize the influence of soil background, and the application on sparse vegetation is obviously improved, but it is greatly affected by spectral and directional characteristics of the canopy. When the visible reflectance and near-infrared reflectance increase in the same proportion, NDVI does not change but DVI changes; DVI is suitable for sparse vegetation, and NDVI is suitable for dense vegetation. The renormalized difference vegetation index (RDVI) (Eq. 11.7) has an approximately linear correlation with FAPAR under any vegetation cover. However, this relationship is greatly affected by the soil background reflectance when the sun illuminates or the sensor observes vertically. Many studies have analyzed the relationship between soil-adjusted vegetation index (SAVI) (Huete, 1988) and the NDVI, DVI, and RDVI. When C is small (<0.15), the SAVI is similar to the NDVI. When C is large (>0.85), the SAVI is similar to the DVI. When $C = \sqrt{NIR + VIS}$, it is equivalent to RDVI.

$$DVI = NIR - VIS \tag{11.6}$$

$$RDVI = (NDVI \cdot DVI)^{1/2} = \frac{NIR - VIS}{\sqrt{NIR + VIS}} \tag{11.7}$$

$$SAVI = \frac{NIR - VIS}{NIR + VIS + C}(1 + C) \tag{11.8}$$

where C is a constant ranging from 0 to 1, the closer to 1, the smaller the soil impact, but the directional impact increases. NIR is the reflectance in the near-infrared band, and VIS is the reflectance in the visible band.

Vegetation index such as NDVI only partially determines the FAPAR value, and FAPAR is also affected by the internal components of the leaf, especially the chlorophyll content. Dawson (2003) showed that the chlorophyll content of vegetation had a great influence on remotely sensed estimates of FAPAR under the same vegetation NDVI, which resulted in the same NDVI value corresponding to a wide range of FAPAR values. High chlorophyll content and an increase in understory vegetation can lead to overestimate FAPAR. Therefore, other factors must be considered when estimating FAPAR, including solar zenith angle, leaf angle distribution type, soil background, etc., and FAPAR and FIPAR (fraction of intercepted PAR) must be distinguished.

11.2.2 MODIS FAPAR product algorithm

In the field of Earth observation, MODIS is a key instrument aboard the Terra (originally known as EOS AM-1) and Aqua (originally known as EOS PM-1) satellites. MODIS has high temporal resolution and acquires global comprehensive information every 1–2 days. MODIS has 36 spectral bands, with spatial resolutions of 250 m, 500 , and 1000 m, transit each

morning and afternoon. The long-term Earth observation data provided by MODIS help to monitor the global dynamics and processes of the Earth's surface and to obtain information on the underlying atmosphere. MODIS' global coverage, multispatial resolution, and multispectral and free product service policies make it an important source of information for studying the atmospheric, oceanic, and terrestrial processes at global scale.

The basic principle of the MODIS FAPAR main algorithm is to describe the spectral and directional characteristics of the canopy using a three-dimensional (3D) radiation transfer model, and then to build a lookup table based on the canopy structure and soil characteristics of a given biome type. Comparing the observed BRF and the stored model BRF in the lookup table, when the observed BRF and model BRF are less than a certain threshold, the corresponding LAI and FAPAR are considered as possible solutions (Knyazikhin et al., 1998b; Myneni, 1997).

In the 3D RT model, the canopy structure is the most important variable in the vegetation canopy. The canopy difference of different vegetation is large. Therefore, three points must be carefully considered when estimating the canopy radiation: (1) the canopy structure of the individual plant or community; (2) the optical characteristics of the vegetation elements (leaves and stems) and the soil background, the former depends on the physiological state of the vegetation, e.g., the water content and pigment content; and (3) the atmospheric conditions, which significantly impact the instantaneous solar radiation. In the original MODIS FAPAR algorithm, global terrestrial vegetation was divided into six categories based on canopy structure: grasses and cereal crops, shrubs, broadleaf crops, savannas, broadleaf forests, and needleleaf forests.

The MODIS algorithm uses a 3D RT model to describe the spectral and directional characteristics of the canopy. Considering the particularity of RT in canopies, the 3D RT model is decomposed into two submodels: (1) the radiation when the background is assumed to be blackbody (black soil) and (2) the radiation when the reflectance is assumed to be anisotropic at the bottom of the canopy, and the canopy reflectance and absorption are considered to be the weighted average of the two. Therefore, the formula for FAPAR at wavelength λ can be expressed as follows:

$$a_\lambda(\Omega_0) = a_{\text{bs},\lambda}(\Omega_0) + a_{S,\lambda}\frac{\rho_{\text{eff}}(\lambda)}{1 - \rho_{\text{eff}}(\lambda)\cdot r_{S,\lambda}}t_{\text{bs},\lambda}(\Omega_0)$$

(11.9)

where $a_\lambda(\Omega_0)$ is the canopy absorption at wavelength λ, Ω_0 is the incident direction of the sun; $a_{\text{bs},\lambda}(\Omega_0)$ and $t_{\text{bs},\lambda}(\Omega_0)$ are the canopy directional absorption and transmittance on black soil, $a_{S,\lambda}$ and $r_{S,\lambda}$ are the canopy absorption and reflectance caused by the anisotropic emission source at the bottom of the canopy, $\rho_{\text{eff}}(\lambda)$ is the effective reflectance of the surface at wavelength λ.

Tian et al. (2000) discussed the reasons why the 3D RT model fails in different regions and situations. They found that the retrieved LAI and FAPAR are effective only when the pixel spectral information falls within the spectral and angular space established by the lookup table. The quality of the retrieval results can be measured by the saturation frequency and the coefficient of variation (standard deviation divided by the mean). The smaller the saturation frequency and the coefficient of variation, the higher the quality of the results. For example, the forest has a high saturation frequency, but its coefficient of variation is small, so its data quality can meet the requirements.

Based on the results obtained by the RT algorithm, when LAI>5, the surface reflectance is insensitive to LAI/FAPAR due to LAI saturation, and this algorithm can only be used at this time. When the uncertainty of the input reflectance data is too large or the model BRF is incorrect due to the model construction error,

the 3D radiation transfer model algorithm fails, and the backup algorithm (LAI/FAPAR-NDVI empirical relationship) is adopted.

MODIS Collection 5 FAPAR products refine the algorithm to improve the quality of FAPAR retrieval. The original 6 vegetation type map were replaced with the new 8 vegetation type map. The broadleaf forest and needleleaf forest types were divided into two subcategories: deciduous forest and evergreen forest. The lookup table algorithm for the FAPAR was also refined. The new stochastic radiation transfer model can better express or demonstrate the inherent spatial heterogeneity of woody vegetation and the canopy structure. The new lookup table parameters will be set to retain the consistency between the model simulated and the measured surface reflectance. It will minimize retrieval anomalies (overestimation of LAI and failure of the retrieval algorithm in moderate or dense vegetation) and inconsistencies in LAI and FAPAR retrieval (in the sparse vegetation area, the LAI retrieval can be correct, while the FAPAR values are overestimated).

11.2.3 JRC_FAPAR product algorithm

The JRC_FAPAR product algorithm was developed by the European Commission Joint Research Center for European vegetation. The resolution of the FAPAR product is 10 km for the world and 2 km for Europe. The JRC_FAPAR algorithm is also based on a physical model to retrieve FAPAR. The FAPAR algorithm was based on the continuous vegetation canopy model (Gobron et al., 2006b), and the 6S model (Vermote et al., 1997). The FAPAR algorithm includes two steps: first, atmospheric correction is performed to eliminate the impact of the atmosphere and angle; second, mathematical methods are used to calculate the FAPAR value.

The JRC_FAPAR algorithm calculates FAPAR based on the adjusted spectral values. The equation is

$$FAPAR = g_0(\rho_{Rred}, \rho_{Rnir})$$

$$= \frac{l_{01}\rho_{Rnir} - l_{02}\rho_{Rred} - l_{03}}{(l_{04} - \rho_{Rred})^2 + (l_{05} - \rho_{Rnir})^2 + l_{06}} \quad (11.10)$$

where the coefficient l_{0m} (m = 1, 2, ..., 6) of the polynomial g_0 have been optimized a priori to force $g_0(\rho_{Rred}, \rho_{Rnir})$ to take on values as close as possible to the FAPAR associated with the plant canopy scenarios used in the training dataset. Once the coefficients are optimized for a specific sensor, then the inputs of the algorithm are the bidirectional reflectance factor values in the blue, red, and near-infrared bands and along different viewing angles.

11.2.4 Four-stream radiative transfer model

In addition to the development of new FAPAR retrieval models suitable for various land cover types, Tao et al. (2016) improved the accuracy of FAPAR estimates by using more accurate model inputs such as LAI and soil background and leaf-scattering albedo (Xiao et al., 2015b). The LAI is calculated by using a hybrid geometric-optic RT model considering the shadowing and multiple scattering in the canopy (Tao et al., 2009; Xu et al., 2009). The algorithm is introduced as follows.

In moderate-resolution images, vegetation pixels are almost continuously distributed across large regions. Therefore, Tao et al. (2016) assume that the land cover is horizontally homogeneous within the targeted surface and develop a four-stream RT model of continuous canopy for FAPAR retrieval. Canopy absorption along the direct- and diffuse-light penetrating paths are calculated separately and summed by using a ratio of scattering light. We denote T_0, T_f, and T_v as

the canopy transmittance along the direct-light penetrating, the diffuse-light penetrating, and the observing paths, respectively, and $\rho_{v,\lambda}$, $\rho_{g,\lambda}$, and $\rho_{c,\lambda}$ as the hemispherical albedo of vegetation, soil background, and leaf, respectively. FAPAR is calculated as the integral of canopy absorption in the upper hemisphere from 400 to 700 nm, in the following manner:

$$FAPAR = (1 - \beta) \int_{400}^{700} \int_0^{\frac{\pi}{2}} \left[(1 - T_0 - 2\rho_{v,\lambda}(\theta)) \right.$$

$$\left. + (1 - T_v(\theta) - 2\rho_{v,\lambda}(\theta)) \frac{T_0\rho_{g,\lambda}}{1 - \rho_{g,\lambda}\rho_{v,\lambda}(\theta)} \right]$$

$$\cos\theta \sin\theta d\theta d\lambda + \beta \int_{400}^{700} \int_0^{\frac{\pi}{2}} \left[(1 - T_f - 2\rho_{v,\lambda}(\theta)) \right.$$

$$\left. + (1 - T_v(\theta) - 2\rho_{v,\lambda}(\theta)) \frac{T_f\rho_{g,\lambda}}{1 - \rho_{g,\lambda}\rho_{v,\lambda}(\theta)} \right]$$

$$\cos\theta \sin\theta d\theta d\lambda$$

$$(11.11)$$

where the canopy transmittance along the direct-light penetrating, the diffuse-light penetrating, and the observing paths is follows:

$$T_{0,f,v} = \exp\left(-\lambda_0 \frac{G_{s,f,v}}{\mu_{s,f,v}} LAI \right) \quad (11.12)$$

and hemispherical albedo of vegetation is as follows:

$$\rho_{v,\lambda}(\theta) = \rho_{c,\lambda}\left[1 - \exp\left(-\lambda_0 \frac{G_v}{\mu_v(\theta)} \Gamma(\phi)LAI \right) \right]$$

$$+ \beta\rho_{c,\lambda}\left[\exp\left(-\lambda_0 \frac{G_v}{\mu_v(\theta)} \Gamma(\phi)LAI \right) \right.$$

$$\left. - \exp\left(-\lambda_0 \frac{G_v}{\mu_v(\theta)} LAI \right) \right]$$

$$(11.13)$$

In Eqs. (11.12) and (11.13), λ_0 is a Nilsson parameter accounting for the vegetation clumping effect; μ_s and $\mu_v(\theta)$ are the cosine values of solar (θ_s) and viewing (θ) zenith angles, respectively; β is the ratio of scattering light; and G_s and G_v are the mean projections of a unit foliage area along the solar and viewing directions, respectively (Liang, 2004; Ross, 1981):

$$G_{s,v} = \frac{1}{2\pi} \int_{2\pi} g_L(\Omega_L) |\Omega_L \cdot \Omega_{s,v}| d\Omega_L \quad (11.14)$$

where $1/2\pi \cdot g_L(\Omega_L)$ is the probability density of a distribution of leaf normals with respect to the upper hemisphere, i.e., leaf angle distribution.

The empirical function $\Gamma(\phi)$ in Eq. (11.13) describes the hot-spot phenomenon, where ϕ accounts for sun−target−sensor position and depends on the angle between solar and viewing directions and the leaf-angle distribution of the canopy.

$$\Gamma(\phi) = \exp\left(\frac{-\phi}{180 - \phi}\right) \quad (11.15)$$

The LAI is assumed known for FAPAR estimation, as a hybrid geometric-optic RT model for LAI retrieval has been previously developed (Tao et al., 2009; Xu et al., 2009). Other important inputs for FAPAR estimation are soil background and leaf-scattering albedo. Overall, the FAPAR estimation model described in Eqs. (11.11)−(11.15) accounts for reflective anisotropic characteristics caused by sun−target−sensor geometry, the vegetation clumping effect, and the hot-spot effect. In consideration of model simplicity and computational efficiency, it neglects reflective anisotropic characteristics caused by leaf and soil background. The parameters of the new FAPAR model are LAI, λ_0, G, $\rho_{c,\lambda}$, $\rho_{g,\lambda}$, and θ_s. The LAI is calculated from a hybrid geometric-optic RT model, θ_s is extracted from satellite data, and the other parameters (λ_0, G, $\rho_{c,\lambda}$ and $\rho_{g,\lambda}$) are from prior knowledge or locally applicable database. For simplicity, the model is referred to as the 4S model. The FAPAR estimated from the model is green FAPAR considering both direct and diffuse radiation. Most of the current

FAPAR products do not consider absorption by diffuse radiation, and no official green FAPAR product including both direct and diffuse radiation is available this far (Tao et al., 2015). Therefore, the study by Tao et al. (2016) serves as a good complement to the current FAPAR products.

11.2.5 GLASS FAPAR algorithm

Xiao et al. (2015) developed a simple scheme to calculate FAPAR values from Global LAnd Surface Satellite (GLASS) LAI product to ensure physical consistency between LAI and FAPAR retrievals.

$$FAPAR = 1 - \tau_{PAR} \quad (11.16)$$

The scheme uses only the transmittance of PAR down to the soil to calculate the approximate FAPAR. The radiation into the vegetation canopy includes direct and diffuse PAR. Therefore, the transmittance of the PAR down to the soil is further expressed as

$$\tau_{PAR} = \tau_{PAR}^{dir} - \left(\tau_{PAR}^{dir} - \tau_{PAR}^{dif}\right) \times f_{skyl} \quad (11.17)$$

where τ_{PAR}^{dir} and τ_{PAR}^{dif} are the fraction of the radiative flux originating from the direct illumination source and the transmitted fraction of the incident diffuse illumination source, respectively; f_{skyl} is the fraction of diffuse sky light and varies with aerosol optical depth, solar zenith angle, band wavelength, and aerosol model type. In our study, f_{skyl} was calculated using a lookup table established from the Second Simulation of the Satellite Signal in the Solar Spectrum (6S) code.

The canopy transmittance is closely related to solar zenith angle, the amount of diffuse radiation, and canopy clumping. If the leaf area index of a canopy is *lai* and the absorptivity of leaves for radiation is *a*, Campbell and Norman (1998) showed that the fraction of the total beam radiation (direct and down scattered) transmitted through the canopy can be approximated using an exponential model,

$$\tau_{PAR}^{dir} = e^{-\sqrt{a} \times k_c(\varphi) \times \Omega \times lai} \qquad (11.18)$$

where Ω is the clumping index, φ is the solar zenith angle, and $k_c(\varphi)$ is the canopy extinction coefficient for PAR. For an ellipsoidal leaf angle distribution, $k_c(\varphi)$ is calculated as follows:

$$k_c(\varphi) = \frac{\sqrt{x^2 + \tan^2(\varphi)}}{x + 1.774 \times (x + 1.182)^{-0.733}} \qquad (11.19)$$

where x is the ratio of average projected areas of canopy elements on horizontal and vertical surfaces. Diffuse radiation comes from all directions. Therefore, the diffuse transmission coefficient, τ_{PAR}^{dif}, can be calculated by integrating the direct transmission coefficient over all illumination directions,

$$\tau_{PAR}^{dif} = 2 \int_0^{\frac{\pi}{2}} \tau_{PAR}^{dir} \sin \varphi \cos \varphi d\varphi \qquad (11.20)$$

For the above scheme, LAI is an important input parameter to estimate FAPAR values. The GLASS LAI product was used to calculate the FAPAR values in this study. The clumping index is another input parameter. Based on the linear relationship between the clumping index and the normalized difference between hotspot and dark-spot indexes, He et al. (2012) employed the MODIS bidirectional reflectance distribution function parameter to derive a global clumping index map at 500 m resolution. In the present study, the MODIS-derived clumping index map was used to calculate canopy transmittance.

The scheme detailed above was used to generate GLASS FAPAR product from the GLASS LAI product.

11.3 FAPAR product intercomparison and validation

Some representative FAPAR products include MISR, MODIS, GLASS, AVHRR, SeaWiFS, MERIS, GEOV1, and JRC FAPAR products.

The most widely used FAPAR products are the AVHRR and MODIS global FAPAR products at 1 km resolution. In addition, the Canadian remote sensing center also established an FAPAR model based on the vegetation index, producing a national FAPAR map using AVHRR data every 10 days with a spatial resolution of 1 km (Chen, 1996). The European Commission Joint Research Center has developed JRC_FAPAR products for vegetation conditions in Europe. JRC_FAPAR has a resolution of 10 km for FAPAR products worldwide and 2 km for Europe.

Satellite FAPAR products have some differences in the definition of their products in terms of the whole canopy or green leaves, direct radiation only or not, and the imaging time. The MISR FAPAR product is the total FAPAR at 10:30 a.m., considering both direct and diffuse radiation absorbed by the whole canopy. The MODIS FAPAR considers only direct radiation, which may result in a smaller value than the MISR FAPAR product. The GLASS FAPAR product corresponds to the total FAPAR at 10:30 a.m. local time. The imaging time of the SeaWiFS sensor is approximately 12:05 p.m. local time, and its FAPAR product corresponds to the black-sky FAPAR (direct radiation only) by green elements. Similarly, the MERIS FAPAR product corresponds to the black-sky FAPAR by green elements at 10 a.m. local time. The GEOV1 FAPAR product corresponds to the instantaneous black-sky FAPAR by green parts around 10:15 a.m. local time. The SeaWiFS, MERIS, and GEOV1 FAPAR products take into account only the absorption by green elements, which may result in lower FAPAR values than the MISR and MODIS FAPAR products, which include the absorption of both green and nongreen elements. Overall, most of the satellite FAPAR products correspond to the instantaneous black-sky FAPAR around 10:15 a.m. which is a close approximation of the daily integrated value obtained by instantaneous FAPAR with the cosine of the solar zenith angle as the weight integral.

In terms of data requirements and processing algorithm, the MODIS FAPAR product uses the lookup table method built on the 3D stochastic RT model for different biomes from MODIS reflectance data (Myneni et al., 2002). MISR applies an RT model with inputs of LAI and soil reflectance without assumptions on biomes (Knyazikhin et al., 1998a). GEOV1 applies a neural network to relate the fused products to the top of canopy SPOT/VEGETATION reflectance (Baret et al., 2013). MERIS uses a polynomial formula based on 1D RT model (Gobron et al., 1999). Similar polynomial formula is used in the SeaWiFS FAPAR product as well (Gobron et al., 2000, 2006a). The main algorithms of MODIS and JRC FAPAR products are the physical model based on the RT. The empirical algorithm is only used as the backup algorithm when the main algorithm fails. The RT model is a relatively mature model based on physical optics and is widely used for large-scale FAPAR retrieval.

In this section, the MODIS, MERIS, MISR, SeaWiFS, and GEOV1 satellite FAPAR products are intercompared globally and over different land cover types in a 1-year period. Specifically, the spatial and seasonal distributions of the five satellite FAPAR products are intercompared globally in Section 11.3.1. The performances of the five satellite FAPAR products over different land cover types are intercompared in Section 11.3.2.

11.3.1 Intercomparison of FAPAR products over the globe

The spatial distribution of the five global FAPAR products during the period July 2005–June 2006 is depicted in Fig. 11.2. The MODIS global FAPAR product generally agrees well with the MISR and GEOV1 FAPAR product, while the MERIS and SeaWiFS FAPAR products agree well with each other. However, the difference between the group of MODIS, MISR, and GEOV1 FAPAR products and the group of MERIS and SeaWiFS FAPAR products is large (>0.1). The results are expected, and the primary reason is that both the SeaWiFS and the MERIS FAPAR products correspond to absorbed fluxes for green leaf single scattering, whereas the MODIS and MISR FAPAR products are based on a priori knowledge of leaf single scattering for each biome. The GEOV1 FAPAR correspond to a fused product which includes MODIS ones.

The seasonal distribution of the five preprocessed 0.5 degrees spatial resolution FAPAR products over the entire globe and the Northern and Southern Hemispheres with the same number of pixels are depicted in the panels of Fig. 11.3. The MODIS FAPAR values remain relatively stable globally from December to March, then increase at an accelerating rate from April to July, and finally decrease from August to the lowest values in December. The trend in the Northern Hemisphere is slightly different, where FAPAR remains relatively stable from January (instead of December globally) to March, then increases from April to July, and finally decreases from August to January (instead of December globally, 1 month longer). The reason is an increase in vegetation FAPAR values from December in the Southern Hemisphere, so that global FAPAR would drop to the lowest value in December even if northern hemispheric FAPAR drops to the lowest value in January. The MERIS, MISR, SeaWiFS, and GEOV1 global FAPAR values have similar trends as the MODIS global FAPAR values. Therefore, satellite FAPAR products agree well both globally and in the Northern Hemisphere in terms of trends. The differences of the mean values of the MODIS, MISR, and GEOV1 FAPAR products at the global scale are very small (<0.05 generally). The difference of the standard deviations of MODIS and MISR is less than 0.02. The mean values of the MERIS and SeaWiFS FAPAR products differ within 0.05 and the standard deviations differ within 0.015. However, the MODIS, MISR, and GEOV1 global FAPAR

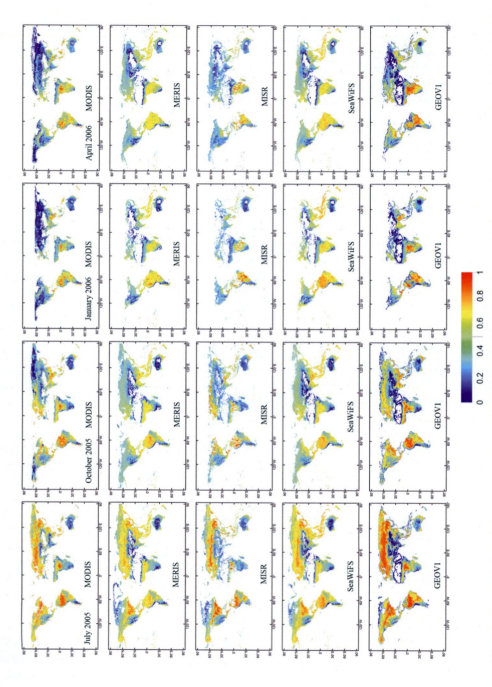

FIGURE 11.2   The MODIS, MERIS, MISR, SeaWiFS, and GEOV1 global FAPAR distributions in Plate-carrée projection during the period July 2005–June 2006 (every 3 months). Note the agreements among the MODIS, MISR, and GEOV1 FAPAR products and between the MERIS and SeaWiFS FAPAR products. However, the MODIS, MISR, and GEOV1 FAPAR values were consistently higher than the MERIS and SeaWiFS FAPAR values (Tao et al., 2015).

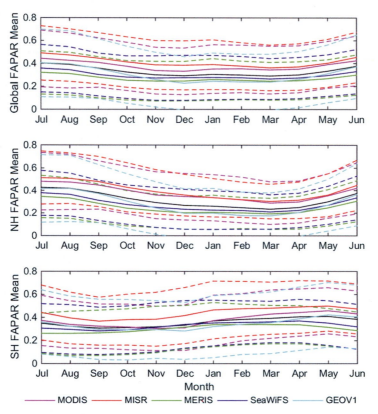

FIGURE 11.3   The global, northern hemispheric, and southern hemispheric mean of quality-controlled MODIS, MISR, MERIS, SeaWiFS, and GEOV1 FAPAR products during the period July 2005–June 2006. The black curve is all five products' mean. The dashed curves correspond to the mean ± standard deviation of each product (Tao et al., 2015).

values are 0.05–0.1 higher than the average of the five products; whereas the MERIS and Sea-WiFS global FAPAR values are 0.05–0.1 lower than the average in terms of magnitudes. Absolute FAPAR values are on average in decreasing order from MISR to MODIS to GEOV1 to Sea-WiFS and MERIS (McCallum et al., 2010).

Compared with the FAPAR trends in the Northern Hemisphere, opposite situations are found in the FAPAR trends in the Southern Hemisphere. The MODIS and GEOV1 southern hemispheric FAPAR remain relatively stable from August to November, then increase to the highest values in May, and finally drop to the lowest values in November. The MISR southern hemispheric FAPAR has similar trend as

the MODIS and GEOV1 southern hemispheric one, except that it drops to the lowest values near September instead of November. The MERIS southern hemispheric FAPAR is slightly different from the MODIS and MISR one. It remains relatively stable from July to September, then increases to the highest values in February, and finally drops to the lowest values near August (3 month variation from MODIS). The SeaWiFS southern hemispheric FAPAR remains relatively stable from July to September, then increases to the highest values in April, and finally drops to the lowest values in September (same as MISR). Overall, southern hemispheric FAPAR remains relatively stable from August to November, then increases to the highest values in April or May, and finally drops to the lowest values between September and November. The increased disparity among products in the Southern Hemisphere is likely a result of fewer vegetation samples there, which is explored in detail for different land covers in Section 11.3.2.

The quality flags of MODIS FAPAR data with nonfill values are analyzed to select maps in high-quality month for further comparisons. The statistics of MODIS Collection 5 FAPAR quality control flags are depicted globally and in the Northern and Southern Hemispheres

(Fig. 11.4). The percentage of the main algorithm retrievals increases in the middle of the growing season and reaches the highest value in September. The percentage of backup retrievals due to bad geometry increases in the winter as expected because of the larger solar zenith angle. This kind of backup retrieval related to bad geometry lasts 6 months, from October to March, both globally and in the Northern Hemisphere and approximately 3 months, from May to July, in the Southern Hemisphere. Overall, the analysis on the MODIS quality flags shows that

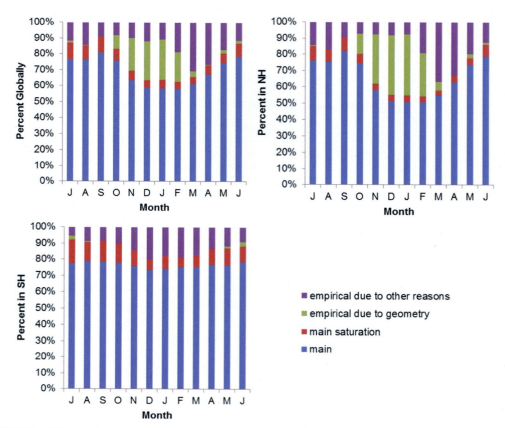

FIGURE 11.4   MODIS collection 5 FAPAR QC statistics globally, in the Northern Hemisphere, and the Southern Hemisphere: the percentage of main algorithm retrievals (blue), the percentage of main algorithm under conditions of saturation (red), the percentage of backup (i.e., NDVI-based) retrievals associated with bad geometry (green), the percentage of pixels using the backup algorithm due to reasons other than geometry (purple). Note the overall increase in high-quality (main algorithm) retrievals during the middle of the growing season (Tao et al., 2015).

the quality of satellite FAPAR products is better in the vegetation growing season than other season.

The difference maps between satellite FAPAR products in July are depicted in Fig. 11.5. The sea/land mask is applied and only pixels with high-quality values from all of the five satellite FAPAR products are included in the difference maps. The MISR FAPAR product exhibits some higher FAPAR values than the MERIS and SeaWiFS FAPAR products at high latitudes, and some slightly lower FAPAR values in the tropical forests near the equator. The difference between the MERIS and SeaWiFS FAPAR products is very small, with a few pixels located along the boundaries of continents. The difference between the MISR and MODIS FAPAR products is quite small as well, with only a few scatters in the boreal forests of Asia and North America. The MISR and MODIS FAPAR products are close to the GEOV1 FAPAR product, except some boundary regions. However, the MODIS FAPAR values are apparently higher than the MERIS and SeaWiFS FAPAR values over the boreal forests and savannahs. The GEOV1 FAPAR product is consistently higher than the MERIS and SeaWiFS FAPAR products over the tropical and boreal forests.

The five global FAPAR datasets are averaged per grid cell and then subtracted from each dataset to obtain the difference to the mean maps (Fig. 11.6). The MODIS, MISR, and GEOV1 FAPAR products have larger values than the average in the boreal and tropical forests and grasslands in the Northern Hemisphere. The GEOV1 FAPAR products are closest to the average of all the products. The MERIS and SeaWiFS FAPAR products have apparently lower than the average values in the forests, savannahs, and grasslands. The differences to

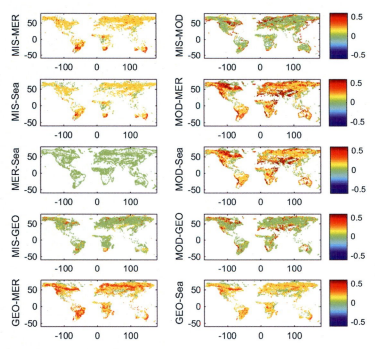

FIGURE 11.5   Global FAPAR difference maps between the MODIS, MISR, GEOV1, MERIS, and SeaWiFS products in July 2005 (MIS: MISR, MER: MERIS, MOD: MODIS, Sea: SeaWiFS, Geo: GEOV1).

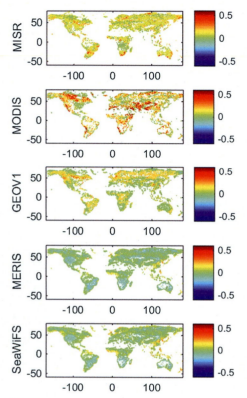

FIGURE 11.6   Maps of the five global FAPAR datasets in July 2005, with the mean of all five products per grid-cell subtracted from each dataset (Tao et al., 2015).

the mean maps are averaged across different latitudes (Fig. 11.7). Their differences are smaller at low and high latitudes but are larger at middle latitudes, especially in the Southern Hemisphere. The possible reason is the saturation of FAPAR values in the tropical forests and the scarcity of vegetation in the high latitudes so that the differences are smaller in these regions.

11.3.2 Intercomparisons over different land cover types

The MODIS global land cover map (MCD12) during the period July 2005–June 2006 is depicted in Fig. 11.8. The vegetated areas are classified by use of the MODIS-derived LAI/

FAPAR scheme into eight land cover types: broadleaf evergreen forest, broadleaf deciduous forest, needleleaf evergreen forest, needleleaf deciduous forest, crop, grass, savannah, and shrubland (Myneni et al., 2002). The MCD12 land cover classification product was resampled into 0.5 degrees using the mode resampling method by selecting the value which appears most often of all the sampled points. Most of the vegetated areas are located in the Northern Hemisphere. The only exception is the broadleaf evergreen forests, the majority of which are located in the Southern Hemisphere, including the northwest part of South America, part of Central Africa, and the southern part of Southeast Asia.

The mean of all five products is averaged globally and in the Northern and Southern Hemispheres during the period July 2005–June 2006 to show their seasonal patterns at the three scales (Fig. 11.9). The trend of northern hemispheric FAPAR is similar to that of global FAPAR, with slight difference in the magnitudes (Fig. 11.9). The explanation is that the majority of the land cover is located in the Northern Hemisphere, resulting in the dominant influence of northern hemispheric FAPAR on global FAPAR. The exceptions are the FAPAR over savannah and broadleaf evergreen forest land covers. The global FAPAR mean over savannah remains almost constant throughout the year, but the northern hemispheric FAPAR mean is a sine curve, with the highest value in September and the lowest value between February and March. There is an opposite trend in the Southern Hemisphere, and the two trends cancel each other out globally. The global FAPAR mean over broadleaf evergreen forest is stabilized throughout the year, but the northern hemispheric FAPAR is a sine curve. In this case, the curve of the global FAPAR mean is similar to the curve in the Southern Hemisphere, because the majority of broadleaf evergreen forests are located in the Southern Hemisphere as noted.

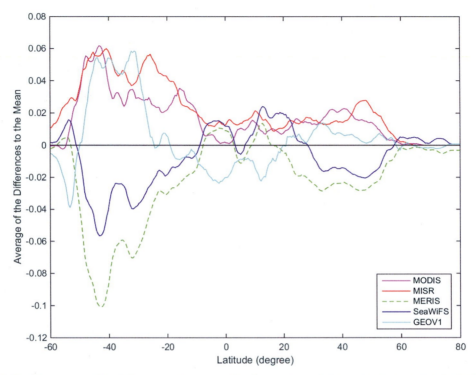

FIGURE 11.7 The average of the difference to the mean of the five products at different latitudes in July 2005. The black line is for reference (Tao et al., 2015).

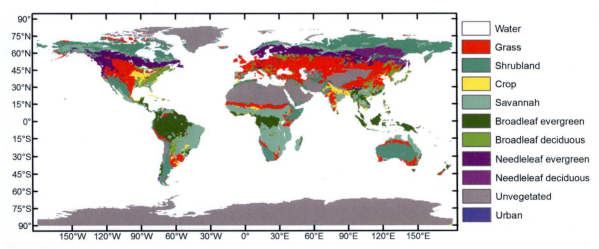

FIGURE 11.8 The resampled MODIS global land cover map (MCD12) at 0.5 degrees during the period July 2005—June 2006. The vegetated areas are classified by use of the MODIS-derived LAI/FAPAR scheme into eight land cover types: broadleaf evergreen forest, broadleaf deciduous forest, needleleaf evergreen forest, needleleaf deciduous forest, crop, grass, savannah, and shrubland. The map also includes the unvegetated, water, and urban area (Tao et al., 2015).

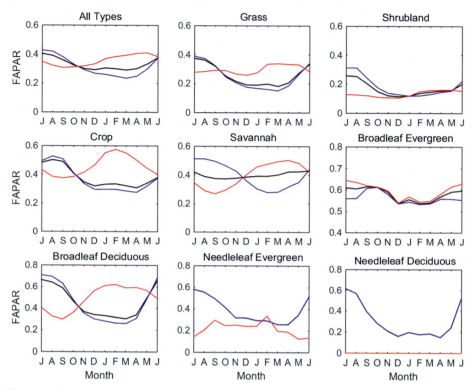

FIGURE 11.9 The global (black), northern hemispheric (blue), and southern hemispheric (red) FAPAR mean of all five products over different land cover types during the period July 2005—June 2006 (Tao et al., 2015).

Compared with the trends of the northern hemispheric FAPAR mean, opposite trends are found in the southern hemispheric FAPAR mean. The opposite relations are very apparent globally, over crop, savannah, grass, broadleaf deciduous forest, and needleleaf evergreen forest. The opposite relations are not apparent over shrubland and broadleaf evergreen forest, where the southern hemispheric FAPAR is stable throughout the year, but the northern hemispheric FAPAR mean has a parabolic shape over shrubland and a sine curve over broadleaf evergreen forest. The global FAPAR curve overlaps with the northern hemispheric FAPAR curve over needleleaf evergreen forests, provided that only a few needleleaf evergreen forests are in the Southern Hemisphere. Barely any needleleaf deciduous forests are in the

Southern Hemisphere. Both the northern hemispheric and the global FAPAR mean have bowl-like shapes over needleleaf deciduous forests throughout the year.

The time series of the mean of the MISR, MODIS, GEOV1, SeaWiFS, and MERIS FAPAR products over different land cover types during the period July 2005—June 2006 are depicted in Fig. 11.10, with the mean of all five products subtracted from each dataset. The MODIS and MISR FAPAR products are approximately 0.05—0.1 higher than the average of the five products, and the MERIS and SeaWiFS FAPAR products are approximately 0.05—0.1 lower than the average of the five products. The GEOV1 FAPAR product has very small difference (<0.05) to the mean over grass, shrubland, crop, and savannah. The deviations to the

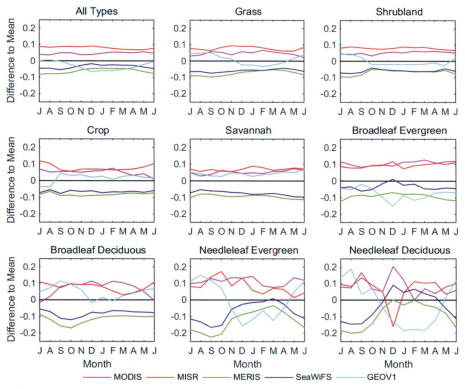

FIGURE 11.10 The time series of the mean of quality controlled MODIS, MISR, MERIS, SeaWiFS, and GEOV1 FAPAR products over different land cover types during the period July 2005—June 2006, with the mean of all five products subtracted from each dataset. The black line is for reference (Tao et al., 2015).

mean for the five products remain stable over grass, shrubland, crops, savannah, and broadleaf evergreen forests throughout the year. However, a different situation occurs over broadleaf deciduous forests, where the deviations are largest in October and smallest in June and July. The deviations of the five products from the average over needleleaf evergreen and needleleaf deciduous forests are largest in September and October, and gradually decrease to the lowest values in March. The GEOV1 FAPAR product has large fluctuations over needleleaf evergreen and needleleaf deciduous forests because of its strong seasonal pattern over the needleleaf forests with a standard deviation of 0.21, compared with standard deviations around 0.11 for other FAPAR products. In such case, it

fluctuates both above and below the average line, although it has similar seasonality as other products as shown in Fig. 11.3. The MISR FAPAR product has a drop in the value over needleleaf deciduous forest in December because of no data. Overall, the differences between the products are consistent throughout the year over most of the land cover types, except over the forests. The possible reason can be traced to the different assumptions in the retrieval algorithms over forests and the large differences between green and total FAPAR products due to tree trunks and branches absorption (Pickett-Heaps et al., 2014). Interestingly, the differences between the products do not fluctuate much in broadleaf evergreen forests over time because FAPAR values remain relatively stable all year long and therefore the differences between the products are small and consistent over broadleaf evergreen forests.

11.3.3 Comparison with FAPAR values derived from high-resolution reference maps

The GLASS, MODIS, GEOV1, and SeaWiFS FAPAR products were compared with high-resolution FAPAR maps to evaluate differences in FAPAR magnitude between the products. 27 high-resolution FAPAR maps over 22 sites from the Validation of Land European Remote sensing Instrument (VALERI) project were collected to validate the accuracy of these FAPAR products. FAPAR ground measurements at the VALERI sites were calculated from digital hemispherical photos. The high-resolution FAPAR maps were derived from the determination of the transfer function between the reflectance values of the high spatial resolution satellite imagery and the FAPAR ground measurements.

The high-resolution FAPAR maps were aggregated over 3×3 km regions centered on the location of the validation sites using spatial averaging to validate the GLASS, MODIS, GEOV1, and SeaWiFS FAPAR products. The characteristics of the validation sites and associated mean values of the high-resolution FAPAR maps over 3×3 km regions centered on the location of the sites are shown in Table 11.1.

Scatterplots of the FAPAR products versus the mean values of the high-resolution FAPAR maps are shown in Fig. 11.11. All the FAPAR products underestimate high FAPAR values and overestimate low FAPAR values compared with the mean values of the high-resolution FAPAR maps. GLASS FAPAR product provide the greatest accuracy ($R^2 = 0.9292$ and RMSE $= 0.0716$) against the mean values of the high-resolution FAPAR maps compared with GEOV1 ($R^2 = 0.8681$ and RMSE $= 0.1085$),

MODIS ($R^2 = 0.8048$ and RMSE $= 0.1276$), and SeaWiFS FAPAR products ($R^2 = 0.7377$ and RMSE $= 0.1635$).

Compared with MODIS, GEOV1, and SeaWiFS FAPAR values, those of GLASS are distributed more closely around the 1:1 line against the mean values of the high-resolution FAPAR maps, showing that GLASS FAPAR product achieves better agreement with the mean values of the high-resolution FAPAR maps across the FAPAR range than the other products.

11.4 Spatiotemporal analysis and applications

The FAPAR estimation results are validated at the site scale, and the method is applied further at the regional scale. Multiple satellite data with different spatial resolutions are used to estimate the FAPAR values for analysis across scales. Multiple satellite data with different spatial resolutions are used to estimate the FAPAR values for analysis across scales. The AmeriFlux sites can be used for temporal validation of FAPAR estimates, due to their continuous FAPAR measurements. The geolocation and land cover information of the AmeriFlux sites are listed in Table 11.2. Validation results show that the estimates have an uncertainty of 0.08 (Tao et al., 2016).

Two study regions covering four AmeriFlux sites in the United States are selected. The MODIS tiles and the MISR and Landsat orbits covering the two study regions are listed in Table 11.3. The temporal resolutions of the MISR, MODIS, and Landsat TM/ETM + reflectance or FAPAR products are 2–9 days, 8 days, and 16 days, respectively. The MISR, MODIS, and Landsat scenes around the four AmeriFlux sites in the vegetation growing season are carefully selected to have close imaging dates and high-quality data without cloud contamination. The imaging dates of the products in each case differ within 4 days (Table 11.3). We assume that the

TABLE 11.1   Characteristics of the 22 validation sites (Xiao et al., 2015a).

Site name	Country	Latitude (°)	Longitude (°)	Biome type	DOY	Year	Mean FAPAR[a]	Uncertainties of FAPAR
Alpilles2	France	43.810	4.715	Broadleaf crops	204	2002	0.399	0.292
Barrax	Spain	39.057	−2.104	Broadleaf crops	194	2003	0.256	0.333
Cameron	Australia	−32.598	116.254	Broadleaf forest	63	2004	0.479	0.109
Concepcion	Chile	−37.467	−73.470	Broadleaf forest	9	2003	0.771	0.197
Counami	French Guyana	5.347	−53.238	Broadleaf forests	269	2001	0.95	0.006
					286	2002	0.887	0.005
Demmin	Germany	53.892	13.207	Broadleaf crops	164	2004	0.741	0.207
Donga	Benin	9.770	1.778	Shrubs	172	2005	0.472	0.159
Fundulea	Romania	44.406	26.583	Grasses and cereal crops	128	2001	0.519	0.370
					160	2002	0.464	0.269
					151	2003	0.374	0.221
Gilching	Germany	48.082	11.320	Grasses and cereal crops	199	2002	0.786	0.201
Gnangara	Australia	−31.534	115.882	Broadleaf forest	61	2004	0.263	0.058
Haouz	Morocco	31.659	−7.600	Shrubs	71	2003	0.489	0.252
Laprida	Argentina	−36.990	−60.553	Savannahs	311	2001	0.837	0.102
					292	2002	0.62	0.040
Larose	Canada	45.380	−75.217	Needleleaf forests	219	2003	0.906	0.080
Larzac	France	43.938	3.123	Savannahs	183	2002	0.349	0.059
Nezer	France	44.568	−1.038	Needleleaf forests	107	2002	0.494	0.269
Plan-de-Dieu	France	44.199	4.948	Broadleaf crops	189	2004	0.223	0.120
Puéchabon	France	43.725	3.652	Broadleaf forests	164	2001	0.601	0.157
Sonian	Belgium	50.768	4.411	Needleleaf forests	174	2004	0.916	0.036
Sud-Ouest	France	43.506	1.238	Grasses and cereal crops	189	2002	0.404	0.258
Turco	Bolivia	−18.239	−68.193	Shrubs	240	2002	0.025	0.013
					105	2003	0.046	0.016
Wankama	Niger	13.645	2.635	Grasses and cereal crops	174	2005	0.073	0.057
Zhangbei	China	41.279	114.688	Grasses and cereal crops	221	2002	0.422	0.143

[a] *The FAPAR ground measurements correspond to the fraction of intercepted PAR.*

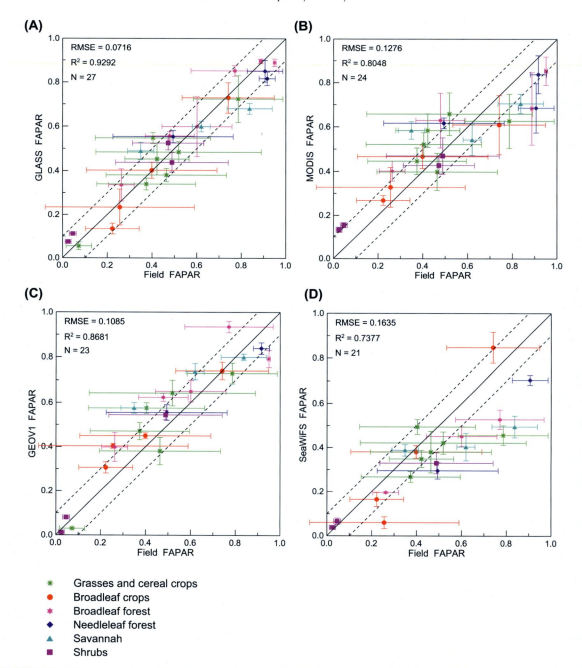

FIGURE 11.11   Scatterplots of (A) GLASS, (B) MODIS, (C) GEOV1, and (D) SeaWiFS FAPAR products versus mean values of the high-resolution FAPAR maps. The R-squared and RMSE values are also shown. N is the number of matched data pairs for each case (Xiao et al., 2015a).

TABLE 11.2   List of the AmeriFlux experimental sites in the United States used in this study.

Site	State	Latitude (°)	Longitude (°)	Land cover
Mead irrigated	Nebraska	41.1651	−96.4766	Crops
Mead irrigated rotation	Nebraska	41.1649	−96.4701	Crops
Mead rainfed	Nebraska	41.1797	−96.4396	Crops
Bartlett	New Hampshire	44.0646	−71.2881	Deciduous broadleaf forests

TABLE 11.3   The spatial coverages and imaging dates of the MODIS, MISR and Landsat data used in the two case studies (Tao et al., 2016).

Case	MODIS tile	MISR orbit	Landsat orbit	MODIS date	MISR date	Landsat date
Case 1	H10V04	P27B58	P28R31	Aug 5−12, 2006	Aug 4, 2006	Aug 3, 2006
Case 2	H12V04	P12B55	P12R29	Aug 5−12, 2005	Aug 8, 2005	Aug 8, 2005

[a]Case 1 covers three sites: Mead Irrigated, Mead Irrigated Rotation, and Mead Rainfed. Case 2 covers the Bartlett site. The "H" and "V" of MODIS tile means horizontal and vertical, respectively. The "P" and "B" of MISR orbit means path and block, respectively. The "P" and "R" of Landsat orbit means path and row, respectively. Tao X., Liang S.L., He T. and Jin H.R., Estimation of fraction of absorbed photosynthetically active radiation from multiple satellite data: Model development and validation, Remot Sens Environ 184, 2016, 539−557.

vegetation remains unchanged within such a short period; therefore, the intercomparison of FAPAR among different sensors is reliable.

The Landsat reflectance data are atmospherically corrected by using the Landsat Ecosystem Disturbance Adaptive Processing System (LEDAPS) preprocessing code (Masek et al., 2006). The missing scan lines in the ETM + -image are filled with values of nearest pixels. The Landsat TM and ETM + surface reflectance scenes are used for estimating FAPAR at a spatial resolution of 30 m. The MISR and MODIS surface reflectance products (MISR L2 and MOD09) are directly used for estimating FAPAR at spatial resolutions of 1 km and 500 m. The MISR and the MODIS FAPAR products (MISR L2 and MOD15) are intended for intercomparison with the FAPAR estimates from this study.

The MODIS FAPAR product uses MCD12 land cover product to distinguish 13 land covers globally. The National Land Cover Database 2006 (NLCD 2006) uses a 16-class land cover classification scheme for Landsat images. A combined land cover classification scheme of the two is used here considering the existing land cover types in the two study regions. Consequently, the MISR, MODIS, and Landsat images are classified into evergreen forest, deciduous forest, urban, grass, crops, barren soil, and water body. The presented 4S model is applied on the surface reflectance and the classified images to estimate the vegetation LAI and FAPAR values. The distributions of the FAPAR estimates from the MISR, MODIS, and Landsat images in Case 1 are shown in Fig. 11.12A−C, respectively. For comparison, the MISR and the MODIS FAPAR products are shown in Fig. 11.12D−E, respectively. The distributions of the FAPAR estimates in Case 2 are shown in Fig. 11.13A−C. For comparison, the MISR and the MODIS FAPAR products are shown in Fig. 11.13D−E, respectively. The MISR FAPAR product is consistently higher (>0.15) than the MODIS FAPAR product in Case 1, and the MODIS and the MISR FAPAR products agree well in Case 2. However, the FAPAR estimates from 4S model are consistent across different scales in both cases. The values have similar distribution patterns across scales, in which the highest values are observed in evergreen forests, followed by deciduous forests,

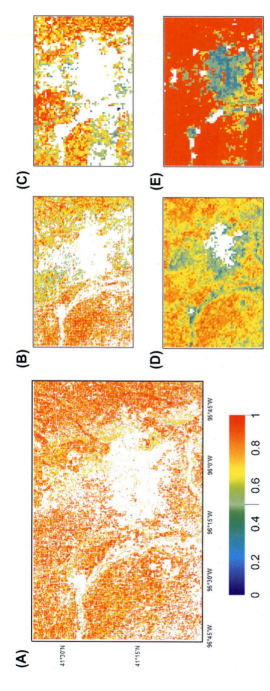

FIGURE 11.12   The FAPAR maps derived from the TM, MODIS, and MISR scenes in the Mead study region in Case 1. (A)–(C) show the TM, MODIS, and MISR FAPAR estimates from 4S model. (D) and (E) show the MODIS and the MISR FAPAR products (Tao et al., 2016).

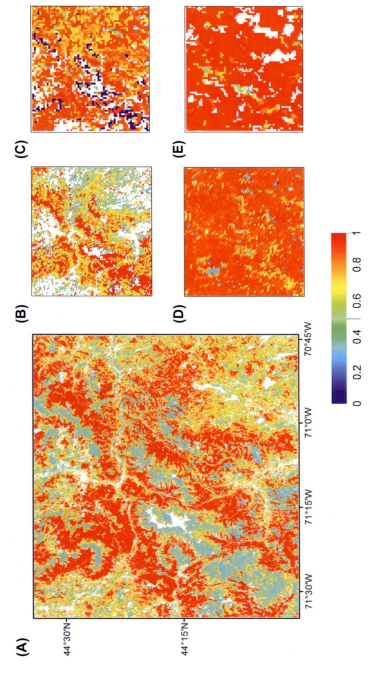

FIGURE 11.13   The FAPAR maps derived from the ETM+, MODIS, and MISR scenes in the Bartlett region in Case 2. (A)–(E) show the ETM+, MODIS, and MISR FAPAR products (Tao et al., 2016). and MISR FAPAR estimates from 4S model. (D) and (E) show the MODIS and MISR FAPAR products (Tao et al., 2016).

crops, and rivers and central urban areas, where the FAPAR estimates are close to zero.

The frequency histograms of the MODIS and the MISR FAPAR products are shown in red and blue bars in Fig. 11.14B and D for Cases 1 and 2, respectively. The MISR FAPAR product has a larger mean (by 0.15) and standard deviation (by 8%) than the MODIS FAPAR product in Case 1, because more pixels with FAPAR values greater than 0.9 are observed in the MISR image than in the MODIS image. The frequency histograms of the MISR and the MODIS FAPAR products agree well in Case 2. The difference between the mean values of the MISR and the MODIS FAPAR products is about 0.05. Overall,

the agreements between the FAPAR products differ in the two regions possibly due to the difference in land cover composition.

The frequency histograms of the FAPAR estimates from the Landsat, MODIS, and MISR reflectance images are shown in green, red, and blue bars in Fig. 11.14A and C for Cases 1 and 2, respectively. Generally, the agreements among the Landsat, MODIS, and MISR FAPAR estimates are reasonably good. The mean values of the FAPAR estimates differ within 0.1, and the standard deviations differ within 0.03 for both cases. Therefore, the FAPAR estimates by the 4S retrieval model have better performance than the MODIS and MISR products regarding

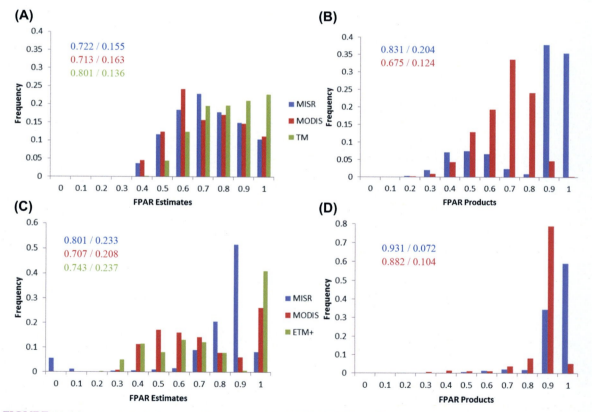

FIGURE 11.14   The FAPAR frequency histograms of the MISR, MODIS, and Landsat scenes in the study region in Case 1 and Case 2. (A) The MISR, MODIS, and ETM + FAPAR estimates from 4S model in Case 1. (B) The MISR and MODIS FAPAR products in Case 1. (C) The MISR, MODIS, and ETM + FAPAR estimates from 4S model in Case 2. (D) The MISR and MODIS FAPAR products in Case 2. The numbers are the regional mean and standard deviation (Tao et al., 2016).

consistency across scales. The comparable results between the estimates from 4S model and the current products indicate that the retrieval algorithms of FAPAR products can partially justify the differences in their data distributions; thus, the FAPAR values from different satellites agree better when using the same algorithm for retrieval (Seixas et al., 2009). Additionally, Tao et al. (2016) provide FAPAR estimates at multiple resolutions ranging from 30 m, 500 m, to 1 km, whereas the existing MODIS and MISR FAPAR products are both at 1 km.

11.5 Summary

This chapter introduces the related concepts of FAPAR, the FAPAR empirical models, the quantitative model, and retrieval method based on the RT, the intercomparison, and validation of the main FAPAR products. Finally, we use AmeriFlux experimental area as an example and present FAPAR retrieval case studies using a physically based model. As described above, the FAPAR retrieval from remotely sensed images has made great progress and there are many mature remote sensing products. The empirical retrieval algorithm serves as an example for generating such a product, but its applications are limited. The physically based retrieval method considers multiple scattering of sunlight in the canopy using the RT mechanism; therefore, it is more universal and useful.

The intercomparison of satellite FAPAR products shows that the seasonality of the products agrees better with each other in the Northern Hemisphere and globally than in the Southern Hemisphere. The differences between the products are consistent throughout the year over most of the land cover types, except over the forests. Possible reasons can be traced back to the different assumptions in the retrieval algorithms over forests and the differences between green FAPAR and total FAPAR products due to tree trunk and branch absorption.

To accurately calculate key parameters such as light energy utilization and provide input parameters for land surface process models, the FAPAR should also be obtained under cloudy conditions. Solar radiation is divided into direct and diffuse radiation by the atmospheric transmission. The directions of the direct and diffuse radiation reaching the surface are different, as are the directions in which they are transmitted within the vegetation. Therefore, the expression should be different in the RT model. At present, most FAPAR retrieval algorithms do not separate the direct solar radiation from diffuse radiation, which inevitably underestimates the effect of diffuse radiation on FAPAR. The 4S model takes into account the absorption of direct and diffuse radiation and is a good complement to existing products. Additionally, it provides FAPAR estimates at three scales of 30 m, 500 m, and 1 km, as a complement to the MODIS and the MISR FAPAR products available at 1 km.

Of course, there are still some areas for improvement in FAPAR products. For example, the MERIS, MODIS, MISR, and GEOV1 FAPAR products have an uncertainty of 0.14 validating with total FAPAR measurements, and 0.09 validating with green FAPAR measurements. At present, the uncertainty of most FAPAR products is within ± 0.1, which cannot meet the threshold accuracy requirement of ± 0.05 proposed by GCOS. Further improvements include reducing the uncertainty of observations and combining multiple observations, taking into account the scale differences between ground observations and moderate-resolution pixels, and so on.

It is also important to improve the accuracy of the model input parameters. Many FAPAR products are based on LAI products, so in addition to developing a more accurate FAPAR model, improving the accuracy of model parameters, especially the accuracy of LAI, is also important to improve the accuracy of FAPAR.

Significant efforts must be devoted to reach Stage 3 of the validation. Product accuracy must be thoroughly assessed, and the uncertainties in the product must be well established via independent measurements made in a systematic and statistically robust way that represents global conditions. Most products currently assume homogeneous landscapes and perform better over homogeneous land cover. Future study would be focused on developing an advanced FAPAR model to improve accuracy over heterogeneous landscapes.

References

Asner, G.P., Wessman, C.A., Archer, S., 1998. Scale dependence of absorption of photosynthetically active radiation in terrestrial ecosystems. Ecol. Appl. 8, 1003–1021.

Asrar, G., Fuchs, M., Kanemasu, E., Hatfield, J., 1984. Estimating absorbed photosynthetic radiation and leaf area index from spectral reflectance in wheat. Agron. J. 76, 300–306.

Asrar, G., Myneni, B.J., Choudhury, B.J., 1992. Spatial heterogeneity in vegetation canopies and remote sensing of absorbed photosyntheticaly active radiation: a modeling study. Remote Sens. Environ. 41, 85–103.

Baret, F., Hagolle, O., Geiger, B., Bicheron, P., Miras, B., Huc, M., Berthelot, B., Nino, F., Weiss, M., Samain, O., Roujean, J.L., Leroy, M., 2007. LAI, fAPAR and fCover CYCLOPES global products derived from VEGETATION - Part 1: principles of the algorithm. Remote Sens. Environ. 110, 275–286.

Baret, F., Weiss, M., Lacaze, R., Camacho, F., Makhmara, H., Pacholcyzk, P., Smets, B., 2013. GEOV1: LAI and FAPAR essential climate variables and FCOVER global time series capitalizing over existing products. Part1: principles of development and production. Remote Sens. Environ. 137, 299–309.

Bonan, G.B., Oleson, K.W., Vertenstein, M., Levis, S., Zeng, X., Dai, Y., Dickinson, R.E., Yang, Z.L., 2002. The land surface climatology of the community land model coupled to the NCAR community climate model. J. Clim. 15, 3123–3149.

Camacho, F., Cemicharo, J., Lacaze, R., Baret, F., Weiss, M., 2013. GEOV1: LAI, FAPAR essential climate variables and FCOVER global time series capitalizing over existing products. Part 2: validation and intercomparison with reference products. Remote Sens. Environ. 137, 310–329.

Campbell, S.G., Norman, J.M., 1998. An Introduction to Environmental Biophysics, second ed. Springer-Verlag, New York, NY.

Carrer, D., Roujean, J.L., Lafont, S., Calvet, J.C., Boone, A., Decharme, B., Delire, C., Gastellu-Etchegorry, J.P., 2013. A canopy radiative transfer scheme with explicit FAPAR for the interactive vegetation model ISBA-A-gs: impact on carbon fluxes. J. Geophys. Res. Biogeosci. 118, 888–903.

Casanova, D., Epema, G.F., Goudriaan, J., 1998. Monitoring rice reflectance at field level for estimating biomass and. LAI Field Crops Research 55, 83–92.

Chen, J.M., 1996. Canopy Architecure and remote sensing of the fraction of photosynthetically active radiation absorbed by boreal forests. IEEE Trans. Geosci. Remote Sens. 34, 1353–1368.

Dawson, T.P., North, P.R.J., Plummer, S.E., Curran, P.J., 2003. Forest ecosystem chlorophyll content: implications for remotely sensed estimates of net primary productivity. Int. J. Remote Sens. 24, 611–617.

Fang, H., Wei, S., Liang, S., 2012. Validation of MODIS and CYCLOPES LAI products using global field measurement data. Remote Sens. Environ. 119, 43–54.

Fensholt, R., Sandholt, I., Rasmussen, M.S., 2004. Evaluation of MODIS LAI, fAPAR and the relation between fAPAR and NDVI in a semi-arid environment using in situ measurements. Remote Sens. Environ. 91, 490–507.

Friedl, M.A., 1997. Examining the effects of sensor resolution and sub-pixel heterogeneity on vegetation spectral indices: implications for biophysical modeling. In: Quattrochi, D.A., Goodchild, M.F. (Eds.), Scale in Remote Sensing and GIS. Lewis, Boca Raton, Fla, pp. 113–139.

GCOS, 2011. Systematic Observation Requirements for Satellite-Based Data Products for Climate, pp. 79–83.

Gobron, N., Pinty, B., Aussedat, O., Chen, J.M., Cohen, W.B., Fensholt, R., Gond, V., Huemmrich, K.F., Lavergne, T., Melin, F., Privette, J.L., Sandholt, I., Taberner, M., Turner, D.P., Verstraete, M.M., Widlowski, J.L., 2006a. Evaluation of fraction of absorbed photosynthetically active radiation products for different canopy radiation transfer regimes: methodology and results using Joint Research Center products derived from SeaWiFS against ground-based estimations. J. Geophys. Res. Atmosp. 111.

Gobron, N., Pinty, B., Taberner, M., Melin, F., Verstraete, M.M., Widlowski, J.L., 2006b. Monitoring the photosynthetic activity of vegetation from remote sensing data. Adv. Space Res. 38, 2196–2202.

Gobron, N., Pinty, B., Verstraete, M., Govaerts, Y., 1999. The MERIS global vegetation index (MGVI): description and preliminary application. Int. J. Remote Sens. 20, 1917–1927.

Gobron, N., Pinty, B., Verstraete, M.M., Widlowski, J.L., 2000. Advanced vegetation indices optimized for up-coming sensors: design, performance, and applications. IEEE Trans. Geosci. Remote Sens. 38, 2489–2505.

Goward, S.N., Huemmrich, K.F., 1992. Vegetation canopy PAR absorptance and the normalized difference vegetation index: an assessment using the SAIL model. Remote Sens. Environ. 39, 119–140.

He, L.M., Chen, J.M., Pisek, J., Schaaf, C.B., Strahler, A.H., 2012. Global clumping index map derived from the MODIS BRDF product. Remote Sens. Environ. 119, 118−130.

Houghton, R.A., 1995. Land-use change and the carbon-cycle. Glob. Chang. Biol. 1, 275−287.

Hu, J.N., Su, Y., Tan, B., Huang, D., Yang, W.Z., Schull, M., Bull, M.A., Martonchik, J.V., Diner, D.J., Knyazikhin, Y., Myneni, R.B., 2007. Analysis of the MISR LA/FPAR product for spatial and temporal coverage, accuracy and consistency. Remote Sens. Environ. 107, 334−347.

Huemmrich, K.F., Privette, J.L., Mukelabai, M., Myneni, R.B., Knyazikhin, Y., 2005. Time-series validation of MODIS land biophysical products in a Kalahari woodland, Africa. Int. J. Remote Sens. 26, 4381−4398.

Huete, A.R., 1988. A soil-adjusted vegetation index (SAVI). Remote Sens. Environ. 25, 295−309.

Huete, A.R., Liu, H.Q., Batchily, K., vanLeeuwen, W., 1997. A comparison of vegetation indices global set of TM images for EOS-MODIS. Remote Sens Environ 59, 440−451.

Kaminski, T., Knorr, W., Scholze, M., Gobron, N., Pinty, B., Giering, R., Mathieu, P.P., 2012. Consistent assimilation of MERIS FAPAR and atmospheric CO2 into a terrestrial vegetation model and interactive mission benefit analysis. Biogeosciences 9, 3173−3184.

Kanniah, K.D., Beringer, J., Hutley, L.B., Tapper, N.J., Zhu, X., 2009. Evaluation of Collections 4 and 5 of the MODIS Gross Primary Productivity product and algorithm improvement at a tropical savanna site in northern Australia. Remote Sens. Environ. 113, 1808−1822.

Knyazikhin, Y., Martonchik, J.V., Diner, D.J., Myneni, R.B., Verstraete, M., Pinty, B., Gobron, N., 1998a. Estimation of vegetation canopy leaf area index and fraction of absorbed photosynthetically active radiation from atmosphere-corrected MISR data. J. Geophys. Res. Atmos. 103, 32239−32256.

Knyazikhin, Y., Martonchik, J.V., Myneni, R.B., Diner, D.J., Running, S.W., 1998b. Synergistic algorithm for estimating vegetation canopy leaf area index and fraction of absorbed photosynthetically active radiation from MODIS and MISR data. J. Geophys. Res. Atmos. 103, 32257−32275.

Liang, S., 2004. Quantitative Remote Sensing of Land Surfaces. John Wiley & Sons, Inc, New York.

Liang, S., Li, X., Wang, J., 2012. Advanced Remote Sensing: Terrestrial Information Extraction and Applications. Academic Press, ISBN 9780123859556.

Liang, S.L., 2007. Recent developments in estimating land surface biogeophysical variables from optical remote sensing. Prog. Phys. Geogr. 31, 501−516.

Martinez, B., Camacho, F., Verger, A., Garcia-Haro, F.J., Gilabert, M.A., 2013. Intercomparison and quality assessment of MERIS, MODIS and SEVIRI FAPAR products over the Iberian Peninsula. Int. J. Appl. Earth Obs. Geoinf. 21, 463−476.

Masek, J.G., Vermote, E.F., Saleous, N.E., Wolfe, R., Hall, F.G., Huemmrich, K.F., Gao, F., Kutler, J., Lim, T.K., 2006. A Landsat surface reflectance dataset for North America, 1990-2000. IEEE Geosci. Remote Sens. Lett. 3, 68−72.

Maselli, F., Chiesi, M., Fibbi, L., Moriondo, M., 2008. Integration of remote sensing and ecosystem modelling techniques to estimate forest net carbon uptake. Int. J. Remote Sens. 29, 2437−2443.

McCallum, A., Wagner, W., Schmullius, C., Shvidenko, A., Obersteiner, M., Fritz, S., Nilsson, S., 2010. Comparison of four global FAPAR datasets over Northern Eurasia for the year 2000. Remote Sens. Environ. 114, 941−949.

Myneni, R.B., Hoffman, S., Knyazikhin, Y., Privette, J.L., Glassy, J., Tian, Y., Wang, Y., Song, X., Zhang, Y., Smith, G.R., Lotsch, A., Friedl, M., Morisette, J.T., Votava, P., Nemani, R.R., Running, S.W., 2002. Global products of vegetation leaf area and fraction absorbed PAR from year one of MODIS data. Remote Sens. Environ. 83, 214−231.

Myneni, R.B., Nemani, R.R., Running, S.W., 1997. Estimation of global leaf area index and absorbed par using radiative transfer models. IEEE Trans. Geosci. Remote Sens. 35, 1380−1393.

Myneni, R.B., Williams, D.L., 1994. On the relationship between FAPAR and NDVI. Remote Sens. Environ. 49, 200−211.

Olofsson, P., Eklundh, L., 2007. Estimation of absorbed PAR across Scandinavia from satellite measurements. Part II: modeling and evaluating the fractional absorption. Remote Sens. Environ. 110, 240−251.

Pickett-Heaps, C.A., Canadell, J.G., Briggs, P.R., Gobron, N., Haverd, V., Paget, M.J., Pinty, B., Raupach, M.R., 2014. Evaluation of six satellite-derived fraction of absorbed photosynthetic active radiation (FAPAR) products across the Australian continent. Remote Sens. Environ. 140, 241−256.

Pinty, B., Clerici, M., Andredakis, I., Kaminski, T., Taberner, M., Verstraete, M.M., Gobron, N., Plummer, S., Widlowski, J.L., 2011. Exploiting the MODIS albedos with the Two-stream Inversion Package (JRC-TIP): 2. Fractions of transmitted and absorbed fluxes in the vegetation and soil layers. J. Geophys. Res. Atmos. 116.

Potter, C.S., Randerson, J.T., Field, C.B., al, a.e, 1993. Terrestrial ecosystem production: a process model based on global satellite and surface data. Glob. Biogeochem. Cycles 7, 811−841.

Prince, S.D., Goward, S.N., 1995. Global primary production: a remote sensing approach. J. Biogeogr. 22, 815−835.

Ross, J., 1981. The Radiation Regime and Architecture of Plant Stands. Dr. W. Junk Publishers, The Hague, Boston and London.

Roujean, J.-L., Breon, F.-M., 1995. Estimating PAR absorbed by vegetation from bidirectional reflectance measurements. Remote Sens. Environ. 51, 375—384.

Ruimy, A., Saugier, B., Dedieu, G., 1994. Methodology for the estimation of terrestrial net primary production from remotely sensed data. J. Geophys. Res. Atmos. 99, 5263—5283.

Seixas, J., Carvalhais, N., Nunes, C., Benali, A., 2009. Comparative analysis of MODIS-FAPAR and MERIS-MGVI datasets: potential impacts on ecosystem modeling. Remote Sens. Environ. 113, 2547—2559.

Sellers, P., 1985. Canopy reflectance, photosynthesis and transpiration. Int. J. Remote Sens. 6, 1335—1372.

Sellers, P.J., Dickinson, R.E., Randall, D.A., Betts, A.K., Hall, F.G., Berry, J.A., Collatz, G.J., Denning, A.S., Mooney, H.A., Nobre, C.A., Sato, N., Field, C.B., Henderson-Sellers, A., 1997. Modeling the exchanges of energy, water, and carbon between continents and the atmosphere. Science 275, 502—509.

Shabanov, N.V., Huang, D., Yang, W.Z., Tan, B., Knyazikhin, Y., Myneni, R.B., Ahl, D.E., Gower, S.T., Huete, A.R., Aragao, L., Shimabukuro, Y.E., 2005. Analysis and optimization of the MODIS leaf area index algorithm retrievals over broadleaf forests. IEEE Trans. Geosci. Remote Sens. 43, 1855—1865.

Steinberg, D.C., Goetz, S.J., Hyer, E.J., 2006. Validation of MODIS F-PAR products in boreal forests of Alaska. IEEE Trans. Geosci. Remote Sens. 44, 1818—1828.

Tao, X., Liang, S., Wang, D.D., 2015. Assessment of five global satellite products of fraction of absorbed photosynthetically active radiation: intercomparsion and direct validation against ground-based data. Remote Sens. Environ. 163, 270—285.

Tao, X., Liang, S.L., He, T., Jin, H.R., 2016. Estimation of fraction of absorbed photosynthetically active radiation from multiple satellite data: model development and validation. Remote Sens. Environ. 184, 539—557.

Tao, X., Yan, B., Wang, K., Wu, D., Fan, W., Xu, X., Liang, S., 2009. Scale transformation of leaf area index product retrieved from multi-resolution remotely sensed data: analysis and case studies. Int. J. Remote Sens. 30, 5383—5395.

Tian, Y., Dickinson, R.E., Zhou, L., Zeng, X., Dai, Y., Myneni, R.B., Knyazikhin, Y., Zhang, X., Friedl, M., Yu II, Wu, W., Shaikh, M., 2004. Comparison of seasonal and spatial variations of leaf area index and fraction of absorbed photosynthetically active radiation from Moderate Resolution Imaging Spectroradiometer (MODIS) and Common Land Model. J. Geophys. Res. Atmos. 109.

Tian, Y., Zhang, Y., Knyazikhin, Y., Myneni, R.B., Glassy, J.M., Dedieu, G., Running, S.W., 2000. Prototyping of MODIS LAI and FPAR algorithm with LASUR and LANDSAT data. IEEE Trans. Geosci. Remote Sens. 38, 2387—2401.

Turner, D.P., Ritts, W.D., Cohen, W.B., Maeirsperger, T.K., Gower, S.T., Kirschbaum, A.A., Running, S.W., Zhao, M.S., Wofsy, S.C., Dunn, A.L., Law, B.E., Campbell, J.L., Oechel, W.C., Kwon, H.J., Meyers, T.P., Small, E.E., Kurc, S.A., Gamon, J.A., 2005. Site-level evaluation of satellite-based global terrestrial gross primary production and net primary production monitoring. Glob. Chang. Biol. 11, 666—684.

Vermote, E., Tanre, D., Deuze, J.L., Herman, M., Morcrette, J.J., 1997. Second simulation of the satellite signal in the solar spectrum: an overview. IEEE Trans. Geosci. Remote Sens. 35, 675—686.

Weiss, M., Baret, F., Garrigues, S., Lacaze, R., 2007. LAI and fAPAR CYCLOPES global products derived from VEGETATION. Part 2: validation and comparison with MODIS collection 4 products. Remote Sens. Environ. 110, 317—331.

Widlowski, J.L., 2010. On the bias of instantaneous FAPAR estimates in open-canopy forests. Agric. For. Meteorol. 150, 1501—1522.

Widlowski, J.L., Taberner, M., Pinty, B., Bruniquel-Pinel, V., Disney, M., Fernandes, R., Gastellu-Etchegorry, J.P., Gobron, N., Kuusk, A., Lavergne, T., Leblanc, S., Lewis, P.E., Martin, E., Mottus, M., North, P.R.J., Qin, W., Robustelli, M., Rochdi, N., Ruiloba, R., Soler, C., Thompson, R., Verhoef, W., Verstraete, M.M., Xie, D., 2007. Third radiation transfer model intercomparison (RAMI) exercise: documenting progress in canopy reflectance models. J. Geophys. Res. Atmos. 112. Art. No. D09111.

Wiegand, C.L., Maas, S.J., Aase, J.K., Hatfield, J.L., Pinter, P.J.J., Jackson, R.D., Kanemasu, E.T., Lapitan, R.L., 1992. Multisite analyses of spectral-biophysical data from wheat. Remote Sens. Environ. 42, 1—21.

Xiao, Z.Q., Liang, S.L., Sun, R., Wang, J.D., Jiang, B., 2015a. Estimating the fraction of absorbed photosynthetically active radiation from the MODIS data based GLASS leaf area index product. Remote Sens. Environ. 171, 105—117.

Xiao, Z.Q., Liang, S.L., Wang, J.D., Xie, D.H., Song, J.L., Fensholt, R., 2015b. A framework for consistent estimation of leaf area index, fraction of absorbed photosynthetically active radiation, and surface albedo from MODIS time-series data. IEEE Trans. Geosci. Remote Sens. 53, 3178—3197.

Xu, X., Fan, W., Tao, X., 2009. The spatial scaling effect of continuous canopy leaves area index retrieved by remote sensing. Sci. China Ser. D Earth Sci. 52, 393—401.

Yang, W.Z., Huang, D., Tan, B., Stroeve, J.C., Shabanov, N.V., Knyazikhin, Y., Nemani, R.R., Myneni, R.B., 2006. Analysis of leaf area index and fraction of PAR absorbed by vegetation products from the terra MODIS sensor: 2000-2005. IEEE Trans. Geosci. Remote Sens. 44, 1829—1842.

Advanced Remote Sensing, Second Edition
https://doi.org/10.1016/B978-0-12-815826-5.00012-X

© 2020 Elsevier Inc. All rights reserved.

Abstract

Fractional vegetation cover (FVC) is an important biophysical parameter describing the Earth's surface system. This chapter summarizes various methods used for the field measurement and remote sensing retrieval of FVC, including visual estimation, sampling and the use of optical measuring instruments, regression modeling, mixed pixel decomposition, and machine learning methods. Some frequently used methods are described in detail, and actual examples and discussions are given in the section describing field measurement. Then, the principal remote sensing—based FVC products and algorithms are briefly introduced, and the temporal and spatial variation of estimated FVC by using remote sensing are analyzed with examples. Finally, the further development of FVC estimation algorithms is prospected.

12.1 Introduction

Fractional vegetation cover (FVC) is generally defined as the ratio of the vertical projection area of above-ground vegetation organs on the ground to the total vegetation area. FVC is an important parameter used to measure surface vegetation cover; additionally, it is an important index for researching the aerosphere, pedosphere, hydrosphere, and biosphere as well as their interactions (Qin et al., 2006; Yin et al., 2016). FVC is also an important biophysical parameter for simulating the exchange between the land surface and the atmospheric boundary level using the soil—vegetation—atmosphere transfer model (Chen et al., 1997). Accurate estimation of the FVC is required for research on land surface processes, climate change, and numerical weather prediction (Zeng et al., 2000). Moreover, FVC is extensively applied in agriculture, forestry, resource and environmental management, land use, hydrology, disaster risk monitoring, and drought monitoring.

In most of these applications, the FVC of healthy vegetation is required, which is also called photosynthesis FVC or green FVC, and the absorption and emission of carbon and water, which are performed by the vegetation and are the foundation for the Earth's biosphere, are emphasized. However, in some applications, the withered and necrotic part of the vegetation is considered (which is nonphotosynthesis FVC), such as studies on water and soil conservation, in which the interception of rainwater by vegetation needs to be determined; in these studies, the focus is on the physical properties of the vegetation.

Field measurement and remote sensing retrieval are two approaches used to obtain FVC accurate values. Field measurement is a conventional approach to extract FVC and includes visual estimation, sampling, and the use of optical measuring instruments (e.g., photography). Remote sensing retrieval can be divided into three methods: the empirical model, the physical model, and machine learning methods. Using an empirical model, FVC is calculated by either a simple statistical model or a regression relationship. Typically, an empirical relationship between the normalized difference vegetation index (NDVI) and the FVC is established, and then the FVC is calculated from the NDVI. The physical model considers the complex canopy radiative transfer process, which involves reflectance, transmittance, and absorption by leaves and other elements. Therefore, the FVC is difficult to directly calculate using the physical model method and must be obtained using a lookup table or by simplifying the retrieval process with other machine learning methods. A machine learning method is a type of retrieval process in which the required knowledge is acquired through sample data training for rapid physical model simulation, thus accomplishing information transmission. It can convert remote sensing data to FVC efficiently; however, a large number of training data are generally needed through other vegetation products, high-resolution data classification, or complex physical model simulation. This chapter introduces the methods used to obtain FVC and the available remote sensing products.

12.2 Field measurements of fractional vegetation cover

In the past, field measurement was the most commonly used method to obtain FVC. However, with the extensive application of remote sensing techniques for monitoring vegetation, field measurement is losing its dominance. Regardless, field measurement continues to play a nonnegligible role by providing basic data for the remote sensing estimation of FVC. Zhou and Robson (2001) believe that an ideal field measurement of FVC should have the following features: (a) instruments with operational ease and economical utilization, (b) available, accurate, and objective land surface observation records, (c) a short measurement duration, and (d) negligible impact from human factors. Most field measurement methods are unable to meet all four requirements, although digital photography is widely applied in the field measurement of FVC because it can satisfy all four requirements simultaneously.

12.2.1 Visual estimation

In visual estimation, FVC is estimated based on the experience of the estimator. This method is characterized by its simplicity and operational ease; however, this method is highly subjective and random because the estimation precision is closely associated with the estimator's experience. The following methods are included:

12.2.1.1 The traditional method

Several sample plots covering a given area are selected based on specific statistical requirements. The FVC of the sample plots is directly estimated based on experience.

12.2.1.2 The digital image method

First, the vegetation within the sample plots is vertically photographed; then, the photos are visually estimated. To improve the estimation precision, reference images are interpreted by personnel in accordance with certain standards, and the average value is taken. In this method, an FVC standard image series must be generated, and the training of survey personnel is required. Images that contain a fractional amount (from 5% up to 95%) of vegetation cover are printed and subsequently coded by survey personnel. Special measurement and calculation software is employed to estimate the FVC. Several colored images are randomly selected each time for the visual estimation of the FVC by trained staff based on standard images. Finally, the results of the visual estimation and those provided by the software are compared to determine the error in visual estimation. Generally, training objectives require that the visual estimation error is less than 10%.

12.2.1.3 The grid method

The grid method is an improvement on traditional visual estimation. In this method, the sample plot is divided into several subquadrats of equal area based on vegetation type. Then, the FVC of each subquadrat is estimated using the traditional method of visual estimation. The mean value of the FVC is taken as the FVC of the sample plot. Research indicates that the grid method is easier to perform and has a higher precision than the traditional method. The grid method for visual estimation is essentially a spatial sampling method that uses equal spacing.

12.2.2 Sampling method

The sampling method is also known as the probability calculation method; in this method, the occurrence probability of vegetation in the sample plot is calculated using measurement methods based on statistical principles. The calculated probability is taken as the FVC of the sample plot. This method has the disadvantages of operational complexity, long measurement duration, a large number of limitations, and low efficiency; however, the method provides high precision. Some sampling methods are listed below.

12.2.2.1 The belt transect sampling method

Two perpendicularly crossed rectangular belt transects are selected. The FVC of the sample plot is defined as the ratio of the plant length that is in contact with the belt transect to the total length of the belt transect (Bonham, 1989). For example, to measure the forest canopy closure, the trees located on the two diagonals of the quadrat are the objects of investigation. The forest canopy closure is defined as the ratio of the number of times that a canopy is visible overhead to the total number of times that the head is raised.

12.2.2.2 The point count sampling method

The basic principle of the point count sampling method is to sample in space, and each sample point only corresponds to a very small space range. There are only two cases of vegetation and nonvegetation for the sample points, and the FVC of the sample square can be obtained through the statistics of multiple sample points. The sample point method is equivalent to estimating the surface (sample square) with points (sample points). According to different measurement methods, there are several representative methods for the sample point method:

12.2.2.2.1 Needle sampling method

Needles are vertically placed in the vegetation, and the FVC is defined as the ratio of the number of needles in contact with leaves to the total number of needles.

12.2.2.2.2 Square frame sampling method

The square frame sampling is composed of two horizontal poles that are aligned up and down and drill 10 holes equally. The observer looks down from the holes in the upper horizontal pole and takes the percentage of the number of holes in the total number of holes observed as the FVC (Lal, 1994).

12.2.2.2.3 Lookup sampling method

This method can be combined with the belt transect sampling method, taking the trees on the two diagonals of the sample plot as the investigation object, walking one step along the diagonals and looking up at the sky once, and taking the percentage of the number of times in which the vegetation can be seen as the FVC. This method is often used to estimate forest canopy density (defined as the proportion of projected canopy area to woodland area, which is close to FVC).

12.2.2.3 The shadow sampling method

The shadow sampling method is also termed the meter-stick method. A meter stick is placed on the land surface parallel to the crop rows. The stick is moved forward a set distance, and the length of shadow on the stick is read; the ratio of shadow length on the stick to the total length of the stick is recorded as the FVC. This method is generally used for row crops, and high noon is considered the optimal time for measurement (Adams and Arkin, 1977).

12.2.2.4 The canopy projection method

The canopy projection of each plant was measured and marked on the drawing paper according to a certain scale. The FVC of the quadrate was obtained according to the total projected area of the plants and the total area of the quadrate. This method is often used to estimate canopy density of tall vegetation such as trees.

12.2.3 Optical measuring instruments

As science and technology has developed, new measuring instruments are now being applied to FVC measurement. For example, electronic equipment is used to record the flux of intercepted sunlight, which is then compared with the level of direct sunlight at that location. From these data, the vegetation gap fraction is obtained. For these measurements, the FVC is calculated as 1—the vertical vegetation gap fraction. Other methods include direct imaging, such as digital photography, which is the most extensively applied method. The proportions of various vegetation types that are present are calculated using image classification. Common measurement techniques include the following:

12.2.3.1 Spatial quantum sensor and traversing quantum sell

Spatial quantum sensor and traversing quantum sell are used to calculate FVC based on the amount of sunlight that is intercepted by vegetation as measured by a sensor. However, these methods have not been extensively applied in practice because specialized sensors are usually needed, resulting in field operation difficulties.

12.2.3.2 Digital photography

Digital photography is performed in a vertical and downward manner. FVC is the ratio of the number of vegetation pixels to the total number of pixels in the digital images. Digital photography is used extensively in the field measurement of FVC because it is free of the subjectivity of other measurements, has high precision and good stability, and is easy to use.

In recent years, some progress has been made in estimating FVC using digital photography. Zhou et al. (1998) acquired FVC digital images using a digital camera to test the consistency of this method. Gitelson et al. (2002) estimated the FVC of wheat in Nebraska, US, using digital photography. Michael et al. (2000) conducted long-term monitoring of the FVC of an arid ecosystem in the United States using an agricultural digital camera with accurate and effective results. Based on a comparison of the various techniques used to obtain field measurements of FVC, White et al. (2000) claimed that digital photography was the easiest and most reliable technique to test and verify the extraction of remote sensing information. Hu et al. (2007) acquired photographs of sample plots in a research area and extracted FVC information from digital images of the area through classification. The information extracted was later used to verify the FVC of the entire research area obtained using remote sensing estimation. With the development of UAV technology, the combination of UAV with digital camera or other sensors has become an extension of ground digital camera photography. Li et al. (2018) proposed an automatic image classification and FVC extraction algorithm based on semi-Gaussian fitting for the problem of mixed pixels in UAV digital photos.

Despite the extensive application of digital photography, which is due to its easy operation and high efficiency, certain problems are encountered in obtaining FVC. Two major issues affect FVC extraction precision, and these factors require certain skills to deal with them. The first issue arises from the methods used for measurement and photography and includes the problem of image edge distortion. The most frequently used solution is to cut the image edge to remove this influence. The other issue, which is more striking and deserves more attention, is the method used to extract FVC information from digital images. Supervised and unsupervised classifications are adopted to solve this problem, and FVC is calculated as the proportion of classified vegetation. All the traditional methods are unable to extract the FVC from digital images rapidly and automatically, thus lowering the practicability of digital photography. However, some researchers have proposed methods that are more convenient to extract the FVC from digital images (Liu and Pattey, 2010; Liu et al., 2012; Song et al., 2015), all of which have good accuracies. Song et al. (2015) showed that the influence of shadows on the FVC extraction algorithm could not be ignored, and the error might reach 0.2, especially when the vegetation is dense and prone to generate heavy shadows (Fig. 12.1).

12.2.3.3 LAI-2000 indirect measurement

LAI-2000 is a more delicately designed measuring instrument and is mainly intended to measure the leaf area index (LAI) (See Section 11.1.2 in Chapter 11). Moreover, LAI-2000 can calculate FVC based on the measured vegetation gap fraction.

LAI-2000 uses a camera with a fisheye lens for observation and imaging, and the image is divided into several rings with variable radii by viewing the zenith angles of 7, 23, 38, 53, and 68 degrees. During FVC measurements, the observed area in the ring with the minimum zenith angle (7 degrees) is approximated as the zenith observation, and the FVC is calculated as 1—the gap fraction for the ring with the minimum zenith angle (7 degrees) (Rautiainen et al., 2005). This measurement shares similar principles with digital photography. In addition, White et al. (2000), when measuring FVC using LAI-2000, first measured the plant area index and then obtained the FVC through conversion. Compared with the method where only the ring with the minimum zenith angle is involved, this method has the advantage of expanding the spatial range for measurement by utilizing a greater number of multiangle observations in a wide field of view using a fisheye lens. However, this method also has the disadvantage of incorporating a greater number of uncertain factors.

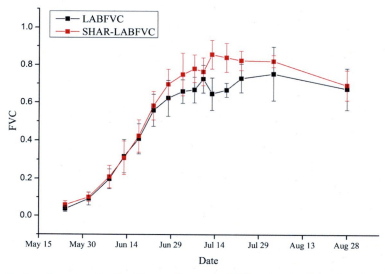

FIGURE 12.1   The shadow impact on fractional vegetation cover (FVC) estimation from digital photography (Song et al., 2015, Fig. 10). LABFVC is an automatic FVC extraction method that does not consider shadow impact, while its improvement version, SHAR-LABFVC, does. In this figure, each point represents the average FVC from 15 samples at the middle reach of Heihe River Basin. LABFVC underestimates FVC significantly when crop is dense.

As with the application of LAI2000 for FVC measurement, early morning and evening are the optimal measuring times; the measurement should not be conducted in direct sunlight. Generally, the use of LAI-2000 is not as convenient as digital photography for measuring FVC.

12.2.4 Examples of field measurement

12.2.4.1 Examples of noninstrumental measurements

Some examples of field measurements of FVC are given below. Depending on the types and features of the vegetation studied, the point sampling, shadow sampling method, and canopy projection method are used.

12.2.4.1.1 Grassland

The point count sampling method is used to measure grassland FVC. In the research plot, 1 m × 1 m subquadrats are selected. Needles are used as markers every 10 cm ($\varphi = 2$ mm), i.e., the needles are successively and vertically inserted in the subquadrat above the grassland at intervals of 10 cm. The points where the needles come into contact with grass are counted, and the points where there are no contacts are not counted. The FVC is the ratio of the number of contact points to the total number of points. The mean value of the measured FVC for three subquadrats at three different positions is recorded as the FVC of the quadrat, as shown in Fig. 12.2.

12.2.4.1.2 Forested land

For tall vegetation, such as trees in forested land, the forest canopy closure is generally used to indicate the cover as expressed in Eq. (12.1):

$$D = \frac{fd}{fe} \times 100\% \qquad (12.1)$$

Here, D is the forest canopy closure (or the FVC of shrub land) expressed as a percentage;

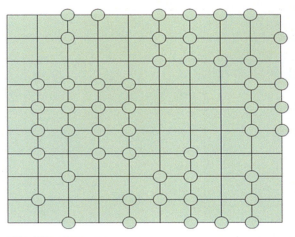

FIGURE 12.2 Schematic diagram for measurement of grassland fractional vegetation cover (FVC). Circles represent sample points, and FVC is 50% in this example.

fe is the area of quadrat (units, m^2); and fd is the vertical projection area of the tree canopy (or grass canopy) in the quadrat (units, m^2).

Forest canopy closure can also be measured using the tree canopy projection method, which is only applicable to forested land. This method is similar to the ellipse method for visual estimation and the grid method for visual estimation; therefore, it not specifically introduced here. A 20 m × 20 m sample plot is typically selected and divided into 5 m × 5 m grids with a measuring tape. The position of each tree in the grid is measured, and the projection length of the canopy of each tree is measured in the north–south and east–west directions using the measuring tape and a compass. The canopy projection is plotted on grid paper on an appropriate scale. The areas of canopy projection and the sample plot are calculated based on the grid; from these data, the forest canopy closure can be derived as shown in Fig. 12.3.

12.2.4.1.3 Shrubbery

The shadow sampling method is usually adopted for measuring the FVC of shrubbery.

FIGURE 12.3　Schematic diagram for measuring the forest canopy closure.

A rope or measuring tape is extended over the quadrat of shrubbery, and the length of that shadow that is cast on the measuring rope is measured. The FVC of the shrubbery is the ratio of the length of shadow cast by the shrub to the total length of the rope or the quadrat. This procedure is repeated three times at three different positions on the quadrat, and the mean value is the FVC of shrubbery in the quadrat. The area of quadrat can be as small as 10 m × 10 m.

12.2.4.2 Examples of digital photography measurement

12.2.4.2.1 Selecting the photography environment

First, illumination conditions must be selected. The vegetation should be photographed on a cloudy day or in the morning or evening when the effect of shadows is minimal. However, because morning dew might also affect the photography, the morning is not recommended. Artificial lighting equipment can be used in the dark during the evening. In short, the spectral difference between the vegetation and the soil background should as strong as possible, and interference from shadows should be prevented. Fig. 12.4 shows images taken with a digital camera in various environments; in (A), a flashlight was used during the night; in (B), the image was taken in direct sunlight; and in (C), the image was taken during a cloudy day.

In field measurements, a long stick with the camera mounted on one end is beneficial to conveniently measure various species of vegetation, enabling a larger area to be photographed with a smaller field of view. The stick can be used to change the camera height; a fixed-focus camera can be placed at the end of the instrument platform at the front end of the

(A)　　　　　　**(B)**　　　　　　**(C)**

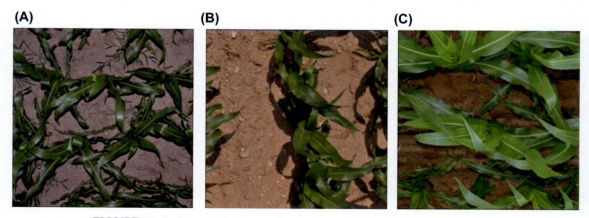

FIGURE 12.4　Images of maize seedlings taken under various lighting conditions.

support bar, and the camera can be operated by remote control. Fig. 12.5 shows a simple observation platform designed by one of the authors during the field measurement of FVC.

The photographic method used depends on the species of vegetation and planting pattern:

• Low crops (<2 m) not in rows

The observation platform is used directly where the height of the installed camera above the canopy of vegetation far exceeds the crown diameter of the vegetation. Sampling is conducted along the diagonals of the quadrat; finally, the arithmetic average is taken. This method is similar to the quadrat sampling method.

• Low crops (<2 m) in rows

In a situation with a small field of view (<30 degrees), rows of more than two cycles should be included in the field of view, and the side of the image should be parallel to the row. If there are no more than two complete cycles, then row spacing and plant spacing are required. The FVC of the entire cycle, i.e., the FVC of the quadrat, can be obtained from the number of rows included in the field of view.

• High crops (>2 m) not in rows

Sampling along the diagonals, as in the quadrat sampling method, can also be conducted. During the photography session, if the sampling points fall on the low vegetation between plants, then the observation platform is used; if the sampling points fall on the tree crown and the observation platform can be maintained above the crown of the vegetation as high as possible, then the observation platform can also be used to obtain photographs. However, if the height is excessively great, then the images should be taken in a bottom-up manner from beneath the crown. Meanwhile, the low vegetation is photographed in a top-down manner beneath the crown to obtain the total FVC of the region near the tree.

• High vegetation in rows (>2 m)

Through the top-down photography of the low vegetation underneath the crown and the bottom-up photography beneath the tree crown,

FIGURE 12.5 Observation platform for fractional vegetation cover (FVC) measurement.

the FVC within the crown projection area can be obtained by weighting the FVC obtained from the two images. Next, the low vegetation between the trees is photographed, and the FVC that does not lie within the crown projection area is calculated. Finally, the average area of the tree crown is obtained using the tree crown projection method. The ratio of the crown projection area to the area outside the projection is calculated based on row spacing, and the FVC of the quadrat is obtained by weighting.

12.2.4.2.2 Fractional vegetation cover extraction from the classification of digital images

Many methods are available to extract the FVC from digital images, and the degree of automation and the precision of identification are important factors that affect the efficiency of field measurements. For example, supervised classification has high precision but low efficiency, whereas unsupervised classification has high efficiency but low precision due to errors of commission and omission. Thus, the defects in these methods restrict their application to a certain extent.

Fig. 12.6 illustrates the results of the automatic and rapid extraction of FVC from digital images (Liu et al., 2012). This method, which is proposed by the authors, has the advantages of a simple algorithm, a high degree of automation, and high precision as well as ease of operation. More rapid classification methods with a higher degree of automation and greater accuracy are required

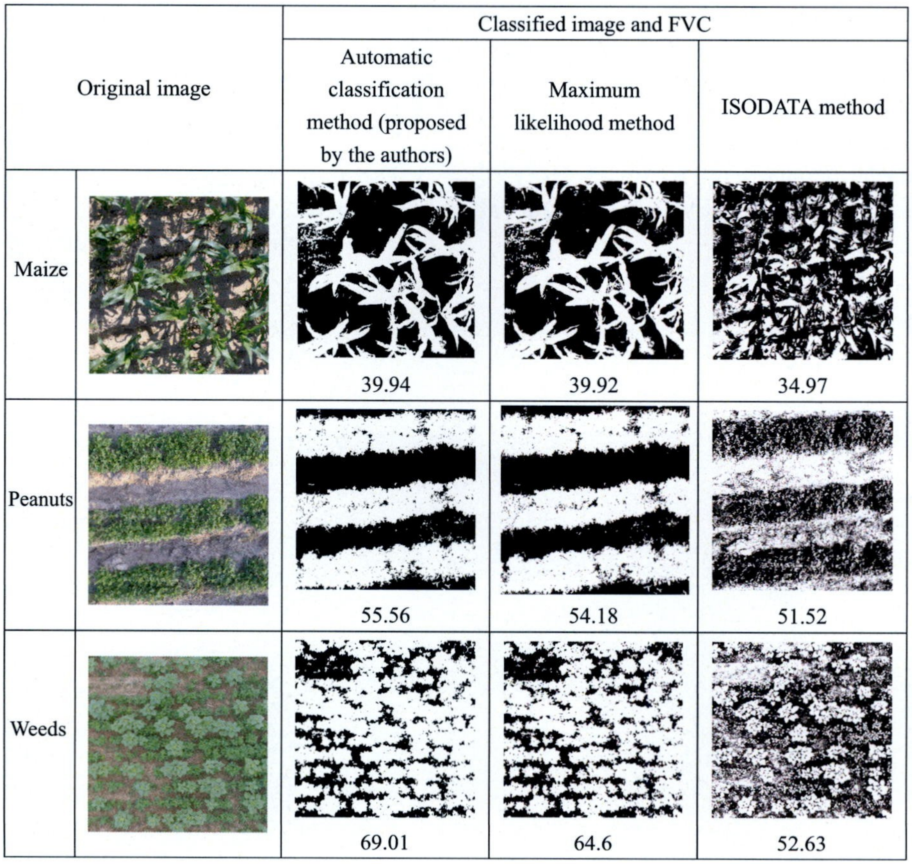

	Original image	Classified image and FVC		
		Automatic classification method (proposed by the authors)	Maximum likelihood method	ISODATA method
Maize		39.94	39.92	34.97
Peanuts		55.56	54.18	51.52
Weeds		69.01	64.6	52.63

FIGURE 12.6 Comparison of the classification results using different classification methods.

to maximize the superiority of digital photography. At the same time, with the development of multispectral and hyperspectral cameras, more spectral information can be used to provide an alternative for digital photos in the extraction of FVC.

12.3 The remote sensing retrieval

The development of remote sensing technology facilitates the acquisition of multitemporal and multiscale data to continually monitor FVC on a large or global scale. Accordingly, many FVC estimation methods are also being developed. The most extensively applicable method involves establishing the relationship between FVC and vegetation index (VI) (such as NDVI) to retrieve FVC. The commonly used remote sensing methods of FVC retrieval are mainly divided into two categories: empirical model methods and pixel decomposition model methods. Additionally, machine learning methods, such as those that employ neural networks, have been adopted by many researchers to estimate FVC.

12.3.1 Regression models

Regression models are also called empirical models and are constructed through the regression of remote sensing data collected using a specific waveband, several wavebands, or a remotely sensed VI to measure FVC. This model can be extended to FVC estimations on a larger scale.

The VI method is most frequently applied. Based on an analysis of the spectral features of vegetation, this method selects the VI that correlates well with FVC and then establishes the conversion relationship between VI and FVC for FVC estimation. Generally, the VI varies depending on area and vegetation species. The

most extensively applied NDVI is expressed using Formula (12.2).

$$NDVI = \frac{\rho_{nir} - \rho_{red}}{\rho_{nir} + \rho_{red}} \quad (12.2)$$

where ρ_{nir} is the reflectance of vegetation in the near-infrared band, and ρ_{red} is the reflectance of vegetation in the visible band. The extensive application of NDVI, which can be used to indicate the growth status of vegetation, stems from the considerable difference in the reflectance of normal vegetation in the visible band and the near-infrared band.

Previous research suggests that FVC is closely correlated with the VI and that the correlation between the two can be either linear or nonlinear. Therefore, the regression model can also be linear or nonlinear, and the regression model method can be subdivided into the linear and nonlinear regression model methods.

12.3.1.1 The linear regression model method

In the linear regression model method, the linear regression of the actual FVC and remotely sensed VI are determined to establish the estimation model for FVC in the research area. The linear regression model of NDVI and FVC provides a simple method to estimate FVC; thus, it has achieved widespread application (Hurcom and Harrison, 1998). For instance, Xiao and Moody (2005) through the linear regression of 60 points selected from a Landsat ETM+ NDVI image and FVC (considered as the actual surface FVC), extracted a high-resolution (0.3 m) true color orthoimage and found a strong linear relationship between NDVI and FVC ($R^2 = 0.89$). They then applied this formula to estimate the FVC of all of the pixels in the Landsat ETM+ image.

For both dense and sparse vegetation, the FVC of a remote sensing pixel can be defined as having a linear relationship with the VI if the influence of multiple scattering is omitted (Hurcom and Harrison, 1998):

$$FVC = a \cdot VI + b \qquad (12.3)$$

where *FVC* is the FVC of the mixed pixel, *VI* is the VI of the mixed pixel, and *a* and *b* are the regression coefficients of FVC and VI, respectively. Fig. 12.7 shows the expression for the linear regression of soil-adjusted VI and FVC established by Choudhury et al. (1994).

Some attempts have been made to grade the value of NDVI such that different grades indicate different FVCs. For instance, Mohammad et al. categorized NDVI after some conversion (NDVI = (NDVI+0.5)*255) into 6 grades, namely, 5, 5—50, 50—100, 100—150, 150—200, and 200—250, which represented the FVC of six situations, namely, 0%, 20%, 40%, 60%, 80%, and 100%, respectively (Mohammad et al., 2002). However, this method of segmentation with discretization of NDVI still utilizes a linear or nonlinear relationship between VI and FVC.

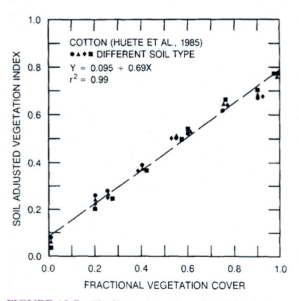

FIGURE 12.7   The linear regression relationship between soil-adjusted vegetation index (VI) and fractional vegetation cover (FVC) (Choudhury et al., 1994, Fig. 12(a)).

12.3.1.2 The nonlinear regression model method

By fitting NDVI and FVC, the nonlinear regression model can be established and applied to calculate the FVC of an entire research area. Research by Carlson and Ripley indicates that for some values of FVC where LAI was 1—3, more clumping of vegetation would result in a better linear correlation between the VI and FVC (Carlson and Ripley, 1997).

Choudhury et al. (1994) found that FVC was related to scaled NDVI in a quadratic manner. Based on this finding, the authors estimated the FVC of a coniferous forest in the US Pacific Northwest. The results indicated that although NDVI is the most commonly applied method, it did not have the strongest correlation with the FVC of trees. Meanwhile, based on remote sensing data from NOAA AVHRR, Choudhury and coworkers estimated the FVC of a coniferous forest using the NDVI and scaled NDVI. They found that the correlation coefficient was 0.55 at a confidence interval of 99% (Choudhury et al., 1994). Gillies and Carlson et al. obtained a quadratic relationship between the FVC and scaled NDVI using various methods and datasets (Gillies et al., 1997; Carlson and Ripley, 1997).

Some researchers have studied the VI method using both the linear regression method and the nonlinear regression method. For instance, Gitelson et al. conducted a regression of three types of VIs, namely the NDVI, Green NDVI, and the visible atmospherically resistant index (VARI) with the FVC of wheat (Fig. 12.8). Nonlinear regression was used for the NDVI and Green NDVI, and linear regression was used for the VARI. A comparative analysis indicated that the VARI was very sensitive to FVC within the range of 0%—10% and that it could minimize the sensitivity to atmospheric influence. It was suggested that the VARI should be utilized in the linear regression model to estimate FVC (Gitelson et al., 2002). Either a

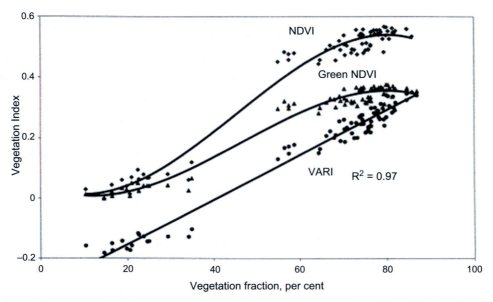

FIGURE 12.8   The expression for the regression relationship between vegetation index (VI) and fractional vegetation cover (FVC) (Gitelson et al., 2002, Fig. 7 (B)).

strong linear relationship (Ormsby et al., 1987) or a nonlinear relationship (Li et al., 2005) was found between the VI and FVC, depending on the specific landscape type. Both the linear and nonlinear models have difficulties with mixed vegetation. Even when the FVC is 100%, the VI depends on the vegetation species that are present due to the differences in their chlorophyll content and canopy structure. It has been reported that even when the actual FVC of different vegetation species is basically the same in a research area, the values of FVC directly calculated from the NDVI can deviate from each other by as much as 40% as a result of the difference in chlorophyll content between natural and artificial vegetation (Glenn et al., 2008).

The estimation precision of FVC can be improved if the features and species of the vegetation are determined by field measurement. Previous researchers have used the average leaf inclination angle corresponding to different vegetation species under various land surface conditions to predict the FVC from the VI (Anderson et al., 1997).

Non-VI forms can also be used to establish an empirical regression relationship with FVC. Graetz et al. conducted a linear regression analysis on measured FVC data from the fifth channel of Landsat MSS. The regression model obtained was later applied to estimate the FVC of sparse grassland (Graetz et al., 1988). North used ATSR-2 remote sensing data in four channels (555, 670, 870, and 1630 nm) to study the linear regression of FVC with LAI. The results indicated that applying the linear regression model based on the data from four channels was superior to the application of VI alone for estimating FVC (North, 2002).

Because of its simplicity, the regression model method has been widely applied in FVC estimations on a regional level with high precision. Nevertheless, it has been demonstrated that the empirical regression model has its own limitations, as it is only applicable to the FVC estimation of specific vegetation species in specific regions. For instance, (Graetz et al., 1988) linear regression model is only applicable to sparse grassland, and his nonlinear regression model was proposed specifically for the study of degraded grassland. In addition, because of its regional limitations, the regression model requires the spatial resolution of remote sensing data. The regression model is not suitable for extensive applications, and the empirical model (on the regional level) might be invalid if it is used to estimate FVC on a large scale.

12.3.2 The linear unmixing model

The linear unmixing model is based on the principle that each pixel in an image is composed of several components, and each pixel contributes separately to the information observed by remote sensors. Through decomposition of the remote sensing information (channel information or VI), the pixel decomposition model can be used to estimate FVC. This model can be either linear or nonlinear; however, thus far, most of the studies on the mixed pixel decomposition model focus on the linear unmixing model. This section, therefore, mainly concentrates on the application of the linear unmixing model to the remote sensing calculation of FVC.

The linear unmixing model is the most extensively applied model among the mixed pixel decomposition models. In this model, pixel information is assumed to result from the linear synthesis of the information of each component. The linear unmixing model involves the assumptions that the photon reaching the sensor acts on only one component and that the components are mutually independent and do not interact. If the photon acts on several components, then nonlinear mixing occurs. In fact, the linear and nonlinear mixing methods are based on a single concept, i.e., linear mixing occurs as an exception from nonlinear mixing when multiple reflections are ignored. To obtain a convenient solution, the number of regional land feature types, and the major land feature types in particular, should not exceed the number of remote sensing channels; otherwise, n+1 unknown variables would

have to be solved using n equations; this constitutes the largest limitation of the linear unmixing model. The linear unmixing model is able to calculate and extract the FVC for each pixel in the image. Suppose that each component contained in a pixel contributes to the pixel information received by the satellite sensors and that the value of the spectral characteristics of the vegetation in each component is taken as a factor, then the linear mixing model is established using the area of this specific component as the weight of this factor, which can be mathematically expressed as (Van der Meer, 1999)

$$R_b = \sum_{i=1}^{n} f_{i,b} r_{i,b} + e_b \tag{12.4}$$

where R_b is the reflectance of the pixel in channel b, f_i is the proportion of the number of subpixel i to that of the mixed pixel, $r_{i,b}$ is the reflectance of subpixel i in channel b, n is the number of subpixels, and e_b is the fitting error of channel b (Van der Meer, 1999).

The proportion of each component in a mixed pixel can be solved using the least squares method. The FVC is then calculated based on the proportion of each vegetation component present. The precision of this solution mainly depends on the selection of each pure component (Lu and Weng, 2004).

Among the various linear unmixing models, the simplest assumes that each pixel is composed of only two components, i.e., vegetation and nonvegetation, and that the spectral information results from linear mixing of the two components. The proportional area of each component in the pixel is the weight of each component. The proportional area of vegetation is the FVC of the pixel, as mathematically expressed using Formulas (12.5) and (12.6) (Gutman and Ignatov, 1998):

$$NDVI = f^* NDVI_v + (1 - f)^* NDVI_s \tag{12.5}$$

and then, $f = \dfrac{NDVI - NDVI_s}{NDVI_v - NDVI_s} \tag{12.6}$

where f is the proportion of vegetation area in the mixed pixel (i.e., FVC), NDVI is the NDVI of the mixed pixel, $NDVI_v$ is the NDVI of the fully covered vegetation, and $NDVI_s$ is the NDVI of bare soil. It is clear from this equation that this dimidiate pixel model is a linear regression model for VI. To obtain the FVC of mixed pixels, the NDVIs of vegetation and of bare soil should be determined. However, the determination of $NDVI_v$ and $NDVI_s$ is affected by many factors, such as soil, vegetation type, and chlorophyll content. Generally, these parameters can be determined through statistical analysis of spatial and temporal NDVI data. Time series NDVI data are analyzed statistically, and the maximum time series NDVI is used as the NDVI of vegetation, whereas the minimum time series $NDVI$ is used as the NDVI of bare soil (Gutman and Ignatov, 1998; Zeng et al., 2000). Some researchers directly select the maximum and minimum NDVIs of the research area as the $NDVI$ values for vegetation and bare soil, respectively (Xiao and Moody, 2005). Qi et al. (2000) used the dimidiate model in their study of the spatial and temporal variation of vegetation in the San Pedro Basin in the US Southwest based on NDVI data; Landsat TM, SPOT 4 VEGETATION, and aerial photography data were also used to study this method, the results indicated that the model can estimate the dynamic variation of vegetation reliably, even without atmospheric correction. Leprieur et al. (1994) used SPOT data after atmospheric correction to calculate NDVI and applied this model to estimate the FVC in the Sahel. Mu et al. (2017) combined multisource remote sensing data to synthesize the NDVI index and produced the FVC product of China—Asian region with good temporal and spatial continuity (http://www.geodoi.ac.cn/doi.aspx?Id=218), which is used in the "Annual Report on Remote Sensing Monitoring of Global Ecosystem and Environment" published by the ministry of science and technology of China.

To some extent, the most critical part of the dimidiate pixel model is to extract the "pure" pixel and realize the transformation from the spectral signal detected by remote sensing to the physical parameters of vegetation. However, there are serious limitations in obtaining $NDVI_v$ and $NDVI_s$ by traditional statistical methods. Firstly, remote sensing pixels fully covered by vegetation or pure bare soil may not exist, for example, in many semiarid or subtropical evergreen forest areas. Secondly, the determination of the maximum or minimum NDVI index using temporal or spatial statistics also needs a judgment threshold, which is largely impacted by human. Fig. 12.9 illustrates the global seasonal FVC estimated by (Gutman and Ignatov, 1998) based on the dimidiate model of FVC and NDVI. Using cluster analysis, the optimal maximum and minimum values of NDVI were selected from the maximum and minimum values of NDVI for each season. Gutman proposed that the FVC estimated using this model has a maximum error of 0.35, which might be derived from the estimation of the maximum and minimum values of NDVI (Montandon and Small, 2008). Song et al. (2017) proposed a method to estimate the NDVI coefficients, $NDVI_v$, and $NDVI_s$ by using the angular information. It combined the gap fraction model with the dimidiate pixel model by using angular VI and then simplified the model with the help of special observation angle (57 degrees zenith angle) and remote sensing LAI products, so as to obtain $NDVI_v$ and $NDVI_s$. Mu et al. (2018) improved this method by using multiangle information to offset the dependence of LAI product and proposed a method to estimate NDVI coefficient only based on multiangle VI.

After considering the surface complexity, the simple dimidiate pixel model is not definitely the most accurate. Jia et al. (2017) compared the dimidiate pixel model using NDVI and EVI, the pixel unmixing model with three and four end-members, the three gradient differential vegetation index (TGDVI) model, and two

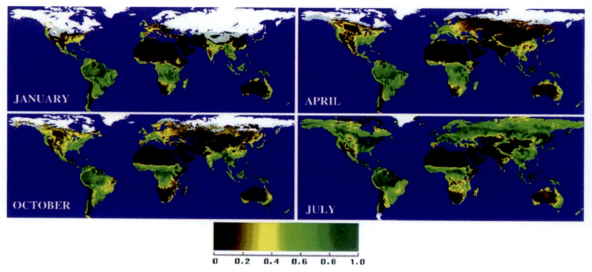

FIGURE 12.9   Seasonal fractional vegetation cover (FVC). Details of snow masks can be found in the report by (Gutman and Ignatov, 1998, Fig. 5).

modified TGDVI models for FVC estimation in an agriculture region and found that the pixel unmixing model with three end-members could achieve better FVC estimation accuracy than other methods. Gutman and Ignatov obtaining FVC using the pixel decomposition model based on the dimidiate pixel model (Gutman and Ignatov, 1998). Depending on the vegetation distribution features located in a mixed pixel, the pixels were divided into uniform and mosaic pixels, and the latter were further classified into dense, nondense, and variable density vegetation pixels. Different FVC models were constructed for different subpixel structures.

Xiao and Moody (2005) estimated the FVC within an area of approximately 4000 km² in New Mexico, US, using various methods. In their report, they presented a comparative analysis on the estimation of FVC by two methods, the mixed pixel decomposition model based on spectral data (SMA3, SMA4, SMA5, and NDVI-SMA) and the linear regression model based on NDVI. The analyses of SMA3, SMA4, and SMA5 included a discussion on whether the sum of the proportions of each component in

each mixed pixel should be constrained to 1. Fig. 12.10 shows the relationship between the FVC values that were estimated using various methods and the actual FVC. Fig. 12.11 shows the FVC values of a research area that was estimated using various methods. The selection of the number of components in mixed pixels and the selection of each end-member, which depends on the vegetation structure and the distribution of vegetation in the research area, are the major factors affecting FVC estimation in the mixed pixel decomposition method.

Yang et al. (2008) described the linear decomposition mixed pixel model as a spectral matching problem: the end-members from images or spectral library according to the preset mixing ratio were used to produce a series of test spectrum, and then the test spectrum and the target spectrum (spectrum that is to be decomposed) were matched, to find the best matching test spectrum with the targets and to obtain the mixing ratio of end-members.

In general, the dimidiate pixel model is the most commonly used model for FVC estimation due to its simple form and the efficiency. Firstly,

12.3.3 Machine learning methods

As computer technology develops, an increasing number of machine learning methods, such as neural networks, support vector machines (SVMs), decision trees, and random forests, are being used in the theoretical study and application of remote sensing technology (Baret et al., 2013; Yang et al., 2016; Wang et al., 2018). SVMs and decision trees are mainly used either for land cover classification or for extracting the vegetation cover from pixels.

12.3.3.1 The neural network method

A neural network is an intelligent computer-based technology that imitates the learning process of human brains and serves as a general computing tool in solving complex problems. A neural network is composed of a series of simple processing units connected by weighting coefficients determined by specific mechanisms, as shown in Fig. 12.12. Through continuous learning with training data, a neural network is capable of outputting optimal results with high computational efficiency.

A neural network can be used to retrieve various features of vegetation from remote sensing data (Baret et al., 1995; Jensen et al., 1999; Foody et al., 2001; Jia et al., 2016). However, these studies usually employ ordinary neural networks or multilayer perceptrons. Other types of neural networks might possess greater potential. However, many issues need to be considered if neural networks are to be applied, as the middle layer of neural networks is a black box. Consequently, it is difficult to control the retrieval of the parameters.

Based on a comparison of multivariable regression analysis, the VI method and the neural network method, Boyd et al. found that a neural network was the most applicable to the FVC estimation of a forest in the US Pacific Northwest. Three types of neural network methods, namely, the multilayer perceptron, the radial basis function, and the generalized

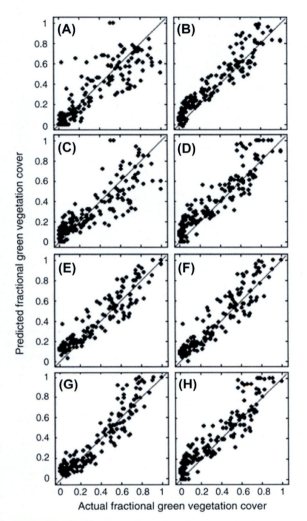

FIGURE 12.10 The relationship between the fractional vegetation cover (FVC) estimated using various methods and the actual FVC: (A) SMA3 without constraint; (B) SMA3 with constraint; (C) SMA4 without constraint; (D) SMA with constraint; (E) a situation based on the expression for regression with NDVI; (F) NDVI–SMA; (G) SMA5 without constraint; (H) SMA5 with constraint; the solid line is 1:1 (Xiao and Moody, 2005, Fig. 4).

estimating the $NDVI_v$ and $NDVI_s$ by using multiangle remote sensing and then applying them to FVC estimation is suitable for large-scale production, and its future development is worthy of attention (Mu et al., 2018).

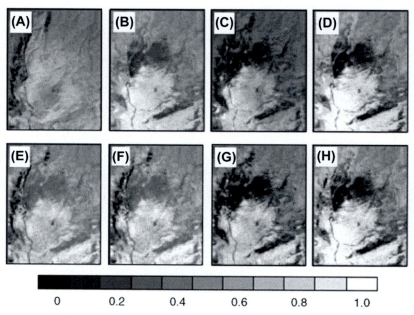

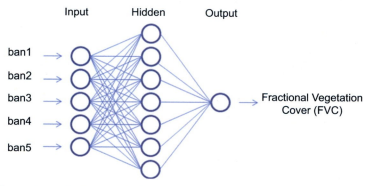

FIGURE 12.11   Estimates of the fractional vegetation cover (FVC) of the research area obtained using various methods: (A) SMA3 without constraint; (B) SMA3 with constraint; (C) SMA4 without constraint; (D) SMA4 with constraint; (3) a situation based on the expression for regression with NDVI; (F) NDVI—SMA; (G) SMA5 without constraint; (H) SMA5 with constraint (Xiao and Moody, 2005, Fig. 5(1)).

FIGURE 12.12   A schematic diagram for the neural network retrieval of fractional vegetation cover (FVC).

regression neural network (GRNN), were also compared. The multilayer perceptron was finally selected to determine the FVC of the research area. Using this method, 40 iterations were performed on six black boxes using a backpropagation algorithm. Within a 99% confidence interval, the results demonstrated that this method

estimated the FVC of the forest with higher precision than the multielement regression analysis or the VI method; the coefficient of determination with respect to the actual FVC of the forest was 0.57 (Boyd et al., 2002). Voorde et al. proposed using a multilayer perceptron to randomly select 3037 pixels from an ETM+

image for training, followed by mixed pixel decomposition, to estimate the FVC of subpixels. Moreover, to compare the results with those estimated using a neural network, regression analysis and the linear unmixing model were also used to estimate the FVC of the research area. Fig. 12.13 shows the estimates of the FVC of the research area obtained using various methods; Fig. 12.14 shows the error in the FVC estimation using various methods (Voorde et al., 2008). Jia et al. (2016) developed an FVC estimation algorithm for the Chinese GF-1 wide field view data based on training back-propagation neural networks (BPNNs) using PROSAIL radiative transfer model simulations. Validation results showed good performance with root mean square error (RMSE) of 0.073. The estimation algorithm provides technical support for the quantitative applications of Chinese high-resolution satellite data.

12.3.3.2 The decision tree method

Natural land features are diverse and constantly change; this change is becoming more

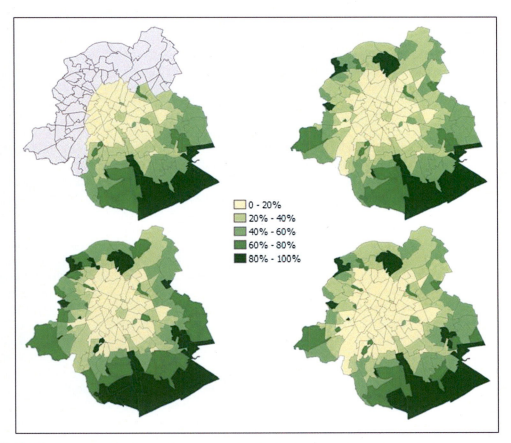

FIGURE 12.13 Estimates of the fractional vegetation cover (FVC) of the research area obtained using various methods. The image in the upper-left corner shows the results of the postclassification processing of high-resolution remote sensing data; the image in the upper-right corner shows the results of the linear regression model with a Landsat ETM+ image; the image in the lower-left corner shows the results of the linear unmixing model based on SVD; and the image in the lower-right corner shows the results of the mixed pixel decomposition model based on a multilayer neural network (Voorde et al., 2008, Fig. 5).

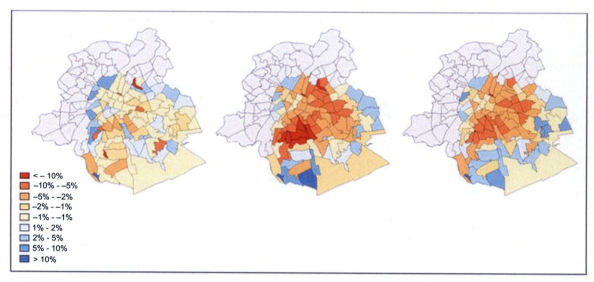

FIGURE 12.14   The error in the estimation of fractional vegetation cover (FVC) using various methods. The image on the left shows the result of the linear regression model; the image in the middle shows the result of the linear unmixing model based on SVD; and the image on the right shows the result of the multilayer perceptron neural network model (Voorde et al., 2008, Fig. 4).

complex due to human and natural factors. Therefore, it is a common phenomenon that one land feature generates different spectra or different land features share the same spectrum in remote sensing. This creates difficulties for identifying and classifying remote sensing images. Therefore, it is necessary to study the inherent rules and associations of apparently disordered and intricate land features; then, based on the rules and associations discovered, a tree structure, or a decision tree, is established. This decision tree functions as the basis for identifying and classifying land features. To ensure the accuracy and objectivity of the classification, decision functions for remote sensing data or nonremote sensing data are constantly involved in the classification process. Other information includes expert knowledge and relevant data (such as boundary conditions and classification parameters) that are used to improve the classification conditions and precision. This involvement of decision functions ensures that the results generated by the classification tree method are reasonable and that these

results can be satisfactorily obtained by forming an optimal logical decision tree.

A decision tree is generally applied to land cover classification and for the FVC estimation of subpixels. MODIS Vegetation Continuous Fields (Hansen et al., 2003) use a decision tree to estimate the FVC of trees and grass on the global scale. Other studies (e.g., Hansen et al., 2002; Huang and Townshend, 2003; Yang et al., 2003; Xu et al., 2005; Gessner et al., 2009) have also used a decision tree to solve problems related to remote sensing subpixels. A decision tree is a parameter-free classification method.

The advantages of the decision tree method, which include the lack of parameterization and the absence of a need to assume the normal distribution or homogeneity of input data, facilitate its extensive application in various fields. Rogan et al. adopted a decision tree that was designed to identify and classify the input data. First, multitemporal spectral mixture analysis was performed to extract four types of end-members of green vegetation (GV), nonphotosynthetic vegetation, shade, and soil. Next, using the designed decision data, the four types of end-members were identified stepwise and classified. When the intercategory deviation was the greatest, a classification threshold was required to separate the two categories at each step. Based on the classification results, four categories of end-members were calculated, and the user precision of the FVC of each end-member was determined to be approximately 76% (Rogan et al., 2002).

12.3.3.3 The random forests regression method

Random forest regression (RFR) is an ensemble algorithm that consists of many regression trees, usually binary decision trees (CART). The average of predicted values from each tree is used as the predictor of RFR. The main advantage of RFR is that it does not overfit with the increasing number of decision trees in a random forest, for it generates limits for generalization errors. Meanwhile, RFR is robust against noise. It can still perform well while data dimension is roughly equal or larger compared with sample size (Breiman, 2001). In addition, RFR is capable of evaluating the importance of different predictors during the process of model establishment, providing reference on band selection for FVC estimation model using multiband remote sensing data. Wang et al. (2018) used RFR models to evaluate the performance of different Sentinel-2 multispectral bands for FVC estimation and established an FVC estimation model using Sentinel-2 data. Results indicated that the three most important Sentinel-2 multispectral bands for FVC estimation are band 4 (red band), band 12 (short wave infrared band), and band 8a (near infrared band). The FVC estimation accuracy appeared to be better using these three bands compared with using all bands.

12.3.3.4 The support vector machines

SVM is a machine learning algorithm based on statistical learning theory that was first proposed by Vapnik. It showed many unique advantages in small sample, nonlinear and high dimensional pattern recognition and can be applied to other machine learning problems such as function fitting (Vapnik, 1995). The principle of SVM is to deal with complicated data classification by solving the optimization problem and finding the optimal classification hyperplane in the high-dimensional feature space. SVM algorithm is mostly used for remote sensing data classification and can achieve better classification results than other algorithms. For example, Su (2009) used SVM to identify vegetation types in semiarid areas and further estimated FVC of research areas based on SVM classification results. Huang et al. (2008) used the SVM method to generate fractional forest cover change products in the study area based on Landsat data. Validation results using high-resolution IKONOS showed the accuracy of the abovementioned products was up to 90%. Yang et al. (2016) compared the effects of support vector regression (SVR), GRNNs, BPNNs, and multivariate adaptive regression splines method on global land surface FVC estimation and found that SVR is comparable to other machines learning algorithms (Fig. 12.15).

However, there is currently no standard criterion for how to select a proper kernel function for a specific problem in SVM algorithm, and there is no unified understanding on how kernel functions affect the classification accuracy. The current method for kernel function selection is to apply different kernel functions separately and select the kernel function with the least classification error. The parameters of the kernel function are also determined by the same method. This selection method is basically based on experience and lacks sufficient theoretical basis. The choice of kernel function has certain influence on the accuracy of SVM algorithm. It is important to make reasonable selection, necessary improvement, correction, and optimization of kernel function.

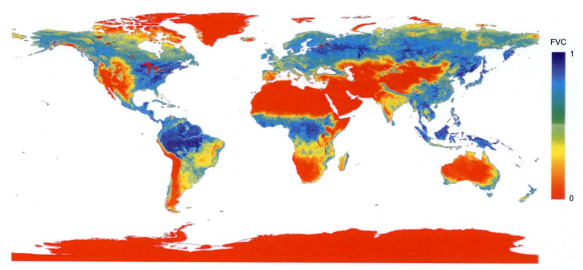

FIGURE 12.15   Global land surface fractional vegetation cover (FVC) maps generated using support vector regression (SVR) (Yang et al., 2016) based on MODIS data (DOY 201, 2003).

12.4 Current remote sensing products

At present, most remote sensing—based FVC products are mainly produced based on the pixel unmixing model and machine learning method, as shown in Table 12.1. And satellite data used in current products mainly come from NOAA/ AVHRR, ADEOS/POLDER, MSG/SEVIRI, ENVISAT/MERIS, SPOT/VEGETATION, Terra-&Aqua/MODIS, and Landsat/TM&ETM+, all of which are polar-orbiting satellite sensors, except MSG/SEVIRI.

AVHRR (onboard NOAA satellite) is an early sensor used in FVC production with the pixel unmixing model. For example, (Gutman and Ignatov, 1998) and Zeng et al. (2000) used the simplest dimidiate pixel model, which required two end-members of fully covered vegetation and pure soil. Defries et al. (1999) divided vegetation end-members into woody plants and herbaceous plants and conducted the decomposition of linearly mixed pixels with three end-members to solve FVC. Dimiceli et al. (2011) produced interannual FVC products (MOD44B) with high spatial resolution (250m)

by using the linear unmixing model with MODIS data. Guan et al. (2012) and Wu et al. (2014) added GIMMS into AVHRR data and extended the time range of each FVC products to 2008 and 2011, respectively, by using the linear unmixing model. Mu et al. (2017) used the MODIS and MERSI (onboard FY3A) data products to produce FVC products with higher temporal resolution (5 days) based on dimidiate pixel model. The SEVIRI sensor is supported by Meteosat Second Generation (MSG) and acquires data at a fixed observation angle with respect to specific observation sites. The sunlight illumination angle varies with time. García-Haro et al. (2005) have produced FVC products in Europe, Africa, and South America since 2005 based on SEVIRI data using the decomposition of mixed pixels.

FVC and LAI products from POLDER use the neural network method, and the radiative transfer model by Kuusk (Lacaze et al., 2003) has been adopted as the training model. LAI products are directly outputted through the neural network method. The FVC products are derived from the LAI products based on the exponential

TABLE 12.1 Current remote sensing fractional vegetation cover (FVC) products (regional scope of complementary products). The FVC products listed in the table were generated at regional or global scale using remote sensing data. The table does not list all the forest coverage products.

Reference	Definition	Name	Sensor	Resolution	Spatial range	Temporal range	Algorithm/ Model
1. Gutman and Ignatov (1998)	Vegetation cover	Global Monthly Greenness Fraction	AVHRR	0.15 degrees; 30 day	Global	1985–1990	The dimidiate pixel model
2. DeFries et al. (1999)	Woody and herb vegetation cover	—	AVHRR	8 km; —	Global	1983, 1993	The linear unmixing model
3. Zeng et al. (2000)	Vegetation cover	—	AVHRR	1 km; 30day	Global	1992–1993	The dimidiate pixel model
4. Roujean and Lacaze (2002); Lacaze et al. (2003)	Vegetation cover	POLDER FVC	POLDER	6 km; 10day	Global	1996–1997 2003	Neural network
5. García-Haro et al., 2005	Vegetation cover	LSA SAF FVC	SEVIRI	3 km; 1day/ 10day	Europe, South and North America, South Africa	2005–now	The linear unmixing model
6. Bacour et al. (2006)	Vegetation cover	TOAVEG FVC	MERIS	0.3 km; 30day/ 10day	Europe	2002–now	Neural network
7. Baret et al. (2007)	Vegetation cover	CYCLOPES FVC	VEGETATION	1 km; 10day	Global	1999–2007	Neural network
8. DiMiceli et al. (2011)	Forest and nonforest vegetation cover	MODIS FVC	MODIS	250 m; 1year	Global	2000–now	The linear unmixing model
9. Guan et al. (2012)	Forest and grassland cover	—	AVHRR GIMMS; QuikSCAT SIR; AMSR-E; TRMM	10 km; 15day	Southeast Africa	1999–2008	The linear unmixing model
10. Baret et al. (2013)	Vegetation over	GEOV1 FVC	VEGETATION	1/112 degrees; 10day	Global	1998–now	Neural network

Continued

TABLE 12.1 Current remote sensing fractional vegetation cover (FVC) products (regional scope of complementary products). The FVC products listed in the table were generated at regional or global scale using remote sensing data. The table does not list all the forest coverage products.—cont'd

Reference	Definition	Name	Sensor	Resolution	Spatial range	Temporal range	Algorithm/Model
11. Sexton et al. (2013)	Forest cover	–	MODIS; TM/ETM+	30 m; 1year	Global	2000, 2005	Linear regression
12. Hansen et al. (2013)	Forest cover	–	ETM+	30 m; 1year	Global	2000–2012	Decision tree
13. Wu et al. (2014)	Vegetation cover	–	AVHRR; GIMMS; MODIS	10 km; 30day	Global	1982–2011	The dimidiate pixel model
14. Jia et al. (2015a)	Vegetation cover	GLASS FVC	MODIS; TM/ETM+	0.5 km; 8day	Global	2000–now	Neural network
15. Xiao et al. (2016)	Vegetation cover	TRAGL FVC	MODIS;	1 km; 8day	Global	2000–14	Gap fraction model and FVC
16. Mu et al. (2017)	Vegetation cover	MuSyQ FVC	MODIS; FY3A/MERSI; FY3B/MERSI	1 km; 5day	China–Asian	2013	The dimidiate pixel model

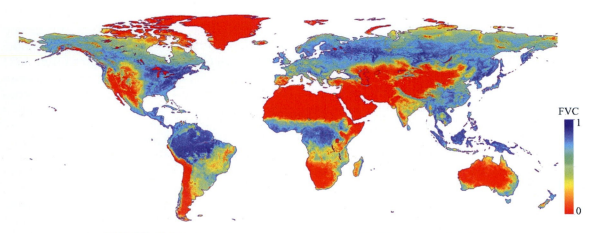

FIGURE 12.16   GLASS fractional vegetation cover (FVC) product on July 2013.

relationship between the two using the following formula: FVC = 1−exp (−0.5 * LAI). Xiao et al. (2016) used the relationship between FVC and LAI and calculated FVC products with the help of the GLASS LAI product, which is an improved version of MODIS LAI product. The FVC product from MERIS onboard ENVISAT, which is able to capture multiangular and multispectral data, is also based on a neural network. Unlike POLDER, FVC products are not directly obtained from LAI products. In addition, FVC products, LAI products, and the fraction of absorbed photosynthetic active radiation products are generated by inputting observations from 13 channels simultaneously. The training of the neural network uses the PROSPECT + SAIL model (Baret et al., 2006), and Bacour et al. (2006) have validated the products. CYCLOPES products, which use SPOT/VEGETATION (VGT) data to generate FVC, also have adopted the neural network method (Baret et al., 2007). To solve the underestimation problem in CYCLOPES FVC products, a series of improved FVC products has been operational (GOEV1, GEOV2, and GEOV3). The new product continues to use the neural network algorithm (Baret et al., 2013, 2016). Research team in Beijing Normal University used machine learning methods combined with MODIS and TM/ETM+ data to produce GLASS FVC products with higher spatial and temporal resolutions (Jia et al., 2015a; Yang et al., 2016). Fig. 12.16 shows the global distribution of GLASS FVC product at 0.5 km spatial resolution. Based on the GLASS FVC product, Liu et al. (2018) developed an FVC estimation algorithm based on machine learning method for VIIRS data, to continue producing 0.5 km resolution time series global FVC product if MODIS sensor failed.

With the development of middle- and high-resolution remote sensing satellite products such as Landsat TM/ETM+, FVC products based on such data have emerged. Sexton et al. (2013) used Landsat TM/ETM+ data (30 m) to downscale MODIS product (MOD44B, 250 m) with the help of empirical model and obtain global FVC with 30m resolution. Hansen et al. (2013) used the decision tree in the machine learning algorithm to produce a 30 m global tree coverage product based on Landsat ETM+ data. This product focuses on forest cover and is mainly used for research on global forest cover change in the 21st century. It is worth noting that although the spatial resolution of this two sets of products has been improved to 30 m, the temporal resolution of these products is only once a year.

Such temporal resolution is difficult to meet the needs of ecological hydrology and urban environment research.

In terms of validation results, the FVC products of SEVIRI and MERIS sensors have good spatial consistency. However, the FVC products of the MERIS sensor systematically underestimate the FVC values (approximately 0.10−0.2) (García-Haro et al., 2008). The same problem also exists for the FVC products of the SEVIRI and VGT sensors (Validation Report of Land Surface Analysis Vegetation products, 2008, URL: http://landsaf.meteo.pt/GetDocument.do?id=301). The FVC products of VGT are higher than those of SEVIRI (by approximately 0.15); therefore, the value of the FVC products of the SEVIRI sensor lies between those of the MERIS and VGT sensors. In their validation report, Fillol et al. (2006) noted that the product value of VGT data was lower than that of high-resolution SPOT data, even after spatial aggregation. Thus, we can infer that the FVC products of the SEVIRI, VGT, and MERIS sensors all underestimate FVC. In addition, Mu et al. (2015) found that the improved first edition of VGT FVC product (GEOV1) existed overestimation in agriculture region, whereas the GLASS FVC product presented better validation performance than the GEOV1 FVC product in the same region (Jia et al., 2018). At present, there is not any unified standard or specification for the evaluation and validation of the remote sensing FVC products, and the evaluation and validation research on different products is not enough.

12.5 Spatiotemporal change analysis of fractional vegetation cover

Based on the time series FVC data, spatiotemporal analysis methods can be used to analyze the temporal and spatial FVC change patterns at regional and global scales, and many researches have been carried out in this field (Wu et al., 2014; Jia et al., 2015b; Vahmani and Ban-Weiss, 2016; Yang et al., 2018). This section takes the spatiotemporal change analysis of fractional forest cover in Northeast China from 1982 to 2011 (Jia et al., 2015b) as an example and demonstrates the impact of long-term time series FVC data on the FVC spatiotemporal variation at regional scale. Northeast China is one of China's main woodland regions. Heilongjiang, Jilin and Liaoning provinces, and the eastern part of Inner Mongolia Autonomous Region contain the largest natural forests in China, mainly distributed in the Daxing'anling, Xiaoxing'anling, and Changbai mountains. Under climate change conditions, the study of long-term time series fractional forest cover change in the northeast region is of great significance for understanding the interaction between forests and climate. The study established fractional forest cover estimation model using neural networks based on land cover data (Liu et al., 2010) from the Landsat interpretation by Chinese Academy of Sciences as the true values and the GIMMS3g NDVI data as the predictor to extract the corresponding training samples. Fractional forest cover in Northeast China during 1982−2011 was estimated using the neural network estimation model.

Results showed that the neural network algorithm has good estimation accuracy for fractional forest cover ($R^2 = 0.81$, RMSE = 11.7%), and the estimation results can be used to analyze the long-term temporal and spatial variation characteristics of fractional forest cover in Northeast China. The spatial patterns of the estimated fractional forest cover in 2000 were shown in Fig. 12.17. In the visual aspects, the spatial distribution of the fractional forest cover had a close agreement with the actual forest cover distribution. The forest cover was mainly distributed in three major mountain regions, including the Daxing'anling Mountains, the Xiaoxing'anling Mountains, and the Changbai Mountains. The

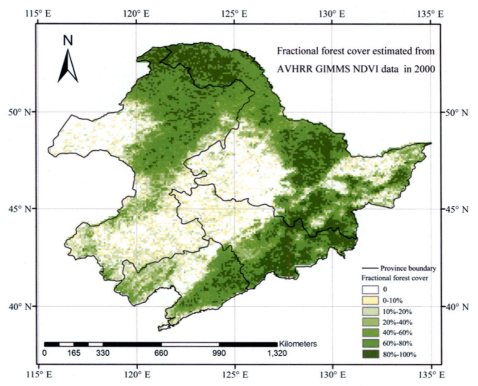

FIGURE 12.17   Fraction forest cover (8-km spatial resolution) estimated using neural networks based on the GIMMS3g normalized difference vegetation index (NDVI) data in 2000 (Jia et al., 2015b).

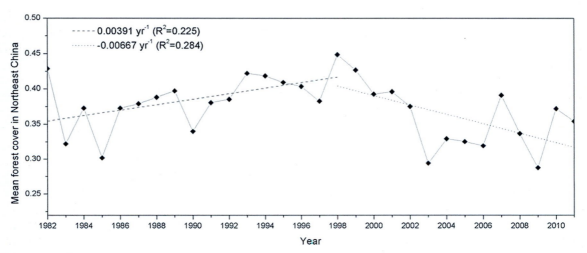

FIGURE 12.18   Interannual variations of the average fractional forest cover over Northeast China in the period of 1982—2011(Jia et al., 2015b).

northwest of Inner Mongolia was the Hulunbuir grassland region, which has few forest land.

Fig. 12.18 shows the interannual variations in the average fractional forest cover in Northeast China during the period of 1982–2011. It can be seen from Fig. 12.18 that there are clearly two distinct periods with opposite trends of change before and after 1998. The fractional forest cover over Northeast China significantly increased by 0.391% year^{-1} from 1982 to 1998, whereas the fractional forest cover decreased by 0.667% year^{-1} from 1998 to 2011.

To facilitate the analysis of the temporal trends in the spatial distribution of the fractional forest cover changes from 1982 to 2011, linear regression analysis was performed for the fractional forest cover in a given year as a function of the years. Fig. 12.19 shows the slope values that resulted from the regression analysis. The linear regression results indicated that the fractional forest cover increased over most mountain regions, which were the main forest cover distribution regions, whereas many plain regions exhibited a decreased forest cover trend. The average regression slope value for the entire Northeast China region was −0.125% year^{-1}, indicating a slow forest cover decrease trend over the study period. The pixels with the largest slopes might be related to human activities or natural disasters, such as afforestation, deforestation, and forest fires. Most pixels (77.63%) had a slope value change between −0.005 and 0.005 year^{-1}, which indicated an insignificant amount of forest cover change (Table 12.2). The obviously decreased pixels (slope < −0.005%, 17.58%) were higher in number than the obviously increased pixels (slope > 0.005, 4.8%). The main increasing forest cover regions were

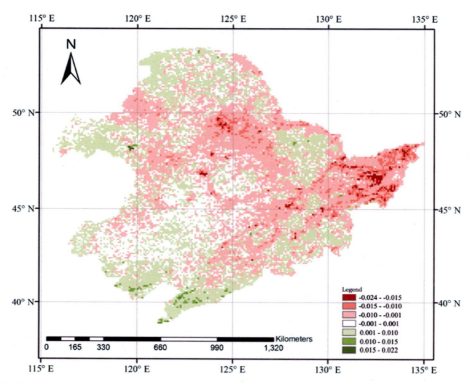

FIGURE 12.19 Spatial distribution of the linear regression slopes for the fractional forest cover changes as a function of year from 1982 to 2011 in Northeast China (Jia et al., 2015b).

TABLE 12.2 Changes in the fractional forest cover from 1982 to 2011 in Northeast China (Jia et al., 2015b).

Regression slopes	Number of pixel	% of total area
<−0.01	1070	5.09%
−0.01 to −0.005	2627	12.49%
−0.005 to 0.005	16,331	77.63%
0.005 to 0.01	908	4.32%
>0.01	100	0.48%

distributed in the remote mountains, whereas a decreasing trend was observed in the edge zones and plain regions.

In summary, neural networks and GIMMS3g NDVI data can be used to estimate the fractional forest cover in Northeast China and achieve satisfactory results. The results indicated that the forest cover in Northeast China tends to decrease slowly during the study period; however, the forest cover change has the opposite tendency before and after 1998. The reasons for the abovementioned forest cover change in Northeast China need to be analyzed along with human activity and climate change data to determine the causes of fractional forest cover change.

12.5.1 Challenges and prospects for fractional vegetation cover estimation

The Earth is a planet brimming with life, where vegetation constitutes the fundamental part of the ecosystem and provides support for the survival of animal life. As a group of living organisms, vegetation has evolved over several billion years, which has given it diversified features. Data extraction would be far less complicated if all vegetation were the same. However, variable features make it difficult to estimate

FVC using remote sensing technology. Currently, no remote sensing method can simultaneously estimate the FVC of all vegetative species, and all remote sensing estimation methods are limited by vegetative species, territory, solar illumination, and soil. In essence, all remote sensing signals are electromagnetic wave, and FVC is a structral parameter of vegetation; thus, converting between the two is crucial for estimating FVC.

Both the empirical and physical methods used in remote sensing have conversion mechanisms for estimating FVC that have advantages and disadvantages. With the empirical method, several empirical coefficients need to be determined regardless of the VI and empirical relationship used. The acquisition of these parameters is related to and restricted by land surface conditions, vegetation types, lighting conditions, and the use of specific sensors. Application of the empirical method is usually limited by changes in these conditions. Using the radiative transfer model, physical method establishes an algorithm framework that considers more factors, thus elucidating the relationship between the optical signals and the physical vegetation parameters. The physical method is more widely applicable, at least in a theoretical sense. Because a large dataset is required and many factors such as time, space, angle, and spectral response need to be considered in the application of satellite observation data, use of the physical method may be restricted due to a lack of data. Moreover, a dilemma exists regarding model selection: if too many parameters need to be estimated in a complex model, then calculations would be difficult; if the model is simplified, then errors would inevitably occur. Thus, two problems, i.e., data and model selection, affect the applicability of the physical method. FVC is estimated more precisely using field measurement than using the remote sensing method on a small scale. However, because of its limitations, field measurement mainly serves as a supplementary validation method for the remote sensing method in most applications.

FVC describes the two-dimensional distribution of vegetation, whereas LAI is a physical quantity describing the vertical distribution of vegetation. Both play an important role in quantitative land surface analysis. With the increase in the quantity and type of remote sensing data, the relevant field data on the global and regional scale are continually established and improved. Therefore, FVC estimation models will become more complex and comprehensive. At the same time, NDVI and other vegetation indices will continue to be applied due to their easy accessibility and will undergo constant development in the foreseeable future.

Some problems are associated with remote sensing—based FVC estimation. For example, remote sensing pixels have a specific spatial scale; when using remote sensing data on a different scale to obtain FVC, the heterogeneity of pixels might bring about mismatches among the results obtained from data of different resolutions. That is, the FVC obtained from high-resolution data differs from that obtained from low-resolution data, thus causing a scale effect. In addition, current FVC estimation is based on the response to chlorophyll content and leaf water content, which results in an FVC value representing GV. Use of the NDVI of vegetation during flowering and fruiting periods would considerably affect FVC determination. Other situations, such as leaf yellowing, water content decline, and discoloration, all have some effect on remote sensing spectral signals, thus resulting in deviations in FVC estimations. Finally, the FVC of different vegetation species in the ecosystem requires further identification and calculation, making this issue even more complex. Therefore, researchers should devote their efforts to improve remote sensing techniques to meet the application requirements of FVC.

FVC is not the most difficult parameter to be estimated by remote sensing techniques.

At present, the precision of results obtained from remote sensing data is approximately 90% or even higher (Jia et al., 2015a; Xiao et al., 2016; Song et al., 2017). Progress in developing software and hardware will increase the precision and efficiency of FVC estimation even further. For field measurement, the digital photography—based extraction method is used most widely. In addition, more and newer types of platforms such as engineering vans and unmanned aircraft will become available to carry digital cameras and to ensure high-precision FVC estimation within a larger scope. VI regression, mixed pixel decomposition, and neural network methods are frequently mentioned in the literature. The neural network method has been widely applied in the algorithms used to generate products because of its capacity to integrate with the physical model, extendibility, and rapidity of calculation.

References

Adams, J.E., Arkin, G.F., 1977. A light interception method for measuring row crop ground cover 1. Soil Sci. Soc. America J. 41 (4), 789—792.

Anderson, M.C., Norman, J.M., Diak, G.R., Kustas, W.P., Mecikalski, J.R., 1997. A two-source time-integrated model for estimating surface fluxes using thermal infrared remote sensing. Remote Sens. Environ. 60 (2), 195—216.

Bacour, C., Baret, F., Béal, D., Weiss, M., Pavageau, K., 2006. Neural network estimation of LAI, fapar, fcover and LAI×Cab, from top of canopy MERIS reflectance data: principles and validation. Remote Sens. Environ. 105 (4), 313—325.

Baret, F., Clevers, J.G.P.W., Steven, M.D., 1995. The robustness of canopy gap fraction estimations from red and near-infrared reflectances: a comparison of approaches. Remote Sens. Environ. 54 (2), 141—151.

Baret, F., Pavageau, K., Béal, D., Weiss, M., Barthelot, B., Regner, P., 2006. Algorithm Theoretical Basis Document for MERIS Top of Atmosphere Land Products (TOAVEG). Report of ESA contract AO/1-4233/02/I-LG.

Baret, F., Hagolle, O., Geiger, B., Bicheron, P., Miras, B., Huc, M., Berthelot, B., Nino, F., Weiss, M., Samain, O., Roujean, J.L., Leroy, M., 2007. LAI, FAPAR, and FCover CYCLOPES global products derived from vegetation. Part 1: principles of the algorithm. Remote Sens. Environ. 110, 305—316.

Baret, F., Weiss, M., Lacaze, R., Camacho, F., Makhmara, H., Pacholcyzk, P., Smets, B., 2013. GEOV1: LAI and FAPAR essential climate variables and FCOVER global time series capitalizing over existing products. Part1: principles of development and production. Remote Sens. Environ. 137, 299–309.

Baret, F., Weiss, M., Verger, A., Smets, B., 2016. Implementing Multi-Scale Agriculture Indicators Exploiting Sentinels ATBD for LAI, FAPAR and FCOVR from PROBA-V Products at 300M Resolution (GEOV3) (ATBD).

Bonham, C.D., 1989. Measurements for Terrestrial Vegetation. Wiley, New York.

Boyd, D.S., Foody, G.M., Ripple, W.J., 2002. Evaluation of approaches for forest cover estimation in the Pacific northwest, USA, using remote sensing. Appl. Geogr. 22, 375–392.

Breiman, L., 2001. Random forests. Mach. Learn. 45 (1), 5–32.

Carlson, T.N., Ripley, D.A., 1997. On the relationship between fractional vegetation cover, leaf area index, and NDVI. Remote Sens. Environ. 62, 241–252.

Chen, T.H., et al., 1997. Cabauw experimental results from the project for inter-comparison of land-surface parameterization schemes. J. Clim. 10 (7), 1194–1215.

Choudhury, B.J., Ahmed, N.U., Idso, S.B., Reginato, R.J., Daughtry, C.S.T., 1994. Relations between evaporation coefficients and vegetation indices studied by model simulations. Remote Sens. Environ. 50 (1), 1–17.

DeFries, R.S., Townshend, J.R.G., Hansen, M.C., 1999. Continuous fields of vegetation characteristics at the global scale at 1-km resolution. J. Geophys. Res. Atmosphere 104, 16911–16923.

DiMiceli, C., Carroll, M., Sohlberg, R., Huang, C., Hansen, M., Townshend, J., 2011. Annual Global Automated MODIS Vegetation Continuous Fields (MOD44B) at 250 m Spatial Resolution for Data Years Beginning Day 65, 2000–2010, Collection 5 Percent Tree Cover. University of Maryland, College Park, MD, USA.

Fillol, E., Baret, F., Weiss, M., Dedieu, G., Demarez, V., Gouaux, P., Ducrot, D., 2006. Cover fraction estimation from high resolution SPOT HRV&HRG and medium resolution SPOT-VEGETATION sensors. Validation and comparison over south-west France. In: Proceedings of Second Recent Advances in Quantitative Remote Sensing Symposium, pp. 659–663.

Foody, G.M., Cutler, M.E., McMorrow, J., Pelz, D., Tangki, H., Boyd, D.S., Douglas, I., 2001. Mapping the biomass of bornean tropical rain forest from remotely sensed data. Glob. Ecol. Biogeogr. 10 (4), 379–387.

García-Haro, F.J., Camacho-de Coca, B., Meliá, J., Martínez, B., 2005a. Operational derivation of vegetation products in the framework of the LSA SAF project. In: EUMETSAT Meteorological Satellite Conference. Dubrovnik (Croatia). 19–23 September. Eumetsat Publ., Darmstad), pp. 247–254. ISBN 92-9110-073-0, ISSN 1011-3932.

García-Haro, F.J., Camacho-de Coca, F., Meliá, J., September 2008. Inter-comparison of SEVIRI/MSG and MERIS/ENVISAT biophysical products over Europe and Africa. In: Proc. of the '2nd MERIS/(A)ATSR User Workshop', Frascati, Italy, pp. 22–26 (ESA SP-666).

Gessner, U., Klein, D., Conrad, C., Schmidt, M., Dech, S., 2009. Towards an automated estimation of vegetation cover fractions on multiple scales: examples of eastern and southern Africa. In: Proceedings of the 33rd International Symposium of Remote Sensing of the Environment, pp. 4–8. Stresa, Italy, May.

Gillies, R.R., Kustas, W.P., Humes, K.S., 1997. A verification of the 'triangle' method for obtaining surface soil water content and energy fluxes from remote measurements of the Normalized Difference Vegetation Index (NDVI) and surface. Inter. J. Remote Sens. 18 (15), 3145–3166.

Gitelson, A.A., Kaufman, Y.J., Stark, R., Rundquist, D., 2002. Novel algorithms for remote estimation of vegetation fraction. Remote Sens. Environ. 80 (1), 76–87.

Glenn, E.P., Huete, A.R., Nagler, P.L., Nelson, S.G., 2008. Relationship between remotely-sensed vegetation indices, canopy attributes and plant physiological processes: what vegetation indices can and cannot tell us about the landscape. Sensors 8, 2136–2160.

Graetz, R.D., Pech, R.P., Davis, A.W., 1988. The assessment and monitoring of sparsely vegetated rangelands using calibrated. Landsat Data Int. J. Remote Sens. 9 (7), 1201–1222.

Guan, K., Wood, E.F., Caylor, K.K., 2012. Multi-sensor derivation of regional vegetation fractional cover in Africa. Remote Sens. Environ. 124, 653–665.

Gutman, G., Ignatov, A., 1998. The derivation of the green vegetation fraction from NOAA/AVHRR data for use in numerical weather prediction models. Int. J. Remote Sens. 19 (8), 1533–1543.

Hansen, M.C., DeFries, R.S., Townshend, J.R.G., Sohlberg, R., Dimiceli, C., Carroll, M., 2002. Towards an operational MODIS continuous field of percent tree cover algorithm: examples using AVHRR and MODIS data. Remote Sens. Environ. 83, 303–319.

Hansen, M.C., DeFries, R.S., Townshend, J., Carroll, M., Dimiceli, C., Sohlberg, R., 2003. Global percent tree cover at a spatial resolution of 500 meters: first results of the MODIS vegetation continuous fields. Algorithm Earth Interact. 7, 1–15.

Hansen, M.C., Potapov, P.V., Moore, R., Hancher, M., Turubanova, S.A., Tyukavina, A., Thau, D., Stehman, S.V., Goetz, S.J., Loveland, T.R., Kommareddy, A., Egorov, A., Chini, L., Justice, C.O., Townshend, J.R.G., 2013. High-resolution global maps of 21st-century forest cover change. Science 342, 850–853.

Hu, Z., He, F., Yin, J., Lu, X., Tang, S., Wang, L., Li, X., 2007. Estimation of fractional vegetation cover based on digital camera survey data and a remote sensing model. J. China Inst. Min. Technol. 17 (1), 116–120.

Huang, C., Townshend, J.R.G., 2003. A stepwise regression tree for nonlinear approximation: applications to estimating subpixel land cover. Int. J. Remote Sens. 24, 75–90.

Huang, C., Song, K., Kim, S., Townshend, J.R.G., Davis, P., Masek, J.G., Goward, S.N., 2008. Use of a dark object concept and support vector machines to automate forest cover change analysis. Remote Sens. Environ. 112, 970–985.

Hurcom, S.J., Harrison, A.R., 1998. The NDVI and spectral decomposition for semi-arid vegetation abundance estimation. Int. J. Remote Sens. 19, 3109–3125.

Jensen, J.R., Qiu, F., Ji, M., 1999. Predictive modelling of coniferous forest age using statistical and artificial neural network approaches applied to remote sensor data. Int. J. Remote Sens. 20, 2805–2822.

Jia, K., Liang, S., Liu, S., Li, Y., Xiao, Z., Yao, Y., Jiang, B., Zhao, X., Wang, X., Xu, S., Cui, J., 2015a. Global land surface fractional vegetation cover estimation using general regression neural networks from MODIS surface reflectance. IEEE Trans. Geosci. Remote Sens. 53, 4787–4796.

Jia, K., Liang, S., Wei, X., Li, Q., Du, X., Jiang, B., Yao, Y., Zhao, X., Li, Y., 2015b. Fractional forest cover changes in Northeast China from 1982 to 2011 and its relationship with climatic variations. IEEE J. Sel. Topics Appl. Earth Observations Remote Sens. 8 (2), 775–783.

Jia, K., Liang, S., Gu, X., Baret, F., Wei, X., Wang, X., Yao, Y., Yang, L., Li, Y., 2016. Fractional vegetation cover estimation algorithm for Chinese GF-1 wide field view data. Remote Sens. Environ. 177, 184–191.

Jia, K., Li, Y., Liang, S., Wei, X., Yao, Y., 2017. Combining estimation of green vegetation fraction in an arid region from Landsat 7 ETM+ data. Remote Sens. 9, 1121.

Jia, K., Liang, S., Wei, X., Yao, Y., Yang, L., Zhang, X., Liu, D., 2018. Validation of global land surface satellite (GLASS) fractional vegetation cover product from MODIS data in an agricultural region. Remote Sens. Lett. 9 (9), 847–856.

Lacaze, R.,P., Richaume, O., Hautecoeur, T., Lalanne, A., Quesney, F., Maignan, P., Bicheron, P., Leroy, M.M., Breon, F.-M., July 2003. Advanced algorithms of the ADEOS2/POLDER2 land surface processing line: application to the ADEOS1/POLDER1 data. In: Proceedings of the XXIIIth IGARSS Symposium, Toulouse, France.

Lal, R., 1994. Soil erosion research methods. CRC Press.

Leprieur, C., Verstraete, M.M., Pinty, B., 1994. Evaluation of the performance of various vegetation indices to retrieve vegetation cover from AVHRR data. Remote Sens. Rev. 10, 265–284.

Li, F., Kustas, W., Preuger, J., Neale, C., Jackson, T., 2005. Utility of remote sensing-based two-source balance model under low- and high-vegetation cover conditions. J. Hydrometeorol. 6, 878–891.

Li, L., Mu, X., Macfarlane, C., Song, W., Yan, G., Yan, K., 2018. A half-Gaussian fitting method for estimating fractional vegetation cover (HAGFVC) of agricultural crops using unmanned aerial vehicle images. Agric. For. Meteorol. 262, 379–390.

Liu, D., Yang, L., Jia, K., Liang, S., Xiao, Z., Wei, X., Yao, Y., Xia, M., Li, Y., 2018. Global fractional vegetation cover estimation algorithm for VIIRS reflectance data based on machine learning methods. Remote Sens. 10, 1648.

Liu, J.G., Pattey, E., 2010. Retrieval of leaf area index from top-of-canopy digital photography over agricultural crops. Agric. For. Meteorol. 150, 1485–1490.

Liu, J.Y., Zhang, Z.X., Xu, X.L., Kuang, W.H., Zhou, W.C., Zhang, S.W., Li, R.D., Yan, C.Z., Yu, D.S., Wu, S.X., Jiang, N., 2010. Spatial patterns and driving forces of land use change in China during the early 21st century. J. Geogr. Sci. 20 (4), 483–494.

Liu, Y., Mu, X., Wang, H., Yan, G., 2012. A novel method for extracting green fractional vegetation cover from digital images. J. Veg. Sci. 23, 406–418.

Lu, D., Weng, Q., 2004. Spectral mixture analysis of the urban landscape in indianapolis with Landsat ETM+ imagery. Photogramm. Eng. Remote Sens. 70, 1053–1062.

Michael, A., White, G.P.A., Ramakrishna, R., Nemani, J., Privette, L., Steven, W.,R., 2000. Measuring fractional cover and leaf area index in arid ecosystems: digital camera, radiation transmittance, and laser Altimetry methods. Remote Sens. Environ. 74, 45–57.

Mohammad, A.A., Shi, Z., Ahmad, Y., Wang, R., 2002. Application of GIS and remote sensing in soil degradation assessments in the Syrian coast. J. Zhejiang Univ. 26 (2), 191–196.

Montandon, L.M., Small, E.E., 2008. The impact of soil reflectance on the quantification of the green vegetation fraction from NDVI. Remote Sens. Environ. 112 (4), 1835–1845.

Mu, X., Huang, S., Ren, H., Yan, G., Song, W., Ruan, G., 2015. Validating GEOV1 fractional vegetation cover derived from coarse-resolution remote sensing images over croplands. IEEE J. Sel. Topics Appl. Earth Observations Remote Sens. 8, 439–446.

Mu, X., Liu, Q., Ruan, G., Zhao, J., Zhong, B., Wu, S., Peng, J., 2017. A 1 km/5 day fractional vegetation cover dataset over China-ASEAN 2013. J. Global Change Data Discov. 1, 45–51.

Mu, X., Song, W., Gao, Z., McVicar, T.R., Donohue, R.J., Yan, G., 2018. Fractional vegetation cover estimation by using multi-angle vegetation index. Remote Sens. Environ. 216, 44–56.

North, P.R.J., 2002. Estimation of FAPAR, LAI, and vegetation fractional cover from ATSR-2 imagery. Remote Sens. Environ. 80, 114–121.

Ormsby, J., Choudry, B., Owe, M., 1987. Vegetation spatial variability and its effect on vegetation indexes. Int. J. Remote Sens. 8, 1301–1306.

Qi, J., Marsett, R.C., Moran, M.S., Goodrich, D.C., Heilman, D.C., Heilman, P., Kerr, Y.H., Zhang, X.X., 2000. Spatial and temporal dynamics of vegetation in the san Pedro River Basin area. Agric. For. Meteorol. 105, 55–68.

Qin, W., Zhu, Q., Zhang, X., Li, W., Fang, B., 2006. Review of vegetation covering and its measuring and calculating method. J. Northwest SCI-TECH Univ. Agric. For. (Nat. Sci. Ed.) 34 (9).

Rautiainen, M., Stenberg, P., Nilson, T., 2005. Estimating canopy cover in Scots pine stands. Silva Fenn. 39, 137–142.

Rogan, J., Franlin, J., Roberts, D.A., 2002. A comparison of methods for monitoring multitemporal vegetation change using thematic mapper imagery. Remote Sens. Environ. 80, 143–156.

Roujean, J.L., Lacaze, R., 2002. Global mapping of vegetation parameters from POLDER multiangular measurements for studies of surface-atmosphere interactions: a pragmatic method and its validation. J. Geophys. Res. 107D, 10129–10145.

Sexton, J.O., Song, X.-P., Feng, M., Noojipady, P., Anand, A., Huang, C., Kim, D.-H., Collins, K.M., Channan, S., DiMiceli, C., Townshend, J.R., 2013. Global, 30-m resolution continuous fields of tree cover: landsat-based rescaling of MODIS vegetation continuous fields with lidar-based estimates of error. Int. J. Digital Earth 6, 427–448.

Song, W., Mu, X., Yan, G., Huang, S., 2015. Extracting the green fractional vegetation cover from digital images using a shadow-resistant algorithm (SHAR-LABFVC). Remote Sens. 7, 10425.

Song, W., Mu, X., Ruan, G., Gao, Z., Li, L., Yan, G., 2017. Estimating fractional vegetation cover and the vegetation index of bare soil and highly dense vegetation with a physically based method. Int. J. Appl. Earth Obs. Geoinf. 58, 168–176.

Su, L., 2009. Optimizing support vector machine learning for semi-arid vegetation mapping by using clustering analysis. J. Photogrammetry Remote Sens. 64, 407–413.

Vahmani, P., Ban-Weiss, G.A., 2016. Impact of remotely sensed albedo and vegetation fraction on simulation of urban climate in WRF-urban canopy model: a case study of the urban heat island in Los Angeles. J. Geophysical Res. Atmospheres 121 (4), 1511–1531.

Van der Meer, F., 1999. Image classification through spectral unmixing. In: Spatial Statistics for Remote Sensing; Stein. Kluwer Academic Publishers, Dordrecht, The Netherlands, pp. 185–193.

Vapnik, V., 1995. The Nature of Statistical Learning Theory. Springer-Verlag, New York, p. 188.

Voorde, T.V.D., Vlaeminck, J., Canters, F., 2008. Comparing different approaches for mapping urban vegetation cover from Landsat ETM+ data: a case study on brussels. Sensors 8, 3880–3902.

Wang, B., Jia, K., Liang, S., Xie, X., Wei, X., Zhao, X., Yao, Y., Zhang, X., 2018. Assessment of sentinel-2 MSI spectral band reflectances for estimating fractional vegetation cover. Remote Sens. 10, 1927.

White, M.A., Asner, G.P., Nemani, R.R., Privette, J.L., Running, S.W., 2000. Measuring fractional cover and leaf area index in arid ecosystems: digital camera, radiation transmittance, and laser Altimetry methods. Remote Sens. Environ. 74, 45–57.

Wu, D., Wu, H., Zhao, X., Zhou, T., Tang, B., Zhao, W., Jia, K., 2014. Evaluation of spatiotemporal variations of global fractional vegetation cover based on GIMMS NDVI data from 1982 to 2011. Remote Sens. 6, 4217.

Xiao, J.F., Moody, A., 2005. A comparison of methods for estimating fractional green vegetation cover within a desert-to-upland transition zone in central New Mexico, USA. Remote Sens. Environ. 98, 237–250.

Xiao, Z., Wang, T., Liang, S., Sun, R., 2016. Estimating the fractional vegetation cover from GLASS leaf area index product. Remote Sens. 8, 337.

Xu, M., Watanachaturaporn, P., Varshney, P.K., Arora, M.K., 2005. Decision tree regression for soft classification of remote sensing data. Remote Sens. Environ. 97, 322–336.

Yang, L., Huang, C., Homer, C.G., Wylie, B.K., Coan, M.J., 2003. An approach for mapping large-area impervious surfaces: synergistic use of landsat-7 ETM+ and high spatial resolution imagery. Can. J. Remote Sens. 29, 230–240.

Yang, W., Chen, J., Songxia, W., Gong, P., Chen, C., 2008. A new spectral mixture analysis method based on spectral correlation matching. J. Remote Sens. 12 (3), 454–461.

Yang, L., Jia, K., Liang, S., Liu, J., Wang, X., 2016. Comparison of four machine learning methods for generating the GLASS fractional vegetation cover product from MODIS data. Remote Sens. 8, 682.

Yang, L., Jia, K., Liang, S., Liu, M., Wei, X., Yao, Y., Zhang, X., Liu, D., 2018. Spatio-temporal analysis and uncertainty of fractional vegetation cover change over northern China during 2001–2012 based on multiple vegetation data sets. Remote Sens. 10, 549.

Yin, J.F., Zhan, X., Zheng, Y., Hain, C.R., Ek, M., Wen, J., Fang, L., Liu, J., 2016. Improving Noah land surface model performance using near real time surface Albedo and green vegetation fraction. Agric. For. Meteorol. 218, 171–183.

Zeng, X.B., Dickinson, R.E., Walker, A., Shaikh, M., DeFries, R.S., Qi, J.G., 2000. Derivation and evaluation of global 1-km fractional vegetation cover data for land modeling. J. Appl. Meteorol. 39, 826–839.

Zhou, Q., Robson, M., 2001. Automated Rangeland vegetation cover and density estimation using ground digital images and a spectral-contextual classifier. Int. J. Remote Sens. 22 (17), 3457–3470.

Zhou, Q., Roberson, M., Pilesjo, R., 1998. On the ground estimation of vegetation cover in Australian Rangelands. Int. J. Remote Sens. 19 (9), 1815–1820.

Further reading

Bartholomé, E., Bogaert, P., Cherlet, M., Defourny, P., Mathoux, P., Vogt, P., March 2002. Rescaling NDVI from the VEGETATION Instrument into Apparent Fraction Cover for Dryland Studies. GLC-2000. First Results Workshop. JRC-Ispra, pp. 18–22.

Brown, M., et al., 1999. Support vector machines for optimal classification and spectral unmixing. Ecol. Modeling 120, 167–179.

Carpenter, G.A., Gopal, S., Macomber, S., Martens, S., Woodcock, C.E., 1999. A neural network method for mixture estimation for vegetation mapping. Remote Sens. Environ. 70, 138–152.

Goodrich, D.C., Heilman, P., Kerr, Y.H., Dedieu, G., Chehbouni, A., Zhang, X.X., 2000. Spatial and temporal dynamics of vegetation in the San Pedro River Basin area. Agric. For. Meteorol. 105 (1–3), 55–68.

Huang, S., Siegert, F., 2006. Land cover classification optimized to detect areas at risk of desertification in North China based on SPOT VEGETATION imagery. J. Arid Environ. 67, 308–327.

Su, L., Choppying, M.J., Rango, A., Martonchik, J.V., Peters, D.P., 2007. Support vector machines for recognition of semi-arid vegetation types using MISR multiangle imagery. Remote Sens. Environ. 107, 299–311.

Walthall, C., Dulaney, W., Anderson, M., Norman, J., Fang, H., Liang, S., 2004. A comparison of empirical and neural network approaches for estimating corn and soybean leaf area index from Landsat ETM+ imagery. Remote Sens. Environ. 92, 465–474.

Vegetation height and vertical structure

Abstract

Vegetation height and vertical structure are important ecosystem parameters because they are highly correlated with the ecological functions and biodiversity of the ecosystem. This chapter addresses the applications of synthetic-aperture radar (SAR) and light detection and ranging (lidar) in the estimation of vegetation height and vertical structures. The first section introduces the field measurement methods. According

© 2020 Elsevier Inc. All rights reserved.

to the size of footprint, the lidar data can be divided into two categories, i.e., small footprint and large footprint lidars. The algorithms used to estimate forest height from these two types of lidar data are different and will be discussed in Sections 2 and 3. Lidars can directly measure vegetation height and vertical structure, but most lidars, especially the space-borne lidars, provide sampling rather than imagery data. Other imagery data, especially the interferometric SAR data (InSAR) are needed for regional mapping of vegetation height and vertical structure. Section 4 is devoted to the application of polarimetric InSAR technique in the estimation of vegetation height and vertical structures. The recent developments of computer vision bring new vitality to photogrammetry and are described in Section 5. Point cloud data can be generated with high density using dense matching technology, which can be used to extract vegetation height information. At last, a brief description of future perspectives is summarized.

Vegetation height and vertical structure are important ecosystem parameters that are highly correlated with the ecological functions and biodiversity of the ecosystem. This chapter addresses the applications of synthetic-aperture radar (SAR) and light detection and ranging (lidar) in the estimation of vegetation height and vertical structures. The first section introduces the field measurement methods. The algorithms used to estimate forest height from lidar data are discussed in Sections 13.2 and 13.3. Section 13.4 is devoted to the application of polarimetric interferometric SAR (PolInSAR) technique in the estimation of vegetation height and vertical structures. The recent developments of computer vision bring new vitality to photogrammetry and are described in Section 13.5. Point cloud data can be generated with high density using dense matching technology, which can be used to extract vegetation height information. At last, a brief description of future perspectives is summarized.

13.1 Field measurement of vegetation height and vertical structure

The vertical structure of vegetation canopy, i.e., the spatial distributions of vegetation mass (Brokaw and Lent, 1999), is a very important

index affecting the mass and energy exchange between landscape and atmosphere as well as the biodiversity of ecosystem.

13.1.1 Height of a single tree

The height of single trees is usually measured by use of hypsometer. Although there are many alternatives, the principal of hypsometer is trigonometry. The height is calculated through the measurement of the other sides and an angle in the triangle composed by tree top, bottom, and the viewer. The angles are measured using the gravity clinometers or gravity sensor. The readings of angles are displayed optically (read it on dial plate) or by electronics (digital numbers). The distance between the tree and the viewer was usually measured using tapes. Ultrasound or laser techniques have been applied in tree height measurement. As an example, how to measure tree height using the *Blume-Leiss* hypsometer is described below.

Blume-Leiss hypsometer (Van Laar and Akça, 2007) is a classic hypsometer used in forestry. From Fig. 13.1, we can see that tree height H can be calculated by

$$H = AB tg\alpha + AE \qquad (13.1)$$

where AB is the horizontal distance, AE is the height of viewer's eyes, and α is the elevation angle.

On the dial plate of *Blume-Leiss* hypsometer, there are several height readings corresponding

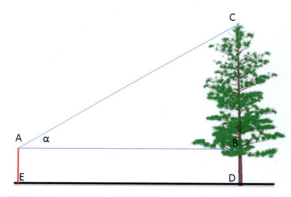

FIGURE 13.1 Tree height measurements using *Blume-Leiss* hypsometer.

to different horizontal distances. The horizontal distance from viewer to tree trunk must be firstly measured. For the convenience of calculation using trigonometry, it is better to set the distance as integers, such as 10, 15, 20, 30 m, and so on. The procedure of measuring a tree height is as follows: press the start button and let the balance move freely; shoot at tree top and wait for 2—3 s until the balance is not moving; press the stop button and read the height in the dial plate. The height plus the height of viewer's eyes AE is the height of the tree.

Ultrasonic or laser hypsometer is frequently employed in the field measurement for remote sensing tasks because of their simplicity. The principle of ultrasonic altimeter is shown in Fig. 13.2 . A is the position of viewer's eyes. B is the tree top. AC is the horizontal line. D is the position of ultrasonic transmitter. E is the position of the tree base. DE is the height of the transmitter and is always set to a constant, such as the breast height. The length of AD is measured by the ultrasonic receiver hold by the viewer. The angles α and β are measured by gravity sensor. According to trigonometry, we have $CD = AD \cdot \sin \alpha$, $AC = AD \cdot \cos \alpha$, and $BC = AC \cdot tg\beta$, so the tree height is $H = BC + CD + DE$.

The principle of laser hypsometer is shown in Fig. 13.3. The lengths of AB and AE are measured by laser range finder, and the angles α and β are measured by gravity sensor. According to trigonometry, $CE = AE \cdot \sin \alpha$ and $BC = AB \cdot \sin \beta$, therefore tree height is $H = BC + CE$.

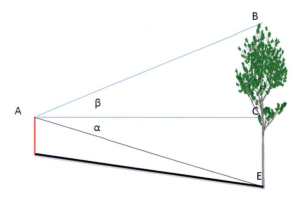

FIGURE 13.3   Principles of laser hypsometer.

Both ultrasonic and laser hypsometers are not perfect. The ultrasonic hypsometer uses ultrasound to measure distance. Its disadvantage is that the transmitting speed can easily be affected by air temperature and moisture content. Its advantage is that ultrasound cannot be stopped by obstacle. It can still work when there are leaves or branches between A and D. Laser hypsometer uses laser range finder to measure distance and does not require a transmitter. But it requires clear sight lines AB and AE, which is sometimes hard to meet under the dense forest circumstance. Some advanced laser hypsometers measure the horizontal distance AC by aiming a portion of the trunk (not to be horizontally) and two angles α and β by aiming the top and bottom; even there are branches and leaves between these points and the viewer.

13.1.2 Relationship between height and diameter at breast height

In field measurement, it is impossible to measure the height of every tree because the top of a tree may be blocked by neighboring trees. The tree height is usually highly correlated with its diameter at breast height (DBH). The relationship between tree height and DBH has the following features:

(1) Tree height increases with the increase of the diameter;

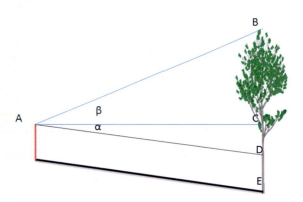

FIGURE 13.2   Principles of ultrasonic hypsometer.

(2) Within a forest canopy at certain diameter class, the probability distribution function is a Gaussian distribution. The number of trees having maximum and minimum heights is very small while that of trees having medium height is big;

(3) The dynamic range of tree height within a diameter class may reach 6–8 m and that within a forest stand is even bigger.
The range is affected by tree species and age. Older forests have narrower dynamic ranges of tree height. Taking the Korean red pine in northeastern China as an example, the coefficient of variation of tree height is 22% in age class III, 15% in age class V, and 7% in age class VII.

(4) The height of majority of trees in a forest stand is close to the mean height of the forest stand.

Certain relationship exists between the arithmetic mean height and diameter at stand level. Taking the diameter of tree classes as horizontal axis and the mean height as vertical axis, the data show a smooth curve representing the relationship between the mean height and corresponding diameter. This curve is called as forest height curve. Various equations have been used to describe the curve, and these equations are called as equations or experimental formula of tree height.

The equations of tree height can be as follows (Li, 2007):

$$h = a_0 + a_1 \log(d) \tag{13.2}$$

$$h = a_0 + a_1(d) + a_2(d^2) \tag{13.3}$$

$$h = a_0 d^{a1} \tag{13.4}$$

$$h = a_0 + \frac{a_1}{d + K} \tag{13.5}$$

$$h = a_0 e^{-a_1/d} \tag{13.6}$$

$$h = a_0 + \frac{a_1}{d} \tag{13.7}$$

where h is tree height, d is diameter at breast height, and log is the common logarithm (base 10). K is a constant. The a_0, a_1, and a_2 are parameters determined by fitting the height–DBH

scatter plot. The best fit will be used as the equation of tree height.

13.1.3 Estimation of average tree height at forest stand level

The height of a forest stand is an index for describing the growing state of the forest stand and also an important indication of the quality of the stand site. Average height is a measurement index for the mean level of tree heights in a forest stand. There are two categories of average height: average height of stand and average top height (HT, mean height of dominant trees). The "average height" commonly used in forestry include conditioned average height, weighted average height, and average top height. These average heights are calculated as follows:

13.1.3.1 Conditioned average height

The height corresponding to the mean diameter of a forest stand (D_g) is called conditioned average height (abbreviation: average height) and expressed as H_D. The height of each diameter class predicted by the forest height curve is called average height of diameter class. To estimate the stand average height in field, the heights of 3–5 trees with diameter similar with the average diameter (D_g) are measured, and the mean value of these trees is taken as the average height of stand.

13.1.3.2 Average height weighted by basal area

The height of a diameter class weighted by corresponding basal area of the class is called weighted average height expressed as $\overline{H}$ (Avery and Burkhart, 1994). The commonly used weighted average height is called Lorey's height, which is calculated by

$$h_L = \frac{\sum\limits_{1}^{n} G_i H_i}{\sum\limits_{1}^{n} G_i} \tag{13.8}$$

where H_i is the height of stem i, G_i is the basal area of stem i, and k is the number of trees.

In multilayered mixed forest, average height should be calculated separately for different layer and species.

With the development of high resolution remote sensing data, it is possible to estimate canopy width of single tree. Pang et al. (2008) proposed an average height weighted by crown area, which can be described as

$$\overline{H_{CW}} = \frac{\sum_{i=1}^{N} h_i \cdot A_i}{\sum_{i=1}^{N} A_i} \quad (13.9)$$

where h_i is tree height and A_i is the area of crown canopy of the tree.

The canopy weighted height is as effective as basal area weighted height in forest inventory. The advantage is that it can be directly measured from remote sensing data.

13.1.3.3 Dominant average height

Besides these average heights described above, the mean height of dominant trees or codominant trees is frequently used in forest inventory. Dominant average height is defined as the mean of dominant trees or codominant trees within a stand. It is expressed as H_T. In practice, 3–5 trees with maximum heights or DBH are taken for calculating dominant average height.

The forest inventory results showed that dominant average height is highly correlated with average height of stand (Li and Lin, 1978). A total of 389 standard forest plots from six provinces of China were analyzed, and a linear relationship between these two heights was obtained (with a correlation coefficient of 0.995):

$$H_1 = 0.233 + 0.828^*H_T \quad (13.10)$$

where H_1 is average height of stand and H_T is dominant average height.

13.2 Small footprint lidar data

13.2.1 Principle of small footprint lidar

The principle of lidar is similar to radar altimeter except the working frequency. Lidar works on optical or near-infrared bands whose frequency is far higher than those of radar altimeters (10,000–100,000 times). Lidar mainly measures the time interval between the transmitting and receiving laser pulses to get the distance from laser transmitter to objects (Bachman, 1979). Its principle can be described as (Baltsavias, 1999)

$$R = (c \cdot t)/2 \quad (13.11)$$

where R is the distance from lidar to an object, c is the speed of light, t is the time for a round-trip from lidar to the object of the laser pulse.

The amplitude of the returned laser pulse can be described by a lidar equation

$$P_R = \frac{P_T G_T}{4\pi R^2} \times \frac{\sigma}{4\pi R^2} \times \frac{\pi D^2}{4} \times \eta_{Atm}^2 \eta_{Sys} \quad (13.12)$$

where P_R is the received power of the returned laser pulse. P_T is the power of transmitted laser pulse. G_T is the gain of transmitting antenna. σ is the cross section of the object. D is the aperture of receiving antenna. η_{Atm} is the one-way atmosphere propagation attenuation coefficient, and η_{sys} is transmission coefficient of the lidar optical system.

Small footprint lidar is usually carried by aircraft. An airborne laser scanner (ALS) system is composed of laser sensor, receiver of Global Position System (GPS), inertial navigation system, or inertial measurement unit. Time stamp is used to synchronize the attitude of aircraft, the lidar, and GPS measurements to get the accurate position of the object (Baltsavias, 1999; Blair et al., 1999; Wynne, 2006).

Small footprint lidar usually uses repeated laser pulses to successively measure the distances to objects (Kirchhof et al., 2008; Wynne, 2006). When laser pulse interacts with leaves, part of the energy is returned back to lidar while remaining energy goes further into lower canopy and ground. Some lidars can only record the first or last returned pulse; others can record both or more returned pulses. Currently, some lidars can record the whole waveform of returned energy as a function of the time delay or distances (Wagner et al., 2006).

The details of forest structure described by small lidar are determined by the sampling frequency of the lidar data. When the sample density is high, such as several, more than 10, or even more pulse shots from a single tree, with enough information of a single tree, it can be used for the estimation parameters of forest structure at both single tree and forest stand level (Lee and Lucas, 2007; Morsdorf et al., 2004). Some lidar system's sampling frequency is so low that there is only one returned pulse from several trees. This type of data can only be used at forest stand level (Anderson et al., 2006; Drake et al., 2002; Lefsky et al., 1999).

The important parameters of individual trees include height, crown width, height of lowest live branch, DBH, and biomass (Bortolot and Wynne, 2005; Brandtberg 2003, 2007). The parameters that can be directly estimated from lidar data are height and crown width (Koch et al., 2006; Popescu et al., 2002, 2003). The rest of the parameters, such as DBH and biomass, needs to be estimated by allometric equations (Maltamo et al., 2004; Popescu, 2007). The parameters of a forest stand are the statistic quantity of the parameters of single trees, such as mean height, basal area per hectare, and forest density (Coops et al., 2007; Donoghue et al., 2007). Table 13.1 summarized the parameters that can be directly or indirectly estimated by lidar data.

The small footprint lidar data are a cloud of three-dimensional points, which is irregularly scattered within a space. Each point has an accurate coordinate (x, y, and z). For bare ground and bare building top surface, the pulse can only be returned once. For vegetations, the laser pulse can be returned more than once because it can penetrate canopies. In dense vegetation, the pulses may be from various places such as from vegetation top, lower branches, understory, and ground surface.

In the processing of small footprint data, the first step is to divide it into ground points and object points, i.e., the pulse returned from ground surface or from vegetation. The data separation is usually implemented by filtering methods. The filtering methods commonly used include local minimum filter, stable linear surface predictions, dynamic surface fitness,

TABLE 13.1　Forest parameters inversed from light detection and ranging (lidar)[a].

Forest parameters	Small footprint lidar system	Large footprint lidar system
Canopy height	Direct retrieval	Direct retrieval
Crown size	Estimated from point cloud	—
Subcanopy topography	Direct retrieval	Direct retrieval
Vertical distribution of intercepted surfaces	Direct retrieval	Direct retrieval
Base area	Modeled using allometric equation	Modeled using allometric equation
Mean stem diameter	Modeled using allometric equation	Modeled using allometric equation
Canopy volume	Estimated from point cloud	Estimated from waveform
Aboveground biomass	Modeled using allometric equation	Modeled using allometric equation
Large tree density	Estimated from point cloud	Estimated from waveform
Canopy density	Estimated from point cloud	Estimated from waveform
LAI	Estimated from point cloud	Estimated from waveform

[a] In most cases, the allometric equation of the specific area is absent, and statistical regression method is used.

mathematical morphology, and irregular triangulation network {Axelsson (2001) #787}. The method based on irregular triangulation network is a stepwise method from coarser to fine scales. The process is similar to the gradually increase of the density of the triangulation network. The ground points are firstly selected from a coarse scale. A triangulation network surface is built using these points. Then the distance and angle from the remaining points to the surface are calculated. If the distance and the angle of a point is less than a threshold, it will be considered as a ground point and taken into the network to form a new network. Otherwise, the point will be discarded. Finally, a digital elevation model (DEM) will be built using the recognized ground points. The height of each point relative to ground surface is calculated and taken as its new z coordinate to form a "normalized" point cloud of canopy, the initial canopy height model (CHM).

13.2.2 Segmentation of single tree and parameters estimation

The dense point cloud data from small footprint lidar have been successfully used in the estimation of structure parameters of single trees, including height, crown width, tree position, and species. The methods often used include watershed method, region growing, curve fitting, vertex clustering, wavelet transform, and the combination of these methods. Table 13.2 (Pang et al., 2008) showed the details

TABLE 13.2 Methods for individual tree crown delineation.

Algorithm	Reference study	Location	Forest type	Lidar system and point density (pts/m²)	Accuracy[a]
Pouring/ watershed	Persson et al., 2002 Koch et al. (2006) Chen et al., 2006	Southern Sweden Southwest of Germany California, USA	Conifer confiner and deciduous blue oak	TopEye; >4 TopoSys; 5–10 ALTM; 9.5	0.98 (0.63 m); 0.58 (0.61 m) 61.7% crowns are correct or satisfactory Crown area 61.3%–68.2%
Region rowing	Hyyppä et al., 2001 Solberg et al., 2015	Southern Finland Southeastern Norway	Conifer	TopoSys; 8–10 ALTM; 5	Standard error of 1.8 m (9.9%) 0.86 (1.4 m); 0.52 (1.1m)
Scale-space theory	Brandtberg et al., (2003)	Eastern USA	deciduous	TopEye; 15	68% (1.1 m); -(-)
Curve fitting	Popescu et al. (2003)	Southeastern USA	Confiner and deciduous	AeroScan; 1.35	-(-); 0.62(1.36 m)
3D clustering	Morsdorf et al. (2004)	Switzerland	Conifer	TopoSys; >10	0.92 (0.61m); 0.20 (0.47 m)
Spatial wavelet	Falkowski et al., 2006	Idaho, USA	Conifer	ALS40; –	0.94 (2.64 m); 0.74 (1.35 m)
Hybrid method	Pang et al. (2008)	California, Washington, Alaska, USA	Conifer	ALTM; 2–5	

[a] *If there are no other notes, the accuracy is in the format of R_h^2 (RMSE$_h$); R_c^2 (RMSE$_c$) for tree height and crown diameter separately. "-" means not available.*

of these methods including reference, research site, forest type, lidar system, point density, and the accuracy of the results.

The main data processing procedure of these segmentation methods includes data classification, height normalization, determination of possible tree top from local maximum, and crown segmentation. Most segmentation methods work on CHM while some directly works on point cloud. Pang et al. proposed a hybrid method to estimate tree height and canopy parameters (Pang et al., 2008). It is based on local maximum, region growing, and polyline fitting. The detailed procedures are as follows:

(1) Apply filtering method to point cloud data to separate ground and vegetation points. DEM are built using ground points. The vegetation points are normalized using the DEM;

(2) Rasterize the normalized dataset, and the maximum height within each pixel is recorded. The CHM is generated after filling the holes where valid lidar returns do not exist;

(3) Apply smooth filter to CHM to further reduce the effect of holes and noise;

(4) Find local maximum as the candidate of tree tops;

(5) Use region-growing method from each candidate tree top to find four sections (profiles) along eight directions;

(6) Use fourth-order polyline to fit each of the four sections using least square method;

(7) Calculate inflexion points of fourth-order polylines, and use these points to calculate crown width;

(8) Average crown widths from eight directions, and use it as the crown width of the tree. The maximum height within a crown is taken as the height of the tree. The position of tree top is the position of the tree;

(9) Process all tree top candidates as described from (5) to (8), then check the overall results. Small trees located within the crown of a big tree will be removed.

Fig. 13.4 is an example of the results from data segmentation. Blue points are the positions of trees. Green circles are the crowns. Red pluses and stars are tree positions from the field measurements.

Fig. 13.5 showed results of tree height estimation using lidar data at Qiliang mountain, Gansu province, China. The horizontal axis is the estimated tree heights using the method described above while the vertical axis is that of field measurements. The correlation coefficient is 0.87.

13.2.3 Estimation of forest parameters at forest stand level

As mentioned above, low-density ALS data may be used to estimate forest parameters at forest stand (or plot) level, especially the forest biomass. Nelson (1997) estimated the basal area, stem volume, and biomass of tropical forest at Costa Rica for this kind of lidar data. The multivariable regression was employed in the research. The results showed that the useful indices derived from lidar data included the mean height of all returns, the mean height of returns from vegetation, and their coefficients of variation. The estimation error would increase if the natural logarithm of independent variables was used in regression. The multivariable regression model with a zero constant and without natural logarithm transformation of indices was the best. The coefficient of determination was 0.4–0.6. Lim and Treitz (2004) estimated the biomass of forests of five species using quartile heights statistically derived from point cloud data. The relationship between the biomass of different forest components (total aboveground biomass, trunk biomass, branch biomass, leaf biomass, bark biomass) and quartile heights was analyzed using linear regression models in logarithmic forms. The results showed that their correlation coefficients were all higher than 0.8. Næsset and Gobakken (2008) studied the relationship between forest biomass and canopy

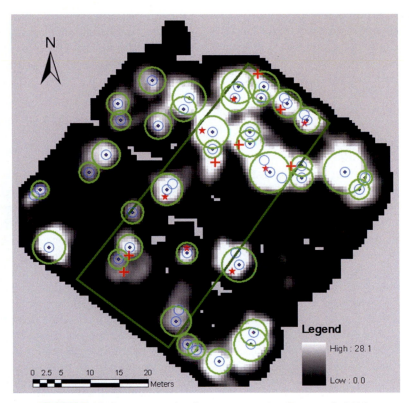

FIGURE 13.4 An example of tree segmentation (Pang et al., 2008).

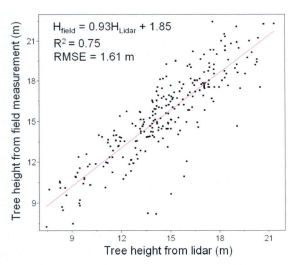

FIGURE 13.5 The results of tree height estimation using light detection and ranging (lidar) data (Pang et al., 2008).

coverage. Two groups of variables derived from lidar data were used in the analysis: one is quartile heights and the other is crown densities. Both the aboveground and belowground biomasses were estimated using 1395 sampling plots within a forest park located at Northern Norway. Sampling plots were classified according to tree species, age class, and site class. A regression model was built using quartile heights and crown density as independent variables and site index and age class as dummy variables. Its coefficient of determination is about 0.7. Zhao et al. (2009) proposed two kinds of scale-invariant models for the estimation of biomass. A linear functional model and an equivalent nonlinear model use lidar-derived canopy height distributions and canopy height quantile functions as predictors, respectively. Results suggested that the models could accurately predict biomass and yield consistent predictive performances across a variety of scales with an R^2 ranging from 0.80 to 0.95 (RMSE: from 14. 3 Mg/ha to 33.7 Mg/ha) among all the fitted models. Latifi et al. (2010) explored the potential of nonparametric prediction and mapping of standing timber volume and biomass. The results showed that the random forest proved to be superior to the nearest neighborhood methods.

In the parameter estimations of forest structures, the 80%–90% percentiles of heights of the first returns or the maximum heights are used to estimate mean forest height or dominant forest height. The prediction models for basal area, stem volume, mean DBH, or stem density use both percentile heights and density variables. These parameters are highly correlated with 20%–30% percentile heights. Some research showed that the dependence of lidar data on forest structure could be affected by locations, species composition, and site conditions (Hall et al., 2005; Holmgren, 2004; Næsset and Gobakken, 2008).

The first return is widely used in the estimation of forest structures because it is relatively stable for different ALS flying at different heights (Næsset and Gobakken, 2008). Generally, the pulses returned from positions of 2m higher than ground surface are considered as returns from vegetation (Nilsson, 1996). Percentile heights present the spatial distribution information of a point cloud (Pang et al., 2008). From the point cloud data of a forest plot, we can calculate heights of percentiles 5% (h_5), 10% (h_{10}), ..., 95 (h_{95}) and maximum height (h_{100}), and then divide the range between lowest height (>2m) and highest heights into 20 intervals. The canopy density is the ratio of the number of points within each interval to the total number of points in the cloud.

Table 13.3 showed the height variables that can be derived from small footprint lidar data. A simple linear regression model can be built for the estimation of forest mean height by using all these variables as independent variables. To consider the nonlinear relationships between these variables and forest biomass, the logarithm form of these parameters are used for biomass estimation:

$$\ln W_i = \beta_0 + \beta_1 \ln h_5 + \beta_2 \ln h_{10} + \ldots$$
$$+ \beta_{19} \ln h_{95} + \beta_{20} \ln h_{max} + \beta_{21} \ln d_5$$
$$+ \beta_{22} \ln d_{10} + \ldots + \beta_{39} \ln d_{95} + \beta_{40} \ln c + \varepsilon$$
$$(13.13)$$

where W_i is the biomass calculated from field data, which can be total aboveground biomass (W_a), trunk biomass (W_s), live branch biomass (W_b), leaf biomass (W_f), underground biomass (W_r), or total biomass (W_t). h_5, h_{10}, ..., h_{95} are percentiles of heights, h_{max} is the maximum height, d_0 is the ratio of the number of all vegetation points to total number of points, d is the ratio of number of points higher than the height of interval n (>2m) (see above on these 20 intervals) to the number of total points in the cloud. c is the ratio of the number of all points higher than 1.8 m to the number of total points. ε is the error of standard distribution [$\varepsilon \sim N(0, \sigma^2)$].

Multivariable linear regression method is the most commonly used method to find the relationships between response and predictor variables, assuming that the relationship between independent and all dependent variables is linear. The variables to be used in the regression model are determined by the stepwise regression based on the change of R^2 due to change of variables (Næsset and Gobakken, 2008). If a variable produces very low F statistics and T test is not significant ($P > 0.1$), it should be removed from the regression. Otherwise, if F statistics is big enough and T test is significant ($P < 0.05$), the variable should be added in the model. The process repeats until all variables in the model do and all variable outside model do not fulfill the requirement. Fig. 13.6 showed the workflow of parameter estimation of forest structures described above.

TABLE 13.3   Light detection and ranging (lidar) matrix.

Variables	Meaning
hmin	Minimum height
hmax	Maximum height
Hrange	Dynamic range of height
hmean	Mean height
hmed	Median height
hvar	Variance of height
hstdv	Standard deviation of height
hskew	Skewness
hkurt	Kurtosis
hcv	Coefficient of variation
Hmad	Mean absolute deviation
ph	Percentile height
Hiqr	Interquartile range
crr	Canopy relief ratio
textureH	Texture of height, standard deviation of returns >0 and ≤ 1m
nPts	Number of lidar returns
nVegPts	Number of lidar vegetation returns
nGrdPts	Number of lidar ground returns
vdensity	Total vegetation density
stratum0	Percentage of ground returns
stratum1	Percentage of vegetation return in $0-1$m
stratum2	Percentage of vegetation return in $1-2.5$m
stratum3	Percentage of vegetation return in $2.5-10$m
stratum4	Percentage of vegetation return in $10-20$m
stratum5	Percentage of vegetation return in $20-30$m
stratum6	Percentage of vegetation return >30 m
pct1	Percentage of 1st returns
pct2	Percentage of 2nd returns
pct3	Percentage of 3rd returns
PCTnotfirst	Percentage of non-1st returns

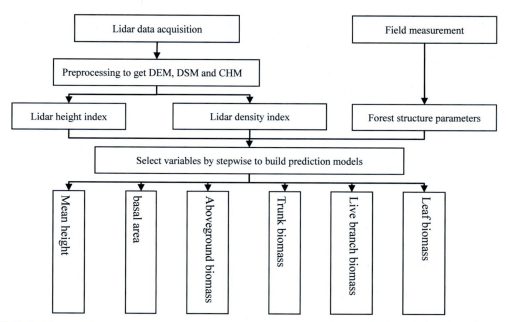

FIGURE 13.6   The workflow of parameters estimation of forest structures using small footprint light detection and ranging (lidar) data.

13.2.4 Large footprint lidar data

13.2.4.1 Principle of large footprint lidar and its application in forestry

Different from the small footprint system that can only record the first or several returned lidar pulses, large footprint lidar system can continuously sample the returned lidar pulse with a given time interval (or step) to form a lidar waveform. The sampling rate determines the details of the information recorded in a waveform. The footprint size of large footprint systems ranges from about 8 −70 m in diameter. Fig. 13.7 illustrates an example of the lidar waveform of a forested area. It indicates the forest structures from tree top to understory vegetation and ground surface. The waveform describes the feature of forest stand rather than a single tree because the footprint is usually greater than a tree crown.

Most of large footprint lidar systems are developed in the United States, such as NASA's airborne systems—Laser Vegetation Imaging Sensor (LVIS) system and Scanning Lidar Imager

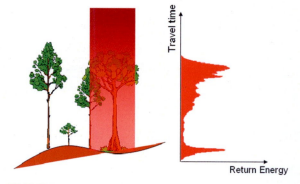

FIGURE 13.7   The light detection and ranging (lidar) waveform from forest.

of Canopies by Echo Recovery (SLICER) system and spaceborne system, Geoscience Laser Altimeter System (GLAS). The size of footprint of LVIS and SLICER is about 8–25m in diameter. LVIS data have been collected over typical forests in the United States, Canada, and Costa Rica. Large footprint lidar data are the direct measurement of vegetation vertical structures and underlying ground. However the extraction of forest structure information depends on our understanding of the relationships between waveform and the forest structure and canopy optical characteristics.

Large footprint lidar systems have been successfully used in the estimation of forest height and biomass. Hyde et al. (2005) examined the extraction of forest structure parameters using LVIS data over mountainous areas. GLAS onboard ICESat (Ice, Cloud, and land Elevation Satellite) launched on January 2002 was the first spaceborne lidar system that can continuously acquires the lidar waveforms returned from atmosphere and terrain objects. It provides a new way to observe the cloud, aerosol, and vegetation structures. It was designed to acquire the thickness and height of cloud to improve the accuracy of short-term weather forecast and to acquire the information of vegetation vertical structures for the assessment of global vegetation distribution and biomass (Zwally et al., 2002). GLAS used non-Doppler, non-interferometric pointing pulse beam. Its footprint size is about 70 m in diameter with footprint intervals about 170 m along flight track (Brenner et al., 2011). GLAS was designed to acquire data continuously for several years. However, after the unexpected death of the first sensor, the data acquisition was adjusted to three times in a year with a period of about 33 days for each period to extend its life time. The three times were at spring (February to March), summer (May to June), and autumn (October to November). This data acquisition strategy lasted till the year 2009. But because of low laser transmitting power, the data acquired after spring of 2007 may not be suitable for vegetation studies. GLAS data have been successfully used in regional estimation of forest structural parameters. The forest height at footprint and regional scales was firstly estimated (Harding and Carabajal, 2005; Lefsky et al., 2005, 2007; Pang et al., 2008; Sun et al., 2008). Lefsky et al. (2005) found that the forest height estimated from GLAS data can be used to accurately estimate forest biomass.

13.2.5 Estimation of forest parameters from lidar waveform data

Large footprint lidar system continuously samples the lidar signals returned from forest canopy top to the ground. It records the reflected energy by various vegetation components (including leaf, trunk, branch, and ticks) at nadir direction. The returned energy at different heights recorded in a lidar waveform is highly correlated with the surface area of reflecting components and can be used to reconstruct the vegetation vertical structure. Lefsky et al. (1999) successfully extracted the vertical distribution of canopy volume using SLICER data. Parker et al. simulated the transmissivity of light within the forest canopy (Parker et al., 2001). The extraction of the vertical distribution of vegetation cross section provides a new tool for vegetation studies and an important parameter for further estimation of other forest parameters such as forest biomass (Drake et al., 2002; Dubayah et al., 1997).

The vertical distribution of forest components change with stand ages (Dubayah et al., 1997; Lefsky et al., 1999). There will be more forest gaps and greater dynamic ranges of ages and heights in mature and overmature forests. On the other hand, even-aged forest stands have more uniform canopies, and a large part of the green crown materials is near the top of the canopy. These spatial structure differences can be clearly identified from lidar waveforms (Lefsky et al., 1999).

The application of ICESat GLAS data in forestry are based on the derivation of various indices from Lidar waveforms and develop correlations between these indices and parameters of forest structures. Lefsky et al. (2005) proposed several indices of GLAS waveform for forestry applications. Other researchers also proposed some indices related to vegetation features (Duncanson et al., 2010; Nelson et al., 2009). Table 13.4 showed the definitions of some of these indexes. More information on application of lidar waveform data will be described in Chapter 15.

TABLE 13.4   The indices derived from large footprint waveform.

Index	Definition	Function
Wavelength	The distance from signal beginning to end	The maximum tree height
Leading edge	The shortest distance between signal beginning to the half energy height of the maximum peak	The variance of forest canopy due to terrain slope
Trailing edge	The shortest distance between signal end to the half energy height of the maximum peak	The variation of ground slope
Wf_variance	The variation of waveform signal	The complexity of terrain
Wf_skew	Waveform slope	Terrain and vegetation
Elevation quartiles	The energy distribution when the waveform is quarterly divided	Elevation variation
Energy quartiles	The energy distribution when the maximum signal amplitude is quarterly divided	Energy variation

13.3 Vegetation canopy height and vertical structure from SAR data

Interferometric SAR (InSAR) is sensitive to the spatial distribution of vegetation components. It can be used to estimate height and vertical profile of vegetation canopy. In this section, we will first introduce the principle of InSAR technique and then summarize the forest height estimation using the difference of penetration depth between short and long wavelength bands. The third part of this section will describe the principle of retrieval of forest height and vertical structure using PolInSAR technique.

13.3.1 Principle of interferometric SAR

The estimation of forest height using SAR data mainly depends on InSAR technique. The use of InSAR technology requires acquiring two backscattering data from the same target from two radar receiving antenna at slightly different locations. These two backscattering signals are from the same target if the distance between the two antennas (baseline) is small and will have similar intensity but different phases. The phase difference of these two signals can be used to infer the height of the target if position of radar locations and the ground

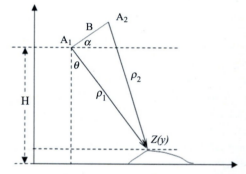

FIGURE 13.8   Principle of interferometric synthetic aperture radar (InSAR).

elevation can be determined. As shown in Fig. 13.8, the two receiving antennas locate at A_1 and A_2 separately. Baseline B is the distance between A_1 and A_2. The angle between horizontal and baseline is baseline angle α. The incidence angle of A_1 is θ and the vertical distance to ground is H. The ranges from A_1 and A_2 to ground are ρ_1 and ρ_2 separately. The elevation of ground point can be expressed as

$$h = H - \rho_1 \cos \theta \qquad (13.14)$$

According to cosine theorem,

$$\rho_2^2 = \rho_1^2 + B^2 - 2\rho_1 B \cos(90^\circ - \theta + \alpha) \qquad (13.15)$$

Therefore,

$$\sin(\alpha - \theta) = \frac{\rho_2^2 - \rho_1^2 - B^2}{2\rho_1 B} \quad (13.16)$$

Suppose the range difference between A_1 and A_2 is expressed as $\delta\rho = \rho_2 - \rho_1$ while phase difference of the two data is φ, then

$$\delta\rho = \frac{\lambda\phi}{2m\pi} \quad (13.17)$$

where $m = 1$ for single pass InSAR mode and $m = 2$ for repeat-pass mode.

Bringing formulas (13.15)–(13.17) into (13.14), we can get

$$h = H - \frac{\left(\frac{\lambda\varphi}{2m\pi}\right) - B^2}{2B\sin(\alpha - \theta) - \frac{\lambda\varphi}{m\pi}}\cos\theta \quad (13.18)$$

Formula (13.18) showed that the ground elevation is correlated with baseline length B, baseline titling angle α, radar height H, and the interferometric phase φ. The data processing steps include the following: (1) estimation of baseline parameters B and α using orbit data; (2) calculation of the interferometric phase through image matching, interferometric phase flattening, and unwrapping; and (3) rebuilding of ground elevation using interferometric phase and baseline parameters.

InSAR technique assumes that the incidence angles from two antennas are nearly the same and there is no change of the target during the acquisition of two images. It is true in single pass mode with two antennas. However, for repeat-pass mode, it is very difficult to guarantee that no ground changes occurred between the two data acquisitions and that the baseline is adequate so incidence angles are almost equal. These requirements will ensure high correlation between two images forming the InSAR data. The correlation or coherence is used to assess the InSAR data quality. The correlation of two single look complex images can be expressed as

$$r = \frac{E(C_1 * C_2)}{\sqrt{E\left(|C_1|^2\right) \cdot E\left(|C_2|^2\right)}} \quad (13.19)$$

where C_1 is the master image while $*C_2$ is the conjunction of the slave image. E is the mathematical expectation. It is obvious that $-1 \leq C \leq 1$. The coherence is defined as

$$\gamma = |r| = \frac{|E(C_1 * C_2)|}{\sqrt{E\left(|C_1|^2\right) \cdot E\left(|C_2|^2\right)}} \quad (13.20)$$

The dynamic range of coherence is 0–1. The coherence of repeat-pass InSAR data over forested area is dominated by volume decorrelation, temporal decorrelation, and thermal decorrelation, in addition to the spatial baseline decorrelation (Zebker and Villasenor, 1992). In practice, the coherence is calculated using pixels within a window of given size N as follows:

$$\gamma = \frac{\sum\limits_{i=1}^{Looks} (C_1 * C_2)}{\sqrt{\sum\limits_{i=1}^{Looks}\left(|C_1|^2\right) \cdot \sum\limits_{i=1}^{Looks}\left(|C_2|^2\right)}} \quad (13.21)$$

13.3.2 Forest height estimation using multifrequency InSAR data

The elevation derived from InSAR data can be used to estimate forest height. As shown in Fig. 13.9, h_{real} is the real forest height from canopy top to ground surface, $h_{effective}$ is the height of scattering phase center of InSAR data. $h_{penetration}$ is the penetration depth of microwave, i.e., the height from canopy top to the scattering phase center. $h_{effective}$ depends on forest structures and SAR system parameters. Generally speaking, the scattering phase center locates at a certain point between canopy top and ground surface. For example, the scattering phase center is located near canopy top in dense forest while it was about half of the canopy height in sparse forest (Hagberg et al., 1995). The different heights of scattering centers at different microwave bands can be used in estimation of top canopy height from multifrequency InSAR data. Neeff et al. (2005) estimated the forest height using the difference of scattering phase centers between X- and

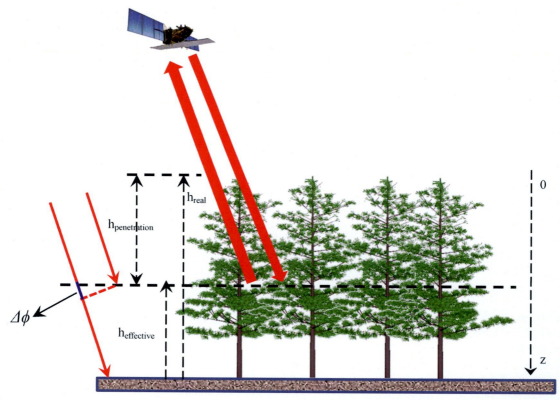

FIGURE 13.9 Sketch map of estimate tree height by interferometric synthetic aperture radar (InSAR) technology (adapted from Floury et al., 1996).

P-bands. The results showed highly correlation between forest canopy height and the difference of scattering phase centers ($R^2 = 0.83$, RMSE = 4.1m).

13.3.3 Retrieval of vegetation vertical structure from PolInSAR data

Polarization is another important attribute of microwave. Polarimetric SAR data are sensitive to the shape and orientation of vegetation components and its dielectric constant. The interaction between vegetation canopy and microwave depends on polarization. Research showed that the cross-polarization return is mainly from volumetric scattering within forest canopy while more scattering from ground surface contributes to copolarization signatures. The combination of polarization and interferometry provides a new possibility for retrieval of vegetation vertical structure. The concept of PolInSAR was first proposed by Cloude and Papathanassion (1998). In their method, the coherence of PolInSAR data was first optimized through polarization combination, and then the coherence decomposition was employed to separate the contributions from different parts of forest canopies for further retrieval of vertical structure. Papathanassiou and Cloude (2001) presented a method for forest height estimation using single baseline PolInSAR data. Assuming that the forest canopy was a randomly distributed vegetation medium with no polarization preference, they derived the explicit model describing relationship between forest structure and complex coherence based on the work (Treuhaft and Siqueira, 2000). The random volume on ground (RVOG) was used to simulate complex multipolarization coherences from initial forest structural parameters, and simulated coherences were compared with those from PolInSAR data. The input forest parameters were then adjusted. The iteration process would continue until the model output matched the real data with a specified accuracy requirement. The adjustable forest parameters in the model include vegetation height, ground phase, canopy attenuation, and the ratio between contributions from canopy and ground at different polarizations. The computation load of the simulation using the proposed model is heavy because it works on six-dimensional space. For reducing the computation load, Cloude and Papathanassiou (2003) proposed a three-step method for the estimation of forest structures. The coherences from polarization combinations in a complex coherence space are first fitted using a line by the least square method. Then the ground phase is estimated from the fitted line. Finally, the forest height and attenuation of forest canopy are estimated through the matching between observation and model simulation on two-dimension space. Cloude (2006) further proposed the concept of polarization coherence tomography (PCT). Given the forest height and ground phase, the vegetation vertical profile function is expanded using Fourier-Legendre series. The vertical profile function can be rebuilt after the expansion coefficients are calculated from PolInSAR data. The following sections will describe these continents in details.

13.3.3.1 The principle of polarimetric SAR interferometry

Section 13.4.1 has described the principle of single polarization interferometry. PolInSAR uses full polarization data, which are always given in the format of scattering matrix. The first step is the vectorization of the scattering matrix. Theoretically, scattering matrix can be vectorized

by any orthogonal basis. For the convenient interpretation of scattering mechanism, it is usually vectorized by Pauli orthogonal basis, which has explicit physical meaning. The vectorization can be expressed as (Cloude and Papathanassiou, 1998)

$$\underline{k} = \frac{1}{2} Trace([S]\psi_p) = \frac{1}{\sqrt{2}} \{s_{hh} + s_{vv}, s_{hh} - s_{vv}, s_{hv} + s_{vh}, i(s_{hv} - s_{vh})\}$$

(13.22)

In reciprocal medium $s_{hv} = s_{vh}$, therefore the vector can be simplified as

$$\underline{k} = \frac{1}{\sqrt{2}} \{s_{hh} + s_{vv}, s_{hh} - s_{vv}, 2s_{hv}\} \quad (13.23)$$

Suppose that the master and slave data are vectorized as $\underline{k}_1$ and $\underline{k}_2$, the conjugate multiplication of them is

$$[T_6] := \left\langle \begin{bmatrix} \underline{k}_1 \\ \underline{k}_2 \end{bmatrix} \begin{bmatrix} \underline{k}_1^{*T} & \underline{k}_2^{*T} \end{bmatrix} \right\rangle = \begin{bmatrix} [T_{11}] & [\Omega_{12}] \\ [\Omega12]^{*T} & [T_{22}] \end{bmatrix}$$

(13.24)

where

$$[T_{11}] = \left\langle \underline{k}_1 \underline{k}_1^{*T} \right\rangle, \quad [T_{22}] = \left\langle \underline{k}_2 \underline{k}_2^{*T} \right\rangle, \quad [\Omega_{12}] = \left\langle \underline{k}_1 \underline{k}_2^{*T} \right\rangle$$

$[T_{11}]$ and $[T_{22}]$ are standard Hermitian coherency matrices that contain the full polarimetric information of master and slave images, respectively (Cloude and Papathanassiou, 1998). $[\Omega_{12}]$ contains not only the polarimetric information but also the interferometric phase information of the different polarimetric channels between InSAR images (Cloude and Papathanassiou, 1998).

As stated previously, the coherence optimization is accomplished by polarization combination. In single polarization interferometry, the master and slave data should be the same polarization. For PolInSAR, master and slave data can be any polarization or polarization combination, for example, the master can be HH + VV while the slave can be HH−VV. A general formula is needed to express the polarization combination.

Suppose w_1 and w_2 are normalized complex vector. Projecting $\underline{k}_1$ and $\underline{k}_2$ onto w_1 and w_2, we can get a new pair of vectors

$$\mu_1 = w_1^{*T}\underline{k}_1, \quad \mu_2 = w_2^{*T}\underline{k}_2 \tag{13.25}$$

Supposing $w_i = \{d_1, d_2, d_3\}^T$, where d_1, d_2, and d_3 are complex number, we can get

$$\mu = \frac{1}{\sqrt{2}}\{d_1(s_{hh} + s_{vv}) + d_2(s_{hh} - s_{vv}) + d_3(2s_{hv})\} \tag{13.26}$$

It is obvious that μ can be any polarization combination determined by the vector w.

Taking μ_1 and μ_2 as master and slave data, the conjugate multiplication of them is

$$[J]: = \left\langle \begin{bmatrix} \mu_1 \\ \mu_2 \end{bmatrix} \begin{bmatrix} \mu_1^* & \mu_2^* \end{bmatrix} \right\rangle = \begin{bmatrix} \mu_1\mu_1^* & \mu_1\mu_2^* \\ \mu_2\mu_1^* & \mu_2\mu_2^* \end{bmatrix}$$

$$= \begin{bmatrix} w_1^{*T}[T_{11}]w_1 & w_1^{*T}[\Omega_{12}]w_2 \\ w_2^{*T}[\Omega_{12}]^{*T}w_1 & w_2^{*T}[T_{22}]w_2 \end{bmatrix} \tag{13.27}$$

The interferometric data are

$$\mu_1\mu_2^* = (w_1^{*T}\underline{k}_1)(w_2^{*T}\underline{k}_2)^{*T} = \underline{w}_1^{*T}[\Omega_{12}]\underline{w}_2 \tag{13.28}$$

Its coherence can be expressed as

$$\gamma = \frac{\left|\langle \underline{w}_1^{*T}[\Omega_{12}]\underline{w}_2 \rangle\right|}{\sqrt{\langle \underline{w}_1^{*T}[T_{11}]\underline{w}_1 \rangle \langle \underline{w}_2^{*T}[T_{22}]\underline{w}_2 \rangle}} \tag{13.29}$$

The coherence optimization is to get the maximum value of γ by selecting w_1 and w_2. This can be achieved through maximizing numerator of Formula (13.29) while keeping the denominator as constant. This in fact is a conditional extremum problem in mathematics with two constraints as

$$\langle \underline{w}_1^{*T}[T_{11}]\underline{w}_1 \rangle = F_1, \quad \langle \underline{w}_2^{*T}[T_{22}]\underline{w}_2 \rangle = F_2 \tag{13.30}$$

We can do this by maximizing the complex Lagrangian defined as

$$L = \underline{w}_1^{*T}[\Omega_{12}]\underline{w}_2 + \lambda_1\left(\underline{w}_1^{*T}[T_{11}]\underline{w}_1 - F_1\right)$$
$$+ \lambda_2\left(\underline{w}_2^{*T}[T_{22}]\underline{w}_2 - F_2\right) \tag{13.31}$$

where λ_1 and λ_2 are Lagrange multipliers. We can solve this maximization problem by setting the partial derivatives of L to zero:

$$\begin{cases} \dfrac{\partial L}{\partial w_1^{*T}} = [\Omega_{12}]\underline{w}_2 + \lambda_1[T_{11}]\underline{w}_1 = 0 \\[2mm] \dfrac{\partial L}{\partial w_2^{*T}} = [\Omega_{12}]^{*T}\underline{w}_1 + \lambda_2^*[T_{22}]\underline{w}_2 = 0 \end{cases} \tag{13.32}$$

It can be transformed into two 3×3 complex eigenvalue problems with common eigenvalues

$$\nu = \lambda_1\lambda_2^*$$

$$\begin{aligned} [T_{22}]^{-1}[\Omega_{12}]^{*T}[T_{11}]^{-1}[\Omega_{12}]\underline{w}_2 &= \nu\underline{w}_2 \\ [T_{11}]^{-1}[\Omega_{12}][T_{22}]^{-1}[\Omega_{12}]^{*T}\underline{w}_1 &= \nu\underline{w}_1 \end{aligned} \tag{13.33}$$

The eigenvectors corresponding to the maximum of eigenvalues ν_{max} should be the solutions of Formula (13.33), i.e., $\underline{w}_{1_{opt}}$ and $\underline{w}_{2_{opt}}$.

13.3.3.2 Mode inversion for forest height estimation

Coherence is preliminarily used to assess the quality of InSAR pairs. Research found that it is also related to the feature of terrain objects. For example, at C band, coherence has inverse relationship with forest biomass/stem volume because more vegetation materials have greater volumetric decorrelation to InSAR data. Models relating coherence to forest structure have been built. Treuhaft and Siqueira (2000) proposed several models that can be applied to different situations. ROV (randomly oriented volume) model only considers the contribution of vegetation. ROVG (randomly oriented volume with an underlying ground surface) considers both the contribution from ground and vegetation. There are two kinds of contribution from ground, i.e., direct backscattering and specular reflection. OV (orientated volume) model considers the vegetation with specified orientation. The following will introduce the ROV and ROVG models in detail because they will be used in forest height estimation from PolInSAR data.

The complex correlation between master and slave images can be expressed as

$$Corss - correlation \equiv \left\langle \hat{p}_1 \vec{E}_{\hat{t}_1}(\vec{R}_1) \hat{p}_2 \vec{E}^*_{\hat{t}_2}(\vec{R}_2) \right\rangle$$

$$(13.34)$$

where $\hat{p}_1$ and $\hat{p}_2$ are the receiving polarizations of master and slave antennae, $\vec{R}_1$ and $\vec{R}_2$ are the ranges of master and slave antennae, $\vec{E}_{\hat{t}_1}(\vec{R}_1)$ and $\vec{E}_{\hat{t}_2}(\vec{R}_2)$ are the signal vectors received at positions $\vec{R}_1$ and $\vec{R}_2$. $\hat{t}_1$ and $\hat{t}_2$ are the polarizations of transmitted and received microwave, and $\langle \rangle$ is the statistical average that is similar to multilook in data processing.

The complex correlation can be further expanded as

master antenna/satellite. In the second line of Formula (13.35), the contributions from vegetation and ground surface are expressed separately. The first item is the complex correlation of signals scattered by the scatterer located at $\vec{R}_{j_v}$ including the backscattering from vegetation and specular reflection from ground. The second item is the complex correlation of signals directly backscattered from ground. For simplicity, the interferometry between the direct backscattering and specular reflection is not considered. However, when their amplitude is comparable, there will be a correlation item between these two terms. The summation in first item is over three dimensions, while in

$$\left\langle \hat{p}_1 \vec{E}_{\hat{t}_1}(\vec{R}_1) \hat{p}_2 \vec{E}^*_{\hat{t}_2}(\vec{R}_2) \right\rangle = \left\langle \left(\sum_{j=1}^{M} \hat{p}_1 \vec{E}_{\hat{t}_1}(\vec{R}_1, \vec{R}_j) \right) \left(\sum_{k=1}^{M} \hat{p}_2 \vec{E}^*_{\hat{t}_2}(\vec{R}_2, \vec{R}_k) \right) \right\rangle$$

$$(13.35)$$

$$= \sum_{j_v}^{M_v} \left\langle \hat{p}_1 \vec{E}_{\hat{t}_1}(\vec{R}_1, \vec{R}_{j_v}) \hat{p}_2 \vec{E}^*_{\hat{t}_2}(\vec{R}_2, \vec{R}_{j_v}) \right\rangle + \sum_{j_g}^{M_g} \left\langle \hat{p}_1 \vec{E}_{\hat{t}_1}(\vec{R}_1, \vec{R}_{j_g}) \hat{p}_2 \vec{E}^*_{\hat{t}_2}(\vec{R}_2, \vec{R}_{j_g}) \right\rangle$$

where $\vec{E}_{\hat{t}_1}(\vec{R}_1, \vec{R}_j)$ is the signal scattered by terrain object located at $\vec{R}_j$ and received by

the second is over two dimensions. It can be further expanded as

$$\left\langle \hat{p}_1 \vec{E}_{\hat{t}_1}(\vec{R}_1) \hat{p}_2 \vec{E}^*_{\hat{t}_2}(\vec{R}_2) \right\rangle = \sum_{j_v=1}^{M_v} \int_{volume} d^3 R_{j_v} P_{vol}(\vec{R}_{j_v}) \left\langle \hat{p}_1 \vec{E}_{\hat{t}_1}(\vec{R}_1; \vec{R}_{j_v}) \hat{p}_2 \vec{E}^*_{\hat{t}_2}(\vec{R}_2; \vec{R}_{j_v}) \right\rangle + \sum_{j_g=1}^{M_g} \int_{surface} d^2 R_{j_g} P_{surf}$$

$$(\vec{R}_{j_g}) \left\langle \hat{p}_1 \vec{E}_{\hat{t}_1}(\vec{R}_1; \vec{R}_{j_g}) \hat{p}_2 \vec{E}^*_{\hat{t}_2}(\vec{R}_2; \vec{R}_{j_g}) \right\rangle$$

$$= \int_{volume} d^3 R \rho_0 W_r^2 \left(\frac{\varphi_1(\vec{R}_1, \vec{R})}{ik_0} - 2|\vec{R}_1 - \vec{R}_0| \right) W_\eta^2(\eta - \eta_0) \times \left\langle \hat{p}_1 \vec{E}_{\hat{t}_1}(\vec{R}_1, w_0; \vec{R}) \hat{p}_2 \vec{E}^*_{\hat{t}_2}(\vec{R}_2, w_0; \vec{R}) \right\rangle +$$

$$\int_{surface} d^2 R \sigma_0 W_r^2 \left(\frac{\varphi_1(\vec{R}_1, \vec{R})}{ik_0} - 2|\vec{R}_1 - \vec{R}_0| \right) W_\eta^2(\eta - \eta_0) \times \left\langle \hat{p}_1 \vec{E}_{\hat{t}_1}(\vec{R}_1, w_0; \vec{R}) \hat{p}_2 \vec{E}^*_{\hat{t}_2}(\vec{R}_2, w_0; \vec{R}) \right\rangle$$

$$(13.36)$$

where $P_{vol}\left(\overrightarrow{R}_{j_v}\right)$ is the probability per unit volume of a scatterer being at $\overrightarrow{R}_{j_v}$ while $P_{surf}\left(\overrightarrow{R}_{j_g}\right)$ is the probability per unit surface area of a surface scattering element being at $\left(\overrightarrow{R}_{j_g}\right)$. w_0 is the central frequency while $k_0 = w_0/c$ is the central wave number of the bandwidth, ρ_0 and σ_0 are the volume density of vegetation and area density of ground surface, respectively, and W_r and W_η are spatial resolutions at range and azimuth directions. $\varphi_1\left(\overrightarrow{R}_1, \overrightarrow{R}\right)$ is the phase of signal $\overrightarrow{E}_{\hat{i}_1}\left(\overrightarrow{R}_1, w_0; \overrightarrow{R}\right)$. Formula (13.36) is the general complex correlation for all models (ROV, RVOG, and OV). The difference between these models is the expansion of the integrals in Formula (13.36).

13.3.3.2.1 Randomly oriented volume model

For randomly oriented vegetation canopy, the signal backscattered by scatterer at $\overrightarrow{R}$ to master antenna/satellite can be expressed as

$$\overrightarrow{E}_{\hat{i}_1}\left(\overrightarrow{R}_1, w_0; \overrightarrow{R}\right) = A^2 F_{b\overrightarrow{R}} \cdot \hat{t}_1 \exp\left[2ik_0|\overrightarrow{R}_1 - \overrightarrow{R}|\right.$$
$$\left. + \frac{4\pi i \rho_0 \left\langle \hat{t} \cdot F_f \cdot \hat{t} \right\rangle (h_v - z)}{k_0 \cos \theta_{\overrightarrow{R}}}\right]$$

(13.37)

where $\theta_{\overrightarrow{R}}$ is the incidence angle from $\overrightarrow{R}_1$ to $\overrightarrow{R}$, A is the reciprocal of slant range, $F_{b\overrightarrow{R}}$ is the backscattering of the scatterer at $\overrightarrow{R}$ while F_f is the forward scattering of the same scatterer. The signal backscattered by scatterer at $\overrightarrow{R}$ to slave antenna/satellite can be expressed in the same way.

Taking them into Formula (13.36) and expanding at $\overrightarrow{R} = \overrightarrow{R}_0$ using Taylor series, we can get

$$\left\langle \hat{p}_1 \overrightarrow{E}_{\hat{i}_1}(\overrightarrow{R}_1) \hat{p}_2 \overrightarrow{E}^*_{\hat{i}_2}(\overrightarrow{R}_2) \right\rangle = A^4 \exp[ik_0(r_1 - r_2)|_0]$$

$$\int_0^{2\pi} W_\eta^2 d\eta \int_{-\infty}^\infty W_r^2 r_0 e^{i\alpha_r r} dr \int_0^{h_v} e^{i\alpha_z z} \rho_0$$

$$\left\langle (\hat{p}_1 \cdot F_b \cdot \hat{t}_1)(\hat{p}_2 \cdot F_b^* \cdot \hat{t}_2) \right\rangle \exp\left[\frac{-8\pi\rho_0 \mathrm{Im}\left\langle \hat{t} \cdot F_f \cdot \hat{t}\right\rangle (h_v - z)}{k_0 \cos \theta_0}\right] dz$$

$$\equiv A^4 e^{i\Phi_0(z_0)} \int_0^{2\pi} W_\eta^2 d\eta \int_{-\infty}^\infty W_r^2 r_0 e^{i\alpha_r r} dr$$

$$\int_0^{h_v} e^{i\alpha_z z} \rho_0 \left\langle (\hat{p}_1 \cdot F_b \cdot \hat{t}_1)(\hat{p}_2 \cdot F_b^* \cdot \hat{t}_2) \right\rangle \exp\left[\frac{-2\sigma_x(h_v - z)}{\cos \theta_0}\right] dz$$

(13.38)

where $r_0 \equiv \left|\overrightarrow{R}_1 - \overrightarrow{R}_0\right|, r_1 \equiv \left|\overrightarrow{R}_1 - \overrightarrow{R}\right|, r_2 \equiv \left|\overrightarrow{R}_2 - \overrightarrow{R}\right|$, h_v is the vegetation height, and $|_0$ means that the distance difference is calculated starting from $\overrightarrow{R} = \overrightarrow{R}_0$. $\varphi_0(z_0) = k_0(r_1 - r_2)|_0$ and $\sigma_x = \frac{4\pi\rho_0 \mathrm{Im}\left\langle \hat{t} \cdot F_f \cdot \hat{t}\right\rangle}{k_0}$. α_z and α_r are the derivations of phase $k_0(r_1 - r_2)$ on vertical and range directions.

Bring Formulas (13–37) and (13–38) into (13–19), we can get the detailed expression of complex coherence as follows:

$$\frac{\left\langle \hat{t}\overrightarrow{E}_{\hat{i}}(\overrightarrow{R}_1) \hat{t}\overrightarrow{E}^*_{\hat{i}}(\overrightarrow{R}_2) \right\rangle}{\sqrt{\left\langle \left|\hat{t}\overrightarrow{E}_{\hat{i}}(\overrightarrow{R}_1)\right|^2 \right\rangle} \sqrt{\left\langle \left|\hat{t}\overrightarrow{E}_{\hat{i}}(\overrightarrow{R}_2)\right|^2 \right\rangle}}$$

$$= \frac{2\sigma_x A_r e^{i\varphi_0(z_0)}}{\cos \theta_0 \left(e^{2\sigma_x h_v/\cos \theta_0} - 1\right)} \int_0^{h_v} \exp\left[\frac{2\sigma_x z'}{\cos \theta_0}\right] dz'$$

(13.39)

The related vegetation structure parameters in (13.39) include (1) vegetation height h_v; (2) ground elevation z_0; and (3) the attenuation coefficient of vegetation layer σ_x.

13.3.3.2.2 ROVG model with specular reflection from ground

Suppose the ground is flat and the parameters describing ground features (elevation and reflecting coefficients) and vegetation are independent. The signal received by single antenna/satellite includes vegetation direct backscattering and the interaction from vegetation to ground and from ground to vegetation. It can be expressed as

$R\left(\theta_{sp1,\overrightarrow{R}}\right)$ is ground reflectance. The explicit expression is

$$\Gamma_{rough} \equiv \exp\left[-2k^2\sigma_H^2\cos\theta_{sp,\overrightarrow{R}}\right], \quad R\left(\theta_{sp1,\overrightarrow{R}}\right) \equiv \begin{pmatrix} R_h\left(\theta_{sp1,\overrightarrow{R}}\right) & 0 \\ 0 & R_v\left(\theta_{sp1,\overrightarrow{R}}\right) \end{pmatrix}$$

(13.41)

where R_h and R_v are Fresnel reflectance coefficients, and σ_H is the standard deviation of surface roughness of Gaussian distribution.

$$\overrightarrow{E}_{\hat{i}_1}(\overrightarrow{R}_1, w_0; \overrightarrow{R}) = A^2 F_{b\overrightarrow{R}} \cdot \hat{t}_1 \exp\left[2ik_0|\overrightarrow{R}_1 - \overrightarrow{R}| + \frac{4\pi i\rho_0\left\langle\hat{t}\cdot F_f\cdot\hat{t}\right\rangle(h_v - z)}{k_0\cos\theta_{\overrightarrow{R}}}\right]$$

$$+A^2 \exp\left[ik_0\left\{\left|\overrightarrow{R}_1 - \overrightarrow{R}_{sp,\overrightarrow{R}}\right| + \left|\overrightarrow{R} - \overrightarrow{R}_{sp,\overrightarrow{R}}\right| + \left|\overrightarrow{R}_1 - \overrightarrow{R}\right|\right\} + \frac{4\pi i\rho_0\left\langle\hat{t}\cdot F_f\cdot\hat{t}\right\rangle h_v}{k_0\cos\theta_{sp1,\overrightarrow{R}}}\right]\Gamma_{rough} \times F_{\overrightarrow{R}_{sp,\overrightarrow{R}}\to\overrightarrow{R}_1}\left\langle R\left(\theta_{sp1,\overrightarrow{R}}\right)\right\rangle_{medg}\cdot\hat{t}_1$$

$$+A^2 \exp\left[ik_0\left\{\left|\overrightarrow{R} - \overrightarrow{R}_1\right| + \left|\overrightarrow{R} - \overrightarrow{R}_{sp,\overrightarrow{R}}\right| + \left|\overrightarrow{R}_{sp,\overrightarrow{R}} - \overrightarrow{R}_1\right|\right\} + \frac{4\pi i\rho_0\left\langle\hat{t}\cdot F_f\cdot\hat{t}\right\rangle h_v}{k_0\cos\theta_{sp1,\overrightarrow{R}}}\right]\Gamma_{rough} \times \left\langle R\left(\theta_{sp1,\overrightarrow{R}}\right)\right\rangle_{medg}\cdot F_{\overrightarrow{R}_1\to\overrightarrow{R}_{sp,\overrightarrow{R}}}\cdot\hat{t}_1$$

(13.40)

where $F_{\overrightarrow{R}_{sp,\overrightarrow{R}}\to\overrightarrow{R}_1}$ is the forward scattering matrix and $F_{\overrightarrow{R}_{sp,\overrightarrow{R}}\to\overrightarrow{R}_1} = F_{\overrightarrow{R}_1\to\overrightarrow{R}_{sp,\overrightarrow{R}}}$, Γ_{rough} is the energy loss due to ground roughness, and

The distance from canopy to ground equals to that from ground to canopy and $\overrightarrow{R}_1 \to \overrightarrow{R}_{sp,R} \to \overrightarrow{R}(x,y,z) \to \overrightarrow{R}_1 = 2\left|\overrightarrow{R}_1 - \overrightarrow{R}(x,y,z)\right|$. Bring Formula (13.40) into (13.36), we can get

$$\left\langle\hat{p}_1\overrightarrow{E}_{\hat{i}_1}(\overrightarrow{R}_1)\hat{p}_2\overrightarrow{E}_{\hat{i}_2}^*(\overrightarrow{R}_2)\right\rangle = \exp[i\varphi_0(z_0)]\exp\left[-\frac{2\sigma_x h_v}{\cos\theta_0}\right]\int_0^{2\pi}W_\eta^2 d\eta\int_{-\infty}^\infty W_r^2 r_0 e^{i\alpha_r r}dr \times \rho_0\left[\left\langle(\hat{p}_1\cdot F_b\cdot\hat{t}_1)(\hat{p}_2\cdot F_b^*\cdot\hat{t}_2)\right\rangle\int_0^{h_v}e^{i\alpha_z z'}\exp\left[\frac{-2\sigma_x z'}{\cos\theta_0}\right]dz'\right.$$

$$\text{Volume} * \text{Volume}$$

$$+\Gamma_{rough}^2\left\langle\left(\hat{p}_1\cdot F_{\overrightarrow{R}_{sp,\overrightarrow{R}}\to\overrightarrow{R}_1}\langle R(\theta_0)\rangle\cdot\hat{t}_1\right)\left(\hat{p}_2\cdot F_{\overrightarrow{R}_{sp,\overrightarrow{R}}\to\overrightarrow{R}_1}\langle R(\theta_0)\rangle\cdot\hat{t}_2\right)\right\rangle\int_0^{h_v}dz'e^{ik_z z'} + \Gamma_{rough}^2\left\langle\left(\hat{p}_1\cdot F_{\overrightarrow{R}_{sp,\overrightarrow{R}}\to\overrightarrow{R}_1}\langle R(\theta_0)\rangle\cdot\hat{t}_1\right)\left(\hat{p}_2\cdot\langle R(\theta_0)\rangle F_{\overrightarrow{R}_1\to\overrightarrow{R}_{sp,\overrightarrow{R}}}\cdot\hat{t}_2\right)\right\rangle\int_0^{h_v}dz'e^{-ik_z z'}$$

$$\text{Ground} - \text{volume} * \text{Ground} - \text{volume} \qquad \text{Ground} - \text{volume} * \text{Volume} - \text{ground}$$

$$+\Gamma_{rough}^2\left\langle\left(\hat{p}_1\cdot\langle R(\theta_0)\rangle F_{\overrightarrow{R}_1\to\overrightarrow{R}_{sp,\overrightarrow{R}}}\cdot\hat{t}_1\right)\left(\hat{p}_2\cdot F_{\overrightarrow{R}_{sp,\overrightarrow{R}}\to\overrightarrow{R}_1}\langle R(\theta_0)\rangle\cdot\hat{t}_2\right)\right\rangle\int_0^{h_v}dz'e^{ik_z z'} + \Gamma_{rough}^2\left\langle\left(\hat{p}_1\cdot\langle R(\theta_0)\rangle F_{\overrightarrow{R}_1\to\overrightarrow{R}_{sp,\overrightarrow{R}}}\cdot\hat{t}_1\right)\left(\hat{p}_2\cdot\langle R(\theta_0)\rangle F_{\overrightarrow{R}_1\to\overrightarrow{R}_{sp,\overrightarrow{R}}}\cdot\hat{t}_2\right)\right\rangle\int_0^{h_v}dz'e^{-ik_z z'}\right]$$

$$\text{Volume} - \text{ground} * \text{Ground} - \text{volume} \qquad \text{Volume} - \text{ground} * \text{Volume} - \text{ground}$$

(13.42)

For reciprocal medium, the items from ground to volume and that from volume to ground should be equal. Therefore, the last four items in Formula (13.42) should be equal. We can get the complex coherence express as

κ_z is the effective vertical interferometric wave number that depends on the imaging geometry and the radar wavelength:

$$\kappa_z = \frac{\kappa \Delta \theta}{\sin \theta_0}, \quad \kappa = \frac{4\pi}{\lambda} \quad (13.46)$$

$$\frac{\left\langle \hat{t}\vec{E}_i(\vec{R}_1)\hat{t}\vec{E}_i^*(\vec{R}_2)\right\rangle}{\sqrt{\left\langle |\hat{t}\vec{E}_i(\vec{R}_1)|^2\right\rangle}\sqrt{\left\langle |\hat{t}\vec{E}_i(\vec{R}_2)|^2\right\rangle}} = A_r e^{i\varphi_0(z_0)} \frac{\left[\frac{\Gamma_{rough}^2 \left\langle R_i(\theta_0)\right\rangle^2 \left\langle \left|\hat{t}\cdot F_{\overrightarrow{R}_{sp,\overrightarrow{R}}\to\overrightarrow{R}_1}\left\langle R(\theta_0)\right\rangle\cdot\hat{t}\right|^2\right\rangle}{\left\langle |\hat{t}\cdot F_b\cdot\hat{t}|^2\right\rangle} h_v \frac{\sin k_z h_v}{k_z h_v}\right]}{\left[\cos\theta_0\left(\frac{e^{2\sigma_x h_v/\cos\theta_0}-1}{2\sigma_x}\right) + 4\frac{\Gamma_{rough}^2\left\langle R_i(\theta_0)\right\rangle^2 \left\langle\left|\hat{t}\cdot F_{\overrightarrow{R}_{sp,\overrightarrow{R}}\to\overrightarrow{R}_1}\left\langle R(\theta_0)\right\rangle\cdot\hat{t}\right|^2\right\rangle}{\left\langle|\hat{t}\cdot F_b\cdot\hat{t}|^2\right\rangle}h_v\right]}$$

$$\equiv A_r e^{i\varphi_0(z_0)} \frac{\left[\int_0^{h_v} e^{i\alpha_z z'} \exp\left[\frac{2\sigma_x z'}{\cos\theta_0}\right]dz' + 4\Delta_t^s h_v \frac{\sin k_z h_v}{k_z h_v}\right]}{\left[\cos\theta_0\left(\frac{e^{2\sigma_x h_v/\cos\theta_0}-1}{2\sigma_x}\right) + 4\Delta_t^s h_v\right]}$$

$$(13.43)$$

It is obvious that the complex coherence in this situation can be fully described by vegetations parameters: (1) vegetation height h_v; (2) ground elevation z_0; (3) attenuation coefficients σ_x; and (4) the contribution of ground relative to those from canopy Δ_i^s.

Papathanassiou and Cloude (2001) simplified the formula (13.43) to

$$\tilde{\gamma} = \exp(i\phi_0)\frac{\tilde{\gamma}_v + m(\vec{w})}{1 + m(\vec{w})} \quad (13.44)$$

where ϕ_0 is the ground phase. m is the ratio of contribution from ground to vegetation volume scattering. $\tilde{\gamma}_v$ is the complex coherence of vegetation canopy, which is correlated to the vegetation attenuation coefficients σ and height h_v as

$$\tilde{\gamma}_v = \frac{I}{I_0}\begin{cases} I = \int_0^{h_v}\exp\left(\frac{2\sigma z'}{\cos\theta_0}\right)\exp(i\kappa_z z')dz' \\ \\ I_0 = \int_0^{h_v}\exp\left(\frac{2\sigma z'}{\cos\theta_0}\right)dz' \end{cases}$$

$$(13.45)$$

where $\Delta\theta$ is the difference of incidence angles from two antennae with respect to the object. It can be seen that complex coherence is correlated with (1) vegetation height h_v; (2) attenuation coefficients σ; (3) effective ratio of ground-to-volume amplitude m, and (4) the phase related to the ground topography φ_0.

For PolInSAR, there are six observations (complex coherence of three polarizations) and six unknowns (φ_0, h_v, σ and m_1, m_2, m_3). The model inversion can be expressed as

$$\begin{bmatrix} h_v \\ \exp(i\varphi_0) \\ \sigma \\ m_1 \\ m_2 \\ m_3 \end{bmatrix} = [M]^{-1}\begin{bmatrix}\tilde{\gamma}_1 \\ \tilde{\gamma}_2 \\ \tilde{\gamma}_3\end{bmatrix} \quad (13.47)$$

where M is a coherence model relating the six observations (three coherences) with six unknown variables. $\tilde{\gamma}_1$, $\tilde{\gamma}_2$, and $\tilde{\gamma}_3$ are the complex coherence for three polarizations. The inversion is a

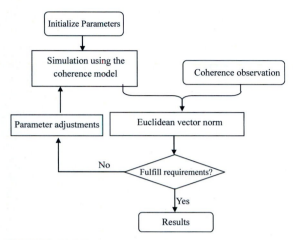

FIGURE 13.10 Flowchart of the model-based inversion for forest height estimation.

nonlinear optimization in six-dimensional space as

$$\min\left(\left\|\begin{bmatrix}\tilde{\gamma}_1\\\tilde{\gamma}_2\\\tilde{\gamma}_3\end{bmatrix} - [M]\begin{bmatrix}h_v\\\exp(i\varphi_0)\\\sigma\\m_1\\m_2\\m_3\end{bmatrix}\right\|\right) \qquad (13.48)$$

where $\|\cdot\|$ indicates norm of the Euclidean vector. The detailed inversion procedure is shown as a flowchart in Fig. 13.10.

The complex coherence is firstly simulated using initial parameters. The distance between simulation and observation is calculated. If the distance is smaller than the requirement, the parameters are output as results. Otherwise, the parameters will be adjusted for new simulation. Repeat the whole process till the results are obtained.

13.3.3.2.3 Three-step method of forest height estimation

Model based method is the nonlinear parameter optimization. Its computation load is heavy and its results can be easily affected by the initial parameters. With these considerations, Cloude

and Papathanassiou (2003) proposed a three-step method.

Transforming formula (13.44), we can get

$$\tilde{\gamma}(\overrightarrow{w}) = \exp(i\varphi_0)\frac{\tilde{\gamma}_v + m(\overrightarrow{w})}{1 + m(\overrightarrow{w})} = \exp(i\varphi_0)$$
$$\left[\tilde{\gamma}_v + \frac{m(\overrightarrow{w})}{1 + m(\overrightarrow{w})}\left(1 - \tilde{\gamma}_v\right)\right]$$

$$(13.49)$$

It can be seen that Formula (13.49) is a line equation in the complex plane. If draw a unit circle within complex plane, the line of (13.49) should intersect with it at two points, and one point should be corresponding to bare ground. According to this analysis, Cloude and Papathanassiou (2003) proposed three-step method for the estimation of forest height. The first step is to get several points in complex plane by polarization combination. These points should be located on the line theoretically. In fact, they will scatter near the line. Therefore, least square method needs to be employed to fit the line and get the intersection points of the line with the unit circle. It is known that the cross-polarization mainly comes from volume scattering. Therefore, the intersection point that is further from the HV polarization point should be the bare ground point. The last step is to calculate vegetation height and attenuation coefficients. $\tilde{\gamma}_v$ can be obtained from the fitted line because the ground phase has been determined. There are two observations (phase and amplitude of $\tilde{\gamma}_v$) and two unknown variables (vegetation height and attenuation coefficients). Model (ROV)-based method can be employed at two-dimensional space to get the vegetation height.

13.3.3.2.4 Polarization coherence tomography

Cloude (Cloude and Papathanassiou, 2003) proposed the concept of PCT to reconstruct the vertical profile of vegetation structure using PolInSAR. Assuming that the vertical profile of vegetation structure is continuous, the vertical profile function can then be expanded using Fourier-Legendre series. If the vegetation height and ground phase are known, the expansion coefficients can be estimated from PolInSAR data, so the vertical profile can be rebuilt. The following section will describe PCT in detail.

Suppose the vertical profile function is $f(z)$, the coherence model can be rewritten as

$$\tilde{\gamma} = \exp(i\varphi_0)\tilde{\gamma}_v = \exp(i\varphi_0)\frac{I}{I_0} \begin{cases} I = \displaystyle\int_0^{h_v} f(z)\exp(ikz)dz \\[2mm] I_0 = \displaystyle\int_0^{h_v} f(z)dz \end{cases}$$

(13.50)

where h_v is vegetation height and φ_0 is ground phase. These two variables are supposed to be known in PCT.

For the expansion using Fourier-Legendre series, we need to make a variable change as follows:

$$\int_0^{h_v} f(z)e^{ikz}dz \xrightarrow{z'=\frac{2z}{h_v}-1} \int_{-1}^1 f(z')e^{ikz'}dz'$$

(13.51)

Numerator and denominator of Formula (13.50) can be rewritten as

$$\int_0^{h_v} f(z)e^{ikz}dz = \frac{h_v}{2}e^{\frac{ikh_v}{2}}\int_{-1}^1 (1+f(z'))e^{\frac{ikh_v}{2}z'}dz'$$

$$\int_0^{h_v} f(z)dz = \frac{h_v}{2}\int_{-1}^1 (1+f(z'))dz'$$

(13.52)

where $f(z')$ can be expanded in $[-1,1]$ as

$$f(z') = \sum_n a_n P_n(z'), \quad a_n = \frac{2n+1}{2}\int_{-1}^1 f(z')P_n(z')dz' \quad (13.53)$$

According to Fourier-Legendre series, the first five items of $P_n(z')$ is

$$P_0(z') = 0; \quad P_1(z') = z'; \quad P_2(z') = \frac{1}{2}(3z'^2-1);$$

$$P_3(z') = \frac{1}{2}(5z'^3-3z'); \quad P_4(z') = \frac{1}{8}(35z'^4-30z'^2+3)$$

(13.54)

Taking (13.53) and (13.52) into (13.50), we can get

$$\tilde{\gamma} = e^{ik_v}\frac{\int_{-1}^1 (1+f(z'))e^{\frac{ik_v}{2}z'}dz'}{\int_{-1}^1 (1+f(z'))dz'} = e^{ik_v}\frac{\int_{-1}^1 \left(1+\sum_n a_n P_n(z')\right)e^{\frac{ik_v}{2}z'}dz'}{\int_{-1}^1 \left(1+\sum_n a_n P_n(z')\right)dz'}$$

$$= e^{ik_v}\frac{(1+a_0)\int_{-1}^1 e^{ik_v z'}dz' + a_1\int_{-1}^1 P_1(z')e^{ik_v z'}dz' + a_2\int_{-1}^1 P_2(z')e^{ik_v z'}dz' + a_3\int_{-1}^1 P_3(z')e^{ik_v z'}dz' + \cdots}{(1+a_0)\int_{-1}^1 dz' + a_1\int_{-1}^1 P_1(z')dz' + a_2\int_{-1}^1 P_2(z')dz' + a_3\int_{-1}^1 P_3(z')dz' + \cdots}$$

$$= e^{ik_v}\frac{(1+a_0)f_0 + a_1 f_1 + a_2 f_2 + a_3 f_3 + \cdots + a_n f_n}{(1+a_0)}$$

(13.55)

Taking Formula (13.54) into (13.55), we can get the explicit forms of first five items in (13.55) as

$$f_0 = \frac{\sin k_v}{k_v}, \quad f_1 = i\left(\frac{\sin k_v}{k_v^2} - \frac{\cos k_v}{k_v}\right),$$

$$f_2 = \frac{3\cos k_v}{k_v^2} - \left(\frac{6-3k_v^2}{2k_v^3} + \frac{1}{2k_v}\right)\sin k_v$$

$$f_3 = i\left(\left(\frac{30-5k_v^2}{2k_v^3} + \frac{3}{2k_v}\right)\cos k_v - \left(\frac{30-15k_v^2}{2k_v^4} + \frac{3}{2k_v^2}\right)\sin k_v\right)$$

(13.56)

$$f_4 = \left(\frac{35\left(k_v^2-6\right)}{2k_v^4} - \frac{15}{2k_v^2}\right)\cos k_v$$

$$+ \left(\frac{35\left(k_v^4-12k_v^2+24\right)}{8k_v^5} + \frac{30\left(2-k_v^2\right)}{8k_v^3} + \frac{3}{8k_v}\right)\sin k_v$$

Formula (13.55) can be further rewritten as

$$\tilde{\gamma} = f_0 + a_{10}f_1 + a_{20}f_2 + \cdots + a_{n0}f_n \quad (13.57)$$

where $a_{n0} = \frac{a_n}{1+a_0}$.

From formula (13.56), it can be seen that the odd items should be real quantity while the even items should be pure imaginary quantity. Therefore, we can get

$$\mathrm{Re}(\tilde{\gamma}) - f_0 = a_{20}f_2 + a_{40}f_4 + \cdots$$
$$\mathrm{Im}(\tilde{\gamma}) = -i(a_{10}f_1 + a_{30}f_3 + \cdots) = a_{10}f_{1i} + a_{30}f_{3i} + \cdots$$

(13.58)

For single baseline PolInSAR data, we can get the first-order solution as

$$
\left.\begin{array}{l}
\widehat{a}_{20} = \dfrac{\mathrm{Re}(\widetilde{\gamma}) - f_0}{f_2} \\[3mm]
\widehat{a}_{10} = \dfrac{\mathrm{Im}(\widetilde{\gamma})}{f_{1i}}
\end{array}\right\} \Rightarrow f(z) = 1 + \widehat{a}_{10} P_1(z)
$$

$$
+ a_{20} P_2(z)
$$

$$
= 1 + \widehat{a}_{10} z + \frac{\widehat{a}_{20}}{2}\left(3z^2 - 1\right), -1 \le z \le 1
$$

$$(13.59)$$

Multibaseline data are needed for higher order solutions. Take dual baselines as example, the equation can be expressed as (x and y denote the two baseline dataset),

$$
\begin{bmatrix}
f_1^x & 0 & f_3^x & 0 \\
0 & f_2^x & 0 & f_4^x \\
f_1^y & 0 & f_2^y & 0 \\
0 & f_3^y & 0 & f_4^y
\end{bmatrix}
\begin{bmatrix}
a_{10} \\
a_{20} \\
a_{30} \\
a_{40}
\end{bmatrix}
=
\begin{bmatrix}
\mathrm{Im}(\widetilde{\gamma}_x) \\
\mathrm{Re}(\widetilde{\gamma}_x) - f_0^x \\
\mathrm{Im}\left(\widetilde{\gamma}_y\right) \\
\mathrm{Re}\left(\widetilde{\gamma}_y\right) - f_0^y
\end{bmatrix}
$$

$$
\Rightarrow [F] \cdot a = g \Rightarrow a = [F]^{-1} g
$$

$$(13.60)$$

It is clear that the first three and five items can be derived using single and dual baseline from PolInSAR data separately. It has to be noted that the expansion coefficients estimated in PCT is a_{n0} rather than a_n because a_0 is unknown. Therefore, the vertical profile obtained from PCT, in fact, is a relative profile.

13.3.4 Forest height from radargrammetry

Radargrammetry is another application of SAR in addition to InSAR. Both of them make use of two SAR images acquired with different views. However, the InSAR reconstruct the terrain elevation using the phase difference of the two complex SAR images, while the radargrammetry depends on the parallax extracted via common points in the two backscattering images. Studies on radargrammetry were initiated in 1960s. Prade (1963) firstly reported the extraction of parallax using two SAR images. The accuracy of parallax measurement depends on the spatial resolution of SAR image. The pixel size of SAR image acquired by European remote sensing (ERS) satellite in range direction is about 26 m, while that of images acquired by Advanced Synthetic Aperture Radar (ASAR) onboard Environmental Satellite (ENVISAT) is 30 m. Therefore, studies on the radargrammetric processing of these SAR images mostly focused on the extraction of DEM other than the forest spatial structure, although what theoretically measured is the terrain elevation including both the forest height and elevation of ground surface. For example, Toutin (2000) reported the extraction of DEM in British Columbia, Canada, using RADARSAT-1 images; Li et al. (2006) explored the terrain and temporal effects on the DEM extraction through the radargrammetric processing of ERS-1/2 images; and d'Ozouville et al. (2008) reported that the accuracy of DEM extracted by the radargrammetric processing of ENVISAT ASAR images was about ±15m.

Spatial resolutions of SAR images acquired by the new satellites, such as TerraSAR-X, COSMO-SkyMed, and Chinese GaoFen-3, have been greatly improved in the last decades. Some studies explored the extraction of DEM using these new SAR images. For example, Nonaka et al. (2009) reported the production of DEM using TerraSAR-X images; Salvini et al. (2015) tested the rebuilding of DEM using COSMO-SkyMed images. In addition to DEM, some studies reported the extraction of digital surface model (DSM). For example, Raggam et al. (2010) reported that there is nearly no difference between the extracted elevation and reference data, while the obvious differences existed over forested areas. Capaldo et al. (2015) reported the potential to extract DSM, based on the evaluation of three sensors, i.e., COSMO-SkyMed, TerraSAR-X, RADARSAT-2, and two radargrammetric processing softwares, including PCI Geomatica and SISAR.

The great improvement of spatial resolution of SAR images makes it possible to measure forest height with the help of ground surface elevation. Therefore, scientists gradually paid attention to the application of radargrammetry over forested areas in recent years. Perko et al. (2011) found that it was possible to extract forest CHM from the DSM of TerraSAR-X subtracted by the elevation of ground surface. Persson and Fransson (2014) examined the extraction of forest aboveground biomass and forest height using the CHM extracted from TerraSAR-X radargrammetric DSM and lidar data. Solberg et al. (2015) reported that the estimation accuracy of forest aboveground biomass using TerraSAR-X radargrammetric DSM was comparable to that from TerraSAR-X InSAR.

13.4 Vegetation canopy height and vertical structure from airborne stereoscopic images

Lidar and InSAR have been widely evaluated in extraction of vegetation canopy height. In addition to these two data types, another one can be used for the measurement of forest height are the stereoscopic images. Studies on stereoscopic images are much earlier than lidar and InSAR. However, the application of stereoscopic data is only limited within the domain of survey and mapping, due to the complicated data processing techniques.

As shown in Fig. 13.11, the two point cloud data are come from CCD camera of Hasselblad-50 and lidar sensor of LMS-Q680i (Pang et al., 2016). Both point cloud data give good characteristics and similar features of canopy surface information. But lidar has more returns from intermediate layers vegetation

and some ground returns beneath of canopy. Photogrammetry has few returns from vertical layers and ground beneath canopy (Xia et al., 2019). So far, it is difficult to get terrain information of forested areas using photogrammetry technology.

The database of Web of Science, searched by "photogrammetry" and "forest" or "stereo" and "forest" as "Title/Keywords/Abstract", shows that the earliest study on the measurement of forest structures was the work of Rhody (1977). The earliest work in the chapter could be traced back to 1959 by Avery using helicopter stereoscopic images. Other earlier studies were carried out by Lyons in 1967, Edwards and Waelti in 1972, Aldred and Sayn-Wittgenstein in 1972, and Aldred and Hall in 1975. They measured tree heights and identified tree species using airborne stereoscopic images. There are two modes to collect airborne stereoscopic images, i.e., cross-track using two camera and along-track using one camera. Rhody (1977) adapted the first modes using two cameras with baseline length of 2.5 and 4.5 m. The overlay is 94.8% and 90.6%, respectively, with a flying height of 70 m aboveground level.

In 1970−1980s, the stereoscopic images is interpreted by the analytical photogrammetry, i.e., common points are visually identified, while the coordinates of common points in object space are mathematically calculated using rigorous equations based on camera parameters, measured photo coordinates, and ground control. In 1990s, the digital photogrammetry becomes popular, i.e., the hardcopy of films is scanned into digital images. For example, Gagnon et al. (1993) scanned films by 300 DPI (dot per inch), 450 DPI and 600 DPI, with corresponding pixel size of 9.3 , 6.2, and 4.6 cm, respectively.

In 2000s, common points can be automatically identified using algorithm of image matching. Therefore, the stereoscopic processing technique becomes much easier. Scientist gradually paid attention to the automatic measurement of forest spatial structures

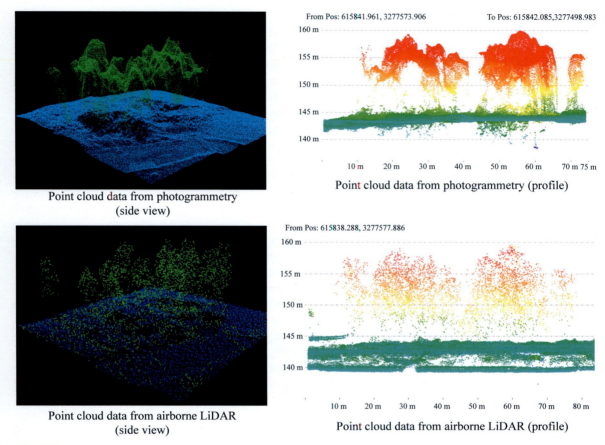

FIGURE 13.11   The comparison of point cloud data from photogrammetry and airborne light detection and ranging (lidar). Blue points are ground returns. Green points are vegetation returns. Point cloud data from photogrammetry (side view). Point cloud data from photogrammetry (profile). Point cloud data from airborne lidar (side view). Point cloud data from airborne lidar (profile).

using stereoscopic images. Gong et al. (2000) reported that the accuracy of automatic measurement of forest heights using digitized films was about 1.5 m. Miller et al. (2000) investigated images acquired with different scales, i.e., 1:20,000, 1:26,000, and 1:10,000, but digitized with a fix resolution of 800 DPI. The errors of height estimation ranges from 1.4 m to 2.2 m. Katsch and Stocker (2000) used images collected with scales of 1:10,000 in Germany and 1:35,000 in South Africa. They reported that the estimation accuracy of forest

height was about −1.1 −1.4 m and −7.5 −1.4 m, respectively.

Studies working on digitized films always talked more about effects of scales or the pixel size in the digitization but less on the forward and side overlaps because they have to use what they had. Overlaps are important factors determining the accuracy of automatic measurement of forest heights. In the period of hardcopy, it is important to use as small overlaps as possible due to the cost of films. The typical forward and side overlaps are about

60% and 30%, respectively, for the mapping of terrain elevations. As the decreasing of image collection cost using digital cameras, some scientists evaluated the estimation of forest height using larger overlaps. For example, Hirschmugl et al. (2007) reported that the accuracy of forest heights was 0.7m, extracted using images collected by UltracamD digital camera with forward overlaps of 90% and pixel size of 15 cm.

The rapidly developed unmanned aerial vehicle (UAV) further reduced the cost of image collection. The application of computer vision makes it possible to make automatic forest measurement using images captured by consumer-grade cameras. The UAV stereoscopic cameras become popular in the studies on forests (Chen et al., 2017; Dandois and Ellis, 2013; Guerra-Hernandez et al., 2017). Ni et al. (2018) systematically evaluated the impacts of forward overlaps and image resolutions on the mapping of three-dimensional structures of forest canopy using UAV stereoscopic imagery and found that finer image resolutions required larger forward overlaps. Ni et al. (2019) synthesized the leaf-on and leaf-off UAV stereo imagery for the inventory of deciduous forests structures. Zhang et al. (2019) demonstrated that forest height extracted from UAV stereo imagery was helpful for the estimation of forest LAI.

This section reviewed the history of measurement of forest spatial structures using stereoscopic images, which was initiated 60 years ago. The platform developed from manned helicopters to UAV; the format of images evolved from hardcopy films to digitized films and finally digital images; the interpretation technique advanced from manual analytical photogrammetry to fully automatic digital photogrammetry. The great decrease of image collection cost, technical difficulty of stereoscopic processing, and the requirement of cameras will finally make the UAV stereoscopic imagery to be practical in forest inventory.

13.5 Future perspectives

This chapter described the field measurement of forest height and its estimation using lidar data and InSAR data. It can be seen from the principle that lidar data are the direct measurement of forest height, especially, the high-density small footprint data that can be used to estimate single tree structure. Large footprint lidar can record the waveform of returned pulse within footprint. It provides detailed vertical distribution of vegetation canopies. However, they both have their own defects. For example, small footprint lidar system is always airborne and can only be used locally. Large footprint lidar systems can be spaceborne but acquire sample data only globally. The estimation of forest parameters is restricted within footprint, and the results are difficult to be expanded onto area especially due to the large interval between flying orbits. InSAR provide another tool for the estimation of forest height. Single polarization needs the underlying ground elevation data or requires dual frequency InSAR data. PolInSAR is a new technique. It can be used to estimate forest height without the use of other data. Therefore, it has great potential in the estimation of forest height over regional area. It should be pointed out that the estimated forest height from PolInSAR is in fact derived from the height of the scattering phase center. The relationship between canopy height and the phase center height can also be affected by forest structures. PCT can only get the relative vertical structure. The forest height mapping in regional or global scale should be conducted by the fusion of lidar and InSAR in the future. lidar point sampling data can be used to calibrate the relationship between scattering phase center and forest height and to transform the relative vertical structures from PCT to absolute structures. If a high-quality DEM is available, the photogrammetry with dense matching features provides an optional way for up story vegetation height information.

References

Anderson, J., Martin, M.E., Smith, M.L., Dubayah, R.O., Hofton, M.A., Hyde, P., Peterson, B.E., Blair, J.B., Knox, R.G., 2006. The use of waveform lidar to measure northern temperate mixed conifer and deciduous forest structure in New Hampshire. Remote Sens. Environ. 105, 248−261.

Avery, T.E., Burkhart, H.E., 1994. Forest Measurements. McGraw-Hill, New York.

Axelsson, P., 2001. Ground estimation of laser data using adaptive TIN-models. Proc. Of OEEPE Workshop on Airborne Laserscanning and Interferometric SAR for Detailed Digital Elevation Models, pp. 185−208.

Bachman, C.G., 1979. Laser Radar Systems and Techniques. Mass: Artech House, Dedham.

Baltsavias, E.P., 1999a. Airborne laser scanning: existing systems and firms and other resources. ISPRS J. Photogrammetry Remote Sens. 54, 164−198.

Baltsavias, E.P., 1999b. Airborne laser scanning: basic relations and formulas. ISPRS J. Photogrammetry Remote Sens. 54, 199−214.

Blair, J.B., Rabine, D.L., Hofton, M.A., 1999. The Laser Vegetation Imaging Sensor: a medium-altitude, digitisation-only, airborne laser altimeter for mapping vegetation and topography. ISPRS J. Photogrammetry Remote Sens. 54, 115−122.

Bortolot, Z.J., Wynne, R.H., 2005. Estimating forest biomass using small footprint LiDAR data: an individual tree-based approach that incorporates training data. ISPRS J. Photogrammetry Remote Sens. 59, 342−360.

Brandtberg, T., 2003. Detection and analysis of individual leaf-off tree crowns in small footprint, high sampling density lidar data from the eastern deciduous forest in North America. Remote Sens. Environ. 85, 290−303.

Brandtberg, T., 2007. Classifying individual tree species under leaf-off and leaf-on conditions using airborne lidar. ISPRS J. Photogrammetry Remote Sens. 61, 325−340.

Brenner, A.C., Zwally, H.J., Bentley, C.R., Csatho, B.M., Harding, D.J., Hofton, M.A., Minster, J.-B., Roberts, L., Saba, J.L., Thomas, H., Yi, D., 2011. GLAS algorithm theoretical basis document version 5.0 - derivation of range and range distributions from laser pulse waveform analysis for surface elevations, roughness, slope, and vegetation heights. http://www.csr.utexas.edu/glas/pdf/WFAtbd_v5_02011Sept.pdf.

Brokaw, N.V.L., Lent, R.A., 1999. Vertical structure. In: Hunter, M.L. (Ed.), Maintaining Biodiversity in Forest Ecosystems. UK Cambridge University Press, Cambridge, pp. 373−399.

Capaldo, P., Nascetti, A., Porfiri, M., Pieralice, F., Fratarcangeli, F., Crespi, M., Toutin, T., 2015. Evaluation and comparison of different radargrammetric approaches for Digital Surface Models generation from COSMO-SkyMed, TerraSAR-X, RADARSAT-2 imagery: analysis of Beauport (Canada) test site. ISPRS J. Photogrammetry Remote Sens. 100, 60−70.

Chen, Q., Baldocchi, D., Gong, P., Kelly, M., 2006. Isolating individual trees in a savanna woodland using small footprint lidar data. Photogramm. Eng. Remote Sens. 72 (8), 923−932.

Chen, S.J., McDermid, G.J., Castilla, G., Linke, J., 2017. Measuring vegetation height in linear disturbances in the boreal forest with UAV photogrammetry. Remote Sens. 9 (12).

Cloude, S.R., 2006. Polarization coherence tomography. Radio Sci. 41 (RS4017), 1−27.

Cloude, S.R., Papathanassiou, K.P., 1998. Polarimetric SAR interferometry. IEEE Trans. Geosci. Remote Sens. 36, 1551−1565.

Cloude, S.R., Papathanassiou, K.P., 2003. Three-stage inversion process for polarimetric SAR interferometry. IEE Proc. - Radar, Sonar Navig. 150, 125−134.

Coops, N.C., Hilker, T., Wulder, M.A., St-Onge, B., Newnham, G., Siggins, A., Trofymow, J.A.T., 2007. Estimating canopy structure of Douglas-fir forest stands from discrete-return LiDAR. Trees Struct. Funct. 21, 295−310.

d'Ozouville, N., Deffontaines, B., Benveniste, J., Wegmueller, U., Violette, S., de Marsily, G., 2008. DEM generation using ASAR (ENVISAT) for addressing the lack of freshwater ecosystems management, Santa Cruz Island, Galapagos. Remote Sens. Environ. 112 (11), 4131−4147.

Dandois, J.P., Ellis, E.C., 2013. High spatial resolution three-dimensional mapping of vegetation spectral dynamics using computer vision. Remote Sens. Environ. 136, 259−276.

Donoghue, D.N.M., Watt, P.J., Cox, N.J., Wilson, J., 2007. Remote sensing of species mixtures in conifer plantations using LiDAR height and intensity data. Remote Sens. Environ. 110, 509−522.

Drake, J.B., Dubayah, R.O., Clark, D.B., Knox, R.G., Blair, J.B., Hofton, M.A., Chazdon, R.L., Weishampel, J.F., Prince, S., 2002. Estimation of tropical forest structural characteristics using large-footprint lidar. Remote Sens. Environ. 79, 305−319.

Dubayah, R., Blair, J.B., Bufton, J.L., Clark, D.B., JaJa, J., Knox, R., Luthcke, S.B., Prince, S., Weishample, J., 1997. The vegetation canopy lidar mission, land satellite information in the next decade II: sources and applications. In: ASPRS Proceedings, pp. 100−112.

Duncanson, L.I., Niemann, K.O., Wulder, M.A., 2010. Estimating forest canopy height and terrain relief from GLAS waveform metrics. Remote Sens. Environ. 114, 138−154.

Falkowski, M.J., Smith, A.M.S., Hudak, A.T., Gessler, P.E., Vierling, L.A., Crookston, N.L., 2006. Automated estimation of individual conifer tree height and crown diameter via two-dimensional spatial wavelet analysis of lidar data. Canadian J. Remote Sens. 32 (2), 153–161.

Floury, N., Le Toan, T., Souyris, J.C., Singh, K., Stussi, N., Hsu, C.C., Kong, J.A., 1996. Interferometry for forest studies. Fringe 96. http://earth.esa.int/workshops/fringe_1996/floury/.

Gagnon, P.A., Agnard, J.P., Nolette, C., 1993. Evaluation of a soft-copy photogrammetry system for tree-plot measurements. Canadian Journal of Forest Research-Revue Canadienne De Recherche Forestiere 23 (9), 1781–1785.

Gong, P., Biging, G.S., Standiford, R., 2000. Use of digital surface model for hardwood rangeland monitoring. J. Range Manag. 53 (6), 622–626.

Guerra-Hernandez, J., Gonzalez-Ferreiro, E., Monleon, V.J., Faias, S.P., Tome, M., Diaz-Varela, R.A., 2017. Use of multi-temporal UAV-derived imagery for estimating individual tree growth in pinus pinea stands. Forests 8 (8).

Hagberg, J.O., Ulander, L.M.H., Askne, J., 1995. Repeat-pass SAR interferometry over forested terrain. IEEE Trans. Geosci. Remote Sens. 33, 331–340.

Hall, S.A., Burke, I.C., Box, D.O., Kaufmann, M.R., Stoker, J.M., 2005. Estimating stand structure using discrete-return lidar: an example from low density, fire prone ponderosa pine forests. For. Ecol. Manag. 208, 189–209.

Harding, D.J., Carabajal, C.C., 2005. ICESat waveform measurements of within-footprint topographic relief and vegetation vertical structure. Geophys. Res. Lett. 32, L21S10.

Hirschmugl, M., Ofner, M., Raggam, J., Schardt, M., 2007. Single tree detection in very high resolution remote sensing data. Remote Sens. Environ. 110 (4), 533–544.

Holmgren, J., 2004. Prediction of tree height, basal area and stem volume in forest stands using airborne laser scanning. Scand. J. For. Res. 19, 543–553.

Hyde, P., Dubayah, R., Peterson, B., Blair, J.B., Hofton, M., Hunsaker, C., Knox, R., Walker, W., 2005. Mapping forest structure for wildlife habitat analysis using waveform lidar: validation of montane ecosystems. Remote Sens. Environ. 96, 427–437.

Hyyppä, J., Kelle, O., Lehikoinen, M., Inkinen, M., 2001. A segmentation-based method to retrieve stem volume estimates from 3-D tree height models produced by laser scanners. IEEE Trans. Geosci. Remote Sens. 39, 969–975.

Katsch, C., Stocker, M., 2000. Automatic determination of stand heights from aerial photography using digital photogrammetric systems. Allg. Forst Jagdztg. 171 (4), 74–80.

Kirchhof, M., Jutzi, B., Stilla, U., 2008. Iterative processing of laser scanning data by full waveform analysis. ISPRS J. Photogrammetry Remote Sens. 63, 99–114.

Koch, B., Heyder, U., Weinacker, H., 2006. Detection of individual tree crowns in airborne lidar data. Photogramm. Eng. Remote Sens. 72, 357–363.

Latifi, H., Nothdurft, A., Koch, B., 2010. Non-parametric prediction and mapping of standing timber volume and biomass in a temperate forest: application of multiple optical/LiDAR-derived predictors. Forestry 83, 395–407.

Lee, A.C., Lucas, R.M., 2007. A LiDAR-derived canopy density model for tree stem and crown mapping in Australian forests. Remote Sens. Environ. 111, 493–518.

Lefsky, M.A., Harding, D., et al., 1999. Surface lidar remote sensing of basal area and biomass in deciduous forests of eastern Maryland, USA. Remote Sens. Environ. 67, 83–98.

Lefsky, M.A., Cohen, W.B., Acker, S.A., Parker, G.G., Spies, T.A., Harding, D., 1999. Lidar remote sensing of the canopy structure and biophysical properties of douglas-fir western hemlock forests. Remote Sens. Environ. 70, 339–361.

Lefsky, M.A., Hudak, A.T., Cohen, W.B., Acker, S.A., 2005. Geographic variability in lidar predictions of forest stand structure in the Pacific Northwest. Remote Sens. Environ. 95, 532–548.

Lefsky, M.A., Keller, M., Pang, Y., De Camargo, P.B., Hunter, M.O., 2007. Revised method for forest canopy height estimation from Geoscience Laser Altimeter System waveforms. J. Appl. Remote Sens. 1, 013537.

Li, F., 2007. Tree Measurement. http://www.jingpinke.com/xpe/portal/22cf354b-1288-1000-887c-5fd719521ae5?start=31&courseID=S0700033&uuid=8a833999-1e4881f5-011e-4881f6ac-008f.

Li, T., Lin, C., 1978. The Designing of Sampling Plot Size for Continuous Forest Inventory Forest Science and Technology, pp. 17–19.

Li, Z.L., Liu, G.X., Ding, X.L., 2006. Exploring the generation of digital elevation models from same-side ERS SAR images: topographic and temporal effects. Photogramm. Rec. 21 (114), 124–140.

Lim, K.S., Treitz, P.M., 2004. Estimation of above ground forest biomass from airborne discrete return laser scanner data using canopy-based quantile estimators. Scand. J. For. Res. 19, 558–570.

Maltamo, M., Eerikäinen, K., Pitkänen, J., Hyyppä, J., Vehmas, M., 2004. Estimation of timber volume and stem density based on scanning laser altimetry and expected tree size distribution functions. Remote Sens. Environ. 90, 319–330.

Miller, D.R., Quine, C.P., Hadley, W., 2000. An investigation of the potential of digital photogrammetry to provide measurements of forest characteristics and abiotic damage. For. Ecol. Manag. 135 (1–3), 279–288.

Morsdorf, F., Meier, E., Kötz, B., Itten, K.I., Dobbertin, M., Allgöwer, B., 2004. LIDAR-based geometric reconstruction of boreal type forest stands at single tree level for forest and wildland fire management. Remote Sens. Environ. 92, 353–362.

Neeff, T., Dutra, L.V., dos Santos, J.R., Freitas, C.C., Araujo, L.S., 2005. Tropical forest measurement by interferometric height modeling and P-band radar backscatter. For. Sci. 51, 585–594.

Nelson, R., 1997. Modeling forest canopy heights: the effects of canopy shape. Remote Sens. Environ. 60, 327–334.

Nelson, R., Ranson, K.J., Sun, G., Kimes, D.S., Kharuk, V., Montesano, P., 2009. Estimating Siberian timber volume using MODIS and ICESat/GLAS. Remote Sens. Environ. 113, 691–701.

Ni, W., Sun, G., Pang, Y., Zhang, Z., Liu, J., Yang, A., Wang, Y., Zhang, D., 2018. Mapping three-dimensional structures of forest canopy using UAV stereo imagery: evaluating impacts of forward overlaps and image resolutions with LiDAR data as reference. Ieee Journal of Selected Topics in Applied Earth Observations and Remote Sensing 11, 3578–3589.

Ni, W.J., Dong, J.C., Sun, G.Q., Zhang, Z.Y., Pang, Y., Tian, X., Li, Z.Y., Chen, E.X., 2019. Synthesis of leaf-on and leaf-off unmanned aerial vehicle (UAV) stereo imagery for the inventory of aboveground biomass of deciduous forests. Remote Sens. 11.

Nilsson, M., 1996. Estimation of tree heights and stand volume using an airborne lidar system. Remote Sens. Environ. 56, 1–7.

Næsset, E., Gobakken, T., 2008. Estimation of above- and below-ground biomass across regions of the boreal forest zone using airborne laser. Remote Sens. Environ. 112, 3079–3090.

Nonaka, T., Hayakawa, T., Griffiths, S., Mercer, B., IEEE, 2009. DEM production utilizing stereo technology of TerraSAR-X data. 2009 IEEE Int. Geosci. Remote Sens. Sympos. 1 (5), 2537–2540.

Pang, Y., Li, Z.Y., et al., 2016. LiCHy: the CAF's LiDAR, CCD and hyperspectral integrated airborne observation system. Remote Sens. 8 (5), 398.

Pang, Y., Lefsky, M., Andersen, H., Miller, M., Sherrill, K., 2008. Automatic tree crown delineation using discrete return LiDAR data and its application in ICEsat vegetation product validation. Can. J. Remote Sens. 34, 471–484.

Papathanassiou, K.P., Cloude, S.R., 2001. Single-baseline polarimetric SAR interferometry. IEEE Trans. Geosci. Remote Sens. 39, 2352–2363.

Parker, G.G., Lefsky, M.A., Harding, D.J., 2001. Light transmittance in forest canopies determined using airborne laser altimetry and in-canopy quantum measurements. Remote Sens. Environ. 76, 298–309.

Perko, R., Raggam, H., Deutscher, J., Gutjahr, K., Schardt, M., 2011. Forest assessment using high resolution SAR data in X-band. Remote Sens. 3, 792–815.

Persson, H., Fransson, J.E.S., 2014. Forest variable estimation using radargrammetric processing of TerraSAR-X images in Boreal forests. Remote Sens 6, 2084–2107.

Persson, Å., Holmgren, J., Söderman, U., 2002. Detecting and measuring individual trees using an airborne laser scanner. Photogramm. Eng. Remote Sens. 68, 925–932.

Popescu, S.C., 2007. Estimating biomass of individual pine trees using airborne lidar. Biomass Bioenergy 31, 646–655.

Popescu, S.C., Wynne, R.H., Nelson, R.F., 2002. Estimating plot-level tree heights with lidar: local filtering with a canopy-height based variable window size. Comput. Electron. Agric. 37, 71–95.

Popescu, S.C., Wynne, R.H., Nelson, R.F., 2003. Measuring individual tree crown diameter with lidar and assessing its influence on estimating forest volume and biomass. Can. J. Remote Sens. 29, 564–577.

Prade, L., 1963. An analytical and experimental study of stereo for radar. Photogramm. Eng. 24 (2), 294–300.

Raggam, H., Gutjahr, K., Perko, R., Schardt, M., 2010. Assessment of the Stereo-Radargrammetric Mapping Potential of TerraSAR-X Multibeam Spotlight Data. IEEE Trans. Geosci. Remote Sens 48, 971–977.

Rhody, B., 1977. New, versatile stereo-camera system for large-scale helicopter photography of forest resources in central-Europe. Photogrammetria 32, 183–197.

Salvini, R., Carmignani, L., Francioni, M., Casazza, P., 2015. Elevation modelling and palaeo-environmental interpretation in the Siwa area (Egypt): Application of SAR interferometry and radargrammetry to COSMO-SkyMed imagery. Catena 129, 46–62.

Solberg, S., Riegler, G., Nonin, P., 2015. Estimating forest biomass from TerraSAR-X stripmap radargrammetry. IEEE Trans. Geosci. Remote Sens. 53, 154–161.

Sun, G., Ranson, K.J., Kimes, D.S., Blair, J.B., Kovacs, K., 2008. Forest vertical structure from GLAS: an evaluation using LVIS and SRTM data. Remote Sens. Environ. 112, 107–117.

Toutin, T., 2000. Evaluation of radargrammetric DEM from RADARSAT images in high relief areas. IEEE Trans. Geosci. Remote Sens. 38, 782–789.

Treuhaft, R.N., Siqueira, P.R., 2000. Vertical structure of vegetated land surfaces from interferometric and polarimetric radar. Radio Sci. 35, 141–177.

Van Laar, A., Akça, A., 2007. Forest Mensuration. Springer, Dordrecht, The Netherlands.

Wagner, W., Ullrich, A., Ducic, V., Melzer, T., Studnicka, N., 2006. Gaussian decomposition and calibration of a novel small-footprint full-waveform digitising airborne laser scanner. ISPRS J. Photogrammetry Remote Sens. 60, 100–112.

Wynne, R.H., 2006. Lidar remote sensing of forest Resources at the scale of management. Photogramm. Eng. Remote Sens. 72, 1310.

Xia, Y., Pang, Y., Liu, L., Chen, B., Dong, B., Huang, Q., 2019. Forest height growth monitoring of cunninghamia lanceolata plantation using multi-temporal aerial photography with the support of high accuracy DEM. Sci. Silvae Sin. 55 (4), 108—121.

Zebker, H.A., Villasenor, J., 1992. Decorrelation in interferometric radar echoes. IEEE Trans. Geosci. Remote Sens. 30, 950—959.

Zhang, D., Liu, J., Ni, W., Sun, G., Zhang, Z., Liu, Q., Wang, Q., 2019. Estimation of forest leaf area index using height and canopy cover information extracted from unmanned aerial vehicle stereo imagery. IEEE J. Sel. Topics Appl. Earth Observ. Remote Sens. 1—11.

Zhao, K., Popescu, S., Nelson, R., 2009. Lidar remote sensing of forest biomass: a scale-invariant estimation approach using airborne lasers. Remote Sens. Environ. 113, 182—196.

Zwally, H.J., Schutz, B., Abdalati, W., Abshire, J., Bentley, C., Brenner, A., Bufton, J., Dezio, J., Hancock, D., Harding, D., Herring, T., Minster, B., Quinn, K., Palm, S., Spinhirne, J., Thomas, R., 2002. ICESat's laser measurements of polar ice, atmosphere, ocean, and land. J. Geodyn. 34, 405—445.

Abstract

Biomass is a critical variable for carbon cycles, soil nutrient allocations, fuel accumulation, and habitat environments in terrestrial ecosystems. A brief introduction is presented in the first section. The ground measurement techniques are described in the second section. It is then followed by various remote sensing methods for biomass estimation using optical remote sensing data (Section 15.3), lidar, and synthetic aperture radar data (Section 15.4). The methods for estimating biomass from multiple data sources are described in Section 15.5. Compared with the last edition, the estimation of biomass using spaceborne stereoscopic images is added in Section 15.4, and the synthesis of multisource remote sensing data is added in Section 15.5.

Advanced Remote Sensing, Second Edition
https://doi.org/10.1016/B978-0-12-815826-5.00014-3

543

© 2020 Elsevier Inc. All rights reserved.

Biomass is a critical variable for carbon cycles, soil nutrient allocations, fuel accumulation, and habitat environments in terrestrial ecosystems. After a brief introduction, the ground measurement techniques are first presented. It is then followed by various remote sensing methods for biomass estimation using optical remote sensing data (Section 14.3), lidar, and synthetic aperture radar (SAR) data (Section 14.4). The methods for estimating biomass from multiple data sources are described in Section 14.5.

14.1 Introduction

Biomass, in general, includes the aboveground and belowground living mass, such as trees, shrubs, vines, roots, and the dead mass of fine and coarse litter associated with the soil. Biomass is often defined in dry weight terms. Aboveground tree biomass, for example, refers to the weight of that portion of the tree found above the ground surface, when oven-dried until a constant weight is reached. Plot-level biomass estimates are typically expressed on a per unit area basis (for example, Mg ha-1 or kg m-2) and are made by summing the biomass values for the individual trees on a plot, then standardizing for the land area covered by that plot.

Biomass governs the potential carbon emission that could be released to the atmosphere due to deforestation, and regional biomass changes have been associated with important outcomes in ecosystem functional characteristics and climate change. The roles and impacts of biomass on carbon cycles, soil nutrient allocations, fuel accumulation, and habitat environments in terrestrial ecosystems have long been recognized. Accurate delineation of biomass distribution at scales from local and regional to global becomes significant in reducing the uncertainty of carbon emission and sequestration, understanding their roles in influencing soil fertility and land degradation or restoration, and understanding the roles in environmental processes and sustainability (Foody et al., 2003).

Because of the difficulty in collecting field data of belowground biomass, most previous research on biomass estimation focused on aboveground biomass (AGB). AGB includes stems, stumps, branches, bark, seeds, and foliage. AGB cannot be directly measured by any sensor from space. Land cover stratification combined with ground sampling is the traditional method to inventory the biomass of a region. Remote sensing data are playing increasingly important roles in forest biomass estimation. Lu (2006) outlined methods for estimating AGB, including field measurements, remote sensing and GIS techniques. Remote sensing techniques have many advantages in AGB estimation over traditional field measurement methods and provide the potential to estimate AGB at different scales. The AGB may be directly estimated using linear or nonlinear regression models and various machine learning algorithms, such as neural network, random forest (RF) (Breiman, 2001; Goetz et al., 2005; Hansen et al., 2008), and MaxEnt (maximum entropy) (Phillips et al., 2006; Saatchi et al., 2011). The data from multisensor are desirable for regional and global biomass mapping, for example, biomass data from field measurements (e.g., FIA—Forest Inventory and Analysis plots; Blackard et al., 2008), lidar (GLAS—Geoscience Laser Altimeter System) (Blackard et al., 2008; Nelson et al., 2009), and image data from Landsat. Moderate Resolution Imaging Spectroradiometer (MODIS) have been used together to perform regional biomass mapping.

In the following sections, we will discuss the allometric method and various remote sensing methods for biomass estimation, then review various algorithms used in biomass estimation using optical remote sensing data. Since the principal and data characteristics of lidar data have been presented in last chapter, and the methods used to infer biomass from remote sensed variables are similar for lidar and radar data, we will first describe possible variables which can be derived from lidar and radar data, then discuss the methods for biomass estimation from these variables.

14.2 Allometric methods

Traditional techniques based on field measurement include destructive sampling methods and nondestructive allometric equations. Destructive sampling of many trees of the dominating species at each site is extremely tedious and not always practical, thus allometric method is practically used.

The basic principle of the allometric equations is that in many organisms, the growth rate of one part of the organism is proportional to that of another. For example, the trunk diameter of a tree is highly correlated with trunk weight. If a range of tree sizes is measured, a regression equation can be derived for predicting tree weight. Because tree diameter is easy to measure but tree weight is much more difficult to determine, this gives a relatively easy way to estimate the standing biomass of forest stands. Allometric equations estimate biomass by regressing a measured sample of biomass against tree variables that are easy to measure in the field (e.g., diameter at breast height, DBH).

Komiyama et al. (2008) reviewed the representative allometric equations for the mangrove forests that are shown in Table 14.1. Note that allometric relationships often show site- or species-dependency and two common equations for mangroves have been proposed and listed at the end of Table 14.1.

Because many biomass allometric equations for many North American tree species have been published, there has been an increasing need for generalized and consistent biomass equations to model the carbon cycle at the national scale. Jenkins et al. (2003) responded to this need for the United States by adopting a so-called generalized regression method. They developed the equation for the total AGB (kg) for trees 2.5 cm DBH and larger hardwood and softwood species in the United States:

$$bm = \exp(\beta_0 + \beta_1 \ln \text{DBH}) \tag{14.1}$$

where β_0 and β_1 are two coefficients given in Table 14.2, and DBH is the diameter at breast height (cm). As cautioned by the authors, this equation may be applied for large-scale analyses of biomass or carbon stocks and trends, but should be used cautiously at very small scales where local equations may be more appropriate.

Lambert et al. (2005) also developed sets of equations based on DBH and on DBH and height for 33 species, groups of hardwood and softwood, and for all species combined, for calculating Canadian national AGB of these trees. The DBH-based equations are as follows:

$$
\begin{aligned}
y_{wood} &= \beta_{wood1} D^{\beta_{wood2}} + e_{wood} \\
y_{bark} &= \beta_{bark1} D^{\beta_{bark2}} + e_{bark} \\
y_{stem} &= \widehat{y}_{wood} + \widehat{y}_{bark} + e_{stem} \\
y_{foliage} &= \beta_{foliage1} D^{\beta_{foliage2}} + e_{foliage} \\
y_{branches} &= \beta_{branches1} D^{\beta_{branchese2}} + e_{branches} \\
y_{crown} &= \widehat{y}_{foliage} + \widehat{y}_{branches} + e_{crown} \\
y_{total} &= \widehat{y}_{wood} + \widehat{y}_{bark} + \widehat{y}_{foliage} + \widehat{y}_{branches} + e_{total}
\end{aligned}
\tag{14.2}
$$

where y_i is the dry biomass compartment i of a living tree (kilograms); i is wood, bark, stem, foliage, branches, crown, and total; $\widehat{y}_i$ is the prediction of y_i; D is the DBH (centimeters); β_{jk} are model parameters with coefficient estimates b_{jk}; j is wood, bark, foliage, and branches; $k = 1$ or 2; and e_i are the error terms.

The DBH and height-based equations are as follows:

$$
\begin{aligned}
y_{wood} &= \beta_{wood1} D^{\beta_{wood2}} H^{\beta_{wood3}} + e_{wood} \\
y_{bark} &= \beta_{bark1} D^{\beta_{bark2}} H^{\beta_{bark3}} + e_{bark} \\
y_{stem} &= \widehat{y}_{wood} + \widehat{y}_{bark} + e_{stem} \\
y_{foliage} &= \beta_{foliage1} D^{\beta_{foliage2}} H^{\beta_{foliage3}} + e_{foliage} \\
y_{branches} &= \beta_{branches1} D^{\beta_{branchese2}} H^{\beta_{branchese3}} + e_{branches} \\
y_{crown} &= \widehat{y}_{foliage} + \widehat{y}_{branches} + e_{crown} \\
y_{total} &= \widehat{y}_{wood} + \widehat{y}_{bark} + \widehat{y}_{foliage} + \widehat{y}_{branches} + e_{total}
\end{aligned}
\tag{14.3}
$$

where H is the height in meters; stem, crown, and total AGBs are obtained by adding their

TABLE 14.1   Allometric equations for various mangroves based on diameter at breast height (DBH) (cm).

Aboveground tree weight (W_{top} in kg)	Belowground tree weight (W_R in kg)
Avicennia germinans	*Avicennia marina*
$W_{top} = 0.140DBH^{2.40}$, $r^2 = 0.97$, $n = 45$, $D_{max} = 4$ cm, Fromard et al. (1998)[a] $W_{top} = 0.0942DBH^{2.54}$, $r^2 = 0.99$, $n = 21$, D_{max}: unknown, Imbert and Rollet (1989)[a]	$W_R = 1.28DBH^{1.17}$, $r^2 = 0.80$, n = 14, $D_{max} = 35$ cm, Comley and McGuinness (2005)
A. marina	*Bruguiera* spp.
$W_{top} = 0.308DBH^{2.11}$, $r^2 = 0.97$, $n = 22$, $D_{max} = 35$ cm, Comley and McGuinness (2005)	$W_R = 0.0188(D^2H)^{0.909}$, r^2: unknown, $n = 11$, $D_{max} = 33$ cm, Tamai et al. (1986) c.f., H = D/(0.025D + 0.583)
Laguncularia racemosa	*Bruguiera exaristata*
$W_{top} = 0.102DBH^{2.50}$, $r^2 = 0.97$, $n = 70$, $D_{max} = 10$ cm, Fromard et al. (1998)[a] $W_{top} = 0.209DBH^{2.24}$, $r^2 = 0.99$, $n = 17$, D_{max}: unknown, Imbert and Rollet (1989)[a]	$W_R = 0.302DBH^{2.15}$, $r^2 = 0.88$, $n = 9$, $D_{max} = 10$ cm, Comley and McGuinness (2005)
Rhizophora apiculata	*Ceriops australis*
$W_{top} = 0.235DBH^{2.42}$, $r^2 = 0.98$, $n = 57$, $D_{max} = 28$ cm, Ong et al. (1995)	$W_R = 0.159DBH^{1.95}$, $r^2 = 0.87$, $n = 9$, $D_{max} = 8$ cm, Comley and McGuinness (2005)
Rhizophora mangle	*R. apiculata*
$W_{top} = 0.178DBH^{2.47}$, $r^2 = 0.98$, $n = 17$, D_{max}: unknown, Imbert and Rollet (1989)[a]	$W_R = 0.00698DBH^{2.61}$, $r^2 = 0.99$, n = 11, $D_{max} = 28$ cm, Ong et al. (1995) c.f., $W_{stilt} = 0.0209DBH^{2.55}$, $r^2 = 0.84$, $n = 41$
Rhizophora spp.	*Rhizophora stylosa*
$W_{top} = 0.128DBH^{2.60}$, $r^2 = 0.92$, $n = 9$, $D_{max} = 32$ cm, Fromard et al. (1998)[a] $W_{top} = 0.105DBH^{2.68}$, $r^2 = 0.99$, $n = 23$, $D_{max} = 25$ cm, Clough and Scott (1989)[a]	$W_R = 0.261DBH^{1.86}$, $r^2 = 0.92$, $n = 5$, $D_{max} = 15$ cm, Comley and McGuinness (2005)
Bruguiera gymnorrhiza	*Rhizophora* spp.
$W_{top} = 0.186DBH^{2.31}$, $r^2 = 0.99$, $n = 17$, $D_{max} = 25$ cm, Clough and Scott (1989)[a]	$W_R = 0.00974(D^2H)^{1.05}$, r^2: unknown, $n = 16$, $D_{max} = 40$ cm, Tamai et al. (1986) c.f., H = D/(0.02D + 0.678)
Bruguiera parviflora	
$W_{top} = 0.168DBH^{2.42}$, $r^2 = 0.99$, $D_{max} = 25$ cm, $n = 16$, Clough and Scott (1989)[a]	
C. australis	
$W_{top} = 0.189DBH^{2.34}$, $r^2 = 0.99$, $n = 26$, $D_{max} = 20$ cm, Clough and Scott (1989)[a]	
Xylocarpus granatum	
$W_{top} = 0.0823DBH^{2.59}$, $r^2 = 0.99$, $n = 15$, $D_{max} = 25$ cm, Clough and Scott (1989)[a]	$W_R = 0.145DBH^{2.55}$, $r^2 = 0.99$, $n = 6$, $D_{max} = 8$ cm, Poungparn et al. (2002)

TABLE 14.1 Allometric equations for various mangroves based on diameter at breast height (DBH) (cm).—cont'd

Aboveground tree weight (W_{top} in kg)	Belowground tree weight (W_R in kg)
Common equation	
$W_{top} = 0.251pD^{2.46}$, $r^2 = 0.98$, $n = 104$, $D_{max} = 49$ cm, Komiyama et al. (2005) $W_{top} = 0.168pDBH^{2.47}$, $r^2 = 0.99$, $n = 84$, $D_{max} = 50$ cm, Chave et al. (2005)	$W_R = 0.199p0.899D^{2.22}$, $r^2 = 0.95$, $n = 26$, $D_{max} = 45$ cm, Komiyama et al. (2005)

D_{stilt}, the weight of prop root of *R. apiculata*.
[a] After Saenger (2002).Table 8.3 on p. 260. D_{max}: the upper range of samples.

TABLE 14.2 Parameters and equations[a] for estimating total aboveground biomass for all hardwood and softwood species in the United States (Jenkins et al., 2003).

		Parameters	
	Species group	β_0	β_1
Hardwood	Aspen/alder/cottonwood/willow	−2.2094	2.3867
	Soft maple/birch	−1.9123	2.3651
	Mixed hardwood	−2.48	2.4835
	Hard maple/oak/hickory/beech	−2.0127	2.4342
Softwood	Cedar/larch	−2.0336	2.2592
	Douglas fir	−2.2304	2.4435
	True fir/hemlock	−2.5384	2.4814
	Pine	−2.5356	2.4349
	Spruce	−2.0773	2.3323
Woodland	Juniper/oak/mesquite	−0.7152	1.7029

[a] For 11 Southeast Asian countries, Yuen et al. (2016) uncovered 402 aboveground and 138 belowground biomass allometric equations for major land covers, such as forest, peat swamp forest, and mangrove forest.

respective compartments ($k = 1$, 2, or 3). All the coefficients are provided in the original paper and omitted here.

More recently, Vargas-Larreta et al. (2017) presented new AGB equations of 17 forest species in the temperate forests of northwestern Mexico. Their equations are based on height and on DBH2 × height.

14.3 Optical remote sensing methods

A sufficient number of field measurements are a prerequisite for developing AGB estimation models and for evaluating the AGB estimation results. Numerous regression models have been developed to estimate AGB in many studies. While these models are accurate at tree, plot, and stand levels, they are limited when considering spatial pattern analysis of AGB across the landscape. However, these approaches are often time consuming, labor intensive, and difficult to implement, especially in remote areas; also, they cannot provide the spatial distribution of biomass in large areas. To scale AGB estimates to the landscape or regional level, the estimates have to be linked with remote sensing data.

The spatial coverage of large area biomass estimates that are constrained by the limited spatial extent of forest inventories or field measurements can be expanded through the use of remotely sensed data. Similarly, remotely sensed data can be used to fill spatial, attributional, and temporal gaps in forest inventory data, thereby augmenting and enhancing estimates of forest biomass and carbon stocks derived from forest inventory data. Such a hybrid approach is particularly relevant for nonmerchantable forests where basic inventory data required for biomass estimation are lacking.

Remotely sensed data have become an important data source for biomass estimation.

Many studies have demonstrated that the surface spectral reflectance of most passive remote sensing sensors is correlated with biomass. For example, Muukkonen and Heiskanen (2005) found that the spectral reflectance of band 1 (green) of Advanced Spaceborne Thermal Emission and Reflection Radiometer (ASTER) showed the highest correlation with the stand volume aboveground tree biomass, branch biomass, and total AGB of all forest vegetation (correlation coefficient r ranged from −0.69 to −0.67) (see Fig. 14.1), and band 3 (NIR) showed the highest correlation with the biomass of the stem and stump system (−0.68) and the biomass of the understory vegetation (−0.36).

The remotely sensed data types and approaches used for biomass estimation have been summarized (Lu, 2006; Patenaude et al., 2005; Lu et al., 2014; Timothy et al., 2016; Galidaki et al., 2017). A variety of remotely sensed data sources have been employed for biomass mapping. For quantifying biomass at local to regional scales, data provided by moderate spatial resolution instruments, such as Landsat

TM (Fazakas et al., 1999; Krankina et al., 2004; Tomppo et al., 2002; Turner et al., 2004; Lu, 2005; Lu et al., 2012; Basuki et al., 2013; Zhang et al., 2014) and ASTER (Muukkonen and Heiskanen, 2005, 2007; Fernández-Manso et al., 2014), are required. At continental- and global-scale biomass mapping, however, the coarse spatial resolution optical sensors, such as the NOAA Advanced Very High Resolution Radiometer (AVHRR) (Dong et al., 2003; Hame et al., 1997) and MODIS (Baccini et al., 2004; Du et al., 2014; Lumbierres et al., 2017; Schucknecht et al., 2017), have been useful due to the good trade-off between spatial resolution, image coverage, and frequency in data acquisition (Lu, 2006).

Biomass estimation over large areas using coarse spatial resolution data has been limited because of the mixed pixels and the huge difference between the support of ground reference data and pixel size of the satellite data (Lu, 2006; Lu et al., 2014). To facilitate the linkage of detailed ground measurements to coarse spatial resolution remotely sensed data (e.g., MODIS,

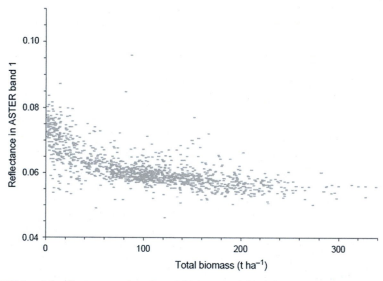

FIGURE 14.1   ASTER band 1 reflectance against the total biomass of the forest stands. *After Muukkonen, P., Heiskanen, J., 2005. Estimating biomass for boreal forests using ASTER satellite data combined with standwise forest inventory data. Remote Sens. Environ. 99 434–447. Fig. 4 on p. 439.*

AVHRR, IRS-WiFS), several studies have integrated multiscale imagery into their biomass estimation methodology and incorporated moderate spatial resolution imagery (e.g., Landsat, ASTER) as an intermediary data source between the field data and coarser imagery (Muukkonen and Heiskanen, 2005, 2007). Research has demonstrated that it is more effective to generate relationships between field measures and moderate spatial resolution remotely sensed data (e.g., Landsat) and then extrapolate these relationships over larger areas using comparable spectral properties from coarser spatial resolution imagery (e.g., MODIS).

Generally, biomass is either estimated via a direct relationship between spectral response and biomass using multiple regression analysis, k-nearest neighbor (kNN), neural networks, or through indirect relationships, whereby attributes estimated from the remotely sensed data, such as leaf area index (LAI), structure (crown closure and height), or shadow fraction, are used in equations to estimate biomass. In this section, we will start to discuss the optical remote sensing methods. Different techniques in microwave and lidar remote sensing will be discussed separately in the following sections.

14.3.1 Using vegetation indices

Many methods may be used to estimate and map forest biomass from optical remotely sensed data. Vegetation indices have, in particular, been used widely. In general, vegetation indices can partially reduce the impacts on reflectance caused by environmental conditions and shadows, thus improve correlation between AGB and vegetation indices, especially in those sites with complex vegetation stand structures.

For example, Dong et al. (2003) correlated seasonally integrated measures of satellite-observed vegetation greenness with ground-based inventory data amounts to a spatial interpolation and extrapolation of the inventory data. They developed the relationship between biomass from the forest inventory and the accumulated normalized difference vegetation index (NDVI) of AVHRR data:

$$B = \frac{1}{\alpha + \beta \left[\dfrac{1}{\text{NDVI}\varphi^2} + \gamma\varphi\right]} \qquad (14.4)$$

in which B is biomass that is a measure of total or above-stump biomass obtained from inventory, NDVI is the cumulative growing season NDVI averaged over a 5-year period before inventory date, φ is the latitude that is the centroid of the area sampled by forest inventory in a province, and α, β, and γ are regression coefficients. The value of these coefficients is estimated using ordinary least squares. For total biomass, they are $\alpha = -0.0377$ (± 0.00977), $\beta = 3809.65$ (± 902.51), and $\gamma = 0.0006$ (± 0.00011) adjusted $R^2 = 0.43$. For above-stump biomass, they are $\alpha = -0.0557$ (± 0.0136), $\beta = 5548.05$ (± 1274.17), and $\gamma = 0.000854$ (± 0.000153) adjusted $R^2 = 0.49$. Values in parentheses are standard errors.

However, Foody et al. (2001) identified four problems with the use of vegetation indices such as the NDVI:

(1) The relationship between the vegetation index and biomass is asymptotic, and this can limit the ability of the index to represent accurate vegetation with a large biomass. NDVI yields poor estimates in areas where there is 100% vegetation cover and, therefore, has limited value in assessing biomass during the peak of seasons due to a saturation level after a certain biomass density or LAI;

(2) It is essential that the remotely sensed data be accurately calibrated, typically to radiance, if the calculated index values are to be correctly interpreted and compared;

(3) The sensitivity of vegetation indices to biomass has been found to vary between environments;

(4) Most vegetation indices fail to use all the spectral data available. Typically, a vegetation index uses only the data acquired in two spectral wavebands, yet the sensor typically acquires the spectral response in several additional wavebands.

Lu (2006) also concluded that the AGB estimation using NDVI from coarse spatial resolution data such as AVHRR and MODIS is very limited because of the common occurrence of mixed pixels and the huge difference between the size of field-measurement data and pixel size in the image, resulting in difficulty in the integration of sample data and remote sensing−derived variables.

One attractive alternative to vegetation indices is the use of multiple regression analysis. This can use all the available spectral data, reducing the need to select a subset of wavebands, and may provide more accurate estimates of biophysical variables than vegetation indices.

14.3.2 Multivariate regression analysis

Stand level biomass is frequently calculated from linear and nonlinear regression models established by species with field measurements. Numerous regression models have been developed to estimate AGB; while these models are accurate at tree, plot, and stand levels, they are limited when considering spatial pattern analysis of AGB across the landscape.

Muukkonen and Heiskanen (2005) undertook a nonlinear regression analysis using ASTER two spectral bands (R_2: band 2; R_3: band 3) as predictors to calculate biomass of boreal forest stands. Their equations are of the following form:

$$y_j = \exp(a + dR_2 + eR_3)(1 + R_2)^b (R_3)^c + \varepsilon_j \tag{14.5}$$

for the forest attributes (y_j) other than the biomass of understory vegetation. The model for the biomass of understory vegetation is

$$y_j = a + bR_2 + cR_3 + \varepsilon_j \tag{14.6}$$

In the models, the $a \sim e$ are the parameters and are listed in Table 14.3, and the ε_j is the error term. Note that R^2 for all vegetation biomass is only 0.56.

The use of regression analysis, however, makes many assumptions about the data that may not be satisfied. In particular, standard regression analysis assumes that a linear relationship exists between the remotely sensed data and the biophysical property, yet in reality most relationships observed may be curvilinear. In addition, multiple regression assumes that the independent variables (the data acquired in the different spectral wavebands by the remote sensor) are uncorrelated. This assumption will be rarely satisfied in remote sensing as the data acquired in different spectral wavebands are commonly very strongly correlated (Mather, 2004). Methods such as correlation coefficient analysis and stepwise regression analysis can be used to identify suitable remote sensing variables that have strong relationships with biomass but weak relationships between the selected remote sensing variables themselves (Lu et al., 2012). Neural networks may, however, be used as a nonparametric alternative to regression.

14.3.3 kNN methods

Forest inventory data often provide the required base data to enable the large area mapping of biomass over a range of scales. However, spatially explicit estimates of AGB over large areas may be limited by the spatial extent of the forest inventory relative to the area of interest (i.e., inventories not spatially exhaustive) or by

TABLE 14.3 Regression equations for estimation of forest parameters from ASTER data.

			β	S.E. of β	R^2	RMSE
Volume		a	24.79	4.28	0.55	70.6
		b	−675.01	0.83		
		c	6.33	1241.93		
		d	588.65	1217.17		
		e	−39.43	2.31		
Age		a	0.46	1.91	0.21	41.4
		b	−907.58	347.93		
		c	−2.24	0.71		
		d	881.45	337.69		
		e	2.11	3.65		
Biomass of trees	Aboveground	a	26.80	2.12	0.56	44.7
		b	−2877.39	581.46		
		c	7.09	0.78		
		d	2739.64	568.81		
		e	−42.73	4.06		
	Stem and stump	a	32.43	2.57	0.59	30.7
		b	−1819.72	639.10		
		c	9.25	0.94		
		d	1725.38	625.26		
		e	−58.78	5.02		
	Branches	a	26.96	2.15	0.57	7.9
		b	−3516.14	544.97		
		c	7.77	0.79		
		d	3352.98	532.98		
		e	−44.98	4.12		
	Foliage	a	26.48	2.6	0.54	4.1
		b	−43.12	1727.76		
		c	8.18	0.93		
		d	−57.02	1694.71		
		e	−45.53	4.74		

(Continued)

TABLE 14.3   Regression equations for estimation of forest parameters from ASTER data.—cont'd

		β	S.E. of β	R^2	RMSE
Biomass of understory vegetation	a	2.90	0.08	0.15	0.5
	b	18.37	2.85		
	c	−6.46	0.41		
Biomass of all vegetation	a	26.29	2.06	0.56	44.8
	b	−2907.02	524.50		
	c	6.90	0.76		
	d	2770.31	512.85		
	e	−41.73	3.96		

After Muukkonen, P., Heiskanen, J., 2005. Estimating biomass for boreal forests using ASTER satellite data combined with standwise forest inventory data. Remote Sens. Environ. 99 434—447., Table 5 on page 441.

the omission of inventory attributes required for biomass estimation. These spatial and attributional gaps in the forest inventory may result in an underestimation of large area AGB. The continuous nature and synoptic coverage of remotely sensed data have led to their increased application for AGB estimation over large areas, although the use of these data remains challenging in complex forest environments.

As discussed previously, some methods for biomass estimation, such as the kNN, rely on the link between ground plots and satellite imagery (Fazakas et al., 1999; Franco-Lopez et al., 2001; Katila, 1999; Katila and Tomppo, 2001; Nelson et al., 2000). Methods such as these are routinely applied in countries where an extensive regional network of field plots is exploited.

14.3.3.1 Overview

The kNN method has become popular for forest inventory mapping applications and is widely used to produce pixel-level estimates of continuous forest variables such as biomass, basal area, and volume (Franco-Lopez et al., 2001; Tomppo et al., 2009).

In the kNN method, forest variables such as biomass for each pixel in a satellite image are estimated as weighted averages of the closest sample plots (the nearest neighbors) in a feature space. The method can be viewed as inverse distance weighting applied to a feature space that are often defined by satellite spectral data acquired in different wavelength bands (Fazakas et al., 1999; Franco-Lopez et al., 2001; Tomppo, 2004; Tomppo et al., 2009).

Tomppo et al. (Tomppo, 1997; Tomppo and Katila, 1991) have led efforts to incorporate the kNN method in forest inventories. The kNN technique has been operationally used by the Finnish National Forest Inventory (NFI) since 1990 (Katila and Tomppo, 2001). Fazakas et al. (1999) applied the kNN method to estimate tree biomass in Sweden using 658 NFI sample plots combined with Landsat TM data and an inverse-squared distance weighting in feature space. The Swedish NFI has been constructing statistical and map products using the kNN technique on a 5-year basis since 2001 and is currently developing an annual product (Reese et al., 2002; Tomppo et al., 2009). Labrecque et al. (2006) used the kNN method with five neighbors and two different weighting functions to estimate AGB in Canada.

The kNN is a nonparametric estimation method in which there are no assumptions about the distributions of the variables involved in the estimation (Franco-Lopez et al., 2001; Hardin, 1994). An essential property of the method is that all forest variables can be estimated at the same time using the same parameters (Tomppo and Katila, 1991; Tomppo et al., 2009). The method also preserves the covariance structure of the variables (Ohmann and Gregory, 2002; Tomppo and Katila, 1991). It can be applied in a straightforward way in different forest conditions and with different remote sensing materials (Franco-Lopez et al., 2001; Katila, 1999).

14.3.3.2 Assumption

The kNN method is based on two important premises (Labrecque et al., 2006). First, the spectral responses of the pixel (pixel values) within a satellite image depend only on the forest condition and not on the geographic location (Fazakas et al., 1999; Franco-Lopez et al., 2001). Second, the ground sample plots are distributed over a large area (e.g., the surface covered by a satellite image) and can be used as ground truth data for estimating the remaining image surface (Franco-Lopez et al., 2001; Tokola et al., 1996).

14.3.3.3 Method description

A general description of the kNN method is described below (Fazakas et al., 1999; Franco-Lopez et al., 2001; Tomppo, 2004).

The estimate of the variable m, i.e., biomass for a given pixel p, is calculated as follows:

$$m_p = \sum_{i=1}^{k} w_{(p_i)p} m_{(p_i)} \qquad (14.7)$$

where $m_{(p_i)}$, $i = 1, ..., k$, is the value of the variable m in sample plot i corresponding to the pixel p_i, which is the ith closest pixel (of "known" pixels) in the feature space to the pixel p, $w_{(p_i)p}$ is the weight of sample plot i to the pixel p and is defined as follows:

$$w_{(p_i)p} = \frac{1}{d_{(p_i)p}^t} \bigg/ \sum_{i=1}^{k} \frac{1}{d_{(p_i)p}^t} \qquad (14.8)$$

$$d_{1p} \leq d_{2p} \leq \cdots \leq d_{kp}$$

where $d_{(p_i)p}$, $i = 1, ..., k$, denotes the distance in the feature space from the pixel p to the pixel p_i representing the sample plot i, $t = 0$, 1, or 2 represents three different weighting functions: (a) equal; (b) inversely proportional to the distance; and (c) inversely proportional to the square of the distance.

Commonly used distance metrics include Euclidean or weighted Euclidean distance (Franco-Lopez et al., 2001; McRoberts et al., 2007; McRoberts, 2012; Reese et al., 2002; Tokola et al., 1996; Tomppo, 1997, 2004) and Mahalanobis distance (Fazakas et al., 1999; Franco-Lopez et al., 2001). Other distance metrics include metrics based on canonical correlation analysis (LeMay and Temesgen, 2005; Temesgen et al., 2003) and canonical correspondence analysis (Ohmann and Gregory, 2002).

The Euclidean distance from the pixel p to the pixel p_i representing the sample plot i is given below:

$$d_{(p_i)p} = \sqrt{\sum_{j=1}^{nf} \left(x_{(p_i),j} - x_{p,j} \right)^2} \qquad (14.9)$$

where $x_{p,j} = $ digital number for the feature j, $nf = $ number of features in the feature space.

The Mahalanobis distance handles the problems with correlated bands and different dynamic ranges between the bands (Fazakas et al., 1999). Several studies showed that the difference between using the Euclidean distance and the Mahalanobis distance in the kNN method is fairly small (Franco-Lopez et al., 2001), at least in forest conditions similar to the ones in northern Sweden. In central Sweden, however, the Mahalanobis distance was considered to be more appropriate as the forest conditions vary more than in the northern parts of Sweden (Fazakas et al., 1999).

Not all the features in the feature space share the same influence in the estimation of biomass for a given pixel. Assuming that there exists a linear combination of features that can provide the best result, additional weights are applied to the original features (Franco-Lopez et al., 2001). The resulting expanded form of Eq. (14.9) is expressed as follows:

$$d_{(p_i)p} = \sqrt{\sum_{j=1}^{nf} a_j^2 \left(x_{(p_i),j} - x_{p,j} \right)^2} \qquad (14.10)$$

where a_j = weighting parameter for the feature j.

This weighting parameter is computed using two methods: (a) the downhill simplex optimization method developed by Nelder and Mead (Nelder and Mead, 1965) and (b) a genetic algorithm (Mitchell, 1998). Tomppo and Halme (Tomppo, 2004) used the *kNN* technique with a genetic algorithm to optimize selection of weights for spectral feature space variables for Landsat TM pixels.

14.3.3.4 *Number of neighbors*

The number of nearest neighbors to employ depends on several factors: the objective of the estimation, the layout and the size of the field plots, size of the pixel, and the variation of the field variables (Franco-Lopez et al., 2001; Katila and Tomppo, 2001).

Usually, there is a rapid early increase in the accuracy of the *kNN* technique as the number of neighbors increases. However, the marginal gains diminish with a large number of neighbors (Franco-Lopez et al., 2001). Studies have shown that no real gain in accuracy was reached by using more than 10 neighbors, and that using a smaller number of k keeps the natural spatial variability and produces results with similar accuracies (Fazakas et al., 1999; Franco-Lopez et al., 2001; Katila and Tomppo, 2001; Reese et al., 2002 also summarized in Labrecque et al., 2006). Using fewer than five neighbors (i.e., $k < 5$), however, has been found to decrease the accuracy, and using too many neighbors ($k > 10$) overgeneralizes the results (Reese et al., 2002).

14.3.4 Artificial neural networks

Artificial neural networks (ANNs) are being increasingly applied in remote sensing as a nonparametric approach for predicting vegetation characteristics (Foody et al., 2001, 2003; Gopal and Woodcock, 1994; Gopal et al., 1999). ANNs are capable of handling the nonnormality, nonlinearity, and collinearity in the system (Haykin and Network, 1994). ANNs are attractive for biomass prediction as they avoid problems in using vegetation indices (e.g., index selection and incomplete use of the data's spectral information content) and multiple regression analyses (e.g., underlying assumptions) and have been used to derive accurate predictions of tropical forest biomass (Foody et al., 2001). Atkinson and Tatnall (1997) identified four important advantages of using ANNs in remote sensing.

1. ANNs perform more accurately than other techniques such as statistical classifiers, particularly when the feature space is complex and the source data have different statistical distributions;
2. ANNs perform more rapidly than other techniques such as statistical classifiers;
3. ANNs are capable of incorporate a priori knowledge and realistic physical constraints into the analysis; and
4. ANNs can incorporate different types of data (including those from different sensors) into the analysis, thus facilitating synergistic studies.

There are many different types of neural networks. The feed-forward multilayer perceptron (MLP) trained by back-propagation algorithm (Rumelhart, 1985) is one of the most commonly used neural networks in remote sensing (Atkinson and Tatnall, 1997; Jensen et al., 1999; Kanellopoulos and Wilkinson, 1997).

14.3.4.1 *Principle*

A typical ANN consists of a number of nodes (or neurons) and three or more layers: one input layer, one output layer, and usually one or two hidden layers (Fig. 14.2). Every node in any layer of the network is connected to all the nodes of the adjacent layer(s). For remotely sensed data, the input layer usually contains DN from spectral bands and the number of nodes equals the number of bands. The hidden nodes calculate a weighted sum of inputs which are then passed through an activation function to produce the node's output value. The output layer presents the network's results, e.g., biomass, basal area, and volume. The number of output nodes equals the number of variables to be estimated. The input data are passed through the connections to the next layer in a feed-forward manner. The back-propagation algorithm refers to passing the errors back to the hidden layer and adjusting weights (Atkinson and Tatnall, 1997; Jensen et al., 1999).

Foody et al. (2001) applied a multilayer perception network to the six nonthermal bands of Landsat TM in a tropical rainforest in Borneo. The MLP provided estimates of biomass that were strongly correlated with those measured in the field ($r = 0.80$). Moreover, these estimates were found to be more strongly correlated with biomass than those derived from 230 conventional vegetation indices, including the widely used NDVI.

In another study, Foody et al. (2003) developed three types of predictive relation for biomass estimation based on vegetation indices, multiple regression, and neural networks. They used ground data and Landsat TM data of moist tropical forests in Brazil, Malaysia, and Thailand, and they also found that neural networks provided stronger relationships between actual and predicted estimates of biomass ($r > 0.71$ at all three sites, significant at the 99% level of confidence).

However, using a network model, forest biomass based on Canada's Forest Inventory was estimated with relatively poor accuracy (RMS = 32 ton/ha) from SPOT VEGETATION reflectance data and terrestrial ecozone (Fraser and Li, 2002).

A brief description of the MLP is described below (Atkinson and Tatnall, 1997; Jensen et al., 1999).

In the MLP algorithm, the input data are passed to the nodes in the next layer in a feed-forward manner. As the data pass from node to node, they is modified by the weights associated with the connection. The receiving node (e.g., j) sums the weighted data from all nodes (e.g., i) to which it is connected in the preceding layer. Formally, the input that a given node j receives is weighted according to the following equation:

$$net_j = \sum_{i=1}^{k} \omega_{ji} o_i \qquad (14.11)$$

$i = 1, 2, \ldots k.$

where k is the number of inputs in the preceding layer, ω_{ji} represents the weight between node i and node j, and o_i is the output from node i.

The output from the single node j is then computed from

$$o_j = f(net_j) \qquad (14.12)$$

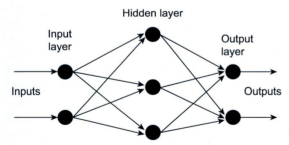

FIGURE 14.2 The multilayer perceptron algorithm. *After Atkinson, P.M., Tatnall, A.R.L., 1997. Introduction neural networks in remote sensing. Int. J. Remote Sens. 18, 699–709. Figure 1 on p. 702.*

where net_j is the weighted sum of inputs before the data passes to the next layer, f is an activation function. The function f is usually a nonlinear sigmoid function defined by (Hagan et al., 1996; Haykin and Network, 1994)

$$f(net_j) = \frac{1}{1 + e^{(-net_j)}} \qquad (14.13)$$

When the data reach the output layer, they form the network output. Then, the network output is compared with the desired output (a set of training data) and the error computed. This error is then back-propagated through the network and the weights of the connections are altered according to the generalized delta rule (Rumelhart, 1985):

$$\Delta\omega_{ji}(n+1) = \eta(\delta_j o_i) + \alpha\Delta\omega_{ji}(n) \qquad (14.14)$$

where n is the nth iteration, η is the learning rate parameter, δ_j is an index of the rate of change of the error, and α is the momentum parameter.

This process of feeding forward input data and back-propagating the error is repeated iteratively until the error of the network as a whole is minimized or reaches an acceptable magnitude. Learning occurs by adjusting the weights in the node. Once learned, the network can recall stored knowledge to perform prediction based on new input data (Haykin and Network, 1994; Lloyd, 1997).

14.3.4.2 Limitations

The training procedure is sensitive to the choice of initial network parameters (Gopal and Woodcock, 1994). It is possible to overtrain a network so that it is able to memorize the training data but is not able to generalize when it is applied to different data (Atkinson and Tatnall, 1997). ANNs have the capacity to predict the output variable but the mechanisms that occur within the network are often ignored. ANNs are often considered as black boxes. Various authors have explored this problem and proposed algorithms to illustrate the role of variables in ANN models (Gevrey et al., 2003).

14.4 Active and stereoscopic remote sensing methods

14.4.1 Lidar data

14.4.1.1 Small-footprint lidar

Discrete-return airborne laser sensors produce a three-dimensional cloud of points depicting the horizontal and vertical distribution of biological material with high spatial resolution. The strong relationships that exist between laser data and AGB can be attributed to the ability of airborne lasers to accurately capture the canopy height and density of the canopy, both of which are highly correlated with biomass. Every point in the point cloud data is a direct measurement of the height of a reflector within the canopy. But these points need to be aggregated to a certain scale to depict the three-dimensional structure of a vegetation canopy and to derive various indices for estimation of canopy height and biomass.

It is essential to separate the echoes in the canopy from the below-canopy echoes. In mature forests, it has been demonstrated that different thresholds to separate canopy and below-canopy echoes can be applied, for example, fixed thresholds of 2 m (e.g., Næsset, 2002b; Nilsson, 1996) or 3 m (Næsset, 2004b). For young forests, Næsset (2011) showed that the different canopy thresholds (0.5, 0.13, 2.0 m) seemed to perform equally well, probably due to that the height accuracy of the ground surface in forested areas determined from airborne laser scanner (ALS) data ranges from 20 to 50 cm (Hodgson and Bresnahan, 2004; Peng and Shih, 2006; Su and Bork, 2006). After the ground points were separated from the last return points, a triangulated irregular network was generated from these terrain ground points to generate a digital terrain model. While the first return points will be used to calculate various indices for biomass estimation since research has shown that properties of the first return echoes tend to be more stable across different ALS instrument configurations and flying altitudes than other echo categories (Næsset and Gobakken, 2008).

During the last decade, airborne scanning lidar has been used for forest management inventories at regional scale in boreal and temperate forests (McRoberts et al., 2010; Næsset, 2004a; Næsset and Gobakken, 2008) and tropical forests (Clark et al., 2004; Nelson, 1997; Nelson et al., 1997). Section 14.2 in this book has described these applications in detail. Table 14.3 lists various height indices which can be derived from point cloud data of ALS. These parameters have been widely used in multivariate regression models for biomass estimation (Næsset and Gobakken, 2008).

14.4.1.2 Large-footprint lidar

Large-footprint lidar systems (Blair et al., 1999) have been developed to provide high-resolution, geo-located measurements of vegetation vertical structure and ground elevations beneath dense canopies. The footprint size is usually greater than a single tree crown, so laser beam will partially pass through the gaps between tree crowns to reach the ground producing a ground peak in the lidar waveform. Over the past decade, several airborne and spaceborne large-footprint lidar systems have been used to make measurements of vegetation. The lidar waveform signature from large-footprint lidar instrument, such as the Scanning LiDAR Imager of Canopies by Echo Recovery (Harding et al., 1998), the Laser Vegetation Imaging Sensor (LVIS) (Blair et al., 1999), and GLAS (Abshire et al., 2005), has been successfully used to estimate the tree height and forest AGB (Baccini et al., 2008; Drake et al., 2002, 2003; Dubayah and Drake, 2000; Hofton and Blair, 2002; Lefsky et al., 1999, 2002, 2005; Simard et al., 2008, 2011; Sun et al., 2011, 2008). Fig. 14.3 shows a GLAS waveform acquired from a forested surface. After estimation of mean and standard deviation of noise, the signal beginning and ending can be determined by a noise threshold (e.g., mean ± 3 standard deviations). If the ground peak can be located, then the percentile energy levels can be located and the heights of these points can be calculated. Some other indices have been also calculated and used in forest stem volume or biomass estimation (Sun et al., 2008; Nelson et al., 2009).

Lidar waveform is a record of returned signals as a function of time delay, or corresponding locations above a reference height. By using a downward Z-axis with $z = 0$ at the top of the canopy, the waveform can be expressed as follows (Sun and Ranson, 2000):

$$P(z) = k \int_{z-r}^{z+r} \rho(x)T(x-z)e^{-x\tau}dx + \varepsilon \quad (14.15)$$

where z is the distance from top canopy, and r is the half width of the transmitted lidar pulse T. $\rho(x)$ $(x = 0, \ldots, h)$ represents the vertical profile of the reflectance of the canopy within the lidar footprint. $T(x)$, $x = -r, \ldots, 0, \ldots, r$, represents a Gaussian-shaped lidar pulse. The k is a constant accounting for all conversion from a target signal to the digital records. The term $e^{-x\tau}$ is the attenuation by the canopy above z in hotspot condition. The ε is the noise term. For a smooth ground surface, ρ will only have one reflectance value at the surface ($z = 0$), and the waveform P will have the same width $2r$ as the lidar transmitted pulse from $z = -r$ to $z = r$. If the transmitted laser pulse has a perfect Gaussian shape, then the lidar waveform $P(z)$ is a summation of multiple Gaussian returns at different heights (time delays). Therefore, a lidar waveform may be decomposed into multiple Gaussian waveforms (Hofton et al., 2000). In GLAS waveform product GLA14, up to six Gaussian peaks were used to approximate the waveform (Brenner et al., 2002). Some waveform indices can be derived from these decompositions (Rosette et al., 2009).

The discrete points from laser scanning systems can be aggregated to create "pseudo" waveforms such as those recorded by large-footprint lidar systems (Muss et al., 2011).

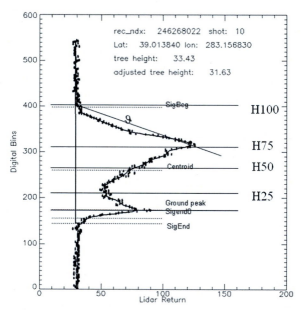

FIGURE 14.3 A GLAS waveform from forests and various waveform indices derived from waveform: SigBeg—signal beginning; SigEnd—signal ending; Ground peak—peak return from ground surface; centroid—centroid of the waveform energy; H25–H100—heights from the ground peak to the points where cumulative waveform energy (upward from SigEnd) reaches 25%, 50%, 75%, and 100% of the total waveform energy.

Under favorable topographic conditions, Popescu et al. (2011) compared the GLAS waveform data with the ALS data within the GLAS footprint and found that GLAS data proved to be accurate in retrieving terrain elevation. Similarly, height metrics are highly correlated with equivalent parameters derived with airborne lidar data and can be used to estimate AGB at footprint level. Table 14.4 is a list of some indices from lidar waveform which can be used for biomass estimation.

Lidar directly measures the heights of components in a vegetation canopy. It is more intuitive to use lidar data for estimation of canopy height than canopy biomass. For example, Lorey's height can be reliably estimated from GLAS waveform data (Lefsky, 2010). The Lorey's height is defined as follows:

$$H_{lorey} = \frac{\sum_{i=1}^{N} BA_i H_i}{\sum_{i=1}^{N} BA_i} \quad (14.16)$$

where BA_i and H_i are the basal area and canopy height of individual trees in a plot. Fig. 14.4 shows that the Lorey's height is highly correlated with biomass at plot sizes equivalent to GLAS-footprint sizes (~70 m diameter) (Saatchi et al., 2011). Lorey's height, or some other canopy heights similarly defined, can be estimated from lidar data first, and then be used for biomass estimation (Kellndorfer et al., 2006; Lefsky, 2010; Saatchi et al., 2011). Some other heights estimated from GLAS waveform were also correlated to biomass (Lefsky et al., 2005; Pflugmacher et al., 2008) by deriving specific allometric equations from field data.

14.4.2 SAR data

14.4.2.1 Backscattering coefficients

Radar, because of its penetration capability and sensitivity to water content in vegetation, is sensitive to the forest spatial structure and

TABLE 14.4 PolSAR attributes used in forest parameter's retrieval (Gonçalves et al., 2011).

Incoherence attributes	Symbol	Equation	Source
Backscattering coefficient[a]	σ_k^0	$10^{\frac{\sigma_k^0[dB]}{10}}, \sigma_k^0[dB] = 20\log_{10}(|S_k|F_j)$	Henderson and Lewis (1998)
Copolarization ratio	R_{co}	$\frac{\sigma_{VV}^0}{\sigma_{HH}^0}$	Henderson and Lewis (1998)
Cross-polarized ratio	R_{cross}	$\frac{\sigma_{HV}^0}{\sigma_{HH}^0}$	Henderson and Lewis (1998)
Total power	P_T	$\sigma_{HH}^0 + \sigma_{VV}^0 + 2\sigma_{HV}^0$	Boerner et al. (1991)
Biomass index	BMI	$\frac{\sigma_{HH}^0 + \sigma_{VV}^0}{2}$	Pope et al. (1994)
Canopy structure index	CSI	$\frac{\sigma_{VV}^0}{\sigma_{VV}^0 + \sigma_{HH}^0}$	Pope et al. (1994)
Volume scattering index	VSI	$\frac{\sigma_{HV}^0}{\sigma_{HV}^0 + (\text{BWI})}$	Pope et al. (1994)

Coherence attributes	Symbol	Equation	Source
HH-VV phase difference[b]	$\Delta\phi$	$\arg(S_{HH}S_{VV}^*)$	Henderson and Lewis (1998)
HH-VV polarimetric coherence[b]	γ	$\frac{|\langle S_{HH}S_{VV}^*\rangle|}{\sqrt{\langle S_{HH}S_{HH}^*\rangle\langle S_{VV}S_{VV}^*\rangle}}$	Henderson and Lewis (1998)
Entropy[c]	H	$H = -\sum_{i=1}^{3} p_i\log_3(p_i), p_i = \frac{\lambda_i}{\sum_{j=1}^{3}\lambda_j}$	Cloude and Pottier (1996)
Anisotropy[c]	A	$\frac{\lambda_2 - \lambda_3}{\lambda_2 + \lambda_3}$	Pottier (1998)
Average alpha angle[c]	$\overline{\alpha}$	$\sum_{i=1}^{3} p_i\alpha_i, p_i = \frac{\lambda_i}{\sum_{j=1}^{3}\lambda_j}$	Cloude and Pottier (1996)
Volume scatter contribution to total power[d]	P_V	$\frac{8f_V}{3}$	Freeman and Durden (1998)
Double-bounce scatter contribution to total power[d]	P_d	$f_d\left(1 + |\alpha|^2\right)$	Freeman and Durden (1998)
Surface scatter contribution to total power[d]	P_S	$f_S\left(1 + |\beta|^2\right)$	Freeman and Durden (1998)

[a] *Expressed in the intensity format, not the more common logarithmic scale (dB). $|S_k|$ is the amplitude response (k = HH, VV, or HV) and F_j is a calibration factor for column j, determined from data collected over the 12 trihedral corner reflectors.*
[b] *arg, S,*, and <7 denote argument function, complex scattering amplitude, complex conjugate, and spatial average, respectively.*
[c] *$\lambda_1 \geq \lambda_2 \geq \lambda_3 \geq 0$ are real eigenvalues of the coherency matrix [T]. α_i is a parameter derived from the eigenvectors of [T] that describes the dominant scattering mechanism. (Cloude and Pottier, 1996).*
[d] *We group P_v with "coherent" attributes because it is often associated with P_d and P_s. But P_v is derived from σ_{HV}^0 and therefore does not depend on polarimetric coherence. f_v, f_d, and f_s are the volume, double-bounce, and surface scatter contributions to the VV cross section. α and β are the second-order statistic, $<S_{HH}S_{VV}^*>$ for double-bounce and surface scattering, respectively, after normalization with respect to the VV term.*

standing biomass. In the past two decades, studies from various campaigns of airborne SAR sensors and launches of spaceborne SAR satellites have demonstrated that multifrequency, multipolarization SAR systems are significant tools for mapping forest types, assessing forest biomass, and monitoring forest changes. Radar data have been used for forest biomass estimation (Dobson et al., 1992; Kasischke et al., 1995; Kurvonen et al., 1999; Ranson et al., 1995, 1997a,b) and canopy height estimation (Hagberg et al., 1995; Kobayashi et al., 2000; Treuhaft et al., 1996). The potential to map forests with different spatial structures and to provide information on forest biomass from polarimetric radar data is limited when forest biomass is high and the structure is complex (Imhoff, 1995; Ranson and Sun, 1997). A consensus is that imaging radar data can be used to estimate AGB in forest ecosystems

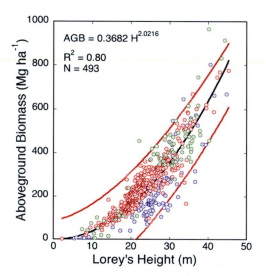

FIGURE 14.4 Allometric relations between Lorey's height and aboveground biomass of calibration plots at spatial scales comparable to GLAS footprints ($AGB = 0.3104H^{2.0608}$, $R^2 = 0.85$, $P < .001$) (Saatchi et al., 2011).

(Kasischke et al., 1995) to a certain biomass level. The utility of imaging radars for this purpose is limited to low-biomass forest stands if a single frequency-polarization system is used (Harrell et al., 1995; Kasischke et al., 1994). The height of the "scattering phase center" retrieved from single-channel interferometric SAR and the coherence of the InSAR images are also correlated to forest biomass (Balzter, 2001; Balzter et al., 2007; Pulliainen et al., 2003; Sun et al., 2011; Treuhaft and Siqueira, 2004).

Mathematically, a scatterer can be characterized by a complex 2×2 scattering matrix:

$$E^{sc} = \begin{bmatrix} S_{hh} & S_{hv} \\ S_{vh} & S_{vv} \end{bmatrix} E^{tr} = [S]E^{tr} \qquad (14.17)$$

where E^{tr} is the electric field vector that was transmitted by the radar antenna, $[S]$ is the 2×2 complex scattering matrix that describes how the scatterer modified the incident electric field vector, and E^{sc} is the electric field vector that is incident on the radar receiving antenna. This scattering matrix is also a function of the radar frequency and the viewing geometry. Once the complete scattering matrix is known and

calibrated, one can synthesize the radar cross section for any arbitrary combination of transmit and receive polarizations (Van Zyl and Kim, 2010). The radar backscattering coefficient is the radar cross section per square meter of the ground surface. Each element of the scattering matrix is a complex number, representing the amplitude and phase of the scattering in a polarization configuration:

$$S_{hv} = a + jb = Ae^{-j\phi} \qquad (14.18)$$

where A and φ are the amplitude and phase of the scattering signal at HV polarization. The cross product $S_{hv}S_{hv}^* = a^2 + b^2 = A^2$ is the backscattering intensity (power) at HV polarization. The radar data (such as PALSAR data), a complex value as shown in Eq. (14.18), can be converted to backscattering coefficient by

$$\sigma^0 = c(a^2 + b^2) + d \qquad (14.19)$$

where c and d are calibration constants (gain and offset) provided with the radar data. The backscattering coefficient is the basic radar signature for biomass estimation.

Most of SAR data available now have been radiometrically calibrated and could also be geocoded. The signal received from a pixel of a radar image is the sum of the signals backscattered from all scatterers within a radar resolution cell. The ground resolution of a radar pixel in range direction depends on the local radar incidence angle. Topographic variations in the image will cause the local incidence angle to be different from that expected for a flat surface with no relief. The distortion of radar backscattering coefficient due to the changes of radar pixel size needs to be corrected before the radar data can be used for prediction of biomass. Slope and aspect can be generated from elevations and then be used to calculate the local incidence angle for a pixel of the radar image:

$$\cos(\vartheta_L) = \sin(s)\cos(\alpha)\sin(s + \phi) + \cos(\vartheta)\sin(\alpha) \qquad (14.20)$$

where θ_L is the local incidence angle, s is local slope, φ is aspect of the slope, θ is radar incidence angle, and α is azimuth angle of the radar look direction. Radiometric distortion due to the illumination areas was corrected using the local incidence angle with an equation of the form used by Kellndorfer et al. (1998).

$$\sigma_{corr}^0 = \sigma^0 \sin(\vartheta_L)/\sin(\vartheta_0) \qquad (14.21)$$

where σ_{corr}^0 is radar backscatter coefficient after correction, σ^0 is original backscatter coefficient and θ_0 is radar incidence angle at image center.

The slope of a pixel not only changes the total surface area of a radar image pixel illuminated by radar beam but also changes the spatial arrangement of scatterers within the medium volume of a pixel. For example, the trees standing on a slope will have different backscattering from the case when these same trees stand on a flat surface (see Fig. 14.5). The incidence angles of airborne radar imagery data can range from 20 to 60°. To reduce the errors in biomass estimation using the backscattering signature, the effect of terrain slope on radar signature needs to be corrected or reduced. The effects of both the terrain and radar incidence angle are target specific, i.e., the changing patterns of radar backscattering due to slope and radar incidence angle depend on the surface type. Fully correction requires knowing the type of the surface and the dependence of backscattering on incidence

angle of the surface type. This is very difficult if not impossible in practice. In practice, we calculate the local incidence angle of each pixel from the radar incidence angle and the slope and aspect of the local terrain surface, develop relationship between backscattering and incidence angle for major land cover types, and then correct or normalize the backscattering coefficients using these relationships according to the type of the image pixel.

The simple backscattering models for vegetation-like media take the form of (Ulaby et al., 1981)

$$\sigma^0(\theta) = \sigma_0 \cos^p(\theta) \qquad (14.22)$$

where θ is the local incidence angle. Both σ_0 and p are polarization dependent. When $p = 1$, the model means that the scattering coefficient (scattering per unit surface area) is dependent on $\cos\theta$, which is the ratio of projected area (normal to the incoming rays) to the surface area. When $p = 2$, the model is based on the Lambert's law for optics. Ulaby et al. (1981) pointed out that although either $p = 1$ or 2 seldom closely approximate the real scattering, sometimes $p = 1$ or 2, or a value between 1 and 2 may be used to represent scattering from vegetation. In a study of biomass estimation in Central Siberia, Sun et al. (2002) used a 3D radar

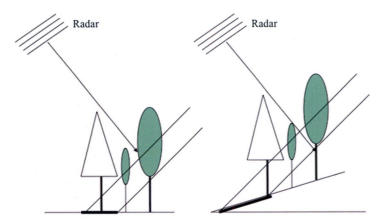

FIGURE 14.5   Terrain changes illumination area of a pixel and the spatial structure of the canopy.

backscatter model (Sun and Ranson, 1995) to simulate the backscattering coefficients from a typical forest stand on a surface with various slopes (10°, 20°, and 30°) and azimuth directions (45° increment from 0 to 360°) at incidence angle of 45°. Local incidence angle was calculated for each case. The model simulation results were fit to this simple model (Eq. 14.22) to estimate σ_0 and p for both L-band HH and HV polarizations. The best fits of the simulated data yield the following two equations:

$$\sigma^0_{hh}(\theta) = 0.361 * \cos^{1.78}\theta \quad r^2 = 0.93 \quad (14.23)$$

$$\sigma^0_{hv}(\theta) = 0.203 * \cos^{1.50}\theta \quad r^2 = 0.95 \quad (14.24)$$

Ideally, equations of this form should be developed for different kind of land cover types. The purpose of the terrain correction is to bring the LHV backscattering coefficients at incidence angle θ to a reference incidence angle θ_0:

$$\sigma^0_{hv}(\theta_0) = \sigma^0_{hv}(\theta)(\cos\theta_0/\cos\theta)^{1.50} \quad (14.25)$$

where θ is the local incidence angle. The local incidence angle can be calculated using elevation data if good DEM data are available. In the study by Sun et al. (2002), the θ was first estimated from HH backscattering using Eq. (14.23):

$$\cos\theta = \left(\sigma^0_{hh}(\theta)/0.361\right)^{1/1.78} \quad (14.26)$$

The real radar data may have different values for 0.361 and 0.203 as shown in Eqs. (14.23) and (14.24). If the actual SAR data are different from the simulated HH data and give a different value of σ^0_i other than the 0.361, the resulting $\cos\theta$ will be as follows:

$$\cos\theta = \left(\sigma^0_{hh}(\theta)/\sigma^0_i\right)^{1/1.78} = \left(\sigma^0_{hh}(\theta)/0.361\right)^{1/1.78}a \quad (14.27)$$

where $a = \left(0.361/\sigma^0_i\right)^{1/1.78}$ and accounts for the difference between σ_0's from the simulation (0.361) and radar image (σ^0_i).

Table 14.4 lists some parameters derived from polarimetric SAR data used for forest volume and biomass estimation (Gonçalves et al., 2011).

The major backscattering components from a forest canopy include (1) volumetric scattering from tree crowns, (2) direct backscattering from ground surface, and (3) double or multiple scattering between trunk-ground and crown-ground (Sun and Ranson, 1995). Various methods of decomposition of polarimetric radar signal into major scattering components (Cloude and Pottier, 1996) have been developed. These components can be used as independent variables for biomass estimation.

14.4.2.2 Interferometric SAR

In Chapter 14, we have shown that the elevation of scattering phase center of a forest stand can be obtained from InSAR data. If elevation of the ground surface is available, the height of the scattering center can be calculated. For example, the elevation difference between SRTM data and USGS NED data was highly correlated to the height indices (RH50) derived from lidar waveform data and had been used in forest biomass estimation models (Kellndorfer et al., 2010; Sun et al., 2011).

The coherence of repeat-pass InSAR data over forested area is dominated by volume decorrelation, temporal decorrelation, and thermal decorrelation in addition to the spatial baseline decorrelation (Zebker and Villasenor, 1992). The volume decorrelation of InSAR data provides additional information on vegetation canopies. ERS-1/2 Tandem coherence was reported to have high potential for the mapping of boreal forest stem volume (i.e, Askne et al., 2003; Santoro et al., 2002; Santoro et al., 2007; Wagner, 2003). It was used with MODIS VCF data in large area forest stem volume mapping (Cartus et al., 2011).

As shown in Chapter 14, the height of the phase center within a forest canopy depends on the structural parameters of the canopy as well as the incident and received signal polarization. Recent studies showed the capabilities of estimation of forest canopy height from InSAR data without the need for a ground DEM

(Cloude and Papathanassiou, 2003; Papathanassiou and Cloude, 2001; Treuhaft et al., 1996; Treuhaft and Siqueira, 2000) using the Polarimetric interferometric Synthetic Aperture Radar (PolInSAR) technique. A simplified radar scattering model was used to invert canopy height or other structural parameters from a set of coherence observations at various polarization configurations. The polarization coherence tomography (PCT) introduced by Cloude (2006) extended conventional polarimetric interferometry by allowing reconstruction of a vertical profile function, physically representing the variation of backscatter as a function of height. The decorrelation caused by factors other than the forest vertical volume distribution such as temporal decorrelation in repeat-pass InSAR data will cause high error in the retrieval of canopy height and vertical profile of scatterer density within a canopy. If multibaseline InSAR data from short repeating interval or single-pass data acquisitions are available, the retrieved canopy height and vertical profile can provide similar information on forest structure as large-footprint lidar waveforms do (Fontana et al., 2010; Luo et al., 2010).

Fig. 14.6 shows the average reflectivity function $f(\underline{w}, z)$ of two typical forest stand derived by PCT algorithm from L-band PolInSAR data. $\underline{w}$ is polarization. The curves are similar to the lidar waveforms. h_2 and h_4 are heights of two points where the reflectivity is close to zero (<0.002). The first peak appears between h_2 and h_4. h_3 is the height of the first peak. h_1 is the height of the other inflection point of the reflectivity function. Between 0 and h_1, there is a second peak. The shape and height of the first peak is highly correlated to AGB. For stand with high biomass (b in Fig. 14.6), the amplitude of the first peak is smaller and taller, and its width $h_4 - h_2$ is larger. The amplitude of second peak is smaller for high biomass stand. The first peak can fit a Gaussian curve very well (red dash lines in Fig. 14.6). Nine indices can be defined from the reflectivity function $\widehat{f}(\underline{w}, z)$ for biomass estimation. These indices are as follows:

$$(1) \quad \widehat{f}(\underline{w}, z); \quad (2) \quad P_2 = \sum_{z=h_2}^{z=h_4} z \cdot \widehat{f}(\underline{w}, z);$$

(3) $1 \big/ \widehat{f}(\underline{w}, h_3)$; (4) and (5) mean and standard deviation of the Gaussian curve used to fit the first peak; (6) $P_6 = 1 \bigg/ \sum_{z=h_2}^{z=h_4} \widehat{f}(\underline{w}, z)$; (7)

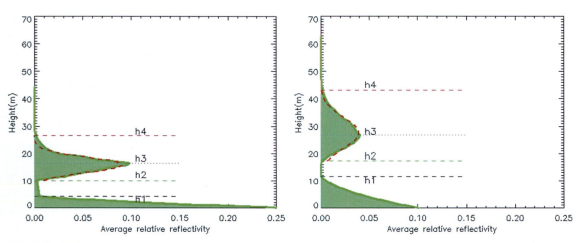

FIGURE 14.6 Average relative reflectivity function of two typical forest stand derived by PCT from PolInSAR data.

$$P_7 = 1 \Big/ \sum_{z=0}^{z=h_1} \widehat{f}(\underline{w}, z); \ (8) \ P_8 = P_6/P_7; \ \text{and} \ (9)$$

$$P_9 = \sum_{z=h_2}^{z=h_3} \widehat{f}(\underline{w}, z) \Big/ \sum_{z=h_3}^{z=h_4} \widehat{f}(\underline{w}, z).$$

The following model for biomass estimation obtained good results:

$$\ln(B) = \ln(b_0') + b_1 \ln(P_1) + b_2 \ln(P_2) + \cdots$$
$$+ b_n \ln(P_n)$$

$$(14.28)$$

where B is biomass, P_i, $i = 1-9$, are nine parameters defined above, and b_i, $i = 0-9$, are regression coefficients.

14.4.3 Spaceborne stereoscopic images

The stereoscopic (or photogrammetric) images were initially used for the extraction of terrain elevations by measuring the parallax contained in two images acquired with different view angles. The accuracy of elevations extracted from stereoscopic images is determined by the spatial resolution as well as the view angle difference of two images. Before 1990s, the spatial resolution of remote sensing images acquired by spaceborne sensors is too low to make accurate measurement of parallax. For example, the multispectral scanner onboard Landsat 1–3 has a nominal spatial resolution of 78 m (Slater, 1979), while that on board Landsat 4–5 is 30 m (Malila et al., 1984). Therefore, most studies have to work on airborne images in last century.

Since the beginning of this century, the spatial resolution of spaceborne remote sensing images has been greatly improved, especially for commercial satellites. For example, Quickbird can acquire 2.44 m multispectral images and 0.61 m panchromatic images (Noguchi and Fraser, 2004); IKONOS can provide 3.28 m multispectral images and 0.82 m panchromatic images (Zhou and Li, 2000). Although they are not specifically designed photogrammetric systems, they can

provide along-track and across-track stereoscopic images by changing attitudes. St-Onge et al. (2008) examined the IKONOS stereo images acquired with the same elevation angle (67.4°) but different azimuth angles (27.8° and 180.52°). The digital surface elevation model was firstly extracted through stereoscopic processing. The forest height model was further derived with the help of ground surface elevation from small-footprint lidar data. The final results demonstrated that the forest AGB estimated using extracted forest height was highly correlated with field measurement with $R^2 = 0.79$.

Spaceborne photogrammetric systems are rapidly growing in the last two decades. The ASTER onboard the Terra satellite are designed for the acquisition of stereoscopic imagery using two cameras pointing to nadir and backward. The images acquired by ASTER have been used for the global digital elevation model (ASTER GDEM) (Ni et al., 2015). The Advanced Land Observing Satellite (ALOS), launched by the Japan Aerospace Exploration Agency in 2006, carried Panchromatic Remote-sensing Instrument for Stereo Mapping (PRISM). The three cameras of ALOS/PRISM, pointing to forward, nadir, and backward with a nominal angle interval of 24.0°, can acquire 2.5 m stereoscopic images. The precise global digital 3D map has been generated using ALOS/PRISM stereoscopic imagery (Tadono et al., 2016). China also launched its first civil spaceborne photogrammetric satellites in 2012, referred to as ZY-3, which has three cameras in the same way as the ALOS/PRISM. The spatial resolution of forward and backward cameras was improved from 3.5 to 2.5 m on the second satellite of ZY-3 launched in 2016.

Compared to most studies focused on the extraction of ground surface elevations, scientists in the field of forest remote sensing gradually paid attention to explore the measurement of forest spatial structure spaceborne stereoscopic imagery in the last decade. Comparing with lidar data, Ni et al. (2015) found that forest height was contained in the ASTER GDEM, although the spatial resolution of ASTER is merely 15 m and the ASTER GDEM is the combined products over 10 years between 2000 and 2010.

Typically, the spaceborne stereoscopic imagery is acquired by pushbroom camera, while the airborne stereoscopic imagery is acquired by frame cameras. One of the biggest differences between them is the field of view and the number of observation angles. Airborne frame cameras have large field of views and thousands of observation angles in both along and across track. For pushbroom camera, the field of view across track is only several degrees while the number of observation angles along track is determined by the number of cameras. Therefore, the automatic identification of common points is much more difficult for pushbroom camera than frame cameras. How to make full use of limited observation angles to get more common points is a critical issue for spaceborne stereoscopic imagery. Ni et al. (2014) analyzed the point cloud extracted from different view combinations and found that they were complementary in both horizontal and vertical distributions. The synthesized point cloud can give better description of forest spatial structures than any of them.

Theoretically, the spaceborne stereoscopic imagery has great potentials for the estimation of forest AGB like lidar data. Current studies only focused on the measurement of forest spatial structures. There is long way to go for the regional mapping of forest AGB using spaceborne stereoscopic imagery. Much effort should be spent on understanding the effects of environmental factors on the distribution of point cloud from stereoscopic imagery, such as the seasonal effects (Montesano et al., 2019), terrain effects, observation geometries (Montesano et al., 2017), and so on. In addition to analysis using as many stereoscopic imagery as possible,

theoretical model is also an important tool to carry out such kind of studies (Ni et al., 2019).

14.5 Synthesis methods of multisource data

Theoretically, forest AGB is directly determined by forest height, stand density, DBH, and species. How to make full use of multisource remote sensing dataset is an important topic.

After the field campaign and acquisition and processing of remote sensing and other environmental data, we have the biomass and corresponding imagery data over a set of points within a study area. Assume B_i, $i = 1, 2, \ldots, N$, is the biomass and X is the data vector. The biomass estimation is to find a prediction model F:

$$\widehat{B} = F(X) \tag{14.29}$$

The objective of developing such model is to minimize the error of the estimation:

$$\text{RMSE} = \sqrt{\sum_{i=1}^{N} \left(\widehat{B}_i - B_i\right)^2 / N} \tag{14.30}$$

The F can be linear or nonlinear regression models, or various machine learning algorithms, such as neural network, K nearest-neighbor, regression tree (RT), RF, and MaxEnt. The spectral responses at optical data, backscattering, and attributes derived from PolSAR and PolInSAR data, various indices from lidar data, and image textures can be directly used as variables for developing prediction models in parametric method such as regression analysis. Most machine learning algorithms, such as MaxEnt, are so-called nonparametric methods, which automatically learn to recognize complex patterns and make intelligent decisions based on data. Data from different sources and with different formats can be used in the process, and the biomass estimation procedure was called spatial modeling of ABG (Saatchi et al.,

2011). Several methods have already been described in the sections dealing with the biomass estimation from optical data. Following sections will be devoted to other methods we have not discussed before.

14.5.1 Regression models

The regression using SAR backscattering intensities was the earliest and most common methods for biomass estimation. Within field sampling plots with certain sizes, all trees above a specified size (DBH greater than 3, 5, 10 cm, etc.) were measured and allometric equations were used to calculate biomass in the plot. Various regression methods (linear and nonlinear, multivariate, stepwise, etc.) were used to find optimal relationships between biomass and selected variables. Various studies have shown that the sensitivity of radar backscattering to biomass depends on radar wavelength and polarization. In early studies (Dobson et al., 1992; Le Toan et al., 1992), the relationships between forest parameters and backscattering of single frequency polarization were investigated. In both studies, logarithm of biomass and backscattering coefficient in dB were used, and polynomial (Dobson et al., 1992) and linear (Le Toan et al., 1992) equations were used. In studies by Ranson et al. (1995; 1997a) and Ranson and Sun (1994, 1997), the cubic root of biomass was used in prediction model. It was also found from these studies that the HV ratio (PHV/CHV, LHV/CHV) suppressed the backscattering abnormality due to terrains or at various edges while retained the relationships with forest biomass.

In a study by Goncalves et al. (2011), the attributes listed in Table 14.4 that were derived from L-band PolSAR data were used to estimate forest stem volume of tropical forest. The linear prediction model took the following form:

$$Y_i = b_0 + b_1 X_1 + b_2 X_2 + \cdots + b_{p-1} X_{p-1} \quad (14.31)$$

where Y_i is the variable to be estimated and X are independent variables. Different transforms might be applied to both dependent and independent variables. Works on forest biomass estimation from SAR data (Beaudoin et al., 1994; Dobson et al., 1992; Kasischke et al., 1995; Le Toan et al., 1992; Ranson and Sun, 1994) have shown that the radar backscattering at HV polarization were clearly correlated with forest AGB up to 100–150 ton/ha for L and P band SAR data. An intrinsic problem for SAR data is the speckle, which can be reduced by various spatial filtering including multilook processing (averaging). It was found that the correlation between biomass and HV backscattering improved when radar signature was averaged within larger plots (Saatchi et al., 2011).

The following model was used by Saatchi et al. (2007) to estimate biomass from multifrequency and multipolarization data:

$$\log(W) = a_0 + a_1 \sigma_{HV}^0 + a_2 \left(\sigma_{HV}^0\right)^2 + b_1 \sigma_{HH}^0$$
$$+ b_2 \left(\sigma_{HH}^0\right)^2 + c_1 \sigma_{VV}^0 + c_2 \left(\sigma_{VV}^0\right)^2$$

$$(14.32)$$

where W is biomass and all radar backscattering coefficient are in decibels. This equation can be considered as a parametric equation including all scattering mechanisms represented by the radar polarimetric measurements (Moghaddam and Saatchi, 1999). As a second-order polynomial, the equation can be referred to as a semiempirical algorithm for estimating forest biomass components.

NASA/JPL's Uninhabited Aerial Synthetic Aperture Radar (UAVSAR) provides high-resolution polarimetric SAR data for use in multiple studies including the retrieval of forest structure parameters. In the study in Howland, Maine, US, UAVSAR data acquired from opposite directions were used for biomass estimation. As other airborne SARs, one disadvantage of UAVSAR image data is the large range of local incidence angles across the image. The methods shown in Eqs. (14.20)−(14.27) were used to reduce the effects of incidence angles. The forests were classified into evergreen forest (EF) and non-EF. Separate correction coefficients were derived. When forest types were not considered the best biomass prediction model selected by stepwise regression

$$B = 27.7346 + 5169.1742\sigma^0_{13}(HV)$$
$$- 1695.2965\sigma^0_{13}(VV) - 874.3129\sigma^0_{1610}(HH)$$
$$+ 4328.7494\sigma^0_{1610}(HV)$$

$$(14.33)$$

with $R^2 = 0.7445$, $N = 57$, and RMSE = 39.99 Mg/ha. When field samples were divided into EF and non-EF, better prediction models could be obtained. For EF, the following model was obtained:

$$B = 17.3313 - 785.3552\sigma^0_{13}(HH)$$
$$+ 6622.0034\sigma^0_{13}(HV) - 1442.1829\sigma^0_{13}(VV)$$
$$- 3230.9985\sigma^0_{10}(HV) + 2291.5723\sigma^0_{10}(VV)$$

$$R^2 = 0.7231, N = 29, \text{RMSE} = 34.63 \text{ Mg/ha}$$

$$(14.34)$$

For non-EF, the model was

$$B = 37.1107 + 5383.5812\sigma^0_{13}(HV)$$
$$- 1607.3955\sigma^0_{13}(VV) - 1231.5441\sigma^0_{1610}(HH)$$
$$+ 4998.7137\sigma^0_{1610}(HV)$$

$$R^2 = 0.7936, N = 28, \text{RMSE} = 36.31 \text{ Mg/ha}$$

$$(14.35)$$

Both models showed smaller RMSE than those models developed without considering the forest type.

14.5.2 Nonparametric algorithms

There are multiple ways of extrapolating the samples of forest biomass data to a gridded map. These include parametric approaches such as the use of regression models and nonparametric approaches such as interpolation, co-kriging, classification, or segmentation (with biomass allocation), decision rule techniques as in RF, MaxEnt, and several other machine learning approaches (Baccini et al., 2008; Kellndorfer et al., 2010; Saatchi et al., 2007, 2011). Two algorithms, i.e., the RF and MaxEnt, are increasingly being used in biomass mapping at complex environments and at regional and global scales because these methods can integrate variables with different statistical distributions and provide more stable and relevant information.

14.5.2.1 Segmentation and biomass allocation

Forest stratification combined with ground sampling is the traditional method to inventory the biomass of a region. In a study by Nelson et al. (2009), the forested area in Central Siberia was stratified into 16 classes using MODIS classification map (evergreen needle, deciduous forest, and mixed forests) and forest fraction cover (four 25% intervals) from MODIS vegetation continuous field product. The biomass samples at GLAS footprints provide statistics of forest stem volumes for each of the 16 classes. This approach was also employed by Lefsky (2010) to develop a global forest height distribution map from GLAS lidar and MODIS data. The MODIS data at 500 m resolution was firstly segmented into spectrally similar patches. The statistics of GLAS lidar sample data that fall into some patches were calculated. The statistics was assigned as the value of similar patches. The methodology is straightforward and can be easily implemented. Because all pixels in a patch have been assigned to the same biomass value, the biomass maps produced with this method are thematic maps. Many data including optical spectral imagery, backscattering from PolSAR data, coherence, and height of scattering phase center from InSAR or PolInSAR data, DEM, and other environmental data (soil, climate, etc.) can be used to improve the quality of segmented patches.

14.5.2.2 Random forest method

Classification and RTs have become increasingly popular within the remote sensing research and applications community (Goetz et al., 2005; Hansen et al., 2008). RT has clear advantages over classical statistical methods. Because no a priori assumptions are made about the nature of the relationships among the response (e.g., biomass) and predictor variables (e.g., remote sensing data), RT allows for the possibility of interactions and nonlinearities among variables (Moore et al., 1991). But part of the output error in a single RT is due to the specific choice of the training dataset. RFs are designed to produce accurate predictions that do not overfit the data (Breiman, 2001, 2002; Liaw and Wiener, 2002). Bootstrap samples are drawn to construct multiple trees with a randomized subset of predictors, hence the name "random" forests. A large number of trees (500–2000) are grown, hence a "forest" of trees. The number of predictors used to find the best split at each node is a randomly chosen subset of the total number of predictors (Prasad et al., 2006).

Kellndorfer et al. (2006) used RF to predict vegetation canopy height (VCH), basal area weighted height, and aboveground dry biomass (ADB). Four variables, i.e., the height of the mean scattering phase center MHSPC, cover type, elevation, and canopy density, were found to be most significant. Because of the mountainous terrain and associated patterns of vegetation zonation in this particular mapping zone, it was not surprising that elevation and cover type had both significant explanatory value in the RT models predicting VCH and ADB. However, slope and aspect did not contribute significantly to the model. Spatial prediction was performed on a per pixel basis using forest cover type, NED elevation, and NLCD 2001 canopy density.

14.5.2.3 Maximum entropy model

Maxent is a general-purpose method for making predictions or inferences from incomplete information. Its origins lie in statistical mechanics (Jaynes, 1957), and it remains an active area of research in diverse areas such as astronomy, portfolio optimization, image reconstruction, statistical physics, and signal processing (Phillips et al., 2006). The idea of Maxent is to estimate a target probability distribution by finding the probability distribution of maximum entropy, subject to a set of constraints that represent incomplete information about the target distribution. The information available about the target distribution often presents itself as a set of real-valued variables, called "features," and the constraints are that the expected value of each feature should match its empirical average (average value for a set of sample points taken from the target distribution). The individual occurrences of a variable's point locality in geographical space (here, biomass sample points) can be approximated with a probability distribution. The best estimation of this probability distribution should satisfy any constrains on the unknown that we are aware of and have maximum entropy (Jaynes, 1957). For biomass estimation from remote sensing data, the environmental constraints are represented by the spectral information from remote sensing image data (MODIS, ALOS, SRTM, Landsat, etc.).

Assuming that the unknown probability distribution of biomass P is defined over a finite set X (the set of pixels sampled within the study area), with the probability value of $P(x)$ for each point x. These probabilities sum to 1 over the space defined by X. The MaxEnt algorithm uses a Bayesian approach to approximate P with a probability distribution $\widehat{P}$ by maximizing the entropy of $\widehat{P}$ as follows:

$$\text{Entropy} = \sum_{x \in X} \widehat{P}(x) \ln\left[\widehat{P}(x)\right] \qquad (14.36)$$

where ln is the natural logarithm. The distribution $\widehat{P}$ with highest entropy is considered the best distribution for inference.

By using features that are continuous real-valued functions of X, such as remote sensing variables, and a set of sample points (training data) of X provided by the field data and lidar sample points, the MaxEnt algorithm employs likelihood estimation procedures to find a probability distribution for all the points in X that have similar statistics as the sample points. MaxEnt assumes a priori a uniform distribution and performs a number of iterations in which the weights (of different features) are adjusted to maximize the average probability of the point localities (also known as the average sample likelihood), expressed as the training gain (Phillips et al., 2004). These weights are then used to compute the MaxEnt distribution over the entire geographic space. MaxEnt has a number of properties that makes it useful for modeling forest biomass over landscapes (Saatchi et al., 2008).

In a study of generating a biomass benchmark map of global tropical forest by Saatchi et al. (2011), MaxEnt model was ran using the locations of inventory plots and GLAS lidar points over forests of all three tropical regions, using 14 remote sensing image layers (5 NDVI, 3 LAI, 4 QSCAT, and 2 SRTM metrics). Total 1877 biomass pixels from inventory plots and 93,188 points randomly selected from 160,918 GLAS lidar points were used as training data and the rest of the GLAS points were used as an independent testing dataset. AGB density from the training data were divided into 11 biomass classes in 25 Mg/ha intervals for biomass ranging from 0 to 100 Mg/ha and 50 Mg/ha intervals for biomass >100 Mg/ha. Biomass classes were used partially to capture the errors associated with the GLAS-derived biomass values and partially because MaxEnt does not run with continuous data. However, biomass ranges for each class can be easily modified as long as there are enough samples (>100

locations) within each class bin. MaxEnt was ran for different biomass ranges to examine the impact on the final AGB map, and Saatchi et al. (2011) found the selected biomass ranges suitable for creating the optimum map. For each interval (biomass class), a MaxEnt model run was performed and 11 continuous probability distribution maps were generated. The values on these maps range from 0 to 100, with 0 as the least suitable pixel for the biomass class and 100 as the most suitable. The continuous probability maps were then combined and converted into a single biomass map by choosing the biomass value associated with the maximum probability weighted mean for each pixel, using the relationship

$$\widehat{B} = \sqrt{\frac{\sum_{i=1}^{N} P_i^n B_i}{\sum_{i=1}^{N} P_i^n}} \tag{14.37}$$

where P_i is the MaxEnt probability estimated for each biomass range class B_i (median value of the range), and $\widehat{B}$ is the predicted value of AGB for each pixel. The power of the probability n is chosen to weight the predicted value toward the maximum probability which is close to the true value when other class probabilities are small. After several iterations and cross validation with the test data, Saatchi et al. (2011) found that for $n = 3$ the distribution of AGB and the cross validation converged to its optimum value.

The predicted probability for each biomass range was to calculate the root mean squared error for an estimation of the uncertainty resulted from the MaxEnt model:

$$\sigma_{\widehat{B}} = \sqrt{\frac{\sum_{i=1}^{N} \left(B_i - \widehat{B}\right)^2 P_i}{\sum_{i=1}^{N} P_i}} \tag{14.38}$$

14.5.2.4 Support vector regression

Support vector regression (SVR) is an application of support vector machine in regression problems. The main idea is, according to the specific sample dataset $[(x1, y1), (x2, y2), ..., (xn, yn)]$, to seek a function $y = f(x)$ which can reflect the samples effectively and to make the cumulative error of the estimated value relative to the samples be the smallest. In general, the error $\varepsilon-$ is measured by the insensitive loss function.

The regression is referred to as linear or nonlinear according to the obtained function. The basic idea of the algorithm is to realize linear regression by constructing a linear decision function in high-dimensional space after dimension upgrading, as shown in the formula,

$$f(x, \omega) = \omega\varphi(x) + b \qquad (14.39)$$

where ω is the weight vector, b is the threshold, and $\varphi(x)$ is a nonlinear mapping function that transfers the original feature space to the high-dimensional space. Nonlinear transformation is usually based on kernel functions. Functions that satisfy Mercer's theorem can be used as kernel functions. Commonly used kernel functions are as follows:

(1) Linear kernel

$$k(x_i, x_j) = x_i \cdot x_j \qquad (14.40)$$

(2) Polynomial kernel

$$k(x_i, x_j) = (x_i \cdot x_j + 1)^d \qquad (14.41)$$

(3) Radial basis kernel (Gauss kernel)

$$k(x_i, x_j) = e^{\left(\gamma\|x_i - x_j\|^2\right)} \qquad (14.42)$$

(4) Sigmoid kernel

$$k(x_i, x_j) = \tanh[\gamma(x_i \cdot x_j + 1)] \qquad (14.43)$$

The standard form of SVR can be obtained by solving optimization problem, as shown in Eq. (14.44) (Vapnik, 1998),

$$\min \frac{1}{2}\|\omega\|^2 + C\sum_{i=1}^{n}(\xi + \xi_i^*)$$

$$s.t. \begin{cases} y_i - f(x_i, \omega) \leq \varepsilon + \xi_i^* \\ f(x_i, \omega) - y_i \leq \varepsilon + \xi_i \\ \xi_i, \xi_i^* \geq 0, \ i = 1, ..., n \end{cases} \qquad (14.44)$$

Its dual form is

$$f(x) = \sum_{i=1}^{l}(\alpha_i^* - \alpha_i)K(x, x_i) + b$$

$$s.t. \begin{cases} 0 \leq \alpha_i \leq C \\ 0 \leq \alpha_i^* \leq C \end{cases} \qquad (14.45)$$

Zhang et al. (2014) estimated forest biomass in GLAS footprints by using stepwise regression, partial least squares regression, and SVR. Ground-measured biomass value was used as reference value; GLAS waveform parameters and MODIS vegetation indices were used as independent variables. The estimation of forest biomass was carried out, and it was found that SVR estimation accuracy was the highest. The addition of MODIS vegetation index improved the accuracy of forest biomass estimation in GLAS footprints. On this basis, using RF method, combined with MODIS vegetation indices and MISR multiangle urban and rural spectro-radiometer data, the biomass of

Northeast China was estimated. The results showed that the estimation accuracy is sufficient when only multitemporal MODIS data are used. Although the additional MISR data improved the accuracy to a certain extent, it is not significant.

Meng et al. (2016) estimated the biomass of a patch of temperate mixed forest located in Little Hinggan Mountains, China. Texture information of high spatial resolution aerial remote sensing images was extracted based on Fourier texture structure classification method (FOTO), and the relationship between texture information and ground measured data was established by SVR. 10-fold cross-validation method was used to verify the optimal model. By comparing the results with the biomass base map obtained by airborne lidar data, it is found that the FOTO texture indices can explain forest biomass well. The R^2 of the model is 0.883 and the RMSE is 34.25 ton/ha. The results showed that the method based on texture parameters has a good potential for estimating other temperate forest biomass.

14.5.3 Multisource remote sensing data

There are many synergistic inversion methods for multisource remote sensing data, which can be categorized as multisource passive optical remote sensing, multisource active remote sensing, and active–passive remote sensing.

Forest biomass estimation using multisource passive optical remote sensing is always based on texture information extracted from high-resolution optical remote sensing data, high-resolution spectral information, and vegetation indices. For example, Nichol et al. (2011) estimated forest biomass in Hong Kong using texture information extracted from two high-resolution optical data such as AVNIR-2 and SPOT-5. It was found that the estimation accuracy was better than that using multispectral

and vegetation indices. Bastin et al. (2014) extracted the texture information of Geoeye-1 and QuickBird-2 by Fourier-based textural ordination (FOTO) method to estimate forest biomass in the Congo region of Africa. The results showed that texture information extracted from optical remote sensing data could greatly reduce the saturation issue at high biomass level.

Lidar-based or field-measured biomass data are always taken as reference data in forest biomass estimation. Sun et al. (2011) investigated the forest biomass mapping by taking Forest biomass based on airborne large-footprint lidar as the dependent variable and SAR data as the independent variable. The results demonstrated the potential of the synthesis of lidar samples and radar imagery for forest biomass mapping. Tsui et al. (2013) estimated forest biomass based on co-kriging method by using benchmark data derived from the airborne small-footprint lidar and spaceborne L-band and C-band SAR data.

For active–passive remote sensing, Hyde et al. (2006) investigated the synergy of the structural information provided by lidar with SAR and passive optical data. They found that lidar is the best choice for estimating forest biomass with single remote sensing data. The addition of ETM+ can improve the estimation accuracy, while the addition of Quickbird and InSAR/SAR information has little effect. Anderson et al. (2008) mapped regional forest biomass using LVIS waveform data and hyperspectral data acquired by Airborne Visible/Infrared Imaging Spectrometer. Boudreau et al. (2008) firstly mapped forest biomass using airborne lidar data with the help of field plot measurements. Then the forest biomass of GLAS footprint was estimated by waveform metrics. Finally, the regional forest biomass map was produced by extrapolating the GLAS estimations based on the image segmentation and fine vegetation classification map (27 categories).

Saatchi et al. (2011) mapped tropical forest biomass using field measurements, GLAS, MODIS, and QuikSCAT. Zhang et al. (2014) compared the three methods (stepwise regression, partial least squares regression, and SVR), using ICESAT GLAS, MODIS, MISR, in the estimation of the forest biomass in Northeast China. The results showed that SVR work better for forest biomass retrieval. Chi et al. (2015) reported the AGB mapping using a combination of GLAS, MODIS data, and ecological zoning data based on RF. Zhang et al (2017) explored the synthesis of SAR and stereo imagery for the estimation of forest AGB. They found that the introduction of forest height from stereo imagery could greatly improve the estimation.

14.6 Future perspective

Understanding of the Earth's carbon cycle is an urgent societal need as well as a challenging intellectual problem. Significant uncertainties remain about what processes cause the observed changes in the atmospheric composition of CO_2. Terrestrial plants contain similar amount of carbon as entire atmosphere does and play important role in carbon cycling. Direct observations of carbon stocks and changes are important in global carbon studies.

Satellite and other remote sensing observations of the Earth system complement the in situ observations and are the only way to provide global data at high spatial and temporal resolutions, making it possible to evaluate global patterns in the carbon system in a manner that would not be feasible using only in situ observations. While airborne sensors, such as Imaging Spectroradiometer, lidar, SAR, are providing high-resolution data in local scale, for continental and global scale, we rely on the data currently available from LANDSAT, MODIS, SRTM, ALOS PALSAR, the GLAS on board Ice, Cloud, and land Elevation Satellite (ICESat), and the data soon to be available from the Landsat

Data Continuity Mission (LDCM), the National Polar-orbiting Operational Environmental Satellite System (NPOESS) Preparatory Project (NPP), and the L-band SAR aboard ALOS-2. In few years, NASA's new lidar on ICESat-2 will provide data to assess carbon stocks, and Soil Moisture Active and Passive (SMAP) will inform linkages between the water and carbon cycles. Other spatial data related to plant growth may also be used, thanks to various intelligent algorithms.

There are numerous methods to link remote sensing data with biomass as we have shown in various chapters in this book. Selection of methods depends on the project scope and the data availability. Although many progresses have been made in developing these algorithms under different environmental conditions, researches will remain active on various issues, such as how to improve the prediction accuracy at complex conditions and provide adequate error estimation of biomass products. The multivariate regression may still be a popular method in biomass estimation; other no-parametric machine learning methods will get more attention, especially for projects at regional and global scales. The inversion algorithms based on theoretical models start to appear either by inversion of canopy parameters from PolInSAR data or using model simulated database as a lookup table for inversion. PolInSAR technology is promising, but it requires Polarimetric InSAR data without (or with negligible) temporal decorrelation in forested area. It is very hard or impossible to acquire this kind of data by the repeat-pass spaceborne SARs.

While many studies showed that the canopy structure, species, and other environmental factors should be considered in development of biomass estimation algorithms, in some cases researchers found that the same prediction model can be used over very different canopy communities. For example, Mitchard et al. (2009) found a widely applicable general relationship exists between AGB and L-band backscatter for woody vegetation with lower-biomass in tropical, temperate, and boreal biomes. The similar findings exist in studies using lidar data. Lefsky et al. (2002) compared the relationships between lidar-derived canopy structure and coincident field measurements of AGB at sites in the temperate deciduous, temperate coniferous, and boreal coniferous biomes and found that a single equation explains 84% of variance in AGB ($P < .0001$) and shows no statistically significant bias in its predictions for any individual site. Investigation of effects of plant species, canopy structure, terrain slope, surface reflectance, roughness, moisture condition, and other factors on biomass prediction algorithms and methods for reducing these effects will be continued. The findings from these studies provide insight for region stratification and accuracy estimation in operational regional or global biomass mapping.

References

Abshire, J.B., Sun, X., Riris, H., Sirota, J.M., McGarry, J.F., Palm, S., Yi, D., Liiva, P., 2005. Geoscience laser altimeter system (GLAS) on the ICESat mission: on-orbit measurement performance. Geophys. Res. Lett. 32.

Anderson, J.E., Plourde, L.C., Martin, M.E., Braswell, B.H., Smith, M.L., Dubayah, R.O., Hofton, M.A., Blair, J.B., 2008. Integrating waveform lidar with hyperspectral imagery for inventory of a northern temperate forest. Remote Sens. Environ. 112 (4), 1856–1870.

Askne, J., Santoro, M., Smith, G., Fransson, J.E.S., 2003. Multi-temporal repeat-pass SAR interferometry of boreal forests. IEEE Trans. Geosci. Remote Sens. 41, 1540–1550.

Atkinson, P.M., Tatnall, A.R.L., 1997. Introduction neural networks in remote sensing. Int. J. Remote Sens. 18, 699–709.

Baccini, A., Friedl, M.A., Woodcock, C.E., Warbington, R., 2004. Forest biomass estimation over regional scales using multisource data. Geophys. Res. Lett. 31, L10501.

Baccini, A., Laporte, N., Goetz, S.J., Sun, M., Dong, H., 2008. A first map of tropical Africa's above-ground biomass derived from satellite imagery. Environ. Res. Lett. 3, 045011.

Balzter, H., 2001. Forest mapping and monitoring with interferometric synthetic aperture radar (InSAR). Prog. Phys. Geogr. 25, 159–177.

Balzter, H., Rowland, C., Saich, P., 2007. Forest canopy height and carbon estimation at Monks Wood national nature reserve, UK, using dual-wavelength SAR interferometry. Remote Sens. Environ. 108, 224–239.

Bastin, J.F., Barbier, N., Couteron, P., Adams, B., Shapiro, A., Bogaert, J., De Canniere, C., 2014. Aboveground biomass mapping of African forest mosaics using canopy texture analysis: toward a regional approach. Ecol. Appl. 24 (8), 1984–2001.

Basuki, T.M., Skidmore, A.K., Hussin, Y.A., Van Duren, I., 2013. Estimating tropical forest biomass more accurately by integrating ALOS PALSAR and Landsat-7 ETM+ data. Int. J. Remote Sens. 34, 4871–4888.

Beaudoin, A., Le Toan, T., Goze, S., Nezry, E., Lopes, A., Mougin, E., Hsu, C.C., Han, H.C., Kong, J.A., Shin, R.T., 1994. Retrieval of forest biomass from SAR data. Int. J. Remote Sens. 15, 2777–2796.

Blackard, J., Finco, M., Helmer, E., Holden, G., Hoppus, M., Jacobs, D., Lister, A., Moisen, G., Nelson, M., Riemann, R., 2008. Mapping U.S. forest biomass using nationwide forest inventory data and moderate resolution information. Remote Sens. Environ. 112, 1658–1677.

Blair, J.B., Rabine, D.L., Hofton, M.A., 1999. The Laser Vegetation Imaging Sensor: a medium-altitude, digitisation-only, airborne laser altimeter for mapping vegetation and topography. ISPRS J. Photogrammetry Remote Sens. 54, 115–122.

Boerner, W.M., Yan, W.L., Xi, A.Q., et al., 1991. On the Basic Principles of Radar Polarimetry - the Target Characteristic Polarization State Theory of Kennaugh, Huynens Polarization Fork Concept, and Its Extension to the Partially Polarized Case. Proceedings of the IEEE 79 (10), 1538–1550.

Boudreau, J., Nelson, R.F., Margolis, H.A., Beaudoin, A., Guindon, L., Kimes, D.S., 2008. Regional aboveground forest biomass using airborne and spaceborne LiDAR in Québec. Remote Sens. Environ. 112 (10), 3876–3890.

Breiman, L., 2001. Random forests. Mach. Learn. 45, 5–32.

Breiman, L., 2002. Using models to infer mechanisms. IMS Wald Lect. 2 (15), 59–71 [online] URL: http://oz.berkeley.edu/users/breiman/wald2002-2.pdf.

Brenner, A.C., Zwally, H.J., Bentley, C.R., Csatho, B.M., harding, D.J., Hofton, M.A., Minster, J.B., Roberts, L.A., Saba, J.L., Thomas, R.H., Yi, D., 2002. GLAS Algorithm Theoretical Basis Document Version 5.0 - Derivation of Range and Range Distributions from Laser Pulse Waveform Analysis for Surface Elevations, Roughness, Slope, and Vegetation Heights. http://www.csr.utexas.edu/glas/pdf/WFAtbd_v5_02011Sept.pdf.

Cartus, O., Santoro, M., Schmullius, C., Li, Z., 2011. Large area forest stem volume mapping in the boreal zone using synergy of ERS-1/2 tandem coherence and MODIS vegetation continuous fields. Remote Sens. Environ. 115, 931—943.

Chave, J., Andalo, C., Brown, S., et al., 2005. Tree allometry and improved estimation of carbon stocks and balance in tropical forests. Oecologia 145, 87—99.

Chi, H., Sun, G.Q., Huang, J.L., Guo, Z.F., Ni, W.J., Fu, A.M., 2015. National forest aboveground biomass mapping from ICESat/GLAS data and MODIS imagery in China. Remote Sens. 7 (5), 5534—5564.

Clark, M.L., Clark, D.B., Roberts, D.A., 2004. Small-footprint lidar estimation of sub-canopy elevation and tree height in a tropical rain forest landscape. Remote Sens. Environ. 91, 68—89.

Cloude, S.R., 2006. Polarization coherence tomography. Radio Sci. 41 (4), RS4017.

Cloude, S.R., Papathanassiou, K.P., 2003. Three-stage inversion process for polarimetric SAR interferometry. IEE Proc. — Radar, Sonar Navig. 150, 125—134.

Cloude, S.R., Pottier, E., 1996. A review of target decomposition theorems in radar polarimetry. IEEE Trans. Geosci. Remote Sens. 34, 498—518.

Clough, B.F., Scott, K., 1989. Allometric relationships for estimating above-ground biomass in six mangrove species. Forest Ecology and Management 27, 117—127.

Dobson, M.C., Ulaby, F.T., LeToan, T., Beaudoin, A., Kasischke, E.S., Christensen, N., 1992. Dependence of radar backscatter on coniferous forest biomass. IEEE Trans. Geosci. Remote Sens. 30, 412—415.

Dong, J., Kaufmann, R.K., Myneni, R.B., Tucker, C.J., Kauppi, P.E., Liski, J., Buermann, W., Alexeyev, V., Hughes, M.K., 2003. Remote sensing estimates of boreal and temperate forest woody biomass: carbon pools, sources, and sinks. Remote Sens. Environ. 84, 393—410.

Drake, J.B., Dubayah, R.O., Clark, D.B., Knox, R.G., Blair, J.B., Hofton, M.A., Chazdon, R.L., Weishampel, J.F., Prince, S., 2002. Estimation of tropical forest structural characteristics using large-footprint lidar. Remote Sens. Environ. 79, 305—319.

Drake, J.B., Knox, R.G., Dubayah, R.O., Clark, D.B., Condit, R., Blair, J.B., Hofton, M., 2003. Above-ground biomass estimation in closed canopy Neotropical forests using lidar remote sensing: factors affecting the generality of relationships. In: Global Ecology and Biogeography. Wiley-Blackwell, pp. 147—159.

Du, L., Zhou, T., Zou, Z., Zhao, X., Huang, K., Wu, H., 2014. Mapping forest biomass using remote sensing and national forest inventory in China. Forests 5 (6), 1267—1283.

Dubayah, R.O., Drake, J.B., 2000. Lidar remote sensing for forestry. J. For. 98, 44—46.

Fazakas, Z., Nilsson, M., Olsson, H., 1999. Regional forest biomass and wood volume estimation using satellite data and ancillary data. Agric. For. Meteorol. 98—99, 417—425.

Fernández-Manso, O., Fernández-Manso, A., Quintano, C., 2014. Estimation of aboveground biomass in Mediterranean forests by statistical modelling of ASTER fraction images. Int. J. Appl. Earth Obs. Geoinf. 31, 45—56.

Fontana, A., Papathanassiou, K.P., Iodice, A., Lee, S.K., 2010. On the Performance of Forest Vertical Structure Estimation via Polarization Coherence Tomography. In: http://ieee.uniparthenope.it/chapter/_private/proc10/22.pdf.

Foody, G.M., Cutler, M.E., McMorrow, J., Pelz, D., Tangki, H., Boyd, D.S., Douglas, I., 2001. Mapping the biomass of Bornean tropical rain forest from remotely sensed data. Glob. Ecol. Biogeogr. 10, 379—387.

Foody, G.M., Boyd, D.S., Cutler, M.E.J., 2003. Predictive relations of tropical forest biomass from Landsat TM data and their transferability between regions. Remote Sens. Environ. 85, 463—474.

Fromard, F., Puig, H., Mougin, E., et al., 1998. Structure, above-ground biomass and dynamics of mangrove ecosystems: new data from French Guiana. Oecologia 115, 39—53.

Franco-Lopez, H., Ek, A.R., Bauer, M.E., 2001. Estimation and mapping of forest stand density, volume, and cover type using the k-nearest neighbors method. Remote Sens. Environ. 77, 251—274.

Fraser, R.H., Li, Z., 2002. Estimating fire-related parameters in boreal forest using SPOT VEGETATION. Remote Sens. Environ. 82, 95—110.

Freeman, A., Durden, S.L., 1998. A three-component scattering model for polarimetric SAR data. IEEE Trans. Geosci. Remote Sens. 36, 963—973.

Comley, B.W.T., McGuinness, K.A., 2005. Above- and belowground biomass, and allometry, of four common northern Australian mangroves. Australian Journal of Botany 53, 431—436.

Galidaki, G., Zianis, D., Gitas, I., Radoglou, K., Karathanassi, V., Tsakiri—Strati, M., Woodhouse, I., Mallinis, G., 2017. Vegetation biomass estimation with remote sensing: focus on forest and other wooded land over the Mediterranean ecosystem. Int. J. Remote Sens. 38, 1940—1966.

Gevrey, M., Dimopoulos, I., Lek, S., 2003. Review and comparison of methods to study the contribution of variables in artificial neural network models. Ecol. Model. 160, 249—264.

Goetz, S.J., Bunn, A.G., Fiske, G.J., Houghton, R.A., 2005. Satellite-observed photosynthetic trends across boreal North America associated with climate and fire disturbance. Proc. Natl. Acad. Sci. U.S.A 102, 13521—13525.

Gonçalves, F.G., Santos, J.R., Treuhaft, R.N., 2011. Stem volume of tropical forests from polarimetric radar. Int. J. Remote Sens. 32, 503—522.

Gopal, S., Woodcock, C., 1994. Theory and methods for accuracy assessment of thematic maps using fuzzy sets. Photogramm. Eng. Remote Sens. 60 (2), 181–188.

Gopal, S., Woodcock, C.E., Strahler, A.H., 1999. Fuzzy neural network classification of global land cover from a 1 AVHRR data set. Remote Sens. Environ. 67, 230–243.

Hagan, M.T., Demuth, H.B., Beale, M.H., University of Colorado, B., 1996. Neural Network Design. PWS Pub.

Hagberg, J.O., Ulander, L.M.H., Askne, J., 1995. Repeat-pass SAR interferometry over forested terrain. IEEE Trans. Geosci. Remote Sens. 33, 331–340.

Hame, T., Salli, A., Andersson, K., Lohi, A., 1997. A new methodology for the estimation of biomass of coniferdominated boreal forest using NOAA AVHRR data. Int. J. Remote Sens. 18, 3211–3243.

Hansen, M.C., Stehman, S.V., Potapov, P.V., Loveland, T.R., Townshend, J.R.G., DeFries, R.S., Pittman, K.W., Arunarwati, B., Stolle, F., Steininger, M.K., Carroll, M., DiMiceli, C., 2008. Humid tropical forest clearing from 2000 to 2005 quantified by using multitemporal and multiresolution remotely sensed data. Proc. Natl. Acad. Sci. U.S.A 105, 9439–9444.

Hardin, P.J., 1994. Parametric and nearest-neighbor methods for hybrid classification: a comparison of pixel assignment accuracy. Photogramm. Eng. Remote Sens. 60, 1439–1448.

Harding, D.J., Blair, J.B., Rabine, D.L., Still, K., 1998. In: SLICER: Scanning Lidar Imager of Canopies by Echo Recovery Instrument and Data Product Description, V. 1.3, NASA's Goddard Space Flight Center, June 2, 1998.

Harrell, P.A., Bourgeau-Chavez, L.L., Kasischke, E.S., French, N.H.F., Christensen Jr., N.L., 1995. Sensitivity of ERS-1 and JERS-1 radar data to biomass and stand structure in Alaskan boreal forest. Remote Sens. Environ. 54, 247–260.

Haykin, S., Network, N., 1994. Neural Networks: A Comprehensive Foundation. Prentice Hall, Upper Saddle River, New Jersey.

Henderson, F.M., Lewis, A.J., 1998. Principles and Applications of Imaging Radar. Manual of Remote Sensing: Volume 2.

Hodgson, M.E., Bresnahan, P., 2004. Accuracy of airborne lidar-derived elevation: empirical assessment and error budget. Photogramm. Eng. Remote Sens. 70, 331–339.

Hofton, M.A., Blair, J.B., 2002. Laser altimeter return pulse correlation: a method for detecting surface topographic change. J. Geodyn. 34, 477–489.

Hofton, M.A., Minster, J.B., Blair, J.B., 2000. Decomposition of laser altimeter waveforms. IEEE Trans. Geosci. Remote Sens. 38, 1989–1996.

Hyde, P., Dubayah, R., Walker, W., Blair, J.B., Hofton, M., Hunsaker, C., 2006. Mapping forest structure for wildlife habitat analysis using multi-sensor (LiDAR, SAR/InSAR, ETM+, Quickbird) synergy. Remote Sens. Environ. 102 (1–2), 63–73.

Imbert, D., Rollet, B., 1989. Phytomasse aerienne et production primaire dans la mangrove du Grand Cul-De-Sac Marin (Guadeloupe, Antillas Francaises). Oecologia Plant 721 (4), 379–396.

Imhoff, M.L., 1995. Radar backscatter and biomass saturation: ramifications for global biomass inventory. IEEE Trans. Geosci. Remote Sens. 33, 511–518.

Jaynes, E.T., 1957. Information theory and statistical mechanics. II. Phys. Rev. 108, 171.

Jenkins, J.C., Chojnacky, D.C., Heath, L.S., Birdsey, R.A., 2003. National-scale biomass estimators for United States tree species. For. Sci. 49, 12–35.

Jensen, J.R., Qiu, F., Ji, M., 1999. Predictive modelling of coniferous forest age using statistical and artificial neural network approaches applied to remote sensor data. Int. J. Remote Sens. 20, 2805–2822.

Kanellopoulos, I., Wilkinson, G.G., 1997. Strategies and best practice for neural network image classification. Int. J. Remote Sens. 18, 711–725.

Kasischke, E.S., Bourgeau-Chavez, L.L., Christensen, N.L., Haney, E., 1994. Observations on the sensitivity of ERS-1 SAR image intensity to changes in aboveground biomass in young loblolly pine forests. Int. J. Remote Sens. 15, 3–16.

Kasischke, E.S., Christensen Jr., N.L., Bourgeau-Chavez, L.L., 1995. Correlating radar backscatter with components of biomass in loblolly pine forests. IEEE Trans. Geosci. Remote Sens. 33, 643–659.

Katila, M., 1999. Adapting Finnish multi-source forest inventory techniques to the New Zealand preharvest inventory. Scand. J. For. Res. 14, 182.

Katila, M., Tomppo, E., 2001. Selecting estimation parameters for the Finnish multisource national forest inventory. Remote Sens. Environ. 76, 16–32.

Kellndorfer, J.M., Pierce, L.E., Dobson, M.C., Ulaby, F.T., 1998. Toward consistent regional-to-global-scale vegetation characterization using orbital SAR systems. IEEE Trans. Geosci. Remote Sens. 36, 1396–1411.

Kellndorfer, J., Walker, W., LaPoint, E., Hoppus, M., Westfall, J., 2006. Modeling height, biomass, and carbon in U.S. Forests from FIA, SRTM, and ancillary national scale data sets. In: IEEE International Conference on Geoscience and Remote Sensing Symposium, 2006. IGARSS 2006, pp. 3591–3594.

Kellndorfer, J.M., Walker, W.S., LaPoint, E., Kirsch, K., Bishop, J., Fiske, G., 2010. Statistical fusion of lidar,

InSAR, and optical remote sensing data for forest stand height characterization: a regional-scale method based on LVIS, SRTM, Landsat ETM+, and ancillary data sets. J. Geophys. Res. 115, G00E08.

Kobayashi, Y., Sarabandi, K., Pierce, L., Dobson, M.C., 2000. An evaluation of the JPL TOPSAR for extracting tree heights. IEEE Trans. Geosci. Remote Sens. 38, 2446–2454.

Komiyama, A., Poungparn, S., Kato, S., 2005. Common allometric equations for estimating the tree weight of mangroves. Journal of Tropical Ecology 21, 471–477.

Komiyama, A., Ong, J.E., Poungparn, S., 2008. Allometry, biomass, and productivity of mangrove forests: a review. Aquat. Bot. 89, 128–137.

Krankina, O.N., Harmon, M.E., Cohen, W.B., Oetter, D.R., Zyrina, O., Duane, M.V., 2004. Carbon stores, sinks, and sources in forests of Northwestern Russia: can we reconcile forest inventories with remote sensing results? Clim. Change 67, 257–272.

Kurvonen, L., Pulliainen, J., Hallikainen, M., 1999. Retrieval of biomass in boreal forests from multitemporal ERS-1 and JERS-1 SAR images. IEEE Trans. Geosci. Remote Sens. 37, 198–205.

Labrecque, S., Fournier, R.A., Luther, J.E., Piercey, D., 2006. A comparison of four methods to map biomass from Landsat-TM and inventory data in western Newfoundland. For. Ecol. Manag. 226, 129–144.

Lambert, M.C., Ung, C.H., Raulier, F., 2005. Canadian national tree aboveground biomass equations. Can. J. For. Res. 35, 1996–2018.

Le Toan, T., Beaudoin, A., Riom, J., Guyon, D., 1992. Relating forest biomass to SAR data. IEEE Trans. Geosci. Remote Sens. 30, 403–411.

Lefsky, M.A., 2010. A global forest canopy height map from the moderate resolution imaging spectroradiometer and the geoscience laser altimeter system. Geophys. Res. Lett. 37, L15401.

Lefsky, M., Harding, D., Cohen, W., Parker, G., Shugart, H., USDA, F., 1999. Surface lidar remote sensing of basal area and biomass in deciduous forests of eastern Maryland, USA. Remote Sens. Environ. 67, 83–98.

Lefsky, M.A., Cohen, W.B., Harding, D.J., Parker, G.G., Acker, S.A., Gower, S.T., 2002. Lidar remote sensing of above ground biomass in three biomes. Glob. Ecol. Biogeogr. 11, 393–399.

Lefsky, M.A., Hudak, A.T., Cohen, W.B., Acker, S.A., 2005. Geographic variability in lidar predictions of forest stand structure in the Pacific Northwest. Remote Sens. Environ. 95, 532–548.

LeMay, V., Temesgen, H., 2005. Comparison of nearest neighbor methods for estimating basal area and stems per hectare using aerial auxiliary variables. For. Sci. 51, 109–119.

Liaw, A., Wiener, M., 2002. Classification and regression by randomForest. R. News 2, 18–22.

Lloyd, R.E., 1997. Spatial Cognition: Geographic Environments. Kluwer Academic Publishers.

Lu, D., 2005. Aboveground biomass estimation using Landsat TM data in the Brazilian Amazon. Int. J. Remote Sens. 26, 2509–2525.

Lu, D., 2006. The potential and challenge of remote sensing-based biomass estimation. Int. J. Remote Sens. 27, 1297–1328.

Lu, D., Chen, Q., Wang, G., Moran, E., Batistella, M., Zhang, M., Laurin, G.V., Saah, D., 2012. Aboveground forest biomass estimation with Landsat and LiDAR data and uncertainty analysis of the estimates. Int. J. Financ. Res. 2012, 16. Article ID 436537.

Lu, D., Chen, Q., Wang, G., Liu, J., Li, G., Moran, E., 2014. A survey of remote sensing-based aboveground biomass estimation methods in forest ecosystems. Int. J. Digit. Earth 9, 1753–8947.

Lumbierres, M., Méndez, P.F., Bustamante, J., Soriguer, R., Santamaría, L., 2017. Modeling biomass production in seasonal wetlands using MODIS NDVI land surface phenology. Remote Sens. 9, 392.

Luo, H., Li, X., Chen, E., Cheng, J., Cao, C., 2010. Analysis of forest backscattering characteristics based on polarization coherence tomography. Sci. China Technol. Sci. 53 (Suppl. 1), 166–175. https://doi.org/10.1007/s11431-010-3242-y.

Malila, W.A., Metzler, M.D., Rice, D.P., Crist, E.P., 1984. Characterization of Landsat-4 Mss and Tm digital image data. IEEE Trans. Geosci. Remote Sens. 22 (3), 177–191.

Mather, P.M., 2004. Computer Processing of Remotely Sensed Images. John Wiley & Sons Ltd, Chichester, West Sussex, England.

McRoberts, R.E., 2012. Estimating forest attribute parameters for small areas using nearest neighbors techniques. For. Ecol. Manag. 272, 3–12.

McRoberts, R.E., Tomppo, E.O., Finley, A.O., Heikkinen, J., 2007. Estimating areal means and variances of forest attributes using the k-Nearest Neighbors technique and satellite imagery. Remote Sens. Environ. 111, 466–480.

McRoberts, R.E., Cohen, W.B., Næsset, E., Stehman, S.V., Tomppo, E.O., 2010. Using remotely sensed data to construct and assess forest attribute maps and related spatial products. Scand. J. For. Res. 25, 340–367.

Meng, S.L., Pang, Y., Zhang, Z.J., Jia, W., Li, Z.Y., 2016. Mapping aboveground biomass using texture indices from aerial photos in a temperate forest of northeastern China. Remote Sens. 8 (3).

Mitchard, E.T.A., Saatchi, S.S., Woodhouse, I.H., Nangendo, G., Ribeiro, N.S., Williams, M., Ryan, C.M., Lewis, S.L., Feldpausch, T.R., Meir, P., 2009. Using satellite radar backscatter to predict above-ground woody biomass: a consistent relationship across four different

African landscapes. Geophys. Res. Lett. 36 https://doi.org/10.1029/2009GL040692. Article Number: L23401.

Mitchell, M., 1998. An Introduction to Genetic Algorithms. The MIT press.

Moghaddam, M., Saatchi, S.S., 1999. Monitoring tree moisture using an estimation algorithm applied to SAR data from BOREAS. IEEE Trans. Geosci. Remote Sens. 37, 901–916.

Montesano, P.M., Neigh, C., Sun, G.Q., Duncanson, L., Van Den Hoek, J., Ranson, K.J., 2017. The use of sun elevation angle for stereogrammetric boreal forest height in open canopies. Remote Sens. Environ. 196, 76–88.

Montesano, P.M., Neigh, C.S.R., Wagner, W., Wooten, M., Cook, B.D., 2019. Boreal canopy surfaces from spaceborne stereogrammetry. Remote Sens. Environ. 225, 148–159.

Moore, D.E., Lees, B.G., Davey, S.M., 1991. A new method for predicting vegetation distributions using decision tree analysis in a geographic information system. J. Environ. Manag. 15, 59–71.

Muss, J.D., Mladenoff, D.J., Townsend, P.A., 2011. A pseudo-waveform technique to assess forest structure using discrete lidar data. Remote Sens. Environ. 115, 824–835.

Muukkonen, P., Heiskanen, J., 2005. Estimating biomass for boreal forests using ASTER satellite data combined with standwise forest inventory data. Remote Sens. Environ. 99, 434–447.

Muukkonen, P., Heiskanen, J., 2007. Biomass estimation over a large area based on standwise forest inventory data and ASTER and MODIS satellite data: a possibility to verify carbon inventories. Remote Sens. Environ. 107, 617–624.

Nelder, J.A., Mead, R., 1965. A simplex method for function minimization. Comput. J. 7, 308.

Nelson, R., 1997. Modeling forest canopy heights: the effects of canopy shape. Remote Sens. Environ. 60, 327–334.

Nelson, R., Oderwald, R., Gregoire, T.G., 1997. Separating the ground and airborne laser sampling phases to estimate tropical forest basal area, volume, and biomass. Remote Sens. Environ. 60, 311–326.

Nelson, R.F., Kimes, D.S., Salas, W.A., Routhier, M., 2000. Secondary forest age and tropical forest biomass estimation using thematic mapper imagery. Bioscience 50, 419–431.

Nelson, R., Ranson, K.J., Sun, G., Kimes, D.S., Kharuk, V., Montesano, P., 2009. Estimating Siberian timber volume using MODIS and ICESat/GLAS. Remote Sens. Environ. 113, 691–701.

Ni, W., Ranson, K.J., Zhang, Z., Sun, G., 2014. Features of point clouds synthesized from multi-view ALOS/PRISM data and comparisons with LiDAR data in forested areas. Remote Sens. Environ. 149, 47–57.

Ni, W.J., Sun, G.Q., Ranson, K.J., 2015. Characterization of AS-TER GDEM elevation data over vegetated area compared with lidar data. Int. J. Digit. Earth 8 (3), 198–211.

Ni, W., Zhang, Z., Sun, G., Liu, Q., 2019. Modeling the stereoscopic features of mountainous forest landscapes for the extraction of forest heights from stereo imagery. Remote Sens. 11 (10).

Nichol, J.E., Sarker, M.L.R., 2011. Improved biomass estimation using the texture parameters of two high-resolution optical sensors. IEEE Trans. Geosci. Remote Sens. 49 (3), 930–948.

Nilsson, M., 1996. Estimation of tree heights and stand volume using an airborne lidar system. Remote Sens. Environ. 56, 1–7.

Noguchi, M., Fraser, C.S., 2004. Accuracy assessment of QuickBird stereo imagery. Photogramm. Rec. 19 (106), 128–137.

Næsset, E., 2002. Predicting forest stand characteristics with airborne scanning laser using a practical two-stage procedure and field data. Remote Sens. Environ. 80, 88–99.

Næsset, E., 2004a. Accuracy of forest inventory using airborne laser scanning: evaluating the first nordic full-scale operational project. Scand. J. For. Res. 19, 554–557.

Næsset, E., 2004b. Practical large-scale forest stand inventory using a small-footprint airborne scanning laser. Scand. J. For. Res. 19, 164–179.

Næsset, E., 2011. Estimating above-ground biomass in young forests with airborne laser scanning. Int. J. Remote Sens. 32, 473–501.

Næsset, E., Gobakken, T., 2008. Estimation of above- and below-ground biomass across regions of the boreal forest zone using airborne laser. Remote Sens. Environ. 112, 3079–3090.

Ohmann, J.L., Gregory, M.J., 2002. Predictive mapping of forest composition and structure with direct gradient analysis and nearest- neighbor imputation in coastal Oregon, U.S.A. Can. J. For. Res. 32, 725–741.

Ong, J.E., 1995. The Ecology of Mangrove Conservation and Management. Hydrobiologia 295, 343–351.

Papathanassiou, K.P., Cloude, S.R., 2001. Single-baseline polarimetric SAR interferometry. IEEE Trans. Geosci. Remote Sens. 39, 2352–2363.

Patenaude, G., Milne, R., Dawson, T.P., 2005. Synthesis of remote sensing approaches for forest carbon estimation: reporting to the Kyoto Protocol. Environ. Sci. Policy 8, 161–178.

Peng, M.H., Shih, T.Y., 2006. Error assessment in two lidar-derived TIN datasets. Photogramm. Eng. Remote Sens. 72, 933–947.

Pflugmacher, D., Cohen, W., Kennedy, R., Lefsky, M., 2008. Regional applicability of forest height and aboveground biomass models for the geoscience laser altimeter system. For. Sci. 54, 647–657.

Phillips, S.J., Dudík, M., Schapire, R.E., 2004. A maximum entropy approach to species distribution modeling. In: Proceedings of the 21st International Conference on Machine Learning. ACMPress, New York, pp. 655—662.

Phillips, S.J., Anderson, R.P., Schapire, R.E., 2006. Maximum entropy modeling of species geographic distributions. Ecol. Model. 190, 231—259.

Pope, K.O., Rey-Benayas, J.M., Paris, J.F., 1994. Radar remote sensing of forest and wetland ecosystems in the Central American tropics. Remote Sensing of Environment 48, 205—219.

Poungparn, S., Komiyama, A., Jintana, V., 2002. A Quantitative analysis on the root system of a mangrove, 12. Xylocarpus granatum Koenig, pp. 35—42. Tropics.

Popescu, S.C., Zhao, K., Neuenschwander, A., Lin, C., 2011. Satellite lidar vs. small footprint airborne lidar: Comparing the accuracy of aboveground biomass estimates and forest structure metrics at footprint level. Remote Sens. Environ. 115, 2786—2797.

Prasad, A.M., Iverson, L.R., Liaw, A., 2006. Newer Classification and Regression Tree Techniques: Bagging and Random Forests for Ecological Prediction. Ecosystems 9 (2006), 181—199. https://doi.org/10.1007/s10021-005-0054-1.

Pulliainen, J., Engdahl, M., Hallikainen, M., 2003. Feasibility of multi-temporal interferometric SAR data for stand-level estimation of boreal forest stem volume. Remote Sens. Environ. 85, 397—409.

Ranson, K.J., Sun, G., 1994. Northern forest classification using temporal multifrequency and multipolarimetric SAR images. Remote Sens. Environ. 47, 142—153.

Ranson, K.J., Sun, G., 1997. An evaluation of AIRSAR and SIR-C/X-SAR images for mapping northern forest attributes in Maine, USA. Remote Sens. Environ. 59, 203—222.

Ranson, K.J., Saatchi, S., Guoqing, S., 1995. Boreal forest ecosystem characterization with SIR-C/XSAR. IEEE Trans. Geosci. Remote Sens. 33, 867—876.

Ranson, K.J., Sun, G., Lang, R.H., Chauhan, N.S., Cacciola, R.J., Kilic, O., 1997a. Mapping of boreal forest biomass from spaceborne synthetic aperture radar. J. Geophys. Res. 102, 29599—29610.

Ranson, K.J., Sun, G., Weishampel, J.F., Knox, R.G., 1997b. Forest biomass from combined ecosystem and radar backscatter modeling. Remote Sens. Environ. 59, 118—133.

Reese, H., Nilsson, M., Sandström, P., Olsson, H., 2002. Applications using estimates of forest parameters derived from satellite and forest inventory data. Comput. Electron. Agric. 37, 37—55.

Rosette, J.A., North, P.R.J., Suárez, J.C., Armston, J.D., 2009. A comparison of biophysical parameter retrieval for forestry using airborne and satellite LiDAR. Int. J. Remote Sens. 30, 5229—5237.

Rumelhart, D.E., 1985. Learning internal representations by error propagation. In: DTIC Document.

Saatchi, S., Halligan, K., Despain, D.G., Crabtree, R.L., 2007. Estimation of forest fuel load from radar remote sensing. IEEE Trans. Geosci. Remote Sens. 45, 1726—1740.

Saatchi, S., Buermann, W., ter Steege, H., Mori, S., Smith, T.B., 2008. Modeling distribution of Amazonian tree species and diversity using remote sensing measurements. Remote Sens. Environ. 112, 2000—2017.

Saatchi, S.S., Harris, N.L., Brown, S., Lefsky, M., Mitchard, E.T.A., Salas, W., Zutta, B.R., Buermann, W., Lewis, S.L., Hagen, S., Petrova, S., White, L., Silman, M., Morel, A., 2011. Benchmark map of forest carbon stocks in tropical regions across three continents. Proc. Natl. Acad. Sci. U.S.A 108, 9899—9904.

Santoro, M., Askne, J., Smith, G., Fransson, J.E.S., 2002. Stem volume retrieval in boreal forests from ERS-1/2 interferometry. Remote Sens. Environ. 81, 19—35.

Santoro, M., Shvidenko, A., McCallum, I., Askne, J., Schmullius, C., 2007. Properties of ERS-1/2 coherence in the Siberian boreal forest and implications for stem volume retrieval. Remote Sens. Environ. 106, 154—172.

Schucknecht, A., Meroni, M., Kayitákire, F., Boureima, A., 2017. Phenology-based biomass estimation to support rangeland management in semi-arid environments. Remote Sens. 9, 463.

Simard, M., Rivera-Monroy, V.H., Mancera-Pineda, J.E., Castañeda-Moya, E., Twilley, R.R., 2008. A systematic method for 3D mapping of mangrove forests based on shuttle radar topography mission elevation data, ICESat/GLAS waveforms and field data: application to Ciénaga Grande de Santa Marta, Colombia. Remote Sens. Environ. 112, 2131—2144.

Simard, M., Pinto, N., Fisher, J.B., Baccini, A., 2011. Mapping forest canopy height globally with spaceborne lidar. J. Geophys. Res. 116 https://doi.org/10.1029/2011JG001708. Article Number: G04021.

Slater, P.N., 1979. Re-Examination of the Landsat Mss. Photogramm. Eng. Remote Sens. 45 (11), 1479—1485.

St-Onge, B., Hu, Y., Vega, C., 2008. Mapping the height and above-ground biomass of a mixed forest using lidar and stereo Ikonos images. Int. J. Remote Sens. 29 (5), 1277—1294.

Su, J., Bork, E., 2006. Influence of vegetation, slope, and lidar sampling angle on DEM accuracy. Photogramm. Eng. Remote Sens. 72, 1265—1274.

Sun, G., Ranson, K.J., 1995. A three-dimensional radar backscatter model of forest canopies. IEEE Trans. Geosci. Remote Sens. 33, 372—382.

Sun, G., Ranson, K.J., 2000. Modeling lidar returns from forest canopies. IEEE Trans. Geosci. Remote Sens. 38, 2617—2626.

Sun, G., Ranson, K.J., Kharuk, V.I., 2002. Radiometric slope correction for forest biomass estimation from SAR data in the Western Sayani Mountains, Siberia. Remote Sens. Environ. 79, 279–287.

Sun, G., Ranson, K.J., Kimes, D.S., Blair, J.B., Kovacs, K., 2008. Forest vertical structure from GLAS: An evaluation using LVIS and SRTM data. Remote Sens. Environ. 112, 107–117.

Sun, G., Ranson, K.J., Guo, Z., Zhang, Z., Montesano, P., Kimes, D., 2011. Forest biomass mapping from lidar and radar synergies. Remote Sens. Environ. 115, 2906–2916.

Tadono, T., Nagai, H., Ishida, H., Oda, F., Naito, S., Minakawa, K., Iwamoto, H., 2016. Generation of the 30 M-Mesh Global Digital Surface Model by Alos Prism. XXIII ISPRS Congr. Comm. IV 41 (B4), 157–162.

Tamai, S., Nakasuga, T., Tabuchi, R., et al., 1986. Standing Biomass of Mangrove Forests in Southern Thailand. Journal of the Japanese forestry society 68, 384–388.

Temesgen, B.H., LeMay, V.M., Froese, K.L., Marshall, P.L., 2003. Imputing tree-lists from aerial attributes for complex stands of south-eastern British Columbia. For. Ecol. Manag. 177, 277–285.

Timothy, D., Onisimo, M., Cletah, S., Adelabu, S., Tsitsi, B., 2016. Remote sensing of aboveground forest biomass: a review. Trop. Ecol. 57, 125–132.

Tokola, T., PitkÄNen, J., Partinen, S., Muinonen, E., 1996. Point accuracy of a non-parametric method in estimation of forest characteristics with different satellite materials. Int. J. Remote Sens. 17, 2333–2351.

Tomppo, E., 1997. Application of remote sensing in Finnish national forest inventory. Application of Remote Sensing in European Forest Monitoring. International Workshop Proceedings. European Commission, CL-NA-17685-EN-C, Vienna, Austria, pp. 375–388 (14-16 October 1996).

Tomppo, E., 2004. Using coarse scale forest variables as ancillary information and weighting of variables in k-NN estimation: a genetic algorithm approach. Remote Sens. Environ. 92, 1–20.

Tomppo, E., Katila, M., 1991. Satellite image-based national forest inventory of finland for publication in the igarss'91 digest. In: Geoscience and Remote Sensing Symposium, 1991. IGARSS '91. Remote Sensing: Global Monitoring for Earth Management., International, pp. 1141–1144.

Tomppo, E., Nilsson, M., Rosengren, M., et al., 2002. Simultaneous use of Landsat-TM and IRS-1C WiFS data in estimating large area tree stem volume and aboveground biomass. Remote Sensing of Environment 82, 156–171.

Tomppo, E.O., Gagliano, C., De Natale, F., Katila, M., McRoberts, R.E., 2009. Predicting categorical forest variables using an improved k-nearest neighbour estimator and landsat imagery. Remote Sens. Environ. 113, 500–517.

Treuhaft, R.N., Siqueira, P.R., 2000. Vertical structure of vegetated land surfaces from interferometric and polarimetric radar. Radio Sci. 35, 141–177.

Treuhaft, R.N., Siqueira, P.R., 2004. The calculated performance of forest structure and biomass estimates from interferometric radar. Waves Random Media 14, 345–358.

Treuhaft, R.N., Madsen, S.N., Moghaddam, M., van Zyl, J.J., 1996. Vegetation characteristics and underlying topography from interferometric radar. Radio Sci. 31, 1449–1485.

Tsui, O.W., Coops, N.C., Wulder, M.A., Marshall, P.L., 2013. Integrating airborne LiDAR and space-borne radar via multivariate kriging to estimate above-ground biomass. Remote Sens. Environ. 139, 340–352.

Turner, D.P., Guzy, M., Lefsky, M.A., Ritts, et al., 2004. Monitoring forest carbon sequestration with remote sensing and carbon cycle modeling. Environmental Management 33, 457–466.

Ulaby, F.T., Moore, R.K., Fung, A.K., 1981. Microwave Remote Sensing: Active and Passive. Volume 1- Microwave remote sensing fundamentals and radiometry. Artech House, Norwood, MA, USA.

Van Zyl, J., Kim, Y., 2010. Synthetic Aperture Radar Polarimetry, JPL Space Science and Technology Series, Series Editor, Joseph H. Yuen. Jet Propulsion Laboratory, California Institute of Technology.

Vargas-Larreta, B., López-Sánchez, C.A., Corral-Rivas, J.J., López-Martínez, J.O., Aguirre-Calderón, C.G., Álvarez-González, J.G., 2017. Allometric equations for estimating biomass and carbon stocks in the temperate forests of north-western Mexico. Forests 8, 269.

Vapnik, V., 1998. Statistical Learning Theory. Wiley, New York, NY.

Wagner, W., 2003. Large-scale mapping of boreal forest in SIBERIA using ERS tandem coherence and JERS backscatter data. Remote Sens. Environ. 85, 125–144.

Yuen, J.Q., Fung, T., Ziegler, A.D., 2016. Review of allometric equations for major land covers in SE Asia: uncertainty and implications for above- and below-ground carbon estimates. For. Ecol. Manag. 360, 323–340.

Zebker, H.A., Villasenor, J., 1992. Decorrelation in interferometric radar echoes. IEEE Trans. Geosci. Remote Sens. 30, 950–959.

Zhang, G., Ganguly, S., Nemani, R.R., White, A.M., Milesi, C., Hashimoto, H., Wang, W., Saatchi, S., Yu, Y., Myneni, B.R., 2014. Estimation of forest aboveground biomass in California using canopy height and leaf area index estimated from satellite data. Remote Sens. Environ. 151, 44–56.

Zhang, Y.Z., Liang, S.L., Sun, G.Q., 2014. Forest Biomass Mapping of Northeastern China Using GLAS and MODIS

Data. IEEE J. Sel. Top. Appl. Earth Obs. Remote Sens. 7 (1), 140−152.

Zhou, G.Q., Li, R., 2000. Accuracy evaluation of ground points from IKONOS high-resolution satellite imagery. Photogramm. Eng. Remote Sens. 66 (9), 1103−1112.

Further reading

Zhang, Z.Y., Ni, W.J., Sun, G.Q., Huang, W.L., Ranson, K.J., Cook, B.D., Guo, Z.F., 2017. Biomass retrieval from L-band polarimetric UAVSAR backscatter and PRISM stereo imagery. Remote Sens. Environ. 194, 331−346.

Advanced Remote Sensing, Second Edition
https://doi.org/10.1016/B978-0-12-815826-5.00015-5

© 2020 Elsevier Inc. All rights reserved.

Abstract

Vegetation, as the principal component of terrestrial ecosystem, plays an important role in sustaining global substance and energy cycle, adjusting carbon balance and alleviating the rise of atmospheric CO_2 concentration and global climate change. Vegetation production of terrestrial ecosystem in particular relates to the process where atmospheric CO_2 is absorbed by plants through photosynthesis and dry matter is accumulated by transforming solar energy to chemical energy. Vegetation production, as a major ecological index to estimate sustainable development of ecosystem, reflects the productivity of terrestrial ecosystem under natural conditions. Vegetation covers over 90% of total terrestrial area and its response to global change is very important in that the adjustment to climate change and the mitigation of the rise of atmospheric CO_2 level mainly depend on vegetation's adjustment and feedback to climate change. Furthermore, about 40% of productivity of terrestrial ecosystem is utilized either indirectly or directly (Vitousek et al., 1997) by being transformed into the food and fuels used by human beings. Thus, terrestrial ecosystem provides the foundation for human survival and sustainable development.

Vegetation production of terrestrial ecosystem has been one of the major subjects for the research on global change. The simulation accuracy of vegetation productivity directly determines the simulation precision of elements in subsequent carbon cycling (such as leaf area index [LAI], litter, soil respiration, and soil carbon). It is also associated with the accurate estimate of the bearing capacity of terrestrial ecosystem for sustainable development of human society. The research on vegetation productivity has undergone several stages of development, including the initial simple statistical model, the later process model driven by remote sensing data, and the current dynamic global vegetation model (DGVM). Remote sensing data, due to their capacity to provide spatially and temporally continuous vegetation change characteristics, occupy a nonsubstitutable position in regional estimation and prediction research. This chapter gives a systematic introduction of various models for estimating vegetation productivity of terrestrial ecosystem, and the role of remote sensing data and their development orientation are emphatically illustrated.

15.1 Concept of vegetation production

The appearance of the concept of vegetation productivity can be dated back to more than 300 years ago (Lieth, 1975), and the thorough and systematic research began in the early 20th century. Demark physiologist P. Boysen-Jeose performed a series of physiological experiments concerned with photosynthesis in 1932, which marked the beginning of phytophysiological research. In his masterwork *Plant Matter Production*, the concepts of gross production and net production, together with their computational formulas, were proposed for the first time. Growth analysis school represented by Watson from the United Kingdom put forward Watson law in 1950s. Japanese ecologists Monsi and Saeki formulated Monsi—Saeki theory on community photosynthesis (Fang et al., 2001). These classical works create solid theoretical basis for the research on ecosystem.

Vegetation production consists of gross primary production (GPP) and net primary production (NPP). The former refers to the rate at which green plants in ecosystem produces organic matters by assimilating carbon dioxide using solar energy through photosynthesis. The latter refers to GPP minus the organic matter consumed by autotrophic respiration of plants. An average of about one half of organic matter in GPP is released into the atmosphere through the respiration of plants, and the other half, which constitutes NPP, produces biomass.

The estimate of vegetation productivity differs considerably in different researches. Earlier International Biological Program (IBP) estimated the global terrestrial NPP of plants to be about 58.8 Pg C/a based on observations and statistical model (Whittaker and Likens, 1975).[1] Cramer et al. (1999) used 16 ecosystem models to estimate global net vegetation primary production, with the results that global average NPP ranged between 39.9 and ~80.5 Pg C/a. The upper limit more than doubled the lower limit (Fig. 15.1).

Currently, Yuan et al. (2010) employed EC-LUE (eddy covariance light use efficiency) model driven by remote sensing data to estimate global average GPP as 125 Pg C/a. This result was close to that (123 Pg C/a) obtained by Beer et al. (2010) based on global eddy data and model estimation. The two were equivalent to the annual average NPP of 62.5 and 61.5 Pg C/a (the ratio of NPP to GPP assumed as 0.5), respectively. Comparison of seven LUE models demonstrated that the global annual mean GPP over the period 2000-2010 ranging from 95.10 to 139.71 Pg C/a with large discrepancy due to different model structures and dominant environmental drivers among different models (Cai et al. 2014).

15.2 Ground observation of vegetation production

Ground observation and model simulation are two approaches to study vegetation production. Accurate ground observation, which is sig-

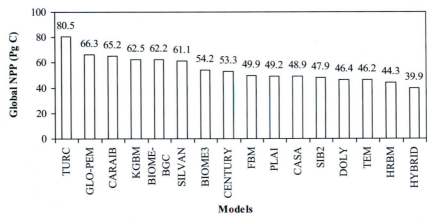

FIGURE 15.1 Average global net primary production of vegetation estimated by 16 ecosystem models (Cramer et al., 1999).

[1] In NPP of global terrestrial vegetation and CO_2 growth rate anomaly. In this figure, atmospheric CO_2 growth rate anomaly is reversely identified.

nificant for improving and validating model simulation, is needed to provide basic data source for parameterization and verification in various vegetation production models. Measurement of vegetation production of terrestrial ecosystem began in 1980s and evolved from the initial biological observation to multiple approaches and methods at present, including phytophysiological approach, chlorophyll measurement, radioactive labeling procedure, eddy covariance, and so on (Fang et al., 2001). This section deals with the principles and measurement procedures for biological observation and eddy covariance, of which the former measures NPP of plants and the latter GPP.

15.2.1 Biological approach

With biological approach, vegetation production is calculated from the measurement of ecosystem biomass. The biomass refers to total organic matter produced by all of the organisms per unit area of ecosystem, including the total weight of root, stem, branch, leaf, flower, fruit, seed, and litter. Comparatively speaking, biomass refers to total organic matter in ecosystem at specific time, while vegetation production indicates the variation in organic matter within a certain period. Due to the disparity of diverse ecosystem in terms of structure and function, the measurement approaches also vary for different ecosystems. This section provides the basic principles and procedures of biological approach for two major terrestrial ecosystems, namely, grassland and forest.

15.2.1.1 Measurement of primary production of vegetation in grassland ecosystem

Plants in grassland ecosystem are mostly herbaceous annuals, and therefore the annual peak value of biomass is NPP of the year. Usually, the most common and ancient harvest method is used to measure NPP of grassland (Odum, 1960). When the peak value of plant growth is reached, the plant community aboveground is harvested within a specific area. Then it is weighed up based on layer and species, to calculate biomass per unit area. For the measurement of underground NPP, soil sampling is adopted; that is, at the start and end of the research period, multiple points are randomly selected within a specific area to collect the root system of plants by soil sampling. The root system, through screening and elimination, is selected and weighed up. The difference between the biomass of root system at the two time moments is the underground primary production of the entire research period.

15.2.1.2 Measurement of primary production of vegetation in forest ecosystem

The measurement of primary production of vegetation in forest ecosystem is relatively complex, as the tree layer, shrub layer, and herbaceous layer are to be measured, respectively.

15.2.1.2.1 Measurement of primary production in tree layer

The primary production of tree layer is estimated indirectly from the measured biomass. In the research period, the biomass of tree layer at t_1 and t_2 is measured, respectively. Between t_1 and t_2, the increase in standing biomass plus the part consumed through plant withering, drying, and nibbling by insects and animals is primary production of vegetation in this period, which can be expressed as

NPP = annual increase in biomass

    + annual litter amount

    + annual plant withering amount

    + annual production of fine root

    + annual consumption by animals and insects

$$\text{(15.1)}$$

where annual increase in biomass is estimated based on tree investigation data on the first measurement (t_1) and the second measurement (t_2) and from the biomass prediction equation established on the first measurement (t_1).

The annual net timber growth and branchwood growth are usually reckoned from the age or growth rate using trunk (branch) analytic method. The age reckoning method refers to the process where the biomass over several years (generally 5–10 years) is used to estimate average annual net growth. It can be expressed as

$$\Delta W = (W - W_n)/n \qquad (15.2)$$

where ΔW is the average annual net growth of timber or branchwood in n years; W and W_n are the standing crop biomass and the biomass n years ago per unit area, respectively; n is the number of growth ring from W_n to W, generally 5–10 years.

Annual net growth of root system is usually reckoned from the same growth rate of aboveground stem according to the following formula

$$\Delta W_r = W_s \times W_r/W_s \qquad (15.3)$$

where ΔW_r is the average annual net growth of root in n years; ΔW_s is the average annual net growth of stem in n years; W_r and W_s are the standing crop biomass of root and stem, respectively; the standing crop mass of leaves divided by the average survival time of leaves is the annual net growth of leaves.

15.2.1.2.2 Measurement of primary production of shrub layer

The biomass of stem and root divided by the average age of basal diameters of multiple shrubs is the annual net production of stem and root of shrub layer; the biomass of branch divided by the average age of basal diameters of branch is the annual net production of branch; the biomass of leaves divided by the average survival time of leaves is the annual net production of leaves. The annual net production of the entire shrub layer is the sum of annual net biomass of roots, stems, branches, and leaves.

15.2.1.2.3 Measurement of primary production of herbaceous layer

Biomass of herbaceous annuals at their peak growth is the annual production of the year. For herbaceous perennials, the sum of biomass difference at different periods is used to calculate annual net production. If the value is negative, then the annual net production is considered as zero. Generally, the total biomass divided by the average age is the annual biomass of herbaceous perennials for domestic researchers.

15.2.2 Eddy covariance

Eddy covariance is a micrometerological method currently popular for direct observation of the exchange between ecosystem and atmosphere in terms of gas, energy, and momentum. Swinbank was the first to publish its principle in 1951. Eddy covariance has the advantages of fewer theoretical assumptions and extensive application scope, and therefore it is highly valued by micro-meteorologists, who consider it as the standard method to determine energy and substance flux. However, the application of eddy covariance is largely limited by its high hardware requirements. Over recent years, with the unceasing progress made in computer acquisition, data processing capacity, and sensors, especially the development and improvement of ultrasonic wind meter and high-performance CO_2 analyzer, eddy covariance is gaining popularity.

Eddy covariance obtains turbulence flux by calculating the covariance of fluctuations of vertical wind velocity and fluctuations of physical quantity to be measured. It is also able to directly measure the carbon, water, and heat flux in plant communities and atmosphere. Under the current technical conditions, minor fluctuations of air mass and energy flux on several time scales (hour, day, season, and year) can be measured. The spatial scope ranging from 100 to 2000 m can be measured with this method. Eddy covariance achieves the observation of carbon and water vapor flux of ecosystem on a direct, precise, and continuous basis and proves itself to be the most efficient method that reveals the interactions between terrestrial biosphere and atmosphere on an ecological scale (Friend et al., 2006; Baldocchi, 2008). According to incomplete statistics, after 20 years of development (especially recent years), observation stations, where eddy covariance technique is employed to study carbon cycling of ecosystem, amount to over 500 globally. Nearly every representative type of terrestrial ecosystem on the earth is covered, and the networks of flux observation on both regional and global scale are established.

Using eddy covariance, the exchange of carbon flux between ecosystem and atmosphere can be continuously observed. With the sampling frequency of 10 Hz, flux observations on the hourly scale are obtained. There are some mature formulas available for indirect calculation of vegetation productivity of ecosystem.

In the nighttime, the photosynthesis stops, and the flux observed represents respiration of ecosystem consisting of autotrophic respiration of plants and soil heterotrophic respiration. Many researchers have found that ecosystem respiration has good correlation with temperature. With adequate water, ecosystem respiration (ER) can be described by Equation Q_{10}:

$$ER = ER_0 \times Q_{10}^{\frac{T-T_0}{T_0}} \qquad (15.4)$$

where ER_0 is ecosystem respiration rate at base temperature (T_0); T is air temperature at the research time moment; Q_{10} is the temperature sensitivity of ER, i.e., the times by which the respiration rate increases for every 10°C higher.

In the daytime, photosynthesis, autotrophic respiration, and soil heterotrophic respiration take place simultaneously. Flux observation is net ecosystem production (NEP), which can be expressed as

$$NEP = GPP - ER_{day} \qquad (15.5)$$

where ER_{day} is ecosystem respiration in the daytime; GPP is vegetation productivity of ecosystem. Inverse operation is needed to obtain GPP with ER_{day}. By virtue of Formula (15.4) and based on daytime ER and temperature actually observed, parameter ER_0 and Q_{10} can be retrieved. From the two parameters and temperature observed at daytime, daytime ecosystem respiration ER_{day} and GPP can be successively calculated. Exiting researches demonstrate that ER_0 and Q_{10} usually change with time and location. Therefore, in actual calculation, 10–15 d is frequently selected as the time window.

15.3 Statistical models based on vegetation index

Statistical model is one of the earliest developed approaches to estimate and simulate regional vegetation productivity. Using remote sensing data (mainly various vegetation indices) and vegetation productivity data through ground measurement, statistical correlation is built and later used to estimate vegetation productivity of other regions. Vegetation index is the key part of statistical model, as previously introduced in Section 11.2, Chapter 11.

Based on remote sensing vegetation index data with spatially and temporally continuous distribution, statistical model is crucial in estimating vegetation productivity on the regional and global scale. Statistical model can be classified into two categories: one is direct establishment of the correlation between vegetation index and vegetation productivity, based on which regional estimation is possible; the other is the establishment of regression parameter vector for regional applications, which is realized through the integrated utilization of vegetation indices and other environmental factors and using regression tree, neural network, and other complex statistical methods.

It is discovered by extensive researches that vegetation index has significantly positive correlation with aboveground primary production of vegetation (Goward et al., 1985). For instance, Paruelo et al. (1997) found that, in prairie in the central part of the United States, observed aboveground vegetation productivity is significantly correlated with NDVI:

$$ANPP = 3803 \times NDVI^{1.9028} \qquad (15.6)$$

This correlation can be utilized to estimate aboveground vegetation productivity of the entire area (Paruelo et al., 1997, 2000). However, it should be noted that this linear correlation cannot be applied for all of the regions. For regions with high vegetation productivity, the correlation between the two gradually declines as vegetation indices become saturated with the development of productivity (Box et al., 1989). Besides, for regions with sparse vegetation, soil background exerts considerable influence on vegetation indices, and as a result, vegetation indices are unable to accurately reflect actual status of vegetation (Huete, 1988). Moreover, for evergreen coniferous forest or broad-leaved forest, seasonal variation as indicated by vegetation indices is far less significant than that of photosynthesis (Gamon et al., 1995). Statistical model based on vegetation indices is not applicable in this area.

Theoretically speaking, vegetation indices are indirect reflection of the characteristics of LAI, which in turn is the key factor determining vegetation productivity. In this sense, statistical model based on vegetation indices and productivity is capable of producing better simulation effects. Vegetation production, however, not only depends on the status of the vegetation itself but also is tremendously affected by environmental factors. Many researches, by integrated utilization of remote sensing data (vegetation index or LAI), meteorological variables, and measured vegetation productivity, and by complex statistical methods such as regression tree and neural network, establish training vectors, which are subsequently used to estimate and simulate regional vegetation productivity (Zhang et al., 2007; Beer et al., 2010). For instance, Beer et al. (2010) selected multiple variables to establish diagnosis model based on advanced statistical methods, to estimate vegetation productivity (Table 15.1).

Zhang et al. (2007), combining NDVI, the starting date, precipitation, temperature, and photosynthetically active radiation of the growth season, and based on GPP retrieval data at five eddy covariance stations, constructed piecewise regression model using regression tree. This model is then applied for the simulation of vegetation productivity in the prairie in the central part of the United States. With regression tree, gradients of various environmental factors are used as classification nodes. Then by taking observations as the targets, the space defined by predictor variables, through recursive partitioning, is divided into categories as homogenous as possible. From the master nodes, the

TABLE 15.1   Elements used to estimate vegetation productivity.

Name	Type	Name	Type
Climate element		**Vegetation structure**	
Mean annual temperature	Split variable	Maximum fAPAR of year	Split variable
Mean annual precipitation sum	Split variable	Minimum fAPAR of year	Split variable
Mean annual climatic water balance	Split variable	Maximum−minimum fAPAR	Split variable
Mean annual potential evapotranspiration	Split variable	Mean annual fAPAR	Split variable
Mean annual sunshine hours	Split variable	Sum of fAPAR over the growing season	Split variable
Mean annual number of wet days	Split variable	Mean fAPAR over the growing season	Split variable
Mean annual relative humidity	Split variable	Growing season length derived from fAPAR	Split variable
Mean monthly temperature	Split variable	Sum of fAPAR × potential radiation of year	Split variable
Mean monthly precipitation sum	Split variable	Maximum of fAPAR × potential radiation of year	Split and regression variable
Mean monthly climatic water balance	Split variable	IGBP vegetation type	Split and regression variable
Mean monthly potential evapotranspiration	Split variable	**Meterological element**	
Mean monthly sunshine hours	Split variable	Temperature	Split and regression variable
Mean monthly number of wet days	Split variable	Precipitation	Split and regression variable
Mean monthly relative humidity	Split variable	Potential radiation	Split and regression variable
		Vegetation status	
		fAPAR	Split and regression variable
		fAPAR × potential radiation	Split and regression variable

Split variable: used for area division and the establishment of regression relationship for each area; regression variable: used for the establishment of regression relationship and the simulation of vegetation productivity; fAPAR: fraction of absorbed photosynthetically active radiation.

predictor variables are successively classified into a series of left and right nodes with hierarchal structure. For each node, the mean value and covariance of predictor variable that falls into this part are calculated, according to which the variable and the value of node classification for each classification node are adjusted.

Statistical model based on vegetation indices is able to estimate regional vegetation productivity to a certain extent. However, many limitations in terms of its applications, due to the defects of the method itself, are still existent. First of all, statistical model, which uses the observations and remote sensing data of vegetation productivity of a specific region to construct correlation, has strong regional applicability. The need for redetermination of empirical parameters when applying such model to other regions has greatly reduced its application universality. Secondly, the time scale which can be simulated by this model is restricted by the time scale adopted for observation data in model establishment. Statistical model does not provide mechanism-based description for the formation process of vegetation productivity, and the correlation built by it is entirely dependent on the observation data collected. Consequently, the time scale of observation data determines the time scale which can be simulated by the model. For instance, statistical models based on average annual production are unable to simulate the variation in vegetation productivity on the monthly scale, with considerably limited application scope. Finally, statistical models cannot be applied for prediction research. As discussed above, statistical model avoids mechanism-based description of phytophysiological process, thus it is unable to reflect the variation of phytophysiological process with climate change. Statistical model is only applicable to assessment of existing productivity.

15.4 Light use efficiency model based on remote sensing data

Light use efficiency model, also called production efficiency model, is the major approach to estimate vegetation production based on remote sensing data. Its basic principle was first proposed by Lieth (1975) when studying the productivity of the cropland. He found (1975) that, under the conditions of sufficient water and field fertility, vegetation production of cropland is only related to absorbed photosynthetically active radiation. As remote sensing technology rapidly develops with extensive applications of remote sensing data at various temporal and spatial scales, light use efficiency model based on remote sensing data has gradually become the mainstream approach to estimate vegetation production.

15.4.1 Principles for light use efficiency model

Light use efficiency model makes simplification and abstraction of photosynthesis in a theoretical sense, with the following several assumptions: under appropriate environmental conditions (temperature, water, and nutrient), the rate of photosynthesis depends on the absorption of effective solar radiation by the leaves, and the solar energy is changed into chemical energy at a fixed ratio (potential optical energy utilization rate) by plants; under actual environmental conditions, potential optical energy utilization rate is affected by water, temperature, and other environmental factors. Therefore, vegetation productivity can be expressed by a series of formulas below:

$$GPP = FPAR \times PAR \times \varepsilon_{max} \times f \qquad (15.7)$$

where FPAR is the fraction of absorbed photosynthetically active radiation; PAR is the incident photosynthetically active radiation (refer to Chapter 6); the product of the two is the photosynthetically active radiation absorbed by plant canopy; ε_{max} is potential light use efficiency; and f is the limitation imposed on light use efficiency by various environmental stresses; the product of the two is light use efficiency under actual environmental conditions.

As mentioned above, vegetation index based on remote sensing data can effectively reflect chlorophyll ratio of the plant canopy, and therefore it is frequently used in calculating FPAR. However, different environmental factors are taken into account for different models. The next section introduces several current major optical energy utilization rate models.

15.4.2 Major light use efficiency model

15.4.2.1 CASA model

CASA (Carnegie Ames Stanford Approach) model (Potter et al., 1993) is one of the earliest developed light use efficiency model. This model directly measures NPP in accordance with the principle of light use efficiency model.

$$NPP = FPAR \times PAR \times \varepsilon_{max} \times f(T1, T2, W)$$
(15.8)

where T1, T2, and W are the limitations imposed by two temperatures and water stress on optical energy utilization rate. ε_{max} is the maximum light use efficiency under ideal conditions, with the value adopted as $0.389\,g\,C/MJ$ (Potter et al., 1993).

In CASA model, the fraction of solar radiation absorbed by plants depends on vegetation type and vegetation cover and its maximum value does not exceed 0.95. Its computational formula is given as

$$FPAR = Min[(SR - SR_{min}) / (SR_{max} - SR_{min}), 0.95]$$
(15.9)

$$SR = (1 + NDVI)/(1 - NDVI)$$
(15.10)

where the value of SR_{min} is set as 1.08; the magnitude of SR_{max} is related to vegetation type, ranging from 4.14 to 6.17.

T1 reflects the limitation imposed by biochemical action of plants on photosynthesis at low and high temperature (Potter et al., 1993; Field et al., 1995), with its computational formula given as

$$T1 = 0.8 + 0.02 \times T_{opt} - 0.0005 \times T_{opt}^2 \quad (15.11)$$

where T_{opt} is the monthly average temperature in the month when NDVI reaches its maximum in a given year for a specific area. When the average temperature of a month is equal to or below $-10°C$, T1 is 0.

T2 shows the decreasing trend of light use efficiency as the environmental temperature either increases or decreases from the optimal temperature (T_{opt}). It can be computed by the following formula:

$$T2 = 1.1814/\left(1 + e^{0.2 \times (T_{opt} - 10 - T)}\right)/$$
$$\times \left(1 + e^{0.3 \times (-T_{opt} - 10 - T)}\right) \quad (15.12)$$

where when the average temperature of a month (T) is $10°C$ higher or $13°C$ lower than the optimal temperature T_{opt}, T2 of this month is one half of T2 when the monthly average temperature T is the optimal temperature T_{opt}.

Water stress factor (w) reflects the influence of effective water utilized by plants on optical energy conversion rate. As the available water in the environment increases, W also increases, ranging between 0.5 (extremely arid condition) and 1 (very wet condition).

$$W = 0.5 + 0.5EET/PET \quad (15.13)$$

where PET is possible evapotranspiration, which is calculated by Thornthwaite formula; estimated evapotranspiration EET is estimated by soil water molecule model. When the monthly average temperature is equal to or below $0°C$, W of this month equals that of the preceding month.

15.4.2.2 CFix model

CFix is a light use efficiency type parametric model driven by temperature (T), PAR, and fPAR (Veroustraete et al., 2002). The model uses the following equations to estimate vegetation GPP on a daily basis:

$$GPP = PAR \times fPAR \times LUE_{wl} \times \rho(T) \times CO_2 fert$$

$$(15.14)$$

where LUE_{wl} is light use efficiency by considering the impact of water stress; $\rho(T)$ is the normalized temperature dependency factor; $CO_2 fert$ is the normalized CO_2 fertilization factor.

Verstraeten et al. (2006) integrated the impact of water limitation on light use efficiency by considering two stomatal regulating factors from soil moisture deficit (F_s) and atmospheric changes (F_a), which were simulated by soil moisture and evaporative fraction (EF), respectively. Because of the difficulties at regional simulations of soil moisture, we only consider the impacts of atmospheric changes on LUE and simplified the regulation equation of water limitation as following (Yuan et al., 2014a):

$$LUE_{wl} = (LUE_{min} + F_a \times (LUE_{max} - LUE_{min}))$$

$$(15.15)$$

where LUE_{wl} was delimited between a maximum (LUE_{max}) and minimum value (LUE_{min}).

CFix model uses a linear equation to describe the relationship between fPAR and NDVI and uses a set of empirical constants according to Myneni and Williams (1994):

$$fPAR = 0.8624 \times NDVI - 0.0814 \quad (15.16)$$

The temperature scalar Ts is described by Wang (1996). CO_2 fertilization effect was calculated following Veroustraete (1994).

$$\rho(T) = \frac{e^{\left(C_1 - \frac{\Delta H_{a,P}}{R_g \times T}\right)}}{1 + e^{\left(\frac{\Delta S \times T - \Delta H_{d,P}}{R_g \times T}\right)}} \quad (15.17)$$

$$CO_2 fert = \frac{[CO_2] - \frac{[O_2]}{2s}}{[CO_2]^{ref} - \frac{[O_2]}{2s}} \frac{K_m \times \left(1 + \frac{[O_2]}{K_0}\right) + [CO_2]^{ref}}{K_m \times \left(1 + \frac{[O_2]}{K_0}\right) + [CO_2]}$$

$$(15.18)$$

where C_1, ΔS, $\Delta H_{a,P}$, $\Delta H_{d,P}$, and R_g at the temperature response equation are 21.77, 704.98 J/K/mol, 52,750 J/mol, 211,000 J/mol, and 8.31 J/K/mol (Veroustraete et al., 2002); the parameter values of s, K_m, K_0, and $[CO_2]^{ref}$ are 2550, 948, 30, and 281 ppm, respectively. The O_2 concentration in the atmosphere $[O_2]$ was set to 209,000 ppm, and CO_2 concentration in the atmosphere $[CO_2]$ was set to be annual mean global carbon dioxide concentration using measurements of weekly air samples from the Cooperative Global Air Sampling network (http://www.esrl.noaa.gov/gmd/ccgg/trends/global.html).

15.4.2.3 CFlux model

The carbon flux model (CFlux) integrates data from multiple sources (Turner et al., 2006; King et al., 2011), and model inputs include daily meteorological data and satellite-derived data (e.g., land cover, stand age, and fPAR). Compared with other LUE models, the CFlux considered the influence of stand age. The model is as following:

$$GPP = PAR \times fPAR \times LUE_{eg} \quad (15.19)$$

$$LUE_{eg} = LUE_{g\_base} \times T_s \\ \times \min(W_s, S_{SWg}) \times S_{SAg} \quad (15.20)$$

$$LUE_{base} = (LUE_{max} - LUE_{cs}) \times S_{CI} + LUE_{cs}$$

$$(15.21)$$

where the scalars for minimum temperature scalar (T_S) and vapor-pressure deficit (VPD) (W_S) are formulated as in MODIS-GPP product (Eq (15.40); Eq (15.41)) ranging from 0 to 1. When the scalar equals 1, there is no temperature (or atmosphere moisture) stress; when the scalar equals 0, the stress is maximum and the LUE is reduced to 0 with no photosynthesis.

Additionally, CFlux introduced the scalar for the influence of soil water (S_{SWg}) is based on the ratio of current soil water content to soil water holding capacity. As it is difficult to obtain high accuracy soil moisture product, we simplified the simulations of soil moisture and used EF to indicate S_{SWg}. S_{CI} cloudiness index scalar varies from 0 on clear days to 1 on fully overcast days and is inferred from the ratio of PAR to potential PAR (Turner et al., 2006). LUEcs is initial LUE for clear sky days. LUEmax is initial LUE for overcast days. In the case of forest cover types, a scalar for the effect of stand age on GPP (S_{SAg}) is implemented to reflect observations of reduced vegetation production in older stands (Van Tuyl et al., 2005; Turner et al., 2006). The age scalar (S_{SAg}) is equal to 1 for nonforest vegetation types.

15.4.2.4 EC-LUE model

EC-LUE model is light use efficiency model based on the carbon flux data measured at the stations using eddy covariance (Yuan et al., 2007, 2010). The general equation can be written as

$$GPP = FPAR \times PAR \times \varepsilon_{max} \times Min(f(T), f(W))$$
$$(15.22)$$

This model takes into account the restriction imposed by temperature (f(T)) and moisture (f(W)) on optical energy utilization rate. It also assumes that the restriction obeys the law of the minimum in ecology, i.e., the final environmental restriction depends on the environmental factor with the most intense stress. Factor of temperature restriction f(T) is calculated by Formula (15.37). When air temperature T_a is below T_{min} or higher than T_{max}, f(T) is zero, which means photosynthesis is stopped. In EC-LUE model, the value of T_{min} and T_{max} is set as 0 and 40°C, respectively. T_{opt} is undetermined parameter.

In EC-LUE model, EF is used to express the restriction imposed by moisture on actual optical

energy utilization rate f (W), and it is calculated by the formula below

$$EF = \frac{LE}{LE + H} \qquad (15.23)$$

where LE and H are the latent heat and sensible heat flux (W/m^2) of the ecosystem, respectively. LE is estimated by the revised RS-PM (remote sensing Penman–Monteith) (Yuan et al., 2010; Chen et al., 2014, 2015).

EF satisfactorily reflects the actual water condition in the ecosystem. With sufficiently available water in the ecosystem, the proportion of latent heat flux in total energy increases, resulting in the increase of EF accordingly, and vice versa. EF has been applied in many researches on local water condition assessment (Kurc and Small, 2004; Zhang et al., 2004; Suleiman and Crago, 2004). When the model was applied on the regional or global scale, however, Yuan et al. (2010), to lower the complexity of the model, postulated that the soil heat flux was negligible, i.e., the sum of latent heat and sensible heat was approximately equal to net radiation. In this way, all of the variables involved in calculating vegetation productivity could be estimated based on remote sensing data and conventional meteorological observation data. Thus, the spatial simulation of vegetation productivity was made easier.

The fraction of absorbed photosynthetically active radiation is calculated by the method proposed by Myneni and Williams (1994).

$$FPAR = a \times NDVI + b \qquad (15.24)$$

where a and b are empirical parameters set as 1.24 and 0.168, respectively. The formula, together with the parameters, has been verified globally in various ecosystems, and their extensive representativeness is proved (Sims et al., 2005).

Yuan et al. (2010) fitted model parameters using data measured at 32 stations in flux networks in America and Europe and conducted verification at other 35 stations. The fitted potential optical energy utilization rate and optimal temperature of photosynthesis was 2.25 g C/MJ and 21°C, respectively. The results showed that EC-LUE model was able to account for 75% and 61% of GPP variation at the fitting stations and verification stations, respectively (Fig. 15.2).

Unlike other light use efficiency models, EC-LUE model has globally unified model parameters. Parameters such as potential light use efficiency and optimal temperature for plant growth do no vary with vegetation type, geographical area, and climate type. This feature greatly enhances the applicability of this model on the regional and global scale. Yuan et al. (2010) simulated the spatial and temporal distribution of global GPP using MERRA reanalysis meteorological data and MODIS data (Fig. 15.3).

15.4.2.5 GLO-PEM

GLO-PEM (Global Production Efficiency Model) was developed by Prince in 1995 (Prince and Goward, 1995). This model involves radiation absorption and utilization by plant canopy, environmental control factors affecting optical energy utilization rate of plants, and autotrophic respiration (Prince and Goward, 1995; Goetz et al., 1999; Cao et al., 2004).

$$NPP = FPAR \times PAR \times \varepsilon_{max} \times f \times Y_g \times Y_m$$
(15.25)

where Y_g is plant growth respiration factor; Y_m is plant maintenance respiration factor.

FAPAR is calculated according to the formula below:

$$FPAR = 1.67 \times NDVI - 0.08 \qquad (15.26)$$

In GLO-PEM, plant autotrophic respiration consists of maintenance respiration and growth respiration, denoted by Y_m (maintenance respiration factor) and Y_g (growth respiration factor), respectively. Growth respiration factor Y_g is set as 0.75. Maintenance respiration factor Y_m is calculated according to the following formula (Hunt, 1994):

$$Y_m = \left\{ 1 - \frac{0.4}{0.75} \times \left(\frac{1000 \times W}{1000 \times W + 50} \right) \right\} \quad (15.27)$$

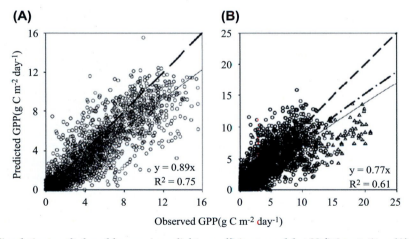

(A)

(B)

y = 0.89x
$R^2 = 0.75$

y = 0.77x
$R^2 = 0.61$

Predicted GPP(g C m^{-2} day^{-1})

Observed GPP(g C m^{-2} day^{-1})

FIGURE 15.2 Simulation results by eddy covariance light use efficiency model at 32 fitting stations (A) and 35 verification stations (B) globally.

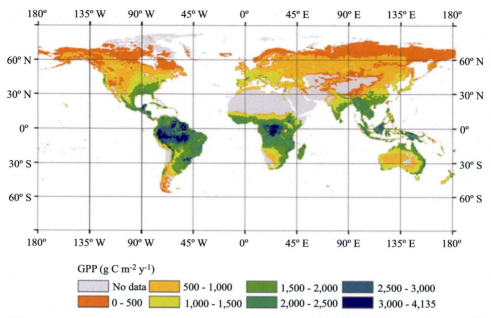

GPP (g C m^{-2} y^{-1})

No data	500 - 1,000	1,500 - 2,000	2,500 - 3,000
0 - 500	1,000 - 1,500	2,000 - 2,500	3,000 - 4,135

FIGURE 15.3   Distribution map of global gross primary production (GPP) simulated by eddy covariance light use efficiency model (mean value from 2003 to 2005).

where W is aboveground viable biomass, calculated as

$$W = 716.61 \times \rho^{-2.6} \quad (15.28)$$

where ρ is the minimum reflectance in the first wave channel of AVHRR of the year.

Depending on photosynthetic pathway, this model has different parameters for C3 and C4 plants. C4 plants refer to those plants which are usually distributed in tropical areas and undergo photosynthesis in which organic acids containing four carbon atoms (oxaloacetic acid) are formed during dark reaction. Maize and sugarcane are common C4 plants. C3 plants are those plants in which the initial product of assimilating carbon dioxide in photosynthesis is 3-phosphoglycerate containing 3 carbon atoms. C4 plants, compared with C3 plants, have higher photosynthetic efficiency, with higher light use efficiency of CO_2. For C4 plants, the potential light use efficiency is set as a constant 2.76 g C/MJ (Collatz et al., 1991), while

that of C3 plants can be calculated by the formula below:

$$\varepsilon_{C3} = 55.2\alpha \quad (15.29)$$

$$\alpha = 0.08\left(\frac{P_i - \Gamma^*}{P_i + 2\Gamma^*}\right) \quad (15.30)$$

where α is quantum efficiency (i.e., CO_2 per unit light quantum which can be fixed); P_i is CO_2 concentration inside the leaves; Γ^* is photosynthetic compensation point for CO_2. Thus,

$$\Gamma^* = \frac{O_i}{2\tau} \quad (15.31)$$

where O_i is O_2 concentration, equaling 20,900 Pa; τ is Michaelis–Menten factor as CO_2 and O_2 concentration change with plant temperature, with its computational formula given as

$$\tau = 2600 \times 0.57^{\left(\frac{T_a - 20}{10}\right)} \quad (15.32)$$

The impact factor of air temperature can be calculated according to the formula below:

$$f(T) = \left[1 + \exp\left(\frac{-220000 + 710(T_a + 273.16)}{8.314(T_a + 273.16)}\right)\right]^{-1}$$
(15.33)

where δ_T is the impact factor of air temperature; T_a is air temperature.

The impact factor of surface soil moisture is calculated using the following formula

$$f(s) = \frac{CSI + 2\Delta\tau}{4\Delta\tau}$$
(15.34)

where $\Delta\tau$ is resistance factor of plant physiological growth under the impact of soil moisture; CSI is cumulative soil humidity index.

The impact factor of saturation VPD on plant growth can be derived from saturation VPD:

$$f(VPD) = 1.2e^{-0.35e_v} - 0.2$$
(15.35)

where e_v is saturation VPD.

After Prince published GLO-PEM model in 1995, Gotze, through modification of the method of calculating autotrophic respiration in GLO-PEM, developed GLO-PEM2 in 1999. With the modified model, no respiration factors are required in calculating autotrophic respiration, and no distinction is made between maintenance respiration and growth respiration. Instead, the total autotrophic respiration rate (R_a) is calculated directly:

$$R_a = \left[0.53 \times \left(\frac{W}{W + 50}\right)\right] \times e^{0.5 \times \left(\frac{T_c - T_a}{25}\right)}$$
(15.36)

where W is aboveground biomass; T_c is average temperature over the years; T_a is air temperature.

Based on GLO-PEM2, Cao et al. (2004) modified the algorithm for the effect of environmental factors on optical energy utilization efficiency of plants in GLO-PEM2. Specifically, restriction equations of three environmental factors about potential optical energy utilization rate were modified.

(1) Effect of air temperature on optical energy utilization rate

$$T_s = \frac{(T_a - T_{min})(T - T_{max})}{(T - T_{min})(T - T_{max}) - (T - T_{opt})^2}$$
(15.37)

where T_a is air temperature; T_{min}, T_{opt}, and T_{max} are the minimum, optimal, and maximum temperature of photosynthesis, respectively; in this model, the maximum and minimum temperature of C3 plants are 50°C and -1°C, respectively. For C4 plants, the maximum and minimum temperature is 50 and 0°C, respectively. Optimal temperature is defined as the average temperature during the growth season over the years.

(2) The impact of surface soil moisture on potential light use efficiency can be calculated by the formula below:

$$\begin{aligned} f(s) &= 1 - 0.05 \times \delta_q \quad 0 < \delta_q \leq 15 \\ f(s) &= 0.25 \qquad\qquad \delta_q \geq 15 \\ \delta_q &= QW(T) - q \end{aligned}$$
(15.38)

where QW(T) is the saturation humidity under specified temperature condition; q is the humidity under the atmosphere.

(3) The impact of saturation VPD on light use efficiency can be calculated using the following formula:

$$f(VPD) = 1 - \exp(0.081 \times (VPD - 83.3))$$
(15.39)

where VPD is the moisture deficit occurring to surface soil above 1 m, and its value is the difference between saturated soil moisture and actual soil moisture.

15.4.2.6 MODIS-GPP product

MODIS-GPP product (MOD-17) is global GPP product using remote sensing data developed according to light use efficiency principle. GPP product has been widely used in the appraisal and application research on various vegetation productions. Piecewise function is used to express the restriction imposed by moisture and temperature on potential light use efficiency.

In the algorithm of MODIS-GPP product, a set of empirical parameters are determined for different vegetation types (details given in Table 15.2). The determination of vegetation types exerts major impact on the precision of GPP product. MODIS-GPP product uses MODIS land use/cover product (MOD12Q1), where 17 categories of land cover classification system of IGBP (International Geosphere-Biosphere

$$
f(TMIN) = \begin{cases} 0 & TMIN < TMIN_{min} \\ \dfrac{TMIN - TMIN_{min}}{TMIN_{max} - TMIN_{min}} & TMIN_{min} < TMIN < TMIN_{max} \\ 1 & TMIN > TMIN_{max} \end{cases} \qquad (15.40)
$$

$$
f(VPD) = \begin{cases} 1 & VPD < VPD_{min} \\ \dfrac{VPD - VPD_{min}}{VPD_{max} - VPD_{min}} & VPD_{min} < VPD < VPD_{max} \\ 0 & VPD > VPD_{max} \end{cases} \qquad (15.41)
$$

where $f(TMIN)$ and $f(VPD)$ are the restrictions imposed by minimum temperature and moisture on potential light use efficiency; TMIN is daily minimum air temperature; VPD is vapor-pressure deficit. $TMIN_{max}$ and $TMIN_{min}$ are the maximum and minimum temperature threshold of photosynthesis, respectively; VPD_{max} and VPD_{min} are minimum and maximum VPD threshold restricting photosynthesis, as shown in Fig. 15.4.

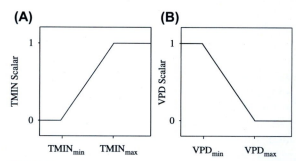

FIGURE 15.4 Curves of restriction imposed by (A) temperature and (B) moisture on potential light use efficiency.

TABLE 15.2 Parameter values of MODIS-GPP product.

Parametric variables	Parameter values					
	ENF	EBF	DNF	DBF	MF	WL
ε_{max}	0.001008	0.001159	0.001103	0.001044	0.001116	0.000800
$TMIN_{min}$	−8.00	−8.00	−8.00	−8.00	−8.00	−8.00
$TMIN_{max}$	8.31	9.09	10.44	7.94	8.50	11.39
VPD_{min}	650	1100	650	650	650	930
VPD_{max}	2500	3900	3100	2500	2500	3100
	Wgrass	Cshrub	Oshrub	Grass	Crop	
ε_{max}	0.000768	0.000888	0.000774	0.000680	0.000680	
$TMIN_{min}$	−8.00	−8.00	−8.00	−8.00	−8.00	
$TMIN_{max}$	11.39	8.60	8.80	12.02	12.02	
VPD_{min}	650	650	650	650	650	
VPD_{max}	3100	3100	3600	3500	4100	

Crop, cropland; *Cshrub*, dense shrub; *DBF*, deciduous broad-leaved forest; *DNF*, deciduous needle-leaved forest; *EBF*, evergreen broad-leaved forest; *ENF*, evergreen needle-leaved forest; *Grass*, grassland; *MF*, mixed forest; *Oshrub*, sparse shrub; *Wgrass*, savanna; *WL*, sparse forest.

Program) are employed (Belward et al. 1999). Existing studies show that this classification image has an average classification precision of 65%−80% for global vegetation types. In areas with smaller spatial heterogeneity, the precision would be higher (Hansen et al., 2000).

In addition to MODIS-FPAR products that drive MODIS-GPP product, other driving data include meteorological elements, namely, daily average and minimum temperature, incident photosynthetically active radiation, and absolute humidity. In actual applications, MODIS-GPP product uses DAO (Data Assimilation Office) reanalysis data to estimate global GPP. This set of reanalysis data employs global circulation model in conjunction with ground and satellite observation data. Land surface meteorological data with the spatial resolution of latitude $1° \times$ longitude $1.25°$ is produced every 6 h.

15.4.2.7 VPM

VPM (vegetation photosynthesis model) divides leaf and forest canopy into photosynthetically active vegetation (PAV) and nonphotosynthetic active vegetation (NAV) (Xiao et al., 2005). The basic expression for VPM is written as

$$GPP = FPAR \times PAR \times \varepsilon_{max} \times f(T) \times f(W) \times f(P)$$

(15.42)

where f(T), f(W), and f(P) are restriction factors of temperature, moisture, and phenology for light use efficiency. Restriction function of temperature can be calculated using Eq. (15.37).

Factor of moisture stress can be calculated as

$$f(W) = \frac{1 + LSWI}{1 + LSWI_{max}}$$ (15.43)

where LSWI is land surface water index; $LSWI_{max}$ is maximum LSWI of the growth season. The value for factor of phenological stress $f(P)$ varies for different vegetation types and requires separate determination.

15.4.2.8 Two-leaf model

The two-leaf light use efficiency (TL-LUE) model was developed to simulate GPP for sunlit and shaded leaves using the MOD17 algorithm separately. And the GPP of total canopy is calculated as

$$GPP = (\varepsilon_{msu} \times APAR_{su} + \varepsilon_{msh} \times APAR_{sh})$$
$$\times f(VPD) \times f(TMIN)$$
$$(15.44)$$

where the TMIN and VPD scalars $f(TMIN)$ and $f(VPD)$ are following MOD17 algorithm according Eqs (15.40) and (15.41). ε_{msu} and ε_{msh} are the maximum LUE of sunlit and shaded leaves, $APAR_{su}$ and $APAR_{sh}$ are the PAR absorbed by sunlit and shaded leaves, respectively.

$$APAR_{sh} = (1 - \alpha) \times$$
$$[(PAR_{dif} - PAR_{dif\_u}) / LAI + C] \times LAI_{sh}$$
$$(15.45)$$

$$APAR_{su} = (1 - \alpha)$$
$$\times \left[\frac{PAR_{dir} \times \cos(\beta)}{\cos(\theta)} + \frac{PAR_{dif} - PAR_{dif\_u}}{LAI} + C\right]$$
$$\times LAI_{su}$$
$$(15.46)$$

where α is the canopy albedo related to vegetation types (Running et al. 2000). $PAR_{dif,u}$ is the diffuse PAR under the canopy and calculated using Eq. (15.47); $(PAR_{dif} - PAR_{dif,u})/LAI$ represents the diffuse PAR on per unit leaf area within the canopy; C quantifies the contribution of multiple scattering of the total PAR to the diffuse irradiance per unit leaf area within the canopy (Eq. (15.48)); β is mean leaf-sun angle and set as $60°$ for a canopy with spherical leaf

angle distribution; and θ is the solar zenith angle. PAR_{dif} and PAR_{dir} are the diffuse and direct components of incoming PAR, respectively (Chen et al., 1999).

$$PAR_{dif\_u} = PAR_{dir} \exp(-0.5\Omega LAI / \cos\theta)$$
$$(15.47)$$

$$C = 0.07\Omega S_{dir}(1.1 - 0.1LAI)\exp(-\cos\theta)$$
$$(15.48)$$

where Ω is the clumping index which is related to land cover, season, and solar elevation angle, and it is set according to vegetation types (Tang et al., 2007).

S_{dif}/S_g is obtained by using the diffuse radiation and total radiation in the meteorological station. For example, the relation between diffuse radiation and total radiation is as follows by analyzing the observed data at Nanjing, Shanghai, Ganzhou, and Nanchang Station:

$$S_{dif}/S_g = 0.7527 + 3.8453R - 16.316R^2$$
$$+ 18.962R^3 - 7.0802R^4$$
$$(15.49)$$

where S_{dif} is the diffuse radiation; S_g is total incoming radiation; R is the clear sky index ($R = S_g/(S_0 * \cos\theta)$); $S_0 = I_0/\rho^2$, I_0 is the solar constant (1367 W/m^2), ρ is the distance between earth and sun.

Assuming PAR accounting for 50% of the incoming solar radiation (Weiss and Norman, 1985; Tsubo and Walker, 2005; Jacovides et al., 2007; Bosch et al., 2009), the diffuse and direct radiation can be converted to the diffuse and direct PAR.

The LAI_{sh} and LAI_{su} in Eqs. (15.45) and (15.46) are the LAI of shaded and sunlit leaves and are computed as (Chen et al., 1999):

$$LAI_{su} = 2 \times \cos(\theta) \times \left(1 - e^{-0.5 \times \Omega \times \frac{LAI}{\cos(\theta)}}\right)$$
$$(15.50)$$

$$LAI_{sh} = LAI - LAI_{su} \qquad (15.51)$$

Sixty-four eddy covariance sites were used to optimize the TL-LUE model to obtain the parameters ε_{msu} and ε_{msh}, which vary with vegetation types.

15.4.3 Disparities among diverse light use efficiency models

Current light use efficiency models follow the same principle, but they employ different model formulas and parameterization schemes, resulting in marked disparities in terms of model parameters and simulation capacity. Firstly, the most important parameter is potential light use efficiency in the model, whose magnitude directly determines the simulation accuracy of vegetation productivity. However, this parameter differs greatly in different models (Table 15.3), with its value ranging from 0.604 to 2.76 g C/MJ APAR in GPP estimation. Generally, potential light use efficiency is retrieved from observed GPP or NPP. Considerable

disparity occurs as a result of different observed values and environmental restriction factors adopted. For instance, EC-LUE model regards the minimum temperature and moisture restriction as representative of the restriction imposed by environmental factors. For other models, however, the product of all environmental factors is adopted as representative of the restriction. As a consequence, significant difference occurs in terms of the order of magnitude of environmental restriction, and hence there is the difference in retrieved potential optical energy utilization rate.

Secondly, fraction of photosynthetically active radiation absorbed by plant canopy also varies enormously for different models. FPAR calculated or adopted by several light use efficiency models are compared, as shown in Fig. 15.5. At Howland station, FPAR adopted by EC-LUE model and MODIS-GPP product

TABLE 15.3 Comparison of potential light use efficiency and environmental restriction factors for different optical energy utilization rate models.

Model	ε_g or ε_n	ε_{max}	Literature
Net primary production			
CASA Model	$\varepsilon_n = \varepsilon_{max} \times f(T1) \times f(T2) \times f(SM)$	0.389	Potter et al. (1993)
Gross primary production			
GLO-PEM	$\varepsilon_n = \varepsilon_{max} \times f(T) \times f(SM) \times f(VPD)$	55.2α[a]	Prince and Goward (1995)
	$\varepsilon_n = \varepsilon_{max} \times f(T) \times f(SM) \times f(VPD)$	2.76[b]	
MODIS-GPP	$\varepsilon_n = \varepsilon_{max} \times f(T) \times f(VPD)$	0.604−1.259[c]	Running et al. (1999)
VPM	$\varepsilon_n = \varepsilon_{max} \times f(T) \times f(W) \times f(P)$	2.208, 2.484[d]	Xiao et al. (2004), 2005
EC-LUE	$\varepsilon_n = \varepsilon_{max} \times Min(f(T) \times f(W))$	2.14[e]	Yuan et al. (2010)
3-PG	$\varepsilon_n = \varepsilon_{max}$	1.8	Landsberg and Waring (1997)
CFix	$\varepsilon_n = \varepsilon_{max}$	1.1	Veroustraete et al. (2002)

NPP = $\varepsilon_n \times$ FPAR$\times$PAR, GPP = $\varepsilon_g \times$FPAR$\times$PAR, where ε_n and ε_g is optical energy utilization rate to calculate NPP and GPP, respectively; ε_{max} is potential light use efficiency; environmental restriction factors include temperature f(T), soil moisture f(SM), water condition of canopy f(W), and phenology f(P).
[a] *Potential light use efficiency of C_3 plants; α is quantum efficiency.*
[b] *Potential light use efficiency of C_4 plants.*
[c] *Potential light use efficiency of 11 vegetation types.*
[d] *Potential light use efficiency of needle-leaved forest and tropical evergreen broad-leaved forest.*
[e] *Applicable for all vegetation types.*

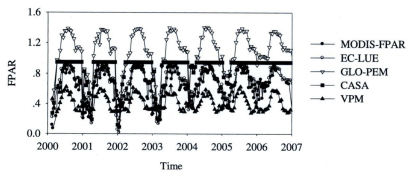

FIGURE 15.5   FPAR calculated or adopted by several light use efficiency models (at Howland station in the United States; latitude, 45.20°N, longitude, −68.74°E; temperate needle-leaved forest).

are very similar, while FPAR of GEO-PEM is far larger than that of other models. For VPM, FPAR is only one half of that in EC-LUE, MODIS-GPP, and CASA. Besides, FPAR calculated by GLO-PEM only shows difference between growth season and nongrowth season, with no difference detected within a specific growth season. This result is inconsistent with actual conditions.

Under the joint action of multiple factors, GPP or NPP simulated by different light use efficiency models also differs considerably. Yuan et al. (2007) used data measured at 28 eddy covariance stations in North America and Europe to compare simulation precision of EC-LUE model and MODIS-GPP product (Fig. 15.6). The results showed that EC-LUE model had higher simulation precision compared with MODIS-GPP product. At regional scale, EC-LUE model did not show systematic deviation. With MODIS-GPP product, however, GPP was overestimated in areas with lower vegetation productivity and was underestimated in areas with higher vegetation productivity. On the global scale, the annual mean values of GPP, respectively, simulated by EC-LUE model and MODIS-GPP product were very similar (Fig. 15.7).

15.4.4 Defects of light use efficiency models

Light use efficiency models are based on clear principle, with simple calculation procedures. Data are acquired using remote sensing technique or measured at meteorological stations, and the simulation precision is high. Due to these advantages, light use efficiency models have become major tools in estimating regional and global vegetation productivity. However, optimal energy utilization rate models still have several shortcomings, which restrict their application in vegetation productivity estimation. This section summarizes the major defects of optical energy utilization rate models and their improvement method to further strengthen their applicability.

15.4.4.1 Difficulty in estimating net primary production

At present, the majority of light use efficiency models directly simulate GPP, and only a small number of them, such as CASA and GLO-PEM, directly simulate NPP. Theoretically speaking, the basic principle based on which the optical energy utilization rate models are built is specifically designed for GPP. To apply this principle

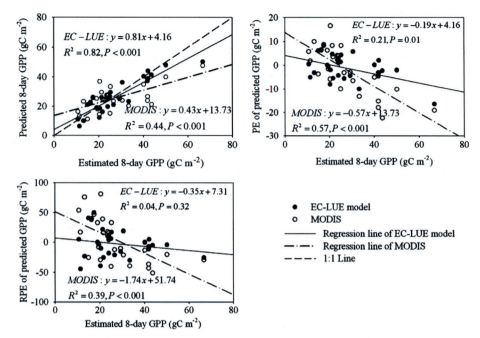

FIGURE 15.6 Comparison of simulation precision of eddy covariance light use efficiency (EC-LUE) model and MODIS-GPP product using data measured at 28 eddy covariance stations in North America and Europe.

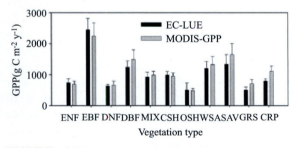

FIGURE 15.7 Simulation results of global gross primary production (GPP) of diverse vegetation types by eddy covariance light use efficiency model and MODIS-GPP product.

in direct simulation of NPP, an assumption is inevitably required: GPP taken up by autotrophic respiration is identical for all ecosystems and geographical areas. Recent research reveals that the ratio of production consumed by autotrophic respiration to GPP is the lowest in area with annual mean temperature of 11°C (Fig. 15.8). This figure increases with the increase in the age of forest for forest ecosystem

(Piao et al., 2010). It is obvious that, in light use efficiency model, calculating NPP based on only a constant ratio of production consumed by autotrophic respiration to GPP would definitely result in systematic error in terms of space.

15.4.4.2 Difference in light use efficiency under the effect of scattering and direct solar radiation

Most of the current light use efficiency models take into account the effect of environmental factors including temperature, moisture, and phenology on potential light use efficiency. Recent studies show that higher ratio of scattered radiation in solar radiation can significantly promote photosynthesis (Gu et al., 2002, 2003; Urban et al., 2007; Alton et al., 2007). Gu et al. (2003) reported that, as a result of increased ratio of scattered radiation caused by eruption of Mount Pinatubo (15.1°N, 121.4°E) in 1991,

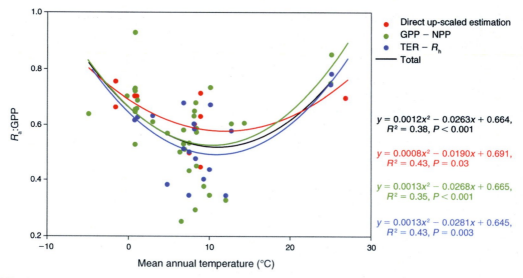

FIGURE 15.8 Variation trend of the ratio of production consumed by autotrophic respiration to gross primary production (GPP) with annual mean temperature.

vegetation productivity of Harvard forest ecosystem in the United States increased by 23% in 1992 and by 8% in 1993. Besides volcanic ash, cloud can also remarkably reduce direct solar radiation, thus greatly increasing the ratio of scattered radiation and the vegetation productivity (Hollinger et al., 1994; Sakai et al., 1996). As scattered radiation increases, the significant increase of light use efficiency of plants can be generally attributed to the following aspects: (1) compared with direct solar radiation, scattered radiation can penetrate plant canopy more easily and thus be absorbed by the leaves of plants on the bottom; (2) with the increased ratio of scattered radiation, the ratio of blue wave radiation to red wave radiation rises; and (3) as blue wavebands are the major wavebands for photosynthetically active radiation, the proportion of photosynthetically active radiation in total radiation also rises significantly (Matsuda et al., 2004; Urban et al., 2007).

However, most light use efficiency models (except TL-LUE model) ignored the influence of diffuse PAR and direct PAR on vegetation photosynthesis. Hence, the influence of this factor fails to be taken into account in simulating regional and global vegetation productivity, leading to simulation error on regional scale. For instance, Yuan et al. (2010) found that vegetation productivity of various ecosystems in Europe was extensively and considerably underestimated, while in North America, no similar errors occurred. The major reason might be that the number of cloudy days was far greater in Europe than in North America, with light use efficiency in European continent higher than that in North America. More emphasis should be placed on developing proper restriction equation to quantitatively describe the effect of scattered radiation and direct solar radiation on optical energy utilization rate.

15.4.4.3 Influence of forest disturbance on GPP estimates

Remote sensing data, usually applied in the calculation of fraction of photosynthetically active radiation absorbed by plant canopy, provide important vegetation information for optical energy utilization rate model. The error of remote sensing data itself would further lead to error in vegetation productivity simulation. For instance, saturation problem related to vegetation index generally results in underestimation of GPP or NPP in areas with dense vegetation. Yuan et al. (2014b) found that light use efficiency model tended to considerably underestimate vegetation productivity in high-latitude cold temperate needle-leaved forest (Table 15.4; Fig. 15.9), and the extent of underestimation increased as the forest age decreased.

The reason might be that the vegetation in tree canopy was sparse, and there was usually moss layer beneath. Therefore, remote sensing signals incorporate the signals of vascular plants such as woody plants and herbaceous plants as well as bryophytes. The photosynthetic capacity of bryophytes is only 1/6 of that of vascular plants (Whitehead and Gower, 2001). If remotely sensed vegetation index is totally regarded as that of vascular plants without any distinction and only the optical energy utilization rate of vascular plants is used to calculate vegetation productivity, overestimation will be inevitable.

Therefore, it is highly necessary to correct the simulation by light use efficiency models, especially for high-latitude areas. Considering the effect of moss in high-latitude area, EC-LUE model is corrected. Computational formula for total GPP of vascular plants (herbaceous and woody plants) and bryophytes can be written as

$$GPP_T = GPP_A + GPP_M \quad (15.52)$$

where GPP_A and GPP_M are GPP of vascular plants and bryophytes, respectively. Based on EC-LUE model, the two can be calculated as

$$GPP_A = PAR \times k\_NDVI \times FPAR \times \varepsilon_{max}$$
$$\times Min(f(t), f(w))$$
$$(15.53)$$

$$GPP_M = PAR_M \times (1 - k\_NDVI) \times FPAR$$
$$\times \varepsilon_{moss} \times Min(f(t), f(w))$$
$$(15.54)$$

TABLE 15.4 The eddy covariance sites used in this study.

ID	Site	Lat	Long	Age	AMT	AP	Period
1	CA-NS1	55.88	−98.48	152	−2.89	500.29	2002−05
2	CA-NS2	55.91	−98.52	72	−2.88	499.82	2001−05
3	CA-NS3	55.91	−98.38	38	−2.87	502.22	2001−05
4	CA-NS95	55.90	−98.21	7	−2.93	498.21	2001−05
5	CA-NS5	55.86	−98.49	21	−2.87	500.34	2001−05
6	CA-NS6	55.92	−98.96	13	−3.08	495.37	2001−05
7	CA-NS7	56.63	−99.95	4	−3.52	483.27	2002−05
8	CA-Oas	53.63	−106.19	83	0.34	428.53	2000−05
9	CA-Obs	53.99	−105.12	111	0.79	405.60	2000−05
10	CA-Ojp	53.92	−104.69	91	0.12	430.50	2000−05
11	CA-Qcu	49.26	−74.04	57	0.13	949.00	2003−05
12	CA-Qfo	49.69	−74.34	100	0.00	961.31	2003−05
13	CA-Sf1	54.48	−105.82	25	−0.15	423.69	2003−05
14	CA-Sf2	54.25	−105.88	13	−0.88	435.12	2001−05
15	CA-Sf3	54.09	−106.01	4	0.08	441.78	2001−05
16	CA-Sj1	53.91	−104.66	8	0.13	430.23	2001−05
17	CA-Sj2	53.95	−104.65	0	0.11	430.33	2003−05
18	CA-Sj3	53.87	−104.65	27	0.13	433.33	2003−05
19	FI-Hyy	61.85	24.28	40	2.18	620.20	2000−03
20	RU-Fyo	56.46	32.92	183	4.91	704.00	2000−06
21	RU-Zot	60.80	89.35	200	−3.27	536.00	2002−04
22	SE-Fla	64.12	19.45	28	0.27	615.98	2001−02
23	SE-Nor	60.08	17.47	100	5.45	561.02	2003
24	TUR	64.12	100.46	102	−9.17	317.00	2004
25	YLF	62.25	129.25	160	−10.40	259.00	2004
26	YPF	62.25	129.65	60	−10.40	259.00	2004

Age, stand age in 2002; *AMT*, annual mean temperature (°C); *AP*, total annual precipitation (mm); *ID*, site label; *Lat*, latitude; *Long*, longitude; *site*, abbreviation of EC site name; *Period*, the study period.

where f_NDVI is the contribution made by vascular plants to the overall NDVI that is remotely sensed. ε_{max} and ε_{moss} are the potential

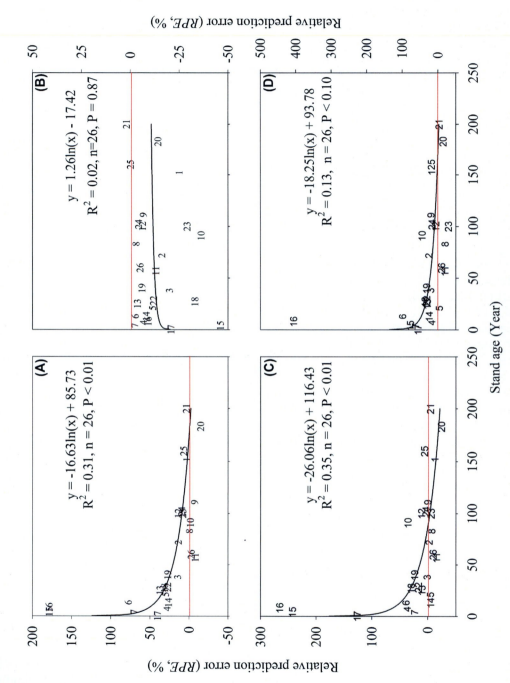

FIGURE 15.9 The RPE of LUE models versus stand age. (A) The correlation relationship between stand age and RPE of eddy covariance light use efficiency (EC-LUE), (B) EC-LUE-2p, (C) MODIS-GPP, and (D) CASA models at all sites. Numbers within the graphs are site IDs (Table 15.4).

light use efficiency of vascular plants and bryophytes, respectively. As vascular plants occupy the top layer of canopy, the photosynthetically active radiation received is the incident photosynthetically active radiation of ecosystem. Photosynthetically active radiation received by bryophytes can be calculated by

$$PAR_M = PAR \times exp(-k \times LAI_A) \quad (15.55)$$

$$LAI_A = k\_NDVI \times LAI \quad (15.56)$$

where k is extinction coefficient of canopy, and its value is set as 0.5.

On this basis and using GPP estimated by eddy covariance technique at 26 boreal forest sites, parameter retrieval is performed with corrected EC-LUE model. The potential light use efficiency of vascular plants is assumed to be $2.14 \, g \, C/m^2/MJ$, and the optimal temperature for photosynthesis is set as 21°C. Based on observation data, k_NDVI and potential light use efficiency of bryophytes are retrieved. At 26 sites, the potential light use efficiency of bryophytes obtained is relatively consistent ($0.6 \, g \, C/m^2/MJ$). Parameter K at first increases with the increase in forest age and then stabilizes (Fig. 15.10). The reason for this is that, at initial growth of forest, the trees and herbaceous plants are sparse, contributing little to remotely sensed vegetation index. With continuous succession of forest ecosystem, the ratio of trees and herbaceous plants as well as their contribution to remote sensing signals increases, until finally they stabilize after reaching certain stages.

15.5 Potential of sun-induced chlorophyll fluorescence for vegetation production estimates

Recently, linking sun-induced chlorophyll fluorescence (SIF) with canopy photosynthesis has received great attention. Light energy absorbed by the leaf chlorophyll molecules has three different pathways: photochemistry,

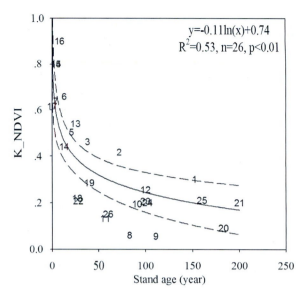

FIGURE 15.10 The contribution of moss to NDVI versus stand age.

nonphotochemical quenching (i.e., heat dissipation), and a small fraction reemitted as SIF (Baker, 2008). Chlorophyll fluorescence is light reemitted by chlorophyll molecules during return from excited to nonexcited states. The spectral extent of chlorophyll fluorescence is around 600–800 nm.

Vegetation photosynthesis closely coupling with chlorophyll emitting fluorescence is the basis for vegetation production estimates using SIF. Unlike the other vegetation indices (e.g., EVI, NDVI), SIF relates more directly to photosynthesis and can provide a direct and quick diagnosis of actual vegetation photosynthesis (Meroni et al., 2009). SIF can be expressed similar as the framework LUE model:

$$SIF = FPAR \times PAR \times LUE_f \times f_{esc} \quad (15.57)$$

where LUE_f is the light use efficiency for SIF, f_{esc} is the fraction of SIF photons escaping from the canopy.

Studies reported the superior performance of SIF over conventional vegetation indices for tracking seasonal variations of photosynthesis across diverse ecosystems (Guanter et al., 2014; Joiner et al., 2014; Sun et al., 2017). However, for the complexity of photosynthesis at different time and spatial scales, the relation between SIF and canopy photosynthesis may vary with vegetation types, canopy conditions, and environmental stresses (e.g., hot wave, water, or radiation).

Recent advances in remote sensing enable the space-based monitoring of SIF globally. Commonly used SIF global measurements include the Greenhouse Gases Observing Satellite (GOSAT), the Global Ozone Monitoring Experiment-2 (GOME-2), the Orbiting Carbon Observatory-2 (OCO-2), and the Scanning Imaging Absorption Spectrometer for Atmospheric ChartographY (SCIAMACHY). Linking SIF with canopy photosynthesis has received great attention, particularly at regional to global scales (Ryu et al., 2019). Numerous studies have investigated the relationship between satellite SIF and canopy photosynthesis (Smith et al., 2018; Sun et al., 2018). Research found there is a universal relationship between SIF from OCO2 and photosynthesis across various biomes (Li et al., 2018). However, in an airborne-based SIF study, the spatial relationships between SIF and canopy photosynthesis varied with vegetation types (Damm et al., 2015).

So whether the relationship between SIF and canopy photosynthesis is linear and universal, whether SIF show better correlation to APAR rather than canopy photosynthesis are still unclear and need to be addressed (Ryu et al., 2019). There are still many uncertainties and many questions to be answered in vegetation production estimates by using SIF.

15.6 Dynamic global vegetation models

Another type of model extensively employed in simulating vegetation productivity of terrestrial ecosystem is DGVM. Models belonging to this type have coupled the major ecological process of terrestrial ecosystem: land surface physical process, physiology of plant canopy, plant phenology, plant succession and competition, and exchange in water, nitrogen, and energy with the atmosphere. The models are able to dynamically simulate vegetation productivity, net ecosystem carbon exchange, soil carbon content, aboveground/underground litter, soil carbon flux, and land surface vegetation structure (such as LAI and vegetation type distribution) (Bonan et al., 2003) (Fig. 15.11). Moreover, the models can also reflect the variation characteristics of ecosystems under various influences such as the increase of CO_2 level, climate change, and various artificial and natural interferences, with irreplaceable application values in the research on the terrestrial ecosystem's response and feedback to global change (Zhuang et al., 2006).

15.6.1 Brief introduction to dynamic global vegetation models

DGVM gives a detailed description of key major physiological process of photosynthesis of leaves. Then according to scaling schemes, vegetation productivity on the scale of ecosystem is simulated. Besides, DGVM includes the mechanism of whole vegetation dynamic change and thus it is able to dynamically simulate the variation characteristics of vegetation with time and space (for instance, variation in plant phenology and LAI at the step length of day). The accurate simulation of photosynthetic rate of leaves is essential to accurate assessment of vegetation productivity.

Photosynthesis refers to the process in which green vegetation absorbs carbon dioxide and water to produce organic matters and release oxygen using solar energy. Photosynthesis process can be represented by the equation below:

$$CO_2 + H_2O \xrightarrow{Radiation} (CH_2O) + O_2 \qquad (15.58)$$

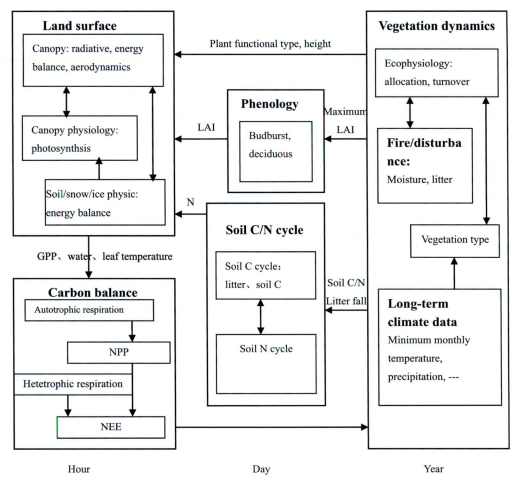

FIGURE 15.11   Block diagram of IBIS model.

Total reaction equation for photosynthesis seems to be a simple redox reaction, but in fact, photosynthesis process involves a series of complex photochemical steps and substance conversion. The complete photosynthesis process can be divided into three major steps: (1) primary reaction (absorption, transmission, and conversion of solar energy); (2) electron transport (the process in which electric energy is converted into active chemical energy); and (3) carbon assimilation (the process in which active chemical energy is converted into stable chemical energy). In DGVM, photosynthesis rate is function of absorbed photosynthetically active radiation, leaf temperature, intercellular carbon dioxide concentration, and Rubisco enzyme activity involved in photosynthesis.

Total photosynthesis production of a forest stand is the sum of photosynthesis production of each leaf. Biochemical process taking place during photosynthesis for each leaf is the principal basis for simulating photosynthesis of forest canopy. As long as the photosynthesis production of a single leaf is simulated, the total photosynthesis production of the forest canopy can be estimated by various scale conversion methods. Among leaf assimilation models, mechanism-based photosynthesis model proposed by Farquhar et al. (1980) has been widely applied. This model determines that instantaneous rate of photosynthesis of C3 plants is the minimum value of the solution of Eqs. (15.59) and (15.60).

$$W_c = V_m \frac{C_i - \Gamma}{C_i + K} \tag{15.59}$$

$$W_j = J \frac{C_i - \Gamma}{4.5C_i + 10.5\Gamma} \tag{15.60}$$

where W_c and W_j are the rate of photosynthesis under the restriction of Rubisco and light, respectively, with the unit of $\mu mol/m/s$; V_m is the maximum value of the rate of carboxylation, with the unit of $\mu mol/m^2/s$; J is the photosynthetic electron transport rate, with the unit of $\mu mol/m^2/s$; C_i is intercellular CO_2 concentration; Γ is the CO_2 compensation point without dark respiration; K is enzyme kinetic function.

Theoretically speaking, the formula above can be applied to any leaf in the canopy for simulating the rate of photosynthesis on the canopy scale. In actual research, however, to expand the rate of photosynthesis of a single leaf to the entire canopy scale is required, which is usually realized using the big-leaf model and double-leaf model.

Big-leaf model assumes that the biochemical process described on the leaf scale can be directly expanded to canopy scale, i.e., the canopy is regarded as a big leaf, and the rate of photosynthesis on the canopy scale is the product of the photosynthetic rate of the leaf and LAI. The principle of big-leaf model is simple, and it includes the basic biochemical process of photosynthesis, with extensive application in various ecosystem models, such as BIOME-BGC model (Hunt and Running, 1992; Kimball et al., 1997; Liu et al., 1997) and SiB2 model (Sellers et al., 1996a,b). However, big-leaf model has one major defect (Chen et al., 1999). On a cloudy day, solar radiation can basically satisfy the optical energy demand of photosynthesis of leaves on the top of the canopy. But it is assumed in big-leaf model that the incident radiation received by the entire canopy equals that on the top of the canopy. In fact, the leaves at the bottom of canopy cannot receive adequate radiation for photosynthesis due to the shelter of other leaves. Therefore, the use of big-leaf model would result in serious overestimation of ecosystem productivity. Chen et al. (1999) used GPP measured by two layers of CO_2 flux to confirm this defect of big-leaf model. Considering the difference in the rate of photosynthesis at different positions of the canopy, Sellers et al. (1992, 1996a,b) gave different weightings to the contribution of leaves to photosynthesis according to their position in the canopy. In this way, the problem arising from low rate of photosynthesis at the bottom of the canopy due to lack of light is overcome.

To avoid this defect of big-leaf model, the total photosynthesis of the canopy can be modeled based on two types of representative leaves: sunshine leaves and shade leaves (DePury and Farquhar, 1997; Wang and Leuning, 1998). The rate of photosynthesis of the two types of leaves can be, respectively, calculated using single-leaf model and then they are multiplied by respective LAI (Norman, 1982).

$$A_{canopy} = A_{sun} \times LAI_{sun} + A_{shade} \times LAI_{shade} \tag{15.61}$$

where A_{sun} and A_{shade} are the rate of photosynthesis of sunshine leaves and shade leaves, respectively; LAI_{sun} and LAI_{shade} are the corresponding LAI. Considering the effect of clumping index (Ω) on radiation mechanism of the canopy, Chen et al. (1999) modified the algorithm for calculating LAI_{sun} and LAI_{shade} by Norman (1982).

$$LAI_{sun} = 2 \times \cos\theta$$
$$\times (1 - \exp(-0.5 \times \Omega \times LAI/\cos\theta))$$
$$LAI_{shade} = LAI - LAI_{sun}$$
$$\tag{15.62}$$

where LAI is leaf area index; θ is solar zenith angle. For needle-leaved forest, Ω is 0.5–0.7, for broad-leaved forest, Ω is 0.7–0.9, and for grass and crops, Ω is 0.9–1.0 (Chen, 1996). Smaller Ω indicates stronger nonrandomness of spatial distribution of vegetation. Clumping index of vegetation changes its reception capacity of incident radiation and therefore is necessary to be taken into account in productivity model. The greater the clumping of vegetation (smaller Ω) is, the more the radiation which is able to penetrate canopy without the interception of leaves there is. Therefore, LAI of sunshine leaves is increased, while that of shade leaves is decreased. The clumping structure of forest canopy implies that the separation of shade leaves from sunshine leaves is very important. As the ratio of shade leaves in the clumped canopy is much larger than that in the canopy with random distribution, shade leaves should be given enough attention in the calculation of forest productivity.

15.6.2 Application of remote sensing data in dynamic global vegetation models

Remote sensing data, which play a significant role in the accurate estimation of vegetation productivity, provide the information related to vegetation growth status in simple empirical models and productivity models. Also in DGVMs, remote sensing data offer important basic data in multiple aspects for the application of the model on regional scale.

15.6.2.1 Land cover map

Land cover map constitutes one of the major basic data sources for the application of DGVMs in estimating vegetation productivity. The parameterization scheme in DGVMs adopts identical set of parameters for the same vegetation type. As shown in Table 15.5, BIOME-BGC model sets key photosynthetic parameter values for each vegetation type, and theses parameters differ considerably across different vegetation types. For instance, the maximum rate of photosynthesis for C3 grassland is nearly two times of that of evergreen needle-leaved forest and shrub. It is obvious that the accuracy of vegetation type classification is directly related to accurate simulation of regional and global vegetation productivity.

Remote sensing technique is the only effective approach available for vegetation type classification. For instance, MODIS land cover product (MOD12) provides annual global land cover information with the resolution of 1 km, laying a solid data foundation for the application of DGVMs.

15.6.2.2 Leaf area index

LAI, as a major decisive factor in estimating vegetation productivity in ecosystem, directly associates photosynthesis of leaves with the rate of photosynthesis of canopy. Therefore, the

TABLE 15.5 Table of regional parameterization scheme in BIOME-BGC model.

Parameter	C3G	C4G	ENF	DNF	EBF	DBF	SHR
Aerodynamic conductance (mm/s)	0.01	0.01	1.0	1.0	1.0	1.0	0.1
Maximum stomatal conductance (mm/s)	5.0	2.0	2.2	4.5	4.0	4.5	3.5
Maximum rate of photosynthesis ($\mu mol/m^2/s$)	6.0	7.5	3.5	5.0	4.0	5.0	3.5
Optimal temperature for photosynthesis (°C)	20	30	20	20	20	20	20

C3G, C3 grassland; *C4G*, C4 grassland; *DBF*, deciduous broad-leaved forest; *DNF*, deciduous needle-leaved forest; *EBF*, evergreen broad-leaved forest; *ENF*, evergreen needle-leaved forest; *SHR*, shrubland.

accurate simulation and retrieval of LAI is essential to vegetation productivity estimation. In fact, LAI simulation is the difficulty in DGVMs, as the simulation involves many complex physiological processes such as carbon allocation strategy and short-term adaptability of leaves to drought and high temperature. In actual applications, simplification of LAI simulation is required. For instance, IBIS model assumes constant LAI in growth season, whose value depends on vegetation productivity of the preceding year (Foley et al., 1996). This simplification apparently cannot meet the actual requirements of LAI simulation and it will result in error in vegetation productivity simulation. Remote sensing technique guarantees reliable direct observation of terrestrial vegetation features. LAI retrieved based on remote sensing data is so far the most accurate LAI information on the regional scale. Many carbon cycle researches, with the purpose of improving the effect of simulation using models, conduct coupling for LAI retrieved from remote sensing data.

15.6.2.3 Model-driven data

For DGVMs to run on the regional scale, numerous meteorological elements are needed as driving variables, including temperature, wind speed, radiation, precipitation, and so on. With the conventional approach, meteorological elements monitored at the weather stations are collected, and spatial interpolation is employed to obtain the spatially continuous driving field of meteorological elements. This method is restricted by the number of weather stations used in interpolation; the greater the number of weather stations is, the closer the spatial driving field obtained to the actual one is. However, what frequently occurs in actual situation is the limited number of weather stations with extremely uneven spatial distribution. Moreover, the spatial interpolation of some elements needs to take into account the effect of topography, latitude, and many other factors. However, these special requirements are difficult to be satisfied in conventional interpolation methods. Thus, the error in spatial driving data in the model constitutes one of the major sources of errors in regional estimation using model.

As remote sensing technique is making rapid progress, the application of remote sensing data in retrieving fields driven by various meteorological elements becomes possible. Chapters 6, 8, and 17 of this book present a detailed introduction of retrieving radiation, temperature, and precipitation from remote sensing data. Current researches show that model-driven field retrieved from remote sensing data is able to overcome the shortcomings of conventional interpolation methods. For instance, data acquired by platform-based radar and meteorological satellite (polar orbiting satellite and geostationary satellite) can be used to retrieve precipitation with higher temporal and spatial resolution and broader observation scope (Michaelides et al., 2009). Sensors carried by LEO satellite can acquire data with high spatial resolution (usually 1 km), while geostationary satellite can achieve long-term continuous observation of a specific region (once per 30 min). The combination of the two can effectively improve the precision of precipitation retrieval.

15.7 Temporal and spatial distribution pattern of global vegetation productivity

Under the influence of climate, soil, and physiological features of vegetation, global vegetation productivity is characterized by salient temporal-spatial variability. Many researchers adopt various productivity models to simulate the variation features of vegetation productivity on daily, monthly, seasonal, yearly, or even longer scales.

On the global scale, the low values of NPP in nearly all of the models are concentrated in the cold and arid areas (Anav et al., 2015). With the increase of temperature and moisture, NPP rises sharply. The simulation results of most of the models show that the highest NPP of biotic community occur in tropical rain forest, with its vegetation productivity accounting for 34% of that of global terrestrial ecosystem (Beer et al., 2010); that of temperate evergreen broad-leaved forest comes the second. Conversely, the minimum value of yearly NPP occurs in arid desert regions as well as high-latitude tundra.

As the terrestrial area of the northern hemisphere constitutes 74% of the global terrestrial area, the seasonal variation of the northern hemisphere dominates that of vegetation productivity of global terrestrial ecosystem. The global vegetation productivity reaches its minimum value in the winter in northern hemisphere, while that in the summer is the maximum (Cramer et al., 1999). The lowest global vegetation productivity estimated by most of the models occurs in February (Cramer et al., 1999). The monthly lowest value of global NPP simulated by the model ranges between $-1.6\,\mathrm{Pg\,C/month}$ (HYBRID) and $4.8\,\mathrm{Pg\,C/month}$ (TURC). In the majority of models, the global highest value simulated occurs in the winter in northern hemisphere (June–August), ranging between $5.6\,\mathrm{Pg\,C/month}$ and $9.1\,\mathrm{Pg\,C/month}$.

On either single-point scale or regional and global scale, vegetation productivity shows significant interannual variation. The researches based on multiple eddy covariance stations worldwide indicate that the interannual variation of vegetation productivity of terrestrial ecosystem differs remarkably with the regions (Yuan et al., 2009). As the interannual variation of vegetation productivity (indicated by standard deviation of GPP over the years) becomes greater with the increase of latitude, that is, GPP of deciduous broad-leaved forest ecosystem in high-latitude regions shows more violent interannual variation. On the contrary, interannual fluctuation in GPP of evergreen needled-leaved forest becomes less violent as the latitude increases. Global vegetation variation directly determined the interannual variation of atmospheric CO_2 level (Fig. 15.12, Zhao and Running, 2010).

Interannual variability of vegetation productivity directly depends on interannual fluctuation of temperature, moisture, and other meteorological elements. The drought events in North America and China in 2000 led to the reduction of NPP of this area. The droughts

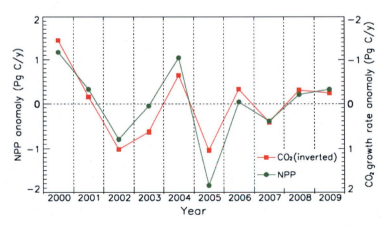

FIGURE 15.12 Interannual variability in net primary production (NPP) of global terrestrial vegetation and CO_2 growth rate anomaly. In this figure, atmospheric CO_2 growth rate anomaly is reversely identified.

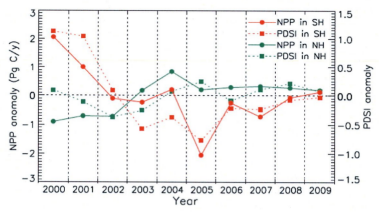

FIGURE 15.13   Interannual variation of net primary production (NPP) of terrestrial vegetation in SH and NP and Palmer drought index in the growth season.

taking place in North America and Australia in 2002 resulted in significant drop of NPP of the regions. Extensive drought and heat wave in Europe in 2003 gave rise to the reduction of vegetation productivity by more than 30%, and large area of forest of this region became the major carbon source (Ciais et al., 2005). Recent studies indicate that, the tropic forest regions, which contribute a large part of global vegetation productivity, have witnessed a considerable drop in vegetation productivity over the past 10 years. For instance, vegetation productivity of Amazon tropical rainforest in the past 10 years decreased by -0.424 Pg C (Zhao and Running, 2010). Regional drought events were the major reason for the reduction of vegetation productivity of this region. Fig. 15.13 shows that NPP of both northern and southern hemispheres showed consistent interannual variation with Palmer drought index.

15.8 Global gross primary production product

In this part, we will demonstrate the global gross primary production product (GLASS-GPP) from the aspects of input data, model algorithm, and model accuracy.

15.8.1 Input data

The input data of GLASS-GPP product include meteorology data, LAI data, and CO_2 concentration data. The meteorology data contain temperature, VPD, diffuse PAR, and direct PAR. The temperature, VPD, diffuse PAR, and direct PAR were derived from MERRA-2 (Modern-Era Retrospective Analysis for Research and Applications) with a spatial resolution of 0.625° in longitude by 0.5° in latitude. The meteorology data were interpolated to 5 km $\times$ 5 km and 500 m $\times$ 500 m globally. The GLASS LAI product (Xiao et al., 2016) was used as the input LAI product.

15.8.2 Brief introduction to global gross primary production product

15.8.2.1 General information

The spatial-temporal information of GLASS-GPP product is as follows:

(1) Spatial extent: global land (except the Antarctica)
(2) Spatial resolution: 1982—2018 (5 km); 2000—2018 (500 m)
(3) Temporal resolution: 8-day

15.8.2.2 Model algorithm description

The revised EC-LUE model simulates terrestrial vegetation GPP as

$$GPP = (\varepsilon_{msu} \times APAR_{su} + \varepsilon_{msh} \times APAR_{sh}) \times C_s$$
$$\times \min(T_s, W_s)$$
$$(15.63)$$

where ε_{msu} and ε_{msh} are the maximum LUE of sunlit and shaded leaves, respectively; $APAR_{su}$ and $APAR_{sh}$ are the PAR absorbed by sunlit and shaded leaves, which can be calculated by Eqs. (15.45) and (15.46), respectively (Chen et al., 1999); C_s, T_s, and W_s represent the downward regulation scalars for the respective effects of atmospheric CO_2 concentration ([CO_2]), temperature (T), and atmospheric water demand (VPD) on LUE of vegetation GPP. min denotes the minimum value of T_s and W_s. The effect of atmospheric CO_2 concentration on GPP is calculated as according to Farquhar et al. (1980) and Collatz et al. (1991).

$$C_s = \frac{C_i - \varphi}{C_i + 2\varphi} \qquad (15.64)$$

$$C_i = C_a \times \chi \qquad (15.65)$$

where φ is the CO_2 compensation point in the absence of dark respiration (ppm), and C_i is CO_2 concentration in the intercellular air spaces of the leaf (ppm), which is the product of atmospheric CO_2 concentration (C_a) and the ratio of leaf internal to ambient CO_2. χ is estimated (Prentice et al., 2014; Keenan et al., 2016)

$$\chi = \frac{\varepsilon}{\varepsilon + \sqrt{VPD}} \qquad (15.66)$$

$$\varepsilon = \sqrt{\frac{356.51K}{1.6\eta^*}} \qquad (15.67)$$

$$K = K_c\left(1 + \frac{P_0}{K_0}\right) \qquad (15.68)$$

$$K_c = 39.97 \times e^{\frac{79.43 \times (T-298.15)}{298.15RT}} \qquad (15.69)$$

$$K_o = 27480 \times e^{\frac{36.38 \times (T-298.15)}{298.15RT}} \qquad (15.70)$$

where K_c and K_o are the Michaelis–Menten coefficient of Rubisco for carboxylation and oxygenation, respectively, expressed in partial pressure units, and P_o is the partial pressure of O_2 (ppm). R is the molar gas constant (8.314 J/mol/K), and η^* is the viscosity of water depending on the air temperature (Korson et al., 1969).

Temperature scalar T_S is calculated by Eq (15.37). When air temperature T_a is below T_{min} or higher than T_{max}, T_S is zero, which means photosynthesis is stopped. In EC-LUE model, the value of T_{min} and T_{max} is set as 0 and 40°C, respectively. T_{opt} is undetermined parameter.

W_s is calculated by the following equations:

$$W_s = \frac{VPD_0}{VPD_0 + VPD} \qquad (15.71)$$

where T_{min}, T_{max}, and T_{opt} are minimum, maximum, and optimum air temperature (°C) for photosynthetic activity, respectively (Yuan et al., 2007). VPD_0 are the half-saturation coefficient of VPD constraint equation (k Pa).

15.8.2.3 Model validation and accuracy

We used 146 eddy covariance flux sites to validate the GLASS-GPP product (Fig. 15.14). These sites distribute global including 10 main terrestrial ecosystem vegetation types.

Fig. 15.15 shows the comparison of the tower-based GPP and simulated GLASS-GPP product at the validation sites with $R^2 = 0.62$.

The distribution of R^2 and relative predictive error (RPE) are shown in Figs. 15.16 and 15.17. For all the 146 validation sites, 41% sites have high accuracy with R^2 over 0.6. But, there are also 24% sites R^2 lower than 0.3. The RPE exhibits normal distribution and over 63% sites with RPE lower than 30%

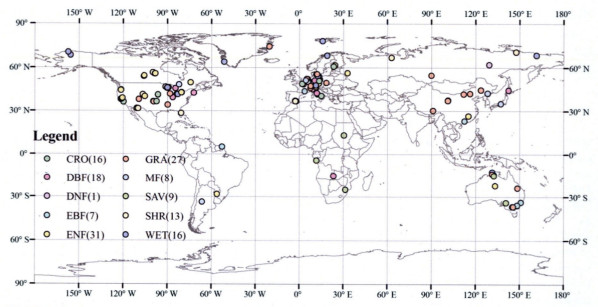

FIGURE 15.14 Site distribution for GLASS-GPP product validation. *CRO*, cropland; *DBF*, deciduous broad-leaved forest; *DNF*, deciduous needle-leaved forest; *EBF*, evergreen broad-leaved forest; *ENF*, evergreen needle-leaved forest; *GRA*, grassland; *MF*, mixed forest; *SAV*, savanna; *SHR*, shrubland; *WET*, wetland.

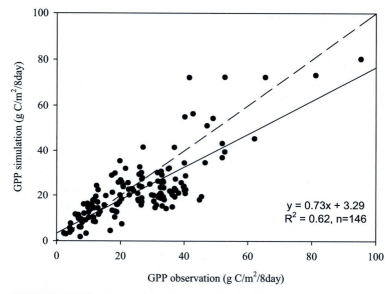

FIGURE 15.15 Comparisons between GLASS-GPP and tower-based GPP.

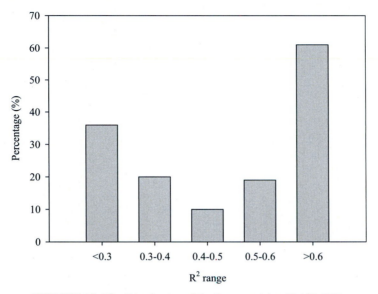

FIGURE 15.16   Distribution of the R-squared for GLASS-GPP.

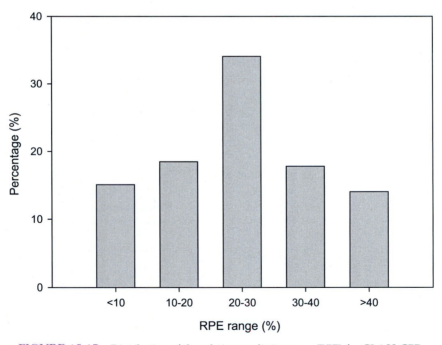

FIGURE 15.17   Distribution of the relative predictive error (RPE) for GLASS-GPP.

15.9 Summary

Vegetation production lays the foundation of the survival and development of human society. As the beginning of global carbon cycle, vegetation production plays vital role in maintaining global atmospheric greenhouse gas concentration and in regulating global climate pattern. Remote sensing data reflect the land surface information with spatial and temporal continuity, providing reliable method and basic data for accurate retrieval and simulation of vegetation productivity of terrestrial ecosystem. On the other hand, light use efficiency models, which are based on light use efficiency principle and make use of vegetation index and LAI retrieved from remote sensing data, have been extensively applied in estimation and simulation studies of vegetation productivity on regional and global scale. Moreover, vegetation distribution types based on remote sensing data, together with meteorological elements such as wind, temperature, and humidity, are also the major driving data for DGVMs. It should also be noted that various estimation methods based on remote sensing data, due to the inherent defects of remote sensing data and the methods themselves, have some shortcomings in estimating vegetation productivity. Further improvement and verification for these models are required.

References

Alton, P.B., North, P.R., Los, S.O., 2007. The impact of diffuse sunlight on canopy light-use efficiency, gross photosynthetic product and net ecosystem exchange in three forest biomes. Glob. Chang. Biol. https://doi.org/10.1111/j.1365-2486.2007.01316.x.

Anav, A., Friedlingstein, P., Beer, C., Ciais, P., Harper, A., Jones, C., Murray-Tortarolo, G., Papale, D., Parazoo, N.C., Peylin, P., Piao, S., Sitch, S., Viovy, N., Wiltshire, A., Zhao, M., 2015. Spatiotemporal patterns of terrestrial gross primary production: a review. Rev. Geophys. 53 (3), 785–818.

Baker, N.R., 2008. Chlorophyll fluorescence: a probe of photosynthesis in vivo. Annu. Rev. Plant Biol. 59, 89–113.

Baldocchi, D.D., 2008. 'Breathing' of the terrestrial biosphere: lessons learned from a global network of carbon dioxide flux measurement systems. Aust. J. Bot. 56, 1–26.

Beer, C., Reichstein, M., Tomelleri, E., Ciais, P., Jung, M., Carvalhais, N., Rödenbeck, C., Arain, M.A., Baldocchi, D., Bonan, G.B., Bondeau, A., Cescatti, A., Lasslop, G., Lindroth, A., Lomas, M., Luyssaert, S., Margolis, H., Oleson, K.W., Roupsard, O., Veenendaal, E., Viovy, N., Williams, C., Woodward, F.I., Papale, D., 2010. Terrestrial gross carbon dioxide uptake: global distribution and covariation with climate. Science. https://doi.org/10.1126/science.1184984.

Belward, A.S., Estes, J.E., Kline, K.D., 1999. The IGBP-DIS global 1-km land-cover data set DISCover: a project overview. Photogramm. Eng. Remote Sens. 65, 1013–1020.

Bonan, G.B., Levis, S., Sitch, S., Vertenstein, M., Oleson, K.W., 2003. A dynamic global vegetation model for use with climate models: concepts and description of simulated vegetation dynamics. Glob. Chang. Biol. 9, 1543–1566.

Bosch, J.L., López, G., Batlles, F.J., 2009. Global and direct photosynthetically active radiation parameterizations for clear-sky conditions. Agric. For. Meteorol. 149 (1), 146–158.

Box, E.O., Holben, B., Kalb, V., 1989. Accuracy of the AVHRR vegetation index as a predictor of biomass, primary productivity and net CO_2 flux. Vegetatio 90, 71–89.

Cai, W.W., Yuan, W.P., Liang, S.L., Liu, S.G., Dong, W.J., Chen, Y., Liu, D., Zhang, H.C., 2014. Large differences in terrestrial vegetation production derived from satellite-based light use efficiency models. Remote Sens. 6 (9), 8945–8965.

Cao, M.K., Prince, S.D., Small, J., Goetz, S.J., 2004. Remotely sensed interannual variations and trends in terrestrial net primary productivity 1981–2000. Ecosystems 7, 233–242.

Chen, J.M., 1996. Canopy architecture and remote sensing of the fraction of photosynthetically active radiation in boreal conifer stands. IEEE Trans. Geosci. Remote Sens. 34, 1353–1368.

Chen, J.M., Liu, J., Cihlar, J., Guolden, M.L., 1999. Daily canopy photosynthesis model through temporal and spatial scaling for remote sensing applications. Ecol. Model. 124, 99–119.

Chen, Y., Xia, J., Liang, S., Feng, J., Fisher, J.B., Li, X., Li, X., Liu, S., Ma, Z., Miyata, A., Mu, Q., Sun, L., Tang, J., Wang, K., Wen, J., Xue, Y., Yu, G., Zha, T., Zhang, L., Zhang, Q., Zhao, T., Zhao, L., Yuan, W., 2014. Comparison of evapotranspiration models over terrestrial ecosystem in China. Remote Sens. Environ. 140, 279–293.

Chen, Y., Yuan, W.P., Xia, J.Z., Fisher, J.B., Dong, W.J., Zhang, X.T., Liang, S.L., Ye, A.Z., Cai, W.W., Feng, J.M., 2015. Using Bayesian model averaging to estimate terrestrial evapotranspiration in China. J. Hydrol. 528, 537–549.

Ciais, P., Reichstein, M., Viovy, N., Granier, A., Ogee, J., Allard, V., Aubinet, M., Buchmann, N., Bernhofer, C., Carrara, A., Chevallier, F., De Noblet, N., Friend, A.D., Friedlingstein, P., Grunwald, T., Heinesch, B., Keronen, P., Knohl, A., Krinner, G., Loustau, D., Manca, G., Matteucci, G., Miglietta, F., Ourcival, J.M., Papale, D., Pilegaard, K., Rambal, S., Seufert, G., Soussana, J.F., Sanz, M.J., Schulze, E.D., Vesala, T., Valentini, R., 2005. Europe-wide reduction in primary productivity caused by the heat and drought in 2003. Nature 437 (7058), 529.

Collatz, G.J., Ball, J.T., Grivet, C., Berry, J.A., 1991. Physiological and environmental regulation of stomatal conductance, photosynthesis and transpiration: a model that includes a laminar boundary layer. Agric. For. Meteorol. 54, 107−136.

Cramer, W., Kicklighter, D.W., Bondeau, A., Moore, B., Churkina, G., Nemry, B., Ruimy, A., Schloss, A.L., 1999. Comparing global models of terrestrial net primary productivity (NPP): overview and key results. Glob. Chang. Biol. 5 (S), 1−15.

Damm, A., Guanter, L., Paul-Limoges, E., van der Tol, C., Hueni, A., Buchmann, N., Eugster, W., Ammann, C., Schaepman, M.E., 2015. Far-red sun-induced chlorophyll fluorescence shows ecosystem-specific relationships to gross primary production: an assessment based on observational and modeling approaches. Remote Sens. Environ. 166, 91−105.

De Pury, D.G.G., Farquhar, G.D., 1997. Simple scaling of photosynthesis from leaves to canopies without the errors of big-leaf models. Plant Cell Environ. 20, 537−557.

Fang, J.Y., Ke, J.H., Tang, Z.Y., Chen, A.P., 2001. Implications and estimations of four terrestrial productivity parameters. Acta Phytoecol. Sin. 25 (4), 414−419 (in Chinese).

Farquhar, C.D., Caemmerers, S., Berry, J.A., 1980. A biochemical model of photosynthetic CO_2 assimilation in leaves of C3 species. Planta 149, 78−90.

Field, C.B., Randerson, J.T., Malmström, C.M., 1995. Global net primary production: combining ecology and remote sensing. Remote Sens. Environ. 51 (1), 74−88.

Foley, J.A., Prentice, I.C., Ramankutty, N., Levis, S., Pollard, D., Sitch, S., Haxeltine, A., 1996. An integrated biosphere model of land surface processes, terrestrial carbon balance, and vegetation dynamics. Glob. Biogeochem. Cycles 10 (4), 603−628.

Friend, A.D., Arneth, A., Kiang, N.Y., 2006. FLUXNET and modelling the global carbon cycle. Glob. Chang. Biol. 12, 1−24.

Gamon, J.A., Field, C.B., Goulden, M.L., Griffin, K.L., Hartley, A.E., 1995. Relationships between NDVI, canopy structure, and photosynthesis in three Californian vegetation types. Ecol. Appl. 5, 28−41.

Goetz, S.J., Prince, S.D., Goward, S.N., Thawley, M.M., Small, J., 1999. Satellite remote sensing of primary production: an improved production efficiency modeling approach. Ecol. Model. 122, 239−255.

Goward, S.A., Tucker, C.J., Dye, D., 1985. North American vegetation patterns observed with the NOAA-7 advanced very high resolution radiometer. Vegetatio 64, 3−14.

Gu, L.H., Baldocchi, D.D., Verma, S.B., Black, T.A., Vesala, T., Falge, E., Dowty, P.R., 2002. Advantages of diffuse radiation for terrestrial ecosystem productivity. J. Geophys. Res. 107 (D6), 4050. https://doi.org/10.1029/2001JD001242.

Gu, L.H., Baldocchi, D.D., Wofsy, S.C., Munger, J.W., Michalsky, J.J., Urbanski, S.P., Boden, T.A., 2003. Response of a deciduous forest to the mount pinatubo eruption: enhanced photosynthesis. Science 299, 2035−2038.

Guanter, L., Zhang, Y., Jung, M., Joiner, J., Voigt, M., Berry, J.A., Frankenberg, C., Huete, A.R., Zarco-Tejada, P., Lee, J.-E., Moran, M.S., Ponce-Campos, G., Beer, C., Camps-Valls, G., Buchmann, N., Gianelle, D., Klumpp, K., Cescatti, A., Baker, J.M., Griffis, T.J., 2014. Global and time-resolved monitoring of crop photosynthesis with chlorophyll fluorescence. Proc. Natl. Acad. Sci. U.S.A 111 (14), E1327−E1333.

Hansen, M.C., DeFries, R.S., Townshend, J.R.G., Sohlberg, R., 2000. Global land cover classification at the 1 km spatial resolution using a classification tree approach. Int. J. Remote Sens. 21, 1331−1364.

Hollinger, D.Y., Kelliher, F.M., Byers, J.N., Hunt, J.E., McSeveny, T.M., Weir, P.L., 1994. Carbon dioxide exchange between an undisturbed old-growth temperate forest and the atmosphere. Ecology 75 (1), 134−150.

Huete, A.R., 1988. A soil adjusted vegetation index (SAVI). Remote Sens. Environ. 25 (3), 295−309.

Hunt, E.R., 1994. Relationship between woody biomass and PAR conversion efficiency for estimating net primary production from NDVI. Int. J. Remote Sens. 15, 1725−1730.

Hunt, E.R., Running, S.W., 1992. Simulated dry matter yields for aspen and spruce stand in the North American Boreal forest. Can. J. Remote Sens. 18, 126−133.

Jacovides, C.P., Tymvios, F.S., Assimakopoulos, V.D., Kaltsounides, N.A., 2007. The dependence of global and diffuse PAR radiation components on sky conditions at Athens, Greece. Agric. For. Meteorol. 143 (3−4), 277−287.

Joiner, J., Yoshida, Y., Vasilkov, A., Schaefer, K., Jung, M., Guanter, L., Zhang, Y., Garrity, S., Middleton, E.M., Huemmrich, K.F., Gu, L., Marchesini, L.B., 2014. The seasonal cycle of satellite chlorophyll fluorescence observations and its relationship to vegetation phenology and ecosystem atmosphere carbon exchange. Remote Sens. Environ. 152, 375−391.

Keenan, T.F., Prentice, I.C., Canadell, J.G., Williams, C.A., Wang, H., Raupach, M., Collatz, G.J., 2016. Recent pause in the growth rate of atmospheric CO_2 due to enhanced terrestrial carbon uptake. Nat. Commun. 7, 13428.

Kimball, J.S., Thornton, P.E., White, M.A., Running, S.W., 1997. Simulating forest productivity and surface-atmosphere carbon exchange in the BOREAS study region. Tree Physiol. 17, 589–599.

King, D.A., Turner, D.P., Ritts, W.D., 2011. Parameterization of a diagnostic carbon cycle model for continental scale application. Remote Sens. Environ. 115 (7), 1653–1664.

Korson, L., Drost-Hansen, W., Millero, F.J., 1969. Viscosity of water at various temperatures. J. Phys. Chem. 73 (1), 34–39.

Kurc, S.A., Small, E.E., 2004. Dynamics of evapotranspiration in semiarid grassland and shrubland ecosystems during the summer monsoon season, central New Mexico. Water Resour. Res. 40, W09305. https://doi.org/10.1029/2004WR003068.

Landsberg, J.J., Waring, R.H., 1997. A generalised model of forest productivity using simplified concepts of radiation-use efficiency, carbon balance and partitioning. For. Ecol. Manag. 95, 209–228.

Li, X., Xiao, J., He, B., Altaf Arain, M., Beringer, J., Desai, A.R., Emmel, C., Hollinger, D.Y., Krasnova, A., Mammarella, I., Noe, S.M., Ortiz, P.S., Rey-Sanchez, A.C., Rocha, A.V., Varlagin, A., 2018. Solar-induced chlorophyll fluorescence is strongly correlated with terrestrial photosynthesis for a wide variety of biomes: first global analysis based on OCO-2 and flux tower observations. Glob. Chang. Biol. 24, 3990–4008.

Lieth, H., 1975. Historical survey of primary productivity research. In: Lieth, H., Whittaker, R.H. (Eds.), Primary Productivity of the Biosphere. Springer-Verlag, New York, pp. 7–16.

Liu, J., Chen, J.M., Cihlar, J., Park, W.M., 1997. A process-based boreal ecosystem productivity simulator using remote sensing inputs. Remote Sens. Environ. 62, 158–175.

Matsuda, R., Ohashi-Kaneko, K., Fujiwara, K., Goto, E., Kurata, K., 2004. Photosynthetic characteristics of rice leaves grown under red light with or without supplemental blue light. Plant Cell Physiol. 45, 1870–1874.

Meroni, M., Rossini, M., Guanter, L., Alonso, L., Rascher, U., Colombo, R., Moreno, J., 2009. Remote sensing of solar-induced chlorophyll fluorescence: review of methods and applications. Remote Sens. Environ. 113, 2037–2051.

Michaelides, S.C., Tymvios, F.S., Michaelidou, T., 2009. Spatial and temporal characteristics of the yearly rainfall frequency distribution in Cyprus. Atmos. Res. 94, 606–615.

Myneni, R.B., Williams, D.L., 1994. On the relationship between FAPAR and NDVI. Remote Sens. Environ. 49, 200–221.

Norman, J.M., 1982. Simulation of microclimates. In: Hatfield, J.L., Thomason, I.J. (Eds.), Biometeorology in Integrated Pest Management. Academic Press, New York, pp. 65–99.

Odum, E.P., 1960. Organic production and turnover in old field succession. Ecology 41, 34–49.

Paruelo, J.M., Epstein, H.E., Lauenroth, W.K., Burke, I.C., 1997. ANPP estimates from NDVI for the central grassland region of the US. Ecology 78, 953–958.

Paruelo, J.M., Oesterheld, M., Di Bella, C.M., Arzadum, M., Lafountaine, J., Cahuepé, M., Rebella, C.M., 2000. Estimation of primary production of subhumid rangelands from remote sensing data. Appl. Veg. Sci. 3, 189–195.

Piao, S.L., Luyssaert, S., Ciais, P., Janssens, I., Chen, A.P., Cao, C., Fang, J.Y., Friedlingstein, P., Luo, Y.Q., Wang, S.P., 2010. Forest annual carbon cost: a global-scale analysis of autotrophic respiration. Ecology 91, 652–657.

Potter, C.S., Randerson, J.T., Field, C.B., Pamela, A.M., Vitousek, P.M., Mooney, H.A., Klooster, S.A., 1993. Terrestrial ecosystem production: a process model based on global satellite and surface data. Glob. Biogeochem. Cycles 7, 811–841.

Prentice, I.C., Dong, N., Gleason, S.M., Maire, V., Wright, I.J., 2014. Balancing the costs of carbon gain and water transport: testing a new theoretical framework for plant functional ecology. Ecol. Lett. 17 (1), 82–91.

Prince, S.D., Goward, S.N., 1995. Global primary production: a remote sensing approach. J. Biogeogr. 22, 815–835.

Running, S.W., Nemani, R., Glassy, J.M., Thornton, P.E., 1999. In: MODIS Daily Photosynthesis (PSN) and Annual Net Primary Production (NPP) Product (MOD17), Algorithm Theoretical Basis Document, Version 3.0. http://modis.gsfc.nasa.gov/.

Running, S.W., Thornton, P.E., Nemani, R., Glassy, J.M., 2000. Global Terrestrial Gross and Net Primary Productivity from the Earth Observing System. Methods in ecosystem science. Springer, New York, NY, pp. 44–57.

Ryu, Y., Berry, J.A., Baldocchi, D.D., 2019. What is global photosynthesis? history, uncertainties and opportunities. Remote Sens. Environ. 223, 95–114.

Sakai, R.K., Fitzjarrald, D.R., Moore, K.E., Freedman, J.M., 1996. How do forest surface fluxes depend on fluctuating light level?. In: Proceedings of the 22nd Conference on Agricultural and Forest Meteorology with Symposium on Fire and Forest Meteorology, vol. 22. American Meteorological Society, pp. 90–93.

Sellers, P.J., Berry, J.A., Collatz, G.J., Field, C.B., Hall, F.G., 1992. Canopy reflectance, photosynthesis, and transpiration. III. A reanalysis using improved leaf models and a new canopy integration scheme. Remote Sens. Environ. 42, 187–216.

Sellers, P.J., Bounoua, L., Collatz, G.J., Randall, D.A., Dazlich, D.A., Los, S.O., Berry, J.A., Fung, I., Tucker, C.J., Field, C.B., Jensen, T.G., 1996a. A revised land surface parameterization (SiB2) for atmospheric GCMs. Part I: model formulation. J. Clim. 9 (4), 676−705.

Sellers, P.J., Randall, D.A., Collatz, G.J., Berry, J.A., Field, C.B., Dazlich, D.A., Zhang, C., Collelo, G.D., Bounoua, L., 1996b. Comparison of radiative and physiological effects of doubled atmospheric CO_2 on climate. Science 271 (5254), 1402−1406.

Sims, D.A., Rahman, A.F., Cordova, V.D., Baldocchi, D.D., Flanagan, L.B., Goldstein, A.H., Hollinger, D.Y., Misson, L., Monson, R.K., Schmid, H.P., Wofsy, S.C., Xu, L.K., 2005. Midday values of gross CO_2 flux and light use efficiency during satellite overpasses can be used to directly estimate eight-day mean flux. Agric. For. Meteorol. 131, 1−12.

Smith, W.K., Biederman, J.A., Scott, R.L., Moore, D.J.P., He, M., Kimball, J.S., Yan, D., Hudson, A., Barnes, M.L., MacBean, N., Fox, A.M., Litvak, M.E., 2018. Chlorophyll fluorescence better captures seasonal and interannual gross primary productivity dynamics across dryland ecosystems of southwestern North America. Geophys. Res. Lett. 45 (2), 748−757.

Suleiman, A., Crago, R., 2004. Hourly and daytime evapotranspiration from grassland using radiometric surface temperatures. Agron. J. 96, 384−390.

Sun, Y., Frankenberg, C., Wood, J.D., Schimel, D.S., Jung, M., Guanter, L., Drewry, D.T., Verma, M., Porcar-Castell, A., Griffis, T.J., Gu, L., Magney, T.S., Köhler, P., Evans, B., Yuen, K., 2017. OCO-2 advances photosynthesis observation from space via solar-induced chlorophyll fluorescence. Science 358.

Sun, Y., Frankenberg, C., Jung, M., Joiner, J., Guanter, L., Köhler, P., Magney, T., 2018. Overview of solar-induced chlorophyll fluorescence (SIF) from the orbiting carbon observatory-2: retrieval, cross-mission comparison, and global monitoring for GPP. Remote Sens. Environ. 209, 808−823.

Tang, S., Chen, J.M., Zhu, Q., Li, X., Chen, M., Sun, R., Zhou, Y., Deng, F., Xie, D., 2007. LAI inversion algorithm based on directional reflectance kernels. J. Environ. Manag. 85 (3), 638−648.

Tsubo, M., Walker, S., 2005. Relationships between photosynthetically active radiation and clearness index at Bloemfontein, South Africa. Theor. Appl. Climatol. 80 (1), 17−25.

Turner, D.P., Ritts, W.D., Styles, J.M., Yang, Z., Cohen, W.B., Law, B.E., Thornton, P.E., 2006. A diagnostic carbon flux model to monitor the effects of disturbance and interannual variation in climate on regional NEP. Tellus B Chem. Phys. Meteorol. 58 (5), 476−490.

Urban, O., Janouš, D., Acosta, M., Czerný, R., Markovâ, I., 2007. Ecophysiological controls over the net ecosystem exchange of mountain spruce stand. Comparsion of the response in direct vs. diffuse solar radiation. Glob. Chang. Biol. 13, 157−168.

Van Tuyl, S., Law, B.E., Turner, D.P., Gitelman, A.I., 2005. Variability in net primary production and carbon storage in biomass across Oregon forests—an assessment integrating data from forest inventories, intensive sites, and remote sensing. For. Ecol. Manag. 209 (3), 273−291.

Veroustraete, F., 1994. On the use of a simple deciduous forest model for the interpretation of climate change effects at the level of carbon dynamics. Ecol. Model. 75, 221−237.

Veroustraete, F., Sabbe, H., Eerens, H., 2002. Estimation of carbon mass fluxes over Europe using the C-Fix model and Euroflux data. Remote Sens. Environ. 83, 376−399.

Verstraeten, W.W., Veroustraete, F., Feyen, J., 2006. On temperature and water limitation of net ecosystem productivity: implementation in the C-Fix model. Ecol. Model. 199 (1), 4−22.

Vitousek, P.M., Mooney, H.A., Lubchenco, J., Melillo, J.M., 1997. Human domination of earth's ecosystems. Science 277, 494−499.

Wang, K.Y., 1996. Canopy CO_2 exchange of Scots pine and its seasonal variation after four-year exposure to elevated CO_2 and temperature. Agric. For. Meteorol. 82 (1−4), 1−27.

Wang, Y.P., Leuning, R., 1998. A two-leaf model for canopy conductance, photosynthesis and partitioning of available energy I: model description and comparison with a multilayered model. Agric. For. Meteorol. 91, 89−111.

Weiss, A., Norman, J.M., 1985. Partitioning solar radiation into direct and diffuse, visible and near-infrared components. Agric. For. Meteorol. 34 (2−3), 205−213.

Whitehead, D., Gower, S.T., 2001. Photosynthesis and light-use efficiency by plants in a Canadian boreal forest ecosystem. Tree Physiol. 21, 925−929.

Whittaker, R.H., Likens, G.E., 1975. The biosphere and man. In: Lieth, H., Whittaker, R. (Eds.), The Primary Productivity of the Biosphere. Springer Verlag, New York, pp. 305−328.

Xiao, X.M., Zhang, Q.Y., Braswell, B., Urbanski, S., Boles, S., Wofsy, S., Moore, B., Ojima, D., 2004. Modeling gross primary production of temperate deciduous broadleaf forest using satellite images and climate data. Remote Sens. Environ. 91, 256−270.

Xiao, X., Zhang, Q., Hollinger, D., Aber, J., Moore, B., 2005. Modeling seasonal dynamics of gross primary production of evergreen needleleaf forest using MODIS images and climate data. Ecol. Appl. 15 (3), 954−969.

Xiao, Z., Liang, S., Wang, J., Xiang, Y., Zhao, X., Song, J., 2016. Long-time-series global land surface satellite leaf area index product derived from MODIS and AVHRR surface reflectance. IEEE Trans. Geosci. Remote Sens. 54 (9), 5301−5318.

Yuan, W.P., Liu, S.G., Zhou, G.S., Zhou, G.Y., Tieszen, L.L., Baldocchi, D., Bernhofer, C., Gholz, H., Goldstein, A.H., Goulden, M.L., Hollinger, D.Y., Hu, Y.M., Law, B.E., Stoy, P.C., Vesala, T., Wofsy, S., 2007. Deriving a light use efficiency model from eddy covariance flux data for predicting daily gross primary production across biomes. Agric. For. Meteorol. 143 (3–4), 189–207.

Yuan, W., Luo, Y., Richardson, A.D., Oren, R., Luyssaert, S., Janssens, I.A., Ceulemans, R., Zhou, X., Gruenwald, T., Aubinet, M., Berhofer, C., Baldocchi, D.D., Chen, J., Dunn, A.L., Deforest, J.L., Dragoni, D., Goldstein, A.H., Moors, E., Munger, J.W., Monson, R.K., Suyker, A.E., Star, G., Scott, R.L., Tenhunen, J., Verma, S.B., Vesala, T., Wofsy, S.C., 2009. Latitudinal patterns of magnitude and interannual variability in net ecosystem exchange regulated by biological and environmental variables. Glob. Chang. Biol. 15, 2905–2920.

Yuan, W.P., Liu, S.G., Yu, G.R., Bonnefond, J.M., Chen, J.Q., Davis, K., Desai, A.R., Goldstein, A.H., Gianelle, D., Rossi, F., Suyker, A.E., Verma, S.B., 2010. Global estimates of evapotranspiration and gross primary production based on MODIS and global meteorology data. Remote Sens. Environ. 114, 1416–1431.

Yuan, W.P., Liu, S.G., Cai, W.W., Liu, S.G., Dong, W.J., Chen, J.Q., Arain, M.A., Blanken, P.D., Cescatti, A., Wohlfahrt, G., Georgiadis, T., Genesio, L., Gianelle, D., Grelle, A., Kiely, G., Knohl, A., Liu, D., Marek, M.V., Merbold, L., Montagnani, L., Panferov, O., Peltoniemi, M., Rambal, S., Raschi, A., Varlagin, A., Xia, J.Z., 2014a. Vegetation-specific model parameters are not required for estimating gross primary production. Ecol. Model. 292, 1–10.

Yuan, W.P., Liu, S.G., Dong, W.J., Liang, S.L., Zhao, S.Q., Chen, J.M., Xu, W.F., Li, X.L., Barr, A., Black, T.A., Yan, W.D., Goulden, M.L., Kulmala, L., Lindroth, A., Margolis, H.A., Matsuura, Y., Moors, E., van der Molen, M., Ohta, T., Pilegaard, K., Varlagin, A., Vesala, T., 2014b. Differentiating moss from higher plants is critical in studying the carbon cycle of the boreal biome. Nat. Commun. 5, 4270. https://doi.org/10.1038/ncomms5270.

Zhang, Y.Q., Liu, C.M., Yu, Q., Shen, Y.J., Kendy, E., Kondoh, A., Tang, C.Y., Sun, H.Y., 2004. Energy fluxes and the Priestley-Taylor parameter over winter wheat and maize in the North China plain. Hydrol. Process. 18, 2235–2246.

Zhang, L., Wylie, B., Loveland, T., Fosnight, E., Tieszen, L.L., Ji, L., Gilmanov, T., 2007. Evaluation and comparison of gross primary production estimates for the Northern Great Plains grasslands. Remote Sens. Environ. 106, 173–189.

Zhao, M., Running, S.W., 2010. Drought-induced reduction in global terrestrial net primary production from 2000 through 2009. Science 329 (5994), 940–943.

Zhuang, Q.L., Melillo, J.M., Sarofim, M.C., Kicklighter, D.W., McGuire, A.D., Felzer, B.S., Sokolov, A.P., Ronald, G., Steudler, P.A., Hu, S.M., 2006. CO_2 and CH_4 exchanges between land ecosystems and the atmosphere in northern high latitudes over the 21st Century. Geophys. Res. Lett. 33, L17403. https://doi.org/10.1029/2006GL026972.

Abstract

Precipitation is a fundamental component of the global water cycle and a key hydrologic variable of the water cycle for meteorology, climatology, and hydrology. Accurate observations of precipitation and its regional and global distributions have long been scientific challenges. The satellite retrieval of precipitation has matured over four decades of development. This chapter briefly introduces surface measurement techniques, the principles of precipitation retrieval from spaceborne sensors, major satellite-based precipitation datasets, and global precipitation climatology. Section 16.2 first introduces surface measurement techniques, including gauge- and radar-based networks. Section 16.3 describes the principles of precipitation retrieval from spaceborne sensors. Section 16.4 introduces major satellite-based precipitation datasets for regional and global applications. Section 16.5 discusses the climatology of global precipitation and associated large climatic events. Section 16.6 summarizes and provides perspectives for the further development of precipitation retrieval in the near future.

16.1 Introduction

Precipitation is a fundamental component of the global water cycle and a key hydrologic variable of the water cycle for meteorology, climatology, and hydrology. Accurate observations of precipitation and its regional and global distributions have long been scientific challenges. The satellite retrieval of precipitation has matured over five decades of development.

© 2020 Elsevier Inc. All rights reserved.

This chapter briefly introduces surface measurement techniques, the principles of precipitation retrieval from spaceborne sensors, major satellite-based precipitation datasets, and global precipitation climatology.

In a physical sense, precipitation is defined as "all liquid or solid phase aqueous particles that originate in the atmosphere and fall to the Earth's surface" (Michaelides et al., 2009). It is a fundamental component of the global water cycle. Precipitation is moisture fluxes from the atmosphere to the earth's surface, and it is a key hydrologic variable of the water cycle for meteorology, climatology, and hydrology. The atmosphere obtains approximately three-quarters of its heat energy from the release of latent heat by precipitation (Kummerow and Barnes, 1998). Precipitation plays an important role in the climate system by connecting the atmosphere and land surface processes.

Rain and snow are the principal forms of precipitation. Unlike other meteorological parameters, rainfall has a high degree of spatiotemporal variability and highly nonnormal behavior. Among atmospheric variables, it is the most difficult to measure. The space—time structure of precipitation produces top-down effects on dynamic terrestrial hydrologic processes such as runoff generation and soil moisture evolution. Thus, accurate observations of precipitation and its regional and global distributions are quite challenging.

This chapter first introduces surface measurement techniques, including gauge- and radar-based networks. Section 16.2 describes the principles of precipitation retrieval from space-borne sensors. Section 16.3 introduces major satellite-based precipitation datasets for regional and global applications. Section 16.4 discusses the climatology of global precipitation and associated large climatic events. Section 16.5 presents conclusions and provides perspectives for the further development of precipitation retrieval in the near future.

16.2 Surface measurement techniques

Rainfall is the most common form of precipitation. Surface rainfall can be measured with a ground instrument. Typically, a rain gauge measures the integrated volume of an ensemble of raindrops, whereas a disdrometer counts and measures drop size at a specific location. Ground-based meteorological radar was invented to monitor precipitation with a high spatiotemporal resolution (Michaelides et al., 2009).

16.2.1 Rain gauge network

Rain gauges are conventionally used to measure rainfall or snowfall in situ. More than 40 rain gauge designs are used throughout the world (Linacre, 1992). The size of the canister opening and the height of the gauge above the ground have yet to be globally standardized (Shelton, 2009). The World Meteorological Organization (WMO) adopted the British design as the standard gauge, which has a cylindrical diameter of 127 mm and a rim located 1 m above the ground (Linacre, 1992). In the United States, the standard rain gauge has a 203-mm diameter opening located 800 mm above the ground.

Existing rain gauges tend to underestimate actual precipitation, with a bias of 9% on average (Groisman and Legates, 1994; Duchon and Essenberg, 2001; Shelton, 2009). The major cause of this inaccuracy is the turbulence over the canister mouth. Other relevant problems include wetting of the internal walls of the gauge, evaporation, splashing into or out of the gauge, and blowing snow (Shelton, 2009).

Sufficiently dense rain gauge networks are often required to characterize spatiotemporal variations in precipitation (Liu, 2003). Most countries have their own gauge networks. WMO now coordinates the Global Telecommunications Network, which distributes rainfall information from participating gauges and forms the largest rain gauge network in the world.

Many factors, including accessibility, ease of maintenance, and topographical aspects, are considered when determining the location of a rain gauge. At the network level, the accuracy of an observation depends on the total number of gauges and their locations. The required minimum density of a rain gauge network is determined by the temporal resolution of the desired precipitation measurements, study objective, geographic configuration of the study region, and economic considerations. The WMO established a minimum density standard for constructing precipitation gauge networks (Table 16.1).

Rodda (1969) summarized the problems and solutions of network design and proposed a useful distinction among three levels of networks. Level I networks acquire information for national planning purposes related to gross estimates of water resources, surveillance of major storms, and national databases. Level II networks supplement level I networks in particular basins or regions and provide extra information for local planning efforts. Level III networks gather operational information for local water management. The networks designed for each level can differ, and the acquired knowledge can be used to classify precipitation networks. In practice, the entire network of rain gauges may contain components from all three levels.

16.2.2 Ground-based radar

Rain gauges measure point precipitation, while meteorological radar systems take snapshots of the drop size distribution (DSD)

of precipitation above the surface of a specified volume (Michaelides et al., 2009). Ground-based radar systems usually have a 1- to 2-km spatial resolution and temporal revisit times of 15–30 min (Shelton, 2009). Single observing points can provide coverage for large areas, with a high spatiotemporal resolution. This feature makes them an attractive alternative to precipitation measurements. Furthermore, composite data from several radar systems extend the observation area, offering spatial and temporal coverages that are better than those obtained with rain gauge networks.

The principal physical basis of precipitation radar (PR) systems is the relationship between range-corrected, backscattered return power and the physical properties of targets in the radar beam (Shelton, 2009) (Fig. 16.1). The radar equation for spherical drops can be expressed as

$$\overline{P_r} = c_r \frac{|K|^2 Z}{r_r^2} \tag{16.1}$$

where $\overline{P_r}$ is the average backscattered power returning to the receiving antenna over several pulses in Watts; r_r is the distance from the target to the radar in Km; $|K|^2$ is the complex index of refraction, representing the physical characteristics of the precipitation particles within the radar resolution volume, with values of 0.93 for water and 0.197 for ice; and c_r is the radar constant, which groups numerical constants and radar

TABLE 16.1 The minimum density for precipitation gauge networks.

Region type	Range of norms for minimum network (km²/gauge)	Range of provisional norms in difficult situations (km²/gauge)
I	600–900	900–3000
II$_a$	100–250	250–1000
II$_b$	25	
III	1500–10,000	

I, Flat regions in temperate, Mediterranean, and tropical zones; *II$_a$*, Mountainous regions in temperate, Mediterranean, and tropical zones; *II$_b$*, Small mountain/islands with irregular precipitation patterns, requiring dense hydrographic networks; *III*, Arid and polar zones.

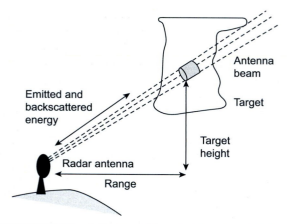

FIGURE 16.1 A diagram of the principles of precipitation radar. *Courtesy of Shelton, M. L., 2008. Hydroclimatology. Cambridge: Cambridge University Press.*

hardware parameters. This last parameter can be expressed as

$$c_r = \frac{P_t G^2 \lambda^2 \theta_H \theta_v c \tau_p \pi^3}{1024 \ln 2 \lambda^2} \qquad (16.2)$$

where P_t is the pulse power of the radar transmitter in Watts; G is the dimensionless antenna gain; λ is the wavelength of the transmitted wave in m; θ_H is the horizontal bandwidth in radians; θ_v is the vertical bandwidth in radians; c is the speed of light $(3 \times 10^8$ m/s); and τ_p is the pulse duration in seconds. Z is the radar reflectivity factor, which represents the intensity of the return signal in relation to the density and size of the targets. It is expressed as

$$Z = \sum_{vol} D_i^6 = \sum n_i D_i^6 = \int_0^\infty N(D) D^6 dD$$

$$(16.3)$$

where D is the drop diameter in mm, and $N(D)$ dD is the number of particles per unit volume in the diameter range from D to $D + dD$.

Both the reflectivity and the rainfall rates are functions of the raindrop size distribution (Stout and Mueller, 1968). Extensive experimental results have shown that

$$Z = aR^b \qquad (16.4)$$

where Z is the radar reflectivity in mm^6/m^3 or log-scale decibel units (dB); R is the rainfall rate in mm/h; and a and b are coefficients, with a ranging from 70 to 500 and b ranging from 1.0 to 2.0. The coefficients are often empirically determined or calibrated using rain gauge data. The $Z-R$ relationship is the principal basis of many retrieval algorithms, but both Z and R are subject to uncertainties. The relationship is scale-dependent, and variations in the relationship depend on the DSD of the precipitation, radar clutter produced by ground echoes, the presence of an enhanced "bright band" related to a melting snow layer, and radar signal attenuation due to heavy rainfall (Morin et al., 2003; Shelton, 2009; Villarini and Krajewski, 2010).

PR systems typically use longer wavelengths of 4–15 cm (e.g., S-band and C-band), and algorithms that have been developed and tested for low-frequency, high-power systems. In recent years, high-frequency, low-power (e.g., X-band) systems have received an increasing amount of attention. Several methods have been developed to correct for rain-path attenuation and estimate DSD parameters for X-band systems (Michaelides et al., 2009).

Radar networks have been established in several countries. The Next-Generation Radar system, which includes 159 high-resolution Doppler weather radar systems, was deployed in the United States (Brown and Lewis, 2005). In Japan, the Automated Meteorological Data Acquisition System (AMeDAS), which includes 20 ground-based radars, was developed by the Japan Meteorological Agency (Makihara et al., 1996). The Canadian weather radar network, which consists of 31 weather radars, spans Canada's most populated regions.

To date, both rain gauge and radar data have been widely applied to a variety of fields, including short-term weather and flood forecasting and hydrologic modeling (Yuter, 2003; Neary et al., 2004). These data are also used for calibration and validation purposes when developing reliable precipitation retrieval algorithms in satellite remote sensing applications (Krajewski, 1987).

16.3 Estimation from satellite data

Surface-based observations of precipitation are mostly performed with rain gauges and, if available, ground radar. However, observations are often impossible over oceans. Furthermore, for some remote and mountainous areas, observations are scarce and generally not uniform.

Satellites can be used to compensate for the limitations of surface-based observations. Since the launch of the first meteorological satellite, the Television and Infrared Observation Satellite (TIROS-1), in April 1960, a plethora of satellite data have been used to extract meteorological information (Kidd, 2001). Passive sensors based on visible (VIS), infrared (IR), and passive microwave (PMW) wavelengths onboard geostationary (GEO) or low earth-orbiting (LEO) satellites have been primarily used to retrieve

precipitation information for approximately 50 years. Launched in 1997, the Tropical Rainfall Measuring Mission (TRMM) was the first spaceborne PR system (Kummerow and Barnes, 1998). On February 28, 2014, the Global Precipitation Measurement (GPM) was launched into space, which is a heritage of the TRMM (Hou et al., 2014). Over 60 empirical and physical retrieval algorithms have been developed for the launched sensors (Ebert and Manton, 1998; Kubota et al., 2009). Levizzani et al. (2007) provided an overview of the state-of-the-art algorithms, sensors, and techniques used in precipitation retrievals from space. This section briefly describes the main types of retrieval algorithms, which use VIS/IR, PMW, active microwave sensors, and various combinations of these technologies (Table 16.2).

16.3.1 VIS/IR algorithms

The primary idea behind the use of VIS/IR data to infer precipitation information is that cold and bright clouds are related to convection. Convective clouds are likely to produce rain. Satellite observations of cloud tops can be used indirectly to derive ground rainfall amounts. More specifically, IR cloud-top temperatures are related to the rainfall probability and intensity on the ground. The most widely used

TABLE 16.2 Classification of satellite-based retrieval of precipitation.

Sensors	Typical algorithms	References
VIS/IR	GPI; GMSRA; OPI; Griffith–Woodley algorithm	Arkin and Meisner (1987); Ba and Gruber (2001); Griffith et al. (1978); Xie and Arkin (1997)
PMW	The Wilheit algorithm; the SSM/I algorithm; GPROF	Wilheit et al. (1977); Ferraro (1997); Kummerow et al. (2001);
Active microwave	The TRMM standard PR algorithm	Iguchi et al. (2000)
Multisensor	CMORPH; TMPA; GSMaP	Joyce et al. (2004); Huffman et al. (2007); Okamoto et al. (2005)

algorithm is the Geostationary Operational Environmental Satellites (GOES) precipitation index (GPI) (Arkin and Meisner, 1987), which can be expressed in millimeters as

$$GPI = r_c F_c t \qquad (16.5)$$

where r_c is a constant (3 mm/h) and t is the length of the period in hours for which F_c is the mean fractional cloudiness, a dimensionless number ranging from 0 to 1. F_c is obtained from the IR pixels associated with brightness temperatures <235K over a spatial domain more than 50×50 km. GPI is simple, but its application is limited to 40°N–40°S, where the major rain systems are convective in nature (Kidd, 2001). Validations showed that while GPI estimates are highly correlated ($R^2 > 0.7$) with gauge data in the Americas, the eastern and central Pacific, and the western Atlantic, biases can be as large as 50%–150% per month (Arkin and Meisner, 1987).

Ba and Gruber (2001) proposed the GOES Multispectral Rainfall Algorithm (GMSRA); this algorithm uses all five VIS/IR bands from the GOES satellite to extract the cloud and precipitation extents and then derives rainfall with corrections to the cloud and precipitation regimes. Validation with gauge-adjusted radar data has demonstrated that the performance of GMSRA is better than that of GPI. The daily estimates have positive biases to the radar data on the order of several millimeters on the global scale. The other major VIS/IR algorithms include the Griffith–Woodley algorithm (Griffith et al., 1978) and the Outgoing Long-wave Radiation (OLR) precipitation index (OPI) (Xie and Arkin, 1997). Ebert and Manton (1998) used the Geostationary Meteorological Satellite (GMS) network and ground radar data to compare retrieval algorithms for rainfall in the equatorial western Pacific Ocean. Among the 16 VIS/IR algorithms, large variations in the magnitude of the retrieved rainfall were observed, while the patterns were similar.

Despite their limited accuracy, these algorithms have been applied to a range of scales and fields including operational meteorology. These algorithms have benefited from the availability of the GEO VIS/IR observations, and they provide an adequate level of temporal precipitation resolution, with high variability in space, time, and intensity.

16.3.2 Passive microwave algorithms

PMW algorithms are more physically direct than the VIS/IR algorithms. A simplified form of microwave radiative transfer can be expressed as (Wilheit et al., 1999)

$$\frac{dR(\theta, \phi)}{ds} = A + S \qquad (16.6)$$

where $R(\theta, \phi)$ is the radiance in the direction specified by the polar angles, θ and ϕ; s is distance in the (θ, ϕ) direction; A denotes the absorption and concomitant emission; and S represents the loss and gain of radiance due to scattering. At PMW wavelengths, precipitation particles are the main source of attenuation of the upwelling radiation. The radiation emitted from atmospheric particles increases the signal received at the sensor. Conversely, scattering induced by the hydrometeors decreases the radiation stream.

Over the ocean, because emissivity at PMW has a narrow range (0.4–0.5), the background radiometric signal is low and constant. Thus, using low-frequency bands (<20 GHz), additional emissions from precipitation can be used to identify and quantify precipitation. Over land, the surface emissivity can be as high as 0.9, and emissions from hydrometeors cannot be measured accurately. Alternatively, scattering caused by ice particles decreases received radiation at high frequencies (>35 GHz). Early research confirmed that PMW radiometry could be used to estimate surface precipitation (e.g., Savage and Weinman, 1975). Alishouse

(1983) found that 18, 19.35, and 37 GHz were particularly useful to extract atmospheric liquid water content, precipitable water, and rainfall rate.

Based on these principles, many approaches have been proposed to retrieve precipitation. The first retrieval algorithm made simple regression steps between surface rainfall rates and the simulated or measured brightness temperatures (Wilheit et al., 1977). Based on a microwave radiative transfer theory, a model is developed between microwave brightness temperatures and rainfall rates. The simulations were compared with the brightness temperature at a wavelength of 1.55 cm for the Electrically Scanning Microwave Radiometer on the Nimbus 5 satellite and the rain rates derived from WSR-57 meteorological radar measurements. Ground-based validations showed that the estimates were accurate within a factor of 2 for the range of 1−25 mm/h over ocean areas.

Ferraro (1997) developed a statistical–physical algorithm that is most widely used for operational uses. The algorithm includes two retrieval techniques: one is based primarily on 85 GHz scattering techniques (Algorithm-85) with the other on a 37 GHz scattering (Algorithm-37) over land and emission only over ocean. Specifically, the Algorithm-37 is described as follows over land

$$SI_{37} = 62.18 + 0.773TB_{19v} - TB_{37v} \quad (16.7a)$$

$$R = 1.3 + 1.46SI_{37} \quad (16.7b)$$

and over ocean

$$Q_{19} = -2.70[\ln(290 - TB_{19v}) - 2.84 \\ - 0.40\ln(290 - TB_{22v})] \quad (16.7c)$$

$$Q_{37} = -1.15[\ln(290 - TB_{37v}) - 2.99 \\ - 0.32\ln(290 - TB_{22v})] \quad (16.7d)$$

$$R = 0.001707(100Q)^{1.7359} \quad (16.7e)$$

where TB_{19v} and TB_{37v} are brightness temperature (K) at band 19 and 37 GHz, respectively;

SI_{37} is the scattering portion of the algorithm; and RR is the rainfall rate (mm/hr). The rainfall rates are calculated from Eq. (16.7b) when $SI_{37} > 5K$ over land. TB_{22v} is the brightness temperature at band 22 GHz (K). Q_{19} and Q_{37} are the liquid water estimates from the 19 and 37 GHz bands, respectively. The rainfall rates are calculated from Eq. (16.7e) when $Q_{19} > 0.60$ mm or $Q_{37} > 0.20$. Alternatively, the Algorithm-85 is described as follows over land

$$SI_L = [451.9 - 0.44TB_{19v} - 1.775TB_{22v} \\ + 0.00575TB_{22v}^2] - TB_{85v} \quad (16.8a)$$

$$R = 0.00513SI_L^{1.9468} \quad (16.8b)$$

and over ocean

$$SI_W = [-174.4 + 0.72TB_{19V} + 2.439TB_{22V} \\ - 0.00504TB_{22V}^2] - TB_{85V} \quad (16.8c)$$

$$R = 0.00188SI_W^{2.0343} \quad (16.8d)$$

where SI_L is the scattering portion of the algorithm over land from the brightness temperature data at bands TB_{19v}, TB_{22v}, and TB_{85v}. The rainfall rates are estimated from Eq. (16.8b) when $SI_L > 10K$ for rainfall between 0.45−35 mm/h. SI_W represents the case over ocean. The rainfall rates are calculated from Eq. (16.8d) when $SI_W > 10K$ for rainfall between 0.20−35 mm/h. Ground evaluation showed that the algorithm using the Special Sensor Microwave Imager (SSM/I) data had a retrieval error of 50% over ocean and 75% over tropical and midlatitude land (Ferraro, 1997).

More complex retrieval algorithms are based on probabilistic theory (Smith et al., 1992; Mugnai et al., 1993). The most widely applied algorithm is the Goddard Profiling Algorithm (GPROF) (Kummerow et al., 2001). This algorithm requires a database of rainfall profiles and the associated brightness temperatures. Given a vertical distribution of hydrometeors represented by **R**, the radiative transfer equation can be used to determine a brightness temperature vector, **Tb**. A Bayesian inversion procedure is performed to find **R** for a given **Tb**. The probability of a particular profile **R** for a given **Tb** is

$$\Pr\left(\mathbf{R}|\mathbf{Tb}\right) = \Pr(\mathbf{R}) \times \Pr\left(\mathbf{Tb}|\mathbf{R}\right) \qquad (16.9)$$

where $\Pr(\mathbf{R})$ is the probability that a certain **R** will be observed, and $\Pr(\mathbf{Tb}|\mathbf{R})$ is the probability of observing **Tb** given a particular rain profile **R**. The first term on the right side of the above equation can be derived using a cloud-resolving model. The strength of the algorithm relies primarily on the accuracy of the profiles in the dataset. Validations with rain gauge and ground radar data indicated that GPROF was biased negatively (positively) by 9% (17%), with a correlation of 0.86 (0.8) for monthly averages over the ocean (land) (Kummerow et al., 2001).

Most existing PMW retrieval algorithms have been optimized for specific satellite sensors. Intercomparisons have revealed that each algorithm has its own strengths and weaknesses and that no algorithm is universally superior. Kummerow et al. (2007) suggested the need for a transparent, parametric algorithm to ensure uniform rainfall products across sensors.

The PMW algorithms are limited to observations currently available from the LEO satellites, typically two per day per satellite. The spatial resolution of the low-frequency bands used over the ocean is 50 × 50 km, and the resolution of the high-frequency bands used over land is at most 10 × 10 km.

16.3.3 Active microwave algorithms

The use of active microwave (radar) observations from space for precipitation retrieval started in November 1997 with the launch of the TRMM (Kummerow and Barnes, 1998), which was the first spaceborne PR. It transmits energy at 13.8 GHz and measures the power reflected by the precipitation particles and earth's surface. It enables three-dimensional precipitation structures over the ocean and land to be captured (Iguchi et al., 2000).

The standard PR algorithm estimates the vertical profiles of attenuation-corrected radar reflectivity values and precipitation rates. The measured apparent radar reflectivity factor $Zm(r)$ is related to the true effective radar reflectivity factor $Ze(r)$ by the following equations (Iguchi et al., 2000; Iguchi, 2007):

$$Z_m(r) = Z_{mt}(r) + \delta_{Z_m}(r) \qquad (16.10)$$

where r is the distance from the radar to the target and $\delta_{Z_m}(r)$ is the measurement error in $Z_m(r)$. $Z_{mt}(r)$ can be expressed as

$$Z_{mt}(r) = Z_e(r)A(r) \qquad (16.11)$$

$$A(r) = e^{0.1\ln(10)\mathrm{PIA}(r)} \qquad (16.12)$$

where $A(r)$ is the attenuation factor. $\mathrm{PIA}(r)$ is the two-way path-integrated attenuation (PIA) in dB. Several factors affect the PIA, which can be described by

$$\mathrm{PIA}(r) = 2\int_0^r (k_p(s) + k_{\mathrm{CLW}}(s) + k_{\mathrm{wv}}(s) + k_{o_2}(s))ds$$

$$(16.13)$$

The two-way attenuation is accounted for by the factor 2 on the right side of this equation. Here, k_P, k_{CLW}, k_{WV}, and k_{o_2} represent the attenuations due to precipitation, cloud liquid water (CLW), water vapor (WV), and molecular oxygen, respectively.

The major attenuation comes from the precipitation particles and is the key to the algorithm. The retrieval algorithm of precipitation profiles performs classifications in two steps. First, Z_e is estimated from the measured vertical profiles Z_m, and this estimation corresponds to the attenuation correction. Next, the estimated Z_e is converted into the precipitation rate (R) using a power law.

Comparisons have demonstrated that the PR retrieval algorithms are accurate on the order of the ground-based rainfall data. Thus, they have been widely used as true values against which other products are compared and evaluated (e.g., Aonashi et al., 2009). However, the attenuation correction and precipitation estimates are affected by uncertainties in the parameters (Iguchi, 2009). The major parameters include the DSD, the phase sate of precipitation particles, their density and shape, the precipitation distribution within the PR footprint, attenuation due to CLW and WV, the freezing height, the surface scattering cross section, and fluctuations of the radar echo signal. The major factors are the phase state of the hydrometeors, the DSD, and the precipitation temperature, followed by the heterogeneity of the precipitation distribution within the radar resolution cell (Iguchi, 2009).

16.3.4 Multisensor algorithms

Comparative studies have shown that the PMW algorithms are more accurate than the VIS/IR algorithms when retrieving instantaneous precipitation data (Kubota et al., 2009). However, the VIS/IR algorithms provided better long-term retrieval due to better temporal sampling of GEO data (e.g., Bauer and Schanz, 1998; Ebert and Manton, 1998). The combined use of multiple sensors may alleviate the deficiencies of single-sensor algorithms. In particular, the combination of active and passive sensors could substantially improve cloud and precipitation retrieval (Michaelides et al., 2009). In the last two decades, many algorithms have been proposed with different combinations of VIS/IR, PMW, and PR data. Here, we discuss three common multisensor algorithms.

Joyce et al. (2004) developed the Climate Prediction Center Morphing (CMORPH) algorithm, which uses motion vectors derived from 30-min interval GEO IR satellite imagery to propagate the relatively high-quality precipitation retrievals from PMW data. IR data are taken from the GOES-8 and -10, the Meteosat (meteorological satellite) -5 and -7, and the GMS-5. PMW data include the TRMM Microwave Imager (TMI), the SSM/I series, the Advanced Microwave Sounding Unit-B (AMSU-B) series, and the Advanced Microwave Scanning Radiometer—Earth Observing System (AMSR-E). The shape and intensity of precipitation are morphed during the time between PMW sensor scans using a linear interpolation. The process yields spatially and temporally complete PMW retrievals, independent of the IR brightness temperature. Kubota et al. (2009) showed that CMORPH achieved better results than other multisensor algorithms in areas around Japan.

Huffman et al. (2007) proposed the TRMM Multisatellite Precipitation Analysis (TMPA) algorithm. It provides a calibration-based sequential scheme for combining precipitation estimates from multiple sensors and gauge analyses at fine scales ($0.25° \times 0.25°$ and every 3 h). The involved sensors include the TRMM TMI, AMSR-E, SSM/I, AMSU, Microwave Humidity Sounders (MHS), and GEO-IR. In the algorithm, the microwave precipitation estimates are calibrated and combined. Next, IR estimates are generated from the calibrated microwave estimates. The two estimates are combined subsequently. Finally, the rain gauge data are incorporated. Overall, the TMPA shows reasonable performance at monthly scales. At finer scales, it reproduces the surface observation—based histogram of precipitation.

Okamoto et al. (2005) developed the Global Satellite Mapping of Precipitation (GSMaP), which uses various attributes of precipitation derived from TRMM data to retrieve hydrometeor profiles from PR, statistical rain/no-rain classification, and scattering algorithms. With polarization-corrected temperatures (PCTs) at 37 and 85.5 GHz, the combined scattering algorithms estimate surface rainfalls. For stronger rainfalls, PCT37 is used, and for weaker rainfalls, PCT85 is used. Monthly surface rainfall amounts are retrieved from six microwave radiometers

(MWRs). GSMaP_MWR is combined with TMI, AMSR-E, AMSR, and SSM/I (Kubota et al., 2007). The combined GSMaP moving vector and Kalman filter approach was proposed to combine PWM and IR data for producing global precipitation maps with high spatial and temporal resolutions (Aonashi et al., 2009; Ushio et al., 2009).

Huffman et al. (2015) proposed the Integrated Multisatellite Retrievals for GPM (IMERG) algorithm. The algorithm combined the precipitation estimates from the various precipitation-relevant satellite PMW sensors and IR rainfall estimates calibrated by MW, then gridded, and intercalibrated to obtain half-hourly $0.1° \times 0.1°$ data. MW sensors include dual-frequency precipitation radar (DPR), PR, the GPM Microwave Imager (GMI), TMI, SSMI/SSMIS (DMSP), AMSR-E (Aqua), AMSU (NOAA), and MHS. IR sensors mainly include GMS, GOES, and Meteosat. GPM-IMERG combined other algorithms of Multisensor Precipitation Estimation. As CMORPH algorithm, PMW data are morphed using the GEO-IR—based cloud motion vectors. Then GEO-IR estimates are used to fill gaps using a Kalman filter when the PMW are too sparse. Following the PERSIANN-CCS (Hong, 2004), at the 60°N—S latitude belt, the IR Tb image is segmented into separable cloud patches using a watershed algorithm, and then cloud patches are classified into a number of groups using an unsupervised clustering analysis. Precipitation is assigned to each cloud patch group based on a training set of PMW precipitation samples. These initial precipitation estimates are then adjusted using coefficients between matched PMW precipitation and cloud-patch precipitation. Finally, at high latitudes, satellite-sounding-based and experimental sounding channel—based algorithms and numerical model—based estimates have the potential to add value. Finally, precipitation gauge analyses are used to adjust the satellite estimates.

In 2007, the Program to Evaluate High-Resolution Precipitation Products was convened at the WMO headquarters (Turk et al., 2008). Sapiano and Arkin (2009) intercompared five high-resolution products including those retrieved from the CMORPH, the TMPA, the Naval Research Laboratory (NRL) blended technique, the National Environmental Satellite, the Data and Information Service Hydro-Estimator, and the Precipitation Estimation from Remotely Sensed Information using Artificial Neural Networks (PERSIANN) (Hong, 2004; Nguyen et al., 2018). The results demonstrate that all the products effectively represented high-resolution precipitation events and adequately reproduced the diurnal cycle of precipitation. CMRPH and TMPA performed best among the multisensor algorithms, with correlations as high as 0.7. Kubota et al. (2009) compared GSMaP, TMPA, CMORPH, PERSIANN, NRL-blended algorithms around Japan with reference to the radar-AMeDAS dataset. Validations showed that GSMaP and CMORPH performed better than the other algorithms. Overall, the estimates were poor for light rainfall during the warm season and for heavy rainfall.

16.4 Global and regional datasets

With the development of surface-based and space-based instruments, many regional and global datasets have been produced. The datasets provide invaluable information for studying precipitation processes and mechanisms. Table 16.3 summarizes the advantages and disadvantages of major precipitation measurements. Various approaches have been proposed to overcome the limitations of given measurements, such as combinations of satellite PMW and IR data. Maggioni et al. (2016) and Sun et al. (2018) provided an overview of the global precipitation datasets based on satellite. Here, we briefly introduce TRMM, GSMaP, Global

TABLE 16.3 Advantages and limitations of major precipitation measurements.

Instruments		Advantages and disadvantages
Surface-based	Rain gauge	Point measurement, type-dependent accuracy, no ancillary information on rain type, etc., available, measures rain accumulation (possibly DSD), operational
	Ground radar	Limited volume measurement, measures possibly polarization and Doppler-spectrum, operational, retrieval accuracy depends on rain gauge calibration
Space-based	Satellite VIS/IR	Available from geostationary satellite with good temporal sampling, high spatial resolution, indirect estimates
	Satellite PMW	Wide-swath measurement, good calibration, TB measures sensitive to hydrometeors, channel selection rather flexible, operational
	Satellite radar	Profile/narrow swath measurement, rather good calibration, measures Z, experimental

Based on Prigent, C., 2010. Precipitation retrieval from space: an overview. Compt. Rendus Geosci. 342, 380–389. Copyright © 2010 Elsevier Masson SAS. All rights reserved.

Precipitation Climatology Project (GPCP), GPM, and CMORPH datasets, all of which are widely used global datasets dedicated to precipitation monitoring.

16.4.1 Tropical Rainfall Measuring Mission

The TRMM is a joint mission between the National Aeronautics and Space Administration (NASA) of the United States and the National Space Development Agency (NASDA) of Japan. The objective of the TRMM is to observe rainfall and energy exchange in tropical and subtropical regions of the world (Kummerow and Barnes, 1998). Two-thirds of all precipitation falls in the tropics, and tropical rainfall is critical to regulating the global hydrological cycle. TRMM data are valuable for understanding, diagnosing, and predicting global cycling mechanisms and the El Nino and Southern Oscillation (ENSO).

The TRMM was designed as an LEO satellite stationed between 35°N and 35°S. It has produced a wealth of detailed information on tropical rainfall. It has a circular orbit of 350 km and an inclination angle of 35° (Kummerow and Barnes, 1998). The satellite includes five scientific measuring instruments (Fig. 16.2):

the Visible and Infrared Scanner (VIRS), PR, the TMI, the Clouds and Earth's Radiant Energy System, and the Lighting Imaging Sensor. The first three sensors have contributed greatly to precipitation retrieval. Table 16.4 lists the major parameters of each instrument.

PR was the first radar designed specifically for rainfall monitoring from space. It provides three-dimensional rainfall structures to achieve quantitative rainfall measurements over land and ocean and to improve the accuracy of TMI measurements by providing rain structure information. TMI is a multichannel, dual-polarized MWR that provides data related to rainfall rates over the oceans. The TMI data, together with the PR data, were the primary dataset for precipitation measurements. The TMI data combined with the PR and VIRS data were also used to derive precipitation profiles. VIRS is a passive, cross-track scanning radiometer that measures scene radiance in five spectral bands in the VIS through IR spectral regions. Comparative analyses of microwave, VIS, and IR data have enabled more precise precipitation estimates than those possible with only VIS and IR data.

The TRMM Science and Data Information System (TSDIS) is responsible for satellite rain data (http://tsdis.gsfc.nasa.gov/). Fig. 16.3 shows the TRMM satellite product processing

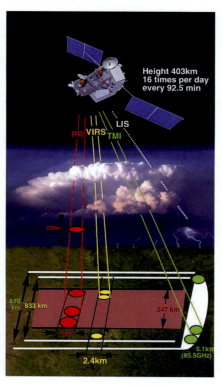

FIGURE 16.2 Configuration of the TRMM. *Source from http://trmm.gsfc.nasa.gov/.*

flow, which includes the retrieval algorithms. V7 algorithms are currently used for processing. The TRMM data were available starting on December 20, 1997. The standard TRMM products can be classified into two groups: satellite and ground validation (GV) products. There are three distinct product levels. The TRMM Level-1 products include VIRS radiance, TMI-calibrated antenna temperature, PR power, and reflectivity. Level-2 products contain geophysical parameters and include TMI rain profiles (2A12), PR surface cross sections (2A21), rain and bright band heights (2A23), TPR rain profiles (2A25), and TRMM Combined Instruments (2B31). There are three TRMM level-2 surface rain rate products (2A12, 2A25, and 2B31). Level-3 and level-4 products are space- and time-averaged products. Table 16.5 details the products.

GV is an essential component of the TRMM. The GV products are primarily used to support the validation of satellite retrieval algorithms. The selected GV radar sites include Darwin, Australia, multiple sites in Florida, Houston, Texas, and the TRMM Large Scale

TABLE 16.4 Major parameters of Visible and Infrared Scanner (VIRS), TRMM Microwave Imager (TMI), and precipitation radar (PR).

Sensor	VIRS	TMI	PR
Frequency/ wavelength	0.63, 1.6, 3.75, 1 0.8, and 12 μm	Dual polarization: 10.65, 19.35, 37, 85.5 GHz; vertical: 21 GHz	Vertical polarization: 13.8 GHz
Scanning mode	Cross track	Conical	Cross track
Ground resolution	2.2 km ([a]2.4 km) at nadir	4.4 km ([a]5.1 km) at 85.5 GHz	4.3 km ([a]5.0 km) at nadir
Swath width	720 km ([a]833 km)	760 km ([a]878 km)	215 km ([a]247 km)
Science applications	Cloud parameters, fire, pollution	Surface rain rate, rain type, distribution, and structure; other atmospheric and oceanic parameters	3D rainfall distribution over both land and oceans and latent heat release into the atmosphere

The TRMM satellite was boosted from an altitude of 350 km to one of 402 km in August 2001.
[a] *Numbers in parentheses represent postboost values.*
Modified from Chiu et al., 2006

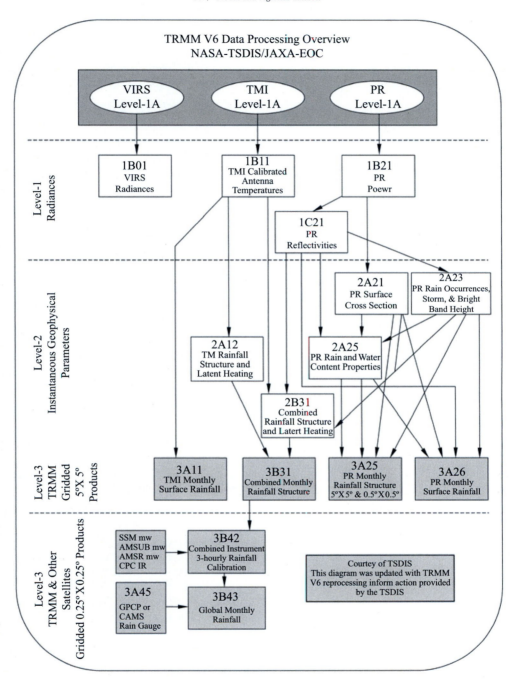

FIGURE 16.3 The TRMM standard satellite product processing flow. *Courtesy of Chiu, L.S., Liu, Z., Rui, H., Teng, W.L., 2006. Tropical rainfall measuring mission data and access tools. In: Qu, J.J., Gao, W., Kafatos, M., Murphy, R.E., Salomonson, V.V. (Eds.), Earth Science Satellite Remote Sensing. vol. 2. Berlin, Springer-Verlag.*

TABLE 16.5   Characteristics of the Tropical Rainfall Measuring Mission (TRMM) standard satellite products.

TSDIS reference	Product name	Product description
1A01	VIRS Raw Data (VIRS)	Reconstructed, unprocessed VIRS (0.63, 1.6, 3.75, 10.8, and 12 μm) data
1A11	TMI Raw Data	Reconstructed, unprocessed TMI (10.65, 19.35, 21, 37, and 85.5 GHz) data
1B01	Visible and Infrared Radiance (VIRS)	Calibrated VIRS (0.63, 1.6, 3.75, 10.8, and 12 μm) radiances at 2.2 km resolution over a 720 km swath
1B11	Microwave Brightness Temperature (TMI)	Calibrated TMI (10.65, 19.35, 21, 37, and 85.5 GHz) brightness temperatures at 5–45 km resolution over a 760 km swath
1B21	Radar Power (PR)	Calibrated PR (13.8 GHz) power at 4 km horizontal and 250 m vertical resolutions over a 220 km swath
1C21	Radar Reflectivity (PR)	Calibrated PR (13.8 GHz) reflectivity at 4 km horizontal and 250 m vertical resolutions over a 220 km swath
2A12	Hydrometeor Profile(TMI)	TMI Hydrometeor (cloud liquid water, precipitation water, cloud ice, precipitation ice) profiles in 14 layers at 5 km horizontal resolution, along with latent heat and surface rain, over a 760 km swath
2A21	Radar Surface Cross-Section (PR)	PR (13.8 GHz) normalized surface cross section at 4 km horizontal resolution and path attenuation (in case of rain), over a 220 km swath
2A23	Radar Rain Characteristics (PR)	Rain type; storm, freezing, and bright band heights; from PR (13.8 GHz) at 4 km horizontal resolution over a 220 km swath
2A25	Radar Rainfall Rate and Profile (PR)	PR (13.8 GHz) rain rate, reflectivity, and attenuation profiles, at 4 km horizontal, and 250 m vertical, resolutions, over a 220 km swath
2B31	Combined Rainfall Profile	Combined PR/TMI rain rate, path-integrated attenuation, and latent heating at 4 km horizontal and 250 m vertical resolutions, over a 220 km swath
3A11	Monthly 5° × 5° Oceanic Rainfall	Rain rate, conditional rain rate, rain frequency, and freezing height for a latitude band from 40°N to 40°S, from TMI
3A12	Monthly 5° × 5° mean 2A12, profile, and surface rainfall	5° × 5° gridded monthly product comprising the mean 2A12 data, calculated vertical hydrometeor profiles, and the mean surface rainfall
3A25	Monthly 5° × 5° and 5° × 5° Spaceborne Radar Rainfall	Total and conditional rain rate, radar reflectivity, path-integrated attenuation at 2, 4, 6, 10, and 15 km for convective and stratiform rain, storm, freezing, and bright band heights, and snow–ice layer depth for a latitude band from 40°N to 40°S, from PR
3A25	Monthly 5° × 5° Surface Rain Total	Rain rate probability distribution at surface, 2, and 4 km for a latitude band from 40°N to 40°S, from PR
3B31	Monthly 5° × 5° Combined Rainfall	Rain rate, cloud liquid water, rain water, cloud ice, graupels at 14 levels for a latitude band from 40°N to 40°S, from PR and TMI
3B46	Monthly 1° × 1° SSM/I Rain	Global rain rate from SSM/I
3B42	3B42: 3-Hour 0.25° × 0.25° TRMM and Other-GPI, Calibration Rainfall	Calibrated geosynchronous IR rain rate using TRMM estimates

TABLE 16.5   Characteristics of the Tropical Rainfall Measuring Mission (TRMM) standard satellite products.—cont'd

TSDIS reference	Product name	Product description
3B43	3B43: Monthly 0.25° × 0.25° TRMM and Other Sources Rainfall	Merged rain rate from TRMM, geosynchronous IR, SSM/I, rain gauges
CSH	CSH: Monthly 0.5° × 0.5° Convective & Stratiform Heating	TRMM Monthly 0.5° × 0.5° Convective/Stratiform Heating (it is a TRMM experimental product)

Courtesy of Chiu, L.S., Liu, Z., Rui, H., Teng, W.L., 2006. Tropical rainfall measuring mission data and access tools. In: Qu, J.J., Gao, W., Kafatos, M., Murphy, R.E., Salomonson, V.V. (Eds.), Earth Science Satellite Remote Sensing. vol. 2. Berlin, Springer-Verlag.

Biosphere-Atmosphere Experiment in Amazonia (TRMM-LBA), Brazil. Characteristics of the GV products are listed in Table 16.6.

16.4.2 Global Satellite Mapping of Precipitation

The GSMaP project, which ran from 2002 until March 2008, was supported by the Japan Science and Technology Agency. Its objectives were to create high-precision and high-resolution global precipitation maps using data derived from PMW radiometers, improve the precipitation physical model and rain rate retrieval algorithms, evaluate the accuracy of various GSMaP productions, and prepare for the coming GPM. Since 2007, the activities of the GSMaP project have been implemented by the Precipitation Measuring Mission Science Team of the Japan Aerospace Exploration Agency (JAXA).

The GSMaP project used multiple datasets from several sensors as inputs to retrieve rainfall rates. The data sources include the MWR datasets observed from the LEO satellites and the infrared (IR) radiometer datasets obtained from the GEO satellites. The PMW sensors include imagers and sounders. The available imagers are TRMM TMI, AMSR, AMSR-E, and SSM/I. The most recent sounder is AMSU-B. In addition, the GEO IR datasets used in the current version of the data processing system are from the Climate Prediction Center (CPC) (Janowiak et al., 2001). The spatial resolution of the data is 0.03635° (4 km at the equator). The latitude ranges from 60°N to 60°S. The temporal resolution is approximately 30 min. Fig. 16.4 shows the structure and flow of the GSMaP project.

Currently, the datasets produced from the GSMaP project can be downloaded from their website at http://sharaku.eorc.jaxa.jp/GSMaP_crest/index.html. The standard version of the GSMaP datasets includes GSMaP_TMI (retrieved from TRMM/TMI algorithm), GSMaP_MWR (retrieved from six spaceborne MWRs), GSMaP_MWR+ (retrieved from six spaceborne MWRs with AMSU-B products), GSMaP_MVK (retrieved from the MWR-GEO IR combined algorithm), GSMaP_MVK+ (retrieved from the MWR-GEO IR combined algorithm with AMSU-B products), and other rainfall estimates from a PMW radiometer. The data from 2006 were produced in a near real-time data processing system developed and operated at the JAXA Earth Observation Research Center (http://sharaku.eorc.jaxa.jp/GSMaP/index.htm). Table 16.7 details the spatial and temporal resolution, data period, data range, and data source of the GSMaP datasets.

The GSMaP datasets have been compared with rain gauge data (Kubota et al., 2007, 2009; Aonashi et al., 2009). Overall, the validation results over the oceans were the best, and the

TABLE 16.6   Characteristics of the Tropical Rainfall Measuring Mission (TRMM) ground validation (GV) products.

TSDIS reference	Product name	Product description
1B51	Radar Reflectivity	Volume scan of radar reflectivity, differential reflectivity, and mean velocity (if available), truncated at 230 km range, at original radar resolution, coordinate, and sampling
1C51	C51:Half-Hourly[a] Calibrated Radar Reflectivity	Volume scan of calibrated radar reflectivity and differential reflectivity (if available) and corresponding QC masks, truncated at 200 km range, at original radar resolution and coordinate
2A52	Half-Hourly[a] Rain Existence	Percent of rain in the radar volume scan
2A53	Half-Hourly[a] 2 km Radar Site Rain Map	Instantaneous rain rate over an area of 300 km × 300 km
2A54	Half-Hourly[a] 2 km Radar Site Rain Type Map	Instantaneous rain type classification (convective, stratiform) over an area of 300 km × 300 km
2A55	Half-Hourly[a] Radar Site 3-D Reflectivity	Instantaneous radar reflectivity and vertical profile statistics over an area of 300 km × 300 km, at 2 km horizontal and 1.5 km vertical resolution
2A56	1 Minute Average and Peak Rain Gauge Rain Rate	Time series of rain gauge rain rates over the radar site rain gauge network
3A53	5-Day 2 km Ground Radar Site Rain Map	Surface rain total from ground radar
3A54	Monthly 2 km Ground Radar Site Rain Map	Surface rain total from ground radar
3A55	Monthly 2 km Ground Radar Site 3-D Rain Map	Vertical profile of reflectivity and contoured frequency by altitude diagrams for stratiform, convective, and anvil rain over land and water

[a] *Volume scans at original radar sampling rates within half an hour of TRMM satellite coincidence.*
Courtesy of Chiu, L.S., Liu, Z., Rui, H., Teng, W.L., 2006. Tropical rainfall measuring mission data and access tools. In: Qu, J.J., Gao, W., Kafatos, M., Murphy, R.E., Salomonson, V.V. (Eds.), Earth Science Satellite Remote Sensing. vol. 2. Berlin, Springer-Verlag.

results over mountainous regions were the worst. The GSMaP_MVK product achieved a score comparable with the CMORPH products (Ushio et al., 2009). Heavy precipitation areas over land and coasts were not easily identified, and precipitation events with more than 10 mm/h were underestimated (Aonashi et al., 2009). Our results show that GSMaP_MWR+ and MVK+ underestimated precipitation over the Poyang Lake Basin of China from January 2003 to December 2006 (Fig. 16.5). The reasons for this discrepancy are unknown.

16.4.3 Global Precipitation Climatology Project

Long-term series of satellite-retrieved global precipitation are not yet available. GPCP was established by the World Climate Research Program in 1986. The aim of the GPCP is to provide area- and time-averaged global precipitation analysis products. The products play an important role in understanding seasonal, annual, and long-term global energy and water cycle variabilities.

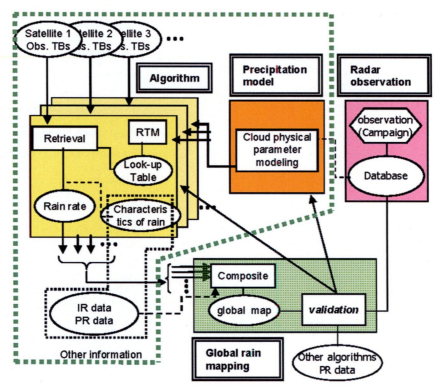

FIGURE 16.4 Structure and flow of the GSMaP project. *Courtesy of Takahashi, N., Iguchi, T., Aonashi, K., Awaka, J., Eito, H., Fujita, M., Hashizume, M., Hirose, M., Inoue, T., Kashiwagi, H., Kozu, T., Satoh, S., Seto, S., Shige, S., Shimizu, S., Takayabu, Y., Okamoto, K., 2004. The global satellite mapping of precipitation (GSMaP) project: Part II algorithm and precipitation model development. 2nd TRMM International Conference. September 6—10, 2004. Tokyo, Japan.*

The GPCP product is a mature dataset of merged IR and PMW retrievals and rain gauge observations (Gruber and Levizzani, 2008). The primary IR data sources are GOES, GMS, Meteosat, and NOAA polar-orbiting satellites. The microwave data source is the SSM/I on DMSP satellites. The station observations are primarily provided by the Global Precipitation Climatology Centre operated by the German Weather Service. The satellite precipitation estimates include SSM/I emission estimates, SSM/I scattering estimates, TOVS (TIROS Operational Vertical Sounder)-based estimates, and GPI and OPI precipitation indices.

The GPCP datasets include three global precipitation sub-datasets (Huffman et al., 2001; Adler et al., 2003; Xie et al., 2003; Adler et al., 2018): the GPCP (Version 2 and 2.3) monthly Satellite-Gauge, the Pentad, and the One-Degree Daily (1DD). The primary website for the GPCP products is the WMO's World Data Center (http://lwf.ncdc.noaa.gov/oa/wmo/wdcamet-ncdc.html) at NOAA's National Climatic Data Center. A backup FTP site for the Version 2 and 1DD products is located at NASA's Goddard Space Flight Center (GSFC). The primary product is 2.5° Version 2 monthly precipitation analyses for the period from 1979

TABLE 16.7 Status of each Global Satellite Mapping of Precipitation (GSMaP) product.

Product	Spatial resolution	Temporal resolution	Data period	Data range	Data source
GSMaP_TMI	0.25°	Hourly, daily, monthly, yearly	1998–2006	±40°	TRMM TMI
GSMaP_MWR	0.25°	Hourly, daily, monthly, yearly	1998–2006	±60°	TRMM/TMI, Aqua/AMSR-E, ADEOS-II/AMSR, DMSP/SSMI-F10,11,13,14,15
GSMaP_MWR+	0.25°	Hourly, daily, monthly	2003–06	±60°	TRMM/TMI, Aqua/AMSR-E, ADEOS-II/AMSR, DMSP/SSMI-F10,11,13,14,15 AMSU-B
GSMaP_MVK	0.1°	Hourly, daily, monthly	2003–06	±60°	TRMM/TMI, Aqua/AMSR-E, ADEOS-II/AMSR, DMSP/SSMI-F13,14,15 GOES-8/10, Meteosat-7/5, GMS
GSMaP_MVK+	0.1°	Hourly, daily, monthly	2003–06	±60°	TRMM/TMI, Aqua/AMSR-E, ADEOS-II/AMSR, DMSP/SSMI-F13,14,15, NOAA/AMSU-B GOES-8/10, Meteosat-7/5, GMS
GSMaP_NRT	0.25°	Hourly, daily, monthly	2007–present	±60°	TRMM/TMI, Aqua/AMSR-E, DMSP/SSMI-F13,14,15, NOAA/AMSU-B GOES-8/10, Meteosat-7/5, GMS, MTSAT

to the present. Both the Pentad and the 1DD products are constrained by monthly analyses. The 1DD dataset was produced on a resolution of 1° from January 1997 to the present. This dataset shows flexibility in computing time and spatial averages. The pentad precipitation analyses are 2.5° × 2.5° analyses from 1979 to the present and use the pentad CMAP (CPC Merged Analysis of Precipitation) estimates. The pentad CMAP merges the IR-based GPI, the SSM/I scattering-based ALG85, the SSM/I emission-based algorithm, the MSU-based method, the OLR-based OPI, and the gauge-based analyses.

Gruber and Levizzani (2008) provided an extensive assessment of the GPCP datasets, which have been applied in global change studies. However, the GPCP still faces challenges. Major issues include determining precipitation over the open ocean, determining solid precipitation, and accurately estimating precipitation in complex terrain areas.

16.4.4 Global Precipitation Measurement

The GPM is a joint mission between the NASA of the United States, the NASDA of Japan, and other international institutes. The objective of the GPM is to improve the global rainfall estimation and advance scientific understanding of the earth's water and energy cycle. The GPM, an outgrowth of TRMM, have the capability to provide physically based retrievals on a global basis with higher resolution and accuracy. It is able to observe weak rainfall (0.5 mm/h) and solid precipitation whose observations are significant in regions in mid- and high-latitudes and mountainous areas.

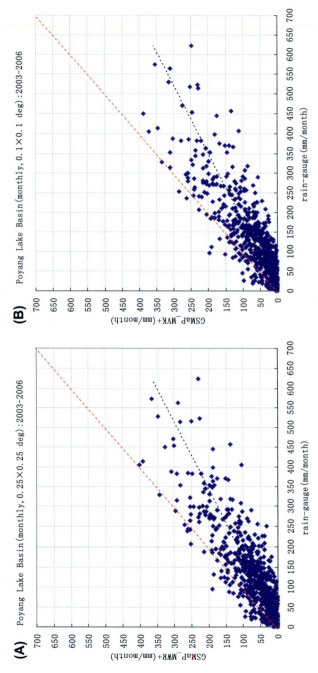

FIGURE 16.5 Rain gauge–based monthly measurements for (A) GSMaP MWR+ and (B) MVK+ datasets over the Poyang Lake Basin of China from January 2003 until December 2006.

TABLE 16.8   Major parameters of GPM Microwave Imager (GMI) and Dual-frequency Precipitation Radar (DPR).

Sensor	TMI	PR
Frequency/ wavelength	Dual polarization: 10.65, 18.7, 36.64, 89.0, 166.0 GHz; Vertical: 23.8, 183.31 ± 3, 183.31 ± 7 GHz	Vertical polarization: 13.8 GHz
Scanning mode	Conical	Cross track
Ground resolution	3.4 km at 183.31 GHz	5.0 km at nadir
Swath width	904 km	245 km
Science applications	Surface rain rate, rain type, distribution, and structure; other atmospheric and oceanic parameters	3D rainfall distribution over both land and oceans, and latent heat release into the atmosphere

GPM Core Observatory (GPMCO) is the heart of the GPM constellation. The GPMCO was designed as an LEO satellite stationed between 65°N and 65°S. It has a circular orbit of 407 km and an inclination angle of 65° (Smith et al., 2007). The satellite includes two scientific measuring instruments: DPR and GMI. The two sensors have contributed greatly to improve precipitation retrieval. Table 16.8 lists the major parameters of each instrument.

DPR was the first dual-frequency radar specifically for rainfall monitoring from space. Similarly to the TRMM, it provides three-dimensional rainfall structures to achieve quantitative rainfall measurements. GMI is a dual-polarized MWR with more multichannels. The GMI data combined with the DPR is able to derive precipitation profiles in higher latitude.

GPM dataset were available starting on March 2014 (https://pmm.nasa.gov/data-access/downloads/gpm). The standard GPM products can be classified into two groups: satellite and GV products. There are three distinct product levels. The GPM Level-1 products include GMI-calibrated antenna temperature, DPR power, and reflectivity. Level-2 products contain surface cross sections, DPR rain profiles, and rain rate based on Level-1 products. Level-3 products are space- and time-averaged global products based on Level-2 products. Table 16.5 details the products.

GSFC is responsible for management of GPM Validation Network (GV). GV started on August 8, 2006. The VN database includes TRMM PR data and coincident ground radar operational meteorological networks. Validation areas are same as TRMM. VN products are processed by VN software (online at http://www.ittvis.com), and the processes include data ingest and preprocessing, the resampling of PR and GR, and the statistical analysis and display of the matching data volumes (Schwaller and Morris, 2011).

16.4.5 Climate Prediction Center Morphing

CPC of the NOAA is responsible for CMORPH products. CMORPH dataset were available starting from December 2012 to October 2017 between 60°N and 60°S. CMORPH includes $0.07° \times 0.07°$, 0.5 h, and $0.25° \times 0.25°$, three hourly products (available at https://climatedataguide.ucar.edu/climate-data/cmorph-cpc-morphing-technique-high-resolution-precipitation-60s-60n). CMORPH first used cloud motion vectors derived from GEO IR satellite imagery to propagate the precipitation retrievals from PMW data.

Many studies evaluated and compared CMORPH with other satellite precipitation products, such as TRMM 3B42 and GSMap (Sapiano and Arkin, 2009; Dinku et al., 2010; Kidd et al., 2012; Li et al., 2013). The results show that CMORPH performed relatively poor over wintery, complex land surfaces with light precipitation. Generally, CMORPH underestimated a rainfall of −6.5−1.4 mm/day.

16.5 Global precipitation climatology

With the aid from satellite remote sensing systems, spatial and temporal variations of precipitation can be predicted at various scales. Because the GPCP dataset spans 25 years, it is commonly used in climatological studies. Here, we illustrate the global precipitation climatology derived from this dataset.

Fig. 16.6 shows the mean annual precipitation produced with the GPCP dataset for the period from 1979 to 2003. The highest annual precipitation levels occurred in the western tropical Pacific, the eastern Pacific Intertropical

Convergence Zone area, with expansions into the Amazon, and over the extreme eastern tropical Indian Ocean. Two high-precipitation zones were identified along the eastern coastal regions of Asia and North America, coinciding with the midlatitude storm tracks associated with the Kuroshio and Gulf Stream currents. In the southern hemisphere, the South Atlantic Convergence Zone displayed local precipitation maxima that extended southeastward from southern Brazil and the South Pacific Convergence Zone to New Guinea.

Fig. 16.7 shows the zonal mean precipitation generated from GPCP datasets for land, ocean, and the globe over the period 1979−2003. The mean global ocean precipitation levels had off-equatorial maxima and relative minima on the equator, reflecting the mean position of the convergence zones in tropics. In contrast, the zonal mean values of the land-only zonal had a single maximum centered on the equator. The zonal mean profiles of both hemispheres in both the ocean- and land-only datasets had

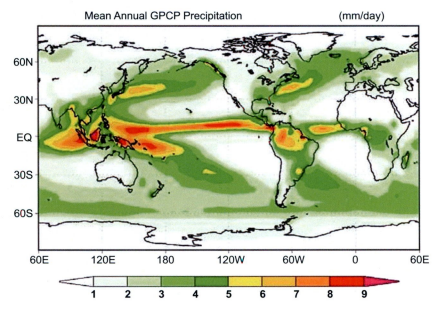

FIGURE 16.6   Mean annual GPCP precipitation for the period 1979−2003. *Courtesy of Gruber, A., Levizzani, V., 2008. Assessment of global precipitation products. WCRP Series Report No. 128 and WMO TD-No.1430, 1−55.*

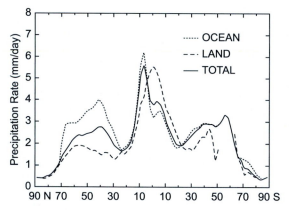

FIGURE 16.7 Zonal mean of GPCP precipitation for land (*dashed line*), ocean (*dotted line*), and globe (*solid line*) over the period 1979–2003. *Courtesy of Gruber, A., Levizzani, V., 2008. Assessment of global precipitation products. WCRP Series Report No. 128 and WMO TD-No.1430, 1–55.*

maxima in the midlatitudes, reflecting the mean position of the storm tracks.

At a seasonal scale, the mean precipitation rate appeared to be highest in the near-equatorial tropical belt for all months of the year (Fig. 16.8). From January to March, the precipitation maximum occurred in the southern hemisphere. From May to early September, the most intense precipitation was in the northern hemisphere. Low precipitation rates in the tropics and subtropics were centered at approximately 20°N from January to April and approximately 15°S from June to September.

Fig. 16.9 shows the variations of monthly precipitation for land (green line), the oceans (blue line), and the globe (black line) over the period 1979–2003. The precipitation rate was

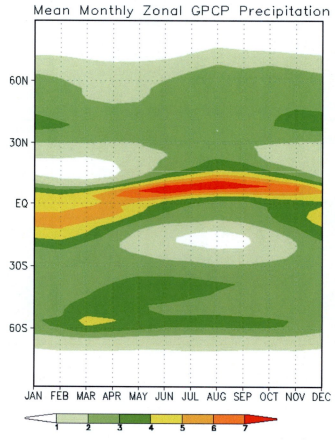

FIGURE 16.8 Annual cycle of zonal precipitation (mm/day) averaged over the period 1979–2003. *Courtesy of Gruber, A., Levizzani, V., 2008. Assessment of global precipitation products. WCRP Series Report No. 128 and WMO TD-No.1430, 1–55.*

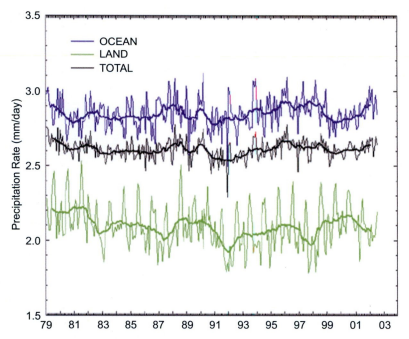

FIGURE 16.9 Monthly precipitation for land (green), ocean (blue), and globe (black) over the period 1979–2003. *Heavy lines* indicate 12-month running means. *Courtesy of Gruber, A., Levizzani, V., 2008. Assessment of global precipitation products. WCRP Series Report No. 128 and WMO TD-No.1430, 1–55.*

higher over the ocean than over the land by approximately 1 mm/day. The land series had larger variations than the ocean, even for the 12-month running mean values. The global total time series were produced by combining the land and ocean precipitation values.

The temporal variations of monthly precipitation may be relevant to the ENSO, which is a quasiperiodic climate pattern characterized by roughly 5-year variations in the surface temperature of the tropical eastern Pacific Ocean. ENSO is often described with the Niño-3.4 sea surface temperature (SST) index. Fig. 16.10 shows the total (top), ocean (middle), and land (bottom) tropical (30°N–30°S) averages of monthly precipitation anomalies. The mean tropical precipitation showed month-to-month variabilities generally within ±0.2 mm/day. The 12-month running mean of global precipitation had a low amplitude of less than 0.1 mm/

day and a low-frequency (2–3 years) variability. However, no significant relationship between ENSO and global precipitation anomalies was observed. Two major volcanic eruptions, El Chichón and Mt. Pinatubo, did not significantly influence the anomalies. Conversely, a significant relationship between the precipitation anomalies and the Niño-3.4 SST index over the oceans or land for both dry and wet conditions was observed. The relationship was strong especially in 1986/87 and 1997/98, two largest ENSO events of the 20th century.

16.6 Summary

Precipitation includes rainfall and snowfall, which are fundamental process in the global water cycle. It has a high degree of spatiotemporal variability, highly nonnormal in behavior. The

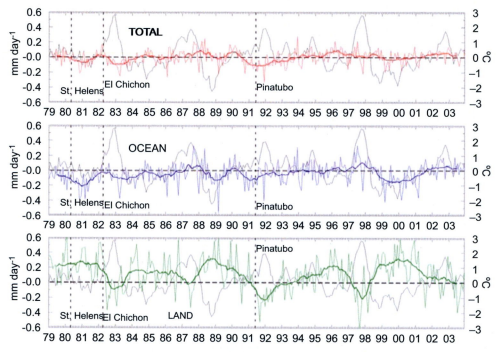

FIGURE 16.10 Tropical (30°N-30°S) averages of monthly precipitation anomalies (mm/day) for total (top), ocean (middle), and land (bottom). Vertical *dashed lines* indicate months with significant volcanic eruptions. The thin *black curves* are the Niño-3.4 SST index (°C). Heavy lines are 12-month running means. *Courtesy of Gruber, A., Levizzani, V., 2008. Assessment of global precipitation products. WCRP Series Report No. 128 and WMO TD-No.1430, 1–55.*

space–time structure of precipitation produces top-down effects on numerous hydrologic processes. Many rain gauge networks and ground-based and space-borne radar techniques have been developed to capture spatial and temporal features of precipitation.

Since 1970s, tens of satellite retrieval algorithms have been proposed on a basis of satellite observation from meteorological sensors and PR. The algorithms use VIS/IR, PMW, active microwave data, or various combinations of these technologies. The GPI algorithm is the most commonly used VIS/IR algorithm. The Ferraro and the GPROF algorithms are the most commonly used PWV algorithms. Combination of active and passive sensors substantially improved cloud and precipitation retrieval and was particularly welcomed in the last two decades.

Fast development of space-based instruments and retrieval approaches fosters to produce many regional and global precipitation datasets. The TRMM, GSMaP, and GPCP datasets are well-known examples of global long-term series precipitation datasets. The global datasets provide invaluable information for studying precipitation processes and mechanisms at regional and global scales.

Global warming and climate change are likely to continue, necessitating a comprehensive understanding of global precipitation for the sustainable development of our changing earth. In addition to surface-based instruments, space-based observations, especially those that use information from the TRMM, have advanced our spatial and temporal understandings of precipitation intensity. As an international effort, the GPM launch in 2014 extends the TRMM measurements to latitudes up to 65°; increases the sampling frequency to 3 h; improves sensitivity to light and heavy rainfalls, with a detectable range from 1 mm/h to more than 200 mm/h; and discriminates between rain and snow.

Long-term global precipitation monitoring is a challenge. Combining observations at different VIS, IR, and microwave wavelengths from passive and active modes and from both low and geostationary orbits can produce hourly samples and long-term precipitation records for the entire globe. The International Precipitation Working Group is focusing on satellite-based quantitative precipitation measurements. With the international efforts, we are approaching an important juncture for the retrieval of global precipitation information from multiple sensors.

References

Adler, R.F., Huffman, G.J., Chang, A., Ferraro, R., Xie, P., Janowiak, J., Rudolf, B., Schneider, B., Curtis, S., Bolvin, D., Gruber, A., Susskind, J., Arkin, P., 2003. The version 2 Global Precipitation Climatology Project (GPCP) monthly precipitation analysis (1979-present). J. Hydrometeorol. 4, 1147–1167.

Adler, R.F., Sapiano, M.R.P., Huffman, G.J., Wang, J.J., Gu, G.J., Bolvin, D., Chiu, L., Schneider, U., Becker, A., Nelkin, E., Xie, P.P., Ferraro, R., Shin, D.B., 2018. The global precipitation climatology project (GPCP) monthly analysis (new version 2.3) and a review of 2017 global precipitation. Atmosphere 9, 14.

Alishouse, J.C., 1983. Total precipitable water and rainfall determinations from the SeaSat scanning multichannel microwave radiometer. J. Geophys. Res. 88, 1929–1935.

Aonashi, K., Awaka, J., Hirose, M., Kozu, T., Kubota, T., Liu, G., Shige, S., Kida, S., Seto, S., Takahashi, N., Takayabu, Y.N., 2009. GSMaP passive microwave precipitation retrieval algorithm: algorithm description and validation. J. Meteorol. Soc. Jpn. 87A, 119–136.

Arkin, P., Meisner, B.N., 1987. The relationship between large-scale convective rainfall and cold cloud over the Western Hemisphere during 1982-84. Mon. Weather Rev. 115, 51–74.

Ba, M.B., Gruber, A., 2001. GOES multispectral rainfall algorithm (GMSRA). J. Appl. Meteorol. 40, 1500–1514.

Bauer, P., Schanz, L., 1998. Outlook for combined TMI-VIRS algorithms for TRMM: lessons from the PIP and AIP projects. J. Atmos. Sci. 55, 1714–1729.

Brown, R.A., Lewis, J.M., 2005. Path to NEXRAD: Doppler radar development at the national severe storms laboratory. Bull. Am. Meteorol. Soc. 86, 1459–1470.

Chiu, L.S., Liu, Z., Rui, H., Teng, W.L., 2006. Tropical rainfall measuring mission data and access tools. In: Qu, J.J., Gao, W., Kafatos, M., Murphy, R.E., Salomonson, V.V. (Eds.), Earth Science Satellite Remote Sensing, vol. 2. Springer-Verlag, Berlin.

Dinku, T., Ruiz, F., Connor, S.J., Ceccato, P., 2010. Validation and intercomparison of satellite rainfall estimates over Colombia. J. Appl. Meteorol. Climatol. 49 (5), 1004–1014.

Duchon, C., Essenberg, G., 2001. Comparative rainfall observations from pit and aboveground rain gauges with and without wind shields. Water Resour. Res. 37, 3253–3263.

Ebert, E.E., Manton, M.J., 1998. Performance of satellite rainfall estimation algorithms during TOGA COARE. J. Atmos. Sci. 55, 1537–1557.

Ferraro, R.R., 1997. Special sensor microwave imager derived global rainfall estimates for climatological applications. J. Geophys. Res. 102, 16715–16735.

Griffith, C.G., Woodley, W.L., Grube, P.G., Martin, D.W., Stout, J., Sikdar, D.N., 1978. Rain estimates from geosynchronous satellite imagery: visible and infrared studies. Mon. Weather Rev. 106, 1153–1171.

Groisman, P., Legates, D., 1994. The accuracy of United States precipitation data. Bull. Am. Meteorol. Soc. 75, 215–228.

Gruber, A., Levizzani, V., 2008. Assessment of global precipitation products. In: WCRP Series Report No. 128 and WMO TD-No.1430, pp. 1–55.

Hong, Y., 2004. Precipitation estimation from remotely sensed information using artificial neural network-cloud classification system. J. Hydrometeorol. 43, 1834–1852.

Hou, A.Y., Kakar, R.K., Neeck, S., Azarbarzin, A.A., Kummerow, C.D., Kojima, M., Oki, R., Nakamura, K., Iguchi, T., 2014. The global precipitation measurement mission. Bull. Am. Meteorol. Soc. 95, 701–722.

Huffman, G.J., Adler, R.F., Morrissey, M., Bolvin, D.T., Curtis, S., Joyce, R., McGavock, B., Susskind, J., 2001. Global precipitation at one-degree daily resolution from multi-satellite observations. J. Hydrometeorol. 2, 36–50.

Huffman, G.J., Alder, R., Bolvin, D.T., Gu, G., Nelkin, E.J., Bowman, K.P., Hong, Y., Stocker, E.F., Wolff, D.B., 2007. The TRMM multisatellite precipitation analysis (TMPA): quasi-global, multiyear, combined-sensor precipitation estimates at fine scales. J. Hydrometeorol. 8, 38–55.

Huffman, G.J., Bolvin, D.T., Braithwaite, D., Hsu, K., Joyce, R., Kidd, C., Nelkin, E.J., Xie, P., 2015. NASA Global Precipitation Measurement Integrated Multi-Satellite Retrievals for GPM (IMERG). Algorithm Theoretical Basis Doc., Version 4.5, 30 pp. Available online at: http://pmm.nasa.gov/sites/default/files/document_files/IMERG_ATBD_V4.5.pdf.

Iguchi, T., Kozu, T., Meneghini, R., Awaka, J., Okamoto, K., 2000. Rain-profiling algorithm for the TRMM precipitation radar. J. Appl. Meteorol. 39, 2038–2052.

Iguchi, T., 2007. Space-borne radar algorithms. In: Levizzani, V., Bauer, P., Turk, F.J. (Eds.), Measuring Precipitation from Space: EURAINSAT and the Future. Springer-Verlag, Berlin.

Iguchi, T., 2009. Uncertainties in the rain profiling algorithm for the TRMM precipitation radar. J. Meteorol. Soc. Jpn. 87A, 1–30.

Janowiak, J., Joyce, R.J., Yahosh, Y., 2001. A real-time global half-hourly pixel-resolution IR dataset and its applications. Bull. Am. Meteorol. Soc. 82, 205–217.

Joyce, R.J., Janowiak, J.E., Arkin, P.A., Xie, P., 2004. CMORPH: a method that produces global precipitation estimates from passive microwave and infrared data at high spatial and temporal resolution. J. Hydrometeorol. 5, 487–503.

Kidd, C., 2001. Satellite rainfall climatology: a review. Int. J. Climatol. 21, 1041–1066.

Kidd, C., Bauer, P., Turk, J., Huffman, G.J., Joyce, R., Hsu, K.L., Braithwaite, D., 2012. Intercomparison of high-resolution precipitation products over northwest europe. J. Hydrometeorol. 13 (1), 67–83.

Krajewski, W., 1987. Cokriging radar-rainfall and rain gage data. J. Geophys. Res. 92, 9571–9580.

Kubota, T., Shige, S., Hashizume, H., Aonashi, K., Takahashi, N., Seto, S., Hirose, M., Takayabu, Y.N., Ushio, T., Nakagawa, K., Iwanami, K., Kachi, M., Okamoto, K., 2007. Global precipitation map using satellite-borne microwave radiometers by the GSMaP project: production and validation. IEEE Trans. Geosci. Remote Sens. 45, 2259–2275.

Kubota, T., Ushio, T., Shige, S., Kida, S., Kachi, M., Okamoto, K., 2009. Verification of high-resolution satellite-based rainfall estimates around Japan using a gauge-calibrated ground-radar dataset. J. Meteorol. Soc. Jpn. 87A, 203–222.

Kummerow, C., Barnes, W., 1998. The tropical rainfall measuring mission (TRMM) sensor package. J. Atmos. Ocean. Technol. 15, 809–817.

Kummerow, C.D., Hong, Y., Olson, W.S., Yang, S., Adler, R.F., McCollum, J., Ferraro, R., Petty, G., Shi, D.B., Wilheit, T.T., 2001. The evolution of the Goddard Profiling Algorithm (GPROF) for rainfall estimation from passive microwave sensors. J. Appl. Meteorol. 40, 1801–1820.

Kummerow, C.D., Masunaga, & H., Bauer, P., 2007. A next-generation microwave rainfall retrieval algorithm for use by TRMM and GPM. In: Levizzani, V., Bauer, P., Turk, F.J. (Eds.), Measuring Precipitation from Space: EURAINSAT and the Future. Springer-Verlag, Berlin.

Levizzani, V., Bauer, P., Turk, F.J., 2007. Measuring Precipitation from Space: EURAINSAT and the Future. Springer, Dordrecht.

Li, Z., Yang, D., Hong, Y., 2013. Multi-scale evaluation of high-resolution multi-sensor blended global precipitation products over the Yangtze river. J. Hydrol. 500 (14), 157–169.

Linacre, E., 1992. Climate Data and Resources: A Reference and Guide. Routledge.

Liu, G., 2003. Satellite remote sensing: precipitation. In: Holton, J.R., Curry, J.A., Pyle, J.A. (Eds.), Encyclopedia of Atmospheric Sciences. Academic Press, London, pp. 1972–1979.

Maggioni, V., Meyers, P.C., Robinson, M.D., 2016. A review of merged high-resolution satellite precipitation product accuracy during the tropical rainfall measuring mission (TRMM) Era. J. Hydrometeorol. 17, 1101–1117.

Makihara, Y., Uekiyo, N., Tabata, A., Abe, Y., 1996. Accuracy of radar-AMeDAS precipitation. IEICE Trans. Commun. E79-B, 751–762.

Michaelides, S., Levizzani, V., Anagnostou, E., Bauer, P., Kasparis, T., Lane, J.E., 2009. Precipitation: measurement, remote sensing, climatology and modeling. Atmos. Res. 94, 512–533.

Morin, E., Krajewski, W.F., Goodrich, D.C., Gao, X., Sorooshian, S., 2003. Estimating rainfall intensities from weather radar data: the scale-dependency problem. J. Hydrometeorol. 4, 782–797.

Mugnai, A., Smith, E.A., Triopli, G.J., 1993. Foundations for statistical-physical precipitation retrieval from passive microwave satellite measurements. Part II: emission-source and generalized weighting-function properties of a time-dependent cloud-radiation model. J. Appl. Meteorol. 32, 17–39.

Neary, V., Habib, E., Fleming, M., 2004. Hydrologic modeling with NEXRAD precipitation in middle Tennessee. J. Hydrol. Eng. 9, 339–349.

Nguyen, P., Ombadi, M., Sorooshian, S., Hsu, K., AghaKouchak, A., Braithwaite, D., Ashouri, H., Thorstensen, A.R., 2018. The PERSIANN family of global satellite precipitation data: a review and evaluation of products. Hydrol. Earth Syst. Sci. 22, 5801–5816.

Okamoto, K., Iguchi, T., Takahashi, N., Iwanami, K., Ushio, T., 2005. The global satellite mapping of precipitation (GSMaP) project. In: 25th IGARSS Proceedings, pp. 3414–3416.

Prigent, C., 2010. Precipitation retrieval from space: an overview. Compt. Rendus Geosci. 342, 380–389. Copyright © 2010 Elsevier Masson SAS. All rights reserved.

Rodda, J.C., 1969. Hydrological network design — needs, problems and approaches. WMO Rep. 12, 1–58.

Sapiano, M.R.P., Arkin, P.A., 2009. An intercomparison and validation of high-resolution satellite precipitation estimates with 3-hourly gauge data. J. Hydrometeorol. 10, 149–166.

Savage, R.C., Weinman, J.A., 1975. Preliminary calculations of the upwelling radiance from rain clouds at 37.0 and 19.35 GHz. Bull. Am. Meteorol. Soc. 56, 1272–1274.

Schwaller, M.R., Morris, K.R., 2011. A ground validation network for the global precipitation measurement mission. J. Atmos. Ocean. Technol. 28 (3), 301–319.

Shelton, M.L., 2009. Hydroclimatology. Cambridge University Press, Cambridge.

Smith, E., Asrar, G., Furuhama, Y., Ginati, A., Kummerow, C., Levizzani, V., Mugnai, A., Nakamura, K., Adler, R., Casse, V., Cleave, M., Debois, M., Durning, J., Entin, J., Houser, P., Iguchi, T., Kakar, R., Kaye, J., Kojima, M., Lettenmaier, D.P., Luther, M., Mehta, A., Morel, P., Nakazawa, T., Neeck, S., Okamoto, K., Oki, R., Raju, G., Shepherd, M., Stocker, E., Testud, J., Wood, E.F., 2007. International global precipitation measurement (gpm) program and mission: an overview. In: Levizzani, V., Bauer, P., Turk, F.J. (Eds.), Measuring Precipitation From Space, Advances In Global Change Research, vol. 28. Springer, Dordrecht.

Smith, E.A., Mugnai, A., Cooper, H.J., Triopli, G.J., Xiang, X., 1992. Foundations for statistical-physical precipitation retrieval from passive microwave satellite measurements. Part I: brightness-temperature properties of a time-dependent cloud-radiation model. J. Appl. Meteorol. 31, 506–531.

Stout, G., Mueller, E.A., 1968. Survey of relationships between rainfall rate and radar reflectivity in the measurement of precipitation. J. Appl. Meteorol. 7, 465–474.

Sun, Q.H., Miao, C.Y., Duan, Q.Y., Ashouri, H., Sorooshian, S., Hsu, K.L., 2018. A review of global precipitation data sets: data sources, estimation, and intercomparisons. Rev. Geophys. 56, 79–107.

Takahashi, N., Iguchi, T., Aonashi, K., Awaka, J., Eito, H., Fujita, M., Hashizume, M., Hirose, M., Inoue, T., Kashiwagi, H., Kozu, T., Satoh, S., Seto, S., Shige, S., Shimizu, S., Takayabu, Y., Okamoto, K., 2004. The global satellite mapping of precipitation (GSMaP) project: Part II algorithm and precipitation model development. In: 2nd TRMM International Conference. September 6-10, 2004. Tokyo, Japan.

Turk, F.J., Arkin, P., Ebert, E.E., Sapiano, M.R.P., 2008. Evaluating high-resolution precipitation products. Bull. Am. Meteorol. Soc. 89, 1911–1916.

Ushio, T., Sasashige, K., Kubata, T., Shige, S., Okamoto, K., Aonashi, K., Inoue, T., Takahashi, N., Iguchi, T., Kachi, M., Oki, R., Morimoto, T., Kawasaki, Z., 2009. A Kalman filter approach to the Global Satellite Mapping of Precipitation (GSMaP) from combined passive microwave and infrared radiometric data. J. Meteorol. Soc. Jpn. 87A, 137–151.

Villarini, G., Krajewski, W.F., 2010. Review of the different sources of uncertainty in single polarization radar-based estimates of rainfall. Surv. Geophys. 31, 107–129.

Wilheit, T.T., Change, A.T.C., Rao, M.S.V., Rodgers, E.B., Theon, J.S., 1977. A satellite technique for quantitatively mapping rainfall rates over oceans. J. Appl. Meteorol. 16, 551–560.

Wilheit, T., Kummerow, C., Ferraro, R., 1999. EOS/AMSR rainfall. Algorithm Theor. Basis Doc. 1–59.

Xie, P., Arkin, P.A., 1997. Global precipitation: a 17 year monthly analysis based on gauge observations, satellite estimates, and predictions. J. Clim. 9, 840–858.

Xie, P., Janowiak, J.E., Arkin, P.A., Adler, R.F., Gruber, A., Ferraro, R.R., Huffman, G.J., Curtis, S., 2003. GPCP pentad precipitation analyses: an experimental dataset based on gauge observations and satellite estimates. J. Clim. 16, 2197–2214.

Yuter, S.E., 2003. Radar: precipitation radar. In: Holton, J.R., Curry, J.A., Pyle, J.A. (Eds.), Encyclopedia of Atmospheric Sciences. Academic Press, London.

Abstract

This chapter provides the theoretical basis for measurement and estimation of terrestrial evapotranspiration. It includes the Monin—Obukhov similarity theory (MOST) and the Penman—Monteith equation. We focus on the application of these theories to satellite remote sensing. Both the MOST and the Penman—Monteith equations are empirical in nature, and they require several assumptions to be met. Further parameterizations and assumptions are necessary to relate the satellite-derived land surface variables to terrestrial evapotranspiration. There is no difference in the physics between the two formulations as the Penman—Monteith equation is derived from MOST. However, there are important differences between the two formulations in their practical application: MOST-like methods use air—surface temperature differences and require accurate estimates of these terms. Therefore, MOST-like methods are sensitive to errors in these differences. This sensitivity is substantially reduced by the Penman—Monteith equation related methods, which use the available energy and stomatal conductance that can be empirically parameterized as a function of the vegetation index (leaf area index), incident radiation, and relative humidity. The accuracy of a given model is usually the first consideration in its development, but the required ancillary data and the method's suitability for different conditions are

© 2020 Elsevier Inc. All rights reserved.

considerations for satellite remote methods as well. Satellite remote sensing can now reliably determine the spatial variability of land surface variables and terrestrial evapotranspiration; however, it still has difficulty in providing long-term, stable estimates.

17.1 Introduction

Terrestrial evapotranspiration (E) is the transfer of water from the soil or the surface of canopies, stems, branches, soils, and paved areas to the atmosphere. This water exchange usually involves the phase change of water from liquid to gas; in the process, energy is absorbed and the surface cooled. The latent heat absorbed by the process, which is the source of the flux to the atmosphere, is λE, where λ is the latent heat of vaporization. In some publications, ET also has the units of W/m^2; under these conditions, ET represents the latent heat flux. λE is required by short-term numerical weather predication models and longer-term climate simulations.

Under the assumption that the energy stored by the canopy is negligible, λE can be calculated as a residual of the available energy at the surface (R_n), the sensible heat flux (H), and the ground heat flux (G) (see Fig. 17.1):

$$R_n = \lambda E + H + G \qquad (17.1)$$

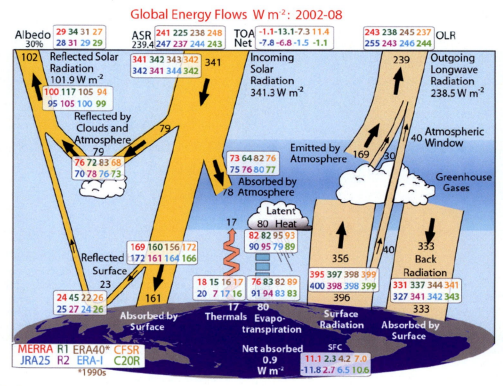

FIGURE 17.1   The background values of radiation or energy flows are based on observations for 2000–05. Superposed, with the key on the lower left, are values from the various reanalyses for the 2002–08 period except for ERA-40, which is for the 1990s, color-coded in W/m^2. On top of the figure, values are given for albedo in percentage (%), absorbed solar radiation ASR, and Net TOA (top of atmosphere) radiation, and the box labeled SFC near the bottom gives the net flux absorbed at the surface. For the 1990s, the latter value is 0.6 W/m^2 (Trenberth et al., 2011).

R_n is determined from the sum of the incident downward and upward shortwave and longwave radiation (see Fig. 17.1):

$$R_n = S_\downarrow - S_\uparrow + L_\downarrow - L_\uparrow$$
$$= S_\downarrow(1-\alpha) + L_\downarrow - L_\uparrow = S_n + L_\downarrow - L_\uparrow \tag{17.2}$$

where $S_\downarrow$ and $S_\uparrow$ are the surface downward and upward shortwave radiation, respectively; $L_\downarrow$ and $L_\uparrow$ are the surface downward and upward longwave radiation, respectively; α is the surface albedo; and S_n is the surface net shortwave radiation.

On average, λE uses approximately 59% (from 48% to 88% for different models) of the surface net energy (Trenberth et al., 2009). The energy required to evaporate a given mass of liquid water corresponds to approximately 600 times the energy required to increase its temperature by 1K and to 2400 times the energy required to increase the temperature of a corresponding mass of air by 1K (Seneviratne et al., 2010). If the terrestrial λE were zero, then the northern hemisphere would be 15–25°C warmer (Shukla and Mintz, 1982).

λE is a major component not only of the surface energy balance but also of the terrestrial water cycle. Therefore, it can be estimated from the surface water balance at basin-wide or continental scale using (see Fig. 17.2):

$$E = P - Q - dw/dt \tag{17.3}$$

where P is the amount of precipitation, Q is the amount of river discharge, and dw/dt is the

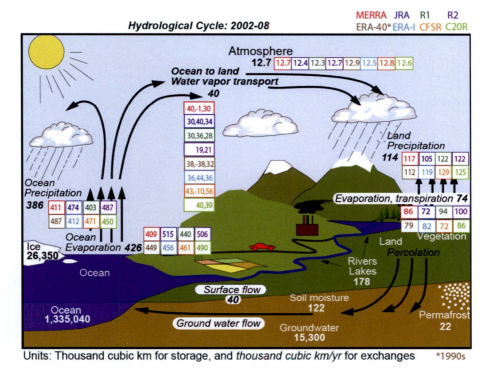

Units: Thousand cubic km for storage, and *thousand cubic km/yr* for exchanges *1990s

FIGURE 17.2 The background figure shows the estimates of the observed hydrological cycle from 2002 to 2008, with units in 1000 km³ for storage and 1000 km³/yr for exchanges. Superposed are values from the eight reanalyses for 2002–08, color-coded as given at top right. The exception is for ERA-40, which is for the 1990s (Trenberth et al., 2011).

change in terrestrial water storage. E accounts for approximately two-thirds of the 700-mm/yr average precipitation that falls over the land (Chahine, 1992; Oki and Kanae, 2006). The global average of the E/P ratio varies from 56% to 82% for eight reanalyses (Fig. 17.2), and the differences at the regional scale are much greater (Fig. 17.3).

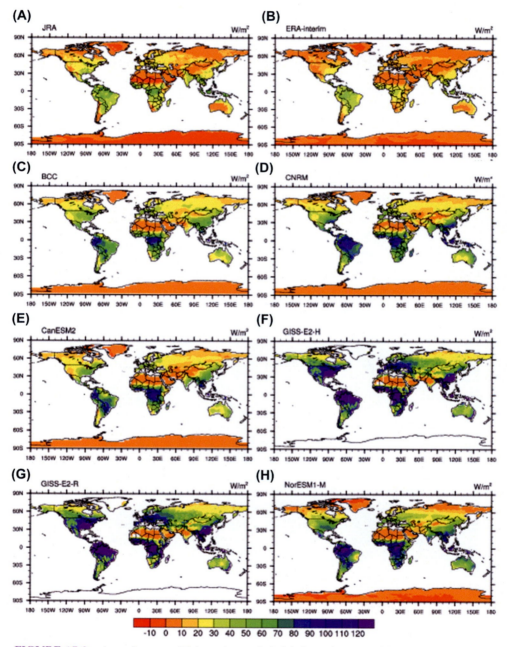

FIGURE 17.3   Annual average λE from six coupled global circulation models and two reanalyses.

FIGURE 17.4 Air flow can be imagined as a horizontal flow of numerous rotating eddies, that is, turbulent vortices of various sizes, with each eddy having horizontal and vertical components. The situation looks chaotic, but vertical movement of the components can be measured from the tower. *Burba, G., Anderson, D., 2008. Introduction to the Eddy Covariance Method: General Guidelines and Conventional Workflow. Available at: http://www.licor.com/env/2010/applications/eddy_covariance.jsp, 141 pp., Copyright LI-COR, Inc., Used By Permission, Copyright LI-COR, Inc., Used By Permission.*

Water and heat are transferred from the land surface to the atmosphere through turbulence, which transports the quantities involved with several orders of magnitude greater efficiency than molecular diffusivity. Turbulence can be assumed to comprise eddies of many different sizes that are superimposed on each other (Fig. 17.4). Large eddies can be thermals of rising warmer air caused by solar heating of the ground during sunny days. Turbulence has been recognized for a long time, but its understanding is still advancing (Lumley and Yaglom, 2001). The Monin–Obukhov similarity theory (MOST) was developed in the 1950s (Monin and Obukhov, 1954) to estimate λE and the sensible heat fluxes (H). MOST is similar to Ohm's Law, which describes the current through a conductor between two points, because MOST relates the turbulence fluxes to the mean gradients of the wind, temperature, and humidity.

A number of methods provide reliable point measurements of λE, e.g., the eddy covariance (*EC*) and the energy balance Bowen ratio (*BR*) tower systems (see Section 19.4). Satellite remote sensing does not provide direct measurement of λE. Rather, most satellite λE algorithms relate the satellite-derived land surface variables to λE using either the MOST or Penman–Monteith equations (Kalma et al., 2008; Li et al., 2009b). The methods that use the surface-air temperature gradient require unbiased T_s retrievals and interpolation of the T_a values from ground-based point measurements (Timmermans et al., 2007). Two approaches have been proposed to reduce the sensitivity of the flux estimates to uncertainties of T_s and T_a: (1) methods that use the temporal variation of T_s (Anderson et al., 1997; Caparrini et al., 2003, 2004b; Norman et al., 2000) and (2) methods that use the spatial variation of T_s (Carlson, 2007; Jiang and Islam, 2001; Wang et al., 2006). The above approaches require input that is not easily obtained. Therefore, many empirical and semiempirical models relate λE to more easily obtained data, including radiation, temperature, the satellite-derived vegetation index, and the water vapor deficit based on meteorological observations (Fisher et al., 2008; Jung et al., 2009; Sheffield et al., 2010; Wang et al., 2010a; Wang and Liang, 2008).

This chapter includes a critical review of the basic theories of λE, including the MOST and Penman–Monteith Equations in Section 17.3. Their assumptions and requirements are discussed in detail to clarify the strengths and weaknesses of the existing remote sensing methods. The chapter also includes a review of the current remote sensing methods in Section 17.4. This review largely focuses on the physical aspects of methods that are typically used, but more exhaustive and complete reviews have been provided by others (Carlson, 2007; Kalma et al., 2008; Li et al., 2009b). A brief introduction to the data used to calibrate and evaluate λE methods (Section 17.5) is also included, as is a discussion of current issues and future prospects (Section 17.6).

17.2 Basic theories of λE

17.2.1 The Monin–Obukhov similarity theory

Within the near surface layer, turbulent fluxes are assumed to be height-independent (for this reason, this layer is termed the constant flux layer). The layer is approximately 10% of the depth of the daytime atmospheric boundary layer (i.e., typically 0.1 km thick during the daytime). MOST relates the turbulent fluxes to the differences between the mean temperature and humidity at two levels in the constant flux layer through its universal similarity functions (Businger et al., 1971; Dyer, 1974). MOST (Monin and Obukhov, 1954) was the starting point for modern micrometeorology, including the development of new measurement devices and the execution of several important experiments. This method of estimating λE and H from time-averaged variables in the surface layer is termed the flux profile method and uses

$$H_p = -\rho C_p u_* \theta_* \qquad (17.4)$$

$$\lambda E_p = -\rho \lambda u_* q_* \qquad (17.5)$$

where the subscript p refers to the flux profile method; C_p is the specific heat at constant pressure; ρ is the air density; and u_*, θ_*, and q_* are the friction velocity, scale potential temperature, and scale humidity, respectively. According to

MOST, the following expressions are valid for the horizontally homogeneous and stationary surface layers:

$$u_* = k \frac{\overline{u_1} - \overline{u_2}}{\ln \frac{z_{m1}}{z_{m2}} - \psi_m(\xi_{m1}, \xi_{m2})} \qquad (17.6)$$

$$\theta_* = k \frac{T_1 - T_2}{\ln \frac{z_{h1}}{z_{h2}} - \psi_h(\xi_{h1}, \xi_{h2})} \qquad (17.7)$$

$$q_* = k \frac{q_1 - q_2}{\ln \frac{z_{q1}}{z_{q2}} - \psi_h(\xi_{q1}, \xi_{q2})} \qquad (17.8)$$

where $\overline{u}$, θ, and q are the mean wind speed, potential temperature, and humidity measurements, respectively. These parameters are mean values over typical averaging times of between 10 min and approximately 1 h. ψ_m and ψ_h are the integral profile functions of the stability, $\xi = z/L$, where L is the Monin–Obukhov length. ψ_m and ψ_h are MOST universal functions (Businger et al., 1971; Dyer, 1974). The subscripts 1 and 2 refer to levels. Under ideal conditions, the theory has an accuracy of approximately 10%–20% (Foken, 2006; Hogstrom and Bergstrom, 1996). Its assumptions of steady, horizontally homogeneous flow (Monin and Obukhov, 1954) are not always applicable.

The temperature and humidity gradients over the surface used by MOST can be very small, especially for forest, irrigated croplands, and grasslands. After the similarity theory was developed for the constant flux layer, much experimental effort was applied to determining its universal functions. The widely used Businger universal functions (Businger et al., 1971) are based on eddy covariance (EC) observations from the 1968 KANSAS experiment. However, some of the results from this experiment have been seriously questioned, in particular, the value of the von Kármán constant (0.35) found by the Kansas group (Businger et al., 1971) because of flow distortion problems caused by the tower, overspeeding of cup anemometers, and the unstable performance of the phase-shift sonic anemometers (Wieringa, 1980). Improved data have led to the reformulation of the Businger universal function (Hogstrom, 1988).

Since the late 1980s, scientists have realized that the sum of H and λE measured by the EC method is generally less than the available energy (Fig. 17.5). The closure problem is always characterized by low turbulent fluxes (Foken, 2008). Some reasons for this underestimation include averaging and coordinate rotation, and these coordinate systems have been recognized and corrected (Finnigan, 2004; Finnigan et al., 2003). Consequently, the MOST universal functions from Eqs. (17.1) and (17.2) must be reformulated to avoid underestimating H and λE.

The universal stability functions are not valid in the roughness sublayer (Sun et al., 1999) of vegetated surfaces (canopy height >0.1 m), and the profile equations must be modified using a universal function that depends on the thickness of this roughness sublayer (Garratt et al., 1996). For tall vegetation or in urban areas, the roughness sublayer may be tens of meters thick, whereas the constant flux layer, for which the

similarity theory is valid, may be shallow. This distinction has particularly important implications for retrieving λE from satellite-derived land surface temperature (T_s) because T_s is the radioactive temperature within the roughness layer of surface.

17.2.2 The Penman–Monteith equation

The Penman equation (Penman, 1948) describes λE from an open water surface. Monteith (Monteith, 1972) modified the Penman equation by including stomatal resistance to make the equation more suitable for terrestrial surfaces. The Penman–Monteith equation was derived from MOST by assuming that surface energy is balanced (Snyder and Paw U, 2002):

$$H = R_n - G - \lambda E \tag{17.9}$$

According to MOST, H and λE have the following forms, depending on the humidity and temperature at the surface and the reference layer:

$$H = \rho C_p \cdot \frac{T_o - T_a}{r_h} \tag{17.10}$$

$$\lambda E = \frac{\rho C_p}{\gamma^*} \cdot \frac{e_s(T_o) - e}{r_h} \tag{17.11}$$

where $\gamma^* = (r_v / r_h) \cdot \gamma$ (γ is the psychrometric constant and r_v is the aerodynamic resistance to water vapor transfer from the surface to the atmosphere), T_o is the aerodynamic air temperature at the surface, T_a is the air temperature above the surface at a reference level, and r_h is the aerodynamic resistance to heat transfer from the surface to the air, which can be calculated using the MOST universal function.

If $\Delta = de_s/dT$ is the derivative of the saturated vapor pressure (e_s) with respect to T_a and then a first-order approximation of $e_s(T_o)$ is

$$e_s(T_o) = e_s(T_a) + \Delta \cdot (T_o - T_a) \tag{17.12}$$

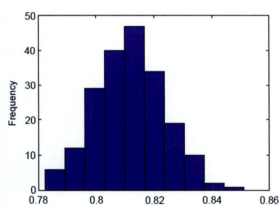

FIGURE 17.5 Histogram of the annual mean energy balance closure at FLUXNET sites. *Cited from Beer, C. et al., 2010. Terrestrial gross carbon dioxide uptake: global distribution and covariation with climate. Science, 329 (5993), 834–838.*

Substituting Eqs. (17.12), (17.10), and (17.9) into Eq. (17.11) and rearranging the terms of Eq. (17.8), we obtain

$$\lambda E = \frac{(R_n - G) + \rho C_p \cdot [e_s(T_a) - e]/r_h}{\Delta + \gamma^*} \quad (17.13)$$

Assuming that $r_v = r_h$, then $\gamma^* = \gamma$, and Eq. (17.13) is the Penman equation. Under the assumption of a big-leaf canopy (Deardorff, 1978), r_v can be separated into the stomatal resistance (r_s) and the aerodynamic resistance (r_h) and $\gamma^* = [(r_s + r_h)/r_h] \cdot \gamma = (1 + r_s/r_h) \cdot \gamma$; then, we obtain the following equation:

$$\lambda E = \frac{(R_n - G) + \rho C_p \cdot [e_s(T_a) - e]/r_h}{\Delta + (1 + r_s/r_h) \cdot \gamma} \quad (17.14)$$

Eq. (17.14) is the Penman—Monteith equation, which is widely regarded as an accurate model (Shuttleworth, 2007). The expression $[e_s(T_a) - e]$ in Eqs. (17.13) and (17.14) is called the water vapor pressure deficit (VPD):

$$e_s(T_a) - e = e_s(T_a) - e_s(T_a) * RH/100$$

$$= e_s(T_a)(1 - RH/100) = VPD$$
$$(17.15)$$

The factors Δ and γ in Eqs. (17.13) and (17.14) depend solely on T_a at a given location (Wang et al., 2006). The Penman—Monteith equation depends on vegetation-specific parameters, e.g., the stomatal conductance (Beven, 1979), which are difficult to measure directly. Therefore, for a long time, the Penman—Monteith equation was regarded as a diagnostic equation to estimate r_s (Alves and Pereira, 2000).

When the water supply is sufficient, the available energy term dominates the equation. Hence, in the case of moist soil, Priestley and Taylor simplified the Penman—Monteith equation to the following (Priestley and Taylor, 1972):

$$\lambda E = \alpha \frac{\Delta}{\Delta + \gamma} \cdot (R_n - G) \quad (17.16)$$

where α is the so-called Priestley—Taylor parameter. The value of α is typically 1.2 to 1.3 (Agam et al., 2010), but it can range from 1.0 from 1.5 (Brutsaert and Chen, 1995; Chen and Brutsaert, 1995). Eq. (17.16) does not consider the roles of the VPD or stomatal resistance that are used in Eqs. (17.13) and (17.14). $\Delta/(\Delta + \gamma)$ is a function of T_a (Fig. 17.6).

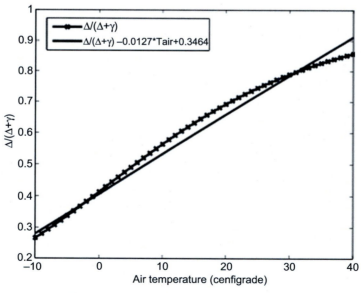

FIGURE 17.6   The scatter plot of $\Delta/(\Delta + \gamma)$ in Eq. (17.16) as a function of air temperature (T_a) (Wang et al., 2006).

17.3 Satellite λE algorithms

Satellite remote sensing does not provide a direct observation of λE; rather, it estimates λE by relating it to land surface variables, such as T_s, the vegetation index (VI), the soil moisture (SM), clouds, and the radiation budget. This method has been elaborated in the previous researches (Wang and Dickinson, 2012; Zhang et al., 2016). The MOST and the Penman–Monteith Equation are the basic formulas for this application. Accordingly, we divide the methods into different categories based on how the methods relate the land surface variables to λE. Tables 17.1 and 17.2 summarize the existing satellite remote sensing methods for estimating λE, along with their strengths and weaknesses.

17.3.1 One-source models

The one-source model was first proposed to estimate λE and H from satellite thermal infrared observations in the 1980s. It uses T_s obtained by the satellite at its view angle to replace the aerodynamic temperature (T_0) in Eq. (17.10) for estimating H (Fig. 17.7). λE is then estimated as a residual of the surface energy budget when the ground heat flux G is known or has been estimated (Clothier et al., 1986; Jacobsen, 1999; Kustas and Daughtry, 1990; Kustas et al., 1993) as in the following equation:

$$G = R_n \cdot (a \cdot VI + b) \qquad (17.17)$$

where VI is the vegetation index, and a and b are constants.

The aerodynamic temperature (T_0) at the surface used in Eq. (17.10) is a "fictitious" temperature at the surface from where the surface turbulent exchanges of water and heat in the constant flux layer originate. The satellite-derived value of T_s is a radiative temperature of the land surface for a homogeneous surface separated from T_o by a roughness layer in which the water vapor and heat exchange primarily depend on molecular diffusion; this diffusion requires much larger gradients than the turbulent exchange in the constant flux layer. During the daytime, T_s can be much greater than T_o,

TABLE 17.1 A summary of estimation ET methods using satellite remote data.

Model	Advantages	Disadvantages	Conditions for best performance
One-source models		Require parameterized excessive resistance, high sensitivity to errors of T_s and T_a; only available for clear sky conditions	Dry and sparse surface with large $T_s - T_a$
Two-source models	Local calibration not requested	High sensitivity to errors of T_s and T_a; only available for clear sky conditions	Dry and sparse surface with large $T_s - T_a$
Ts-VI models	Low sensitivity to errors of T_s, does not require T_a or wind speed	Relationship between ET and T_s complicated with temperature and energy control on ET, key parameters; only available for clear sky conditions	Middle latitude where soil moisture determines ET
Empirical models	Simple, low requirements of accuracy of T_s, T_a	Most models need local calibration	Depends on calibration data
Penman–Monteith equation	Simple, low requirements of accuracy of T_s, T_a	Need local calibration	Depends on calibration data
Assimilation method	Temporal integrated estimation	High sensitivity to errors of T_s and T_a	Dry and sparse surface with large $T_s - T_a$

TABLE 17.2 Validation results of remote sensing techniques for estimating evaporation.

Method	Source	Validation	Surface type	RMSE (W/m²)	R²
One-source method	Kustas and Daughtry (1990)	Kustas and Daughtry (1990)	Furrowed cotton	24—85 (10%—25%)	
One-source method	Su (2002)	Su et al. (2005)	Corn	47	0.89
One-source method	Su (2002)	Su et al. (2005)	Soybean	40	0.84
One-source method	Su (2002)	McCabe and Wood (2006)	Corn and soybean	99	0.66
One-source method	Su (2002)	McCabe and Wood (2006)	Corn and soybean	68	0.77
One-source method	Su (2002)	Su et al. (2007)	Grassland, crops, (rain) forest	44 (25%)	
Two-source method	Norman et al. (1995)	Kustas and Norman (1999)	Furrowed cotton	37—47 (12%—15%)	
Two-source method	Norman et al. (1995)	Norman et al. (2000)	Shrub, rangeland, pasture, salt cedar	105 (27%)	
Two-source method	Norman et al. (1995)	French et al. (2005)	Corn and soybean	94 (26%)	
Two-source method	Norman et al. (1995)	Li et al. (2006)	Corn and soybean	50—55 (10%—15%)	
Two-source method	Anderson et al. (1997); Norman et al. (2000)	Norman et al. (2003)	Wheat, pasture	40—50 (20%)	
Two-source method	Norman et al. (2000)	Norman et al. (2000)	Shrub,	65 (17%)	
Two-source method	Anderson et al. (1997); Norman et al. (2000)	Anderson et al., (2007a,b)	Water, forest, woodland, shrub, grassland, crops, bare, built-up	58 (25%)	
Triangle method	Bastiaanssen et al. (1998a)	French et al. (2005)	Corn and soybean	55 (15%)	
Triangle method	Roerink et al. (2000)	Verstraeten et al. (2005)	Forests	35 (24%)	
Triangle method	Carlson et al. (1995)	Gillies et al. (1997)	Tallgrass prairie, grassland, steppe shrub	25—55 (10%—30%)	0.80—0.90
Triangle method	Jiang and Islam (2001)	Jiang and Islam (2001)	Mixed farming, forest, tall and short grass	85 (30%)	0.64
Triangle method	Jiang and Islam (2001)	Jiang and Islam (2001)	Mixed farming, forest, tall and short grass	50 (17%)	0.9

TABLE 17.2 Validation results of remote sensing techniques for estimating evaporation.—cont'd

Method	Source	Validation	Surface type	RMSE (W/m^2)	R^2
Triangle method	Jiang and Islam (2001)	Jiang and Islam (2003)	Mixed farming, forest, tall and short grass	59 (15%)	0.79
Triangle method	Jiang and Islam (2001)	Batra et al. (2006)	Mixed farming, forest, tall and short grass	51—56 (22%—28%)	0.77—0.84
Triangle method	Nishida et al. (2003)	Nishida et al. (2003)	Forest, corn, soybean, wheat, shrub, rangeland, tall grass	45	0.86
Empirical method	Wang et al. (2007b)	Wang et al. (2007b)	Forest, grassland, cropland	32 (36%)	0.81
Trad and climate data	McVicar and Jupp (2002)	McVicar and Jupp (2002)	Cropping	88—72 (27%—28%)	0.52—0.81
Trad and compl. approach	Venturini et al. (2008)	Venturini et al. (2008)	Rangelands, pasture, wheat	34 (15%)	0.79
Assimilation of Trad	Caparrini et al. (2004b)	Caparrini et al. (2004b)	Tallgrass prairie	56	
Assimilation of Trad	Caparrini et al. (2004b)	Caparrini et al. (2004b)	Tallgrass prairie	20	0.96

Modified from Kalma, J.D., McVicar, T.R., McCabe, M.F., 2008. Estimating land surface evaporation: a review of methods using remotely sensed surface temperature data. Surv. Geophys. 29 (4—5), 421—469.

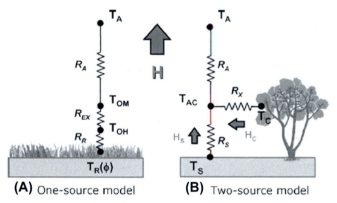

FIGURE 17.7 Schematic of resistance network for the one-source model and two-source model formulations for computing sensible heat flux, H (Kustas and Anderson, 2009).

especially for bare soil or sparsely vegetated surfaces (Chehbouni et al., 1996; Friedl, 2002). Therefore, directly replacing T_o with T_s may result in a significant overestimation of H. The universal stability functions that are used to calculate the aerodynamic r_h in Eq. (17.10) are not valid in the roughness layer (Sun et al., 1999). The system is more complicated for heterogeneous vegetation, for which T_s is a view angle—dependent mixture of the canopy and soil radiative temperatures.

This scenario raises serious doubts about the utility of the satellite-derived T_s for estimating λE (Hall et al., 1992; Shuttleworth, 1991). Its overestimation can be corrected by adding an excess heat exchange resistance into Eq. (17.10), $r_{ex} = kB^{-1} = \ln(z_m/z_h)$, which is a function of the roughness height for the momentum transfer (z_m) and the roughness height for the heat transfer (z_h) (Blümel, 1999; Kustas and Anderson, 2009; Su et al., 2001; Verhoef et al., 1997). Over sparse vegetation, kB^{-1} can be large and variable. This parameter is a function of the structural characteristics of the vegetation (e.g., the leaf area index, LAI), the level of water stress, the view angle of the radiometer, and climatic conditions (Blümel, 1999; Kustas and Anderson, 2009; Lhomme and Chehbouni, 1999; Lhomme et al., 2000; Su et al., 2001; Verhoef et al., 1997; Yang et al., 2009).

A widely used one-source model is the Surface Energy Balance Algorithm for Land (SEBAL) (Bastiaanssen et al., 1998a,b). This model requires only field information about shortwave atmospheric transmittance, surface temperature, and vegetation height as well as empirical relationships for different geographical regions and times of image acquisition. The empirical relationships used in the model require local calibration (Teixeira et al., 2009a,b).

Another well-known one-source model is the Surface Energy Balance System (SEBS), which estimates H and λE from satellite data and routinely available meteorological data (Su, 2002). This model has been widely used with high-resolution satellite data (Landsat and ASTER) and with modest resolution data (MODIS).

17.3.2 Two-source models

Two-source models have been proposed to improve the accuracy of λE estimates from satellite remote sensing data, especially over sparse surfaces. A typical widely used two-source model was first proposed by Norman et al. (1995). This model and its improvements are reviewed here. This model divides the surface into soil and vegetation components, each of which transfers H and λE into the atmosphere above its surface (Fig. 17.7). The satellite-derived surface directional radiometric temperature, $T_s(\theta_v)$, is considered to be a composite of the soil (T_{soil}) and canopy temperatures (T_{veg}) (Fig. 17.7):

$$\varepsilon_g \sigma [T_s(\theta_v)]^4 = [1 - f(\theta_v)]\varepsilon_s \sigma T_{soil}^4 + f(\theta_v)\varepsilon_c \sigma T_{veg}^4$$

(17.18)

where the Stefan–Boltzmann constant $\sigma = 5.6697 \times 10^{-8}$ Wm^{-2}/K^4, $f(\theta_v)$ is the vegetation cover fraction at the view zenith angle θ_v, and ε_g, ε_s, and ε_c are the surface emissivities of the composite surface, soil, and vegetation, respectively. If $\varepsilon_t = \varepsilon_s = \varepsilon_c$, then

$$T_{rad}(\theta_v) = \left\{ f(\theta_v)T_{veg}^4 + [1 - f(\theta_v)]T_{soil}^4 \right\}^{\frac{1}{4}}$$

(17.19)

H and λE are also divided into soil and vegetation components as follows:

$$H_{veg} = \rho C_p \frac{T_{veg} - T_a}{r_{hc}}$$

(17.20)

$$H_{soil} = \rho C_p \frac{T_{soil} - T_a}{r_{hv}}$$

(17.21)

where r_{hc} and r_{hs} are the aerodynamic resistances from the canopy and the soil to the atmosphere, respectively.

The canopy λE (λE_{veg}) has been estimated using the Priestley–Taylor equation (Priestley and Taylor, 1972) (Eq. 17.16). The two-source model iterates to obtain T_{soil} and T_{veg} from the satellite-derived value of T_s using an initial value of 2.0 for the Priestley–Taylor parameter (Anderson et al., 2008; Kustas and Anderson, 2009). This initial high value of the Priestley–Taylor parameter overestimates λE_{veg} and gives a negative soil evaporation value (λE_{soil}), which is a nonphysical solution. Therefore, the effective Priestley–Taylor parameter is obtained by iterative reduction until λE_{soil} approaches zero, producing a final Priestley—Taylor constant and, hence, a final T_r and T_c. λE and H are then calculated using these estimates.

This modeling scheme has been demonstrated to provide reasonable estimates of λE over a wide range of climatic and vegetation cover conditions (Anderson et al., 2004; Li et al., 2008b; Norman et al., 1995, 2003). However, this scheme may be limited by its accuracy in the partitioning of λE between λE_{soil} and λE_{veg} because soil evaporation is usually a significant component of the total λE and is not zero (Anderson et al., 2008). The above iteration may result in an overestimation of λE_{veg}. In a recent study, the canopy evaporation term was replaced by (Anderson et al., 2008):

$$\lambda E_c = \lambda \frac{e_s(T_c) - e}{r_s + r_{hc}} \qquad (17.22)$$

Both the one- and two-source models are sensitive to the use of temperature differences to estimate H, as mentioned in Section 17.4.1. For example, Timmermans et al. (2007) showed that over subhumid grassland and semiarid rangeland, a ± 3K error in T_s results in an average

error of approximately 75% in H for a typical one-source model (Bastiaanssen et al., 1998a, 1998b) and an averaged error of approximately 45% for the two-source model (Norman et al., 1995). The methods require unbiased T_s retrievals and a value of T_a that is interpolated from ground-based point measurements. Attempts to estimate the spatial variability of T_a at regional scales with remote sensing suggest an uncertainty of 3—4K (Goward et al., 1994; Prince et al., 1998). The uncertainties associated with the T_s retrievals are on the order of several K (Wang et al., 2007a; Wang and Liang, 2009). Consequently, except in areas containing low vegetation cover, the derived values of $T_s - T_a$ may be comparable to its uncertainty (Caselles et al., 1998; Norman et al., 2000).

Consequently, several methods have used a time series of satellite-derived values of T_s to reduce the sensitivity of the flux estimation to errors (Anderson et al., 1997, 2007a, 2007b; Mecikalski et al., 1999; Norman et al., 2000); however, considerable sensitivity to temperature errors (in T_s and T_a) remain. Most validation experiments have considered sparse grasslands, croplands, or shrubs (Table 17.3). For these

TABLE 17.3 A summary of the satellite methods that provide an estimate of global terrestrial λE.

Method type	Source	Validation and application	Surface type	RMSE (W/m², relative error)	R²
Empirical P-M	Zhang et al. (2009)	Zhang et al. (2010a)	82 FLUXNET stations	12.5—14.6	0.8—0.84
Empirical P-M	Zhang et al. (2009)	Zhang et al. (2010a)	261 river basins globally	14.8	0.8
Empirical P-M	Wang et al. (2010a)	Wang et al. (2010a,b)	64 globally distributed FLUXNET sites	17.0 (25%)	0.88
Empirical method	Jung et al. (2009)	Jung et al. (2010)	112 river basins	11.8	0.92
Empirical P-M	Leuning et al. (2008)	Zhang et al. (2008)	120 catchments in Australia	6.4	0.67
Empirical method	Wang and Liang (2008)	Wang and Liang (2008)	12 sites in the United States	28.6 (daytime)	0.85
Empirical P-M	Fisher et al. (2008)	Fisher et al. (2008)	16 FLUXNET sites	15.2 (28%)	0.90
Empirical P-M	Mu et al. (2007)	Sheffield et al. (2010)	19 FLUXNET sites	27.3	0.70—0.76

surfaces, $T_s - T_a$ is relatively large; therefore, this term has a relatively low sensitivity to temperature errors. No capability to predict λE and H over forests where the value of $T_s - T_a$ is small has been established. Furthermore, the parameterizations of aerodynamic resistance may also cause large deviations in the modeling of H (Liu et al., 2007; van der Kwast et al., 2009).

17.3.3 Ts-VI space methods

In contrast to the one- and two-source models, Ts-VI methods use the spatial variation of T_s to partition R_n into λE and H. If a sufficiently large number of satellite pixels are present, then the shape of the pixel envelope resembles a triangle or trapezoid in the T_s-VI space (Fig. 17.8). The T_s-VI method was first introduced by Price

(1990). Its principle is simple (Gao and Long, 2008; Wang et al., 2006): a surface that absorbs solar radiation experiences surface heating during the daytime. However, the changes in T_s for wet surfaces are relatively small because more energy is expended on the λE of a wet surface and such surfaces have higher thermal inertia. The greater cooling effect of λE and the thermal inertia of wet surfaces produce small or varying values of T_s. Therefore, the warm edge of the T_s-VI space has the lowest evaporative fraction (EF) or λE, and the cold edge of the space represents the highest EF or λE (Murray and Verhoef, 2007; Nemani and Running, 1997; Verhoef et al., 1996). The T_s-VI method is based on an interpretation of the image (pixel) distribution in T_s-VI space (Carlson, 2007).

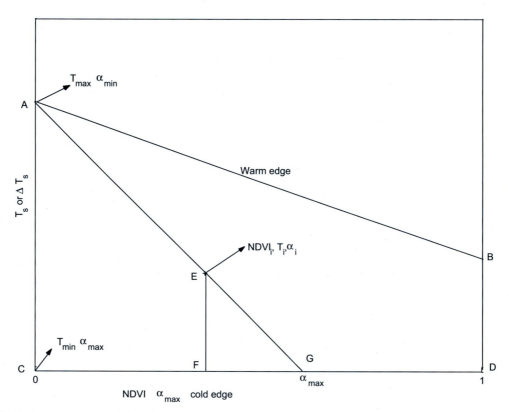

FIGURE 17.8   A schematic plot for the linear interpretation of the Priestley–Taylor parameter using T_s-VI space (Wang et al., 2006).

Fig. 17.8 is a schematic plot of the linear interpretation of the Priestley–Taylor parameter in T_s-VI space (Eq. 17.16). The trapezoid ABCD represents the spatial variation of T_s-VI; CD is the "cold edge" and AB is the "warm edge" of the spatial variation. The linear interpolation of T_s-VI was first proposed by Jiang and Islam (2001) to parameterize α. For a pixel i at a point E in T_s-VI space, A and E can be connected, and AE can be extended to G. Because α at point A is equal to $\alpha_{\min}$ and the "cold edge" has a maximum of $\alpha_{\max}$, the length of AG is $\alpha_{\max} - \alpha_{\min}$, and the length of AE is $\alpha_i - \alpha_{\min}$. Because the triangle EFG is similar to the triangle ACG, the following equation can be derived:

$$\frac{|EF|}{|AC|} = \frac{|EG|}{|AG|} \qquad (17.23)$$

Eq. (17.23) can be rewritten as

$$\frac{T_i - T_{\min}}{T_{\max} - T_{\min}} = \frac{(\alpha_{\max} - \alpha_{\min}) - (\alpha_i - \alpha_{\min})}{\alpha_{\max} - \alpha_{\min}} \qquad (17.24)$$

α_i at (T_i, NDVI_i) is equal to

$$\alpha_i = \frac{T_{\max} - T_i}{T_{\max} - T_{\min}}(\alpha_{\max} - \alpha_{\min}) + \alpha_{\min} \qquad (17.25)$$

where T is T_s in T_s-VI space. Therefore, λE is parameterized as follows (Wang et al., 2006):

$$\lambda E = (R_n - G) \cdot \frac{\Delta}{\Delta + \gamma}$$
$$\cdot \left[\frac{T_{\max} - T_i}{T_{\max} - T_{\min}}(\alpha_{\max} - \alpha_{\min}) + \alpha_{\min} \right] \qquad (17.26)$$

The values of $T_{\max}$ and $T_{\min}$ are derived from a visual inspection of the T_s-VI space (Fig. 17.2). $\alpha_{\max}$ is the maximum Priestley–Taylor parameter without surface water stress, and $\alpha_{\min}$ is often assumed to be zero, corresponding to the energy fraction used for evaporating the driest bare soil. Because the T_s-VI method makes use of only T_s spatial information, its requirement for T_s accuracy is low. Batra et al. (Batra et al., 2006) demonstrated that λE (or EF) can be

accurately derived from satellite brightness temperature observations (without the use of atmospheric and emissivity corrections to convert T_s). Similarly, other valid methods interpret the spatial variation of T_s, such as the albedo-T_s method (Sobrino et al., 2007) and EVI-T_s(Helman et al., 2015).

Because of its simplicity, this method has been widely accepted. It has been reviewed in detail by Petropoulos et al. (2009) and Carlson (2007). Some well-known, one-source models, such as SEBS and METRIC (mapping evapotranspiration at high resolution with internalized calibration) (Allen et al., 2007a,b; Tittebrand et al., 2005), also rely on T_s-VI space to estimate λE.

However, some model simulations have shown that the Priestley–Taylor parameter α is not linear as Eq. (17.25) would suggest; rather, it is highly curved (Carlson, 2007; Mallick et al., 2009). In addition, as reported by Wang et al. (2006), the information used in T_s-VI space is the change of T_s after the absorption of solar radiation. The dependence of λE on T_s or changes of T_s is further complicated by the series of processes that are mixed together, i.e., the cooling effect of λE; the change in thermal inertia with vegetation, soil moisture, and soil texture; and the impact of soil moisture, T_a, and available energy on the λE process (Nemani et al., 2003; Wang and Liang, 2008).

A key assumption of the T_s-VI method is that λE is negatively correlated with temperature because of the cooling effect from λE and the thermal inertia effect of SM (Fig. 17.2). However, observations and model simulations have shown that this assumption is not always true. At high latitudes and in cold areas, temperature is a major controlling factor for λE and is generally positively correlated with λE (Iwasaki et al., 2010; Nemani et al., 2003). Therefore, the T_s-VI method is suitable for use only during the growing season (when the range of VI is sufficiently large) in middle latitude regions where soil moisture, rather than T_a or available energy, is the key factor controlling λE.

The following additional sources of uncertainty affect the operational application of the T_s-VI method to the estimation of λE (Carlson, 2007; Mallick et al., 2009). (1) The triangle may not be fully determined if the area of interest does not include the full range of land surface types and conditions. (2) The main weakness of the triangle method is that it requires some subjectivity in identifying the warm edge and the dense vegetation and bare soil extremes. Identification is more easily obtained from high-resolution imagery. Tang et al. proposed an algorithm to automatically determine the edges of the T_s-VI triangle (Tang et al., 2010). Finally, (3) T_s and NDVI depend on the surface type due to differences in aerodynamic resistance, which is not considered in this method.

17.3.4 Empirical models

Because satellites can provide only limited information pertaining to λE, a major task in remote sensing is to identify key factors that influence the processes involved and their parameterization from satellite data. Long-term continuous measurements collected by globally distributed projects (as reviewed by Shuttleworth (2007)) provide an opportunity to identify these factors, especially with the data obtained from the FLUX-NET and Atmospheric Radiation Measurements (ARM) projects (see also Section 17.5).

By analyzing the long-term surface measurements of λE collected by the ARM project, Wang et al. (2006) and Wang et al. (2007b) found that the dominant parameters controlling λE are the surface net radiation (R_n), the temperature (either as air temperatures (T_a) or land surface temperature (T_s)), and the vegetation cover, which is quantified using vegetation indices (VI) (Fig. 17.9). Therefore, a method was developed to parameterize the daily value of λE using R_n, T_a, or T_s and VI from satellite data (Fig. 17.10) (Wang et al., 2007b):

$$\lambda E = R_n(a_0 + a_1 \cdot T + a_2 \cdot VI) \qquad (17.27)$$

In its simplest form, Eq. (17.27) expresses the dependence of the variation of λE on the vegetation and is consistent with the Priestley–Taylor equation (Eq. 17.16) while incorporating the influence of vegetation control on λE (through the stomatal conductance). A similar equation was proposed to relate the vegetation fraction to λE (Anderson and Goulden, 2009; Choudhury, 1994; Kim and Kim, 2008; Schuttemeyer et al., 2007):

$$\lambda E = VF \cdot \alpha \frac{\Delta}{\Delta + \gamma} \cdot (R_n - G) \qquad (17.28)$$

where VF is a linear function of VI. Eqs. (17.27) and (17.28) do not include the effects of soil water stress on λE. These effects were parameterized by adding a dependence on the diurnal temperature range (Wang and Liang, 2008):

$$\lambda E = R_n(b_0 + b_1 \cdot T + b_2 \cdot VI + b_3 \cdot DTR) \qquad (17.29)$$

A comparison of Eq. (17.29) with the land model simulations showed that it works well on a global scale, at spatial resolutions of several kilometers to hundreds of kilometers (Wang and Liang, 2008). A relative humidity-based Priestley–Taylor parameter is derived from ARM measurements (Yan and Shugart, 2010). The above empirical methods seek to relate λE to vegetation and key environmental control factors based on the analysis of ground measurements. The so-called crop coefficient methods (Gordon et al., 2005; Rana and Katerji, 2000) are similar methods that empirically link λE with vegetation conditions and potential λE.

The microwave emissivity difference vegetation index (EDVI), which is defined as the difference between the microwave land surface emissivities at 19 and 37 GHz, was also used to estimate λE (Becker and Choudhury, 1988; Li et al., 2009a; Min and Lin, 2006). The EDVI values represent the physical properties of crown vegetation, such as the vegetation water content of crown canopies.

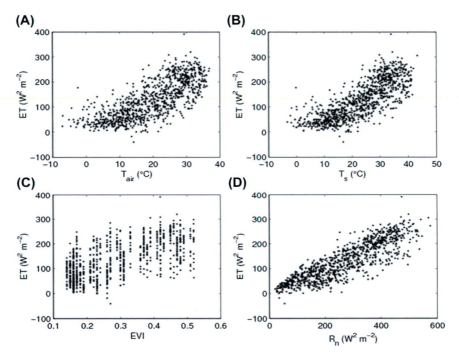

FIGURE 17.9 Scatterplots of λE (or ET) as a function of (A) the daytime-averaged air temperature (T_{air}), (B) daytime-averaged land surface temperature (T_s), (C) enhanced vegetation index (EVI), and (D) surface net radiation (R_n) at a site in the Southern Great Plains (Wang et al., 2007b).

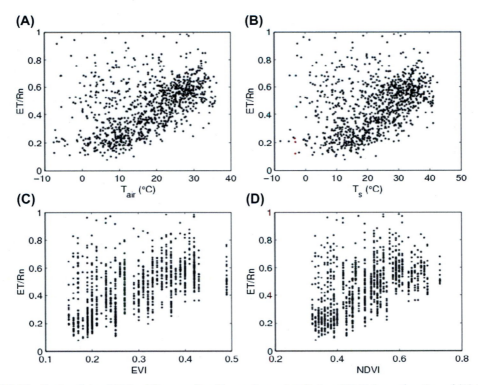

FIGURE 17.10 Scatterplots of λE (or ET) normalized by surface net radiation (ET/R_n) as a function of (A) the daytime-averaged air temperature (T_{air}), (B) the daytime-averaged land surface temperature (T_s), (C) the enhanced vegetation index (EVI), and (D) the normalized difference vegetation index (NDVI) at a site in the Southern Great Plains (Wang et al., 2007b).

Some methods that use artificial neural networks or support vector machine techniques have been used to relate satellite retrievals, such as R_s, T_s, T_a, VI, and land cover, with λE measurements and to apply their training results spatially. The methods are naturally empirical, although their application in other areas is limited because the methods do not provide explicit formulas to follow (Jung et al., 2009; Lu and Zhuang, 2010; Yang et al., 2006).

The above simple empirical methods have accuracy comparable to that of more complicated models, as shown in the comparison between the λE models (Jiménez et al., 2011; Kalma et al., 2008). But one study shows that the empirical methods show a wide range of difference in simulating λE, and the more detailed models considering net solar radiation will significantly improve the simulation accuracy (Carter and Liang, 2018). This simplicity allows global application of the model (Wang and Liang, 2008), although further improvements are needed (Wang et al., 2010a) (see Table 17.3).

Empirical methods often require local calibration. To reduce this requirement, some universal empirical methods have been proposed that are suitable for different types of land cover (Jung et al., 2009; Wang et al., 2010a; Wang and Liang, 2008). However, experiments have shown that the composition of tree species strongly influences λE; for example, the value of λE of deciduous species is much greater than that of coniferous species (Margolis and Ryan, 1997; Yuan et al., 2010), and physiological limitations to transpiration in boreal conifers reduce λE and increase H over large regions, even when soil water is abundant (Margolis and Ryan, 1997). Therefore, land cover–dependent coefficients may be a better choice, such as the tabulated maximum values of g_s for various land covers used in land surface models (Sellers et al., 1997).

17.3.5 The empirical Penman–Monteith equation

The Penman–Monteith equation is widely used in land process modeling to estimate vegetation transpiration or soil evaporation because it requires only conventional observations as input and does not require input from two levels. This equation has been recommended by the Food and Agriculture Organization (FAO) of the United Nations for estimating λE (Allen et al., 1998). Its usefulness for the satellite remote sensing of λE has not been fully recognized until recently, most likely because it does not explicitly use a satellite-derived T_s and its parameterization of r_s is complex.

Most Penman–Monteith-like methods empirically relate satellite-derived variables with r_s. Cleugh et al. (2007) directly related the stomatal conductance (the inverse of stomatal resistance) to the satellite-derived LAI using the Penman–Monteith equation (Eq. 17.14) to estimate λE. The model was modified with adjustments to the stomatal conductance based on environmental controls, and the algorithm was applied globally (Mu et al., 2007):

$$g_s = g_{s,\mathrm{min}} \cdot LAI \cdot f(T_{\mathrm{min}}) \cdot f(VPD) \qquad (17.30)$$

Similarly, Wang et al. (2010a) used VI and the relative humidity deficit ($RHD = 1 - RH/100$) to parameterize the stomatal conductance as follows:

$$g_s \approx (1 - RH/100) \cdot (a_0 + a_1 \cdot VI) \qquad (17.31)$$

Eq. (17.30) linearly relates g_s to LAI; consequently, it may overestimate g_s when LAI is greater than 3 or 4. The equation may be improved by using the vegetation indices in Eq. (17.31) (Bucci et al., 2008; Glenn et al., 2007; Lu et al., 2003; Suyker and Verma, 2008). Leuning et al. (2008) replaced Eq. (17.30) with a biophysical two-parameter model that requires local calibration, a requirement that limits its application. Eq. (17.30) is also improved by simulations of the variable infiltration capacity model (Sheffield et al., 2010).

17.3.6 Assimilation methods and temporal scaling up

Remotely sensed data are acquired instantaneously and provide the spatial variation of land surface variables that relate to λE. However, some of these variables can be accurately estimated only from satellite optical and thermal observations under clear sky conditions, and such retrievals are impossible under cloudy conditions (i.e., albedo, VI, and T_s) (Baroncini et al., 2008). The satellite retrievals must be temporally and spatially interpolated to provide temporally and spatially continuous values. Land data assimilation provides a physical solution to this problem by merging satellite measurements with estimates from land surface models (Li et al., 2009b; Reichle, 2008).

A typical assimilation method proposed by Caparrini et al. (2003) is discussed here. This method assimilates the satellite value of T_s into the following force-restore equation:

$$\frac{dT_s}{dt} = \frac{2\sqrt{\pi\omega}}{P}\left[R_n - \frac{1}{1-EF}H\right] - 2\pi\omega(T_s - T_d)$$

(17.32)

$$H = \rho \cdot C_p \cdot C_{HN} \cdot \left[1 + 2\left(1 - e^{10RiB}\right)\right] \cdot U \cdot (T_s - T_a)$$

(17.33)

where C_{HN} is the bulk transfer coefficient for heat under neutral conditions, R_{iB} is the bulk Richardson number, and U is the wind speed. This variational assimilation algorithm uses a bulk transfer model (a one-source model) and $T_s - T_a$ to parameterize H, and it connects T_s, the radiation components, and the ground heat flux G. The variational problem is solved by obtaining an adjoint state model for Eq. (17.32) and utilizing the model to efficiently search for values of

C_{HN} and EF that minimize the root mean square difference between the predictions of T_s obtained from Eq. (17.33) and the observed T_s values (Boni et al., 2001a; Castelli et al., 1999; Crow and Kustas, 2005).

These approaches have a number of advantages over purely diagnostic approaches, as shown in Sections 17.4.1–17.4.4. Most importantly, they provide flux estimates that are continuous in time using a physically realistic force-restore prognostic equation to interpolate between the sparse T_s observations (Boni et al., 2001b).

The above variational approach demonstrates promise for λE and H retrievals at dry and lightly vegetated sites. However, its results suggest that a simultaneous retrieval of both EF and C_{HN} by this variational approach will be difficult for wet or heavily vegetated land surfaces (Crow and Kustas, 2005). The single-source nature of the variational approach hampers the physical interpretation of the retrieved C_{HN} (Crow and Kustas, 2005). Consequently, the algorithm was modified to incorporate the two-source model concept by dividing the EF into soil and canopy components (Caparrini et al., 2004b). The impact of precipitation on λE was explicitly incorporated using the antecedent precipitation index (Caparrini et al., 2004a):

$$EF = a + b\frac{\arctan(K \cdot API)}{\pi}$$

(17.34)

K is another parameter that must be optimized using the variational approach. These improvements add additional parameters to be solved from the satellite T_s observations. However, incorporating Eq. (17.34) does not change the bulk transfer model of H and depends on $T_s - T_a$; therefore, the algorithm is more accurate for dry and lightly vegetated sites where $T_s - T_a$ is relative large, and the sensitivity of the algorithm to errors of T_s or T_a is relative small. The algorithm's requirement for accurate estimates of both T_a and T_s at pixel scale hampers its wide application.

Recently, satellite-based remotely sensed T_s was assimilated into a coupled atmosphere/ land global data assimilation system (Jang et al., 2010) with explicit accounting for biases in the model state. However, some studies have reported that T_s assimilation does not substantially improve λE simulations (Bosilovich et al., 2007). One likely reason for this lack of improvement is the fact that the satellite-derived value of T_s represents only a surface skin temperature (Crow and Wood, 2003; Reichle, 2008), whereas the land surface model uses the temperature of a surface layer, whose thickness is approximately 5—20 cm for various models (Tsuang et al., 2009).

17.4 Observations for algorithm calibration and validation

Estimating λE from satellite-derived instantaneous variables requires instantaneous observations to calibrate and validate the methods. Currently, observations collected by the eddy covariance technique (EC), energy balance BR tower systems, and scintillometer systems meet this requirement. The measurements are briefly discussed here.

17.4.1 Eddy covariance technique

The EC technique measures H and λE through statistical covariance between the heat and moisture variation and vertical velocity using rapid response sensors at frequencies typically equal to or greater than 10 Hz (Fig. 17.11). It was first developed in the 1950s by scientists from CSIRO in Australia (Garratt and Hicks, 1990; Hogstrom and Bergstrom, 1996). Now, it is regarded as the best method to direct measure H and λE and widely accepted in many important boundary layer experiments (Aubinet et al., 1999; Baldocchi et al., 1996, 2001; Wilson et al., 2002), in particular FLUXNET (Fig. 17.12). The typical error of λE is about 5%—20% or 20—50/Wm^2 (Foken, 2008; Vickers et al., 2010).

The EC method suffers from an energy closure problem, i.e., $H + \lambda E < R_n - G$. If the energy closure ratio is defined as $R = (H + \lambda E)/(R_n - G)$, it has values about 0.8 (Wilson et al., 2002) (see also Fig. 17.5). Several reasons for this energy closure problem have been reported and corrected (Foken, 2008; Foken et al., 2006), including averaging and coordinate rotation and coordinate systems (Finnigan, 2004; Finnigan et al., 2003; Fuehrer and Friehe, 2002; Gockede et al., 2008; Massman and Lee, 2002;

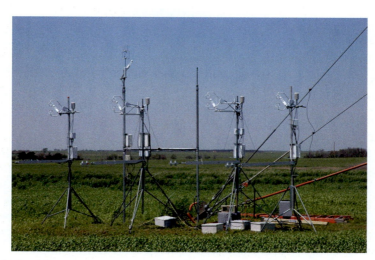

FIGURE 17.11   Multiple eddy covariance systems were installed at the Southern Great Plains Central Facility to collect data and compare latent and sensible heat fluxes. *From http://www.flickr.com/photos/armgov/4621337701/in/photostream. Image courtesy of the U.S. Department of Energy Atmospheric Radiation Measurement (ARM) user facility.*

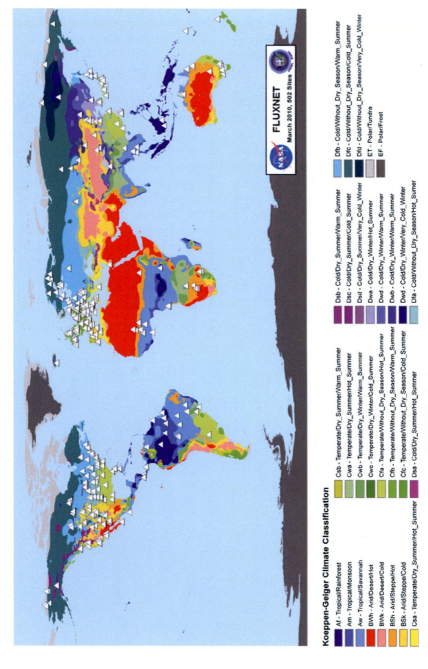

FIGURE 17.12   A map of Global FLUXNET sites.

Mauder et al., 2008). These corrections substantially increase λE and H estimates and improve energy closure ratio R (Finnigan et al., 2003; Kanda et al., 2004; Oncley et al., 2007) but the closure ratio R is still less than unit after these corrections. Foken (2008) argue that the eddy covariance technique can only measure small eddies, but large eddies in the lower boundary layer provide an important contribution to the energy balance, but do not touch the surface or are steady state and therefore cannot be measured with the eddy covariance method. Given our current (limited) understanding of the nature of the energy imbalance (Foken, 2008), it may be better to preserve the BR and closes the energy balance on a larger time scale (Wohlfahrt et al., 2009).

17.4.2 Energy balance Bowen ratio method

The energy balance BR method uses simultaneous measurements of vertical air temperature and humidity gradients to partition the energy balance (Fig. 17.13) (Bowen, 1926). Under moist conditions, the BR approach for determining λE from plant communities can provide good results, but this method may be less accurate under very dry conditions or with considerable advection of energy in moist conditions (Angus and Watts, 1984).

The primary challenge in implementing the BR method is the difficulty measuring the small gradients of temperature and humidity over surfaces with efficient turbulent transfer, especially forests, irrigated croplands, and grasslands. There are several assumptions or requirements for using the BR method (Angus and Watts, 1984; Kanemasu et al., 1992). The first is that the turbulent transfer coefficients for heat and water vapor are identical; this assumption is known to hold for conditions that are not too far from neutral, but it is not valid in very strong lapse (or inversion) conditions (Angus and Watts, 1984; Blad and Rosenber, 1974). The two levels at which the temperature and humidity

FIGURE 17.13   An energy balance Bowen ratio (EBBR) system installed at the Southern Great Plains. Flux estimates are calculated from observations of net radiation, soil surface heat flux, and the vertical gradients of temperature and relative humidity. *From http://www.flickr.com/photos/armgov/4857704844/. Image courtesy of the U.S. Department of Energy Atmospheric Radiation Measurement (ARM) user facility*

are measured must be within a constant flux layer. The second requirement for the BR method is that the surface energy must be closed; this requirement can be met for homogeneous surfaces or periods that are daily or longer. The components of the radiation and ground heat fluxes measured in BR systems are point measurements, especially for the upward radiation components. However, the turbulent fluxes are related to the landscape, are affected by its heterogeneities, and fetch in the upwind direction when the air flow over the surface is extensive (at least 100 times the maximum height of the measurement) (Wiernga, 1993).

17.4.3 The scintillometer method

One important issue with the use of EC and BR measurements to calibrate and evaluate satellite-derived values of λE is the mismatch between the estimate scales. EC and BR measurements usually represent a small scale of several hundreds of meters, including the upwind fetch; however, satellite λE estimates usually represent pixel averages covering kilometers, which may produce a substantial difference in the estimates, even when both estimates are accurate (Li et al., 2008a; McCabe and Wood, 2006). This issue has been addressed using scintillometers to take the measurements (Fig. 17.14). Recently, the use of

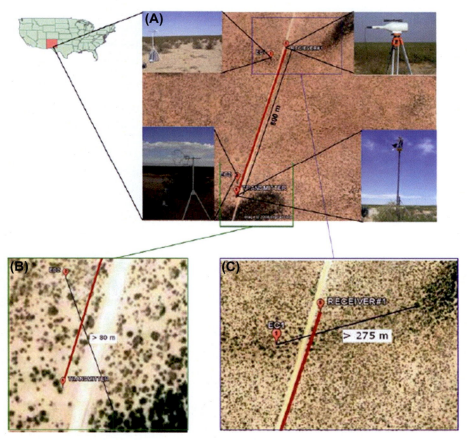

FIGURE 17.14 A Layout of the deployed large aperture scintillometer and two eddy-covariance systems in the experimental site near Las Cruces, New Mexico, USA (Zeweldi et al., 2010).

scintillometers to determine H and λE has become widely accepted because of their ability to quantify energy distributions at the landscape scale. They can calculate H (Moene et al., 2009; Solignac et al., 2009) over distances of several kilometers. Scintillometers are cost-effective and simple to operate (De Bruin, 2009).

However, scintillometer does not provide a direct measurement of λE and its sign (i.e., positive or negative) (Lagouarde et al., 2002). Scintillometer relies on the validity of MOST for calculating surface fluxes (de Bruin et al., 1993; Solignac et al., 2009). Zhang et al. (2010b) showed that the overestimation of H using scintillometer (Chehbouni et al., 2000; Hoedjes et al., 2007; Lagouarde et al., 2002; Von Randow et al., 2008) is associated with a higher frictional velocity estimation by MOST. Furthermore, studies have reported significant differences of up to 21% between six Kipp & Zonen large-aperture scintillometers (Kleissl et al., 2008, 2009).

17.4.4 Terrestrial water budget method

Regional mean λE values can be calculated from the following terrestrial water budget equation:

$$\lambda E = P - R - \Delta S \qquad (17.35)$$

where P is the precipitation, R is the runoff, and ΔS is the change in terrestrial water storage. ΔS is generally assumed to be zero at annual or longer time scales, and the long-term variability of the global or regional λE can be calculated by $P - R$.

However, human activities such as reservoir construction have drastically altered regional water cycles in the past decades (Chao et al., 2008; Haddeland et al., 2014; Mateo et al., 2014; Vörösmarty et al., 1997). These newly built reservoirs can alter the regional water balance by allowing more water to be stored on the land surface, which lead to more pronounced terrestrial water storage changes and decreased runoff (Biemans et al., 2011; Jaramillo and Destouni,

2015). Therefore, the ΔS caused by reservoirs should be considered in the surface water balance equation. Mao et al. (2016) quantifies this reservoir effect on the estimated λE at the national and basin scale from 1997 to 2014 in China. The results show that the reservoir total storage capacity in China has increased by 0.38×10^{12} m^3 from 1997 to 2014. If this change in ΔS is not considered, a significant increase in the calculated λE of 4.2% per decade is derived. However, after this change in ΔS is taken into account, the calculated trend of λE decreases to almost zero for the period 1997–2014 (Fig. 17.15), which is consistent with the negligible changes in the determining factors for λE, including precipitation, surface incident solar radiation, and air temperature. Therefore, the λE datasets considering the reservoir impact will provide a better reference product to evaluate the satellite retrieved λE (Mao and Wang, 2017).

17.5 The spatiotemporal characteristics of global and regional λE

Many global λE products have been developed and used to study the spatiotemporal λE characteristics based on the above theories and various satellite retrieval methods. Cleugh et al. (2007) proposed a method using gridded meteorological data and Penman–Monteith equation to calculate the λE at 1 km × 1 km resolution with the vegetation stomatal conductance calculated from LAI of Moderate Resolution Imaging Spectroradiometer (MODIS). Furthermore, this method was improved by Mu et al. (2007) and Mu et al. (2011), which not only can be applied to global simulation but also the obtained spatial pattern of λE agrees well with that of the MODIS global terrestrial gross and net primary production.

The improvement of remote sensing λE simulation provides key information for studying global hydrological cycle, energy cycle, and environment change. Fisher et al. (2008)

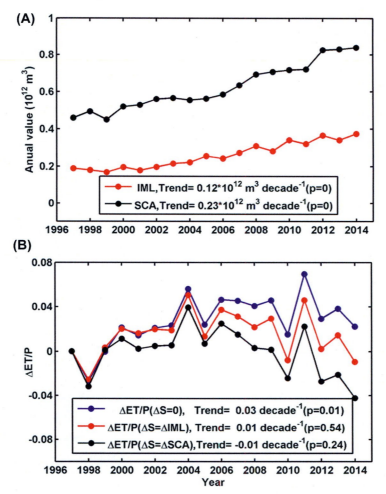

FIGURE 17.15   (A) Time series of the water storage capacity of all reservoirs (SCA) and water impoundment in the medium-sized and large reservoirs (IML) from 1997 to 2014 in China. (B) Time series of $\Delta ET/P$ by assuming (1) $\Delta S = 0$, (2) $\Delta S = \Delta$IML, and (3) $\Delta S = \Delta$SCA. ΔET for each year from 1997 to 2014 was calculated from $\Delta ET = \Delta P - \Delta R - \Delta S$. P and R were the observed annual precipitation and runoff for each year. Trend and its P values by t-test method are also shown. For comparison purpose, the unit of SCA and IML of m^3/yr is converted to mm/yr by dividing by China territorial area including land area and water area, i.e., $9.6 \times 1012\ m^2$.

developed a method to convert the potential λE from Priestley—Taylor to actual λE, then calculated the global monthly λE dataset from 1986 to 93 at 1km × 1 km resolution, and found that the simulated λE were improved in water limited area. Jung et al. (2010) obtained global λE data from 1982 to 2008 using a global monitoring network, meteorological and remote sensing observations, and a machine learning algorithm and found a decreasing trend during this time

period which should be contributed to limited soil moisture. Zhang et al. (2010a) calculated global λE from 1998 to 2008 by separating the three components with the Penman—Monteith method calculating the vegetation transpiration and soil evaporation and Priestley—Taylor method calculating water evaporation, and the obtained global λE spatiotemporal variation agrees with that from the water balance method. Miralles et al. (2011) estimated the global λE at

$0.25° \times 0.25°$ resolution at high accuracy using various satellite products and Priestley—Taylor method.

The spatiotemporal characteristic of λE plays a significant role in understanding land—atmosphere interaction, improving region water resource and land resource regulation (Raupach, 2001), and monitoring the region drought (McVicar and Jupp, 1998). Global λE distribution demonstrates a spatial pattern that higher in tropical area and lower in high latitude area, but in specific regions, the different models indicate much

difference (Yuan et al., 2010). Jiménez et al., (2011) made a comparison of 12 λE products and found that all the datasets could reproduce the spatial pattern that corresponds to the climate regions and geographic features, but the absolute value shows a wide range of difference, especially in rain forest. Mueller et al. (2011) evaluated various observation-based λE datasets and found a similar pattern, while IPCC AR4 product shows much difference with other products (Fig. 17.16).

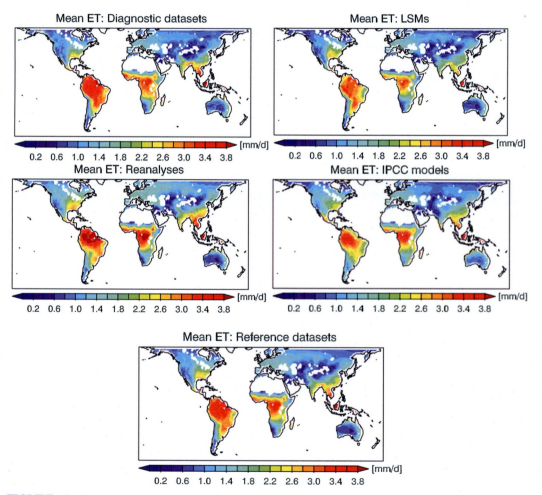

FIGURE 17.16   Spatial pattern of mean global land λE values from difference sources (Mueller et al., 2011).

The global hydrological cycle will be intensified under global warming (Huntington, 2006), but the direct observed evidence of increasing λE is still lacking, and most studies are performed in an indirect way. Gedney et al. (2006) found a decreasing λE trend from 1960s calculated by a mechanistic land surface model and optimal fingerprinting statistical techniques and attributed the decrease to the suppression of plant transpiration due to CO_2-induced stomatal closure. Piao et al. (2007) pointed out that global λE shows decreasing trend from 1901 to 99, which is attributed to the land use factor when considering CO_2 as a plant fertilizer, which can lead to the increase in foliage area.

Wang et al. (2010b) detected an increasing λE trend from 1982 to 2002, which is mainly controlled by surface solar radiation and wind speed. Jung et al. (2010) provided an estimate of global terrestrial λE from 1982 to 2008 using a machine learning method with meteorological and satellite data as input and concluded that global λE increased from 1982 to 1997 but decreased since 1998. Wang and Dickinson (2012) doubted the reliability of this study by declaring that this study excluded the impact from incident solar radiation, which plays a key role in moist regions. However, the later studies by Mueller et al. (2013) and Zeng et al. (2014) still found a decreasing trend from 1982 to 2009, which was consistent with Jung et al. (2010).

Zeng et al. (2012) developed a spatial regression model by integrating precipitation, temperature, and vegetation index and detected an increasing trend at $1.10 \, \text{mm/yr}^2$ of global λE during 1982–2009. During this period, Mao et al. (2015) also found similar increasing trends using multiple satellite datasets and land surface models and attributed it to climate change, CO_2 concentrations, and nitrogen deposition. Another study which investigated the variation of three λE components (vegetation transpiration, soil evaporation, and vaporization of intercepted rainfall from vegetation) from 1981 to 2012 shows that the enhanced vegetation transpiration and increased intercepted rainfall lead to the increase of global λE (Zhang et al,. 2016). One study even indicated stable global λE trend in the past 60 years by using the more reliable models which considered the energy, humidity, and wind speed meantime (Sheffield et al., 2012).

Considering the debates on the long-term variability of global λE, the reliable estimation of global and regional λE by integrating the high accurate satellite products and models will be the important sources of large-scale and long-term λE datasets and will be better used for investigating spatiotemporal characteristic of global λE.

17.6 Conclusions and discussion

This chapter has discussed the basic theories for measuring and estimating λE, including the MOST and the Penman–Monteith equation, and reviewed how the theories have been applied to the satellite remote sensing field. The MOST and Penman–Monteith equation methods depend on turbulence; therefore, they are partly empirical, and assumptions are needed. Further parameterizations and assumptions are required to relate the satellite-derived land surface variables to λE because satellites do not supply direct measurements of λE. No substantial difference exists between the two formulas because the Penman–Monteith equation was derived from the MOST. In practical application, however, there is a large difference between the two types of methods. MOST-type methods use $T_s - T_a$; therefore, they require accurate estimates of T_s and T_a. This sensitivity is substantially reduced using the Penman–Monteith-type methods that require the available energy and stomatal conductance, which are usually parameterized as a function of the vegetation index or LAI.

Satellite instantaneous retrieval of T_s, which is a key parameter that controls both available energy and λE, is available only under clear sky conditions and for limited periods. The data assimilation method can be used to fill temporal and spatial gaps. As a result, data assimilation is a useful tool for integrating satellite-derived instantaneous λE estimates over daily, monthly, or longer times.

The obtaining of satellite-derived instantaneous value of λE requires instantaneous observations to calibrate and validate the methods. Currently, observations collected using the eddy covariance technique (EC), the energy balance BR system, and the scintillometer system meet this requirement. The EC method directly measures the turbulence and uses a statistical method to calculate λE; however, it usually underestimates the value. The BR technique uses point measurements of the available energy (surface net radiation minus soil heat flux) and the observed BR that is affected by the upwind fetch. This approach makes the energy balance requirement of the BR technique a problem because of the different scales of the available energy and the BR.

An important issue with using EC or BR measurements for calibrating and evaluating satellite-derived λE values is the mismatch in the estimate scales. EC and BR measurements usually are of a smaller scale than that of satellite λE retrievals (Li et al., 2008b; McCabe and Wood, 2006). This issue has been addressed using scintillometer measurements because scintillometers can quantify H and λE at a scale of several kilometers (Solignac et al., 2009).

Significant progress has been made toward the satellite derivation of λE in recent years, with experiments showing systematic agreement between ground-based estimates and aircraft measurements over several test areas (French et al., 2005; Kustas et al., 2005; McCabe and Wood, 2006; Su et al., 2005). Evaluation studies have shown that satellite-derived λE values

have accuracies of approximately 15%—30% (Kalma et al., 2008), whereas the accuracy of EC measurements is 5%—20% (Foken, 2008).

A number of methods require ancillary data to obtain the spatial variability of λE, for example, near-surface air temperature and wind speed. This requirement hinders their application. Methods have been provided to reduce the need for ancillary data. Some methods rely solely on satellite-derived variables. Other methods have attempted to reduce the sensitivity of the method to the ancillary data, such that the ancillary data can be interpolated from weather station observations without substantially reducing their accuracy. The model's accuracy is usually the first consideration in choosing a method. However, the ancillary data requirement and the method's suitability over different conditions are also considerations when using satellite-based remote methods. The balance of the accuracy of the methods with their capability for practical application is also an issue to be considered in the evaluation of satellite-based remote sensing methods for λE. Simple empirical methods are of comparable accuracy to the more complicated models, as demonstrated by comparisons between various global λE models (Jiménez et al., 2011; Kalma et al., 2008).

The evaluation and comparison of different methods remains a significant issue. Currently, most existing methods are developed and calibrated using local data. The capability of these methods to describe other land cover types and climates must be evaluated. The lack of standard reference data also limits the critical evaluation of remote sensing methods.

Satellite remote sensing has an obvious advantage for monitoring spatial variability. However, the ability of satellite-based remote sensing to provide long-term stable estimates of λE is controversial. Most satellite-based remote sensing methods for λE are limited to their ancillary data requirement; thus, they are not applicable for obtaining long-term estimates.

Acknowledgments

This study is funded by the National Basic Research Program of China (2012CB955302), the National Natural Science Foundation of China (41175126), and the State Key Laboratory of Earth Surface Processes and Resource Ecology.

References

Agam, N., et al., 2010. Application of the Priestley-Taylor approach in a two-source surface energy balance model. J. Hydrometeorol. 11 (1), 185–198.

Allen, R.G., Pereira, L.S., Raes, D., Smith, M., 1998. Crop evapotranspiration: Guidelines for computing crop water requirements, FAO Irrig Drain. Pap. 56, Food and Agric. Organ. of the United Nations, Rome.

Allen, R.G., et al., 2007a. Satellite-based energy balance for mapping evapotranspiration with internalized calibration (METRIC) — Applications. J. Irrig. Drain. Eng, 133 (4), 395–406.

Allen, R.G., Tasumi, M., Trezza, R., 2007b. Satellite-based energy balance for mapping evapotranspiration with internalized calibration (METRIC) — Model. J. Irrig. Drain. Eng, 133 (4), 380–394.

Alves, I., Pereira, L.S., 2000. Modelling surface resistance from climatic variables? Agric. Water Manag. 42 (3), 371–385.

Anderson, M.C., Norman, J.M., Diak, G.R., Kustas, W.P., Mecikalski, J.R., 1997. A two-source time-integrated model for estimating surface fluxes using thermal infrared remote sensing. Remote Sens. Environ. 60 (2), 195–216.

Anderson, M.C., et al., 2008. A thermal-based remote sensing technique for routine mapping of land-surface carbon, water and energy fluxes from field to regional scales. Remote Sens. Environ. 112 (12), 4227–4241.

Anderson, M.C., Norman, J.M., Mecikalski, J.R., Otkin, J.A., Kustas, W.P., 2007a. A climatological study of evapotranspiration and moisture stress across the continental United States based on thermal remote sensing: 1. Model formulation. J. Geophys. Res. Atmos. 112 (D10), D10117.

Anderson, M.C., Norman, J.M., Mecikalski, J.R., Otkin, J.A., Kustas, W.P., 2007b. A climatological study of evapotranspiration and moisture stress across the continental United States based on thermal remote sensing: 2. Surface moisture climatology. J. Geophys. Res. Atmos. 112 (D11), D11112.

Anderson, M.C., et al., 2004. A multiscale remote sensing model for disaggregating regional fluxes to micrometeorological scales. J. Hydrometeorol. 5 (2), 343–363.

Anderson, R.G., Goulden, M.L., 2009. A mobile platform to constrain regional estimates of evapotranspiration. Agric. For. Meteorol. 149 (5), 771–782.

Angus, D.E., Watts, P.J., 1984. Evapotranspiration — how good is the Bowen ratio method? Agric. Water Manag. 8 (1–3), 133–150.

Aubinet, M., et al., 1999. Estimates of the annual net carbon and water exchange of forests: the EUROFLUX methodology. In: Advances in Ecological Research. Academic Press, pp. 113–175.

Baldocchi, D., et al., 2001. FLUXNET: a new tool to study the temporal and spatial variability of ecosystem scale carbon dioxide, water vapor, and energy flux densities. Bull. Am. Meteorol. Soc. 82 (11), 2415–2434.

Baldocchi, D., Valentini, R., Running, S., Oechel, W., Dahlman, R., 1996. Strategies for measuring and modelling carbon dioxide and water vapour fluxes over terrestrial ecosystems. Glob. Chang. Biol. 2 (3), 159–168.

Baroncini, F., Castelli, F., Caparrini, F., Ruffo, S., 2008. A dynamic cloud masking and filtering algorithm for MSG retrieval of land surface temperature. Int. J. Remote Sens. 29 (12), 3365–3382.

Bastiaanssen, W.G.M., Menenti, M., Feddes, R.A., Holtslag, A.A.M., 1998a. A remote sensing surface energy balance algorithm for land (SEBAL) — 1. Formulation. J. Hydrol. 213 (1–4), 198–212.

Bastiaanssen, W.G.M., et al., 1998b. A remote sensing surface energy balance algorithm for land (SEBAL) — 2. Validation. J. Hydrol. 213 (1–4), 213–229.

Batra, N., Islam, S., Venturini, V., Bisht, G., Jiang, J., 2006. Estimation and comparison of evapotranspiration from MODIS and AVHRR sensors for clear sky days over the Southern Great Plains. Remote Sens. Environ. 103 (1), 1–15.

Becker, F., Choudhury, B.J., 1988. Relative sensitivity of normalized difference vegetation index (NDVI) and microwave polarization difference index (MPDI) for vegetation and desertification monitoring. Remote Sens. Environ. 24 (2), 297–311.

Beer, C., et al., 2010. Terrestrial gross carbon dioxide uptake: global distribution and covariation with climate. Science 329 (5993), 834–838.

Beven, K., 1979. Sensitivity analysis of the Penman-Monteith actual evapotranspiration estimates. J. Hydrol. 44 (3–4), 169–190.

Biemans, H., Haddeland, I., Kabat, P., Ludwig, F., Hutjes, R.W.A., Heinke, J., von Bloh, W., Gerten, D., 2011. Impact of reservoirs on river discharge and irrigation water supply during the 20th century. Water Resour. Res. 47 (3), W03509.

Blümel, K., 1999. A Simple formula for estimation of the roughness length for heat transfer over partly vegetated surfaces. J. Appl. Meteorol. 38 (6), 814–829.

Blad, B.L., Rosenber, N.J., 1974. Lysimetric calibration of Bowen ratio-energy balance method for evapotranspiration estimation in central great plains. J. Appl. Meteorol. 13 (2), 227–236.

Boni, G., Castelli, F., Entekhabi, D., 2001a. Sampling strategies and assimilation of ground temperature for the estimation of surface energy balance components. IEEE Trans. Geosci. Remote Sens. 39 (1), 165–172.

Boni, G., Entekhabi, D., Castelli, F., 2001b. Land data assimilation with satellite measurements for the estimation of surface energy balance components and surface control on evaporation. Water Resour. Res. 37 (6), 1713–1722.

Bosilovich, M.G., Radakovich, J.D., da Silva, A., Todling, R., Verter, F., 2007. Skin temperature analysis and bias correction in a coupled land-atmosphere data assimilation system. J. Meteorol. Soc. Jpn. 85A, 205–228.

Bowen, I.S., 1926. The ratio of heat losses by conduction and by evaporation from any water surface. Phys. Rev. 27 (6), 779.

Brutsaert, W., Chen, D., 1995. Desorption and the two stages of drying of natural Tallgrass Prairie. Water Resour. Res. 31 (5), 1305–1313.

Bucci, S.J., et al., 2008. Controls on stand transpiration and soil water utilization along a tree density gradient in a Neotropical savanna. Agric. For. Meteorol. 148 (6–7), 839–849.

Burba, G., Anderson, D., 2008. Introduction to the Eddy Covariance Method: General Guidelines and Conventional Workflow. Available at: http://www.licor.com/env/2010/applications/eddy_covariance.jsp, 141 pp., Copyright LI-COR, Inc, Used By Permission.

Businger, J.A., Wyngaard, J.C., Izumi, Y., Bradley, E.F., 1971. Flux-profile relationships in atmospheric surface layer. J. Atmos. Sci. 28 (2), 181–189.

Caparrini, F., Castelli, F., Entekhabi, D., 2003. Mapping of land-atmosphere heat fluxes and surface parameters with remote sensing data. Boundary-Layer Meteorol. 107 (3), 605–633.

Caparrini, F., Castelli, F., Entekhabi, D., 2004a. Estimation of surface turbulent fluxes through assimilation of radiometric surface temperature sequences. J. Hydrometeorol. 5 (1), 145–159.

Caparrini, F., Castelli, F., Entekhabi, D., 2004b. Variational estimation of soil and vegetation turbulent transfer and heat flux parameters from sequences of multisensor imagery. Water Resour. Res. 40 (12), W12515.

Carlson, T., 2007. An overview of the "triangle method" for estimating surface evapotranspiration and soil moisture from satellite imagery. Sensors 7 (8), 1612–1629.

Carlson, T.N., Gillies, R.R., Schmugge, T.J., 1995. An interpretation of methodologies for indirect measurement of soil water content. Agric. For. Meteorol. 77 (3–4), 191–205.

Carter, C., Liang, S., 2018. Comprehensive evaluation of empirical algorithms for estimating land surface evapotranspiration. Agric. For. Meteorol. 256–257, 334–345.

Caselles, V., Artigao, M.M., Hurtado, E., Coll, C., Brasa, A., 1998. Mapping actual evapotranspiration by combining landsat TM and NOAA-AVHRR images: application to

the Barrax area, Albacete, Spain. Remote Sens. Environ. 63 (1), 1–10.

Castelli, F., Entekhabi, D., Caporali, E., 1999. Estimation of surface heat flux and an index of soil moisture using adjoint-state surface energy balance. Water Resour. Res. 35 (10), 3115–3125.

Chahine, M.T., 1992. The hydrological cycle and its influence on climate. Nature 359 (6394), 373–380.

Chao, B.F., Wu, Y.H., Li, Y.S., 2008. Impact of artificial reservoir water impoundment on global sea level. Science 320 (5873), 212–214.

Chehbouni, A., Lo Seen, D., Njoku, E.G., Monteny, B.M., 1996. Examination of the difference between radiative and aerodynamic surface temperatures over sparsely vegetated surfaces. Remote Sens. Environ. 58 (2), 177–186.

Chehbouni, A., et al., 2000. Estimation of heat and momentum fluxes over complex terrain using a large aperture scintillometer. Agric. For. Meteorol. 105 (1–3), 215–226.

Chen, D.Y., Brutsaert, W., 1995. Diagnostics of land surface spatial variability and water vapor flux. J. Geophys. Res. Atmos. 100 (D12), 25595–25606.

Choudhury, B.J., 1994. Synergism of multispectral satellite-observations for estimating regional land-surface evaporation. Remote Sens. Environ. 49 (3), 264–274.

Cleugh, H.A., Leuning, R., Mu, Q., Running, S.W., 2007. Regional evaporation estimates from flux tower and MODIS satellite data. Remote Sens. Environ. 106 (3), 285–304.

Clothier, B.E., et al., 1986. Estimation of soil heat flux from net radiation during the growth of alfalfa. Agric. For. Meteorol. 37 (4), 319–329.

Crow, W., Kustas, W., 2005. Utility of assimilating surface radiometric temperature observations for evaporative fraction and heat transfer coefficient retrieval. Boundary-Layer Meteorol. 115 (1), 105–130.

Crow, W.T., Wood, E.F., 2003. The assimilation of remotely sensed soil brightness temperature imagery into a land surface model using Ensemble Kalman filtering: a case study based on ESTAR measurements during SGP97. Adv. Water Resour. 26 (2), 137–149.

De Bruin, H.A.R., 2009. Time to think: reflections of a pre-pensioned scintillometer researcher. Bull. Am. Meteorol. Soc. 90 (5), ES17–ES26.

de Bruin, H.A.R., Kohsiek, W., Hurk, B.J.J.M., 1993. A verification of some methods to determine the fluxes of momentum, sensible heat, and water vapour using standard deviation and structure parameter of scalar meteorological quantities. Boundary-Layer Meteorol. 63 (3), 231–257.

Deardorff, J.W., 1978. Efficient prediction of ground surface temperature and moisture, with inclusion of a layer of vegetation. J. Geophys. Res. 83 (C4), 1889–1903.

Dyer, A.J., 1974. A review of flux-profile relationships. Boundary-Layer Meteorol. 7 (3), 363–372.

Finnigan, J.J., 2004. A re-evaluation of long-term flux measurement techniques – part II: coordinate systems. Boundary-Layer Meteorol. 113 (1), 1–41.

Finnigan, J.J., Clement, R., Malhi, Y., Leuning, R., Cleugh, H.A., 2003. A re-evaluation of long-term flux measurement techniques – Part I: averaging and coordinate rotation. Boundary-Layer Meteorol. 107 (1), 1–48.

Fisher, J.B., Tu, K.P., Baldocchi, D.D., 2008. Global estimates of the land-atmosphere water flux based on monthly AVHRR and ISLSCP-II data, validated at 16 FLUXNET sites. Remote Sens. Environ. 112 (3), 901–919.

Foken, T., 2006. 50 years of the Monin-Obukhov similarity theory. Boundary-Layer Meteorol. 119 (3), 431–447.

Foken, T., 2008. The energy balance closure problem: an overview. Ecol. Appl. 18 (6), 1351–1367.

Foken, T., Wimmer, F., Mauder, M., Thomas, C., Liebethal, C., 2006. Some aspects of the energy balance closure problem. Atmos. Chem. Phys. 6, 4395–4402.

French, A.N., et al., 2005. Surface energy fluxes with the advanced Spaceborne thermal emission and Reflection radiometer (ASTER) at the Iowa 2002 SMACEX site (USA). Remote Sens. Environ. 99 (1–2), 55–65.

Friedl, M.A., 2002. Forward and inverse modeling of land surface energy balance using surface temperature measurements. Remote Sens. Environ. 79 (2–3), 344–354.

Fuehrer, P.L., Friehe, C.A., 2002. Flux corrections revisited. Boundary-Layer Meteorol. 102 (3), 415–457.

Gao, Y.C., Long, D., 2008. Intercomparison of remote sensing-based models for estimation of evapotranspiration and accuracy assessment based on SWAT. Hydrol. Process. 22 (25), 4850–4869.

Garratt, J.R., Hess, G.D., Physick, W.L., Bougeault, P., 1996. The atmospheric boundary layer – advances in knowledge and application. Boundary-Layer Meteorol. 78 (1–2), 9–37.

Garratt, J.R., Hicks, B.B., 1990. Micrometeorological and PBL experiments in Australia. Boundary-Layer Meteorol. 50 (1), 11–29.

Gedney, N., Cox, P., Betts, R., et al., 2006. Continental runoff: a quality-controlled global runoff data set (reply). Nature 444, E14–E15.

Gillies, R.R., Kustas, W.P., Humes, K.S., 1997. A verification of the 'triangle' method for obtaining surface soil water content and energy fluxes from remote measurements of the Normalized Difference Vegetation Index (NDVI) and surface e. Int. J. Remote Sens. 18 (15), 3145–3166.

Glenn, E.P., Huete, A.R., Nagler, P.L., Hirschboeck, K.K., Brown, P., 2007. Integrating remote sensing and ground methods to estimate evapotranspiration. Crit. Rev. Plant Sci. 26 (3), 139–168.

Gockede, M., et al., 2008. Quality control of CarboEurope flux data – part 1: coupling footprint analyses with flux data quality assessment to evaluate sites in forest ecosystems. Biogeosciences 5 (2), 433–450.

Gordon, L.J., et al., 2005. Human modification of global water vapor flows from the land surface. Proc. Nat. Acad. Sci. U.S.A 102 (21), 7612–7617.

Goward, S.N., Waring, R.H., Dye, D.G., Yang, J.L., 1994. Ecological remote-sensing at Otter – satellite macroscale observations. Ecol. Appl. 4 (2), 322–343.

Haddeland, I., Heinke, J., Biemans, H., Eisner, S., Flörke, M., Hanasaki, N., Konzmann, M., Ludwig, F., Masaki, Y., Schewe, J., Stacke, T., Tessler, Z.D., Wada, Y., Wisser, D., 2014. Global water resources affected by human interventions and climate change. Proc. Natl. Acad. Sci. U.S.A 111 (9), 3251–3256.

Hall, F.G., Huemmrich, K.F., Goetz, S.J., Sellers, P.J., Nickeson, J.E., 1992. Satellite remote sensing of surface energy balance: success, failures, and unresolved issues in FIFE. J. Geophys. Res. 97 (D17), 19061–19089.

Helman, D., Givati, A., Lensky, I., 2015. Annual Evapotranspiration Retrieved from Satellite Vegetation Indices for the Eastern Mediterranean at 250 M Spatial Resolution.

Hoedjes, J.C.B., Chehbouni, A., Ezzahar, J., Escadafal, R., De Bruin, H.A.R., 2007. Comparison of large aperture scintillometer and eddy covariance measurements: can thermal infrared data Be used to capture footprint-induced differences? J. Hydrometeorol. 8 (2), 144–159.

Hogstrom, U., 1988. Non-Dimensional wind and temperature profiles in the atmospheric surface-layer – a re-evaluation. Boundary-Layer Meteorol. 42 (1–2), 55–78.

Hogstrom, U., Bergstrom, H., 1996. Organized turbulence structures in the near-neutral atmospheric surface layer. J. Atmos. Sci. 53 (17), 2452–2464.

Huntington, T.G., 2006. Evidence for intensification of the global water cycle: review and synthesis. J. Hydrol. 319, 83–95.

Iwasaki, H., Saito, H., Kuwao, K., Maximov, T.C., Hasegawa, S., 2010. Forest decline caused by high soil water conditions in a permafrost region. Hydrol. Earth Syst. Sci. 14 (2), 301–307.

Jacobsen, A., 1999. Estimation of the soil heat flux/net radiation ratio based on spectral vegetation indexes in high-latitude Arctic areas. Int. J. Remote Sens. 20 (2), 445–461.

Jang, K., et al., 2010. Mapping evapotranspiration using MODIS and MM5 four-dimensional data assimilation. Remote Sens. Environ. 114 (3), 657–673.

Jaramillo, F., Destouni, G., 2015. Local flow regulation and irrigation raise global human water consumption and footprint. Science 350 (6265), 1248–1251.

Jiang, L., Islam, S., 2001. Estimation of surface evaporation map over Southern Great Plains using remote sensing data. Water Resour. Res. 37 (2), 329–340.

Jiang, L., Islam, S., 2003. An intercomparison of regional latent heat flux estimation using remote sensing data. Int. J. Remote Sens. 24 (11), 2221–2236.

Jiménez, C., et al., 2011. Global inter-comparison of 12 land surface heat flux estimates. J. Geophys. Res. 116, 3–25.

Jung, M., Reichstein, M., Bondeau, A., 2009. Towards global empirical upscaling of FLUXNET eddy covariance observations: validation of a model tree ensemble approach using a biosphere model. Biogeosciences 6 (10), 2001–2013.

Jung, M., et al., 2010. Recent decline in the global land evapotranspiration trend due to limited moisture supply. Nature 467 (7318), 951–954.

Kalma, J.D., McVicar, T.R., McCabe, M.F., 2008. Estimating land surface evaporation: a review of methods using remotely sensed surface temperature data. Surv. Geophys. 29 (4–5), 421–469.

Kanda, M., Inagaki, A., Letzel, M.O., Raasch, S., Watanabe, T., 2004. LES study of the energy imbalance problem with Eddy covariance fluxes. Boundary-Layer Meteorol. 110 (3), 381–404.

Kanemasu, E.T., et al., 1992. Surface flux measurements in FIFE – an overview. J. Geophys. Res. Atmos. 97 (D17), 18547–18555.

Kim, S., Kim, H.S., 2008. Neural networks and genetic algorithm approach for nonlinear evaporation and evapotranspiration modeling. J. Hydrol. 351 (3–4), 299–317.

Kleissl, J., et al., 2008. Large aperture scintillometer intercomparison study. Boundary-Layer Meteorol. 128 (1), 133–150.

Kleissl, J., Watts, C.J., Rodriguez, J.C., Naif, S., Vivoni, E.R., 2009. Scintillometer intercomparison study-continued. Boundary-Layer Meteorol. 130 (3), 437–443.

Kustas, W., Anderson, M., 2009. Advances in thermal infrared remote sensing for land surface modeling. Agric. For. Meteorol. 149 (12), 2071–2081.

Kustas, W.P., Daughtry, C.S.T., 1990. Estimation of the soil heat-flux net-radiation ratio from spectral data. Agric. For. Meteorol. 49 (3), 205–223.

Kustas, W.P., Daughtry, C.S.T., Van Oevelen, P.J., 1993. Analytical treatment of the relationships between soil heat flux/net radiation ratio and vegetation indices. Remote Sens. Environ. 46 (3), 319–330.

Kustas, W.P., Hatfield, J.L., Prueger, J.H., 2005. The soil moisture-atmosphere coupling experiment (SMACEX): background, hydrometeorological conditions, and preliminary findings. J. Hydrometeorol. 6 (6), 791–804.

Kustas, W.P., Norman, J.M., 1999. Evaluation of soil and vegetation heat flux predictions using a simple two-source model with radiometric temperatures for partial canopy cover. Agric. For. Meteorol. 94 (1), 13–29.

Lagouarde, J.P., Bonnefond, J.M., Kerr, Y.H., McAneney, K.J., Irvine, M., 2002. Integrated sensible heat flux measurements of a two-surface composite landscape using scintillometry. Boundary-Layer Meteorol. 105 (1), 5–35.

Leuning, R., Zhang, Y.Q., Rajaud, A., Cleugh, H., Tu, K., 2008. A simple surface conductance model to estimate regional evaporation using MODIS leaf area index and the Penman-Monteith equation. Water Resour. Res. 44 (10), W10419.

Lhomme, J.P., Chehbouni, A., 1999. Comments on dual-source vegetation-atmosphere transfer models. Agric. For. Meteorol. 94 (3–4), 269–273.

Lhomme, J.P., Chehbouni, A., Monteny, B., 2000. Sensible heat flux-radiometric surface temperature relationship over sparse vegetation: parameterizing B-1. Boundary-Layer Meteorol. 97 (3), 431–457.

Li, F., et al., 2006. Comparing the utility of microwave and thermal remote-sensing constraints in two-source energy balance modeling over an agricultural landscape. Remote Sens. Environ. 101 (3), 315–328.

Li, F.Q., Kustas, W.P., Anderson, M.C., Prueger, J.H., Scott, R.L., 2008a. Effect of remote sensing spatial resolution on interpreting tower-based flux observations. Remote Sens. Environ. 112 (2), 337–349.

Li, R., Min, Q.L., Lin, B., 2009a. Estimation of evapotranspiration in a mid-latitude forest using the microwave emissivity difference vegetation index (EDVI). Remote Sens. Environ. 113 (9), 2011–2018.

Li, S., et al., 2008b. A comparison of three methods for determining vineyard evapotranspiration in the and desert regions of northwest China. Hydrol. Process. 22 (23), 4554–4564.

Li, Z.L., et al., 2009b. A review of current methodologies for regional evapotranspiration estimation from remotely sensed data. Sensors 9 (5), 3801–3853.

Liu, S.M., Lu, L., Mao, D., Jia, L., 2007. Evaluating parameterizations of aerodynamic resistance to heat transfer using field measurements. Hydrol. Earth Syst. Sci. 11 (2), 769–783.

Lu, H., Raupach, M.R., McVicar, T.R., Barrett, D.J., 2003. Decomposition of vegetation cover into woody and herbaceous components using AVHRR NDVI time series. Remote Sens. Environ. 86 (1), 1–18.

Lu, X.L., Zhuang, Q.L., 2010. Evaluating evapotranspiration and water-use efficiency of terrestrial ecosystems in the conterminous United States using MODIS and AmeriFlux data. Remote Sens. Environ. 114 (9), 1924–1939.

Lumley, J.L., Yaglom, A.M., 2001. A century of turbulence. Flow, Turbul. Combust. 66 (3), 241–286.

Mallick, K., Bhattacharya, B.K., Patel, N.K., 2009. Estimating volumetric surface moisture content for cropped soils using a soil wetness index based on surface temperature and NDVI. Agric. For. Meteorol. 149 (8), 1327–1342.

Mao, J., et al., 2015. Disentangling climatic and anthropogenic controls on global terrestrial evapotranspiration trends. Environ. Res. Lett. 10 (9), 094008.

Mao, Y., Wang, K., Liu, X., et al., 2016. Water storage in reservoirs built from 1997 to 2014 significantly altered the calculated evapotranspiration trends over China. J. Geophys. Res. Atmos. 121, 10,097-010,112.

Mao, Y., Wang, K., 2017. 'Comparison of evapotranspiration estimates based on the surface water balance, modified Penman-Monteith model, and reanalysis data sets for continental China. J. Geophys. Res. Atmos. 122, 3228–3244, 2016JD026065.

Margolis, H.A., Ryan, M.G., 1997. A physiological basis for biosphere-atmosphere interactions in the boreal forest: an overview. Tree Physiol. 17 (8–9), 491–499.

Massman, W.J., Lee, X., 2002. Eddy covariance flux corrections and uncertainties in long-term studies of carbon and energy exchanges. Agric. For. Meteorol. 113 (1–4), 121–144.

Mateo, C.M., Hanasaki, N., Komori, D., Tanaka, K., Kiguchi, M., Champathong, A., Sukhapunnaphan, T., Yamazaki, D., Oki, T., 2014. Assessing the impacts of reservoir operation to floodplain inundation by combining hydrological, reservoir management, and hydrodynamic models. Water Resour. Res. 50, 7245–7266.

Mauder, M., et al., 2008. Quality control of CarboEurope flux data − Part 2: inter-comparison of eddy-covariance software. Biogeosciences 5 (2), 451–462.

McCabe, M.F., Wood, E.F., 2006. Scale influences on the remote estimation of evapotranspiration using multiple satellite sensors. Remote Sens. Environ. 105 (4), 271–285.

McVicar, T.R., Jupp, D.L., 1998. The current and potential operational uses of remote sensing to aid decisions on drought exceptional circumstances in Australia: a review. Agric. Syst. 57, 399–468.

McVicar, T.R., Jupp, D.L.B., 2002. Using covariates to spatially interpolate moisture availability in the Murray-Darling Basin: a novel use of remotely sensed data. Remote Sens. Environ. 79 (2–3), 199–212.

Mecikalski, J.R., Diak, G.R., Anderson, M.C., Norman, J.M., 1999. Estimating fluxes on continental scales using remotely sensed data in an atmospheric and exchange model. J. Appl. Meteorol. 38 (9), 1352–1369.

Min, Q.L., Lin, B., 2006. Remote sensing of evapotranspiration and carbon uptake at Harvard Forest. Remote Sens. Environ. 100 (3), 379–387.

Miralles, D.G., Holmes, T.R.H., De Jeu, R.A.M., et al., 2011. Global land-surface evaporation estimated from satellite-based observations. Hydrol. Earth Syst. Sci. 15, 453–469.

Moene, A.F., Beyrich, F., Hartogensis, O.K., 2009. Developments in scintillometry. Bull. Am. Meteorol. Soc. 90 (5), 694–698.

Monin, A.S., Obukhov, A.M., 1954. Basic laws of turbulent mixing in the ground layer of the atmosphere (in Russian). Tr. Geofiz. Inst. Akad. Nauk SSSR 151, 163–187.

Monteith, J.L., 1972. Solar radiation and productivity in tropical ecosystems. J. Appl. Ecol. 9, 747–766.

Mu, Q., Zhao, M., Running, S.W., 2011. Improvements to a MODIS global terrestrial evapotranspiration algorithm. Remote Sens. Environ. 115 (8), 1781–1800.

Mu, Q., Heinsch, F.A., Zhao, M., Running, S.W., 2007. Development of a global evapotranspiration algorithm based on MODIS and global meteorology data. Remote Sens. Environ. 111 (4), 519–536.

Mueller, B., Seneviratne, S.I., Jimenez, C., et al., 2011. Evaluation of global observations-based evapotranspiration datasets and IPCC AR4 simulations. Geophys. Res. Lett. 38, L06402.

Mueller, B., Hirschi, M., Jimenez, C., et al., 2013. Benchmark products for land evapotranspiration: LandFlux-EVAL multi-data set synthesis. Hydrol. Earth Syst. Sci. 17, 3707–3720.

Murray, T., Verhoef, A., 2007. Moving towards a more mechanistic approach in the determination of soil heat flux from remote measurements − II. Diurnal shape of soil heat flux. Agric. For. Meteorol. 147 (1–2), 88–97.

Nemani, R., Running, S., 1997. Land cover characterization using multitemporal red, near-IR, and thermal-IR data from NOAA/AVHRR. Ecol. Appl. 7 (1), 79–90.

Nemani, R.R., et al., 2003. Climate-driven increases in global terrestrial net primary production from 1982 to 1999. Science 300 (5625), 1560–1563.

Nishida, K., Nemani, R.R., Running, S.W., Glassy, J.M., 2003. An operational remote sensing algorithm of land surface evaporation. J. Geophys. Res. 108 (D9), 4270.

Norman, J.M., et al., 2003. Remote sensing of surface energy fluxes at 10(1)-m pixel resolutions. Water Resour. Res. 39 (8), 1221.

Norman, J.M., Kustas, W.P., Humes, K.S., 1995. Source approach for estimating soil and vegetation energy fluxes in observations of directional radiometric surface temperature. Agric. For. Meteorol. 77 (3–4), 263–293.

Norman, J.M., Kustas, W.P., Prueger, J.H., Diak, G.R., 2000. Surface flux estimation using radiometric temperature: a dual temperature-difference method to minimize measurement errors. Water Resour. Res. 36 (8), 2263–2274.

Oki, T., Kanae, S., 2006. Global hydrological cycles and world water resources. Science 313 (5790), 1068–1072.

Oncley, S.P., et al., 2007. The energy balance experiment EBEX-2000. Part I: overview and energy balance. Boundary-Layer Meteorol. 123 (1), 1–28.

Penman, H.L., 1948. Natural evaporation from open water, bare soil and grass. Proc. R. Soc. Lond. Ser A Math. Phys. Sci. 193 (1032), 120–145.

Petropoulos, G., Carlson, T.N., Wooster, M.J., Islam, S., 2009. A review of T-s/VI remote sensing based methods for the retrieval of land surface energy fluxes and soil surface moisture. Prog. Phys. Geogr. 33 (2), 224—250.

Piao, S., Friedlingstein, P., Ciais, P., et al., 2007. Changes in climate and land use have a larger direct impact than rising CO(2) on global river runoff trends. Proc. Natl. Acad. Sci. U.S.A 104, 15242—15247.

Price, J.C., 1990. Using spatial context in satellite data to infer regional scale evapotranspiration. IEEE Trans. Geosci. Remote Sens. 28 (5), 940—948.

Priestley, C.H.B., Taylor, R.J., 1972. On the assessment of surface heat flux and evaporation using large-scale parameters. Mon. Weather Rev. 100 (2), 81—92.

Prince, S.D., Goetz, S.J., Dubayah, R.O., Czajkowski, K.P., Thawley, M., 1998. Inference of surface and air temperature, atmospheric precipitable water and vapor pressure deficit using advanced very high-resolution radiometer satellite observations: comparison with field observations. J. Hydrol. 212—213, 230—249.

Rana, G., Katerji, N., 2000. Measurement and estimation of actual evapotranspiration in the field under Mediterranean climate: a review. Eur. J. Agron. 13 (2—3), 125—153.

Raupach, M., 2001. Combination theory and equilibrium evaporation. Q. J. R. Meteorol. Soc. 127, 1149—1181.

Reichle, R.H., 2008. Data assimilation methods in the Earth sciences. Adv. Water Resour. 31 (11), 1411—1418.

Roerink, G.J., Su, Z., Menenti, M., 2000. S-SEBI: a simple remote sensing algorithm to estimate the surface energy balance. Phys. Chem. Earth — Part B Hydrol., Oceans Atmos. 25 (2), 147—157.

Schuttemeyer, D., Schillings, C., Moene, A.F., De Bruin, H.A.R., 2007. Satellite-based actual evapotranspiration over drying semiarid terrain in West Africa. J. Appl. Meteorol. Climatol. 46 (1), 97—111.

Sellers, P.J., et al., 1997. Modeling the exchanges of energy, water, and carbon between continents and the atmosphere. Science 275 (5299), 502—509.

Seneviratne, S.I., et al., 2010. Investigating soil moisture-climate interactions in a changing climate: a review. Earth Sci. Rev. 99 (3—4), 125—161.

Sheffield, J., Wood, E.F., Munoz-Arriola, F., 2010. Long-term regional estimates of evapotranspiration for Mexico based on downscaled ISCCP data. J. Hydrometeorol. 11 (2), 253—275.

Sheffield, J., Wood, E.F., Roderick, M.L., 2012. Little change in global drought over the past 60 years. Nature 491, 435—438.

Shukla, J., Mintz, Y., 1982. Influence of land-surface evapotranspiration on the earth's climate. Science 215 (4539), 1498—1501.

Shuttleworth, W.J., 1991. Insight from large-scale observational studies of land/atmosphere interactions. Surv. Geophys. 12 (1), 3—30.

Shuttleworth, W.J., 2007. Putting the 'vap' into evaporation. Hydrol. Earth Syst. Sci. 11 (1), 210—244.

Snyder, R.L., Paw U, K.T., 2002. Penman-Monteith Equation Derivation. Available: http://biomet.ucdavis.edu/Evapotranspiration/PMDerivation/PMD.htm.

Sobrino, J.A., Gomez, M., Jimenez-Munoz, C., Olioso, A., 2007. Application of a simple algorithm to estimate daily evapotranspiration from NOAA-AVHRR images for the Iberian Peninsula. Remote Sens. Environ. 110 (2), 139—148.

Solignac, P.A., et al., 2009. Uncertainty analysis of computational methods for deriving sensible heat flux values from scintillometer measurements. Atmos. Meas.Tech. 2 (2), 741—753.

Su, H., McCabe, M.F., Wood, E.F., Su, Z., Prueger, J.H., 2005. Modeling evapotranspiration during SMACEX: comparing two approaches for local- and regional-scale prediction. J. Hydrometeorol. 6 (6), 910—922.

Su, H., Wood, E.F., McCabe, M.F., Su, Z., 2007. Evaluation of remotely sensed evapotranspiration over the CEOP EOP-1 reference sites. J. Meteorol. Soc. Jpn. 85A, 439—459.

Su, Z., 2002. The Surface Energy Balance System (SEBS) for estimation of turbulent heat fluxes. Hydrol. Earth Syst. Sci. 6 (1), 85—99.

Su, Z., Schmugge, T., Kustas, W.P., Massman, W.J., 2001. An evaluation of two models for estimation of the roughness height for heat transfer between the land surface and the atmosphere. J. Appl. Meteorol. 40 (11), 1933—1951.

Sun, J.L., Massman, W., Grantz, D.A., 1999. Aerodynamic variables in the bulk formulation of turbulent fluxes. Boundary-Layer Meteorol. 91 (1), 109—125.

Suyker, A.E., Verma, S.B., 2008. Interannual water vapor and energy exchange in an irrigated maize-based agroecosystem. Agric. For. Meteorol. 148 (3), 417—427.

Tang, R., Li, Z.-L., Tang, B., 2010. An application of the Ts-VI triangle method with enhanced edges determination for evapotranspiration estimation from MODIS data in arid and semi-arid regions: implementation and validation. Remote Sens. Environ. 114 (3), 540—551.

Teixeira, A., Bastiaanssen, W.G.M., Ahmad, M.D., Bos, M.G., 2009a. Reviewing SEBAL input parameters for assessing evapotranspiration and water productivity for the Low-Middle Sao Francisco River basin, Brazil Part A: calibration and validation. Agric. For. Meteorol. 149 (3—4), 462—476.

Teixeira, A., Bastiaanssen, W.G.M., Ahmad, M.D., Bos, M.G., 2009b. Reviewing SEBAL input parameters for assessing evapotranspiration and water productivity for the Low-Middle Sao Francisco River basin, Brazil Part B: application to the regional scale. Agric. For. Meteorol. 149 (3—4), 477—490.

Timmermans, W.J., Kustas, W.P., Anderson, M.C., French, A.N., 2007. An intercomparison of the surface energy balance algorithm for land (SEBAL) and the two-source energy balance (TSEB) modeling schemes. Remote Sens. Environ. 108 (4), 369—384.

Tittebrand, A., Schwiebus, A., Berger, F.H., 2005. The influence of land surface parameters on energy flux densities derived from remote sensing data. Meteorol. Z. 14 (2), 227—236.

Trenberth, K.E., Fasullo, J.T., Kiehl, J., 2009. Earth's global energy budget. Bull. Am. Meteorol. Soc. 90 (3), 311—324.

Trenberth, K.E., Fasullo, J.T., Mackaro, J., 2011. Atmospheric moisture transports from ocean to land and global energy flows in reanalyses. J. Clim. 24 (18), 4907—4924.

Tsuang, B.-J., et al., 2009. A more accurate scheme for calculating Earth's skin temperature. Clim. Dyn. 32 (2), 251—272.

van der Kwast, J., et al., 2009. Evaluation of the surface energy balance system (SEBS) applied to ASTER imagery with flux-measurements at the SPARC 2004 site (Barrax, Spain). Hydrol. Earth Syst. Sci. 13 (7), 1337—1347.

Venturini, V., Islam, S., Rodriguez, L., 2008. Estimation of evaporative fraction and evapotranspiration from MODIS products using a complementary based model. Remote Sens. Environ. 112 (1), 132—141.

Verhoef, A., De Bruin, H.A.R., Van Den Hurk, B.J.J.M., 1997. Some practical Notes on the parameter kB-1 for sparse vegetation. J. Appl. Meteorol. 36 (5), 560—572.

Verhoef, A., van den Hurk, B.J.J.M., Jacobs, A.F.G., Heusinkveld, B.G., 1996. Thermal soil properties for vineyard (EFEDA-I) and savanna (HAPEX-Sahel) sites. Agric. For. Meteorol. 78 (1—2), 1—18.

Verstraeten, W.W., Veroustraete, F., Feyen, J., 2005. Estimating evapotranspiration of European forests from NOAA-imagery at satellite overpass time: towards an operational processing chain for integrated optical and thermal sensor data products. Remote Sens. Environ. 96 (2), 256—276.

Vickers, D., Gockede, M., Law, B.E., 2010. Uncertainty estimates for 1-h averaged turbulence fluxes of carbon dioxide, latent heat and sensible heat. Tellus Ser. B Chem. Phys. Meteorol. 62 (2), 87—99.

Von Randow, C., Kruijt, B., Holtslag, A.A.M., de Oliveira, M.B.L., 2008. Exploring eddy-covariance and large-aperture scintillometer measurements in an Amazonian rain forest. Agric. For. Meteorol. 148 (4), 680—690.

Vörösmarty, C.J., Sharma, K.P., Fekete, B.M., Copeland, A.H., Holden, J., Marble, J., Lough, J.A., 1997. The storage and aging of continental runoff in large reservoir systems of the world. Ambio 26, 210—219.

Wang, K., Dickinson, R.E., Wild, M., Liang, S., 2010a. Evidence for decadal variation in global terrestrial evapotranspiration between 1982 and 2002: 1. Model development. J. Geophys. Res. 115 (D20), D20112.

Wang, K., Dickinson, R.E., Wild, M., Liang, S., 2010b. Evidence for decadal variation in global terrestrial evapotranspiration between 1982 and 2002: 2. Results. J. Geophys. Res. 115 (D20), D20113.

Wang, K., Liang, S., 2008. An improved method for estimating global evapotranspiration based on satellite estimation of surface net radiation, vegetation index, temperature, and soil moisture. J. Hydrometeorol. 9 (4), 712—727.

Wang, K., et al., 2007a. Evaluation and improvement of the MODIS land surface temperature/emissivity products using ground-based measurements at a semi-desert site on the western Tibetan Plateau. Int. J. Remote Sens. 28 (11), 2549—2565.

Wang, K.C., Li, Z.Q., Cribb, M., 2006. Estimation of evaporative fraction from a combination of day and night land surface temperatures and NDVI: a new method to determine the Priestley-Taylor parameter. Remote Sens. Environ. 102 (3—4), 293—305.

Wang, K.C., Liang, S.L., 2009. Evaluation of ASTER and MODIS land surface temperature and emissivity products using long-term surface longwave radiation observations at SURFRAD sites. Remote Sens. Environ. 113 (7), 1556—1565.

Wang, K.C., Wang, P., Li, Z.Q., Cribb, M., Sparrow, M., 2007b. A simple method to estimate actual evapotranspiration from a combination of net radiation, vegetation index, and temperature. J. Geophys. Res. Atmos. 112 (D15), D15107.

Wang, K., Dickinson, R.E., 2012. A review of global terrestrial evapotranspiration: observation, modeling, climatology, and climatic variability. Rev. Geophys. 50 (2), RG2005.

Wieringa, J., 1980. A revaluation of the Kansas mast influence on measurements of stress and cup anemometer overspeeding. Boundary-Layer Meteorol. 18 (4), 411—430.

Wiernga, J., 1993. Representative roughness parameters for homogeneous terrain. Boundary-Layer Meteorol. 63 (4), 323—363.

Wilson, K., et al., 2002. Energy balance closure at FLUXNET sites. Agric. For. Meteorol. 113 (1—4), 223—243.

Wohlfahrt, G., et al., 2009. On the consequences of the energy imbalance for calculating surface conductance to water vapour. Agric. For. Meteorol. 149 (9), 1556—1559.

Yan, H., Shugart, H.H., 2010. An air relative-humidity-based evapotranspiration model from eddy covariance data. J. Geophys. Res. Atmos. 115, D16106.

Yang, F.H., et al., 2006. Prediction of continental-scale evapotranspiration by combining MODIS and AmeriFlux data through support vector machine. IEEE Trans. Geosci. Remote Sens. 44 (11), 3452—3461.

Yang, K., Chen, Y.Y., Qin, J., 2009. Some practical notes on the land surface modeling in the Tibetan Plateau. Hydrol. Earth Syst. Sci. 13 (5), 687—701.

Yuan, W., et al., 2010. Impacts of precipitation seasonality and ecosystem types on evapotranspiration in the Yukon River Basin, Alaska. Water Resour. Res. 46 (2), W02514.

Yuan, W., Liu, S., Yu, G., et al., 2010. Global estimates of evapotranspiration and gross primary production based on MODIS and global meteorology data. Remote Sens. Environ. 114, 1416–1431.

Zeng, Z., Wang, T., Zhou, F., et al., 2014. A worldwide analysis of spatiotemporal changes in water balance-based evapotranspiration from 1982 to 2009. J. Geophys. Res. Atmos. 119, 2013JD020941.

Zeng, Z., Piao, S., Lin, X., et al., 2012. Global evapotranspiration over the past three decades: estimation based on the water balance equation combined with empirical models. Environ. Res. Lett. 7, 014026.

Zeweldi, D.A., et al., 2010. Intercomparison of sensible heat flux from large aperture scintillometer and eddy covariance methods: field experiment over a homogeneous semi-arid region. Boundary-Layer Meteorol. 135 (1), 151–159.

Zhang, K., et al., 2009. Satellite based analysis of northern ET trends and associated changes in the regional water balance from 1983 to 2005. J. Hydrol. 379 (1–2), 92–110.

Zhang, K., Kimball, J.S., Nemani, R.R., Running, S.W., 2010a. A continuous satellite-derived global record of land surface evapotranspiration from 1983 to 2006. Water Resour. Res. 46 (9), W09522.

Zhang, X.D., Jia, X.H., Yang, J.Y., Hu, L.B., 2010b. Evaluation of MOST functions and roughness length parameterization on sensible heat flux measured by large aperture scintillometer over a corn field. Agric. For. Meteorol. 150 (9), 1182–1191.

Zhang, Y., Peña-Arancibia, J.L., McVicar, T.R., Chiew, F.H.S., Vaze, J., Liu, C., Lu, X., Zheng, H., Wang, Y., Liu, Y.Y., Miralles, D.G., Pan, M., 2016. Multi-decadal trends in global terrestrial evapotranspiration and its components. Sci. Rep. 6, 19124.

Zhang, Y.Q., Chiew, F.H.S., Zhang, L., Leuning, R., Cleugh, H.A., 2008. Estimating catchment evaporation and runoff using MODIS leaf area index and the Penman-Monteith equation. Water Resour. Res. 44 (10), W10420.

Soil moisture contents

Abstract

Soil moisture content (SMC) is an important parameter in various applications. This chapter reviews conventional techniques for in situ point measurements and then introduces the basic principles, sensors, and inversion methods of microwave remote sensing in both the passive and active modes. Various optical and thermal IR methods are also reviewed. A list of SMC products is given at the end.

© 2020 Elsevier Inc. All rights reserved.

18.1 Introduction

Soil moisture generally refers to the amount of water stored in the spaces (pores) between soil particles in the unsaturated soil zone, also termed the vadose zone. Surface soil moisture (SSM) refers to the water content within the upper 5 cm of soil, whereas root-zone soil moisture is the water that is available to plants, which is generally considered to be in the upper 200 cm of soil. In practice, often only a fraction of soil moisture is relevant or measureable.

The ability to determine the spatial and temporal distribution of soil moisture would be of significant help in understanding the Earth as an integrated system. The volume of soil moisture is small compared with other components of the hydrologic cycle; nonetheless, it is of fundamental importance to many hydrological, biological, and biogeochemical processes (Legates et al., 2011; Wang et al., 2019). Soil moisture is a key variable in controlling the exchange of water and heat energy between the land surface and the atmosphere through evaporation and plant transpiration. Through its impact on the partitioning of the incoming energy in the latent and sensible heat fluxes, soil moisture has several additional effects on climate processes, in particular on air temperature, boundary layer stability, and, in some instances, precipitation.

Soil moisture can affect air temperature. Whenever soil moisture limits the total energy used by latent heat flux, more energy is available for sensible heating, inducing an increase of near surface air temperature. Increased temperature leads to a higher vapor pressure deficit and evaporative demand and thus to a potential increase in evapotranspiration despite the dry conditions, possibly leading to a further decrease in soil moisture. Soil moisture can also impact precipitation, mostly in an indirect way, by influencing boundary layer stability and precipitation formation, rather than in terms of the absolute moisture input resulting from modified evapotranspiration. For example, the additional precipitated water falling over wet soils may originate from oceanic sources, but the triggering of precipitation may itself be the result of enhanced instability induced by the wet or dry soil conditions (Seneviratne et al., 2010).

Soil moisture information is also valuable to a wide range of governmental and commercial organizations concerned with weather and climate, runoff potential and flood control, soil erosion and slope failure, reservoir management, early warning of droughts, irrigation scheduling, crop yield forecasting, geotechnical engineering, and water quality. For example, it is very important to minimize water stress for plants and avoid overirrigation when scheduling irrigation. A tendency to underirrigate can result in stress on the plant root water-uptake mechanism and a decrease in photosynthesis and cell expansion of the plant. Overirrigation will result in poor utilization of resources and high surface runoff, as well as possible erosion, leaching of nutrients and pesticides, and low system efficiency in general. Additionally, timely information on soil moisture is also used by the military to accurately plan infantry and vehicular movements in remote areas.

Soil moisture is estimated by in situ measurements, remote sensing techniques, or by land surface or hydrologic modeling. In situ measurements are mostly suitable for field-scale studies. Remote sensing of soil moisture can be advantageous owing to its continuous spatial coverage, as shown in Fig. 18.1, although the estimation accuracy still needs to be improved. Most studies agree that the penetration depth for microwave sensing is between 0.1 and 0.2 times the wavelength, where the longest wavelengths (L-band) are approximately 21 cm. A combined data assimilation approach with remotely sensed estimates and land surface models (LSM) or soil—vegetation—atmosphere transfers (SVAT) models, calibrated with in situ measurements

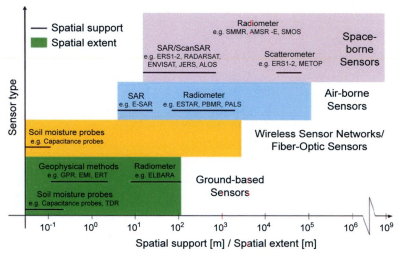

FIGURE 18.1   Support scale of soil moisture observations or measurements obtained from ground-based sensors (*ELBARA*, L band radiometer; *EMI*, electromagnetic induction; *GPR*, ground penetrating radar; *TDR*, time domain reflectometry), wireless sensor networks, airborne sensors (*E-SAR*, experimental airborne SAR; *ESTAR*, electronically scanned thinned aperture radiometer; *PBMR*, L band push broom microwave radiometer; *PALS*, passive and active L/S band sensor; *SAR*, synthetic aperture radar), and spaceborne sensors (*ALOS*, Advanced Land Observing Satellite; *AMSR-E*, Advanced Microwave Scanning Radiometer; *ENVISAT*, Environmental Satellite; *ERS1−2*, European Remote Sensing Satellite 1−2; *JERS*, Japanese Environmental Remote Sensing; *SMMR*, Scanning Multichannel Microwave; *SMOS*, Soil Moisture and Ocean Salinity Satellite) (Vereecken et al., 2008).

to quantify errors and uncertainties, is the most promising approach for soil moisture estimation involving large areas, particularly for total moisture stored in the root zone and the vertical profile of soil moisture.

18.2 Conventional SMC measurement techniques

Measurement at the local scale (~ 0.01 m^2) is performed in situ, using sensors of different sizes and shapes. The website www.sowacs.com provides a list of commercially available sensors for measuring soil moisture content (SMC). Among the most common devices for SMC measurement at the local scale are those based on dielectric measurements. One of the main advantages of using in situ sensors is that they can be connected to data loggers, automatically retrieve SMC data in real time, and provide

detailed time series. Various methods are available (Lekshmi et al., 2014), and a general, nonexhaustive list of methods to determine soil moisture at the local scale is given in Table 18.1.

All SMC measurement methods are either direct or indirect. As outlined by Bittelli (2011), direct or contact-based methods directly measure the amount of water, for instance, by measuring its weight as a fraction of the total soil weight (gravimetric method). Examples of contact methods include capacitance sensors, time-domain reflectometry, electrical resistivity measurements, heat pulse sensors, fiber optic sensors, and destructive sampling (e.g., gravimetric methods). The thermogravimetric method starts by taking a soil sample in the field, which is immediately weighed and then subsequently dried in an oven for 24 h at 105°C, to determine the mass of the dry soil. The difference in weight (the weight of liquid water) is expressed as fraction of the soil solid weight,

TABLE 18.1 An overview of methods to measure in situ soil moisture content (SMC) (Verstraeten et al., 2008).

Methods	Example	Description
Gravimetric	Oven-drying	Standard method, destructive sampling
Nuclear	Neutron scattering	Fast neutrons emitted from a radioactive source are slowed down by hydrogen atoms in the soil
	Gamma attenuation	The scattering and absorption of gamma rays are related to the density of matter in their path
	Nuclear magnetic resonance	Soil water is subjected to both a static and an oscillating magnetic field at right angles to each other
Electromagnetic	Resistive sensor	Soil resistivity depends on the soil electrical properties and moisture
	Capacitive sensor	Using the dielectric constant by measuring capacitance between two electrodes implanted in the soil
	Time-domain reflectometer	Propagation of electromagnetic signals. Velocity and attenuation depend on soil properties: water content and electrical conductivity
	Frequency domain	An oscillator detects changes in soil dielectric properties linked to variations in soil water content
Tensiometric	Soil matrix tension	Measures the soil matrix potential (capillary tension)
Hydrometric	Thermal inertia	Relationship between moisture in porous materials and the relative humidity. Because thermal inertia of a porous medium depends on moisture, soil surface temperature is indicative
Heat dissipation	Heat pulse	Rising or cooling of temperature in a porous block is measured after a heat pulse
Feel and appearance	Manual	Soil moisture interpretation chart based on texture classification and manual squeezing of soil samples

called gravimetric water content. These measurement techniques typically provide spatially and temporally highly resolved measurements. However, this measurement method is usually destructive because the soil sample is removed from the field to be analyzed in the laboratory and is also a time-consuming and impractical way of measuring SMC in the field.

An indirect or contact-free method measures another variable that is affected by the amount of soil water and then relates changes in this variable to changes in SMC; such methods include passive microwave radiometers, synthetic aperture radars (SAR), scatterometers, and thermal methods. For instance, dielectric sensors exploit changes in soil dielectric properties as a function of SMC. Dielectric

measurement takes advantage of the differences in dielectric permittivity values of the different soil phases (solid, liquid, and gas). Liquid water has a dielectric permittivity of about 80 (depending on temperature, electrolyte solution, and frequency), air has a dielectric permittivity of about 1, and the solid soil phase has a permittivity of 4–16. This variability makes dielectric permittivity of soil very sensitive to variation in SMC. The direct gravimetric method is the reference method for SMC measurement (and commonly used for calibration of indirect methods), but nowadays, the majority of commercial sensors are based on indirect methods.

Although it is possible to install distributed wireless sensor networks to obtain data across a field or a watershed, SMC field measurements are generally insufficient to provide the near-continuous spatial coverage needed for some applications. As an alternative to in situ SMC measurement, remote sensing can be used to generate vast amounts of information about the Earth's surface at sufficiently high spatial resolutions with coverage that is sufficiently extensive to be appropriate for various applications. Currently, there are three main remote sensing methods in use: the first two methods based on the optical/thermal IR and microwave information have attracted the most attention; the third method detects changes in the gravity potential field above the soil, which are related to changes in the density of the soil, and thus SMC. At present, SMC measurements using the third method are only possible at very large scales, i.e., 600–1000 km. In the following section, we will describe the first two methods.

18.3 Microwave remote sensing methods

Research on soil moisture remote sensing began in the mid-1970s. Subsequent research has investigated many diverse paths. Microwave remote sensing using satellites has become the primary remote sensing technique for measurement of SMC at the regional and global scales. Microwave remote sensing involves two techniques: passive and active. Passive techniques rely on a radiometer to observe natural microwave emission, often in terms of brightness temperature, of the land surface, whereas active remote techniques use radar to measure the backscattered/reflected power from the surface compared with the transmitted signal. Quantitative measurements of surface soil layer moisture have been most successful using passive remote sensing methods.

The theory behind the microwave remote sensing of soil moisture (Karthikeyan et al., 2017a) is based on the large contrast between the dielectric properties of liquid water (~ 80) and dry soil (<4). The dielectric constant of wet soil is usually less than 35, as shown in Fig. 18.2. As the moisture content increases, the dielectric constant of the soil–water mixture increases, and this change is detectable by microwave sensors.

Vegetation and surface roughness reduce the sensitivity of microwave observations to variations in soil moisture. These effects become more pronounced as the frequency increases; hence, low frequencies in the L-band range (1–2 GHz) are preferred for soil moisture sensing. Furthermore, at lower microwave frequencies such as the L-band, soil microwave emission originates from deeper in the soil profile (a few centimeters), resulting in a more representative measurement of moisture conditions below the surface crust or skin layer.

18.3.1 Passive microwave remote sensing

Previous research has shown that passive microwave remote sensors can be effectively used to monitor SSM over land surfaces. These sensors measure the intensity of a soil's microwave emission and, hence, dielectric properties, using brightness temperatures. However, the

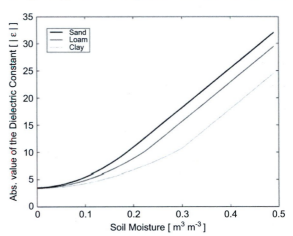

FIGURE 18.2 Comparison of the soil dielectric constant and soil moisture for typical sand, loam, and clay soils (de Jeu et al., 2008).

effects of vegetation cover, soil temperature, snow cover, topography, and soil surface roughness also play a significant role in the microwave emission recorded from the land surface. Other parameters such as soil texture, bulk soil density, and atmospheric effects have a smaller (second-order) influence. Many approaches have been developed to determine soil moisture from microwave radiometric measurements where each of the various effects contributing to the surface microwave emission is taken into account.

Passive systems are typically characterized by broad spatial coverages and high temporal resolutions but coarse spatial resolutions ($\sim 10-30$ km). Consequently, passive systems are more suitable for climatic studies at regional to global scales. In the following section, we will introduce the basic principles of passive microwave remote sensing, followed by a review of satellite sensors and two inversion methods.

18.3.1.1 Basic principles

Microwave radiometers measure emitted microwave radiation, expressed in terms of brightness temperature, for vertical or horizontal polarization. When the temperature of the emitting layer is known, the emissivity and the reflectivity can be calculated. Over a bare soil surface, the soil dielectric constant can be derived from the reflectivity, after correcting for soil roughness effects. Soil moisture can then be estimated from the soil dielectric constant using dielectric mixing models that account for the soil characteristics (e.g., texture, structure, density).

The emission of microwave energy is commonly referred to as the microwave brightness temperature (T_B):

$$T_B = t(H, \theta)[RT_{sky} + (1 - R)T_{surf}] + T_{atm}(H, \theta)$$

$$(18.1)$$

where $t(H, \theta)$ is the atmospheric transmittance, T_{sky} is atmospheric downward thermal emission, $T_{atm}(H, \theta)$ is the upwelling thermal emission of

the atmosphere received by the sensor, T_{surf} is the soil surface thermometric temperature, and R is the smooth surface reflectivity.

Passive microwave measurements at frequencies as low as 1.4 GHz measure only soil moisture (w_S) at shallow soil depths (approximately 2–5 cm). Therefore, for rather smooth soil surfaces, the soil microwave reflectance can be approximated using the Fresnel equation of a plane surface. For horizontally (h) and vertically (v) polarized waves at off-nadir incidence angle (θ), the Fresnel reflectance is given by

$$\begin{cases} R_h = \left| \dfrac{\cos \theta - \sqrt{d - \sin^2 \theta}}{\cos \theta + \sqrt{d - \sin^2 \theta}} \right|^2 \\[4mm] R_v = \left| \dfrac{d \cos \theta - \sqrt{d - \sin^2 \theta}}{d \cos \theta + \sqrt{d - \sin^2 \theta}} \right|^2 \end{cases} \quad (18.2)$$

where d is the complex dielectric constant of the emitter. For soils, d is mainly determined from the SMC and, to a lesser extent, from the textural and structural properties of soil.

For typical soil moisture applications using longer microwave wavelengths at low altitude, temperature contributions from the atmosphere and the sky can be neglected. Thus, the brightness temperature of an emitter of microwave radiation is related to the physical temperature of the soil surface through the emissivity, such that

$$T_B = (1 - R_{h,v})T_{surf} = \varepsilon_{h,v}T_{surf} \quad (18.3)$$

where ε is surface emissivity, as determined by the Fresnel reflectance. Note that the above Fresnel equation is suitable for smooth surfaces. However, in general, several other factors that influence soil emission readings should also be taken into account, such as surface roughness.

Volumetric soil moisture w_S can be considered as a monotonically decreasing function of the emissivity ε_h of bare soil. If the soil roughness conditions do not change much during the observations, this function can be well approximated as a linear equation:

$$\varepsilon_h = a_0 + a_1 w_s \qquad (18.4)$$

This simple relationship for bare soils proves to be valid for a wide range of soil moisture and roughness conditions, provided that sufficient ground data are available to calibrate the coefficients a_0 and a_1.

In the presence of vegetation, the canopy brightness can be computed as a function of three main surface variables: surface soil moisture w_s, vegetation optical depth τ that is related to vegetation water content, and canopy temperature. Therefore, several measurement data are required to differentiate between the effects of these three variables. These data can be obtained from measurements for several configurations of the sensor in terms of polarization, view angle, and frequency (Jackson, 2008).

18.3.1.2 Satellite sensors

Eight passive microwave satellite missions have been widely utilized for soil moisture estimation: the Scanning Multichannel Microwave Radiometer (SMMR) onboard Nimbus-7 (1978–87), the Special Sensor Microwave/Imager (SSM/I) onboard Defense Meteorological Satellite Program (DMSP) (1987–2007), the microwave imager onboard Tropical Rainfall Measuring Mission (TRMM) (1997–2015), the WindSAT mission onboard Coriolis (2003–12), the Advanced Microwave Scanning Radiometer–Earth Observing System (AMSR-E) onboard Aqua satellite (2002–11), the Advanced Microwave Scanning Radiometer 2 (AMSR2) onboard the GCOM-W satellite (2012–present), along with two dedicated satellite missions, the Soil Moisture Ocean Salinity (SMOS) mission (Kerr et al., 2016) and the Soil Moisture Active Passive (SMAP) mission (Chan et al., 2016; McColl et al., 2017; Zhang et al., 2019).

Frequency selection is important in microwave remote sensing of soil moisture. Lower frequencies provide greater sensitivity to changes in soil moisture across a wider range of vegetation cover conditions. Low-frequency radiation is better suited to soil moisture monitoring because it can more easily penetrate the vegetation layer to sense moisture; at higher frequencies (>15 GHz), the corrections required for atmospheric effects strongly limit the all-weather capabilities of microwave instruments. Multifrequency measurements can be useful to distinguish soil contribution from that of vegetation. At L-band ($f \sim 1.4$ GHz), soil contribution is dominant for most low vegetation covers. As frequency increases, the screening effect of vegetation increases. Thus, at 5 GHz, for low vegetation cover, the soil and vegetation contributions are similar in magnitude, while at 10 GHz, the vegetation effects become dominant.

Several multifrequency spaceborne microwave window channel radiometers have significant soil moisture sensitivity and comparable channel configurations in the 6–37 GHz frequency range. These include the SMMR, TRMM Microwave Imager (TMI), AMSR-E, and the WindSat radiometer.

Although sensors in the L-band (1–2 GHz) provide the most reliable SSM estimates, these radiometers are operating at higher frequencies due to the technical challenges of constructing L-band antennae long enough (4–6 m) to provide adequate resolution and complex processing to obtain well-calibrated L-band brightness temperature data. In 2009, the European Space Agency (ESA) launched its SMOS mission, which uses an L-band (1.4 GHz) SAR with a spatial resolution of 50 km and has higher soil moisture sensitivity than heritage radiometers (Kerr et al., 2010). Because of the reliance on lower microwave frequencies (e.g., L- and C-bands), the temporal resolution is 1–2 days. SMOS is the first satellite devoted to soil moisture remote sensing, and then it was followed by the NASA SMAP satellite, which, by combining active and passive sensors within

18. Soil moisture contents

the L-band, provides soil moisture products at high spatial resolutions (10 km active and 40 km passive) (Entekhabi et al., 2010) (http://smap.jpl.nasa.gov/). Comparative specifications for the sensors are provided in Table 18.2.

18.3.1.3 Inversion algorithms

A number of inversion algorithms have been proposed to estimate SSM. Besides many statistical methods, most physically based algorithms use a radiative transfer (RT) model. Various RT models have different characteristics, but they all include three modules: a dielectric model relating SSM to the dielectric constants; a surface roughness model accounting for surface scattering effects; and a vegetation layer model accounting for the vegetation attenuation effects. Some RT models also include an atmospheric module. In the following section, we will introduce two physically based inversion methods, although more models have been proposed in the literature (Shi et al. 2002, 2005).

TABLE 18.2 Specifications of the typical microwave sensors for soil moisture content (SMC).

Instrument	Satellite	Duration	Frequency (GHz)	Footprint size (km)	Incidence angle at earth surface (°)	Temporal resolution (Days)	Coverage
SMMR	Nimbus 7	Oct. 1978−Aug. 1987	6.6	148 × 95	50.2	2	Global
			10.7	91 × 59			
SSM/I	DMSP	Aug. 1987 − Dec. 2007	19.3	70 × 45	53.1	1	Global
TMI	TRMM	Dec. 1997−Apr. 2015	10.7	59 × 36	35	1	N40° to S40°
AMSR-E	Aqua	Jun. 2002−Sep. 2011	6.9	76 × 44	55	1	Global
			10.7	49 × 28			
WindSAT	Coriolis	Feb. 2003−Jul. 2012	6.8	60 × 40	53.5	1	Global
			10.7	38 × 25	49.9		
AMSR2	GCOM-W	Jul. 2012−present	6.9	62 × 35	55	1	Global
			10.7	42 × 24			
MIRAS	SMOS	Jan. 2010−present	1.4	35 − 50	0 − 55	1−3	Global
SMAP	SMAP	Mar. 2015−present	1.4	47 × 39	40	1−3	85.044°N
							85.044°S
ERS	ERS-1/2	Jul. 1991−Sep. 2011	5.3	50 × 50	16 − 50	2−7	Global
ASCAT	MetOp-A/B	Jan. 2007−present	5.225	25 × 25	25 − 65	1−2	Global
ESA−CCI −COMBI	Various	Oct. 1978−Dec. 2014	Various	25 × 25	Various	1	Global

18.3.1.3.1 AMSE-R instrument algorithm

A major issue in using passive microwaves is that the effects of soil moisture and vegetation water content on microwave emission are inversely related: a decrease in vegetation water content and an increase in soil moisture have the same effect on the signal and conversely. Another issue concerns the strong effects of temperature on daytime measurements (ascending orbit). A steep temperature gradient in the top soil layers makes it difficult for soil moisture inversion in these conditions. To alleviate this problem, only data from descending passes (i.e., nighttime) are usually used.

AMSR-E C-band (6.9 GHz) and X-band (10.7 GHz) channels are suitable for soil moisture remote sensing. Various approaches have been considered for retrieving w_s from the AMSR-E measurements. These approaches differ primarily in the methods used to correct for the effects of soil texture, roughness, vegetation, and surface temperature. A common assumption is that, over most land areas, at the AMSR-E footprint scale, the effects of variability in soil texture, roughness, and single-scattering albedo on the observations are relatively small compared with the effect of variability in soil moisture (Njoku et al., 2003).

The method developed by Njoku and Li (1999) and Njoku et al. (2003) is the official AMSR-E soil moisture science team contribution. The AMSE-R algorithm is an iterative least-squares minimization algorithm. The retrieval algorithm is based on an RT model that relates parameters describing the surface and atmosphere to the observed brightness temperatures. The algorithm simultaneously retrieves M geophysical variables ($j = 1-M$) from measurements at brightness temperature channels N ($i = 1-N$), where N should be greater than M for stable retrievals. The cost function is the weighted sum of squared differences between observed T_i^{obs} and computed brightness temperatures $\Phi_i(x)$,

$$\chi^2(x) = \sum_{i=1}^{N} \left(\frac{T_i^{obs} - \Phi_i(x)}{\sigma_i} \right)^2 \tag{18.5}$$

where σ_i represents the measurement noise standard deviation in channel i. The efficient Levenberg–Marquardt algorithm is used to search for the set of variables x^* that minimizes χ^2. The basic procedure has been widely used.

For the forward function $\Phi_i(x)$, the following expresses the brightness temperature of a homogeneous vegetated surface

$$T_{b_p} = \Phi(x) = T_s \{ \varepsilon_{sp} e^{-\tau_c} + (1 - \omega_p) \left[1 - e^{-\tau_c} \right] \\ \times \left[1 + r_{sp} e^{-\tau_c} \right] \} \tag{18.6}$$

where the soil emissivity ε_{sp} and reflectivity are related by $r_{sp} \varepsilon_{sp} = 1 - r_{sp}$, and ω_p is the vegetation single-scattering albedo. Multiple scattering in the vegetation layer is ignored, and the soil and vegetation temperatures T_s are assumed to be approximately equal. A quasi-specular soil surface and no reflection at the air–vegetation boundary are assumed; τ_c is the vegetation opacity along the viewing path:

$$\tau_c = b_p w_c / \cos \theta \tag{18.7}$$

where θ is the incidence angle, w_c is the vegetation columnar water content, and b_p is a coefficient that depends on frequency and vegetation type.

Soil moisture influences the soil reflectivity and emissivity (ε_{sp} and r_{sp}) through the effect of moisture on the soil dielectric constant. Eq. (18.2) defines the formula for the smooth surfaces. Surface roughness is modeled theoretically using a rough surface spectrum with two parameters: the rms surface height and the horizontal correlation length. In practice, for fixed viewing angle sensors such as AMSR-E, an empirical formulation has been found useful for relating the reflectivity of a rough soil surface to that of the equivalent smooth surface:

$$r_{sp} = [(1 - QR_p + QR_q)]\exp(-h) \qquad (18.8)$$

where p and q represent either of the orthogonal polarization states (V or H), and Q and h are roughness parameters. In several studies, Q has been found to be small (0.1) and, at L-band and C-band frequencies, may be approximated as zero. The influence of soil texture is also significant, although of less importance.

Brightness temperature polarization ratios (PRs) have often been used to study soil moisture and vegetation effects when polarized off-nadir measurements are available. A commonly used PR is given by the expression

$$PR = \frac{T_{b_v} - T_{b_h}}{T_{b_v} + T_{b_h}} \qquad (18.9)$$

PRs effectively normalize the surface temperature, leaving a quantity that is dependent primarily on soil moisture and vegetation. As frequency increases, the PR becomes more dependent on vegetation and roughness and less on soil moisture. This forms one basis for a multichannel (polarization and frequency) approach for separating the effects of temperature, vegetation, and moisture. The off-nadir constant incidence angle (54.8°) observations of AMSR-E are suitable for implementing the PR approach.

The parameter set to be estimated include w_s, T_s, ω_p, b_p, h, and soil texture. Soil texture, b_p, and h are predetermined based on model fitting in conjunction with a land surface classification map. Thus, the iteratively estimated parameter set becomes $x = \{w_s, T_s, \omega_p\}$. The algorithm can also be configured to iterate on computed and observed PRs at 6.9, 10.6, and 18 GHz instead of (or in addition to) brightness temperatures. Convergence will normally occur, except where the model does not adequately represent the surface emission (e.g., in densely vegetated areas

or where the model calibration to the observations is in error). In such cases, the retrievals will yield high values of minimized χ^2. Retrieval flags are used to indicate the quality of the retrieval results based on the number of iterations required and the residual value, among other criteria.

Initially, the algorithm described above was developed for 6.9–10.7 GHz frequencies. Because of radio frequency interference affecting C-band data over large regions, the 10.7 GHz data were used instead. Land surface parameters such as soil moisture, vegetation water content, and surface temperature are also provided as AMSR-E products.

The SMC products from AMSR-E using this algorithm will be presented in Section 18.7.2.

18.3.1.3.2 Land Parameter Retrieval Model (LPRM)

Another recently developed retrieval approach is based on the microwave polarization difference index, defined as Eq. (18.9). The algorithm developed by VU University Amsterdam in collaboration with NASA (VUA-NASA) can be used for all bands in the passive microwave domain (Owe et al., 2008), allowing data collected by different satellites to be combined.

The VUA-NASA algorithm uses the Land Parameter Retrieval Model (LPRM), requiring horizontal (H) and vertical (V) polarization, C-band brightness temperatures (T_b), and V polarization Ka-band T_b, from which soil surface temperature is estimated. The vegetation optical depth (dimensionless, an indicator of vegetation density) and soil dielectric constant are derived simultaneously. The soil moisture (m^3/m^3) is solved from the dielectric constant using the mixing model proposed by Wang and Schmugge (1980).

The upwelling radiation from the land surface, as observed from above the canopy, may be expressed in terms of the radiative brightness temperature and is given as a simple RT equation (Owe et al., 2001)

$$T_{b_p} = T_s \varepsilon_{sp} \Gamma_p + (1 - \omega_p) T_c (1 - \Gamma_p)$$
$$+ (1 - \varepsilon_{s,p})(1 - \omega_p) T_c (1 - \Gamma_p) \Gamma_p \quad (18.10)$$

where p is the state of polarization (H or V), Γ is canopy transmittance, and $\varepsilon_p = 1 - R_p \exp(-h \cos^2 \theta)$. This method assumes that the roughness effect is small. Other terms are the same as described previously. The first term of Eq. (18.10) defines the radiation from the soil, as attenuated by the overlying vegetation. The second term accounts for the upward radiation directly from the vegetation, while the third term defines the downward radiation from the vegetation, reflected upward by the soil and again attenuated by the canopy. The transmissivity Γ is related to the optical depth τ as $\Gamma = \exp(-\tau_c / \cos \theta)$.

Canopy optical depth can be related to the brightness temperature PR as

$$\tau_c = C_1 \ln(C_2 * PR + C_3) \quad (18.11)$$

where C_i are coefficients as a function of the absolute value of the soil dielectric constant k

$$C_i = P_{i,1} k^N + P_{i,2} k^{N-1} + \cdots P_{i,N} k + P_{i,N+1} \quad (18.12)$$

where N is the degree of the polynomial, and $P_{j,i}$ are the polynomial coefficients predetermined from extensive simulations.

By substituting Eqs (18.11) into (18.10), the optical depth is eliminated, and the vegetation term in the RT equation is now expressed as a function of the PR (T_{b_v} and T_{b_H}) and the soil dielectric constant k. The remaining term in the RT Eq. (18.10) is the soil emissivity $\varepsilon_{s,p}$. As H-polarization has the greatest sensitivity to soil moisture, we solve it using T_{b_H}. The emissivity of the soil is calculated from the Fresnel Eq. (18.2), where the only unknown is the dielectric constant of the soil. We now have both the canopy optical depth and the soil emissivity defined in terms of the soil dielectric constant.

Next, the model uses a nonlinear iterative procedure, the Brent method, in a forward approach to solve the RT equation in horizontal polarization by optimizing the dielectric constant.

Once convergence of the modeled and observed horizontal brightness temperatures is achieved, the model uses a global database of soil physical properties (Rodell et al., 2004) together with a soil dielectric mixing model (Wang and Schmugge, 1980) to solve for the SSM. The VUA-NASA model uses either the 6.9 GHz or the 10.7 GHz channel for soil moisture retrieval.

18.3.2 Active microwave remote sensing

18.3.2.1 Basic principles

Active microwave sensors can provide soil moisture estimates at the high spatial resolution and large coverage required for many applications. The most common active-microwave imaging configuration is the SAR, which transmits a series of pulses as the radar antenna traverses the scene. These pulses are then processed together to simulate a very long aperture capable of high surface resolution.

The magnitude of the SAR backscatter coefficient (σ) is related to w_s through the contrast of the dielectric constants of bare soil and water. The basic principles can be illustrated by an empirical model (Ulaby et al., 1996). As illustrated in Fig. 18.3, the radar backscatter from a vegetated surface is composed of three contributions:

$$\sigma = t^2 \sigma_{soil} + \sigma_{veg} + \sigma_{multi}$$
$$= \exp(-2\tau_c) \sigma_{soil} + \sigma_{veg} + \sigma_{multi} \quad (18.13)$$

where σ_{soil} is the backscatter contribution of the bare soil surface, t^2 is the two-way attenuation of the vegetation layer, σ_{veg} is the direct backscatter contribution of the vegetation layer, and σ_{multi} represents multiple scattering involving the vegetation elements and the ground surface. For densely vegetated targets, $t^2 \approx 0$ and σ are

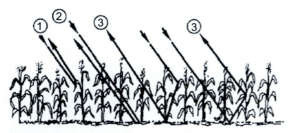

FIGURE 18.3 Backscattering contributions of a canopy over a soil surface (Ulaby et al., 1996): (1) is the direct backscattering from plants; (2) direct backscattering from soil (includes a two-way attenuation by plants); and (3) plant–soil multiple scattering.

determined largely by volumetric scattering from the vegetation canopy. For sparsely vegetated targets, $t^2 \approx 1$ and the second and third terms in Eq. (18.13) are negligible; in that case, σ is determined by the soil roughness and moisture content.

For bare soil, σ_{soil} has a functional relation with w_s, where

$$\sigma_{soil} = f(R, w_s) \qquad (18.14)$$

and R is a surface roughness term. Considering this, many algorithms using single-wavelength, single-polarization SAR to estimate w_s follow a standard two-step approach, where the first step is to estimate and remove the signal owing to backscatter from the vegetation canopy, and thus $\sigma \cong \sigma_{soil}$; the second step is to determine the relationship between σ_{soil} and w_s, based on the assumption that the surface roughness adds a signal to the backscatter intensity that can be treated as an offset. Thus, for a target of uniform R,

$$w_s = a + b\sigma_{soil} \qquad (18.15)$$

where a and b are regression coefficients determined primarily from field experiments, which encompass the target-invariant R and the scene-invariant λ, θ_i, polarization, and calibration. Therefore, Eq. (18.15) is only valid for a given sensor, land use, and soil type and for targets in which t^2, σ_{veg}, and σ_{multi} are known or negligible.

Experimental evidence supports the conclusion that radar signals can penetrate through vegetation, particularly in the longer wavelength segment of the microwave band ($\lambda > 5\,cm$). Although σ exhibits a good response to w_s for each vegetation type, the response curves have different slopes and intercepts, which indicates the need to determine the vegetation parameters.

Another major limitation is that the sensitivity of radar backscatter to R can be much greater than the sensitivity to w_s. For example, Oh et al. (1992) stated that the primary cause of backscatter variation in radar image scenes was surface roughness and, secondarily, moisture content. Thus, it is imperative that surface roughness and topography be accounted for in any operational approach.

18.3.2.2 Satellite sensors

There are primarily two types of radar currently being used for SWC retrieval: SAR and scatterometers.

Investigations into the potential of radars for soil moisture retrieval began in the 1960s and gained momentum in the 1990s due to the launch of several satellites that carried a SAR on board. Earlier-generation SAR sensors operating in C-band (e.g., ERS-1 and RADARSAT-1) provided SMC retrievals based on a single polarization (ERS-1 VV polarization; RADARSAT-1 HH polarization) and, in the case of ERS-1, a 23° incidence angle.

However, it is not possible to accurately retrieve SMC information with only single-frequency, single-incidence angle, single-pass SAR data, without a priori information, as a range of factors impact the backscattered signal from soil. The more recent SAR sensors (e.g., ENVISAT-ASAR, RADARSAT-2, ALOS, and TerraSAR-X) have tried to address this problem by operating at several frequencies, polarizations, and incidence angles, thus allowing acquisitions to be chosen as a function of the required parameter, to minimize the effects of other soil surface characteristics. Additionally, some of the new-generation sensors (ALOS, RADARSAT-2, and TerraSAR-X) are capable of operating in polarimetric mode (four polarizations simultaneously).

There have been five operational SAR satellite systems with frequencies suitable for soil moisture retrieval: ESA ERS-1/2 C-band SAR, ESA ENVISAT (ERS-3) C-band ASAR (Advanced SAR), the Canadian C-band RADARSAT-1/2, the Japanese L-band ALOS-PALSAR (Advanced Land Observing Satellite Phased Array type L-band SAR, JERS-2), and the German X-band Terra-SAR. These SAR systems can provide resolutions from 10 to 100 m over a swath width of 50−500 km.

Note that a significant limitation of SAR for watershed-scale applications is that sun-synchronous satellites can provide only weekly repeat coverage and even longer for the same orbital path (e.g., ERS-1 has a scheduled repeat pass every 35 days for the same orbital path).

Scatterometers are microwave radar sensors that measure the normalized radar cross section of the surface, scanned from an airplane or a satellite. They were primarily developed for measurement of near-surface winds over the ocean, based on the fact that wind determines small-scale changes of the sea surface, affecting the sea surface roughness and, therefore, the backscattering properties. In addition to their original purpose, now scatterometers are also being used for polar ice studies, vegetation coverage, and SWC measurements.

Spaceborne scatterometers, flown on a series of European and US satellites, and used operationally for retrieval of wind data over the oceans, have been found to be useful for soil moisture monitoring over land. While all US scatterometers have been operated in Ku-band (around 14 GHz), Europe relies on C-band scatterometers. Because of their longer wavelength, European scatterometers are better suited to soil moisture retrieval than the US scatterometers. The wind scatterometers on the European Remote Sensing (ERS) satellites have been performing continuous active microwave measurements at C-band (5.3 GHz) between 1991 and 1996 (ERS-1) and since 1996 (ERS-2). Their 20-year continuity has been ensured since 2006 by the Advanced Scatterometer (ASCAT) on the Meteorological Operational satellite (METOP). METOP/ASCAT provided near real-time soil moisture products since 2007 to 2014, which have been widely used (Brocca et al., 2017).

18.3.2.3 Inversion methods

There are many algorithms, from empirical to physical, that are used to estimate SMC from radar data (Moran et al., 2004; Wang and Qu, 2009). Empirical algorithms are generally derived from experimental measurements to establish useful regression between soil moisture and backscatter coefficients. Quite often, the backscatter coefficients are transformed into some form of indices before carrying out the regression analysis. For example, Shoshany et al. (2000) used the normalized backscatter moisture index (NBMI), defined below:

$$\text{NBMI} = \frac{\sigma_{t1} - \sigma_{t2}}{\sigma_{t1} + \sigma_{t2}} \quad (18.16)$$

where σ_{t1} and σ_{t2} are the backscatter coefficients at two different times.

Most methods have been very effective for estimating SMC over bare soils, but challenges remain over vegetated surfaces, although many studies on removing the impacts of vegetation have been reported (Baghdadi et al., 2017; Bao et al., 2018).

18.4 Optical and thermal infrared remote sensing methods

The use of optical sensors for soil moisture assessment has not been very popular (Zhang and Zhou, 2016). Efforts to directly relate soil reflectance to moisture have achieved success only when models are fit for specific soil types in the absence of vegetation cover. This is due partly to the fact that optical remote sensing measures the reflectance or emittance from only the top millimeter (s) of the surface. Another difficult situation is that, while at low moisture levels, increasing moisture content led to a decrease in soil reflectance, the opposite was true at higher moisture levels (Liu et al., 2002). That is, increasing moisture content led to an increase in soil reflectance, albeit determined by much poorer regression results. In addition to moisture content, as discussed earlier, soil reflectance measurements are also strongly affected by the soil composition, physical structure, and observation conditions, resulting in poor predictors of soil moisture on combined soil type samples.

Thus, the use of optical reflectance as a *direct* measure of soil moisture over a large region is greatly constrained, but it has shown some success in combination with thermal remote sensing based on radiative temperature T_R measurements (Zhang and Zhou, 2016). Variations in T_R of bare soils were found to be highly correlated with variations in w_s. However, there is not a universal relationship between surface wetness and soil temperature because of the influence of rapidly varying factors (e.g., wind speed, soil texture, incoming solar radiation, vegetation condition, leaf area index).

Many approaches combine soil temperature and vegetation index (VI) to estimate SMC, including the triangle method, temperature-vegetation contextual approach (TVX), surface temperature—VI space (T_s/NDVI), temperature—vegetation dryness index (TVDI), moisture index, and the VI/T_{rad} relationship. Some of the methods used are described below.

18.4.1 The triangle method

The triangle method is based on an interpretation of the pixel distribution in surface radiant temperature (T_R) and fractional vegetation

coverage (F_r) space (Carlson, 2007; Wang et al., 2018). If a sufficiently large number of pixels are present, excluding cloud, surface water, and outliers, the shape of the pixel envelope resembles a triangle (Fig. 18.4). Basically, a triangle emerges because the range of surface radiant temperature decreases as the vegetation cover increases, its narrow vertex attesting to the narrow range of surface radiant temperature over dense vegetation.

SVAT simulations also confirm the existence of the triangular patterns and can guide the estimate of soil moisture conditions. For example, Carlson (2007) defined the soil surface moisture availability (M_o), the ratio of soil water content to that at field capacity (ranging from 0 for completely dry soil to 1 for the completely wet soil), and then generated a set of polynomials from the model simulations:

$$M_0 = \sum_{i=1}^{i=3} \sum_{j=1}^{j=3} a_{ij} T^{*i} F_r^j \qquad (18.17)$$

where the subscripts i and j pertain to the modeled surface radiant temperature

$$T^* = \frac{T_R - T_{min}}{T_{max} - T_{min}} \qquad (18.18)$$

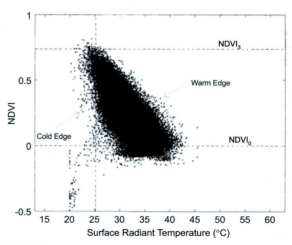

FIGURE 18.4   Scatter plots of NDVI versus surface radiant temperature (Gillies et al., 1997).

where T_{max} and T_{min} are the maximum and minimum values of T_R, and F_r is the fractional vegetation cover determined from NDVI:

$$F_r = \left(\frac{\text{NDVI} - \text{NDVI}_{min}}{\text{NDVI}_{max} - \text{NDVI}_{min}} \right)^2 \qquad (18.19)$$

The coefficients for these parameters are given in Table 18.3. Note that Eq. (18.17) would not precisely fit every situation but is probably sufficiently accurate (and certainly very convenient to use) for many applications where a suitable SVAT model is not available. These polynomials require relatively little expense in computer time or human resources in processing large images.

Lambin and Ehrlich (1996) provided an excellent explanation of the VI and LST space in terms of evaporation, transpiration, and fractional vegetation coverage (see Fig. 18.5) based on previous studies. As they explained, variations in surface brightness temperature are highly correlated with variations in surface water content over base soil. Thus, points A and B in Fig. 18.5 represent dry bare soil (low VI, high LST) and moist bare soil, respectively. As the fractional vegetation cover increases, surface temperature decreases due to several biophysical mechanisms. Point C corresponds to continuous vegetation canopies with a high resistance to evapotranspiration (high VI, relatively high LST), which may result from low soil water availability. Point D corresponds to continuous vegetation canopies with low resistance to evapotranspiration (high VI, low LST), which may occur on well-watered surfaces. The upper envelope of observations (A—C) in the VI—LST space represents the low-evapotranspiration line (i.e., dry conditions). The lower envelop, B—D, represents the line of potential evapotranspiration (wet conditions).

18.4.2 The trapezoid method

The distribution of surface temperature minus air temperature at a particular time was found to form a trapezoid when plotted against percentage vegetation cover (Sadeghi et al., 2017; Tian et al., 2019), as illustrated in Fig. 18.6. The upper left of the trapezoid corresponds to a well-watered crop at 100% cover and the upper right to a nontranspiring crop at 100% cover (points 1 and 2, respectively). These two points are the same as the upper and lower limits for the standard Crop Water Stress Index and can be estimated using the same techniques. The lower portion of the trapezoid (bare soil) is bounded by a wet and dry soil surface. Details on calculating the soil corners are not presented here but are explained by Moran et al. (1994).

With corners of the trapezoid, the Water Deficit Index (WDI) for a measured percentage of vegetation coverage becomes

$$\text{WDI} = \frac{\Delta T - \Delta T_{L13}}{\Delta T_{L24} - \Delta T_{L13}} \qquad (18.20)$$

where ΔT is the measure of surface temperature minus air temperature at a particular percentage vegetation coverage, ΔT_{L13} is surface temperature minus air temperature, determined by the line from points 1 to 3 for the percentage coverage of interest ("wet" line), and ΔT_{L24} is the temperature difference on the line formed between points 2 and 4 ("dry" line). Graphically, the WDI can be viewed as the ratio of the distances A—C to A—B in the previous figure.

TABLE 18.3 Coefficients of the polynomial relationship for M_o between T^* and F_r.

a_{ij}	$j = 0$	$j = 1$	$j = 2$	$j = 3$
$i = 0$	2.058	−1.644	0.850	−0.313
$i = 1$	−6.490	1.112	−3.420	−0.062
$i = 2$	7.618	3.494	10.869	4.831
$i = 3$	−3.190	−3.871	−6.974	−16.902

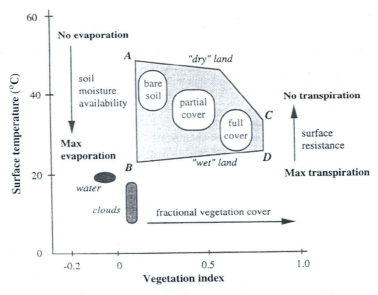

FIGURE 18.5   Relationship between vegetation index and surface temperature (Lambin and Ehrlich, 1996).

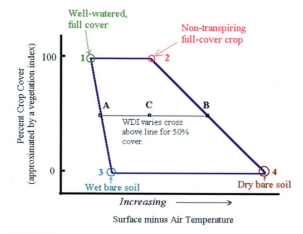

FIGURE 18.6   Relationship between vegetation and surface minus air temperature (Moran et al., 1994).

18.4.3 Temperature–vegetation dryness index

Sandholt et al. (2002) defined a simplified land surface dryness index (TVDI) based on an empirical parameterization of the relationship between surface temperature (T_s) and NDVI.

The index is conceptually and computationally straightforward:

$$\text{TVDI} = \frac{T_s - T_{\min}}{a + b\text{NDVI} - T_{\min}} \qquad (18.21)$$

where $T_{\min}$ is the minimum surface temperature in the triangle, defining the wet edge, T_s is the observed surface temperature at the given pixel, NDVI is the observed normalized difference vegetation index, and a and b are parameters defining the dry edge modeled as a linear fit to data ($T_{\text{smax}} = a + b\text{NDVI}$), where T_{smax} is the maximum surface temperature observation for a given NDVI. The parameters a and b are estimated on the basis of pixels from an area large enough to represent the entire range of surface moisture contents, from wet to dry, and from bare soil to fully vegetated surfaces.

The uncertainty of TVDI is larger for high NDVI values, where the TVDI isolines are closely spaced. The simplification of representing the T_s/NDVI space with a triangle rather than a trapezoid (e.g., Moran et al., 1994) adds uncertainty to TVDI at high NDVI values.

Similarly, the "wet edge" is modeled as a horizontal line in TVDI, as opposed to the sloping wet edge used in the trapezoid approach, which may lead to over estimation of TVDI at low NDVI values.

Sandholt et al. (2002) reported regression coefficients of 0.70 when comparing TVDI results for a study site in Senegal to those from a distributed hydrological model.

Although much progress has been made, the returns from optical/IR sensors remain equally sensitive to soil types, and it is difficult to decouple the two signatures. In addition, the soil moisture estimates derived from optical/IR sensors require surface micrometeorological and atmospheric information that is not routinely available. Controlled experiments continue to show that the optical/IR approach has the potential to sense soil moisture, but implementation, particularly from space, has not been accomplished so far.

18.4.4 The thermal inertia method

This method is based on the principle that soil moisture is directly proportional to thermal inertia (TI) (Lu et al., 2018). For a homogeneous material, TI can be expressed by

$$TI = \sqrt{\rho c K} \qquad (18.22)$$

where K is the thermal conductivity (Jm/ s/ K), ρ is the bulk density (kg/m^3), and c is the specific heat capacity (Jkg/K) of the material. *TI* is expressed in TI units (W·s$^{1/2}$/m^2/K). The TI represents the measure of the material's resistance to externally imposed temperature changes, meaning that, for a given incoming heat flux, the variation of soil temperature is inversely proportional to its TI, which is in turn influenced by soil water content due to its high specific heat.

Table 18.4 shows typical values of TI for different surface types (Sobrino and Cuenca, 1999). Water bodies have a higher *TI* than dry

TABLE 18.4 Thermal inertia values for some typical materials.

Material	TI (TIU)	Material	TI (TIU)
Water and clouds	5000	Granite	2200
Ice	2000	Bush	2000
Snow	150	Grass	2100
Dry sand	590	Corn	2700
Wet sand	2500	Alfalfa	2900
Dry clay	550	Oats	2500
Wet clay	2200	Wood land	4200
Shale	1900		

soils and rocks and exhibit lower diurnal temperature fluctuation. When soil water content increases, *TI* also shows a proportional increase, thereby reducing the diurnal temperature fluctuation range. Because water has a higher TI value than dry soil, a change in moisture content will cause a change in soil TI. Thus, soil moisture may be obtained based on the relationships between TI and soil moisture if TI can be derived from remotely sensed data.

One of the simplest approximations of *TI* is apparent thermal inertia (*ATI*). For an area with uniform solar energy, we obtain the following relationship:

$$ATI = \frac{a(1 - \alpha)}{T_{day} - T_{night}} \qquad (18.23)$$

where a is an experimental coefficient related to soil type, α is the albedo of the land surface, and T_{day} and T_{night} are the land surface temperatures for daytime and night, respectively. SMC can be estimated using a linear model, logarithmic model, or exponential model, established by regression methods. For example, Verstraeten et al. (2008) mapped SMC (w_t) from Meteosat imagery using the following relationship:

$$w_t = SMSI(t)(w_{max} - w_{min}) + w_{min} \qquad (18.24)$$

$$\text{SMSI}(t) = \frac{\sum_t \text{SMSI}_0(t_i)e^{-\left(\frac{t-t_i}{T}\right)}}{\sum_t e^{-\left(\frac{t-t_i}{T}\right)}} \qquad (18.25)$$

where soil moisture saturation index (SMSI) is calculated from the *ATI*:

$$\text{SMSI}_0(t) = \frac{\text{ATI}(t) - \text{ATI}_{\min}}{\text{ATI}_{\max} - \text{ATI}_{\min}} \qquad (18.26)$$

where t_i is the time at which a discrete change of $SMSI_0$ takes place (day).

Although *AIT* has been used to map SMC, it is warned that ATI should not be used in regions that have variability in surface moisture. Sobrino and El Kharraz (1999) developed a formula to map TI using four AVHRR images: an image at 14:30 h (nominal hour) is used to estimate the surface albedo; others at 2:30 and 14:30 h are used to estimate the diurnal surface temperature range, while images at 7:30, 14:30, and 19:30 h are used to determine the phase difference. TI is expressed as

$$\text{TI} = \frac{(1-\alpha)S_0 C_t}{\Delta T \sqrt{\omega}}(Y_1 + Y_2) \qquad (18.27)$$

$$Y_1 = \frac{A_1[\cos(\omega t_2 - \delta_1) - \cos(\omega t_1 - \delta_1)]}{\sqrt{1 + \frac{1}{b} + \frac{1}{2b^2}}} \qquad (18.28)$$

$$Y_2 = \frac{A_2[\cos(2\omega t_2 - \delta_2) - \cos(2\omega t_1 - \delta_2)]}{\sqrt{2 + \frac{\sqrt{2}}{b} + \frac{1}{2b^2}}} \qquad (18.29)$$

where α is the surface albedo, S_0 is the solar constant (typical value: 1367 W/m²), $\omega = 7.272$ 10^{-5} rad/s is the angular velocity of rotation of the Earth, C_t is the atmospheric transmittance in the visible spectrum (typical value of 0.75 was assumed), t_1 and t_2 are the times of diurnal and nocturnal satellite overpasses,

respectively (around 2:30 and 14:30 h), $b = \tan(\delta_1)/[1 - \tan(\delta_1)]$, $\delta_2 = \arctan[\sqrt{2}b/(1-+\sqrt{2}b)]$, and δ_1 is the phase difference given by

$$\delta_1 = \arctan(\xi) + (2m+1)\pi, m = 0,1,2,\ldots$$

$$\xi = \frac{(T_j - T_k)[\cos(\omega t_i) - \cos(\omega t_j)] - (T_i - T_j)[\cos(\omega t_j) - \cos(\omega t_k)]}{(T_i - T_j)[\sin(\omega t_j) - \sin(\omega t_k)] - (T_j - T_k)[\sin(\omega t_i) - \sin(\omega t_j)]}$$
$$(18.30)$$

where T_i, T_j, and T_k are the actual local solar times of the three different satellite passes that correspond to three surface temperatures $T_i \equiv T(t_i)$, $T_j \equiv T(t_j)$, and $T_k \equiv T(t_k)$, respectively. A_1 and A_2 are the first and second coefficients of Fourier series, respectively,

$$A_1 = \frac{2}{\pi}\sin\delta\sin\lambda\sin\psi$$
$$+ \frac{1}{2\pi}\cos\delta\cos\lambda[\sin(2\psi) + 2\psi] \qquad (18.31)$$

$$A_2 = \frac{\sin\delta\sin\lambda}{\pi}\sin(2\psi)$$
$$+ \frac{2\cos\delta\cos\lambda}{3\pi}[2\sin(2\psi)\cos\psi$$
$$- \cos(2\psi)\sin\psi]$$

where $\psi = \arccos(\tan\delta\tan\lambda)$, λ is the latitude of the study area, and δ is the solar declination (both in radians),

$$\delta = (0.006918 - 0.399912\cos\Gamma$$
$$+ 0.070257\sin\Gamma - 0.006758\cos 2\Gamma$$
$$+ 0.000907\sin 2\Gamma - 0.002697\cos 3\Gamma$$
$$+ 0.00148\sin 3\Gamma)(180/\pi)$$
$$(18.32)$$

where $\Gamma = 2\pi(d_n - 1)/365.25$ is the day angle, and d_n is the day of the year.

Cai et al. (2007) developed a method suitable for MODIS data. Because it involves solving two nonlinear equation sets, a lookup table approach has been suggested in their method.

It should be pointed out that a TI model derived from heat conductivity for soil moisture monitoring is only suitable for bare soil or sparsely vegetated areas.

18.5 Estimation of soil moisture profile

The main limitations associated with remote sensing are that only SSM can be retrieved (a few centimeters at most), while it is root-zone soil moisture that is relevant for most climate applications. Root-zone soil moisture constitutes an important variable for hydrological and weather forecast models. It plays a vital role in the regulation of water and energy budgets at the soil—vegetation—atmosphere interface through evaporation processes of the uppermost surface soil layer and plant transpiration. If the initialization of this variable is not accurate, significant drifts of the temporal evolution of the surface state variables may develop and may consequently cause a degradation of the weather forecast reliability (Sabater et al., 2007).

Under dry conditions, SSM might become "decoupled" from the root-zone soil moisture, limiting any approach to derive root-zone soil moisture from surface observations. We have discussed various techniques (above) to estimate near-SSM. This quantity is physically related to root-zone soil moisture through diffusion processes, and both surface and root-zone soil layers are commonly simulated by LSMs. Many investigators have demonstrated the feasibility of estimating soil moisture profiles by assimilating time series of SSM into dynamic models to estimate root-zone soil moisture profiles.

Some assumptions need to be made on the vertical distribution of soil moisture within the soil profile to predict root-zone soil moisture from measured SSM. A number of studies have applied representations of the one-dimensional Richards equation. The ability to retrieve the profile to the extent of the root zone is determined largely by the specifications of the error covariances in the data assimilation framework, and the errors in both the observations and the model must be sufficiently small for the combination approach to be successful. Assimilation methods differ in terms of flexibility,

performance, and computational effort. Sabater et al. (2007) compared two types of Kalman filters and two variational schemes with respect to their potential in deriving root-zone soil moisture from near-SSM data. In general, all four data assimilation methods provided satisfactory results.

Improved retrieval of soil moisture profiles is often accomplished by including additional information besides measured soil moisture data (Baldwin et al., 2017; Vereecken et al., 2008). This may include state variables (e.g., temperature, LAI), fluxes, and spatial attributes.

18.6 Comparison of different remote sensing techniques

Although optical and thermal data can be used to estimate soil moisture conditions, as discussed in Section 18.4, a disadvantage of optical and thermal RS approaches is their limited soil surface penetration depth (Table 18.5). Additionally, optical RS suffers from a high perturbation by clouds and vegetation and a significant signal perturbation by the Earth's atmosphere.

Microwave remote sensing has the potential to facilitate direct measurement of soil moisture. It also has the advantage of providing all-weather observations and penetration into the vegetation canopy for soil moisture sensing. Considering the large differences in footprint size and temporal sampling, it becomes evident that the different microwave sensors provide very different information. While radiometers and scatterometers allow regular monitoring of the large-scale atmosphere-related soil moisture component, SARs allow access to small-scale land surface—related patterns, albeit very infrequently. An important limitation of passive microwave remote sensing is the perturbation of the signal by surface roughness and vegetation biomass, its irregular revisit frequency, and,

TABLE 18.5   Comparison of the advantages and limitations of different remote sensing techniques (Wang and Qu, 2009).

Spectrum domain		Properties observed	Advantages	Limitations
Optical		Soil reflection	Fine spatial resolution broad coverage	Limited surface penetration cloud contamination many other noise sources
Thermal infrared		Surface temperature	Fine spatial resolution broad coverage physical well understood	Limited surface penetration cloud contamination perturbed by meteorological conditions and vegetation
Microwave	Passive	Brightness temperature dielectric properties soil temperature	Low atmospheric noise moderate surface penetration physical well understood	Low spatial resolution perturbed by surface roughness and vegetation
	Active	Backscatter coefficient dielectric properties	Low atmospheric noise moderate surface penetration high spatial resolution physical well understood	Limited swath width perturbed by surface roughness and vegetation

quite importantly, water bodies (e.g., coastlines) are major obstructions in terrestrial applications.

Many studies have been reported to estimate SMC by integrating visible thermal data with microwave data, particularly for removing the effects of vegetation cover. For example, Bao et al. (2018) presented a new methodology for retrieving SSM under conditions of partial vegetation cover based on the synergy between Sentinel-1 SAR and Landsat Operational Land Image data. Bousbih et al. (2018) presented a technique for the mapping of soil moisture and irrigation, at the scale of agricultural fields, based on the synergistic interpretation of multitemporal Sentinel-1 and Sentinel-2 optical and SAR data. Amazirh et al. (2018) presented a synergistic method combining Sentinel-1 microwave and Landsat-7/8 (L7/8) thermal data.

18.7 Available datasets and spatial and temporal variations

18.7.1 Ground point measurements

There are a growing number of in situ soil moisture networks, typically run by universities or national and regional organizations.

However, globally, the number of meteorological networks and stations measuring soil moisture, particularly on a continuous basis, is still limited, and the data they provide lack standardization of technique and protocol. To overcome many of these limitations, the International Soil Moisture Network (ISMN; https://ismn.geo.tuwien.ac.at/en/) was initiated to serve as a centralized data-hosting facility where globally available in situ soil moisture measurements from operational networks and validation campaigns are collected, harmonized, and made available to users (Dorigo et al., 2011) (Fig. 18.7). ISMN includes the in situ datasets in the Global Soil Moisture Data Bank (Robock et al., 2000) that has been closed.

Incoming soil moisture data are automatically transformed into common volumetric soil moisture units and checked for outliers and implausible values. In addition to soil water measurements at different depths, the ISMN database includes important metadata and meteorological variables (e.g., precipitation and soil temperature). The database is queried via a graphical user interface, and data selected for download are provided according to common output standards for data and metadata. The

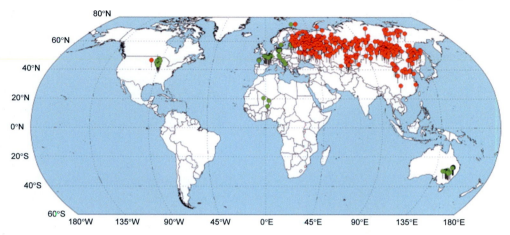

FIGURE 18.7   Map of the distribution of networks and stations in the ISMN as of May 2011 (Dorigo et al., 2011).

database spans from 1952 until the present, although most datasets originated during the last decade.

It is worthy to mention the in situ measurements from various field campaigns, such as the Soil Moisture Experiments series, organized at the Southern Great Plains, USA; the Canadian Experiment for Soil Moisture in 2010 (CanEx-SM10) from May 31, 2010, to June 17, 2010, over agricultural and forested sites located in Saskatchewan, Canada; the Soil Moisture Active Passive Validation Experiments in 2008 and 2012 at the Eastern Shore of Maryland and Delaware, USA; and the Soil Moisture Active Passive Experiments series in Australia.

18.7.2 Microwave remote sensing

Multiple SMC products have been generated from microwave remote sensing data, which will be briefly introduced below.

18.7.2.1 AMSR-E/Aqua daily L3 surface soil moisture

This gridded AMSR-E Level-3 land surface product includes daily measurements of SSM and vegetation/roughness water content interpretive information, as well as brightness temperatures and quality control variables. Ancillary data include time, geolocation, and quality assessment. Input brightness temperature data, corresponding to a 56-km mean spatial resolution, are resampled to a global cylindrical 25-km Equal-Area Scalable Earth Grid Version 2.0 (EASE-Grid 2.0) cell spacing. Data are stored in HDF-EOS format and are available from June 19, 2002, to October 3, 2011, via FTP from the National Snow and Ice Data Center at https://nsidc.org/data/ae_land3/versions/2. The basic algorithm for deriving this dataset was described in 18.3.1.3.1.

18.7.2.2 VUA-NASA soil moisture products

The Land Surface Parameter Model (LPRM), described in Section18.3.1.3.2, that was developed by researchers at NASA and the Department of Hydrology and Geoenvironmental Sciences at Vrije Universiteit (VUA, Amsterdam, the Netherlands) (Owe et al., 2008) has been used to derive soil moisture products from multiple sensors shown in Table 18.6; they are available at https://www.geo.vu.nl/~jeur/lprm/.

The AMSR-E/Aqua level 3 global monthly Surface Soil Moisture Averages and Standard Deviations at 1° by 1° using the same algorithm is also available at the Goddard Earth Sciences

TABLE 18.6   A list of available VUA-NASA soil moisture products.

Satellite	Years	Products
SMMR	1978–87	LST, soil moisture (C- and X-band), vegetation optical depth (C- and X-band)
SSM/I	1987-present	LST, soil moisture (Ku-band), vegetation optical depth (Ku-band)
TRMM-TMI	1998-present	LST, soil moisture (X-band), vegetation optical depth (X-band)
AMSR-E	June 2002–October 2011	LST, soil moisture (C- and X-band), vegetation optical depth (C- and X-band),

Data and Information Services Center, formerly known as the Goddard Distributed Active Archive Center at http://mirador.gsfc.nasa.gov/cgi-bin/mirador/presentNavigation.pl?tree=project&project=NEESPI&dataGroup=Soil%20Moisture.

18.7.2.3 Scatterometer-derived soil moisture product from the Vienna University of Technology

Global coarse-resolution soil moisture data (25–50 km) are derived from backscatter measurements acquired with scatterometers aboard the ERS-1 and ERS-2 satellites (1991–present) and the three MetOp satellites (2006–20). Two different product types are derived:

- Level 2 products representing the SMC within a thin soil surface layer (<2 cm) during the time of overflight of the satellite.
- Level 3 products representing the water content in the soil profile, regularly sampled in space and time.

The Level 2 SSM data are derived using a change detection method that relies on the multi-incidence observation capabilities of the ERS and METOP scatterometers to model the effects of vegetation phenology. The surface soil moisture values are scaled between 0 and 1, representing zero soil moisture and saturation, respectively. Retrieval is not possible over tropical forest, which affects about 6.5% of the land surface area.

The Level 3 soil moisture products are obtained by a modeling approach using the ASCAT Level 2 and other data sources as input. A typical Level 3 product is an estimate of the SMC for different soil layers and different temporal and spatial sampling characteristics, tailored to the needs of specific user groups. Our standard Level 3 product is the so-called Soil Water Index that is a measure of the profile SMC obtained by filtering the surface soil moisture time series with an exponential function. They are available at http://hsaf.meteoam.it/.

18.7.2.4 Soil Moisture and Ocean Salinity

The SMOS satellite has a single payload instrument, the Microwave Imaging Radiometer with Aperture Synthesis (MIRAS), a dual-polarized L-band 2D passive interferometer radiometer operating at 1.4 GHz (21 cm). Soil moisture product has been generated from the radiometer observations at 1.4 GHz (21 cm) and L-band. Global maps of soil moisture values are produced at 3-day, 10-day, and monthly temporal resolutions at 25-km EASE-Grid 2.0 from January 2010 to April 2015, which are available at http://www.catds.fr/Products/Available-products-from-CPDC.

18.7.2.5 Soil Moisture Active and Passive

The SMAP mission concept was to utilize the respective advantages offered by using both passive and active sensors in retrieving soil moisture and to obtain accurate soil moisture observations with high spatial and temporal resolutions. In this regard, SMAP has two components, an active SAR (1.2 GHz with VV, HH, and HV polarizations) and a passive radiometer at L-band frequency (1.41 GHz with, V, H, the third and the fourth Stokes' parameters), and achieves global coverage every 2—3 days. Global daily products of soil moisture have been generated from SMAP from 31 March 2015 to present at the 36-km EASE-Grid 2.0 and further interpolated to the 9-km EASE-Grid 2.0. The products are available at https://nsidc.org/data/SPL3SMP/versions/5 and https://nsidc.org/data/SPL3SMP_E/versions/2, respectively.

18.7.2.6 ESA soil moisture ECV products

The global Essential Climate Variable (ECV) soil moisture dataset has been generated using active and passive microwave spaceborne instruments (see Fig. 18.8) and covers the (nearly) 40 year period from 1978 to 2018 (Dorigo et al., 2017). The dataset consists of three products: an active dataset based on observations from the C-band scatterometers on board of ERS-1, ERS-2, MetOp-A, and MetOp-B; a passive dataset based on passive microwave observations from Nimbus 7 SMMR, DMSP SSM/I, TRMM TMI, Aqua AMSR-E, Coriolis WindSat, GCOM-W1 AMSR2, and SMOS; and a combined dataset. It provides daily surface soil moisture with a spatial resolution of 0.25°. Soil moisture uncertainties are now available globally. The current version is V04.4, freely available at www.esa-soilmoisture-cci.org/dataregistration.

18.7.3 LSM estimates with observation-based forcing

LSM estimates, driven with observation-based data, can be a useful alternative to derive global soil moisture datasets. Two such products are the Global Land Data Assimilation System dataset (GLDAS) (Rodell et al., 2004) and Global Soil Wetness Project 2 (GSWP-2) (Dirmeyer et al., 2006). There are two versions: GLDAS-1 and

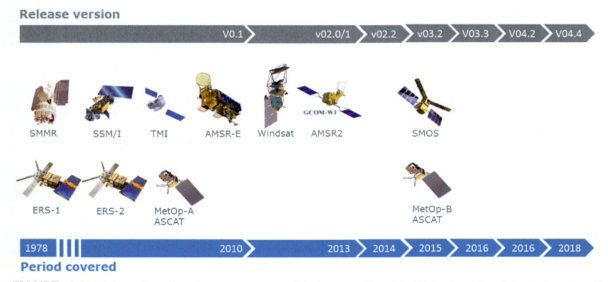

FIGURE 18.8 Active and passive microwave sensors used for the generation of the ECV soil moisture datasets. *From https://www.esa-soilmoisture-cci.org/node/93. Used with permission from Prof. Wouter Dorigo, The Vienna University of Technology.*

TABLE 18.7   Vertical soil layers used in different models in the global or North America Land Data Assimilation System.

Land data Assimilation Systems	Models	Vertical layers (meters)
GLDAS	CLM (10 layers)	0–0.018, 0.018–0.045, 0.045–0.091, 0.091–0.166, 0.166–0.289, 0.289–0.493, 0.493–0.829, 0.829–1.383, 1.383–2.296, and 2.296–3.433
	Mosaic (3 layers)	0–0.02, 0.02–1.50, and 1.5–3.50
	NOAH (4 layers)	0–0.1, 0.1–0.4, 0.4–1.0, and 1.0–2.0
	VIC (3 layers)	0–0.1, 0.1–0.4, 0.4–1.0, and 1.0–2.0
NLDAS	Mosaic (6 layers)	0–10, 0–40, 0–100, 0–200, 10–40, and 40–200

GLDAS-2. The outputs of SMC in GLDAS-1 are from four LSMs with different layer depths, including Community Land Model, Mosaic (MOS), variable infiltration capacity (VIC), and Noah (NOAH). The GLDAS-2 only involves the NOAH model. More details about the GLDAS data can be found from https://disc.sci.gsfc.nasa.gov/datasets?keywords=GLDAS.

The Global Soil Wetness Project 2 dataset (GSWP-2) (Dirmeyer et al., 2006), available for the time period 1986–95 (a previous dataset for 1987–88 is also available) (Dirmeyer et al., 1999), was generated from an environmental modeling research activity of the Global Land/Atmosphere System Study and the International Satellite Land Surface Climatology Project (ISLSCP). A major product of GSWP-2 will be a multimodel land surface analysis for the ISLSCP Initiative II period. Information on this dataset is available at http://cola.gmu.edu/gswp/

One of the biggest advantages of the LSM estimates is to provide the average layer soil moisture, which is the depth-averaged amount of water present in a specific soil layer beneath the surface. In the Land Data Assimilation Systems (LDAS), the number of vertical levels for soil moisture is model-specific (see Table 18.7). Data are available at http://disc.gsfc.nasa.gov/hydrology/data-holdings/parameters/average_layer_soil_moisture_hsb.shtml.

18.8 Conclusions

SMC at the top surface layer can be estimated from optical, thermal, and microwave remote sensing data, but the operational products are mainly derived from passive microwave remote sensing data. The vertical profiles of SMC can also be estimated by assimilating remotely sensed data products into a dynamic process model (Baldwin et al., 2017).

Simplified large-scale SWC modeling at the regional and global scales is necessary and remains a challenge, due to the complexity of acquiring the knowledge of soil hydraulic properties necessary for reliable modeling.

None of the available sensor configurations is able to perfectly fulfill all requirements, nor is there an ideal retrieval algorithm. Rather, the combination of sensor/algorithm has to be optimized to derive accurate soil moisture data. A major challenge for the future will be the simultaneous use of different measurement technologies and the development of a framework that optimally combines the information at different scales (Dorigo et al., 2017), especially the identification of efficient techniques for downscaling of satellite data (Peng et al., 2017).

The accuracy of remotely sensed soil moisture products is determined by the sensor characteristics and the retrieval algorithms used. Many validation efforts have been conducted, but methods and reference data for validation have varied considerably between different studies. It is therefore difficult to assess the relative accuracy of the different sensors/algorithms and provide reliable accuracy estimates (Karthikeyan et al., 2017b). This demonstrates the urgent need to introduce standards and common datasets for the validation of remotely sensed soil moisture data to compare the accuracy of different soil moisture products for different vegetation zones and climatic conditions.

The Gravity Recovery and Climate Experiment satellite, launched in 2002, allows derivation of the Earth's gravitational field (approximately monthly temporal resolution) related to seasonal variations in terrestrial water storage (soil moisture, groundwater, snow, surface water, and ice cover). Even after determining the terrestrial water storage (Chapter 21 discusses some of the technical issues), however, disaggregation of the data in separate estimates of the individual terrestrial water storage components is necessary to estimate the soil moisture at such a coarse spatial resolution (e.g., 500–1000 km).

References

Amazirh, A., Merlin, O., Er-Raki, S., Gao, Q., Rivalland, V., Malbeteau, Y., Khabba, S., Escorihuela, M.J., 2018. Retrieving surface soil moisture at high spatio-temporal resolution from a synergy between Sentinel-1 radar and Landsat thermal data: a study case over bare soil. Remote Sens. Environ. 211, 321–337.

Baghdadi, N., El Hajj, M., Zribi, M., Bousbih, S., 2017. Calibration of the water cloud model at C-band for winter crop fields and grasslands. Remote Sens. 9, 969.

Baldwin, D., Manfreda, S., Keller, K., Smithwick, E.A.H., 2017. Predicting root zone soil moisture with soil properties and satellite near-surface moisture data across the conterminous United States. J. Hydrol. 546, 393–404.

Bao, Y., Lin, L., Wu, S., Deng, K.A.K., Petropoulos, G.P., 2018. Surface soil moisture retrievals over partially vegetated areas from the synergy of Sentinel-1 and Landsat 8 data using a modified water-cloud model. Int. J. Appl. Earth Obs. Geoinf. 72, 76–85.

Bittelli, M., 2011. Measuring soil water content: a review. HortTechnology 21, 293–300.

Bousbih, S., Zribi, M., El Hajj, M., Baghdadi, N., Lili-Chabaane, Z., Gao, Q., Fanise, P., 2018. Soil moisture and irrigation mapping in A semi-arid region, based on the synergetic use of sentinel-1 and sentinel-2 data. Remote Sens. 10, 1953.

Brocca, L., Crow, W.T., Ciabatta, L., Massari, C., de Rosnay, P., Enenkel, M., Hahn, S., Amarnath, G., Camici, S., Tarpanelli, A., Wagner, W., 2017. A review of the applications of ASCAT soil moisture products. Ieee J. Sel. Top. Appl. Earth Obs. Remote Sens. 10, 2285–2306.

Cai, G., Xue, Y., Hu, Y., Wang, Y., Guo, J., Luo, Y., Wu, C., Zhong, S., Qi, S., 2007. Soil moisture retrieval from MODIS data in Northern China Plain using thermal inertia model. Int. J. Remote Sens. 28, 3567–3581.

Carlson, T., 2007. An overview of the "triangle method" for estimating surface evapotranspiration and soil moisture from satellite imagery. Sensors 7, 1612–1629.

Chan, S.K., Bindlish, R., O'Neill, P.E., Njoku, E., Jackson, T., Colliander, A., Chen, F., Burgin, M., Dunbar, S., Piepmeier, J., Yueh, S., Entekhabi, D., Cosh, M.H., Caldwell, T., Walker, J., Wu, X.L., Berg, A., Rowlandson, T., Pacheco, A., McNairn, H., Thibeault, M., Martinez-Fernandez, J., Gonzalez-Zamora, A., Seyfried, M., Bosch, D., Starks, P., Goodrich, D., Prueger, J., Palecki, M., Small, E.E., Zreda, M., Calvet, J.C., Crow, W.T., Kerr, Y., 2016. Assessment of the SMAP passive soil moisture product. IEEE Trans. Geosci. Remote Sens. 54, 4994–5007.

de Jeu, R.A.M., Wagner, W., Holmes, T.R.H., Dolman, A., van de Giesen, N., Friesen, J., 2008. Global soil moisture patterns observed by space borne microwave radiometers and scatterometers. Surv. Geophys. 29, 399–420.

Dirmeyer, P.A., Dolman, A., Sato, N., 1999. The pilot phase of the global soil wetness project. Bull. Am. Meteorol. Soc. 80, 851–878.

Dirmeyer, P.A., Gao, X.A., Zhao, M., Guo, Z.C., Oki, T.K., Hanasaki, N., 2006. GSWP-2 — multimodel analysis and implications for our perception of the land surface. Bull. Am. Meteorol. Soc. 87, 1381–1397.

Dorigo, W., Wagner, W., Albergel, C., Albrecht, F., Balsamo, G., Brocca, L., Chung, D., Ertl, M., Forkel, M., Gruber, A., Haas, E., Hamer, P.D., Hirschi, M., Ikonen, J., de Jeu, R., Kidd, R., Lahoz, W., Liu, Y.Y., Miralles, D., Mistelbauer, T., Nicolai-Shaw, N., Parinussa, R., Pratola, C., Reimer, C., van der Schalie, R., Seneviratne, S.I., Smolander, T., Lecomte, P., 2017. ESA CCI soil moisture for improved earth system understanding: state-of-the art and future directions. Remote Sens. Environ. 203, 185–215.

Dorigo, W.A., Wagner, W., Hohensinn, R., Hahn, S., Paulik, C., Xaver, A., Gruber, A., Drusch, M., Mecklenburg, S., van Oevelen, P., Robock, A., Jackson, T., 2011. The International Soil Moisture Network: a data hosting facility for global in situ soil moisture measurements. Hydrol. Earth Syst. Sci. 15, 1675–1698.

Entekhabi, D., Njoku, E.G., O'Neill, P.E., Kellogg, K.H., Crow, W.T., Edelstein, W.N., Entin, J.K., Goodman, S.D., Jackson, T.J., Johnson, J., 2010. The soil moisture active passive (SMAP) mission. Proc. IEEE 98, 704–716.

Gillies, R., Cui, J., Carlson, T., Kustas, W., Humes, K., 1997. Verification of the "triangle" method for obtaining surface soil water content and energy fluxes from remote measurements of NDVI and surface radiant temperature. Int. J. Remote. Sens. 18, 3145–3166.

Jackson, T., 2008. Passive microwave remote sensing for land applications. In: Liang, S. (Ed.), Advances in Land Remote Sensing: System, Modeling, Inversion and Application. Springer, New York, pp. 9–18.

Karthikeyan, L., Pan, M., Wanders, N., Kumar, D.N., Wood, E.F., 2017a. Four decades of microwave satellite soil moisture observations: Part 1. A review of retrieval algorithms. Adv. Water Resour. 109, 106–120.

Karthikeyan, L., Pan, M., Wanders, N., Kumar, D.N., Wood, E.F., 2017b. Four decades of microwave satellite soil moisture observations: Part 2. Product validation and inter-satellite comparisons. Adv. Water Resour. 109, 236–252.

Kerr, Y.H., Al-Yaari, A., Rodriguez-Fernandez, N., Parrens, M., Molero, B., Leroux, D., Bircher, S., Mahmoodi, A., Mialon, A., Richaume, P., 2016. Overview of SMOS performance in terms of global soil moisture monitoring after six years in operation. Remote Sens. Environ. 180, 40–63.

Kerr, Y.H., Waldteufel, P., Wigneron, J.P., Delwart, S., Cabot, F., Boutin, J., Escorihuela, M.J., Font, J., Reul, N., Gruhier, C., 2010. The SMOS mission: new tool for monitoring key elements of the global water cycle. Proc. IEEE 98, 666–687.

Lambin, E.F., Ehrlich, D., 1996. The surface temperature - vegetation index space for land cover and land-cover change analysis. Int. J. Remote Sens. 17, 463–487.

Legates, D.R., Mahmood, R., Levia, D.F., DeLiberty, T.L., Quiring, S.M., Houser, C., Nelson, F.E., 2011. Soil moisture: a central and unifying theme in physical geography. Prog. Phys. Geogr. 35, 65–86.

Lekshmi, S.U.S., Singh, D.N., Baghini, M.S., 2014. A critical review of soil moisture measurement. Measurement 54, 92–105.

Liu, W., Baret, F., Gu, X., Tong, Q., Zheng, L., 2002. Relating soil surface moisture to reflectance. Remote Sens. Environ. 81, 238–246.

Lu, Y.L., Horton, R., Zhang, X., Ren, T.S., 2018. Accounting for soil porosity improves a thermal inertia model for estimating surface soil water content. Remote Sens. Environ. 212, 79–89.

McColl, K.A., Alemohammad, S.H., Akbar, R., Konings, A.G., Yueh, S., Entekhabi, D., 2017. The global distribution and dynamics of surface soil moisture. Nat. Geosci. 10, 100–104.

Moran, M., Peters-Lidard, C., Watts, J., McElroy, S., 2004. Estimating soil moisture at the watershed scale with satellite-based radar and land surface models. Can. J. Remote Sens. 30, 805–826.

Moran, S.M., Clarke, T.R., Inoue, Y., Vidal, A., 1994. Estimating crop water deficit using the relationship between surface-air temperature and spectral vegetation index. Remote Sens. Environ. 49, 246–263.

Njoku, E.G., Jackson, T.J., Lakshmi, V., Chan, T.K., Nghiem, S.V., 2003. Soil moisture retrieval from AMSR-E. IEEE Trans. Geosci. Remote Sens. 41, 215–229.

Njoku, E.G., Li, L., 1999. Retrieval of land surface parameters using passive microwave measurements at 6-18 GHz. IEEE Trans. Geosci. Remote Sens. 37, 79–93.

Oh, Y., Sarabandi, K., Ulaby, F.T., 1992. An empirical model and an inversion technique for radar scattering from bare soil surfaces. IEEE Trans. Geosci. Remote Sens. 30, 370–381.

Owe, M., de Jeu, R., Holmes, T., 2008. Multisensor historical climatology of satellite-derived global land surface moisture. J. Geophys. Res. 113, F01002.

Owe, M., de Jeu, R., Walker, J., 2001. A methodology for surface soil moisture and vegetation optical depth retrieval using the microwave polarization difference index. IEEE Trans. Geosci. Remote Sens. 39, 1643–1654.

Peng, J., Loew, A., Merlin, O., Verhoest, N.E., 2017. A review of spatial downscaling of satellite remotely sensed soil moisture. Rev. Geophys. 55, 341–366.

Robock, A., Vinnikov, K., Srinivasan, G., Entin, J., Hollinger, S., Speranskaya, N., Liu, S., Namkhai, A., 2000. The global soil moisture data bank. Bull. Am. Meteorol. Soc. 81, 1281–1299.

Rodell, M., Houser, P.R., Jambor, U., Gottschalck, J., Mitchell, K., Meng, C.J., Arsenault, K., Cosgrove, B., Radakovich, J., Bosilovich, M., Entin, J.K., Walker, J.P., Lohmann, D., Toll, D., 2004. The global land data assimilation system. Bull. Am. Meteorol. Soc. 85, 381–394.

Sabater, J.M., Jarlan, L., Calvet, J.C., Bouyssel, F., De Rosnay, P., 2007. From near-surface to root-zone soil moisture using different assimilation techniques. J. Hydrometeorol. 8, 194–206.

Sadeghi, M., Babaeian, E., Tuller, M., Jones, S.B., 2017. The optical trapezoid model: a novel approach to remote sensing of soil moisture applied to Sentinel-2 and Landsat-8 observations. Remote Sens. Environ. 198, 52–68.

Sandholt, I., Rasmussen, K., Andersen, J., 2002. A simple interpretation of the surface temperature/vegetation index space for assessment of surface moisture status. Remote Sens. Environ. 79, 213–224.

Seneviratne, S.I., Corti, T., Davin, E.L., Hirschi, M., Jaeger, E.B., Lehner, I., Orlowsky, B., Teuling, A.J., 2010. Investigating soil moisture-climate interactions in a changing climate: a review. Earth Sci. Rev. 99, 125–161.

Shi, J., Chen, K., Li, Q., Jackson, T.J., O'Neill, P.E., Tsang, L., 2002. A parameterized surface reflectivity model and estimation of bare-surface soil moisture with L-band radiometer. IEEE Trans. Geosci. Remote Sens. 40, 2674–2686.

Shi, J., Jiang, L., Zhang, L., Chen, K.S., Wigneron, J.P., Chanzy, A., 2005. A parameterized multifrequency-polarization surface emission model. IEEE Trans. Geosci. Remote Sens. 43, 2831–2841.

Shoshany, M., Svoray, T., Curran, P., Foody, G.M., Perevolotsky, A., 2000. The relationship between ERS-2 SAR backscatter and soil moisture: generalization from a humid to semi-arid transect. Int. J. Remote Sens. 21, 2337.

Sobrino, J., El Kharraz, M., 1999. Combining afternoon and morning NOAA satellites for thermal inertia estimation 1. Algorithm and its testing with Hydrologic Atmospheric Pilot Experiment-Sahel data. J. Geophys. Res. 104, 9445–9453.

Sobrino, J.A., Cuenca, J., 1999. Angular variation of emissivity for some natural surfaces from experimental measurements. Appl. Optic. 38, 3931–3936.

Tian, J., Deng, X.Z., Su, H.B., 2019. Intercomparison of two trapezoid-based soil moisture downscaling methods using three scaling factors. Int. J. Digit. Earth 12, 485–499.

Ulaby, F.T., Dubois, P.C., vanZyl, J., 1996. Radar mapping of surface soil moisture. J. Hydrol. 184, 57–84.

Vereecken, H., Huisman, J.A., Bogena, H., Vanderborght, J., Vrugt, J.A., Hopmans, J.W., 2008. On the value of soil moisture measurements in vadose zone hydrology: a review. Water Resour. Res. *44*.

Verstraeten, W.W., Veroustraete, F., Feyen, J., 2008. Assessment of evapotranspiration and soil moisture content across different scales of observation. Sensors 8, 70–117.

Wang, C., Fu, B.J., Zhang, L., Xu, Z.H., 2019. Soil moisture-plant interactions: an ecohydrological review. J. Soils Sediments 19, 1–9.

Wang, J.R., Schmugge, T.J., 1980. An empirical model for the complex dielectric permittivity of soils as a function of water content. IEEE Trans. Geosci. Remote Sens. 18, 288–295.

Wang, L., Qu, J.J., 2009. Satellite remote sensing applications for surface soil moisture monitoring: a review. Front. Earth Sci. China 3, 237–247.

Wang, S., Garcia, M., Ibrom, A., Jakobsen, J., Koppl, C.J., Mallick, K., Looms, M.C., Bauer-Gottwein, P., 2018. Mapping root-zone soil moisture using a temperature-vegetation triangle approach with an unmanned aerial system: incorporating surface roughness from structure from motion. Remote Sens. 10, 28.

Zhang, D.J., Zhou, G.Q., 2016. Estimation of soil moisture from optical and thermal remote sensing: a review. Sensors 16, 29.

Zhang, R., Kim, S., Sharma, A., 2019. A comprehensive validation of the SMAP Enhanced Level-3 Soil Moisture product using ground measurements over varied climates and landscapes. Remote Sens. Environ. 223, 82–94.

Advanced Remote Sensing, Second Edition
https://doi.org/10.1016/B978-0-12-815826-5.00019-2

© 2020 Elsevier Inc. All rights reserved.

Abstract

The amount of snow affects the energy, water, and carbon balance on the Earth's surface and the interactions between the land and the atmosphere. It is a key parameter for the numerical weather prediction and the long-term climate research. Therefore, it is very important to monitor snow water equivalent (SWE). This chapter describes the methods to measure SWE using both the ground-based and the remote sensing techniques, and introduces the spatiotemporal distribution characteristics of the SWE and its application in the hydrological, meteorological, biological and economical fields. Microwave remote sensing has a strong physical basis to detect snow depth and SWE, due to the penetration ability of microwave and its sensitivity to the snow volume scattering. Therefore, we focus on the theories and algorithms using the passive and active microwave remote sensing. The optical remote sensing utilizes the snow cover fraction to build empirical relationship with SWE. The SWE can also be "reconstructed" by summing the snowmelt calculated by the snow cover fraction from the last to the current snow days. These methods are also briefly introduced, as an extension to the microwave remote sensing.

This chapter describes the methods to conduct the snow water equivalent (SWE) measurement, which include both the ground-based in situ approaches and the remote sensing techniques, and briefly introduces the spatiotemporal distribution characteristics of the remotely sensed SWE and its application. Microwave remote sensing has a stronger physical basis to detect snow depth and SWE compared with the optical remote sensing, due to its penetration ability and sensitivity to the snow volume scattering. Therefore, we focus on the theories and algorithms using the passive and active microwave remote sensing. The optical remote sensing utilizes the snow cover fraction to build empirical

relationship with snow depth or to estimate snowmelt to "reconstruct" SWE. These methods are also briefly introduced, as an extension to the microwave remote sensing. The chapter is organized as follows. Section I introduces the ground-based SWE measurement methods. Section II introduces the snow microwave remote sensing theory. Section III introduces the passive and active remote sensing techniques for the SWE estimation. Section IV provides a brief overview of the optical remote sensing applied for the snow depth or SWE calculation. Section V is the summary.

Snow is one of the most active natural elements on the Earth's surface in the winter. About 3/4 global terrestrial freshwater resources exist in the form of ice and snow. Every winter, about 46 million km^2 land in the northern hemisphere (NE) is covered by seasonal snow (Brown and Robinson, 2011). Snow has a high albedo and low thermal conductivity. Therefore, the snow amount strongly affects the energy balance on the Earth's surface and the interactions between the land and the atmosphere. It is a key parameter for the numerical weather prediction and the long-term climate research (Flanner et al., 2011). Seasonal snow accumulates in the winter and melts in the spring. It leads to a delay of several months between the precipitation and the runoff and influences the water, energy, and carbon cycles. 1/6 of the worlds' population relies on the snowmelt for water demand (Barnett et al., 2005). The changes in snow cover can result in large economical and social effects. Therefore, regular monitoring of SWE is of critical importance and high priority.

SWE (*W*) represents the equivalent height of liquid water when a unit column of snow cover completely melts. It can be expresses as (Ulaby et al., 1986):

$$W = \int_0^d \rho_s dz / \rho_w \qquad (19.1)$$

where ρ_s is snow density, and ρ_w is the density of liquid water (1 g/cm^3 or 1000 kg/m^3).

If the snow layer is uniform, then Eq. (19.1) can be simplified as

$$W = \rho_s d / \rho_w \qquad (19.2)$$

In Eq. (19.2), mm or cm is often utilized as the unit for SWE (W).

After the SWE is defined, from the next section, we will start to introduce the in situ, remote sensing SWE estimation methods and the SWE applications.

19.1 Snow water equivalent ground measurement method

SWE on the ground can be measured by the direct and indirect ways. In the direct ways,

the SWE is measured by weighing the mass of a vertical column of the snow cover within a unit section area. The Mt. Rose federal sampler and the snow pillow are widely used tools. The Mt. Rose sampler is a metal tube with a sharpe edge at the bottom. As shown in Fig. 19.1A, by pressing the federal sampler into the snow surface until it reaches the ground, a column of snow can be held in the sampler and lifted for weighing. SWE is determined by the mass divided by the density of liquid water (1 g/cm^3) and the cross-sectional area of the sampler. The snow pillow (Fig. 19.1B) is an automatic tool utilized in the SNOWTEL (SNow TElemetry) network in the Western United States. It is a pressure-sensing pillow made of stainless steel, aluminum, or synthetic rubber, filled with an antifreeze solution (for example, glycol) to measure the weight of the snow resting on it from the snowfalls. The measured weight is converted into SWE in the field and transmitted to the data collection center. As shown in Eq. (19.1), the SWE can also be calculated by multiplying the snow depth and the average

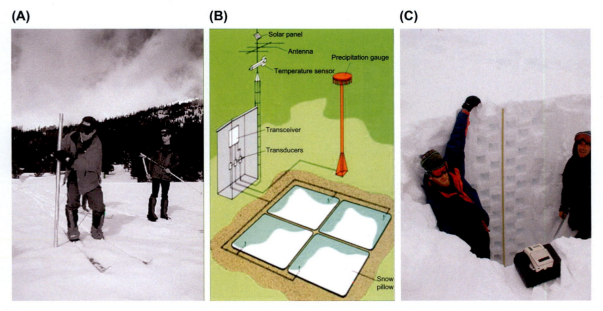

(A)   **(B)**   **(C)**

FIGURE 19.1 In situ snow water equivalent measurement method: (A) the Mt. Rose federal sampler (photographed by Randy Julander); (B) snow pillow (from the website of the Natural Resources Conservation Service, USDA, https://www.wcc.nrcs.usda.gov/factpub/sect_4b.html); and (C) snowpit and snow density measurement (From the National Snow and Ice Data Center, photographed during the Cold Land Processes Field Experiment campaign). *(C) Image/photo courtesy of the National Snow and Ice Data Center, University of Colorado, Boulder.*

snow density measured by the snow cutter from a snowpit (Fig. 19.1C). Digging a pit is time consuming, but the density measurement can work together with other measurements (for example, snow stratigraphy and snow grain size) at the same site.

One indirect method to measure SWE is to measure the attenuation of the water molecules (ice or liquid) to the gamma radiation. The natural source of the gamma radiation comes from the decay of potassium (^{40}K), thorium, and uranium in the Earth (Glynn et al., 1988). The gamma detector compares the amount of gamma radiation (in counts per hour) emitted from the ground before and after the snowfall and converts the attenuation to the SWE using a coefficient determined from the calibration processes. In the Nordic Snow Radar Experiment (NoSREx) (Lemmetyinen et al., 2016), an experimental Gamma Water Instrument based on this method was utilized to automatically measure the real-time SWE (see Fig. 19.2).

19.2 Snow microwave scattering and emission modeling

Dry snow is a mixture of ice and air; the underlying is often rough soil surfaces. In the interactions of microwaves with the terrestrial snow, the volume scattering of snow particles, the surface scattering from the snow—ground interface, snow layer boundary and air—snow interface, and the interaction between snow volume and rough soil surface need to be considered.

Microwave snow forward models are expected to accurately simulate the microwave scattering and emission characteristics of snow covered terrain. The forward models mainly include three parts under radiative transfer concept (Pan et al., 2016):

(1) Snow microstructure representations: The representation of snow microstructure is critical because it partly dominates the choice of formulation to compute the scattering coefficient. Several models use a collection of discrete ice spheres to characterize the snow microstructure

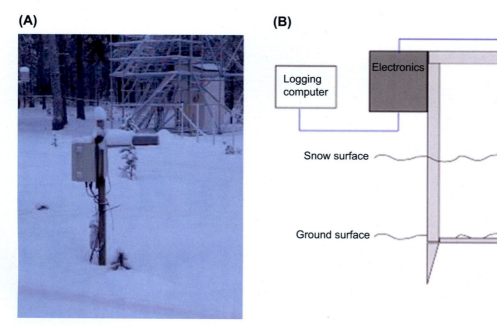

FIGURE 19.2 The Gamma Water Instrument (GWI) installed in the Finnish Meteorological Institute to measure SWE: (A) the instrument photo and (B) the schematics (Lemmetyinen et al., 2011).

(the Helsinki University of Technology snow emission model, HUT; dense media radiative transfer model based on quasicrystalline approximation, DMRT-QCA). Models, such as the microwave emission model for layered snowpacks (MEMLS) and bicontinuous model, treat snow as a continuous medium.

(2) Scattering properties calculations: For microwave spectrum, the snow grain size is comparable with incident wavelengths, thus scattering effect dominates the interaction of microwave and snow. The sensitivity of brightness temperature and backscattering coefficient to scattering is much stronger than absorption. According to the complexity of calculating scattering coefficient, the models can be classified into three categories: semiempirical models, analytical models, and numerical models. Semiempirical models are based on limited field observations, and the scattering coefficient is usually expressed as function of snow grain size and incident wave frequency. Analytical models are derived from Maxwell equations after some simplification and approximations. In numerical models, the scattering properties are directly calculated using numerical electromagnetic computations, and the snow microstructure is usually treated as the bicontinuous medium (Ding et al., 2010).

(3) The solution of radiative transfer equation (RTE): Usually, the microwave scattering and emission of snow-covered terrain are modeled using RTE, in which the snow volume scattering, soil and snow surface scattering, and the interaction between snow volume and soil surface are included. The RTE is a partial differential/integral equation; various methods were proposed to solve it. By considering the complexity and the number of discrete directions of fluxes, the methods of solution can be classified into three categories: two streams (such as HUT model), six streams (such as MEMLS model), and user-defined number of streams (such as DMRT model).

In passive remote sensing, the radiometers only receive the emitted microwave radiation from the snow and soil medium. In active remote sensing, radars emit concentrated pluses into the ground and measure the backscattered signals from the snow and the snow. Their difference is illustrated in Fig. 19.3, with major sources of radiation drawn in colored arrows.

In the following, commonly used snow emission and scattering models are introduced, such as the HUT model, the MEMLS model, the DMRT-QCA model, and DMRT-bicontinuous model. Each model can be used for snowpack with multiple layers. These models can be classified into three categories according to their strategy in calculating the scattering properties: semiempirical models, analytical models, and numerical models.

19.2.1 Semiempirical models

The semiempirical models rely on experimental measurements, thus extensive electromagnetic computations are avoided. The models have simple forms, strong practicability, and have already been successfully used in the research of passive microwave snow parameters inversions. HUT model and MEMLS are two typical semiempirical models.

19.2.1.1 Helsinki University of Technology model

HUT model (Pulliainen et al., 1999) is a semiempirical model based on radiative transfer. The scattering phase function of snow is simplified as a forward-scattering function. The extinction coefficient of snow is estimated according to experiments (Hallikainen et al., 1987) as

$$\kappa_e = 0.0018 f^{2.8} D^2 \qquad (19.3)$$

where κ_e is the extinction coefficient (dB), f is the frequency (GHz), and d_0 is the snow particle diameter (mm). The absorption coefficient of snow is calculated by dry snow dielectric constant, with the real part calculated based on Mätzler (1987) and the imaginary part calculated using Polder—van Santen mixture dielectric constant model (Hallikainen and Ulaby, 1986).

(A)

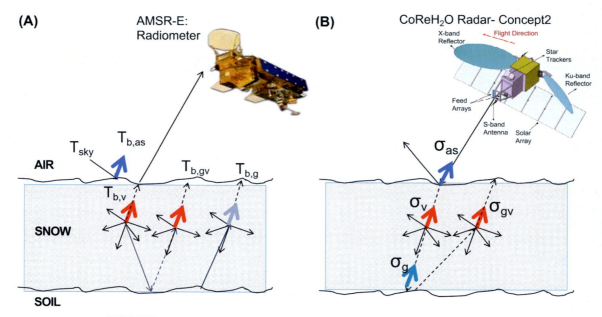

(B)

FIGURE 19.3   The passive (left) and active (right) microwave radiative transfer.

In the HUT model, the scattering phase function is assumed to be forward concentrated, and the relationship between extinction coefficient and frequency and particle size is empirical. These aspects are the short end of the HUT model. Roy et al. (2004) fitted another equation for the extinction coefficient referring to Rayleigh scattering theory and the experiment data as

$$\kappa_e = 2\left(f^4 D^6\right)^{0.2} \qquad (19.4)$$

The model considers most of the scattered radiation intensity to be concentrated in the forward direction and introduces an empirical parameter q to define the total forward scattered incoherent intensity. Then, the emission of snow medium with thickness d' can be described as

$$\frac{\partial T_B(d',\theta)}{\partial d'} = \kappa_a \sec \theta T_s + \sec \theta (q\kappa_s - \kappa_e) T_B(d',\theta)$$

$$(19.5)$$

where T_s is the physical snow temperature. κ_a, κ_s, and κ_e are absorption coefficient, scattering coefficient, and extinction coefficient. θ is the propagation direction.

Considering the multiple reflections between air—snow and snow—soil boundaries, the brightness temperature observed from snow surface T_B (d^+, θ) at an incidence angle of θ can be calculated as (Pulliainen et al., 1999)

$$T_B(d^+,\theta) = \frac{\gamma_{as}}{1 - \Gamma_{as}\Gamma_{ss}e^{-2(\kappa_e - q\kappa_s)\sec\,\theta\cdot d}} T_B(d^-,\theta)$$

$$(19.6)$$

where $T_B(d^-, \theta)$ is the emitted brightness temperature below the snow—air boundary and can be described as

$$T_B(d^-,\theta) = \gamma_{ss}T_{soil}e^{-(\kappa_e - q\kappa_s)\sec\,\theta\cdot d}$$

$$+\left(1 + \Gamma_{ss}e^{-(\kappa_e - q\kappa_s)\sec\,\theta\cdot d}\right)$$

$$\frac{\kappa_a T_{snow}}{\kappa_e - q\kappa_s}\left(1 - e^{-(\kappa_e - q\kappa_s)\sec\,\theta\cdot d}\right)$$

$$(19.7)$$

$$\equiv T_{B,soil} + T_{B,snow}$$

where γ_{ss} and Γ_{ss} are the transmissivity and the reflectivity of the snow—soil boundary. γ_{as} and Γ_{as} are the transmissivity of the air—snow boundary.

The HUT model is widely used in current passive microwave snow depth inversion algorithms because of its simplicity and robustness. The empirical formula of scattering coefficient calculation is valid for a wide frequency range of 18–90 GHz. Recently, a study found that the HUT model underestimates brightness temperature for deep snow. It is suggested that the HUT model should be used for snow depth smaller than 50 cm (Pan et al., 2016).

19.2.1.2 Microwave emission model for layered snowpack model

MEMLS (Wiesmann and Mätzler, 1999) uses exponential correlation length p_{ec} to represent the snow microstructure. There are two strategies to calculate volume scattering coefficient. One is the semiempirical approach. The scattering coefficient γ_s is calculated using snow density and correlation length, which is determined according to experiment data as

19.2.2 Analytical models

In the analytical models, the scattering properties of snow are expressed by analytical formulas. Early models are based on traditional RTE. The snow layer is modeled as composed of independent scattering spheres or spheroids. The scattering field of single particle is calculated using Rayleigh scattering theory, and the overall collective scattering magnitude is the sum of scattering for all particles. This independent scattering theory ignores the collective scattering effect of snow particles. For microwave remote sensing of snow, multiple scattering particles exists within one wavelength, and the permittivity of ice particle is obviously higher than background permittivity, thus the independent scattering theory and the traditional radiative transfer theory are not appropriate.

To consider the dense medium effect of snow, Tsang et al. (1985) proposed DMRT (dense me-

$$\gamma_s = \left(9.2\frac{p_{ec}}{1\text{ mm}} - 1.23\frac{\rho}{1\text{ g/cm}^3} + 0.54\right)^2 \left(\frac{f}{50\text{ GHz}}\right)^{2.5}, 0.05\text{ mm} \leq p_{ec} \leq 0.3\text{ mm}$$

$$\gamma_s = \left(3.16\frac{p_{ec}}{1\text{ mm}} + 295\left[\frac{p_{ec}}{1\text{ mm}}\right]^{2.5}\right)^2 \left(\frac{f}{50\text{ GHz}}\right)^{2.5}, p_{ec} < 0.05\text{ mm}$$

(19.8)

MEMLS can also use the physically based improved Born approximation to obtain the scattering coefficient (Mätzler and Wiesmann, 1999).

MEMLS solves the RTE based on six-stream approximation, and the total, coherent, and incoherent reflection on the snow layer interface are considered. The "trapped" radiation due to total reflection is considered in the MEMLS model. To solve the RTE, the six-stream formulation is modified to two-stream form. The MEMLS model considers the multilayer structure of snow and is applicable for 5–100 GHz frequency range. Recent research has found that MEMLS performs better than the HUT model for deep snow (Pan et al., 2016).

dia radiative transfer) for microwave remote sensing of snow. DMRT and its application in forward model for snow-covered terrain are described in detail in Tsang and Kong (2001). The traditional RTE is derived based on equation of energy conservation, in which the coherence of scattering field is ignored, and only the superposition of intensity is considered. The traditional RTE (independent scattering) is not valid for dense medium, in which the scattering particles are densely distributed. The coherence of scattering by particles makes the traditional radiative transfer to overestimate the scattering.

The Foldy–Lax equations of multiple scattering are used to study scattering from the dense medium. Analytical methods, including effective field approximation (EFA), the quasicrystalline approximation (QCA) approximation, the QCA with the Percus—Yevick pair distribution function (QCA-CP), etc., are approximations of the multiple scattering equations. The EFA makes truncation at the first equation of the hierarchy. The EFA is valid for a sparse concentration of particles. The QCA and QCA-CP are high-order approximation than the EFA. DMRT-QCA and DMRT-QCA-CP are applicable to densely distributed moderate size particles and scattering particles far less than the wavelength, respectively.

Considering the size distribution of snow particles, Tsang et al. (2003) modeled the snow medium as composed by densely distributed particles with size distribution. Tan et al. (2004, 2005) extended this model to sticky particles. QCA-DMRT has very different physical features from classical independent scattering: (1) the extinction increases with fraction volume and then decreases; (2) the scattering coefficient has weaker frequency dependence than the fourth power as predicted by Rayleigh scattering theory; and (3) the phase matrix shows more forward scattering, and this is not the same as the equally scattered forward and backward as predicted by the Mie and Rayleigh theory.

By using DMRT-QCA model, snow with sticky, nonsticky particles, or with size distribution can be modeled. discrete-ordinate method can be used to solve DMRT equation, in which both QCA and QCA-CP can be used. By using discrete-ordinate method, the number of streams can be much greater than the "six stream" as used by MEMLS model. This is important for active microwave remote sensing because the incident wave is azimuth dependent.

19.2.3 Numerical models

In previous models, simplified analytical expressions were derived to avoid time-consuming computations. But the computer hardware develops rapidly in these years so that the computation problem can be less considered. Using parallel computers, large-scale numerical computation of electromagnetic scattering can be completed in relatively short time period. In numerical models, the snow structure is first simulated using computer simulations, then the snow scattering properties are calculated by numerical electromagnetic computations. These kinds of models are well summarized in Tsang et al. (2000a,b, 2013).

The real snow particles in nature are irregular and nonspherical. Ding et al. (2010) proposed a bicontinuous medium for snow microwave scattering and emission simulation. In this model, the level-cut of Gaussian random field is used for snow microstructure simulation (Berk, 1991), and the simulated microstructure is very similar to real snow microstructures. This avoids unrealistic assumptions in snow particle shape. Then the discrete dipole approximation is used for the scattering field calculations. Coherent and incoherent fields are separated by many realizations, with the incoherent field used for scattering phase matrix and scattering coefficient calculations. In bicontinuous model, the input parameters include ζ, b, and the volume fraction. The ζ and b parameter are not easy to measure in practice. In Xiong et al. (2012), the bicontinuous parameters are determined by correlation function fit using measured snow two-dimensional section images (see Fig. 19.4).

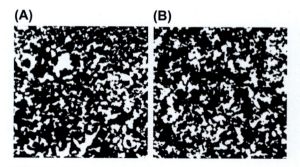

(A)　　　　　**(B)**

FIGURE 19.4　(A) Snow slices measured experimentally and (B) simulated bicontinuous structures.

In numerical models, the cross-polarization elements of phase matrix are not equal to zero so that they are different with analytical models such as DMRT-QCA. In bicontinuous model, the cross-polarization also arises from the irregular feature of microstructure, and the collective scattering effect is also considered through numerical computations. When combined with multiple scattering radiative transfer solution, the bicontinuous model can predict stronger cross-polarization signal, which is underestimated in DMRT-QCA.

The semiempirical model, analytical theoretical model, and numerical calculation model play important roles in microwave remote sensing studies and applications. The semiempirical model is relatively simple. Parts of its parameters are obtained from experimental measurements. The model simulation is relatively stable and practical. The semiempirical models have already been successfully applied in the retrievals of snow parameters. The analytical models are represented by the scattering theory of dense medium. This kind of models makes several assumptions for approximating the multiple scattering equations. The calculation is moderate and the accuracy is higher than the semiempirical model. Numerical models are closest to the actual scattering problems with few assumptions. But numerical models require time-consuming calculations.

In conclusion, HUT model is suitable for simulations of shallow snow and MEMLS for simulation of both shallow snow and deep snow. Numerical models as well as analytical models can simulate active and passive microwave radiation and scattering characteristics, but the numerical models perform better in the simulation of cross-polarization. At present, the numerical models, such as DMRT-bicontinuous model, are the most time-consuming models. With the continuous development of computer capability in future, the calculation problem of numerical models can be better solved. However, at the current stage it is still necessary to parameterize the numerical models to perform remote sensing inversion.

19.3 Microwave snow water equivalent retrieval techniques

19.3.1 Snow water equivalent inversion techniques using passive microwave remote sensing

The long-term satellite passive microwave brightness temperature data of almost 40 years from the 1980s to present is of great value for snow parameter estimation. The National Snow and Ice Data Center (NSIDC) of United States has archived daily passive microwave data since 1979, including the Scanning Multichannel Microwave Radiometer (SMMR), the Special Sensor Microwave/Image (SSM/I), the Tropical Rainfall Measuring Mission Microwave Imager, and the Advanced Microwave Scanning Radiometer for Earth Observing System (AMSR-E). As a continuation of the legacy of AMSR-E, the Advanced Microwave Scanning Radiometer 2 (AMSR2) on the Global Change Observation Mission 1-Water (GCOM-W1) satellite was launched in May 2012 by the Japan Aerospace Exploration Agency (JAXA). From 2008, China launched a series of polar-orbit Chinese meteorological satellites (FY-3A/B/C/D), which carried the Microwave Radiation Imager (MWRI). The MWRI passive microwave brightness temperature data is distributed by the National Satellite Climate Center. Table 19.1 lists the passive microwave sensors that are currently available for SWE inversion.

The microwave radiation from the snow cover mainly comprises the radiation from snow and the radiation from underlying surfaces (Chang, 1987). The low-frequency emission is mainly affected by the properties of the soil under the snow. In contrast, high-frequency emission is sensitive to SWE and snow particle size caused by the volume scattering (Hofer and Matzler, 1980). Because the scattering effect is greater for high frequencies than for low frequencies, the brightness temperature of dry snow decreases with increasing frequency (Ulaby et al., 1981). Therefore, we can use the brightness temperature difference between high

TABLE 19.1 Passive microwave sensors for snow water equivalent inversion[1,2].

Passive microwave sensor	SMMR	AMSR-E	MWRI	AMSR-2	SMM/I	Special sensor microwave imager/sounder (SSMIS)
Satellite platform	Nimbus-7 Pathfinder	Aqua	FY-3B, C, D	GCOM-W1 (Global Change Observation Mission for Water-1)	DMSP-F8/F11/F13*	DMSP-F17*/F18
Operation time	1978.10.25– 1987.8.20	2002.06.01– 2011.10.04	2010.11.18– present (FY3B) 2013.09.29– present (FY3C) 2017.11.15– present (FY-3D)	2012.08.10–present	1987.6.18–1991.12.31 (85 GHz decommissioned in February 1989) (F8) 1991.12.03–2000.05 (F11) 1995.05–2009.11 (F13)	2006.11.04– 2017.09.30 (F17) 2009.10– present (F18)
Altitude/km	955	705	836	700	856 (F8) 853 (F11) 850 (F13)	850 (F17) 830 (F18)
Orbital inclination angle/°	99.1	98.2	98.75	98.2	98.8	98.8
Spatial content	Global	Global	Global	Global	Global	Global
Frequency (GHz) IFOV/km	6.6 GHz: 148×95 10.7 GHz: 91×59 18 GHz: 55×41 21 GHz: 46×30 37 GHz: 27×18	6.9 GHz: 75×43 10.7 GHz: 51×29 18.7 GHz: 27×16 23.8 GHz: 31×18 36.5 GHz: 14×8 89.0 GHz: 6×4	10.65 GHz: 85×51 18.7 GHz: 50×30 23.8 GHz: 45×27 36.5 GHz: 30×18 89 GHz: 15×9	6.9 GHz: 62×35 7.3 GHz: 62×35 10.65 GHz: 42×24 18.7 GHz: 22×14 23.8 GHz: 19×11 36.5 GHz: 12×7 89 GHz: 5×3	19.3 GHz: 70×45 22.2V: 60×40 37 GHz: 38×30 85.5 GHz: 16×14	19.3 GHz: 70×45 22.2V: 60×40 37 GHz: 38×30 91.7 GHz: 16×13
Orbit width/km	780	1445	1400	1450	1394	1700
Polarization	H/V	H/V	H/V	H/V	H/V	H/V
Local ascending (A) and descending (D) overpass time	A: 12:00 p.m. D: 0:00 a.m.	A: 1:30 p.m. D: 1:30 a.m.	3B A: 1:30 p.m. D: 1:30 a.m. 3C A: 1:30 a.m. D: 1:30 p.m. 3D A: 2:00 p.m. D: 2:00 a.m.	A: 1:30 p.m. D: 1:30 a.m.	F8 A: 6:20 a.m. D: 6:20 p.m. F11 A: 6:20 p.m. D: 6:20 a.m. F13 A: 6:00 p.m. D: 6:00 a.m.	A: 6:00 p.m. D: 6.00 a.m.
Observation angle (degree)	50.3	55	53	55	53.4 (F8) 52.8 (F11) 53.4 (F13)	53.1

[1] *The information listed here was extracted from the NSIDC; CMA; WMO OSCAR websites.*
[2] *The DMSP F10, F14, F15, F16 satellites also carry the passive microwave sensors, but their local overpass time has drifts and is not as stable as the F13 and F17 satellites. Therefore, details of these satellites are not listed here.*

and low frequencies to retrieve snow depth and SWE.

The brightness temperature signal observed by orbiting satellite for the snow-covered terrain is also affected by emission from the other snow types. Because of the complex signal characteristics of the land surface, there is currently no effective fully physically based inversion method for SWE. Current algorithms include the semiempirical algorithms, iterative algorithms, look-table method, data assimilation method, and machine learning techniques.

19.3.1.1 Semiempirical algorithms

19.3.1.1.1 Static algorithms

Static algorithms were the earliest and most common SWE retrieval algorithm in the past. It is essentially a linear regression between the brightness temperature gradient and the snow depth (or SWE). Various fitted coefficients have been used by different researchers. The general form of a static algorithm can be expressed as follows:

$$SWE = A + B \cdot \Delta T_B \text{ (mm)} \quad (19.9)$$

where A and B are the offset (mm) and slope (mm/K), respectively. Because the equations were fitted for different regions using different experimental data, different A and B can be obtained. ΔT_B is the brightness temperature difference between a high frequency (37 or 85 GHz) and a low frequency (18 or 19 GHz), which can be in either vertically or a horizontally polarized.

19.3.1.1.1.1 The Chang (1987) algorithm (the NASA algorithm) Chang et al. (1982, 1987) calculated snow brightness temperature based on the RTE for a homogeneous snowpack with a snow density of 300 kg/m³ and a snow grain size of 0.3 mm. By substituting the simulated results into Eq. (19.9), a SWE algorithm for the SMMR sensor was derived as

$$SWE = 4.8 \times \left(T_{B,18H} - T_{B,37H}\right) \text{ (mm)} \quad (19.10)$$

where SWE is snow water equivalent (mm); $T_{B,18H}$ and $T_{B,37H}$ are the brightness temperatures at 18 and 37 GHz, respectively, in a horizontal polarization. Armstrong and Brodzik (2001) demonstrated that when the $(Tb18H-Tb37H)$ brightness temperature gradient algorithm is used for SSM/I, A should be changed to 5 K due to the differing features of the SSM/I and SMMR sensors.

The Chang algorithm is typical among the earliest semiempirical algorithms for SWE and snow depth. From Chang's simulation, this algorithm can be used only when the SWE is less than 300 mm. When the SWE is greater, the brightness temperature difference becomes saturated. Th algorithm used a constant grain size for simulation; therefore, it is not adaptable from different snow grain sizes.

19.3.1.1.1.2 The Foster et al. (1997) algorithm (the NASA 96 algorithm) Boreal forest (or taiga) is widely distributed over North America and Northern Eurasia. Compared with the nearby tundra or prairie regions, the accumulated snow in this region is often deeper and melts later in the spring because the trees reduce the incidence of solar radiation on the snow cover. The presence of vegetation greatly complicates the accurate estimation of snow depth (SWE) using satellite data. Because vegetation can attenuate the microwave emission emitted from snowpack under the canopy, the brightness temperature difference will be reduced. Therefore, an algorithm that does not consider the forest canopy will underestimate the SWE for these regions.

Foster et al. (1997) considered the effect of the forest fraction f and revised the Chang algorithm (1987) to obtain the NASA 96 algorithm for the global SWE inversion. Because the forest emissivities are closer at 18 and 37 GHz, the brightness temperature difference consists almost entirely of the contributions from forest-free regions, represented by an areal fraction $(1-f)$. The adjusted algorithm is

$$SWE = \frac{4.8 \times (T_{B,18H} - T_{B,37H})}{1 - f} \text{ (mm)} \quad (19.11)$$

Figs. 19.5 displays the relationship between the SWE and brightness temperature difference for different forest fractions, graphically illustrating the adjustment for forest fraction.

In Central Eurasia, because the winter air temperature in is low, it is easier for the snow to increase the grain size or form a depth hoar layer. Therefore, the algorithm uses a 0.4 mm grain size and revised the coefficient B for this region as

$$SWE = \frac{2.3 \times (T_{B,18H} - T_{B,37H})}{1 - f} \text{ (mm)} \quad (19.12)$$

The Foster et al. (1997) algorithm accounted for the forest-covered land surface. It reduced the SWE inversion error from 50% to 15% in North America and significantly increased the SWE retrieval accuracy for Europe in the spring (Foster et al., 1997).

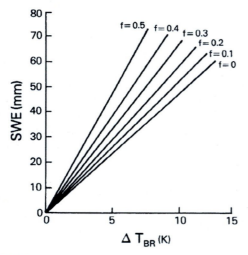

FIGURE 19.5 Relationship between SWE and brightness temperature difference for different forest fractions (calculated results) (Foster et al., 1997). *From Foster, J.L., Chang, A.T.C., Hall, D.K. Rango, A. 1991. Derivation of snow water equivalent in boreal forests using microwave radiometry. Arctic 44 (5), 147–152.*

19.3.1.1.1.3 The Foster et al. (2005) algorithm

Globally, the snow cover has significant spatial and seasonal property divergence. For example, the Siberian snow tends to have depth hoar, larger grain size, and lower average density than that in North America. Sturm et al. (1995) classified seasonal snow into six classes: tundra, taiga, alpine, maritime, prairie, and ephemeral. Each class is defined by a unique ensemble of textural and stratigraphic characteristics, including the sequence of snow layers and the thickness, density, crystal morphology, and grain size of each layer. Table 19.2 lists the mean density and grain size of the six snow classes (Tedesco and Kim, 2006).

The results from different radiative transfer models indicate that the snow brightness temperature is strongly related to snow grain size and SWE. In general, a larger snow grain size can lead a stronger scattering and a larger brightness temperature difference, which could provide similar brightness temperature signal to a smaller grain size but larger snow depth (at least for a single band) (Chang et al., 1982, 1987; Kelly et al., 2003). Several researchers have demonstrated that the grain size is critical for determining the snow volume scattering characteristics. Neither the Chang et al. (1987) nor the Foster et al. (1997) algorithms adequately considered the snow grain size and snow-type differences.

Subsequently, Foster et al. (2005) employed two dynamic parameters to represent the effects

TABLE 19.2   Snow mean density and grain size for the six seasonal snow types (Tedesco and Kim, 2006).

	Tundra	Alpine	Taiga	Maritime	Prairie	Ephemeral
Snow density (kg/m^3)	380	250	260	350	300	300
Snow grain size (mm)	1.0	0.6	1.3	0.7	0.8	0.5

of forest fraction and snow grain size. He also emphasized the importance of prior knowledge of snow classes and of a land cover–type database. A new SWE inversion equation was thereby established:

$$SWE = F_t c_t \left(T_{B,19H} - T_{b,37H} \right) \qquad (19.13)$$

Here, F_t is the adjustment coefficient for forest fraction, and c_t is the adjustment coefficient for the effect of snow grain size; these coefficients are defined as follows:

$$F_t = 1/(1 - \varepsilon) \qquad (19.14)$$

$$c_t = (1 - \gamma)c_0 \qquad (19.15)$$

where ε is the mean relative error of the estimated SWE for different forest fraction. The forest fraction is acquired from the International Geosphere-Biosphere Program (IGBP) 1 km global land cover map. The algorithm varies the coefficient B from 3.3 to 5.8 mm/K.

Using observations collected over seven winters in Canada, the Foster et al. (2005) algorithm was shown to reflect the snow accumulation and melting phases well. In most regions, the known bias of the Chang algorithm has been corrected. However, large errors are still found for alpine and maritime snow.

19.3.1.1.1.4 The Derksen et al. (2005) algorithm (the Canada algorithm) Beside forest, different land cover background also affects the snow properties and brightness temperature and results in the differences in the A and B values in the general equation. Hallikainen (1984) demonstrated that B varies with vegetation type and

fitted a B of 2.9 mm/K for swamps and 5.4 mm/K for forests.

Derksen et al. (2005) developed the following formula for the SWE estimation in the boreal forest/tundra region in northern Canada:

$$SWE = \sum_{j=1}^{n} F_j (A + B \cdot \Delta T_B)_j \, (mm) \qquad (19.16)$$

where F_j is the percentage of open prairie, coniferous forest, deciduous forest, and sparse forest in every SSM/I grid cell obtained from IGBP 1 km land cover data; A and B are fitted for each type individually. The total SWE is computed by summing the SWE values from each land type weighted by the coverage fraction. The inversion equations for different land types use the Meteorological Service of Canada algorithm, which includes different terrain conditions based on brightness temperature differences between 19 and 37 GHz in vertical polarization. The detailed coefficients for each land type are listed in Table 19.3.

Derksen et al. (2005) validated the algorithm using measurements from a dedicated snow survey campaign from November 2003 to March 2004 in Manitoba, Canada, which has tundra-type snow according to the classification system of Sturm et al. (1995). The snow parameters were observed along a path 500 km long, running from Thompson (boreal forest) to Gillam (sparse boreal forest) to Churchill (open tundra). The results show that most of the retrieved SWE are within ±20 mm of the ground observations, except for a 50% underestimation in the tundra region.

TABLE 19.3 The Meteorological Service of Canada (MSC) algorithms for different land cover types (Derksen et al., 2005).

Land cover type	SMMR algorithm	SSM/I algorithm
Open prairie	SWE=−20.7−2.59[(37V−18V)/19]	SWE=−20.7−2.59[(37V−19V)/18]
Coniferous forest	SWE= 16.81− 1.96(37V−18V)	SWE= 16.81− 1.96(37V−19V)
Deciduous forest	SWE= 33.5− 1.97(37V−18V)	SWE= 33.5− 1.97(37V−19V)
Sparse forest	SWE=−1.95−2.28(37V−18V)	SWE=−1.95−2.28(37V−19V)

The Derksen et al. (2005) algorithm considers the local vegetation type more explicitly than the Foster et al. (2005) algorithm but not adequately with respect to the growth of snow grain size. The Derksen et al. (2005) algorithm provides a method to establish a general algorithm with high accuracy for regions with complex land surfaces. This algorithm underestimated the SWE in tundra region in Canada. The results of SWE retrieval in boreal forest are also not very accurate for deep snow in April.

19.3.1.1.1.5 The snow depth estimation algorithms in China SWE retrieval algorithms for China have been studied during the past two decades. Chao et al. (1993) adjusted the coefficients of Chang's algorithm (Chang et al., 1987) for Western China, using SSMR brightness temperature. Using digital elevation model (DEM) data, they classified five areas: high mountains, plateau, low mountains, rolling hills, and basin. Che et al. (2008) modified Chang's algorithm using 1980–81 ground observations for SMMR and 2003 for SSM/I. They considered influences of vegetation, wet snow, precipitation, cold desert, and frozen ground on snow depth estimation. Root mean square errors (RMSE) of their estimates were 6.22 and 5.22 cm for SSMR and SSM/I, respectively. Sun (2006) and Chang et al. (2009) developed an empirical snow depth algorithm using AMSR-E, by incorporating MODIS 8-day snow-covered fraction data (MYD10A2). In the work of Sun (2006), China was divided into three snow regions. The semiempirical snow depth retrieval algorithms for these regions were developed individually. In the Xinjiang region, snow fraction was integrated into the regression of snow depth retrieval. Through validation, estimated error decreased when snow fraction was included. Following the work of Sun (2006), Chang et al. (2009) improved snow depth retrieval techniques for four different land types, using MODIS land cover product MOD12Q1 V004. The brightness temperature difference between 89 and 18.7 or 36.5 GHz is also included in snow depth algorithms to detect shallow snow. This algorithm is valid with a snow layer thicker than 3 cm. Its RMSE is small, 5.6 cm. However, this algorithm is not suitable for monitoring snow cover in real time, as it requires 8-day MODIS snow cover data (MYD10A2). To estimate real-time snow depth, Jiang et al. (2014) proposed a suit of semiempirical mixed pixel snow depth algorithms for FY3B-MWRI based on 7-year (2002–09) observations of brightness temperature (AMSR-E) and snow depth (from meteorological stations) in China.

A 1-km resolution land use/land cover map from the Data Center for Resources and Environmental Sciences, Chinese Academy of Sciences, is used to determine fractions of four mainland cover types (grass, farmland, bare soil, and forest). Pure pixel snow depth retrieval algorithms (land cover fraction greater than 85%) are initially developed using AMSR-E brightness temperature data. Each grid-cell snow depth was estimated as the sum of snow depths from each land cover algorithm weighted by percentages of land cover types within each grid cell. All frequencies, 10.7, 18.7, 36.5, and 89 GHz at both polarizations, were used for the regressions of empirically derived algorithms. These regressions are as follows:

$$SD_{farmland} = -4.235 + 0.432 \times d18h36h + 1.074 \times d89v89h$$
$$SD_{grass} = 4.320 + 0.506 \times d18h36h - 0.131 \times d18v18h + 0.183 \times d10v89h - 0.123 \times d18v89h$$
$$SD_{baresoil} = 3.143 + 0.532 \times d36h89h - 1.424 \times d10v89v + 1.345 \times d18v89v - 0.238 \times d36v89v$$
$$SD_{forest} = 11.128 - 0.474 \times d18h36v - 1.441 \times d18v18h + 0.678 \times d10v89h - 0.649 \times d36v89h$$

$$(19.17)$$

Then the estimated snow depth with mixed pixels method is

$$SD = f_{grass} \times SD_{grass} + f_{barren} \times SD_{barren} + f_{forest}$$
$$\times SD_{forest} + f_{farmland} \times SD_{farmland}$$

$$(19.18)$$

Through evaluation of this algorithm using station measurements from 2006, RMSE of snow depth retrieval is about 5.6 cm, and the bias is within ±5 cm. The snow depth retrieval algorithms performed well for shallow snow but underestimated the depth of deep snow.

Regional semiempirical static algorithms are developed for specific areas, which have been calibrated at a local scale and may be capable of providing a reasonable, stable, and accurate SD estimation. However, it is hard to be used globally because of both the complex snow conditions and the lack of data for regression.

19.3.1.1.2 Dynamic algorithms

As the snow season progresses, the snow cover condition, such as grain size, density, snow depth, and layered characteristics, changes. The most typical examples of the changes are related to the snow grain size and the formation of depth hoar. To represent the temporal and spatial variations of snow's physical characteristics, the slope B in the generation inversion equation based on brightness temperatures difference is allowed to vary dynamically with time and space. Thus, the algorithm is called the dynamic algorithm. The key point of the dynamic algorithm is to consider the influence of internal metamorphism in the snowpack (mainly the growth of grain size) on SWE retrieval.

19.3.1.1.2.1 The temperature gradient index dynamic algorithm The TGI dynamic algorithm developed by Josberger and Mognard (2002) explains temporal and spatial variations of the internal characteristics of a snowpack, especially the changes in snow grain size. They found from the ground observations in the US Northern Great Plains that the brightness temperature gradient continued to increase after the snow

depth had reached a maximum and started to decrease. This discrepancy with the static algorithm principle is likely due to variations in grain size in a snowpack.

The principle of grain size growth in dry snowpack is based on constructive metamorphism, specifically temperature gradient metamorphism. The thermal isolation of snow results in a higher ground temperature compared with the snow surface temperature, which leads to a large temperature gradient in the snow cover (°C/m). This temperature gradient leads to a vapor pressure gradient, forcing the water vapor to move upward through gaps in the snowpack. The vapor from the warmer bottom of the snow layer condenses at the cold surfaces of ice particles in overlying layers, thereby increasing the snow grain size and the pore size where ice is deposited. If the temperature gradient in the snow cover is very large (more than 10°C/m) or the snowpack temperature is near 0°C, depth hoar with large ice particles soon forms.

Josberger and Mognard (2002) defined a temperature gradient index (TGI) to express the cumulative temperature gradient in snowpack:

$$TGI = \frac{1}{C} \int \frac{T_{ground} - T_{air}}{D(t)} dt \qquad (19.19)$$

where C is an empirical coefficient (20°C/m), T_{ground} is the temperature of the snow–soil interface, T_{air} is the temperature of the air–snow interface, and $D(t)$ is the snow depth as a function of time. TGI starts to accumulate from the beginning of the snow season and excludes the negative differences between T_{ground} and T_{air}. In this case, TGI becomes a cumulative index that indicates the trend of grain size growth.

Assuming T_{ground} to be 0°C and discretizing Eq. (19.19), we obtain

$$TGI = \sum (-T_{air} / D) / C \qquad (19.20)$$

Josberger and Mognard (2002) obtained the following linear relationship between the brightness temperature gradient (SG) and TGI at several observation sites by discarding data from the wet snow period at the end of the snow season:

$$SG = (T_{B,19H} - T_{B,37H}) = \alpha TGI + \beta \quad (19.21)$$

Differentiating this formula, we obtain the snow depth $D(t)$ at moment t:

$$D(t) = \frac{\alpha(T_{ground} - T_{air})}{C(dSG/dt)} \quad (19.22)$$

The TGI algorithm represents physical processes in snowpack evolution. However, it also has several problems. Firstly, this algorithm can only be used for cold snow cover in Alaska, Siberia, and Canada and is valid only if T_{ground} is above T_{air}. Secondly, the algorithm applies a differential equation as a function of time; thus, daily fluctuations in brightness temperature introduce errors. Smoothing of the time series brightness temperature is required when the algorithm is used for daily SWE retrieval. The same reprocessing procedure is also required for air temperature. Thirdly, the use of the change in brightness temperature dSG/dt in the denominator means that a very small SG change can yield abnormally large snow depth. Fourthly, this algorithm does not correctly predict the actual snow depth after melting begins or in cases when early season snowfall melts after hitting the ground. Finally, the coefficient α is not the same between years due to differences in the metamorphic processes in each year's snowpack.

19.3.1.1.2.2 The Kelly et al. (2003) dynamic algorithm The dynamic algorithm developed by Kelly et al. (2003) adopts a semiempirical model to describe the temporal change of snow grain size and snow density:

$$r(t) = r_\infty - (r_\infty - r_0)\exp(-kt) \text{ (mm)} \quad (19.23)$$

$$mv(t) = mv_\infty - (mv_\infty - mv_{s0})\exp(-lt) \text{ (\%)} \quad (19.24)$$

where t is time (day); $r(t)$ is snow grain size (mm); r_0 is the initial snow grain size (mm), set at 0.2 mm; r_∞ is the upper limit of snow grain

size (mm), set at 1.0 mm; k is an empirical coefficient, set at 0.01/day; $mv(t)$ is the volumetric proportion of snow particles in the snowpack (snow volume fraction) (%); mv_{s0} is the snow volume fraction of fresh snow right after the snowfall (%), which is a function of air temperature; and mv_∞ is the maximum snow volume fraction (%), which equals $mv_{s0} + 27\%$. The empirical coefficient l is set at 0.007/day.

Setting the snow temperature at 260 K and applying DMRT model to relate brightness temperature difference and snow depth, the following quadratic polynomial regression equation was obtained:

$$SD = b(\Delta T_b)^2 + c\Delta T_b \text{ (cm)} \quad (19.25)$$

where ΔT_B is the brightness temperature difference between 19 and 37 GHz in vertical polarization. b and c are empirical coefficients related to the grain size (gs in mm) and snow volume fraction (mv in %).

$$b = 0.898(gs/mv)^{-3.716} \quad (19.26)$$

$$c = 1.060(gs/mv)^{-1.915} \quad (19.27)$$

The brightness temperature at 37 GHz becomes saturated when the snow depth reaches 50–100 cm (depends on grain size). Kelly et al. (2003) established a threshold for the brightness temperature difference, $satK$. When the brightness temperature difference is greater than $satK$, it is replaced by $satK$ to ensure that the retrieval process is in the valid range of the DMRT model.

$$satK = 15.09(gs/mv) - 5.79(K) \quad (19.28)$$

This algorithm is a semiempirical dynamic algorithm that combines the parameterized DMRT model with snow grain size and snow density. The changes in grain size and density are calculated with time and then inserted into the regression equation to compute the snow depth.

The algorithm considers the nonlinearity between snow depth and brightness temperature difference caused by the effects of grain size and snow density (see Fig. 19.6).

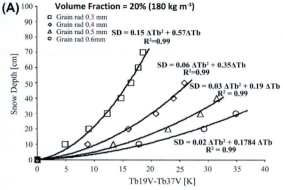

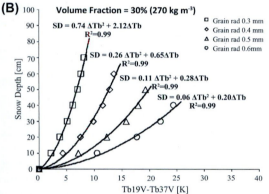

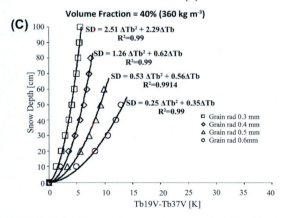

FIGURE 19.6 Relationships between the DMRT simulated brightness temperature difference (Tb19V—Tb37V) and snow depth for snow volume fractions of 20% (A), 30% (B), and 40% (C) (Kelly et al., 2003).

The semiempirical grain size and snow density are fixed. However, the actual snow grain size growth rate changes according to time and geographical location. Confined by these factors, the comparison with ground truth data from global WTO-GTS sites in 1992—95 and 2000—01 indicates that the improvement of this algorithm compared with the Chang static algorithm was limited. Although its mean error is closer to zero, its RMSE is larger than the Chang static algorithm. In addition, its performance is less stable than the Foster et al. (2005) algorithm.

19.3.1.1.2.3 Later development of the TGI algorithm: a combined static and dynamic algorithm

Grippa et al. (2004) used air temperature data from the NCEP reanalysis dataset and the IIASA (International Institute of Applied System Analysis) permafrost temperature as the land surface temperature to calculate the monthly average snow depth in Siberia with the TGI algorithm:

$$SD(t) = \alpha \frac{T_{ground} - T_{air}}{dSG/dt} \tag{19.29}$$

where $SD(t)$ is the snow depth and SG is the brightness temperature difference at 19 and 37 GHz in horizontal polarization. In late winter, when the snow grains grow slowly, the brightness temperature difference (SG) also changes slowly, resulting in large errors in snow depth retrieval from the denominator of dSG/dt. Therefore, when dSG/dt is smaller than 1 K/5 days, they adjusted the coefficient in the Chang static algorithm for every pixel to match the last retrieved snow depth using the TGI dynamic algorithm:

$$a = \frac{SD(t_{last})}{T_{B,19H} - T_{B,37H}} \tag{19.30}$$

where a is similar to B in the static algorithm with the original A set as zero. Grippa et al. (2004) used a spatially varied static algorithm to compensate for the shortcomings of the dynamic algorithm and created a combined static and dynamic algorithm. The primary goal of Grippa et al. (2004) was to serve the climate and ecosystem change studies. They used monthly averaged meteorological data to validate the estimated monthly averaged SWE; the performance was proven to be reasonable.

Biancamaria et al. (2008) used the snow depth data from the Global Soil Wetness Project—Phase 2 to test and verify this combined algorithm in high-latitude West Siberia and the US Great Plains. If the change in the brightness temperature difference is always less than 1 K/5 day over the whole snow season, external multiyear average global snow depth data from USAF/ETAC were used to calculate a. The results showed that the TGI algorithm yields better results when applied to a long-term climate change scale (monthly) and has particularly good accuracy from January to March. The performance of this algorithm is better in Eurasia ($R^2 = 0.65$) than in North America ($R^2 = 0.29$); however, it cannot be used in some areas, such as southern coniferous forest regions.

Although the TGI algorithm still has problems, such as its preferential use for monthly SWE retrieval and a need for prior knowledge to fill in the gaps, the advantage of this algorithm in Siberia is the consideration of the influence of the permafrost layer in the temperature gradient, which yields a snow depth that is more reasonable with regard to the spatial distribution.

19.3.1.1.2.4 Kelly (2009) dynamic algorithm This algorithm is used to estimate snow depth for AMSR-E and AMSR-2 instruments, which considered the sensitivity of different frequencies to snow depth. It calculates the shallow snow depth using the 19- and 37-GHz and the deep snow depth using the 19- and 10-GHz. Different with the static algorithm, the scaling factor between SWE and brightness temperature gradient (B) was calculated from daily brightness temperature measurements to include the sensitivity from the snow grain size and snow density. Furthermore, the algorithm considered

the forest and calculates the SD as the sum of a forested component and a nonforested component,

$$SD = ff(SD_f) + (1 - ff) * (SD_o)[cm] \qquad (19.31)$$

where ff is the forest fraction ranging from 0 to 0.75. Forested areas are identified as ff, whereas the nonforested component is identified as (1-ff). SD is calculated for both the forested and the nonforested components, respectively, as

$$SD_f = {}^1/_{\log_{10}(pol_{37})} * (Tb_{19V} - Tb_{37V}) / (1 - fd * 0.6) \qquad (19.32)$$

$$SD_o = \left[{}^1/_{\log_{10}(pol_{37})} * (Tb_{10V} - Tb_{37V}) \right]$$
$$+ \left[{}^1/_{\log_{10}(pol_{19})} * (Tb_{10V} - Tb_{19V}) \right] \qquad (19.33)$$

where fd is the forest density, and pol_{19} and pol_{37} are the polarization difference at 19 and 37 GHz, respectively.

The snow grain size has a significant effect on the snow depth and SWE inversion (Tedesco and Narvekar, 2010). Tedesco et al. (2016) replaced $1/\log_{10}(pol_{37})$ and $1/\log_{10}(pol_{19})$ by equations of two factors: a grain size and a permafrost indicator calculated from a climatologically averaged 10 GHz, vertical polarization brightness temperature at 2002—2009. The grain size was fitted using neural network to match the brightness temperature at 36.5 GHz and at both 18.7 and 36.5 GHz, respectively.

19.3.1.2 Physically based statistical algorithm

Jiang et al. (2007) adopted the matrix doubling approach to consider multiple scattering in the snowpack vector RTE. In this model, the DMRT model based on Mie scattering is used to describe extinction and emission in snow cover, and the AIEM model is used to establish boundary conditions. Because of the complex form of the matrix doubling model when considering multiple scattering, it can only provide numerical solutions to radiation transfer equations that cannot be used directly for SWE retrieval.

Therefore, Jiang et al. (2007) developed a parameterized model by comparing the present model, including multiple scattering, with the form of the zeroth-order model. The model is suitable for snowpacks with optical thicknesses $\tau \leq 2$ but requires further development for snow cover in a half-infinite space. Based on this parameterized model, Jiang et al. (2011) developed a model-based statistical SWE retrieval algorithm. Assuming that the surface temperature and snow temperature are known, this algorithm uses the relationship between ground surface emissivities at different frequencies and polarizations to remove the disturbance from the underlying ground from the total satellite signal, after which the signal from the snowpack is extracted for SWE retrieval. Jiang et al. (2011) established a simulated database of snow signals with a wide range of several combinations of model required snow parameters. Using the method described above, the snowpack radiation (A) and attenuation properties (B) can be calculated. The SWE has a good regression relationship with these two parameters and can be estimated using the following relationship:

$$swe \approx \exp\left(a + b \cdot A + c \cdot A^2 + d \cdot \log(-\log(B))\right)$$
(19.34)

where a, b, c, and d are coefficients fitted from the simulated database ($a = -4.945,399$, $b = 3.917$, 603, $c = 0.6,443,147$, $d = 0.2,471,171$). A and B are only influenced by the snow properties. A and B will be independent of polarizations, if the snow particles are assumed to be spherical and randomly distributed. They can be calculated by dual-polarization brightness temperature at two frequencies:

$$B = \frac{E_v^t(f_1) - E_h^t(f_1)}{E_v^t(f_2) - E_h^t(f_2)}$$
(19.35)

$$A = E_p^t(f1) - B \cdot E_p^t(f1)$$
(19.36)

The retrieval algorithm was first tested in the synthetic test using the simulated dataset. Fig. 19.7 compares the SWE retrieved from simulated brightness temperature with the model

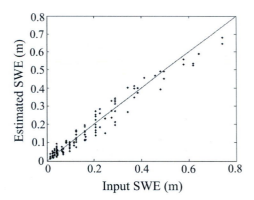

FIGURE 19.7 Comparison of model input SWE with retrieved SWE. *From Jiang, L., Shi, J., Tjuatja, S., et al., 2011. Estimation of snow water equivalence using the polarimetric scanning radiometer from the cold land processes experiments (CLPX03). IEEE Geosci. Remote Sens. Lett. 8 (2), 359 -364. © 2011, IEEE.*

inputs (as true values). RMSE was 32.8 mm, which showed that SWE may be retrieved using the parameterized model.

Airborne data from the Cold Land Processes Field Experiment in 2003 in Northern Colorado and southern Wyoming (Cline et al., 2002) were used to validate this algorithm. To compare the performance of different algorithms, Jiang et al. (2011) compared the retrieved SWE using this algorithm with the operational AMSR-E algorithm (Chang et al., 1997). The results showed that both methods overestimate SWE. However, the accuracy of the newly developed physically based statistical algorithm is slightly higher than the AMSR-E algorithm. Regardless, the new algorithm can only be used for sparsely vegetated areas and requires further development.

19.3.1.3 Iterative algorithms

Model iteration algorithms are based on a forward model. It iteratively adjusts the snow depth/SWE, starting from an initialized value, until the cost function between the simulated and measured brightness temperatures is minimized.

This method is adaptable to consider the nonlinearity between SWE and bright temperature differences but has a high computational cost. To reduce the cost, usually only the snow depth and grain size were iterated, whereas other snow parameters were fixed and determined according to snow features in different regions.

A well-developed example of this algorithm is to use the HUT model as the forward iteration model (Pulliainen et al., 1999). Pulliainen et al. (1999) used the HUT snow emission model to retrieve SWE by iteration. They used a least square method (LSM) with a constraint condition to iteratively fit the model-simulated brightness temperature with observations and outputted the cost-minimized SWE.

Butt and Kelly (2008) and Butt (2009) employed the algorithm based on the HUT model to retrieve snow depth in England and compared it with the Chang algorithm. Unlike the seasonal snow on the continent, snowpack in the United Kingdom appears mostly in Scotland with a short lifetime and a small coverage fraction and is strongly influenced by air temperature fluctuations. The snowpack in the United Kingdom is classified as ephemeral snow and maritime snow in the Sturm et al. (1995) system. The biases of the HUT algorithm in January, February, and March of 1995 were −0.59 cm, 1.89 cm and 1.64 cm, respectively, whereas the Chang algorithm would underestimate snow depth.

19.3.1.4 Lookup table algorithms

The algorithm is a method by building lookup table on inputs and outputs of the snow forward model. The lookup table is used to retrieve the snow parameters by searching the simulated brightness temperature combinations that are close to the satellite observations. Dai et al. (2012) used MEMLS to build a lookup table algorithm based on prior knowledge. Three lookup tables using one, two, and three snow layers were built, and the brightness temperature at 10, 18, and 36 GHz were simulated. A lot of prior knowledge was used to ensure the accuracy of the lookup table, which includes the snow vertical stratigraphy, snow grain size, and snow density from in situ measurements, the relationship

between the number of snow layers and the snow depth, the snow temperature fitted as a function of the daily minimal air temperature, and the snow depth. The prior knowledge of the snow density and grain size was updated monthly. It used the 18.7 and 36.5 GHz temperature difference to determine which number of snow layers should be used. It then searched the corresponding brightness temperature lookup table and found the group of snow parameters to calculate the snow depth/SWE output. Che et al. (2016) established an improved lookup table in forested areas by inputting forest transmissivity and snow properties. It achieved an RMSE of 4.5 cm in the Northeast forest of China.

The key of this algorithms is the building of the lookup table. It can potentially fail if the relationship between the brightness temperature and the snow depth is not one to one. Many studies illustrated that prior knowledge can limit lookup table to achieve a one-to-one correspondence and improve the accuracy. However, the errors from snow forward model and the absence of prior knowledge limit its application for large areas.

19.3.1.5 Machine learning algorithms

Passive microwave SWE retrieval faces a number of challenges, including the coarse spatial resolution, the sensitivity to a large number of parameters (snow temperature, snow grain size, snow density, and stratigraphy), the influence of forest attenuation, the so-called saturation effect after snow depth exceeds a certain threshold, and the atmospheric effect. These factors result in complex nonlinear relationship between snow depth and brightness temperature difference. Many studies have started to use the machine learning techniques, including the supported vector machines (SVMs), the artificial neural networks (ANN), random forests (RF), and Markov Chain Monte Carlo (MCMC), to improve the snow parameter estimation accuracy.

The ANN model is a first-order mathematical approximation to the human's nervous system, which has been widely used to solve various nonlinear problems. The BP neural network can accurately simulate the nonlinear relationships between the multifrequency, dual-polarization brightness temperature, and land surface parameters. As applied for the inversion problem, it does not require a priori knowledge of the land surface condition and can overcome the limitations of the present empirical models; they are thus powerful tools for solving the parameter retrieval problems.

Davis et al. (1993) used a neural network to train the snow signal simulated by the DMRT model, which considers the multiple scattering. They retrieved four snow parameters (mean snow particle size, snow density, snow temperature, and snow depth) using five brightness temperature measurements. Santi et al. (2012) used the ANN to retrieve soil moisture and snow depth at the same time.

However, the neural network still has the following problems. Firstly, the inputs to train the network should be independent or have little relation between each other. Otherwise, there can be large errors in the trained results. Secondly, a suitable training method and neural network structure is the key to the success of the training. These two factors restrict the application of the neural network inversion algorithm.

SVM is now widely used in image classification to address the nonlinear issues. Forman and Reichle (2015) compared the ability of SVM and ANN to model the multifrequency brightness temperature and found the SVM outperformed the ANN. Xiao et al. (2018) used the station snow depth, classified them into three snow stages and four land cover types, and established 12 nonlinear regression methods based on the support vector regression (SVR) to estimate snow depth using PM brightness temperature, latitude, longitude, and elevation as inputs. The results showed this algorithm performs better than the Chang algorithm, the spectral polarization difference algorithm, the ANN algorithm, and the linear regression algorithm. It also showed that the SVR snow-depth algorithm can alleviate the saturation effect and be used for deep snow.

Compared with ANN, the Decision Tree (DT) algorithm has many advantages. The model logic can be visualized as a set of if–then rules. DTs can utilize categorical data, and once the model has been developed, the classification is extremely rapid because no further complex mathematics is required. RF is a DT algorithm introduced by Breiman and Cutler in 2001. It is an ensemble classifier, which uses and summarizes a large number of DTs to overcome the weaknesses of a single DT. Moreover, the input variables are not necessarily uncorrelated. It does not have to select representative variables and has a high computational speed. Bair et al. (2018) used the ParBal (Parallel Energy Balance) model to estimate the SWE and uses it as the truth to train the RF model based on the spaceborne satellite brightness temperature. The results showed that RF model had a smaller RMSE and bias compared with the ANN method.

Durand and Liu (2012) and Pan et al. (2017) developed the Bayesian-based algorithm for SWE Estimation–Passive Microwave (BASE-PM). The core of BASE-PM is to estimate SWE with four frequencies brightness temperatures using the MCMC method. This algorithm considered multiple snow layers and allowed incorporating the prior knowledge of grain size, density, and temperature etc., into the retrieval. Pan et al. (2017) demonstrated that at least two snow layers are required to explain the sensitivity of high-frequency (37 and 89 GHz) brightness temperature to the top new snow layer of small grain size. MCMC is a strong machine to conduct a synergy retrieval using multiple frequencies or active and passive remote sensing at the same time. The forward model was iteratively used as a black box, and it requires no further understanding of the forward model details for the algorithm developer.

However, its shortcoming is obvious: the computational cost is high. To obtain a stable posterior SWE distribution, Pan et al. (2017) conducted 20,000 iterations. Except for its limitation, MCMC is a good tool to test the possibility of retrieval for a parameter, better than the sensitivity test when the number of influencing parameters is large, also it can be used to test "how accurate" the prior knowledge needs to be.

The nonlinear machine learning methods based on the prior knowledge is capable of describing the complex relationships between the snow parameters and the PMW brightness temperature. It increased the retrieval accuracy to some extent. However, some machine learning algorithms reply on the training data; the algorithm itself is a black box without clear physical explanations. Therefore, the use of the machine learning algorithm needs to be very careful.

19.3.1.6 Data assimilation methods

Data assimilation is a scientific method to introduce the observations into a snow process model. Early approaches to merge observations into snow process models were very simple. Liston et al. (1999) used a direct insertion method to update snow depth; they directly substituted of the simulated snow depth of a coupled land atmosphere model with the observations. Sun et al. (2004) applied the extended Kalman filter to assimilate the observed SWE into the NSIPP (NASA Seasonal-to-Interannual Project) watershed land surface model. The submodel for snowpack in the land surface model is the Lynch-Stieglitz (1994) model. The snow model runs both single- and three-layer models simultaneously. Each layer includes the SWE, snow depth, and thermal parameters. To realize the transition from no cover to complete snow cover, the model at first used a single-layer model with a snow cover fraction for SWE smaller than 13 mm. When snow begins to accumulate, the single-layer model is assumed to expand horizontally at first, from no snow (a snow fraction of 0) to fully snow covered (a snow fraction of 1). Once the full snow coverage is reached (with a SWE of 13 mm and a snow fraction of 1), the snowpack begins to grow vertically. A uniform snow density and snow temperature were used in the single-layer model. After running the three-layer mode, different values of these physical parameters were calculated in each layer.

Andreadis and Lettenmaier (2006) applied an ensemble Kalman filter (EnKF) to assimilate remotely sensed snow observation into the Variable Infiltration Capacity (VIC) meso-scale hydrologic model. The MODIS snow cover product for 1993–2003 was used to update the SWE simulated by the VIC model, using a simple snow depletion curve to relate snow cover and SWE. The VIC is a meso-scale hydrology model that has been used for numerous river basins on continents with diverse climates. The results showed that mean absolute error in snow cover fraction estimation is 0.128 for the open loop and 0.106 when the MODIS snow cover area was assimilated. Andreadis and Lettenmaier (2006) also tried to assimilate AMSR-E SWE. However, because the AMSR-E SWE product cannot be used for deep snow, assimilating it reduced the land surface model accuracy.

Durand and Margulis (2006) assimilated the SSM/I and AMSR-E brightness temperature and broadband albedo into the simple snow–atmosphere–soil (SAST) model, a component of the Simplified Simple Biosphere (SSiB3) model, to evaluate the feasibility of improving SWE estimation accuracy. SAST is a three-layer snow model of intermediate complexity based on the energy balance. The model considered the snowpack compaction due to recrystallization, overburden, and snow melting. Considering the sensitivity of passive microwaves to snow grain size, Durand and Margulis (2006) incorporated the dynamic snow grain size growth model of SNTHERM (SNow THERmal Model) into SSiB3. When the ensemble Kalman smoother (EnKS) was applied, the effects of vegetation and atmosphere were included in the radiative transfer model to build a better observation operator. The RMSE of SWE estimated by Durand and Margulis (2006) was 2.95 cm. In addition, they also found that among all the SSM/I and AMSR-E frequencies from 6.925 to 89 GHz, 10.65 GHz provided the most information for SWE estimation.

Che et al. (2014) combined the multilayer MEMLS model with snow module of CLM to study the snow parameter assimilation. Tedesco et al. (2010) expanded the linear regression and the iterative algorithms, by updating the fitting coefficients A, B or the effective grain size of the HUT model every several days to match the in situ observed or the CLSM simulated snow depth. Li et al. (2017) retrieved SWE with Ensemble Kalman Batch Smoother (EnBS). EnBS has an advantage over EnKF and EnKS because it utilizes time series of brightness temperature observations and can partly overcome saturation of brightness temperature to snow depth.

The algorithm of the current GlobSnow SWE product (Takala et al., 2011) used the HUT model as the observation operator. It fitted the effective grain size at the stations when the snow depth is measured and interpolated it to the nearby pixels as a prior to fit the snow depth (Pulliainen, 2006) using a Bayesian approach. It considered the forest canopy transmissivity and the variation in the snow density. The cost function is

$$F_{\cos t} = \frac{1}{\sigma^2}\left\{\left[T_{B,HUT}^{19V}(SD, d_0) - T_{B,HUT}^{37V}(SD, d_0)\right]\right.$$

$$\left. - \left[T_{B,obsr}^{19V} - T_{B,obsr}^{37V}\right]\right\}^2 + \frac{1}{\lambda^2}\left[d_0 - \widehat{d_0}\right]^2$$

$$(19.37)$$

where $T_{B,HUT}$ is the simulated brightness temperature, $T_{B,obsr}$ is satellite observation, $\widehat{d_0}$ is original snow grain size, and λ is the standard deviation of snow grain size. σ is the standard deviation of satellite brightness temperature. The spatially continuous background field of the prior snow grain size $\widehat{d_0}$ is interpolated using a kriging technique.

To get SWE from snow depth, snow density was obtained by a statistical model:

$$\rho_{hi}, DOY_i = (\rho_{\max} - \rho_0)$$

$$\times [1 - \exp(-k_1 \times h_i - k_2 \times DOY_i)] + \rho_0$$

$$(19.38)$$

where $\rho_{\max}$ is maximal snow density, ρ_0 is original snow density, and k_1 and k_2 vary with snow class (Sturm et al., 2010).

The Globsnow system takes account of the effect of forest on snow depth. The empirical forest transmissivity model by Kruopis et al. (1999) is used to correct brightness temperature. The model calculates the transmissivity of the vegetation using stem volume:

$$t(f, V) = t(f, V_{high}) + [1 - t(f, V_{high})] \cdot e^{-0.035V}$$

$$(19.39)$$

$$t(f, V_{high}) = 0.42 + [1 - 0.42] \cdot e^{-0.028f} \quad (19.40)$$

where V is forest stem volume and f is frequency.

The Globsnow SWE product assimilated the station snow depth observations and thus has a high accuracy over other operational snow depth products at global scale (Takala et al., 2011, 2017; Luojus et al., 2013; Metsämäki et al., 2015; Larue et al., 2017).

19.3.1.7 The mixed-pixel problem in the passive microwave SWE retrieval

The current satellite-borne passive microwave radiometer has inadequate spatial resolution to support the hydrological application requirements (Foster et al., 1997; Kelly et al., 2003). Although in certain area and season, the SWE may be accurately estimated, and the satellite SWE product still has biased in temporal and spatial scales. Besides the systematical biases, current algorithms underestimated the SWE for deep snow (Derksen et al., 2005). The spatial heterogeneity of snow properties and the different land cover types inside a single pixel, i.e., the mixed pixel problem, challenges the satellite SWE retrieval. Because the spatial resolution of the satellite radiometer is low, the pixel is a mixture of several land cover types, and the measurements can have the contributions from the bare soil, vegetation, water body, etc. The emission properties, temperature, and area fraction of different land cover types influence the satellite signals. However, most algorithm considered the snow as a pure pixel, which will reduce the retrieval accuracy. Therefore, the mixed pixel problem is one of the key problems to be solved.

Derksen (2008) and Lemmetyinen et al. (2009) had proved that the lake area fraction has a good relationship with the snow cover brightness temperature. The lake fraction also has a strong influence to the current SWE algorithms based on brightness temperature difference. Based on the previous work, Lemmetyinen et al. (2011) utilized the HUT model to simulate the brightness temperature in the lake rich areas. Based on a linear combination assumption, it used the lake fraction to correct the old iterative retrieval algorithm. The results indicated an improvement of the retrieval accuracy when the lake was considered. This is because the emissivity of water and lake ice is very different with the snow. For mixed pixels with other land types, likewise, we need to use the emission model for different land cover types to analyze their influences on the SWE retrieval accuracy.

The mixed-pixel problem can be alleviated by using brightness temperature downscaling methods, such as the enhanced resolution reconstruction method (Long and Daum, 1998; Long and Brodzik, 2016; Santi et al., 2010) and the mixed pixel decomposition method (Gu et al., 2014, 2016, 2018; Liu et al., 2018), to improve the spatial resolution. However, these downscaling methods also have errors that can affect the snow depth retrieval. Therefore, the mixed pixel problem of the passive microwave remote sensing is unsolved and requires for further studies.

19.3.2 Active snow water equivalent inversion algorithms

Satellite passive microwave observations are available every 1—3 days for large-scale snow parameter retrieval. However, because of low spatial resolution (tens of kilometers), these data cannot provide detailed snow distribution at small scale. In active microwave remote sensing, the radiation is first transmitted by the antenna and thus has much higher power compared to the passive mode. The active remote sensing has the potential to map SWE at higher spatial resolution and improve the applicability for local applications.

In early research on radar snow measurement, important efforts were made using Space-borne Imaging Radar-C and X-band Synthetic Aperture Radar (SIR-C/X-SAR) C-band and X-band data to retrieve snow parameters (Shi and Dozier, 2000a,b). Presently, many SAR sensors, such as PALSAR (the Phased Array type L-band Synthetic Aperture Radar, L-band, 7—44 m resolution) onboard ALOS (http://www.eorc.jaxa.jp/ALOS), China's HJ-1 and GF-3, the ESA's ERS-1/2, Envisat ASAR, Canada's RADARSAT-1/2, and Germany's Terra SAR-X, provide a multitude of data at different frequencies, incidence angles, and polarizations that can be used for model validation and parameter inversion.

Generally, inversion algorithms for SWE using active microwave remote sensing data can be classified into two categories: inversion algorithms based on physical backscattering models of snow-covered terrain and algorithms based on interference SAR measurements. The algorithms based on physical backscattering models are developed by analyzing the scattering intensity compositions and their different responses to polarizations and frequencies; and the total backscattering intensity is commonly decomposed into surface scattering and volume scattering (including a volume—surface interaction term) in the algorithm. For example, the soil surface parameters and snow density can be estimated using L-band data, and other snow parameters (e.g., SWE and grain size) can be inverted by multipolarization and high-frequency SAR measurements. The radar frequencies selected should have sufficiently high sensitivity to snow parameters while avoiding problems related to decreased penetration depth with increased frequency. The X-to Ku-band microwave is generally deemed as ideal for SWE retrievals. Algorithms based on interference SAR are developed using the phase shift of radar waves caused by snow layer; thus, these algorithms are suitable for low-frequency (long-wavelength) data such as L- and C-band data.

19.3.2.1 Snow water equivalent inversion algorithm based on physical backscattering models

Radar backscattering coefficients measured at a given incidence angle θ_i over seasonally snow-covered terrain can generally be expressed as a four-component model:

$$\sigma_{pq}^t(f) = \sigma_{pq}^a(f) + \sigma_{pq}^v(f) + \sigma_{pq}^{gv}(f)$$
$$+ T_p T_q L_p L_q \sigma_{pq}^g(f) \qquad (19.41)$$

The total backscattering intensity consists of surface backscattering on the air—snow and snow—soil interfaces, volume scattering from the snow layer, and an interaction term between the snow and underlying soil. These four terms are represented by the superscripts t, a, g, v, and gv in Eq. (19.41), respectively. T is the power transmissivity at the air—snow interface, and L is the snow attenuation factor, $L_p = e^{-\tau_p/\cos\theta_r}$. Forward simulations for dry snow conditions with Eq. (19.41) require 13 ground and snow parameters. The six snowpack parameters include snow depth, density, ice particle size, size variation, stickiness, and temperature. In addition to the dielectric constant of the ground, the six surface roughness parameters include RMS height, correlation length, and the two-parameter correlation functions at the air—snow and snow—ground interfaces. The relative importance of each scattering component in Eq. (19.41) depends on the frequency, polarization, incident geometry of the sensor, and snow properties.

19.3.2.1.1 Snow water equivalent retrieval algorithm based on multifrequency (L/C/X) radar observations

Based on the frequency dependence of SAR measurements of snow and ground properties, Shi and Dozier (2000b) developed a multifrequency (L-, C-, and X-bands) and dual-polarization (VV and HH) technique to estimate SWE and applied SIR-C/X-SAR image data. The names, frequencies, and wavelengths of the microwave bands are listed in Table 19.4. This technique uses L-band measurements to estimate snow density and soil dielectric and roughness

TABLE 19.4   Names, frequencies, and wavelengths of the microwave bands.

Band	Frequency (GHz)	Wavelength (cm)	Band	Frequency (GHz)	Wavelength (cm)
Ka	27—40	0.75—1.1	S	2—4	7.5—15
K	18—27	1.1—1.67	L	1—2	15—30
Ku	12—18	1.67—2.4	P	0.3—1	30—100
X	8—12	2.4—3.75	UHF	0.03—0.3	100—1000
C	4—8	3.75—7.5	VHF	0.003—0.03	1000—10,000

properties. The C-band and X-band measurements are used to minimize the effects of the underlying soil surface backscattering and to estimate snow depth and ice particle sizes. This technique requires measurements at all three SIR-C/X-SAR frequencies.

Snow volume scattering at L-band is negligible because ice particles are much smaller in size compared with L-band wavelengths. Under these conditions, we can simplify the backscattering model by considering a dry and homogeneous snowpack over a bare soil or rock surface, and Eq. (19.41) becomes

$$\sigma_{pq}^t(k_0, \theta_i) = T_p(\theta_i)T_q(\theta_i)\sigma_{pq}^g(k_1, \theta_r) \quad (19.42)$$

where θ_i and θ_r are the radar wave incidence angle at the air—snow interface and the propagation angle in snowpack layer, respectively. k_1 represents the radar wave propagation number in snowpack for the coherent component.

When the electromagnetic wave passes through the snowpack, instead of directly striking the ground, the following phenomena occur:

- Because of refraction within the snow, the incidence angle at the snow—ground interface is reduced.
- The wavelength at the snow—ground interface is shorter because snow is dielectrically thicker than air.
- The snow layer reduces the dielectric contrast above the snow—ground interface, which in turn reduces the reflectivity at the snow—ground interface.
- The power loss at the air—snow interface reduces the total energy reaching the snow—ground interface.

The first two factors result in changes in the sensor observation configuration. The shift in wave number ($k_1 = k_0\sqrt{\varepsilon_s}$) is a function of snow density. If the range of snow densities is limited to $100-550 \text{ kg/m}^3$, then the L-band wavelength during propagation in snow ranges from 21 to 16 cm (for comparison, this wavelength is 24 cm in air). Because the surface roughness effect depends on the scale of the roughness relative to the incident radar wavelength, the shortening of the incident wavelength for high snow densities makes the soil surface appear rougher than the case without snow. This effect increases the surface backscattering signal and is especially strong for a small roughness surface. However, this effect, which is caused by the wavelength shift, becomes smaller when the surface is rougher or the incident frequency is higher.

Snell's Law describes the relationship between the incidence and refraction angles at an interface. The incidence angle at the snow—ground interface depends only on the incidence angle at the snow surface and the dielectric constant of the snow. For a given snow density, however, a larger incidence angle at the snow surface results in a greater change in the refractive angle in the snow layer. Therefore, a greater increase in the backscattered power at larger incidence angles is expected.

Furthermore, the increase of VV-polarization scattering caused by snow is smaller than that of HH for the same snow and soil properties. The VV-polarization backscattering declines with incidence angle more slowly than that of the HH polarization when the surfaces are not very rough. Therefore, the changes in the VV polarization are smaller than those in the HH polarization. However, for a very rough surface, the angular dependence is smaller than that for a smooth surface, and the difference between the VV and HH polarizations is smaller.

Although a dry snowpack does not absorb or scatter a radar signal at low frequencies, it affects the magnitude of the backscattering from the underlying soil and the relationship between the HH and VV polarizations. The magnitude of the effect depends on the radar incidence angle, snow density, and roughness and dielectric properties of soil. Snow is more likely to enhance the backscattering magnitude of a smooth soil than a rough soil surface. These factors enable the development of an algorithm for inferring snow density using L-band SAR measurements.

Based on the above understanding of the effects of snow density on radar backscattering at L-band, Shi and Dozier (2000a) developed an algorithm to estimate snow density by characterizing the dependence of the surface backscattering on both the incidence angle and the wavelength. This development was accomplished by establishing a backscattering coefficient database using the IEM model (Fung, 1994) over a wide range of incidence angles, dielectric and roughness conditions, and incident wave numbers corresponding to a range of snow densities from 100 to 550 kg/m³. The relationships between the HH and VV backscattering signals and a wide range of surface dielectric and roughness properties were then characterized for each incidence angle and wave number using a regression analysis:

$$\log_{10}\left[\sqrt{\sigma_{hh}^v} + \sqrt{\sigma_{vv}^v}\right] = a(\theta_r, k_1)\log_{10}(\sigma_{hh}^g + \sigma_{vv}^g)$$

$$+ c(\theta_r, k_1)\log_{10}(\sigma_{hh}^g) + d(\theta_r, k_1)\log_{10}\left(\frac{\sigma_{hh}^g}{\sigma_{vv}^g}\right)$$

$$+ e(\theta_r, k_1)\log_{10}\left(\frac{\sigma_{hh}^g}{\sigma_{vv}^g}\right)^2$$

$$(19.43)$$

The above formula describes the relationship between the surface backscattering coefficients σ_{hh}^g and σ_{vv}^g at a given incidence angle and wave number for a wide range of random rough surfaces. The formula form minimizes its sensitivity to the surface dielectric and roughness properties by maximizing its sensitivity to the incidence angle and wave number. The coefficients a, b, c, d, and e in Eq. (19.43) depend only on the incidence angle and the wave number at

the ground surface. These values are given in Shi and Dozier (2000a). By introducing Eq. (19.45) into Eq. (19.69), an algorithm for the estimation of snow density using only δ_{hh} and δ_{vv} SAR measurements can be derived:

$$\log_{10}\left[\frac{\sqrt{\sigma_{hh}^t}}{T_{hh}(\theta_i, \varepsilon_s)} + \frac{\sqrt{\sigma_{vv}^t}}{T_{vv}(\theta_i, \varepsilon_s)}\right] = a(\theta_r, k_1)$$

$$+ b(\theta_r, k_1)\log_{10}\left[\frac{\sigma_{hh}^t}{T_{hh}^2(\theta_i, \varepsilon_s)} + \frac{\sigma_{vv}^t}{T_{vv}^2(\theta_i, \varepsilon_s)}\right]$$

$$+ c(\theta_r, k_1)\log_{10}\left[\frac{\sigma_{hh}^t}{T_{hh}^2(\theta_i, \varepsilon_s)}\right]$$

$$+ d(\theta_r, k_1)\log_{10}\left[\frac{\sigma_{hh}^t T_{vv}^2(\theta_i, \varepsilon_s)}{\sigma_{vv}^t T_{hh}^2(\theta_i, \varepsilon_s)}\right]$$

$$+ e(\theta_r, k_1)\log_{10}\left[\frac{\sigma_{hh}^t T_{vv}^2(\theta_i, \varepsilon_s)}{\sigma_{vv}^t T_{hh}^2(\theta_i, \varepsilon_s)}\right]^2$$

$$(19.44)$$

where T_{pp} depends on the polarization pp, the incidence angle θ_i at the air–snow interface, and the dielectric constant of the snowpack ε_s. Here, ε_s is the only unknown; θ_i can be calculated from the combination of orbital data and DEM. Therefore, for a given L-band, with SAR measurements of σ_{hh}^t and σ_{vv}^t, we can estimate the snow dielectric constant numerically by varying the values of the coefficients a, b, c, d, and e to find the root of Eq. (19.45). This process does not require a priori knowledge of the dielectric and roughness properties of the soil under the snow. Furthermore, snow density can be estimated from Looyenga's semiempirical dielectric formula (Looyenga, 1965):

$$\varepsilon_s = 1.0 + 1.5995\rho_s + 1.861\rho_s^3 \qquad (19.45)$$

Three snow density maps were derived by this technique using the SIR-C/X-SAR L-band data from its first mission in April 1994 for the Mammoth Mountain study area. To reduce the effect of image speckle on the estimation of snow density, the Stokes matrixes were averaged in azimuth and slant range to form multilook imagery; radiometric and terrain corrections were then performed. The snow-free pixels were also masked based on the snow classification results derived from SIR-C/X-SAR image data (Shi and Dozier, 1997).

Fig. 19.8 compares the field measurements (the x-axis) with SIR-C derived snow density (the y-axis). The RMSEs are 42 kg/m^3 and 13% for the absolute and relative errors, respectively. The estimated snow density should represent the mean value of the top and bottom layers of the snowpack because the backscattering signal is mainly controlled by two factors: (1) the snow density at the top layer near the surface, which mainly affects the power transmissivity at the air−snow interface; and (2) the snow density at the bottom layer, which controls the incidence angle and wave number at the snow−ground interface.

In addition to the inversion of snow density from the L-band data, SWE can also be retrieved from the C- and X-band data. The basic ideas behind this algorithm are as follows:

(1) To reduce the effect of the surface backscattering signal from the air−snow interface: For SWE retrieval, the

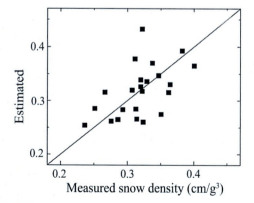

FIGURE 19.8 Comparison of the field measurements (x-axis) with the snow densities (g/cm^3) derived from the SIR-C/X-SAR L-band image data (y-axis). *From Shi, J., Dozier, J., 2000a. Estimation of snow water equivalence using SIR-C/X-SAR, part I: inferring snow density and subsurface properties. IEEE Trans. Geosci. Remote Sens. 38 (6), 2465−2474,© 2000, IEEE*

backscattering signals from the air−snow interface $\sigma_{pq}^a(f)$ are generally considered the "noise" because they are typically smaller than the other scattering contributions in Eq. (19.67) and contain no information on SWE. Their effect on the retrieval of snow properties can be reduced using the signal generated by the estimated snow density and typical roughness parameters.

(2) To develop semiempirical models to characterize the snow−ground interaction terms: These models represent the snow−ground interaction components more realistically than the formulas developed under the assumption of independent scattering. Natural surfaces in alpine regions are quite rough. Therefore, a significant contribution to the snow−ground interactions from noncoherent components is expected. The semiempirical models are developed explicitly in terms of the snowpack volume scattering albedo, optical thickness, ground reflectivity, and surface RMS height.

(3) To develop a semiempirical model to characterize the relationships between the ground surface backscattering components at C- and X-bands: With the estimation of the ground dielectric constant and RMS height derived from the L-band observations, the relationship between the ground surface backscattering at C- and X-bands can be accurately described. Thus, the number of unknowns in the ground-surface backscattering formulation can be reduced to one.

(4) To parameterize the relationships between the snowpack extinction properties at C- and X-bands: Because the extinction properties are highly correlated, we developed analytical forms for the extinction relationships at C- and X-bands. Using these relationships, the number of unknowns in the snowpack backscattering components can be reduced to two: the volume scattering albedo ω and the optical thickness τ.

In the procedures above, the number of unknowns has been reduced to three: τ, ω, and the surface component $\sigma_{vv}^s(X)$. Thus, these three unknowns can be solved numerically for each pixel using Eq. (19.41) and three SAR measurements: $\sigma_{vv}^t(C)$, $\sigma_{hh}^t(C)$, and $\sigma_{vv}^t(X)$. To estimate the snow depth d, the effects of the extinction coefficient κ_e and the snow depth d on the optical thickness τ must be separated. In other words, we must determine the absolute value of the extinction coefficient, which can be accomplished by first estimating the absorption coefficient:

$$\kappa_a(X) = 1.334 + 1.2182 \log(V_s)$$
$$- 3.4217 \log\left(\frac{\tau(X)[1 - \omega(X)]}{\tau(C)[1 - \omega(C)]}\right) \quad (19.46)$$

where V_s is the volume fraction of ice, which can be derived from the estimated snow density in each pixel from the L-band measurements. $\tau[1 - \varpi] = \kappa_a d = \tau_a$, which is the absorption component of the optical thickness for each frequency after removing the scattering effects. $\frac{\tau(X)[1-\varpi(X)]}{\tau(C)[1-\varpi(C)]} = \frac{\kappa_a(X)}{\kappa_a(C)}$ represents the snowpack temperature information (Shi and Dozier, 2000b). Finally, snow depth can be estimated as

$$d = \frac{\tau(X)[1 - \varpi(X)]}{\kappa_a(X)} \quad (19.47)$$

Fig. 19.9 compares the field measurements with the snow depths estimated from the SIR-C/X-SAR image data for the Mammoth Mountain study area. The algorithm reproduces the overall trend of observed snow depth, with an RMSE of 34 cm.

Moreover, it is possible to estimate the optically equivalent particle size, which is defined as the particle size at which the extinction properties from a natural snow volume equals to those obtained from an ideal snow volume with uniform distributions of ice particles. This optically equivalent particle size can be estimated as

$$\bar{r}_s = \left(\frac{0.01\kappa_s(X)}{2V_s S_f(2.8332 + 6.6143 V_s)}\right)^{\frac{1}{3}} \quad (19.48)$$

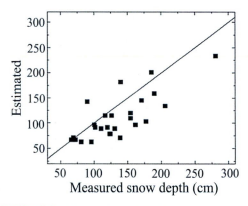

FIGURE 19.9 Comparison of the field measurements (x-axis) with the snow densities (g/cm³) derived from the SIR-C/X-SAR L-band image data (y-axis). *From Shi, J., Dozier, J., 2000a. Estimation of snow water equivalence using SIR-C/X-SAR, part II: inferring snow depth and particle size. IEEE Trans. Geosci. Remote Sens. 38 (6), 2475−2488,© 2000, IEEE.*

where κ_s is the scattering coefficient, which can be obtained from the estimated albedo ω, the optical thickness τ, and snow depth in each pixel.

A sensitivity analysis indicated that the C-band SAR measurements were affected mainly by the ground surface properties. The contributions of C-band signals from a typical snowpack are approximately 30% and 15% at HH and VV polarizations, respectively. Thus, the estimation of snow depth using C-band SAR measurements requires a technique to accurately remove the ground backscattering component. At X-band, the fraction of the signal from the snowpack itself is approximately 60%. Thus, we expect that this measurement is much more sensitive to snowpack properties, and the inversion of snow parameters is more reliable when using X-band or high-frequency SAR data.

19.3.2.1.2 Snow water equivalent retrieval algorithm based on X- and Ku-band radar observations

The selection of suitable radar frequencies for observing snowpack properties by means of remote sensing has to consider the microwave capability of penetrating a snowpack with substantial thickness, taking into account the two-way propagation path; on the other hand, the snow scattering signal is supposed to be strong enough for inversion. Accordingly, Ku-band (10.9–22 GHz) and X-band (5.75–10.9 GHz) are deemed as optimal for snow retrievals. According to regulation by the International Telecommunication Union, active remote sensing within these bands is confined to the following frequencies: 8.55–8.65 GHz, 9.50–9.80 GHz, 13.25–13.75 GHz, and 17.20–17.30 GHz. For the satellite mission COld REgions Hydrology High-resolution Observatory (CoReH$_2$O), a candidate mission of the ESA Earth Explorer Program (ESA SP-1324/2; Rott et al., 2010), a dual frequency SAR operating at Ku-band (17.2 GHz) and X-band (9.6 GHz) frequencies has been proposed. Microwave with shorter wavelengths (e.g., Ka-band) does not provide sufficient penetration and the signal is dominated by microstructure (grain size and shape) of the top snow layers. The radar return of microwave at longer wavelengths (e.g., C-band, L-band) is dominated by backscattering of ground below a dry snowpack (Rott et al., 2018). The Water Cycle Observatory Mission satellite was proposed for water cycle monitoring (Shi et al., 2014). A dual-frequency scatterometer (DPS, X- and Ku-band) is planned to achieve high temporal and spatial resolution for SWE observations.

To retrieve SWE from X- and Ku-band radar observations, first a parameterization scheme of snow volume backscattering is developed to describe the relationship between single scattering albedo, optical thickness of snow, and snow volume backscattering (Du et al., 2010). The relationship between snow optical thickness and single scattering albedo at X- and Ku-bands is established by snow scattering model simulations. The SWE is finally estimated through an iteratively minimized cost function with prior constraints. The estimation of geophysical parameters from satellite data is hampered by uncertainties of the complicated geosystem and mutual influences of parameters in this system. It is very popular to use statistical methods to retrieve geophysical variables, such as LSM. In this approach, the target parameter is retrieved through a process iteratively minimizing a cost function that describes the difference between the simulated and observed signals. The cost function used for SWE retrieval from X- and Ku-band radar is

$$F = \sum_{i=1}^{4} \frac{\left[\sigma_i^{meas} - \sigma_i^{model}(x_1, x_2)\right]^2}{2\mathrm{var}_i^2} + \sum_{i=1}^{2} \frac{\left[x_i - \overline{x}_i\right]^2}{2\lambda_i^2} \quad (19.49)$$

where i represents the number of the measurement channel (X- and Ku-bands at VV and VH polarizations), x_i denotes albedo and optical thickness, respectively, σ_i^{meas} denotes the measured backscattering signals, and σ_i^{model} denotes the modeled backscattering, which is the function of albedo and optical thickness. var_i^2 is the expected error variance of the measurements, which can be obtained from the sensor configuration. Through optimization process of the cost function, the single scattering albedo and optical thickness of snow can be solved.

The snow optical thickness is a product of snow depth and the extinction coefficient, which is closely related to SWE. The absorption part of snow optical thickness τ_a is the product of snow absorption coefficient and snow depth, which is linearly related to SWE. τ_a can be calculated by the optical thickness and albedo retrieved in the previous step as shown below:

$$\tau_a(fre) = (1 - \omega_{fre}) \cdot \tau_{fre} = k_a(fre)d \quad (19.50)$$

where k_a is the absorption coefficient of snow and can be empirically described as

$$k_a = V_s k_0 \frac{\xi''}{\xi} \left| \frac{3\xi}{\xi_i + 2\xi} \right|^2 \quad (19.51)$$

where, k_0 is the radar wave number, ξ is the dielectric constant of the background ($\xi = 1$ for air and ξ_i $i = 3.15$ for ice), ξ'' is the imaginary part of the dielectric constant of ice that is mainly determined by snow temperature, and V_s is volume fraction of ice (snow density $\rho_s = 0.917V_s$).

FIGURE 19.10 Image of the NoSREx experimental site (Lemmetyinen et al., 2016). There is a tower for SnowScat implementation in the left and the observation area is in the middle of the scene.

From the above two equations, SWE can be expressed in the following equation:

$$SWE = \tau_a(fre) \cdot \frac{0.917}{0.339 k_0 \xi''_{ice}(fre)} \qquad (19.52)$$

where ξ''_{ice} is the imaginary part of dielectric constant of ice, which is determined by the snow temperature.

Validation against 2-year measurements, using the SnowScat instrument from the NoSREx, shows that the estimated SWE using the presented algorithm has a RMSE of 16.59 mm for the winter of 2009−10 and 19.70 mm for the winter of 2010−11. Fig. 19.10 shows the experiment details. Fig. 19.11 shows the SWE retrieval validation results.

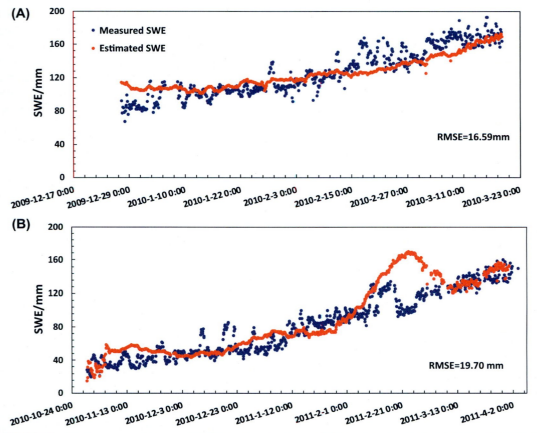

FIGURE 19.11 The time series of observed SWE (black) and estimated SWE (red) during (A) 2009/12/27−2010/03/19 and (B) 2010/10/29−2011/04/04 (Cui et al., 2016).

19.3.2.2 *Estimation of SWE and its variation by repeat-pass interferometric SAR*

The inversion algorithm discussed above is physically based; however, it requires accurate radar backscattering models and is very complex. Another promising technique for SWE estimation is to use the repeat-pass SAR interferometric measurements. The interferometric phase shift caused by snow has been proved effective for direct estimation of SWE or relative changes in SWE (Guneriussen et al., 2001). Fig. 19.12 explains how this phase shift

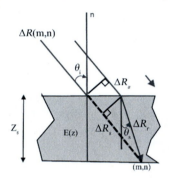

FIGURE 19.12   The geometric process of repeat-pass SAR interferometric measurements (dashed: no snow, solid: snow) (Guneriussen et al., 2001).

is produced. This technique has been applied to low-frequency interferometric SAR measurements in the C and (Guneriussen et al., 2001) and the L-band.

Considering a layer of dry snow, radar backscattering is dominated by scattering from the snow–ground interface at low frequencies (C-band or lower). The repeat-pass interferometric phase Φ consists of the following contributions:

$$\Phi = \Phi_{flat} + \Phi_{topo} + \Phi_{atm} + \Phi_{snow} + \Phi_{noise}$$

(19.53)

where Φ_{flat} and Φ_{topo} are the phase differences due to changes in the relative distance between the satellite and the target for a flat landscape and for a convoluted topography, respectively. Φ_{atm} results from changes in atmospheric propagation, Φ_{noise} is phase noise, and Φ_{snow} is the two-way propagation difference in the snowpack relative to air as a result of the refraction of radar waves in a dry snowpack. As shown in Fig. 19.13, the existence of snow changes the microwave penetration process compared with that under snow-free conditions. When snowpack volume scattering is neglected, the snow phase

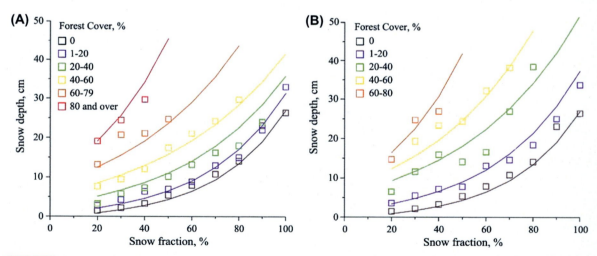

FIGURE 19.13   In situ snow depth versus the satellite-derived snow fraction statistics for different forest types on the US Great Plains from 2000 to 2003 (Romanov et al., 2007): (A) Deciduous forest, (B) coniferous forest. Snow depth is averaged within 10% snow fraction bins.

term Φ_{snow} for a uniform layer of snow of depth d can be written as (Guneriussen et al., 2001)

$$\Phi_{snow} = -2\kappa_i \cdot d \cdot \left(\cos \theta_i - \sqrt{\varepsilon_s - \sin^2 \theta_i} \right)$$

$$(19.54)$$

For ERS SAR with an incidence angle of 23 degrees, the phase shift caused by SWE changes can be approximated by a linear relationship as follows (Guneriussen et al., 2001):

$$\Phi_{snow} = -2\kappa_i \cdot 0.87 \cdot \Delta SWE \qquad (19.55)$$

This estimation indicates that at the ERS wavelength, one fringe is equivalent to a 32.5 mm SWE, and for the L-band (wavelength, 24 cm; incidence angle, 23 degrees), one fringe corresponds to an SWE of 138 mm.

The coherence is determined by several factors:

$$\gamma_{total} = \gamma_{thermal} \cdot \gamma_{surface} \cdot \gamma_{volume} \cdot \gamma_{temporal} \qquad (19.56)$$

where $\gamma_{thermal}$ depends on the signal-to-noise ratio, and its contribution to the decorrelation of dry snow-covered ground is usually small. For dry snow, the wave number shifts mainly affect the terms $\gamma_{surface}$ and γ_{volume}. The temporal decorrelation $\gamma_{temporal}$ has a major effect on the incoherence of the repeat-pass SAR data over dry snow-covered areas. This decorrelation mainly results from changes in snowpack properties due to snowfall or snow drift, which change the microwave propagation path in the snowpack. Temporal decorrelation caused by snowfall or wind redistribution is the major limitation of this method.

More studies on snow depth estimation using interferometry methods can be found in Li et al. (2016), Deeb et al. (2017), and Leinss et al. (2015). Different frequencies from X- to Ku-band were used. New techniques such as airborne Ka-band interferometric SAR were also used for snow depth remote sensing. Assuming wet snow surface and negligible penetration depth,

the elevation difference of snow-free and wet snow surface estimated from Ka-band InSAR can be used for snow depth estimation (Moller et al., 2017).

19.4 Optical remote sensing techniques

Unlike microwave remote sensing, the penetration depth of visible light into snowpack rarely exceeds a few centimeters. Thus, there is no direct physical relationship between the snow depth and the snow reflectivity. However, for most types of terrain surfaces, owing to the influences of the vegetation and topography, the fraction of the land surface covered by the snow increases with increasing snow depth. The surface reflectance changes until the snow depth reaches a value that masks the surface completely. This relationship between snow depth and the surface reflectance of fractional snow cover is significant for shallow to moderate snowpacks, providing a way to retrieve snow depth from the reflectance in the visible spectral band. At the same time, this relationship is more valid in the snow ablation period, see more details related to the snow depletion curve in the chapter of Molotch et al. (2014).

Using the visible band to estimate snow depth or SWE includes two steps: the first step is to calculate the snow cover fraction and the second step is to calculate SWE from the snow cover fraction.

19.4.1 Snow cover fraction estimation using subpixel decomposition method

The snow has a high reflectivity at the visible bands. Therefore, the apparent reflectance of a pixel increases when the snow cover fraction increases. Romanov and Tarpley (2004) established a simple linear mixing method used for the GOES geostationary satellite, where the satellite-observed reflectance is a linear combination of two end-members: a completely snow-covered and a completely snow-free pixel. The snow fraction F is calculated as

$$F = \frac{R - R_{land}}{R_{snow} - R_{land}} \qquad (19.57)$$

where R is the satellite-observed visible reflectance at a single band, and R_{snow} and R_{land} are the reflectances of a completely snow-covered and a completely snow-free land surface, respectively. R_{land} in every pixel is determined from satellite observations during snow-free seasons. R_{snow} is assumed to be independent of the location and is determined from reflectance statistics from geostationary satellites signals (e.g., GOES) in several target areas during the winter.

The polar-orbit satellite has a higher resolution and more bands. The snow has a high reflectivity at the visible (VIS) bands but a lower reflectivity at the near-infrared (NIR) bands; the plant and the soil are on the contrary. The VIS and the NIR band can build a Normalized Difference Snow Index (NDSI), which can be used to establish an empirical relationship with the snow fraction area. For the MODIS sensor, an example snow cover fraction equation is written as (Rittger et al., 2013):

$$F = \begin{cases} -0.01 + 1.45\text{NDSI}, & \text{where NDSI} = \dfrac{R_4 - R_6}{R_4 + R_6} \text{ for Terra} \\ -0.64 + 1.91\text{NDSI}, & \text{where NDSI} = \dfrac{R_4 - R_7}{R_4 + R_7} \text{ for Aqua} \end{cases}$$
$$(19.58)$$

where R_4, R_6, and R_7 are the reflectance at the fourth, sixth and the seventh bands of MODIS. F was calculated when $0.1 \leq \text{NDSI} \leq 1.0$.

Painter et al. (2009) developed a mixed-pixel decomposition algorithm for the TM and the MODIS sensors. The algorithm is called as the MODSCAG (MODIS Snow Covered-Area and Grain size) retrieval algorithm. It uses the spectrum of the snow cover, soil, and vegetation of different types as end members. Especially for the snow, the spectrum library was built

considering different snow grain size. Therefore, it can retrieve the snow cover fraction and the snow grain size at the snow surface at the same time.

19.4.2 The empirical algorithm to estimate snow depth

Romanov and Tarpley (2004) developed a snow depth algorithm using GOES data for the US Great Plains. According to the relation between the GOES-based snow fraction and the snow depths derived from ground-based stations from 1999 to 2003, an exponential-shape function was derived to estimate snow depth as follows:

$$D = \exp(aF) - 1 \qquad (19.59)$$

where D is the snow depth (cm), F is the snow-covered fraction (%), and a is an empirical parameter. The value of a was fitted as 0.0333 by Romanov and Tarpley (2004).

Fig. 19.13 shows the statistical relationship between snow fractions and snow depths for different forest cover fractions. For areas with forest, Romanov and Tarpley (2007) modified Eq. (19.59) by adding an adjustment factor b:

$$D = \exp(aF + b) - 1 \qquad (19.60)$$

where $F(\%)$ is the snow fraction, and a and b are fitted as a third-order polynomial of the tree-cover fraction. For mixed coniferous and deciduous forests, the snow depth formula is

$$D = (D_c f_c + D_d f_d)/(f_c + f_d) \qquad (19.61)$$

where f_c and f_d are the fractions of coniferous and deciduous forest, respectively, and D_c and D_d are the snow depth retrieval equations for pure coniferous and deciduous forests.

This algorithm relies on the relationship between the snow depth and snow fraction; therefore, it cannot give a meaningful estimate when the snow cover is 100%. In other words, when F is 100% in Eqs. (19.60) and (19.61), a maximum snow depth of about 30 cm is returned. Romanov and Tarpley (2007) concluded that when using a subpixel snow fraction to calculate snow depths based on GOES data, the error for snow depth below 30 cm is 30%, but this error increases to 50% when the snowpack depth increases to 30−50 cm. In forested areas, this algorithm was suggested for the areas with deciduous tree cover below 80% or coniferous tree cover below 50%.

19.4.3 The SWE reconstruction algorithm combined with the snowmelt model

The SWE reconstruction algorithm uses the estimated snowmelt from the energy balance of a pure snow cover and the snow cover fraction to retrieve the peak SWE and the SWE during the ablation period. It sums up the accumulated snowmelt from the last day of the snow season when the snow completely melts. Molotch et al. (2014) mentioned several examples of this algorithm. A simple model to calculate daily snowmelt is (Molotch and Margulis, 2008)

$$M = (T_d * a_r + R_n * M_q) \times F \qquad (19.62)$$

where F is the snow cover fraction, T_d is the mean daily air temperature that exceeds 0°C, R_n is the daily net radiation. a_r and M_q are the energy transfer coefficients.

The SWE in the ablation period can be calculated as

$$SWE_{day,i} = \sum_{j=day,i}^{day,e} M_j \qquad (19.63)$$

where day,i is the current day; day,e is the day when snow completely melts; M_j is the snowmelt on the jth day. $SWE_{day,i}$ is the SWE on the day,i, which was summed from day,e reversely back to day,i. It can be seen that this method is more likely to be used for historical SWE reanalysis, not for predictions. Its accuracy is higher than the empirical SD algorithm in Romanov and Tarpley (2004). It can achieve a 24% relative error even for thick mountainous snow.

19.5 Snow water equivalent product and applications

19.5.1 Snow water equivalent products

The SWE products were calculated from satellite observations and archived in several public scientific data centers in the world to support the various hydrological, meteorological, climatological, and economical applications.

Examples of the continental satellite SWE products are listed as follows:

(1) GlobSnow Northern Hemisphere SWE product from 1979 to present produced by the European Space Agency (ESA) (http://www.globsnow.info/);
(2) the NASA AMSR-E/Aqua daily global SWE product from 2002 to 2011 (https://nsidc.org/data/ae_dysno);
(3) the NASA global monthly SWE climatology product calculated for the SMMR and SSM/I sensors from 1978 to 2007 (https://nsidc.org/data/nsidc-0271);
(4) the NOAA (National Oceanic and Atmospheric Administration) GCOM-W1 AMSR-2 SWE product from 2012 to present (http://www.ospo.noaa.gov/Products/atmosphere/gpds/index.html);
(5) the JAXA AMSR-2 daily snow depth product from 2012 to present (http://suzaku.eorc.jaxa.jp/GCOM_W/data/data_w_dpss.html);
(6) the FY3B (Fengyun-3B) MWRI daily SWE product from 2014 to present (http://satellite.nsmc.org.cn/PortalSite/Default.aspx).

These products were calculated solely based on the satellite passive microwave brightness temperature datasets and have a resolution of 25 km or 0.25 degrees.

In 2016, the NASA NSIDC published a MEaSUREs Calibrated Enhanced-Resolution Passive Microwave Daily EASE-Grid 2.0 Brightness Temperature dataset (http://nsidc.org/data/nsidc-0630), which reproduced the SMMR, AMSR-E, and DMSP series SSM/I and SSMI/S brightness temperature data from 1978 to present, and provides multiband brightness temperature of 3.125—12.5 km resolution. It gives the potential to improve the current resolution of the SWE product.

In the US mainland, the SNODAS (Snow Data Assimilation System) product from 2003 to present (https://nsidc.org/data/G02158) is available with a resolution of 1 km.

More SWE satellite and data assimilation products can be found in the Snow Dataset Inventory in the Global Cryosphere Watch (http://globalcryospherewatch.org/reference/snow_inventory.php).

19.5.2 Snow spatiotemporal distribution characteristics

Snow has very strong seasonal characteristics. In NH, in general, the snow starts to accumulate in every November, peaks in late March, and starts to melt in April. In detail, these dates that represent the snow season vary in different regions: in the low-latitude or low-elevation areas, snow falls later, melts earlier, and the snow duration is shorter; in the high-latitude or high-elevation areas, snow falls earlier, melts later, and the snow duration is longer. Fig. 19.14 shows the monthly SWE distribution in 2012—13 from the GlobSnow product (Version 2.0). As an example, it shows the snow cover was first found on November 1, 2012, in the Northern Canada and Siberia. It expanded to the entire Canada and northern part of Euroasia on January 1, 2013. On February 1, it was close to

the maximum coverage. On March 1, it maintained the maximum coverage. Until April 1, the south edge of the snow cover started to retreat.

The snow cover onset, snow melt onset, the snow duration, snow cover area, and the peak SWE are the important parameters to describe the seasonal characteristics of the local snow. Usually, the snow cover onset is related to the decreasing of the air temperature to below freezing that allows a transfer from rainfall to snowfall. The peak SWE represents the total snowfall amount in the entire winter (with a small percent as the sublimation and evaporation excluded). It is influenced by both the air temperature and the precipitation, in other words, how much precipitation is in a solid form. The length of the snow season was determined by the peak SWE, air temperature, and radiation. Thicker snow melts relatively slower. However, once the snow temperature reaches 0°C forced by the air temperature and radiation, it will melt rapidly. The continental snow cover area is a comprehensive result of the spatial and temporal distributions of the snow cover and snow melt onset days.

Seasonal snow is also sensitive to climate change and shows significant interannual difference. From NH snow cover product published by NOAA (Fig. 19.15), it shows the NH snow cover in January was stable and even slightly increased from 1967 to 2018. However, the NH snow cover in March decreased significantly with a speed of 1.15% per 10 years. The March snow cover after 1990 is lower than before 1990.

Allchin and Dery (2017) summarized the snow-dominated duration (SDD, i.e., snow duration) from 1971 to 2014. See Fig. 19.16. Results showed that the average snow duration in NH is between 4 and 48 weeks. Of the entire seasonal snow area, 23.3% has a significantly shortened snow duration, and 5.4% has a significantly lengthened snow duration. The areas

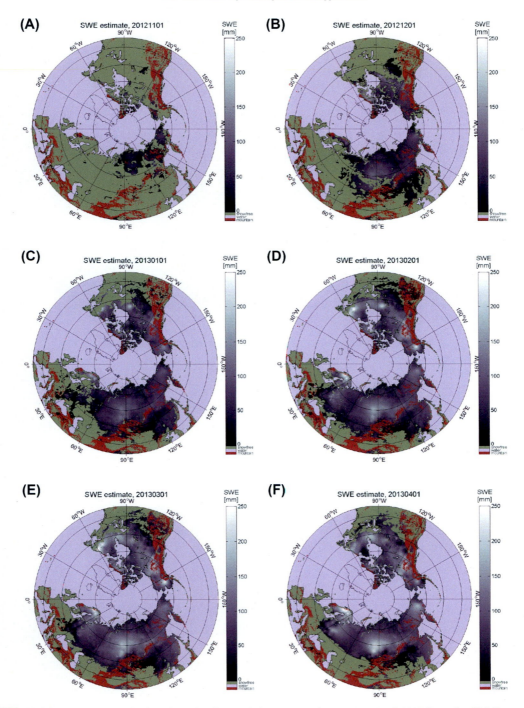

FIGURE 19.14 The SWE on the first day of each month from November 2012 to April 2013 from the GlobSnow product quick view (http://www.globsnow.info/).

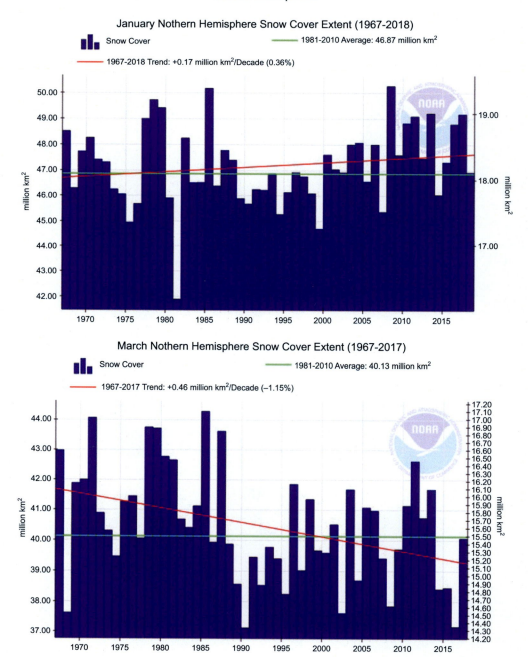

FIGURE 19.15   January and March snow cover area from 1967 to 2017 published by NOAA (unit: million km²). In situ snow depth versus the satellite-derived snow fraction statistics for different forest types on the US Great Plains from 2000 to 2003 (Romanov et al., 2007): (A) Deciduous forest, (B) coniferous forest. Snow depth is averaged within 10% snow fraction bins.

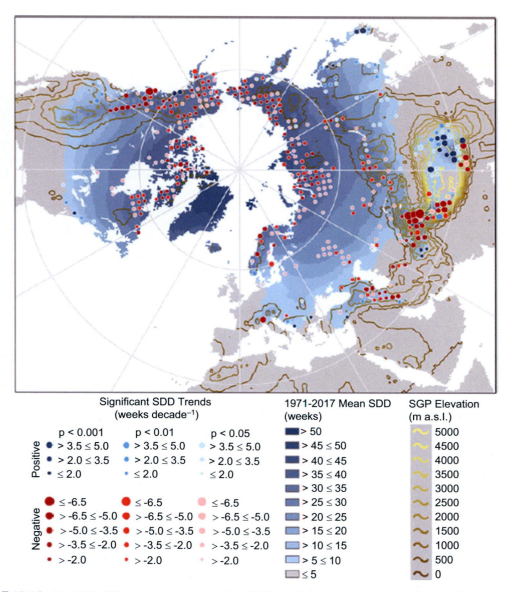

FIGURE 19.16 The 1971—2014 average snow season length (shown by background color) and the significance of changing trends (positive marked by blue points, negative marked by red points) (Allchin and Dery, 2017).

that show a shorter snow season locate at the south—central Eurasia, the Alborz and Zagros ranges of the northwest Iran, the eastern front edges of the Himalayas, the Rocky Mountains of western North America, and throughout the circumpolar regions of North America, Scandinavia, and Russia. The only areas with extended snow duration are found in the northern and eastern parts of the Tibetan Plateau, Japan, and near the central Pacific coast of Russia.

Mudryk et al. (2015) compared multiple SWE products from the remote sensing, land surface modeling, and data assimilation reanalysis. It showed that the multiyear average total snow mass in NH is between 2.4×10^{15} and 4.3×10^{15} kg and its peak was found between late February and mid-March. Figs. 19.17 shows the spatial distributions of the average SWE of five products and their difference.

Luojus et al. (2011) analyzed the monthly snow mass trend from 1980 to 2010 using the GlobSnow product. It showed that the total snow mass in NH had a significant decreasing trend. The total snow mass was relatively stable in Euroasia but decreased in North America. This result agrees with Allchin and Dery (2017).

19.5.3 Snow water equivalent Product application

19.5.3.1 Hydrological applications

The monitoring of the peak SWE in mountainuous area provides valuable database to predict the runoff and the flood risk in the next spring. Every month, institutes such as the National Resources Conservation Service and the Northwest River Forecast Center in the United States publish the SWE observed from SNOTEL network and combine it with hydro station observations to provide runoff prediction, water supply assess, and flood risk potential (probability of exceeding flood storage) for snow-dominated watersheds. See https://www.wcc.nrcs.usda.gov/wsf/index.html and https://www.nwrfc.noaa.gov/snow/index.html?version=20171004v2. Real-time SWE satellite retrievals are not highly involved in this process due to the intensive SWE observation network in western US mountains. Limited by the spatial and temporal combined resolution of the microwave sensors, the applications of the satellite SWE products in the hydraulic field is limited. Li et al. (2017) uses the snow process model to assimilate the passive microwave brightness temperature at 36.5 GHz and modeled the SWE, snowmelt, and the contribution of snowmelt to the total runoff in the US western mountains. See Fig. 19.18. This is an example when passive microwave was used for runoff estimation.

19.5.3.2 Meteorological applications

A great example of the passive microwave remote sensing applied for the meteorological

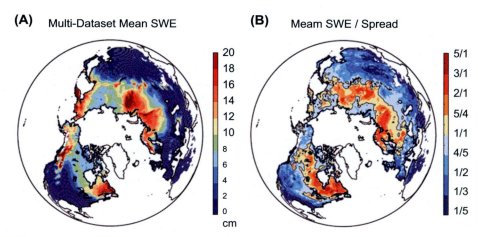

(A) Multi-Dataset Mean SWE

(B) Meam SWE / Spread

FIGURE 19.17   Comparison of multiple SWE datasets for February—March over the 1981—2010 period: (A) the mean SWE of all datasets; (B) the ratio of mean to the spread (|maximum-minimum|) of the datasets (Mudryk et al., 2015).

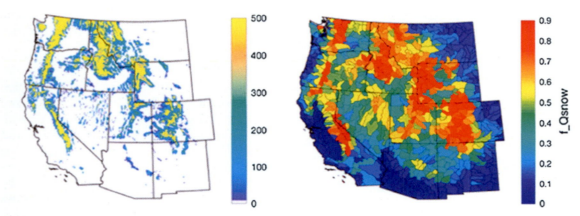

FIGURE 19.18 Data assimilation algorithm estimated average peak SWE on April 1 (left) and contribution of snowmelt to the runoff (right) from 1950 to 2005.

applications is the Interactive Multisensor Snow and Ice Mapping System product. Every day, a 1-km snow cover map is produced for the National Oceanic and Atmospheric Administration in the United States to provide land surface boundary conditions for daily weather forecasting. When the visible images are obscured by the clouds, the gaps have to be filled by satellite microwave observations. To be specific, the SWE products from SSM/I, AMSR-E, and AMSR-2 sensors are utilized and converted to snow-covered area when SWE is above zero. The FY2 satellites launched by the National Satellite Meteorological Center of the China Meteorological Administration operate well to provide continuous all-weather SWE distribution maps for meteorological services.

Based on SWE predictions calculated from the historical satellite data archives, the change of snow accumulation can be estimated and used for global change analysis. See the work carried out for Eurasia (Zhang and Ma, 2018) and Tibetan Plateau (Sun et al., 2014). A good example of the snow impact on the climate is related to the Himalayas snow cover and the Indian monsoon. Early in 1982, Dey and Kumar found there is an apparent relationship between the Eurasian spring snow cover and the advance

period of the Indian summer monsoon (Fig. 19.19). The new research of Senan et al. (2016) concluded it as follows: in high snow years of Himalayan-Tibetan Plateau region, the onset of the Indian summer monsoon was delayed by about 8 days, which can result in a reduced rainfall over the Indian subcontinent.

19.5.3.3 Biological applications

The SWE estimates can also be used for phenological analysis. Trujillo et al. (2002) found

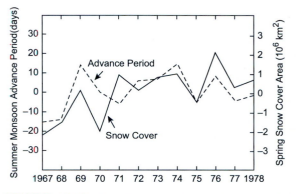

FIGURE 19.19 Year to year variation of mean spring (March to May) snow cover departure over Eurasia (south of 70°N) and the corresponding departure of the advance period of the Indian summer monsoon (Dey and Kumar, 1982).

the increased snow accumulation resulted in an increased forest greenness in the next winter in Sierra Nevada Mountains (Fig. 19.20) because the effect of increased water supply overtakes the coldness from the longer snow duration.

Pulliainen et al. (2017) found that the snow clearance day (SCD), extracted from the day when passive microwave SD decreased to 0 cm, has a close relationship ($R^2 = 0.57$) with the sprint recovery (SR) day of the taiga forest, observed by 10 flux towers from 1996 to 2014 in the world. The SR day is determined by the day when the forest gross primary production (GPP) first reaches 15% of the maximum GPP in summer. The earlier snowmelt onset results in earlier recovery of the taiga forest and increases its accumulative spring GPP. Based on this relationship, they transferred the climatological SCD trend from 1979 to 2014 to the phenological SR and accumulative spring GPP trends. The calculated result shows that, from 1979 to 2014, the SR day of the entire NH taiga forest advanced 8.1 days in average (2.3 days per 10 years) and GPP increased $29\,\mathrm{g\,C\,m^{-2}}$ ($8.4\,\mathrm{g\,C\,m^{-2}}$ per 10 years) from January to June. The global circulation and land surface biological modeling also proved the sensitivity of GPP to SCD, and from its statistics, the NH taiga forest GPP increased $13.1\,\mathrm{g\,C\,m^{-2}}$ per 10 years.

19.5.3.4 Economical applications

The research to study the snow accumulation and snowmelt timing will help the optimized design for hydroelectric power plants at high altitudes or in snow-dominated watershed in arid regions. The influence of climate change to the amount and the temporal distribution of snowmelt runoff is also a critical topic concerned by the energy industry (Koch et al., 2011; Dematteis et al., 2015).

The snow accumulation process and the speed will also influence the snow-related tourism and recreation. Natural snowfall is very important to the operation of the snow restores. At the same time, to prevent uncontrolled snow slides, the depth of snow on the slope, its wetness, and the grain size of the bottom layer need to be monitored. Using SAR and ground-based lidar system to map the distribution and the change of snow depth on the slopes can help locate potential avalanche site. See Eckerstorfer et al. (2016) for details. Fig. 19.21 is an example of the changed elevation observed by the ground-based radar.

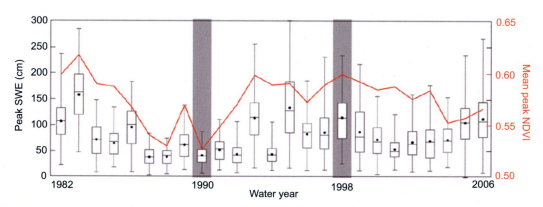

FIGURE 19.20   The relationship between peak SWE in winter and the peak NDVI in summer (Trujillo et al., 2012).

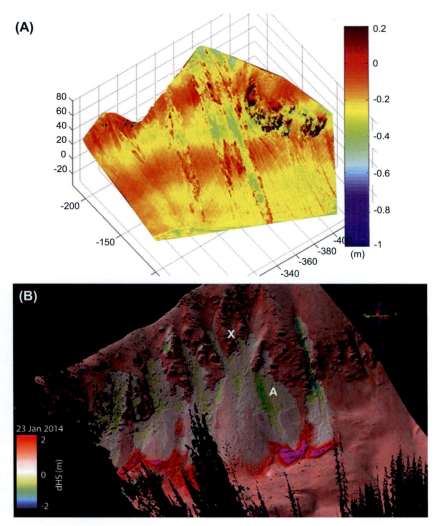

FIGURE 19.21   TLS scanning of the mass gain and mass loss of snow on slope from (A) a natural and (B) an artificially triggered avalanche (Eckerstorfer et al., 2016). *A) From Prokop, A., 2008. Assessing the applicability of terrestrial laser scanning for spatial snow depth measurements. Cold Reg. Sci. Technol. 54 (3), 155–163. (B) From Deems, J.S., Gadomski, P.J., Vellone, D., Evanczyk, R., LeWinter, A.L., Birkeland, K.W., Finnegan, D.C., 2015. Mapping starting zone snow depth with a ground-based lidar to assist avalanche control and forecasting. Cold Reg. Sci. Technol. 120, 197–204. doi:10.1016/j.coldregions.2015.09.002*

19.6 Summary

In summary, the current methods in common use to acquire snow depth and SWE from remote sensing data are based on microwave observations. Although we can build regression relationships between visible reflectance and snow depth, because visible light has short wavelengths and cannot penetrate snowpack to directly probe deep snow cover, their physical relationships remain unclear. Furthermore, remote sensing in visible bands is incapable of obtaining snow depth information under all weather conditions because clouds and rain can affect the signal. Thus, regional hydrological models, atmospheric circulation models, and climate models still adopt passive microwave remote sensing products when snow depth or SWE is required as an input.

At present, the passive microwave sensors providing snow depth and SWE products are SSM/I and AMSR-E. The AMSR-E algorithm used for the SWE product is based on studies in North America, Northern Europe, and Siberia. These areas have long, cold winters with widely distributed deep snow. However, in China, most snow regions are located at middle latitudes, and winter snowfall mostly forms a seasonal, thin, and uneven snowpack that is also subjected to complex terrain and deep hoar. For the above reasons, large errors are found in the AMSR-E SWE product (Chang et al., 1997) in China.

Because current SWE algorithms in passive microwave remote sensing mainly use semiempirical methods that were developed from limited experimental data, these algorithms only work well for the same set of data as the algorithm was constructed for and require further validation if they are to be used in other regions. Jiang et al. (2011) applied a parameterized snow emission model to build a physically based inversion algorithm. However, this model can only be used for pure snow pixels, and there are still many problems in global applications, for example, the difficulty of dealing with mixed pixels in passive microwave remote sensing. Furthermore, vegetation cover should be included when addressing the mixed-pixel problem. In addition to plants, the existence of other types of surface features can affect the accuracy of SWE inversions based on coarse, passive microwave pixels at a scale of tens of kilometers. For the same types of surfaces, emissions in different bands are believed to have certain relationships; thus, we can attempt to build statistical relationships between, e.g., MODIS thermal infrared brightness temperatures and AMSR-E passive microwave brightness temperatures using these relationships to eliminate disturbing signals from other types of surfaces in mixed pixels. Currently, the AQUA satellite loads the MODIS and AMSR-E sensors simultaneously, thus providing simultaneous data for validation. In addition, problems also arise when applying theoretical inversion algorithms to practical uses. Shallow snow usually occurs in early and late winter. When the snow depth is less than or equal to 10 cm, shallow snow is almost transparent in the microwave bands. The 89 GHz sensor on AMSR-E can be used to detect this type of shallow snow; however, a consideration of atmospheric effects must be included at this frequency. Therefore, a method for quantitative atmospheric corrections must be developed in the future.

In addition, the inversion algorithm should consider the impact of terrain in mountainous areas. The signal observed by a passive microwave radiometer consists of contributions from the atmosphere and the land surface, and each component of these contributions is influenced by the topography. Topography has two main effects on passive remote sensing:

First, the length of the route by which microwaves pass through the atmosphere between the land surface and the sensor mainly depends on the altitude of the land surface emitting radiation; thus, there are atmospheric effects related to the altitude of the imaged region.

Second, most of the current discussion focuses on the effects of surface topography such as slopes, ridges, valleys, and the surrounding topography. The radiation from these surfaces is influenced not only by the atmosphere but also by the interactions among these surfaces, called cross-radiation, which increases the total effective radiation observed from a land surface.

The main problem when passive microwave remote sensing is used in mountainous areas is the coarse spatial resolution of the radiometers. The spatial resolutions of current radiometers are often in tens of kilometers, and the pixel size is sometimes larger than the scale of terrain features. Thus, the signals from these sensors cannot reflect local changes in snow parameters due to terrain complexity. Therefore, to obtain snow depth over complex terrain in mountainous areas is a challenge for low-resolution passive microwave sensors.

Compared with passive microwave remote sensing snow research, there are as yet no highly developed snow products using spaceborne active microwave remote sensing radar because the observing frequencies of the existing spaceborne radar systems are generally no higher than the X-band. Low-frequency radar is not sensitive to dry snowpack and is not helpful in the inversion of snow depth and related parameters. Newly promoted regional and global snow monitoring projects using radar, such as the COld REgions Hydrology High-resolution Observatory (CoReH$_2$O) fostered by ESA and the Snow and Cold Land Processes project fostered by the National Aeronautics and Space Administration, have adopted research frames in higher frequency SAR (the X- and Ku-bands) with multipolarization observation. These instruments are more capable of detecting snow depth and SWE and promise to improve the low spatial resolution of passive microwave remote sensing. The development of snow equivalence inversion algorithms for high-frequency radar observations and, in particular, the technology to separate snow signals from underlying soil signals and to correct for the influence of terrain, are important issues for active microwave remote snow sensing in the future.

References

Allchin, M.I., Dery, S.J., 2017. A spatio-temporal analysis of trends in Northern Hemisphere snow-dominated area and duration, 1971–2014. Ann. Glaciol. 58, 21–35.

Andreadis, K.M., Lettenmaier, D.P., 2006. Assimilation remotely sensed snow observations into a macroscale hydrology model. Adv. Water Resour. 29, 872–886.

Armstrong, R.L., Brodzik, M.J., 2001. Recent Northen Hemisphere snow extent: a comparison of data derived from visible and microwave satellite sensors. Geophys. Res. Lett. 28, 3673–3676.

Bair, E.H., Abreu Calfa, A., Rittger, K., Dozier, J., 2018. Using machine learning for real-time estimates of snow water equivalent in the watersheds of Afghanistan. Cryosphere 12, 1579–1594.

Barnett, T.P., Adam, J.C., Lettenmaier, D.P., 2005. Potential impacts of a warming climate on water availability in snow-dominated regions. Nature 438, 303–309.

Berk, N.F., 1991. Scattering properties of the lev-eled-wave model of random morphologies. Phys. Rev. A 44 (8), 5069.

Biancamaria, S., Mognard, N.M., Boone, A., et al., 2008. A satellite snow depth multi-year average derived from SSM/I for the high latitude regions. Remote Sens. Environ. 112, 2557–2568.

Brown, R., Robinson, D., 2011. Northern Hemisphere spring snow cover variability and change over 1922–2010 including an assessment of uncertainty. The Cryosphere 5 (1), 219–229.

Butt, M.J., 2009. A comparative study of Chang and HUT models for UK snow depth retrieval. Int. J. Remote Sens. 30, 6361–6379.

Butt, M.J., Kelly, R., 2008. Estimation of snow depth in the UK using the HUT snow emission model. Int. J. Remote Sens. 29 (14), 4249–4267.

Chang, A.T.C., Foster, J.L., Hall, D.K., et al., 1982. Snow water equivalent estimation by microwave radiometry. Cold Reg. Sci. Technol. 5 (3), 259–267.

Chang, A.T.C., Foster, J.L., Hall, D.K., et al., 1987. Nimbus-7 SMMR derived global snow cover parameters. Ann. Glaciol. 9, 39–44.

Chang, A.T.C., Grody, N., Tsang, L., et al., November 1997. Algorithm Theoretical Basis Document (ATBD) for AMSR-E Snow Water Equivalent Algorithm. NASA/GSFC.

Chang, S., Shi, J., Jiang, L., Zhang, L., Yang, H., 2009. Improved snow depth retrieval algorithm in China area using passive microwave remote sensing data. In: Proceedings of IEEE International Geoscience and Remote Sensing Symposium, 2. IGARSS, pp. II614–II617, 2009 IEEE International.

Chao, M., Li, P., Robinson, D.A., et al., 1993. Evaluation and Preliminary application of microwave remote sensing SMMR-derived snow cover in western China. Environ. Remote Sens. 8 (4), 260–269.

Che, T., Li, X., Jin, R., Armstrong, R., Zhang, T.J., 2008. Snow depth derived from passive microwave remote-sensing data in China. Ann. Glaciol. 49, 145–154.

Che, T., Li, X., Jin, R., Huang, C., 2014. Assimilating passive microwave remote sensing data into a land surface model to improve the estimation of snow depth. Remote Sens. Environ. 143, 54–63. https://doi.org/10.1016/j.rse.2013.12.009.

Che, T., Dai, L., Zheng, X., et al., 2016. Estimation of snow depth from passive microwave brightness temperature data in forest regions of northeast Chin. Remote Sens. Environ. 183, 334–349.

Cline, D., Armstrong, R., Davis, R., Elder, K., Liston, G., 2002. CLPX-Ground: ISA Snow Pit Measurements. CO: National Snow and Ice Data Center. Digital Media.

Cui, Y., Xiong, C., Lemmetyinen, J., Shi, J., Jiang, L., Peng, B., et al., 2016. Estimating snow water equivalent with backscattering at x and ku band based on absorption loss. Remote Sens. 8 (6), 505.

Dai, L., Che, T., Wang, J., et al., 2012. Snow depth and snow water equivalent estimation from AMSR-E data based on a priori snow characteristics in Xinjiang, China. Remote Sens. Environ. 127, 14–29.

Davis, D.T., Chen, Z., Tsang, L., et al., 1993. Retrieval of snow parameters by iterative inversion of a neural network. IEEE Trans. Geosci. Remote Sens. 31, 842–851.

Deeb, E.J., Marshall, H.P., Forster, R.R., Jones, C.E., Hiemstra, C.A., Siqueira, P.R., 2017. Supporting NASA SnowEx remote sensing strategies and requirements for L-band interferometric snow depth and snow water equivalent estimation. In: International Geoscience and Remote Sensing Symposium (IGARSS), 2017–July, pp. 1395–1396.

Dematteis, N., Davide, M., Cassardo, C., 2015. Application of Climate Downscaled Data for the Design of Micro-Hydroelectric Power Plants. Engineering Geology for Society and Territory − Volume 1. Springer International Publishing, pp. 205–208.

Derksen, C., 2008. The contribution of AMSR-E 18.7 and 10.7 GHz measurements to improved boreal forest snow water equivalent retrievals. Remote Sens. Environ. 112, 2700–2709.

Derksen, C., Walker, A., Goodison, B., 2005. Evaluation of passive microwave snow water equivalent retrievals across the boreal forest/tundra transition of western Canada. Remote Sens. Environ. 96 (3–4), 315–327.

Dey, B., Kumar, O.S.R.U.B., 1982. An apparent relationship between Eurasian spring snow cover and the advance period of the Indian summer monsoon. J. Appl. Meteorol. 21, 1929–1932.

Ding, K., Xu, X., Tsang, L., 2010. Electromagnetic scattering by bicontinuous random microstructures with discrete permittivities. IEEE Trans. Geosci. Remote Sens. 48 (8), 3139–3151.

Du, J., Shi, J., Rott, H., 2010. Comparison between a multi-scattering and multi-layer snow scattering model and its parameterized snow backscattering model. Remote Sens. Environ. 114 (5), 1089–1098.

Durand, M.T., Liu, D., 2012. The need for prior information in characterizing snow water equivalent from microwave brightness temperatures. Remote Sens. Environ. 126 (4), 248–257.

Durand, M., Margulis, S.A., 2006. Feasibility test of multifrequency radiometric data assimilation to estimate snow water equivalent. J. Hydrometeorol. 7 (3), 443–457.

Eckerstorfer, M., Bühler, Y., Frauenfelder, R., Malnes, E., 2016. Remote sensing of snow avalanches: recent advances, potential, and limitations. Cold Reg. Sci. Technol. 121, 126–140.

Flanner, M.K., Shell, M., Barlage, D., Perovich, Tschudi, M., 2011. Radiative forcing and albedo feedback from the Northern Hemisphere cryosphere between 1979 and 2008. Nat. Geosci. 4 (3), 151–155.

Forman, B.A., Reichle, R.H., 2015. Using a support vector machine and a land surface model to estimate large-scale passive microwave brightness temperatures over snow-covered land in north America. IEEE J. Sel. Top. Appl. Earth Obs. Remote Sens. 8 (9), 4431–4441.

Foster, J.L., Chang, A.T.C., Hall, D.K., 1997. Comparison of snow mass estimation from a prototype passive microwave algorithm, a revised algorithm and a snow depth climatology. Remote Sens. Environ. 62 (2), 132–142.

Foster, J.L., Sun, C., Walker, J.P., Kelly, R., Chang, A., Dong, J., Powell, H., 2005. Quantifying the uncertainty in passive microwave snow water equivalent observations. Remote Sens. Environ. 94, 187–203.

Fung, A.K., 1994. Microwave Scattering and Emission Models and Their Applications. Artech House, Boston.

Glynn, J., Carrol, T., Holman, P., Grasty, R., 1988. An airborne gamma ray snow survey of a forest covered area with a deep snowpack. Remote Sens. Environ. 26 (2), 149–160.

Grippa, M., Mognard, N., Le Toan, T., et al., 2004. Siberia snow depth climatology derived from SSM/I data using a combined dynamic and static algorithm. Remote Sens. Environ. 93, 30–41.

Gu, L., Ren, R., Zhao, K., Li, X., 2014. Snow depth and snow cover retrieval from fengyun3b microwave radiation imagery based on a snow passive microwave unmixing method in northeast China. J. Appl. Remote Sens. 8 (1), 084682.

Gu, L., Ren, R., Li, X., 2016. Snow depth retrieval based on a multifrequency dual-polarized passive microwave unmixing method from mixed forest observations. IEEE Trans. Geosci. Remote Sens. 54 (99), 1–13.

Gu, L., Ren, R., Li, X., et al., 2018. Snow depth retrieval based on a multifrequency passive microwave unmixing method for saline-alkaline land in the Western Jilin Province of China. IEEE J. Sel. Top. Appl. Earth Obs. Remote Sens. 11 (7), 2210-22.

Guneriussen, T., Hogda, K.A., Johnsen, H., et al., 2001. InSAR for estimation of changes in snow water equivalent of dry snow. IEEE Trans. Geosci. Remote Sens. 39 (10), 2101–2108.

Hallikainen, M.T., 1984. Retrieval of snow water equivalent from Nimbus-7 SMMR data: effect of land-cover categories and weather conditions. IEEE J. Ocean. Eng. OE-9, 372–376.

Hallikainen, M.T., Ulaby, F.T., 1986. Abdelrazik M. Dielectric properties of snow in 3 to 37 GHz range. IEEE Trans. Antennas Propag. 34 (11), 1329–1340.

Hallikainen, M.T., Ulaby, F.T., Van Deventer, T.E., 1987. Extinction behavior of dry snow in the 18-to 90-GHz range. IEEE Trans. Geosci. Remote Sens. 25 (6), 737–745.

Hofer, R., Matzler, C., 1980. Investigations on snow parameters by radiometry in the 3- to 60-mm wavelength region. J. Geophy. Res. Oceans 85 (C1), 453–460.

Looyenga, H., 1965. Dielectric constants of heterogeneous mixtures. Physica 31 (3), 401–406.

Jiang, L., Shi, J., Tjuatja, S., et al., 2007. A parameterized multiple-scattering model for microwave emission from dry snow. Remote Sens. Environ. 111 (2–3), 357–366.

Jiang, L., Shi, J., Tjuatja, S., et al., 2011. Estimation of snow water equivalence using the polarimetric scanning radiometer from the cold land processes experiments (CLPX03). IEEE Geosci. Remote Sens. Lett. 8 (2), 359–363.

Jiang, L.M., Wang, P., Zhang, L.X., Hu, Y., Yang, J., 2014. Improvement of snow depth retrieval for FY3B-MWRI in China. Sci. China Earth Sci. 57 (6), 1278–1292.

Josberger, E.G., Mognard, N.M., 2002. A passive microwave snow depth algorithm with a proxy for snow metamorphism. Hydrol. Process. 16 (8), 1557–1568.

Kelly, R., 2009. The AMSR-E snow depth algorithm: description and initial results. J. Remote Sens. Soc. Jpn 29 (1), 307–317.

Kelly, R.E., Chang, A.T.C., Tsang, L., et al., 2003. A prototype AMSR-E global snow area and snow depth algorithm. IEEE Trans. Geosci. Remote Sens. 41, 230–242.

Koch, F., Prasch, M., Bach, H., Mauser, W., Appel, F., Weber, M., 2011. How will hydroelectric power generation develop under climate change scenarios? a case study in the upper danube basin. Energies 4 (10), 1508–1541.

Kruopis, N., Praks, J., Arslan, A.N., et al., 1999. Passive microwave measurements of snow-covered forest areas in EMAC'95. IEEE Trans. Geosci. Remote Sens. 37, 2699–2705.

Larue, F., Royer, A., De Sève, D., Langlois, A., Roy, A., Brucker, L., 2017. Validation of globsnow-2 snow water equivalent over eastern Canada. Remote Sens. Environ. 194, 264–277.

Leinss, S., Wiesmann, A., Lemmetyinen, J., Hajnsek, I., 2015. Snow water equivalent of dry snow measured by differential interferometry. IEEE J. Sel. Top. Appl. Earth Obs. Remote Sens. 8 (8), 3773–3790. https://doi.org/10.1109/JSTARS.2015.2432031.

Lemmetyinen, J., Derksen, C., Pulliainen, J., Strapp, W., Toose, P., Walker, A., Tauriainen, S., Plhlflyckt, J., Karna, J., Hallikainen, M.T., 2009. A comparison of airborne microwave brightness temperatures and snow-pack properties across the boreal forests of Finland and western Canada. IEEE Trans. Geosci. Remote Sens. 47, 965–978.

Lemmetyinen, J., Kontu, A., Rautiainen, K., Vehvilainen, J., Pulliainen, J., 2011. Technical assistance for the deployment of an X- to ku-band scatterometer during the NoS-REx experiment: final report. In: ESTEC Contract: No. 22671/09/NL/JA. November 4, 2011, Technique Report Prepared by Finnish Meteorological Institute.

Lemmetyinen, J., Kontu, A., Pulliainen, J., et al., 2016. Nordic Snow Radar Experiment. Geoscientific Instrumentation, Methods and Data Systems, pp. 1–23.

Li, H., Xiao, P., Feng, X., He, G., Wang, Z., 2016. Monitoring snow depth and its change using repeat-pass interferometric SAR in Manas River Basin. In: International Geoscience and Remote Sensing Symposium (IGARSS), 2016–November, pp. 4936–4939.

Li, D.Y., Wrzesien, M.L., Durand, M., Adam, J., Lettenmaier, D.P., 2017. How much runoff originates as snow in the western United States, and how will that change in the future? Geophys. Res. Lett. 44, 6163–6172.

Liston, G.E., Pielke, R.A., Greene, E.M., 1999. Improving first-order snow-related deficiencies in a regional climate model. J. Geophys. Res. 104 (D16), 19559–19567.

Liu, X., Jiang, L., Wang, G., et al., 2018. Using a linear unmixing method to improve passive microwave snow depth retrievals. IEEE J. Sel. Top. Appl. Earth Obs. Remote Sens. 11 (11), 4414-29.

Long, D.G., Brodzik, M.J., 2016. Optimum image formation for spaceborne microwave radiometer products. IEEE Trans. Geosci. Remote Sens. 54 (5), 2763–2779.

Long, D.G., Daum, D.L., 1998. Spatial resolution enhancement of SSM/I data. IEEE Trans. Geosci. Remote Sens. 36 (2), 407-17.

Luojus, K., Pulliainen, J., Takala, M., Lemmetyinen, J., Derksen, C., Metsamaki, S., Bojkov, B., 2011. Investigating hemispherical trends in snow accumulation using globsnow snow water equivalent data. In: 2011 Ieee International Geoscience and Remote Sensing Symposium (Igarss), pp. 3772–3774.

Luojus, K., Pullianinen, J., Takala, M., Lemmetyinen, J., Kangwa, M., Smolander, T., Derksen, C., Pinnock, S., 2013. ESA Globsnow: Algorithm Theoretical Basis Document-SWE-Algorithm. Technical Report. European Space Agency (ESA).

Lynch-Stieglitz, M., 1994. The development and validation of a simple snow model for the GISS GCM. J. Clim. 7, 1842–1855.

Mätzler, C., 1987. Applications of the interaction of microwaves with the natural snow cover. Remote Sens. Rev. 2, 259–391.

Mätzler, C., Wiesmann, A., 1999. Extension of the microwave emission model of layered snowpacks to coarse-grained snow. Remote Sen. of Environ. 70 (3), 317–325.

Metsämäki, S., Pulliainen, J., Salminen, M., Luojus, K., Wiesmann, A., Solberg, R., Böttcher, K., Hiltunen, M., Ripper, E., 2015. Introduction to GlobSnow Snow Extent products with considerations for accuracy assessment. Remote Sens. Environ. 156, 96–108.

Moller, D., Andreadis, K.M., Bormann, K.J., Hensley, S., Painter, T.H., 2017. Mapping snow depth from ka-band interferometry: proof of concept and comparison with scanning lidar retrievals. IEEE Geosci. Remote Sens. Lett. 14, 886–890. https://doi.org/10.1109/LGRS.2017.2686398.

Molotch, N.P., Margulis, S.A., 2008. Estimating the distribution of snow water equivalent using remotely sensed snow cover data and a spatially distributed snowmelt model: a multi-resolution, multi-sensor comparison. Adv. Water Resour. 31, 1503–1514.

Molotch, N.P., Durand, M.T., Guan, B., Margulis, S.A., Davis, R.E., 2014. Snow cover depletion curves and snow water equivalent reconstruction. In: Remote Sensing of the Terrestrial Water Cycle. John Wiley & Sons, Inc, Hoboken, NJ.

Mudryk, L.R., Derksen, C., Kushner, P.J., Brown, R., 2015. Characterization of northern hemisphere snow water equivalent datasets, 1981–2010. J. Clim. 28, 8037–8051.

Painter, T.H., Rittger, K., McKenzie, C., Slaughter, P., Davis, R.E., Dozier, J., 2009. Retrieval of subpixel snow covered area, grain size, and albedo from MODIS. Remote Sens. Environ. 113, 868–879.

Pan, J., Durand, M., Sandells, M., Lemmetyinen, J., Kim, E.J., Pulliainen, J., Kontu, A., Derksen, C., 2016. Differences between the HUT snow emission model and MEMLS and their effects on brightness temperature simulation. IEEE Trans. Geosci. Remote Sens. 54, 2001–2019.

Pan, J., Durand, M.T., Jagt, B.J.V., et al., 2017. Application of a Markov chain Monte Carlo algorithm for snow water equivalent retrieval from passive microwave measurements. Remote Sens. Environ. 192, 150–165.

Pulliainen, J., 2006. Mapping of snow water equivalent and snow depth in boreal and sub-arctic zones by assimilating space-borne microwave radiometer data and ground-based observations. Remote Sens. Environ. 101 (2), 257–269.

Pulliainen, J.T., Grandell, J., Hallikainen, M.T., et al., 1999. HUT snow emission model and its applicability to snow water equivalent retrieval. IEEE Trans. Geosci. Remote Sens. 37, 1378–1390.

Pulliainen, J., Aurela, M., Laurila, T., Aalto, T., Takala, M., Salminen, M., Kulmala, M., Barr, A., Heimann, M., Lindroth, A., Laaksonen, A., Derksen, C., Mäkelä, A., Markkanen, T., Lemmetyinen, J., Susiluoto, J., Dengel, S., Mammarella, I., Tuovinen, J.-P., Vesala, T., 2017. Early snowmelt significantly enhances boreal springtime carbon uptake. Proc. Natl. Acad. Sci. 114 (42), 11081–11086.

Rittger, K., Painter, T.H., Dozier, J., 2013. Assessment of methods for mapping snow cover from MODIS. Adv. Water Resour. 51, 367–380.

Romanov, P., Tarpley, D., 2004. Estimation of snow depth over open prairie environments using GOES Imager observations. Hydrol. Process. 18, 1073–1087.

Romanov, P., Tarpley, D., 2007. Enhanced algorithm for estimating snow depth from geostationary satellites. Remote Sens. Environ. 108, 97–110.

Rott, H., Yueh, S.H., Cline, D.W., Duguay, C., Essery, R., Haas, C., Hélière, F., Kern, M., Macelloni, G., Malnes, E., Nagler, T., Pulliainen, J., Rebhan, H., Thompson, A., 2010. Cold regions hydrology high-resolution observatory for snow and cold land processes. Proc. IEEE 98 (5), 752–765.

Rott, H., Shi, J., Xiong, C., Cui, Y., 2018. Snow properties from active remote sensing instruments. Compr. Remote Sens. 237–257.

Roy, V.K., Goita, A., Royer, A.E., Walker, B., Goodison, 2004. Snow water equivalent retrieval in a Canadian boreal environment from microwave measurements using the HUT snow emission model. IEEE Trans. Geosci. Remote Sens. 42 (9), 1850−1858.

Santi, E., 2010. An application of the SFIM technique to enhance the spatial resolution of spaceborne microwave radiometers. Int. J. Remote Sens. 31 (10), 2419-28.

Santi, E., Pettinato, S., Paloscia, S., Pampaloni, P., Macelloni, G., Brogioni, M., 2012. An algorithm for generating soil moisture and snow depth maps from microwave spaceborne radiometers. Hydrol. Earth Syst. Sci. 16, 3659−3676.

Senan, R., Orsolini, Y.J., Weisheimer, A., Vitart, F., Balsamo, G., Stockdale, T.N., Dutra, E., Doblas-Reyes, F.J., Basang, D., 2016. Impact of springtime Himalayan-Tibetan Plateau snowpack on the onset of the Indian summer monsoon in coupled seasonal forecasts. Clim. Dyn. 47, 2709−2725.

Shi, J., Dong, X., et al., 2014. WCOM: the science scenario and objectives of a global water cycle observation mission. In: Geoscience and Remote Sensing Symposium (IGARSS), pp. 3646−3649.

Shi, J., Dozier, J., 1997. Mapping seasonal snow with SIR-C/X-SAR in mountainous areas. Remote Sens. Environ. 59, 294−307.

Shi, J., Dozier, J., 2000a. Estimation of Snow Water Equivalence Using SIR-C/X-SAR, Part I: inferring snow density and subsurface properties. IEEE Trans. Geosci. Remote Sens. 38 (6), 2465−2474.

Shi, J., Dozier, J., 2000b. Estimation of snow water equivalence using SIR-C/X-SAR, Part II: inferring snow depth and particle size. IEEE Trans. Geosci. Remote Sens. 38 (6), 2475−2488.

Sturm, M., Holmgren, J., Liston, G.E., 1995. A seasonal snow cover classification system for local to global applications. J. Clim. 8 (5), 1261−1283.

Sturm, M., Taras, B., Liston, G.E., et al., 2010. Estimating snow water equivalent using snow depth data and climate classes. J. Hydrometeorol. 11 (6), 1380−1394.

Sun, C., Walker, J.P., Houser, P.R., 2004. A methodology for snow data assimilation in a land surface model. J. Geophys. Res. 109, D08108.

Sun, Z., Shi, J., Jiang, L., Yang, H., Zhang, L., 2006. Development of snow depth and snow water equivalent algorithm in Western China using passive microwave remote sensing data. Advances in Earth Science 12, 1363−1369.

Sun, Y., Huang, X., Wang, W., Feng, Q., Li, H., Liang, T., 2014. Spatio-temporal changes of snow cover and snow water equivalent in the Tibetan Plateau during 2003−2010. J. Glaciol. Geocryol. 36 (6), 1337−1344.

Takala, M., Luojus, K., Pulliainen, J., et al., 2011. Estimating northern hemisphere snow water equivalent for climate research through assimilation of space-borne radiometer data and ground-based measurements. Remote Sens. Environ. 115 (12), 3517−3529.

Takala, M., Ikonen, J., Luojus, K., et al., 2017. New snow water equivalent processing system with improved resolution over Europe and its applications in hydrology. IEEE J. Sel. Top. Appl. Earth Obs. Remote Sens. 10 (2), 428−436.

Tan, Y., Li, Z., Tsang, L., Chang, A.T.C., Li, Q., 2004. Modeling passive and active microwave remote sensing of snow using DMRT theory with rough surface boundary conditions. In: Proceedings of IEEE Geoscience and Remote Sensing Symposium, IGARSS 2004. IEEE International, vol. 3, pp. 1842−1844.

Tan, Y., Li, Z., Tse, K.K., Tsang, L., 2005. Microwave model of remote sensing of snow based on dense media radiative transfer theory with numerical Maxwell model of 3D simulations (NMM3D). In: Proceedings of IEEE Geoscience and Remote Sensing Symposium, IGARSS 2005, IEEE International, vol. 1, pp. 578−581.

Tedesco, M., Jeyaratnam, J., 2016. A new operational snow retrieval algorithm applied to historical AMSR-E brightness temperatures. Remote Sens. 8, 1037.

Tedesco, M., Kim, E.J., 2006. Intercomparison of electromagnetic models for passive microwave remote sensing of snow. IEEE Trans. Geosci. Remote Sens. 44 (10), 2654−2666.

Tedesco, M., Narvekar, P.S., 2010. Assessment of the NASA AMSR-E SWE product. IEEE J. Sel. Top. Appl. Earth Obs. Remote Sens. 3 (1), 141-59.

Tedesco, M., Reichle, R., Löw, A., Markus, T., Foster, J.L., 2010. Dynamic approaches for snow depth retrieval from spaceborne microwave brightness temperature. IEEE Trans. Geosci. Remote Sens. 48, 1955−1967. https://doi.org/10.1109/TGRS.2009.2036910.

Tsang, L., Kong, J.A., Shin, R.T., 1985. Theory of Microwave Remote Sensing. Wiley Interscience.

Tsang, L., Chen, C.T., Chang, A.T.C., Guo, J., Ding, K.H., 2000a. Dense media radiative transfer theory based on quasicrystalline approximation with application to passive microwave remote sensing of snow. Radio Sci. 35 (3), 731−749.

Tsang, L., Kong, J.A., Ding, K.H., 2000b. Scattering of Electromagnetic Waves. Theories and Applications. Wiley-Interscience, New York.

Tsang, L., Kong, J.A., 2001. In: Scattering of Electromagnetic Waves, Volume 3. John Wiley & Sons, Inc. Advanced Topics.

Tsang, L., Ding, K.H., Chang, A.T.C., 2003. Scattering by densely packed sticky particles with size distributions and applications to microwave emission and scattering from snow. In: Proceedings of IEEE Geoscience and Remote Sensing Symposium, IGARSS 2003, IEEE International, vol. 4, pp. 2844–2846.

Tsang, L., Ding, K., Huang, S., Xu, X., 2013. Electromagnetic computation in scattering of electromagnetic waves by random rough surface and dense media in microwave remote sensing of land surfaces. In: Proceedings of IEEE Geoscience and Remote Sensing Symposium, vol. 101, pp. 255–279, 2.

Ulaby, F.T., Moore, R.K., Fung, A.K., 1981. Microwave Remote Sensing, Active and Passive: Volume 1, Microwave Remote Sensing Fundamentals and Radiometry. Artech House, MA.

Ulaby, F.T., Moore, R.K., Fung, A.K., 1986. Microwave Remote Sensing, Active and Passive: Volume 3, from Theory to Applications. Artech House, MA.

Wiesmann, A., Mätzler, C., 1999. Microwave emission model of layered snowpacks. Remote Sens. Environ. 70, 307–316.

Xiao, X., Zhang, T., Zhong, X., et al., 2018. Support vector regression snow-depth retrieval algorithm using passive microwave remote sensing data. Remote Sens. Environ. 210, 48–64.

Xiong, C., Shi, J.C., Brogioni, M., Tsang, L., 2012. Microwave snow backscattering modeling based on two-dimensional snow section image and equivalent grain size. In: Geoscience and Remote Sensing Symposium (IGARSS).

Zhang, Y., Ma, N., 2018. Spatiotemporal variability of snow cover and snow water equivalent in the last three decades over Eurasia. J. Hydrol. 238–251.

Further reading

Brogioni, M., Macelloni, G., Palchetti, E., et al., 2009. Monitoring snow characteristics with ground-based multifrequency microwave radiometry. IEEE Trans. Geosci. Remote Sens. 47 (11), 3643–3655.

Chen, C., Tsang, L., Guo, J., Chang, A.T.C., Ding, K., 2003. Frequency dependence of scattering and extinction of dense media based on three-dimensional simulations of Maxwell's equations with applications to snow. IEEE Trans. Geosci. Remote Sens. 41 (8), 1844–1852. https://doi.org/10.1109/TGRS.2003.811812.

ESA, 2012. Report for Mission Selection: CoReH2O, ESA SP-1324/2 (3 Volume Series). European Space Agency, Noordwijk, The Netherlands.

Girotto, M., Margulis, S.A., Durand, M., 2014. Probabilistic SWE reanalysis as a generalization of deterministic SWE reconstruction techniques. Hydrol. Process. 28 (12), 3875–3895.

Goodison, B., Walker, A., 1994. Canadian development and use of snow cover information from passive microwave satellite data. In: Choudhury, B., Kerr, Y., Njoku, E., Pampaloni, P. (Eds.), Passive Microwave Remote Sensing of Land-Atmosphere Interactions. VSP BV, Utrecht, pp. 245–262.

Hutengs, C., Vohland, M., 2016. Downscaling land surface temperatures at regional scales with random forest regression. Remote Sens. Environ. 178, 127–141.

Kleindienst, H., 2000. Integrated system for water resources assessment – a tool for optimised operation of hydroelectric power plants. Politics, (in 1998) 1–14.

Kurvonen, L., Hallikainen, M.T., 1997. Influence of Landcover category on brightness temperature of snow. IEEE Trans. Geosci. Remote Sens. 35 (2), 367–377.

Lax, M., 1952. Multiple scattring of waves II. The effective field in dense systems. Phys. Rev. 85 (4), 621–629.

Lemmetyinen, J., Pulliainen, J., Rees, A., Kontu, A., Qiu, Y., Derksen, C., 2010. Multiple-layer adaptation of HUT snow emission model: comparison with experimental data. IEEE Trans. Geosci. Remote Sens. 48, 2781–2794.

Li, D., Durand, M., 2015. Large-scale high-resolution modeling of microwave radiance of a deep maritime alpine snowpack. IEEE Trans. Geosci. Remote Sens. 53 (5), 2308–2322.

Painter, T., Berisford, F., Boardman, J., Boardman, K., Deems, J., Gehrke, F., Hedrick, A., Joyce, M., Laidlaw, R., Marks, D., Mattmann, C., McGurk, B., Ramirez, P., Richardson, M., Skiles, M., Seidel, F., Winstral, A., 2016. The Airborne Snow Observatory: fusion of scanning lidar, imaging spectrometer, and physically-based modeling for mapping snow water equivalent and snow albedo. Remote Sens. Environ. 184, 139–152.

Rango, A., Martine, J., Chang, A.T.C., Foster, J.L., et al., 1989. Average areal water equivalent of snow in a mountain basin using microwave and visible satellite data. IEEE Trans. Geosci. Remote Sens. 27 (6), 740–745.

Stankov, B., Cline, D., Weber, B., et al., 2008. High-resolution airborne polarimetric microwave imaging of snow cover during the NASA cold land processes experiment. IEEE Trans. Geosci. Remote Sens. 46 (11), 3672–3693.

Tedesco, M., Pulliainen, J., Takala, M., et al., 2004. Artificial neural network-based techniques for the retrieval of SWE and snow depth from SSM/I data. Remote Sens. Environ. 90 (1), 76–85.

Tedesco, M., Reichle, R., Loew, A., Markus, T., Foster, J., 2009. Dynamic approaches for snow depth retrieval from spaceborne microwave brightness temperature. IEEE Trans. Geosci. Remote Sens. 48, 1955–1967.

Trujillo, E., Molotch, N.P., Goulden, M., Kelly, A., Bales, R.C., 2012. Elevation-dependent influence of snow accumulation on forest greening. Nat. Geosci. 5.

Tsang, L., 1992. Dense media radiative transfer theory for dense discrete random media with particles of multiple sizes and permittivities. Prog. Electromagn. Res. 6 (5), 181–225.

Tsang, L., Kong, J.A., 1980. Multiple scattering of electromagnetic waves by random distribution of discrete scatterers with coherent potential and quantum mechanical formulism. J. Appl. Phys. 57 (7), 3465–3485.

Tsang, L., Kong, J.A., 1992. Scattering of electromagnetic waves from a dense medium consisting of correlated Mie scatterers with size distributions and applications to dry snow. J. Electromagn. Waves Appl. 6, 265–286.

Tsang, L., Kong, J.A., 2001. Scattering of Electromagnetic Waves. Volume 3: Advanced Topics. John Wiley & Sons, Inc.

Tsang, L., Pan, J., Liang, D., Li, Z., Cline, D.W., Tan, Y., 2007. Modeling active microwave remote sensing of snow using dense media radiative transfer (DMRT) theory with multiple-scattering effects. IEEE Trans. Geosci. Remote Sens. 45 (4), 990–1004.

Wiesmann, A., Matzler, C., Weise, T., 1998. Radiometric and structural measurements of snow samples. Radio Sci. 33 (2), 273–289.

Xu, L., Dirmeyer, P., 2013. Snow-atmosphere coupling strength. Part II: albedo effect versus hydrological effect. J. Hydrometeorol. 14, 404–418.

Abstract

Monitoring water storage and its variation is important to understanding local hydrological processes and the global water cycle, which sustains all life on Earth. The development of satellite remote sensing techniques has benefited the retrieval of terrestrial water storage and its variation, which has emerged as a new discipline. Focusing on terrestrial water storage, this chapter describes major retrieval approaches: the water balance—based approach, the surface parameter—based approach, and the Gravity Recovery and Climate Experiment—based approach. Accurate estimates of terrestrial water storage and its variation are still being developed.

20.1 Introduction

Surface water (as rivers, lakes, reservoirs, wetlands, and inundated areas) represents less than 1% of the total amount of water on Earth. However, it is one of the most important components in the terrestrial water and is essential for both human beings and ecosystems (Frappart et al., 2005). Terrestrial surface water plays a major role in global climate variability and the hydrological cycles (Calmant et al., 2008). The volume of water storage is an important parameter in

© 2020 Elsevier Inc. All rights reserved.

monitoring the quantity of surface water resources. The surface water storage variations also affect their physical, chemical, and biological processes. Therefore, accurate and timely monitoring of water storage variations in surface water is essential for the effective management of water allocation, ecosystem services, and for a better understanding of the impact of climate change and human activities on the terrestrial water resources (Birkett, 1995). However, our knowledge of water storage variations on terrestrial land is yet insufficient (Alsdorf and Lettenmaier, 2003).

The volume of surface water storage cannot be measured directly. The traditional approach to monitor surface water storage relies on in situ measurements of water levels combined with accurate bathymetric data (Furnans and Austin, 2008). However, availability of continuous in situ measurement has been limited in most regions of the Earth. This approach is also time-consuming, labor-intensive, and costly. Therefore, it is a challenge to provide reliable surface water volume information given the absence of continuous and public in situ hydrological measurements.

In the past few decades, remote sensing techniques offer the great potentials to large-scale hydrological observations. It enables to retrieve hydrological variables and parameters from space. The retrievals include precipitation (Chapter 17), evapotranspiration (Chapter 18), soil moisture (Chapter 19), and water surface area and stage (Smith and Pavelsky, 2009). Especially, the Gravity Recovery and Climate Experiment (GRACE) satellite provides a novel way to estimate water storage at global scale (Alsdorf and Lettenmaier, 2003; Schmidt et al., 2008). These developments have benefited the retrieval of terrestrial water storage and its variations. It emerges as a new discipline in remote sensing applications.

Focusing on terrestrial water storage, this chapter describes three major retrieval approaches. Section 20.2 introduces water balance—based approach using the hydrologic components retrieved from remote sensing. Section 20.3 depicts surface parameter—based approach, in which advantages of satellite retrieval and hydraulic functions are described. Section 20.4 reviews the GRACE instrument and the principle of retrieving water storage. The last section makes remarks on future prospects of this new field.

20.2 Water balance—based estimation

In general, the volume of water storage in lakes or reservoirs is dependent on the balance between water inputs and outputs (Fig. 20.1). The water inputs consist of direct precipitation, surface runoff of the tributaries, and seepage, whereas the water outputs include evaporation, groundwater outflow, and surface water discharge (Crétaux and Birkett, 2006). Therefore, the simple water balance equation for a lake or reservoir can be written as:

$$\frac{dS}{dt} = P - ET - Q_s - Q_g + \varepsilon \qquad (20.1)$$

where S is total water storage in a lake or reservoir (m^3), and t is time in hour. dS/dt is the change in lake or reservoir water volume over the time period dt. P is the precipitation over the surface of the lake or reservoir (m^3 h^{-1}). ET is the actual evapotranspiration from the lake or reservoir (m^3 h^{-1}). Q_s and Q_g is the surface runoff and groundwater runoff, respectively (m^3 h^{-1}). ε represents the accumulated errors from all components and other anthropogenic factors such as human water use. Water storage is the residual of the four components on the right side of Eq. (20.1).

Water balance analysis method is an established approach of assessing changes in water volume, which seems simple enough. However, to close the water balance equation, all components should be measured or estimated independently. In this respect, the accuracy of the estimated water storage depends on the accuracy

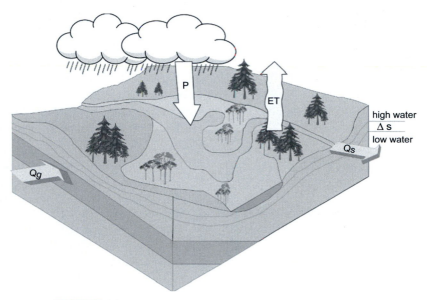

FIGURE 20.1   Conceptual picture of water balance over a lake.

of each components. At present, there remain significant uncertainties to estimate the precipitation and evapotranspiration by using remote sensing (Roads et al., 2003; Kutoba et al., 2009). The largest uncertainty generally comes from *ET* estimates due to the limited temporal sampling of qualified satellite images and the large uncertainties in existing retrieval approach (Chapter 18). The large uncertainties restrict this water balance–based approach from practical applications. Whatever, the approach is simple in principle. It may become operational once the retrieval approaches of the relevant four components mature with desired accuracy.

20.3 Surface parameter–based estimation

With the improved remote sensing techniques, numerous surface parameters including surface area and water level can be retrieved from satellite data. Given the parameter retrievals, water storage can be easily estimated with a digital elevation model (DEM) of the area of interest. Here, we briefly introduce the principle of the approach and its applications.

20.3.1 Principles

The volume of water surface storage cannot be measured by any single satellite sensor data directly. Instead, it can be inferred indirectly by estimating the satellite-derived surface extent and water level. The principle of this method is as follows:

$$S = \int_0^h A(h)dh = \int_0^A h(A)dA \qquad (20.2)$$

where S is the volume of water storage (m³), h is the water level of terrestrial water bodies (m), and A is the surface water area (m²). A and h can be a function of the other. A straight forward approach to estimating S is to measure A and/or h. If the DEM is known, S can be estimated from either A or h. This method provides accurate volume estimations when detailed topographic data are available (Fig. 20.2).

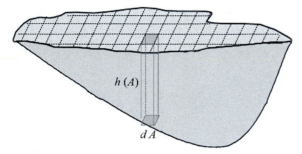

FIGURE 20.2   Principle diagram of A-H method to calculate water storage.

In practical situation, topographic data are usually nonexistent or unavailable for given surface water bodies. Therefore, it is generally difficult to determine the absolute water volume of a lake or reservoir from space (Cretaux et al., 2005; Song et al., 2014). However, the determination of absolute volume of water storage is not fundamental when it is easier to calculate water volume variations. Instead, water storage variations dS, rather than S itself, becomes the central variables in the current studies (Cretaux et al., 2005). To estimate this variable, measurements of both surface water extent (A) and water level (h) at different dates are needed. Through multitemporal satellite data, water volume variations can be measured directly by estimating the variations of the surface extent and water level. A can be determined from VIS/IR/PMW satellite data (Jain et al., 2005) and h can be obtained from satellite radar altimetry. Then, a relationship between A and h may be established empirically. The empirical relationship allows dS to be retrievable from either A or h. Without the relationship, multitemporal observations of A or h can also be multiplied to produce dS. Satellite approaches for monitoring water extent (A) and water level (h) are described in detail below.

20.3.2 Satellite-derived water surface area

Satellite retrieval algorithms of water surface area have long-time been developed (Hallberg

et al., 1973). With the development of remote sensing technology, it has become necessary to delineate the water surface extent from Earth observing satellites. At present, a variety of satellite sensor data are available for water surface area delineation at a range of spatial and temporal resolutions (Birkett, 2000; Alsdorf and Lettenmaier, 2003). The typical satellite instruments include

(1) Optical imagery, such as National Oceanic and Atmospheric Administration (NOAA), Advanced Very High–Resolution Radiometer (AVHRR), Moderate Resolution Imaging Spectroradiometer (MODIS), Satellite pour l'Observation de la Terre (SPOT), and Landsat Thematic Mapper (TM)/Enhanced Thematic Mapper+;

(2) Active microwave imagery, such as RADARSAT, Advanced Synthetic Aperture Radar (ASAR), Phased Array Type L-Band SAR (PALSAR), and European Space Agency (ERS) scatterometer.

(3) Passive microwave imagery, such as Special Sensor Microwave/Imager (SSM/I), Advanced Microwave Scanning Radiometer-Earth Observing System (AMSR-E), and TRMM Microwave Imager.

Some of these sensors are listed in Table 20.1 for quick reference.

Based on these satellite data, many accepted methods for the delineation of surface water extent have been proposed. According to different satellite sensor characteristics, these methods can be categorized into four basic types as follows.

20.3.2.1 Optical satellite sensors

Optical satellite sensors have most frequently been employed in surface water extent delineation research (Huang et al., 2018). The parts of the electromagnetic spectrum covered by these sensors include the visible blue, green, and red wavelengths, as well as emissivity data through infrared wavelengths. In general, in contrast to bare soil and vegetation, water body is highly

TABLE 20.1 Overview of main satellite sensors for water surface delineation.

Sensor type	Satellite mission	Operation period	Spatial resolution	Temporal resolution	Main algorithm
Optical satellite sensors	NOAA/AVHRR	1978—present	1100 m	0.5 days	Single band method; Multiband index; Image classification; Decision tree method; Neural network method
	MODIS	1999—present	250 m	0.5 days	
	Sentinel-3 OLCI	2016—present	300 m	2 days	
	Landsat	1972—present	30 m	16 days	
	SPOT	1986—present	2.5—20 m	26 days	
	ASTER	1999—present	15—90 m	16 day	
	HJ_1A/1B CCD	2008—present	30 m	4 days	
Active microwave sensors	ENVISAT ASAR	2002—2012	30—1000m	35 days	Visual interpretation; Image classification; Histogram thresholding; Image texture analysis; Multitemporal change detection methods
	RADARSAT	1995—present	50—100 m	24 days	
	ERS	1991—2011	25 m	35 days	
	ALOS PALSAR	2006—2012	10—100m	46 days	
	TerraSAR-X	2007—present	1—16 m	11 days	
	TANDEM-X	2010—present	3 m	11 days	
Passive microwave sensors	DMSP SSM/I	1987—present	25 km	1—2 days	Brightness temperature differences method
	AQUA AMSR-E	2002—present	25 km	1—2 days	
	TRMM TMI	1997—2014	25 km	1—2 days	

absorptive in the infrared bands and slightly more reflective in the visible bands (Fig. 20.3).

According to this principle, in the past 40 years, a number of approaches have been proposed to derive water surface area from optical satellite imagery (Alsdorf and Lettenmaier, 2003; Gao, 2015). The typical ones include single-band approach (Hallberg et al., 1973; Smith and

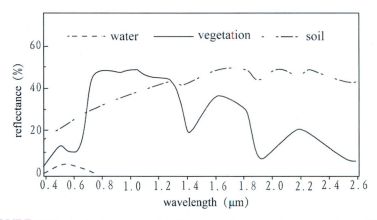

FIGURE 20.3 Reflectance characteristics comparison of water, vegetation, and soil.

Pavelsky, 2009), supervised, and unsupervised image classification approach (Davranche et al., 2010; Jin et al., 2017; Berhane et al., 2018), decision tree approach (Acharya et al., 2016; Olthof, 2017), artificial neural network approach (Jiang et al., 2018), density slicing approach (Bennett, 1987; Frazier and Page, 2000; Jain et al., 2005), spectral analysis, and multiband water index approach (Hui et al., 2008; Jain et al., 2005; Tulbure et al., 2016).The multiband water index is generally derived from an arithmetic operation (e.g., ratio, difference, and normalized difference) of two or more spectral bands, such as normalized difference vegetation index (NDVI) (Ji et al., 2009), normalized difference water index (NDWI) (McFeeters, 1996), and modified normalized difference water index (Xu, 2006). These indices are described as follows:

$$NDVI = \frac{\rho_{NIR} - \rho_{RED}}{\rho_{NIR} + \rho_{RED}} \quad (20.3)$$

$$NDWI = \frac{\rho_{GREEN} - \rho_{NIR}}{\rho_{GREEN} + \rho_{NIR}} \quad (20.4)$$

$$MNDWI = \frac{\rho_{GREEN} - \rho_{MIR}}{\rho_{GREEN} + \rho_{MIR}} \quad (20.5)$$

where ρ_{NIR}, ρ_{RED}, ρ_{GREEN} and ρ_{MIR} indicate the reflectance values in the near-infrared, red, green, and mid-infrared bands of remote sensing imagery, respectively. From the index histogram, surface water bodies can be separated from their surrounding land cover types with a simple segmentation algorithm based on selected optimal threshold (Otsu, 1979).

According to these methods, a considerable amount of research has been conducted on the surface water delineation in the past decade (Huang et al., 2018). The multiband water index method has been demonstrated to be more efficient and much more commonly used in various lakes, rivers, and inundated area (Jain et al., 2005; Hui et al., 2008; Wu and Liu, 2015b). The extracted results could be as accurate as up to 90% (Birkett, 2000). For example, Duane Nellis et al. (1998) used Landsat TM images to observe temporal and spatial variations in Tuttle Creek

Reservoir in Manhattan. Lu et al. (2011) evaluated the potential of the HJ-1A/B imagery on water body monitoring and proposed an integrated water mapping method (NDVI-NDWI index). Rokni et al. (2014) modeled the spatiotemporal changes of Lake Urmia in the period 2000–13 using the multitemporal Landsat imagery. More recently, based on MODIS images from 2000 to 2011, Wu and Liu (2015b) investigated the spatial–temporal distribution and changing processes of surface inundation in the Poyang Lake (the largest freshwater lake in China). Fig. 20.4 shows the Poyang Lake's intra-annual and interannual variations in average inundation area. The results were retrieved from MODIS reflective bands using the NDWI approach. It disclosed the large fluctuations of the lake at seasonal scale (Fig. 20.4A) and the decreasing trend of lake surface at annual scale in the last 10 years (Fig. 20.4B).

20.3.2.2 Active microwave sensors

Retrievals of water surface area have been made successfully with optical sensors, but their routine applications are of limitation in that the optical imagery can be easily affected by clouds, smoke, floating emergent vegetation, and inundated forests (Alsdorf and Lettenmaier, 2003). Radar remote sensing offers an opportunity to routinely acquire surface water extent. It can penetrate cloud cover and can acquire data during day and night (Schumann and Moller, 2015). Synthetic aperture radar (SAR) is an advanced active microwave imagery system that emits microwave pulses at an oblique angle toward the target (Grimaldi et al., 2016). Examples include Japanese Earth Resources Satellite (JERS)-1 SAR, RADARSAT, ASAR, PALSAR, and so on.

As a specular reflector, open water has a relatively smooth surface. Microwave energy is reflected away from the sensor, resulting in low backscatter in the SAR imagery (Bates et al., 2013). However, terrestrial land surfaces reflect the energy in many directions and generally appear as high backscatter features. Such

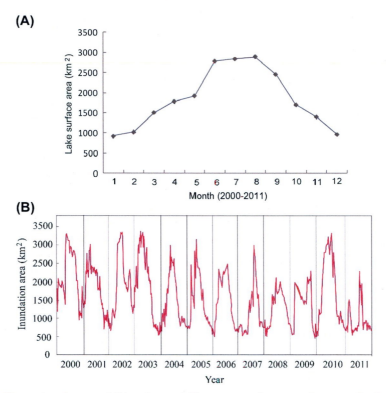

FIGURE 20.4 Satellite-retrieved seasonal (A) and annual (B) variations of water surface areas for Poyang Lake, the largest fresh lake in China.

differences make inundation extent easy to be delineated using many methods, and commonly used methods include simple visual interpretation, change vector analysis, supervised classification, image histogram thresholding, image texture algorithms and various multitemporal change detection methods (Schumann and Moller, 2015; Giustarini et al., 2016).

Recent scientific literature is full of articles describing methodologies using SAR imagery for surface water extent mapping. For example, Wang (2004) studied seasonal change in inundation extent of North Carolina and South Carolina, USA, with JERS-1 SAR data by using a decision tree classification method. Voormansik et al. (2014) developed a supervised classification algorithm to produce high-resolution maps of

the flooded area from the TerraSAR-X images in forested regions in Estonia. Eilander et al. (2014) proposed a new Bayesian approach to delineate surface areas of small reservoirs in Ghana using the SAR imagery. Matgen et al. (2011) introduced an automated hybrid methodology for SAR-based water extent mapping that combines thresholding, region growing, and change detection, benefitting from the respective strengths of the specific processes. More recently, Pradhan et al. (2016) proposed an efficient methodology that is based on rule-based classification to recognize and map flooded areas by using TerraSAR-X imagery. However, active microwave remote sensing has its own deficiency. For example, it can only provide a limited number of images available per year in some regions, making the technique unsuitable for monitoring inundation variations in large water bodies (Gao, 2015; Pham-Duc et al., 2017). In addition, SAR-based water surface areas are likely to be confused by wind roughening of water surface or emergent vegetation, and the image processing is complicated (Smith and Alsdorf, 1998).

20.3.2.3 Passive microwave sensors

Similar to active microwave measurements, passive microwave imagery can also reveal the presence of water surface despite cloud cover. In addition, passive microwave sensors can offer the advantage of higher revisit frequency of the available data collection (Hamilton et al., 2002; De Groeve, 2010).Theoretically, because of the different thermal inertia and dielectric properties of terrestrial land and water, the observed microwave radiation has a much lower brightness temperature for water bodies than for other land features (Grimaldi et al., 2016). Based on this principle, water surface extent can be efficiently detected, and detailed techniques have been proposed by many researchers. To estimate the water surface extent, Basist et al. (1998) proposed a Basin Wetness Index (BWI) based on the correlation between the decrease of emissivity and the brightness temperature differences. Sippe et al. (1998) determined inundation variations for the Amazon floodplain in Brazil based on an analysis of the 37 GHz polarization difference observed by the scanning multichannel microwave radiometer. Hamilton et al. (2002) applied a similar method to the South America. Likewise, Temimi et al. (2005)investigated the flood inundation area over the Mackenzie River Basin based on the use of a water surface fraction derived from SSM/I passive microwave images.

However, the spatial resolutions of passive microwave imagery are very coarse (~25 km) due to the large angular beams of such systems. Therefore, the potential of using passive microwave imagery for water surface mapping is thus limited to very large scale (Bates et al., 2013). In addition, the quantitative estimates of subpixel inundation extent is very difficult particularly when the flooded area is vegetated (Prigent et al., 2001; Papa et al., 2006).

20.3.2.4 Combination of multisatellite sensors

In summary, surface water area and extent can be measured with a variety of satellite sensors, but each of which has its strengths and weaknesses. Some of the high spatial resolution optical sensors make it possible to accurately detect and delineate the water body information (Smith, 1997). However, the routine water monitoring with high spatial resolution optical data is difficult due to narrow scanning coverage and the long return period between successive satellites overpasses (Alsdorf and Lettenmaier, 2003). High temporal resolution multispectral data including MODIS and AVHRR have therefore been widely used to conduct routine inundation monitoring in mesoscale (Brakenridge and Anderson, 2006), but the resolution is relatively coarse. The overall uncertainty of these measurements is $\sim6\%-13\%$ for small water bodies (Bryant and Rainey, 2002), which identified coarse spatial resolution imagery's inability to detect small inundated regions. In addition, active microwave remote sensing data such as SAR can penetrate clouds and thick forest canopies but perform poorly on water with wind waves or roughened surfaces from emergent vegetation. Passive microwave observations have too coarse spatial resolutions, which restrict their potential values in regional applications (Temimi et al., 2007).

To overcome the disadvantages of each satellite sensor, the methodology of combing the observations from multisource remote sensing (e.g., optical, active microwave, and passive microwave sensor data) to measure the water surface extent was gradually being recognized in recent years. For example, Townsend and Walsh (1998) combined SAR with Lansat TM imagery and a DEM to derive potential areas of inundation within the lower Roanoke River floodplain, North Carolina. Töyrä et al. (2002) evaluated the use of radar (RADARSAT) and visible/infrared satellite imagery (SPOT) for mapping the extent of flooded wetland areas. Prigentet al. (2001,

2007) developed global floodplain and wetland inundation extent datasets with a suite of satellite observations, including passive and active microwave along with visible and infrared measurements. Likewise, Adhikari et al. (2010) used 250m resolution MODIS and other sensor data to map flooding in near real time and from this compile an archive of global floods from 1998 to 2008.

20.3.3 Satellite-derived water level

There are several accepted methods for retrieving water levels using various remote sensing data. These methods can be summarized in three categories: the method of water level/area relationship, land—water contact method based on DEM data, and radar altimetry method.

20.3.3.1 Water level/area relationship method

Establishing the water level/area relationship is a simple approach that uses the water level/area relationship to estimate the water level. For this method, a number of satellite images of inundation areas are needed to establish empirical rating curve relating water level to surface water area (Smith, 1997). This method was more commonly applied because water surface area is relatively easily to measure directly from satellite sensors with a high accuracy. For example, Al-Khudhairy et al. (2001) used multitemporal Landsat TM imagery and simultaneous ground-based measurements of water levels to establish statistical relationships between water area and water level and estimated historical water levels in the North Kent Marshes. Pan and Nichols (2013) successfully obtained 16 lake levels of Lake Champlain in Vermont through the constructed inundation area/lake level rating curves from satellite-measured inundation areas. However, to estimate lake level based on satellite measured inundation area, we first need the inundation area—lake level rating curve. The problem associated with this approach, as pointed out by Smith (1997), is that we often do

not have observed data to construct the inundation area—lake level curves. In addition, the derived relationships between water level and area are essentially empirical, and the transferability from one hydrogeomorphological setting to another has not been proven (Wu and Liu, 2015a).

20.3.3.2 Land—water contact method

Water level can also be estimated using the land—water contact method, which uses water inundation extent in combination with high-resolution topographic maps to derive water levels (Smith, 1997; Matgen et al., 2007; Schumann et al., 2008). This principle relies on the combination of the position of the water—land boundaries and a database of bottom topographic points (Fig. 20.5). Each water surface has contact line with the land at a corresponding water level. Thus, the challenge is to assign bottom topographic values to the pixels corresponding to the border of the water extent. The process can be described as follows. First, the water body from satellite images was transformed into water polygons. Next, the water polygons were used to generate vertices as points form the outer edge of water—land boundary pixel. Then, the height values along those water—land borders were extracted from the matched bottom topographic data. Finally, the spatial distribution of water level can then be constructed with the spatial interpolation method from the height values of the water—land borders.

Several studies have attempted to monitor water level variation by the land—water contact method. For example, Raclot (2006) used this method to extract water levels from aerial photographs on floodplains, which have been shown to be sufficiently precise (mean ± 15 cm). Matgen et al. (2007) retrieved water level maps combining SAR-derived inundation areas with high-precision topographic data, and a root mean squared error of 41 cm was obtained. More recently, Wu and Liu (2015a) illustrated the use of MODIS images combined with topographic data to characterize complex water level

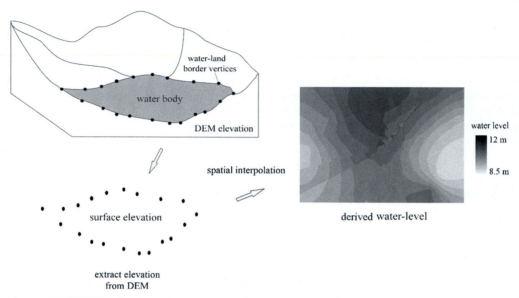

FIGURE 20.5   Principle diagram of land—water contact method to derive water level.

variation in Poyang Lake. An error analysis was conducted to assess the derived water level relative to gauge data. Validation results demonstrated that this method can capture spatial patterns and seasonal variations in water level fluctuations. However, the absolute accuracy of the resulting map depends too much on DEM uncertainties and errors both in the horizontal and vertical directions (Zwenzner and Voigt, 2008).

20.3.3.3 Satellite altimetry method

Satellite radar/laser altimetry is a promising technique for directly detecting water levels of open water bodies from space (Frappart et al., 2006). Satellite altimeters transmit a series of pulses toward the terrestrial surface in the nadir direction and receive the echo reflected by the surface (Duan and Bastiaanssen, 2013).The two-way travel time of radar (or laser) is measured and used to calculate the distance between the satellite and the target surface, called "range". The shape of the reflected signal, known as the "waveform," represents the power distribution of accumulated echoes as the radar pulse hits

the surface (Calmant et al., 2008).The principle of satellite altimetry is shown in Fig. 20.6.

The water surface height can then be determined by the difference between the altitude of satellite orbit and the range observation, which is based on the following equation:

$$H = Alt - R - T_E \qquad (20.6)$$

where "Alt" is the satellite elevation above a reference ellipsoid provided by a precise orbit solution. "R" is the measured range. "T_E" indicates various instrument and geophysical corrections, including atmospheric refraction, tidal effects, and so on (Birkett, 1995; Fu and Cazenave, 2011).

At present, several altimetry satellites have been launched since the early 1990s. The most commonly used altimeters for measuring the height of terrestrial water bodies are Topex/Poseidon (T/P) (1992—2002), ERS-1 (1992—2005), ERS-2 (1995—2003), Envisat (2002—12), Jason-1 (2002—08), Jason-2 (2008—present), CryoSat-2 (2010—present), SARAL (2013—present), and Sentinel-3 (2015—present). Different satellite altimetry sensors are flying at different orbits,

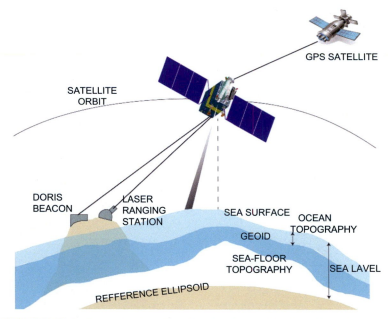

FIGURE 20.6   Principle of satellite altimetry. *Source: http://oceantopo.jpl.nasa.gov.*

resulting in the different spatiotemporal coverage of the surface water bodies. For example, T/P altimeter has a 10-days orbital cycle (temporal resolution) and 350 km intertrack spacing at the equator, while for CryoSat-2, the revisit cycle is 369 days (with the subcycle of 28 days) allowing a very dense coverage of the terrestrial surface. Table 20.2 provides a summary of the main instrument characteristics for past, current, and future satellite altimeter missions.

So far, these satellite altimeter data have been widely used to monitor water levels in rivers (e.g., Tourian et al., 2016; Biancamaria et al., 2017; Pham et al., 2018; Huang et al., 2018), lakes (e.g., Crétaux and Birkett, 2006; Crétaux et al., 2015; Song et al., 2015; Jiang et al., 2017), reservoirs (e.g., Birkett and Beckley, 2010; Troitskaya et al., 2012; Duan and Bastiaanssen, 2013; Avisse et al., 2017), and floodplains (e.g., Dettmering et al., 2016; Yuan et al., 2017; Ovando et al., 2018). The altimetry technique has shown great potential in land surface hydrology research as these data are freely available worldwide, and it

is the essential information source for most water bodies in remote regions (Crétaux and Birkett, 2006). Fig. 20.7 shows the T/P altimeter—derived relative stage variations of the Amazon River near the Manaus (Birkett et al., 2002). The T/P measurements (triangles) agreed quite well with in situ observations (solid line).

However, there are also few limitations for monitoring some smaller inland water bodies (Crétaux et al., 2015). For one thing, waveform over inland water bodies are generally contaminated by land features, resulting in degraded accuracy or altimeter losing lock. To address this issue, the processing of current satellite altimetry waveform for inland water bodies still remains challenging (Liu et al., 2016). For another, each satellite altimeter is still limited by its long revisit periods (10—35 days) and coarser intertrack spatial resolution (Table 20.2). This limitation causes temporal or spatial gaps during each altimeter overpass (Crétaux et al., 2015; Biancamaria et al., 2017). To overcome this problem, it is expected to be continuously

TABLE 20.2 Overview of main satellite sensors for water surface delineation.

Altimeter sensor	Frequency	Operation period	Spatial resolution	Temporal resolution
ERS-1	Ku	1991–1995	80 km	35 days
TOPEX/Poseidon	Ku	1992–2005	315 km	10 days
ERS-2	Ku	1995–2003	80 km	35 days
Jason-1	Ku	2001–2013	315 km	10 days
Envisat	Ku	2002–2012	70 km	35 days
ICESat-1	Laser	2003–2010	170 m	91 days
Jason-2	Ku	2008–present	315 km	10 days
CryoSat-2	Ku	2010–present	7.5 km	369 days (subcycle: 28 days)
HY-2A	Ku	2011–present	315 km	14 days
SARAL/AltiKa	Ka	2013–present	80 km	35 days
Sentinel-3	Ku	2016–present	104 km	27 days
Jason-3	Ku	2016–present	315 km	10 days
ICESat-2	Laser	2017–present	170 m	91 days
Jason-CS	Ku	Launch 2020	315 km	10 days
SWOT	Ka	Launch 2021	120 km wide swath with a ±10 km gap	15–25 days

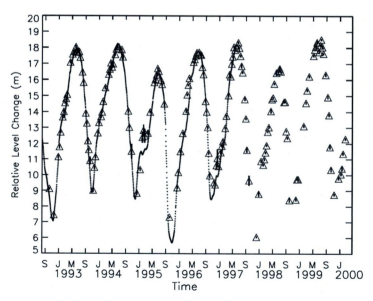

FIGURE 20.7 Relative stage changes of the Amazon River near Manaus. *Courtesy of Birkett et al., 2002.*

updated in the next few years with missions such as Jason-CS and SWOT (Surface Water Ocean Topography) (see Table 20.2), which will provide a high-resolution water surface height measurements for lakes and rivers as narrow as 100m with accuracy of 10 cm (Sulistioadi, 2013). In addition, combinations of multisatellite altimetry dataset are likely to be a better way to increase the spatiotemporal resolution of the derived water level (Calmant et al., 2008).

In recent years, the combination from different altimeters for better spatiotemporal sampling of water level has been investigated by many researchers (Frappart et al., 2006; Birkett et al., 2011; Schwatke et al., 2015). Some global surface water level databases have also been developed by combining multisatellite altimeter observations. At present, three main global lakes' and rivers' water level databases based on different altimetry are operationally accessible. They are Global Reservoir and Lake Monitoring (GRLM) database by the US Department of Agriculture (USDA), River Lake Hydrology (RLH) database by European Space Agency (ESA), and Hydroweb database by the French Space Agency Center National d'Etudes Spatiales' (CNES), which are listed in Table 20.3.

Fig. 20.8 shows the global distribution of the lakes and reservoirs from Hydroweb database. For almost 150 lakes and reservoirs, monthly level variations can be freely provided by this database (Crétaux et al., 2011). This database is based on merged T/P, Jason, ERS-2, Envisat, and Geosat. Follow-on data are provided by ESA, the National Aeronautics and Space Administration (NASA), and CNES data centers. In a longer perspective, the Hydroweb database will integrate data from future missions (Jason-3, Jason-CS, Sentinel-3A/B) and finally will serve for the design of the SWOT mission.

20.3.4 Applications

Multisatellite retrievals of water surface area and water levels from either an altimetry altimeter or in situ data are available to generate

TABLE 20.3 List of global reservoir and lake level databases based on multialtimeter data.

Database	Altimeter data source	Period	URL
GRLM	T/P, Jason-1/2, Envisat	1992–present	http://www.pecad.fas.usda.gov/cropexplorer/global_reservoir/
RLH	Envisat, Jason-2	2002–present	http://tethys.eaprs.cse.dmu.ac.uk/RiverLake/shared/main
Hydroweb	T/P, Jason-1/2, 2ERS-2, GFO, Envisat	1992–present	http://www.legos.obs-mip.fr/soa/hydrologie/hydroweb/

water storage variations. At present, many researchers have combined these two parameters to calculate lake and reservoir water storage changes worldwide (Gao et al., 2012; Song et al., 2013; Zhang et al., 2014; Crétaux et al., 2016). For example, Smith and Pavelsky (2009) calculated water storage variations in nine lakes of the Peach—Athabasca Delta, Canada, by using in situ water levels and satellite-derived surface areas. Duan and Bastiaanssen (2013) estimated water volume variations for three lakes and reservoirs by using four operational satellite altimetry databases combined with satellite imagery dada. Besides lakes and reservoirs, several studies also performed to reconstruct seasonal or interannual variations of water storage at large floodplains, such as Mekong River Basin and the Negro River Basin (Frappart et al., 2005, 2006, 2008). The Negro River Basin is the tributary carrying the largest discharge to the Amazon River. Fig. 20.9 displays the monthly variations of water storage in the basin estimated. It used water stage data from the T/P radar altimetry and inundation areas derived from multisatellite for the 1993—2000 periods (black line).The monthly variation from GRACE satellite was included for comparison, and the data were

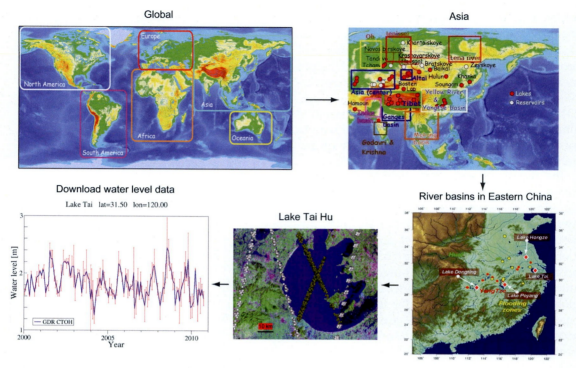

FIGURE 20.8   Schema of use of Hydroweb database.

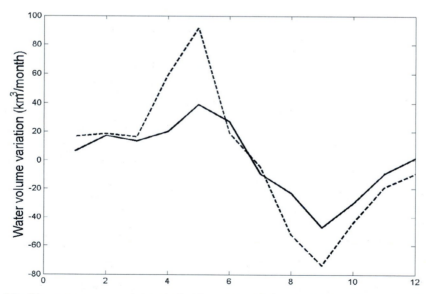

FIGURE 20.9   Monthly water storage variations in the Negro River Basin estimated from Topex/Poseidon (T/P) altimetry and multisatellite—derived inundation areas over 1993—2000 period (*black line*) and monthly variation from GRACE averaged over 2003—05 (*black dotted line*).

averaged over 2003—05 (black dotted line). The overall results showed good agreement.

It has been recognized that water storage accommodated lateral area variability or vertical stage adjustment varied significantly from one to other. Remote sensing estimates based on "area only" or "stage only" may lose important information of water storage variations. For example, inferring just from satellite-based areas may not work well in environments where small changes yield little surface area change but yield great variations in water storage (Alsdorf and Lettenmaier, 2003). A and h should be measured simultaneously for reliable estimation of storage variations.

Recent work has demonstrated that Interferometric Synthetic Aperture Radar (InSAR) provides an intrinsic, image-based direct measurement to yield $\delta h / \delta t$ (Alsdorf et al., 2000; Smith, 2002), and the accuracy could be up to centimeter scale. InSAR uses the phase values from two radar images (Smith, 2002). For water is highly reflective, microwave pulses from off-nadir imaging SAR reflect away from the antennae unless intercepted by vegetation. In this way, subtle changes of water level can be mapped. With water surface area determined from the radar imagery, water storage variations were then estimated from the simultaneous measurement.

20.4 GRACE-based estimation

In addition to the water balance—based and the surface parameter—based approaches, water storage can also be estimated with the data from the GRACE satellite (e.g., Schmidt et al., 2008). This approach is novel and of great potential in monitoring global water storage. We hereby introduce the GRACE satellite, the principle of the GRACE-based approach, the GRACE dataset, and its application.

20.4.1 GRACE satellite

GRACE satellite mission, launched in March 17, 2002, from Plesetsk, Russia, was implemented through a collaborative endeavor by NASA in the United States and Deutsche Forschungsanstalt für Luft- und Raumfahrt in Germany (Tapley and Reigber, 2002).With twin co-orbiting satellites at a low and near polar orbit, separated by a mean intersatellite distance of approximately 220 km, they measure continuous variations in the ranges of relative motion. Exploiting the differential observation, the underlying gravity field can be derived with high resolution (Fig. 20.10). It was originally designed to operate for 5 years and terminated in October 2017. It was expected to provide the Earth's gravity with a ground resolution of 300—400 km, on an order of 150s for 1 month and 160s for the whole 5 years (Tapley et al., 2004).Fig. 20.10 shows the GRACE flight configuration, which involves science goals, deputy teams of mission systems, and orbit parameters.

20.4.2 Principles

Temporal variations of the Earth's gravity mainly result from surface redistribution of the water inside and among outer fluid envelopes of the Earth (including oceans, atmosphere, cryosphere, and hydrosphere). The GRACE provides the measures of vertically integrated total water storage change. Over the continents, it involves in river basins, surface reservoirs, soil moisture, and groundwater (Longuevergne et al., 2010). The change in the total hydrological signature from GRACE is often written as a combination of different components (Moore and Williams, 2014), namely

$$\Delta TWS = \Delta SWE + \Delta SME + \Delta SWS + \Delta CWE + \Delta GWS$$

(20.7)

where ΔTWS is the change in total water storage, ΔSWE is the ice and snow storage water equivalent, ΔSME is the soil moisture equivalent, ΔSWS is the surface water storage, ΔCWE is the water equivalent in the canopy, and ΔGWS is the groundwater storage. Therefore, the surface water storage (ΔSWS) can be investigated if the

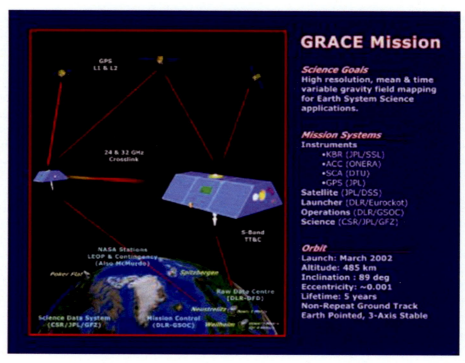

FIGURE 20.10   Flight configuration of GRACE. *From http://www2.csr.utexas.edu/grace/mission/, Used with permission from The University of Texas Center for Space Research.*

other components in Eq. (20.7) are negligible or estimated through land models or other remote sensing data.

GRACE measures spatiotemporal change of the Earth's gravity field at monthly interval. These monthly gravity fields are provided as sets of spherical harmonic coefficients, which can be inverted to global and regional estimates of vertically integrated total water storage change (ΔTWS) at a spatial resolution of few hundred kilometers or more. Based on the changes of spherical harmonic coefficients, we can estimate the water storage variations over a certain region (Wahr et al., 1998):

$$\Delta\sigma(\theta, \lambda) = \frac{a\rho_a}{3} \sum_{n=0}^{\infty} \sum_{m=0}^{n} \frac{2n+1}{1+k'_n} \times$$
$$\left[\Delta\overline{C}_{nm} \cos(m\lambda) + \Delta\overline{S}_{nm} \sin(m\lambda) \right] \overline{P}_{nm}(\cos\theta)$$
$$(20.8)$$

where $\Delta\sigma(\theta, \lambda)$ denotes the changes in surface areal density. θ and λ denote spherical coordinates with respect to an Earth-fixed frame. $\rho_a = 5517 \ \text{kg m}^{-3}$, which is the average density of the Earth. $\Delta\overline{C}_{nm}$ and $\Delta\overline{S}_{nm}$ are the corresponding changes of the harmonic coefficients with an order of m and degree of n. $\overline{P}_{nm}(\cos\theta)$ are fully normalized Legendre polynomials. The water mass variation can be expressed by equivalent water depth (ρ_w/ρ_a, where $\rho_a = 1000 \ \text{kg m}^{-3}$). The detailed expressions of the relevant variables can be found in Wahr et al. (1998).

20.4.3 GRACE dataset and applications

Since GRACE's launch March 17, 2002, the official GRACE Science Data System (SDS) continuously releases gravity solutions from three different processing centers: the University of Texas at Austin Center for Space Research, the GeoforschungsZentrum Potsdam, and the Jet Propulsion Laboratory. The provided time series consist of monthly and long-term mean sets of spherical harmonic coefficients of the global gravity potential of the Earth (Schmidt et al., 2008). These can be easily transferred into any gravity functional or surface mass anomalies in the space domain as required for the individual application. The GRACE data are divided into three levels. Level 1 data represent the raw data, collected from satellites, calibrated, and time-tagged in a nondestructive (or reversible) sense. It includes the intersatellite range, range rate, range acceleration, the nongravitational accelerations from each satellite, the pointing estimates, and the orbits. Level 2 data are the monthly gravity field estimates in form of spherical harmonic coefficients. All the Level 2 and

accompanying Level 1B products are released to the public. Since mission launch, several releases have been published by SDS. The most recent model generation provided by SDS is RL06 (published on April 26, 2018). Level 3 data are mass anomalies, and the products are available through several groups. Fig. 20.11 shows the data flow of the GRACE mission.

Since its launch, the GRACE satellite data have been widely used to study water storage at various scales over different parts of the globe (Wahr et al., 2004; Han et al., 2005; Chambers, 2006; Schmidt et al., 2008; Ramillien et al., 2008). Frappart et al. (2008) determined spatiotemporal variations of water volume over the basin of the Negro River using combined observations from GRACE, T/P, and multisatellite inundation dataset. Alsdorf et al. (2010) combined GRACE satellite data with other remote

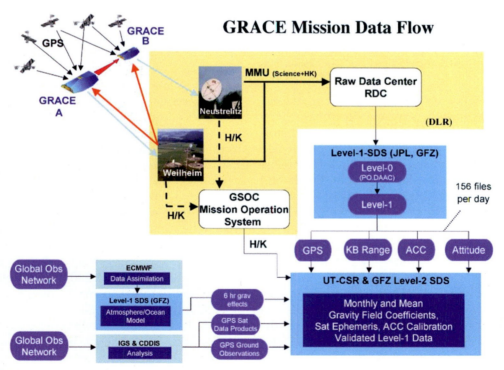

FIGURE 20.11 Data flow of Gravity Recovery and Climate Experiment (GRACE) mission. *From http://www2.csr.utexas.edu/ grace/asdp.html, Used with permission from The University of Texas Center for Space Research*

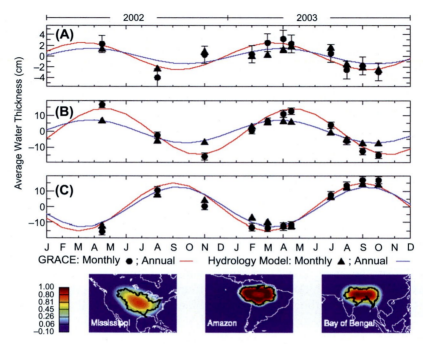

FIGURE 20.12 Gravity Recovery and Climate Experiment (GRACE)—based water variability (*dots*) within (A) the Mississippi River Basin, (B) the Amazon River basin, and (C) a drainage system flowing into the Bay of Bengal. The *triangles* represent the results from hydrological model. The best-fitting signals are in contrast for Gravity Recovery and Climate Experiment (GRACE)—derived results (blue) and model prediction (red).

sensing imagery to successfully explore seasonal water storage on the Amazon floodplain. Singh et al. (2012) investigated interannual water storage changes in the Aral Sea by using the GRACE satellite gravimetry in combination with satellite altimetry and optical imagery data. Panda and Wahr (2016) used 129 monthly gravity solutions from GRACE satellite to characterize spatiotemporal evolution of water storage changes in the Ganges River Basin of India. These studies have demonstrated the usefulness of the GRACE to study the water balance of large river basins, through combination with water stage data, precipitation data, and some other variables in terrestrial branch. It is demonstrated that the characteristics of GRACE restrict its meaningful application to study areas not smaller than 200,000 km^2 (Singh et al., 2012), which is a big limitation for hydrological study of many lakes

and reservoirs with relatively smaller surface areas. In addition, the GRACE-derived results can also be used for validation of other results from indirect estimation or model simulation. Fig. 20.12 shows an example of water storage variations with respect to model-based results (Wahr et al., 2004).

On the global scale, Schmidt et al. (2008) estimated the variations in global water storage and explored the spatiotemporal distribution of surface mass anomalies from GRACE dataset. Andersen and Hinderer (2005) estimated global gravity field changes using 15 monthly gravity field solutions from the GRACE twin satellites. The results demonstrated that GRACE is capable of capturing the changes in groundwater on interannual scales with an accuracy of 9 mm water thickness on spatial scales longer than 1300 km. More recently, Long et al. (2017)

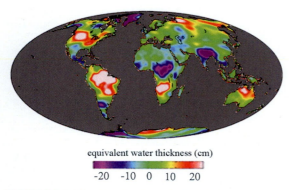

equivalent water thickness (cm)

-20 -10 0 10 20

FIGURE 20.13 The distribution of global water storage variation in May 2011.

analyzed the spatiotemporal variability in global total water storage using multiple GRACE products and global hydrological models. Fig. 20.13 shows the distribution of global water storage variation in May 2011 through Level-2 RL05 GRACE monthly field data (water abundance is represented by red, and water deficit is represented by blue).

20.5 Summary

Satellite retrieval of water storage is an immature field but with rapid development. While remote sensing techniques have made great achievements in a variety of disciplines, accurate estimation of terrestrial water storage and its variations is facing several challenging tasks.

First, terrestrial DEM is yet generally unavailable at the present. With international research efforts, the Shuttle Radar Topography Mission (SRTM) obtained DEMs on a near-global scale with a high spatial resolution of 30 m. However, there are void values over numerous land water surfaces. This is due that the DEM was undetectable with SRTM in these areas. Without reliable DEM, it is often impossible to obtain water storage over the areas.

Second, the GRACE satellite often has difficulty in reliably estimating water storage over land. Its coarse spatial resolution limits its applications to basins greater than 200,000 km^2 at an altitude of 500 m (Rodell and Famiglietti, 2001). The detecting signal is not sensitive to small basins or areas. In addition, the monthly observation may not capture short flood event. Large rivers and lakes may have large variations of water storage in a few days during a specific flood event.

Third, simultaneous estimation of surface area and water stage is under study. The altimeters currently in operation are solely limited to measures of elevation profile. Accurate retrieval of terrestrial water storage requires simultaneous estimation of surface area and water stage. This will be implemented with the SWOT Wide-Swath Altimeter, which was scheduled to be launched into space in 2021. In overall, we believe a promising prospect of satellite retrieval of global terrestrial water storage.

References

Acharya, T.D., Lee, D.H., Yang, I.T., Lee, J.K., 2016. Identification of water bodies in a landsat 8 oli image using a j48 decision tree. Sensors 16 (7), 1075.

Adhikari, P., Hong, Y., Douglas, K.R., Kirschbaum, D.B., Gourley, J., Adler, R., Brakenridge, G.R., 2010. A digitized global flood inventory (1998–2008): compilation and preliminary results. Nat. Hazards 55 (2), 405–422.

AI-Khudhairy, D.H.A., Leemhuis, C., Hoffmann, V., Calaon, R., Shepherd, I.M., Thompson, J.R., Gasca-Tucker, D.L., 2001. Monitoring wetland ditch water levels in the North Kent Marshes, UK, using Landsat TM imagery and ground-based measurements. Hydrol. Sci. J. 46 (4), 585–597.

Alsdorf, D.E., Lettenmaier, D.P., 2003. Tracking fresh water from space. Science 301 (5639), 1491–1494.

Alsdorf, D.E., Melack, J.M., Dunne, T., Mertes, L.A., Hess, L.L., Smith, L.C., 2000. Interferometric radar measurements of water level changes on the Amazon flood plain. Nature 404 (6774), 174.

Alsdorf, D., Han, S.C., Bates, P., Melack, J., 2010. Seasonal water storage on the Amazon floodplain measured from satellites. Remote Sens. Environ. 114 (11), 2448–2456.

Andersen, O.B., Hinderer, J., 2005. Global inter-annual gravity changes from GRACE: early results. Geophys. Res. Lett. 32 (1).

Avisse, N., Tilmant, A., Müller, M.F., Zhang, H., 2017. Monitoring small reservoirs' storage with satellite remote sensing in inaccessible areas. Hydrol. Earth Syst. Sci. 21 (12), 6445.

Basist, A., Grody, N.C., Peterson, T.C., Williams, C.N., 1998. Using the Special Sensor Microwave/Imager to monitor land surface temperatures, wetness, and snow cover. J. Appl. Meteorol. 37 (9), 888–911.

Bates, P.D., Neal, J.C., Alsdorf, D., Schumann, G.J.P., 2013. Observing global surface water flood dynamics. In: The Earth's Hydrological Cycle. Springer, Dordrecht, pp. 839–852.

Berhane, T.M., Lane, C.R., Wu, Q., Autrey, B.C., Anenkhonov, O.A., Chepinoga, V.V., Liu, H., 2018. Decision-tree, rule-based, and random forest classification of high-resolution multispectral imagery for wetland mapping and inventory. Remote Sens. 10 (4), 580.

Bennett, M.W.A., 1987. Rapid monitoring of wetland water status using density slicing. Proceedings of the Fourth Australasian Remote Sensing Conference 682–691.

Biancamaria, S., Frappart, F., Leleu, A.S., Marieu, V., Blumstein, D., Desjonquères, J.D., Valle-Levinson, A., 2017. Satellite radar altimetry water elevations performance over a 200 m wide river: evaluation over the Garonne River. Adv. Space Res. 59 (1), 128–146.

Birkett, C.M., 2000. Synergistic remote sensing of Lake Chad: variability of basin inundation. Remote Sens. Environ. 72 (2), 218–236.

Birkett, C.M., 1995. The contribution of TOPEX/POSEIDON to the global monitoring of climatically sensitive lakes. J. Geophys. Res. Oceans. 100 (C12), 25179–25204.

Birkett, C.M., Beckley, B., 2010. Investigating the performance of the Jason-2/OSTM radar altimeter over lakes and reservoirs. Mar. Geod. 33 (S1), 204–238.

Birkett, C.M., Mertes, L.A.K., Dunne, T., Costa, M.H., Jasinski, M.J., 2002. Surface water dynamics in the Amazon Basin: application of satellite radar altimetry. J. Geophys. Res. 107 (D20).

Brakenridge, R., Anderson, E., 2006. MODIS-based flood detection, mapping and measurement: the potential for operational hydrological applications. In: Transboundary Floods: Reducing Risks through Flood Management. Springer, Dordrecht, pp. 1–12.

Birkett, C., Reynolds, C., Beckley, B., et al., 2011. From research to operations: The USDA global reservoir and lake monitor. Coastal altimetry. Springer, Berlin, Heidelberg, pp. 25–29.

Bryant, R.G., Rainey, M.P., 2002. Investigation of flood inundation on playas within the Zone of Chotts, using a time-series of AVHRR. Remote Sens. Environ. 82 (2–3), 360–375.

Calmant, S., Seyler, F., Cretaux, J.F., 2008. Monitoring continental surface waters by satellite altimetry. Surv. Geophys. 29 (4–5), 247–269.

Chambers, D.P., 2006. Evaluation of new GRACE time-variable gravity data over the ocean. Geophys. Res. Lett. 33 (17).

Crétaux, J.F., Birkett, C., 2006. Lake studies from satellite radar altimetry. Compt. Rendus Geosci. 338 (14–15), 1098–1112.

Crétaux, J.F., Jelinski, W., Calmant, S., Kouraev, A., Vuglinski, V., Bergé-Nguyen, M., et al., 2011. SOLS: a lake database to monitor in the Near Real Time water level and storage variations from remote sensing data. Adv. Space Res. 47 (9), 1497–1507.

Crétaux, J.F., Biancamaria, S., Arsen, A., Bergé-Nguyen, M., Becker, M., 2015. Global surveys of reservoirs and lakes from satellites and regional application to the Syrdarya river basin. Environ. Res. Lett. 10 (1), 015002.

Crétaux, J.F., Abarca-del-Río, R., Berge-Nguyen, M., et al., 2016. Lake volume monitoring from space. Surv. Geophys. 37 (2), 269–305.

Crétaux, J.F., Kouraev, A.V., Papa, F., et al., 2005. Evolution of sea level of the big Aral Sea from satellite altimetry and its implications for water balance. J. Great Lakes Res. 31 (4), 520–534.

Davranche, A., Lefebvre, G., Poulin, B., 2010. Wetland monitoring using classification trees and SPOT-5 seasonal time series. Remote Sens. Environ. 114 (3), 552–562.

Dettmering, D., Schwatke, C., Boergens, E., Seitz, F., 2016. Potential of ENVISAT radar altimetry for water level monitoring in the Pantanal wetland. Remote Sens. 8 (7), 596.

De Groeve, T., 2010. Flood monitoring and mapping using passive microwave remote sensing in Namibia. Geomatics Nat. Hazards Risk. 1 (1), 19–35.

Duan, Z., Bastiaanssen, W.G.M., 2013. Estimating water volume variations in lakes and reservoirs from four operational satellite altimetry databases and satellite imagery data. Remote Sens. Environ. 134, 403–416.

Eilander, D., Annor, F.O., Iannini, L., van de Giesen, N., 2014. Remotely sensed monitoring of small reservoir dynamics: a Bayesian approach. Remote Sens. 6 (2), 1191–1210.

Frappart, F., Seyler, F., Martinez, J.M., Leon, J.G., Cazenave, A., 2005. Floodplain water storage in the Negro River basin estimated from microwave remote sensing of inundation area and water levels. Remote Sens. Environ. 99 (4), 387–399.

Frappart, F., Calmant, S., Cauhopé, M., Seyler, F., Cazenave, A., 2006. Preliminary results of ENVISAT RA-2-derived water levels validation over the Amazon basin. Remote Sens. Environ. 100 (2), 252–264.

Frappart, F., Papa, F., Famiglietti, J.S., Prigent, C., Rossow, W.B., Seyler, F., 2008. Interannual variations of river water storage from a multiple satellite approach: a case study for the Rio Negro River basin. J. Geophys. Res. 113 (D21).

Frazier, P.S., Page, K.J., 2000. Water body detection and delineation with Landsat TM data. Photogramm. Eng. Remote Sens. 66 (12), 1461–1468.

Furnans, J., Austin, B., 2008. Hydrographic survey methods for determining reservoir volume. Environ. Model. Softw. 23 (2), 139–146.

Satellite altimetry and earth sciences: a handbook of techniques and applications. In: Fu, L.L., Cazenave, A. (Eds.), 2011. International Geophysics Series, 69. Academic Press, San Diego.

Gao, H., 2015. Satellite remote sensing of large lakes and reservoirs: from elevation and area to storage. Wiley Interdiscip. Rev. 2 (2), 147–157.

Gao, H., Birkett, C., Lettenmaier, D.P., 2012. Global monitoring of large reservoir storage from satellite remote sensing. Water Resour. Res. 48 (9).

Giustarini, L., Hostache, R., Kavetski, D., Chini, M., Corato, G., Schlaffer, S., Matgen, P., 2016. Probabilistic flood mapping using synthetic aperture radar data. IEEE Trans. Geosci. Remote Sens. 54 (12), 6958–6969.

Grimaldi, S., Li, Y., Pauwels, V.R., Walker, J.P., 2016. Remote sensing-derived water extent and level to constrain hydraulic flood forecasting models: opportunities and challenges. Surv. Geophys. 37 (5), 977–1034.

Hallberg, G.R., Hoyer, B.N.E., Rango, A., 1973. Application of ERTS-1 Imagery to Flood Inundation Mapping.

Hamilton, S.K., Sippel, S.J., Melack, J.M., 2002. Comparison of inundation patterns among major South American floodplains. J. Geophys. Res. 107 (D20).

Han, S.C., Shum, C.K., Jekeli, C., et al., 2005. Improved estimation of terrestrial water storage changes from GRACE. Geophys. Res. Lett. 32 (7).

Huang, C., Chen, Y., Zhang, S., Wu, J., 2018. Detecting, extracting, and monitoring surface water from space using optical sensors: a review. Rev. Geophys. 56 (2), 333–360.

Huang, Q., Long, D., Du, M., Zeng, C., Li, X., Hou, A., Hong, Y., 2018. An improved approach to monitoring Brahmaputra River water levels using retracked altimetry data. Remote Sens. Environ. 211, 112–128.

Hui, F., Xu, B., Huang, H., Yu, Q., Gong, P., 2008. Modelling spatial-temporal change of Poyang Lake using multitemporal Landsat imagery. Int. J. Remote Sens. 29 (20), 5767–5784.

Jain, S.K., Singh, R.D., Jain, M.K., Lohani, A.K., 2005. Delineation of flood-prone areas using remote sensing techniques. Water Resour. Manag. 19 (4), 333–347.

Ji, L., Zhang, L., Wylie, B., 2009. Analysis of dynamic thresholds for the normalized difference water index. Photogramm. Eng. Remote Sens. 75 (11), 1307–1317.

Jiang, L., Nielsen, K., Andersen, O.B., Bauer-Gottwein, P., 2017. Monitoring recent lake level variations on the Tibetan Plateau using CryoSat-2 SARIn mode data. J. Hydrol. 544, 109–124.

Jiang, W., He, G., Long, T., Ni, Y., Liu, H., Peng, Y., Wang, G., 2018. Multilayer perceptron neural network for surface water extraction in landsat 8 OLI satellite images. Remote Sens. 10 (5), 755.

Jin, H., Huang, C., Lang, M.W., et al., 2017. Monitoring of wetland inundation dynamics in the Delmarva Peninsula using Landsat time-series imagery from 1985 to 2011. Remote Sens. Environ. 190, 26–41.

Kubota, T., Ushio, T., Shige, S., Kida, S., Kachi, M., Okamoto, K.I., 2009. Verification of high-resolution satellite-based rainfall estimates around Japan using a gauge-calibrated ground-radar dataset. J. Meteorol. Soc. Jpn. 87 (II), 203–222.

Liu, K.T., Tseng, K.H., Shum, C.K., Liu, C.Y., Kuo, C.Y., Liu, G., Shang, K., 2016. Assessment of the impact of reservoirs in the upper Mekong River using satellite radar altimetry and remote sensing imageries. Remote Sens. 8 (5), 367.

Long, D., Pan, Y., Zhou, J., Chen, Y., Hou, X., Hong, Y., Longuevergne, L., 2017. Global analysis of spatiotemporal variability in merged total water storage changes using multiple GRACE products and global hydrological models. Remote Sens. Environ. 192, 198–216.

Longuevergne, L., Scanlon, B.R., Wilson, C.R., 2010. GRACE Hydrological estimates for small basins: Evaluating processing approaches on the High Plains Aquifer, USA. Water Resour. Res. 46 (11).

Lu, S., Wu, B., Yan, N., Wang, H., 2011. Water body mapping method with HJ-1A/B satellite imagery. Int. J. Appl. Earth Obs. Geoinf. 13 (3), 428–434.

Matgen, P., Schumann, G., Henry, J.B., Hoffmann, L., Pfister, L., 2007. Integration of SAR-derived river inundation areas, high-precision topographic data and a river flow model toward near real-time flood management. Int. J. Appl. Earth Obs. Geoinf. 9 (3), 247–263.

Matgen, P., Hostache, R., Schumann, G., Pfister, L., Hoffmann, L., Savenije, H.H.G., 2011. Towards an automated SAR-based flood monitoring system: lessons learned from two case studies. Phys. Chem. Earth Parts A/B/C 36 (7–8), 241–252.

McFeeters, S.K., 1996. The use of the Normalized Difference Water Index (NDWI) in the delineation of open water features. Int. J. Remote Sens. 17 (7), 1425–1432.

Moore, P., Williams, S.D.P., 2014. Integration of altimetric lake levels and GRACE gravimetry over Africa: inferences for terrestrial water storage change 2003—2011. Water Resour. Res. 50 (12), 9696—9720.

Nellis, M.D., Harrington Jr., J.A., Wu, J., 1998. Remote sensing of temporal and spatial variations in pool size, suspended sediment, turbidity, and Secchi depth in Tuttle Creek Reservoir, Kansas: 1993. Geomorphology 21 (3—4), 281—293.

Olthof, I., 2017. Mapping seasonal inundation frequency (1985—2016) along the St-John river, new Brunswick, Canada using the landsat archive. Remote Sens. 9 (2), 143.

Otsu, N., 1979. A threshold selection method from gray-level histograms. IEEE Trans. Syst., Man Cybern. 9 (1), 62—66.

Ovando, A., Martinez, J.M., Tomasella, J., Rodriguez, D.A., von Randow, C., 2018. Multi-temporal flood mapping and satellite altimetry used to evaluate the flood dynamics of the Bolivian Amazon wetlands. Int. J. Appl. Earth Obs. Geoinf. 69, 27—40.

Pan, F., Nichols, J., 2013. Remote sensing of river stage using the cross-sectional inundation area-river stage relationship (IARSR) constructed from digital elevation model data. Hydrol. Process. 27 (25), 3596—3606.

Panda, D.K., Wahr, J., 2016. Spatiotemporal evolution of water storage changes in India from the updated GRACE-derived gravity records. Water Resour. Res. 52 (1), 135—149.

Papa, F., Prigent, C., Durand, F., et al., 2006. Wetland dynamics using a suite of satellite observations: A case study of application and evaluation for the Indian Subcontinent. Geophys. Res. Lett. 33 (8).

Pham, H.T., Marshall, L., Johnson, F., Sharma, A., 2018. Deriving daily water levels from satellite altimetry and land surface temperature for sparsely gauged catchments: a case study for the Mekong River. Remote Sens. Environ. 212, 31—46.

Pham-Duc, B., Prigent, C., Aires, F., 2017. Surface water monitoring within Cambodia and the Vietnamese Mekong Delta over a year, with sentinel-1 SAR observations. Water 9 (6), 366.

Pradhan, B., Tehrany, M.S., Jebur, M.N., 2016. A new semiautomated detection mapping of flood extent from TerraSAR-X satellite image using rule-based classification and taguchi optimization techniques. IEEE Trans. Geosci. Remote Sens. 54 (7), 4331—4342.

Prigent, C., Matthews, E., Aires, F., Rossow, W.B., 2001. Remote sensing of global wetland dynamics with multiple satellite data sets. Geophys. Res. Lett. 28 (24), 4631—4634.

Prigent, C., Papa, F., Aires, F., Rossow, W.B., Matthews, E., 2007. Global inundation dynamics inferred from multiple satellite observations, 1993—2000. J. Geophys. Res. 112 (D12).

Raclot, D., 2006. Remote Sensing of Water Levels on Floodplains: A Spatial Approach Guided by Hydraulic Functioning. Internation.

Ramillien, G., Famiglietti, J.S., Wahr, J., 2008. Detection of continental hydrology and glaciology signals from GRACE: a review. Surv. Geophys. 29 (4-5), 36—374.

Roads, J., Chen, S.C., Kanamitsu, M., 2003. US regional climate simulations and seasonal forecasts. J. Geophys. Res. 108 (D16).

Rodell, M., Famiglietti, J.S., 2001. An analysis of terrestrial water storage variations in Illinois with implications for the Gravity Recovery and Climate Experiment (GRACE). Water Resour. Res. 37 (5), 1327—1339.

Rokni, K., Ahmad, A., Selamat, A., Hazini, S., 2014. Water feature extraction and change detection using multitemporal Landsat imagery. Remote Sens. 6 (5), 4173—4189.

Schmidt, R., Flechtner, F., Meyer, U., Neumayer, K.H., Dahle, C., König, R., Kusche, J., 2008. Hydrological signals observed by the GRACE satellites. Surv. Geophys. 29 (4—5), 319—334.

Schumann, G.J.P., Moller, D.K., 2015. Microwave remote sensing of flood inundation. Phys. Chem. Earth Parts A/B/C 83, 84—95.

Schumann, G., Matgen, P., Cutler, M.E.J., Black, A., Hoffmann, L., Pfister, L., 2008. Comparison of remotely sensed water stages from LiDAR, topographic contours and SRTM. ISPRS J. Photogram. Remote Sens. 63 (3), 283—296.

Schwatke, C., Dettmering, D., Bosch, W., Seitz, F., 2015. DAHITI—an innovative approach for estimating water level time series over inland waters using multi-mission satellite altimetry. Hydrol. Earth Syst. Sci. 19 (10), 4345.

Singh, A., Seitz, F., Schwatke, C., 2012. Inter-annual water storage changes in the Aral Sea from multi-mission satellite altimetry, optical remote sensing, and GRACE satellite gravimetry. Remote Sens. Environ. 123, 187—195.

Sippe, S.J., Hamilton, S.K., Melack, J.M., Novo, E.M.M., 1998. Passive microwave observations of inundation area and the area/stage relation in the Amazon River floodplain. Int. J. Remote Sens. 19 (16), 3055—3074.

Smith, L.C., 1997. Satellite remote sensing of river inundation area, stage, and discharge: a review. Hydrol. Process. 11 (10), 1427—1439.

Smith, L.C., 2002. Emerging applications of interferometric synthetic aperture radar (InSAR) in geomorphology and hydrology. Ann. Assoc. Am. Geogr. 92 (3), 385—398.

Smith, L.C., Alsdorf, D.E., 1998. Control on sediment and organic carbon delivery to the Arctic Ocean revealed with space-borne synthetic aperture radar: Ob'River, Siberia. Geology 26 (5), 395—398.

Smith, L.C., Pavelsky, T.M., 2009. Remote sensing of volumetric storage changes in lakes. Earth Surf. Process. Landforms 34 (10), 1353—1358.

Song, C., Huang, B., Ke, L., 2013. Modeling and analysis of lake water storage changes on the Tibetan Plateau using multi-mission satellite data. Remote Sens. Environ. 135, 25–35.

Song, C., Huang, B., Ke, L., 2014. Inter-annual changes of alpine inland lake water storage on the Tibetan Plateau: Detection and analysis by integrating satellite altimetry and optical imagery. Hydrol. Processes 28 (4), 2411–2418.

Song, C., Huang, B., Ke, L., 2015. Heterogeneous change patterns of water level for inland lakes in High Mountain Asia derived from multi-mission satellite altimetry. Hydrol. Process. 29 (12), 2769–2781.

Sulistioadi, Y.B., 2013. Satellite Altimetry and Hydrologic Modeling of Poorly-Gauged Tropical Watershed. The Ohio State University.

Tapley, B., Reigber, C., April 2002. The GRACE Mission, Status and Future Plans. EGS XXVII General Assembly, Nice, pp. 21–26.

Tapley, B.D., Bettadpur, S., Watkins, M., Reigber, C., 2004. The gravity recovery and climate experiment: mission overview and early results. Geophys. Res. Lett. 31 (9).

Temimi, M., Leconte, R., Brissette, F., Chaouch, N., 2005. Flood monitoring over the Mackenzie River Basin using passive microwave data. Remote Sens. Environ. 98 (2–3), 344–355.

Temimi, M., Leconte, R., Brissette, F., Chaouch, N., 2007. Flood and soil wetness monitoring over the Mackenzie River Basin using AMSR-E 37 GHz brightness temperature. J. Hydrol. 333 (2–4), 317–328.

Tourian, M.J., Tarpanelli, A., Elmi, O., Qin, T., Brocca, L., Moramarco, T., Sneeuw, N., 2016. Spatiotemporal densification of river water level time series by multimission satellite altimetry. Water Resour. Res. 52 (2), 1140–1159.

Townsend, P.A., Walsh, S.J., 1998. Modeling floodplain inundation using an integrated GIS with radar and optical remote sensing. Geomorphology 21 (3–4), 295–312.

Töyrä, J., Pietroniro, A., Martz, L.W., Prowse, T.D., 2002. A multi-sensor approach to wetland flood monitoring. Hydrol. Process. 16 (8), 1569–1581.

Troitskaya, Y., Rybushkina, G., Soustova, I., Balandina, G., Lebedev, S., Kostianoy, A., 2012. Adaptive retracking of Jason-1 altimetry data for inland waters: the example of

the Gorky Reservoir. Int. J. Remote Sens. 33 (23), 7559–7578.

Tulbure, M.G., Broich, M., Stehman, S.V., Kommareddy, A., 2016. Surface water extent dynamics from three decades of seasonally continuous Landsat time series at subcontinental scale in a semi-arid region. Remote Sens. Environ. 178, 142–157.

Voormansik, K., Praks, J., Antropov, O., Jagomagi, J., Zalite, K., 2014. Flood mapping with TerraSAR-X in forested regions in Estonia. IEEE J. Selected Top. Appl. Earth Obs. Remote Sens. 7 (2), 562–577.

Wahr, J., Molenaar, M., Bryan, F., 1998. Time variability of the Earth's gravity field: hydrological and oceanic effects and their possible detection using GRACE. J. Geophys. Res. 103 (B12), 30205–30229.

Wahr, J., Swenson, S., Zlotnicki, V., Velicogna, I., 2004. Time-variable gravity from GRACE: first results. Geophys. Res. Lett. 31 (11).

Wang, Y., 2004. Seasonal change in the extent of inundation on floodplains detected by JERS-1 Synthetic Aperture Radar data. Int. J. Remote Sens. 25 (13), 2497–2508.

Wu, G., Liu, Y., 2015a. Combining multispectral imagery with in situ topographic data reveals complex water level variation in China's largest freshwater lake. Remote Sens. 7 (10), 13466–13484.

Wu, G., Liu, Y., 2015b. Capturing variations in inundation with satellite remote sensing in a morphologically complex, large lake. J. Hydrol. 523, 14–23.

Xu, H., 2006. Modification of normalised difference water index (NDWI) to enhance open water features in remotely sensed imagery. Int. J. Remote Sens. 27 (14), 3025–3033.

Yuan, T., Lee, H., Jung, H.C., Aierken, A., Beighley, E., Alsdorf, D.E., Kim, D., 2017. Absolute water storages in the Congo River floodplains from integration of InSAR and satellite radar altimetry. Remote Sens. Environ. 201, 57–72.

Zhang, S., Gao, H., Naz, B.S., 2014. Monitoring reservoir storage in South Asia from multisatellite remote sensing. Water Resour. Res. 50 (11), 8927–8943.

Zwenzner, H., Voigt, S., 2008. Improved estimation of flood parameters by combining space based SAR data with very high resolution digital elevation data. Hydrol. Earth Syst. Sci. Discuss. 5 (5).

High-level land product integration methods

Abstract

More and more satellite products of land variables are becoming available. Multiple products of the same variable usually have various uncertainties and are incompatible in spatial and temporal coverage and resolution. They usually are inconsistent with one another. It is essential to provide the end user with a set of consistent data thorough integrating various data sources. This is what high-level product integration does. We start this chapter with a discussion of the problems in land remote sensing products. We then give a brief review of methods used to address these issues and also introduce a simple integration strategy for single-point univariate cases in Section 21.1. In Sections 21.2—21.4, we discuss three types of commonly used product integration methods: geostatistical methods, multiresolution methods, and empirical orthogonal function—based methods. In Section 21.5, we summarize the chapter and provide an outlook on the future work.

Advanced Remote Sensing, Second Edition
https://doi.org/10.1016/B978-0-12-815826-5.00021-0

© 2020 Elsevier Inc. All rights reserved.

21.1 Introduction

Data integration, also known as data fusion in some literature, is not a new terminology in remote sensing. Chapter 4 was particularly written on this topic, where remotely sensed data from multiple sources are fused into a new dataset to take advantage of the features of each data source. A typical example of data fusion is to fuse high spatial resolution panchromatic imagery that lacks spectral information with multispectral imagery that has relatively coarser spatial resolution. The objective of such practice is to improve the spatial resolution of multispectral data (commonly known as "pansharpening"), that is, to generate a new dataset with both high spatial resolution and ample spectral resolution. Another example is to combine the higher revisiting frequency of moderate or coarse resolution data (e.g., Moderate Resolution Imaging Spectroradiometer [MODIS]) with less frequent but higher spatial resolution images (e.g., Landsat ETM+ data) to obtain the observations with both high spatial and temporal resolutions (e.g., see Gao et al., 2006). As the two examples show, this type of data fusion methods introduced in Chapter 4 focuses on fusing the low-level remote sensing data, such as radiance or reflectance. Here, in this chapter, we also try to integrate multiple data sources to generate an improved dataset superior to each of the data source. However, the methods discussed in this chapter directly integrate high-level products of some land variable and generate an improved product of the same variable. The purpose of data integration here is different from that of data fusion in Chapter 4. Thus, the theory and approaches are also quite different.

With development of terrestrial remote sensing, more and more products of various land variables are available to the user community. Let us take the example of the variable of leaf area index (LAI). There are a couple of global satellite products generated by different agencies (see Section 11.5). These LAI products are retrieved from different sensors with various algorithms and usually have variable temporal and spatial resolutions. Besides these remote sensing products, land variables are also measured through various regional and global networks or field campaigns. However, some major problems with current land products limit their applications in understanding land dynamics:

First, there are discontinuities and gaps due to instrumental malfunction, cloud contamination, and other factors (Fang et al., 2007). For instance, the gaps in the MODIS LAI product can be as high as 80% in the winter at high latitudes due to snow coverage. In the wet seasons, the retrieval rates of land variables are usually small or the data quality is relatively low even if data are retrievable because of the contamination of clouds and cloud shadows.

Second, different satellite data generate the products of the same variable with different uncertainties. Fig. 21.1 shows the discrepancies among three common satellite LAI products, including the Carbon cYcle and Change in

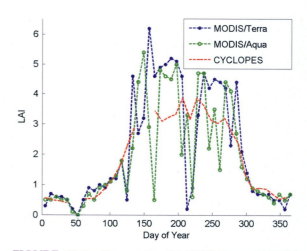

FIGURE 21.1 Time series of MODIS/Terra, MODIS/Aqua, and CYCLOPES LAI products in 2003 at a temperate forest pixel (39.04°N, 79.86°W). *From Wang, D., Liang, S., 2011. Integrating MODIS and CYCLOPES leaf area index products using empirical orthogonal functions. IEEE Trans. Geosci. Remote Sens. 49, 1513–1519, ©2011, IEEE.*

Land Observational Products from an Ensemble of Satellites (CYCLOPES) LAI, MODIS/Terra, and MODIS/Aqua LAI. Significant differences between MODIS/Terra and MODIS/Aqua LAI products are noticeable, although they are estimated using the same inversion algorithm from the same sensor. Phenology recognition from time series data is difficult due to large uncertainties. These uncertainties may produce substantial errors in driving various numerical models.

Third, the products of the same variable are typically incompatible in spatial and temporal resolution, map projection, and spatial ground coverage. Most sensors have the capability of mapping, while some types of sensors such as light detection and ranging (lidar) can only observe the surface at sparse sampling points. For the sensors with mapping ability, the swath widths vary significantly. For example, MODIS has a swath of more than 2000 km, while Multiangle Imaging SpectroRadiometer (MISR)'s swath is as narrow as 400 km. The difference between geometric and orbital properties among various sensors increases the difficulty in applying and incorporating multiple-source data.

Fourth, the use of multiple data sources (e.g., multiple products and other ancillary knowledge) provides additional information that generates more accurate land products. For example, lidar accurately estimates vegetation vertical parameters at sparse sampling points whereas multiangular sensors map less accurately. There is, however, no formal mechanism in the instrument inversion algorithms that can integrate multiple data sources effectively.

In a word, the accuracy of current land products cannot satisfy the systematic requirements for climate study and other applications. Generally, there are two types of approaches to address these problems (Wang and Liang, 2011). One is to continue developing sophisticated remote sensing inversion methods, by incorporating prior and/or background

knowledge (Liang, 2007). The other way is to improve the quality of exiting products through various postprocessing methods. Data assimilation, or combining the uncertainties of remotely sensed measurements with physically based dynamic models, is an effective postprocessing method, which has been extensively used in meteorology and oceanography (Liang and Qin, 2008). Land data assimilation is an emerging field, and many issues remain. As part of the data assimilation scheme, special efforts are needed to integrate multiple remote sensing products and other ancillary data. The integrated products and their uncertainties could be used as initial (background) values and background error matrix in the data assimilation formulation. This chapter will focus on the product integration techniques, whereas Chapter 24 of this book will introduce the theory and applications of data assimilation in land remote sensing.

21.1.1 Overview of product integration methods

Product integration aims to improve estimation of one variable by combing multiple observations of this variable. However, the existing data usually have some missing values. This feature determines that product integration methods are closely linked with gap-filling methods. Because most of product integration methods work also for a single dataset, gap-filling methods could be taken as special cases of product integration methods in some sense. The discontinuity or inconsistency in scientific data records is a universal phenomenon, both for in situ measurements (Falge et al., 2001; Ooba et al., 2006) and for satellite observations (Fang et al., 2008; Moody et al., 2005), existing in land surface datasets (Fang et al., 2008; Moody et al., 2005) and also existing in atmospheric products (Zhang et al., 2007) and in oceanic data archives (Pottier et al., 2008). Spatially, the discontinuity or inconsistency prevents forming an integrated map and spatial analysis; temporally, it limits the ability of making time series analysis and obtaining trend and change information. There are numerous investigations on developing algorithms to build spatially and temporally continuous scientific datasets and to improve the quality of these products (Beckers and Rixen, 2003; Buermann et al., 2002; Chen et al. 2004, 2006; Fang et al., 2008; Gregg and Conkright, 2001; Gu et al., 2006; Kwiatkowska and Fargion, 2003; Sellers et al., 1994). Generally, we could classify them into temporal methods, spatial methods, and spatiotemporal methods according to the information methods used.

Temporal curve fitting methods have been extensively used to smooth or interpolate land products. Temporal methods are able to analyze the time series at the mean time filling gaps. These methods would be good choices when investigating the trend or phenology of vegetation activities (Piao et al., 2006; Sakamoto et al., 2005; Zhang et al. 2003, 2006). A good discussion of this type of techniques could be found in Chapter 3. Here, we focus on product integration methods using spatial information or both spatial and temporal information, as incorporation of more information expects to achieve better accuracy. For example, Borak and Jasinski (2009) compared several methods of interpolating LAI and found that the incorporation of spatial information would improve results over methods using only temporal information. One simple approach of utilizing spatial information is a weighted spatial average, averaging observations within a spatial window, and the weights are empirically given by the distance among observations (Cressman, 1959). This method is simple but useful. Based on Cressman's approach, Fang et al. (2008) designed a temporal spatial filter to generate spatially and temporally continuous LAI data. Different from Cressman's approach, optimal interpolation (OI) uses more strict way to calculate the weights

by considering the covariance among points and measurement errors (Gandin, 1965). This method is also called Kriging in geostatistics. Recently, Christakos (2000) developed a new set of modern geostatistics theory with the core method called Bayesian maximum entropy (BME). Unlike the "traditional" geostatistics, BME makes no restrictive modeling assumption such as linearity and normality (Serre and Christakos, 1999). BME is different from the linear interpolation used in OI in that a flexible form is incorporated. BME has been successfully applied to solve many problems. For example, Christakos et al. (2004) employed a nonlinear estimator based on BME to combine both the satellite ozone product and the empirical relationships between ozone and tropopause pressure to produce high spatial resolution ozone products with greater accuracy.

Both OI and BME need to inverse a matrix with the dimension proportional to the square of the number of data used. Besides, BME involves high-dimensional integral. The computational cost limits their applications for large datasets. Multiresolution Tree (MRT) employs the Kalman filter on a tree-structured data and is computationally effective (Fieguth et al., 1995). Besides, MRT is able to integrate data with different spatial resolutions.

Besides geostatistical methods, the EOFs based method is another choice to fill gaps and reduce noises. Compared with the abovementioned methods, EOF-based methods require little input on measurement errors and covariance structure (Beckers and Rixen, 2003). In these methods, the leading eigenvector components of the covariance matrix are used to reconstruct the original matrix to achieve the goal of filling data gaps and reduce errors.

In the following sections, we will introduce integration methods based on OI, BME, MRT, and EOFs respectively. Before this, let us start the product integration theory with a simplified example.

21.1.2 A toy model of product integration

We have multiple observations of one variable and each of these observed values contains uncertainties. We want to estimate the "true" value of the variable from these observations. In this example, we do not consider the spatio-temporal relationship between observations made at different time over various locations.

Without loss of generality, supposing that we have two observations x_1 and x_2 of one variable x at the same space–time point from two satellite products with errors e_1 and e_2, the "true" x value at this space–time coordinate is x_t (Fig. 21.2). Then we have two measurement equations:

$$x_1 = x_t + e_1 \qquad (21.1)$$

$$x_2 = x_t + e_2 \qquad (21.2)$$

We assume the error terms follow Gaussian distribution with variance of σ_1^2 and σ_2^2, respectively. Usually, we also assume both of the satellite products are unbiased, that is

$$E(x_1) = E(x_2) = x_t \qquad (21.3)$$

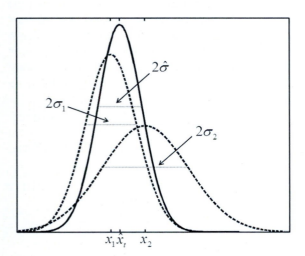

FIGURE 21.2 Illustration of the toy model. The predicted value $\widehat{x}_t$ is the weighted average of x_1 and x_2.

If this is not the case, we need to conduct some kinds of preprocessing such as calibration to remove the systematic difference (relative bias) among the products.

Because the two observations are independently retrieved from different sensors using different algorithms, the errors of the two products are assumed to be uncorrelated, that is

$$Cov(e_1, e_2) = 0 \qquad (21.4)$$

According to the Gauss–Markov theorem, the best estimator of x_t takes the linear combination of measurements:

$$\widehat{x}_t = a_1 x_1 + a_2 x_2 \qquad (21.5)$$

where $a_1 + a_2 = 1$, as we are talking about unbiased measurements and a_1 and a_2 are given by the least squares method. The best estimator here means the one that minimizes the mean squared estimation error:

$$
\begin{aligned}
E\left[(\widehat{x}_t - x_t)^2\right] &= E\left[(a_1 x_1 + a_2 x_2 - x_t)^2\right] \\
&= E\left[(a_1 x_1 + a_2 x_2 - (a_1 + a_2)x_t)^2\right] \\
&= E\left[(a_1 e_1 + a_2 e_2)^2\right]
\end{aligned}
$$

$$(21.6)$$

Given the fact that the errors of the two products are uncorrelated, we have

$$
\begin{aligned}
E\left[(\widehat{x}_t - x_t)^2\right] &= E(a_1^2 e_1^2) + E(a_2^2 e_2^2) \\
&= a_1^2 \sigma_1^2 + a_2^2 \sigma_2^2
\end{aligned}
\qquad (21.7)
$$

This constrained minimization problem could be easily solved by introducing a Lagrange multiplier λ to convert the original problem to an unconstrained optimization problem:

$$
\begin{aligned}
\operatorname*{argmin}_{a_1, a_2, \lambda}(\Lambda) = \operatorname*{argmin}_{a_1, a_2, \lambda} \Big[&a_1^2 \sigma_1^2 + a_2^2 \sigma_2^2 \\
&+ \lambda(a_1 + a_2 - 1) \Big]
\end{aligned}
\qquad (21.8)
$$

Calculate the partial derivative with a_1, a_2, and λ:

$$\frac{\partial \Lambda}{\partial a_1} = 2\sigma_1^2 a_1 + \lambda = 0$$

$$\frac{\partial \Lambda}{\partial a_2} = 2\sigma_2^2 a_2 + \lambda = 0 \qquad (21.9)$$

$$\frac{\partial \Lambda}{\partial \lambda} = a_1 + a_2 - 1 = 0$$

By solving this system of linear equations, we have

$$a_1 = \frac{\sigma_2^2}{\sigma_1^2 + \sigma_2^2}, \quad a_2 = \frac{\sigma_1^2}{\sigma_1^2 + \sigma_2^2} \qquad (21.10)$$

We achieve this result by following the classic Gaussian statistics. Actually, we will have the same result if we use the Bayesian theorem. In the Bayesian theorem, we could treat the probability distribution function of the first measurement x_1 as a prior distribution of x_t. Then given the second measurement x_2, we update the distribution and obtain the final posterior distribution. For the derivation of the parameters a_1 and a_2 using Bayesian theorem, the reader could refer to a tutorial on Bayesian data fusion by Wikle and Berliner (2007).

So, starting from either theory, for a Gaussian process, the coefficients will be the same (Lorenc, 1986; Wikle and Berliner, 2007). Intuitively, it is also easy to understand that a better estimate of the variable given two unbiased and uncorrelated observations would be the weighted average of the two observations and the weights would be determined by their relative "accuracy." According to Eq. (21.10), the observation with a smaller error will contribute more to the result through a greater weight.

Although it looks simple, this toy model serves as the foundation of many other complex product integration and data assimilation methods. This method could also be used to solve practical problems. Gu et al. (2006) method of improving MODIS LAI is based on this simple one-point OI. However, in product integration of land products, usually there is no reliable observation available in the same spatiotemporal point of estimation. Information from temporally and spatially adjacent points is needed to improve the estimation. Accordingly, temporal or spatial methods with the ability of interpolating are needed.

21.2 Geostatistics methods

This section will discuss two geostatistics-based integration methods. The first one is based on traditional geostatistics approach, OI, and the other is based on modern geostatistics method, BME. Before we go into the details of these two methods, we first introduce some basic terminologies and concepts used in geostatistics.

21.2.1 Introduction to stochastic process

Like other statistics, geostatistics also deals with stochastic (random) processes. Here, we will give a brief introduction to stochastic process. The interested readers could refer to some textbooks on random field for further details (e.g., Christakos, 1992). Stochastic process means a collection of random variables L_i over probability space (Ω, F, P). Specifically, the index in geostatistics usually refers to spatial location (x_i, y_i).

To describe the probability and other properties of L_i, we define two functions.

The first one is cumulative distribution function (CDF) F:

$$F_{L_i}(x) = \text{Probability}(L_i \leq x) \qquad (21.11)$$

The derivative of CDF defines another function of L_i probability density function (pdf) f:

$$f_{L_i}(x) = \frac{dF_{L_i}(x)}{dx} \qquad (21.12)$$

Given *pdf*, the mean of this stochastic process can be calculated by

$$\mu(L_i) = \int x \cdot f_{L_i}(x)dx \qquad (21.13)$$

For two points i and j, their covariance is defined by

$$C_L(i,j) = \int\int (L_i - \mu(L_i)) \cdot (L_j - \mu(L_j)) f_{L_i,L_j}(x,x\prime)dxdx\prime \qquad (21.14)$$

where $f_{L_i,L_j}(x,x\prime)$ is multivariate *pdf* of random variable L_i and L_j. The covariance function must be nonnegative definite to keep it permissible. For a covariance function to be nonnegatively definite, it must satisfy

$$\sum_{i=1}^{n}\sum_{j=1}^{n} a_i a_j C_L(i,j) \geq 0 \quad \text{given any } a \qquad (21.15)$$

For the sake of simplicity, the random field usually is assumed to have the property of stationarity. A strict stationarity is defined by the multivariate CDF. The CDF is invariant with the spatial lag. In most cases, only weak stationary (wide sense stationary) is required, which has the following properties:

$$\mu_L(i) = c \qquad (21.16)$$

$$C_L(i,j) = C_L\left(\sqrt{(x_i - x_j)^2 + (y_i - y_j)^2}\right) \qquad (21.17)$$

This means the mean is a constant and the covariance is only the function of distance between the two points. According to its definition, a stationary random field cannot contain nonconstant trend. One important task of geostatistics is to remove large-scale trend T and get homogenous zero-mean small-scale variation V:

$$L(x_i, y_i) = T(x_i, y_i) + V(x_i, y_i) \qquad (21.18)$$

where T is assumed to be deterministic and V is stochastic. V is the real target of geostatistics.

21.2.2 Optimal interpolation

OI, also known as objective analysis, is the statistical method of spatial interpolation on the Gaussian process. OI was invented in the field of numerical weather prediction by Gandin (1965). In geostatistics, a very similar set of theories, Kriging, was independently developed almost at the same period (Matheron, 1963). OI is designed for spatial interpolation; however, it is natural to add time as another dimension of space and extend OI to the spatiotemporal domain. For some variables, this extension is plausible. These variables evolve in both spatial and temporal domains and have both spatial and temporal dependency. For example, insolation can be modeled through a spatiotemporal covariance matrix and predicted using Kriging in the spatiotemporal domain (Huang et al., 2007). However, some variables show significantly variable properties in spatial and temporal domains. Uz and Yoder (2004) found that temporal correlation barely exists in oceanic chlorophyll concentration although spatial correlations were found, so Pottier et al. (2006) used only spatial dependency in merging MODIS and SeaWIFS chlorophyll products.

Equations used in OI/Kriging could have varied forms according to assumptions on whether mean is known or fixed. For example, simple Kriging handles cases where a known constant mean exists, ordinary Kriging does not predecide the mean but requires the mean to be fixed, whereas universal Kriging allows the mean to change linearly with some covariates. In the practice of remote sensing product integration, detrending is usually the first step of preprocessing. So we mainly deal with the data with known means.

Recall the toy model mentioned in Section 21.1.2. We used the approach of least squares to obtain a linear combination of multiple measurements of the same variable at the same point. OI is also based on the Gauss–Markov theorem, so it also pursues the best linear unbiased predictor of a group of measurements. However, in the geostatistics field, interpolation is needed as the measurements happen at points different from the points to be predicted. These existing measurements and the value to be predicted are spatially correlated. We usually assume the covariance is only the function of distance.

The problem setting is that given the mean and covariance function of this Gaussian process, we have observations $x(p_i)$ of the variable x at a set of space time points p_i, and we want to predict the value $\widehat{x}(p_0)$ of this variable at a given point (Fig. 21.3)

$$\widehat{x}(p_0) = \mu(p_0) + \sum \lambda_i \cdot [x(p_i) - \mu(p_i)] \quad (21.19)$$

through minimizing the error $[\widehat{x}(p_0) - x(p_0)]^2$. There are a lot of well-written textbook on OI/Kriging (e.g., Cressie, 1993), please refer to them for the deduction process of the parameters λ_i (Bretherton et al., 1976):

$$\lambda_i = \sum_{i=1}^{N} \sum_{j=1}^{N} C(p_i, p_j)^{-1} C(p_0, p_i) \quad (21.20)$$

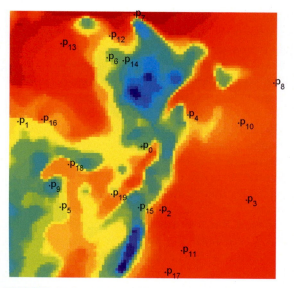

FIGURE 21.3   With regard to a second-order stationary field, optimal interpolation gives the best prediction at p_0 from the observations at p_i.

21.2.2.1 Application of optimal interpolation in product integration

OI was not originally designed for product integration. However, the interpolating ability makes it perfectly suitable for handling the problem of missing values in land remote sensing. Because OI does not require the existing measurements from one single product, it is natural to apply OI over multiple products. In recent practice, OI was extensively used to reconstruct satellite products, for example, sea surface temperature (Reynolds and Smith, 1994), sea surface height (Le Traon et al., 1998), chlorophyll (Pottier et al., 2006), precipitation (Sapiano et al., 2008), etc.

Generally, steps of applying OI in integrating satellite products include the following:

1. Detrend: Remove large-scale trends to obtain residues with the properties of second-order stationary. This preprocessing step is not only used in the geostatistics methods. The other methods also usually work on homogeneous fields only. However, it is not an easy task to estimate the trends (Johannesson and Cressie, 2004). This detrending step is relatively subjective. Multiple years' climatology (mean and variance) is usually used in this step.
2. Estimate covariance functions: Commonly used spatial function include exponential, Gaussian, spherical, etc. (Fig. 21.4). If treating spatial covariance and temporal covariance separately, space—time covariance functions could be simply the combination (e.g., addition, multiplying) of two functions, which are called separable covariance functions. Besides, we also have nonseparable space—time covariance functions, which are still hot research topics (Cressie and Huang, 1999; Gneiting, 2002; Stein, 2005).
3. Evaluate measurement errors: Besides covariance functions, measurement errors are another vital parameters in applying OI. Under the assumption of no measurement errors, Kriging does not change the points

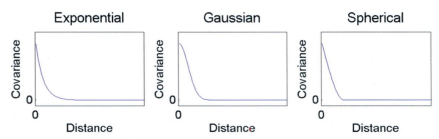

FIGURE 21.4 Illustrations of three common covariance functions. The three figures use the same sill and range parameters but have different shapes.

where measurement exists and only interpolate over points where no measurement is available. For the case with measurement errors, OI will interpolate over points without measurements and smooth over points with measurements. The smoothness depends on the relative ratio between measurement errors and standard deviation of the variable itself. Comprehensive validation is the ideal way to assign the value of measurement errors. However, the available in situ measurements of most land variables are spatially or temporally limited. Sometime, an alternative way needs to be figured out.

4. After completing all these preprocessing steps, we could run the OI code now. But we need to keep in mind that OI involves the inversion of covariance matrix, and the computational cost of matrix inversion exponentially depends on the dimension of the matrix. It is wise to limit the number of neighbor points used in OI and balance accuracy and efficiency.

21.2.2.2 A case study

LAI, one-sided leaf area per unit of ground area (Chen and Black, 1992), is a required key input for various ecosystem productivity and land process models. Numerous efforts have been made to inverse LAI from optical satellite data. Among existing global LAI products, MODIS (Myneni et al., 2002) and CYCLOPES

(Baret et al., 2007) are two representatives, which have continuous and relatively long time coverage, moderate spatial (around 1 km), and temporal (a couple of weeks) resolutions. Both have been validated extensively, see, e.g, Yang et al. (2006) and Weiss et al. (2007). However, neither of these two products could meet the requirement of 0.5 for climate study specified by Global Climate Observing System (GCOS, 2006).

Facing these problems, Wang and Liang (2014) applied OI to integrate MODIS and CYCLOPES LAI products. In their paper, the multiple years' LAI climatology was used as the background information to remove the large-scale trend in the two LAI datasets. A revised locally adjusted cubic spline capping method was applied to smooth the climatology to reduce errors (Chen et al., 2006). Fig. 21.5 shows the LAI climatology from July 20 to July 27 calculated from MODIS/Terra and MODIS/Aqua LAI products. The random error is reduced through the smoothing procedure and the smoothed climatology is more likely to represent the natural phenology curve. The authors used a linear LAI measurement model to represent different types of LAI measurement errors:

$$L_M = a_M \cdot L_t + b_M + e_M \qquad (21.21)$$

where L_M is the MODIS estimated LAI, and L_t is the true LAI. a_M is the ratio of dynamic range between the estimation and actual LAI values, b_M is the estimation bias, and e_M is the random

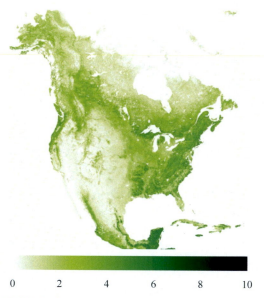

| | | | | | |
|0|2|4|6|8|10|

FIGURE 21.5 North America LAI climatology calculated from multiple years' MODIS data from July 20 to July 27 after temporal smoothing. *From Wang, D., Liang, S., 2011. Integrating MODIS and CYCLOPES leaf area index products using empirical orthogonal functions. IEEE Trans. Geosci. Remote Sens. 49, 1513–1519, ©2011, IEEE.*

residue error. Based on this linear measurement model, a normalization scheme was carefully designed to account for the relative difference between MODIS LAI and CYCLOPES LAI:

$$Ls\prime = \frac{L_M - \mu_M}{a_M \sigma_t} \tag{21.22}$$

where μ_M is the product mean calculated directly from multiyear's MODIS LAI time series and σ_t^2 is the variance of true LAI. With regard to the covariance function, the exponential functions were used to fit both the spatial and temporal covariance. A nested exponential covariance model is chosen to account for the spatial dependency at both short distance and long distances:

$$C(s,t) = \left[c_1 \exp\left(-\frac{3s}{a_{s1}} \right) + c_2 \exp\left(-\frac{3s}{a_{s2}} \right) \right] \exp\left(-\frac{3t}{a_t} \right) \quad s > 0, t > 0 \tag{21.23}$$

$$C(s,t) = c_{Nugget} \quad s = 0, t = 0$$

Fig. 21.6 shows the fitted and calculated covariance for data from the Beltsville Agriculture Research Center (BARC). Together with the measurement errors from validation results, these parameters are used to drive OI codes and obtain integrated LAI results both in time series and validation sites.

Fig. 21.7 shows a time series of integrated results over an agricultural cropland. The integration process removed large fluctuations within the original LAI products, producing a smoother time series. The discrepancies between the two LAI datasets are mitigated through the product integration, that is, a consistent time series was derived from the incompatible two data sources. Fig. 21.8 shows the error budgets with the product integration. One merit of the OI method is its ability to provide estimation errors with estimated values. We could clearly see that the integrated results have larger uncertainties in summer when original LAI products have bigger errors. We could also tell the incorporation of Aqua LAI after year 2002 reduces the estimation uncertainties.

The authors also validated their integrated results using high-resolution LAI reference maps. The integration process reduced the estimation bias and also improved the accuracy in terms of RMSE and R^2.

21.2.3 Bayesian maximum entropy

OI is a linear predictor on the Gaussian process. A new epistemological spatiotemporal mapping method, BME, has been recently developed. Uncertain observations are treated as soft data in BME, and the errors are rigorously considered. Unlike the "traditional" geostatistics, Kriging, BME makes no restrictive modeling assumption such as linearity and normality (Serre and Christakos, 1999). BME is different from the linear interpolation used in Kriging in that a flexible form is incorporated. Under some scenarios, BME is simplified to Kriging.

(A)

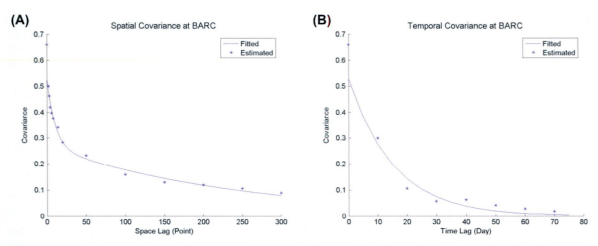

(B)

FIGURE 21.6   Estimated and fitted covariance at the Beltsville Agriculture Research Center. (A) Spatial covariance and (B) temporal covariance (Wang and Liang, 2014).

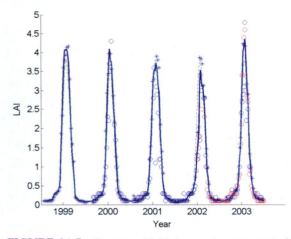

FIGURE 21.7   Five years' LAI time series at an agricultural site. The *solid line* is integrated results. *Stars* are CYCLOPES LAI, *green circles* are MODIS/Terra LAI, and *red circles* are MODIS/Aqua LAI. *From Wang, D., Liang, S., 2014. Improving LAI mapping by integrating MODIS and CYCLOPES LAI products using optimal interpolation. IEEE J. Select. Topics Appl. Earth Obser. Remote Sens. 7, 445–457, © 2014,IEEE.*

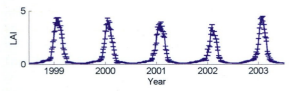

FIGURE 21.8   Five years' time series of integrated leaf area indices (LAIs) and their errors at this agricultural site (Wang and Liang, 2014).

Christakos (2000) monograph systematically describes the BME theory. BME treats spatiotemporal stochastic processes in a different way from traditional geostatistics (e.g., Kriging) by incorporating physical knowledge into the spatiotemporal analysis instead of using a "pure inductive" framework (Christakos, 1990; Serre and Christakos, 1999). Under the BME's framework, "traditional" geostatistical methods are a special case of BME.

BME has been successfully applied to solve many problems. For example, Christakos et al. (2004) employed a nonlinear estimator based on BME to combine both the satellite ozone product and the empirical relationships between ozone and tropopause pressure to produce high spatial resolution ozone products with greater accuracy. Douaik et al. (2005) used BME to map soil salinity and found that the BME approach produced less biased and more accurate predictions than the traditional Kriging approach. Kolovos et al. (2002) combined site-specific observations and stochastic partial differential equations into an assimilation of the advection-reaction process.

BME is capable of incorporating both general knowledge and site-specific information. The general knowledge could be either physical laws or statistical moments. Both are expressed in the form of teleologic equations. There are generally three stages for applying BME: *prior*, *meta-prior*, and *posterior* (Christakos, 2000; Serre and Christakos, 1999):

In the *prior stage*, the general knowledge, related with a g_α function, is expressed by the representation of a G-operator:

$$\int d\chi_{map} G(g_\alpha) f_G(\chi_{map}) = 0 \quad (\alpha = 1, ..., N) \tag{21.24}$$

The prior *pdf* $f_G(\chi_{map})$ is obtained by maximizing the informative entropy constrained by this equation. χ_{map} represents all the data points, including observed data χ_{data} and the data χ_k to be predicted at unobserved points k. In the *meta-prior stage*, the available site-specific data can be divided into true (hard) data χ_{hard} or uncertain (soft) data χ_{soft}. The uncertainties with soft data will be considered explicitly in a rigorous way at this stage. In the *posterior stage*, the posterior *pdf* $f_K(\chi_k)$ of predicted points χ_k is given by:

$$f_K(\chi_k) = A^{-1} \int_D d\chi_{soft} f_G(\chi_{map}) \tag{21.25}$$

where $A = \int_D d\chi_{data} f_G(\chi_{data})$ is the normalization coefficient. When the physical knowledge is the statistical moments of underlying stochastic process, the G-operator of the prior stage takes the form of multivariate Gaussian function and the posterior *pdf* leads to (Serre and Christakos, 1999)

$$f_K(x_k) = A^{-1} \varphi\left(x_k; B_{k|h} x_{hard}, c_{k|h}\right)$$
$$\times \int dx_{soft} f_S(x_{soft}) \varphi\left(x_{soft}; B_{s|kh} x_{kh}, c_{s|kh}\right) \tag{21.26}$$

where $\varphi(x, \overline{x}, c)$ is the multivariate Gaussian function of variable x with the mean $\overline{x}$ and covariance c; $c_{k|h}$ is the covariance matrix of predicted points conditional to hard data; $c_{s|kh}$ is the covariance matrix of soft data points conditional

to predicted points and hard data points; $B_{k|h}$ is defined as $c_{k,h} c_{h,h}^{-1}$ and $B_{s|kh}$ is defined as $c_{s,kh} c_{kh,kh}^{-1}$.

Based on the posterior *pdf*, the estimates $\overline{x}_k$ and the errors of estimation σ_k^2 can be expressed as

$$\overline{x}_k = \int x_k f_K(x_k) dx_k \tag{21.27}$$

$$\sigma_k^2 = \int (x_k - \overline{x}_k)^2 f_K(x_k) dx_k \tag{21.28}$$

Under this framework, all the information is integrated into the Bayesian inference to maximize the information. The results are expected to be more informative and accurate. These properties of BME can address the problems the land remote sensing community is faced with and can make BME ideally suitable for product integration of satellite LAI products.

21.2.3.1 Application of Bayesian maximum entropy to product integration

The steps of applying of BME in product integration are similar to those of applying OI. However, the way BME treats measurement errors are different from OI. OI assumes the normal distribution of measurement errors and use the variance to fully characterize them. In BME, inaccurate measurements are treated as soft hard, which could take a variety of forms. For example, soft data could be in the form of intervals, the lower and upper limit of one measurement. BME also take soft data in *pdf* forms. This makes the representation of measurement errors very flexible.

De Nazelle et al. (2010) applied BME to integrate in situ ozone measurements with ozone predictions from a simulation model of air quality to generate ozone maps in North Carolina, USA. In this example, field measurements of ozone are available only at sparse locations (46 sites in North Carolina). Kriging can be used as an interpolation method to map ozone from the measurements. In the paper, BME integrated another piece of information, ozone data with a resolution of 4 km predicted by the model, to provide a better estimation of ozone maps. Modeled ozone here was used as the soft input of BME, and the uncertainties with the model simulation were considered by a normal distribution truncated at zero (Fig. 21.9) (De Nazelle et al., 2010). Cross-validations showed that the integration of the field measurements with the model simulation improved the accuracy of the ozone prediction, especially over location where field measurements are far from the site.

21.3 Multiresolution tree

When applying OI or BME on a large dataset, the computational cost will be obstacle due to the inefficient inversion of large covariance functions. One way to solve this problem is through a scale-recursive filter of MRT-based data structure. Chou developed this method is his dissertation (1991). Based on Dr. Chou's work, Fieguth further improved this method in his PhD work (Fieguth, 1995) and successfully interpolated and smoothed satellite ocean surface height data (Fieguth et al. 1995, 1998). With the popularization of this method, this multiresolution estimation framework has been attempted to solve a couple of practical problems, such as mapping temperature (Menemenlis et al., 1997), assimilating soil moisture (Parada and Liang, 2004), and generating aerosol maps at different resolutions(Huang et al., 2002). The data fusion ability is another big merit of MRT in addition to the computational efficiency. MRT is able to handle observations with different spatial resolutions and integrate them together in a resolution-consistent way. This across-resolution product integration does not

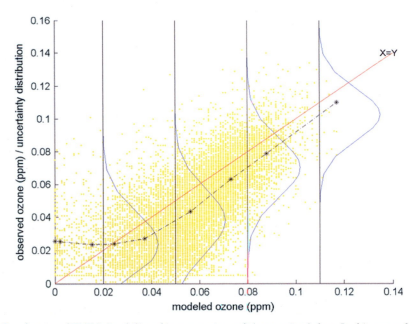

FIGURE 21.9   One feature of BME is its ability of incorporating soft (uncertainty) data. In this example, the ozone concentrations predicted by the model are used in form of pdf (*blue lines*) as the soft input of BME. *The figure is copied from De Nazelle, A., Arunachalam, S., Serre, M.L., 2010. Bayesian maximum entropy integration of ozone observations and model predictions: an application for attainment demonstration in North Carolina. Environ. Sci. Technol. 44, 5707–5713.*

only improve the mapping accuracy by incorporating more information but also produce self-consistent results throughout all resolutions. The properties perfectly satisfy the requirement of land remote sensing integration, where some data cover larger area and also with larger errors, while other data have smaller footprints but with better accuracy. This data fusion ability of MRT has been demonstrated by several researches. Slatton et al. (2001) extended the mapping ability of accurate lidar height data by combing them with large-scale interferometric synthetic aperture radar data. de Vyver and Roulin (2009) used this approach to integrate two remotely sensed precipitation datasets with different spatial resolutions. Tao et al. (2018) recently improved the quality and integrity of fraction of absorbed photosynthetically active radiation using the MRT data integration approach.

21.3.1 Methodology

The core part of MRT is the Kalman filter. The Kalman filter was designed to update measurements and model parameters temporally according to dynamic models in time domain. The multiresolution OI actually carries out the Kalman filter in space domain. Chou (1991) applies this approach to a tree-structured spatial data (Fig. 21.10). Similar to the regular Kalman filter in time domain, the MRT also needs two groups of equations. One is the observation equation, describing how the variable measured y is linked with the variable interested x:

$$y(s) = Hx(s) + v(s) \qquad (21.29)$$

where H is the observation matrix, and $v(s)$ is the normally distributed measurement errors: $N(0, R(s))$.

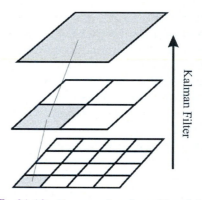

FIGURE 21.10 Framework of multiresolution tree method. *From Wang, D., Liang, S., 2010. Using multiresolution tree to integrate MODIS and MISR-L3 LAI products. In, 2010 IGRASS, Honululu HI, pp. 1027–1030, © 2010,IEEE.*

Besides, we also need an equation to describe the state transfer model from coarse resolution to fine resolution (Luettgen (1993)):

$$x(s) = A(s)x(ps) + w(s) \qquad (21.30)$$

where, $x(s)$ and $x(ps)$ are the variable of interest, respectively, at scale s and its parent scale ps. $w(s)$ represents the state transfer noise with Gaussian distribution $N(0, Q(s))$. A is the state transition matrix from parent to children. Similarly, we also have a state transfer model from fine resolution to coarser resolution.

$$x(ps) = F(s)x(s) + w\prime(s) \qquad (21.31)$$

The state transition matrix F here could be calculated by (Luettgen, 1993)

$$F(s) = P(ps)A(s)P^{-1}(s) \qquad (21.32)$$

where $P(s)$ is the variance at scale s.

To fully utilize both coarse resolution and fine resolution data, the Kalman filter is carried out from children to parents and the Kalman smoother is applied from parent to children. From child scale s to parent scale ps, we have two information sources: the state transfer model prediction $\hat{x}(ps|s)$ using observations up to scale p and observation at scale ps $y(ps)$. The Kalman filter links the two sources by (Luettgen, 1993)

$$\hat{x}(ps|ps) = \hat{x}(ps|s) + K(ps)(y(ps) - H\hat{x}(ps|s)) \tag{21.33}$$

where $\hat{x}(ps|ps)$ is the estimator of scale ps by incorporating observations up to scale ps, and $K(s)$ is the Kalman gain and is given by

$$K(ps) = P(ps|s)HV^{-1}(ps) \tag{21.34}$$

and $V(s)$ is the innovation covariance given by

$$V(ps) = HP(ps|s)H^T + R(ps) \tag{21.35}$$

After the Kalman filter reaches the root of the tree, the Kalman smoother is carried out from parent ps to children s to obtain the final estimator $\hat{x}\prime(s)$ (Luettgen, 1993):

$$\hat{x}\prime(s) = \hat{x}(s|s) + J(s)(\hat{x}\prime(ps) - \hat{x}(ps|s)) \tag{21.36}$$

The critical step in using MRT is to assign the model parameters. Usually, an identity matrix is used to represent the state transition matrix (Tzeng et al., 2005). The measurement matrix usually takes the form of identity matrix for the cases where the variable measured is the same as the variable interested. The measurement error is dependent on sensor characteristics and the retrieval algorithm, which can be obtained through validation. However, the acquisition of the state transfer noise is still a research topic. Huang et al. (2002) calculated the variance parameter from the covariance function. Some use a $1/f^\mu$-like stochastic model, as many natural phenomena display a self-similar property (Fieguth et al. 1995, 1998; Fieguth and Willsky, 1996). Kannan et al. (2000) used an expectation—maximization algorithm to estimate the parameters. de Vyver and Roulin (2009) directly calculated the parameters from the average radar measurement.

21.3.2 A case study with leaf area index

The product integration ability of MRT in land remote sensing was demonstrated by combining MODIS and MISR L3 LAI (Wang and Liang, 2010). Because of its narrow swath, MISR is less frequent to cover the entire earth surface than MODIS. MISR LAI data are aggregated to 0.5° to generate L3 monthly data to enhance the surface coverage. To collocate with MODIS data, MISR L3 data are resampled and reprojected to MODIS's sinusoidal map projection. Validation results are used to characterize the measurement errors of MODIS and MISR LAI products, whereas the state transfer noises are empirically calculated using aggregated MODIS data.

The paper show the integrated LAI map over one MODIS tile (Fig. 21.11). The MRT integration algorithm filled all the data gaps and removed spurious extremes, resulting in a smoother LAI anomaly map. It is also worthy to mention that the inconsistency between different resolutions is also removed through the cross-scale integration process.

21.3.3 A case study with albedo

The MRT-based approach was also recently applied to integrate albedo retrieved from MISR, MODIS, and Landsat Thematic Mapper/Enhanced Thematic Mapper Plus (ETM+) data (He et al., 2014). The MODIS and MISR official albedo products were used in the study. The Landsat albedo was retrieved from the Landsat L1T data through three major steps: atmospheric correction, BRDF modeling, and narrow-to-broadband conversion. The study area is the North Central region of the United States, covered by a variety of surface types, such as cropland, grassland, forest, and inland water. Field albedo measurements at eight AmeriFlux

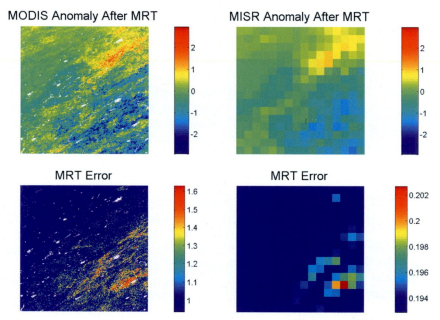

FIGURE 21.11   MRT-integrated LAI anomaly and its error at MODIS and MISR scales. *From Wang, D., Liang, S., 2010. Using multiresolution tree to integrate MODIS and MISR-L3 LAI products. In, 2010 IGRASS, Honululu HI, pp. 1027−1030, © 2010, IEEE.*

sites in the study area were collected to validate the satellite albedo data. The Landsat albedo agrees best with the field data with a relative error of 15% due to its fine spatial resolution.

The MODIS and MISR albedo products were first converted to the UTM map projection of the Landsat data. Selection of Landsat data with cloud coverage <30% was done to minimize the contamination of residue cloud and cloud shadow. Multiday mean of MISR albedo was used to reduce the effects of missing MISR data. A method of spatial moving window was used to remove the trends of each satellite albedo dataset. Landsat uses a window size of 15 pixels, whereas MODIS and MISR use 3 pixels. The time series of the three albedo data before and after applying MRT was displayed in Fig. 21.12. MISR and Landsat albedo maps contain substantial data gaps. The gaps in MISR are mainly results of the strict criteria in the MISR retrieval algorithm, while those in the Landsat ETM+ data are caused by the instrumental issue of scan line corrector. The maps demonstrated

MRT could fill data gaps with the information from the other scales and generate consistent maps across various scales. The validation suggested the MRT-based data integration process also improved the albedo accuracy and reduced relative RMSE almost to half its original value (He et al., 2014).

21.4 Empirical orthogonal function–based methods

EOF analysis is one of the most extensively used methods in geosciences (Hannachi et al., 2007; Preisendorfer, 1988). EOF analysis sometimes is also called principal components analysis. Singular spectrum analysis (SSA) also belongs to the EOF family but deals with the temporal correlations of short and noisy time series (Vautard and Ghil, 1989). Multichannel SSA (MSSA) may handle multivariate time series, and these different channels can be the same variable at different spatial locations, so

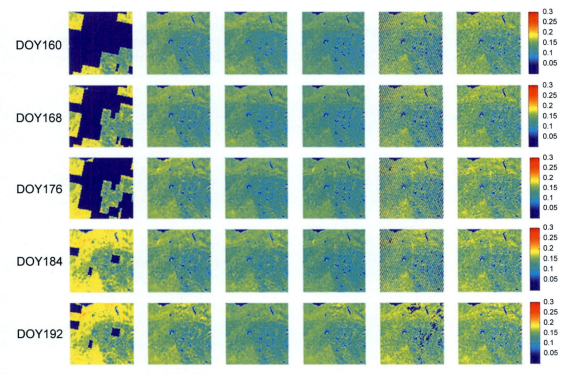

FIGURE 21.12 Time series of MISR albedo before and after MRT, MODIS albedo before and after MRT, and Landsat albedo before and after MRT. *From He, T., Liang, S., Wang, D., Shuai, Y., Yu, Y., 2014. Fusion of satellite land surface albedo products across scales using a multiresolution tree method in the North Central United States. IEEE Trans. Geosci. Remote Sens. 52, 3428–3439, ©2014, IEEE.*

MSSA has the ability of processing both temporal and spatial information. Similarly, an extended version of EOF deals with both spatial and temporal correlation by using a sliding window to incorporate the lagged information in the time domain (Weare and Nasstrom, 1982). In the atmospheric literature, this technique is called extended EOF (EEOF), which can be treated as the synonym of MSSA.

One major application of EOF methods is to analyze multivariate geophysical data. Recently, algorithms based on EOFs are also developed to solve the problem of missing value in geophysical datasets (e.g., Liu et al., 2005 Zhang et al., 2007). Among all these algorithms, the Data Interpolating Empirical Orthogonal Functions (DINEOF) developed by Beckers and Rixen (2003) is a representative one. Different from geostatistics methods, DINEOF needs less parameter input and is called "parameter free." A similar method is also interpreted in the perspective of SSA by Kondrashov and Ghil (2006). Alvera-Azcarate et al. (2005) applied this method to reconstruct sea surface temperature map from gappy satellite SST data in the area of the Adriatic Sea and showed DINEOF is able to generate realistic data even when missing data percentage is as high as 80%. In another paper, Alvera-Azcarate et al. (2007) tested this method to reconstruct multiple datasets and found that the incorporation of other data (e.g., chlorophyll) produced significantly better results than the univariate reconstruction of SST.

21.4.1 Introduction to Data Interpolating Empirical Orthogonal Functions method

First, we need organize the satellite data in the form a matrix. Say the number of pixels in one image is P and the counts of the images is N, we have a $P \times N$ matrix $\mathbf{L}$:

$$\mathbf{L} = \begin{pmatrix} l_{1,1} & l_{1,2} & \cdots & l_{1,N} \\ l_{2,2} & l_{2,2} & \cdots & l_{2,N} \\ \vdots & \vdots & \ddots & \vdots \\ l_{P,1} & l_{P,2} & \cdots & l_{P,N} \end{pmatrix} \tag{21.37}$$

where $l_{s,t}$ is the pixel at location s and time t. Then we will have the covariance matrix:

$$\mathbf{C_L} = \mathbf{LL}^T \tag{21.38}$$

Given this covariance matrix, we could obtain a group of orthogonal vectors $\mathbf{u}$ by solving the eigenvalue problem (Hannachi et al., 2007):

$$\mathbf{C_L u} = \lambda^2 \mathbf{u} \tag{21.39}$$

where λ_k^2 is the kth eigenvalue, which represents the variance the kth vector $\mathbf{u}_k$ could explain. The vector $\mathbf{u}_k$ is all called EOFs (spatial domain). For each $\mathbf{u}_k$, we could obtain a corresponding $\mathbf{v}_k$:

$$\mathbf{v}_k = \mathbf{L}^T \mathbf{u}_k \tag{21.40}$$

$\mathbf{v}_k$ is usually called PC (temporal domain). The original data matrix $\mathbf{L}$ could be exactly reconstructed by using all the EOFs and PCs:

$$\mathbf{L} = \mathbf{u}\lambda^2 \mathbf{v}^T \tag{21.41}$$

However, we are usually more interested in the first leading EOFs and PCs, as combination of them will explain most of the variance of data. The last EOFs and PCs are believed to contain lots of noises and other unwanted variations. The basic idea of data reconstruction using EOFs is straightforward: just using the leading components.

Besides the approach of solving the eigenvalue problem of covariance $\mathbf{C_L}$, the same results could also be achieved by singular value decomposition (SVD) of $\mathbf{L}$ to obtain directly u and v. This is where the name of a similar method SSA comes from. Different from EOF analysis, SSA is working on a time series of a single variable. Say we choose only one pixel from the time series of images $l_{0,t}$, $t \in [1, N]$:

$$\begin{pmatrix} l_{0,1} & l_{0,2} & \cdots & l_{0,N} \end{pmatrix} \tag{21.42}$$

We will build a matrix $\mathbf{L}'$ from this vector for SSA by sliding a window of the length W:

$$\mathbf{l}' = \begin{pmatrix} l_{0,1} & l_{0,2} & \cdots & l_{0,N-W+1} \\ l_{0,2} & l_{0,3} & \cdots & l_{0,N-W+2} \\ \vdots & \vdots & \ddots & \vdots \\ l_{0,W} & l_{0,W+1} & \cdots & l_{0,N} \end{pmatrix} \tag{21.43}$$

Then we could apply SVD on this matrix $\mathbf{l}'$ and obtain its EOFs and PCs. We could reconstruct original univariate time series by manipulating the EOFs and PCs of $\mathbf{l}'$. This is the idea of SSA (Ghil et al., 2002). We could see that EOF analysis and SSA are two closely related but different methods. The two methods can be linked by extending SSA to a multivariate case called MSSA and expanding EOF to EEOF analysis. Thus, both methods work on the same matrix with dimension of $WP \times (N - W + 1)$:

$$\mathbf{L}' = \begin{pmatrix} l_{1,1} & l_{2,1} & \cdots & l_{W,1} & l_{1,2} & \cdots & l_{W,P} \\ l_{2,1} & l_{3,1} & \cdots & l_{W+1,1} & l_{2,2} & \cdots & l_{W+1,P} \\ \vdots & \vdots & \ddots & \vdots & \vdots & \ddots & \vdots \\ l_{N-W+1,1} & l_{N-W+2,1} & \cdots & l_{N,1} & l_{N-W+1,2} & \cdots & l_{N,P} \end{pmatrix} \tag{21.44}$$

We have introduced the basic schema of EOF methods. This simplified procedure cannot be directly used to reconstruct gappy data. Sophisticated algorithms are needed to estimate missing values from existing values. Most of these algorithms use iterative loops to fill missing values gradually until they converge. Different algorithms have various details about how to transfer information from existing points to data gaps (Beckers and Rixen, 2003; Kondrashov and Ghil, 2006; Schoellhamer, 2001; Zhang et al., 2007). Here, we briefly review the steps used in DINEOF (Beckers and Rixen, 2003) (Alvera-Azcarate et al., 2005):

1. Calculate the mean of the data matrix and fill the missing values with mean. Then remove the mean from the whole data and obtain "demeaned" residues.
2. Carry out a double-loop iteration on the "demeaned" residues.
   a. Start the data filling with the inner loop: carry out EOF analysis on the mean-filled matrix and use the first k EOFs and PCs to reconstruct the matrix. Use the reconstructed values to replace the missing values to form a new matrix.
   b. Repeat the inner loop (procedure 2.a) on the new matrix until the reconstructed values converge.
   c. The convergence of inner loop invokes the outer loop, where the number of EOFs and PCs used in reconstruction increase from 1 to a predefined value.
3. The outer loop stops when k reaches the predefined value. Then add back the mean and obtain the final reconstructed matrix.

21.4.2 Application of DINEOF in product integration

DINEOF is originally developed for reconstructing one product. Although Alvera-Azcarate et al. (2007) used it in a multivariate case and reconstructing products of multiple variables, the original DINEOF is still not ready for integration of land remote sensing products. The first problem is its heavy computational cost. Unlike oceanographic variables that DINEOF was designed for, land products usually have much higher spatial resolutions. To use the MODIS LAI product as an example, the magnitude for the data over North America alone will be 10^8. It is inefficient and impractical to manipulate such a huge matrix at one time. The second problem is that multiple products

of the same variable usually have different temporal resolution. Without temporal interpolation, it is hard to form all the products in one matrix. Wang and Liang (2011) proposed a hierarchical EOF (HEOF)—based method to solve these problems. EOF analysis is carried out on both coarse resolution aggregated data and fine resolution small subdatasets. Results from coarse resolution are used as prior knowledge in filtering the fine resolution data. To account for the incompatible temporal resolution of multiple products, this approach is first run as a univariate case on each product individually. The reconstructed products are interpolated to a common temporal resolution to match with each other. Then, EOF is run as a multivariate case on all interpolated products. The intermediate data are averaged to obtain the final integration results.

21.4.2.1 A case study on leaf area index

We also use LAI as example to show the ability of EOF methods in integrating land remote sensing products (Wang and Liang, 2011). Although one advantage of EOF-based methods over geostatistical methods is that the former need less parameters, we have assigned two parameters, including the number of leading components in reconstructing and the window length, before running the integration algorithm. EOF with varied parameters is used to reconstruct LAI and compare with "true" values. The relative errors of the reconstructed datasets are then calculated to determine the optimum parameters. The results of cross-validation are shown in Fig. 21.13. The incorporation of temporal information (where window length is larger than 1) could greatly reduce the estimation error of EOF analysis. For a fixed window length, using more EOF modes in reconstruction first reduces and then increases the estimation errors.

In their paper, two runs of HEOF analysis were used to handle the incompatibility of temporal resolutions between MODSI and CYCLOPES LAI data. First, EOF analysis is

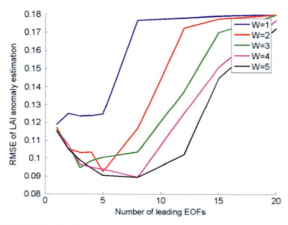

FIGURE 21.13   Relative errors of cross-validation using CYCLOPES leaf area index (LAI) anomaly, as the function of window length W and the number of leading components. *From Wang, D., Liang, S., 2011. Integrating MODIS and CYCLOPES leaf area index products using empirical orthogonal functions. IEEE Trans. Geosci. Remote Sens. 49, 1513–1519,* ©2011, IEEE

applied on MODIS LAI only. The gap-filled and error-reduced results from this step are interpolated to match with CYCLOPES data in terms of temporal resolution. To keep consistent with

Baret et al.'s (2007) smoothing strategy in producing CYCLOPES LAI data, Wang and Liang (Wang and Liang, 2011) use a Gaussian function with a standard deviation of 12.7 $f(t)$ to smooth the EOF-processed MODIS data $L(t_i)$:

$$\widehat{L}(t_C) = \sum_{i=1}^{N} L(t_i) \frac{f(t_i - t_C)}{\sum\limits_{j=1}^{N} f(t_j - t_C)} \qquad (21.45)$$

Then the multivariate EOF analysis is applied on the combined matrix of interpolated MODIS data and CYCLOPES data. Weighted average is employed to merge the output MODIS and CYCLOPES data and obtain the final results. MODIS and CYCLOPES LAI data of peak growing season 2001 around BARC are used to test this algorithm. The integrated images are free of gaps and smoother than the original LAI maps (Fig. 21.14).

High-resolution LAI reference maps are used to validate the algorithm. Validation shows that product integration improved the quality of original LAI products, especially when compared with MODIS data. In addition to the

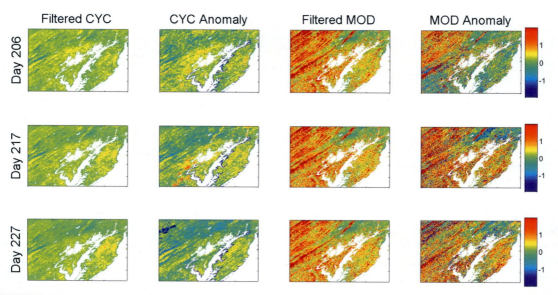

FIGURE 21.14   EOF results on three consecutive maps from July 25 to August 15, 2001, around BARC. Each row shows the data at 1 day. The four columns, respectively, are the filtered and original CYCLOPES anomaly and the filtered and original MODIS anomaly. *From Wang, D., Liang, S., 2011. Integrating MODIS and CYCLOPES leaf area index products using empirical orthogonal functions. IEEE Trans. Geosci. Remote Sens. 49, 1513–1519,* ©2011, IEEE

increase of R^2 and decrease of RMSE, the estimation bias is also significantly reduced to -0.08 from $+0.28$ (MODIS) and -0.20(CYCLOPES).

21.5 Summary

This chapter introduces three types of methods for integrating land products. With regard to gaps and large uncertainties of current land remote sensing products, data assimilation may be a choice to address these issues. Nevertheless, independent observational data without incorporating physical dynamic models are essential to drive and validate all kinds of physical models. However, integration of land remote sensing products is still a relatively new research topic. This chapter examined four methods based on OI, BME, MRT, and EOFs, respectively, to integrate multiple land products and improve their integrity and accuracy.

These methods have their own advantages and shortcomings. Geostatistical methods have solid mathematical foundations. However, one big shortcoming of this type of methods is their heavy computational cost. There are numerical efforts trying to improve their efficiency. One is to introduce predictive process model, which has a lower dimension (Banerjee et al., 2008). One is approximating covariance matrix, through sparse matrix technique (Barry and Pace, 1997), tapering (Furrer et al., 2006), spectral domain (Fuentes, 2007), or wavelet basis functions (Nychka et al., 2002). Another possible choice is using random effect model, expressing covariance matrix in the forms of limited basic function with much smaller dimension (Cressie and Johannesson, 2008). In this chapter, we introduced MRT with significantly improved efficiency. Nevertheless, the parameters in MRT are hard to estimate. Especially, the selection of state transfer matrix across resolutions is relatively subjective. Besides, MRT could not directly handle multiple products with the same resolution, although it has the advantage of integrating products with varied resolutions. The EOF-based method requires no prior knowledge on covariance function and measurement errors and is "parameter-free" and easy to implement (Beckers and Rixen, 2003). However, EOFs methods are more empirical than statistical, and the results usually lack statistical meanings.

This chapter is by no means a review of existing product integration methods. Instead, we aim to provide the readers basic ideas of some common methods. For instance, we cover little on the hierarchical Bayesian space—time models (Wikle et al., 1998). The types of models have been used to model various variables: ocean winds (Wikle et al., 2001), bird population (Wikle, 2003), air pollution (Cocchi et al., 2007), ozone (McMillan et al., 2005), surface temperature (Lu et al., 2018; Zhu et al., 2019), and so on. Interested readers could refer to specific literature for further details on new development of product integration methods.

References

Alvera-Azcarate, A., Barth, A., Beckers, J.M., Weisberg, R.H., 2007. Multivariate reconstruction of missing data in sea surface temperature, chlorophyll, and wind satellite fields. J. Geophys. Res. Oceans 112.

Alvera-Azcarate, A., Barth, A., Rixen, M., Beckers, J.M., 2005. Reconstruction of incomplete oceanographic data sets using empirical orthogonal functions: application to the Adriatic Sea surface temperature. Ocean Model. 9, 325—346.

Banerjee, S., Gelfand, A.E., Finley, A.O., Sang, H., 2008. Stationary process approximation for the analysis of large spatial datasets. J. R. Stat. Ser. B Stat. Methodol. 70, 825—848.

Baret, F., Hagolle, O., Geiger, B., Bicheron, P., Miras, B., Huc, M., Berthelot, B., Nino, F., Weiss, M., Samain, O., Roujean, J.L., Leroy, M., 2007. LAI, fAPAR and fCover CYCLOPES global products derived from VEGETATION — Part 1: principles of the algorithm. Remote Sens. Environ. 110, 275—286.

Barry, R.P., Pace, R.K., 1997. Kriging with large data sets using sparse matrix techniques. Commun. Stat. Simulat. Comput. 26, 619—629.

Beckers, J.M., Rixen, M., 2003. EOF calculations and data filling from incomplete oceanographic datasets. J. Atmos. Ocean. Technol. 20, 1839—1856.

Borak, J.S., Jasinski, M.F., 2009. Effective interpolation of incomplete satellite-derived leaf-area index time series for the continental United States. Agric. For. Meteorol. 149, 320–332.

Bretherton, F., Davis, R., Fandry, C., 1976. A technique for objective analysis and design of oceanographic experiments applied to MODE-73. Deep Sea Res. 23, 559–582.

Buermann, W., Wang, Y.J., Dong, J.R., Zhou, L.M., Zeng, X.B., Dickinson, R.E., Potter, C.S., Myneni, R.B., 2002. Analysis of a multiyear global vegetation leaf area index data set. J. Geophys. Res. Atmos. 107.

Chen, J., Jonsson, P., Tamura, M., Gu, Z.H., Matsushita, B., Eklundh, L., 2004. A simple method for reconstructing a high-quality NDVI time-series data set based on the Savitzky-Golay filter. Remote Sens. Environ. 91, 332–344.

Chen, J.M., Black, T.A., 1992. Defining Leaf area index for non-flat leaves. Plant Cell Environ. 15, 421–429.

Chen, J.M., Deng, F., Chen, M.Z., 2006. Locally adjusted cubic-spline capping for reconstructing seasonal trajectories of a satellite-derived surface parameter. IEEE Trans. Geosci. Remote Sens. 44, 2230–2238.

Chou, K., 1991. A stochastic modeling approach to multiscale signal processing. In: Dept. EECS: MIT.

Christakos, G., 1990. A Bayesian maximum entropy view to the spatial estimation problem. Math. Geol. 22, 763–777.

Christakos, G., 1992. Random Field Models in Earth Sciences. Academic Press.

Christakos, G., 2000. Modern Spatiotemporal Geostatistics. Oxford University Press, New York.

Christakos, G., Kolovos, A., Serre, M.L., Vukovich, F., 2004. Total ozone mapping by integrating databases from remote sensing instruments and empirical models. IEEE Trans. Geosci. Remote Sens. 42, 991–1008.

Cocchi, D., Greco, F., Trivisano, C., 2007. Hierarchical space-time modelling of PM10 pollution. Atmos. Environ. 41, 532–542.

Cressie, N., 1993. Statistics for Spatial Data. Wiley, New York.

Cressie, N., Huang, H.C., 1999. Classes of nonseparable, spatio-temporal stationary covariance functions. J. Am. Stat. Assoc. 94, 1330–1340.

Cressie, N., Johannesson, G., 2008. Fixed rank kriging for very large spatial data sets. J. R. Stat. Ser. Soc. B Stat. Methodol. 70, 209–226.

Cressman, G.P., 1959. An operational objective analysis system. Mon. Weather Rev. 87, 367–374.

De Nazelle, A., Arunachalam, S., Serre, M.L., 2010. Bayesian maximum entropy integration of ozone observations and model predictions: an application for attainment demonstration in North Carolina. Environ. Sci. Technol. 44, 5707–5713.

de Vyver, H.V., Roulin, E., 2009. Scale-recursive estimation for merging precipitation data from radar and microwave cross-track scanners. J. Geophys. Res. Atmos. 114.

Douaik, A., Van Meirvenne, M., Toth, T., 2005. Soil salinity mapping using spatio-temporal kriging and Bayesian maximum entropy with interval soft data. Geoderma 128, 234–248.

Falge, E., Baldocchi, D., Olson, R., Anthoni, P., Aubinet, M., Bernhofer, C., Burba, G., Ceulemans, R., Clement, R., Dolman, H., Granier, A., Gross, P., Grunwald, T., Hollinger, D., Jensen, N.O., Katul, G., Keronen, P., Kowalski, A., Lai, C.T., Law, B.E., Meyers, T., Moncrieff, H., Moors, E., Munger, J.W., Pilegaard, K., Rannik, U., Rebmann, C., Suyker, A., Tenhunen, J., Tu, K., Verma, S., Vesala, T., Wilson, K., Wofsy, S., 2001. Gap filling strategies for defensible annual sums of net ecosystem exchange. Agric. For. Meteorol. 107, 43–69.

Fang, H., Liang, S., Townshend, J.R., 2007. Spatially and temporally continuous LAI data sets based on an integrated filtering method: examples from North America. Remote Sens. Environ. 112, 75–93.

Fang, H.L., Liang, S.L., Townshend, J.R., Dickinson, R.E., 2008. Spatially and temporally continuous LAI data sets based on an integrated filtering method: examples from North America. Remote Sens. Environ. 112, 75–93.

Fieguth, P., Menemenlis, D., Ho, T., Willsky, A., Wunsch, C., 1998. Mapping Mediterranean altimeter data with a multiresolution optimal interpolation algorithm. J. Atmos. Ocean. Technol. 15, 535–546.

Fieguth, P.W., 1995. Application of multiscale estimation to large multidimensional imaging and remote sensing problems. In: Department of Electrical Engineering and Computer Science: Massachusetts Institute of Technology.

Fieguth, P.W., Karl, W.C., Willsky, A.S., Wunsch, C., 1995. Multiresolution optimal interpolation and statistical analysis of TOPEX/POSEIDON satellite altimetry. IEEE Trans. Geosci. Remote Sens. 33, 280–292.

Fieguth, P.W., Willsky, A.S., 1996. Fractal estimation using models on multiscale trees. IEEE Trans. Signal Process. 44, 1297–1300.

Fuentes, M., 2007. Approximate likelihood for large irregularly spaced spatial data. J. Am. Stat. Assoc. 102, 321–331.

Furrer, R., Genton, M.G., Nychka, D., 2006. Covariance tapering for interpolation of large spatial datasets. J. Comput. Graph. Stat. 15, 502–523.

Gandin, L.S., 1965. Objective Analysis of Meteorological Fields. Israel Program for Scientific Translations, Jerusalem.

Gao, F., Masek, J., Schwaller, M., Hall, F., 2006. On the blending of the Landsat and MODIS surface reflectance: predicting daily Landsat surface reflectance. IEEE Trans. Geosci. Remote Sens. 44, 2207–2218.

GCOS, 2006. Systematic observation requirements for satellite-based products for climate. In: Geneva, Switzerland.

Ghil, M., Allen, M.R., Dettinger, M.D., Ide, K., Kondrashov, D., Mann, M.E., Robertson, A.W., Saunders, A., Tian, Y., Varadi, F., Yiou, P., 2002. Advanced spectral methods for climatic time series. Rev. Geophys. 40.

Gneiting, T., 2002. Nonseparable, stationary covariance functions for space-time data. J. Am. Stat. Assoc. 97, 590—600.

Gregg, W.W., Conkright, M.E., 2001. Global seasonal climatologies of ocean chlorophyll: blending in situ and satellite data for the Coastal Zone Color Scanner era. J. Geophys. Res. Oceans 106, 2499—2515.

Gu, Y.X., Belair, S., Mahfouf, J.F., Deblonde, G., 2006. Optimal interpolation analysis of leaf area index using MODIS data. Remote Sens. Environ. 104, 283—296.

Hannachi, A., Jolliffe, I.T., Stephenson, D.B., 2007. Empirical orthogonal functions and related techniques in atmospheric science: a review. Int. J. Climatol. 27, 1119—1152.

He, T., Liang, S., Wang, D., Shuai, Y., Yu, Y., 2014. Fusion of satellite land surface albedo products across scales using a multiresolution tree method in the north Central United States. IEEE Trans. Geosci. Remote Sens. 52, 3428—3439.

Huang, H.C., Cressie, N., Gabrosek, J., 2002. Fast, resolution-consistent spatial prediction of global processes from satellite data. J. Comput. Graph. Stat. 11, 63—88.

Huang, H.C., Martinez, F., Mateu, J., Montes, F., 2007. Model comparison and selection for stationary space-time models. Comput. Stat. Data Anal. 51, 4577—4596.

Johannesson, G., Cressie, N., 2004. Finding large-scale spatial trends in massive, global, environmental datasets. Environmetrics 15, 1—44.

Kannan, A., Ostendorf, M., Karl, W.C., Castanon, D.A., Fish, R.K., 2000. ML parameter estimation of a multiscale stochastic process using the EM algorithm. IEEE Trans. Signal Process. 48, 1836—1840.

Kolovos, A., Christakos, G., Serre, M.L., Miller, C.T., 2002. Computational Bayesian maximum entropy solution of a stochastic advection-reaction equation in the light of site-specific information. In: Water Resources Research, p. 1318.

Kondrashov, D., Ghil, M., 2006. Spatio-temporal filling of missing points in geophysical data sets. Nonlinear Process Geophys. 13, 151—159.

Kwiatkowska, E.J., Fargion, G.S., 2003. Application of machine-learning techniques toward the creation of a consistent and calibrated global chlorophyll concentration baseline dataset using remotely sensed ocean color data. IEEE Trans. Geosci. Remote Sens. 41, 2844—2860.

Le Traon, P.Y., Nadal, F., Ducet, N., 1998. An improved mapping method of multisatellite altimeter data. J. Atmos. Ocean. Technol. 15, 522—534.

Liang, S., 2007. Recent developments in estimating land surface biogeophysical variables from optical remote sensing. Prog. Phys. Geogr. 31, 501—516.

Liang, S., Qin, J., 2008. Data assimilation methods for land surface variable estimation. In: Liang, S. (Ed.), Advances in Land Remote Sensing: System, Modeling, Inversion and Applications. Springer, pp. 319—339.

Liu, H.Q., Pinker, R.T., Holben, B.N., 2005. A global view of aerosols from merged transport models, satellite, and ground observations. J. Geophys. Res. Atmos. 110.

Lorenc, A.C., 1986. Analysis methods for numerical weather prediction. Q. J. R. Meteorol. Soc. 112, 1177—1194.

Lu, N., Liang, S.L., Huang, G.H., Qin, J., Yao, L., Wang, D.D., Yang, K., 2018. Hierarchical Bayesian space-time estimation of monthly maximum and minimum surface air temperature. Remote Sens. Environ. 211, 48—58.

Luettgen, M., 1993. Image processing with multiscale stochastic models. In: Department of Electrical Engineering and Computer Science. Massachusetts Institute of Technology.

Matheron, G., 1963. Principles of geostatistics. Econ. Geol. 58, 1246—1266.

McMillan, N., Bortnick, S.M., Irwin, M.E., Berliner, L.M., 2005. A hierarchical Bayesian model to estimate and forecast ozone through space and time. Atmos. Environ. 39, 1373—1382.

Menemenlis, D., Fieguth, P., Wunsch, C., Willsky, A., 1997. Adaptation of a fast optimal interpolation algorithm to the mapping of oceanographic data. J. Geophys. Res. Oceans 102, 10573—10584.

Moody, E.G., King, M.D., Platnick, S., Schaaf, C.B., Gao, F., 2005. Spatially complete global spectral surface albedos: value-added datasets derived from terra MODIS land products. IEEE Trans. Geosci. Remote Sens. 43, 144—158.

Myneni, R.B., Hoffman, S., Knyazikhin, Y., Privette, J.L., Glassy, J., Tian, Y., Wang, Y., Song, X., Zhang, Y., Smith, G.R., Lotsch, A., Friedl, M., Morisette, J.T., Votava, P., Nemani, R.R., Running, S.W., 2002. Global products of vegetation leaf area and fraction absorbed PAR from year one of MODIS data. Remote Sens. Environ. 83, 214—231.

Nychka, D., Wikle, C., Royle, J.A., 2002. Multiresolution models for nonstationary spatial covariance functions. Stat. Model. 2, 315—331.

Ooba, M., Hirano, T., Mogami, J.I., Hirata, R., Fujinuma, Y., 2006. Comparisons of gap-filling methods for carbon flux dataset: a combination of a genetic algorithm and an artificial neural network. Ecol. Model. 198, 473—486.

Parada, L.M., Liang, X., 2004. Optimal multiscale Kalman filter for assimilation of near-surface soil moisture into land surface models. J. Geophys. Res. Atmos. 109.

Piao, S.L., Fang, J.Y., Zhou, L.M., Ciais, P., Zhu, B., 2006. Variations in satellite-derived phenology in China's temperate vegetation. Glob. Chang. Biol. 12, 672—685.

Pottier, C., Garcon, V., Larnicol, G., Sudre, J., Schaeffer, P., Le Traon, P.Y., 2006. Merging SeaWiFS and MODIS/Aqua ocean color data in North and Equatorial Atlantic using weighted averaging and objective analysis. IEEE Trans. Geosci. Remote Sens. 44, 3436–3451.

Pottier, C., Turiel, A., Garcon, V., 2008. Inferring missing data in satellite chlorophyll maps using turbulent cascading. Remote Sens. Environ. 112, 4242–4260.

Preisendorfer, R., 1988. Principal Component Analysis in Meteorology and Oceanography. Elsevier.

Reynolds, R.W., Smith, T.M., 1994. Improved global sea surface temperature analyses using optimal interpolation. J. Clim. 7, 929–948.

Sakamoto, T., Yokozawa, M., Toritani, H., Shibayama, M., Ishitsuka, N., Ohno, H., 2005. A crop phenology detection method using time-series MODIS data. Remote Sens. Environ. 96, 366–374.

Sapiano, M.R.P., Smith, T.M., Arkin, P.A., 2008. A new merged analysis of precipitation utilizing satellite and reanalysis data. J. Geophys. Res. Atmos. 113.

Schoellhamer, D.H., 2001. Singular spectrum analysis for time series with missing data. Geophys. Res. Lett. 28, 3187–3190.

Sellers, P.J., Tucker, C.J., et al., 1994. A global 1° by 1° NDVI data set for climate studies. Part 2: the generation of global fields of terrestrial biophysical parameters from the NDVI. Int. J. Remote Sens. 15, 3519–3545.

Serre, M.L., Christakos, G., 1999. Modern geostatistics: computational BME analysis in the light of uncertain physical knowledge — the Equus Beds study. Stoch. Environ. Res. Risk Assess. 13, 1–26.

Slatton, K.C., Crawford, M.M., Evans, B.L., 2001. Fusing interferometric radar and laser altimeter data to estimate surface topography and vegetation heights. IEEE Trans. Geosci. Remote Sens. 39, 2470–2482.

Stein, M.L., 2005. Space-time covariance functions. J. Am. Stat. Assoc. 100, 310–321.

Tao, X., Liang, S., Wang, D., He, T., Huang, C., 2018. Improving satellite estimates of the fraction of absorbed photosynthetically active radiation through data integration: methodology and validation. IEEE Trans. Geosci. Remote Sens. 56, 2107–2118.

Tzeng, S.L., Huang, H.C., Cressie, N., 2005. A fast, optimal spatial-prediction method for massive datasets. J. Am. Stat. Assoc. 100, 1343–1357.

Uz, B.M., Yoder, J.A., 2004. High frequency and mesoscale variability in SeaWiFS chlorophyll imagery and its relation to other remotely sensed oceanographic variables. Deep Sea Res. Part II Top. Stud. Oceanogr. 51, 1001–1017.

Vautard, R., Ghil, M., 1989. Singular spectrum analysis in nonlinear dynamics, with applications to paleoclimatic time-series. Physica D 35, 395–424.

Wang, D., Liang, S., 2010. Using multiresolution tree to integrate MODIS and MISR-L3 LAI products. In: 2010 IGRASS, pp. 1027–1030 (Honululu, HI).

Wang, D., Liang, S., 2011. Integrating MODIS and CYCLOPES leaf area index products using empirical orthogonal functions. IEEE Trans. Geosci. Remote Sens. 49, 1513–1519.

Wang, D., Liang, S., 2014. Improving LAI mapping by integrating MODIS and CYCLOPES LAI products using optimal interpolation. IEEE J. Sel. Top. Appl. Earth Obs. Remote Sens. 7, 445–457.

Weare, B.C., Nasstrom, J.S., 1982. Examples of extended empirical orthogonal function analyses. Mon. Weather Rev. 110, 481–485.

Weiss, M., Baret, F., Garrigues, S., Lacaze, R., 2007. LAI and fAPAR CYCLOPES global products derived from VEGETATION. Part 2: validation and comparison with MODIS collection 4 products. Remote Sens. Environ. 110, 317–331.

Wikle, C.K., 2003. Hierarchical Bayesian models for predicting the spread of ecological processes. Ecology 84, 1382–1394.

Wikle, C.K., Berliner, L.M., 2007. A Bayesian tutorial for data assimilation. Phys. D Nonlinear Phenom. 230, 1–16.

Wikle, C.K., Berliner, L.M., Cressie, N., 1998. Hierarchical Bayesian space-time models. Environ. Ecol. Stat. 5, 117–154.

Wikle, C.K., Milliff, R.F., Nychka, D., Berliner, L.M., 2001. Spatiotemporal hierarchical Bayesian modeling: tropical ocean surface winds. J. Am. Stat. Assoc. 96, 382–397.

Yang, W.Z., Tan, B., Huang, D., Rautiainen, M., Shabanov, N.V., Wang, Y., Privette, J.L., Huemmrich, K.F., Fensholt, R., Sandholt, I., Weiss, M., Ahl, D.E., Gower, S.T., Nemani, R.R., Knyazikhin, Y., Myneni, R.B., 2006. MODIS leaf area index products: from validation to algorithm improvement. IEEE Trans. Geosci. Remote Sens. 44, 1885–1898.

Zhang, B.L., Pinker, R.T., Stackhouse, P.W., 2007. An empirical orthogonal function iteration approach for obtaining homogeneous radiative fluxes from satellite observations. J. Appl. Meteorol. Climatol. 46, 435–444.

Zhang, X.Y., Friedl, M.A., Schaaf, C.B., 2006. Global vegetation phenology from Moderate Resolution Imaging Spectroradiometer (MODIS): evaluation of global patterns and comparison with in situ measurements. J. Geophys. Res. Biogeosci. f.

Zhang, X.Y., Friedl, M.A., Schaaf, C.B., Strahler, A.H., Hodges, J.C.F., Gao, F., Reed, B.C., Huete, A., 2003. Monitoring vegetation phenology using MODIS. Remote Sens. Environ. 84, 471–475.

Zhu, Y.X., Kang, E.L., Bo, Y.C., Zhang, J.Z., Wang, Y.X., Tang, Q.X., 2019. Hierarchical Bayesian model based on robust fixed rank filter for fusing MODIS SST and AMSR-E SST. Photogramm. Eng. Remote Sens. 85, 119–131.

Data production and management system

Advanced Remote Sensing, Second Edition
https://doi.org/10.1016/B978-0-12-815826-5.00022-2

© 2020 Elsevier Inc. All rights reserved.

Abstract

This chapter describes the production and management of high-level remote sensing data products. The data production system provides task management, algorithm execution, data quality inspection, and system monitoring functions. An overview of the ground segment system is presented. For a very long time, Earth observation agencies have played a major role in remote sensing data management. The Google Earth Engine system consists of four main parts—a data management system, computing engine system, programming interface, and user interface system—which dramatically accelerates the research efficiency for remote sensing scientists.

This chapter describes the production and management of high-level remote sensing data products. The data production system provides task management, algorithm execution, data quality inspection, and system monitoring functions. The data management system is responsible for data archiving, data storage, and data distribution.

The first section presents an overview of the ground segment system, followed by an introduction to data production and management system, with Global LAnd Surface Satellite (GLASS) and Center for Resources Satellite Data and Applications (CRESDA) systems as two exemplars. As the very recent progress in this field, cloud computing—enabled online analytics is illustrated in Section 22.3.

22.1 Remote sensing ground system

The remote sensing satellite ground system receives and processes data obtained from various aerospace observations, such as satellites. Among many Earth observation data centers and providers (Lu et al., 2011), two typical remote sensing ground systems are described below: Earth Observation System (EOS) built by National Aeronautics and Space Administration (NASA) and European remote sensing (ERS) satellite system by European Space Agency (ESA).

22.1.1 NASA's Earth Observation System Data and Information System

From the perspective of the history of global remote sensing development, NASA first began the construction of a global-scale EOS in 1991. This EOS system includes satellite manufacturing, payload design, satellite control, data product production, and data sharing. The EOS has been a world-leading pioneering system for more than several decades.

NASA's Earth Observing System Data and Information System (Ramapriyan et al., 2010) is shown in Fig. 22.1. It manages NASA's geoscience data from a variety of sources such as satellites, aircraft, and field measurements. Overall, the system is divided into two parts: the Mission Operations and the Science Operations.

The Mission Operations mainly includes components such as the observation satellites, relay satellite, satellite ground control station, satellite broadcasting ground receiving station, polar ground receiving station, ground pretreatment system, and ground control system. It provides command and control of satellite operations, scheduling, data reception, and initial (level 0) processing of data processing capabilities.

The Science Operations focuses on building the following capabilities: generating high-level (Level 1—4) scientific data products for EOS tasks, archiving, and distributing data products from EOS and other satellite missions, as well as aircraft and on-site measurement activities. The production of scientific data products is carried out in a distributed system of many interconnected nodes, including the Scientific Investigator-led Processing System, abbreviated as SIPS, and the Distributed Active Archive Center (DAAC). The latter focuses on the production, archiving, and distribution of Earth science data products. DAAC serves a large and diverse user community by providing the ability to search and access scientific data products and professional services. It includes the data processing system of the load development team and the scientific analysis team, different

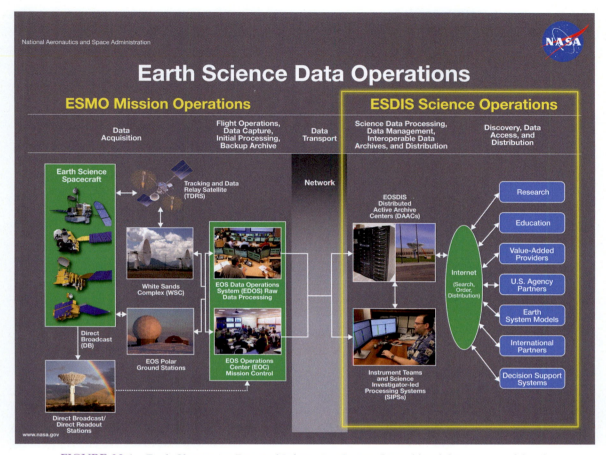

FIGURE 22.1 Earth Observation Data and Information System (https://earthdata.nasa.gov/about).

scientific data center systems, and data sharing distribution systems. The connection between the observing task and the scientific application is done through a dedicated data and service network.

The main facilities of NASA's DAAC are shown in Fig. 22.2. These agencies are responsible for retaining data from NASA's Earth observation missions, including various satellite data and scientific surveys and field collections, as well as data processing, archiving, recording, and distribution. The difference between 12 centers is that the scientific fields and processed data are different from each other. For example, the Alaska Satellite Facility DAAC is responsible for the acquisition, processing, archiving, and distribution of synthetic aperture.

These centers work closely together to follow the same standards in data product organization, data product description, metadata coding, and multidata product retrieval. Together they implement reliable and powerful data services, including assistance in selecting and acquiring data, accessing data processing and visualization tools, and technical support.

22.1.2 European remote sensing satellite ground system

ESA's two ERS satellites, ERS-1 and ERS-2, were launched into the same orbit in 1991 and 1995, respectively. They were the most sophisticated Earth observation satellite ever developed in Europe. Their payloads include SAR, radar

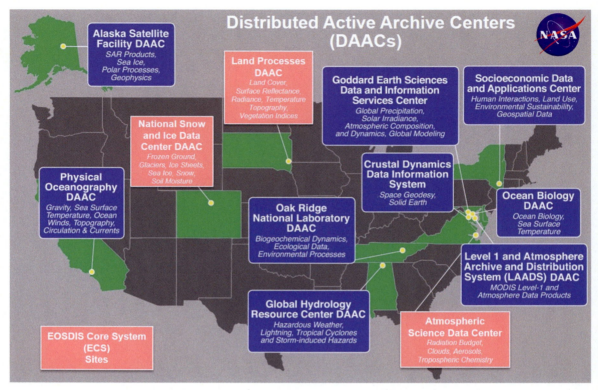

FIGURE 22.2   Physical location of NASA's DAACs (https://earthdata.nasa.gov/about).

altimeters, and instruments that measure ocean surface temperatures and wind fields. The ground system structure diagram of the ERS satellite is shown in Fig. 22.3.

The diagram on the right side of the receiving antenna is the receiving system for receiving ERS satellites. The locations of the receiving stations are Fucino, Italy; Maspalomas, Spain; Gatineau, western Quebec, Canada; Prince Albert, Saskatchewan, Canada; and Kiruna, Sweden. These receiving stations can receive data in real time or obtain recent historical data from the onboard tape drive. Within 3 h of data reception, standard data preprocessing can be completed, and the preprocessed data can be transmitted to the main data processing node and the main user unit through the data transmission network.

The top four Processing and Archiving Facilities (PAFs) in the figure represent four data processing and archiving centers. Similar to NASA's DAACs, these PAFs are clearly defined and take on the task of processing and archiving specific data products.

As shown in Fig. 22.4, the D-PAF Center is primarily responsible for the precision and geocoded data processing of SAR and its higher level of height measurement products and precise orbit calculations. The F-PAF Center is responsible for wind scatterers and radar altimeters for marine and related products. I-PAF processes SAR and L-band radiometer data for the Mediterranean region. UK-PAF handles ice and land products.

22.2 Data production system

The production system usually consists of data processing subsystem, data archive and

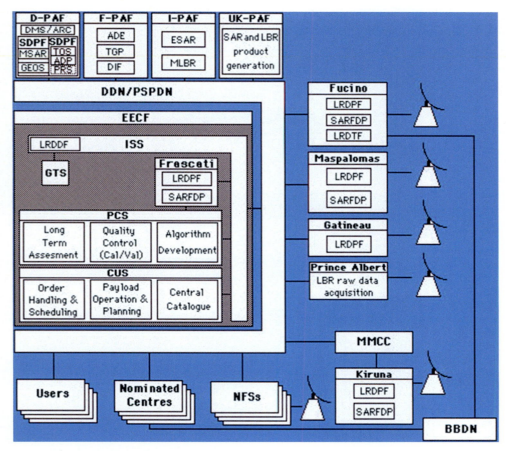

FIGURE 22.3 Ground segment of European remote sensing satellites (https://earth.esa.int/web/guest/missions/esa-operational-eo-missions/ers/ground-segment).

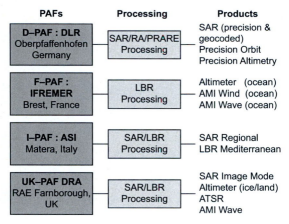

FIGURE 22.4 Processing units in the ground segment of European remote sensing (https://earth.esa.int/web/guest/missions/esa-operational-eo-missions/ers/ground-segment).

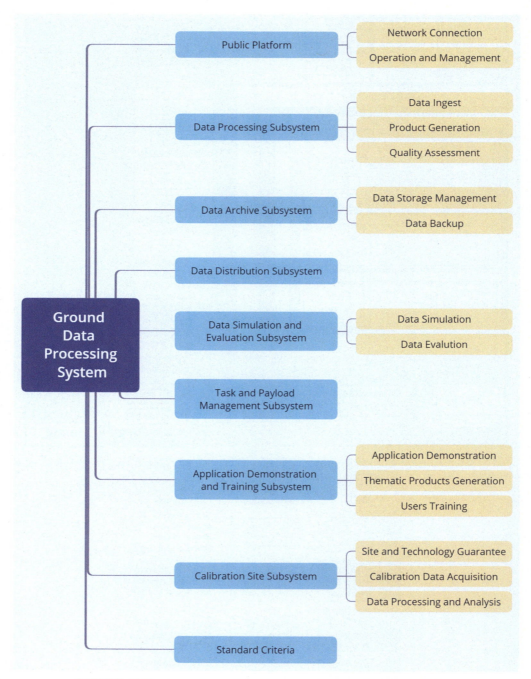

FIGURE 22.5   Composition of the data production system built in CRESDA.

information management subsystem, task and payload management subsystem, data distribution subsystem, data simulation and evaluation subsystem, application demonstration and training subsystem, and calibration site subsystem. One public platform may be utilized to provide network connection and system operation support. Fig. 22.5 shows the composition of the data production system built in the China CRESDA.

Another exemplar is the GLASS product production system. China launched the 863 key projects entitled "Generation and application of global products of essential land variables." The central component of this project is the development of GLASS product production system to produce long-term land observation products, for example, LAI, shortwave broadband albedo, longwave broadband emissivity,

incident short radiation, and photosynthetically active radiation.

Fig. 22.6 illustrates the network structure of GLASS data production system.

Fig. 22.7 shows a schematic diagram of the hardware configuration of GLASS data production system. It includes a data processing server and a production management server. The production management server primarily comprises a task management server, a resource monitoring server, a production scheduling server, and a quality inspection server. The data management system hardware mainly comprises a database management server, a mass data storage system, and a product distribution server. These servers are interconnected via a 10-gigabit switch or 1-gigabit switch.

The detailed hardware configuration of GLASS production system in China is listed in Table 23.1.

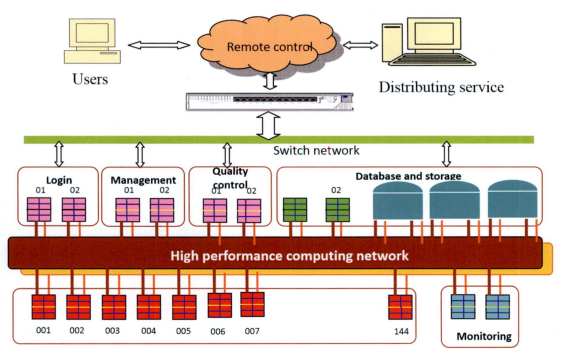

FIGURE 22.6 Network structure of GLASS data production system. *Excerpted from Zhao, X., Liang, S., Liu, S., Yuan, W., Xiao, Z., Liu, Q., Cheng, J., Zhang, X., Tang, H., Zhang, X., 2013. The Global Land Surface Satellite (GLASS) remote sensing data processing system and products. Remote Sens. 5 (5), 2436—2450, https://doi.org/10.3390/rs5052436.*

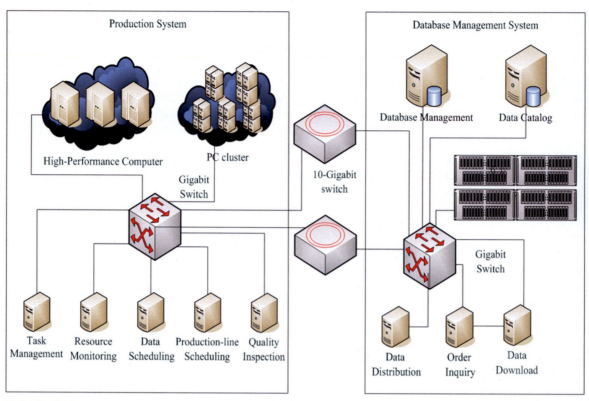

FIGURE 22.7 Schematic diagram of the production and data management system network. *Excerpted from Liang, S., Zhao, X., Liu, S., Yuan, W., Cheng, X., Xiao, Z., Zhang, X., Liu, Q., Cheng, J., Tang, H., Qu, Y., Bo, Y., Qu, Y., Ren, H., Yu, K., Townshend, J., 2013. A long-term Global LAnd Surface Satellite (GLASS) data-set for environmental studies. Int. J. Digit. Earth 6, 5−33.*

TABLE 23.1 Hardware configuration of Global LAnd Surface Satellite (GLASS) data production and management system.

Number	Name	Major equipment parameters
1	Computation server	144 computation nodes with a double quad-core CPU for each node; 1-gigabit/10-gigabit Ethernet switch; InfiniBand switch
2	Storage system	SAS storage: capacity 200 TB; 15K rpm SAS hard disk; IOPS no less than 650,000; and 2 RAID cards for each IO node, totaling 24
		SATA storage: capacity 500TB; SATA 7200 rpm hard drive; IOPS no less than 650,000; 2 RAID cards for each IO node, totaling 24
		Mobile disk: capacity 500TB; SATA 7200 rpm hard drive, mainly used for data download and backup
3	Database management server	HP ProLiant ML150 server, 2; quad-core Intel Xeon E5506, 2; 4GB DDR3 RAM; 2 * 500GB hot swap; SATA hard drive; gigabit server adaptor, 2; Smart Array P410RAID controller; 460W power supply
4	Distribution server	Website/mail server, 2; database server, 2; FTP server, 2; map distribution server, 1; user service disk array, 1; firewall, 1
5	Switch	Activated 48-port InfiniBand switch, 2; totally activated 48-port gigabit Ethernet switch

22.2.1 Production task management

Task management refers to task list—based realization and management of the production of various products. A task list refers to the list of production steps formulated according to specific needs, which mainly involve such requirements as production type, data used, and production time and algorithm version. Because of the large number of products and the limited computation nodes, unified management and distribution of various tasks is required. During the operation of the production system, monitoring and management of the computation nodes and computation tasks are required, including task list formulation, task list checking, task list execution, task list resetting, task list cancellation, task list priority setting, display of task list execution status, and display of computational resource status. Fig. 22.8 shows a flowchart of the execution of production task management, with its major functions described below.

22.2.1.1 Task list formulation

The task list for data preprocessing and production must first be formulated. The task list

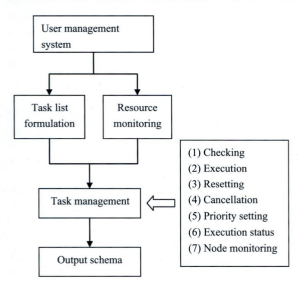

FIGURE 22.8 Task management flowchart.

generation interface is filled out according to the user's needs. Its contents involve various aspects of the product as specified by the user, including time range, spatial range, parameter type, production mode, and auxiliary parameter selection. The task list submitted by the user is stored in the system, and a task list document is then generated.

22.2.1.2 Task list checking

The validity and feasibility of the execution of the task list should be checked. The status of the task list (i.e., whether it is deficient in data or executable) is set according to whether the required data are available. The system, based on the data required by the task list, conducts data preparation and inspection for the working area. If the conditions are not met, then the system will send instructions for data repreparation; if the data for the working area are ready, the system will assess the computation quantity, establish the dispatch list, and submit the task list for execution.

22.2.1.3 Task list execution

The executable task list is dispatched to a computation node for the execution of a specific production task.

22.2.1.4 Task list resetting

Task lists with a data deficiency or production failure can be submitted for rechecking after the necessary production conditions are met. The task list is ready for execution after task list rechecking.

22.2.1.5 Task list cancellation

The execution queue status of task lists can be canceled, causing these task lists to receive awaiting-execution status. Unless the task execution is initiated, these task lists will not be executed.

22.2.1.6 Task list priority setting

Priority can be set for various production task lists. Tasks with higher priorities are executed first.

22.2.1.7 Display of task list execution status

The possible task list execution statuses include new task list, checking passed, failure to pass checking, cancellation, awaiting execution, production in progress, and operation error.

22.2.1.8 Display of computational resource status

The usage status of the CPU at each computation node can be displayed.

22.2.2 High-performance computing

In this chapter, high-performance computing (HPC) refers to either a computing system with multiple processors or a computing environment with several computers in a cluster to support the computation of a massive data volume using a wide range of various retrieval algorithms to generate the advanced products, such as the LAI, albedo, reflectance, and radiation products.

Many types of HPC systems exist, and these differ in scope, ranging from large-scale clusters of standard computers to highly specialized hardware. Many cluster-based HPC systems adopt high-performance network interconnections, such as InfiniBand and/or Myrinet network interconnections. For basic network topology and structure, a simple bus topology can be used. In a high-performance environment, a mesh network system allows shorter latency between host computers, thus improving the overall network performance and transmission rate.

Because of their superiority in improving computation efficiency and increasing computation scale, the parallel computing techniques associated with HPC are adopted for some

retrieval algorithms when more computing time is needed.

One of the most popular high-performance parallel computing schemes is the message passing interface (MPI); this interface exchanges work requests among the nodes via message passing and thus provides a simple method for work creation. A number of processors (herein referring to individual nodes instead of individual CPUs) may be required during the development process. Labor division in an HPC environment depends on both the application and scale of the environment. If the ongoing work assignment relies on multiple steps and computations, then the parallel and sequential features of the HPC environment will play an important role in the speed and flexibility of the grid. After the assigned work is completed, a message is sent to every node to instruct them to execute the assigned work. Related work is put into an HPC unit and is then sent to each node, with the expectation that the nodes will return their results simultaneously. The results delivered from each node are returned to the application program of the host computer via another message provided by the MPI. All of the messages are then received by the application program, thus completing the work (Pakin et al., 2009).

Of all the retrieval algorithms for global land surface products, the GLASS LAI retrieval algorithm uses the year as the minimum time range. Its minimum task unit contains the data from an entire year, thus leading to a heavy computational burden. Therefore, parallel decomposition and execution methods are more applicable for this algorithm. Based on the features of the LAI retrieval algorithm, a high-efficiency parallel algorithm is established on the hardware platform of the existing high-performance parallel computer system with software support. In this algorithm, the LAI program is ported into the parallel environment, and the computational efficiency is increased. As shown in Fig. 22.9, the program is based on the MPI standard and

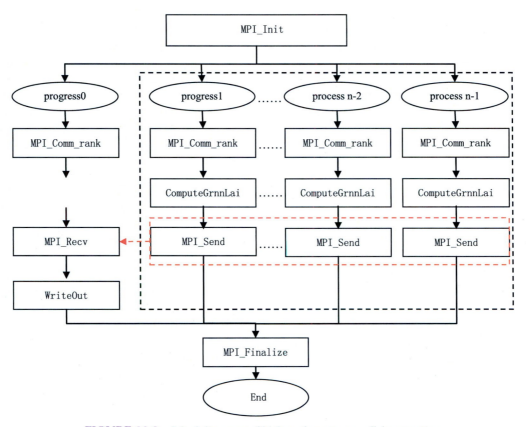

FIGURE 22.9   Scheduling map of high-performance parallel computing.

exploits large-scale, multiple-node parallel computing. Parallel computational tasks are segmented based on a coarse resolution. Process0 is responsible for distributing, receiving, and summarizing data, while the other processes are concerned with computation. All of these processes are synchronized. As the number of processes increases, the run time of the program decreases considerably. When 16 processes are responsible for a parallel computing task, the computation time can be reduced by over 90%. No strict linear relationship exists between the computational efficiency and the number of parallel nodes. This phenomenon is associated with the cost of memory allocation, interprocess communication, and internode communication.

22.2.3 Data quality inspection

Quality inspection refers to the quality inspection of various high-level products and the generation of quality control information or documents for the reference of users at various levels. There are many data quality inspection methods, including automatic inspection and visual inspection (Hubert, 1961).

Automatic inspection of product quality is automatically achieved using a quality inspection program that generates pixel-by-pixel data quality designations. Automatic inspection methods include the following: (1) time continuity test, which is based on long time series profiles for each land surface type at a typical data collection station, and products' capacities for characterizing seasonal and interannual variations are tested and (2) spatial consistency test, which checks that the regional land surface status remains stable during a given period and that the biophysical and geophysical parameters characterizing the land surface status show a certain spatial pattern. Regarding the product dataset, a priori knowledge can first be used to test the spatial distributions of various seasons and time images.

In addition to automatic inspection, visual inspection is needed for each data product tile, which requires an independent quality inspector. With the use of remote sensing data visualization software, the quality inspector can open each data product tile and browse data products through magnification and histogram stretching. With the aid of statistical tools or statistical information on the products, the data product quality is inspected by checking whether the image features of the data products are consistent with the background field. Therefore, various types of quality defects that might influence the usage of the data products can be eliminated.

Efficient quality inspection can be achieved by constructing a product quality verification platform and identifying an association among production, quality analysis, and validation. This platform provides a communication channel with which the user can oversee production, quality analysis, and validation tests, thereby enabling the association of various information. Fig. 22.10 is the flowchart of product quality inspection. The platform primarily involves the following three aspects.

22.2.3.1 Construction of quality inspection database

Global land surface characteristic parameter products are provided by the production model; additionally, the quality inspection database includes predefined control documents used for quality analysis, ground synchronous measurement data used for authenticity inspection, and a priori knowledge corresponding to parameters and historical data.

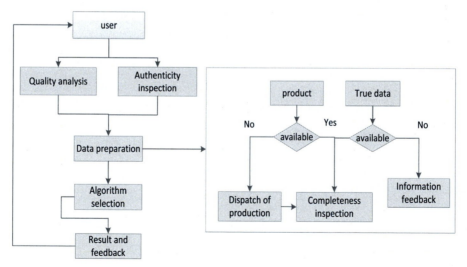

FIGURE 22.10 Quality inspection flowchart.

22.2.3.2 *Algorithm module integration*

The quality inspection platform needs to provide an algorithm interface between the quality analysis module and the authenticity inspection module. Integrating the various algorithm modules used for product quality control and the scientific design of the inspection platform module interface as well as the web interface–based integration of the database interface, quality analysis interface, and authenticity inspection interface will facilitate related operation and usage processes for users and administrators. After the administrator enters the quality inspection platform and triggers the relevant interfaces, this platform will perform quality analysis or tests of the remote sensing data products. The administrator will subsequently receive the report of the inspection results via the feedback mechanism.

22.2.3.3 *Construction of user feedback mechanism*

A corresponding representation method should be selected for each different inspection result. Through the formulation of a unified feedback standard, the collation and reprocessing of the inspection results obtained from each algorithm module can be realized. Feedback regarding inspection results can be transmitted in a timely manner to a user or an administrator.

22.2.4 System monitoring

With a layered monitoring system structure, the system can simultaneously achieve the management of computational tasks running in a distributed environment, task list optimization, and automatic production. The system is divided into three parts: a web server module, a monitoring service registration module, and a computation node monitoring agency. Via the web server, users can check the execution status at each computational node and the task operating status at each computational node. If a computational node breaks down, an email will automatically be sent to the administrator so that the node failure can be solved immediately.

A distributed monitoring system is built on a cross-platform, extensible, and HPC system, which requires excellent capabilities for such specifications as CPU utilization rate, RAM, HDD utilization rate, I/O load, and network flow. This module can monitor a node's operation status, thereby providing considerable capacity for the reasonable adjustment and allocation of system resources as well as the improvement of the system's overall performance.

The requirements of the system monitoring module include the monitoring and management of computational nodes and computational tasks during system operation. Fig. 22.11 shows a flowchart of the monitoring system. Real-time

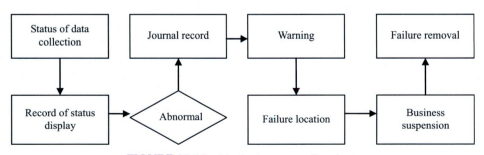

FIGURE 22.11   Monitoring system flowchart.

business monitoring is performed primarily based on the status of the computational nodes and the execution status of tasks. The execution status of the computational resources is displayed on a real-time basis, and a warning is dispatched when an abnormal situation occurs. Administrators will respond to emergencies on receiving this warning.

22.2.5 Data management

The purpose of remote sensing data management is to provide long-term data services and provide data retrieval and retrieval methods to facilitate researchers to obtain remote sensing data they need and to further analyze and realize data value. Remote sensing data product management refers to the management of remote sensing data obtained after the preprocessing, processing, production, and other steps described in other chapters of this book.

Remote sensing data products generally fall into two categories: datasets (or data collection) and data (or granule). Among them, the dataset refers to a set of data with the same characteristics produced according to a certain processing flow. For example, "MODIS/Terra Land Surface Temperature/Emissivity Daily L3 Global 1 km SIN Grid V006" is a dataset. It also has a short name, MOD11A1. This dataset is a gridded data product. This dataset provides ground temperature and emissivity over a range of 1200 km × 1200 km per 1 km × 1 km (referred to as "spatial resolution") over a certain day (referred to as "time resolution").

The following is an example of two data (granule) under the dataset:

MOD11A1.A2019131.h24v06.006. 2019132083759.hdf
MOD11A1.A2019131.h25v06. 006.2019132083805.hdf

These two data, as can be seen from the file name, belong to MOD11A1 dataset, and the rules for file names are consistent. In fact, the 2019131 part of the two file names indicates that they are all data on the 131st day of 2019 (corresponding to the previous day mentioned, that is, each data under this dataset correspond to a specific day). H24v06 and h25v06 correspond to two fixed grid areas on the surface of the earth, as shown in the two highlighted polygon boxes in Fig. 22.12.

The number of data contained in each dataset varies greatly. For example, the "Landsat 7 Enhanced Thematic Mapper Plus (ETM+) Collection 1 V1" dataset contains images from every scene from April 15, 1999, to the present, with a total of 2,696,876 images. Each of the scene data covers a range of 185 km × 170 km on the ground, corresponding to an observation period of about 20 min. The "Global Annual PM2.5 Grids from MODIS, MISR, and SeaWiFS Aerosol Optical Depth with GWR, 1998—2016" dataset is a global product that covers data for each of the 19 years from 1998 to 2016.

22.2.6 Product data management

Dataset information is the key to realizing the management of remote sensing data products. Because each dataset represents a set of data that have been processed through the same process flow, the difference between datasets can be large, for instance, some data products refer to the original image and some are global air quality data. However, the difference between the data in the same dataset is often limited to observation time or spatial coverage.

Dataset management needs to first ensure the relative stability and consistency of the dataset name. On this basis, the information of the dataset can be centrally maintained and effectively managed in the form of a searchable directory. For example, NASA has established the Global Change Master Directory (GCMD) system to provide detailed information on data change systems related to global change research, avoiding the repetitive establishment of many isolated

FIGURE 22.12 Spatial coverage information for two granules in MOD11A1 collection (https://earthdata.nasa.gov/).

data directories by multiple organizations and strengthening global changes at the national level. NASA's remote sensing dataset information is maintained in a standard, uniform format and is regularly updated. Another example is the International Data Directory Network (IDN) of the Intergovernmental Committee on Earth Observations (CEOS). The members of the CEOS are the Earth observation departments of major aerospace providers, such as NASA, NOAA, and the USGS in the United States, CNES in France, JAXR in Japan, the National Meteorological Administration, and the National Remote Sensing Center in China. The IDN provides registration, maintenance, and query capabilities for standard dataset information for national Earth observation departments. Among them, the standard described by the IDN

dataset information is the key factor to accomplish information sharing of remote sensing Earth observation dataset at the international level.

In addition, the use of digital object unique identifiers (DOIs) for each dataset has become an important trend. For example, Global Annual PM2.5 Grids from MODIS, MISR, and SeaWiFS Aerosol Optical Depth with GWR, 1998—2016 dataset, DOI logo is "10.7927/H4ZK5DQS." This identifier, along with the URL address of the DOI service, can be concatenated into a unique URN, i.e., https://dx.doi.org/10.7927/H4ZK5DQS, which becomes the unique identifier for the dataset. It is worth noting that this URN is also a valid URL. If you open it in your browser, you can point to the home page of the dataset on the NASA Social Data and Application Center website.

The storage of data (or granule) is often managed according to a preorganized directory structure, and the data of the same product dataset are put together. The naming of data generally adopts certain agreed rules to achieve consistency between different products. For example, for the piece of data of the MOD11A1 dataset introduced earlier, MOD11-A1.A2019131.h24v06.006.2019132083759.hdf, the MOD11A1 in the file name indicates the short name of the dataset. A means AM, which means the data come from Terra. 2019131, represents data acquisition time, which is the 131st day in the year of 2019. h24v06 stands for the position of the ground grid proposed by MODIS. 006 represents the version number of the dataset. 2019132083759 is for the production time of the data expressed in the YYYY-DDD-HH-MM-SS format.

22.2.7 Product metadata management

In addition to the above dataset information and data file management, an important aspect of product management is metadata management. Metadata are data about data. Conceptually, it refers to a type of descriptive information used to describe important information such as the content, quality, presentation, spatial reference system, management method, and access location of the data itself. Metadata of data in different fields differ greatly in their content and expression.

For remote sensing scientific data, metadata can be either for a dataset/data collection or for a granule. At present, there is already a relatively mature standardization organization management method. Typical standards include the Directory Exchange Format (DIF) developed by NASA, the Geospatial Metadata Content Standard (CSDGM) developed by the American Geographic Data Council, and the ISO 19115, 19115-2, and 19139 geographic information developed by the ISO Geographic Information Standards Committee.

Among them, DIF is a metadata content standard for remote sensing datasets. It specifies the metadata information that needs to be provided to describe a dataset. The main ones are Directory Entry Identifier, Directory Entry Title, Start and Stop Dates, Sensor Name, Source Name, Investigator, Technical Contact, Author, Data Center (Name, Contact Person, and Dataset ID), Originating Center, Campaign or Project Name, Storage Medium, Parameter Measured, Discipline Keywords, Location Keywords, General Keywords, Coverage, Revision Date, Science Review Date, Future Review Date, Reference, Quality, and Summary. Each item in the DIF specification has a detailed and accurate description. The aforementioned GCMD system and the IDN of the CEOS all adopt the DIF standard to manage the information of remote sensing data products.

The CSDGM provides a set of specialized terminology definitions that define the content organization of metadata for geospatial data. Two versions were released before and after the CSDGM. The first version is for geospatial data, and the second version is further extended for remote sensing data. Currently, it contains nearly 500 items of metadata content such as the following types of information: identification, data quality, spatial data organization, spatial reference system, entity and attribute, distribution, metadata reference, reference, time period, and contact.

ISO 19115 and 19115-2 are two ISO geographic information metadata standards. They draw on the DIF and CSDGM to a large extent. Among them, 19115 is for geospatial data, while 19115-2 is for remote sensing data. 19139 is the coding standard for these two metadata content standards, which stipulates that the metadata contents of 19115 and 19115-2 are organized and managed in XML.

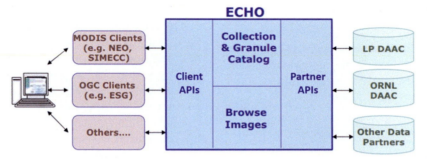

FIGURE 22.13   NASS's metadata clearinghouse (https://wiki.earthdata.nasa.gov/display/echo/Before+you+begin).

The organization and management of remote sensing product data and its metadata information can provide the function of searching and distributing remote sensing data products. Fig. 22.13 shows the system architecture of NASA's Earth Observation Metadata System, ECHO (Bai et al., 2007). In the figure, the right side shows the data centers responsible for data production, such as LP DAAC. All remote sensing product data are produced and managed in these data centers. In the process of producing data, the data center synchronously completes the production of the corresponding metadata information and sends the metadata information to the ECHO system through the Partner API interface of the ECHO system. The ECHO management maintains metadata information for all datasets (shown in the collection) and data (shown by Granule in the figure), as well as all preview images. ECHO further provides remote sensing data product retrieval services, which provide various spatial data services and systems displayed on the left side through the client API interface, and supports the retrieval function of remote sensing data products. Currently, the ECHO system has been upgraded to CMR. The remote sensing data retrieval system listed on the left side of the figure currently has updated versions such as Reverb and EarthData.

22.3 Cloud computing—based integration of data management and analytics

For a very long time, Earth observation agencies have played a major role in remote sensing data management. The main reason is that only these agencies are capable of receiving, processing, and management of remote sensing satellite data. However, just in recent years, commercial companies represented by Google have begun to enter the field of remote sensing data, value-added services. They build large-scale storage and computing environments, copy the current open-sharing remote sensing data products shared by Earth observation agencies, and purchase high-resolution commercial satellite data. On this basis, Google Inc., in conjunction with Carnegie Mellon University, NASA, USGS, and TIME, launched the Google Earth Engine system (Gorelick et al., 2017).

Google Earth Engine is a cloud computing platform designed to enable planetary-scale satellite image analytics. The platform leverages Google's computing infrastructure to optimize parallel processing of geospatial data. It provides web-based interactive data analytics environment for petabytes of satellite imagery. It further enables JavaScript and Python programming API interfaces.

22.3.1 Components of the Google Earth Engine system

The Google Earth Engine system consists of four main parts: a data management system, computing engine system, programming interface, and user interface system

22.3.1.1 Data management system

The Google Earth Engine System integrates nearly 40 years of Earth observation data, including complete Landsat 4, 5, 7, and 8 satellite imagery, MODIS global data, and other remote sensing and vector data products. All data are preprocessed, georegistered, and can be used directly. At the same time, users can also upload raster and vector data for free sharing to others or for their own use.

22.3.1.2 Calculation engine system

The computational framework of the Google Earth Engine provides a very efficient calculation method that provides great convenience for large-scale data processing. Analytical tasks are submitted by users worldwide in real time, automatically implementing parallel analysis in Google's high-performance data centers. After the calculation is complete, the results are typically stored in the cache so that requests for the same data analysis do not have to be recalculated.

22.3.1.3 Programming interface

Google Earth Engine provides a programming interface that supports complex geospatial data analysis, including overlay analysis, map calculation, matrix calculation, image processing, classification, change detection, time series analysis, raster vector conversion, image statistics, and more. Users can write complex scripts

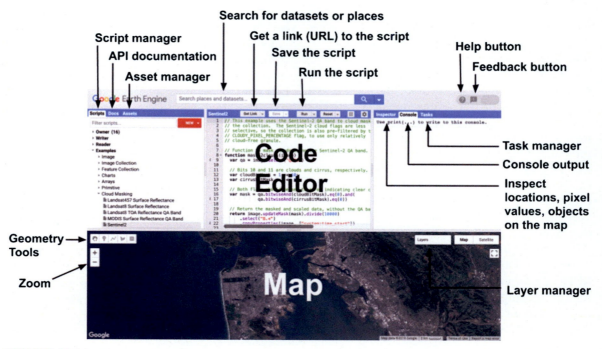

FIGURE 22.14   Diagram of components of the Earth Engine Interface (https://developers.google.com/earth-engine/playground).

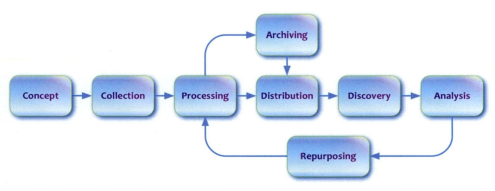

FIGURE 22.15   Data life cycle (http://www.ddialliance.org/training/why-use-ddi).

to perform complex geographic analysis and output the results in charts, maps, and pictures, which is very convenient for programmers to achieve data customization analysis.

22.3.1.4 User interface system

As shown in Fig. 22.14, the Google Earth Engine's Code Editor is a web-based integrated development environment for the JavaScript API that handles complex geospatial workflows simply and quickly. It has a user-friendly interface that allows users to edit code, run programs, output visualizations, search datasets, add geographic features, find help files, and more.

22.4 Summary

A typical remote sensing data process pipeline includes data reception and preprocessing, data processing and product production, data organization and management, data sharing and distribution, and analysis and utilization. As an important type of scientific data, the life cycle of remote sensing data can also be captured by Fig. 22.15.

Among these stages, concept, collection, processing, archiving, and distribution can be understood as observation design, data reception, data processing and product production, data storage management, and data sharing. For remote sensing satellite data, these responsibilities are usually fulfilled by various types of Earth observation agencies. Discovery, analysis, and repurposing can be understood as data retrieval, data analysis, and result integration. These responsibilities are generally done by researchers. Therefore, this general life cycle represents a "data provider—data demander" two-party interaction pattern, where the distribution stage is the key to link two parties together.

The new paradigm, represented by Google Earth Engine, provides an integration of remote sensing data management and analytics. Through Google Earth Engine's Code Editor, researchers do not have to download the massive remote sensing data, archive them, and finish the analytics work in their local computing environment. Instead, they just need to define the area of interest, select the data to be analyzed, and program the processing workflow. All the tasks will be automatically executed by the Google Earth Engine. The analytics results will be presented to researchers at the end. Therefore, it dramatically accelerates the research efficiency for remote sensing scientists.

References

Bai, Y., Di, L., Chen, A., Liu, Y., Wei, Y., 2007. Towards a geospatial catalogue federation service. Photogramm. Eng. Remote Sens. 73 (6), 699–708.

Gorelick, N., Hancher, M., Dixon, M., Ilyushchenko, S., Thau, D., Moore, R., 2017. Google earth engine: planetary-scale geospatial analysis for everyone. Remote Sens. Environ. 202, 18–27.

Hubert, L.F., 1961. TIROS I: Camera Attitude Data, Analysis of Location Errors, and Derivation of Correction for Calibration. U.S. Dept. of Commerce, Weather Bureau.

Liang, S., Zhao, X., Liu, S., Yuan, W., Cheng, X., Xiao, Z., Zhang, X., Liu, Q., Cheng, J., Tang, H., Qu, Y., Bo, Y., Qu, Y., Ren, H., Yu, K., Townshend, J., 2013. A long-term Global LAnd Surface Satellite (GLASS) data-set for environmental studies. Int. J. Digit. Earth 6, 5–33.

Lu, X.F., Cheng, C.Q., Gong, J.Y., Guan, L., 2011. Review of data storage and management technologies for massive remote sensing data. Sci. China Technol. Sci. 54, 3220–3232.

Pakin, S., Lang, M., Kerbyson, D.J., 2009. The reverse-acceleration model for programming petascale hybrid systems. IBM J. Res. Dev. 53 (5), 721–735.

Ramapriyan, H.K., Behnke, J., Sofinowski, E., Lowe, D., Esfandiari, M.A., 2010. Evolution of the earth observing system (EOS) data and information system (EOSDIS). In: Standard-Based Data and Information Systems for Earth Observation. Springer, Berlin, Heidelberg, pp. 63–92.

Zhao, X., Liang, S., Liu, S., Yuan, W., Xiao, Z., Liu, Q., Cheng, J., Zhang, X., Tang, H., Zhang, X., 2013. The Global Land Surface Satellite (GLASS) remote sensing data processing system and products. Remote Sens. 5 (5), 2436–2450.

Further reading

Wang, X.M., Wang, W.Y., Mao, W., He, Y.C., 2018. Design and implementation of S/X/Ka LO remote sensing satellite data receiving system. Space Electron. Technol. 15 (1), 105–110.

Wulder, M.A., White, J.C., Loveland, T.R., Woodcock, C.E., Belward, A.S., Cohen, W.B., Fosnight, E.A., Shaw, J., Masek, J.G., Roy, D.P., 2016. The global Landsat archive: status, consolidation, and direction. Remote Sens. Environ. 185, 271–283.

Abstract

Urbanization is the inevitable phenomenon of social and economic development. The rapid growth of global urbanization not only leads to complex land cover changes but also affects the vegetation coverage, the surface albedo, the surface roughness of the city, and the surrounding area and then affects the circulation of the urban hydrological and ecological systems and changes the urban environment and the near ground climate. Remote sensing provides broad, precise, impartial, and easily available information for quantifying the scope of the city, monitoring the process of urban expansion, and analyzing the impact of urbanization. This chapter aims to describe how to use remote sensing data products to map urban areas and expansion areas (23.2), monitor urban ecological environment (23.3), and analyze the impact of urbanization on vegetation phenology, net primary productivity, surface parameters (vegetation coverage, surface albedo, and surface temperature) and air quality (23.4).

Advanced Remote Sensing, Second Edition
https://doi.org/10.1016/B978-0-12-815826-5.00023-4

© 2020 Elsevier Inc. All rights reserved.

23.1 Introduction

Urbanization is defined by the United Nations as movement of people from rural to urban areas with population growth equating to urban migration. People move into cities to seek economic opportunities. The global proportion of urban population rose dramatically from 13% (220 million) in 1900, to 29% (732 million) in 1950, to 49% (3.2 billion) in 2005 (DESA, 2006). The same report projected that the urban population is likely to rise to 60% (4.9 billion) by 2030.

Driven in part by the doubling of the global population over the last 50 years and an unprecedented era of economic development, the urban footprint in developed and developing cities continues to grow and intensify. While the developed world's cities are experiencing urban renewal and suburban sprawl, the developing World's megacities are being formed by the in-migration of rural populations in search of work.

Although information is available for some cities, the true extent and rate of urban land expansion worldwide remains unknown. The resulting impact on biodiversity, local climate, and global change processes is an area of research (Huang et al., 2018; Mahmoud et al., 2016; Xie et al., 2018a,b), while geographic information systems and other quantitative data tools on urbanization will assist decisions on smart growth, equitable development, infrastructure management, and public health. Public and governmental actions will remain the most significant factor in mitigating urbanization's negative externalities.

23.2 Urban area monitoring

23.2.1 Mapping urban areas

Information about urban extent is important for monitoring the progress of urbanization, optimizing distribution of resources and transportation, and ensuring sustainable urban growth. The spatial extent of urban areas has been mapped successfully for a long time using remote sensing data. Table 23.1 lists 10 global urban or urban-related maps (Schneider et al., 2009).

The technologies used for identifying urban areas have evolved greatly over the past several decades. The traditional supervised classifications (e.g., maximum likelihood) are still popular, but more complex classifications have also been developed (He et al., 2019), including Support Vector Machine (SVM), decision tree, wavelet transform classifications, artificial neural networks classification, linear spectral unmixing, support vector regression, decision tree regression, and vegetation—impervious surface—soil models. In terms of the classification unit, the classification methods can be divided into three main groups, namely, subpixel, pixel-based, and object-oriented methods. Pixel-based methods are the most popular in mapping urban areas. However, as urban land cover classes are often spectrally mixed and spatially complex, the subpixel methods are gradually receiving more and more attention, especially when identifying urban areas in lower resolution imagery. Urban areas also exhibit distinctive texture information because this is the land use type most influenced by human planning, which allows successful use of object-oriented image processing in urban identification. Object-oriented classification methods are more suitable for medium-resolution imagery.

In the following parts of this section, we will show one example using optical remote sensing data to identify urban areas and another example using nighttime remotely sensed light data to identify urban areas.

23.2.1.1 Mapping by optical remote sensing

Schneider et al. (2009) produced a global urban map using Moderate Resolution Imaging Spectroradiometer (MODIS) satellite data. MODIS satellite data used in this study spanned 1 year and comprised 8-day composites of the seven land bands and the enhanced vegetation index (EVI) from February 18, 2001, to February 17, 2002. The spatial resolution of their data was 500 m.

TABLE 23.1   Ten global urban or urban-related maps listed in order of increasing global urban extent.

Abbreviation	Map (citation)	Definition of urban or urban-related feature	Resolution	Extent (km²)
VMAP	Vector Map Level Zero (Danko, 1992)	Populated places	1:1 mil	276,000
GLC2000	Global Land Cover 2000 (Bartholome and Belward, 2005)	Artificial surfaces and associated areas	988 m	308,000
GlobCover	GlobCover v2.2 (ESA, 2008)	Artificial surfaces and associated areas (urban areas >50%)	309 m	313,000
HYDE	History Database of the Global Environment v3 (Goldewijk, 2005)	Urban area (built-up areas, cities)	9000 m	532,000
IMPSA	Global Impervious Surface Area (Elvidge et al., 2007)	Density of impervious surface area	927 m	572,000
MODIS 500 m	MODIS Urban Land Cover 500 m (Schneider et al., 2010)	Areas dominated by built environment (>50%), including nonvegetated, human-constructed elements, with minimum mapping unit >1 km²	463 m	657,000
MODIS 1 km	MODIS Urban Land Cover 1 km (Schneider et al., 2003)	Urban and built-up areas	927 m	727,000
GRUMP	Global Rural–Urban Mapping Project (CIESIN, 2004)	Urban extent	927 m	3,524,000
Lights	Nighttime Lights v2 (Elvidge et al., 2001; Imhoff et al., 1997)	Nighttime illumination intensity	927 m	NA
LandScan	LandScan 2005 (Bhaduri et al., 2002)	Ambient (average over 24 h) global population distribution	927 m	NA

DOE, Department of Energy; *MODIS*, Moderate Resolution Imaging Spectroradiometer; *NASA*, National Aeronautics and Space Administration; *NOAA*, National Oceanographic and Atmospheric Administration.

To map global urban area, they stratified the global landscape into 16 quasihomogeneous areas that they termed urban ecoregions in terms of climate, vegetation trends, gross domestic product, and regional differences in city structure, organization, and historic development. In each ecoregion, they customized the training samples and postprocessing. The method used to classify the urban area was a supervised decision tree algorithm (C4.5) in conjunction with boosting. The C4.5 algorithm is a nonparametric classifier and shows excellent performance for coarse resolution datasets with complex, nonlinear relationships between features and classes. Boosting is an ensemble classification technique and is used to improve class discrimination by training each new model instance to emphasize the training instances that previous models misclassified. The training samples used were a set of urban samples selected from 182 cities distributed across the globe and 1860 sites collected across 17 land cover classes by manual interpretation of Landsat and Google Earth imagery. In the postprocessing, they excluded water pixels from the urban map using the MODIS 500-m water mask, and they also verified misclassification by hand editing. Fig. 23.1 shows two urbanized regions on the new MODIS 500-m global urban map.

They collected Landsat-based urban maps for 140 cities to evaluate the accuracy of their newly produced global urban map. The Landsat-based urban maps had 30-m spatial resolution. They first resampled the Landsat-based urban maps

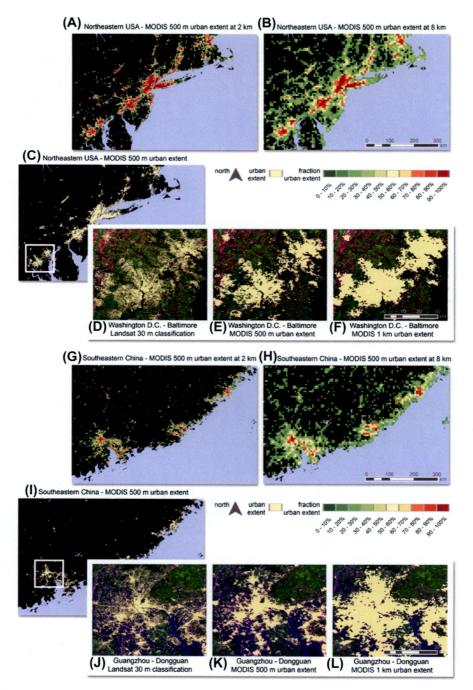

FIGURE 23.1 An illustration of the new MODIS 500 m global urban map for two urbanized regions: the Northeastern United States (A)—(F) and Southeastern China (G)—(L). In addition to the binary map (C), (I), the map has been aggregated to 2 and 8 km resolution for display (A), (B), (G), and (H). The inset maps (shown in rows) include a Landsat-based classification (30 m resolution) (Schneider and Woodcock, 2008), the new map of urban extent from MODIS 500 m data, and the previous version of the MODIS-based map (1 km resolution) (Schneider et al., 2009).

to 500-m binary maps and then used those maps to assess the accuracy of six global urban maps, namely, Global Impervious Surface Area (IMPSA), GlobCover v2.2, Global Land Cover 2000 (GLC2000), the urban map from the Global Rural—Urban Mapping Project available at http://sedac.ciesin.columbia.edu/gpw, MODIS 1-km urban map (Schneider et al., 2003), and the newly produced global urban map. Based on their evaluation, the new map has the highest accuracy among six global urban maps. The overall accuracy of their new map is 93% (kappa = 0.65).

23.2.1.2 Mapping by nighttime remotely sensed light data

Cao et al. (2009) mapped urban areas from DMSP-OLS nighttime light data and SPOT VGT data for 25 cities in Eastern China (Fig. 23.2). The data used in their study included the DMSP-OLS nighttime light data for 2000 from the National Oceanic and Atmospheric Administration/National Geophysical Data Center, the SPOT VGT NDVI global 10-day

composite product from April to September 2000, and 6 Landsat ETM+ images for 1999—2000.

Their method comprised the following steps. (1) They first coregistered the DMSP-OLS and SPOT NDVI images, resampled them to 1-km spatial resolution, and stacked them together. (2) They chose both urban and nonurban training samples from the stacked image. The urban training samples were pixels with OLS DN greater than 30. The nonurban training samples were from water pixels and the pixels where NDVI values were greater than 0.4 and OLS DNs were less than 30. (3) They used the training samples collected in the last step to train SVM and classify the urban and nonurban areas on DMSP-OLS and SPOT NDVI images using an SVM-based region-growing algorithm developed by the authors. In the SVM-based region-growing algorithm, all unknown pixels were labeled by a loop procedure. First, the urban training samples were used as seed pixels, and the unknown pixels within a 3 × 3 window of each seed were classified by SVM. After the first loop, all pixels labeled as urban were applied as

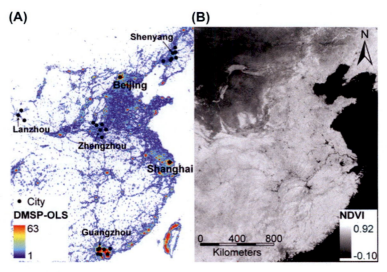

FIGURE 23.2 Study area and data: (A) DMSP-OLS image in 2000 and 25 selected cities and (B) SPOT VGT NDVI (MVC of April—September 2000) (Cao et al., 2009).

new training samples to train SVM and classify unknown pixels. The above steps were repeated many times and stopped when no new urban pixel was identified. Finally, they masked out the pixels with NDVI greater than 0.6 from the urban map produced by the SVM-based region-growing algorithm to produce the final urban map. (4) They also used global-fixed and local-optimized thresholding methods to produce another two urban maps. Those two maps were further used for comparison with the urban map produced by their own method. (5) They extracted the urban areas in ETM+ images using a maximum likelihood classifier and aggregated 28.5-m classification results to 1-km urban maps. The 1-km urban maps from ETM+ images were further used as reference maps to validate the urban maps derived from the DMSPOLS and SPOT NDVI images.

The accuracy of the urban map derived by their new method was comparable to that derived using the local-optimized threshold. However, the authors indicated that their method was a semiautomatic procedure and could avoid tedious trial-and-error procedures.

23.2.2 Monitoring urban growth

Urban expansion is the physical growth of urban areas as a consequence of socioeconomic development. The rapid development of global urbanization leads to a complex process of landscape transformation, which further alters the structure and function of ecosystems. The tremendous increase in energy consumption in urban areas contributed most of the greenhouse gases, changing the urban thermal environment. The large amount of air pollutants emitted in urban areas not only contributes to local climate change but also impacts on human health. Increases in impervious surfaces modify urban soil hydrological conditions and the land surface radiation budget (Wang et al., 2018). Urban invasion to cropland and forestland decreases

productivity and even species biodiversity. Therefore, information on urban extent, progress of urbanization, and the spatial distribution of various urban land uses are extremely important to help researchers understand how human activity impacts on the Earth system and for planners to make sustainable urban development policies.

The morphology and evolution of cities caused by urban sprawl have long been hot topics in geography, and the analysis of spatiotemporal characteristics of urban expansion have been receiving more and more attention because of the enhanced feasibility of data from remote sensing technology. Detection of urbanization from satellite data relies on various changes in detection technologies. Generally, the change detection technologies can be divided into three categories, namely, (1) visual analysis and digitizing changed areas, (2) classification methods, and (3) postclassification methods. Classification methods include algebra methods (such as image differencing, image regression, image ratioing, vegetation index differencing, change vector analysis, and background subtraction), transformation methods (e.g., principal component, tasseled cap, Gram—Schmidt, and chi-square analyses), and classification by various classifiers. Algebra methods and transformation methods require the determination of thresholds to identify changed areas, and they can provide changed/unchanged information but no information on from-to changes. Classification methods usually stack up multistage satellite images of the study area and then choose the change and no-change training samples and suitable classifiers to identity the changed area. Postclassification methods usually require two steps, namely, (1) making classifications based on images that cover the same area but correspond to different years and (2) comparing the thematic maps from the classifications and making land cover change metrics. Postclassification methods can detect detailed from-to changes.

The most popular and straightforward change detection method in urbanization detection is the postclassification comparison method. For example, Ji et al. (2006) used multistage Landsat images to monitor urbanization in the Kansas City metropolitan area using the postclassification comparison method. The data used for their study area included six Landsat images covering the area. The images were collected for years 1972, 1979, 1985, 1992, 1999, and 2001. The 1972, 1979, and 1985 images were 28.5-m spatial resolution, and the others were 30-m spatial resolution. All images were rectified and georeferenced to the UTM projection. They identified four classes—built-up area, forestland, nonforest vegetation, and water body—in each image using the maximum likelihood classification method. The nonforest vegetation class included grasslands, brush land, and cropland. The classified images for years 1972, 1979, and 1985 were resampled to produce new maps with 30-m spatial resolution using the nearest neighbor interpolation. After that, they collected 200 validation samples (50 for each class) per image and assessed the accuracy of classifications and analyzed the characteristics of urban expansion in Kansas City.

Fig. 23.3 shows their classification results. The authors found that the significant increase of built-up land in Kansas was mainly at the expense of nonforest vegetation cover during the past 30 years. The areas of nonforest vegetation and forestland became more fragmented.

23.3 Urban ecological environment monitoring

23.3.1 Urban vegetation monitoring

Green vegetation can regulate urban heat island (UHI) effects and urban environmental quality and affect material circulation and energy flow in urban ecosystems through selective solar radiation reflection or absorption, water transpiration regulation, and suspended solids and harmful atmospheric gases adsorption (Gunawardena et al., 2017; Small, 2001; Vieira et al., 2018). Global climate change and sustained urban expansion has increased urban ecological pressure. Monitoring is particularly necessary as a means to evaluate urban vegetation's role for urban dwellers' lifestyle and urban ecosystem services.

In this section, we introduce a case study of using mixed pixel decomposition to obtain urban vegetation fraction. Song (2005) proposed a Bayesian spectral mixture analysis (BSMA) model by considering the impact of end-member variability on derivation of subpixel vegetation fractions in an urban environment. In contrast to traditional spectral mixture analysis (SMA) methods, the proposed method did not consider end-member spectral signatures as constants but explicitly considered end-member uncertainty and used end-member spectral signature distribution probabilities for unmixing. The BSMA model can be expressed as

$$p(c|\text{NDVI}_m) = \frac{f(\text{NDVI}_m|c) \times \pi(c)}{\int_0^1 f(\text{NDVI}_m|u)\pi(u)du} \quad (23.1)$$

where c and u are subpixel vegetation fraction and nonvegetation fraction, respectively (ranging from 0.0 to 1.0). NDVI_m is the NDVI values of nonvegetaion pixels. $p(c|\text{NDVI}_m)$ is the conditional probability density function of c at a given NDVI_m. $f(\text{NDVI}_m|c)$ and $f(\text{NDVI}_m|u)$ are the conditional probability density function of NDVI_m at a given vegetation fraction and nonvegetation fraction, respectively. $\pi(c)$ and $\pi(u)$ are prior probability density function of subpixel vegetation fractions and nonvegetaion fractions, respectively.

$$f'(c|DN_m) = \frac{f(DN_m|c) \times p(c)}{\int_0^1 f(DN_m|u)p(u)du} \quad (23.2)$$

where $f(DN_m|c)$ and $f(DN_m|u)$ are the conditional probability density function of DN_m at a given vegetation fraction and nonvegetation

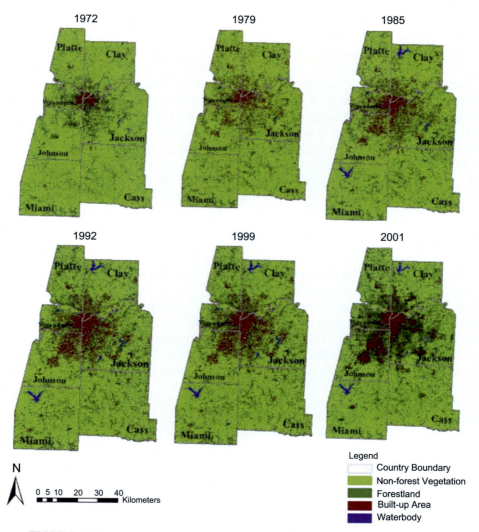

FIGURE 23.3   Classified satellite images for the six detection years (Ji et al., 2006).

fraction, respectively. $p(c)$ and $p(u)$ are the probability of subpixel end-member vegetation and nonvegetation fractions obtained from the first spectral measurement, respectively.

The study included two parts: simulation and real remote sensing data experiments. The basic steps of the simulation experiment were to first generate three 100×100 pixel simulated mixture images and then use BSMA and traditional SMA models to perform mixed pixel analysis on the simulated images. The first two simulated image cases contained only vegetation and nonvegetation end-members.

The first simulated image assumed that vegetation end-member spectral signatures distribution followed Gaussian distribution with mean NDVI = 0.3 and standard deviation SD = 0.12 and nonvegetation end-member spectral signatures distribution followed Gaussian distribution with mean NDVI = −0.1 and SD = 0.08.

The second simulated image assumed end-member spectral signatures followed skewed probability density distribution functions, with vegetation and nonvegetation end-members NDVI $= -0.1$ to -0.3 and $0.1-0.7$ and peak $= -0.20$, and 0.5, respectively.

The third simulated image includes three end-members (vegetation, high albedo, and low albedo). Vegetation end-members have low albedo in the red wave band and high albedo in the near-infrared band, and low/high albedo end-members have low/high albedo in both red wave band and near-infrared band.

Interpixel variation of vegetation end-member fractions was assumed to follow Gaussian distribution with mean $= 0.4$ and SD $= 0.15$; the interpixel variation of the low albedo end-member fractions was assumed to follow Gaussian distribution with mean $= 0.3$ and SD $= 0.1$; the remaining fractions were assumed to match the high albedo end-members. Simulation details were provided in the original paper (Song, 2005).

The real remote sensing data experiment used 4 m resolution Ikonos and 30 m resolution Landsat 7 ETM+ images from Bangkok and Thailand, obtained on November 27, 2002, and November 16, 1999, respectively. The main processing steps were as follows:

(1) The Ikonos image was classified as vegetation or nonvegetation using an unsupervised classification method and then used as the vegetation/nonvegetation end-member map.
(2) The end-member map was degraded from 4 to 30 m spatial resolution and used as a reference for accuracy validation.
(3) DN values of Ikonos and Landsat images were converted to reflectance reflectivity, and then NDVI were calculated.
(4) Ikonos data were degraded from 4 to 30 m spatial resolution, and the NDVI images were unmixed using BSMA and traditional SMA. For the BSMA unmixing process,

end-member spectral values were the entire probability distribution of all end-members. In the traditional SMA unmixing process, end-member spectral values were the mean and mode of end-member signature distributions.

(5) BSMA and traditional SMA unmixing of the Landsat NDVI image was performed similarly.
(6) Unmixing accuracies obtained by the different methods were compared.

Figs. 23.4−23.6, show the unmixed results of the simulated images, and Figs. 23.7 and 23.8 show unmixed results for the two real images. SMA using the end-member signature distribution mode produced the poorest mapping, whereas SMA using the end-member signature distribution mean was similar to, and in some cases better than, that the BSMA method. However, BSMA can provide estimation uncertainty, which SMA cannot. Unmixing uncertainty of areas with high or low NDVI values is small, whereas it can be very large for areas with intermediate NDVI. The study also showed that traditional SMA chose the purest pixels or extremes in the feature space, which may differ from end-member signature means and produce significant unmixing bias. This conclusion helps guide end-member selection.

23.3.2 Estimation of carbon storage and sequestration by urban forests

In recent years, human activities have greatly increased greenhouse gas emissions (Arrieta et al., 2018; Karelin et al., 2017), land desertification (Guo et al., 2017; Zhao et al., 2017), and decreased biodiversity (Fuller et al., 2017). Global environmental change problems, characterized by climate warming, reduced water resources, food shortages, and frequent natural disasters, have gradually become a research focus of international concern (Paudyal et al., 2018; Rossati, 2017; Shrestha et al., 2017;

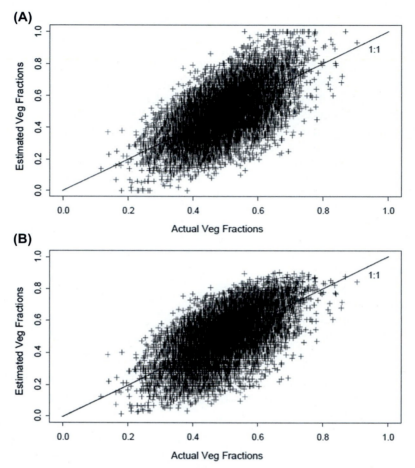

FIGURE 23.4   Results of spectral mixture analysis for subpixel vegetation fractions from the first simulated image. (A) SMA with the means of end-member signature distributions (RMSE = 0.12); (B) BSMA with the entire signature distribution (RMSE = 0.11) (Song, 2005).

Solomon et al., 2007; Xie et al., 2018a,b). Urban vegetation is an important green infrastructure in cities and has attracted increase attention for its carbon storage function in the background of global change (He et al., 2016). Urban vegetation has important roles in absorbing and fixing CO_2 by photosynthesis and mitigating climate change (Eric et al., 2018; Mac Dowell et al., 2017; Mchale et al., 2007). Shade and wind shelter also reduce building energy consumption and carbon emissions (Pataki et al., 2006). Therefore, research on the impact of changes in urban

forest carbon stocks is important to help understand the impact of urbanization and regional carbon budgets in the process of rapid urbanization, and it can provide policy support for sustainable urban development and planning (Puhlick et al., 2017; Ren et al., 2012).

Urban forest reduction of CO_2 has been largely studied in the United States (Rowntree and Nowak, 1991; Nowak, 1993, 1994). However, although Chinese research started late, studies on urban forest carbon storage are increasing. Liu and Li (2012) used Shenyang,

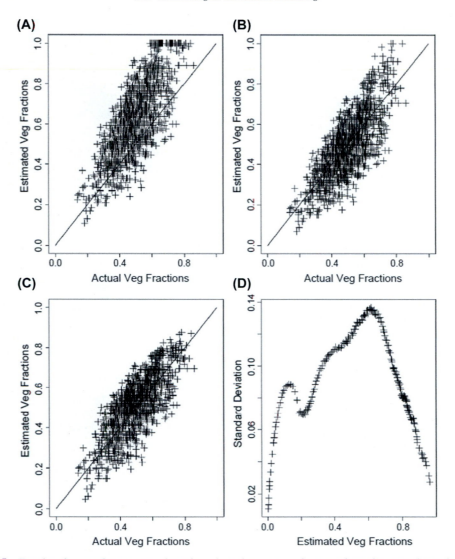

FIGURE 23.5 Results of spectral mixture analysis for subpixel vegetation fractions from the second simulated image: (A) SMA with the mode of end-member signature distributions (RMSE = 0.18); (B) SMA with the means of end-member signature distributions (RMSE = 0.11); (C) BSMA with the entire signature distribution (RMSE = 0.10); and (D) the standard deviation of unmixed vegetation fractions using BSMA (Song, 2005).

a heavy industrial city, as an example to study urban carbon storage and carbon sequestration quantitatively through the biomass equation using field survey data and urban forest data obtained from high-resolution QuickBird images. Their study mainly included six steps:

(1) Data preprocessing of QuickBird image data collected on August 19, 2006. Urban forests were extracted from the images by manual visual interpretation and divided into five categories: ecological and public welfare forest (EF), attached forest (AF), landscape

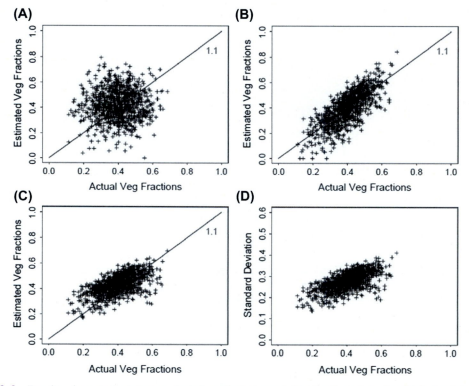

FIGURE 23.6   Results of spectral mixture analysis for subpixel vegetation fractions from the third simulated image: (A) SMA with the mode of end-member signature distributions; (B) SMA with the means of end-member signature distributions; (C) BSMA with the entire signature distribution; and (D) the standard deviation of unmixed vegetation fractions using BSMA (Song, 2005).

and relaxation forest (LF), production and management forest (PF), and road forest (RF).

(2) A total of 213 plots were field surveyed in September and October 2006. The plot number of each urban forest category was identified in terms of the area of each forest category and urban forest structure type differences, including species diversity and tree diameter at breast height (DBH) distribution. The results are detailed in Table 23.2.

(3) Biomass equations were used to calculate the dry biomass of each investigated tree. The biomass equation was selected because it contains only the DBH parameter, which

was close to the study area and suitable for tree species in the study area. Tree pruning was not common in Shenyang, with pruned trees mainly being old trees in the AF and RF. Therefore, they applied 20% biomass reduction for trees with DBH >30 cm. They then multiplied dry biomass by 0.5 to get carbon stocks and calculated carbon density for each plot (t/ha).

(4) Annual DBH growth rates of deciduous and evergreen trees were determined based on DBH measurements of 186 trees located in Shenyang from 2003 to 06, and sequestration was calculated based on this annual growth rate and adjusted depending on tree health. Adjustment factors for excellent, fair, poor,

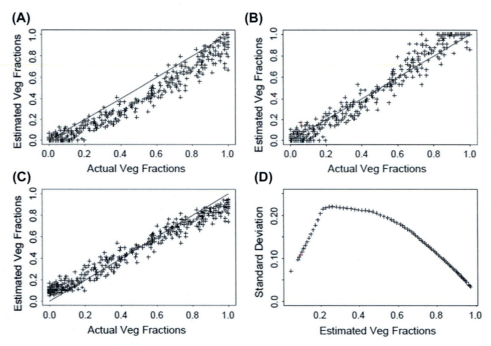

FIGURE 23.7 Results of spectral mixture analysis for subpixel vegetation fractions from the degraded Ikonos image: (A) SMA with the modes of end-member signature distributions (RMSE = 0.11 when the subpixel vegetation fractions are between 0.0 and 1.0, and RMSE = 0.13 when the subpixel vegetation fractions are between 0.1 and 0.9); (B) SMA with the means of end-member signature distributions (RMSE = 0.07 when the subpixel vegetation fractions are between 0.0 and 1.0, and RMSE = 0.09 when the subpixel vegetation fractions are between 0.1 and 0.9); (C) BSMA with the entire signature distribution(-RMSE = 0.09 when the subpixel vegetation fractions are between 0.0 and 1.0, and RMSE = 0.07 when the subpixel vegetation fractions are between 0.1 and 0.9); and (D) the standard deviation of unmixed vegetation fractions using BSMA (Song, 2005).

critical, dying, and dead were set to 1, 1, 0.76, 0.42, 0.15, and 0, respectively. They then calculated carbon sequestration (t/ha/y) for each plot.

(5) Carbon storage, sequestration, and number of trees were calculated for the entire study area based on tree densities, carbon density, and sequestration calculated for the survey plots.

(6) The value of carbon storage and sequestration by Shenyang urban forest was evaluated in terms of the cost of afforestation (RMB 273.3/t C), and the ratios for carbon storage, carbon sequestration, and urban carbon emissions were calculated for the urban forests in the study area. The

calculation method for urban carbon emissions was as follows. Shenyang City combustion volume 2004–06 was first calculated based on statistical data and then different fossil fuel consumption types were converted into standard coal using appropriate conversion factors. Standard coal volume was then multiplied by 2.877 to convert to CO_2 and finally multiplied by 0.2727 to convert to carbon.

The study reached the following conclusions:

(1) There was 101 km^2 of urban forest in the study area, accounting for 22.28% of the total area. Ecological public welfare and public

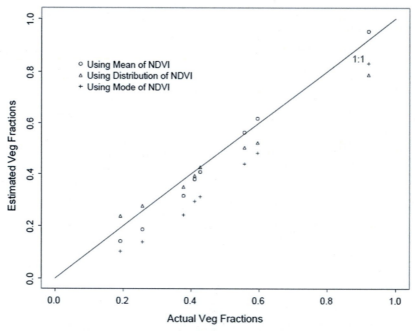

FIGURE 23.8   Validation of window-based spectral mixture analysis. The estimated vegetation fractions are the average vegetation fractions of the 40 × 40 pixel window of the Landsat ETM+ spatial resolution (30 × 30 m). The actual vegetation fractions are the average vegetation fractions in a 300 × 300 pixel window at Ikonos multispectral image spatial resolution (4 × 4 m) (Song, 2005).

TABLE 23.2   The number and area of survey sites in each city (mean ± standard deviation) (Liu and Li, 2012).

Urban forest type	Number of surveyed plots	Average area of plots(ha)
EF	30	0.25 ± 0.15
AF	46	0.20 ± 0.09
LF	70	0.23 ± 0.14
PF	14	0.33 ± 0.14
RF	49	0.31 ± 0.21
Total	213	0.25 ± 0.16

welfare forests were the main forest types, accounting for approximately half the area.

Table 23.3 shows the detailed results.

(2) Urban forest carbon storage = 337,000 t, with EF and AF accounting for 41% and 34%, respectively. Average carbon density = 33.22 t/ha, with afforested forests having the highest carbon density, followed by RFs, landscape forests, and forest production and management. Urban forest annual carbon sink = 28,820 t. Similar to carbon storage, most of carbon was segregated by ecological and affiliated forests. Carbon absorption rate = 2.84 t/ha/y, ranging over 1.16–4.78 t/ha/y for different urban forest types, as shown in Table 23.4.

(3) Carbon storage value in urban forests was 92.02 M yuan, and urban forest carbon absorption was 7.88 M yuan. Annual average carbon emissions 2004–06 from burning fossil fuels in Shenyang amounted

TABLE 23.3 Urban forest area and tree number of different urban forest types ($\pm$SE) (Liu and Li, 2012).

Urban forest type	Area (km^2)	Tree density (number/ha)	Estimated tree number (10^4)
EF	47.1	653 ± 117	307 ± 54.88
AF	22.7	375 ± 55	85.1 ± 12.38
LF	12.3	502 ± 42	61.9 ± 5.17
PF	10.9	905 ± 133	98.5 ± 14.52
RF	8.4	279 ± 47	23.4 ± 3.96
Total	101	569 ± 90	576 ± 90.9

TABLE 23.4 Carbon density and total carbon storage and sequestration by urban forests ($\pm$SE) (Liu and Li, 2012).

	Carbon density (t/ha)		Total (10^3t)	
Types	Storage	Sequestration (/yr)	Storage	Sequestration (/yr)
EF	29.25 ± 4.18	2.45 ± 0.32	137.63 ± 19.68	11.55 ± 1.50
AF	50.17 ± 5.50	4.78 ± 1.54	113.81 ± 12.48	10.85 ± 3.40
LF	33.65 ± 2.94	2.47 ± 0.25	41.49 ± 3.62	3.05 ± 0.31
PF	13.17 ± 3.71	1.16 ± 0.28	14.34 ± 4.04	1.26 ± 0.31
RF	34.95 ± 4.76	2.51 ± 0.28	29.41 ± 4.01	2.11 ± 0.24
Total	33.22 ± 4.32	2.84 ± 0.58	336.68 ± 43.82	28.82 ± 5.84

to 11.16 Mt. Carbon stored in urban forests was equivalent to 3.02% of annual carbon emissions from fossil fuel combustion, and carbon absorption could offset 0.26% of annual carbon emissions in Shenyang.

23.4 Study on the impact of urbanization

23.4.1 The effects of urbanization on vegetation growth season

Vegetation is an important part of the terrestrial ecosystem. It plays an important role in reducing greenhouse effect, improving local climate, preventing soil erosion, regulating river flow, and reducing environmental pollution. Relationships between vegetation phenology and climate have become the focus of global environmental change. On one hand, vegetation phenology is an extremely sensitive indicator of climate change (Yang et al., 2017). On the other hand, vegetation itself regulates climate through the exchange of energy, water, and carbon with the atmosphere and the Earth's surface (Zhou et al., 2016). Urbanization changes the urban climate (such as the heat island effect) (Morris et al., 2017), urban vegetation environment (Lu et al., 2017), and urban vegetation itself (Guida-Johnson et al., 2017).

White et al. (2002) analyzed urbanization effects on vegetation growth season for broadleaved deciduous forest regions of the United States using AVHRR data. The data and the main research steps used in their study are as follows.

(1) Extraction of the study area: They divided continental United States into $1° \times 1°$ grids and extracted urban and deciduous broadleaved forests (DBF) pixels from the 1 km resolution land cover classification map produced by Hansen et al. (2000). DBF included deciduous broadleaf forest, mixed forest, and woodland. They calculated DBF coverage for the grid and extracted grids with DBF coverage >50% and urban pixel count $\geq$50. DBF pixels with mean elevation difference between DBF pixel elevation and urban pixels >100 m were removed in each

1° grid. Consequently, the study area was limited to the area east of latitude 100°W and north of latitude 32°N.

(2) Growing season parameter estimation: They collected 2-week AVHRR datasets from the Earth resources observing system data center (EDC) 1990—99 and identified and removed cloud pixels using the method developed by Thornton (1998). Then they extracted the beginning of the growing season (SOS), ending time (EOS), and growing season length (GSL) for each year using the method developed by White et al., (1999) and calculated 9-year average SOS, EOS, and GSL pixel by pixel. Differences between cities and DBFs for each 1° grid were denoted as ΔSOS, ΔEOS, and ΔGSL, respectively. Negative ΔSOS indicated that growth season advances and positive ΔEOS indicated growth season delays in urban areas, both of which could lead to urban growth season extension.

(3) Fractional vegetation coverage (FC): 1 km resolution pixels labeled as DBF or urban type were grouped in the 1° grid. NDVI was extracted for these pixels in July and maximum and minimum values ($NDVI_{max}$ and $NDVI_{min}$, respectively) identified. FC was calculated for each pixel (Formula 23.3) and mean FC was calculated for urban and DBF areas for each grid.

$$FC = \frac{NDVI - NDVI_{min}}{NDVI_{max} - NDVI_{min}} \qquad (23.3)$$

where NDVI is the NDVI value in July; the maximum NDVI refers to area that DBF completely covers; and the minimum value of the NDVI refers to either bare areas or completely urbanized area.

(4) Seasonal NDVI amplitude ($NDVI_{amp}$): $NDVI_{amp} = NDVI_{max} - NDVI_{min}$ was calculated for each pixel for each year, averaged over the study period, and mean

urban and DBF $NDVI_{amp}$ calculated for each 1° grid.

(5) Auxiliary data collection and processing: Average annual maximum temperature, minimum temperature, average temperature, water pressure, precipitation, and shortwave (SW) radiation data were collected for 1983 to 1999 at 1 km scale from continental climate data, and average values for various meteorological indicators were calculated for each grid. Urban pixels were from the 1 km resolution land cover classification map produced by Hansen et al. (2000), and urbanization proportion within each grid calculated as the number of urban pixels divided by the number of DBF plus urban pixels. Population density and total population (CIESIN 2000) were collected at 2.5′ spatial resolution and aggregated to 1° resolution. CO_2 emission was also collected at 1° resolution from data produced by Andres et al. (1996).

(6) Generalized additive model: Explanatory variables were screened, as shown in Table 23.5, and a generalized additive model constructed suitable explanatory variables and dependent variables (SOS and EOS) to explain factors that influenced growing season differences between urban and DBF areas.

Table 23.6 summarizes the results of their study. Average urban area SOS was 5.7 days earlier, EOS was 2.6 days later, and the entire growing season was extended by 7.6 days. ΔSOS was largest between DBF and cities in the midlatitude zone (35°—40°), most differences = 8 days, with some differences > 15 days. Urban area FC and $NDVI_{amp}$ are lower than those in rural areas for every 1° grid, and average city FC was 62% lower than average DBF. $ΔNDVI_{amp}$ was very small in Southern, Central, and Eastern parts of the United States and large in other regions.

TABLE 23.5 Explanatory variables in stepwise generalized additive model (White et al., 2002).

Variable	Smoothed	Linear
Maximum temperature[a]	X	
Minimum temperature[a]	X	
Mean temperature[a]	X	
Precipitation[a]	X	X
Vapor pressure[a]	X	
Shortwave radiation[a]	X	
Elevation[b]	X	X
Elevation[b]	X	X
Latitude[b]	X	
Longitude[b]	X	
Urban fractional cover[c]		X
DBF fractional cover[c]		X
Fractional cover[c]		X
Urban NDVI amplitude[c]		X
DBF NDVI amplitude[c]		X
NDVI amplitude[c]		X
CO_2[d]	X	X
Population density[e]	X	X
Population[e]	X	X
Percent urbanization[f]	X	X

Input variables used to construct models predicting ΔSOS and ΔEOS. The X in the smoothing column indicates that the corresponding variable is fitted using a smooth spline. The X in the linear function column indicates that the corresponding variable is fitted with a linear function. The choice of smoothness and linearity is based on visual inspection of scatter plots. For visually unclear variables, smooth spline interpolation and linear variables are used at the same time.

[a] *Extract only F pixels from urban and deciduous broad-leaved forests; see Thornton et al. (1997) for calculations.*

[b] *Extract only pixels from urban and deciduous broad-leaved forests; auxiliary data from satellite datasets of the Earth resources observing system data center.*

[c] *Extracts only urban and deciduous broad-leaved forest pixels; normalized differential vegetation index data from the Earth resources observing system data center.*

[d] *1° × 1° grid, data from Andres et al. (1996).*

[e] *Reaggregates to 1° × 1° (CIESIN 2000).*

[f] *From Hansen et al. (2000).*

However, the study failed to construct a suitable model for ΔEOS.

A general additive ΔSOS model was constructed based on annual mean temperature, urban vegetation coverage, and ΔNDVI$_{amp}$ for urban and DBF areas. Fig. 23.9A shows that SOS impact is greatest for urban average temperature = 13°C. The effect of temperature on SOS decreases, and the direction of influence changes as temperature increases or decreases. ΔSOS decreases with the reduction of city FC and ΔNDVI$_{amp}$ (Fig. 23.9B,C).

Thus, the study showed that urbanization had significant impact on the beginning and end of the vegetation growth season for BDFs of the Eastern United States, and that although urbanization has extended the growing season of urban vegetation, urban vegetation productivity may decrease.

Differences in phenology between the three indicators: difference between urban and DBF and P value at the beginning of season (ΔSOS, days); difference between urban and DBF at the end of season (ΔEOS, days) and P value; and length of growing season (ΔGSL, number of days) difference between city and DBF and P value. The P value of HS indicates that it is a highly significant P value < .001. Land cover indicators: n Urban and deciduous forest (n DBF). Stepwise GAM predicts the variable ΔSOS generated by the response variable: the annual mean temperature intervening day temperature (T_{avg}, °C) calculated from 1983 to 1999; the seasonal amplitude of the normalized difference vegetation index (NDVI$_{amp}$) and the difference between urban and DBF; and urban vegetation coverage (Urban FC).

23.4.2 Impact of urbanization on net primary productivity

Net primary productivity (NPP) is the amount of solar energy converted to chemical energy through the process of photosynthesis.

TABLE 23.6   Statistics for each $1° \times 1°$ grid (White et al., 2002).

Latitude	Longitude	ΔSOS	ΔSOS P value	ΔEOS	ΔEOS P value	ΔGSL	ΔGSL P value	n Urban	n DBF	T_{avg}	$\Delta NDVI_{amp}$	Urban FC
31	−94	−0.1	0.948	2.1	0.042	2.2	0.308	68	546	18.7	−0.032	0.44
31	−89	−7.7	HS	4.5	0.002	12.3	HS	60	323	18.3	−0.032	0.11
32	−94	−3.3	HS	1.2	0.036	5.1	HS	121	362	18.0	−0.049	0.27
32	−93	−4.3	HS	2.6	HS	7.6	HS	153	686	17.9	−0.047	0.22
32	−85	−3.7	HS	3.1	HS	7.1	HS	77	563	17.4	−0.049	0.34
32	−84	−7.2	HS	3.4	HS	7.9	HS	134	209	17.9	−0.071	0.14
33	−92	−0.5	0.561	0.3	0.781	−2.0	0.314	62	828	17.0	−0.021	0.19
33	−87	−9.0	HS	2.1	HS	12.6	HS	90	709	16.5	−0.062	0.21
33	−86	−8.0	HS	3.5	HS	10.9	HS	514	029	16.4	−0.057	0.34
33	−83	−3.5	HS	5.4	HS	6.2	HS	128	251	17.0	−0.041	0.29
33	−82	−6.4	HS	10.3	HS	16.4	HS	137	569	16.9	−0.064	0.27
34	−92	−1.3	0.016	1.8	HS	−1.1	0.287	242	103	16.5	−0.016	0.22
34	−87	−8.4	HS	2.1	HS	9.5	HS	121	728	15.4	−0.054	0.26
34	−86	−3.9	HS	4.3	HS	9.2	HS	236	359	15.4	−0.085	0.19
34	−85	−6.6	HS	2.7	HS	8.9	HS	128	277	15.2	−0.057	0.19
34	−84	−7.5	HS	8.5	HS	16.9	HS	128	921	15.3	−0.086	0.20
34	−83	−1.5	0.153	3.1	HS	5.3	HS	61	198	15.7	−0.070	0.32
35	−88	−7.9	HS	0.4	0.448	6.9	HS	55	345	15.1	−0.053	0.10
35	−87	−18.5	HS	−0.4	0.322	16.8	HS	54	267	14.4	−0.101	0.08
35	−86	−7.3	HS	1.5	HS	11.7	HS	138	170	14.5	−0.078	0.11
35	−85	−5.3	HS	1.9	HS	7.4	HS	230	524	14.5	−0.101	0.25
35	−84	−9.1	HS	1.7	HS	11.9	HS	113	447	14.4	−0.059	0.19
35	−83	−10.8	HS	0.3	0.543	11.0	HS	120	607	14.0	−0.118	0.24
35	−82	−8.2	HS	1.3	0.009	9.2	HS	84	250	13.9	−0.101	0.13
35	−81	−7.5	HS	2.3	HS	8.0	HS	251	4019	15.1	−0.052	0.14
35	−79	−6.6	HS	3.6	HS	9.2	HS	103	7275	15.6	−0.038	0.23
36	−87	−13.1	HS	−2.6	HS	11.6	HS	76	6112	14.1	−0.091	0.13
36	−83	−18.6	HS	3.0	HS	21.7	HS	89	3834	13.1	−0.125	0.19
36	−82	−10.0	HS	1.4	HS	10.1	HS	201	3179	13.1	−0.109	0.18
36	−80	−8.9	HS	4.2	HS	13.2	HS	241	3741	14.0	−0.070	0.18

TABLE 23.6 Statistics for each $1° \times 1°$ grid (White et al., 2002).—cont'd

Latitude	Longitude	ΔSOS	ΔSOS P value	ΔEOS	ΔEOS P value	ΔGSL	ΔGSL P value	n Urban	n DBF	T_{avg}	$\Delta NDVI_{amp}$	Urban FC
36	−79	−8.0	HS	2.6	HS	9.5	HS	289	6214	14.1	−0.056	0.19
36	−78	−7.5	HS	2.0	HS	9.1	HS	108	6446	14.4	−0.048	0.24
37	−81	−9.2	HS	3.0	HS	14.4	HS	62	7642	11.4	−0.110	0.19
37	−80	−18.6	HS	8.5	HS	29.1	HS	72	1517	12.1	−0.133	0.11
37	−79	−10.0	HS	2.4	HS	12.5	HS	159	348	13.3	−0.082	0.21
37	−77	−8.6	HS	6.3	HS	14.6	HS	448	406	14.2	−0.062	0.26
37	−76	−11.3	HS	13.8	HS	24.3	HS	147	3727	14.6	−0.100	0.26
38	−82	−8.5	HS	0.7	0.045	9.1	HS	107	7151	12.0	−0.107	0.21
38	−81	−5.2	HS	−0.5	0.151	4.2	HS	71	6016	12.0	−0.088	0.21
38	−78	−8.9	HS	2.4	HS	11.4	HS	107	3713	12.7	−0.091	0.14
38	−77	−7.2	HS	4.5	HS	11.5	HS	588	5865	13.3	−0.075	0.30
39	−82	−20.3	HS	2.9	HS	22.1	HS	179	5798	11.0	−0.185	0.18
39	−81	−14.3	HS	2.0	0.005	14.5	HS	51	6753	11.1	−0.119	0.15
39	−80	−6.5	HS	1.3	0.002	7.9	HS	72	7497	10.7	−0.091	0.26
39	−79	−9.1	HS	2.5	0.002	12.7	HS	50	700	10.8	−0.136	0.12
39	−78	−11.0	HS	5.6	HS	17.7	HS	56	4228	11.2	−0.114	0.15
40	−80	−2.1	HS	0.1	0.558	1.9	HS	668	4663	10.2	−0.069	0.36
40	−79	−4.2	HS	0.8	0.001	5.8	HS	625	3799	9.8	−0.091	0.31
40	−78	−2.7	HS	0.5	0.151	3.3	HS	157	5133	9.2	−0.079	0.30
40	−77	−7.8	HS	2.4	0.001	9.8	HS	57	3857	10.3	−0.140	0.11
40	−76	−7.0	HS	1.9	HS	8.1	HS	275	2823	10.5	−0.138	0.19
40	−75	−7.5	HS	1.9	HS	9.2	HS	893	2962	10.9	−0.126	0.31
41	−80	−9.3	HS	2.0	HS	11.4	HS	368	5583	9.1	−0.103	0.19
41	−79	0.1	0.861	0.5	0.272	−0.4	0.667	85	6785	8.2	−0.046	0.25
41	−78	−4.5	HS	2.4	HS	7.0	HS	66	5804	7.7	−0.081	0.23
41	−76	−5.5	HS	1.3	0.052	6.2	HS	50	1695	9.2	−0.114	0.17
41	−75	−1.8	HS	0.5	0.103	2.3	0.004	210	1897	8.8	−0.151	0.20
41	−74	−1.8	HS	2.2	HS	3.7	HS	343	3029	9.8	−0.094	0.22
41	−73	−2.8	HS	1.5	HS	4.3	HS	573	3611	10.2	−0.104	0.35
41	−72	−8.3	HS	2.9	HS	11.1	HS	449	3616	10.0	−0.123	0.24

(Continued)

TABLE 23.6   Statistics for each $1° \times 1°$ grid (White et al., 2002).—cont'd

Latitude	Longitude	ΔSOS	ΔSOS P value	ΔEOS	ΔEOS P value	ΔGSL	ΔGSL P value	n Urban	n DBF	T_avg	ΔNDVI_amp	Urban FC
41	−71	−4.0	HS	0.8	0.010	4.3	HS	284	2882	9.9	−0.134	0.24
41	−70	−0.0	1.000	−1.5	0.150	−0.9	0.782	81	1269	10.0	−0.052	0.26
42	−79	−2.8	HS	0.1	0.932	2.1	0.092	57	1622	8.5	−0.077	0.25
42	−78	−5.0	HS	1.4	HS	8.5	HS	365	981	8.5	−0.206	0.22
42	−77	−1.9	0.003	−0.2	0.771	1.8	0.160	78	2093	8.2	−0.068	0.22
42	−76	−5.7	HS	2.4	HS	8.5	HS	154	2216	8.3	−0.105	0.30
42	−75	2.5	HS	−4.8	HS	−7.2	HS	69	2257	7.6	−0.114	0.28
42	−73	1.8	HS	−3.9	HS	−5.8	HS	221	3246	8.5	−0.106	0.20
42	−72	0.6	0.312	−0.1	0.708	−1.5	0.065	308	1833	8.6	−0.095	0.21
42	−71	−0.1	0.721	−1.7	HS	−1.3	0.001	1188	4015	9.1	−0.094	0.43
42	−70	0.2	0.736	−2.4	HS	−2.4	0.016	207	963	9.7	−0.073	0.43
43	−91	1.8	0.033	−0.7	0.391	−2.9	0.073	50	5864	7.6	−0.084	0.21
43	−86	−1.5	0.072	1.0	0.021	2.9	0.011	55	2190	8.0	−0.152	0.11
43	−77	−8.4	HS	3.8	HS	12.5	HS	341	1545	9.1	−0.097	0.34
43	−76	−7.9	HS	4.0	HS	12.0	HS	154	3724	8.4	−0.146	0.21
43	−75	−2.0	0.001	1.0	0.250	2.5	0.039	90	2250	7.6	−0.123	0.21
43	−73	1.0	0.220	−3.1	0.001	−4.5	0.006	67	2769	7.8	−0.082	0.28
43	−70	6.3	HS	−3.3	HS	−8.8	HS	73	2935	7.7	−0.073	0.39
44	−92	−3.2	HS	2.1	HS	6.9	HS	99	4438	7.0	−0.113	0.21
44	−91	−0.4	0.547	−2.1	0.002	−0.3	0.838	53	6220	6.9	−0.131	0.21
44	−89	−6.9	HS	5.4	HS	13.8	HS	59	5786	6.4	−0.114	0.16
44	−88	−6.4	HS	5.4	HS	12.9	HS	182	4260	6.7	−0.188	0.17
44	−70	4.2	HS	0.3	0.692	−4.4	0.019	52	3059	6.5	−0.077	0.32
46	−92	4.1	HS	3.5	HS	0.4	0.658	73	3485	4.8	−0.112	0.37
Mean		−5.7		2.0		7.6		190	4269	11.9	−0.091	0.23
Standard deviation		5.1		2.9		7.0		198	1870	3.5	−0.037	0.08
Median		−6.5		2.0		8.3		121	4123	11.2	−0.091	0.21
CV (%)		89		144		92		104	44	30	41	35
Minimum		−20.3		−4.8		−8.8		50	607	4.8	−0.206	0.08
Maximum		6.3		13.8		29.1		188	8267	18.7	−0.016	0.44

The latitude and longitude of the table correspond to the lower right corner of each grid. ΔSOS and ΔSOS P values are the differences between the start time of urban and DBF growing season and the significance test P value; ΔEOS and ΔEOS P values are the differences between the end of the growing season of urban and DBF. The P value was tested for significance; the P value for HS was significantly different ($P < .001$). n Urban indicates the number of pixels in the city, and n DBF indicates the number of pixels in the DBF. The input variables for the progressive generalized additive model used to predict ΔSOS include the annual mean temperature calculated from 1983 to 99, the NDVI amplitude difference (ΔNDVI_amp), and urban coverage for urban and DBFs.

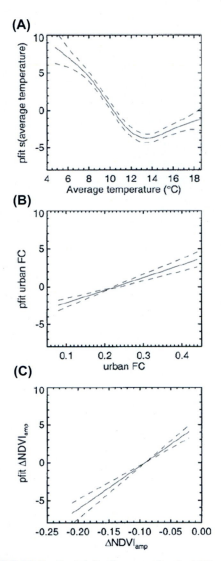

FIGURE 23.9 Partial fit of a generalized additive model for predicting differences between the city and the DBF SOS. (A) Average annual temperature in 1982–99; (B) urban vegetation coverage (FC); and (C) amplitude difference of normalized vegetation index in cities and DBF ($\Delta NDVI_{amp}$) (White et al., 2002).

It is an integrative descriptor of ecosystem functioning and resources and a key variable in assessing the effects of land use changes on ecosystem conditions. The estimation methods from remote sensing and modeling have been presented in Chapter 16. Urban sprawl has rapidly transformed landscapes and profoundly altered biodiversity and ecosystem functioning. Understanding the effects of urbanization on NPP is initially important for the estimation of urbanization impacts on ecosystem functioning and the development of sustainable urban landscapes (Zhou et al., 2017).

Some efforts have been made to quantify the impact of urbanization on NPP in previous studies (Buyantuyev and Wu, 2009; Lu et al., 2010; Milesi et al., 2003; Yu et al., 2009; Guan et al., 2019; Peng et al., 2016; Wen et al., 2019). For example, Imhoff et al. (2004) studied the impact of urbanization on NPP in the United States. The data used in their study included the DMSP-derived urban map data produced by Imhoff et al. (1997), the global monthly composite of the NDVI data set from the AVHRR, a vegetation map derived by Hansen et al. (2000), monthly solar radiation (Bishop and Rossow, 1991), soil texture (Zobler, 1986), and temperature and precipitation (Shea, 1986). The AVHRR data were collected from April 1992 to March 1993 and had a 1-km spatial resolution.

Their method included three main steps. (1) They first divided the landscape of the United States into three types, namely, urban, periurban, and nonurban land in terms of the DMSP-derived urban map (Fig. 23.10). Urban areas were the areas with high levels of nighttime light emission. Periurban areas were those with substantial but unstable illumination. Nonurban lands were the areas with no or little observed illumination. (2) They used the Carnegie-Ames-Stanford Approach (CASA) terrestrial carbon model to estimate NPP and produced a map of total annual NPP for the United States at 1-km spatial resolution (Fig. 23.11). More details about the CASA model can be found in Chapter 16. (3) They compared monthly rates of NPP and annual total carbon production for urban, periurban, and nonurban lands over the whole United States.

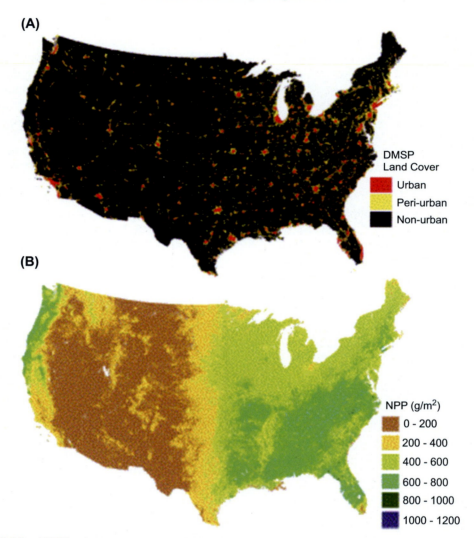

FIGURE 23.10   (A) Urbanization map generated from nighttime satellite images from the Defense Meteorological Satellite's Operational Linescan System (DMSP/OLS) collected from October 1994 to March 1995. Red (urban), yellow (periurban), black (nonurban). (B) Simulated total annual NPP for the United States at 1×1 km horizontal resolution (Imhoff et al., 2004).

Their results indicate that urbanization had negative impact on NPP. The overall annual reduction in NPP caused by urbanization in the United States was estimated to be 0.04 pg, which is 1.6% of the periurban input.

23.4.3 The influence of urbanization on land surface parameters and environment

Urbanization is an inevitable phenomenon of social and economic development and not only changes land use/cover types but also affects underlying surface properties, such as green vegetation cover (Fan et al., 2017; Jenerette et al., 2007), surface albedo (Sailor, 1995), surface roughness (Christen et al., 2007) etc., along with urban hydrological and ecological circulation (Grimm et al., 2008; Kalantari et al., 2017), changing the urban environment and near-surface climate.

Hou et al. (2014) used Landsat TM/ETM+ images with 30 m resolution and MODIS

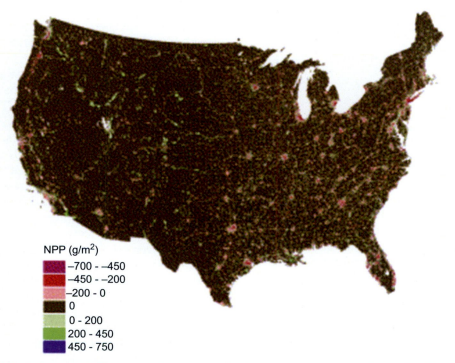

NPP (g/m²)

■	−700 - −450
■	−450 - −200
■	−200 - 0
■	0
■	0 - 200
■	200 - 450
■	450 - 750

FIGURE 23.11 Difference in NPP showing the total annual reduction (negative) or gain (positive) in the rates of NPP (g/m²) (POST-urban−PRE-urban) (Imhoff et al., 2004).

Collection 5 datasets to analyze change trends, intensity, and spatial patterns for vegetation cover, urbanization impact on vegetation coverage, and vegetation coverage effects on surface albedo in Guangzhou, China. The main processes of the study were as follows:

(1) Landsat images were preprocessed, including radiation calibration and atmospheric correction, to reduce noise due to water vapor and air pollution particles in the atmosphere.

(2) Green vegetation fraction (GVF) was calculated for the Landsat images using (Gutman and Ignatov, 1998)

$$\text{NDVI} = \frac{\rho(band4) - \rho(band3)}{\rho(band4) + \rho(band3)} \quad (23.4)$$

$$\text{GVF} = \frac{\text{NDVI} - \text{NDVI}_{min}}{\text{NDVI}_{max} - \text{NDVI}_{min}} \quad (23.5)$$

where NDVI is the normalized difference vegetation index calculated by using the red and near-infrared bands; $\rho(band4)$ and $\rho(band3)$ are the spectral reflectance of near-infrared band and red band for Landsat images, respectively; NDVI_{min} and NDVI_{max} represent the minimum and maximum value of NDVI, respectively.

(3) Total SW albedo α_{short} was calculated for the Landsat TM/ETM+

$$\alpha_{short} = 0.356\alpha_1 + 0.130\alpha_3 + 0.373\alpha_4 + 0.085\alpha_5 + 0.072\alpha_7 - 0.0018, \quad (23.6)$$

MODIS images

$$\alpha_{short} = 0.160\alpha_1 + 0.291\alpha_2 + 0.243\alpha_3 + 0.116\alpha_4$$
$$+ 0.112\alpha_5 + 0.081\alpha_7 - 0.0015,$$

(23.7)

visible

$$\alpha_{visible} = 0.443\alpha_1 + 0.317\alpha_2 + 0.240\alpha_3, \quad (23.8)$$

and near-infrared

$$\alpha_{NIR} = 0.693\alpha_4 + 0.212\alpha_5 + 0.116\alpha_7 - 0.003.$$

(23.9)

Albedos were calculated for Landsat TM/ETM+ following the Liang (2001) algorithm based on different band combinations. In the formula, α_i (i = 1, 2, ..., 7) represents the spectral reflectance of the *ith* band.

(4) Surface albedo and vegetation coverage change trends and seasonal variations as relationships between them were compared.

Fig. 23.12 shows GVF and surface albedo changes for different Guangzhou regions. Urbanization Guangzhou accelerated year by year 1990−2000 with GVF decreasing in each subregion except Conghua, while surface albedo increased. However, albedo changes in the visible and near-infrared bands were not significant.

Fig. 23.13 shows the relationships between SW, visible, and near-infrared band albedos and GVF for the Guangzhou area. The relationship between surface albedo and GVF changes for different GVF threshold settings. The appropriate GVF threshold was ~0.21. When GVF was <0.21, there was no significant change in visible band reflectance with increasing of GVF (Fig. 23.13A), whereas SW and NIR reflectance increased significantly (Fig. 23.13B,C). When GVF was >0.21, SW, visible, and near-infrared albedo decreased significantly with increasing GVF. Regions with GVF <0.21 were mainly distributed in urban and water bodies areas (Fig. 23.14A), and surface albedo was also low in urban suburbs and rural areas in the north (Fig. 23.14B). For GVF < threshold, the surface was mainly a mixture of vegetation, impervious layer, and soil, and reflectivity increased with increasing GVF. When GVF > threshold, albedo started to decline because land cover type became dense vegetation.

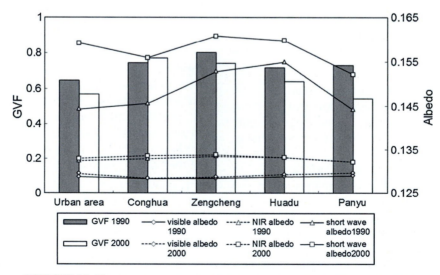

FIGURE 23.12　The change trend of GVF and surface albedo (Hou et al., 2014).

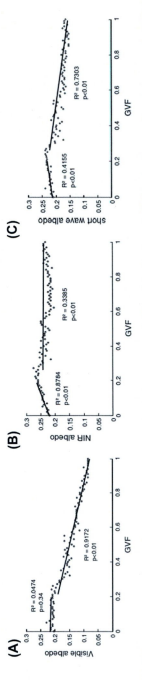

FIGURE 23.13 Relationship between near (A) infrared, (B) shortwave, and (C) visible albedo and GVF in the Greater Guangzhou area in 2000 (Hou et al., 2014).

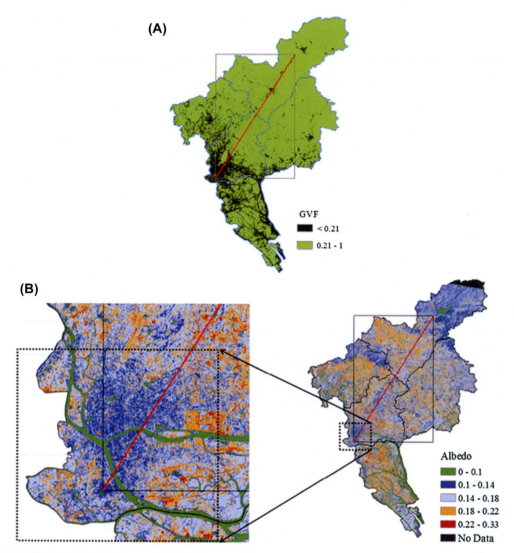

FIGURE 23.14　Distribution of (A) GVF and (B) shortwave albedo from Landsat ETM+ (Hou et al., 2014).

23.4.4 Urban heat island effects

UHIs occur as a result of the phenomenon whereby urban areas are warmer than the surrounding rural areas (Sobstyl et al., 2018). UHIs are more obvious at nighttime than during the day. UHIs are caused by multiple factors, such as changes in the thermal properties of surface materials, lack of evapotranspiration in urban areas, the urban canyon effect, and building heat blocked by buildings (Deilami et al., 2018; Wu and Zhang, 2018; Zhou and Chen, 2018). Fig. 23.15 shows the surface energy balance over a rural area and an urban area during both daytime and nighttime.

The characteristics of the UHI effect have been studied extensively. Traditionally, the UHI

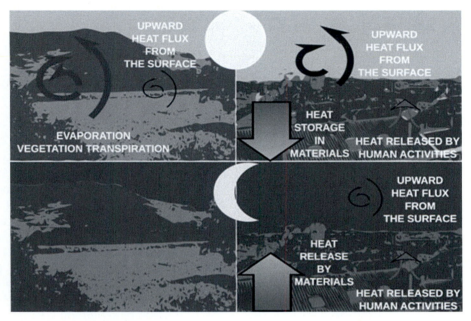

FIGURE 23.15 Sketch of surface energy balance (SEB) over a rural area (left) and an urban land use (right) during the day (upper panels) and the night (lower panels). During the day, the rural SEB is dominated by evaporation of water from the surface, while the urban SEB is dominated by the accumulation of heat in materials and air heating. During the night, while few exchanges occur over rural areas, the heat accumulated in material and released by human activities delays the air cooling (Hidalgo et al., 2008).

intensity for a given urban location is studied by comparing the temperature difference between a given urban site (multiple urban sites) and a carefully selected nearby nonurban (reference) site (or multiple nonurban sites) from meteorological observations. However, the representativeness of observational sites is quite important for accurately estimating UHI intensity, and the UHI intensity is influenced by many factors, such as land cover type, urban size, urban population, climate background, latitude, longitude, and urban parameters (e.g.,xbuilt-up ratio, green surface ratio, sky view factor, etc.). Choosing representative observation sites is challenging. Besides, with fast urban expansion, the nonurban sites might be merged into the urban area, which impedes continuous UHI examination.

With the advent of thermal remote sensing technology, remote observation of UHIs became possible using satellite and aircraft platforms. The first satellite-based study was conducted by Rao (1972). Subsequently, many studies of UHIs based on remote sensing technology have been reported (Chen et al., 2006; Hung et al., 2006; Jenerette et al., 2007; Yuan and Bauer, 2007). Some researchers successfully reviewed the use of thermal remote sensing in the study of UHIs (Arnfield, 2003; Roth et al., 1989; Voogt and Oke, 2003; Weng, 2009).

The most popular parameter in UHI study using remote sensing technology is land surface temperature (LST). LST modulates the air temperature of the lower layer of the urban atmosphere and is a primary factor in exploring surface radiation and energy exchange, the internal climate of buildings, the spatial structure of urban thermal patterns and their relation to urban surface characteristics, surface—air temperature relationships, and human comfort in cities.

Various LST products are being produced from different satellite data (see Chapter 8), and some of them have been used in studies of the urban thermal environment and its dynamics. For example, Imhoff et al. (2010) used the MODIS LST products to explore the relationship between UHIs and urban development intensity, size, and ecological setting in 38 of the most populous cities in the continental United States.

Their methods included the following several steps. (1) They used the terrestrial ecoregions map developed by Olson et al. (2001) to stratify 38 cities into 8 groups. (2) They used the impervious surface area (ISA) data from the land cover map of the United States (NLCD2001) to identify urban areas of each city, estimated their size, and then stratified 38 cities into 5 groups in term of ISA density. (3) They extracted LST and NDVI for each individual city from MODIS LST and VI products. The pixels with NDVI smaller than 0.1 were excluded from the analysis. (4) On the basis of the topographic data, they calculated the mean elevation of each city and masked out the areas outside a ±50-m window from mean elevation. This step is to avoid the effects of elevation difference on the analysis of UHIs. (5) They compared the LST difference between different ISA groups defined in step 2 under each biome and also compared the average LST differences between the urban core area and the rural area for all the cities in each biome, for summer time and winter time. (6) They further computed the relationships between the fractional area of impervious surfaces and the UHI and the relationship between the total size of the urban area and the amplitude of the UHI.

In light of their results, they concluded that the fraction of ISA is highly correlated to LST for all cities in the continental United States, in all biomes except deserts and xeric shrublands (Fig. 23.16). The highest LST difference between urban and rural areas occurred in summer at midday. The ecological background plays a significant role in modulating the diurnal and seasonal UHI.

23.4.5 Impact of urbanization on air quality

Air pollution has harmful effects on human health (An et al., 2018; Guo et al., 2018; Peng et al., 2002; Vopham et al., 2018; Wang and Mauzerall, 2006) and natural ecosystems (Bobbink et al., 1998; Duan et al., 2017; Karnosky et al., 2007; Munzi et al., 2017; Taylor et al., 1994; Dirnböck et al., 2017). Air pollution is not only impacted by climatic conditions but also influences the climate (Doherty et al., 2017; Kinney and Patrick, 2018; Xu, 2001). Air pollutants modify the properties of cloud condensation nuclei (Duan et al., 2018; Wiedensohler et al., 2009), inhibit precipitation (Rosenfeld, 2000; Zhao et al., 2006), cause atmospheric haze (Ma et al., 2010; Wang et al., 2017), affect radiative forcing (Jacobson, 2001; Qian and Giorgi, 2000), reduce visibility (Guo et al., 2004; Qiu and Yang, 2000), and lead to meteorological disasters (Larssen and Carmichael, 2000). Urban expansion has resulted in tremendous increases in energy consumption and emissions of air pollutants. Air pollution has become one of the top environmental concerns.

Aerosols, ozone (O_3), nitrogen dioxide (NO_2), carbon monoxide (CO), and sulfur dioxide (SO_2) are common pollutants in surface air. Aerosols are colloid suspensions of fine solid particles or liquid droplets in the air and play an important role in the global climate balance. Aerosols influence climate by altering radiative forcing directly (e.g., the scattering of solar radiation and the absorption/emission of terrestrial radiation) and indirectly (e.g., effects of aerosols on cloud properties). Ozone in the atmosphere absorbs long-wave infrared radiation emitted from the Earth's surface and ozone near the ground damages plants, reducing their ability to sink carbon dioxide from the atmosphere. NO_2 has a long

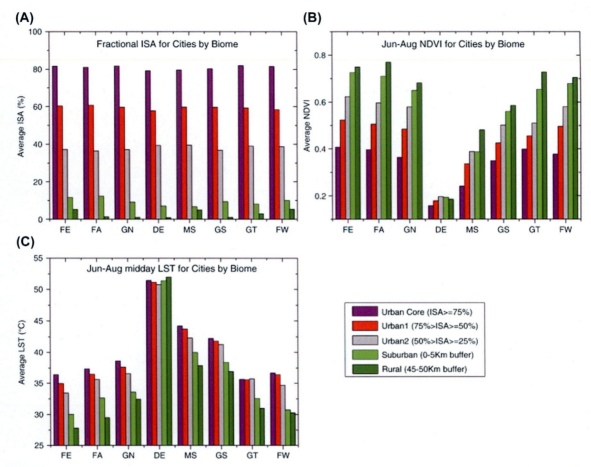

FIGURE 23.16   Average remotely sensed parameters of the five zones for cities in each of the eight biomes defined in this study. Panel A represents the fractional imperviousness area (%), Panel B shows the average summer (June—August) MODIS NDVI for each zone in each group, and Panel C shows the average summer daytime (around 1:30 p.m. local time) land surface temperature derived from MODIS product (Imhoff et al., 2010).

atmospheric lifetime (about 120 yr), and the heat trapping effects of NO_2 are 310 times more powerful than carbon dioxide. Sulfur dioxide emissions are a precursor to acid rain.

Satellite remote sensing of trace gases and aerosols for air quality applications can be traced back to 1976, when Lyons and Husar first showed a large area of haze covering the Midwest of the United States in a GOES image. Since that time, satellite remote sensing of air quality has evolved dramatically. Table 23.7

summarizes the currently available Earth observation instruments that measure trace gases in the lower troposphere. Fig. 23.17 shows an example of tropospheric NO_2 columns retrieved from the SCIAMACHY.

Martin (2008) reviewed satellite remote sensing—based studies of air quality and grouped these studies into three categories, namely (1) those analyzing the events that affect air quality; (2) those examining surface air quality; and (3) those estimating emissions. Among remote

TABLE 23.7   Nadir sensors measure trace gases in the lower troposphere (less than 6 km), which can be used for estimating surface sources and sinks. The local equatorial crossing time (LECT) denotes the sunlit portion of the complete orbit (Palmer, 2008).

Nadir sensors (launch year)	Orbit (LECT)	Horizontal resolution (vertical degrees of freedom)	Approximate repeat time	Gases measured
Aqua satellite AIRS (2002)	LEO (1330)	13.5 km diameter circle (1 d.f.)	Daily	CO, CO_2, O_3, CH_4
Aura satellite OMI (2003)	LEO (1338)	13×24 km^2 (column)	Daily	O_3, HCHO, NO_2, SO_2
TES		5×8 km^2 (1–2 d.f.)	6 days	O_3, CH_4, CO
Envisat satellite (2002) SCIAMACHY	LEO (1000)	30×60 km^2 (column)	6 days	O_3, HCHO, NO_2, SO_2, CO, CO_2, CH_4
MetOp satellite (2007) IASI	LEO (0930)	12 km diameter circle (1–3 d.f.)	12 h	O_3, CO, CH_4, BrO, SO_2
GOME-2		40×80 km^2 (column)	Daily	O_3, HCHO, NO_2, BrO, SO_2
Terra satellite (1999) MOPITT	LEO (1030)	22×22 km^2 (1–2 d.f.)	3 days	CO
ERS-2 satellite (1995) GOME	LEO (1030)	40×320 km^2 (column)	3 days	O_3, HCHO, NO_2, BrO, SO_2
To be launched OCO (2008)	LEO (c1315)	1.3×2.3 km^2 (column)	16 days	CO_2
GOSAT (2008)	LEO (c1300)	10 km diameter circle (1–2 levels)	3 days	CO_2, CH_4
SENTINEL 4/5 (c2018)	GEO (all)	To be determined	n.a.	TBD

sensing—based studies, aerosol pollution has been paid most attention in studies of air quality from satellite observations. For example, Ramachandran (2007) analyzed aerosol optical depth (AOD) and fine-mode fraction (FMF) variations over four Indian cities, namely, Chennai (13.04°N, 80.17°E, south), Mumbai (18.90°N, 72.81°E, west), Kolkata (22.65°N, 88.45°E, east), and New Delhi (28.56°N, 77.11°E, north). The data used in his study included the daily mean MODIS Terra—derived Level 3 $1° \times 1°$ grid AOD and FMF, as well as in situ AOD measurements at Kanpur from the Aerosol Robotic Network. All these data were collected from 2001 to 2005. They first validated MODIS-derived AOD and FMF data by comparing them with the in situ measurements at Kanpur and then analyzed the interannual and intraannual variations in MODIS-derived AOD and FMF for the period 2001—05 over all four study sites. They also explored the relationship between AOD and climate factors (rainfall and relative humidity). Figs. 23.18—23.20, show part of their results. They found that MODIS-derived AODs and FMFs compared well with in situ measurements over Kanpur. The AODs showed obvious seasonal variation in all the four studied sites. FMF variation over Kolkata was different from that over Chennai, Mumbai, and New Delhi, where FMFs were low during the premonsoon and summer seasons. AODs showed a positive correlation with rainfall.

(A)

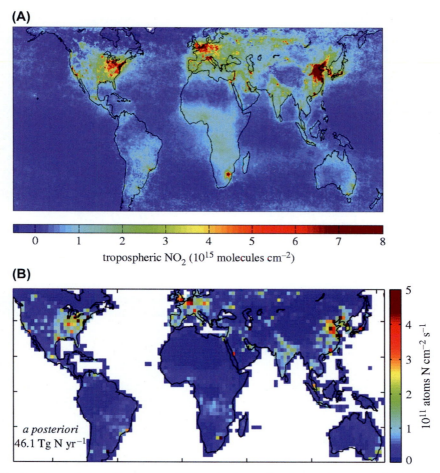

tropospheric NO$_2$ (10^{15} molecules cm^{-2})

(B)

a posteriori
46.1 Tg N yr^{-1}

FIGURE 23.17   (A) Tropospheric NO$_2$ columns (1015 molecules/cm^2) retrieved from the SCIAMACHY for May 2004–April 2005 and (B) associated NO$_x$ emissions (10^{11} atoms N/cm^2/s). NO$_2$ columns, filtered for cloud radiance less than 0.5, are averaged on a 0.4° × 0.4°. The NO$_x$ emissions, averaged on a 2° × 2.5° grid, are determined through inverse modeling of the NO$_2$ columns using the GEOS-Chem chemistry transport model. *Figure courtesy of Randall Martin, Dalhousie University, from Martin et al., 2006. Evaluation of space-based constraints on global nitrogen oxide emissions with regional aircraft measurements over and downwind of eastern North America. J. Geophy. Res. doi: http://10.1029/2005JD006680.*

Kolkata had the lowest variation in annual mean FMF, and New Delhi had the highest variation in annual mean FMF.

23.5 Summary

This chapter focuses on the research cases of remote sensing data and products in urban area mapping, urban expansion monitoring, urban ecological environment survey, and urbanization impact assessment. However, with the acceleration of global urbanization, the urban landscape is changing with each passing day, and the problems caused by urban development are becoming increasingly prominent. How to coordinate the relationship between urbanization and ecological

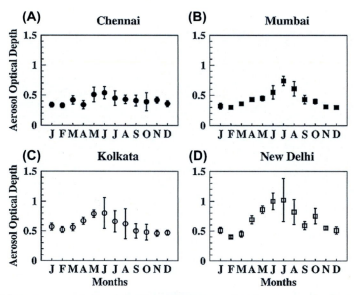

FIGURE 23.18   Monthly mean aerosol optical depths at 550 nm obtained by averaging the data from 2001 to 05 over: (A) Chennai, (B) Mumbai, (C) Kolkata, and (D) New Delhi. *Vertical bars* indicate ± 1 s from the mean (Ramachandran, 2007).

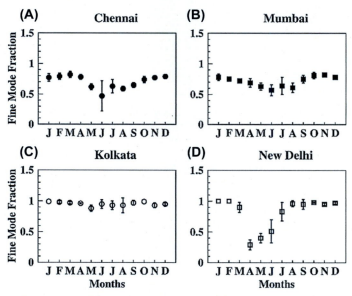

FIGURE 23.19   The 2001–05 averaged intraannual fine-mode fraction values at (A) Chennai, (B) Mumbai, (C) Kolkata, and (D) New Delhi. *Vertical bars* denote ± 1 s from the mean (Ramachandran, 2007).

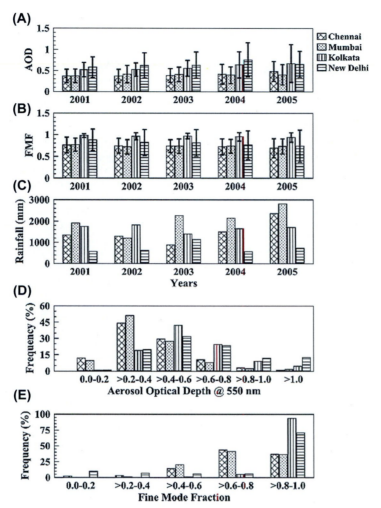

FIGURE 23.20 Annual averages of (A) aerosol optical depths, (B) fine-mode fraction, and (C) rainfall for Chennai, Mumbai, Kolkata, and New Delhi from 2001 to 05. Frequency distribution of (D) aerosol optical depths and (E) fine-mode fraction of aerosols over Chennai, Mumbai, Kolkata, and New Delhi obtained from daily mean data during 2001–05 (Ramachandran, 2007).

environment is a common concern of people from all walks of life. The demand of remote sensing in urban research is deepening, and the information elements are becoming more comprehensive and detailed. It has changed from two-dimensional plane thematic information to three-dimensional spatial analysis, from explanatory study of urban phenomena, patterns and functions to diagnostic study of urban problems, and urban functions. It is an inevitable trend in the future to synthesize remote sensing data and other data sources and develop new information processing methods.

References

An, R., Ji, M., Yan, H., Guan, C., 2018. Impact of ambient air pollution on obesity: a systematic review. Int. J. Obes. 42, 1.

Andres, R.J., Gregg, M., Inez, F., Elaine, M., 1996. Distribution of carbon dioxide emissions from fossil fuel consumption and cement manufacture, 1950–1990. Glob. Biogeochem. Cycles 10, 419–430.

Arnfield, A.J., 2003. Two decades of urban climate research: a review of turbulence, exchanges of energy and water, and the urban heat island. Int. J. Climatol. 23, 1–26.

Arrieta, E.M., Cuchietti, A., Cabrol, D., González, A.D., 2018. Greenhouse gas emissions and energy efficiencies for soybeans and maize cultivated in different agronomic zones: a case study of Argentina. Sci. Total Environ. 625, 199–208.

Bartholome, E., Belward, A.S., 2005. GLC2000: a new approach to global land cover mapping from earth observation data. Int. J. Remote Sens. 26, 1959–1977.

Bhaduri, B., Bright, E., Coleman, P., Dobson, J., 2002. LandScan: locating people is what matters. Geoinfomatics 5, 34–37.

Bishop, J.K.B., Rossow, W.B., 1991. Spatial and temporal variability of global surface solar irradiance. J. Geophys. Res. Oceans 96, 16839–16858.

Bobbink, R., Hornung, M., Roelofs, J.G.M., 1998. The effects of air-borne nitrogen pollutants on species diversity in natural and semi-natural European vegetation. J. Ecol. 86, 717–738.

Buyantuyev, A., Wu, J., 2009. Urbanization alters spatiotemporal patterns of ecosystem primary production: a case study of the Phoenix metropolitan region, USA. J. Arid Environ. 73, 512–520.

Cao, X., Chen, J., Imura, H., Higashi, O., 2009. A SVM-based method to extract urban areas from DMSP-OLS and SPOT VGT data. Remote Sens. Environ. 113, 2205–2209.

Chen, X.L., Zhao, H.M., Li, P.X., Yin, Z.Y., 2006. Remote sensing image-based analysis of the relationship between urban heat island and land use/cover changes. Remote Sens. Environ. 104, 133–146.

Christen, A., Gorsel, E.V., Vogt, R., 2007. Coherent structures in urban roughness sublayer turbulence. Int. J. Climatol. 27, 1955–1968.

CIESIN, 2004. Global Rural-Urban Mapping Project(-GRUMP) Alpha Version: Urban Extents. http://sedac.ciesin.columbia.edu/gpw.

CIESIN, 2000. Gridded population of the world (GPW). Version 2. Palisades (NY). Center for International Earth Science Information Network (Columbia University), International Food Policy Research Institute, and World Resources Institute.

Danko, D.M., 1992. The digital chart of the world project. Photogramm. Eng. Remote Sens. 58, 1125–1128.

Deilami, K., Kamruzzaman, M., Liu, Y., 2018. Urban heat island effect: a systematic review of spatio-temporal factors, data, methods, and mitigation measures. Int. J. Appl. Earth Obs. Geoinf. 67, 30–42.

DESA, 2006. World Urbanization Prospects: The 2005 Revision. Pop. Division, Department of Economic and Social Affairs, UN. http://www.un.org/esa/population/publications/WUP2005/2005wup.htm.

Dirnböck, T., Djukic, I., Kitzler, B., Kobler, J., Moldijkstra, J.P., Posch, M., Reinds, G.J., Schlutow, A., Starlinger, F., Wamelink, W.G.W., 2017. Climate and air pollution impacts on habitat suitability of austrian forest ecosystems. PLoS One 12 (9), e0184194.

Doherty, R.M., Heal, M.R., O'Connor, F.M., 2017. Climate change impacts on human health over europe through its effect on air quality. Environ. Health 16 (S1), 118.

Duan, K., Sun, G., Zhang, Y., Yahya, K., Wang, K., Madden, J.M., Caldwell, P.V., Cohen, E.C., McNulty, S.G., 2017. Impact of air pollution induced climate change on water availability and ecosystem productivity in the conterminous United States. Clim. Change 140 (2), 259–272.

Duan, L., Wang, Y., Xie, X., Li, M., Tao, J., Wu, Y., Cheng, T., Zhang, R., Liu, Y., Li, X., He, Q., Gao, W., Wang, J., 2018. Influence of pollutants on activity of aerosol cloud condensation nuclei (ccn) during pollution and post-rain periods in guangzhou, southern China. Sci. Total Environ. 642, 1008–1019.

Elvidge, C.D., Imhoff, M.L., Baugh, K.E., Hobson, V.R., Nelson, I., Safran, J., Dietz, J.B., Tuttle, B.T., 2001. Nighttime lights of the world: 1994–1995. ISPRS J. Photogram. Remote Sens. 56, 81–99.

Elvidge, C.D., Tuttle, B.T., Sutton, P.S., Baugh, K.E., Howard, A.T., Milesi, C., Bhaduri, B.L., Nemani, R., 2007. Global distribution and density of constructed impervious surfaces. Sensors 7, 1962–1979.

Eric, D.M., Duarte André, G., Way, D.A., 2018. Plant Carbon Metabolism and Climate Change: Elevated Co_2, and Temperature Impacts on Photosynthesis, Photorespiration and Respiration (New Phytologist).

ESA, 2008. The Ionia Globcover Project the GlobaCover Portal. http://ionia1.esrin.esa.int.

Fan, C., Myint, S.W., Kaplan, S., Middel, A., Zheng, B., Rahman, A., Huang, H.P., Brazel, A., Blumberg, D.G., 2017. Understanding the impact of urbanization on surface urban heat islands—a longitudinal analysis of the oasis effect in subtropical desert cities. Remote Sens. 9 (7), 672.

Fuller, R.J., Williamson, T., Barnes, G., Dolman, P.M., Mac Nally, R., 2017. Human activities and biodiversity opportunities in pre-industrial cultural landscapes: relevance to conservation. J. Appl. Ecol. 54, 459–469.

Goldewijk, K.K., 2005. Three centuries of global population growth: a spatial referenced population (density) database for 1700–2000. Popul. Environ. 26, 343–367.

Grimm, N.B., Faeth, S.H., Golubiewski, N.E., Redman, C.L., Wu, J., Bai, X., 2008. Global change and the ecology of cities. Science 319, 756.

Guan, X.B., Shen, H.F., Li, X.H., Gan, W.X., Zhang, L.P., 2019. A long-term and comprehensive assessment of the urbanization-induced impacts on vegetation net primary productivity. Sci. Total Environ. 669, 342–352.

Guida-Johnson, B., Faggi, A.M., Zuleta, G.A., 2017. Effects of urban sprawl on riparian vegetation: is compact or dispersed urbanization better for biodiversity? River Res. Appl. 33, 959–969.

Gunawardena, K.R., Wells, M.J., Kershaw, T., 2017. Utilising green and bluespace to mitigate urban heat island intensity. Sci. Total Environ. 584–585, 1040–1055.

Guo, J., Rahn, K.A., Zhuang, G.S., 2004. A mechanism for the increase of pollution elements in dust storms in Beijing. Atmos. Environ. 38, 855–862.

Guo, Q., Fu, B., Shi, P., Cudahy, T., Zhang, J., Xu, H., 2017. Satellite monitoring the spatial-temporal dynamics of desertification in response to climate change and human activities across the Ordos Plateau, China. Remote Sens. 9 (6), 525.

Guo, X., Zhao, L., Chen, D., Jia, Y., Zhao, N., Liu, W., Cheng, S., 2018. Air quality improvement and health benefit of pm2.5 reduction from the coal cap policy in the Beijing–Tianjin–Hebei (BTH) region, China. Environ. Sci. Pollut. Control Ser. 25, 32709–32720.

Gutman, G., Ignatov, A., 1998. The derivation of the green vegetation fraction from NOAA/AVHRR data for use in numerical weather prediction models. Int. J. Remote Sens. 19, 1533–1543.

Hansen, M.C., Defries, R.S., Townshend, J.R.G., Sohlberg, R., 2000. Global land cover classification at 1km spatial resolution using a classification tree approach. Int. J. Remote Sens. 21, 1331–1364.

He, C., Liu, Z., Gou, S., Zhang, Q., Zhang, J., Xu, L., 2019. Detecting global urban expansion over the last three decades using a fully convolutional network. Environ. Res. Lett. 14 (3).

He, C., Zhang, D., Huang, Q., Zhao, Y., 2016. Assessing the potential impacts of urban expansion on regional carbon storage by linking the lusd-urban and invest models. Environ. Model. Softw 75, 44–58.

Hidalgo, J., Masson, V., Baklanov, A., Pigeon, G., Gimeno, L., 2008. Advances in urban climate modeling. Trends Dir Clim. Res. 354–374.

Hou, M., Hu, Y., He, Y., 2014. Modifications in vegetation cover and surface albedo during rapid urbanization: a case study from south China. Environ. Earth Sci. 72, 1659–1666.

Huang, C.W., Mcdonald, R.I., Seto, K.C., 2018. The importance of land governance for biodiversity conservation in an era of global urban expansion. Landsc. Urban Plan. 173, 44–50.

Hung, T., Uchihama, D., Ochi, S., Yasuoka, Y., 2006. Assessment with satellite data of the urban heat island effects in Asian mega cities. Int. J. Appl. Earth Obs. Geoinf. 8, 34–48.

Imhoff, M.L., Bounoua, L., DeFries, R., Lawrence, W.T., Stutzer, D., Tucker, C.J., Ricketts, T., 2004. The consequences of urban land transformation on net primary productivity in the United States. Remote Sens. Environ. 89, 434–443.

Imhoff, M.L., Lawrence, W.T., Stutzer, D.C., Elvidge, C.D., 1997. A technique for using composite DMSP/OLS "city lights" satellite data to map urban area. Remote Sens. Environ. 61, 361–370.

Imhoff, M.L., Zhang, P., Wolfe, R.E., Bounoua, L., 2010. Remote sensing of the urban heat island effect across biomes in the continental USA. Remote Sens. Environ. 114, 504–513.

Jacobson, M.Z., 2001. Strong radiative heating due to the mixing state of black carbon in atmospheric aerosols. Nature 409, 695–697.

Jenerette, G.D., Harlan, S.L., Brazel, A., Jones, N., Larsen, L., Stefanov, W.L., 2007. Regional relationships between surface temperature, vegetation, and human settlement in a rapidly urbanizing ecosystem. Landsc. Ecol. 22, 353–365.

Ji, W., Ma, J., Twibell, R.W., Underhill, K., 2006. Characterizing urban sprawl using multi-stage remote sensing images and landscape metrics. Comput. Environ. Urban Syst. 30, 861–879.

Kalantari, Z., Ferreira, C.S.S., Walsh, R.P.D., Ferreira, A., José, D., Destouni, G., 2017. Urbanization development under climate change: hydrological responses in a peri-urban mediterranean catchment. Land Degrad. Dev. 28.

Karelin, D.V., Goryachkin, S.V., Zamolodchikov, D.G., Dolgikh, A.V., Zazovskaya, E.P., Shishkov, V.A., 2017. Human footprints on greenhouse gas fluxes in cryogenic ecosystems. Dokl. Earth Sci. 477 (2), 1467–1469.

Karnosky, D.F., Skelly, J.M., Percy, K.E., Chappelka, A.H., 2007. Perspectives regarding 50 years of research on effects of tropospheric ozone air pollution on US forests. Environ. Pollut. 147, 489–506.

Kinney, Patrick, L., 2018. Interactions of climate change, air pollution, and human health. Curr. Environ. Health Rep. 5 (1), 179–186.

Larssen, T., Carmichael, G.R., 2000. Acid rain and acidification in China: the importance of base cation deposition. Environ. Pollut. 110, 89–102.

Liang, S., 2001. Narrowband to broadband conversions of land surface albedo I : algorithms. Remote Sens. Environ. 76, 213–238.

Liu, C., Li, X., 2012. Carbon storage and sequestration by urban forests in Shenyang, China. Urban For. Urban Green. 11, 121–128.

Lu, D.S., Xu, X.F., Tian, H.Q., Moran, E., Zhao, M.S., Running, S., 2010. The effects of urbanization on net primary productivity in southeastern China. Environ. Manag. 46, 404–410.

Lu, Y., Coops, N.C., Hermosilla, T., 2017. Chronicling Urbanization and Vegetation Changes Using Annual Gap Free Landsat Composites from 1984 to 2012. Urban Remote Sensing Event, IEEE.

Ma, J., Chen, Y., Wang, W., Yan, P., Liu, H., Yang, S., Hu, Z., Lelieveld, J., 2010. Strong air pollution causes widespread haze-clouds over China. J. Geophys. Res. Atmos. 115.

Mac Dowell, N., Fennell, P.S., Shah, N., Maitland, G.C., 2017. The role of CO_2 capture and utilization in mitigating climate change. Nat. Clim. Chang. 7 (4), 243–249.

Mahmoud, I.M., Alfred, D., Christopher, C., Michael, T., Halilu, S.A., 2016. Analysis of settlement expansion and urban growth modelling using geoinformation for assessing potential impacts of urbanization on climate in abuja city, Nigeria. Remote Sens. 8, 220.

Martin, R.V., 2008. Satellite remote sensing of surface air quality. Atmos. Environ. 42, 7823–7843.

Mchale, M.R., Mcpherson, E.G., Burke, I.C., 2007. The potential of urban tree plantings to be cost effective in carbon credit markets. Urban For. Urban Green. 6, 49–60.

Milesi, C., Elvidge, C.D., Nemani, R.R., Running, S.W., 2003. Assessing the impact of urban land development on net primary productivity in the southeastern United States. Remote Sens. Environ. 86, 401–410.

Morris, K.I., Chan, A., Morris, K.J.K., Ooi, M.C.G., Oozeer, M.Y., Abakr, Y.A., 2017. Impact of urbanization level on the interactions of urban area, the urban climate, and human thermal comfort. Appl. Geogr. 79, 50–72.

Munzi, S., Ochoa-Hueso, R., Gerosa, G., Marzuoli, R., 2017. Emerging directions on air pollution and climate change research in mediterranean basin ecosystems. Environ. Sci. Pollut. Control Ser. 24 (34), 26155–26159.

Nowak, D.J., 1993. Atmospheric carbon reduction by urban trees. J. Environ. Manag. 37, 207–217.

Nowak, D.J., 1994. Atmospheric Carbon Dioxide Reduction by Chicago's Urban Forest. General Technical Report Ne.

Olson, D.M., Dinerstein, E., Wikramanayake, E.D., Burgess, N.D., Powell, G.V.N., Underwood, E.C., D'Amico, J.A., Itoua, I., Strand, H.E., Morrison, J.C., Loucks, C.J., Allnutt, T.F., Ricketts, T.H., Kura, Y., Lamoreux, J.F., Wettengel, W.W., Hedao, P., Kassem, K.R., 2001. Terrestrial ecoregions of the worlds: a new map of life on Earth. Bioscience 51, 933–938.

Palmer, P.I., 2008. Quantifying sources and sinks of trace gases using space-borne measurements: current and future science. Philos. Trans. R. Soc. A Math. Phys. Eng. Sci. 366, 4509–4528.

Pataki, D.E., Alig, R.J., Fung, A.S., Golubiewski, N.E., Kennedy, C.A., Mcpherson, E.G., 2006. Urban ecosystems and the north American carbon cycle. Glob. Chang. Biol. 12, 2092–2102.

Paudyal, N., Khadka, R.B., Khadka, R., 2018. Agricultural perspectives of climate change induced disasters in Doti, Nepal. J. Agric. Environ. 14.

Peng, C.Y., Wu, X.D., Liu, G., Johnson, T., Shah, J., Guttikunda, S., 2002. Urban air quality and health in China. Urban Stud. 39, 2283–2299.

Peng, J., Shen, H., Wu, W., Liu, Y., Wang, Y., 2016. Net primary productivity (npp) dynamics and associated urbanization driving forces in metropolitan areas: a case study in beijing city, China. Landsc. Ecol. 31 (5), 1077–1092.

Puhlick, J., Woodall, C., Weiskittel, A., 2017. Implications of land-use change on forest carbon stocks in the eastern United States. Environ. Res. Lett. 12 (2), 024011.

Qian, Y., Giorgi, F., 2000. Regional climatic effects of anthropogenic aerosols? The case of Southwestern China. Geophys. Res. Lett. 27, 3521–3524.

Qiu, J.H., Yang, L.Q., 2000. Variation characteristics of atmospheric aerosol optical depths and visibility in North China during 1980–1994. Atmos. Environ. 34, 603–609.

Ramachandran, S., 2007. Aerosol optical depth and fine mode fraction variations deduced from Moderate Resolution Imaging Spectroradiometer (MODIS) over four urban areas in India. J. Geophys. Res. Atmos. 112.

Rao, P.K., 1972. Remote sensing of urban heat islands from an environmental satellite. Bull. Am. Meteorol. Soc. 53, 647.

Ren, Y., Yan, J., Wei, X., Wang, Y., Yang, Y., Hua, L., 2012. Effects of rapid urban sprawl on urban forest carbon stocks: integrating remotely sensed, gis and forest inventory data. J. Environ. Manag. 113, 447–455.

Rosenfeld, D., 2000. Suppression of rain and snow by urban and industrial air pollution. Science 287, 1793–1796.

Rossati, A., 2017. Global warming and its health impact. Int. J. Occup. Environ. Med. 8 (1), 7–20.

Roth, M., Oke, T.R., Emery, W.J., 1989. Satellite-derived urban heat islands from 3 coastal cities and the utilization of such data in urban climatology. Int. J. Remote Sens. 10, 1699–1720.

Rowntree, R.A., Nowak, D.J., 1991. Quantifying the role of urban forests in removing atmospheric carbon dioxide. J. Arboric. 17, 269–275.

Sailor, D.J., 1995. Simulated urban climate response to modifications in surface albedo and vegetation cover. Appl Meteorol 34, 1694–1704.

Schneider, A., Woodcock, C.E., 2008. Compact, dispersed, fragmented, extensive? A comparison of urban growth in twenty-five global cities using remotely sensed data, pattern metrics and census information. Urban Stud. 45, 659–692.

Schneider, A., Friedl, M.A., Potere, D., 2009. A new map of global urban extent from MODIS satellite data. Environ. Res. Lett. 4.

Schneider, A., Friedl, M.A., McIver, D.K., Woodcock, C.E., 2003. Mapping urban areas by fusing multiple sources of coarse resolution remotely sensed data. Photogramm. Eng. Remote Sens. 69, 1377–1386.

Schneider, A., Friedl, M.A., Potere, D., 2010. Mapping global urban areas using MODIS 500-m data: new methods and datasets based on 'urban ecoregions'. Remote Sens. Environ. 114, 1733–1746.

Shea, D., 1986. Climatological Atlas:1950-1979, Surface Air Temperature, Precipitation, Sea Level Pressure, and Sea Surface Temperature. NCAR, Boulder.

Shrestha, N.K., Du, X., Wang, J., 2017. Assessing climate change impacts on fresh water resources of the athabasca river basin, Canada. Sci. Total Environ. 601–602, 425–440.

Small, C., 2001. Estimation of urban vegetation abundance by spectral mixture analysis. Int. J. Remote Sens. 22, 1305–1334.

Sobstyl, J.M., Emig, T., Qomi, M.J.A., Ulm, F.J., Pellenq, R.J.M., 2018. Role of city texture in urban heat islands at nighttime. Phys. Rev. Lett. 120 (10), 108701.

Solomon, S.D., Qin, D., Manning, M., Chen, Z., Marquis, M., Avery, K.B., 2007. Climate Change 2007: The Physical Science Basis. Working Group I Contribution to the Fourth Assessment Report of the IPCC. Computational Geometry, pp. 1–21.

Song, C., 2005. Spectral mixture analysis for subpixel vegetation fractions in the urban environment: how to incorporate endmember variability? Remote Sens. Environ. 95, 248–263.

Taylor, G.E., Johnson, D.W., Andersen, C.P., 1994. Air-pollution and forest ecosystems — a regional to global perspective. Ecol. Appl. 4, 662–689.

Thornton, P.E., 1998. Regional Ecosystem Simulation: Combining Surface- and Satellite-Based Observations to Study Linkages between Terrestrial Energy and Mass Budgets. Thesis University of Montana, p. 1015.

Thornton, P.E., Running, S.W., White, M.A., 1997. Generating surfaces of daily meteorological variables over large regions of complex terrain. J Hydrol 190, 214–251.

Vieira, J., Matos, P., Mexia, T., Silva, P., Lopes, N., Freitas, C., 2018. Green spaces are not all the same for the provision of air purification and climate regulation services: the case of urban parks. Environ. Res. 160, 306–313.

Voogt, J.A., Oke, T.R., 2003. Thermal remote sensing of urban climates. Remote Sens. Environ. 86, 370–384.

Vopham, T., Bertrand, K.A., Tamimi, R.M., Laden, F., Hart, J.E., 2018. Ambient pm2.5 air pollution exposure and hepatocellular carcinoma incidence in the United States. Cancer Causes Control 29 (7), 563–572.

Wang, P., Zheng, H., Ren, Z., Zhang, D., Zhai, C., Mao, Z., 2018. Effects of urbanization, soil property and vegetation configuration on soil infiltration of urban forest in Changchun, northeast China. Chin. Geogr. Sci.

Wang, Q., Sun, Y., Xu, W., Du, W., Zhou, L., Tang, G., Chen, C., Cheng, X., Zhao, X., Ji, D., Han, T., Wang, Z., Li, J., Wang, Z., 2017. Vertically-resolved characteristics of air pollution during two severe winter haze episodes in urban Beijing, China. Atmos. Chem. Phys. Discuss. 1–28.

Wang, X.P., Mauzerall, D.L., 2006. Evaluating impacts of air pollution in China on public health: implications for future air pollution and energy policies. Atmos. Environ. 40, 1706–1721.

Wen, Y.Y., Liu, X.P., Bai, Y., Sun, Y., Yang, J., Lin, K., Pei, F.S., Yan, Y.C., 2019. Determining the impacts of climate change and urban expansion on terrestrial net primary production in China. J. Environ. Manag. 240, 75–83.

Weng, Q.H., 2009. Thermal infrared remote sensing for urban climate and environmental studies: methods, applications, and trends. ISPRS J. Photogram. Remote Sens. 64, 335–344.

White, M.A., Nemani, R.R., Thornton, P.E., Running, S.W., 2002. Satellite evidence of phenological differences between urbanized and rural areas of the eastern United States deciduous broadleaf forest. Ecosystems 5 (3), 260–273.

White, M.A., Running, S.W., Thornton, P.E., 1999. The impact of growing-season length variability on carbon assimilation and evapotranspiration over 88 years in the eastern us deciduous forest. Int. J. Biometeorol. 42, 139–145.

Wiedensohler, A., Cheng, Y.F., Nowak, A., Wehner, B., Achtert, P., Berghof, M., Birmili, W., Wu, Z.J., Hu, M., Zhu, T., Takegawa, N., Kita, K., Kondo, Y., Lou, S.R., Hofzumahaus, A., Holland, F., Wahner, A., Gunthe, S.S., Rose, D., Su, H., Poeschl, U., 2009. Rapid aerosol particle growth and increase of cloud condensation nucleus activity by secondary aerosol formation and condensation: a case study for regional air pollution in northeastern China. J. Geophys. Res. Atmos. 114.

Wu, Z.J., Zhang, Y.X., 2018. Spatial variation of urban thermal environment and its relation to green space patterns: implication to sustainable landscape planning. Sustainability 10, 2249.

Xie, W., Huang, Q., He, C., Zhao, X., 2018a. Projecting the impacts of urban expansion on simultaneous losses of ecosystem services: a case study in Beijing, China. Ecol. Indicat. 84, 183–193.

Xie, W., Xiong, W., Pan, J., Ali, T., Cui, Q., Guan, D., Meng, J., Mueller, N.D., Lin, E., Davis, S.J., 2018b. Decreases in global beer supply due to extreme drought and heat. Nat. Plants 4 (11), 964–973.

Xu, Q., 2001. Abrupt change of the mid-summer climate in central east China by the influence of atmospheric pollution. Atmos. Environ. 35, 5029–5040.

Yang, B., He, M., Shishov, V., Tychkov, I., Vaganov, E., Rossi, S., Ljungqvist, F.C., Brauning, A., Griebinger, J., 2017. New perspective on spring vegetation phenology and global climate change based on Tibetan Plateau tree-ring data. Proc. Natl. Acad. Sci. U. S. A 114 (27), 6966.

Yu, D., Shao, H.B., Shi, P.J., Zhu, W.Q., Pan, Y.Z., 2009. How does the conversion of land cover to urban use affect net primary productivity? A case study in Shenzhen city, China. Agric. For. Meteorol. 149, 2054–2060.

Yuan, F., Bauer, M.E., 2007. Comparison of impervious surface area and normalized difference vegetation index as indicators of surface urban heat island effects in Landsat imagery. Remote Sens. Environ. 106, 375–386.

Zhao, C.S., Tie, X.X., Lin, Y.P., 2006. A possible positive feedback of reduction of precipitation and increase in aerosols over eastern central China. Geophys. Res. Lett. 33.

Zhao, Y., Ding, G., Gao, G., Peng, L., Cui, X., 2017. Regionalization for aeolian desertification control in the mu us sandy land region, China. J. Desert Res. 37 (4), 635–643.

Zhou, D., Zhao, S., Zhang, L., Liu, S., 2016. Remotely sensed assessment of urbanization effects on vegetation phenology in China's 32 major cities. Remote Sens. Environ. 176, 272–281.

Zhou, X., Chen, H., 2018. Impact of urbanization-related land use land cover changes and urban morphology changes on the urban heat island phenomenon. Sci. Total Environ. 635, 1467–1476.

Zhou, Y., Xing, B., Ju, W., 2017. Assessing the impact of urban sprawl on net primary productivity of terrestrial ecosystems using a process-based model—a case study in Nanjing, China. IEEE J. Sel. Top. Appl. Earth Obs. Remote Sens. 8 (5), 2318–2331.

Zobler, L., 1986. A world soil file for global climate modeling. NASA Tech. Memo. 87802.

CHAPTER

24

Remote sensing application in agriculture

Abstract

Agriculture is the most important land use activity in the world. Agriculture not only affects the change of land cover but also has a profound impact on the sustainable development of social economy, food security, water and environment, ecosystem services, climate change, and carbon cycle. The area, location, status, and conversion information of farmland are important to understand how human activities affect the biosphere, hydrosphere, atmosphere, and lithosphere, as well as the simulation of carbon nitrogen cycle and the formulation of sustainable agricultural development policies. This chapter aims to describe how to use remote sensing data products for agricultural research, including extracting farmland information (24.2), detection of farmland change, estimation of grain yield (24.3), monitoring of agricultural disasters (drought, 24.4), monitoring crop residue (24.5), analysis of influence of farmland changes on surface parameters and environment (24.6), and study of the impact of climate change on farmland ecosystem (24.7).

Advanced Remote Sensing, Second Edition
https://doi.org/10.1016/B978-0-12-815826-5.00024-6

© 2020 Elsevier Inc. All rights reserved.

24.1 Introduction

Agriculture is the predominant land use activity on the planet. Globally, cropland area was estimated at 265 Mha in 1700, 1471 Mha in 1990 (Goldewijk, 2001), and about 1.5—1.8 Bha at the end of the millennium (Ramankutty and Foley, 1999). It is estimated that around 46.5 million km^2 of national vegetation had been converted to agricultural land (one-third for cropland and two-third for pasture) (Ramankutty et al., 2008).

Agriculture is the largest component of anthropogenic water use. Irrigated agriculture consumes about 84% of the water used by humans globally (Shiklomanov, 2000). It is estimated that the global total crop water use was 6685 km^3/yr during the time period 1998—2002; blue water use was 1180 km^3/yr; green water use of irrigated crops was 919 km^3/yr; and green water use of rainfed crops was 4586 km^3/yr (Siebert and Doll, 2010). Blue water refers to water in lakes, reservoirs, rivers, ice caps, and groundwater, and green water refers to effective rainfall.

Agriculture also alters the C and N cycles. Agriculture contributes 52% of global anthropogenic methane emissions and 84% of global nitrous oxide emissions (Smith et al., 2008). It is estimated that the global technical mitigation potential from agriculture by 2030 will be approximately 5500—6000 Mt CO$_2$-eq./year (Smith et al., 2008), and the C sink capacity of global agricultural and degraded soil is 50% —66% of the historic carbon loss of 42—78 GtC (Lal, 2004).

Therefore, the impacts of agriculture go far beyond changes in land cover; agriculture has implications for social economy, food security, water and environment sustainability, ecosystem services, climate change, and the carbon cycle (Foley et al., 2005; Khan and Hanjra, 2009; Lal, 2004, 2007; Paustian et al., 1997). Information on acreage, location, status, and transformation of cropland is crucial for an understanding of how human activity impacts on the biosphere, hydrosphere, atmosphere, and pedosphere, for modeling the C and N cycles and for guiding policies for sustainable agricultural development.

Various remote sensing data products have been widely used in the extraction of farmland spatial distribution (Zhu et al., 2014; Steele-Dunne et al., 2017), crop type identification (Wardlow et al., 2007; Pena-Barragan et al., 2011; Sonobe et al., 2018), growth monitoring (Gitelson, 2004; Shafian et al., 2018), crop phenological monitoring (Sakamoto et al., 2005), yield estimation (Xin et al., 2013; Chlingaryan et al., 2018), crop disaster monitoring (Xin et al., 2013; Chlingaryan et al., 2018), crop response to climate change (Lobell et al., 2012; Brown et al., 2012), and others (Zheng et al., 2014; Begue et al., 2018).

24.2 Cropland information extracting

24.2.1 Cropland mapping

Information about the distribution of cropland is important for land management and trade decisions, and it is also needed to estimate crop stress and productivity as well as other relative variables such as irrigation requirements and so on (Monfreda et al., 2008; Wu et al., 2010). Traditionally, crop areas are reported based on census data that cannot provide geographical distribution information. Besides, the process is tedious, time-consuming, and costly. Remote sensing has proven to be an effective tool to estimate crop distribution for a wide range of end users including government agencies, farmers, and modelers (Biradar et al., 2009; Frolking et al., 2002; Gumma et al., 2011; Karkee et al., 2009; Murthy et al., 2003; Ozdogan, 2010; Pena-Barragan et al., 2011; Wardlow et al., 2007).

The technologies used for identifying cropland from satellite data have evolved from simple unsupervised approaches to various complex supervised classifications (e.g., maximum likelihood classification, support vector machine (SVM), decision tree, and wavelet transform classification, artificial neural networks classification, and so on), from pixel-based methods to subpixel (e.g., linear spectral unmixing, support vector regression, and decision tree regression) and object-oriented methods and from exploring spectral differences to examining crop phenology differences in time series data.

The data used for cropland mapping depends on the purpose and extent of the study, as well as data availability and consumption. Local and regional studies work best with high resolution (e.g., IKONOS, Quickbird, TM, ETM, SPOT), national and continental studies work best with medium-resolution imagery (e.g., MODIS), and global studies are more likely to use low-resolution imagery. Optical, hyperspectral, and radar imageries have all proven to be useful for crop monitoring. It is commonly accepted that the classification results from higher resolution data are usually more accurate than those from lower resolution data, and the results from multitemporal images are more reliable than those from a single image.

For example, Lobell and Asner (2004) mapped cropland using MODIS vegetation index (VI) data in two agricultural regions, i.e., the Yaqui Valley (YV) in Northwest Mexico and the Southern Great Plains (SGP) in the United States. The data used in their study included Landsat data and MODIS 16-day 250-m VI composite products (NDVI, EVI, and quality assessment flags).

The method used to map cropland distribution in their study is known as probabilistic temporal unmixing. Fig. 24.1 shows the entire procedure. They first constructed time series of red and near-infrared (NIR) reflectance based on the MODIS NDVI composite products and chose image end-members from the reflectance time series with the help of Landsat classification. Instead of defining end-members with a single spectrum, they defined them as a set of spectra. In YV, the end-member set included 52

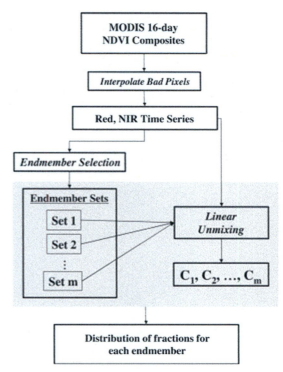

FIGURE 24.1 Outline of probabilistic temporal unmixing (PTU) algorithm. The steps within the *gray box*, namely, the selection of an end-member from each set and the calculation of end-member fractions, are repeated many (50) times to derive distributions of end-member fractions that reflect the uncertainty associated with end-member variability (Lobell and Asner, 2004).

end-members of wheat, 28 end-members of maize, and 46 end-members of uncropped land. In SGP, the end-member set included 22 end-members of wheat, 24 end-members of pasture, and 42 end-members of summer crops. For each pixel, they repeatedly carried out spectral unmixing analysis using randomly selected end-members from end-member sets and produced several fraction maps. Finally, they built up a probability distribution of fractions of each end-member. Fig. 24.2 shows the mean and standard deviation images of wheat fractions derived using their method for two study areas. This study provides a reference for crop identification in areas where mixed pixel problems are prominent.

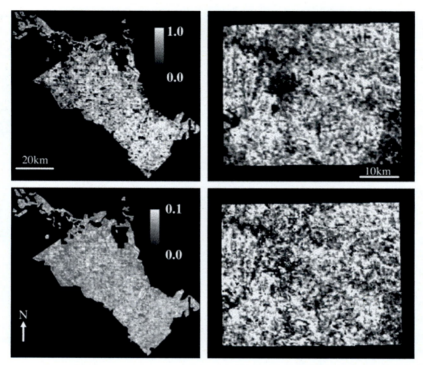

FIGURE 24.2   Mean (top) and standard deviation (bottom) images of wheat fractions from PTU, for YV (left side) and SGP (right side) (Lobell and Asner, 2004).

24.2.2 Monitoring cropland change

Besides cropland extent, information on cropland changes and their drivers is very important for assessing food and water security and guiding policies on sustainability (Liu et al., 2005a; Yan et al., 2009). Traditionally, the information on crop area change is derived by comparing the census data for many years. However, this method cannot provide geographical distributions. As a result, remote sensing plays a more and more important role in monitoring the transition of cropland (Amissah-Arthur et al., 2000; Doygun, 2009; Zhao et al., 2004; Zomeni et al., 2008).

Methods for monitoring cropland change include various change detection technologies that are commonly applied in LCLUC detection and have been summarized in Section 23.2.2 of this chapter. The most popular and straightforward change detection method is a postclassification comparison method for cropland. In Section 2.2, we have shown a case study in which a postclassification comparison method was used to detect urbanization. Here, we show a case study in which the changed areas are directly distinguished from the unchanged areas by multistage classification. For example, Kuemmerle et al. (2008) monitored change in Carpathian cropland by classification of multitemporal composites. The data used in their study included field measurement data, 16 Quickbird images available from Google Earth, TM and ETM+ images (path/row 186/

26) for October 2, 1986; July 27, 1988; June 10, 2000; and August 2, 2000. Field measurement was conducted in the summer of 2004, spring of 2005, and spring of 2006, and a total of 481 ground truth points were collected by field measurement.

Their method included the following steps. (1) They preprocessed Landsat images by geometrical and atmospheric rectifications and then masked out forests, water bodies, and built-up areas from Landsat images from 1988. They further masked out areas with altitudes higher than 1000 m from all four Landsat images and then stacked the four masked images into one multitemporal dataset. (2) They digitized 1171 plots using Quickbird images with the help of field measurement data. The 1171 plots plus 481 field measurements were used as ground truth data for training and validation purposes. (3) They divided all ground truth data into three classes, namely, unchanged area, fallow land, and reforested land, and used 1079 ground truth points to train the classifier of SVM and classify the multitemporal dataset. (4) They used the remaining 573 ground truth samples to validate the accuracy of classification.

Fig. 24.3 shows their farmland abandonment map. The overall accuracy of their farmland abandonment map is 90.9% and the kappa is 0.82. During the period from 1988 to 2000, abandoned farmland in the border triangle of Poland, Slovakia, and Ukraine covered 1285 km^2, 12.5% of which was converted to forests.

24.2.3 Agricultural irrigation

Irrigated agriculture consumes about 84% of the water used by humans globally (Shiklomanov, 2000). It affects hydrological processes (Rosenberg, 2000; Shibuo et al., 2007), local climate (Adegoke et al., 2007; Boucher et al., 2004; Lobell et al., 2006b; Sen Roy et al., 2007; Stohlgren et al., 1998; Tilman et al., 2001; Trenberth, 2004), and environmental variables such as soil salinity (Metternicht and Zinck, 2003) and soil quality depletion (Asadi et al., 2008; Dobermann and Oberthiir, 1997; Liu et al., 2005b). As a result, accurate information on the extent of irrigation is needed in many areas of research, such as water exchange between the land surface and the atmosphere (Boucher et al., 2004; Gordon et al., 2005; Ozdogan et al., 2006), climate change, irrigation water requirements (Döll and Siebert, 2002), water resources management (Vörösmarty et al., 2005), hydrological modeling, and agricultural planning.

Several efforts have been made to map irrigated areas globally. One is the US Geological Survey (USGS) Global Land Cover Map (Loveland et al., 2000) produced on the basis of 1-km Advanced Very High–Resolution Radiometer (AVHRR) observations between April 1992 and September 1993. It includes four irrigated land classes, namely, irrigated grassland, rice paddies and fields, hot irrigated cropland, and cool irrigated cropland. The USGS map provides a general classification scheme, so the irrigated classes used as subsets of the general classification scheme are less accurate.

In addition, the FAO and the University of Frankfurt (FAO/UF) have published several versions of a global map of irrigated areas (FAO map) (Döll and Siebert, 1999; Siebert and Döll, 2001; Siebert et al., 2005b). The latest version is a global dataset of monthly irrigated and rainfed crop areas around the year 2000 (MIRCA2000) (Portmann et al., 2010) (Fig. 24.4). This dataset describes the monthly growing areas of 26 irrigated and rainfed crops including wheat, rice, maize, barley, rye, millet, sorghum, soybeans, sunflower, potatoes, cassava, sugar cane, sugar beet, oil palm, rape seed/canola, groundnuts/peanuts, pulses, citrus, date palm, grapes/vines, cocoa, and coffee, as well as related crop calendars for 402 spatial units. The spatial resolution of this map is 5' × 5'.

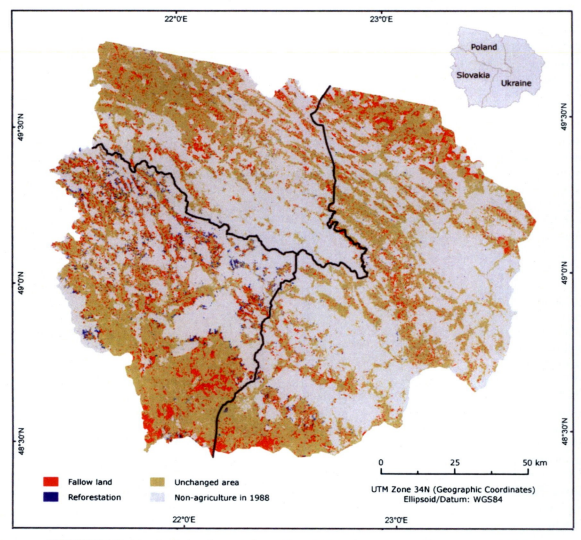

FIGURE 24.3　Farmland abandonment from 1986 to 2000 in the study area (Kuemmerle et al., 2008).

The International Water Management Institute (IWMI) also produced a global map of irrigated areas (IWMI map) (Thenkabail et al., 2006, 2008, 2009). The IWMI map has 10-km grid resolution and was produced using 20 years of AVHRR data and other additional data, including SPOT-VEGETATION, Japanese Earth Resources Satellite, and Landsat GeoCover 2000 data. The area statistics are reported as annualized irrigated area and total area available for irrigation. IWMI's method offers two advantages; it considers irrigation type and irrigation intensity information and uses subpixel decomposition techniques to derive the irrigated fraction within a pixel (Thenkabail et al., 2007).

In addition to global-scale data, irrigated area studies at other scales have been reported (Beltran and Belmonte, 2001; Biggs et al., 2006;

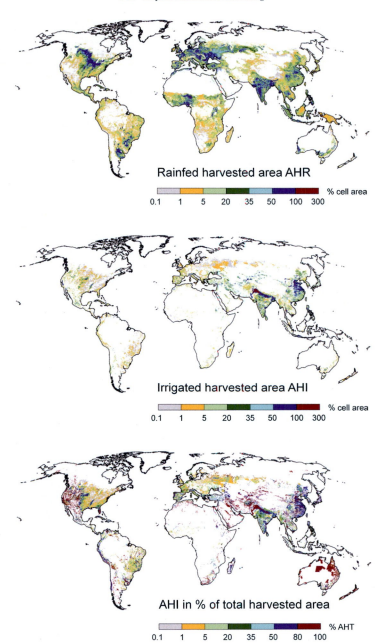

FIGURE 24.4 Global distribution of (top) rainfed harvested area (AHR) and (middle) irrigated harvested area (AHI) in percent of grid cell area and (bottom) AHI in percent of total harvested area (AHT), for 1998–2002 (Portmann et al., 2010).

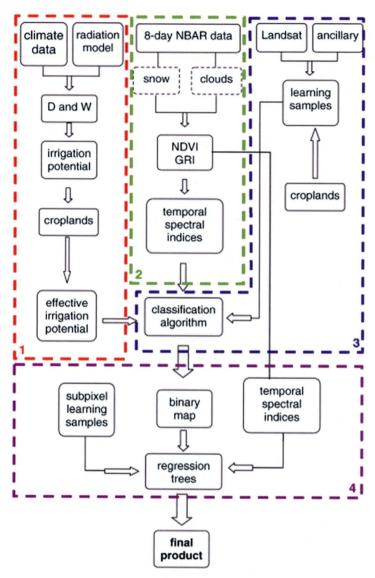

FIGURE 24.5   Flowchart of the major steps in the proposed mapping algorithm. Each dashed box with a number refers to the processing step in the proposed irrigation mapping procedure (Ozdogan and Gutman, 2008).

Boken et al., 2004; Dheeravath et al., 2010; El-Magd et al., 2003; Ozdogan and Gutman, 2008; Thenkabail et al., 2005; Wriedt et al., 2009; Zhu et al., 2014). For example, Ozdogan and Gutman (2008) produced a national irrigation map of the United States using two tree-based models (decision trees and regression trees). Decision trees were used to distinguish irrigated pixels from nonirrigated pixels, while regression trees were used to estimate irrigation fraction within each irrigated pixel identified by decision trees.

Fig. 24.5 shows their whole procedures. They first calculated the radiative dryness index (D) using mean annual net radiation, annual precipitation, and the latent heat of vaporization and then used it further to calculate a water availability parameter (W). They found a linear relationship between W and the fractional irrigated area, from which they calculated the irrigation potential and referred to it as the effective irrigation potential. Second, they excluded the cloud, snow, and nonagricultural pixels from the MODIS time series data and calculated NDVI and GRI using MODIS reflectance data. GRI equals NIR band reflectance divided by green band reflectance. Third, they built two training sets from Landsat images. One set consisted of irrigated/nonirrigated polygons that were used to train the decision tree classifier (C4.5) and classify the input dataset into two categories, namely, irrigated and nonirrigated areas. Input datasets included GRI, NDVI, and the

effective irrigation potential. Another dataset consisted of the irrigation fraction in each 500-m pixel that was used to train the regression tree. To get the irrigation fraction, they classified each Landsat image into irrigated and non-irrigated classes and aggregated these maps up to 500 m. Finally, they used the regression tree method to retrieve the irrigation fraction in each irrigated pixel identified in the former step.

Fig. 24.6 shows their irrigation map. They collected validation samples from the high-resolution Landsat images and calculated the error matrix. Accuracy assessment suggested that the map was highly accurate in the Western United States but less accurate in the Eastern United States. They also compared their map to a global irrigation map made by the FAO, and their map shows more detail. The irrigated area estimated from their map is also highly correlated with the irrigated agriculture statistical datasets.

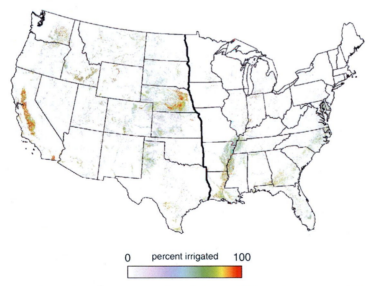

0   percent irrigated   100

FIGURE 24.6   Spatial distribution of irrigation in the United States c.2001 mapped form MODIS and ancillary data using the proposed procedure. The thick *vertical line* separates the east and west portions of the country individually selected for accuracy assessment (Ozdogan and Gutman, 2008).

24.3 Crop yield prediction

Crop yield estimation methods based on remote sensing can be generally divided into three categories: statistical analysis of remote sensing (Shanahan et al., 2001; Panda et al., 2010; Liu and Kogan, 2002; Dempewolf et al., 2014), production efficiency model (Bastiaanssen and Ali, 2003; Lobell et al., 2003; Tao et al., 2005; Liu et al., 2010; Peng et al., 2014; Yuan et al., 2016), and crop growth model (Mo et al., 2005; Fang et al., 2008; Moriondo et al., 2007; Padilla et al., 2012; Wang et al., 2013, 2014). The first method develops a relationship model between remote sensing bands (band combinations or various remote sensing indices, etc.) and crop yield (or yield components). The second method suggests crop yields under nonstressed conditions correlate linearly with the amount of absorbed photosynthetically active radiation. This method first estimates crop aboveground dry matter using remote sensing data and then converts this into crop yield. The third method applies satellite data to calibrate physiologically based crop models and then uses crop models to simulate physical crop growth processes and finally estimate yield.

24.3.1 Rice yield prediction by using NOAA-AVHRR NDVI and historical rice yield data

Since the 1980's, NDVI time series has been used for crop yield forecast. The most popular method is to build the empirical relationship between NDVI and crop yield (Mkhabela et al., 2005; Huang et al., 2014; Mashaba et al., 2017). However, different studies use different NDVI variables, such as original NDVI, average or accumulated NDVI over the growth period, average or accumulated NDVI over key growing stages, etc. Although NDVI time series can reflect interannual crop yield variation, crop yield changes caused by technical development or management improvement cannot be detected by NDVI. Therefore, Huang et al. (2013) used long-term historical yield data and a time series of AVHRR NDVI composite imagery to estimate paddy rice yield of Heilongjiang, Sichuan, Jiangxi, Hunan, and Guangxi provinces in China. Their basic concept was to first decompose historical rice yield into trend and remotely sensed yields, where trend yield refers to the component regulated by agricultural technology and remotely sensed yield refers to the component regulated by natural environmental conditions, such as temperature, precipitation, pets, disease, etc. Independent models were built to estimate the trend and remotely sensed yields, and then final crop estimation was calculated from the sum of trend and remotely sensed yield. The data used in their study mainly included time series 15 day AVHRR maximum NDVI composite imagery with 8 km spatial resolution from July 1981 to December 2006 and historical rice yield data from the China Statistical Year Book 1979—2009. The main processing steps were as follows:

(1) Five-year moving average and linear regression models were employed to fit historical rice yield data for each province, and they found the result detrended by linear regression performed better than that detrended by a 5-year moving average for Heilongjiang, Henan, Jiangxi, and Sichuan, and performed worse for Guangxi. Therefore, they used trend yield estimates from the linear regression method for Heilongjiang, Henan, Jiangxi, and Sichuan and from the 5-year moving average for Guangxi for further analysis.

(2) Remotely sensed yield was calculated using the difference between historical yield data and the trend yield estimated in step 1.

(3) Remotely sensed yield was used as the dependent variable and NDVI as independent variables to build remotely sensed yield prediction models through

stepwise regression, selecting remotely sensed yield prediction models with the highest correlation coefficients for each province. Table 24.1 shows NDVI variables used in their study and Table 24.2 shows the best model chosen for each province.

(4) Trend yield estimates from step 1 and remotely sensed yield estimates from step 3 were summed to obtain the final yield estimation.

Fig. 24.7 compares observed and predicted rice yields for Heilongjiang, Hunan, Jiangxi, Sichuan, and Guangxi provinces form 1982 to 2004. Table 24.3 shows validation results for predicted rice yields form 2005 to 06. Validation results indicate that relative error between predicted and observed rice yield is ~5.85%. Thus, their proposed method would be suitable for rice yield prediction at provincial level. The authors also argued that their method has potential to predict crop yield for other counties and other crop types.

24.3.2 A production efficiency model—based method for satellite estimates of corn and soybean yields

Photosynthesis is the main material source of crop yield formation, but only a part of sunlight can be utilized by leaf photosynthesis. Therefore, the concept of vegetation productivity has been proposed to describe the ability of plants to absorb CO_2 in the atmosphere through photosynthesis, converting light to chemical energy and hence accumulating organic dry matter. Vegetation productivity can be divided into gross primary (GPP) and net primary (NPP) productivity. GPP refers to the total amount of organic carbon fixed by photosynthesis by green plants for unit time and unit area, whereas NPP is the residual GPP from organic matter consumed by plant autotrophic respiration. GPP and NPP are highly correlated with crop biomass; hence, they are often used to study crop yield estimation.

Considering the difference of radiation use efficiency (RUE) of various crops and the mixed pixel problem, Xin et al. (2013) improved the crop GPP calculation method and used it to estimate maize and soybean yields in Midwestern United States. Their data included 8-day 1 km leaf area index (LAI) and fraction of photosynthetically active radiation product (MOD15A2) and vegetation productivity product (MOD17A2) over the period of 2009—11. Statistical data describing corn and soybean yields and harvested areas in each US county were provided by the Quick Stats database from the National Agricultural Statistics Service (NASS) of the US Department of Agriculture. Crop type classification maps with 30 m spatial resolution were sourced from the NASS cropland data layer (CDL) program and irrigation maps with 250 m resolution from the USGS early warning program. The main processing included the following steps:

(1) Relatively pure corn and soybean MODIS pixels were identified using CDL.
(2) Crop type-specific RUE values for corn and soybeans were used to estimate yields rather than the biome-wide value in the MOD17 algorithm. Corn and soybean RUE used in their study were 3.35 g/MJ PAR and 1.44 g/MJ PAR, respectively.
(3) The GPP for each MODIS pixel was divided into contributions from crops and from other vegetation types, and Formula 24.1—24.6 were used to calculate the contribution from crops in a given MODIS pixel.
(4) Crop GPP was converted to crop yield by Formula 24.7.
(5) Corn and soybean yield estimation accuracy was validated for 12 states in Midwest United States (North Dakota, South Dakota, Nebraska, Kansas, Minnesota, Iowa, Missouri, Michigan, Wisconsin, Illinois, Indiana, and Ohio).

TABLE 24.1   NVDI variables and their calculation formulas (Huang et al., 2013).

Number	NDVIs	Description of formulas
1	$NDVI_{maxb1}$	The first biweekly NDVI before $NDVI_{max}$
2	$NDVI_{maxb2}$	The second biweekly NDVI before $NDVI_{max}$
3	$NDVI_{maxb3}$	The third biweekly NDVI before $NDVI_{max}$
4	$NDVI_{maxb4}$	The fourth biweekly NDVI before $NDVI_{max}$
5	$NDVI_{max}$	The maximum NDVI during the growth period
6	$NDVI_{maxa1}$	The first biweekly NDVI after $NDVI_{max}$
7	$NDVI_{maxa2}$	The second biweekly NDVI after $NDVI_{max}$
8	$NDVI_{maxb4-b3}$	$(NDVI_{maxb4} + NDVI_{maxb3})/2$
9	$NDVI_{maxb4-b2}$	$(NDVI_{maxb4} + NDVI_{maxb3} + NDVI_{maxb2})/3$
10	$NDVI_{maxb4-b1}$	$(NDVI_{maxb4} + NDVI_{maxb3} + NDVI_{maxb2} + NDVI_{maxb1})/4$
11	$NDVI_{maxb4-max}$	$(NDVI_{maxb4} + NDVI_{maxb3} + NDVI_{maxb2} + NDVI_{maxb1} + NDVI_{max})/5$
12	$NDVI_{maxb4-a1}$	$(NDVI_{maxb4} + NDVI_{maxb3} + NDVI_{maxb2} + NDVI_{maxb1} + NDVI_{max} + NDVI_{maxa1})/6$
13	$NDVI_{maxb4-a2}$	$(NDVI_{maxb4} + NDVI_{maxb3} + NDVI_{maxb2} + NDVI_{maxb1} + NDVI_{max} + NDVI_{maxa1} + NDVI_{maxa2})/7$
14	$NDVI_{maxb3-b2}$	$(NDVI_{maxb3} + NDVI_{maxb2})/2$
15	$NDVI_{maxb3-b1}$	$(NDVI_{maxb3} + NDVI_{maxb2} + NDVI_{maxb1})/3$
16	$NDVI_{maxb3-max}$	$(NDVI_{maxb3} + NDVI_{maxb2} + NDVI_{maxb1} + NDVI_{max})/4$
17	$NDVI_{maxb3-a1}$	$(NDVI_{maxb3} + NDVI_{maxb2} + NDVI_{maxb1} + NDVI_{max} + NDVI_{maxa1})/5$
18	$NDVI_{maxb3-a2}$	$(NDVI_{maxb3} + NDVI_{maxb2} + NDVI_{maxb1} + NDVI_{max} + NDVI_{maxa1} + NDVI_{maxa2})/6$
19	$NDVI_{maxb2-b1}$	$(NDVI_{maxb2} + NDVI_{maxb1})/2$
20	$NDVI_{maxb2-max}$	$(NDVI_{maxb2} + NDVI_{maxb1} + NDVI_{max})/3$
21	$NDVI_{maxb2-a1}$	$(NDVI_{maxb2} + NDVI_{maxb1} + NDVI_{max} + NDVI_{maxa1})/4$
22	$NDVI_{maxb2-a2}$	$(NDVI_{maxb2} + NDVI_{maxb1} + NDVI_{max} + NDVI_{maxa1} + NDVI_{maxa2})/5$
23	$NDVI_{maxb1-max}$	$(NDVI_{maxb1} + NDVI_{max})/2$
24	$NDVI_{maxb1-a1}$	$(NDVI_{maxb1} + NDVI_{max} + NDVI_{maxa1})/3$
25	$NDVI_{maxb1-a2}$	$(NDVI_{maxb1} + NDVI_{max} + NDVI_{maxa1} + NDVI_{maxa2})/4$
26	$NDVI_{max-a1}$	$(NDVI_{max} + NDVI_{maxa1})/2$
27	$NDVI_{max-a2}$	$(NDVI_{max} + NDVI_{maxa1} + NDVI_{maxa2})/3$
28	$NDVI_{maxa1-a2}$	$(NDVI_{maxa1} + NDVI_{maxa2})/2$

TABLE 24.2 Results of the stepwise regression models for remotely sensed rice yield using AVHRR-derived NDVI measures as independent variables (Huang et al., 2013).

Study areas	Model	R	F-test value	RMSE
Heilongjiang	$Y_{RS} = -849.158 + 0.137\text{NDVI}_{\text{maxa1}}$	0.42^a	4.508	361.99
Hunan	$Y_{RS} = -1240.690 + 0.229m\text{NDVI}_{\text{maxb1}-\text{a2}}$	0.69^b	19.342	114.57
Jiangxi	$Y_{RS} = -1553.145 + 0.261m\text{NDVI}_{\text{maxb1}-\text{max}}$	0.46^b	5.689	166.38
Sichuan	$Y_{RS} = -1495.515 + 0.403m\text{NDVI}_{\text{maxb4}-\text{b3}}$	0.73^b	24.238	207.07
Guangxi	$Y_{RS} = -1832.285 + 1.138m\text{NDVI}_{\text{maxb4}-\text{b3}}$ $+ 0.214\text{NDVI}_{\text{maxa2}} - 1.315m\text{NDVI}_{\text{maxb4}-\text{b2}}$ $+ 0.307_{\text{maxb2}-\text{b1}}$	0.92^b	25.103	87.70

[a] Significant at 0.05 level.
[b] Significant at 0.01 level, $n = 23$.

$$\text{GPP}_{\text{total,DOY}_n} = \text{GPP}_{\text{crop,DOY}_n} + \text{GPP}_{\text{other,DOY}_n} \quad (24.1)$$

$$\text{GPP}_{other,\text{DOY}_n} = \text{MR}_{\text{other,DOY}_n}/(1 - \text{CUE}) \quad (24.2)$$

$$\text{MR}_{\text{other,DOY}_n} = \text{Leaf\_MR}_{\text{other}} + \text{Froot\_MR}_{\text{other}} \quad (24.3)$$

$$\text{Leaf\_MR}_{\text{other}} = \text{LAI}_{\text{other}}/\text{SLA} \times \text{leaf\_mr\_base} \times Q10\_mr^{[(Tavg-20.0/10.0)]} \quad (24.4)$$

$$\text{Froot\_MR}_{\text{other}} = \text{LAI}_{\text{other}}/\text{SLA} \times \text{froot\_leaf\_ratio} \times \text{froot\_mr\_base} \times Q10\_mr^{[(Tavg-20.0/10.0)]} \quad (24.5)$$

$$\text{LAI}_{\text{other}} \approx \max\left(\text{LAI}_{\text{total.DOY } n=89:N_0}\right) \quad (24.6)$$

$$\text{Yield} = \sum_{n>N_0}^{n=N_1} \text{GPP}_{\text{crop,DOY}_n} \times \frac{\text{HI}}{(1 + \text{RS})} \times \frac{1}{1 - \text{MC}} \quad (24.7)$$

where $\text{GPP}_{total,\text{DOY}_n}$ refers to the GPP value for a specific day of year (DOY)$_n$. $\text{GPP}_{\text{crop,DOY}_n}$ and $\text{GPP}_{other,\text{DOY}_n}$ are the contribution from crops (i.e., corn or soybeans) and the contribution from other vegetation types for the same time period, respectively. $\text{MR}_{\text{other,DOY}_n}$ is the maintenance respiration of other vegetation (Kg C/day). It includes the maintenance respiration of the leaves of other vegetation (Leaf_MR$_{\text{other}}$) and the maintenance respiration of fine roots of other vegetation (Froot_MR$_{\text{other}}$). CUE is the carbon use efficiency. LAI$_{\text{other}}$ (m$^2 \cdot$leaf$\cdot$m^{-2} ground area) is the estimated LAI of vegetation other than crops. SLA (projected leaf area m^2/kg$\cdot$leaf C) is the specific leaf area for a given pixel. LAI$_{\text{total.DOY } n=89:N_0}$ is the LAI value between April (DOY 89) and before mid-May (DOY N0). leaf_mr_base and froot_mr_base (kg$\cdot$C$\cdot$kg/C/day$\cdot$20 C) are the maintenance respiration of leaves and fine roots per unit mass at 20 C, respectively. froot_leaf_ratio is the ratio of the fine root to leaf mass. $Q10\_mr$ is an exponent shape parameter that controls respiration as a function of temperature. T_{avg} is the average daily temperature. Yield is estimated crop yield. HI is the harvest index. RS is the root to shoot ratio. MC is the grain moisture content. In their study, LAI was obtained from MOD15d. Tavg was estimated by using meteorological data from the NASA Data Assimilation Office. HI of corn and soybean are 0.53 and 0.42, respectively. RS of corn and soybean are 0.18 and 0.15, respectively. MC of corn and soybean are 0.11 and 0.10, respectively.

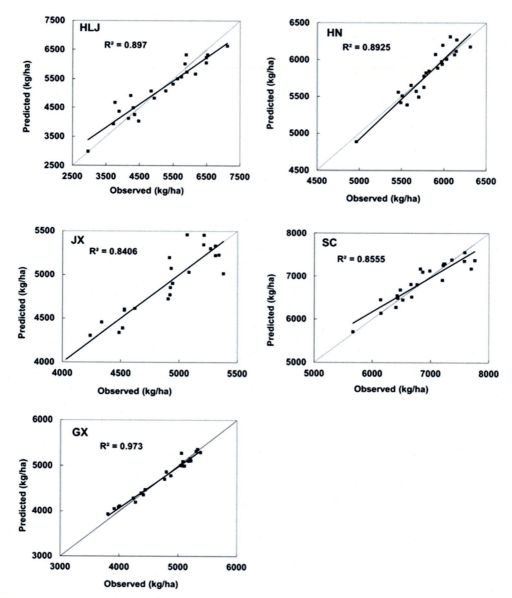

FIGURE 24.7   Observed versus predicted yields of rice (kg/ha) for the provinces of Heilongjiang (HLJ), Hunan (HN), Jiangxi (JX), Sichuan (SC), and Guangxi (GX) over the period 1982–2004 (Huang et al., 2013).

Other biome-specific coefficients are obtained from a Biome Parameter Look-Up Table from a general ecosystem model (Running et al., 2000).

Figs. 24.8 and 24.9 compare crop yields estimated from MODIS data and reported by NASS for rainfed counties and states in Midwestern United States, respectively, and Table 24.4 compares estimated crop yield errors from the authors' approach with those from the original MOD17 GPP. Rainfed corn and soybean

TABLE 24.3 Observed and predicted rice yields (independent test) (Huang et al., 2013).

Provinces	Year	Observed (kg/ha)	Predicted (kg/ha)	Relative error (%)
Heilongjiang	2005	6795.7	6780.7	−0.22
	2006	6261.3	6897.8	10.17
Hunan	2005	6050.3	6337.5	4.75
	2006	6141.3	6441.2	4.88
Jiangxi	2005	5328.2	5545.9	4.09
	2006	5475.1	5634.9	2.92
Sichuan	2005	7213.0	8018.4	11.17
	2006	6420.7	7680.3	19.62
Guangxi	2005	4953.0	5028.98	1.53
	2006	5088.0	5053.44	−0.68

yields estimated using the authors' method was highly correlated with the NASS survey data. Crop yield estimate accuracy by the authors' method was higher than that estimated using the original MOD17 GPP product.

24.4 Drought monitoring of crop

Drought is a phenomenon of water deficiency caused by water imbalance. Droughts are generally divided into four categories: meteorological droughts, agricultural droughts, hydrological droughts, and economic and social droughts. Agricultural ecosystems are most directly and severely affected by droughts. Drought indices based on meteorological stations are highly accurate and can be extended to the surface by interpolation. However, in areas lacking sufficient weather stations, they are impossible to accurately monitor crop drought conditions. Compared with meteorological stations, remote sensing provides large-scale real-time monitoring. Therefore, drought indices based on remote sensing data are widely used for crop drought monitoring.

Drought indices derived from remote sensing data began to emerge in the 1990s, and common indices include NDVI, vegetation condition index (VCI), vegetation water supply index, and temperature vegetation dryness index. However, it is difficult for a single index to fully capture

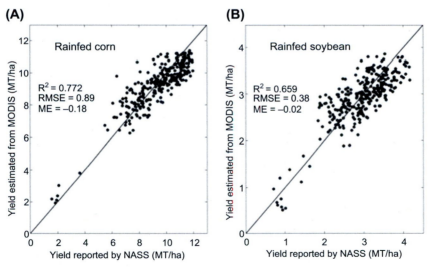

FIGURE 24.8 Comparisons between crop yields estimated from MODIS data and reported by the NASS for rainfed counties in the Midwestern United States. The *black line* is the 1:1 line. (A) Rainfed corn and (B) rainfed soybean (Xin et al., 2013).

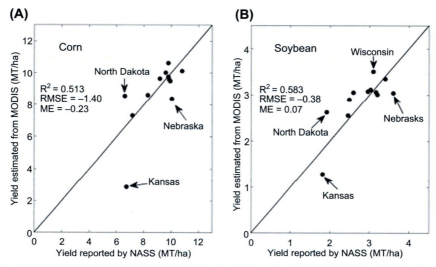

FIGURE 24.9 Comparisons between crop yields estimated from MODIS data and reported by the NASS for states in the Midwestern United States. The *black line* is the 1:1 line. (A) Corn and (B) soybean (Xin et al., 2013).

agricultural drought information. Domestic and international studies have proposed integrating various remote sensing data to construct comprehensive drought indices and improve the accuracy of drought monitoring. For example, Liu and Kogan (2002) proposed a vegetation health index (VHI) by linearly combining TCI and VCI, and Rhee et al. (2010) proposed a drought monitoring index suitable for both dry and humid areas by linearly weighting land surface temperature (LST), NDVI, and tropical rainfall measuring mission (TRMM) precipitation. Zhang and Jia (2013) used microwave data to monitor droughts and constructed weights by

enumeration to construct a series of drought indices suitable for different regions and timescales. Hao and Aghakouchak (2014) studied several drought monitoring indices and proposed a multivariate drought monitoring index by integrating precipitation and soil moisture (SM) information and verified their proposed index for drought monitoring.

This section introduces two crop drought monitoring cases based on remote sensing data. The first case uses the study of Patel et al. (2012) as an example to introduce using single remote sensing data (e.g., MODIS) to monitor agricultural drought. The second case uses the

TABLE 24.4 Statistics between crop yields estimated from our approach using MODIS data and reported by the NASS for rainfed counties for each year from 2009 to 2011.

	Corn			Soybean		
Year	R^2	RMSE (MT/ha)	ME (MT/ha)	R^2	RMSE (MT/ha)	ME (MT/ha)
2009	0.55 (0.15)	1.21 (5.52)	−0.60 (−5.39)	0.50 (0.35)	0.38 (0.86)	−0.07 (0.77)
2010	0.54 (0.22)	1.17 (4.65)	−0.14 (−4.38)	0.73 (0.53)	0.30 (0.97)	−0.09 (0.89)
2011	0.77 (0.46)	0.89 (4.56)	−0.18 (−4.28)	0.66 (0.53)	0.38 (1.06)	−0.02 (0.95)

For comparison, values in parentheses are statistics using the standard MOD17 products (Xin et al., 2013).

study of Rhee et al. (2010) as an example to introduce a method to monitor agricultural drought using multisensor remote sensing data.

24.4.1 Analysis of agricultural drought using vegetation temperature condition index

Patel et al. (2012) analyzed agricultural drought conditions in Gujarat, India, 2000–04, based on the vegetation temperature condition index (VTCI) calculated from the MODIS data. The data used in their study included 8-day NDVI data and surface temperature data from MODIS during cloudless periods 2000–04 along with maximum and minimum temperature and rainfall data from 20 weather stations. The main processing included the following steps:

(1) NDVI and MODIS surface temperature data during cloudless periods (day 241–297) over 2000–04 was used to construct NDVI-Ts spaces, and then wet and dry edges were extracted. VTCI was calculated by Formula 24.8–24.10 based on NDVI-Ts space.

$$VTCI = \frac{LST_{NDVIimax} - LST_{NDVIi}}{LST_{NDVIimax} - LST_{NDVIimin}} \quad (24.8)$$

$$LST_{NDVIimax} = a + bNDVI_i \quad (24.9)$$

$$LST_{NDVIimin} = a' + b'NDVI_i \quad (24.10)$$

where the $LST_{NDVIimax}$ and $LST_{NDVIimin}$ are maximum and minimum LSTs of pixels, which have the same NDVI value in a study region on each DOY or period of image, respectively. LST_{NDVIi} denotes actual LST of one pixel whose NDVI value is $NDVI_i$. Coefficients a, b, a', and b' can be estimated from an area large enough where SM at surface layer should span from wilting point to field capacity at pixel level.

(2) Crop moisture index (CMI) was estimated using the 20 meteorological sites (Palmer, 1968).

(3) Historical production was detrended (Larson et al., 2004), and abnormal outputs were calculated,

$$DYa_i = \left(\frac{Ya_i}{Yt_i} - 1\right) \times 100 \quad (24.11)$$

where Ya_i is detrended yield anomaly for the ith year, and Ya_i and Yt_i are actual and time trend-based yield of ith year.

(4) The relationship between VTCI and CMI and abnormal outputs was explored (Table 24.5), and VTCI drought threshold was divided according to the CMI, where 0.45–1.0, 0.45–0.38, 0.38–0.31, and 0.31–0.0 refer to normal, light, mid, and severe drought, respectively.

(5) They used 0.45 VTCI threshold = 0.45, where VTCI < 0.45 was classified as arid and VTCI > 0.45 was classified as areas without drought. Accumulated drought days in the study period were counted and then compared with actual crop yields to verify

TABLE 24.5 Correlation coefficient of VTCI with yield and detrended yield anomalies (Patel et al., 2012).

	Food grains				Oilseeds	
	2000		**2002**		**2000**	**2002**
DOY	Yield	Anomaly	Yield	Anomaly	Anomaly	Anomaly
	0.414	0.036	0.46	0.342	0.033	0.206
249	0.589	0.233	0.54	0.451	−0.079	0.275
257	0.542	0.273	0.49	0.330	−0.087	0.221
265	0.451	0.150	0.38	0.189	−0.044	0.237
273	0.559	0.186	0.42	0.192	−0.075	0.315
281	0.436	0.138	0.39	0.212	−0.043	0.282
289	0.509	0.150	0.30	0.166	−0.025	0.319
297	0.464	0.214	0.39	0.241	0.114	0.331

(A)   **(B)**   **(C)**

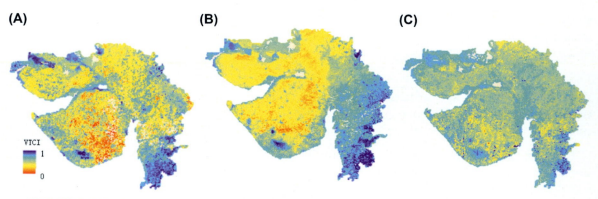

FIGURE 24.10   Spatial pattern of VTCI on DOY 249 in year (A) 2000, (B) 2002, and (C) 2004 (Patel et al., 2012).

drought duration monitoring performance using VTCI images.

Fig. 24.10 shows VTCI spatial distribution in drought (2000 and 2002) and drought-free (2004) years. VTCI was generally lower in 2000 and 2002 than 2004, VTCI was mostly >0.5 in 2004, and droughts were mainly distributed in midwestern and northern regions. Areas with higher VTCI indicate better irrigation, and the southern regions had higher VTCI because cover type was forest.

Fig. 24.11 shows 2002 drought classification in the study area. Most of the study area experienced severe drought in 2002, mainly distributed in the central and northern regions, with only a proportion of the study area experiencing moderate and light drought. Drought situation can be more directly described by the classification of drought level.

Fig. 24.12 shows accumulated drought days during the study period. The drought situation was the most serious in 2002, followed by 2000. Crop yield monitoring results show good agreement with actual crop yield. Thus, VTCI can be used to distinguish drought and drought-free years and further analyze drought levels and drought duration.

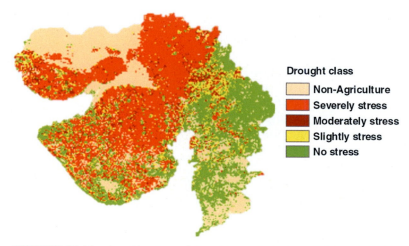

Drought class
- Non-Agriculture
- Severely stress
- Moderately stress
- Slightly stress
- No stress

FIGURE 24.11   Spatial pattern of drought severity in 2002 (Patel et al., 2012).

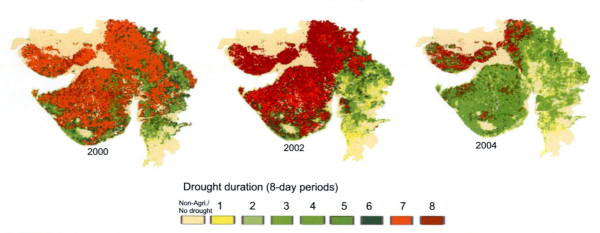

FIGURE 24.12   Spatial pattern of drought duration obtained from VTCI for drought (2000 and 2002) and wet (2004) years (Patel et al., 2012).

24.4.2 Monitoring agricultural drought using multisensor remote sensing data

Rhee et al. (2010) used LST, VI, and precipitation data from the TRMM satellite to construct a remote sensing scaled drought condition index (SDCI) and compared it with previous drought indices: Z index, PDSI, and standardized precipitation index (SPI). SDCI drought monitoring accuracy was superior to the previous indices, and monitoring results were consistent with the US drought monitor map.

Data used in their study included monthly total precipitation and monthly mean temperature 1971–2009 obtained from the Applied Climate Information System for weather stations, MODIS 8-day LST (MOD11A2, collection v005), monthly VI (NDVI, MOD13A3, collection v005), 8-day surface reflectance data (MOD09A1, collection v005), and TRMM monthly rainfall 2000–09. They also used the National Drought Monitor (USDM) provided by the United States Drought and Disaster Reduction Center and historical crop yield statistics obtained from NASS

(http://www.nass.usda.gov) 1981–2000. The main processing included the following steps:

(1) MODIS surface reflectance data were employed to calculate normalized multiband drought index (NMDI) and normalized difference water index (NDWI). Then, the normalized difference drought index (NDDI) was calculated by using NDVI and NDWI, and VHI (Unganai and Kogan, 1998) was calculated using NDVI and LST.

(2) Correlations were calculated for VI, LST, and precipitation data with 3 month SPI, and then three groups of weights were defined in terms of these correlations to construct the composite index (CI), as shown in Table 24.6.

(3) Correlation coefficients were analyzed between the CI and common previous drought indices (Z index, PDSI, 3 and 6 month SPI) at meteorological sites, selecting the optimal drought index individually for the different situations, which was called SDCI.

(4) SDCI monitoring performance for agricultural drought conditions was

TABLE 24.6 Formulas for remote sensing variables. ρ represents the spectral reflectance.

Remote sensing variable	Formula
NDVI(500 m)	$(\rho_{band2} - \rho_{band1})/(\rho_{band2} + \rho_{band1})$
NMDI	$(\rho_{band2} - (\rho_{band6} - \rho_{band7}))/(\rho_{band2} + (\rho_{band6} - \rho_{band7}))$
NDWI	$\left(\rho_{band2} - \rho_{band5(or\ 6\ or\ 7)}\right)\Big/\left(\rho_{band2} + \rho_{band5(or\ 6\ or\ 7)}\right)$
NDDI	$(NDVI - NDWI)/(NDVI + NDWI)$
Scaled LST	$(LST_{max} - LST)/(LST_{max} - LST_{min})$
Scaled NDVI(=VCI)	$(NDVI - NDVI_{min})/(NDVI_{max} - NDVI_{min})$
Scaled NMDI	$(NMDI_{max} - NMDI)/(NMDI_{max} - NMDI_{min})$ for the arid region
	$(NMDI - NMDI_{min})/(NMDI_{max} - NMDI_{min})$ for the humid region
Scaled NDWI	$(NDWI - NDWI_{min})/(NDWI_{max} - NDWI_{min})$
Scaled NDDI	$(NDDI_{max} - NDDI)/(NDDI_{max} - NDDI_{min})$
Scaled TRMM	$(TRMM - TRMM_{min})/(TRMM_{max} - TRMM_{min})$
VHI	$(1/2)$scaled LST $+ (1/2)$scaled NDVI
CI1	$(1/3)$scaled LST $+ (1/3)$scaled TRMM $+ (1/3)$scaled VI
CI2	$(1/4)$scaled LST $+ (2/4)$scaled TRMM $+ (1/4)$scaled VI
CI3	$(2/5)$scaled LST $+ (2/5)$scaled TRMM $+ (1/5)$scaled VI

The vegetation component shown as VI is one of NDVI, NMDI, NDWI, and NDDI, and the scaled TRMM uses one of the 1-, 3-, 6-, 9-, and 12-month timescales (Rhee et al., 2010).

assessed by comparing year to year SDCI changes with those for VHI and Z-index.

(5) The May 2000–May 2009 SDCI map was compared with USDM maps of the arid zone in May, and the relationship between SDCI and crop yields analyzed to further verify SDCI for crop drought monitoring.

The main study conclusions were as follows:

(1) Correlation coefficients between the comprehensive drought index and the in situ drought indices show that LST, TRMM, and NDVI provided the best combination of CO_2 weights in arid regions, whereas LST, TRMM, and NDDI6 worked best with CO_2 weights in wet regions.

(2) Comparing yearly changes of SDCI, VHI, and Z-index (Fig. 24.13) show that SDCI was consistent with Z-index. The VHI trend was slightly different from those of SDCI and Z-index because it did not use direct precipitation information.

(3) Comparing SDCI and USDM (Fig. 24.14) shows that for arid regions, droughts evident in 2000, 2002, 2006, 2007, and 2009 USDI maps were successfully monitored using SDCI, where extreme drought conditions in Carolina in 2008 (humid region) were not caught by SDCI. Overall, SDCI and USDM provided comparable monitoring.

(4) Correlations between yearly crop yield and SDCI for each month from May to September show that for arid areas, only May SDCI and yearly cotton yield were significantly correlated. June and July SDCI had significant correlation with yearly corn

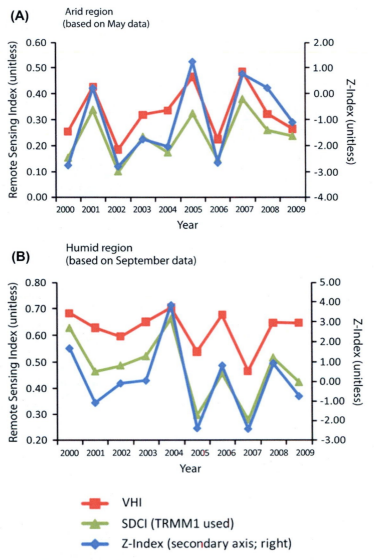

FIGURE 24.13 Year-to-year changes of VHI, SDCI (TRMM1 used), and Z-Index averaged over (A) the arid region and (B) the humid region (Rhee et al., 2010).

yield, and September SDCI had significant correlation with yearly soybean yield. TRMM precipitation timescale was shown to affect SDCI applicability.

The authors concluded that integrating VI, surface temperature, and precipitation data into a comprehensive drought index could better monitor agricultural drought, and monitoring accuracy would continue to improve increased use of remote sensing data.

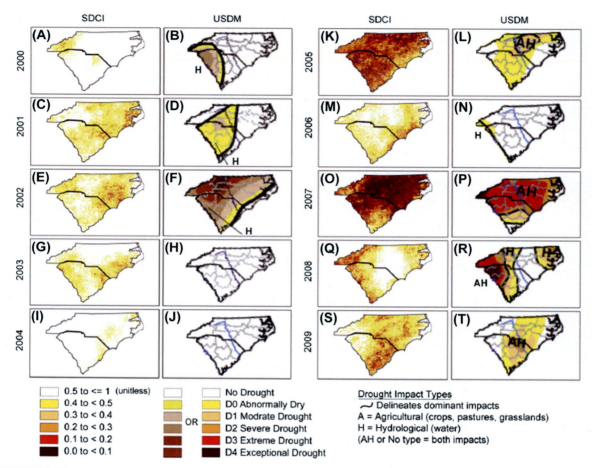

FIGURE 24.14   Year-to-year changes of SDCI (TRMM1 used) and the USDM maps in the humid region for September from 2000 to 2009. (A), (C), (E), (G), (I), (K), (M),(O), (Q), (S) refers to the SDCI of september from 2000 to 2009. (A), (C), (E), (G), (I), (K), (M), (O), (Q), (S) refers to the SDCI of september from 2000 to 2009.

24.5 Crop residue monitoring

24.5.1 Crop residue cover

Crop residues are materials left on cultivated land after the crop has been harvested. Retention of crop residues after harvesting is considered to be an effective antierosion measure. Crop residues can improve soil structure, increase organic matter content in the soil, reduce evaporation, and help fix CO_2 in the soil. Good residue management practices on agricultural lands have many positive impacts on soil quality. Besides, crop residues can be used in biofuel production. Information on residue cover guides polices for promoting beneficial management practices and helps the estimation of soil carbon.

Traditional methods of residue cover measurement, such as line-point transects or photographic techniques, are inefficient in large-scale investigations, and their accuracy is impacted by operator bias and sampling representatives. A satellite-based approach is an efficient and less costly way to measure residue cover (Daughtry et al., 1996; Sullivan et al., 2004).

Currently, both optical and microwave remote sensing have been used to estimate crop residues; however, both of these face some challenges. The biggest challenge to using optical remote sensing to detect crop residues is the ability to distinguish crop residues from bare soil. The spectra of crop residues and soils are often similar and differ only in amplitude for a certain wavelength. Moreover, the spectral reflectance of soils is affected by many factors, such as organic matter, moisture, texture, particle size distribution, iron oxide content, and surface roughness. The spectral reflectance of crop residues is affected by degree of decomposition, water content, and harvesting time. Therefore, there is no uniform relationship between crop residue spectra and soil spectra. In microwave images, the difference between soil and crop residues lies in different backscatter (McNairn and Brisco, 2004). However, radar backscatter is influenced by surface roughness, soil status, SM, and crop residue distributions (McNairn et al., 2002).

So far, some indices have been developed for detecting crop residues, such as the brightness index, normalized difference index, normalized difference tillage index, normalized difference senescence index, normalized difference residue index, soil adjusted corn residue index, modified soil adjusted corn residue index, crop residue index multiband, cellulose absorption index (CAI), lignin cellulose absorption index, and shortwave infrared normalized difference residue index (Bannari et al., 2006; Biard and Baret, 1997). All of the above indices try to intensify the spectral difference between soil and crop residues and usually use visible (400−700 nm), NIR (700−1200 nm), and shortwave infrared (1200−2500 nm) bands (Fig. 24.15). Among these, CAI was reported to have the highest accuracy, followed by LCP (Serbin et al., 2009). This is because crop residues are primarily composed of cellulose, hemicelluloses, and lignin and the absorption characteristics of crop residues are highly correlated to the absorption characteristics of cellulose and lignin. The CAI can capture the cellulose absorption at 2101 nm.

$$\text{CAI} = 100 \left(\frac{R_{2031} + R_{2211}}{2} - R_{2101} \right) \quad (24.12)$$

where R is the reflectance, and the subscripts 2031, 2101, and 2211 denote 11-nm-wide bands centered at 2031, 2101, and 2211 nm, respectively.

Besides spectral indices, classification methods (such as linear spectral mixture analysis, LSMA) have also been used to measure crop residue by remote sensing in recent years (Bannari et al., 2006; Zhang et al., 2011a). For example, Pacheco and McNairn (2010) mapped crop residue on SPOT and Landsat images. They extracted end-members of three different residues (corn, soybean, and grain) and two soil types (clay and loam) directly from the image based on the ground data and then used the LSMA model in the PCI Geomatica software to derive the crop residue fractions. They compared the percentage residue cover estimated by SMA with that observed by ground measurements via the coefficient of determination (R^2) and the root mean square error (RMSE). Fig. 24.16 shows one of their results. R^2 for the estimated and measured crop residue was between 0.58 and 0.78, and the RMSEs were between 17.29% and 20.74%.

Using radar images to estimate crop residues is usually conducted by building a linear relationship between crop residue coverage and backscattering coefficient (Zhang et al., 2011a). However, the simple linear relationship cannot accurately represent the complex relationship between crop residue and radar backscatter and consequently cannot produce ideal results. For example, McNairn et al. (2001) studied different crop type, residue moisture content, and residue amount using radar backscatter

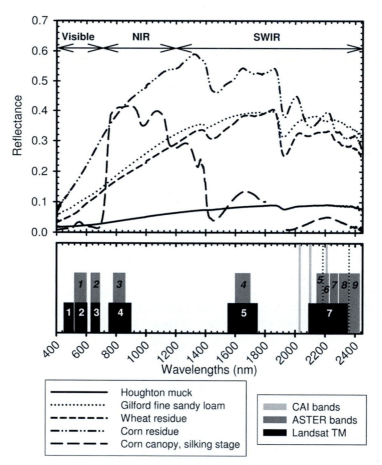

FIGURE 24.15   Visible, near-infrared (NIR), and shortwave-infrared reflectance spectra for two soils, two crop residues, and live corn canopy, with ranges of Landsat Thematic Mapper (TM), Advanced Spaceborne Thermal Emission and Reflection Radiometer (ASTER), and cellulose absorption index (CAI) bands. Wheat residues were acquired from a strip-tilled field in June, ~9 mo after harvest. Corn residues were acquired from a field of standing corn residue stubble in May, ~8 mo after harvest. Corn canopy was at the silking stage (R1) (Serbin et al., 2009).

and found that radar backscatter increased with increasing amount and moisture content of residue. McNairn et al. (2002) examined the sensitivity of linear polarizations and polarimetric parameters of polarimetric synthetic aperture radars to crop residues and found that the polarimetric parameters vary depending on the type and amount of residue cover.

24.5.2 Crop residue burning

Crop residue burning is a common type of land management in cultivated land. It can help control pests and weeds and remove extra residues before planting. It also emits substantial amounts of particulate matter and other pollutants into the atmosphere. The burned areas

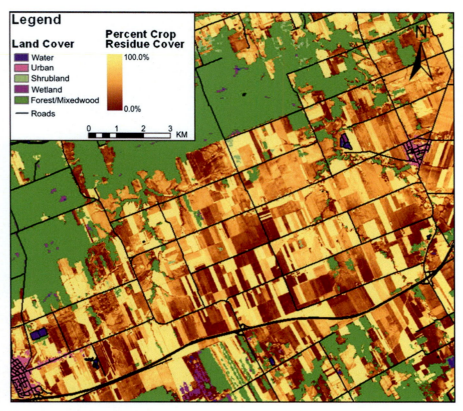

FIGURE 24.16   Percent crop residue cover map over the Casselman/St. Isidore study site derived from spectral mixture analysis on a SPOT image acquired on November 9, 2007 (Pacheco and McNairn, 2010).

also affect the hydrological and ecological environments.

Satellite remote sensing provides the only means of monitoring vegetation burning at regional to global scales and has been used to monitor fire globally for more than two decades. Currently, the NASA (MODIS) on the Terra and Aqua satellites has been used to systematically generate a suite of global MODIS fire products including the active fire product and burned area product (http://modis-fire.umd.edu/index.html). The algorithms for MODIS fire products have been continuously improved and can be found in references (Giglio et al., 2003; Roy et al., 2005, 2008).

The MODIS active fire product is derived by a contextual algorithm, in which thresholds are applied to the observed middle-infrared and thermal infrared brightness temperature and then false fire pixels are deleted by comparing the brightness temperature of the detected pixel with those of neighboring pixels (Giglio et al., 2003). Fig. 24.17 shows an example of the density of MODIS 1-km active fire detections per 10 km grid in the contiguous United States (CONUS). The MODIS burned area product is developed using a detection method in which the predicted bidirectional reflectance (BRF) in a pixel is estimated by a bidirectional reflectance model and compared with observed BRF. If the difference

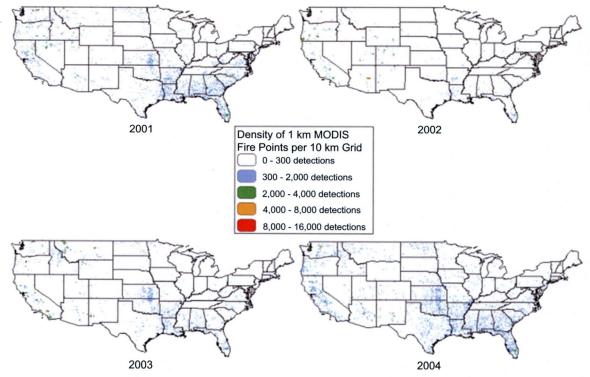

FIGURE 24.17   Density of MODIS 1 km Active Fire Detections (MOD14) per 10 km grid in the contiguous United States (2001–2004) (McCarty et al., 2007).

between predicted and observed BRF exceeds a certain threshold in a pixel, the pixel will be labeled as burned (Roy et al., 2005). The MODIS burned area product is distributed as a monthly gridded 500-m product in Hierarchical Data Format and sinusoidal equal area projection.

The MODIS fire product focuses not only on agricultural land cover but is also used to monitor agricultural burning together with other data (McCarty et al., 2007, 2008, 2009; Zhang et al., 2011b). For example, McCarty et al. (2009) monitored crop residue burning in the CONUS using multiple MODIS products including the 500-m MODIS 8-day surface reflectance product (MOD09A1), the 1-km MODIS active fire product (TERRA/AQUA, MOD14/MYD14), and the MODIS 1-km Land Cover

Dataset (MOD12) during growing seasons from 2003 to 2007.

Fig. 24.18 shows their workflow for monitoring crop residue burning. Their method included two parts. The first part was to detect burned pixels using the Normalized Burn Ratio (dNBR). However, dNBR method is more suitable for burned areas bigger than 20 ha. For burned areas smaller than 20 ha, the 1-km MODIS active fire product was used; this is the second part of the workflow. Then, they combined the burning maps from these two parts to generate the final burning map. The dNBR method included the following steps. (1) They generated a dNBR map for each tile of the MODIS surface reflectance product covering the study area. The method for calculating the

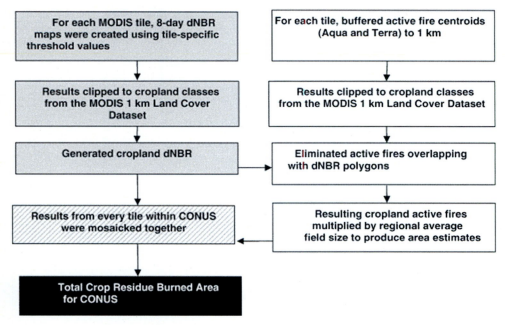

FIGURE 24.18 Workflow for calculating the crop residue burned area for the CONUS (McCarty et al., 2009).

dNBR can be found in Lopez Garcia and Caselles (1991). (2) Corresponding to each MODIS tile, they built a crop mask using the MODIS 1-km Land Cover Dataset, used the crop mask to get rid of the noncrop area from the dNBR map, and finally produced a cropland dNBR map. (3) For each cropland dNBR map, they set a different burn threshold based on ground measurements, and pixels in the dNBR map with values greater than the burn threshold were recognized as burned pixels. The ground measurements included a total of 296 GPS data points that can well represent the difference in cropping systems, soil properties, irrigation activities, and residue burning frequencies of CONUS. According to their estimation, 1,239,000 ha of croplands are burned each year in the CONUS. Fig. 24.19 shows the annual burned area in CONUS and two magnified burned areas in California and Louisiana.

24.6 The impact from cropland

24.6.1 Irrigation impacts on land surface parameters

Like land cover and land use change, land management also can have a big impact on the climate system, but it is not yet of too much concern. Irrigation is one of most important land management practices by which people try to grow crops in dry areas or increase food production.

It is reported that agricultural irrigation accounts for 84% of global water use by the world's population (Shiklomanov, 2000) and has grown rapidly over the past 200 years. The irrigated area was estimated to be about 8 Mha around 1800, 47 Mha around 1900 (Shiklomanov, 2000), and 274 Mha around 2000 (Siebert et al., 2005a). Considering the tremendous increase of

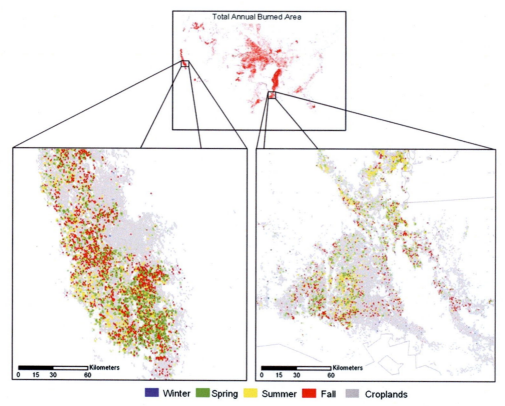

FIGURE 24.19   Seasonal cropland burned area for subregions of California and Louisiana for years 2003–07; for mapping purposes, active fire detections were not calibrated into area for display purposes and remain as original point shapefiles but are symbolized as squares; seasons defined as winter: January–March; spring: April–June; summer: July–September; and Fall: October–December (McCarty et al., 2009).

irrigated area and the potential impact of irrigation on the climate system, it may have contributed to the formation of current climate system and may impact our future climate system.

So far, there have been some reports about the impact of irrigation on near-surface air temperature (Bonfils and Lobell, 2007; Kueppers et al., 2007; Lobell and Bonfils, 2008; Mahmood et al., 2006), energy fluxes (Devries, 1959; Douglas et al., 2006), groundwater (Kendy et al., 2004), water vapor (Boucher et al., 2004), and precipitation (Barnston and Schickedanz, 1984; Lee et al., 2009; Lohar and Pal, 1995; Moore and Rojstaczer, 2001; Segal et al., 1998) based on climate observations and modeling studies. Observational studies usually make comparisons between pre- and postirrigation temperature trends in irrigated areas (Adegoke et al., 2003; Mahmood et al., 2004) or between irrigated and nonirrigated areas (Christy et al., 2006; Segal et al., 1998). Modeling studies usually compare the outputs from different models (regional or global, coupled or uncoupled) with and without irrigation, for example, fixing a high value of SM (Kanamaru and Kanamitsu, 2008; Lobell et al., 2006a), imposing a fixed amount of evapotranspiration (ET) from irrigated areas (Boucher et al., 2004; Segal et al., 1998), and designing an irrigation model based on the balance between water demand and supply (De Rosnay et al.,

2003; Haddeland et al., 2006) throughout the growing season.

Both observational and modeling studies are facing some challenges (Bonfils and Lobell, 2007; Lobell and Bonfils, 2008). For example, meteorological observations essentially provide point measurements, which usually do not represent area means. It is difficult to clearly distinguish the impact of irrigation on climate from the impact of other factors as the characteristics of irrigated sites such as land cover type, altitude, latitude, and longitude, distance from urban/ocean areas, and black carbon concentration may vary considerably. The results from modeling rely heavily on input parameters associated with four key aspects of irrigation, namely, where to irrigate, when to irrigate, how much to irrigate, and how to irrigate (e.g., rain, spray, drip, and rate), causing over- or underestimation.

Remote sensing observation is a promising tool and could provide land parameter information on a large scale, including SM, albedo, LST, vegetation cover, and so on. It could be a valid method for determining the impact of irrigation on the local surface climate—especially in those regions where direct observations are limited or obscured by other factors, such as urbanization in China. It can also be integrated into models to better represent reality. However, the studies based on remote sensing observations are rare so far.

Zhu et al. (2011) used satellite observations to analyze the impact of irrigation on land surface biogeophysical parameters in Jilin Province, China. The parameters used in their study included albedo, LST, NDVI, SM, and ET. The surface albedo dataset was from the (MODIS) albedo product (MCD43C3: Albedo 16-Day L3 Global 0.05Deg CMG). The LST dataset was from the MODIS/Aqua LST/Emissivity Monthly L3 Global 0.05Deg CMG (MYD11C3) products. The NDVI dataset was from the MODIS Vegetation Indices Monthly L3 Global 0.05Deg CMG (MOD13C2) product. SM data

were from the Advanced Microwave Scanning Radiometer-Earth Observing System L3 Surface Soil Moisture products provided by the National Snow and Ice Data Center (Njoku, 2005). ET was calculated using a statistical equation (Wang and Liang, 2008). To evaluate land surface parameters under different irrigation intensities, they compare land surface parameters between cultivated areas featuring a high percentage of irrigated land and those with a low irrigation percentage. Fig. 24.20 shows the comparison results. They found that highly irrigated areas always corresponded to a lower albedo and LST and higher SM, NDVI, and ET over the study period of 2000–08. Their study proved that satellite observations are sufficiently valid to determine the impact of irrigation on land surface parameters and provide another viable method for understanding the impact of irrigation on local climate, especially in those regions where direct observations are limited or obscured by other factors, such as urbanization in China.

24.6.2 Impacts of cropland on surface temperature

The role of intensive agriculture in modifying surface climate has been documented by various studies based on both modeling and observational methods. LST is generally defined as the skin temperature of the ground. LST is a key parameter in the physics of land surface processes, combining surface–atmosphere interactions and energy fluxes between the atmosphere and the ground. Ge (2010) used MODIS LST to study the impacts of intensive agriculture on surface temperature. His study area was the winter wheat belt in the North American Great Plains. The data used in this study included MODIS land cover products (MOD12C1) with the International Geosphere-Biosphere Programme (IGBP) classification scheme from 2001 (http://modis-land.gsfc.nasa.gov/landcover.htm) and monthly Terra

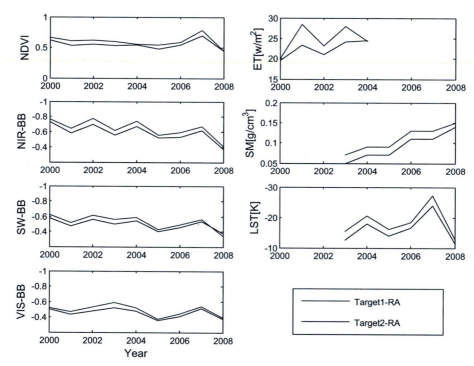

FIGURE 24.20   Comparison of land surface parameters between highly and lowly irrigated areas. The pixels with an irrigation percentage smaller than 10 were classified as "reference" areas (RA). The pixels with an irrigation percentage greater than 30 and 50 were denoted as "target1" and "target2" areas, respectively. NIR-BB, SW-BB, and VIS-BB are black-sky near-infrared, shortwave, and visible band albedo, respectively. ET, evapotranspiration; *LST*, land surface temperature; *SM*, soil moisture. *From Zhu, X., Liang, S., Pan, Y., Zhang, X., 2011. Agricultural irrigation impacts on land surface characteristics detected from satellite data products in Jilin Province, China. IEEE J. Select. Topics Appl. Earth Obser. Remote Sens. 4, 721–729, ©2011, IEEE.*

and Aqua LST data (version 5) with 0.05° spatial resolution from August 2002 to July 2008 (https://wist.echo.nasa.gov/wist-bin/api/ims.cgi?mode=MAINSRCH&JS=1).

His method involved two main steps. First, he identified the winter wheat and grassland pixels in his study area by using the MODIS land cover product (Figs. 24.21) and then he compared the daytime, nighttime, and diurnal LSTs of winter wheat with those of grassland from August 2002 to July 2008. Figs. 24.22 and 24.23 are parts of his results. Fig. 24.22 shows LST difference between wheat and grassland observed by MODIS Aqua. Fig. 24.23 is averaged LST diurnal difference between winter wheat and grassland.

In term of his results, the author concluded that the wheat field has a cool anomaly in the growing season and warm anomaly when bare soil is exposed and that the temperature over the wheat field is more uniform than that of surrounding areas. His study indicated the potential advantages of using satellite observations in land–climate interaction research.

24.6.3 Impact of crop residue burning

Burning crop residue can be an important source of particulate and trace gas emissions that affect both air quality and public health (Badarinath et al., 2006; Zhang et al., 2011b). In recent years, satellite-based approaches have been employed to study the impact of residue burning on near surface atmosphere and land surface characteristics (Badarinath et al., 2009;

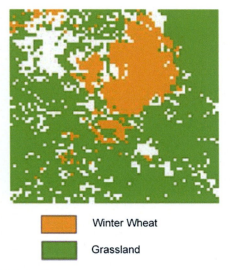

Winter Wheat

Grassland

FIGURE 24.21   Distribution of winter wheat and grassland in the study area (Ge, 2010).

Serbin et al., 2009). For example, McCarty et al. (2009) studied the impact of burning crop residue using multiple satellite products in the Indo-Gangetic Plains (IGP), which is one of the world's largest and most intensively cultivated areas.

The data used in their study included MODIS (AOD) data (available at http://g0dup05u.ecs.nasa.gov/Giovanni), Indian remote sensing satellite (IRS)-P4 ocean color monitor and CO data from the troposphere instrument (available at http://eosweb.larc.nasa.gov/PRODOCS/mopitt/table_mopitt. html), and aerosol index (AI) from the ozone monitoring instrument flown on the EOS Aura spacecraft and MODIS fire products (available at http://maps.geog.umd.edu/activefire_html/). These data were acquired in November 2007, which is a typical crop residue burning period in the IGP. Besides, a major Indian festival known as Diwali was celebrated on November 9, 2007.

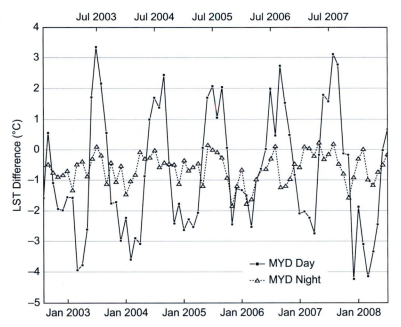

FIGURE 24.22   LST difference between wheat and grassland (LSTwheat−LSTgrassland) from August 2002 to July 2008 observed by MODIS Aqua (MYD) (Ge, 2010).

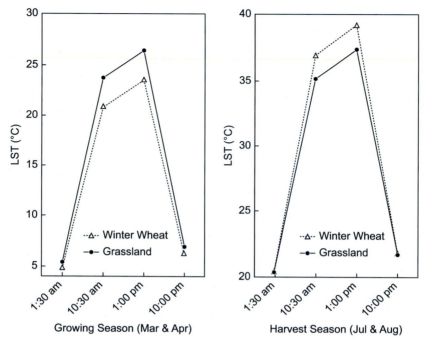

FIGURE 24.23   Averaged LST diurnal cycles for winter wheat and grassland from August 2002 to July 2008. The left panel is for 2 months (March and April) in the growing season; the right panel is for 2 months (July and August) in harvest season (Ge, 2010).

To estimate impacts of crop residue burning on AOD, AI, and CO in IGP, they compared the AOD, AI, and CO observed during the Diwali period with those observed after Diwali. Fig. 24.24 shows a comparison example. They also conducted back trajectory analysis via the HYSPLIT model. Their results indicated that crop residue burning and fireworks led to an increase of 30% in AOD at 550 nm and influenced the AI and CO over the Arabian Sea.

24.7 Response of crops to climate change

Climate is the basic environment for human survival, and climate change has brought many adverse effects on the global environment and human life. Agriculture is very sensitive to climate change, and many previous studies have shown changes to agricultural planting

areas (Newman, 1980), planting systems, crop growth periods (Wang et al., 2004), crop photosynthesis efficiency, water use efficiency, crop yield (Parry and Swaminathan 1992; Terjung et al., 1989), etc., due to climate changes. This section introduces two case studies of crop response to climate change based on remote sensing.

24.7.1 Effects of extreme heat on wheat growth

Wheat is one of the world's major food crops and has the most extensive planting area globally. Depending on the different planting periods, it can be divided into winter and spring wheat. Ground temperature increases through the wheat growing season, reaching maximum during the grain filling stage. Temperature

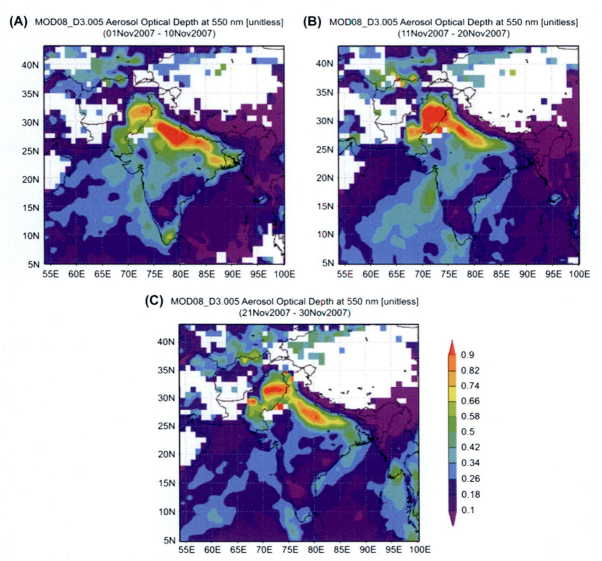

FIGURE 24.24 MODIS derived aerosol optical depth at 550 nm during (A) November 1–10, 2007, (B) November 11–20, 2007, and (C) November 21–30, 2007 (Badarinath et al., 2009).

during this latter period can affect grain filling speed, leaf senescence speed, and wheat yield (Wardlaw and Wrigley, 1994; Wardlaw and Moncur, 1995; Alkhatib and Paulsen, 1999).

Lobell et al. (2012) used satellite data in Northern India to monitor wheat senescence rates at temperatures >34°C. The main experimental data and steps were as follows:

(1) VI products based on MODIS satellite data were employed to extract wheat green-up and senescence dates in IGP in India and estimate the green season length (GSL).

(2) Long-term average monthly maximum and minimum temperatures from high resolution climatology maps from the WorldClim database (http://worldclim.org) and Global

Summary of the Day data from the National Climate Data Center (http://www.ncdc. noaa.gov/cgi-bin/res40.pl?page=gsod. html) were used to estimate daily minimum and maximum temperatures over 1 km grid cells and then calculate 2° days (DDs): one called GDD, where base temperature was 0°C and maximum 30°C and the other called EDD, where base temperature was 34°C.

(3) Rainfall could be related to extreme heat, mitigating moisture stress and therefore delaying senescence. Therefore, gridded rainfall data was collected from NASA (http://power.larc.nasa.gov/).

(4) Depending on the green-up period, they divided the study area into three groups and established regression equations for GDD, EDD, rainfall, and GSL for each group

$$GSL = \beta_0 + \beta_G GDD + \beta_E EDD + \beta_R RAIN$$
(24.13)

(5) The established regression equations and two crop models (CERES-Wheat and APSIM) were used to simulate extreme heat effects on the crop yield and explore the impact of

climate change on crop growing season. To ensure that the selected sites covered the possible extreme heat range, sites were divided into three groups depending on planting date and selected corresponding to 5th, 50th, and 90th percentiles of EDD in each group for analysis. GDD and EDD were calculated after temperature rises (assumed to be 1, 2, 3, and 4°C), and GSL was estimated from the regression Eq. (24.13). Under conditions without nitrogen and water stress, CERES-Wheat and APSIM models were used to simulate temperature increase effects on growth season length and yield, respectively.

(6) Regression model and crop model simulation estimates were compared.

All three simulation models indicated that shortening of GSL in the study area was related to sowing date (Fig. 24.25A). The shortening length of the growing season simulated by CERES-Wheat and APSIM was smaller than that estimated from the MODIS regression model, especially for late seeding wheat. For example, when the temperature rises by 2°C and the sow date is November 25, the shortening

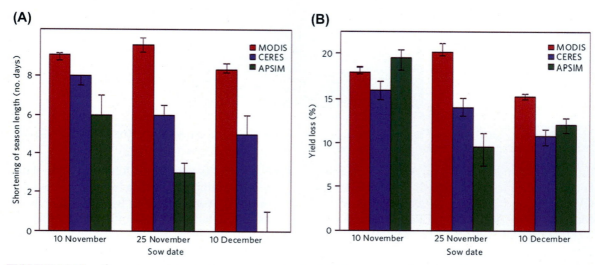

FIGURE 24.25 Comparison of simulation results based on MODIS data and crop-based models. (A) Shortening length of wheat growing season at different sowing dates and (B) loss of wheat yield at different sowing dates (Lobell et al., 2012).

lengths predicted by MODIS Regression, CERES-Wheat, and APSIM were 9, 6, and 3 d, respectively. They estimated MODIS yield loss using the CERES simulation relationship between GSL shortening and yield loss. Comparing CERES and APSIM, larger yield losses were predicted from MODIS regression (Fig. 24.25B). Both CERES and APSIM models may have underestimated yield loss caused by temperature increase; however, the outcomes verified that crop phenological information extracted based on satellite data can be used to evaluate model performance.

24.7.2 Effects of changes in humidity and temperature on crops

Climate is the key factor for agricultural production, as soil temperature and moisture are the basis for crop growth. Climate changes, such as temperature increase, precipitation changes, and light condition changes, directly affect food yields. For example, temperature changes in temperate and tropical regions have significant effects on agricultural production (Piao et al., 2007). Water resource shortages lead to reduced crop yields from rainfed agriculture (Jeong et al., 2010).

Brown et al. (2012) investigated crop phenology responses to climate change globally from 1981 to 2006. The data used in this study included AVHRR NDVI datasets from the NASA Global Inventory Monitoring and Modeling Systems (GIMMS) July 1981–December 2006 with 8 km resolution; 3-hourly gridded meteorological data with a 1° resolution from the Global Land Data Assimilation System, global crop distribution map with $5'' \times 5''(\sim 10 \times 10$ km) latitude-longitude grid produced by Monfreda et al. (2008); rainfed cereal production (country level) 1982–2006 from United Nation Food and Agriculture Organization, and US wheat yields 1982–2006

from NASS. The main processing included the following steps:

(1) Data preprocessing: The GIMMS data were regridded from native Albers equal area 8 km resolution to 0.08° resolution. Average daily temperature was calculated by aggregating 3 h temperatures, subtracting base temperature (5°C) from average daily temperature to obtain GDD. Average daily temperature was not used for calculating GDD if it was <5°C. Finally, the global crop distribution map was regridded from 5″ to 0.08° resolution and the rainfed crop distribution map extracted.

(2) Model building and phenology prediction: The method proposed by White et al. (2009) was used to estimate growing season start and end dates and GSL based on NDVI data. A quadratic regression model (24.14) was fitted for growing season NDVI and used to estimate NDVI peak height and position. See Brown and de Beurs (2008) for further details regarding the quadratic regression model.

$$\text{NDVI} = \alpha + \beta x + \gamma x^2 \qquad (24.14)$$

where x is either accumulated growing degree days (AGDD) or accumulated relative humidity (Arhum); α is NDVI at SOS, and β and γ determine GSL. Peak position is estimated by model parameters. Peak NDVI is determined based on the peak position. They defined two 18-month crop growing periods: October to March (cycle 1) and April to September (cycle 2). AGDD and Arhum were calculated by summing daily mean GDD and relative humidity over the study periods. Two models were built for each cycle using AGDD and Arhum, respectively, i.e., four models in total, called AGDD-Cycle1, AGDD-Cycle2, Arthum-Cycle1, and Arthum-Cycle2.

(3) Statistical analysis. Based on the phenology prediction above the last step, interannual

variation and temporal trends were analyzed using phenology estimates for SOS, peak period, and GSL, as well as relationships between crop phenology variation and agricultural production 1982–2006. Peak period refers to the period from SOS to date when peak NDVI occurs.

Fig. 24.26A shows SOS trends 1982–2006, where positive indicates later and negative indicates earlier SOS. West African Sahel, Southeast Asia, and North America have been experiencing later SOS, and Europe generally shows earlier SOS. Fig. 24.26B shows GSL trends 1982–2006, where positive indicates longer and negative indicates shorter GSL. Approximately 27% of cereal crop areas have experienced longer GSL 1982, with average lengthening 2.3 days.

Fig. 24.27 shows peak growing period trends, where positive indicates later and negative indicates earlier peak timing. More areas have experienced later peak timing estimated from Arhum than estimated from AGDD. Arhum results

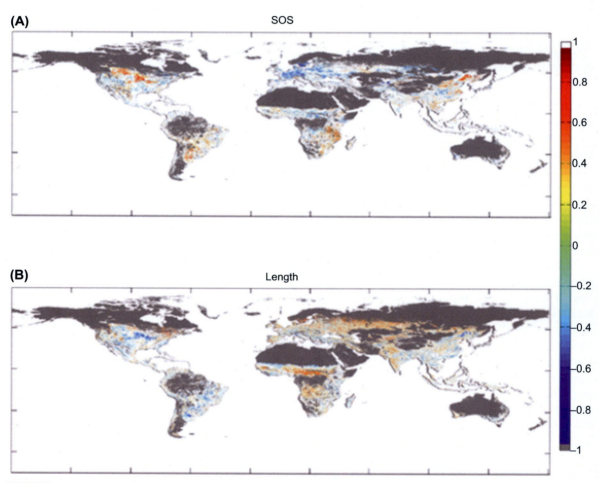

FIGURE 24.26   Significant trends of (A) start of season and (B) length of growing season over the 26 years in cropping regions, given by the regression coefficient of the parameter versus time. Significance is measured by P value of less than 0.1 (Brown et al., 2012).

(A)
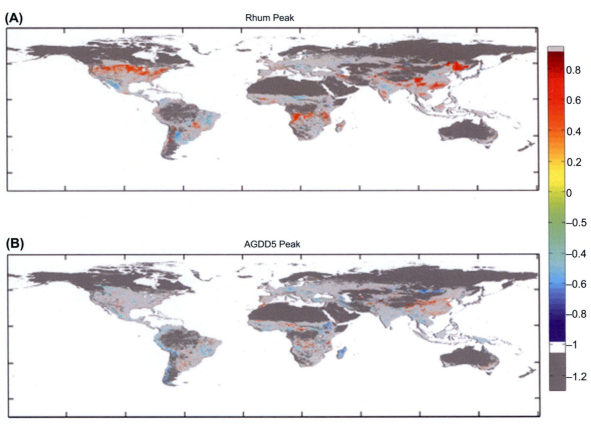

FIGURE 24.27   Significant trends of peak period for (A) relative humidity model and (B) accumulated growing degree days for 26 years in cropping regions, given by the regression coefficient of the parameter versus time. Significance is measured by *P* value of less than 0.1 (Brown et al., 2012).

show that Eastern Europe and Eastern United States have significantly later peak timing, whereas dry regions of south Asia, India, and arid southwest in North and South America have significantly earlier peak timing. AGDD results show peak timing has a large positive trend across Asia.

Fig. 24.28 shows the correlations between agricultural production and peak growing season positions estimated from AGDD-Cycle1, AGDD-Cycle2, Arthum-Cycle1, and Arthum-Cycle2. More than 25% of the pixels showed significant correlation between agricultural production and peak growing season position variance in 75 countries. Correlations between

agricultural production and peak NDVI estimated from AGDD-Cycle1 and Arthum-Cycle1 are stronger than for AGDD-Cycle2 and Arthum-Cycle2.

24.8 Summary

This chapter focuses on remote sensing data and products in the field information extraction, crop yield estimation, drought monitoring, land use management of farmland on the impact of surface parameters and climate change on crop growth. However, remote sensing research in agriculture is in-depth and extensive, and

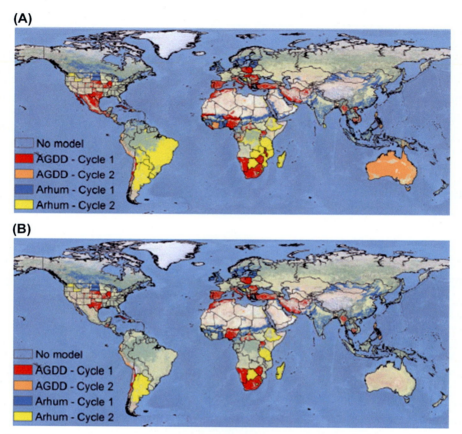

FIGURE 24.28   (A) Overview map showing which model results in phenological metrics that best correlates with production statistics. For each cropland pixel in each country/state or region, the model type that reveals the most correlated pixels with production is shown. Countries that were part of the former Soviet Union and countries that were part of former Yugoslavia are omitted. Countries that do not reveal a significant model are omitted as well. (B) Same as above but only the countries for which at least 25% of the pixels show a significant behavior are show (Brown et al., 2012).

gradually toward application, such as precision agriculture, agricultural insurance monitoring and evaluation, agricultural engineering monitoring, agricultural policy effect evaluation, etc. On the other hand, the emerging new sensors (hyperspectral remote sensing data, fluorescence remote sensing, polarization remote sensing, and UAV remote sensing) provide more abundant data resources for agricultural research and application. The newly proposed and developed technologies (artificial intelligence, deep learning, and large data) provide a technical approach for agricultural remote sensing information extraction and information inversion. The new application demand is bound to promote the further development of remote sensing technology and products, and the further development of agricultural remote sensing technology will also promote the intelligence and automation of modern agricultural development.

References

Adegoke, J.O., Pielke, R., Carleton, A.M., 2007. Observational and modeling studies of the impacts of agriculture-related land use change on planetary boundary layer processes in the central US. Agric. For. Meteorol. 142, 203–215.

Adegoke, J.O., Pielke, R.A., Eastman, J., Mahmood, R., Hubbard, K.G., 2003. Impact of irrigation on midsummer surface fluxes and temperature under dry synoptic conditions: a regional atmospheric model study of the U.S. high plains. Mon. Weather Rev. 131, 556–564.

Alkhatib, K., Paulsen, G.M., 1999. High-temperature effects on photosynthetic processes in temperate and tropical cereals. Crop Sci. 39, 119–125.

Amissah-Arthur, A., Mougenot, B., Loireau, M., 2000. Assessing farmland dynamics and land degradation on Sahelian landscapes using remotely sensed and socioeconomic data. Int. J. Geogr. Inf. Sci. 14, 583–599.

Asadi, S.S., Azeem, S., Prasad, A.V.S., Anji Reddy, M., 2008. Analysis and mapping of soil quality in Khandaleru catchment area using remote sensing and GIS. Curr. Sci. 95, 391–396.

Badarinath, K.V.S., Chand, T.R.K., Prasad, V.K., 2006. Agriculture crop residue burning in the Indo-Gangetic Plains – a study using IRS-P6 AWiFS satellite data. Curr. Sci. 91, 1085–1089.

Badarinath, K.V.S., Kharol, S.K., Sharma, A.R., Prasad, V.K., 2009. Analysis of aerosol and carbon monoxide characteristics over Arabian Sea during crop residue burning period in the Indo-Gangetic Plains using multi-satellite remote sensing datasets. J. Atmos. Sol. Terr. Phys. 71, 1267–1276.

Bannari, A., PacheCo, A., Staenz, K., McNairn, H., Omari, K., 2006. Estimating and mapping crop residues cover on agricultural lands using hyperspectral and IKONOS data. Remote Sens. Environ. 104, 447–459.

Barnston, A.G., Schickedanz, P.T., 1984. The effect of irrigation on warm season precipitation in the southern great plains. J. Clim. Appl. Meteorol. 23, 865–888.

Bastiaanssen, W.G.M., Ali, S., 2003. A new crop yield forecasting model based on satellite measurements applied across the Indus Basin, Pakistan. Agric. Ecosyst. Environ. 94, 321–340.

Begue, A., Arvor, D., Bellon, B., Betbeder, J., de Abelleyra, D., Ferraz, R.P.D., Lebourgeois, V., Lelong, C., Simoes, M., Veron, S.R., 2018. Remote sensing and cropping practices: a review. Remote Sens. 10, 99.

Beltran, C.M., Belmonte, A.C., 2001. Photogrammetric irrigated crop area estimation using Landsat TM imagery in La Mancha, Spain. Eng. Remote Sens. 67, 1177–1184.

Biard, F., Baret, F., 1997. Crop residue estimation using multiband reflectance. Remote Sens. Environ. 59, 530–536.

Biggs, T.W., Thenkabail, P.S., Gumma, M.K., Scott, C.A., Parthasapadhi, G.R., Turral, H.N., 2006. Irrigated area mapping in heterogeneous landscapes with MODIS time series, ground truth and census data, Krishna Basin, India. Int. J. Remote Sens. 27, 4245–4266.

Biradar, C.M., Thenkabail, P.S., Noojipady, P., Li, Y.J., Dheeravath, V., Turral, H., Velpuri, M., Gumma, M.K., Gangalakunta, O.R.P., Cai, X.L., Xiao, X.M., Schull, M.A., Alankara, R.D., Gunasinghe, S., Mohideen, S., 2009. A global map of rainfed cropland areas (GMRCA) at the end of last millennium using remote sensing. Int. J. Appl. Earth Obs. Geoinf. 11, 114–129.

Boken, V.K., Hoogenboom, G., Kogan, F.N., Hook, J.E., Thomas, D.L., Harrison, K.A., 2004. Potential of using NOAA-AVHRR data for estimating irrigated area to help solve an inter-state water dispute. Int. J. Remote Sens. 25, 2277–2286.

Bonfils, C., Lobell, D., 2007. Empirical evidence for a recent slowdown in irrigation-induced cooling. Proc. Natl. Acad. Sci. U.S.A 104, 13582–13587.

Boucher, O., Myhre, G., Myhre, A., 2004. Direct human influence of irrigation on atmospheric water vapour and climate. Clim. Dyn. 22, 597–603.

Brown, M.E., de Beurs, K., 2008. Evaluation of multi-sensor semi-arid crop season parameters based on NDVI and rainfall. Remote Sens. Environ. 112, 2261–2271.

Brown, M.E., Beurs, K.M.D., Marshall, M., 2012. Global phenological response to climate change in crop areas using satellite remote sensing of vegetation, humidity and temperature over 26 years. Remote Sens. Environ. 126, 174–183.

Chlingaryan, A., Sukkarieh, S., Whelan, B., 2018. Machine learning approaches for crop yield prediction and nitrogen status estimation in precision agriculture: a review. Comput. Electron. Agric. 151, 61–69.

Christy, J.R., Norris, W.B., Redmond, K., Gallo, K.P., 2006. Methodology and results of calculating central California surface temperature trends: evidence of human-induced climate change? J. Clim. 19, 548–563.

Daughtry, C.S.T., McMurtrey, J.E., Chappelle, E.W., Hunter, W.J., Steiner, J.L., 1996. Measuring crop residue cover using remote sensing techniques. Theor. Appl. Climatol. 54, 17–26.

De Rosnay, P., Polcher, J., Laval, K., Sabre, M., 2003. Integrated parameterization of irrigation in the land surface model ORCHIDEE. Validation over Indian Peninsula. Geophys. Res. Lett. 30.

Dempewolf, J., Adusei, B., Beckerreshef, I., Hansen, M., Potapov, P., Khan, A., Barker, B., 2014. Wheat yield forecasting for Punjab province from vegetation index time series and historic crop statistics. Remote Sens. 6, 9653–9675.

Devries, D.A., 1959. The influence of irrigation on the energy balance and the climate near the ground. J. Meteorol. 16, 256—270.

Dheeravath, V., Thenkabail, P.S., Chandrakantha, G., Noojipady, P., Reddy, G.P.O., Biradar, C.M., Gumma, M.K., Velpuri, M., 2010. Irrigated areas of India derived using MODIS 500 m time series for the years 2001—2003. ISPRS J. Photogram. Remote Sens. 65, 42—59.

Dobermann, A., Oberthiir, T., 1997. Fuzzy mapping of soil fertility- a case study on irrigated riceland in the Philippines. Geoderma 77, 317—339.

Döll, P., Siebert, S., 1999. A Digital Global Map of Irrigated Areas. Report A9901, vol. 3. Center for Environmental Systems Research, University of Kassel, Kurt Wolters Strasse, Kassel, Germany, p. 34109.

Döll, P., Siebert, S., 2002. Global modeling of irrigation water requirements. Water Resour. Res. 38, 8.1—8.10.

Douglas, E.M., Niyogi, D., Frolking, S., Yeluripati, J.B., Pielke, R.A., Niyogi, N., Vorosmarty, C.J., Mohanty, U.C., 2006. Changes in moisture and energy fluxes due to agricultural land use and irrigation in the Indian Monsoon Belt. Geophys. Res. Lett. 33.

Doygun, H., 2009. Effects of urban sprawl on agricultural land: a case study of Kahramanmara AY, Turkey. Environ. Monit. Assess. 158, 471—478.

El-Magd, Abou, I., Tanton, T.W., 2003. Improvements in land use mapping for irrigated agriculture from satellite sensor data using a multi-stage maximum likelihood classification. Int. J. Remote Sens. 24, 4197—4206.

Fang, H., Liang, S., Hoogenboom, G., Teasdale, J., Cavigelli, M., 2008. Corn-yield estimation through assimilation of remotely sensed data into the CSE-CERES-Maize model. Int. J. Remote Sens. 29, 3011—3032.

Foley, J.A., DeFries, R., Asner, G.P., Barford, C., Bonan, G., Carpenter, S.R., Chapin, F.S., Coe, M.T., Daily, G.C., Gibbs, H.K., Helkowski, J.H., Holloway, T., Howard, E.A., Kucharik, C.J., Monfreda, C., Patz, J.A., Prentice, I.C., Ramankutty, N., Snyder, P.K., 2005. Global consequences of land use. Science 309, 570—574.

Frolking, S., Qiu, J.J., Boles, S., Xiao, X.M., Liu, J.Y., Zhuang, Y.H., Li, C.S., Qin, X.G., 2002. Combining remote sensing and ground census data to develop new maps of the distribution of rice agriculture in China. Glob. Biogeochem. Cycles 16.

Ge, J.J., 2010. MODIS observed impacts of intensive agriculture on surface temperature in the southern Great Plains. Int. J. Climatol. 30, 1994—2003.

Giglio, L., Descloitres, J., Justice, C.O., Kaufman, Y.J., 2003. An enhanced contextual fire detection algorithm for MODIS. Remote Sens. Environ. 87, 273—282.

Gitelson, A.A., 2004. Wide dynamic range vegetation index for remote quantification of biophysical characteristics of vegetation. J. Plant Physiol. 161, 165—173.

Goldewijk, K.K., 2001. Estimating global land use change over the past 300 years: the HYDE Database. Glob. Biogeochem. Cycles 15, 417—433.

Gordon, L.J., Steffen, W., Jonsson, B.F., Folke, C., Falkenmark, M., Johansen, A., 2005. Human modification of global water vapor flows from the land surface. Proc. Natl. Acad. Sci. U.S.A 102, 7612—7617.

Gumma, M.K., Nelson, A., Thenkabail, P.S., Singh, A.N., 2011. Mapping rice areas of South Asia using MODIS multitemporal data. J. Appl. Remote Sens. 5.

Haddeland, I., Lettenmaier, D.P., Skaugen, T., 2006. Effects of irrigation on the water and energy balances of the Colorado and Mekong river basins. J. Hydrol. 324, 210—223.

Hao, Z., Aghakouchak, A., 2014. A nonparametric multivariate multi-index drought monitoring framework. J. Hydrometeorol. 15, 89—101.

Huang, J., Wang, H., Dai, Q., Han, D., 2014. Analysis of NDVI data for crop identification and yield estimation. IEEE J. Sel. Top. Appl. Earth Obs. Remote Sens. 7, 4374—4384.

Huang, J., Wang, X., Li, X., Tian, H., Pan, Z., 2013. Remotely sensed rice yield prediction using multi-temporal NDVI data derived from NOAA'S-AVHRR. PLoS One 8, e70816.

Jeong, S.J., Ho, C.H., Brown, M.E., Kug, J.S., Piao, S., 2010. Browning in desert bound- aries in Asia in recent decades. J. Geophys. Res. Atmos. 116, D02103.

Kanamaru, H., Kanamitsu, M., 2008. Model diagnosis of nighttime minimum temperature warming during summer due to irrigation in the California central valley. J. Hydrometeorol. 9, 1061—1072.

Karkee, M., Steward, B.L., Tang, L., Aziz, S.A., 2009. Quantifying sub-pixel signature of paddy rice field using an artificial neural network. Comput. Electron. Agric. 65, 65—76.

Kendy, E., Zhang, Y.Q., Liu, C.M., Wang, J.X., Steenhuis, T., 2004. Groundwater recharge from irrigated cropland in the North China plain: case study of Luancheng county, Hebei province, 1949-2000. Hydrol. Process. 18, 2289—2302.

Khan, S., Hanjra, M.A., 2009. Footprints of water and energy inputs in food production-Global perspectives. Food Policy 34, 130—140.

Kogan, F.N., 1998. Global drought and flood-watch from NOAA polar-orbitting satellites. Adv. Space Res. 21, 477—480.

Kuemmerle, T., Hostert, P., Radeloff, V.C., van der Linden, S., Perzanowski, K., Kruhlov, I., 2008. Cross-border comparison of post-socialist farmland abandonment in the Carpathians. Ecosystems 11, 614—628.

Kueppers, L.M., Snyder, M.A., Sloan, L.C., 2007. Irrigation cooling effect: regional climate forcing by land-use change. Geophys. Res. Lett. 34.

Lal, R., 2004. Soil carbon sequestration impacts on global climate change and food security. Science 304, 1623—1627.

Lal, R., 2007. Anthropogenic influences on world soils and implications to global food security. In: Sparks, D.L. (Ed.), Advances in Agronomy, vol. 93. Elsevier Academic Press Inc, San Diego, pp. 69–93.

Larson, D.W., Jones, E., Pannu, R.S., Sheokand, R.S., 2004. Instability in Indian agriculture-a challenge to the green revolution technology. Food Policy 29, 257–273.

Lee, E., Chase, T.N., Rajagopalan, B., Barry, R.G., Biggs, T.W., Lawrence, P.J., 2009. Effects of irrigation and vegetation activity on early Indian summer monsoon variability. Int. J. Climatol. 29, 573–581.

Liu, W.T., Kogan, F., 2002. Monitoring Brazilian soybean production using NOAA/AVHRR based vegetation condition indices. Int. J.Remote Sens. 23 (6), 1161–1179.

Liu, J.Y., Liu, M.L., Tian, H.Q., Zhuang, D.F., Zhang, Z.X., Zhang, W., Tang, X.M., Deng, X.Z., 2005a. Spatial and temporal patterns of China's cropland during 1990-2000: an analysis based on Landsat TM data. Remote Sens. Environ. 98, 442–456.

Liu, J., Pattey, E., Miller, J.R., Mcnairn, H., Smith, A., Hu, B., 2010. Estimating crop stresses, aboveground dry biomass and yield of corn using multi-temporal optical data combined with a radiation use efficiency model. Remote Sens. Environ. 114, 1167–1177.

Liu, W.H., Zhao, J.Z., Ouyang, Z.Y., Leif, S., Liu, G.H., 2005b. Impacts of sewage irrigation on heavy metal distribution and contamination in Beijing. China 31, 805–812.

Liu, X., Zhu, X., Pan, Y., Li, S., Liu, Y., Ma, Y., 2016. Agricultural drought monitoring: progress, challenges, and prospects. J. Geogr. Sci. 26, 750–767.

Lobell, D.B., Asner, G.P., 2004. Cropland distributions from temporal unmixing of MODIS data. Remote Sens. Environ. 93, 412–422.

Lobell, D.B., Bonfils, C., 2008. The effect of irrigation on regional temperatures: a spatial and temporal analysis of trends in California, 1934–2002. J. Clim. 21, 2063–2071.

Lobell, D.B., Asner, G.P., Ortiz-Monasterio, J.I., Benning, T.L., 2003. Remote sensing of regional crop production in the Yaqui valley, Mexico: estimates and uncertainties. Agric. Ecosyst. Environ. 94, 205–220.

Lobell, D.B., Bala, G., Duffy, P.B., 2006b. Biogeophysical impacts of cropland. Management changes on climate. Geophys. Res. Lett. 33, L06708.

Lobell, D.B., Bala, G., Bonfils, C., Duffy, P.B., 2006a. Potential bias of model projected greenhouse warming in irrigated regions. Geophys. Res. Lett. 33.

Lobell, D.B., Sibley, A., Ortizmonasterio, J.I., 2012. Extreme heat effects on wheat senescence in India. Nat. Clim. Chang. 2, 186–189.

Lohar, D., Pal, B., 1995. The effect of irrigation on premonsoon season precipitation over south-West Bengal, India. J. Clim. 8, 2567–2570.

Lopez Garcia, M.J., Caselles, V., 1991. Mapping burns and natural reforestation using Thematic Mapper data. Geocarto Int. 31–37.

Loveland, T.R., Reed, B.C., Brown, J.F., Ohlen, D.O., Zhu, J., Yang, L., Merchant, J.W., 2000. Development of a global land cover characteristics database and IGBP DISC over from 1-km AVHRR data. Int. J. Remote Sens. 21, 1303–1330.

Mahmood, R., Foster, S.A., Keeling, T., Hubbard, K.G., Carlson, C., Leeper, R., 2006. Impacts of irrigation on 20th century temperature in the northern Great Plains. Glob. Planet. Chang. 54, 1–18.

Mahmood, R., Hubbard, K.G., Carlson, C., 2004. Modification of growing-season surface temperature records in the northern Great Plains due to land-use transformation: verification of modelling results and implication for global climate change. Int. J. Climatol. 24, 311–327.

Mashaba, Z., Chirima, G., Botai, J.O., Combrinck, L., Munghemezulu, C., Dube, E., 2017. Forecasting Winter Wheat Yields Using MODIS NDVI Data for the Central Free State Region, vol. 113, pp. 11–12.

McCarty, J.L., Justice, C.O., Korontzi, S., 2007. Agricultural burning in the Southeastern United States detected by MODIS. Remote Sens. Environ. 108, 151–162.

McCarty, J.L., Korontzi, S., Justice, C.O., Loboda, T., 2009. The spatial and temporal distribution of crop residue burning in the contiguous United States. Sci. Total Environ. 407, 5701–5712.

McCarty, J.L., Loboda, T., Trigg, S., 2008. A hybrid remote sensing approach to quantifying crop residue burning in the United States. Appl. Eng. Agric. 24, 515–527.

McNairn, H., Brisco, B., 2004. The application of C-band polarimetric SAR for agriculture: a review. Can. J. Remote Sens. 30, 525–542.

McNairn, H., Duguay, C., Boisvert, J., Huffman, E., Brisco, B., 2001. Defining the sensitivity of multi-frequency and multi-polarized radar backscatter to post-harvest crop residue. Can. J. Remote Sens. 27, 247–263.

McNairn, H., Duguay, C., Brisco, B., Pultz, T.J., 2002. The effect of soil and crop residue characteristics on polarimetric radar response. Remote Sens. Environ. 80, 308–320.

Metternicht, G.I., Zinck, J.A., 2003. Remote sensing of soil salinity: potentials and constraints. Remote Sens. Environ. 85, 1–20.

Mkhabela, M.S., Mkhabela, M.S., Mashinini, N.N., 2005. Early maize yield forecasting in the four agro-ecological regions of Swaziland using NDVI data derived from NOAA'S-AVHRR. Agric. For. Meteorol. 129, 1–9.

Mo, X., Liu, S., Lin, Z., Xu, Y., Xiang, Y., Mcvicar, T.R., 2005. Prediction of crop yield, water consumption and water use efficiency with a svat-crop growth model using remotely sensed data on the north China plain. Ecol. Model. 183, 301–322.

Monfreda, C., Ramankutty, N., Foley, J.A., 2008. Farming the planet: 2. Geographic distribution of crop areas, yields, physiological types, and net primary production in the year 2000. Glob. Biogeochem. Cycles 22.

Moore, N., Rojstaczer, S., 2001. Irrigation-induced rainfall and the great plains. J. Appl. Meteorol. 40, 1297–1309.

Moriondo, M., Maselli, F., Bindi, M., 2007. A simple model of regional wheat yield based on NDVI data. Eur. J. Agron. 26, 266–274.

Murthy, C.S., Raju, P.V., Badrinath, K.V.S., 2003. Classification of wheat crop with multi-temporal images: performance of maximum likelihood and artificial neural networks. Int. J. Remote Sens. 24, 4871–4890.

Newman, J.E., 1980. Climate Changeimpacton the growing season of the North American corn belt. Biometeorology 7, 128–142.

Njoku, E.G., 2005. AMSR-E/Aqua Daily L3 Surface Soil Moisture, Interpretive Parms, & QC EASE-Grids,Jan 2005 to Dec 2006. National Snow and Ice Data Center.Digital media, Boulder, CO, USA. Availabe at: https://wist.echo.nasa.gov/api/.

Ozdogan, M., 2010. The spatial distribution of crop types from MODIS data: temporal unmixing using independent component analysis. Remote Sens. Environ. 114, 1190–1204.

Ozdogan, M., Gutman, G., 2008. A new methodology to map irrigated areas using multi-temporal MODIS and ancillary data: an application example in the continental US. Remote Sens. Environ. 112, 3520–3537.

Ozdogan, M., Salvucci, G.D., Anderson, B.C., 2006. Examination of the Bouchet-Morton complementary relationship using a mesoscale climate model and observations under a progressive irrigation scenario. J. Hydrometeorol. 7, 235–251.

Pacheco, A., McNairn, H., 2010. Evaluating multispectral remote sensing and spectral unmixing analysis for crop residue mapping. Remote Sens. Environ. 114, 2219–2228.

Padilla, F.L.M., Maas, S.J., González-Dugo, M.P., Mansilla, F., Rajan, N., Gavilán, P., Dominguez, J., 2012. Monitoring regional wheat yield in southern Spain using the grami model and satellite imagery. Field Crop. Res. 130, 145–154.

Palmer, W.C., 1968. Keeping Track of Crop Moisture Conditions, Nationwide: The New Crop Moisture Index.

Panda, S.S., Ames, D.P., Panigrahi, S., 2010. Application of vegetation indices for agricultural crop yield prediction using neural network techniques. Remote Sens. 2, 673–696.

Parry, M.L., Swaminathan, M.S., 1992. Effects of Climate Changes on Food Production. Cambridge University Press, Cambridge.

Patel, N.R., Parida, B.R., Venus, V., Saha, S.K., Dadhwal, V.K., 2012. Analysis of agricultural drought using vegetation temperature condition index (VTCI) from Terra/MODIS satellite data. Environ. Monit. Assess. 184, 7153–7163.

Paustian, K., Andren, O., Janzen, H.H., Lal, R., Smith, P., Tian, G., Tiessen, H., Van Noordwijk, M., Woomer, P.L., 1997. Agricultural soils as a sink to mitigate CO_2 emissions. Soil Use Manag. 13, 230–244.

Pena-Barragan, J.M., Ngugi, M.K., Plant, R.E., Six, J., 2011. Object-based crop identification using multiple vegetation indices, textural features and crop phenology. Remote Sens. Environ. 115, 1301–1316.

Peng, D.L., Huang, J.F., Li, C.J., Liu, L.Y., Huang, W.J., Wang, F.M., Yang, X.H., 2014. Modelling paddy rice yield using MODIS data. Agric. For. Meteorol. 184, 107–116.

Piao, S., Friedlingstein, P., Ciais, P., Viovy, N., Demarty, J., 2007. Growing season extension and its impact on terrestrial carbon cycle in the Northern Hemisphere over the past 2 decades. Glob. Biogeochem. Cycles 21, GB3018.

Portmann, F.T., Siebert, S., Doll, P., 2010. MIRCA2000-Global monthly irrigated and rainfed crop areas around the year 2000: a new high-resolution data set for agricultural and hydrological modeling. Glob. Biogeochem. Cycles 24, 24.

Ramankutty, N., Foley, J.A., 1999. Estimating historical changes in global land cover: croplands from 1700 to 1992. Glob. Biogeochem. Cycles 13, 997–1027.

Ramankutty, N., Evan, A.T., Monfreda, C., Foley, J.A., 2008. Farming the planet: 1. Geographic distribution of global agricultural lands in the year 2000. Glob. Biogeochem. Cycles 22.

Rhee, J., Im, J., Carbone, G.J., 2010. Monitoring agricultural drought for arid and humid regions using multi-sensor remote sensing data. Remote Sens. Environ. 114, 2875–2887.

Rosenberg, D.M., 2000. Global-scale environmental effects of hydrological alterations. Bioscience 50, 746–751.

Roy, D.P., Boschetti, L., Justice, C.O., Ju, J., 2008. The collection 5 MODIS burned area product – global evaluation by comparison with the MODIS active fire product. Remote Sens. Environ. 112, 3690–3707.

Roy, D.P., Jin, Y., Lewis, P.E., Justice, C.O., 2005. Prototyping a global algorithm for systematic fire-affected area mapping using MODIS time series data. Remote Sens. Environ. 97, 137–162.

Running, S.W., Thornton, P.E., Nemani, R., Glassy, J.M., 2000. Global Terrestrial Gross and Net Primary Productivity from the Earth Observing System. Methods in Ecosystem Science. Springer, New York.

Sakamoto, T., Yokozawa, M., Toritani, H., Shibayama, M., Ishitsuka, N., Ohno, H., 2005. A crop phenology detection method using time-series MODIS data. Remote Sens. Environ. 96, 366–374.

Segal, M., Pan, Z., Turner, R.W., Takle, E.S., 1998. On the potential impact of irrigated areas in North America on summer rainfall caused by large-scale systems. J. Appl. Meteorol. 37, 325–331.

Sen Roy, S., Mahmood, R., Niyogi, D., Lei, M., Foster, S.A., Hubbard, K.G., Douglas, E., Pielke, R., 2007. Impacts of the agricultural Green Revolution – induced land use changes on air temperatures in India. J. Geophys. Res. Atmos. D21108.

Serbin, G., Daughtry, C.S.T., Hunt, E.R., Brown, D.J., McCarty, G.W., 2009. Effect of soil spectral properties on remote sensing of crop residue cover. Soil Sci. Soc. Am. J. 73, 1545–1558.

Shafian, S., Rajan, N., Schnell, R., Bagavathiannan, M., Valasek, J., Shi, Y., Olsenholler, J., 2018. Unmanned aerial systems-based remote sensing for monitoring sorghum growth and development. PLoS One 13, e0196605.

Shanahan, J.F., Schepers, J.S., Francis, D.D., Varvel, G.E., Wilhelm, W.W., Tringe, J.M., Schlemmer, M.R., Major, D.J., 2001. Use of Remote-Sensing Imagery to Estimate Corn Grain Yield, vol. 93, pp. 583–589.

Shibuo, Y., Jarsjo, J., Destouni, G., 2007. Hydrological responses to climate change and irrigation in the Aral Sea drainage basin. Geophys. Res. Lett. 34, L21406.

Shiklomanov, I.A., 2000. Appraisal and assessment of world water resources. Water Int. 25, 11–32.

Siebert, S., Döll, P., 2001. A Digital Global Map of Irrigated Areas-An Update for Latin America and Europe. Kassel World Water Series 4. Center for Environmental Systems Research, University of Kassel, Germany, p. 14 (pp + Appendix).

Siebert, S., Doll, P., 2010. Quantifying blue and green virtual water contents in global crop production as well as potential production losses without irrigation. J. Hydrol. 384, 198–217.

Siebert, S., Doll, P., Hoogeveen, J., Faures, J.M., Frenken, K., Feick, S., 2005a. Development and validation of the global map of irrigation areas. Hydrol. Earth Syst. Sci. 9, 535–547.

Siebert, S., Feick, S., Döll, P., Hoogeveen, J., 2005b. Global Map of Irrigation Areas Version 3.0. University of Frankfurt (Main), Germany, and FAO, Rome, Italy.

Smith, P., Martino, D., Cai, Z., Gwary, D., Janzen, H., Kumar, P., McCarl, B., Ogle, S., O'Mara, F., Rice, C., Scholes, B., Sirotenko, O., Howden, M., McAllister, T., Pan, G., Romanenkov, V., Schneider, U., Towprayoon, S., Wattenbach, M., Smith, J., 2008. Greenhouse gas mitigation in agriculture. Philos. Trans. R. Soc. Biol. Sci. 363, 789–813.

Sonobe, R., Yamaya, Y., Tani, H., Wang, X., Kobayashi, N., Mochizuki, K., 2018. Crop classification from Sentinel-2-derived vegetation indices using ensemble learning. J. Appl. Remote Sens. 12, 026019.

Steele-Dunne, S.C., McNairn, H., Monsivais-Huertero, A., Judge, J., Liu, P., Papathanassiou, K., 2017. Radar remote sensing of agricultural canopies: a review. IEEE J. Sel. Top. Appl. Earth Obs. Remote Sens. 10, 2249–2273.

Stohlgren, T.J., Chase, T.N., Pielke, R.A.S., Kittel, T.G., Baron, J.S., 1998. Evidence that local land use practices influence regional climate, vegetation, and stream flow patterns in adjacent natural areas. Glob. Chang. Biol. 4, 495–504.

Sullivan, D.G., Shaw, J.N., Mask, P.L., Rickman, D., Guertal, E.A., Luvall, J., Wersinger, J.M., 2004. Evaluation of multispectral data for rapid assessment of wheat straw residue cover. Soil Sci. Soc. Am. J. 68, 2007–2013.

Tao, F., Yokozawa, M., Zhang, Z., Xu, Y., Hayashi, Y., 2005. Remote sensing of crop production in China by production efficiency models: models comparisons, estimates and uncertainties. Ecol. Model. 183, 385–396.

Terjung, W.H., Ji, H.Y., Hayes, J.T., 1989. Actual and potential yield for rain-fed and irrigation maize in China. Int. J. Biometeorol. 28, 115–135.

Thenkabail, P.S., Biradar, C.M.T., Noojipady, P.L., Vithanage, J., Dheeravath, V., Velpuri, M., Schull, M., Cai, X.L., Dutta, R., 2006. An Irrigated Area Map of the World (1999) Derived from Remote Sensing. Research Report 105. International Water Management Institute, Colombo, Sri Lanka.

Thenkabail, P.S., Biradar, C.M., Noojipady, P., Cai, X.L., Dheeravath, V., Li, Y.J., Velpuri, M., Gumma, M., Pandey, S., 2007. Sub-pixel area calculation methods for estimating irrigated areas. Sensors 7, 2519–2538.

Thenkabail, P.S., Biradar, C.M., Noojipady, P., Dheeravath, V., Li, Y.J., Velpuri, M., Gumma, M., Gangalakunta, O.R.P., Turral, H., Cai, X.L., Vithanage, J., Schull, M.A., Dutta, R., 2009. Global irrigated area map (GIAM), derived from remote sensing, for the end of the last millennium. Int. J. Remote Sens. 30, 3679–3733.

Thenkabail, P.S., Biradar, C.M., Noojipady, P., Dheeravath, V., Li, Y.J., Velpuri, M., Reddy, G.P.O., Cai, X.L., Gumma, M., Turral, H., Vithanage, J., Schull, M., Dutta, R., 2008. A global irrigated area map (GIAM) using remote sensing at the end of the last millennium. Int.Water Manag. Inst. 63.

Thenkabail, P.S., Schull, M., Turral, H., 2005. Ganges and Indus river basin land use/land cover(LULC) and irrigated area mapping using continuous streams of MODIS data. Remote Sens. Environ. 95, 317–341.

Tilman, D., Fargione, J., Wolff, B., D'Antonio, C., Dobson, A., Howarth, R., Schindler, D., Schlesinger, W.H., Simberloff, D., Swackhamer, D., 2001. Environmental change forecasting agriculturally driven global. Science 292, 281.

Trenberth, K.E., 2004. Rural land-use change and climate. Nature 427, 213.

Unganai, L.S., Kogan, F.N., 1998. Drought monitoring and corn yield estimation in Southern Africa from AVHRR data. Remote Sens. Environ. 63, 219–232.

Vörösmarty, C.J., Douglas, E.M., Green, P.A., Revenga, C., 2005. Geospatial indicators of emerging water stress: an application to Africa. Ambio 34.

Wang, H., Zhu, Y., Li, W., Cao, W., Tian, Y., 2014. Integrating remotely sensed leaf area index and leaf nitrogen accumulation with rice grow model based on particle swarm optimization algorithm for rice grain yield assessment. J. Appl. Remote Sens. 8, 083674.

Wang, J., Li, X., Lu, L., Fang, F., 2013. Estimating near future regional corn yields by integrating multi-source observations into a crop growth model. Eur. J. Agron. 49, 126–140.

Wang, K.C., Liang, S.L., 2008. An improved method for estimating global evapotranspiration based on satellite determination of surface net radiation, vegetation index, temperature, and soil moisture. J. Hydrometeorol. 9, 712–727.

Wang, R., Zhang, Q., Wang, Y., Yang, X., Han, Y., Yang, Q., 2004. Response of corn to climate warming in arid areas in northwest China. Acta Bot. Sin. 46, 1387–1392.

Wardlaw, I.F., Wrigley, C.W., 1994. Heat tolerance in temperate cereals: an overview. Funct. Plant Biol. 21, 695–703.

Wardlaw, I.F., Moncur, L., 1995. The response of wheat to high temperature following anthesis. I. the rate and duration of kernel filling. Aust. J. Plant Physiol. 22, 391–397.

Wardlow, B.D., Egbert, S.L., Kastens, J.H., 2007. Analysis of time-series MODIS 250 m vegetation index data for crop classification in the US Central Great Plains. Remote Sens. Environ. 108, 290–310.

White, M.A., de Beurs, K., Didan, K., Inouye, D., Richardson, A., Jensen, O., 2009. Intercomparison, interpretation and assessment of spring phenology in North America estimated from remote sensing for 1982 to 2006. Glob. Chang. Biol. 15, 2335–2359.

Wriedt, G., van der Velde, M., Aloe, A., Bouraoui, F., 2009. A European irrigation map for spatially distributed agricultural modelling. Agric. Water Manag. 96, 771–789.

Wu, W.B., Yang, P., Tang, H.J., Zhou, Q.B., Chen, Z.X., Shibasaki, R., 2010. Characterizing spatial patterns of phenology in cropland of China based on remotely sensed data. Agric. Sci. China 9, 101–112.

Xin, Q., Gong, P., Yu, C., Yu, L., Broich, M., Suyker, A.E., Ranga, B.M., 2013. A production efficiency model-based method for satellite estimates of corn and soybean yields in the midwestern us. Remote Sens. 5, 5926–5943.

Yan, H.M., Liu, J.Y., Huang, H.Q., Tao, B., Cao, M.K., 2009. Assessing the consequence of land use change on agricultural productivity in China. Glob. Planet. Chang. 67, 13–19.

Yuan, W., Chen, Y., Xia, J., Dong, W., Magliulo, V., Moors, E., Olesen, J.E., Zhang, H., 2016. Estimating crop yield using a satellite-based light use efficiency model. Ecol. Indicat. 60, 702–709.

Zhang, A., Jia, G., 2013. Monitoring meteorological drought in semiarid regions using multi-sensor microwave remote sensing data. Remote Sens. Environ. 134, 12–23.

Zhang, M., Li, Q.Z., Meng, J.H., Wu, B.F., 2011a. Review of crop residue fractional cover monitoring with remote sensing. Spectrosc. Spectr. Anal. 31, 3200–3205.

Zhang, X.Y., Kondragunta, S., Quayle, B., 2011b. Estimation of biomass burned areas using multiple-satellite-observed active fires. IEEE Trans. Geosci. Remote Sens. 49, 4469–4482.

Zhao, G.X., Lin, G., Warner, T., 2004. Using Thematic Mapper data for change detection and sustainable use of cultivated land: a case study in the Yellow River delta, China. Int. J. Remote Sens. 25, 2509–2522.

Zheng, B., Campbell, J.B., Serbin, G., Galbraith, J.M., 2014. Remote sensing of crop residue and tillage practices: present capabilities and future prospects. Soil Tillage Res. 138, 26–34.

Zhu, X., Liang, S., Pan, Y., Zhang, X., 2011. Agricultural irrigation impacts on land surface characteristics detected from satellite data products in Jilin Province, China. IEEE J. Sel. Top. Appl. Earth Obs. Remote Sens. 4, 721–729.

Zhu, X., Zhu, W., Zhang, J., Pan, Y., 2014. Mapping irrigated areas in China from remote sensing and statistical data. IEEE J. Sel. Top. Appl. Earth Obs. Remote Sens. 7, 4490–4504.

Zomeni, M., Tzanopoulos, J., Pantis, J.D., 2008. Historical analysis of landscape change using remote sensing techniques: an explanatory tool for agricultural transformation in Greek rural areas. Landsc. Urban Plan. 86, 38–46.

Further reading

Hazaymeh, K., Hassan, Q.K., 2016. Remote sensing of agricultural drought monitoring: a state of art review. Aims Environ. Sci. 3, 604–630.

Khanal, S., Fulton, J., Shearer, S., 2017. An overview of current and potential applications of thermal remote sensing in precision agriculture. Comput. Electron. Agric. 139, 22–32.

Liu, W.T., Kogan, F., 2002. Monitoring Brazilian soybean production using NOAA/AVHRR based vegetation condition indices. Int. J. Remote Sens. 23 (6), 1161–1179.

Mahlein, A.K., Oerke, E.C., Steiner, U., Dehne, H.W., 2012. Recent advances in sensing plant diseases for precision crop protection. Eur. J. Plant Pathol. 133, 197–209.

25

Forest cover changes: mapping and climatic impact assessment

Abstract

As the largest terrestrial ecosystem on Earth, forests are affected by climate change and human activities. In turn, forests influence the climate at local, regional, and global scales through physical, chemical, and biological processes. Forest cover change and its climatic effects are of considerable interest in land cover and land-use change studies. Remote sensing is a powerful tool used for monitoring forest cover change as it provides abundant spatial and temporal information. In this chapter, we describe how to apply remotely sensed data to detect and map the location, extent, and variability of forest cover change (Section 25.2) and quantify the climatic effects resulting from forest

© 2020 Elsevier Inc. All rights reserved.

cover change (Section 25.3). Several examples, including deforestation in the Amazon and forest cover change in China, are given (Section 25.4) to better illustrate the causes and effects of forest cover change.

25.1 Introduction

Forests are ecosystems characterized by more or less dense and extensive tree cover. A common definition of a "forest" is *"land of at least 0.5 ha covered by trees higher than 5 m and with a canopy cover of more than 10%, or by trees able to reach these thresholds, and predominantly under forest land use,"* devised by the United Nations Food and Agricultural Organization's (FAO) Forest Resource Assessment (FRA) in 2000. This definition excludes land that is mainly under agricultural or urban land uses. Indeed, this definition is essentially a land use–based definition. According to FAO FRA (2015), forests cover 3999 M-ha globally, which is equivalent to 31% of total land area or 0.6 ha for every person on the planet (FAO, 2015). Forests are increasingly being managed for a variety of uses and values, including productive, protective, and socioeconomic functions.

- Forests and trees outside forests provide a wide range of wood and nonwood forest products. Many products are extracted from forests, ranging from wood for timber and fuel to food (berries, mushrooms, edible plants, bushmeat), fodder, and other nonwood forest products. One-third of the world's forests are used primarily for the production of wood and nonwood forest products.
- The world's forests have many protective functions, ranging from soil and water conservation and avalanche control to sand dune stabilization, desertification control, and coastal protection. About 11% of the world's forests have been designated for the conservation of biological diversity. The overall proportion of forests designated for protective functions increased from 8% in 1990 to 9% in 2005.
- Forests provide a wide range of economic and social benefits to humankind, including contributions to the overall economy—for example, through employment and processing and trade of forest products and energy—and investments in the forest sector and the hosting and protection of sites and landscapes of high cultural, spiritual, or recreational value. In Europe, about 2.4% of all forests are used for recreation, tourism, education, and conservation of cultural and spiritual sites.

Forest change dynamics are illustrated as in Fig. 25.1. A reduction in forest area can occur through either deforestation or natural disasters. Deforestation is the permanent removal of forest cover by people and the conversion of the land for other uses, such as agriculture or infrastructure. An increase in forest area occurs in two ways—either through afforestation, i.e., planting trees on land that was not previously forested, or through the natural expansion of forests, e.g., on abandoned agricultural land. Forest change also includes reforestation, where a forest is cut down but replanted, and natural regeneration, where the forest grows back on its own within a relatively short period.

Characterizing and mapping forest cover is essential for planning and managing natural forest resources (e.g., conservation, and/or development). Detecting changes in forest via satellite observations has a long tradition, beginning with Landsat satellites in the 1980s and 1990s. Satellite remote sensing provides a viable

FIGURE 25.1   Forest change dynamics (FAO, 2006).

means for identifying, mapping, assessing, and monitoring forest cover at a range of spatial and temporal scales. Remote sensors offer a quick and low cost way to map thousands of acres per day to monitor deforestation, logging, and other disturbances, as well as the recovery of forests. Meanwhile, repeated and consistent observations from satellites allow us to investigate the local effects of forests from a global perspective at high resolution, which can improve our understanding of forestry policy outcomes. In the following sections, we will introduce how remote sensing data are used to map forest change and quantify the corresponding climatic effects.

25.2 Mapping forest change

Complex social and economic forces contribute to large-scale changes in forest cover throughout the world, where these changes vary in different regions. For example, forests are being cleared in the tropics and in the boreal forests in Canada and Siberia but are regrowing on abandoned agricultural lands in Europe and temperate North America. Such changes alter feedback to the climate system at local to global scales. One goal of large-area forest cover mapping is to produce globally consistent characterizations that have local relevance and utility, that is, reliable information across many spatial and temporal scales. Earth observation satellite data are used to map global forest loss and gain because of its advantages in offering the abundant temporal and spatial information.

25.2.1 Change detection based on forest cover mapping

The detection of forest change, including forest disturbances, forest degradation, reforestation, and afforestation, is a key aspect in sustainable forest management. Thus, reliable maps of forest cover change have become

increasingly important. Forest change mapping by means of remote sensing has a long tradition. The spatial and temporal characteristics of remote sensors are very important when choosing sources for assessing and mapping land cover change. While very high spatial resolution sensors (<4 m multispectral pixel size) may have better change-detection capabilities because of their much more smaller footprints, their wall-to-wall coverage is generally more cost- and time-consuming and even impossible to obtain large areas. Medium- to coarse-resolution sensors (>60 m multispectral pixel size) have lower cost and time requirements when obtaining wall-to-wall data coverage, but they are very limited in their ability to detect small area disturbances. High-resolution satellite systems (5–30 m multispectral pixel size) can be considered a good compromise, offering high enough spatial resolution and a large enough footprint for cost-efficient large-scale land cover change monitoring. Nowadays, in addition to the continuous Landsat data, which have been publicly available without restrictions from the US Geological Survey since 1972, more and more optical high-resolution data have become available (e.g., Sentinel-2). Therefore, change-detection methods based on high-resolution data have drawn more attention recently.

Surface reflectance values or indices generated from remote sensing optical imaging bands are usually applied for land cover change mapping. Some examples includes photosynthetic active radiation, the fraction of absorbed photosynthetic active radiation, index variables such as normalized difference vegetation index (NDVI), enhanced vegetation index (EVI), soil adjusted vegetation index, topographical variables (e.g., slope, aspect, height), and texture variables (e.g., homogeneity, object size and shape, connectivity).

Land cover mapping is roughly divided into two main categories: first, classical image-to-image change detection and second, time series analyses.

- Classical image-to-image change detection

This method is currently the most commonly used method for land cover change mapping. It requires at least two images, of which the first must be from the start of a monitoring period and the second at the end. Changes are then classified based on the change of indicators (e.g., NDVI, normalized burnt ratio, etc.) between the two dates. This approach is also often applied two or more times, each time comparing two images or classifications. In the case of forest cover change mapping, a forest mask is necessary to focus on changes in the forest areas only, and it must be as up-to-date as possible. However, this method does not detect change in a strict sense, as only one remote sensing image is involved. Moreover, the geometric and radiometric calibrations are highly important among the standard preprocessing for this method.

- Time series analysis—based change detection

A typical remote sensing time series consists of three components: a long-term directional trend component, a seasonal component, and a residual component (Kuenzer et al., 2015). With regard to forest monitoring, the residual component, which has to be differentiated from residual noise, is of the highest interest, while the seasonal component is also very important. A series of images continuously over a period of time is required for this approach. Thus, there is a need for substantially more and regular image acquisitions of the area of interest. Although low- to medium-resolution satellite imagery (e.g., AVHRR, MODIS) was frequently used in various time series analysis methods over the last few decades, high-resolution satellite data (e.g., Landsat) is now preferred because of their spatial resolution. Specifically, there are four subcategories of this approach: (a) threshold-based change detection, (b) curve fitting, (c) trajectory fitting, and (d) trajectory segmentation. Fig. 25.2 gives a schematic comparison of the four methods.

(a) Threshold-based change detection

This method uses a thresholding procedure to separate forests from nonforests, or intact forests from changed forests, in a time series. An automated forest change mapping algorithm (e.g., vegetation change tracker [VCT]) was developed based on the likelihood of a pixel being forested. The VCT algorithm is presented in Section 25.2.2. The main drawback of threshold-based time series analyses is that the thresholds are empirically determined and thus they cannot be directly transferred to other study areas, which are characterized by different vegetation types, densities, or change patterns.

(b) Curve fitting

Curve fitting is often used for monitoring forest dynamics. A pixel-wise trend function based on a regression model is fitted between spectral variables and time, where the spectral variables can be single-band reflectance or derived indices. A slope that is significantly different from zero determines the presence or absence of a trend. The trend is expressed as the slope of the regression curve and provides information on the magnitude of change over time, while the sign of the slope can be used to separate different change processes, such as an increase or decrease of crown cover. Linear and quadratic coefficients are used to analyze forest cover change trends, where the coefficients of quadratic curvature have been found to be especially important for understanding degradation and recovery. A known shortcoming of single-curve fitting is that standard underlying statistical assumptions, such as normality and equal variance, generally need to be met, and violation of these assumptions can result in an inadequate representation of the data by the fitted function.

(c) Trajectory fitting

Trajectory fitting can be interpreted as "*a supervised change-detection method with idealized trajectories representing training signatures specific to*

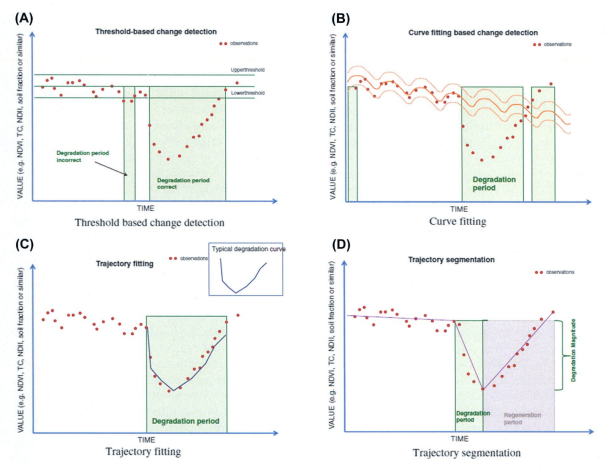

FIGURE 25.2   Schematic illustration of different time series analysis methods (Hirschmugl et al., 2017).

different degradation types" (Kennedy et al., 2007). This requires the definition of hypothesized change trajectories, which are based on the expected spectral—temporal behavior of the selected variables. A major limitation of the trajectory fitting approach is that the shape of the typical degradation curve must be known from some reference data, and the method will only work properly if the observed spectral trajectory matches one of the predefined typical degradation curves.

(d) Trajectory segmentation

The trajectory can be decomposed into a series of straight-line segments to capture broad features of the trajectory as well as subtrends. The first phase of segmentation is the determination of the vertex years that define the endpoints of the segments. In the second phase, the best straight-line trajectory is fit through the vertices using either point-to-point or regression lines. The advantage of this approach is that the straight-line segments allow for the detection of abrupt events such as disturbances and longer-duration processes such as regrowth. Another advantage is that a typical curve of degradation is not required because the data themselves determine the shape of the trajectory. A main drawback of the method is that seasonal effects caused by phenology are not taken into account.

The LandTrendR is one of the most popular algorithms of this kind of method (Kennedy et al., 2010).

25.2.2 Techniques using temporal landsat imagery

As mentioned in Section 25.2.1, temporal information, especially the high-resolution satellite data, is particularly useful for characterizing land cover and change processes. To minimize potential omission errors that may arise from using temporally sparse observations, dense Landsat time series stacks (LTSS) can be very useful. To improve the efficiency of land cover change analyses using LTSS, Huang et al. (2010) developed the VCT algorithm for mapping forest change through the simultaneous analysis of all images in an LTSS.

The VCT is based on the spectral—temporal characteristics of land cover and forest change processes. It involves two major steps (Fig. 25.3). The first step consists of individual image masking and normalization, in which each image in an LTSS is analyzed separately to mask water, cloud, and cloud shadow and to identify some forest samples. The identified forest samples are then used to calculate several indices as measures of forest likelihood. The second step performs the time series analysis. Once the first step is completed, the indices and masks derived for all images of an LTSS are used to form time series trajectories and produce forest change products.

Specifically, the first step includes the following major processes, namely, the creation of a land—water mask, identification of forest samples, calculation of forest indices, and masking of cloud and cloud shadow. After the first step, temporal interpolation is used to derive interpolated values for pixels flagged as cloud, shadow, or other bad observations. The resultant masks and indices are then used to determine change and nonchange classes and to derive a suite of attributes that characterize the mapped changes.

The VCT algorithm can handle a certain level of bad observations in individual images. Those bad observations typically do not leave gaps or footprints in the derived yearly forest disturbance maps (Fig. 25.4). The VCT achieves such a level of resistance to bad observations through (1) the automatic masking of clouds and cloud shadows, (2) temporally interpolating for identified bad observations, (3) requiring consecutive observations when determining forest and change, and (4) simultaneously considering the entire temporal domain of each LTSS. Of course, unflagged, consecutive bad observations may result in false changes, and excessive bad observations may completely fool the VCT algorithm. Therefore, it is still necessary to select high-

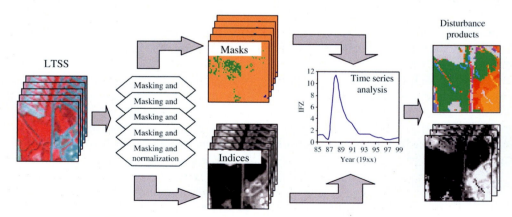

FIGURE 25.3   Flowchart of the VCT algorithm (Huang et al., 2010).

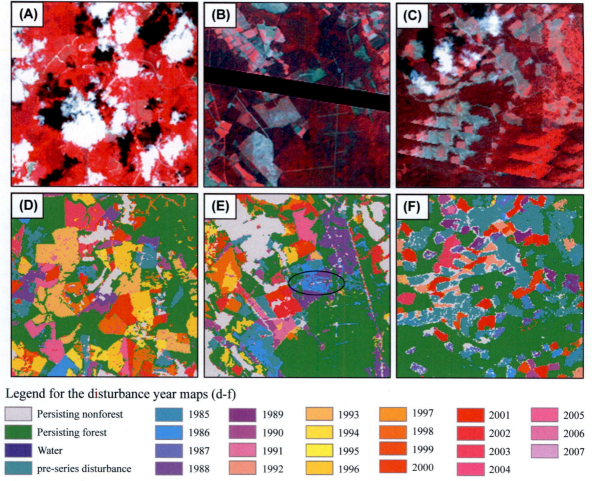

Legend for the disturbance year maps (d-f)

☐ Persisting nonforest	☐ 1985	☐ 1989	☐ 1993	☐ 1997	☐ 2001	☐ 2005
☐ Persisting forest	☐ 1986	☐ 1990	☐ 1994	☐ 1998	☐ 2002	☐ 2006
☐ Water	☐ 1987	☐ 1991	☐ 1995	☐ 1999	☐ 2003	☐ 2007
☐ pre-series disturbance	☐ 1988	☐ 1992	☐ 1996	☐ 2000	☐ 2004	

FIGURE 25.4 Bad observations such as (A) cloud/shadow contamination, (B) missing scan lines, or (C) duplicate scan lines in individual images leave little or no signs in the yearly disturbance maps produced by the VCT ((D), (E), and (F), produced with (A), (B), and (C), respectively, as part of the inputs). When preceding a 1988 disturbance, however, the missing scan line in (B) caused the 1988 disturbance to be mapped as a 1986 disturbance (circled in (E)). The images in (A–C) are shown with bands 4, 3, and 2 in red, green, and blue (Huang et al., 2010).

quality images for use in the LTSS to produce satisfactory disturbance products.

The VCT has been widely tested. Visual assessment of the yearly disturbance products revealed that most of them are reasonably reliable. Design-based accuracy assessments have revealed that overall accuracies of around 80% are achieved for disturbances mapped at the individual-year level. Average user's and producer's accuracies of the disturbance classes are around 70% and 60%, respectively (Huang et al., 2010).

25.2.3 MODIS vegetation continuous fields products

The MODIS science team has produced the vegetation continuous fields (VCFs) and the vegetative cover conversion (VCC) products.

- MODIS VCF product

The VCF product improved on traditional discrete land cover classifications by providing global subpixel estimates of landscape components at 500-m spatial resolution. Estimates are provided for percentage tree cover, herbaceous cover, and bare cover. Vegetative cover is further partitioned into leaf type (broad leaf and needle leaf) and longevity (evergreen and deciduous). By providing continuous percentage cover estimates, users are able to define their own vegetation class boundaries without constraints imposed by a priori definitions inherent in discrete land cover classifications. VCFs are produced on an annual time steps, allowing for the analysis of changes in vegetation density through time.

The technical details of the VCF algorithm are described by Hansen et al. (2002). It is an automated procedure that employs a regression tree algorithm. A regression tree is a nonlinear, distribution-free algorithm that is highly suited to handling the complexity of global spectral land cover signatures. Training data are used as the dependent variable, which are predicted by the independent variables in the form of the annual MODIS metrics. Outputs from the regression tree are further modified by stepwise regression and bias adjustment. The derivation of tree cover in this way creates the possibility of using subsequent depictions to measure change. An example is shown in Fig. 25.5.

The MODIS VCF algorithm was improved in 2011 by DiMiceli et al. (2011). The early MODIS VCF algorithms (Collection 3) were developed using a semiautomated process where the regression trees were created using machine learning software. In the new approach, the training data are completely updated using Landsat Geocover data and have been revised and refined using the plethora of fine and ultrafine-resolution data available through the NASA science data purchase, Google Earth, among many others. The model creation process was also improved by creating 30 independent regression trees, and later the 30 independent results were yield with MODIS data, and then the final VCF is produced by averaging the 30 independent results. Moreover, new and improved data-mining software without human intervention is also used. Fig. 25.6 shows the image pairs; the image on left is the old 500 m product, and the image on right is the new 250 m product. Both are shown in a 250 m grid to emphasize the improved in spatial detail, where the updated VCF product (Collection 5) has better accuracy.

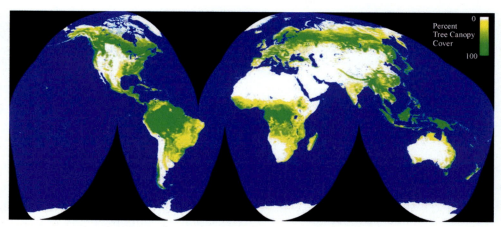

FIGURE 25.5   Global VCF percentage tree cover for 2001 (interrupted Goode homolosine projection) (Hansen et al., 2003).

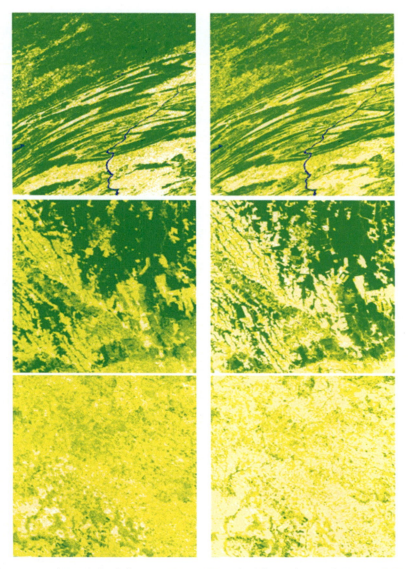

FIGURE 25.6 Image pairs showing the Collection 3500 m VCF on the left and the new Collection 5250 m VCF on the right. Darker green indicates denser tree cover (Townshend et al., 2006).

- MODIS VCC product

The VCC product provides quarterly identification of land cover change hot spots at a spatial resolution of 250 m and an annual scale. Daily MODIS data are assimilated to collect the best possible set of cloud-free, near-nadir observations from which to assess change. Changes of interest include deforestation, flooding, and burning. The dataset is intended to identify any areas undergoing rapid land cover change that should be further delineated using fine-resolution observations from instruments such as Landsat and

IKONOS. The technical details of the VCC algorithm are described by Zhan et al. (2000) and briefly summarized in Table 25.1.

Zhan et al. (2002) applied the VCC method to the imagery shown in Fig. 25.7 to assess the areas burned by Idaho—Montana wildfires. The fire perimeter polygons generated by the United States Department of Agriculture Forest Service (USFS) from ground observations, helicopter GPS flights, and airborne infrared imaging were used for validation. The burned area detection results are shown in Fig. 25.8, where green

pixels are correctly identified burned areas, blue are incorrectly labeled burned areas, and red are undetected burned areas. The polygonwise analysis demonstrates that most of the five change-detection methods and the integrated measure of change identified burned areas very well.

Similar to the VCF product, the VCC product (Collection 5) has also evolved over time. The original method employed five different change-detection methods, which were combined to yield a single end result. In contrast,

TABLE 25.1 MODIS VCC procedure (Zhan et al., 2000).

Name of method	Criteria used	Implementation
Red-NIR space partitioning	Based on partitioning of red-NIR space between cover types at a given time of year and latitude. Identification of whether pixel value changes from subspace of one cover type to a second.	Implemented using LUTs depicting cover type for a given pair of red and NIR reflectance values for each month and latitudinal zone at times 1 and 2.
Red-NIR space change vector	Using the red-NIR space, based on the angle and magnitude of the vector defined by the pixel value at time 1 to that at time 2. The direction and magnitude of change are defined as $$\Theta = \arctan(\Delta\rho_{red}/\Delta\rho_{NIR})$$ $$A = \sqrt{(\Delta\rho_{red})^2 + (\Delta\rho_{NIR})^2}$$ where Θ = change angle; A = change magnitude; ρ_{red} = red reflectance; and ρ_{NIR} = NIR reflectance.	Implemented using LUTs depicting the angle and magnitude couplets associated with all possible land cover changes for each pair of months and latitudinal zone.
Modified delta space thresholding	Uses space defined by differences in pixel values for times 1 and 2 for the red and NIR values of each pixel (no change occurs at the origin). Type of conversion defined by angle and distance from origin and the initial state of the pixel.	Implemented using LUTs defining the initial cover type at time 1 and the sector of the delta space in which the pixel lies at time 2 for each latitudinal zone.
Texture	Uses coefficient of variation of the NDVI within a 3×3 kernel at times 1 and 2. Conversion flagged when change exceeds a threshold.	Implemented using LUTs defining threshold value for change between each pair of cover types for each month, latitudinal zone, and initial state of the pixel.
Linear feature	Compute the mean of the absolute difference of the pixel value for each neighbor pixel in a 3×3 kernel. A threshold determines whether a linear feature is present.	Implemented using rule identifying whether linear feature exists in time 2 when absent in time 1.
Integrated measure of change		Voting method: conversion confirmed where three out of five flag conversion methods.

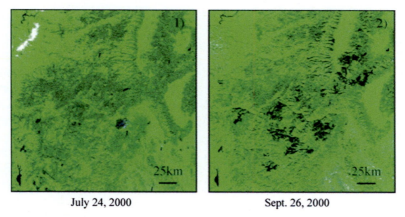

July 24, 2000   Sept. 26, 2000

FIGURE 25.7   MODIS 250 m image acquired at the start of the Idaho—Montana wildfires on July 24, 2000 (left), and immediately after the fires on September 26, 2000 (right). In these images, the MODIS Level 1B band 1 reflectance is shown using both red and blue, and band 2 reflectances are shown using green in RGB visualization. The dark purple areas in the middle of the September image are the areas burned by wildfires (Zhan et al., 2002).

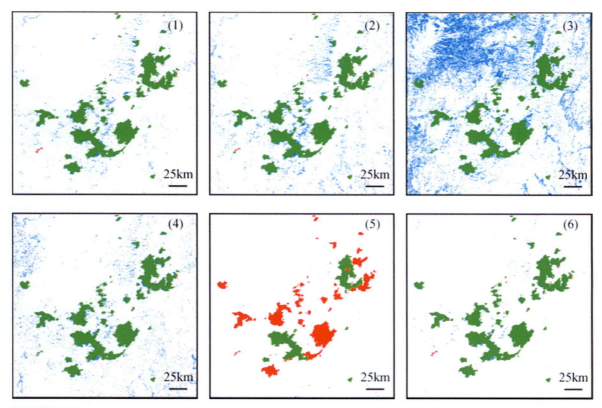

FIGURE 25.8   Results from the VCC change-detection algorithms. The green color indicates burned areas identified by both the VCC algorithms and USFS polygons. The red color shows omission errors, and the blue color shows commission errors. Algorithms for the maps are (1) the red—NIR space partitioning method, (2) the red—NIR space change vector method, (3) the modified delta space thresholding method, (4) the texture change method, (5) the linear feature method, and (6) the integration of the five methods. The black bars are 25 km in length (Zhan et al., 2002).

the current method uses separate methods for determining different types of change. The method for determining deforestation is derived from the original space-partitioning method (see Table 25.1), which relies on decision tree classification to determine antecedent vegetation conditions that are compared with current vegetation conditions. Changes due to burning are derived using the different normalized burn ration (van Wagtendonk et al., 2004) with two scenes imaged a year apart. Flooding is determined similarly to deforestation using decision tree classification to determine where water exists in any given scene, which is then compared with a static map of water extent to determine areas that may be inundated (Townshend et al., 2006).

25.2.4 FAO FRA 2010 remote sensing survey

The primary source for global information of forest resources to date is FAO FRA. These data, supplied by contributing member countries, are the current reference data for global forest change. FAO FRA collects and collates national statistics on forest resources in 5—10 yr intervals, and the remote sensing survey was implemented aiming at complementing the assessment since FRA 1990. In FRA's 2010 survey, results from the global remote sensing survey were included as an independent means of collecting comparable time series data on the state of the worlds' forests. The next global remote sensing forest survey is expected in 2020.

In the FAO FRA 2010, the systematic sampling design was based on each longitude and latitude intersection, as illustrated below (Fig. 25.9), where the reduced intensity above 60N is due to the curvature of the Earth. The assessment covers the whole land surface of the Earth and consists of about 13,500 samples, of which about 9000 are outside deserts and areas with permanent ice. The area covered at each

sample site is 10×10 km, providing a sampling intensity of about 1% of the global land surface. For each sample plot, Landsat images, dating from around 1990, 2000, and 2005, were interpreted and classified using an automated supervised approach. Nearly, 7 million polygons were analyzed in each time interval to detect forest area, forest gains, and forest losses that were ≥ 5 ha in size.

The survey shows the total forest area in 2005 was 3.69 Bha, a decrease from 3.726 Bha in 2000 and 3.767 Bha in 1995. However, there were important regional differences in forest loss and gain, as shown in Fig. 25.10.

25.3 Qualifying the climatic effects of forest change

The Intergovernmental Panel on Climate Change (IPCC) concluded that there may be significant regional transitions associated with shifts in forest locations and composition in the United States due to climate change. Climate change will likely alter the geographic distribution of North American forests, including regionally important tree species, such as New England sugar maples and boreal forests in Alaska. Climate change effects on forests are likely to include changes in forest health and productivity and changes in the geographic range of certain tree species. These effects could in turn alter timber production, outdoor recreational activities, water quality, wildlife, and rates of carbon storage.

On the other hand, forest change can also impact the climate system through several processes (Fig. 25.11) (Chapin et al., 2008) such as (1) the emission of greenhouse gases, which cause an imbalance in the Earth's energy budget at the top of the atmosphere, (2) altering of albedo (the proportion of solar radiation that the Earth's surface reflects back to space), which influences the amount of heat transferred from ecosystems to the atmosphere, (3) altering

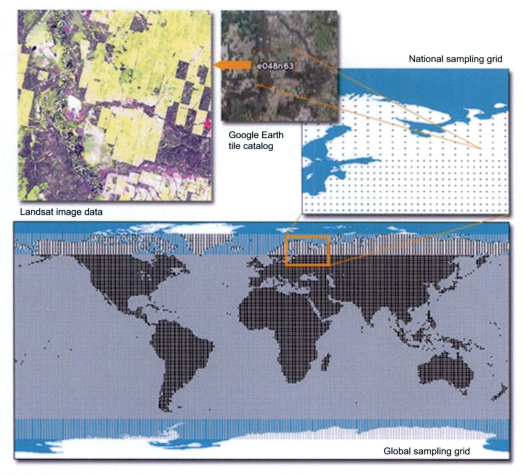

FIGURE 25.9   The sampling scheme (http://geonetwork4.fao.org/geonetwork/srv/en/fra.samplingGrid).

evapotranspiration (ET; i.e., evaporation from the Earth's surface plus that from leaves), which cools the Earth's surface and provides moisture to form clouds and drive atmospheric mixing, (4) altering longwave radiation, which depends on the Earth's surface temperature and cloudiness, (5) changes in the production of aerosols (small particles that scatter and absorb light), and (6) changes in the Earth's surface roughness, which determines the strength of coupling between the atmosphere and the surface and therefore, the efficiency of water and energy exchange. The first process is usually referred

to as a biogeochemical effect, and the rest are biogeophysical effects.

Three major balance systems are illustrated in Fig. 25.11: carbon balance, energy balance, and water balance. Carbon balance is the difference between CO_2 uptake by ecosystems (photosynthesis) and CO_2 loss to the atmosphere by respiration. Energy balance is the balance between incoming solar radiation, the proportion of this incoming solar radiation that is reflected (albedo), and the transfer of the absorbed radiation to the atmosphere as sensible heat (warming the near-surface air), ET (cooling the surface), and

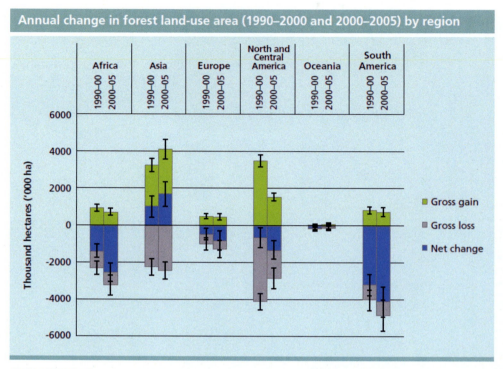

FIGURE 25.10   Regional differences in the rate of changes of forest areas worldwide (FAO, 2010).

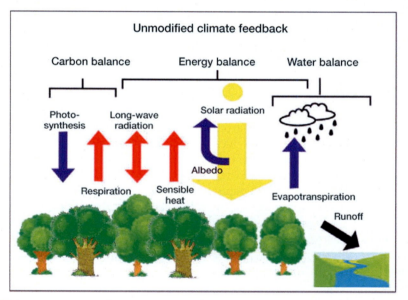

FIGURE 25.11   The major categories of ecosystem—climate interactions (each shown by the arrows beneath the brackets). Arrows show the direction of mass or energy transfer. Cooling effects on the surface climate are shown by *blue arrows* and warming effects by *red arrows* (Chapin et al., 2008).

longwave radiation. Water balance between the ecosystem and atmosphere is the difference between precipitation inputs and water via ET; the remaining water leaves the ecosystem as runoff. Each of these ecosystem—atmosphere exchanges influences the Earth's climate, as shown in Fig. 25.11. Both ET and sensible heat cool the surface, but surface cooling is more pronounced when ET predominates. Other interactions with the climate system are influenced by ecosystems (but not shown in this diagram), which include the effects of particulates, CH_4, N_2O, ozone, and reflectance by clouds (Chapin et al., 2008).

Forests can sequester C, but this leads to other important biogeophysical changes. Forests often have lower surface albedos than those of the ecosystems they replace, and so they absorb more solar radiation. They can also affect other biogeophysical surface properties, including surface roughness, which influences the exchange of energy and mass between the land surface and the atmosphere. Fig. 25.12 shows the impacts of forest on surface energy balance. Forests have greater heat fluxes than nonforest ecosystems, resulting from their greater surface roughness. Tropical rainforests have large latent heat fluxes that result in the development of clouds, which reflect solar radiation back to space. Temperate and boreal forests have major seasonal variations in energy fluxes and can reduce seasonal cooling by masking snow. Forests and trees must be recognized as prime regulators within the water, energy, and carbon cycles. In the following sections, we will briefly introduce the climatic effects of forest change of three factors, namely, greenhouse gases, temperature, and precipitation.

25.3.1 Greenhouse gases

Forests can sequester C through photosynthesis. Afforestation leads to C accumulation in living biomass, coarse woody debris, and soil organic carbon, with the relative importance of accumulation in these pools varying considerably across different biomes (Table 25.2). Tropical forests exhibit high globally averaged C storage and uptake per unit area, cover the greatest amount of land, and are responsible for the highest level of net cooling of any biome. Tropical deforestation currently accounts for over 90% of net C emissions resulting from global land-use change; therefore, avoiding tropical deforestation reduces anthropogenic C emissions from land-use change (Anderson et al., 2011). Bala et al. (2007) showed that complete tropical deforestation could increase global land temperatures by 0.9°C. In addition, transpiration is also closely connected to many other ecological and biogeochemical processes, such as the nitrogen cycle.

Water vapor is another natural, important, and abundant greenhouse gas, which is thought to be inextricably linked to vegetation greenness. Water vapor comprises 0.25% of the mass of the atmosphere—equivalent to just 2.5 cm of liquid over the entire Earth (atmospheric water in the form of liquid droplets and ice adds less than one-hundredth to this miniscule total; Sheil, 2018). Water vapor plays a major role as a dominant feedback variable in the hydrological cycle and in radiative forcing, which impacts both global and local weather as well as climate. Vegetation, especially forests, affect water vapor mainly through ET processes. Observational evidence (Jiang and Liang, 2013) has shown that increased vegetation greenness in summer increased the amount of atmospheric water vapor over Northern China during 1982—2008 and that vegetation accounts for as much as 30% of the total water vapor variance, especially occurred in water-limited areas in the middle and west of China (Fig. 25.13B).

25.3.2 Temperature

Forests influence the local temperature, including the land surface temperature (LST) and air temperature (2 m height from the ground), through two major biophysical factors:

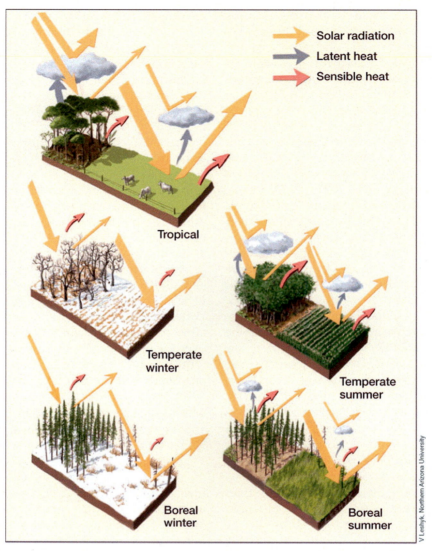

FIGURE 25.12   Effects of forest and nonforest ecosystems on surface energy fluxes in tropical, temperate winter, temperate summer, and boreal summer scenarios (Anderson et al., 2011).

TABLE 25.2   Area and C storage in vegetation of select biomes (Anderson and Goulden, 2011).

	Area (millions km²)	Total C (gigatons)	C per unit area (kg/m²)
Tropical forests	17.5	553	31.6
Temperate forests	10.4	292	28.1
Boreal forests	13.7	395	28.8
Crops	13.5	15	1.1
Tropical grasslands	27.6	326	11.8
Temperature grasslands	15.0	182	12.3

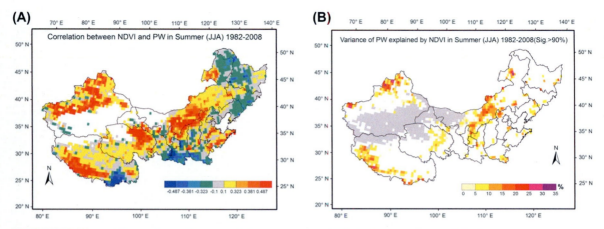

FIGURE 25.13 (A) Pearson's correlation result between summer NDVI and water vapor and (B) the explained water vapor variance determined from the NDVI in summer over Northern China from 1982 to 2008. Northern China is delineated by the *black line*, whereas the white areas represent insignificant regions (<90% confidence level) (Jiang and Liang, 2013).

albedo and ET. With lower albedo, forests absorb more shortwave radiation that can potentially lead to a warming effect. However, this net energy gain is offset by a greater latent heat loss via high ET in forests that result in a net cooling effect, where the resultant temperature is determined by the sum of the two effects. In the decade 2003−12, variations in forest cover generated a mean biophysical warming on land corresponding to about 18% of the global biogeochemical signal due to CO_2 emission from land-use change (Alkama and Cescatti, 2016). Fig. 25.14 shows the surface temperature distribution of forests and other landscapes.

The effects of forests on temperature have distinctive latitudinal patterns, ranging from strong cooling in the tropics to moderate cooling in temperate regions and warming in high latitudes. The strength of albedo warming generally increases with latitude, while the strength of cooling by ET decreases. It has been observed that the transitional latitude that separates the cooling/warming of LST is around 45°N, while the air temperature effect of forests is similar but with a transitional latitude of 35°N (Li et al., 2015). Forests north of 45°N tend to be warming, such as boreal forests, as albedo

warming completely surpasses the negligible ET cooling, and forests south of 35°N tend to be cooling, such as tropical forests, as high ET cooling completely offsets albedo warming. Temperate forests between 35°N and 45°N tend to be moderately cooling, as the different biophysical effects exhibit comparable strengths and the net effect is more susceptible to other factors that can alter the relative role of the biophysical effects.

Forests change also affects the diurnal asymmetric features of LST by different biophysical drivers. For example, China's land cover transition from forest to cropland, which tends to decrease LST by 0.53 K during the day and by 0.07 K at night (Zhang and Liang, 2018). These features exhibit considerable spatial and seasonal variations, as Fig. 25.15 shows, where deforestation results in daytime warming and nighttime cooling. The strongest daytime warming is in the tropics, while the strongest nighttime cooling is observed in the boreal zones over most regions of the world. Daytime patterns of diurnal asymmetry are explained by differences in the latent heat flux and absorbed solar radiation. At night, the LST is closely related to energy stored during the daytime

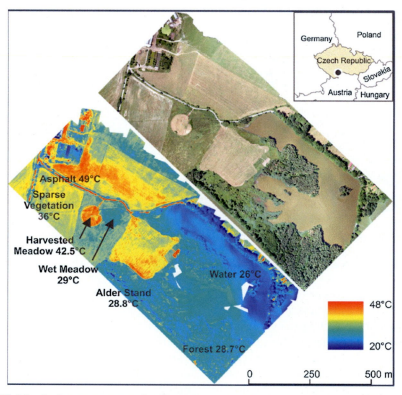

FIGURE 25.14   Surface temperature distribution in a mixed landscape with forests (Ellison et al., 2017).

and to the near-surface atmospheric boundary layer. When afforestation occurs, it is likely that the increase in surface heat capacity (e.g., as a result of an increase in soil moisture) may result in more daytime heat storage, and thus more nighttime heating, and that the increase in air humidity (e.g., as a result of enhanced daytime ET) near the surface and the enhancement of boundary layer cloud information may result in more downward longwave radiation received from the atmosphere that reduces the outgoing longwave radiation from the planted forests. This longwave radiative imbalance has a stronger effect during the nighttime, when the boundary layer is thinner and more stable, than during the daytime. In addition, the nighttime warming effect could be magnified as a result of reduced atmospheric turbulence from a more stable stratification over trees.

Moreover, net forest effects on regional and global climate warming and cooling are also influenced by the background climate (e.g., snow and rainfall) and human management practices (e.g., irrigation). For example, the interplay between vegetation and surface biophysics is amplified up to five times under extreme warm—dry and cold—wet years (Forzieri et al., 2017).

25.3.3 Precipitation

Forests play a large role in regulating the fluxes of atmospheric moisture and rainfall patterns over land. Forests exert a strong control on the ET process, which can affect the amount of water vapor in the atmosphere. It has been verified that forests evaporate more moisture into the atmosphere than other types of

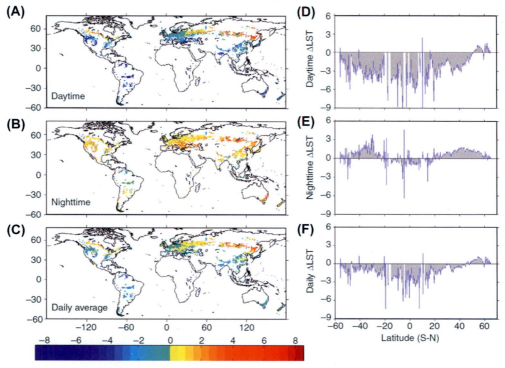

FIGURE 25.15   Annual LST difference of forest minus open land. Effect of forest on temperature is represented by ΔLST (forest minus open land), where positive (negative) values indicate a warming (cooling) effect of forests. (A–C) Spatial pattern (average on 1 × 1 degree grids) and (D–F) corresponding latitudinal dependence of ΔLST for daytime, nighttime, and daily averages (*blue line* denotes 95% confidence interval [CI] estimated by the Student's t-test). Latitude bars with CI out of display range are not drawn (Li et al., 2015).

vegetation. Under a sufficient water supply, the higher capture of advective energy in forest canopies translates into higher evaporative water losses than those achieved by shorter herbaceous canopies. Under drier conditions, forests tend to have deeper roots than herbaceous plants and hence can maintain higher ET than grasslands when the water supply declines. Fig. 25.16 shows comparisons of the monthly ET between forests and grasslands in central Argentina.

In addition to affect atmospheric moisture directly, forests also play an even more extensive role in transporting moisture. The biotic pump

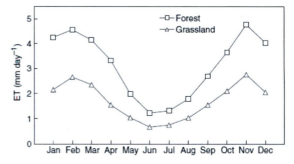

FIGURE 25.16   Seasonal ET patterns of forests and native grasslands in central Argentina, south hemisphere (Nosetto et al., 2005).

theory, proposed by Makarieva and Gorshkov (2007), suggests that the atmospheric circulation that brings rainfall to continental interiors is driven and maintained by large, continuous areas of forests beginning from coasts. The theory, which is controversial, explained that through transpiration and condensation, forests actively create low pressure regions that draw in moist air from oceans, thereby generating prevailing winds capable of carrying moisture and sustaining rainfall far within continents. Afterward, some studies found that the biotic pump theory also works for interior forest (Jiang and Liang, 2013). Fig. 25.17 gives detailed

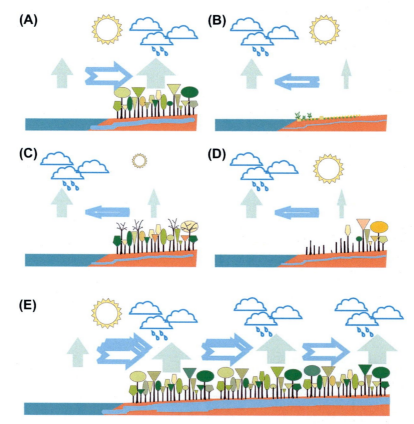

FIGURE 25.17   Makarieva and Gorshkov's "biotic pump." Atmospheric volume reduces at a higher rate over areas with more intensive evaporation (*solid vertical arrows*, widths denotes relative flux). The resulting low pressure draws in additional moist air (*open horizontal arrows*) from areas with weaker evaporation. This leads to a net transfer of atmospheric moisture to the areas with the highest evaporation. (A) Under full sunshine, forests maintain higher evaporation than oceans and thus draw in moist ocean air. (B) In deserts, evaporation is low, and air is drawn toward the oceans. (C) In seasonal climates, solar energy may be insufficient to maintain forest evaporation at rates higher than those over the ocean during a winter dry season and the oceans draw air from the land. However, in summer, high forest evaporation rates are reestablished (as in panel A). (D) With forest loss, the net evaporation over the land declines and may be insufficient to counterbalance that from the ocean: air will flow seaward, and the land becomes arid and unable to sustain forests. (E) In wet continents, continuous forest cover maintaining high evaporation allows large amounts of moist air to be drawn in from the coast. Not shown in diagrams: dry air returns at higher altitudes, from wetter to drier regions, to complete the cycle and internal recycling of rain contributes significantly to continental-scale rainfall patterns. *Source: Adapted from ideas presented in Makarieva, A., Gorshkov, V, 2007. Biotic pump of atmospheric moisture as driver of the hydrological cycle on land. Hydrol. Earth Syst. Sci. 11, 1013–1033.*

explanations of the biotic pump theory. However, the biotic pump theory is still in controversy.

The impacts of forest-derived ET can be seen in satellite observations of rainfall: over most of the tropics, air that passes over forests for 10 days typically produces at least twice as much rain as air that passes over sparse vegetation. A 10% rise in relative humidity can lead to 2–3 times the amount of precipitation. Satellite observations further suggest that European forests have a major influence on cloud formation and thus sunshine/shade dynamics and rainfall. Further, forests contribute to the intensification of rainfall through the biological particles they release into the atmosphere, which include fungal spores, pollen, and bacterial debris. Atmospheric moisture condenses when air becomes sufficiently saturated with water and much more readily when suitable surfaces, provided by aerosol particles (condensation nuclei), are present. Moreover, some plants are particularly effective in facilitating freezing cloud droplets, which is very crucial in the formation of rain in temperate regions.

The formation of precipitation over continents is the product of complex processes that are only partly influenced but not controlled by vegetation. Even though, deforestation has already reduced vapor flows derived from forests by almost 5% (an estimated 3000 km^3 per year of a global terrestrial derived total of 67,000 km^3), with little sign of slowing, and has been implicated as contributing to declining rainfall in various regions (including the Sahel, West Africa, Cameroon, central Amazonia, and India), as well as weakening monsoons.

Modeling is the popular method when analyzing and predicting the climatic effects of forest change. However, more observational evidences are urgently needed to verify and even complete the conclusions derived from the models. Remote sensing can provide the necessary comprehensive observations at various tempospatial resolutions, and it has been applied more and more in the study of this topic. Section 25.4 cites several cases on the climatic effects caused by forest cover change using remotely sensed data.

25.4 Case studies

The climate response to forest cover change depends on the different biomes and where it occurs, and there are still lots of uncertainties regarding coupling between forest change and the atmosphere. Remote sensing is a useful tool to explore and evaluate the climate effects of forest cover change. From satellite observations and other properties inferred from the observations, we can obtain the quantitative evidence of the magnitude and spatial extent of the impact of forest change. Recently, more and more parameters that characterize the surface properties contained in remotely sensed products, such as vegetation indices, skin temperature, albedo, radiation fluxes, ET, GPP, and various atmospheric products, have been developed and utilized in related studies. We give several cases including deforestation in Amazon and forest change in China to illustrate the application of remotely sensed data on this topic.

25.4.1 Deforestation in the Amazon Basin

Humans have been part of the vast forest—river system of the Amazon basin for many thousands of years, but expansion and intensification of agriculture, logging, and urban footprints during the past few decades have been unprecedented. Tropical forests are under immense pressure. The human population of the Brazilian Amazon region increased from 6 million in 1960 to 25 million in 2010, and while tropical forests once covered 3.6 billion hectares, almost a third have been lost as a result of deforestation. Of the remaining area, 46% is fragmented, 30% degraded, and only 24% is in a mature and relatively undisturbed state (Hirschmugl et al., 2017).

25.4.1.1 Deforestation in the Amazon Basin

Detailed spatially and temporal information on global-scale forest change is vital. Hansen et al. (2013) mapped global tree cover extent, loss, and gain for the period from 2000 to 2012 based on 30-m Landsat data, where losses were allocated annually. They concluded that the global forest loss of 2.3 million km^2 was due to disturbance and gain, while 0.8 million km^2 of new forest was established over this period. The tropics experienced the greatest total forest loss and the highest ratio of loss and gain (3.6 for >50% of tree cover), which indicates the prevalence of deforestation dynamics. The results are depicted in Fig. 25.18. The tropics are the only climate domain to exhibit a statistically significant trend in annual forest loss, with an estimated increase in loss of 2101 km^2/year. Tropical rainforest ecozones totaled 32% of the global forest cover loss, nearly half of which occurred in South American rainforests.

Of all countries globally, Brazil exhibits the largest decline in annual forest loss, with a peak of over 40,000 km^2/year in 2003–2004 and a low point of under 20,000 km^2/year in 2010–2011. Brazil is a global exception in terms of forest change, implementing policies to dramatically reduce Amazon Basin deforestation. Fig. 25.19 shows the converging rates of forest disturbance in Brazil. Although the short-

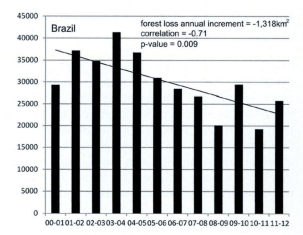

FIGURE 25.19   Annual forest loss total for Brazil from 2000 to 2012. The forest loss annual increment is the slope of the estimated trend line of change in annual forest loss (Hansen et al., 2013).

term decline of Brazilian deforestation is well documented, changing legal frameworks governing Brazilian forests could reverse this trend.

25.4.1.2 The drivers of deforestation

Road paving is a key economic activity that can stimulate deforestation. Further clearing occurs along networks of unofficial roads that result from the interacting interests of colonist farmers and loggers, where the latter minimize

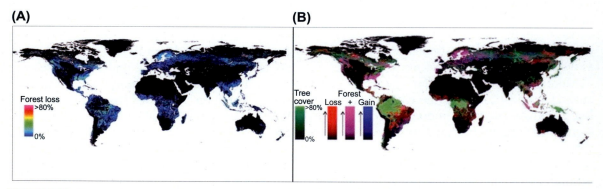

FIGURE 25.18   A color composite of tree cover in green, forest loss in red (A), forest gain in blue, and forest loss and gain in magenta is shown in (B), with loss and gain enhanced for improved visualization (Hansen et al., 2013).

their costs by buying the right to log private lands. Although practices vary widely across the regions, most small land holders (<200 ha) have kept more than 50% of their land as some combination of mature and secondary forests.

Secondly, international and national demands for cattle and livestock feed are increasingly driving land-use change. The direct conversion of forests to croplands in 2003, mostly by large land holders, represented 23% of the deforestation in forest and Cerrado regions of the state of Mato Grosso. Although cattle pasture remains the dominant use of cleared land, the growing importance for the larger and faster conversion to cropland, mostly for soybean export, has defined a trend of forest loss in Amazonia since the early 2000s.

Although selective logging is not an immediate land-use change, it often leads to deforestation. From 1999 to 2003, the total area annually logged in the Amazon basin was similar in magnitude to that of deforested areas. Logged areas are accessible by logging roads and are likely to be cleared within only a few years of initial disturbance, and those that are not cleared have a high risk of burning. On the other hand, reduced-impact logging has been demonstrated to be economically viable, while causing only modest and transient effects on carbon storage and water exchange. Moreover, the expansion of protected areas has also played an important role in reducing deforestation in the Brazilian Amazon.

25.4.1.3 Deforestation alters the energy and water balance

Incoming air from the Atlantic Ocean provides about two-thirds of the moisture that forms precipitation over the Amazon basin. The remainder is supplied through recycling of ET, primarily driven by the deep-rooted Amazon trees. Long-term research of the Amazon has suggested that the basin produces some 9.4 trillion liters of water vapor per year through the process of ET from plant and soil,

where a large portion of this water vapor—about 3.4 trillion liters per year—appears to be transported to South America's southern regions. This means that, in an undisturbed ecosystem, the forest and atmosphere are basically recycling the same water, as well as exporting a massive amount of water vapor to distant basins.

In Amazonia, the replacement of forest by pastures has led to an increase of albedo from approximately 0.13 to approximately 0.18 and a decrease in net radiation of approximately 11% at the surface. A large number of observational and modeling studies have suggested that deforestation causes two main changes in the energy and water balance of the Amazon basin, as follows: first, by partitioning of the net radiation that is absorbed by the land surface changes, with a decrease in the latent heat flux and an increase in the sensible heat flux, primarily because deforestation results in less vegetation being available to transpire water to the atmosphere, and second, replacing the dark rainforest (low albedo) with more reflective pasturelands or crops (high albedo) results in a decrease in solar radiation absorbed by the land surface.

Atmospheric convection and precipitation are driven by the fluxes of energy and water from the land surface. Where clearings for cattle pastures extend for tens of kilometers outward from a road, the air above the deforested areas warms up more quickly and tends to rise and draw moist air from the surrounding forest, creating so-called vegetation breezes. This decreases rainfall over the forest while increasing cloudiness, rainfall, and thunderstorms over the pasture. Heterogeneous deforestation at large scales (hundreds to thousands of km^2) leads to more complex circulation changes, with suppressed rainfall over core clearings, particularly at the beginning and the end of the wet season, and unchanged or increased rainfall over large remnant forest patches. These changes also affect water and light availability and C uptake by the remaining forests, but these effects have not been yet well quantified. A framework

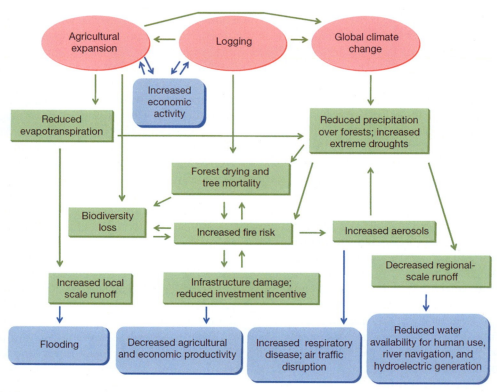

FIGURE 25.20 Interactions between global climate, land use, fire, hydrology, ecology, and human dimensions. Forcing factors are indicated with *red ovals*, processes are indicated by *green boxes and arrows*, and the consequences of human society are indicated by *blue boxes with rounded corners* (Davidson et al., 2012).

for understanding the linkages between natural variability, drivers of change, response and feedbacks in the Amazon basin is presented in Fig. 25.20.

25.4.1.4 Deforestation case

Silverio et al. (2015) collected a combination of MODIS products (MOD43A3, MOD11A2, MOD08E3) and weather station data covering the Xingu region of southern Amazonia to quantify the individual and combined impacts of the three most widespread land-use transitions (LUTs) (forest-to-pasture, forest-to-cropland and pasture-to-cropland) on the major components of the surface energy balance: LST, net radiation (R_{net}), latent heat flux (ET), and sensible heat flux (H). According to the study,

approximately 12% (18,838 km^2) of the Xingu region's forests were converted into croplands (3347 km^2; 2.4%) or pasturelands (15,491 km^2; 9.6%), decreasing the region's forest cover from 61% to 49%. Meanwhile, some pastures (4962 km^2) were converted to croplands.

Generally speaking, the surface energy budget, hydrological cycle, and LSTs in Xingu have been significantly altered by the three LUTs, particularly transition involving forest clearing. According to the results shown in Fig. 25.21, when a given unit area of forest is converted to cropland or pasture, R_{net} decreases by 18% and 21%, ET decreases by 32% and 24%, H increases by 6% and 9%, and LST increases by 0.2 and 0.07°C, respectively. Regarding to the LST, all three LUTs during the 2000s caused

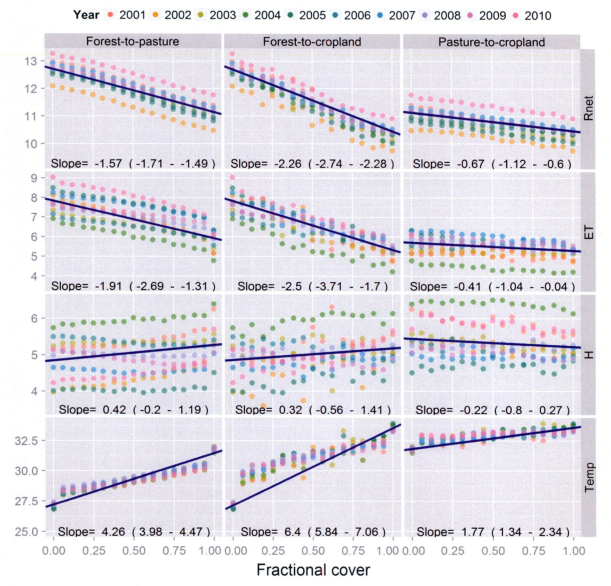

FIGURE 25.21 Changes in the net radiation (Rnet), latent heat (ET), sensible heat (H), and LST (Temp) as a function of fractional change in land cover, estimated from yearly remotely sensed data (*dots*) using linear regressions (*solid lines*). The slopes in parentheses indicate the minimum and maximum slopes calculated for each year from 2001 to 2010. Temperature is in °C and other variables are in MJ·m-2d-1 (Silverio et al., 2015).

a mean basin-wide increase in the Xingu region of 0.3°C, where the forest-to-pasture transitions had the greatest accumulative impact in increasing the LST by 0.2°C (Fig. 25.22A).

To evaluate the role of protected forest areas in mitigating regional climates, the LST changes inside and outside the Xingu Indigenous Park (XIP, protected forest area) were compared

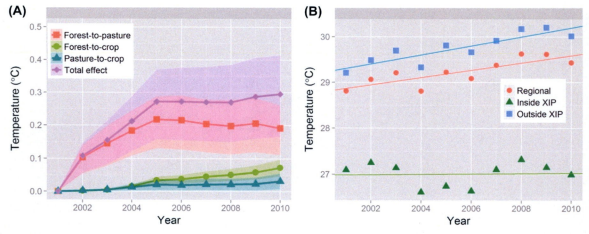

FIGURE 25.22   Effects of LUTs on the daytime LST (from MOD11A2). (A) Effect on temperature in the entire upper Xingu region during the 2000s for each of the three LUTs (shaded areas represent the first and third quartiles). (B) Trends in the annual mean surface temperature for the entire upper Xingu basin (regional), inside and outside the XIP (Silverio et al., 2015).

during the 2000s (Fig. 25.22B). The LST inside the XIP was 1.9°C cooler than the upper Xingu basin average outside the XIP in 2001, and the difference increased to 2.5°C by 2010 due to warming outside the XIP driven by LUTs. In a word, this study provided observational evidence to that illustrated how important forests are in terms of stabilizing the climate in the Amazon.

25.4.2 Forest disturbance in China

China is a vast country with a diverse physical environment, which contains almost all the main forest vegetation types of the northern hemisphere. However, because of war, traditional farming styles, and other historical reasons, as well as rapid economic growth in recent decades, China is facing various environmental problems, including desertification, sandstorms, soil water erosion, and land degradation. To mitigate environmental degradation, afforestation and reforestation have been the subject of nationwide efforts in China since the 1950s. China now has the largest area of forest plantations in the world, and vegetation greening has been observed in many areas of China.

25.4.2.1 Historical forest cover change

China has a poor base for forestry before the foundation of the People's Republic of China in 1949, and it is estimated that Chinese forests covered a mere 8.6% of all the land in 1949. After that, although the government increasingly addressed the protection of forest resources, the degradation of forests caused by unsound exploitation, forest fires, pests, and diseases are still serious threats. Rapid population growth coupled with the development of agriculture, industry, and construction, the overexploitation of forest resources, and subsequent cultivation on steep slopes have led to the deterioration of forest ecosystems and a reduction in biodiversity. As a result of these changes, China has faced a series of hazards and disasters including soil erosion, desertification, dust storms, elevated levels of greenhouse gas emissions, and severe damage to wildlife habitat.

To address these concerns and to improve its environmental conditions, the Chinese government has launched several major forestry ecosystem projects since the 1970s, particularly in the "Three-North" region (Northeast, Northwest, And North China) (see Fig. 25.23A), where

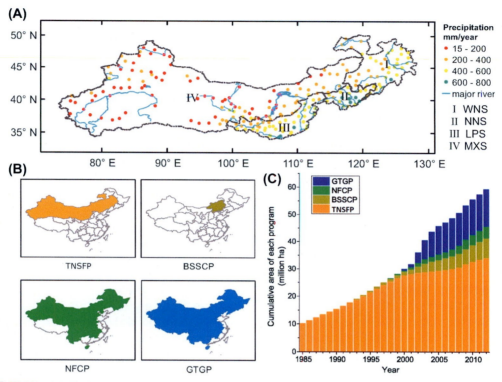

FIGURE 25.23 (A) The location of the "Three-North" region. The dots represent the annual mean precipitation measured by weather stations beginning in 1951. (B) The spatial extent of the four major ecological restoration programs in China since 1978. (C) The cumulative afforested areas for each ecological restoration program from 1985 to 2012. The TNSFP began in 1978, but data regarding the annual planted areas could not be obtained until 1986; therefore, the total planted area from 1978 to 1985 is shown for 1985. The GTGP and NFCP are countrywide restoration programs, and only provinces in our study area, i.e., Heilongjiang, Jilin, Liaoning, Inner Mongolia, Beijing, Tianjin, Hebei, Shanxi, Shaanxi, Ningxia, Gansu, Xinjing, and Qinghai, are included in our statistics (Zhang et al., 2016).

the annual precipitation for most areas in this regions is less than 600 mm. The series of projects (Fig. 25.23B) include the "Three North Shelterbelt Forest Program", which was initiated in 1978 and aims at planting protective forests in arid and semiarid areas, the subsequent "Beijing-Tianjin Sand Source Control Programs", which started in the 2000s and aims to protect against sandstorms through afforestation, the "Natural Forest Conservation Program", which began in 1990 and aims at protecting natural forests through logging bans, and the "Grain for Green Program", which started in 2000 and targeted the conversion of farmland into forests and grasslands.

Among all the ecosystem projects, TNSFP is the most famous one. As mentioned above, this program was started in 1978 and is scheduled to complete in 2050. It involves 13 provinces, autonomous regions, and municipalities in the "Three-North" region, which are some of the most environmentally vulnerable and water-limited regions in China. Its goal is to prevent desertification and dust storms and to improve hydrological and climatic conditions in this region. The total planned area is nearly 42% of the total area of China, and a total of 30.6 M ha of afforestation has been conducted at a total cost of ¥4 billion to date.

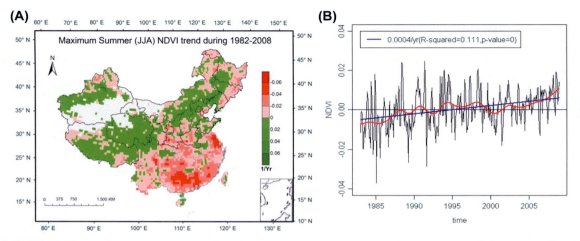

FIGURE 25.24   (A) Linear regression trends in the maximum summer (JJA, June–July–August) NDVI over the Chinese mainland from 1982 to 2008. (B) Anomalies in the monthly NDVI from 1982 to 2008 over the entire TNSFP region.

Various satellite data were used to investigate the trend of greenness in the "Three-North" region to understand the effectiveness of ecological restoration programs during the past three decades. Jiang et al. (2015) evaluated changes in the greenness in China between 1982 and 2008 by using the GIMMS NDVI. Fig. 25.24A shows that the greenness increases mainly occurred in the "Three-North" region. A time series of anomalies in the NDVI in the TNSFP region is shown in Fig. 25.24B, where the trend is also significantly positive. Additionally, according to the China Forestry Yearbook, it is claimed that forest cover has increased dramatically and the total forest cover in China reached 20.36% in 2008. Moreover, acceleration in the amount of afforestation areas was found during the past three decades.

However, changes in vegetation vary in different environments (Qiu et al., 2017). As revealed from field survey photos, vegetation gain and bare soil decline can be simultaneously acquired by tree and grass growth (Fig. 25.25A and B). According to the field survey shown in Fig. 25.25, site A was barren around 10 years ago, but now, it is fully covered by trees and grass. In the second group, site C (Fig. 25.25C),

small trees are growing in the mountains, but there is little grass, and the ground is only partially covered, thus only a partially favorable effect on vegetation growth was obtained in site C. Even though no trends in vegetation change were exhibited in the TNSFP region, diminishing bare soil cover is also very important for reducing sandstorms, as site D shows (Fig. 25.25D).

25.4.2.2 Forestry influence in China

(1) De(re)forestation and climate warming in Heilongjiang Province

Gao and Liu (2012) established a quantitative association between de(re)forestation and climate warming at regional and local scales in Heilongjiang Province, located in Northeast China, through remote sensing and geographic information systems. The study region contains the largest remnant virgin forest in China, where large-scale deforestation started in the 1920s. In recent decades, forest protection and reforestation have been implemented.

Gao and Liu (2012) used a land-use map of Northeast China generated from the visual interpretation of aerial photographs taken in 1958,

(A)

(B)

(C)

(D)

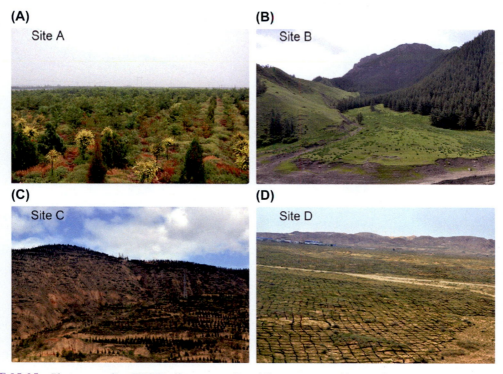

FIGURE 25.25   Photos revealing TNSFP efficacy regarding: (A) vegetation and bare soil in site A and (B) site B, (C) merely on vegetation in site C, and (D) bare soil decline in site D (Qiu et al., 2017).

two digital land cover maps of the province derived from the visual interpretation of Landsat TM/Enhanced TM Plus (ETM+) images recorded in 1980 and 2000, and temperature data collected from 28 meteorological stations. First, deforestation was detected by overlaying consecutive land cover maps in ArcGIS, and then the trends in the mean annual temperature for each weather station were calculated using a linear regression during the same period. Fig. 25.26 shows a drastic reduction of forests in Heilongjiang during 1958–2008; meanwhile, the annual temperature fluctuated widely around a mean of 1.5°C without any apparent trend or change from the 1960s to the mid-1980's, and since then, it has started to rise, reaching the first maxima of 3.56°C in 1990 and stayed at an unprecedentedly high level until 2000. Apparently, the warming trend has accelerated over the last two decades. This pattern of warming is also reminiscent in Fig. 25.27. The annual temperature anomalies of whole China hovered around −0.5°C. After considering other influencing factors, the author concluded that the deforestation accounted for the higher rise in decadal temperature at the regional scale.

The local effects of de(re)forestation on climate change was studied by examining the amount of de(re)forestation within a 25-km radius from weather stations during 1958–80 and 1980–2000, irrespective of the type of forest cover. Fig. 25.28 illustrates that climate warming exhibits a generally negative relationship with deforestation. After a statistical analysis, Gao and Liu (2012) also found that more forest cover is conducive to a slower rise in the decadal temperature (Fig. 25.29).

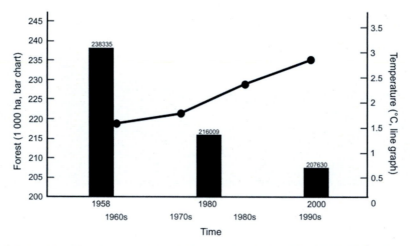

FIGURE 25.26   Relationship of forest cover and decadal temperature during 1958–2000 in Heilongjiang Province (Gao and Liu, 2012).

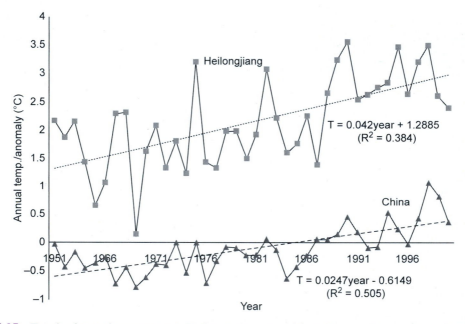

FIGURE 25.27   Trends of annual temperature in Heilongjiang, averaged from the annual temperature observed at 28 stations (*top line*) and annual temperature anomalies in China during 1961–2000 (bottom line, source: Ren et al., 2005) (Gao and Liu, 2012).

This study showed an inverse relationship between the decadal temperature and observed deforestation. In addition, it also indicated that the rise in decadal temperature is also affected by the amount of initial forest cover.

(2) Radiative forcing by four forest disturbances over Northeastern China

Radiative forcing is often used to assess the climatic impact of disturbances per unit area

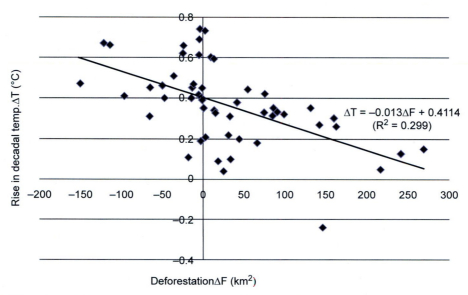

FIGURE 25.28   Relationship between deforestation during 1958–80 and the rise in decadal temperature from the 1960s to the 1970s and that during 1980–2000 and the rise in decadal temperature from the 1980s to the 1990s (Gao and Liu, 2012).

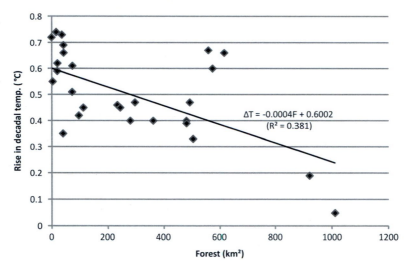

FIGURE 25.29   Relationship between the average forest cover during 1980–2000 and the rise in decadal temperature from the 1980s to the 1990s (Gao and Liu, 2012).

and to compare anthropogenic and natural drivers of climate change. Zhang and Liang (2014) quantitatively estimated the climatic effects of forest disturbances due to changes in forest biomass and surface albedo in terms of radiative forcing over Northeastern China by using multiple remotely sensed products from 2000 to 2010. Four types of forest disturbances were considered: fires, insect damage, logging, and afforestation/reforestation. Two types of surface radiative forcings were considered: CO_2-driven radiative forcing related to the loss of forest biomass carbon to the atmosphere and albedo-driven radiative forcing related to changes in land cover properties. The net forcing was defined as the sum of the albedo-driven and CO_2-driven radiative forcing.

First, the MODIS global disturbance index (MGDI) algorithm based on annual maximum MODIS LST data and annual maximum MODIS EVI data were used to detect the locations of forest disturbances. Second, the MODIS fire products, VCF product, and the calculated MGDI were combined to determine the four disturbances pixel-by-pixel. Third, radiative forcing from changes in albedo caused by the four disturbances was estimated using Global LAnd Surface Satellite albedo data and the Global Energy and Water Exchanges Surface Radiation Budget (SRB) data. Lastly, CO_2-driven radiative forcing due to changes in forest biomass was also calculated, where the forest biomasses were mapped by using a random forests model with field data, geoscience laser altimeter system, and MODIS surface reflectance data firstly, and the final net forcing was obtained.

Table 25.3 summarizes the affected areas of the four forest disturbance during the period 2000–10. The "instantaneous" and "decadal" radiative forcings (albedo-driven, CO_2-driven, and net) are also shown. The results indicate that the radiative forcings resulting from changes in the albedo and CO_2 release are of the same order of magnitude. The "instantaneous" net forcings of albedo-driven radiative forcing and CO_2-driven radiative forcing are 0.53 ± 0.08, 1.09 ± 0.14, 2.23 ± 0.27, and 0.14 ± 0.04 W/m^2 for fire, insect damage, logging, and afforestation and reforestation, respectively. Over a decadal, the estimated net forcings are 2.24 ± 0.11, 0.20 ± 0.31, 1.06 ± 0.41, and -0.47 ± 0.07 W/m^2, respectively.

The trajectories of the CO_2-driven, albedo-driven, and net radiative forcings are illustrated in Fig. 25.30. It was found that the trajectories of the net forcings are different with time for these disturbances, and the CO_2-driven forcing is relatively stable compared with the albedo-driven forcing. During the first 4 or 5 years after disturbances, the net forcings are positive for fires, insect damage, and logging, but negative for afforestation/reforestation. The mechanisms behind each disturbance are different and complicated.

(3) Impact of vegetation change on the local surface climate over North China

Jiang et al. (2015) sought observational evidences to explore the local monthly climate impact of vegetation change caused by a series of ecological projects in the "Three-North" region in China. Remotely sensed data (GIMMS NDVI) and ground meteorological measurements, including the mean air temperature (T, units of °C), air maximum temperature (T_{max}, units of °C), air minimum temperature (T_{min}, units of

TABLE 25.3 Summary of disturbance information and associated radiative forcing for logging, afforestation/reforestation, fire, and insect damage.

Disturbance type	Logging	Afforestation/reforestation	Fire	Insect damage
Annual area disturbed (millions of hectares)	0.15	2.89	1.26	0.57
Instantaneous albedo-driven radiative forcing (W·m^{-2})	-0.22 ± 0.19	0.62 ± 0.03	-0.92 ± 0.06	0.65 ± 0.12
Instantaneous CO_2-driven radiative forcing (W·m^{-2})	2.45 ± 0.08	-0.48 ± 0.01	1.44 ± 0.02	0.43 ± 0.03
Decadal albedo-driven radiative forcing (W·m^{-2})	-1.48 ± 0.33	0.56 ± 0.44	-0.52 ± 0.08	-0.56 ± 0.28
Decadal CO_2-driven radiative forcing (W·m^{-2})	2.53 ± 0.08	-1.03 ± 0.03	2.76 ± 0.08	0.76 ± 0.03

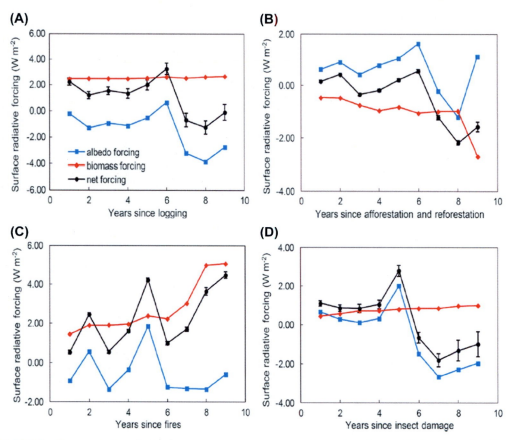

FIGURE 25.30 Changes in the albedo during forest recovery since a forest disturbance. (A) Logging, (B) afforestation/reforestation, (C) fire, and (D) insect damage (Zhang and Liang, 2014).

°C), diurnal temperate range (DTR, units of °C), precipitation (P, units of mm), relative humidity (RH, units of %), wind speed (W, units of m/s), and sunshine hours (SH, units of h, stands for cloud cover), were used in a bivariate Granger Causality test model during the growing season (May to September) from 1982 to 2011. By considering the diversity of the natural environment in China, the study region was divided into four climatic zones (Northeast, eastern arid/semiarid, western arid/semiarid, and North China), where the most significant difference among these zones is the amount of precipitation therein.

First, the changes in vegetation in the TNSFP area were evaluated with remotely sensed data.

Fig. 25.31 shows the range of annual maximum NDVIs between 1982 and 2011 for the four climatic zones. Generally speaking, vegetation increased but with some fluctuations for most areas in the TNSFP region during the past three decades, which is consistent with the result of Fig. 25.24, and a rapid-increase period occurred in all climatic zones from 2007 to 2011. Granger causality tests between the NDVIs and the different climatic variables for the four climatic zones during 1982–2011 were implemented in the study of Jiang et al. (2015). The Granger causality test can be used to detect any causal relationships (e.g., forcing and feedbacks) exclusively from vegetation observations in

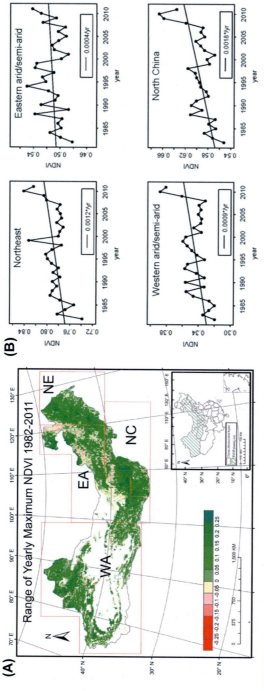

FIGURE 25.31   (A) Range of yearly maximum NDVIs during 1982—2011 in the TNSFP region. (B) Annual average NDVI time series profiles within four climatic regions (Northeast, eastern arid/semiarid, western arid/semiarid, and North China) for 1982 to 2011.

terms of predictability, and it has been applied in various applications, including vegetation—atmosphere interactions. In this study, the Granger causality links were tested and detected between the NDVIs and each of the climatic factors. To explore this further, the test results were further divided into thermal effects (T, T_{min}, T_{max}, and DTR), hydrological effects (P, RH, and SH), and wind effect (W).

Based on the results, it can be concluded that arid/semiarid areas are the most sensitive to interactions between vegetation changes and the local climate. The changes in the NDVIs had a more evident causal link with the minimum temperature and also a significant influence on the hydroclimate in these water-limited regions, but an insignificant effect on reducing wind erosion. Furthermore, most of the temperature and hydrological effects highlighted by the NDVIs were mixed, suggesting that ecological restoration projects should be developed carefully and should take into account both the short-term and long-term effects of vegetation. Moreover, NDVI changes in Northeastern China had a significantly negative influence on the air temperate but not on any other climatic variable. The test results in Northern China were not as objective as the other zones due to the rapid urbanization occurring therein.

25.5 Conclusions

Forests, which cover one-third of the Earth's land surface, provide ecological, economic, social, and esthetic services to natural systems and humankind. The maintenance of the health of forests is a necessary precondition of this globally preferential state. Forest cover change, including practices such as afforestation (the planting of trees on land where they have not recently existed), reforestation, deforestation, and forest degradation, is a potentially important climate change mitigation strategy. Forest cover change affects the climate through associated biophysical and biochemical changes that modify the water cycle and surface energy balance and also through alterations to the carbon balance. A call to action on forests is emerging on many fronts. An in-depth understanding of the causes and consequences of forest cover change and how forest conversion affects regional and global climate and their background mechanisms is essential for more reasonable climate change adaption and mitigation policies.

Monitoring and mapping forests and forest change in a consistent and robust manner over large areas is made possible with satellite data. Remote sensing technology enables observations, identification, mapping, assessment, and monitoring of forest cover at a range of spatial, temporal, and thematic scales and has facilitated forest cover change impact studies in recent past years.

Remote sensing is a powerful tool used for monitoring and mapping forest cover change. Detecting changes in forest cover already has a long tradition starting with Landsat data in the 1980s and 1990s (Running and Bauer, 1996; Singh, 1989). Various satellite optical data, including coarse (e.g., AVHRR and MODIS) and high spatial resolution data (e.g., Landsat), have been used in land cover mapping. Coarse spatial resolution data are suitable for large-scale cover mapping because of their big footprint and frequent revisit period, but they are not recommended for use in thematic class mapping due to their low local accuracy (Frey and Smith, 2007; Fritz et al., 2010). High-resolution optical data are used more often in forest change mapping as they provide information describing heterogeneous and dynamic regions, and data with higher spatial resolutions are available than ever before, which can be used to build time series.

The technologies used for forest change mapping have also evolved from image-to-image classical change detection to time series analyses. The image-to-image change detection produces the maps that relate changes between two dates, which typically lack information regarding the underlying processes and do not enable insights regarding the rate or persistence of the detected changes. Time series of satellite observations have been widely used in forest change mapping (Kennedy et al., 2010; Schroeder et al., 2014; Shimabukuro et al., 2014), where the approach can be further divided into four subcategories: (a) threshold-based change detection, (b) curve fitting, (c) trajectory fitting, and (d) trajectory segmentation.

Satellite data play a more and more important role in forest change impact studies. Over the past few decades, most of our understanding on how forest affects climate have come from climate modeling, in which uncertainties are unavoidable, especially those relating to land cover classification and biophysical process modeling or parameterization (Mahmood et al., 2010; Pielke et al., 2011). Therefore, solid observational evidence is badly needed to better understand of the climate effects resulting from forest change. Consecutive time series of advanced satellite observations provide new insights into forest change and its climatic impacts. For example, the contrasting responses of regional climates to widespread greening on Earth were revealed by using satellite observations (Forzieri et al., 2017). In addition, the multiple remote sensing approaches can bring together more reliable interpretations and attributions.

Studies of forest change and its impact using remote sensing data and technologies have already been successful applied in various applications, but some, change-detection methods, and different definitions of the same category. Current validations of the satellite products are also not sufficient. Moreover, the current available time periods of satellite products are not long enough for most vegetation—climate interaction studies. Futhermore, spatial-scale problems have not been well addressed. However, remote sensing is an irreplaceable tool for forest change and its subsequent impact studies, and it will remain an active research topic for the near future.

References

Alkama, R., Cescatti, A., 2016. Biophysical climate impacts of recent changes in global forest cover. Science 351, 600—604.

Anderson, R.G., Canadell, J.G., Randerson, J.T., Jackson, R.B., Hungate, B.A., Baldocchi, D.D., Ban-Weiss, G.A., Bonan, G.B., Caldeira, K., Cao, L., Diffenbaugh, N.S., Gurney, K.R., Kueppers, L.M., Law, B.E., Luyssaert, S., O'Halloran, T.L., 2011. Biophysical considerations in forestry for climate protection. Front. Ecol. Environ. 9, 174—182.

Anderson, R.G., Goulden, M.L., 2011. Relationships between climate, vegetation, and energy exchange across a montane gradient. J. Geophys. Res. 116, G01026.

Bala, G., Caldeira, K., Wickett, M., Phillips, T.J., Lobell, D.B., Delire, C., Mirin, A., 2007. Combined climate and carbon-cycle effects of large-scale deforestation. Proc. Natl. Acad. Sci. U.S.A. 104, 6550—6555.

Chapin, F.S., Randerson, J.T., McGuire, A.D., Foley, J.A., Field, C.B., 2008. Changing feedbacks in the climate—biosphere system. Front. Ecol. Environ. 6, 313—320.

Davidson, E.A., de Araujo, A.C., Artaxo, P., Balch, J.K., Brown, I.F., Bustamante, M.M.C., Coe, M.T., DeFries, R.S., Keller, M., Longo, M., Munger, J.W., Schroeder, W., Soares-Filho, B.S., Souza, C.M., Wofsy, S.C., 2012. The Amazon basin in transition. Nature 481, 321—328.

DiMiceli, C.M., Carroll, M., Sohlberg, R.A., Huang, C., Hansen, M.C., Townshend, J.R.G., 2011. Annual Global Automated MODIS Vegetation Continuous Fields (MOD44B) at 250 m Spatial Resolution for Data Years Beginning Day 65, 2000—2010, Collection 5 Percent Tree Cover. University of Maryland, College Park, MD, USA.

Ellison, D., Morris, C.E., Locatelli, B., Sheil, D., Cohen, J., Murdiyarso, D., Gutierrez, V., van Noordwijk, M., Creed, I.F., Pokorny, J., Gaveau, D., Spracklen, D.V., Tobella, A.B., Ilstedt, U., Teuling, A.J., Gebrehiwot, S.G., Sands, D.C., Muys, B., Verbist, B., Springgay, E., Sugandi, Y., Sullivan, C.A., 2017. Trees, forests and water: cool insights for a hot world. Glob. Environ. Chang. Hum. Policy Dimens. 43, 51—61.

FAO, 2006. Global forest resources assessment 2005: progress towards sustainable forest management. In: Forestry Paper 147. Rome.

FAO, 2010. Global forest resources assessment 2010. In: Rome Food and Agrigulture Organization of the United Nations.

FAO, 2015. Global forest resources assessment 2015. In: Rome: Food and Agriculture Organization of the U. N.

Forzieri, G., Alkama, R., Miralles, D.G., Cescatti, A., 2017. Satellites reveal contrasting responses of regional climate to the widespread greening of Earth. Science 356, 1180.

Frey, K.E., Smith, L.C., 2007. How well do we know northern land cover? Comparison of four global vegetation and wetland products with a new ground-truth database for West Siberia. Glob. Biogeochem. Cycles 21.

Fritz, S., See, L., Rembold, F., 2010. Comparison of global and regional land cover maps with statistical information for the agricultural domain in Africa. Int. J. Remote Sens. 31, 2237–2256.

Gao, J., Liu, Y., 2012. De(re)forestation and climate warming in subarctic China. Appl. Geogr. 32, 281–290.

Hansen, M.C., DeFries, R.S., Townshend, J.R.G., Carroll, M., Dimiceli, C., Sohlberg, R.A., 2003. Global percent tree cover at a spatial resolution of 500 meters: first results of the MODIS vegetation continuous fields algorithm. Earth Interact. 7, 1–15.

Hansen, M.C., DeFries, R.S., Townshend, J.R.G., Sohlberg, R., Dimiceli, C., Carroll, M., 2002. Towards an operational MODIS continuous field of percent tree cover algorithm: examples using AVHRR and MODIS data. Remote Sens. Environ. 83, 303–319.

Hansen, M.C., Potapov, P.V., Moore, R., Hancher, M., Turubanova, S.A., Tyukavina, A., Thau, D., Stehman, S.V., Goetz, S.J., Loveland, T.R., Kommareddy, A., Egorov, A., Chini, L., Justice, C.O., Townshend, J.R.G., 2013. High-resolution global maps of 21st-century forest cover change. Science 342, 850–853.

Hirschmugl, M., Gallaun, H., Dees, M., Datta, P., Deutscher, J., Koutsias, N., Schardt, M., 2017. Methods for mapping forest disturbance and degradation from optical Earth observation data: a review. Curr. For. Rep. 3, 32–45.

Huang, C.Q., Coward, S.N., Masek, J.G., Thomas, N., Zhu, Z.L., Vogelmann, J.E., 2010. An automated approach for reconstructing recent forest disturbance history using dense Landsat time series stacks. Remote Sens. Environ. 114, 183–198.

Jiang, B., Liang, S.L., 2013. Improved vegetation greenness increases summer atmospheric water vapor over Northern China. J. Geophys. Res. Atmosphere 118, 8129–8139.

Jiang, B., Liang, S.L., Yuan, W.P., 2015. Observational evidence for impacts of the vegetation changes in the three-north region in China on local surface climate using Granger Causality test. J. Geophys. Res. − Biogeosci. 120, 1–12.

Kennedy, R.E., Cohen, W.B., Schroeder, T.A., 2007. Trajectory-based change detection for automated characterization of forest disturbance dynamics. Remote Sens. Environ. 110, 370–386.

Kennedy, R.E., Yang, Z., Cohen, W.B., 2010. Detecting trends in forest disturbance and recovery using yearly Landsat time series: 1. LandTrendr — temporal segmentation algorithms. Remote Sens. Environ. 114, 2897–2910.

Kuenzer, C., Dech, S., Wagner, W., 2015. Remote sensing time series revealing land surface dynamics: status quo and the pathway ahead. Remote Sens. Time Ser. 1–24 (Springer).

Li, Y., Zhao, M.S., Motesharrei, S., Mu, Q.Z., Kalnay, E., Li, S.C., 2015. Local cooling and warming effects of forests based on satellite observations. Nat. Commun. 6.

Mahmood, R., Pielke, R.A., Sr, Hubbard, K.G., Niyogi, D., Bonan, G., Lawrence, P., McNider, R., McAlpine, C., Etter, A., Gameda, S., Qian, B., Carleton, A., Beltran-Przekurat, A., Chase, T., Quintanar, A.I., Adegoke, J.O., Vezhapparambu, S., Conner, G., Asefi, S., Sertel, E., Legates, D.R., Wu, Y., Hale, R., Frauenfeld, O.W., Watts, A., Shepherd, M., Mitra, C., Anantharaj, V.G., Fall, S., Lund, R., Trevino, A., Blanken, P., Du, J., Chang, H.-I., Leeper, R.E., Nair, U.S., Dobler, S., Deo, R., Syktus, J., 2010. Impacts of land use/land cover change on climate and future research priorities. Bull. Am. Meteorol. Soc. 91, 37–46.

Makarieva, A., Gorshkov, V., 2007. Biotic pump of atmospheric moisture as driver of the hydrological cycle on land. Hydrol. Earth Syst. Sci. 11, 1013–1033.

Nosetto, M.D., Jobbagy, E.G., Paruelo, J.M., 2005. Land-use change and water losses: the case of grassland afforestation across a soil textural gradient in central Argentina. Glob. Chang. Biol. 11, 1101–1117.

Pielke, R.A., Pitman, A., Niyogi, D., Mahmood, R., McAlpine, C., Hossain, F., Goldewijk, K.K., Nair, U., Betts, R., Fall, S., Reichstein, M., Kabat, P., de Noblet, N., 2011. Land use/land cover changes and climate: modeling analysis and observational evidence. Wiley Interdiscipl. Rev. Clim. Change 2, 828–850.

Qiu, B.W., Chen, G., Tang, Z.H., Lu, D.F., Wang, Z.Z., Chen, C.C., 2017. Assessing the Three-North Shelter Forest Program in China by a novel framework for characterizing vegetation changes. ISPRS J. Photogrammetry Remote Sens. 133, 75–88.

Ren, G.Y., Xu, M.Z., 2005. Changes of Surface Air Temperature in China During 1951-2004. Climat. Environ. Res. 10 (4), 717–727.

Running, T., Bauer, M., 1996. Change detection in forest ecosystems with remote sensing digital imagery. Remote Sens. Rev. 13, 207–234.

Schroeder, T.A., Healey, S.P., Moisen, G.G., Frescino, T.S., Cohen, W.B., Huang, C., Kennedy, R.E., Yang, Z., 2014. Improving estimates of forest disturbance by combining observations from Landsat time series with US Forest Service Forest Inventory and Analysis data. Remote Sens. Environ. 154, 61–73.

Sheil, D., 2018. Forests, atmospheric water and an uncertain future: the new biology of the global water cycle. Forest Ecosyst. 5.

Shimabukuro, Y.E., Beuchle, R., Grecchi, R.C., Achard, F., 2014. Assessment of forest degradation in Brazilian

Amazon due to selective logging and fires using time series of fraction images derived from Landsat ETM+ images. Remote Sens. Lett. 5, 773–782.

Silverio, D., Brando, P.M., Macedo, M.N., Beck, P.S.A., Bustamante, M., Coe, M.T., 2015. Agricultural expansion dominates climate changes in southeastern Amazonia: the overlooked non-GHG forcing. Environ. Res. Lett. 10.

Singh, A., 1989. Digital change detection techniques using remotely-sensed data. Int. J. Remote Sens. 10, 989–1003.

Townshend, J., Hansen, M., Carroll, M., Dimiceli, C., Sohlberg, R., Huang, C., 2006. In: User Guide for the MODIS Vegetation Continuous Field Product Collection 5 Version 1. University of Maryland, College Park, Maryland.

van Wagtendonk, J.W., Root, R.R., Key, C.H., 2004. Comparison of AVIRIS and Landsat ETM+ detection capabilities for burn severity. Remote Sens. Environ. 92, 397–408.

Zhan, X., Defries, R., Townshend, J.R.G., Dimiceli, C., Hansen, M., Huang, C., Sohlberg, R., 2000. The 250 m global land cover change product from the moderate resolution imaging spectroradiometer of NASA's Earth observing system. Int. J. Remote Sens. 21, 1433–1460.

Zhan, X., Sohlberg, R.A., Townshend, J.R.G., DiMiceli, C., Carroll, M.L., Eastman, J.C., Hansen, M.C., DeFries, R.S., 2002. Detection of land cover changes using MODIS 250 m data. Remote Sens. Environ. 83 (1-2), 336–350.

Zhang, Y., Peng, C.H., Li, W.Z., Tian, L.X., Zhu, Q.Q., Chen, H., Fang, X.Q., Zhang, G.L., Liu, G.M., Mu, X.M., Li, Z.B., Li, S.Q., Yang, Y.Z., Wang, J., Xiao, X.M., 2016. Multiple afforestation programs accelerate the greenness in the 'Three North' region of China from 1982 to 2013. Ecol. Indicat. 61, 404–412.

Zhang, Y.Z., Liang, S.L., 2014. Surface radiative forcing of forest disturbances over northeastern China. Environ. Res. Lett. 9.

Zhang, Y.Z., Liang, S.L., 2018. Impacts of land cover transitions on surface temperature in China based on satellite observations. Environ. Res. Lett. 13.

Index

Note: 'Page numbers followed by "*f*" indicate figures and "*t*" indicate tables'.